I0887967

SINGLE-PARTICLE CRYO-ELECTRON MICROSCOPY

The Path Toward Atomic Resolution

Selected Papers of Joachim Frank with Commentaries

Series in Structural Biology

ISSN: 2424-8398

Series Editor: Anders Liljas *(Lund University, Sweden)*

Published

Vol. 10 *Single-Particle Cryo-Electron Microscopy: The Path Toward Atomic Resolution/ Selected Papers of Joachim Frank with Commentaries*
edited by Joachim Frank

Vol. 9 *Structure and Mechanism in Protein Science: A Guide to Enzyme Catalysis and Protein Folding*
by Alan Fersht

Vol. 8 *Textbook of Structural Biology (Second Edition)*
by Anders Liljas, Lars Liljas, Miriam-Rose Ash, Göran Lindblom, Poul Nissen and Morten Kjeldgaard

Vol. 7 *Curiosity and Passion for Science and Art: S-layer Proteins of Bacteria and Archaea*
by Uwe B. Sleytr

Vol. 6 *Structure and Action of Molecular Chaperones: Machines that Assist Protein Folding in the Cell*
by Lila M. Gierasch, Arthur L. Horwich, Christine Slingsby, Sue Wickner and David Agard

Vol. 5 *The Struggles and Dreams of Robert Langer*
by Robert Langer

Vol. 4 *From a Grain of Salt to the Ribosome: The History of Crystallography as seen Through the Lens of the Nobel Prize*
edited by Ivar Olovsson, Anders Liljas and Sven Lidin

Vol. 3 *Selected Papers of Michael G. Rossmann with Commentaries: The Development of Structural Biology*
by Michael G. Rossmann

Vol. 2 *Found in Translation: Collection of Original Articles on Single-Particle Reconstruction and the Structural Basis of Protein Synthesis*
by Joachim Frank

Vol. 1 *Structural Aspects of Protein Synthesis (Second Edition)*
by Anders Liljas and Måns Ehrenberg

Series in Structural Biology — Vol. 10

SINGLE-PARTICLE CRYO-ELECTRON MICROSCOPY

The Path Toward Atomic Resolution

Selected Papers of Joachim Frank with Commentaries

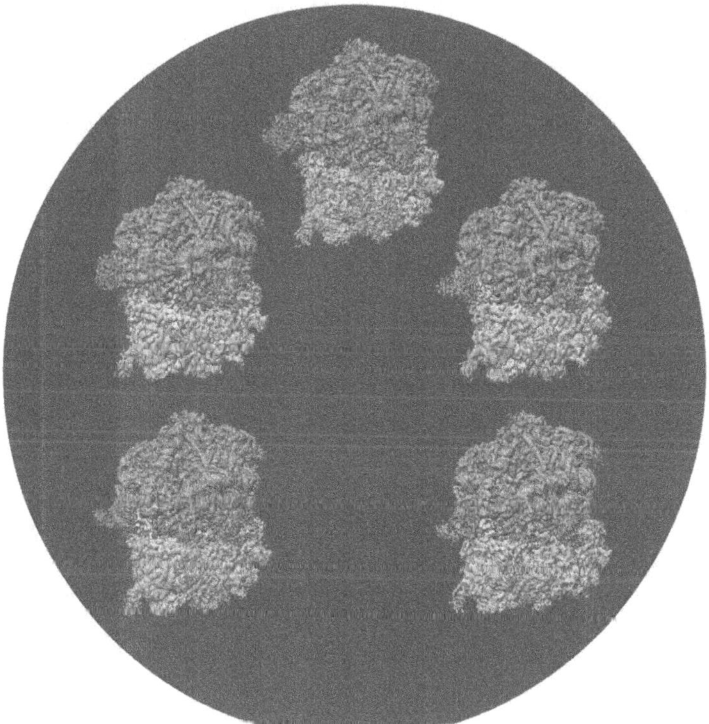

Editor

Joachim Frank

Columbia University, USA

World Scientific

NEW JERSEY • LONDON • SINGAPORE • BEIJING • SHANGHAI • HONG KONG • TAIPEI • CHENNAI • TOKYO

Published by

World Scientific Publishing Co. Pte. Ltd.

5 Toh Tuck Link, Singapore 596224

USA office: 27 Warren Street, Suite 401-402, Hackensack, NJ 07601

UK office: 57 Shelton Street, Covent Garden, London WC2H 9HE

Library of Congress Cataloging-in-Publication Data

Names: Frank, J. (Joachim), 1940– author, editor.

Title: Single-particle cryo-electron microscopy : the path toward atomic resolution : selected papers of Joachim Frank with commentaries / editor, Joachim Frank, Columbia University, USA.

Description: New Jersey : World Scientific, [2018] | Series: Series in structural biology ; Vol. 10 | Includes bibliographical references.

Identifiers: LCCN 2017057652| ISBN 9789813234857 (hardcover : alk. paper) | ISBN 9813234857 (hardcover : alk. paper)

Subjects: LCSH: Electron microscopy--Technique. | Electron microscopy--Methodology. | Microscopy--Technique.

Classification: LCC QH212.E4 F73 2018 | DDC 570.28/25--dc23

LC record available at https://lccn.loc.gov/2017057652

British Library Cataloguing-in-Publication Data

A catalogue record for this book is available from the British Library.

Copyright © 2018 by World Scientific Publishing Co. Pte. Ltd.

All rights reserved. This book, or parts thereof, may not be reproduced in any form or by any means, electronic or mechanical, including photocopying, recording or any information storage and retrieval system now known or to be invented, without written permission from the publisher.

For photocopying of material in this volume, please pay a copying fee through the Copyright Clearance Center, Inc., 222 Rosewood Drive, Danvers, MA 01923, USA. In this case permission to photocopy is not required from the publisher.

For any available supplementary material, please visit
http://www.worldscienti ic.com/worldscibooks/10.1142/10844#t=suppl

to Carol

Contents

I

INTRODUCTION AND COMMENTARIES

As with many biologists trained as physicists, I entered biology through the "back door," by initially focusing on developing or perfecting a tool, then demonstrating its utility in a particular area of biology, and finally allowing myself to be drawn into this area, after realizing that it was in fact possible to make useful contributions there. Reading Horace Freeland Judson's book "The Eighth Day of Creation" (*Judson, 1996*) made me aware of the fact that I'm in good company, as Jim Watson and Francis Crick were complete outsiders when they started to approach one of the most vexing problems of biology.

In the beginning of my career, nothing could have prepared me for the extraordinary, exciting and intellectually challenging 50-year long journey that started in 1967. It started, quite prosaically, with the analysis of carbon film images affected by drift, and culminated in the reconstruction of an important eukaryotic ribosome at 2.5Å resolution, at the time this collection is being compiled. What is more, I came to realize that the whole trajectory of a molecular machine such as the ribosome can be followed by classifying and visualizing molecules contained in a single sample — an ensemble of molecules all encountered in different stages of a highly complex repetitive process.

In the selection of papers for this collection I have tried to highlight important methodological contributions to the development of the single-particle reconstruction method, and in particular, the knowledge gained in the application of this method to the ribosome and the structural basis of protein synthesis. The focus and scope of this project is on single-particle reconstruction, and therefore omits any mention of my lab's work related to Electron Tomography.

This narrative, far from aiming to be exhaustive, is meant to provide pointers to the selected papers in the context of the overall work of my lab. Page restrictions, and the restriction in the number of color reproductions posed some limitations on the number of papers related to protein synthesis. Nevertheless I hope the volume will be useful for novices to the field of cryo-EM and old-timers alike.

1. Image Formation, Optical Diffraction, and Digital Image Processing

When I started my dissertation in the lab of **Walter Hoppe** at the Max-Planck Institute in Munich (Fig. 1a), in 1967, he suggested I work on the three-dimensional (3D) reconstruction of biological macromolecules. An X-ray crystallographer by training and vocation, with initial contributions to molecular replacement to his name, he had developed an interest in electron microscopy. In theory, recovery of the 3D object from a series of projections is facilitated by two tenets: (1) The approximate validity of the linear transfer theory in bright field transmission electron microscopy, stating that the intensity in micrographs recorded as optical

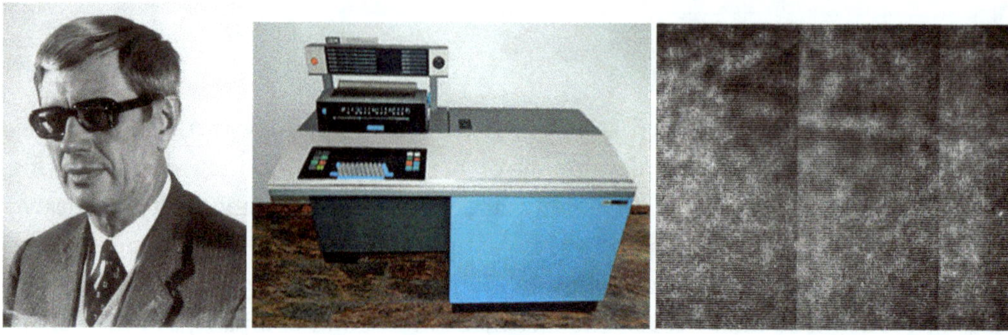

(a) (b) (c)

Fig. 1. The Hoppe Lab. **(a)** Walter Hoppe (1917–1986). **(b)** The IBM 1130 (1968). **(c)** Digitized image of fatty acid synthetase, rendered by overprinting of characters (from J. Frank, Ph.D. thesis 1970).

density on film or, more lately, as a direct electron count, is proportional to the projection of the electrostatic (Coulomb) potential; (2) The known Fourier theorem linking projections to central sections of the three-dimensional Fourier transform of the object they originate from.

At that time no suitable programs were available to perform 3D reconstructions from projections. DeRosier and Klug's 1968 *Nature* paper on the 3D reconstruction of the bacteriophage tail had yet to appear. I was supposed to re-write programs used in X-ray crystallography, just — as my mentor made it sound — by adding in the Fourier phases that are readily available with an imaging instrument. However, I would soon discover that I had steep mountains to climb. I had to start at the very bottom, by familiarizing myself with the uncanny properties of electron micrographs, their corruption by noise, and with the strange behavior of the phase contrast transfer function.

At the time, we scanned electron micrographs recorded on film with a custom-built microdensitometer, which stored the information on rolls of punched paper-tape. Computers were massive, filling a good part of a room, though they were extremely primitive compared to the microchip in today's cell phones. During the time in the Hoppe Lab I worked with an IBM 1130 (Fig. 1b), equipped with 16k words of memory and a speed of 455 kHz. The computer was connected with a card reader, a punch-tape reader, a line printer, and a digital plotter. To display images, one had to use overprinting of characters on the line printer (Fig. lc). Workstations with gray scale image display did not exist until the 1980s. Later on, for larger-scale computations, a main frame became available at the Max-Planck Institute for Nuclear Physics in Garching, requiring a subway ride north for each job to start.

As has been recounted at another place (*Mossman, 2007*), the analysis of drift and its manifestation in the Fourier transform of the image was a catalyst in my first forays into digital image processing. When I entered the field, **Fritz Thon**'s classic analysis of the contrast transfer function (CTF) by optical diffraction had just been published (*Thon, 1965; 1966*). "Thon rings" is now a household word among electron microscopists. The optical diffraction patterns show a footprint of the CTF modulating the more or less uniform "white" background of the Fourier transform expected for an amorphous carbon film. From this pattern we can "read" the defocus, astigmatism, and range of resolution of the instrument. In fact, at the time when Fourier transformation in the computer was prohibitively time-consuming, optical diffraction was the way to go in order to get instant information about aberrations in the microscope and their effects on the images.

By a fortuitous circumstance I came into the possession of a stack of micrographs recorded on film and taken on an unstable electron microscope. It was therefore almost by reflex that I decided to examine them in the optical diffractometer. Here I discovered that the expected Thon ring pattern was disrupted in a striking way (Frank, 1969). The pattern was either limited to one single central stripe, or it was modulated by periodic fringes. I was able to link the observed patterns to mechanical vibrations or gradual movements of the specimen stage in the first case, and to a sudden jump ("jump drift" as I called it) of the specimen stage during the exposure in the second case. In the theoretical analysis, the periodic modulation turned out to be equivalent to Young's fringes, familiar to the physicist from light-optical interference experiments.

I was intrigued by the realization that the cosine-modulated, fringe-producing term visible in the Fourier transform of the image affected by "jump drift" was a term directly related to the cross-correlation function of the two images superimposed. This result told me that two images of the same carbon film taken in two separate shots in the microscope could be aligned with the help of the two-dimensional cross-correlation function to within a few Angstroms. Moreover, by locating the centroid of the cross-correlation peak one could obtain alignment precision in the order of an Angstrom. To me this was an exciting revelation since it foretold the feasibility of highly accurate digital alignment of micrographs — or parts thereof — originating from the same object. Obviously alignment of micrographs with high precision is a critical prerequisite for many image processing purposes, such as averaging, classification, contrast transfer function correction, and 3D reconstructions from projections.

Another, related realization that sprang from the observation of the "jump drift" phenomenon was that the extent of Young's fringes could be regarded as a measure of *potential resolution*. That is, in comparing two independent images of the same object, I had in fact separated signal from noise since only the Fourier components common to both images — the signal — will be modulated. This realization was the seed later on for the method of estimating resolution by comparing half-set averages or half-set reconstructions using some measure (Euclidan distance, differential phase residual, Fourier ring or shell correlation) in Fourier space. The fact that the CTF inverts the phases of some of the zones is captured in the term "potential resolution" — i.e., the best resolution, but achievable only when the CTF is entirely corrected for.

Part of the subsequent research during my time as a postdoc, first in the United States sponsored by a Harkness Fellowship (1970–1972), then in the Cavendish Lab in Cambridge, England (1973–1975), addressed the way partial coherence affects the CTF, and the way the CTF can be corrected. At the Jet Propulsion Laboratory, with access to image processing facilities of the highest caliber, I was able to make use of an algorithm proposed earlier by Peter Schiske (*Schiske*, 1968), and demonstrated the ability to separate amplitude from phase contrast of a defocus series showing DNA on carbon (Frank, 1972a).

Partial coherence is an illumination condition created by an extended source, as opposed to a point source for the fully coherent case. Partial coherence limits the resolution obtainable in EM bright field imaging. At that time, sources with high brightness such as field emission guns — which allowed a point source to be approximated, were not available in transmission electron microscopes, and partial coherence was still a serious bottleneck in achieving high resolution. The calculations necessary to predict the effects of partial coherence in EM image formation are quite complicated, making the effects of particular choices of parameters such as source size and defocus difficult to gauge. As a remedy I developed an approximate expression — the envelope function — which greatly simplified the task as it had a simple analytical form (Frank, 1973). Using this expression for the envelope function along with the CTF, I was able to explain the observed oscillations in the signal-to-noise ratio of electron micrographs, measured via cross-correlation, as a function of defocus (Frank and Al-Ali, 1975). During the time I was at the Cavendish Lab, I greatly benefited from interactions with **Peter Hawkes**, an expert in Electron Optics, who later moved to the Laboratory of Electron Optics of Centre National de la Recherche Scientifique (CEMES, Toulouse, France).

2. Development of Methods for 2D Averaging and 3D Reconstruction of Biological Macromolecules

Research in the Hoppe Lab, after the computer programs were finally in place, had centered on 3D reconstruction of negatively stained molecules from tilt series, by electron tomography. The development of electron tomography was Hoppe's pioneering contribution to electron microscopy. However, the direction of the work was rightly criticized because of the questionable relevance of images obtained with accumulated exposures of more than 1000 electrons per square Angstrom. Under those conditions, molecules are essentially left as cinder, giving no useful structural information. Spending some of my time as a postdoc in the lab of **Robert M. Glaeser** (Donner Lab, Berkeley, CA) in 1971, I witnessed his and **Ken Taylor**'s experiments on radiation damage of L-valine crystals exposed to the electron beam, and it was in this lab that I developed an appreciation for the extreme sensitivity of biological molecules.

I therefore became convinced that valid 2D averages or 3D reconstructions of a biological macromolecule could only be obtained by averaging or merging images from multiple realizations, or "copies," of a molecule. In this way, the exposure could be reduced by spreading it over the large number of copies (Frank, 1975b). This paper makes reference to Unwin and Henderson's seminal article on the structure of bacteriorhodopsin reconstructed from low-dose micrographs of the purple membrane, a paper that was published earlier that year (*Unwin and Henderson*, 1975). Indeed, the single-particle approach I proposed is equivalent to Fourier filtering of a 2D crystal, except for the fact that in the former case the images of the repeating unit must be located and aligned explicitly while they are being located automatically in the latter case. The formulation of conditions — in terms of minimum dose, contrast, and object size — for reliable alignment of low-dose electron images by cross-correlation were therefore an important topic of research during my time at the Cavendish Lab. The results of this analysis (published later, in collaboration with **Owen Saxton**: Saxton and Frank, 1977) encouraged me to put all my energy into this single-particle imaging approach to structure research with the electron microscope.

After accepting the offer of a Senior Research Scientist position at the New York State Department of Health, Division of Labs and Research (later renamed *Wadsworth Center*) in Albany, in 1975, I focused my efforts on the development of a modular image processing suite (SPIDER) that contained all necessary operations for aligning and averaging images of single molecules (Frank *et al.*, 1981a), a crucial prerequisite for the ultimate goal of reconstructing then in three dimensions. At that time, **William Goldfarb** (Fig. 2a), hired as my assistant who helped me set up the lab in Albany, greatly contributed to the development and testing of the programs (Figs. 2a,b). Later on, **Ardean Leith** (Fig. 5), a Research Scientist at the Wadsworth

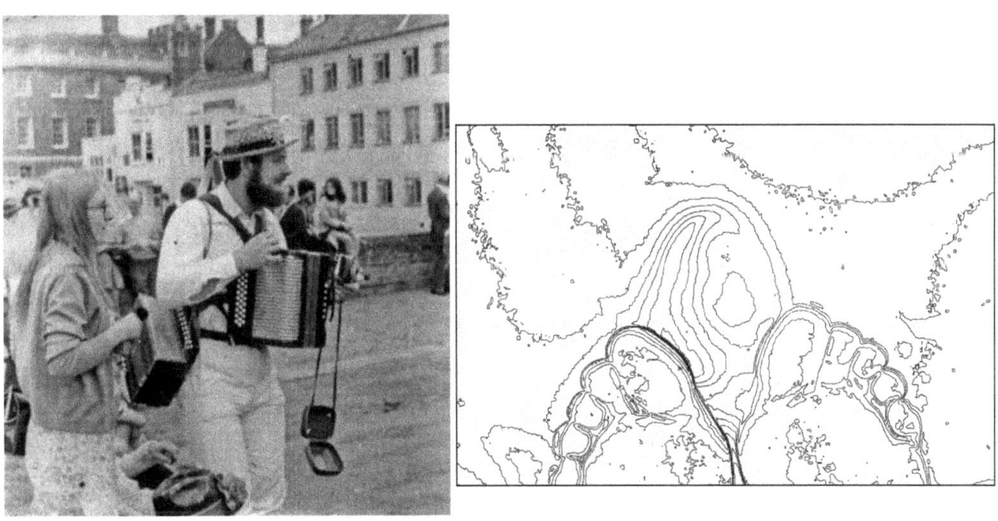

(a) (b)

Fig. 2. Testing the newly developed image processing suite SPIDER. **(a)** William Goldfarb, as Morris Dancer in Cambridge, England (~1975). This image was repeatedly used to test contrast enhancement options in SPIDER. **(b)** Testing the contour plot option in SPIDER.

Center, took on a major role in the maintenance and further development of SPIDER. With the help of this suite of Fortran programs, the first success (in terms of obtaining "clean," noise-free 2D averages of aligned molecules having a preferred orientation) was achieved with negatively stained molecules of glutamine synthetase (Frank *et al.*, 1978; Kessel *et al.*, 1980) and ribosomes (Frank *et al.*, 1981b). **Adriana Verschoor** (Figs. 3, 4 — I will say more about her later in the section on Eukaryotic ribosomes) assisted me with the image processing. Collaborators on those studies were **David Eisenberg** (UCLA, Los Angeles, California), **Tim Baker** (Purdue University, Indiana), **Martin Kessel** (Hadassah Medical Center, Jerusalem) (Figs. 3, 4), and **Miloslav Boublik** (Roche, Nutley, New Jersey) — who all supplied samples and much encouragement. At that time Martin Kessel was spending a Sabbatical year in my lab.

However, the heterogeneity of images posed a serious obstacle to this approach. Together with **Marin van Heel** (Fig. 3), a student visiting from The Netherlands, I hit upon a solution to this problem, inspired by an approach used by **Jean-Pierre Bretaudiere** (Fig. 3), a scientist in Laboratory Medicine at the Wadsworth Center, in analyzing blood samples of patients in a database of the New York State Health Department. (This episode is contained in a bio sketch on the occasion of my election to the National Academy of Sciences (*Mossman, 2007*)). Van Heel brought with him a stack of electron micrographs of negatively stained *Limulus polyphemus* hemocyanin. In these micrographs the molecule appeared in "flip" and "flop" versions (i.e., projections related by flipping the molecule over on the grid) and also had different "rocking" positions. The gallery of hemocyanin images thus

Fig. 3. At the first meeting of the newly minted Gordon Research Conference on 3D Electron Microscopy, in New Hampshire, 1985. From left to right: Marin van Heel, Jean-Pierre Bretaudiere, Adriana Verschoor, Bruce McEwen, Joachim Frank, Terry Wagenknecht, Michael Radermacher, and Martin Kessel.

Fig. 4. At the Gordon Research Conference on 3D Electron Microscopy, in New Hampshire 1988. Standing (left to right): Harald Rose, Carmen Mannella, Martin Kessel, Bruce McEwen, Adriana Verschoor, J.F., Terry Wagenknecht, Nicolas Boisset, Pawel Penczek, Jose-Maria Carazo, Weiping (Mark) Liu. Sitting: Suman Srivastava, Neng Yu Zhang, Michael Radermacher, Bob Grassucci, and Xiaowei Guo.

presented a paragon for the problem of heterogeneity. By equating aligned images with patients, and arrays of pixels with the vectors describing the sets of various measurements in a blood sample, we showed that Correspondence Analysis — a variant of Principal Component Analysis — could be readily used to sort out these images into different groups based on their shape (van Heel and Frank, 1981; Frank and van Heel, 1982; Frank, 1982a; Bretaudiere and Frank, 1986). This method of dimensionality reduction, following standard methods of multivariate data analysis, proved to be of great value for the whole emerging field of single-particle electron microscopy since classification of molecule images was greatly simplified in the low-dimensional subspace. The use of other classification methods such as fuzzy classification was later explored by **Jose-Maria Carazo** (Fig. 4), who joined my group as a postdoc for a year in 1987 (see Carazo *et al.*, 1990).

The next step was to find a method for the reconstruction of a molecule in three dimensions based on a set of its projections in random directions. Three-dimensional (3D) reconstruction of an asymmetric object from its randomly oriented projections poses a fundamental problem that is quite different, for instance, from the problem of reconstructing a highly symmetric virus. The reason is that (1) the individual projections themselves contain no clues (except for one common line for each pair; see below) on their placement in the three-dimensional coordinate framework, and (2) the absence of symmetries necessitates collection of a much larger dataset than required for a virus, amounting to a 60-fold increase over a virus with icosahedral symmetry.

Two methods for the determination of orientations of projections, referred to as "*ab initio* methods," were proposed. One is called *random-conical reconstruction* method, which was originally mentioned in the concluding section of an earlier paper (Frank *et al.*, 1978a). This method was brought to fruition by my postdoc **Michael Radermacher** (Figs. 3, 4) (Radermacher *et al.*, 1987b), who had already been working in his dissertation project in Walter Hoppe's lab on the problem of 3D reconstruction of an object from its projections in arbitrary directions — in contrast to the regular angular spacing that had always been assumed in the mathematical treatment of 3D reconstruction. The idea behind the random-conical approach is that two images of the same field of molecules are taken, one with tilt, the other without. It is easy to see that from such a tilt pair, the azimuth and tilt of each molecule can be determined. Among the first structures visualized with the help of the new method, apart from the ribosome, was the ryanodine receptor (Radermacher *et al.*, 1994). The second method, proposed by **Marin van Heel** (Fig. 3) under the name of ***angular reconstitution*** at the same time the random-conical method was first demonstrated, is based on the alignment of common lines shared by projections in Fourier space (*van Heel, 1987*). As already mentioned above, only one line is shared between any two projections of a completely asymmetric structure.

First studies of ribosomes samples with the random-conical reconstruction method, first negatively stained (Wagenknecht *et al.*, 1988a) and later ice-embedded (Wagenknecht *et al.*, 1988b; 1989), were carried out by **Terry Wagenknecht** (Figs. 3–5), a postdoc who had obtained his Ph.D. working with **David DeRosier** (Brandeis University, Waltham, MA). He also started the ground-breaking work on the structure of the calcium release channel/ryanodine receptor, in collaboration with **Sydney Fleischer** (Vanderbilt University, Nashville, TN), resulting in the first two-dimensional averages of this huge molecule (Wagenknecht *et al.*, 1989b). (Much later the structure of the closed calcium release channel would be solved by us in collaboration with Andy Marks and Wayne Hendrickson (also at Columbia University) and by two competing groups; see Zalk *et al.*, 2015; Des Georges *et al.*, 2016; see Section 4.2).

The methodology of 3D reconstruction was further advanced by a new postdoc, **Pawel Penczek** (Fig. 4) who arrived at my lab with a strong background in signal processing (Penczek *et al.*, 1992; Frank *et al.*, 1992; Penczek *et al.*, 1994; Penczek *et al.*, 1997). The main improvements and breakthroughs reported in these articles were reference-free alignment of images, merging of random-conical reconstructions, angular refinement, and CTF correction via Wiener filtering. Random-conical reconstructions suffer from artifacts related to the lack of information in a conical domain in Fourier space. The remedy, described in Penczek *et al.* (1992), is to merge random-conical data related to different view classes of the same molecule, followed by angular refinement.

A group of structures to which the newly developed reconstruction methods were applied with great success were hemocyanins, which we explored in collaboration with **Jean Lamy** (University of Tours, France) and his student, the late **Nicolas Boisset,** who joined my lab for a year as a postdoc (Fig. 4). During his short life as a scientist (he succumbed to cancer in 2008), Nicolas made exceptional contributions to the study of hemocyanins (Boisset *et al.*, 1990; 1992; 1993a; 1993b; 1994; 1995), as well as to the teaching of the new technique of single-particle reconstruction — some of his slides live on in my lectures and in the lectures of other colleagues in the cryo-EM field. Hemocyanins are hetero-oligomers composed of closely similar monomers, and beginning with our earlier success with the sorting of the *Limulus* hemocyanin images taken by van Heel, they continued to present instructional examples for the application of multivariate data analysis techniques.

3. Focus Shifting from Methodology to Biology: the Ribosome and the Mechanism of Translation

3.1. *Discoveries related to ribosome structure*

The Bacterial Ribosome

A breakthrough in cryo-EM of asymmetric objects was achieved with the reconstruction of the *Escherichia coli* ribosome at 25Å resolution (Fig. 5b; Frank *et al.*, 1995a) by my graduate student **Jun Zhu** from data he collected at the Max-Planck Institute for Medical Research in Heidelberg, where I spent a Sabbatical with **Rasmus Schröder** in **Ken Holmes'** department. Details of image processing leading to this reconstruction, including the use of Wiener filtering for CTF correction, are contained in subsequent papers (Zhu *et al.*, 1997; Penczek *et al.*, 1997). Polypeptide tunnel and mRNA channel were revealed, as well as the complex topology of the intersubunit space. Intersubunit bridges were identified and named for the first time (Frank *et al.*, 1995b). These results marked a turning point in my career because of the realization that the newly emerging technique — cryo-EM combined with single-particle reconstruction — was able to make significant contributions to the study of the mechanism of translation. It was evidently the only method that permitted visualization of molecules in functionally meaningful states as it did not require the formation of crystals, which trap the molecule in a state that is not necessarily representative of its functional repertoire of states. Earlier considerations (Saxton and Frank, 1977) had shown that the larger the molecular complex, the bigger this advantage would become because alignment accuracy is reciprocally related to the particle's diameter.

Already in 1991, when my group obtained the first cryo-EM map of the *E. coli* ribosome, at a mere 45Å resolution, I actively started to recruit postdocs with a background in biochemistry, or — specifically — ribosome biochemistry. **Suman Srivastava** (Fig. 4) and **Rajendra Agrawal** (Figs. 7a, 8b), who had both trained with

Debi Prasad Burma (Banaras Hindu University, Varanasi, India) joined my lab in the early 1990s. Later additions were **Christian Spahn** (Fig. 6), from Knud Nierhaus' lab in Berlin, **Mikel Valle** (Fig. 8), **Jayati Sengupta**, **Urmila Rawat** (Fig. 8), **Sukhjit Kaur, Kakoli Mitra, Greg Allen, Derek Taylor** (Fig. 7b), **Ning Gao**, and **Xabier Agirrezabala** (Fig. 7c). In short order we were able to visualize tRNAs (Agrawal *et al.*, 1996) and EF-G (Agrawal *et al.*, 1998, 1999a) bound to the ribosome. A closer analysis of cryo-EM reconstructions of ribosomes with and without bound EF-G gave rise to the discovery of the ratchet-like motion that the ribosome performs during mRNA-tRNA translocation (Frank and Agrawal, 2000). It was an exciting discovery; in fact it was the first one showing evidence of a large conformational change related to ribosomal function. Many subsequent studies by my group and others followed up on this finding and its implications for the mechanism of translocation. A few papers related to it are included in this compilation, to be mentioned further below.

(a) (b)

Fig. 5. The late 90s (Wadsworth Center, 1998). **(a)** Upper (left to right): Manjuli Sharma, Montserrat Samso. Middle: Bruce McEwen, Terry Wagenknecht, Mike Marko. Lower: J.F., Yanhong Li, Amy Heagle, Doryen Bubeck, Ardean Leith. **(b)** With Jun Zhu (1996).

(a) (b)

Fig. 6. The new Millenium. **(a)** Christian Spahn and Roland Beckmann, at Ribosome Conference in Rensselaerville (2000). **(b)** Bimal Rath, J.F., Elizabeth Cavosie, and Christian Spahn (2002).

(a) (b) (c)

Fig. 7. **(a)** Raj Agrawal; **(b)** Derek Taylor; **(c)** Xabier Agirrezabala.

The cryo-EM structure of the *E. coli* ribosome was refined to 11.5Å from about 130,000 single-particle images (Gabashvili *et al.*, 2000), just at the time the X-ray structures of the subunits were published (*Ban et al., 2000*; *Schluenzen et al., 2000*; *Wimberly et al., 2000*). As part of the effort toward the phasing of the X-ray structure of the large ribosomal subunit of the archaea bacterium *Haloarcula marismortui* in **Thomas Steitz**' lab (Yale University, New Haven, CT), we obtained a 18-Å cryo-EM reconstruction of this subunit (Ban *et al.*, 1998, Penczek *et al.*, 1999). Incidentally, Penczek's study marked the first time Fourier amplitudes from cryo-EM and X-ray crystallography of ribosomes were quantitatively compared. We later obtained a cryo-EM density map at higher resolution (6.7Å) for an *E. coli* ribosome complex to which the ternary complex formed by EF-Tu, aminoacyl-tRNA, and GDP was bound, stabilized in the presence of the antibiotic kirromycin (LeBarron *et al.*, 2008). This density map was subsequently used to build an atomic model by flexible fitting of a structure available in the Protein Data Bank (Trabuco *et al.*, 2008; Villa *et al.*, 2009).

These structural studies on ribosome complexes related to the elongation cycle went hand in hand with the exploration of ribosome binding states during initiation (collaborators **Venki Ramakrishnan**, University of Utah, Salt Lake City, and **Mans Ehrenberg**, Uppsala University, Sweden; McCutcheon *et al.*, 1999; Allen *et al.*, 2005), termination (in collaboration with **Mans Ehrenberg**; Rawat *et al.*, 2003; Rawat *et al.*, 2006; Gao *et al.*, 2007a) and recycling (in collaboration with **Akira Kaji**; Agrawal *et al.*, 2004; Gao *et al.*, 2005; 2007b). In addition, a number of studies focused on Sec61-ribosome interaction and the path of the polypeptide exiting the ribosome (in collaboration with **Nenad Ban**, University of Zurich; Mitra *et al.*, 2005; 2006a; 2006b; Mitra and Frank, 2006).

The Eukaryotic Ribosome

The first forays into the study of eukaryotic ribosomes with cryo-EM were made by **Adriana Verschoor** (Figs. 3, 4), a graduate student and later postdoc who had been with my group from the start (Verschoor and Frank, 1990; Verschoor *et al.*, 1996). In collaboration with **Guenter Blobel** (Rockefeller University, New York), we achieved the first visualization of the translocon bound to the 80S ribosome from yeast, showing the alignment of its central pore with the exit of the polypeptide tunnel of the ribosome (Beckmann *et al.*, 1997). **Roland Beckmann** (Fig. 6a), a student of **Guenter Blobel**, received training in cryo-EM during several visits to my lab at the Wadsworth Center, and later collaborated with my postdoc **Christian Spahn** who joined the lab soon after. This work was followed by another study at higher resolution (Beckmann *et al.*, 2001; Spahn *et al.*, 2001a). At the same time, we succeeded in visualizing the binding of IRES (Internal Ribosome Entry Site) from the hepatitis C virus to the 40S ribosomal subunit, in collaboration with **Jennifer Doudna** (Yale University, New Haven, CT) (Spahn *et al.*, 2001b). The results of this study were remarkable as they implicated conformational changes in the subunit induced by the binding of IRES, which were required for processing the viral mRNA. Binding of another viral IRES — of grasshopper paralysis virus to the human ribosome — was later studied by Christian Spahn, shortly before he left my group (Spahn *et al.*, 2004b).

The binding of eEF2, the equivalent to the translocase EF-G in the 80S ribosome, was the subject of several studies on 80S ribosomes from yeast (Gomez-Lorenzo *et al.*, 2000; Spahn *et al.*, 2004a; Sengupta *et al.*, 2004; 2008) and a closely related thermophilic fungus, *Thermomyces lanuginosus* (Taylor *et al.*, 2007) in collaboration with **Poul Nissen** (University of Aarhus, Denmark). With the improved resolution, an atomic model was built showing the interaction of the 80S ribosome with eEF2 trapped by sordarin (Taylor *et al.*, 2009).

At one of the Science conferences of the Howard Hughes Medical Institute, I met the late **Mariano Levin** (Chagas Institute, Buenos Aires) who proposed collaboration on the structure of the ribosome from *Trypanosoma cruzi*. This collaboration resulted in the first (12Å) reconstruction of a kinetoplastid ribosome (Gao *et al.*, 2005) and led to the discovery of the structure of unusually large rRNA expansion segments in the 40S subunit. This study also spawned a long-term interest, in our lab, in the alternate-host specialization of ribosomes from kinetoplastid parasites. A much higher resolution, 2.5Å, would later be achieved in our lab for the large subunit of the *T. cruzi* ribosome with the help of the direct electron detection camera; see Section 4 below. In collaboration with **Susan Madison-Antennucci** (Wadsworth Center, Albany, New York), we visualized the ribosome of *Trypanosoma brucei*, yielding a 5.5-Å reconstruction (Hashem *et al.*, 2013a). This map allowed us to build an atomic model by flexible fitting via MDFF, making use of Marat Yusupov's X-ray structure of the ribosome from yeast (*Ben-Shem et al., 2010*), as well as homology modeling of proteins and *ab initio* modeling of several rRNA expansion segments in collaboration with **Eric Westhof** (CNRS Strasbourg, France).

For years, the structural basis of **Eukaryotic translation initiation** remained elusive as relevant structures such as the 43S pre-initiation complex containing initiator tRNA, mRNA and factors eIF1, eIF2, and eIF3 proved to be unstable. My long-term collaboration with **Tatyana Pestova** and **Chris Hellen** (SUNY

Downstate Medical Center, Brooklyn) finally bore fruit (Hashem *et al.*, 2013b; Des Georges *et al.*, 2015) mainly as the result of their discovery of a new initiation factor, the helicase DHX29, which apparently contributes toward stabilizing the 43S complex. Still, a multitude of complexes was encountered in the sample, all with different combinations of ligands, and the feat of computational isolation of a nearly complete 43S complex representing just 4.5% of the heterogeneous mix in Hashem *et al.* (2013b) would not have been possible without the progress in classification techniques; in particular the introduction of RELION, as will be expounded in the next section.

3.2. *The need for additional tools*

3D Variance Analysis and Classification. We learned early on in the development of the single-particle analysis method that many of the studies of ribosomal complexes are adversely affected by sample heterogeneity, due to the fact that (i) different molecule species coexist in the sample; or (ii) the same molecule occurs in different binding states, or (iii), as in the case of the pre-translocational ribosome (Agirrezabala *et al.*, 2008; 2012; *Julian et al., 2008*), the molecule equilibrates between two or more states that are chemically identical but possess different conformations. Any compositional, stoichiometric, and conformational heterogeneity obviously violates the main assumption of single-particle reconstruction: that all molecules imaged must be identical and their positions in the frozen sample must be related by rigid-body transformations. This realization called for the development of two new types of tools: one is for the *estimation of three-dimensional variance*, which shows where exactly the regions of high variability are located within the molecule, and the other is a tool for *classification*, to identify and separate the different subpopulations, in order to separately reconstruct them. (To avoid confusion, I must note that this task is different from the task of classifying particles solely according to the views they present, under the assumption that they all originate from the same structure — a problem that was solved in the early 1980s, as described in Section 2, and is now referred to as "2D classification").

Estimation of three-dimensional variance analysis poses a peculiar problem since in the method of single-particle reconstruction, each particle is represented by one projection only while the estimation of 3D variance calls for multiple 3D maps. This problem was first addressed in a *tour de force* study by one of my graduate students, **Weiping Liu** (Fig. 4), leading to two papers published in the *Journal of the Optical Society* (Liu and Frank, 1995; Liu *et al.*, 1995a). It was later addressed in a different way through the bootstrap reconstruction method (Penczek *et al.*, 2005) and an algebraic method developed by **Hstau Liao**, a former graduate student of **Gabor Herman** (University of Philadelphia, PA) (Liao *et al.*, 2015).

So-called supervised classification methods can perform classification of images quite rapidly but they are handicapped by the need to come up with the correct three-dimensional references (or templates). An example of this approach is contained in the article by Valle *et al.* (2002); see particularly the supplement of that paper. Unsupervised methods are evidently preferable, as they identify clusters in the dataset on the basis of proximity relationships within the data, not according to similarity to known models. We collaborated with **Jose Maria Carazo** (Centro Nacional de Biotecnología, Universidad Autónoma de Madrid, Spain) (Fig. 4) and **Gabor Herman** in the demonstration of a novel maximum-likelihood approach to classification (Scheres *et al.*, 2007), which we later used in the investigation of intermediate states during translocation (Agirrezabala *et al.*, 2012). In a revised form, incorporating a Bayesian approach, this program is known as RELION and is now widely used (*Scheres, 2012*).

The power of the new classification programs has profoundly changed the game in structure research by cryo-EM. Instead of trying to obtain a single structure, by optimizing the conditions favorable for stability of a particular complex and by purification, it now often proves easier to sort out the different species of molecules co-existing in the sample, so that they can be reconstructed separately. In the end, one obtains an exhaustive inventory of all the most populated states present in the sample. I have called this approach, first practiced by a group at The Scripps Institute (*Mulder et al., 2010*) and by Fischer and coworkers in Göttingen, Germany (*Fischer et al., 2010*), "Story in a Sample" (Frank, 2010). For this approach to work, one must collect very large datasets, and employ newer, fully automated procedures in the workflow. In the whole emerging picture of single-particle cryo-EM, however, the identification of discrete structures representing most populated states still falls short of the technique's true potential: the depiction of the entire multidimensional ensemble of states in continuum: see separate section below regarding a novel technique that we introduced in 2014 in collaboration with **Abbas Ourmazd** (University of Wisconsin, Milwaukee, WI).

Flexible Fitting. Starting with the publication of the first X-ray structures of the bacterial ribosome and its subunits, in the year 2000, it became possible to interpret cryo-EM density maps in terms of the underlying atomic structures. At first, crude interpretations of cryo-EM maps were obtained by rigid-body fitting. We then started to look for methods of flexible fitting, by deforming and otherwise modifying the X-ray structure — while maintaining the integrity of chemical bonds and rules of stereochemistry — such that the resulting structure would best explain the observed density map. Over the next decade, we participated in three collaborations on flexible fitting, each with a different fitting method, in which the ribosome was taken on as a challenge: (1) real-space refinement, in collaboration with **Michael Chapman** (Florida State University, Tallahassee). The challenge of applying real-space refinement to the gigantic structure of the ribosome was taken on by **Haixiao Gao**, a postdoc with physics background who joined my group at the beginning of the new century (Gao *et al.*, 2003; Gao and Frank, 2005); (2) normal-mode analysis, in collaboration with **Charles Brooks III** (The Scripps Institute,

La Jolla, CA) (Tama *et al.*, 2003; *Tama et al., 2004*); and (3) molecular dynamics flexible fitting "MDFF"), in collaboration with **Klaus Schulten** (University of Illinois at Urbana–Champaign, IL) (Trabuco *et al.*, 2008). We have employed this MDFF tool, now available as a module in Klaus Schulten's program suite VMD, in much of our subsequent structural work.

(a)

(b)

Fig. 8. **(a)** With Christian Spahn, Doryen Bubeck, Urmila Rawat, and Mikel Valle (2003). **(b)** With Rajendra Agrawal at the RNA Conference in Banff (2005), posing as 30S and 50S ribosomal subunits.

Fig. 9. Roof of William Black Building, Columbia University Medical Center, Manhattan (2012). Upper (left to right): Robert Langlois, Harry Kao, Amedee des Georges, Yaser Hashem, J.F., Cheri Hampton, Drew Kennedy, and Tornubari Barinee. Middle: Wen Li, Ming Sun, Bob Grassucci, Sanchaita Das, Melissa Thomas, Amy Jobe, Bingxin Shen, and Nadia Severina. Lower: Nam Ho, Nesh Sharma, Bo Chen, and Hstau Liao.

Flexible fitting, initially by real-space refinement, made it possible to derive atomic models of the pre-translocational ribosome in the two conformations related by the ratchet-like movement, which we later (Frank *et al.*, 2007) termed *macrostates I and II* (researchers employing single-molecule FRET used the term global state I and II). The first application of flexible fitting via MDFF was in the interpretation of the reconstruction mentioned earlier: the 6.7-Å density map of the *E. coli* ribosome bound with the aa-tRNA•EF-Tu•GTPNP ternary complex in the presence of the antibiotic kirromycin (Trabuco *et al.*, 2008; Villa *et al.*, 2009).

Characterization of Domain Motions. The discovery of domain motions and their importance in the functional cycle of translation (see next section) pointed to the need for a quantitative characterization of these motions using reproducible descriptions of domains and defined conventions for coordinate transformations. In collaboration with **Klaus Schulten**'s group we developed such a tool, based on a tensor representation of individual domains such as small subunit body, small subunit head, etc. (Agirrezabala *et al.*, 2011) and applied it to the study of mRNA-tRNA translocation (Agirrezabala *et al.*, 2012). Recently this tool was further developed and integrated with VMD (Maji *et al.*, 2016).

3.3. *Discoveries shedding light on ribosome dynamics*

A series of papers, some of which were already mentioned above, relate to studies that shed light on dynamic events in the elongation cycle. It was cryo-EM, as it presents no obstacles to motions of molecules, that first showed dramatic changes in conformation. The first group of papers was related to mRNA-tRNA translocation. Evidence of conformational changes began to emerge when the resolution crossed the 15-Å mark (Malhotra *et al.*, 1998; Agrawal *et al.*, 1998; 1999). During translocation, and upon binding with EF-G, the bacterial ribosome undergoes a large conformational change, characterized as a ratchet-like intersubunit motion (Frank and Agrawal, 2000; Valle *et al.*, 2003c; Frank *et al.*, 2007). Following the discovery, by Harry Noller's group using FRET methods, that the transition between the two macro states is spontaneous, requiring no elongation factor to be present, we examined a pre-translocational ribosome complex by cryo-EM (in collaboration with **Rachel Green**, Johns Hopkins University, Baltimore, MD) and indeed found the ribosomes coexisting in two different conformations: "unrotated," with tRNAs in the classic A/A, P/P configurations, and "rotated," with tRNAs in the hybrid A/P, P/E configuration (Agirrezabala *et al.*, 2008).

These findings — confirmed by Rodnina's group (*Julian et al., 2008*) — led to the formulation of a "population shift" hypothesis: that the ribosome performs the rotation required for translocation under the influence of the thermal environment, as a Brownian machine, and that EF-G will bind only at a point when its structure fits the ribosome conformation (Frank and Gonzalez, 2010; Frank, 2012). **Ruben Gonzalez'** (Department of Chemistry, Columbia University) experiments indeed showed that the rate of the rotations has the predicted temperature-dependence

(*Wang et al., 2011*). These results gave credence to **Alexander Spirin**'s (Institute of Protein Research Russian Academy of Sciences, Puschino, Russia) earlier view of the ribosome as a molecular machine buffeted by thermal energy (*Spirin, 2002*). Moreover, with the emergence of better classification techniques it was also possible to distinguish intermediate states of the ratchet-like motion (Agirrezabala *et al.*, 2012). These states apparently correspond to local energy minima created by different binding constellation as the subunits rotate with respect to one another. It was fascinating to realize that the ribosome, and by inference many other large molecular machines, conduct their work with the help of thermal energy in the environment, and that evolutionary pressure has optimized molecular architecture with regard to dynamics, not just structure *per se*.

The discovery of intermediate states in an equilibrating sample of the pre-translocational ribosome complex (Agirrezabala *et al.*, 2012) seemingly posed a paradox as such states (characterized by intermediate constellations of ligands/domains labeled with fluorophores) had never been observed by single-molecule FRET (smFRET) in a wild-type ribosome. This paradox was resolved with theoretical calculations which showed that the time-resolution of smFRET was likely insufficient to resolve the expected signals from transient states (Kinz-Thompson *et al.*, 2015).

The second group of papers showing evidence of dynamic features of the ribosome relate to tRNA selection and decoding. During this process, aminoacyl-tRNA enters the ribosome in concert with EF-Tu and GTP as a ternary complex. A reconstruction of the aa-tRNA•EF-Tu•GTPNP complex bound to the ribosome, in the presence of the antibiotic kirromycin, showed that the aa-tRNA is in an unusual, strongly kinked and twisted conformation (Valle *et al.*, 2002; 2003b; Li *et al.*, 2008). Comparison of this bound aa-tRNA with a fully accommodated tRNA suggested that the accommodation of the aa-tRNA, after departure of EF-Tu, could be explained as the relaxation of a spring originally in a high-energy state (Valle *et al.*, 2003b; Frank *et al.*, 2005), and that the stability of the kinked, twisted conformation might set the threshold for proofreading (Yarus *et al.*, 2003). A higher-resolution density map of the ternary complex-bound ribosome, at 6.7Å resolution (LeBarron *et al.*, 2008), was subsequently used in the first demonstration of the flexible fitting method MDFF (Trabuco *et al.*, 2008) and the resulting structure served as the basis for a model for the mechanism by which GTP hydrolysis is triggered (Villa *et al.*, 2009).

4. The "Resolution Revolution"

Revolutions are historical events with profound consequences, resulting in a new order, new insights, or a radically new way to approach a problem. We are taught not to use the word lightly. There is, however, widespread agreement that the introduction of direct electron detecting cameras can be counted as nothing but revolutionary (*Faruqi and Henderson, 2007*).

Following this event, two methods of 3D visualization are now in competition in the world of structure research of biological macromolecules: X-ray crystallography and single-particle cryo-EM. The former is well established and by now has provided more than 100,000 atomic models, all publicly available in the Protein Data Bank. X-ray crystallography, of course, requires crystals, and the ability of this technique to give information on the full range of a molecule's functional conformations is limited. Cryo-EM, on the other hand, does not require crystals, and it is able to provide a full inventory of conformations and compositions co-existing in the sample. However, until recently, the full potential of cryo-EM could not be reached because of the poor performance of electronic detection media. The best resolution for an asymmetric structure, prior to the introduction of the new cameras, was in the range of 5–6Å — see for instance our *T. brucei* ribosome reconstruction at 5.5Å by Hashem *et al.* (2013a). The large discrepancy between the resolution expected and the resolution achieved in the experiment for a given number of particles was brought out in a drastic way in a study by Richard Henderson (1995). According to Henderson's theoretical calculations, 10,000 particles should suffice to obtain a resolution of 3Å for the ribosome, while in practice much lower resolutions were being achieved with that number (see for instance Malhotra *et al.*, 1998, where only 15Å was obtained using about three times as many particles). Thus, Henderson's study pointed out the most serious bottleneck in the flow of information from object to recorded image. All this changed with the advance of direct electron detection cameras, so that current numbers of particles are much better in line with Henderson's predictions.

4.1. *Impact on ribosome research*

In our own ribosome research, the ability to see atomic detail in the interactions between mRNA, tRNA and factors and the ribosome is critical for understanding translation at each step. Previously, in attempts to interpret our cryo-EM maps, we had to rely on flexible fitting or docking X-ray structures deposited in the Protein Data Bank. Three studies reflect the recent progress in our lab and the new normal in the field. The first article (Li *et al.*, 2015) shed light on the interaction between EF-G and the ribosome during mRNA-tRNA translocation, with reconstructions in the 3.6 to 4.2Å range. When using a mutation on the switch II loop of EF-G, which slows down GTP hydrolysis, EF-G was found to bind to the ribosome in two different constellations, one to the unrotated and one to the rotated form. The resolution readily allowed atomic models to be built, which help explain the mechanism of GTPase activation on EF-G during mRNA-tRNA translocation.

The second article (Sun *et al.*, 2015) reported a study of ribosomes purified from a cell extract of the human malaria parasite *Plasmodium falciparum*. It is a good example of the "Story in the Sample" idea as multiple states of the elongation cycle were encountered, which could be analyzed for domain motions and compared with current knowledge from other eukaryotes.

The third article (Liu *et al.*, 2016a) describes a 2.5-Å reconstruction of the *Trypanosoma cruzi* ribosome's large subunit, the highest resolution we have obtained thus far in all of our single-particle studies. Importantly, we see Mg^{2+} and Zn^{2+} ions and even water molecules in the cryo-EM map. In addition, we can make out a large number of rRNA modifications. At this resolution, according to *Wang and Moore (2016)*, the electrostatic (Coulomb) potential distribution obtained by EM differs from the charge density distribution obtained by X-ray crystallography in significant ways, showing off modification sites more clearly. With the greatly improved resolution, it was possible to model large stretches of the sugar-phosphate backbone of rRNA and Cα chains of proteins (Liu *et al.*, 2016b).

4.2. *Foray into channel structures and their gating mechanisms*

Yifan Cheng's (University of California at San Francisco) publication of a 3.4-Å reconstruction of the TRPV channel (*Liao et al., 2013*) marked a huge step in the application of cryo-EM to small membrane-bound channel structures. In the wake of it many X-ray crystallographers started to seek collaborations with cryo-EM labs, or started to import the technique into their own labs. In this way I became involved in an exciting project at Columbia University conducted in the lab of **Alexander Sobolevsky**, the structural characterization of the AMPA receptor, an important glutamate receptor engaged in fast synaptic signal transmission in the central nervous system (Twomey *et al.*, 2016, 2017a, 2017b).

I mentioned earlier that the Ryanodine receptor/calcium release channel was one of the early applications of single-particle reconstruction in my lab while I was still at the Wadsworth Center (Wagenknecht *et al.*, 1989; Radermacher *et al.*, 1994). Fortuitously, a recent collaboration with two labs at Columbia University Medical Center, **Andrew Marks** (Department of Physiology) and **Wayne Hendrickson** (Department of Biochemistry and Molecular Biophysics), gave me an opportunity to contribute to the elucidation of this enormous channel structure (Zalk *et al.*, 2015) and its gating mechanism (des Georges *et al.*, 2016; Dashti *et al.*, 2017) at near-atomic resolution.

5. Adding the Dimension of Time

5.1. *Time-resolved cryo-EM*

In standard single-particle cryo-EM, all molecules in the sample are in equilibrium, and the inventory of structures resulting from classification remain the same. If we wish to follow a reaction over time, then we need to bring the two reactants together at a defined time point and then take samples of the evolving mixture (i.e., reactant1, reactant2, short-lived intermediate(s), product) in defined intervals as function of time. However, with few exceptions, the total reaction time is quite short (in the order of tens to hundreds of milliseconds) compared to the time needed both for depositing the sample on the grid and blotting the grid (in the order of seconds). Time-resolved cryo-EM, therefore, requires a special apparatus for fast mixing, reacting the mixture for a defined time, and fast deposition of the products on the grid. To this end, a microfluidic chip was developed and successfully tested in a collaboration between scientists at the Wadsworth Center and Rensselaer Polytechnic Institute (*Lu et al., 2009*; *Shaikh et al., 2014*).

After moving to Columbia University, in 2008, I assembled a team in my group to import the novel technology and improve on it with the help of **Howard White** (Eastern Virginia Medical School, VA) and, at a later stage, the Engineering Department of Columbia University. My graduate student **Bo Chen** deserves special credit for the rapid implementation of the apparatus, for the success of the first experiments, and for training new students who became part of the team. Three papers came out of these studies so far: Chen *et al.* (2015) studied Mg^{2+}-induced association of ribosomal subunits, a process in the 500 ms range. Fu *et al.* (2016) represented the first successful time-resolved study of a biological process with the microfluidic chip device: the trapping of a short-lived intermediate in the ribosome recycling process. Finally, Feng *et al.* (2017) introduced a newly designed PDMS-based sprayer, which will be integrated with a full PDMS-based microfluidic chip but is also promising in its own right, as a method for fast sample deposition on the grid without the need for blotting.

5.2. *The Free-Energy Landscape: mapping continuous changes in conformation*

There is evidence that many molecules possess conformations in a continuous range; in this case the standard methods of maximum-likelihood classification (Scheres *et al.*, 2007; *Scheres, 2012*; *Lyumkis et al., 2013*), which is based on the assumption of discrete states, merely pick out classes grouped around local maxima of the distribution of conformations occupied. In search of a true ensemble approach, capable of mapping a continuum of states, I started a collaboration with **Abbas Ourmazd** (University of Wisconsin, Milwaukee). Ourmazd's approach is based on a technique called *manifold embedding*: after initial sorting of single-particle images into bins representing projection directions, each bin contains samples from all co-existing states. As such the dataset in the bin forms a *manifold* in multidimensional space. If conformations occur in a continuum, then the manifold can be tracked by a technique called diffusion mapping or manifold embedding. Next, the data are represented in a low-dimensional system of eigenvectors. Consolidation of all manifolds from all projection directions results in a continuous map of state occupancies, which can be converted into a free-energy landscape (Dashti *et al.*, 2014; Frank and Ourmazd, 2016). Remarkably, the application to a data set of ribosome images results in an energy landscape containing a closed-loop low-energy pathway (i.e., a path of high occupancies), and reconstructions from data taken along this pathway exhibit conformational changes known from the elongation cycle of translation (Dashti *et al.*, 2014). We subsequently applied this data analysis to images of the Ryanodine receptor/calcium release channel, bound or unbound with ligands, which were obtained earlier in our lab (des Georges *et al.*, 2016). The resulting energy landscapes for the bound or unbound channel allowed us to trace the most likely pathway for ligand binding (Dashti *et al.*, 2017) — exciting new information unavailable by maximum likelihood methods.

6. Outlook

This brief account of the work spanning a period of half a century shares with high school history textbooks an element of breathlessness, an effect that is unavoidable when years of developments in concepts, techniques, and results are compacted in a few sentences. What I hope will come across is the sense of increasing momentum and excitement in the field, just as the technology of electron microscopy and computation were developing at ever-accelerating pace. It is exhilarating to see the field flourish in the hands of students and young colleagues and to see it come close to the point — thanks to the most recent advances in detector technology — where the structure of asymmetric molecular complexes, even some small membrane-associated ones, can be solved *ab initio* entirely by cryo-EM.

I would like to add that in view of the latest developments, summarized in Section 5, I can foresee a time when deep learning and ensemble-based methods of data analysis, such as the manifold embedding technique described in the last section, will transform single-particle cryo-EM in a profound way: into a tool to study kinetics at atomic resolution.

7. Acknowledgments

I would like to thank **Anders Liljas** for commissioning me to write this book, and for his critical reading of this Summary. Many thanks go to **Melissa Thomas-Baum** (Buckyball Design, New York) for help with the illustrations and to **Masgan Saidi**, my dedicated Administrative Coordinator, for assisting me with the compilation of the literature.

The research I have summarized here has been supported for almost 40 years by the National Institute of Health (NIGMS). During part of this time, from 1998 to 2017, I also received support from the Howard Hughes Medical Institute. My research would not have been possible without the dedicated work of many gifted students and staff members, and the support of patient, congenial collaborators. It was not possible to mention all students and staff members in this short account, but I wish to single out **Bob Grassucci** (Fig. 4) for his outstanding and dedicated support in his role as Lab Manager and Electron Microscope Specialist since 1986. I acknowledge the inspirations and encouragement I received from my colleagues at the Wadsworth Center in Albany during the many years I worked there. I wish to say a special thanks to **Carmen Mannella** (Fig. 4), who gave me his unwavering support and counsel over many years.

8. Comprehensive List of Original Publications

References for all original, peer-reviewed papers (status 3/20/13), ordered by year. Those reproduced in this collection are marked by underlining of first author. Excluded are all invited reviews, book chapters, and books.

Note: The list also includes some references to the work of other groups, provided for context (marked in italics).

__Thon__, F. (1965). (in German) Z. Naturforsch. __20a__: 154–155.

__Thon__, F. (1966). (in German) Zur Defokussierungsabhängigkeit des Phasenkontrastes bei der electronenemikroskopischen Abbildung. [On the defocus dependence of phase contrast in electron microscopic imaging]. Z. Naturforsch. __21a__: 476–478.

__Schiske__, P. (1968) Fourth European Conference on Electron Microscopy, Rome, Vol I, pp. 145–146.

Frank, J. (1969). (in German) Nachweis von Objektbewegungen im lichtoptischen Diffraktogramm von elektronenmikroskopischen Aufnahmen [Detection of object movements in optical diffractograms of electron micrographs]. *Optik* **30**: 171–178. [Translated version of paper included.]

Hoppe, W., Langer, R., Frank, J., and Feltynowski, A. (1969). (in German) Bilddifferenzverfahren in der Elektronenmikroskopie. [Image difference methods in electron microscopy]. *Naturwissenschaften* **56**: 267–272.

Langer, R., Formanek, S., Frank, J., Martin, H.H., and Hoppe, W. (1969). (in German) [Structure of the murein sacculus of *Spirillum serpens.* Quantitative electron microscopy studies at high resolution]. Hoppe Seyler's *Z. Physiol. Chem.* **350**: 1173–1174.

Frank, J. (1972). A study of heavy-light atom discrimination in bright-field electron microscopy using the computer. *Biophys. J.* **12**: 484–511.

Frank, J. (1972). Observation of the relative phases of electron microscopic phase contrast zones with the aid of the optical diffractometer. *Optik* **35**: 608–612.

Frank, J. (1973). A remark on phase determination in electron microscopy. *Optik* **38**: 582–584.

Frank, J. (1973). The envelope of electron microscopic transfer functions for partially coherent illumination. *Optik* **38**: 519–539.

Frank, J. (1974). Radiation damage assessment from electron images using digital correlation methods. *J. Phys. D: Appl. Phys.* **7**: L75–L78.

Frank, J. (1974/75). Phasenbestimmung und Berücksichtigung der Linsenfehler in der Elektronenmikroskopie; eine Klarstellung. [Phase determination and accounting for lens aberrations in electron microscopy — a clarification]. *Optik* **41**: 90–91.

Frank, J. (1975). A practical resolution criterion in optics and electron microscopy. *Optik* **43**: 25–34.

Frank, J. (1975). Averaging of low exposure electron micrographs of non-periodic objects. *Ultramicroscopy* **1**: 159–162.

Frank, J. (1975). Controlled focusing and stigmating in the conventional and scanning transmission electron microscope. *J. Phys. E: Sci. Instrum.* **8**: 582–587.

Frank, J. (1975). Partial coherence and efficient use of electrons in bright field electron microscopy. *Optik* **43**: 103–109.

Frank, J., and Al-Ali, L. (1975). Signal-to-noise ratio of electron micrographs obtained by cross-correlation. *Nature* **256**: 376–379.

***Unwin**, P.N., and Henderson, R. (1975). Molecular structure determination by electron microscopy of unstained crystalline specimens. J. Mol. Biol. **94**: 425–440.*

Frank, J. (1976). Determination of source size and energy spread from electron micrographs using the method of Young's fringes. *Optik* **44**: 379–391.

Saxton, W.O., and Frank, J. (1977). Motif detection in quantum noise-limited electron micrographs by cross-correlation. *Ultramicroscopy* **2**: 219–227.

Wade, R.H., and Frank, J. (1977). Electron microscopic transfer functions for partially coherent axial illumination and chromatic defocus spread. *Optik* **49**: 81–92.

Frank, J., Goldfarb, W., Eisenberg, D., and Baker, T.S. (1978). Reconstruction of glutamine synthetase using computer averaging. *Ultramicroscopy* **3**: 283–290.

Frank, J., McFarlane, S.C., and Downing, K.H. (1978). A note on the effect of source size and defocus spread in bright field electron microscopy. *Optik* **52**: 49–60.

Frank, J., Goldfarb, W., Eisenberg, D., and Baker, T.S. (1979). Addendum to "reconstruction of glutamine synthetase using computer averaging". *Ultramicroscopy* **4**: 247.

Al-Ali, L., and Frank, J. (1980). Resolution estimation in electron microscopy. *Optik* **56**: 31.

Kessel, M., Frank, J., and Goldfarb, W. (1980). Averages of glutamine synthetase molecules as obtained with various skin and electron dose conditions. *J. Supramol. Struct.* **14**: 405–422.

Zingsheim, H.P., Neugebauer, D.C., Barrantes, F.J., and Frank, J. (1980). Structural details of membrane-bound acetylcholine receptor from *Tropedo marmorata. Proc. Natl. Acad. Sci. USA* **77**: 952–956.

Frank, J., Shimkin, B., and Dowse, H. (1981a). SPIDER — A modular software system for image processing. *Ultramicroscopy* **6**: 343–358.

Frank, J., Verschoor, A., and Boublik, M. (1981b). Computer averaging of electron micrographs of 40S ribosomal subunits. *Science* **214**: 1353–1355.

van Heel, M., and Frank, J. (1981). Use of multivariate statistics in analysing the images of biological macromolecules. *Ultramicroscopy* **6**: 187–194.

Frank, J. (1982). Methods for studying the dynamic behavior of biological macromolecules. *Ultramicroscopy* **9**: 3–8.

Frank, J. (1982). New methods for averaging non-periodic objects and distorted crystals in biologic electron microscopy. *Optik* **63**: 67–89.

Frank, J., and van Heel, M. (1982). Correspondence analysis of aligned images of biological particles. *J. Mol. Biol.* **161**: 134–137.

Frank, J., Verschoor, A., and Boublik, M. (1982). Multivariate statistical analysis of ribosome electron micrographs. L and R lateral views of the 40S subunit from HeLa cells. *J. Mol. Biol.* **161**: 107–133.

Lamy, J., Sizaret, P.Y., Frank, J., Verschoor, A., Feldmann, R., and Bonaventura, J. (1982). Architecture of *Limulus polyphemus* hemocyanin. *Biochemistry* **21**: 6825–6833.

Mannella, C., and Frank, J. (1982). Effects of divalent metal ions and chelators on the structure of the outer mitochondrial membranes from *Neurospora crassa. Biophys. J.* **37**: 3–4.

Sizaret, P.Y., Frank, J., Lamy, J., Weill, J., and Lamy, J.N. (1982). A refined quaternary structure of *Androctonus australis* hemocyanin. *Eur. J. Biochem.* **127**: 501–506.

Zingsheim, H.P., Barrantes, F.J., Frank, J., Hänicke, W., and Neugebauer, D.C. (1982). Direct structural localization of two toxin-recognition sites on an ACh receptor protein. *Nature* **299**: 81–84.

Zingsheim, H.P., Neugebauer, D.C., Frank, J., Hänicke, W., and Barrantes, F.J. (1982). Dimeric arrangement and structure of the membrane-bound acetylcholine receptor studied by electron microscopy. *EMBO J.* **1**: 541–547.

Mannella, C.A., Colombini, M., and Frank, J. (1983). Structural and functional evidence for multiple channel complexes in the outer membrane of *Neurospora crassa* mitochondria. *Proc. Natl. Acad. Sci. USA* **80**: 2243–2247.

Frank, J., and Wagenknecht, T. (1984). Automatic selection of molecular images from electron micrographs. *Ultramicroscopy* **12**: 169–176.

Hänicke, W., Frank, J., and Zingsheim, H.P. (1984). Statistical significance of molecule projections by single-particle averaging. *J. Microsc.* **133**: 223–238.

Mannella, C.A., and Frank, J. (1984). Electron microscopic stains as probes of the surface charge of mitochondrial outer membrane channels. *Biophys. J.* **45**: 139–141.

Radermacher, M., and Frank, J. (1984). Representation of three-dimensionally reconstructed objects in electron microscopy by surfaces of equal density. *J. Microsc.* **136**: 77–85.

Verschoor, A., Frank, J., Radermacher, M., Wagenknecht, T., and Boublik, M. (1984). Three-dimensional reconstruction of the 30S ribosomal subunit from randomly oriented particles. *J. Mol. Biol.* **178**: 677–698.

Wagenknecht, T., and Frank, J. (1984). Localization of lipoyl-bearing domains in the alpha-ketoglutarate dehydrogenase multienzyme complex. *Biochemistry* **23**: 3383–3389.

Woodcock, C.L., and Frank, J. (1984). Nucleosome mass distribution using image averaging. *J. Ultrastruct. Res.* **89**: 295–302.

Hui, S.W., and Frank, J. (1985). Analysis of the irregular planar distribution of proteins in membranes. *J. Microsc.* **137**: 293–303.

Kessel, M., Radermacher, M., and Frank, J. (1985). The structure of the stalk surface layer of a brine pond microorganism: correlation averaging applied to a double layered lattice structure. *J. Microsc.* **139**: 63–74.

Kim, C.H., Radermacher, M., Kessel, M., Frank, J., and King, T.E. (1985). Three-dimensional reconstruction of cytochrome oxidase vesicle crystals prepared by cholate solubilization. *J. Inorg. Biochem.* **23**: 163–169.

Lamy, J., Billiald, P., Sizaret, P.Y., Cave, G., Frank, J., and Motta, G. (1985). Approach to the direct intramolecular localization of antigenic determinants in *Androctonus australis* hemocyanin with monoclonal antibodies by molecular immunoelectron microscopy. *Biochemistry* **24**: 5532–5542.

Parsons, D.F., Marko, M., Radermacher, M., and Frank, J. (1985). Shape changes and polarization of cells migrating through tissue. A high-voltage electron microscope and computer graphics study of serial thick sections. *Tissue Cell* **17**: 491–510.

Radermacher, M., and Frank, J. (1985). Use of nonlinear mapping in multivariate image analysis of molecule projections. *Ultramicroscopy* **17**: 117–126.

Subirana, J.A., Munoz-Guerra, S., Aymami, J., Radermacher, M., and Frank, J. (1985). The layered organization of nucleosomes in 30 nm chromatin fibers. *Chromosoma* **91**: 377–390.

Verschoor, A., Frank, J., and Boublik, M. (1985). Investigation of the 50S ribosomal subunit by electron microscopy and image analysis. *J. Ultrastruct. Res.* **92**: 180–189.

Bretaudiere, J.P., and Frank, J. (1986). Reconstitution of molecule images analysed by correspondence analysis: a tool for structural interpretation. *J. Microsc.* **144**: 1–14.

Mannella, C.A., and Frank, J. (1986). Effect of succinylation on images of negatively stained arrays of mitochondrial outer membrane channels. *J. Ultrastruct. Mol. Struct. Res.* **96**, 31–40. PMID: 2445865; PMCid.

Mannella, C.A., Ribeiro, A., and Frank, J. (1986). Structure of the channels in the outer mitochondrial membrane: Electron microscopic studies of the periodic arrays induced by phospholipase A2 treatment of the *Neurospora* membrane. *Biophys. J.* **49**: 307–317.

McEwen, B.F., Radermacher, M., Rieder, C.L., and Frank, J. (1986). Tomographic three-dimensional reconstruction of cilia ultrastructure from thick sections. *Proc. Natl. Acad. Sci. USA* **83**: 9040–9044.

Radermacher, M., Wagenknecht, T., Verschoor, A., and Frank, J. (1986). A new 3D reconstruction scheme applied to the 50S ribosomal subunit of *E. coli*. *J. Microsc.* **141**: RP1–2.

Tsuji, M., Frank, J., and Manley, R.S.J. (1986). Image analysis in the electron microscopy of cellulose protofibrils. II: Digital correlation methods. *Colloid Polym. Sci.* **264**: 89–96.

Verschoor, A., Frank, J., Wagenknecht, T., and Boublik, M. (1986). Computer-averaged views of the 70S monosome from *Escherichia coli*. *J. Mol. Biol.* **187**: 581–590.

Mannella, C.A., Frank, J., and Delihas, N. (1987). Interrelatedness of 5S RNA sequences investigated by correspondence analysis. *J. Mol. Evol.* **24**: 228-235.

Mannella, C.A., Ribeiro, A.J., and Frank, J. (1987). Cytochrome c binds to lipid domains in arrays of mitochondrial outer membrane channels. *Biophys. J.* **51**: 221–226.

Radermacher, M., Wagenknecht, T., Verschoor, A., and Frank, J. (1987). Three-dimensional structure of the large ribosomal subunit from *Escherichia coli*. *EMBO J.* **6**: 1107–1114.

Radermacher, M., Wagenknecht, T., Verschoor, A., and Frank, J. (1987). Three-dimensional reconstruction from a single-exposure, random conical tilt series applied to the 50S ribosomal subunit of *Escherichia coli*. *J. Microsc.* **146**: 113–136.

***van Heel**, M. (1987). Angular reconstitution:* a posteriori *assignment of projection directions fior 3D reconstruction. Ultramicroscopy* ***21**: 111–124.*

Boisset, N., Frank, J., Taveau, J.C., Billiald, P., Motta, G., Lamy, J., and Sizaret, P.Y. (1988). Intramolecular localization of epitopes within an oligomeric protein by immunoelectron microscopy and image processing. *Proteins* **3**: 161–183.

Carazo, J.M., and Frank, J. (1988). Three-dimensional matching of macromolecular structures obtained from electron microscopy: an application to the 70S and 50S E. coli ribosomal particles. *Ultramicroscopy* **25**: 13–22.

Carazo, J.M., Wagenknecht, T., Radermacher, M., Mandiyan, V., Boublik, M., and Frank, J. (1988). Three-dimensional structure of 50S *Escherichia coli* ribosomal subunits depleted of proteins L7/L12. *J. Mol. Biol.* **201**: 393–404.

Frank, J., Bretaudiere, J.P., Carazo, J.M., Verschoor, A., and Wagenknecht, T. (1988). Classification of images of biomolecular assemblies: a study of ribosomes and ribosomal subunits of *Escherichia coli*. *J. Microsc.* **150**: 99–115.

Frank, J., Chiu, W., and Degn, L. (1988). The characterization of structural variations within a crystal field. *Ultramicroscopy* **26**: 345–360.

Saito, A., Inui, M., Radermacher, M., Frank, J., and Fleischer, S. (1988). Ultrastructure of the calcium release channel of sarcoplasmic reticulum. *J. Cell Biol.* **107**: 211–219.

Wagenknecht, T., Frank, J., Boublik, M., Nurse, K., and Ofengand, J. (1988a). Direct localization of the tRNA–anticodon interaction site on the *Escherichia coli* 30S ribosomal subunit by electron microscopy and computerized image averaging. *J. Mol. Biol.* **203**: 753–760.

Wagenknecht, T., Grassucci, R., and Frank, J. (1988b). Electron microscopy and computer image averaging of ice-embedded large ribosomal subunits from *Escherichia coli*. *J. Mol. Biol.* **199**: 137–147.

Carazo, J.M., Wagenknecht, T., and Frank, J. (1989). Variations of the three-dimensional structure of the *Escherichia coli* ribosome in the range of overlap views. An application of the methods of multicone and local single-cone three-dimensional reconstruction. *Biophys. J.* **55**: 465–477.

Unser, M., Trus, B.L., Frank, J., and Steven, A.C. (1989). The spectral signal-to-noise ratio resolution criterion: computational efficiency and statistical precision. *Ultramicroscopy* **30**: 429–433.

Verschoor, A., Zhang, N.Y., Wagenknecht, T., Obrig, T., Radermacher, M., and Frank, J. (1989). Three-dimensional reconstruction of mammalian 40S ribosomal subunit. *J. Mol. Biol.* **209**: 115–126.

Wagenknecht, T., Carazo, J.M., Radermacher, M., and Frank, J. (1989). Three-dimensional reconstruction of the ribosome from *Escherichia coli*. *Biophys. J.* **55**: 455–464.

Wagenknecht, T., Grassucci, R., Frank, J., Saito, A., Inui, M., and Fleischer, S. (1989). Three-dimensional architecture of the calcium channel/foot structure of sarcoplasmic reticulum. *Nature* **338**: 167–170.

Boisset, N., Taveau, J.C., Lamy, J., Wagenknecht, T., Radermacher, M., and Frank, J. (1990). Three-dimensional reconstruction of native *Androctonus australis* hemocyanin. *J. Mol. Biol.* **216**: 743–760.

Carazo, J.M., Rivera, F.F., Zapata, E.L., Radermacher, M., and Frank, J. (1990). Fuzzy sets-based classification of electron microscopy images of biological macromolecules with an application to ribosomal particles. *J. Microsc.* **157**: 187–203.

Verschoor, A., and Frank, J. (1990). Three-dimensional structure of the mammalian cytoplasmic ribosome. *J. Mol. Biol.* **214**: 737–749.

Frank, J., Penczek, P., Grassucci, R., and Srivastava, S. (1991). Three-dimensional reconstruction of the 70S *Escherichia coli* ribosome in ice: the distribution of ribosomal RNA. *J. Cell. Biol.* **115**: 597–605.

Woodcock, C.L., McEwen, B.F., and Frank, J. (1991). Ultrastructure of chromatin. II. Three-dimensional reconstruction of isolated fibers. *J. Cell Sci.* **99**(Pt 1): 107–114.

Barnard, D.P., Turner, J.N., Frank, J., and McEwen, B.F. (1992). A 360 degrees single-axis tilt stage for the high-voltage electron microscope. *J. Microsc.* **167**: 39–48.

Boisset, N., Grassucci, R., Penczek, P., Delain, E., Pochon, F., Frank, J., and Lamy, J.N. (1992). Three-dimensional reconstruction of a complex of human alpha 2-macroglobulin with monomaleimido Nanogold (Au1.4nm) embedded in ice. *J. Struct. Biol.* **109**: 39–45.

Frank, J., Penczek, P., and Liu, W. (1992). Alignment, classification, and three-dimensional reconstruction of single particles embedded in ice. *Scanning Microsc. Suppl.* **6**: 11–20.

Frank, J., and Radermacher, M. (1992). Three-dimensional reconstruction of single particles negatively stained or in vitreous ice. *Ultramicroscopy* **46**: 241–262.

Morgan, D.G., Grant, R.A., Chiu, W., and Frank, J. (1992). Patch averaging of electron images of GP3*I crystals with variable thickness. *J. Struct. Biol.* **108**: 245–256.

Penczek, P., Radermacher, M., and Frank, J. (1992). Three-dimensional reconstruction of single particles embedded in ice. *Ultramicroscopy* **40**: 33–53.

Radermacher, M., Wagenknecht, T., Grassucci, R., Frank, J., Inui, M., Chadwick, C., and Fleischer, S. (1992). Cryo-EM of the native structure of the calcium release channel/ryanodine receptor from sarcoplasmic reticulum. *Biophys. J.* **61**: 936–940.

Srivastava, S., Verschoor, A., and Frank, J. (1992). Eukaryotic initiation factor 3 does not prevent association through physical blockage of the ribosomal subunit-subunit interface. *J. Mol. Biol.* **226**: 301–304.

Wilson, C.J., Mastronarde, D.N., McEwen, B., and Frank, J. (1992). Measurement of neuronal surface area using high-voltage electron microscope tomography. *Neuroimage* **1**: 11–22.

Boisset, N., Penczek, P., Pochon, F., Frank, J., and Lamy, J. (1993a). Three-dimensional architecture of human alpha 2-macroglobulin transformed with methylamine. *J. Mol. Biol.* **232**: 522–529.

Boisset, N., Radermacher, M., Grassucci, R., Taveau, J.C., Liu, W., Lamy, J., Frank, J., and Lamy, J.N. (1993b). Three-dimensional immunoelectron microscopy of scorpion hemocyanin labeled with a monoclonal Fab fragment. *J. Struct. Biol.* **111**: 234–244.

Frank, J., Chiu, W., and Henderson, R. (1993). Flopping polypeptide chains and Suleika's subtle imperfections: analysis of variations in the electron micrograph of a purple membrane crystal. *Ultramicroscopy* **49**: 387–396.

McEwen, B.F., Arena, J.T., Frank, J., and Rieder, C.L. (1993). Structure of the colcemid-treated PtK1 kinetochore outer plate as determined by high voltage electron microscopic tomography. *J. Cell Biol.* **120**: 301–312.

Boisset, N., Penczek, P., Pochon, F., Frank, J., and Lamy, J. (1994). Three-dimensional reconstruction of human alpha 2-macroglobulin and refinement of the localization of thiol ester bonds with monomaleimido nanogold. *Ann. NY Acad. Sci.* **737**: 229–244.

Lambert, O., Boisset, N., Penczek, P., Lamy, J., Taveau, J.C., Frank, J., and Lamy, J.N. (1994). Quaternary structure of *Octopus vulgaris* hemocyanin. Three-dimensional reconstruction from frozen-hydrated specimens and intramolecular location of functional units Ove and Ovb. *J. Mol. Biol.* **238**: 75–87. PMID: 8145258; PMCid:

Mannella, C.A., Marko, M., Penczek, P., Barnard, D., and Frank, J. (1994). The internal compartmentation of rat-liver mitochondria: tomographic study using the high-voltage transmission electron microscope. *Microsc. Res. Tech.* **27**: 278–283.

Penczek, P.A., Grassucci, R.A., and Frank, J. (1994). The ribosome at improved resolution: new techniques for merging and orientation refinement in 3D cryo-electron microscopy of biological particles. *Ultramicroscopy* **53**: 251–270.

Radermacher, M., Rao, V., Grassucci, R., Frank, J., Timerman, A.P., Fleischer, S., and Wagenknecht, T. (1994). Cryo-electron microscopy and three-dimensional reconstruction of the calcium release channel/ryanodine receptor from skeletal muscle. *J. Cell Biol.* **127**: 411–423.

Boisset, N., Penczek, P., Taveau, J.C., Lamy, J., and Frank, J. (1995). Three-dimensional reconstruction of *Androctonus australis* hemocyanin labeled with a monoclonal Fab fragment. *J. Struct. Biol.* **115**: 16–29.

Frank, J., and Penczek, P. (1995). On the correction of the contrast transfer function in biological electron microscopy. *Optik* **98**: 125–129.

Frank, J., Zhu, J., Penczek, P., Li, Y., Srivastava, S., Verschoor, A., Radermacher, M., Grassucci, R., Lata, R.K., and Agrawal, R.K. (1995a). A model of protein synthesis based on cryo-electron microscopy of the *E. coli* ribosome. *Nature* **376**: 441–444.

Frank, J., Verschoor, A., Li, Y., Zhu, J., Lata, R.K., Radermacher, M., Penczek, P., Grassucci, R., Agrawal, R.K., and Srivastava, S. (1995b). A model of the translational apparatus based on a three-dimensional reconstruction of the *Escherichia coli* ribosome. *Biochem. Cell Biol.* **73**: 757–765.

Lata, K.R., Penczek, P., and Frank, J. (1995). Automatic particle picking from electron micrographs. *Ultramicroscopy* **58**: 381–391.

Liu, W., Boisset, N., and Frank, J. (1995). Estimation of variance distribution in three-dimensional reconstruction. II. Applications. *J. Opt. Soc. Am.* A **12**: 2628–2635.

Liu, W., and Frank, J. (1995). Estimation of variance distribution in three-dimensional reconstruction. I. Theory. *J. Opt. Soc. Am.* A **12**: 2615–2627.

Liu, Y., Penczek, P.A., McEwen, B.F., and Frank, J. (1995). A marker-free alignment method for electron tomography. *Ultramicroscopy* **58**: 393–402.

Penczek, P., Marko, M., Buttle, K., and Frank, J. (1995). Double-tilt electron tomography. *Ultramicroscopy* **60**: 393–410.

Srivastava, S., Verschoor, A., Radermacher, M., Grassucci, R., and Frank, J. (1995). Three-dimensional reconstruction of mammalian 40S ribosomal subunit embedded in ice. *J. Mol. Biol.* **245**: 461–466.

Agrawal, R.K., Penczek, P., Grassucci, R.A., Li, Y., Leith, A., Nierhaus, K.H., and Frank, J. (1996). Direct visualization of A-, P-, and E-site transfer RNAs in the *Escherichia coli* ribosome. *Science* **271**: 1000–1002.

Frank, J., Radermacher, M., Penczek, P., Zhu, J., Li, Y., Ladjadj, M., and Leith, A. (1996). SPIDER and WEB: processing and visualization of images in 3D electron microscopy and related fields. *J. Struct. Biol.* **116**: 190–199.

Judson, *H.F. (1996). The Eighth Day of Creation. Cold Spring Harbor Laboratory Press, 2013.*

Lata, K.R., Agrawal, R.K., Penczek, P., Grassucci, R., Zhu, J., and Frank, J. (1996). Three-dimensional reconstruction of the *Escherichia coli* 30S ribosomal subunit in ice. *J. Mol. Biol.* **262**: 43–52.

Marabini, R., Vaquerizo, C., Fernandez, J.J., Carazo, J.M., Engel, A., and Frank, J. (1996). Proposal for a new distributed database of macromolecular and subcellular structures from different areas of microscopy. *J. Struct. Biol.* **116**: 161–165.

Penczek, P.A., Zhu, J., and Frank, J. (1996). A common-lines based method for determining orientations for N > 3 particle projections simultaneously. *Ultramicroscopy* **63**: 205–218.

Verschoor, A., Srivastava, S., Grassucci, R., and Frank, J. (1996). Native 3D structure of eukaryotic 80s ribosome: morphological homology with E. coli 70S ribosome. *J. Cell Biol.* **133**: 495–505.

Beckmann, R., Bubeck, D., Grassucci, R., Penczek, P., Verschoor, A., Blobel, G., and Frank, J. (1997). Alignment of conduits for the nascent polypeptide chain in the ribosome-Sec61 complex. *Science* **278**: 2123–2126.

Li, Y., Leith, A., and Frank, J. (1997). Tinkerbell — a tool for interactive segmentation of 3D data. *J. Struct. Biol.* **120**: 266–275.

Penczek, P.A., Zhu, J., Schröder, R., and Frank, J. (1997). Three dimensional reconstruction with contrast transfer compensation from defocus series. *Scanning Microsc.* **11**: 147–154.

Rath, B.K., Marko, M., Radermacher, M., and Frank, J. (1997). Low-dose automated electron tomography: a recent implementation. *J. Struct. Biol.* **120**: 210–218.

Wadzack, J., Burkhardt, N., Jünemann, R., Diedrich, G., Nierhaus, K.H., Frank, J., Penczek, P., Meerwinck, W., Schmitt, M., Willumeit, R., and Stuhrmann, H.B. (1997). Direct localization of the tRNAs within the elongating ribosome by means of neutron scattering (proton-spin contrast-variation). *J. Mol. Biol.* **266**: 343–356.

Zhu, J., Penczek, P.A., Schröder, R., and Frank, J. (1997). Three-dimensional reconstruction with contrast transfer function correction from energy-filtered cryoelectron micrographs: procedure and application to the 70S *Escherichia coli* ribosome. *J. Struct. Biol.* **118**: 197–219.

Agrawal, R.K., Penczek, P., Grassucci, R.A., and Frank, J. (1998). Visualization of elongation factor G on the *Escherichia coli* 70S ribosome: the mechanism of translocation. *Proc. Natl. Acad. Sci. USA* **95**: 6134–6138.

Ban, N., Freeborn, B., Nissen, P., Penczek, P., Grassucci, R.A., Sweet, R., Frank, J., Moore, P.B., and Steitz, T.A. (1998). A 9Å resolution X-ray crystallographic map of the large ribosomal subunit. *Cell* **93**: 1105–1115.

Frank, J., and Agrawal, R.K. (1998). The movement of tRNA through the ribosome. *Biophys. J.* **74**: 589–594.

Malhotra, A., Penczek, P., Agrawal, R.K., Gabashvili, I.S., Grassucci, R.A., Jünemann, R., Burkhardt, N., Nierhaus, K.H., and Frank, J. (1998). *Escherichia coli* 70S ribosome at 15Å resolution by cryo-electron microscopy: localization of fMet-tRNA^{fMet} and fitting of L1 protein. *J. Mol. Biol.* **280**: 103–116.

Samso, M., Radermacher, M., Frank, J., and Koonce, M.P. (1998). Structural characterization of a dynein motor domain. *J. Mol. Biol.* **276**: 927–937.

Verschoor, A., Warner, J.R., Srivastava, S., Grassucci, R.A., and Frank, J. (1998). Three-dimensional structure of the yeast ribosome. *Nucl. Acids Res.* **26**: 655–661.

Agrawal, R.K., Heagle, A.B., Penczek, P., Grassucci, R.A., and Frank, J. (1999). EF-G-dependent GTP hydrolysis induces translocation accompanied by large conformational changes in the 70S ribosome. *Nat. Struct. Biol.* **6**: 643–647.

Agrawal, R.K., Penczek, P., Grassucci, R.A., Burkhardt, N., Nierhaus, K.H., and Frank, J. (1999). Effect of buffer conditions on the position of tRNA on the 70S ribosome as visualized by cryoelectron microscopy. *J. Biol. Chem.* **274**: 8723–8729.

Frank, J., Heagle, A.B., and Agrawal, R.K. (1999). Animation of the dynamical events of the elongation cycle based on cryoelectron microscopy of functional complexes of the ribosome. *J. Struct. Biol.* **128**: 15–18.

Gabashvili, I.S., Agrawal, R.K., Grassucci, R., and Frank, J. (1999). Structure and structural variations of the *Escherichia coli* 30S ribosomal subunit as revealed by three-dimensional cryo-electron microscopy. *J. Mol. Biol.* **286**: 1285–1291.

Gabashvili, I.S., Agrawal, R.K., Grassucci, R., Squires, C.L., Dahlberg, A.E., and Frank, J. (1999). Major rearrangements in the 70S ribosomal 3D structure caused by a conformational switch in 16S ribosomal RNA. *EMBO J.* **18**: 6501–6507.

McCutcheon, J.P., Agrawal, R.K., Philips, S.M., Grassucci, R.A., Gerchman, S.E., Clemons, W.M., Jr., Ramakrishnan, V., and Frank, J. (1999). Location of translational initiation factor IF3 on the small ribosomal subunit. *Proc. Natl. Acad. Sci. USA* **96**: 4301–4306.

Penczek, P., Ban, N., Grassucci, R.A., Agrawal, R.K., and Frank, J. (1999). Haloarcula marismortui 50S subunit-complementarity of electron microscopy and X-Ray crystallographic information. *J. Struct. Biol.* **128**: 44–50.

Spahn, C.M., Grassucci, R.A., Penczek, P., and Frank, J. (1999). Direct three-dimensional localization and positive identification of RNA helices within the ribosome by means of genetic tagging and cryo-electron microscopy. *Structure* **7**: 1567–1573.

Agrawal, R.K., Spahn, C.M., Penczek, P., Grassucci, R.A., Nierhaus, K.H., and Frank, J. (2000). Visualization of tRNA movements on the *Escherichia coli* 70S ribosome during the elongation cycle. *J. Cell Biol.* **150**: 447–460.

Frank, J., and Agrawal, R.K. (2000). A ratchet-like inter-subunit reorganization of the ribosome during translocation. *Nature* **406**: 318–322.

Gabashvili, I.S., Agrawal, R.K., Spahn, C.M., Grassucci, R.A., Svergun, D.I., Frank, J., and Penczek, P. (2000). Solution structure of the *E. coli* 70S ribosome at 11.5 Å resolution. *Cell* **100**: 537–549.

Gomez-Lorenzo, M.G., Spahn, C.M., Agrawal, R.K., Grassucci, R.A., Penczek, P., Chakraburtty, K., Ballesta, J.P., Lavandera, J.L., Garcia-Bustos, J.F., and Frank, J. (2000). Three-dimensional cryo-electron microscopy localization of EF2 in the *Saccharomyces cerevisiae* 80S ribosome at 17.5 Å resolution. *EMBO J.* **19**: 2710–2718.

Spahn, C.M., Penczek, P.A., Leith, A., and Frank, J. (2000). A method for differentiating proteins from nucleic acids in intermediate-resolution density maps: cryo-electron microscopy defines the quaternary structure of the *Escherichia coli* 70S ribosome. *Structure* **8**: 937–948.

VanLoock, M.S., Agrawal, R.K., Gabashvili, I.S., Qi, L., Frank, J., and Harvey, S.C. (2000). Movement of the decoding region of the 16S ribosomal RNA accompanies tRNA translocation. *J. Mol. Biol.* **304**: 507–515.

Wriggers, W., Agrawal, R.K., Drew, D.L., McCammon, A., and Frank, J. (2000). Domain motions of EF-G bound to the 70S ribosome: insights from a hand-shaking between multi-resolution structures. *Biophys. J.* **79**: 1670–1678.

***Ban**, N., Nissen., P., Hansen, J., Moore, P.B., and Steitz, T.A. (2000). The complete atomic structure of the large ribosomal subunit at 2.4 Å resolution. Science **289**: 905–920.*

***Schlünzen**, F., Tocilj, A., Zarivach, R., Harms, J., Gluehmann, M., Janell, D., Bashan, A., Bartels, H., Agmon, I., Franceschi, F., and Yonath, A. (2000). Structure of functionally activated small ribosomal subunit at 3.3 Å resolution. Cell **102**: 615–623.*

*Wimberly, B.T., Brodersen, D.E., Clemons, Jr, W.M., Morgan-Warren, R.J., Carter, A.P., Vonrhein, C., Hartsch, T., and Ramakrishnan, V. (2000). Structure of the 30S ribosomal subunit. Nature **407**: 327–339.*

Agrawal, R.K., Linde, J., Sengupta, J., Nierhaus, K.H., and Frank, J. (2001). Localization of L11 protein on the ribosome and elucidation of its involvement in EF-G-dependent translocation. *J. Mol. Biol.* **311**: 777–787.

Beckmann, R., Spahn, C.M., Eswar, N., Helmers, J., Penczek, P.A., Sali, A., Frank, J., and Blobel, G. (2001). Architecture of the protein-conducting channel associated with the translating 80S ribosome. *Cell* **107**: 361–372.

Gabashvili, I.S., Gregory, S.T., Valle, M., Grassucci, R., Worbs, M., Wahl, M.C., Dahlberg, A.E., and Frank, J. (2001). The polypeptide tunnel system in the ribosome and its gating in erythromycin resistance mutants of L4 and L22. *Mol. Cell* **8**: 181–188.

McEwen, B.F., and Frank, J. (2001). Electron tomographic and other approaches for imaging molecular machines. *Cur. Opin. Neurobiol.* **11**: 594–600.

Sengupta, J., Agrawal, R.K., and Frank, J. (2001). Visualization of protein S1 within the 30S ribosomal subunit and its interaction with messenger RNA. *Proc. Natl. Acad. Sci. USA* **98**: 11991–11996.

Spahn, C.M., Beckmann, R., Eswar, N., Penczek, P.A., Sali, A., Blobel, G., and Frank, J. (2001a). Structure of the 80S ribosome from *Saccharomyces cerevisiae* — tRNA-ribosome and subunit–subunit interactions. *Cell* **107**: 373–386.

Spahn, C.M., Kieft, J.S., Grassucci, R.A., Penczek, P.A., Zhou, K., Doudna, J.A., and Frank, J. (2001b). Hepatitis C virus IRES RNA-induced changes in the conformation of the 40S ribosomal subunit. *Science* **291**: 1959–1962.

Spahn, C.M., Blaha, G., Agrawal, R.K., Penczek, P., Grassucci, R.A., Trieber, C.A., Connell, S.R., Taylor, D.E., Nierhaus, K.H., and Frank, J. (2001c). Localization of the ribosomal protection protein Tet(O) on the ribosome and the mechanism of tetracycline resistance. *Mol. Cell* **7**: 1037–1045.

Gao, H., Spahn, C.M., Grassucci, R.A., and Frank, J. (2002). An assay for local quality in cryo-electron micrographs of single particles. *Ultramicroscopy* **93**: 169–178.

Hsieh, C.E., Marko, M., Frank, J., and Mannella, C.A. (2002). Electron tomographic analysis of frozen-hydrated tissue sections. *J. Struct. Biol.* **138**: 63–73.

***Spirin**, A. S. (2002). Ribosome as a molecular machine. FEBS Lett. **514**: 2–10.*

Stagg, S.M., Valle, M., Agrawal, R.K., Frank, J., and Harvey, S.C. (2002). Problems with the transorientation hypothesis. *RNA* **8**: 1093–1094.

Valle, M., Sengupta, J., Swami, N.K., Grassucci, R.A., Burkhardt, N., Nierhaus, K.H., Agrawal, R.K., and Frank, J. (2002). Cryo-EM reveals an active role for aminoacyl-tRNA in the accommodation process. *EMBO J.* **21**: 3557–3567.

Gao, H., Sengupta, J., Valle, M., Korostelev, A., Eswar, N., Stagg, S.M., Van Roey, P., Agrawal, R.K., Harvey, S.C., Sali, A., Chapman, M.S., and Frank, J. (2003). Study of the structural dynamics of the *E. coli* 70S ribosome using real-space refinement. *Cell* **113**: 789–801.

Gomez-Lorenzo, M.G., Valle, M., Frank, J., Gruss, C., Sorzano, C.O., Chen, X.S., Donate, L.E., and Carazo, J.M. (2003). Large T antigen on the simian virus 40 origin of replication: a 3D snapshot prior to DNA replication. *EMBO J.* **22**: 6205–6213.

Moccia, R., Chen, D., Lyles, V., Kapuya, E., E, Y., Kalachikov, S., Spahn, C.M., Frank, J., Kandel, E.R., Barad, M., and Martin, K.C. (2003). An unbiased cDNA library prepared from isolated *Aplysia* sensory neuron processes is enriched for cytoskeletal and translational mRNAs. *J. Neurosci.* **23**: 9409–9417.

Rath, B.K., Hegerl, R., Leith, A., Shaikh, T.R., Wagenknecht, T., and Frank, J. (2003). Fast 3D motif search of EM density maps using a locally normalized cross-correlation function. *J. Struct. Biol.* **144**: 95–103.

Rawat, U.B., Zavialov, A.V., Sengupta, J., Valle, M., Grassucci, R.A., Linde, J., Vestergaard, B., Ehrenberg, M., and Frank, J. (2003). A cryo-electron microscopic study of ribosome-bound termination factor RF2. *Nature* **421**: 87–90.

Shaikh, T.R., Hegerl, R., and Frank, J. (2003). An approach to examining model dependence in EM reconstructions using cross-validation. *J. Struct. Biol.* **142**: 301–310.

Tama, F., Valle, M., Frank, J., and Brooks, C.L., 3rd (2003). Dynamic reorganization of the functionally active ribosome explored by normal mode analysis and cryo-electron microscopy. *Proc. Natl. Acad. Sci. USA* **100**: 9319–9323.

Valle, M., Gillet, R., Kaur, S., Henne, A., Ramakrishnan, V., and Frank, J. (2003). Visualizing tmRNA entry into a stalled ribosome. *Science* **300**: 127–130.

Valle, M., Zavialov, A., Li, W., Stagg, S.M., Sengupta, J., Nielsen, R.C., Nissen, P., Harvey, S.C., Ehrenberg, M., and Frank, J. (2003). Incorporation of aminoacyl-tRNA into the ribosome as seen by cryo-electron microscopy. *Nat. Struct. Biol.* **10**: 899–906.

Valle, M., Zavialov, A., Sengupta, J., Rawat, U., Ehrenberg, M., and Frank, J. (2003). Locking and unlocking of ribosomal motions. *Cell* **114**: 123–134.

Yarus, M., Valle, M., and Frank, J. (2003). A twisted tRNA intermediate sets the threshold for decoding. *RNA* **9**: 384–385.

Agrawal, R.K., Sharma, M.R., Kiel, M.C., Hirokawa, G., Booth, T.M., Spahn, C.M., Grassucci, R.A., Kaji, A., and Frank, J. (2004). Visualization of ribosome-recycling factor on the *Escherichia coli* 70S ribosome: functional implications. *Proc. Natl. Acad. Sci. USA* **101**: 8900–8905.

Gao, H., Valle, M., Ehrenberg, M., and Frank, J. (2004). Dynamics of EF-G interaction with the ribosome explored by classification of a heterogeneous cryo-EM dataset. *J. Struct. Biol.* **147**: 283–290.

Halic, M., Becker, T., Pool, M.R., Spahn, C.M., Grassucci, R.A., Frank, J., and Beckmann, R. (2004). Structure of the signal recognition particle interacting with the elongation-arrested ribosome. *Nature* **427**: 808–814.

Nilsson, J., Sengupta, J., Frank, J., and Nissen, P. (2004). Regulation of eukaryotic translation by the RACK1 protein: a platform for signalling molecules on the ribosome. *EMBO Rep.* **5**: 1137–1141.

Rath, B.K. and Frank, J. (2004). Fast automatic particle picking from cryo-electron micrographs using a locally normalized cross-correlation function: a case study. *J. Struct. Biol.* **145**: 84–90.

Sengupta, J., Nilsson, J., Gursky, R., Spahn, C.M., Nissen, P., and Frank, J. (2004). Identification of the versatile scaffold protein RACK1 on the eukaryotic ribosome by cryo-EM. *Nat. Struct. Mol. Biol.* **11**: 957–962.

Spahn, C.M., Gomez-Lorenzo, M.G., Grassucci, R.A., Jorgensen, R., Andersen, G.R., Beckmann, R., Penczek, P.A., Ballesta, J.P., and Frank, J. (2004). Domain movements of elongation factor eEF2 and the eukaryotic 80S ribosome facilitate tRNA translocation. *EMBO J.* **23**: 1008–1019.

Spahn, C.M., Jan, E., Mulder, A., Grassucci, R.A., Sarnow, P., and Frank, J. (2004). Cryo-EM visualization of a viral internal ribosome entry site bound to human ribosomes: the IRES functions as an RNA-based translation factor. *Cell* **118**: 465–475.

Tama, F. *Miyashita*, O. and *Brooks, C.L. III (2004). Normal mode based flexible fitting* of *high-resolution structure into low-resolution experimental data* from *cryo-EM. J. Struct. Biol.* ***147* *(2004)*:** *315–326.*

Adiga, U., Baxter, W.T., Hall, R.J., Rockel, B., Rath, B.K., Frank, J., and Glaeser, R. (2005). Particle picking by segmentation: a comparative study with SPIDER-based manual particle picking. *J. Struct. Biol.* **152**: 211–220.

Allen, G.S., Zavialov, A., Gursky, R., Ehrenberg, M., and Frank, J. (2005). The cryo-EM structure of a translation initiation complex from *Escherichia coli*. *Cell* **121**: 703–712.

Datta, P.P., Sharma, M.R., Qi, L., Frank, J., and Agrawal, R.K. (2005). Interaction of the G' domain of elongation factor G and the C-terminal domain of ribosomal protein L7/L12 during translocation as revealed by cryo-EM. *Mol. Cell* **20**: 723–731.

Gao, H., Ayub, M.J., Levin, M.J., and Frank, J. (2005). The structure of the 80S ribosome from *Trypanosoma cruzi* reveals unique rRNA components. *Proc. Natl. Acad. Sci. USA* **102**: 10206–10211.

Gao, H., and Frank, J. (2005). Molding atomic structures into intermediate-resolution cryo-EM density maps of ribosomal complexes using real-space refinement. *Structure* **13**: 401–406.

Gao, N., Zavialov, A.V., Li, W., Sengupta, J., Valle, M., Gursky, R.P., Ehrenberg, M., and Frank, J. (2005). Mechanism for the disassembly of the posttermination complex inferred from cryo-EM studies. *Mol. Cell* **18**: 663–674.

Halic, M., Becker, T., Frank, J., Spahn, C.M., and Beckmann, R. (2005). Localization and dynamic behavior of ribosomal protein L30e. *Nat. Struct. Mol. Biol.* **12**: 467–468.

Lei, J., and Frank, J. (2005). Automated acquisition of cryo-electron micrographs for single particle reconstruction on an FEI Tecnai electron microscope. *J. Struct. Biol.* **150**: 69–80.

Mitra, K., Schaffitzel, C., Shaikh, T., Tama, F., Jenni, S., Brooks, C.L., 3rd, Ban, N., and Frank, J. (2005). Structure of the E. coli protein-conducting channel bound to a translating ribosome. *Nature* **438**: 318–324.

Hsieh, C.E., Leith, A., Mannella, C.A., Frank, J., and Marko, M. (2006). Towards high-resolution three-dimensional imaging of native mammalian tissue: electron tomography of frozen-hydrated rat liver sections. *J. Struct. Biol.* **153**: 1–13.

Kaur, S., Gillet, R., Li, W., Gursky, R., and Frank, J. (2006). Cryo-EM visualization of transfer messenger RNA with two SmpBs in a stalled ribosome. *Proc. Natl. Acad. Sci. USA* **103**: 16484–16489.

Li, W., Sengupta, J., Rath, B.K., and Frank, J. (2006). Functional conformations of the L11-ribosomal RNA complex revealed by correlative analysis of cryo-EM and molecular dynamics simulations. *RNA* **12**: 1240–1253.

Marko, M., Hsieh, C., Moberlychan, W., Mannella, C.A., and Frank, J. (2006). Focused ion beam milling of vitreous water: prospects for an alternative to cryo-ultramicrotomy of frozen-hydrated biological samples. *J. Microsc.* **222**: 42–47.

Mitra, K., and Frank, J. (2006). A model for co-translational translocation: ribosome-regulated nascent polypeptide translocation at the protein-conducting channel. *FEBS Lett.* **580**: 3353–3360.

Mitra, K., Frank, J., and Driessen, A. (2006a). Co- and post-translational translocation through the protein-conducting channel: analogous mechanisms at work? *Nat. Struct. Mol. Biol.* **13**: 957–964.

Mitra, K., Schaffitzel, C., Fabiola, F., Chapman, M.S., Ban, N., and Frank, J. (2006b). Elongation arrest by SecM via a cascade of ribosomal RNA rearrangements. *Mol. Cell* **22**: 533–543.

Penczek, P.A., Frank, J., and Spahn, C.M. (2006). A method of focused classification, based on the bootstrap 3D variance analysis, and its application to EF-G-dependent translocation. *J. Struct. Biol.* **154**: 184–194.

Penczek, P.A., Yang, C., Frank, J., and Spahn, C.M. (2006). Estimation of variance in single-particle reconstruction using the bootstrap technique. *J. Struct. Biol.* **154**: 168–183.

Rawat, U., Gao, H., Zavialov, A., Gursky, R., Ehrenberg, M., and Frank, J. (2006). Interactions of the release factor RF1 with the ribosome as revealed by cryo-EM. *J. Mol. Biol.* **357**: 1144–1153.

Slagter-Jager, J.G., Allen, G.S., Smith, D., Hahn, I.A., Frank, J., and Belfort, M. (2006). Visualization of a group II intron in the 23S rRNA of a stable ribosome. *Proc. Natl. Acad. Sci. USA* **103**: 9838–9843.

Baxter, W.T., Leith, A., and Frank, J. (2007). SPIRE: the SPIDER reconstruction engine. *J. Struct. Biol.* **157**: 56–63.

Faruqi, *A.R., and Henderson, R. (2007). Electronic detectors for electron microscopy. Curr. Opin. Struct. Biol.* ***17****: 549–555.*

Frank, J., Gao, H., Sengupta, J., Gao, N., and Taylor, D.J. (2007). The process of mRNA-tRNA translocation. *Proc. Natl. Acad. Sci. USA* **104**: 19671–19678.

Mossman, *K. (2007). Profile of Joachim Frank. Proc. Natl. Acad. Sci. USA* ***104****: 19668–19670.*

Fu, J., Gao, H., and Frank, J. (2007). Unsupervised classification of single particles by cluster tracking in multi-dimensional space. *J. Struct. Biol.* **157**: 226–239.

Gao, H., Zhou, Z., Rawat, U., Huang, C., Bouakaz, L., Wang, C., Cheng, Z., Liu, Y., Zavialov, A., Gursky, R., Sanyal, S., Ehrenberg, M., Frank, J., and Song, H. (2007). RF3 induces ribosomal conformational changes responsible for dissociation of class I release factors. *Cell* **129**: 929–941.

Gao, N., Zavialov, A.V., Ehrenberg, M., and Frank, J. (2007). Specific interaction between EF-G and RRF and its implication for GTP-dependent ribosome splitting into subunits. *J. Mol. Biol.* **374**: 1345–1358.

Gillet, R., Kaur, S., Li, W., Hallier, M., Felden, B., and Frank, J. (2007). Scaffolding as an organizing principle in trans-translation. The roles of small protein B and ribosomal protein S1. *J. Biol. Chem.* **282**: 6356–6363.

LeBarron, J., Mitra, K., and Frank, J. (2007). Displaying 3D data on RNA secondary structures: coloRNA. *J. Struct. Biol.* **157**: 262–270.

Li, W., and Frank, J. (2007). Transfer RNA in the hybrid P/E state: correlating molecular dynamics simulations with cryo-EM data. *Proc. Natl. Acad. Sci. USA* **104**: 16540–16545.

Marko, M., Hsieh, C., Schalek, R., Frank, J., and Mannella, C. (2007). Focused-ion-beam thinning of frozen-hydrated biological specimens for cryo-electron microscopy. *Nat. Methods* **4**: 215–217.

Nilsson, J., Sengupta, J., Gursky, R., Nissen, P., and Frank, J. (2007). Comparison of fungal 80S ribosomes by cryo-EM reveals diversity in structure and conformation of rRNA expansion segments. *J. Mol. Biol.* **369**: 429–438.

Scheres, S.H., Gao, H., Valle, M., Herman, G.T., Eggermont, P.P., Frank, J., and Carazo, J.M. (2007). Disentangling conformational states of macromolecules in 3D-EM through likelihood optimization. *Nat. Methods* **4**: 27–29.

Taylor, D.J., Nilsson, J., Merrill, A.R., Andersen, G.R., Nissen, P., and Frank, J. (2007). Structures of modified eEF2 80S ribosome complexes reveal the role of GTP hydrolysis in translocation. *EMBO J.* **26**: 2421–2431.

Yang, C., Penczek, P.A., Leith, A., Asturias, F.J., Ng, E.G., Glaeser, R.M., and Frank, J. (2007). The parallelization of SPIDER on distributed-memory computers using MPI. *J. Struct. Biol.* **157**: 240–249.

Agirrezabala, X., Lei, J., Brunelle, J.L., Ortiz-Meoz, R.F., Green, R., and Frank, J. (2008). Visualization of the hybrid state of tRNA binding promoted by spontaneous ratcheting of the ribosome. *Mol. Cell* **32**: 190–197.

Frank, J. (2008). Under the rug, or a plea for the lighter touch. *J. Struct. Biol.* **163**: 270.

Julián, *P., Konevega, A.L., Scheres, S.H.W., Lázaro, M., Gil, D., Wintermeyer, W., Rodnina, M.V., and Valle, M. (2008). Structure of ratcheted ribosomes with tRNAs in hybrid states. Proc. Natl. Acad. Sci. USA* ***105****: 16924–16927.*

LeBarron, J., Grassucci, R.A., Shaikh, T.R., Baxter, W.T., Sengupta, J., and Frank, J. (2008). Exploration of parameters in cryo-EM leading to an improved density map of the *E. coli* ribosome. *J. Struct. Biol.* **164**: 24–32.

Li, W., Agirrezabala, X., Lei, J., Bouakaz, L., Brunelle, J.L., Ortiz-Meoz, R.F., Green, R., Sanyal, S., Ehrenberg, M., and Frank, J. (2008). Recognition of aminoacyl-tRNA: a common molecular mechanism revealed by cryo-EM. *EMBO J.* **27**: 3322–3331.

Renken, C., Hsieh, C.E., Marko, M., Rath, B., Leith, A., Wagenknecht, T., Frank, J., and Mannella, C.A. (2008). Structure of frozen-hydrated triad junctions: a case study in motif searching inside tomograms. *J. Struct Biol.* **165**: 53–63.

Sengupta, J., Nilsson, J., Gursky, R., Kjeldgaard, M., Nissen, P., and Frank, J. (2008). Visualization of the eEF2-80S ribosome transition-state complex by cryo-electron microscopy. *J. Mol. Biol.* **382**: 179–187.

Shaikh, T.R., Trujillo, R., LeBarron, J.S., Baxter, W.T., and Frank, J. (2008). Particle-verification for single-particle, reference-based reconstruction using multivariate data analysis and classification. *J. Struct. Biol.* **164**: 41–48.

Trabuco, L.G., Villa, E., Mitra, K., Frank, J., and Schulten, K. (2008). Flexible fitting of atomic structures into electron microscopy maps using molecular dynamics. *Structure* **16**: 673–683.

Baxter, W.T., Grassucci, R.A., Gao, H., and Frank, J. (2009). Determination of signal-to-noise ratios and spectral SNRs in cryo-EM low-dose imaging of molecules. *J. Struct. Biol.* **166**: 126–132.

Lu, *Z., Shaikh, T.R., Barnard, D., Meng, X., Mohamed, H., Yassin, A., Mannella, C.A., Agrawal, R.K., Lu, T.M., and Wagenknecht, T. (2009). Monolithic microfluidic mixing-spraying devices for time-resolved cryo-electron microscopy. J. Struct. Biol.* ***168****: 388–395.*

Shi, Y., Giammartino, D. C. D., Taylor, D.J., Sarkeshik, A., Rice, W. J., Yates III, J.R., Frank, J., and Manley, J. L. (2009). Molecular architecture of the human pre-mRNA 3′ processing complex. *Mol. Cell* **33**: 365–376.

Sorci, M., Grassucci, R.A., Hahn, I., Frank, J., and Belfort, G. (2009). Time-dependent insulin oligomer reaction pathway prior to fibril formation: cooling and seeding. *Proteins* **77**: 62–73.

Taylor, D., Devkota, B., Huang, A.D., Topf, M., Narayanan, E., Sali, A., Harvey, S.C., and Frank, J. (2009). Comprehensive molecular structure of the eukaryotic ribosome. *Structure* **17**: 1591–1604.

Villa, E., Sengupta, J., Trabuco, L.G., LeBarron, J., Baxter, W.T., Shaikh, T.R., Grassucci, R.A., Nissen, P., Ehrenberg, M., Schulten, K. (2009). Ribosome-induced changes in elongation factor Tu conformation control GTP hydrolysis. *Proc. Natl. Acad. Sci. USA* **106**: 1063–1068.

Ben-Shem, *A., Jenner, L., Yusupova, G., and Yusupov, M. (2010). Crystal structure of the Eukaryotic ribosome. Science* ***330****: 1203–1209.*

Fischer, *N., Konevega, A.L., Wintermeyer, W., Rodnina, M.V., and Stark, H. (2010). Ribosome dynamics and tRNA movement by time-resolved electron cryomicroscopy. Nature* ***466****: 329–333.*

Frank, J. (2010). The ribosome comes alive. *Israel J. Chem.* **50**: 95–98.

Frank, J., and Gonzalez, R.L. (2010). Structure and dynamics of a processive Brownian motor: the translating ribosome. *Ann. Rev. Biochem.* **79**: 381–412.

Fu, J., Hashem, Y., Wower, I., Lei, J., Liao, H. Y., Zwieb, C., Wower, J., and Frank, J. (2010). Visualizing the transfer-messenger RNA as the ribosome resumes translation. *EMBO J.* **29**: 3819–3825.

Huang, T., S.T., Gupta, K., Contreras-Martin, L.M., Grassucci, R.A., Van Duyne, G.D., Frank, J., Belfort, M. (2010). The group II intron ribonucleoprotein precursor is a large, loosely packed structure. *Nucl. Acids Res.* **39**: 2845–2854.

Liao, H., and Frank, J. (2010). Classification by bootstrapping in single particle methods. Proceedings / IEEE International Symposium on Biomedical Imaging: from nano to macro *IEEE International Symposium on Biomedica* pp. 169–172.

***Mulder**, A.M., Yoshoika, C., Beck, A.H., Bunner, A.E., Milligan, R.A., Potter, C.S., Carragher, B., and Williamson, J. R. (2010). Visualizing ribosome biogenesis: parallel assembly pathways for the 30S subunit. Science **330**: 673–677.*

Reblova K, R.F., Li W, Gao H, Frank J, Sponer J (2010). Dynamics of the base of ribosomal A-site finger revealed by molecular dynamics simulations and cryo-EM. *Nucl. Acids Res.* **38**: 1325–1340.

Sengupta, J., Bussiere, C., Pallesen, J., West, M., Johnson, A., and Frank, J. (2010). Characterization of the nuclear export adaptor protein Nmd3 in association with the 60S ribosomal subunit. *J. Cell Biol.* **7**: 1079–1086.

Agirrezabala, X., Schreiner, E., Trabuco, L.G., Lei, J., Ortiz-Meoz, R. F., Schulten, K., Green, R., and Frank, F. (2011). Structural insights into cognate vs. near-cognate discrimination during decoding. *EMBO J.* **30**: 1497–1507.

Fu, J., Munro, J.B., Blanchard, S.C., Frank, J. (2011). Cryo-EM structures of the ribosome complex in intermediate states during tRNA translocation. *Proc. Natl. Acad. Sci. USA* **108**: 4817–4821.

Fu, J., Hashem, Y., Wower, J., and Frank, J. (2011). tmRNA on its way through the ribosome: Two steps of resume, and what next? *RNA Biol.* **8**: 586–590.

Langlois, R., Frank, J. (2011). A clarification of the terms used in comparing semi-automated particle selection algorithms in cryo-EM. *J. Struct. Biol.* **175**: 348–352.

Langlois, R., Pallesen, J., and Frank, J. (2011). Reference-free particle selection enhanced with semi-supervised machine learning for cryo-electron microscopy. *J. Struct. Biol.* **175**: 353–361.

Li, W., Trabuco, L.G., Schulten, K., Frank, F. (2011). Molecular dynamics of EF-G during translocation. *Proteins* **79**: 1478–1486.

***Wang**, B., Ho, J., Fei, J., Gonzalez, Jr., R.L., and Lin, Q. (2011). A microfluidic approach for investigating the temperature dependence of biomolecular activity with single-molecule resolution. Lab. Chip **11**: 274–281.*

Agirrezabala, X., Liao, H., Schreiner, E., Fu, J., Ortiz-Meoz, R.F., Schulten, K., Green, R., and Frank, J. (2012). Structural characterization of mRNA-tRNA translocation intermediates. *Proc. Natl. Acad. Sci. USA* **109**: 6094–6099.

Bussiere, C., Hashem, Y., Arora, S., Frank, J., Johnson, A.W. (2012). Integrity of the P-site is probed during maturation of the 60S ribosomal subunit. *J. Cell Biol.* **197**: 747–759.

***Scheres**, S.H.W. (2012). RELION: Implementation of a Bayesian approach to cryo-EM structure determination. J. Struc. Biol. **180**: 519–530.*

Taylor, D., Unbehaun, A., Li, W., Das, S., Lei, J., Liao, H., Grassucci, R.A., Pestova, T., and Frank, J. (2012). Cryo-EM structure of the mammalian eukaryotic release factor eRF1–eRF3-associated termination complex. *Proc. Natl. Acad. Sci. USA* **109**: 18413–18418.

des Georges, A., Hashem, Y., Unbehaun, A., Grassucci, R.A., Taylor, D., Hellen, C.U.T., Pestova, T.V., and Frank, J. (2013). Structure of the mammalian ribosomal pre-termination complex associated with eRF1•eRF3•GDPNP. *Nucleic Acids Res.* **42**: 3409–3418.

Hashem, Y., des Georges, A., Fu, J., Buss, S.N., Jossinet, F., Jobe, A., Zhang, Q., Liao, H.Y., Grassucci, R.A., Bajaj, C., Westhof, E., Madison-Antenucci, S., and Frank, J. (2013a). High-resolution cryo-EM structure of *Trypanosoma brucei* ribosome. *Nature* **494**: 385–389.

Hashem, Y., des Georges, A., Dhote, V., Langlois, R., Liao, H.L., Grassucci, R.A., Hellen, C.U.T., Pestova, T.V., and Frank, J. (2013b). Structure of the mammalian ribosomal 43S preinitiation complex bound to the scanning factor DHX29. *Cell* **153**: 1108–1119.

Hashem, Y., des Georges, A., Dhote, V., Langlois, R., Liao, H.Y., Grassucci, R.A., Pestova, T.V., Hellen, C.U., and Frank, J. (2013c). Hepatitis-C-virus-like internal ribosome entry sites displace eIF3 to gain access to the 40S subunit. *Nature* **503**: 539–543.

Li, W., Atkinson, G.C., Thakor, N.S., Allas, U., Lu, C., Chan, K., Tenson, T., Schulten, K., Wilson, K.S., Hauryliuk, V., and Frank, J. (2013). Mechanism of tetracycline resistance by ribosomal protection protein Tet(O). *Nature Commun.* **4**: #1477; doi: 10.1038/ncomms2470.

***Liao**, M., Cao, E., Julius, D. and Cheng, Y. (2013) Structure of the TRPV1 ion channel determined by electron cryo-microscopy. Nature **504**: 107–112.*

Sharma, G., Pallesen, J., Das, S., Grassucci, R., Langlois, R., Hampton, C.M., Kelly, D.F., des Georges, A., and Frank, J. (2013). Affinity grid-based cryo-EM of PKC binding to RACK1 on the ribosome. *J. Struct. Biol.* **181**: 190–194.

***Lyumkis**, D., Brilot, A.F., Theobald, D.L., and Grigorieff, N. (2013). Likelihood-based classification of cryo-EM images using FREALIGN. J Struct. Biol. **183**: 377–388.*

Pallesen, J., Hashem, Y., Korkmaz, G., Koripella, R., Huang, C., Ehrenberg, M., Sanyal, S., and Frank, J. (2013). Cryo-EM visualization of the ribosome in termination complex with apo-RF3 and RF1. *eLife* doi: 10.7554/eLife.00411.

Boël, G., Smith, P.C., Ning, W., Englander, M.T., Chen, B., Hashem, Y., Testa, A.J., Fischer, J.J., Wieden, H.-J., Frank, J., Gonzalez, R.L., and Hunt, J.F. (2014). The ABC-F protein EttA gates ribosome entry into the translation elongation cycle. *Nat. Struct. Mol. Biol.* **21**: 143–151.

Chen, B., Boël, G., Hashem, Y., Ning, W., Fei, J., Wang, C., Gonzalez, R.L., Hunt, J.F., and Frank, J. (2014). EttA regulates translation by binding to the ribosomal E site and restricting ribosome-tRNA dynamics. *Nat. Struct. Mol. Biol.* **21**: 152–159.

Chen, B., Shen, B., and Frank, J. (2014). Particle migration analysis in iterative classification of cryo-EM single-particle data. *J. Struct. Biol.* **188**: 267–273.

Dashti, A., Schwander, P., Langlois, R., Fung, R., Li, W., Hosseinizadeh, A., Liao, H.Y., Pallesen, J., Sharma, G., Stupina, V.A., Simon, A.E., Dinman, J.D., Frank, J., and Ourmazd, A. (2014). Trajectories of the ribosome as a Brownian nanomachine. *Proc. Natl. Acad. Sci. USA* **111**: 17492–17497.

Langlois, R., Pallesen, J., Ash, J.T., Ho, D.N., Rubinstein, J.L., and Frank, J. (2014). Automated particle picking for low-contrast macromolecules in cryo-electron microscopy. *J. Struct. Biol.* **186**: 1–7.

***Shaikh**, T.R., Yassin, A.S., Lu, Z., Barnard, D., Meng, X., Lu, T.-M., Wagenknecht, T., and Agrawal, R.K. (2014). Initial bridges between two ribosomal subunits are formed within 9.4 milliseconds, as studied by time-resolved cryo-EM. Proc. Natl. Acad. Sci. **111**: 9822–9827.*

Chen, B., Kaledhonkar, S., Sun, M., Shen, B., Lu, Z., Barnard, D., Lu, T., Gonzalez, R.L., and Frank, J. (2015). Structural dynamics of ribosome subunit association studied by mixing-spraying time-resolved cryo-EM. *Structure* **23**: 1097–1105.

des Georges, A., Dhote, V., Kuhn, L., Hellen, C.U., Pestova, T.V., Frank, J., and Hashem, Y. (2015). Structure of mammalian eIF3 in the context of the 43S preinitiation complex. *Nature* **525**: 491–495.

Kinz-Thompson, C.D., Sharma, A.K., Frank, J., Gonzalez, R.L., and Chowdhury, D. (2015). Quantitative connection between ensemble thermodynamics and single-molecule kinetics: a case study using cryo-EM and smFRET investigations of the ribosome. *J. Phys. Chem. B* **119**: 10888–10901.

Li, W., Liu, Z., Koripella, R.K., Langlois, R., Sanyal, S., and Frank, J. (2015). Activation of GTP hydrolysis in mRNA-tRNA translocation by Elongation Factor G. *Science Advances* **1**: e1500169.

Liao, H.Y., Hashem, Y., and Frank, J. (2015). Efficient estimation of three-dimensional covariance and its application in the analysis of heterogeneous samples in cryo-electron microscopy. *Structure* **23**: 1129–1137.

Marabini, R., Carragher, B., Chen, S., Chen, J., Cheng, A., Downing, K.H., Frank, J., Grassucci, R.A., Heymann, J.B., Jiang, W., Jonic, S., Liao, H.Y., Ludtke, S.J., Patwari, S., Piotrowski, A.L., Quintana, A., Sorzano, C.O.S., Stahlberg, H., Vargas, J., Voss, N.R., Chiu, W., and Carazo, J.M. (2015). CTF challenge: result summary. *J. Struct. Biol.* **190**: 348–359.

Sun, M., Li, W., Blomqvist, K., Das, S., Hashem, Y., Dvorin, J.D., and Frank, J. (2015). Dynamical features of the Plasmodium falciparum ribosome during translation. *Nucleic Acids Res.* **43**: 10515–10524.

Zalk, R., Clarke, O.B., des Georges, A., Grassucci, R.A., Reiken, S., Mancia, F., Hendrickson, W.A., Frank, J., and Marks, A.R. (2015). Structure of a mammalian ryanodine receptor. *Nature* **517**: 44–49.

Chen, Y., Clarke, O.B., Kim, J., Stowe, S., Kim, Y.-K., Assur, Z., Cavalier, M., Godoy-Ruiz, R., von Alpen, D.C., Manzini, C., Blaner, W.S., Frank, J., Quadro, L., Weber, D.J., Shapiro, L., Hendrickson, W.A., and Mancia, F. (2016). Structure of the STRA6 receptor for retinol uptake. *Science* **353**. DOI: 10.1126/science.aad8266.

des Georges, A., Clarke, O.B., Zalk, R., Yuan, Q., Condon, K.J., Grassucci, R.A., Hendrickson, W.A., Marks, A.R., and Frank, J. (2016). Structural basis for gating and activation of RyR1. *Cell* **167**: 145–157.

Frank, J., and Ourmazd, A. (2016). Continuous changes in structure mapped by manifold embedding of single-particle data in cryo-EM. *Methods* **100**: 61–67.

Fu, Z., Kaledhonkar, S., Borg, A., Sun, M., Chen, B., Grassucci, R.A., Ehrenberg, M., and Frank, J. (2016). Key intermediates in ribosome recycling visualized by time-resolved cryo-electron microscopy. *Structure* **24**: 2092–2101.

Liu, Z., Gutierrez-Vargas, C., Wei, J., Grassucci, R.A., Ramesh, M., Espina, N., Sun, M., Tutuncuoglu, B., Madison-Antenucci, S., Woolford, J.L., Tong, L., and Frank, J. (2016). Structure and assembly model for the *Trypanosoma cruzi* 60S ribosomal subunit. *Proc Natl Acad Sci USA* **113**: 12174–12179.

Liu, Z., Gutierrez-Vargas, C., Wei, J., Grassucci, R.A., Sun, M., Espina, N., Madison-Antenucci, S., Tong, L., and Frank, J. (2016). Determination of the ribosome structure to a resolution of 2.5Å by single-particle cryo-EM. *Protein Science* **26**: 82–92.

Twomey, E.C., Velshanskaya, M.V., Grassucci, R.A., Frank, J., and Sobolevsky, A.I. (2016). Elucidation of AMPA receptor–stargazin complexes by cryo–electron microscopy. *Science* **353**: 83–86.

***Wang**, J. and Moore, P.B. (2016). On the interpretation of electron microscopic maps of biological macromolecules. Protein Sci. **26**: 122–129.*

Dashti, A., Ben Hail, D., Mashayekhi, G., Schwander, P., des Georges, A., Frank, J., and Ourmazd, A. (2017). Conformational dynamics and energy landscapes of ligand binding in RyR1. *Biorxiv* doi: https://doi.org/10.1101/167080.

Feng, X., Fu, Z., Kaledhonkar, S., Jia, Y., Shah, B., Jin, A., Liu, Z., Sun, M., Chen, B., Grassucci, R.A., Ren, Y., Jiang, H., Frank, J., and Lin, Q. (2017). A fast and effective microfluidic spraying-plunging method for high-resolution single-particle cryo-EM. *Structure* **25**: 663–670.

Frank, J. (2017). The mechanism of translation. *F1000Research* **6**: 198.

Maji, S., Shahoei, R., Schulten, K., and Frank, J. (2017). Quantitative characterization of domain motions in molecular machines. *J. Phys. Chem. B* **121**: 3747-3756.

Malyutin, A.G., Musalgaonkar, S., Patchett, S., Frank, J., and Johnson, A.W. (2017). Nmd3 is a structural mimic of eIF5A, and activates the cpGTPase Lsg1 during 60S ribosome biogenesis. *EMBO J.* **36**: 854–868.

Twomey, E.C., Yelshanskaya, M.V., Grassucci, R.A., Frank, J., and Sobolevsky, A.I. (2017a). Structural bases of desensitization in AMPA receptor-auxiliary subunit complexes. *Neuron* **94**: 569–580.

Twomey, E.C., Yelshanskaya, M.V., Grassucci, R.A., Frank, J., and Sobolevsky, A.I. (2017b). Channel opening and gating mechanism in AMPA-subtype glutamate receptors. *Nature* doi: 10.1038/nature23479.

II

REPRINTS OF SELECTED ARTICLES, IN CHRONOLOGICAL ORDER

Optik

Volume 30 (1969)

WISSENSCHAFTLICHE VERLAGSGESELLSCHAFT MBH STUTTGART

Detection of Object Movement in the Optical Diffractograms of Electron Micrographs

Joachim Frank

Department of X-ray Crystallography at the Max-Planck Institut für Eiweiss und Lederforschung (Protein Research), Munich, and Department of Structure Research, Institute of Physical Chemistry, Technical University

Received on July 7, 1969

Abstract

[original text] **Detection of object movement in the optical diffractograms of electron micrographs.** Movements of the objects during an electron microscopic exposure cause characteristic modulations of the Fourier transform of the photographic density distribution, from which the nature of the movements can in turn be reconstructed. Modulations of this type were observed in optical diffractograms of a number of micrographs of amorphous carbon films and could be assigned to particular movements.

[starting here: translation]

1. Introduction

It is known from the work of Thon [1] that the wave aberrations prevailing during an exposure in the electron microscope can be determined from the optical diffraction pattern [of the micrograph] if a carbon film is used as object. The image of a statistical object1) essentially documents the electron microscopic parameters (voltage, wavelength, defocus, astigmatic defocus difference), though in an encoded way. The "decoding" [of this information] is carried out by taking the Fourier transform, for instance in the light-optical diffractometer.

We will show that beyond this information, the light-optical diffractogram contains additional clues regarding the behavior of the object during the exposure in the electron microscope. Specifically we will focus on the movements of the object, considered rigid during that time span. This study was prompted by [the analysis of] micrographs of carbon films which showed a strong periodic modulation.

2. General Description of [Object] Movements in Fourier Space.

We define $I(\mathbf{r})$ as the image intensity (in the following "image" for short) which hits the film during the exposure time for a resting, unchanging object. The coordinate system $\{\mathbf{r}\}$ refers to the resting film. We now assume that the image changes over time, for one due to radiation damage, and second due to the movements of the object (for instance, the thermal drift of the stage) and due to changes of the parameters describing the imaging process (for instance drift of the lens current, changes

[1] an object, in other words, in whose phase structure all spatial frequencies are distributed with equal probability ("white scatterer")]

of the astigmatism due to the buildup of contamination, etc.). If these changes occur over a time comparable to the exposure time T, we have

$$I = I(\mathbf{r}', t) \text{ with } \mathbf{r}' = \alpha(t)\,\mathbf{r}.$$

Here $\alpha(t)$ stands for a general time-dependent coordinate transformation. The recorded optical density will be proportional to (see [2])

$$S(\mathbf{r}') = \int_0^T I(\mathbf{r}', t)dt \tag{1}$$

Despite the blurring of the image that is due to the superposition of the component images the time dependence [of the process] can be inferred in simple cases. To this end we can use procedures known from both X-ray Crystallography and Statistical Optics.

We restrict ourselves to the case of translational movements of a rigid object, ignoring all other changes of the object. Let us assume that, on account of the movement of the object, the image moves by $\mathbf{q}(t)$ relative to its starting position, and as a consequence (1) is rewritten in the form

$$S(\mathbf{r}) = \int_0^T I(\mathbf{r}) \circ \delta(\mathbf{r} - \mathbf{q}(t))\, dt = T\, I(r) \circ 1/T \int_0^T \delta(\mathbf{r} - \mathbf{q}(t))dt, \tag{2}$$

using the notation $\mathrm{a}(\mathbf{r}) \circ \mathrm{b}(\mathbf{r}) = \int \mathrm{a}(\mathbf{r} - \mathbf{r}')\mathrm{b}(\mathbf{r}')\, d\mathbf{r}'$ for the convolution operation. Fourier-transformation of (2) and application of the convolution theorem yields

$$\mathfrak{F}\{S(\mathbf{r})\} = T\,\mathfrak{F}\{I(\mathbf{r})\} \cdot 1/T \int_0^T e^{-2\pi i \mathbf{k}\mathbf{q}(t)}\, dt \;=\; \mathfrak{F}\{T\, I(\mathbf{r})\}\, M(\mathbf{k}) \tag{3}$$

[A side issue:] Instead of the optical density [of the film], the transparency $e^{-S(\mathbf{r})}$ is the quantity effectively being Fourier-transformed when the film is placed into the optical diffractometer. However, when the density differences $\Delta S(\mathbf{r})$ are sufficiently small, linearization around the average density S_0 is possible, leading to

$$e^{-S(\mathbf{r})} \approx (1 + \Delta S(\mathbf{r}))\, e^{-S_0} = S'(\mathbf{r}).$$

As the additive term produces only an irrelevant modification to the zero term of the Fourier transform, we will not distinguish in the following between $S(\mathbf{r})$ and $S'(\mathbf{r})$.

The intensity obtained in the diffractogram becomes therefore proportional to

$$\mathfrak{F}\{S\} \cdot \mathfrak{F}^*\{S\} = \mathfrak{F}\{T\,I\}\, \mathfrak{F}^*\{T\,I\} \cdot M(\mathbf{k})M^*(\mathbf{k}) \tag{4}$$

Now when a carbon film is used as object, the diffractogram shows a distribution approximately given by

$$|\mathfrak{F}\{S\}|^2 = \sin^2 \gamma(\mathbf{k}), \text{ [except for a falloff term not mentioned here]}$$

where $$\gamma(\mathbf{k}) = \gamma(\theta,\varphi) = \frac{\pi}{2\lambda} C_s\, \theta^4 - \frac{\pi}{\lambda}\theta^2 [\Delta f + \tfrac{1}{2} \Delta f_A \sin 2\varphi]$$

modulated by a [two-dimensional] function MM* whose shape depends on the function $\mathbf{q}(t)$. $\mathbf{k}$ and (θ,φ) are Cartesian and polar coordinates in the plane of the diffractogram. Further definitions are as follows:

C_s = spherical aberration constant of the objective lens
Δf = defocus
Δf_A = focus difference due to axial astigmatism
λ = wavelength of the electron

The approximation lies in the replacement of density differences by proportional differences in transparency, described above, and the assumption of a linear relationship between [scattered wave] amplitudes and intensities in the image plane of the electron microscope, which is known to hold in good approximation for bright field imaging of weak phase objects.

3. Special Movements

Without loss of essentials we can restrict ourselves to translations along a straight line. Let the image move into the x-direction,

$$\mathbf{q}(t) = \{x(t), 0\}.$$

We consider three special cases:
(a) *Jump drift* during the exposure. After $t = T_1$: the image has the position $x = 0$ until time T_1, then jumps into a new position $x = \xi$ and stays there till the end of the exposure $t = T$.

$$x(t) = \begin{matrix} 0 \; for \; 0 \leq t \leq T_1 \\ \xi \; for \; T_1 \leq t \leq T \end{matrix} \tag{5}$$

With the introduction of $\eta_1 = T_1/T; \eta_2 = T_2/T; T = T_1 + T_2; \eta_1 + \eta_2 = 1$ (6)
we obtain

$$M(\mathbf{k}) = M(k_x) = \eta_1 + (1 - \eta_2)e^{2\pi i k_x \xi} \tag{7}$$

$$MM^* = 1 + \eta_1 \eta_2 [\cos(2\pi k_x \xi) - 1] \tag{8}$$

The modulating function thus creates a pattern of equidistant stripes whose period is inversely proportional to the size of the movement, and whose amplitude depends upon the dwell times in the two positions.

We call now the ratio of periodic to constant portions of MM* the degree of modulation:

$$G = \frac{2\eta_1\eta_2}{1 - 2\eta_1\eta_2} \tag{9}$$

Since G can be determined from the diffractogram if – from a control experiment – the undisturbed distribution $|F\{I\}|^2$ is known, the dwell times T_1 and T_2 can be determined.

G becomes largest when $\eta_1 = \eta_2$, hence $G = 1$, and (8) now becomes

$$MM^* = \tfrac{1}{2}\,[1+ \cos(2\pi k_x \xi)] \qquad (8')$$

Fig. 1a-c depicts diffractograms of a single micrographs of a carbon film showing a modulation by $sin^2\,\gamma(\mathbf{k})$ and an additional modulation which can be explained by assuming a sudden jump (a translation) of the object during the exposure. Fig. 2 shows the electron micrograph of a contaminated graphite foil along with its diffractogram. (The contamination has the effect that the Fourier transform is dominated by a strong continuous contribution from an amorphous structural component; this is why neither the image nor the diffractogram can be distinguished from the case of an amorphous object.) One recognizes in the micrograph a preferred orientation, which runs in the direction orthogonal to the direction of the stripes in the diffractogram (Fig. 2b). Finally, Fig. 2c represents the theoretically expected intensity distribution in the diffractogram.

That we are not dealing with properties of the object can be demonstrated by examining other micrographs of the same object taken at different defocusses, which lack the stripes.

As mentioned at another place (5), one obtains the same periodic modulation if one takes two images of the same area, and brings these first into perfect registration in the diffractometer, then applies a small shift.

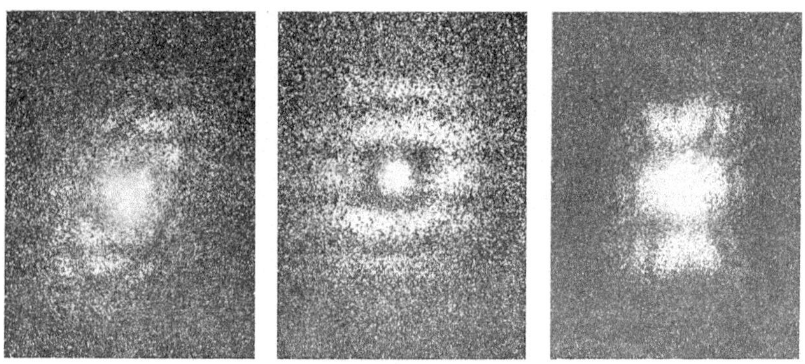

a b c

Fig. 1. Examples for the modulation of the diffractogram due to jump-drift. ($U = 80kV$, $V_{elopt} = 220{,}000\,x$). Object: amorphous carbon film.
(a) $\xi = 20$ Å, $\Delta f = 4100$ Å, $\Delta f_A = 2100$ Å
(b) $\xi = 17$ Å, $\Delta f = 2000$ Å, $\Delta f_A = 200$ Å
(c) $\xi = 7$ Å, $\Delta f = 1900$ Å, $\Delta f_A = 2100$ Å

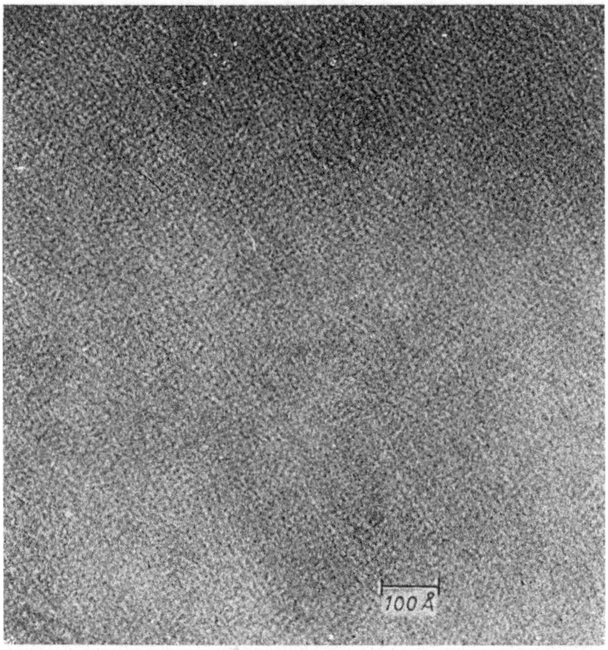

2a

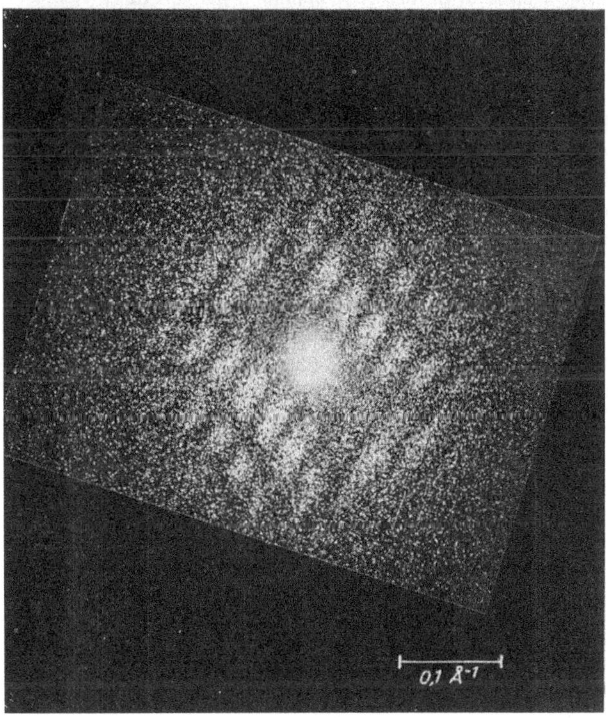

2b

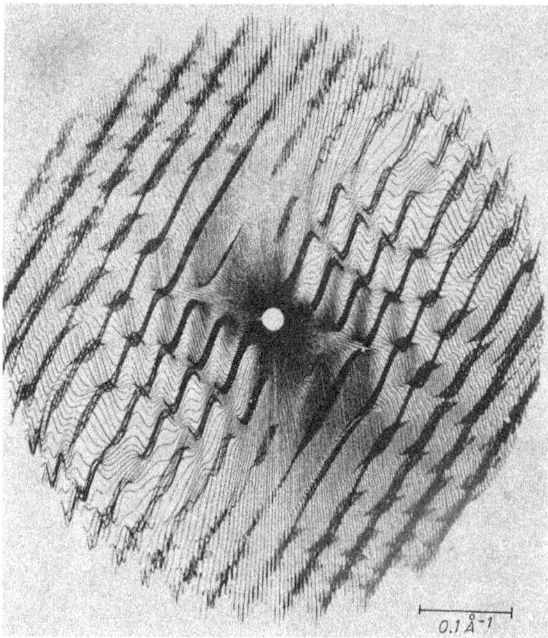

2c

Fig. 2. Further example for the detection of jump-drift. *(U = 80kV, V_{elopt} = 160,000 x).* *Object: contaminated graphite film.*
(a) Electron micrograph with $\Delta f \approx 1200Å$, $\Delta f_A \approx 3800Å$.
(b) Light-optical diffractogram of (a) displayed in matching orientation; ξ= 30Å.
(c) Representation of the expected intensity distribution using a plotter linked to an IBM 1130 computer. MM $\sin^2 \gamma(\mathbf{k})$ was computed using the parameters in (b), in terms of U, V_{elopt}, C_s, Δf, Δf_A, ξ, and for $\theta \leq 1.2 \times 10^{-2}$.*

b) *Continuous drift:* Due to a steady drift, let the object move with constant velocity v during the exposure of length T.

$$x_D(t) = v\,t \tag{10}$$

For such a case of blurring by linear motion (see also ref. 6), some procedures for recovery of the undisturbed image have already been given (7).

We now obtain with (10)

$$M(k_x) = \frac{1}{T}\int_0^T e^{2\pi i k_x v t}dt = \frac{\sin(\pi k_x v T)}{\pi k_x v T}, \tag{11}$$

up to a phase factor and

$$MM^* = \left(\frac{\sin(\pi k_x \xi)}{\pi k_x \xi}\right)^2; \quad \xi = vT. \tag{12}$$

Because of the rapid decrease of the amplitudes, this distribution is not as easy to observe as in the case of (a). However, the frequent observation that the Fourier plane is covered with the characteristic pattern of $\sin^2\gamma$ only within a narrow, sharply delineated stripe must be attributed to continuous drift. Fig. 3 shows a few examples for this effect, in comparison with the diffraction pattern for a normal, undisturbed exposure. The pronounced central stripe corresponds to the principal maximum of expression (12).

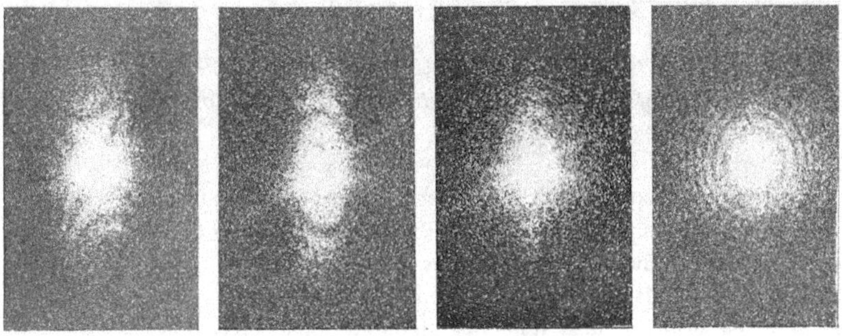

Fig. 3. Examples for modulation of the diffractogram in the case of continuous drift (U = 80 kV, V_{elopt} = 220,000 x). Object: amorphous carbon film.

(c) *Periodic oscillations:* This is another interesting case, which occurs when the object itself, or its image, oscillates relative to the photographic plate.

$$x_D(t) = \frac{\xi}{2}\cos(\omega t). \tag{13}$$

One obtains

$$M(k_x) = \frac{1}{T}\int_0^T e^{i[\pi k_x \xi]\cos(\omega t)}dt \tag{14}$$

$$= \frac{1}{T}\int_0^{wt=2\pi N} e^{i[\pi k_x \xi]\cos(\omega t)}d(\omega t) = J_0(\pi k_x \xi).$$

(Bessel function of zero order)

and

$$MM^* = J_0^2(\pi k_x \xi) \tag{15}$$

From the positions of the zeros of J_0, or the positions of the minima of $J_0{}^2$, respectively, the amplitude of the oscillations can be determined. For instance, the first zero occurs for an amplitude of ξ = 2Å at the (electron microscope's) resolution limit of 3 Å.

The term (14) also occurs in X-ray crystallography in the context of thermal vibrations. Since the amplitudes of thermal vibrations are quite small relative to the resolution achieved in electron microscopy, the term only accounts for a very small decrease of Fourier amplitudes. For an amplitude of 0.2 Å (the mean amplitude of oscillations [of an atom] at room temperature), the decrease is in the order of 1%, or 2% in quadrature.

However, oscillations that occur during the electron microscopical exposure due to mechanical vibrations of the column should be detectable for sufficiently large amplitudes.

178 J. Frank

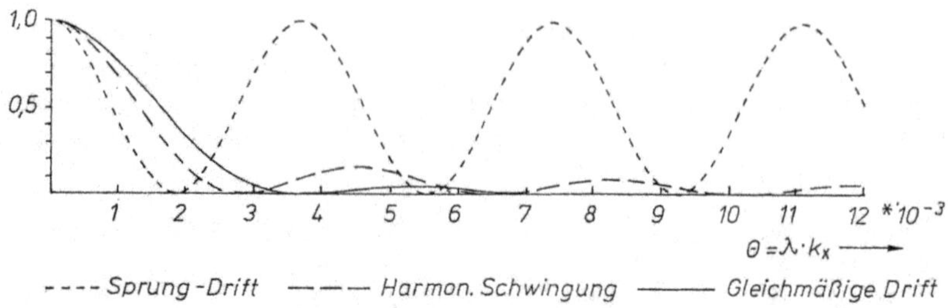

Fig. 4. Comparison of the modulation functions MM for jump-, oscillating, and continuous drift with the amplitude ξ = 10 Å.*

Figure 4 gives a comparison of modulation functions for the three cases considered, with the same amplitude parameter ξ = 10 Å. Qualitatively, the three cases can be distinguished with the help of the following criteria:

(a) *jump drift:* equidistant minima, same amplitude throughout;
(b) *continuous drift:* equidistant minima, amplitude falling off rapidly;
(c) *harmonic oscillation:* non-equidistant minima, amplitude falling off.

Acknowledgments

I would like to thank Professor Dr. W. Hoppe for stimulating discussions. Part of the micrographs analyzed in this study were kindly supplied by Drs. Langer and Feltinowski in our institute. I thank both for discussions, as well.

This work has been supported by the *Deutsche Forschungsgemeinschaft*, and the Fonds of the Chemical Industry and the *Badische Anilin- und Soda-Fabrik*.

References

[1] F. Thon, Z. Naturforschg. *20a* (1965) 154.
[2] R.C. Valentine, Advances in optical and Electron Microscopy I, ed.: Barer, Casslett, Academic Press, London/N.Y. 1966, 180.
[3] O. Scherzer, J. Appl. Phys. *20* (1949) 20.
[4] W. Hoppe, Acta Cryst. *10* (1957) 750.
[5] W. Hoppe, R. Langer, J. Frank, A. Feltynowski, Naturwiss. 56 (1969) 267.
[6] E.L. O'Neill, "Introduction to Statistical Optics," Addison-Wesley Publ. Comp., Reading, Mass. 1963, 27.
[7] J.L. Harris, J. Opt. Soc. Amer. *56* (1966) 569.
[8] R. Hosemann, S.N. Bagchi, "Direct Analysis of Diffraction by Matter," North-Holland Publ. Comp., Amsterdam 1962, p. 271, 286.

Optik

Sonderdruck 38. Band (1973) Seiten 519–536

WISSENSCHAFTLICHE VERLAGSGESELLSCHAFT MBH STUTTGART

The Envelope of Electron Microscopic Transfer Functions for Partially Coherent Illumination

By *Joachim Frank**

Max-Planck-Institut für Biochemie, Strukturforschung I, D-8033 Martinsried bei München

Received 7 May 1973

Abstract

It is known that the linear transfer theory of optical image formation remains valid for partially coherent illumination, provided that the object contrast is small. Starting from the results of the theory of coherence, the properties of the amplitude and phase contrast transfer function are examined. For not too large source sizes, each transfer function can be represented as a product of the corresponding coherent transfer function and an envelope function. Such a product representation is convenient both for theoretical analysis and for the purpose of object restoration from images taken with partially coherent illumination. It is shown that the envelope function is closely related to the phase coherence function. The envelope functions are explicitly given for sources with disk-shaped and Gaussian intensity distributions. Electron microscopic image formation is then treated as a special case of the hitherto general analysis. The electron microscopic envelope functions are calculated for a Gaussian source, using generalized variables. In a second order approximation, the product representation can be retained, but it has to be modified to allow for a phase shift of the coherent transfer function. A comparison of exact and approximate computation of the partially coherent phase contrast transfer function shows that the formula derived in this paper works satisfactorily within the range of illumination apertures used in practice in high resolution experiments.

Inhalt

Die Enveloppe der elektronenmikroskopischen Übertragungsfunktionen für partiell kohärente Beleuchtung. Die lineare Übertragungstheorie der optischen Bildentstehung verliert bekanntlich nicht ihre Gültigkeit für partiell kohärente Beleuchtung, falls der Objektkontrast klein ist. Ausgehend von den Ergebnissen der Theorie der Kohärenz werden die Eigenschaften der Amplituden- und Phasen-Übertragungsfunktionen untersucht. Für nicht zu große Beleuchtungsquellen können die Übertragungsfunktionen in Produktform dargestellt werden, mit der entsprechenden kohärenten Übertragungsfunktion und einer Enveloppe als Faktoren. Eine solche Produktdarstellung besitzt Vorteile sowohl für die theoretische Analyse als auch für die praktische Berücksichtigung der partiellen Kohärenz bei der Restoration der Objektfunktion. Es wird gezeigt, daß zwischen der Enveloppe und der Phasenkohärenzfunktion eine enge Beziehung besteht. Die Enveloppen werden explizit angegeben für Quellen mit scheiben- und gaußförmiger Intensitätsverteilung. Anschließend wird die elektronenmikroskopische Abbildung als Spezialfall der bis dahin allgemeinen Betrachtung behandelt. Die elektronenmikroskopi-

* Present address: Cavendish Laboratory, University of Cambridge.

schen Enveloppen wurden für gaußförmige Quellenverteilungen verschiedener Halbwertsbreite berechnet, unter Zuhilfenahme von generalisierten Variablen. Auch in zweiter Näherung kann die Produktdarstellung noch beibehalten werden, allerdings muß eine Phasenverschiebung der kohärenten Übertragungsfunktion berücksichtigt werden. Wie ein Vergleich von exakter und näherungsweiser Berechnung der Phasenkontrast-Übertragungsfunktion zeigt, arbeitet die hergeleitete Näherungsformel zufriedenstellend im Bereich der Beleuchtungsaperturen, die in Höchstauflösungsexperimenten zur Anwendung kommen.

1. Introduction

For partially coherent illumination, the relation between the object transmission and the image intensity cannot generally be described by the linear transfer theory, i.e. as a linear superposition of elementary images derived from weighted object points. But *Menzel* [1], *Hauser* [2] and *Slansky* [3] have shown that the linear theory still holds provided that the object contrast is small. This fact is of great importance for the practicability of computational object restoration from electron micrographs [4-8] since evaluations based on a coherent illumination model can give only crude results if the micrographs are taken under the illumination conditions of practical electron microscopy. In particular, the high resolution information of the image is not optimally used if the finite angular extension of the source is ignored.

Stroke dna *Halioua* [9] have already proposed holographic methods for deconvolution of partially coherent micrographs which are based on the results of the work of *Menzel*, *Hauser* and *Slansky* cited above.

The properties of the electron microscopic transfer functions for partially coherent illumination have been examined in detail by *Hanszen* and *Trepte* [10] and by *Hirt* and *Hoppe* [11]. Inspection of the results of these studies shows that the positions of the zeros of the partially coherent transfer[1] functions nearly coincide with the positions of the zeros of their coherent counterparts, at least for illumination angles commonly used in electron microscopy. This suggests that it may be possible to represent the partially coherent transfer functions in the form

$$\mathrm{B}_{\mathrm{pc}} = \mathrm{B}_{\mathrm{c}}\, \mathrm{E}_{\alpha} \tag{1}$$

where B_c is the corresponding transfer function for coherent illumination, and E_α is an envelope function which reflects the influence of the finite illumination aperture α.

A representation of the form (1) is advantagous for the theoretical analysis of partially coherent image formation. Furthermore, it is convenient for the purpose of object restoration since it permits the determination of the parameter(s) of a model source function from the Fourier transform of a carbon film micrograph with the aid of a simple least squares algorithm, like that with which the parameters of the pupil function are obtained [4, 12].

[1] For expediency, the terms "coherent transfer function" and "partially coherent transfer function" will be used instead of "transfer function for coherent/partially coherent illumination", respectively.

Finally it should be mentioned that the influence of fluctuations of voltage and lens current (and, in a modified way, the energy spread of the electron beam) allows a product representation in a similar form as (1) [13]. It is therefore possible, within the range of validity of (1), to summarize all resolution limiting effects on the transfer functions in a simple equation

$$B_{eff} = B_c \, E_{res}$$

where B_c is the ideal transfer function for coherent illumination, zero energy spread and absence of voltage and current fluctuations, E_{res} is the envelope function resulting from the combined action of the resolution limiting effects, and B_{eff} is the effective transfer function.

2. Contrast Transfer Functions for Partially Coherent Illumination and Low Object Contrast

This chapter will serve to put the further discussion into the context of the theory of coherence [14, 15] and to define some expressions to be used later on. (The low contrast limit is treated in detail in [1–3, 10, 16].)

We start with the general expression for the Fourier transform of the image intensity as obtained by *Born* and *Wolf* [17] in the chapter dealing with coherence theory:

$$I(\mathbf{k}) = \int T(\mathbf{k}' + \mathbf{k}, \mathbf{k}') \, F(\mathbf{k}' + \mathbf{k}) \, F^*(\mathbf{k}') \, d\mathbf{k}' \tag{2}$$

where $F(\mathbf{k})$ is the Fourier transform of the object transmission function $\tilde{o}(\mathbf{r})$

$$F(\mathbf{k}) = \int \tilde{o}(\mathbf{r}) \exp(2\pi i \mathbf{k}\mathbf{r}) \, d\mathbf{r} \tag{3}$$

and $T(\mathbf{k}', \mathbf{k}'')$ is the so-called *transmission cross coefficient*. We may regard (2) as the autocorrelation integral of the object transform but with the integrand weighted, according to the vector displacement in the cross product, by a cross coefficient. The transmission cross coefficient is defined in the following way [18]:

$$T(\mathbf{k}', \mathbf{k}'') = \int s(\mathbf{q}) \, t(\mathbf{q} + \mathbf{k}') \, t^*(\mathbf{q} + \mathbf{k}'') \, d\mathbf{q}. \tag{4}$$

Here $s(\mathbf{q})$ is the normalized intensity distribution of the *effective source* (definition below) that illuminates the object, and $\mathbf{q}$ is the coordinate of the source plane, reduced according to the scale factors between the source plane and the back focal plane of the objective lens. $t(\mathbf{k})$ is the pupil function of the objective lens. $T(\mathbf{k}', \mathbf{0})$ and $T(\mathbf{0}, \mathbf{k}'')$ can be regarded as weighted autocorrelation integrals of the pupil function where the weighting function is the intensity distribution of the effective source.

In the derivation of (2) and (4), the source is considered as consisting of incoherently emitting point sources. The term "effective source" was introduced by *Hopkins* [14]. If, for an arbitrary illumination system, the *phase-coherence function* $\Gamma(\mathbf{r}_1 - \mathbf{r}_2)$ [19] is known between all pairs of points in the object plane, then the object plane may be regarded as illuminated by a *single* source which has the intensity distribution

$$\mathrm{s}(\mathbf{q}) = \int \Gamma(\mathbf{r}) \exp(2\pi i \mathbf{q}\mathbf{r})\, d\mathbf{r} \tag{5}$$

and is termed the effective source associated with the illumination system.

For later use the representation of Γ is given in terms of $\mathrm{s}(\mathbf{q})$ as well:

$$\Gamma(\mathbf{r}_1 - \mathbf{r}_2) = \int \mathrm{s}(\mathbf{q}) \exp(-2\pi i \mathbf{q}(\mathbf{r}_1 - \mathbf{r}_2))\, d\mathbf{q}. \tag{6}$$

The normalization condition is

$$\int \mathrm{s}(\mathbf{q})\, d\mathbf{q} = \Gamma(\mathbf{0}) = 1. \tag{7}$$

Now let us consider an object transmission function of the form

$$\tilde{\mathrm{o}}(\mathbf{r}) = 1 + \mathrm{o}(\mathbf{r}) \tag{8}$$

where

$$\mathrm{o}(\mathbf{r}) = \mathrm{o}_1(\mathbf{r}) + i\mathrm{o}_2(\mathbf{r}) \tag{9}$$

is the complex object variation[2]. Let $\mathrm{O}(\mathbf{k})$ be the Fourier transform of $\mathrm{o}(\mathbf{r})$. Then we have according to (3) and (8)

$$\mathrm{F}(\mathbf{k}) = \delta(\mathbf{k}) + \mathrm{O}(\mathbf{k}) \tag{10}$$

where δ is the Dirac delta funktion. Forming $\mathrm{F}(\mathbf{k}' + \mathbf{k})$ and $\mathrm{F}(\mathbf{k}')$ from (10) we obtain as Fourier coefficient of the image intensity

$$\begin{aligned} \mathrm{I}(\mathbf{k}) &= \int \mathrm{T}(\mathbf{k}' + \mathbf{k}, \mathbf{k}') \{\delta(\mathbf{k}' + \mathbf{k}) + \mathrm{O}(\mathbf{k}' + \mathbf{k})\} \{\delta(-\mathbf{k}') + \mathrm{O}^*(\mathbf{k}')\}\, d\mathbf{k}' \\ &= \mathrm{T}(\mathbf{k}, \mathbf{0})\, \delta(\mathbf{k}) \\ &\quad + \{\mathrm{T}(\mathbf{0}, -\mathbf{k})\, \mathrm{O}^*(-\mathbf{k}) + \mathrm{T}(\mathbf{k}, \mathbf{0})\, \mathrm{O}(\mathbf{k})\} \\ &\quad + \int \mathrm{T}(\mathbf{k}' + \mathbf{k}, \mathbf{k}')\, \mathrm{O}(\mathbf{k}' + \mathbf{k})\, \mathrm{O}^*(\mathbf{k}')\, d\mathbf{k}'. \end{aligned} \tag{11}$$

The first term is the zero beam term and will produce the background intensity in the image. The second term is linear in the object spectrum, and the third term (the only term we would expect for dark field illumination) is quadratic in $\mathrm{O}(\mathbf{k})$. For small object contrast,

$$|\mathrm{o}(\mathbf{r})| \ll 1, \tag{12}$$

and the second term makes the predominant contribution to the image contrast. We see that in this case there still is a linear relation between the Fourier coefficients of the object transform and those of the image intensity transform, a fact first pointed out by *Menzel* [1]. This means that no loss of

---

[2] Representation of the object transmission in this form is convenient for the electron microscopic application since the Fourier transforms of $\mathrm{o}_1(\mathbf{r})$ and $\mathrm{o}_2(\mathbf{r})$ can be interpreted as the structure factors for anomalous and normal elastic scattering, respectively (cf. [5]). In the optical literature (cf. [2]) it is more common to regard (8), for $|\mathrm{o}(\mathbf{r})| \ll 1$, as the first order approximation of the object transmission of a pure phase object $\exp(i\mathrm{o}_2(\mathbf{r}))$ (for $\mathrm{o}_1(\mathbf{r}) = 0$) or, alternatively, of a purely absorbing object $\sqrt{1 + 2\,\mathrm{o}_1(\mathbf{r})}$ (for $\mathrm{o}_2(\mathbf{r}) = 0$).

information occurs due to partial coherence, at least for moderate angular source sizes.

We shall restrict our further discussion to the linear term, assuming that condition (12) is fulfilled. The linear term is

$$I_2(\mathbf{k}) = T(\mathbf{k}, \mathbf{0})\, O(\mathbf{k}) + T(\mathbf{0}, -\mathbf{k})\, O^*(-\mathbf{k}). \tag{13}$$

The transmission cross coefficients in (13) are obtained from the defining equation (4) as

$$T(\mathbf{k}, \mathbf{0}) = \int s(\mathbf{q})\, t(\mathbf{q}+\mathbf{k})\, t^*(\mathbf{q})\, d\mathbf{q} \tag{14}$$

$$T(\mathbf{0}, -\mathbf{k}) = \int s(\mathbf{q})\, t(\mathbf{q})\, t^*(\mathbf{q}-\mathbf{k})\, d\mathbf{q}\ (= T^*(\mathbf{k}, \mathbf{0}) \quad \text{for symmetric sources}) \tag{15}$$

Note in passing that the symmetry property of T guarantees that $I_2(\mathbf{k})$ is the Fourier transform of a real function for any $O(\mathbf{k})$:

Since $$T^*(-\mathbf{k}, \mathbf{0}) = T(\mathbf{0}, -\mathbf{k}) \tag{16}$$

and $$T^*(\mathbf{0}, -\mathbf{k}) = T(\mathbf{k}, \mathbf{0}) \tag{17}$$

it follows from (13) that $I_2(\mathbf{k}) = I_2{}^*(-\mathbf{k})$ (Friedel's Law).

Now let us split up $O(\mathbf{k})$ according to

$$O(\mathbf{k}) = O_1(\mathbf{k}) + i\, O_2(\mathbf{k}) \tag{18}$$

where $O_{1,2}(\mathbf{k})$ are the Fourier transforms of $o_{1,2}(\mathbf{r})$, respectively. By observing that both $O_1(\mathbf{k})$ and $O_2(\mathbf{k})$ have the Friedel symmetry, we obtain from (13)

$$I_2(\mathbf{k}) = O_1(\mathbf{k})\, \{T(\mathbf{k}, \mathbf{0}) + T(\mathbf{0}, -\mathbf{k})\} \quad + i\, O_2(\mathbf{k})\, \{T(\mathbf{k}, \mathbf{0}) - T(\mathbf{0}, -\mathbf{k})\}. \tag{19}$$

The pupil function has the form

$$t(\mathbf{k}) = a(\mathbf{k}) \exp(-2\pi i \chi(\mathbf{k})) \tag{21}$$

where $a(\mathbf{k})$ is the aperture function of the exit pupil

$$a(\mathbf{k}) = \begin{cases} 1 & |\mathbf{k}| \leq k_0 \\ 0 & \text{elsewhere.} \end{cases} \tag{22}$$

We assume that the aperture function is symmetric: $a(\mathbf{k}) = a(-\mathbf{k})$. $\chi(\mathbf{k})$ is the wave aberration function[3] which describes the deviation of the wave front from the Gaussian reference sphere. With (21) as pupil function in expressions (14) and (15), and with a symmetric source, the linear term (19) of the image transform becomes

$$\begin{aligned} I_2(\mathbf{k}) - 2\, O_1(\mathbf{k}) & \int s(\mathbf{q})\, a(\mathbf{q})\, a(\mathbf{q}+\mathbf{k}) \cos\{2\pi[\chi(\mathbf{q}+\mathbf{k}) - \chi(\mathbf{q})]\}\, d\mathbf{q} \\ & + 2\, O_2(\mathbf{k}) \int s(\mathbf{q})\, a(\mathbf{q})\, a(\mathbf{q}+\mathbf{k}) \sin\{2\pi[\chi(\mathbf{q}+\mathbf{k}) - \chi(\mathbf{q})]\}\, d\mathbf{q} \\ & =_{\text{def}} O_1(\mathbf{k})\, B_1(\mathbf{k}) + O_2(\mathbf{k})\, B_2(\mathbf{k}) \end{aligned} \tag{23}$$

[3] In earlier papers [4, 5], $\gamma(\mathbf{k}) = 2\pi\chi(\mathbf{k})$ was erroneously called the "wave aberration function", in disagreement with its definition in the optical literature.

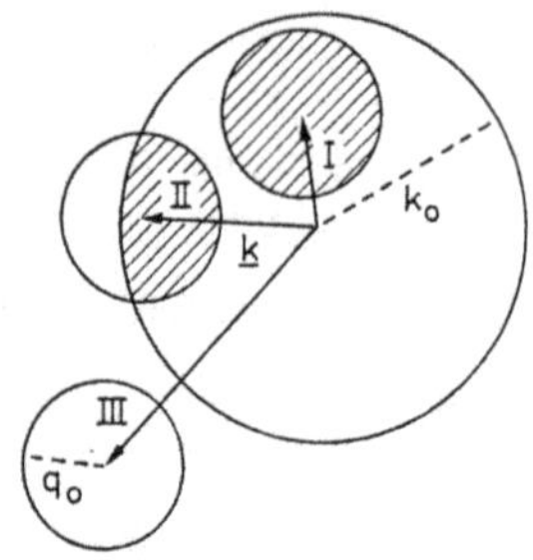

Fig. 1. The area of integration of expression (23) (= shaded area in this drawing) for various spatial frequencies **k**.
Position I: $|\mathbf{k}| < k_0 - q_0$, area of integration is the full source area.
Position II: $k_0 - q_0 < |\mathbf{k}| < k_0 + q_0$, area of integration is a continuous function of **k**.
Position III: $k_0 + q_0 < |\mathbf{k}|$, area of integration is zero.

where $B_{1,2}(\mathbf{k})$ are the *amplitude* and *phase contrast transfer functions*, respectively, *for partially coherent illumination.* While the term $a(\mathbf{q})$ can be omitted for source apertures q_0 smaller than the exit pupil, the term $a(\mathbf{q} + \mathbf{k})$ has the effect that the full integration is performed only as long as

$$|\mathbf{q}| + |\mathbf{k}| < k_0. \tag{24}$$

For larger values of **k**, the area of integration becomes dependent on **k** until, for $|\mathbf{k}| - q_0 > k_0$, the area of integration vanishes. This behavior can be visualized by placing the source of radius q_0 onto the exit pupil (radius k_0) and considering various positions **k** of its center with respect to the center of the exit pupil (see fig. 1). The effective area of integration of (23) is then the common area of source and exit pupil. For the purpose of the electron microscopic application, the source size is small compared with the size of the limiting aperture of the exit pupil, and condition (24) is valid save for **k** - values close to the limiting aperture.

There is therefore no need to maintain the distinction implied in the above reasoning through the following discussion. The aperture term in the integrand, $a(\mathbf{q})\, a(\mathbf{q} + \mathbf{k})$ will be omitted below, and the integration is extended over the full source size. This means that from now on, the argument is valid only for **k**-values compatible with (24).

3. The Product Representation and the Properties of the Envelope Function

If the discussion is further restricted to a spatial frequency region where $|\mathbf{k}| \gg q_0$, the wave aberration at $\mathbf{k} + \mathbf{q}$ may be obtained by expanding $\chi(\mathbf{k})$ in the neighborhood of **k** and ignoring terms of higher than first order in **q**:

$$\chi(\mathbf{k} + \mathbf{q}) \approx \chi(\mathbf{k}) + \mathbf{q}\nabla\chi(\mathbf{k}) + \ldots \tag{25}$$

where $\nabla\, \chi(\mathbf{k})$ stands for $\left(\frac{\partial\chi(\mathbf{k})}{\partial k_x}, \frac{\partial\chi(\mathbf{k})}{\partial k_y}\right)$.

It can be assumed that $\chi(\mathbf{q})$ does not contain terms of lower than second order in $\mathbf{q}$. Therefore $\chi(\mathbf{q})$ may be omitted in (23), to the same accuracy as (25).

We now insert (25) into (23) and make use of the addition theorems for the trigonometric functions:

$$B_{\frac{1}{2}}(\mathbf{k}) = 2 \,{\cos \atop \sin}\, \{2\pi\chi(\mathbf{k})\}\ E(\mathbf{k}) \pm 2 \,{\sin \atop \cos}\, \{2\pi\chi(\mathbf{k})\}\ \hat{E}(\mathbf{k}) \tag{26}$$

where

$$E(\mathbf{k}) = \int \cos\{2\pi\mathbf{q}\nabla\chi(\mathbf{k})\}\ s(\mathbf{q})\, d\mathbf{q} \tag{27}$$

and

$$\hat{E}(\mathbf{k}) = \int \sin\{2\pi\mathbf{q}\nabla\chi(\mathbf{k})\}\ s(\mathbf{q})\, d\mathbf{q}. \tag{28}$$

We note that $\hat{E}(\mathbf{k})$ vanishes if the source function is symmetric (which has been assumed earlier), for the argument of the sine is an odd function of $\mathbf{q}$. The remainder of (26) is the desired representation: The transfer functions $B_{1,2}(\mathbf{k})$ are obtained as a product of the transfer functions for coherent illumination, $2\cos 2\pi\chi(\mathbf{k})$ and $2\sin 2\pi\chi(\mathbf{k})$, respectively, and an envelope function, $E(\mathbf{k})$.

Regarded as a function of $\nabla\chi(\mathbf{k})$, the envelope function (27) is the Fourier transform of the source. That is to say, the integral (27) transforms from the $\mathbf{q}$-space into the space of gradients of the wave aberration function. Since the Fourier transform of the source function is, according to (6), the phase-coherence function, we obtain

$$E(\mathbf{k}) = \Gamma(\nabla\chi(\mathbf{k})). \tag{29}$$

Note that, from (6), we know that $\Gamma(\mathbf{r})$ is real for a real symmetric source function.

The result (29) can be stated in the following way: *The slope of the wave aberration surface as a function of the spatial frequency is expressed by a vector* $\mathbf{g}(\mathbf{k})$ *whose dimension is length. The envelope of the contrast transfer functions is the phase-coherence function* $\Gamma(\mathbf{r}_1 - \mathbf{r}_2)$ *taken at* $\mathbf{r}_1 - \mathbf{r}_2 = \mathbf{g}(\mathbf{k})$.

According to its definition, the phase-coherence function describes the degree of correlation between the wave amplitudes at two different points. The relation (29) seems to reveal a new aspect of the phase-coherence function: In spatial frequency space it describes the attenuation of the transfer functions as a function of the gradient of the wave aberration function. It is interesting to study some properties of $E(\mathbf{k})$ in more detail. These properties follow from the equations (6) and (7).

(1) For coherent illumination, $\Gamma = \text{const} = 1$ for any argument vector. Therefore $E(\mathbf{k}) = 1$: the envelope leaves the coherent transfer function unchanged.

(2) It follows from $\Gamma(\mathbf{0}) = 1$ that $E(\mathbf{k}) = 1$ wherever $\nabla\chi(\mathbf{k})$ vanishes. In particular, this means that coherent and partially coherent transfer functions are identical in spatial frequency regions where $\chi(\mathbf{k}) = \text{const}$.

(3) For incoherent illumination where $\Gamma(\mathbf{r}) = \delta(\mathbf{r})$, equation (29) yields formally

$$\mathrm{E}(\mathbf{k}) = \delta(\nabla\chi(\mathbf{k})) \tag{30}$$

i. e. transfer of information is confined to regions of the wave aberration function where its gradient is zero.

Although the condition $|\mathbf{k}| \gg q_0$ and hence (29) are not valid for incoherent illumination, the result (30) gives an idea of the behavior of the transfer functions for increasing source sizes: Regions of the transfer functions where $\chi(\mathbf{k}) = \text{const}$ are least attenuated as the source size increases.

The possibility of representing the partially coherent transfer function in product form (1) implies that the image point spread function $p_{pc}(\mathbf{r})$ can be written as a convolution product of the coherent point spread function $p_c(\mathbf{r})$ and a function $b(\mathbf{r})$ that describes the blurring effect due to the partially coherent illumination.

$$p_{pc}(\mathbf{r}) = \int p_c(\mathbf{r}')\, b(\mathbf{r}' - \mathbf{r})\, d\mathbf{r}' \tag{31}$$

where

$$p_c(\mathbf{r}) = \int B(\mathbf{k}) \exp(-2\pi i\mathbf{k}\mathbf{r})\, d\mathbf{k} \tag{32}$$

and

$$b(\mathbf{r}) = \int E(\mathbf{k}) \exp(-2\pi i\mathbf{k}\mathbf{r})\, d\mathbf{k}\,. \tag{33}$$

It follows from the above discussion that $b(\mathbf{r}) = \delta(\mathbf{r})$ for coherent illumination, and that the integral (31) reproduces $p_c(\mathbf{r})$.

The expression (31) may be compared with Slansky's representation of the partially coherent point spread function as a conjugate product [3] of the coherent point spread function and a quasi-point spread function. But unlike (31), Slansky's representation does not involve any approximation but follows directly from our equations (13) and (14).

For sources with rotational symmetry, the envelope is

$$\mathrm{E}(\mathbf{k}) = \Gamma(|\nabla\chi(\mathbf{k})|) \tag{34}$$

We will consider two particular sources with rotational symmetry, a disk of radius q_0 and a Gaussian distribution with q_0 as halfwidth.

1) Disk source

$$s(|\mathbf{q}|) = \begin{cases} \dfrac{1}{\pi q_0{}^2} & |\mathbf{q}| \le q_0 \\ 0 & \text{elsewhere}\,. \end{cases} \tag{35}$$

By introducing polar coordinates

$$q_x = q\cos\varphi,\quad q_y = q\sin\varphi \quad \text{where} \quad q = |\mathbf{q}| \tag{36}$$
$$x = \varrho\cos\varphi',\quad y = \varrho\sin\varphi',$$

the Fourier integral (6) becomes a Fourier-Bessel integral [20]

$$\Gamma(\varrho) = \frac{1}{\pi q_0^2} \int_0^{2\pi} \int_0^{q_0} \exp(-2\pi i q \varrho \cos(\varphi - \varphi'))\, q \, dq \, d\varphi$$

$$= \frac{2}{q_0^2} \int_0^{q_0} J_0(2\pi q \varrho)\, q \, dq = 2 \frac{J_1(2\pi q_0 \varrho)}{2\pi q_0 \varrho} \tag{37}$$

where J_0 and J_1 are the Bessel functions of zeroth and first order. Therefore, the envelope function can be expressed as

$$E_{disk}(\mathbf{k}) = 2 \frac{J_1(2\pi q_0 |\nabla\chi(\mathbf{k})|)}{2\pi q_0 |\nabla\chi(\mathbf{k})|}. \tag{38}$$

2) Gaussian source

$$s(|\mathbf{q}|) = \frac{1}{\pi q_0^2} \exp\left(-\frac{q^2}{q_0^2}\right). \tag{39}$$

The phase-coherence function is

$$\Gamma(\varrho) = \frac{2}{q_0^2} \int_0^{\infty} \exp\left(-\frac{q^2}{q_0^2}\right) J_0(2\pi q \varrho)\, q \, dq = \exp(-\pi^2 \varrho^2 q_0^2). \tag{40}$$

Hence, the envelope function[4] has the form

$$E_{gauss}(\mathbf{k}) = \exp(-\pi^2 q_0^2 |\nabla\chi(\mathbf{k})|^2). \tag{41}$$

4. Application to Electron Microscopy

So far, no specific assumptions have been made about the properties of the wave aberration function, except that it is an even function of the pupil coordinate **k**, and that it does not employ orders lower than the second in **k**. Now we shall specify the wave aberration function of the objective lens of the electron microscope, in the presence of defocus Δz, axial astigmatism Δz_a and third order spherical aberration C_s [21]. It is convenient at this point to use a polar coordinate system (ϑ, φ) with $\vartheta = \frac{|\mathbf{k}|}{\lambda}$. ϑ is the scattering angle that corresponds to the spatial frequency **k** and the wavelength λ. For a

[4] In this case, the envelope function and the phase-coherence function are not exactly equal. This is due to the aperture truncation in (23) which was ignored in later expressions. However, the error is not appreciable for small sources since the area under the gaussian function decreases rapidly as the argument is increased beyond the halfwidth.

particular choice of the polar coordinate axis for which the effect of the axial astigmatism disappears at $\varphi = 0$, we have

$$\chi(\vartheta, \varphi) = -\frac{1}{2\lambda}\left(\Delta z + \frac{\Delta z_a}{2}\sin 2\varphi\right)\vartheta^2 + \frac{1}{4\lambda} C_s \vartheta^4. \tag{42}$$

The gradient of χ has, in polar coordinates, the components $\chi_{,\vartheta} = \frac{\partial \chi}{\partial \vartheta}$, $\chi_{,\varphi} = \frac{1}{\vartheta}\frac{\partial \chi}{\partial \varphi}$. For sources with rotational symmetry, only the magnitude of this gradient, $\sqrt{\chi_{,\vartheta}^2 + \chi_{,\varphi}^2}$, has to be known. We obtain from (42)

$$|\nabla \chi(\mathbf{k})| = \lambda \sqrt{\left(\frac{\partial \chi}{\partial \vartheta}\right)^2 + \left(\frac{1}{\vartheta}\frac{\partial \chi}{\partial \varphi}\right)^2} \tag{43}$$

$$= \vartheta \sqrt{\Delta z_{eff}^2(\varphi) + \tfrac{1}{4}\Delta z_a^2 \cos^2 2\varphi - 2 C_s \vartheta^2 \Delta z_{eff}(\varphi) + C_s^2 \vartheta^4} =_{def} P(\vartheta, \varphi)$$

where $\Delta z_{eff}(\varphi) =_{def} \Delta z + \frac{\Delta z_a}{2}\sin 2\varphi$.

The envelope function can now be written in the form

$$E(\vartheta, \varphi) = \Gamma(P(\vartheta, \varphi)). \tag{44}$$

In particular, if the axial astigmatism is compensated, equation (43) yields

$$P(\vartheta, \varphi) = P(\vartheta) = C_s \vartheta^3 - \Delta z \vartheta \tag{45}$$

and

$$E(\vartheta, \varphi) = E(\vartheta) = \Gamma(C_s \vartheta^3 - \Delta z \vartheta) \tag{46}$$

If Δz is positive, the function $P(\vartheta)$ has a second zero besides $\vartheta = 0$ at $\vartheta = \sqrt{\frac{\Delta z}{C_s}}$. It is therefore a remarkable feature of the electron microscopic envelope function $E(\vartheta)$ for *positive* defocus that it contains a section where it leaves the coherent transfer function unchanged. This behavior can clearly be distinguished in plots of the results of exact computations (see for instance [11]).

The specific form of (46) for disk-shaped and Gaussian source distributions is easily obtained from (38) and (41), respectively[5].

$$E_{disk}(\vartheta) = 2\,\frac{J_1(2\pi q_0(C_s \vartheta^3 - \Delta z \vartheta))}{2\pi q_0(C_s \vartheta^3 - \Delta z \vartheta)} \tag{47}$$

and

$$E_{gauss}(\vartheta) = \exp\{-\pi^2 q_0^2 (C_s \vartheta^3 - \Delta z \vartheta)^2\}. \tag{48}$$

Some calculations were done to examine the behavior of $E_{gauss}(\vartheta)$, as the Gaussian profile was considered a fairer model of a physical source than the disk profile with its sharp cut-off. For evaluation, equation (48) was cast into

[5] A formula similar to (48) has been mentioned by *Welton* [22].

a different form where it does not contain the instrumental constant C_s and is valid for *any* objective lens.

Following *Hanszen* and *Trepte* [23], we introduce the generalized coordinate

$$\tilde{k} = \sqrt[4]{C_s \lambda^3}\, k \tag{49}$$

and the generalized defocus

$$\tilde{\Delta z} = \frac{\Delta z}{\sqrt{C_s \lambda}}. \tag{50}$$

Similarly, we define the generalized source size $\tilde{q}_0 = \sqrt[4]{C_s \lambda^3}\, q_0$. (51)

With these new variables, the wave aberration function and its gradient have the generalized form (for $\Delta z_a = 0$)

$$\chi(\tilde{k}) = -\tfrac{1}{2} \tilde{\Delta z}\, \tilde{k}^2 + \tfrac{1}{4} \tilde{k}^4 \tag{52}$$

$$\nabla \chi(\tilde{k}) = -\tilde{\Delta z}\, \tilde{k} + \tilde{k}^3. \tag{53}$$

Instead of (48), we now have

$$E_{gauss}(\tilde{k}) = \exp\{-\pi^2 \tilde{q}_0^2 (\tilde{k}^3 - \tilde{\Delta z}\, \tilde{k})^2\} \tag{54}$$

The generalized source halfwidth is chosen in such a way that it corresponds to angular sizes of $2\alpha = 1 \times 10^{-3}$ and 2×10^{-3} rad for $C_s = 1.35$ mm and $\lambda = 0.037$ Å. With these parameters the generalized spatial frequency range $0 \le \tilde{k} \le 2$ corresponds to the range of scattering angles $0 \le \vartheta \le 1.45 \times 10^{-2}$ rad. As generalized defocus parameter, the values $\sqrt{1}$, $\sqrt{3}$ and $\sqrt{5}$ were examined (Generally, the defocus values $\sqrt{2n+1}$, $n = 0, 1, 2 \ldots$ produce broad transfer bands in the phase contrast transfer function, which are most desirable for practical work).

The exact evaluation was performed by numerical integration of the integral (23) in the generalized form

$$B_2(\tilde{k}) = \frac{2}{\pi \tilde{q}_0^2} \int_0^\infty \exp\left(-\frac{\tilde{q}^2}{\tilde{q}_0^2}\right) \int_0^{2\pi} \sin 2\pi\, \{\chi(\sqrt{\tilde{k}^2 + \tilde{q}^2 - 2\tilde{k}\tilde{q}\cos\varphi}) - \chi(\tilde{q})\}\, d\varphi\, \tilde{q}\, d\tilde{q}$$

with $\chi(\tilde{k})$ as defined in (52).

The results are shown in fig. 2. For $\tilde{q}_0 = 0.069$, the agreement between exact and approximate curves is excellent throughout the spatial frequency range considered, except for slight phase shifts at frequencies $\tilde{k} > 1$.

For $\tilde{q}_0 = 0.138$, the discrepancies already appear at $\tilde{k} > 0.75$ and are more severe. Beyond $\tilde{k} = 1.5$ the approximation fails completely to represent the phase and amplitude of the transfer functions.

In short, the results imply that the first order approximation works well within the range $0 < \tilde{k} < 2$ for sources smaller than 0.069, and fairly well within the range $0 < \tilde{k} < 1.5$ for sources smaller than 0.138. It will be shown

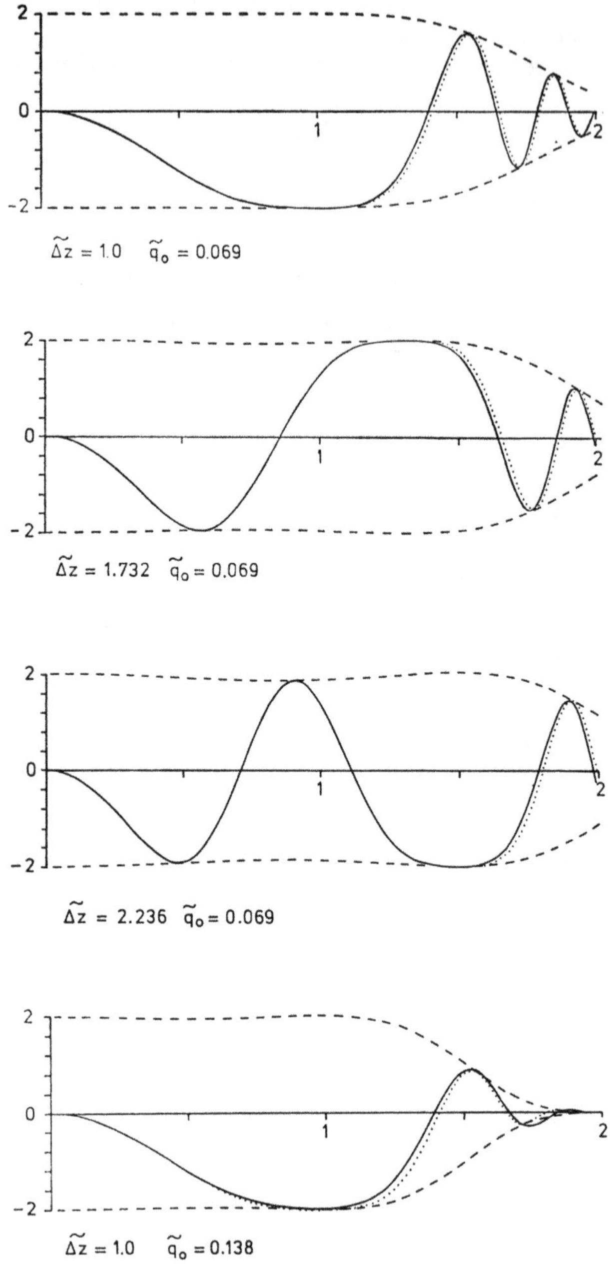

Fig. 2. Partially coherent transfer functions for Gaussian source distributions, as a function of the generalized spatial frequency $\tilde{k}$ (horizontal axis). Generalized

in the next chapter that the agreement can be improved considerably if the approximation is extended to the second order.

5. Second Order Approximation

The ordinary product representation (1) of the transfer function becomes invalid as soon as the second order term in the expansion of $\chi(\mathbf{k} + \mathbf{q})$ (cf. (25)) becomes sizable. The results of the first order computations discussed in the previous chapter suggest that the main effect of this term is to produce a *phase shift* in the argument of the coherent transfer function. Hence, the convenient product representation can be maintained in a modified form, provided that it is possible to express this phase shift in an analytic form.

The second order term in (25) is a quadratic form which may be written as the scalar product

$$\frac{1}{2!}\chi_{,,\mathbf{k}} \cdot (\mathbf{qq}) \tag{55}$$

of the tensors

$$\chi_{,,\mathbf{k}} = \begin{pmatrix} \dfrac{\partial^2\chi}{\partial k_x{}^2} & \dfrac{\partial^2\chi}{\partial k_x\,\partial k_y} \\ \dfrac{\partial^2\chi}{\partial k_y\,\partial k_x} & \dfrac{\partial^2\chi}{\partial k_y{}^2} \end{pmatrix} \tag{56}$$

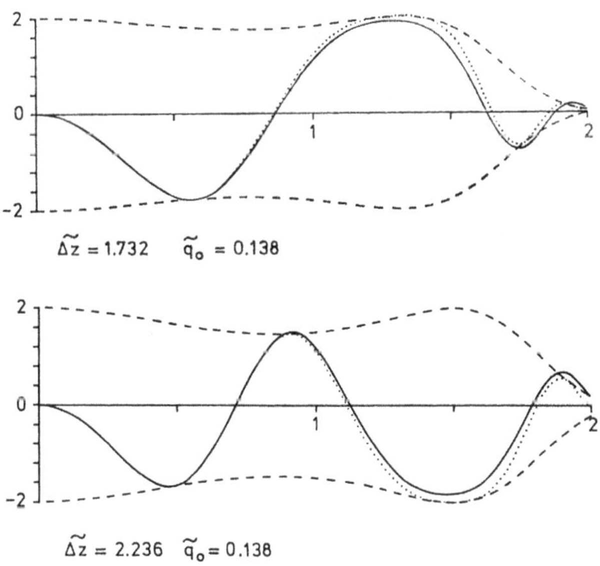

source halfwidths are 0.069 and 0.138; generalized defocus values $\tilde{\Delta}z = \sqrt{1}, \sqrt{3}$, and $\sqrt{5}$. Broken line shows $\pm 2E(\tilde{k})$, solid line exact computation, dotted line first order approximation.

and

$$(\mathbf{qq}) = \begin{pmatrix} q_x{}^2 & q_x\, q_y \\ q_y\, q_x & q_y{}^2 \end{pmatrix}. \tag{57}$$

$\chi_{,,k}$ is a tensor which is related to the curvature of the wave aberration surface.

The full second order term in the argument of the sine or cosine function in (23) is

$$\varepsilon(\mathbf{k}, \mathbf{q}) = \pi\chi_{,,\mathrm{k}} \cdot (\mathbf{qq}) - 2\pi\chi(\mathbf{q})\,. \tag{58}$$

Now we obtain the following expressions for the envelope functions $E(\mathbf{k})$ and $\hat{E}(\mathbf{k})$ instead of (27) and (28):

$$\begin{aligned} E_2(\mathbf{k}) &= \int \cos\{2\pi\mathbf{q}\nabla\chi(\mathbf{k}) + \varepsilon(\mathbf{k}, \mathbf{q})\}\, s(\mathbf{q})\, d\mathbf{q} \\ &\approx \int (1 - \tfrac{1}{2}\varepsilon^2) \cos\{2\pi\mathbf{q}\nabla\chi(\mathbf{k})\}\, s(\mathbf{q})\, d\mathbf{q} - \int \varepsilon(\mathbf{k}, \mathbf{q}) \sin\{2\pi\mathbf{q}\nabla\chi(\mathbf{k})\}\, s(\mathbf{q})\, d\mathbf{q} \\ &\approx E(\mathbf{k}) \end{aligned} \tag{59}$$

$$\begin{aligned} \hat{E}_2(\mathbf{k}) &= \int \sin\{2\pi\mathbf{q}\nabla\chi(\mathbf{k}) + \varepsilon(\mathbf{k}, \mathbf{q})\}\, s(\mathbf{q})\, d\mathbf{q} \\ &\approx \int (1 - \tfrac{1}{2}\varepsilon^2) \sin\{2\pi\mathbf{q}\nabla\chi(\mathbf{k})\}\, s(\mathbf{q})\, d\mathbf{q} + \int \varepsilon(\mathbf{k}, \mathbf{q}) \cos\{2\pi\mathbf{q}\nabla\chi(\mathbf{k})\}\, s(\mathbf{q})\, d\mathbf{q} \\ &= \tilde{E}(\mathbf{k})\,. \end{aligned} \tag{60}$$

In (59) and (60) we have written

$$\cos\varepsilon \approx 1 - \tfrac{1}{2}\varepsilon^2; \qquad \sin\varepsilon \approx \varepsilon$$

and have finally ignored the term containing ε^2. We have also made use of the fact that $\varepsilon(\mathbf{k}, \mathbf{q})$ is an even function of $\mathbf{q}$. Returning to (26) where we replace $\hat{E}(\mathbf{k})$ by $\hat{E}_2(\mathbf{k})$ we see that an additive term appears in the transfer functions

$$B_{\frac{1}{2}}(\mathbf{k}) = 2 \begin{matrix}\cos\\ \sin\end{matrix} \{2\pi\chi(\mathbf{k})\}\, E(\mathbf{k}) \mp 2 \begin{matrix}\sin\\ \cos\end{matrix} \{2\pi\chi(\mathbf{k})\}\, \tilde{E}(\mathbf{k}) \tag{61}$$

which is the product of the "complementary" coherent transfer function and the function

$$\tilde{E}(\mathbf{k}) = \int \varepsilon(\mathbf{k}, \mathbf{q}) \cos\{2\pi\mathbf{q}\nabla\chi(\mathbf{k})\}\, s(\mathbf{q})\, d\mathbf{q} \tag{62}$$

It will be shown that $\tilde{E}(\mathbf{k})$ can be derived analytically for the electron microscopic transfer functions if the axial astigmatism is compensated and if illumination by a Gaussian source (39) is assumed.

For $\Delta z_a = 0$ in (42), equation (58) assumes the simple form

$$\begin{aligned} \varepsilon(k, q) &= 2\pi \left(\frac{1}{2} \frac{\partial^2 \chi}{\partial k^2} q^2 - \chi(q) \right) \\ &= 2\pi \left(\frac{3}{2} C_s \lambda^3 k^2 q^2 - \frac{1}{2} \lambda \Delta z q^2 + \frac{1}{2} \lambda \Delta z q^2 - \frac{1}{4} \lambda^3 C_s q^4 \right) \\ &\approx 3\pi C_s \lambda^3 k^2 q^2. \end{aligned} \tag{63}$$

Now according to (62), in polar coordinates,

$$\begin{aligned}\tilde{E}(k) &= 3\pi C_s \lambda^3 k^2 \frac{2}{q_0^2} \int_0^\infty q^2 \exp\left(-\frac{q^2}{q_0^2}\right) J_0(2\pi q |\nabla\chi(k)| q\, dq \\ &= 3\pi C_s \lambda^3 k^2 (q_0^2 - \pi^2 |\nabla\chi(k)|^2 q_0^4) \exp(-\pi^2 q_0^2 |\nabla\chi(k)|^2) \\ &= 3\pi C_s \lambda^3 k^2 q_0^2 (1 - \pi^2 |\nabla\chi(k)|^2 q_0^2) E_{gauss}(k). \qquad (64)\end{aligned}$$

(The envelope function for a Gaussian source, $E_{gauss}(k)$, was derived earlier., cf. (41)).

The integral (64) was evaluated by differentiating the identity

$$\int \exp\left(-\frac{|\mathbf{q}|^2}{q_0^2}\right) \exp(2\pi i \nabla\chi\mathbf{q})\, d\mathbf{q} = \pi q_0^2 \exp(-\pi^2 q_0^2 |\nabla\chi|^2) \qquad (65)$$

with respect to q_0^2, and introducing polar coordinates. It is now possible to specify the range of validity of the first order approximation treated in the previous chapter: The relative error due to the second order term,

$$e(k) =_{def} \frac{\tilde{E}(k)}{E(k)} = 3\pi \tilde{k}^2 \tilde{q}^2 (1 - \pi^2 |\nabla\chi(\tilde{k})|^2 \tilde{q}_0^2) \qquad (66)$$

has to be small compared with unity. For $\tilde{k} \leq 1.38$ (i.e. $\vartheta \leq 10^{-2}$ rad for $C_s = 1.35$ mm and $\lambda = 0.037$ Å) this means

$$\tilde{q}_0 < 2.4\sqrt{p} \times 10^{-2} \text{ (illumination angle } 2\alpha < 3.5\sqrt{p} \times 10^{-4} \text{ rad)}$$

if a relative error of p% is acceptable.

Since the correction term appears with $\frac{\pi}{2}$ phase shift with respect to the coherent transfer function, its relative contribution becomes sizable where the latter has zeros. Its main effect is therefore to shift the zeros of the coherent transfer function and the first order envelope function still represents the heights of the maxima and minima of the partially coherent transfer function fairly well.

For

$$|e(k)|^2 \ll 1 \quad (\tilde{q}_0 < 9\sqrt[4]{p} \times 10^{-2} \text{ rad} \quad \text{with } k, C_s, \lambda \text{ and the meaning of p as above})$$

it is possible to represent the resulting transfer functions as single trigonometric functions, the phase of which is shifted by an amount depending on the spatial frequency. From (61) and (66) we obtain

$$\begin{aligned}B_{\frac{1}{2}}(k) &= E(k)\left[2 \begin{matrix}\cos\\ \sin\end{matrix} \{2\pi\chi(k)\} \mp 2e(k) \begin{matrix}\sin\\ \cos\end{matrix} \{2\pi\chi(k)\}\right] \\ &\approx E(k)\left[2 \begin{matrix}\cos\\ \sin\end{matrix} \{2\pi\chi(k) + \arcsin e(k)\}\right]. \qquad (67)\end{aligned}$$

The phase shift is positive or negative (i.e. the zeros are shifted in the direction of decreasing or increasing k), depending on whether $\pi^2|\nabla\chi(k)|^2 q_0^2 < 1$ or > 1, respectively.

The form of equation (67) suggests the introduction of a complex *effective wave aberration function*

$$\chi_{eff}(k) = \chi(k) + \frac{1}{2\pi} \arc\sin e(k) - \frac{i}{2} \pi q_0^2 |\nabla\chi(k)|^2, \tag{68}$$

with the imaginary part representing the dissipative effect of the partially coherent illumination on the contrast transfer functions[6]. Within the limits of the low contrast approximation, image formation with partially coherent illumination and a pupil function $t(k) = a(k) \exp(-2\pi i \chi(k))$ is equivalent to image formation with coherent illumination where t(k) is replaced by the *effective pupil function*

$$t_{eff}(k) = a(k) \exp(-2\pi i \chi_{eff}(k)).$$

Expression (68) was derived assuming compensated astigmatism and a Gaussian source distribution. In the general case it will not be possible to perform the integration (62) analytically, but it is still possible to define an effective wave aberration function. Its general form is

$$\chi_{eff}(\mathbf{k}) = \chi(\mathbf{k}) + \frac{1}{2\pi} \arc\sin e(\mathbf{k}) - \frac{i}{2\pi} \ln \Gamma(\nabla\chi(\mathbf{k})) \tag{69}$$

where

$$e(\mathbf{k}) = \frac{\int \varepsilon(\mathbf{k}, \mathbf{q}) \cos(2\pi \mathbf{q} \nabla\chi(\mathbf{k}))\, s(\mathbf{q})\, d\mathbf{q}}{\Gamma(\nabla\chi(\mathbf{k}))}$$

with $\varepsilon(\mathbf{k}, \mathbf{q})$ as defined in (58).

The range of validity of (67), first line, can only be examined by estimating the cumulative contribution of the higher than second order terms in the expansion of the wave aberration function. For expediency, it will rather be demonstrated how well the approximation works for two source half-widths, by comparing the result of an exact evaluation of the integral (23) with that of the approximation.

The approximate representation was obtained from (67), first line, by computing

$$B'_2(\tilde{k}) = 2 \exp(-\pi^2 \tilde{q}_0^2 |\nabla\chi(\tilde{k})|^2) [\sin\{2\pi\chi(\tilde{k})\} + 3\pi \tilde{k}^2 \tilde{q}_0^2 (1 - \pi^2 |\nabla\chi(\tilde{k})|^2 \tilde{q}^2_0) \cos\{2\pi\chi(\tilde{k})\}].$$

The results are shown in fig. 3 and may be compared with the corresponding curves in fig. 2. For $\tilde{q}_0 = 0.069$ the approximate curve coincides with

[6] A simple consideration shows that a complex wave aberration function does not preserve the number of particles. Therefore this description cannot be generally valid; it is useful only in order to describe the behavior of the *linear* term of the image transform, $I_2(\mathbf{k})$ (cf. (13)).

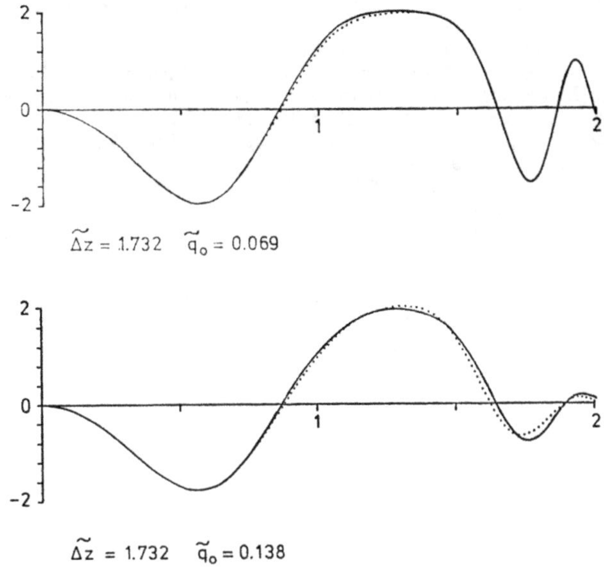

Fig. 3. Partially coherent transfer functions for Gaussian source distribution. Generalized source halfwidths as in fig. 2, for $\Delta\tilde{z} = \sqrt{3}$. Solid line exact computation, dotted line second order approximation.

the exact curve within the accuracy of the drawing pen. The accuracy of the $\tilde{q}_0 = 0.138$ curve has also improved when compared with the first order approximation, and the remaining deviations are only small.

Hence, in its second order modification, the product representation applies within $0 < \tilde{k} < 2$ for source sizes up to 0.138 at least.

As for source distributions other than Gaussian, it may be expected that the behaviour of the approximation as a function of spatial frequency an source size is very similar to that for the Gaussian distribution considered here.

Acknowledgments

This work was supported by a Research Grant of the Deutsche Forschungsgemeinschaft.

I would like to thank Prof. Dr. *W. Hoppe* for kindly providing the facilities of the Abteilung Strukturforschung I. The computations were done using the computer Siemens 4004/150 of the Max-Planck-Institut für Biochemie. I wish to express my thanks to Dr. *P. W. Hawkes* for his helpful discussion.

References

[1] *Menzel, E.*, Optik *15* (1958) 460.
[2] *Hauser, H.*, Optica Acta *9* (1962) 121.
[3] *Slansky, S.*, Optica Acta *9* (1962) 277.

[4] *Frank, J.*, *Bussler, P. H.*, *Langer, R.* and *Hoppe, W.*, Ber. Bunsenges. Phys. Chem. *74* (1970) 1105.
[5] *Frank, J.*, Biophys. J. *12* (1972) 484.
[6] *Frank, J.*, Use of Anomalous Scattering for Element Discrimination. In: Image Processing and Computer-aided Design in Electron Optics. (P. W. Hawkes, ed.) Academic Press, 1973, 196–211.
[7] *Bussler, P. H.*, *Feltynowski, A.* and *Hoppe, W.*, Proc. 5th Europ. Congr. El. Micr., Manchester 1972, 626.
[8] *Hoppe, W.*, *Bussler, P. H.*, *Feltynowski, A.*, *Hunsmann, N.* and *Hirt, A.* Some Experience with Computerized Image Reconstruction Methods. In: Image Processing and Computer-aided Design in Electron Optics. (P. W. Hawkes, ed.) Academic Press, 1973, 92–126.
[9] *Stroke, W.* and *Halioua, M.*, Optik *35* (1972) 489.
[10] *Hanszen, K.-J.* and *Trepte, L.*, Optik *33* (1971) 182.
[11] *Hirt, A.* and *Hoppe, W.*, Proc. 5th Europ. Congr. El. Micr., Manchester 1972, 12.
[12] *Bussler, P. H.* (1973) to be published.
[13] *Hanszen, K.-J.* and *Trepte, L.*, Optik *32* (1971) 519.
[14] *Hopkins, H. H.*, Proc. Roy. Soc. *A217* (1953) 408.
[15] *Born, M.* and *Wolf, E.*, Principles of Optics, Pergamon Press 1959, p. 490 ff.
[16] *Fischer, B.* and *Menzel, E.*, Optik *21* (1964) 70.
[17] *Born, M.* and *Wolf, E.*, see [15] p. 527, eq. 49.
[18] *Born, M.* and *Wolf, E.*, see [15] p. 527, eq. 47.
[19] *Hopkins, H. H.*, Proc. Roy. Soc. *A208* (1951) 263.
[20] *Jahnke, E.*, *Emde, F.* and *Lösch, F.*, Tables of Higher Functions. B. G. Teubner Verlagsgesellschaft, Stuttgart 1960, p. 131 ff.
[21] *Scherzer, O.*, J. Appl. Phys. *20* (1949) 20.
[22] *Welton, T. A.* (1971) private communication.
[23] *Hanszen, K.-J.* and *Trepte, L.*, Optik *33* (1971) 166.

Note added in proof

(Received 19 October 1973)

Eq. (63) is in error and has to be replaced by

$$\varepsilon(\mathbf{k}, \mathbf{q}) = \pi C_S \lambda^3 [\mathbf{k}^2 \mathbf{q}^2 + 2(\mathbf{k}\,\mathbf{q})^2]$$

which is consistent with eq. (55). As a consequence, the integration in eq. (64) becomes more involved but can still be carried out analytically. The result is that the term 1 in the bracket of eq. (64), last line, has to be replaced by $\frac{2}{3}$. As it is apparent from fig. 3 which was calculated using the erroneous expression, the effect of the error cannot be large, but may nevertheless account for part of the small deviations of the computed second order curves from the exact curves. I would like to thank Dr. *G. Ade* for suggesting this correction.

Ultramicroscopy 1 (1975) 159–162
© North-Holland Publishing Company

SHORT NOTE

AVERAGING OF LOW EXPOSURE ELECTRON MICROGRAPHS OF NON-PERIODIC OBJECTS

Joachim FRANK *
The Cavendish Laboratory, Free School Lane, Cambridge CB2 3RQ, UK

Received 20 October 1975

The investigation concerns the possibility of extending to non-periodic objects the low exposure averaging techniques recently proposed for non-destructive electron microscopy of periodic biological objects. Two methods are discussed which are based on cross-correlation and are in principle suited for solving this problem.

1. Introduction

Recent work on low exposure techniques combined with averaging [1–3] (called 'SNAP shot techniques' in [3]) shows that information can be retrieved from periodic biological objects at higher than conventionally available resolutions [4]. Unwin and Henderson [2] were able to achieve 7 Å image resolution, by reducing the exposure to less than 1 electron/Å^2. Although a number of interesting biological materials exist in the native state as (or can be induced to form) periodic arrays suitable for study by electron microscopy, many others exist only as isolated units or in the form of disordered aggregates. We will investigate how the averaging techniques could be extended to this general case. Of all the possible irregular specimens which are being investigated in biological high resolution microscopy, we are interested in those which form identical particles, sufficiently well separated on the microscope grid so as not to overlap. In order to avoid any three-dimensional complications, we will assume that these particles are flat, with preferential attachment occurring on one side so that all projections in the direction of the axial electron beam are identical.

We note that averaging has been applied to dark field images of molecules labelled with heavy atoms and lying isolated on a carbon film, for removing the noise due to the structure of the supporting film [5, 6]. In these applications, the contrast of the individual marker atom image to be superposed is sufficient for straightforward alignment. However, the requirement of subminimum exposure poses a new problem: the alignment of features that are only faintly visible on a noisy background.

* Present address: State of New York Dept. of Health, 120 New Scotland Avenue, Albany, New York 12201, USA.

2. Averaging of a motif repeated on an irregular lattice

2.1. Formulation of the problem

In order to consider this problem, we first assume an arrangement of repeats of a motif (e.g. a virus particle) where each repeat has the same orientation but appears at irregular positions $\boldsymbol{r}_j$. Such an arrangement can be denoted in convolution form:

$$i(\boldsymbol{r}) = m(\boldsymbol{r}) \circ g(\boldsymbol{r}) + n(\boldsymbol{r}) , \qquad (1)$$

with the irregular point structure $g(\boldsymbol{r}) = \Sigma_{j=1}^{N} \delta(\boldsymbol{r} - \boldsymbol{r}_j)$, and an additive uncorrelated noise term. (For the limits of additive formulation, see ref. [7].)

The idea of SNAP techniques is that a large number of noisy repeats of $m(\boldsymbol{r})$ are averaged to form a noise-free representation of the motif. We immediately see the unique position of the periodic arrangement in that it forces a common orientation on all repeats, and that it makes them appear on a regular lattice, leaving alignment as a trivial problem. So trivial, indeed, that it does not even surface in the usual Fourier treatment [8,9] unless data from different experiments have to be combined, in which case a phase factor must be attached to the Fourier transforms.

In contrast, we have here to consider the position vectors $\boldsymbol{r}_j$ in (1) as unknowns which must be determined from the image.

2.2. Matched filtering

The general problem of accurate alignment of unperiodic patterns can be solved by cross-correlation techniques [7,10]; however, the present problem requires a special treatment.

By applying the idea of matched filtering [11], one could proceed to determine the lattice as follows. From theoretical considerations or some experimental evidence the most probable motif structure $m_0(\boldsymbol{r})$ ('prototype') may be known. Cross-correlation of $m_0(\boldsymbol{r})$ with the experimental image (1) gives

$$m_0(\boldsymbol{r}) \otimes i(\boldsymbol{r}) = [m_0(\boldsymbol{r}) \otimes m(\boldsymbol{r})] \circ \sum_{j=1}^{N} \delta(\boldsymbol{r} - \boldsymbol{r}_j)\,, \qquad (2)$$

i.e. the cross-correlation function of the prototype with the motif $m_0 \otimes m$ appears in each position $\boldsymbol{r}_j$. Ideally, if the prototype happened to be identical with the motif sought, this cross-correlation function would show a sharp peak, of halfwidth corresponding to the image resolution, indicating the position of the origin of $m(\boldsymbol{r})$ with respect to the position of each repeat of the motif with precision.

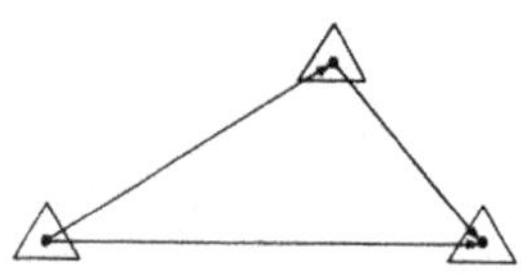

Fig. 1a. Simple structure consisting of three repeats of a triangular motif. Underlying point structure marked by arrows.

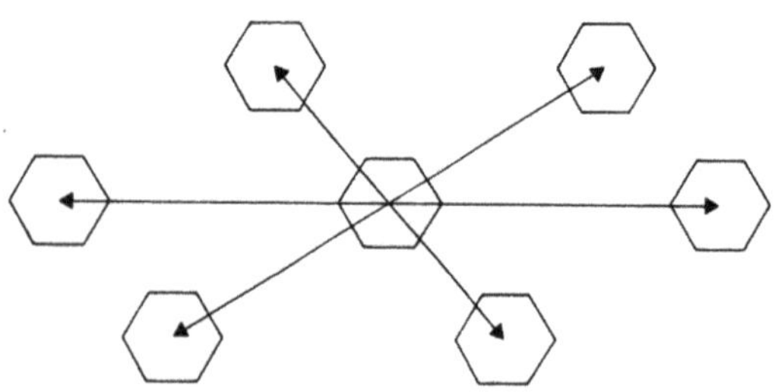

Fig. 1b. Patterson function of the structure in a.

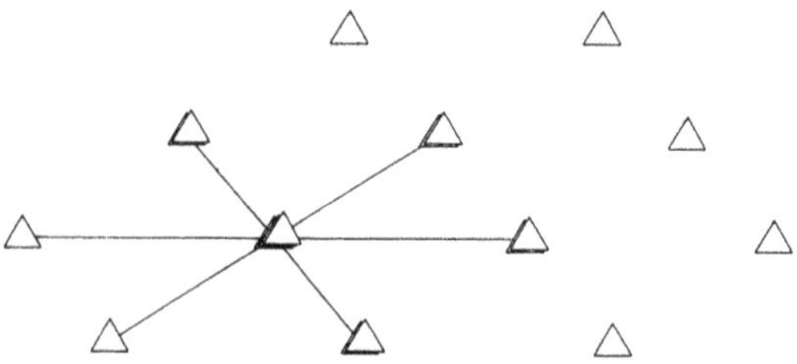

Fig. 1c. The structure a is placed on each point of the Patterson function of its point lattice (endpoints of arrows in b). The number of overlaps in each position is schematically indicated. The "reconstruction" is formed at the centre of the Patterson function, in this case by superposition of three repeats.

In practice, of course, any degree of knowledge about the motif structure is possible – from complete certainty to total ignorance. Generally, deviations of the motif from the prototype affect the correlation peak in two ways: first, the peak is reduced in height and second, it is blurred out since deviations occur most likely in the high resolution range. For unfortunate choices of the prototype pattern, the correlation peak may in fact be so broad and insignificant that the locations of the repeats cannot be established. To give an extreme example, it will not be of much help to know that the motif is circular with approximate radius R, because the auto-correlation between two circular discs of that radius is a function whose halfwidth is in the order of $2R$.

Once the lattice is established, the noisy image is superposed N-times on itself, shifted by $\boldsymbol{r}_j$, and the average is computed. The average of the repeats is an enhanced version of the motif structure, whereas the average of the noise portion levels out to a constant background term if N is sufficiently large. To be accurate, the condition for incoherent noise superposition is that the lengths of the lattice vectors $\boldsymbol{r}_j$ and all possible difference vectors $\boldsymbol{r}_j - \boldsymbol{r}_k$ are larger than the auto-correlation radius of the noise function.

2.3. Auto-correlation superposition

The other approach follows from the idea that the repeats present in (1) can act as their own prototypes. In the simple case where the signal-to-noise ratio is sufficiently large, the repeats of the motif can be approximately located in the image by eye. An image area containing a single repeat is then selected and cross-correlated against the whole image. The result is essentially the same as in (2): the autocorrelation function appears superposed on all lattice positions and its peak thus marks the lattice.

We now assume the more complicated case where, due to low signal-to-noise ratio, the repeats cannot be clearly distinguished from the noise background.

By auto-correlating the image one obtains

$$i(\boldsymbol{r}) \otimes i(\boldsymbol{r}) = [m(\boldsymbol{r}) \otimes m(\boldsymbol{r})] \circ [g(\boldsymbol{r}) \otimes g(\boldsymbol{r})] + n(\boldsymbol{r}) \otimes n(\boldsymbol{r}) . \quad (3)$$

Here the cross-terms between noise and signal are omitted because of the assumption of uncorrelated noise made at the beginning. Apart from the noise auto-correlation $n \otimes n$ appearing in the centre, this is a convolution product of the auto-correlation function of the motif with that of the irregular lattice $g(\boldsymbol{r})$. Following the practice of X-ray structure analysis, we call the auto-correlation function of the point structure 'Patterson function' [12]. We have here the complication that $m \otimes m$ appears repeated on the Patterson function of the lattice rather than on the lattice itself. This function contains points at each possible difference position $\boldsymbol{r}_{jk} = \boldsymbol{r}_j - \boldsymbol{r}_k$ (cf. fig. 1a, b). For an irregular lattice, the number of these points is $N(N-1)+1$ if one counts the multiple zero difference vector $j = k$ only once and observes that each non-zero difference vector appears twice, with opposite directions. The proposal is now that the image should be superposed $N(N-1)+1$ times on itself, shifted by the Patterson vectors found in (3).

The resulting pattern (fig. 1c) will contain single overlaps lying on one half of the Patterson function, and a number of repeats without overlap at the periphery. However, most interesting for our purpose is the fact that exactly N overlaps occur at the origin of the Patterson function, with each position of the irregular lattice contributing only once, so that here the desired average version of the motif is built up. Another way of looking at this result is the following: since the Patterson function of the lattice contains the lattice itself, shifting by all of its vectors assures that all lattice points come to an overlap at one point. Apart from the 'signal' position, our reconstruction contains a noise portion from $N(N-2)+1$ unsuccessful overlaps which tends to be levelled out for large N.

One difficulty has to be mentioned here*: with an increasing number of repeats, the number of difference vectors in the irregular point structure increases, and so does the number of Patterson vectors. As a result, the auto-correlation discs which according to (3) are centred on the end points of these vectors will with increasing probability come to a partial overlap, thus obscuring the Patterson function sought and preventing accurate superposition. One can show by simple geometrical considerations that overlap will be unlikely if the extension of the selected field is large compared to that of a single particle.

This second approach therefore seems to solve the problem of averaging over N noisy occurrences of a motif on an irregular lattice in a general way, although the assumption of equal orientation made at the beginning is a most unsatisfactory restriction. Nevertheless, we may have situations where biological structures appear in irregular intervals in a background matrix with preferred orientation, or where a crystalline supporting film favours a distinct orientation of the support particles.

2.4. *The case of arbitrary relative orientation*

Following is an outline for a strategy that may be successful in the case of arbitrary relative orientation. The approximate location of the repeats can be found either by inspection or by application of matched filtering with a circularly symmetric prototype function that has the expected radius and radial distribution of the motif. One repeat so located is then rotationally correlated with each of the others. Since the exact position of the origins is unknown at this stage, rotational correlation has to be done by using translation-invariant methods [10,13]. Once the relative orientations of the $N-1$ repeats is found with respect to the reference repeat, these can be rotated so that an image with oriented repeats is produced, which can then be treated by the methods outlined above.

3. Conclusion

As in the case of periodic objects, averaging can be used to build up a high resolution image from noise-

* I would like to thank Prof. W. Hoppe for drawing my attention to this point.

contaminated repeats of a structure lying in random positions but common orientation. Non-destructive electron microscopy which has recently proved successful in the determination of an ordered membrane structure [2] can therefore in principle be extended to biological material in disordered form. The main complication introduced by the random positions is that they are not revealed by Fourier transformation, but have to be determined prior to the averaging procedure. Two methods have been discussed in this context: matched filtering and a new method, called auto-correlation superposition.

While matched filtering requires a certain amount of a-priori knowledge about the structure to be located, the other method makes no such assumption, and entirely relies on the correlation between any pair of repeats.

The question of noise sensitivity which has not been answered here makes a detailed theoretical investigation of the effects of size, shape, resolution, signal-to-noise ratio and nature of the noise on the position-finding algorithms necessary. This will be done in a later study. It is already obvious, however, that the exposure reduction that can be achieved in the case of non-periodic specimens is limited by the requirement of some high-resolution similarity between any pair of repeats which is necessary for accurate positioning, whereas no similar limitation seems to exist for periodic specimens.

As a concluding note to interested biologists, it would be desirable to know at this stage if some interesting biological material occurs in disordered yet oriented form, or can be brought into this form by the use of special supporting films, since this would greatly affect the motivation for extended theoretical studies on these lines.

Acknowledgements

I gratefully acknowledge valuable suggestions made by Drs. H.P. Erickson, R.M. Glaeser and P.N.T. Unwin. This work was supported by a S.R.C. Grant.

References

[1] R.M. Glaeser, I. Kuo and T.F. Budinger, 29th Ann. Proc. Electron Microscopy Soc. Amer. (1971) 466.
[2] P.N.T. Unwin and R. Henderson, J. Mol. Biol. (1975), in press.
[3] I.A.M. Kuo and R.M. Glaeser, Ultramicroscopy 1 (1975) 53.
[4] A brief discussion of these techniques is also contained in a report by M. Beer, J. Frank, K.-J. Hanszen, E. Kellenberger and R.C. Williams, Quart. Rev. Biophys. 7 (1975) 211.
[5] F.P. Ottensmeyer, E.E. Schmidt, T. Jack and J. Powell, J. Ultrastr. Res. 40 (1972) 546.
[6] F.P. Ottensmeyer, E.E. Schmidt and A.J. Olbrecht, Science 179 (1973) 175.
[7] J. Frank, Computer Processing of Electron Micrographs, in 'Advanced Techniques in Biological Microscopy', ed. J.K. Koehler (Springer Verlag, Heidelberg, 1973) p. 215.
[8] H.P. Erickson and A. Klug, Ber. Bunsenges. Phys. Chem. 74 (1970) 1129.
[9] U. Aebi, P.R. Smith, J. Dubochet, C. Henry and E. Kellenberger, J. Supermol. Str. 1 (1973) 498.
[10] R. Langer, J. Frank, A. Feltynowski and W. Hoppe, Ber. Bunsenges. Phys. Chem. 74 (1970) 1120.
[11] See, for instance, H.C. Andrews, "Computer Techniques in Image Processing" (Academic Press New York/London, 1970) p. 56 ff.
[12] R. Hosemann and S.N. Bagchi, 'Direct Analysis of Diffraction by Matter' (North-Holland Publishing Company, Amsterdam, 1962).
[13] W.O. Saxton, Eight Int. Congr. El. Micr. Canberra 1974, I, p. 314.

Signal-to-noise ratio of electron micrographs obtained by cross correlation

J. Frank & L. Al-Ali
Cavendish Laboratory, Cambridge, UK

The signal-to-noise ratio of electron micrographs can be determined by two-dimensional digital cross correlation even though neither signal nor noise can be analysed separately. Such measurements suggest how best to make use of the electron microscope.

HIGH resolution electron micrographs contain a large amount of detail that is not related to the object structure and is commonly referred to as noise. The relative amount of significant information compared with noise in the image can be expressed in terms of the ratio of signal variance to noise variance, or signal-to-noise (s/n) ratio. The s/n ratio is an important criterion for the quality of the electron micrograph, and for the imaging process by which it is obtained. Unfortunately, no exactly defined electron microscopic model object is available from which the relative portion of the signal variance in the image could be inferred; hitherto, therefore, the s/n ratio has been a merely theoretical concept. Using a cross-correlation technique, we have studied the defocus dependence of the s/n ratio for a carbon film micrograph and found a characteristic behaviour which closely follows the predictions of linear contrast transfer theory, if one takes into account the effects of partial coherence and energy spread and assumes purely elastic scattering.

Our results support the view that, for thin biological objects and bright field illumination, elastic scattering plays the dominant role in contrast formation. They also stress the importance of the underfocus range for practical work and suggest a way of achieving accurate reproducible focusing of instrument with on-line image readout.

The technique

It has been suggested[1] that the sample cross-correlation coefficient could be used for estimating the s/n ratio of band-limited stochastic time functions, if two different versions of the desired signal disturbed by uncorrelated noise are available, each as a series of N time samples taken at a rate equal to twice the bandwidth. The sample cross-correlation coefficient of two series x_i, y_i with zero lag is defined as

$$r_N = \frac{\langle (x_i - \langle x_i \rangle)(y_i - \langle y_i \rangle) \rangle}{\{\langle (x_i - \langle x_i \rangle)^2 \rangle \langle (y_i - \langle y_i \rangle)^2 \rangle\}^{1/2}}$$

where the angle brackets denote averaging. Assuming stationary Gaussian zero-mean statistics for both signal and noise one obtains the s/n estimate

$$\begin{aligned} \alpha &= \exp[-2/(N-3)](r_N/(1-r_N)+\tfrac{1}{2})-\tfrac{1}{2} \\ &\approx r_N/(1-r_N) \quad \text{for large } N \end{aligned} \tag{1}$$

The formula for large N can be shown to be identical with the formula that generally relates the s/n ratio of two continuous series to their cross-correlation coefficient, irrespective of the statistical distribution of noise and signal. In image processing application where N is of the order of 10,000, the Gaussian assumption can therefore be dropped.

We have adopted this method of s/n measurement for the quantitative analysis of electron micrographs of a thin amorphous carbon film, and found excellent agreement between the defocus dependence of α obtained from two-dimensional digital correlation, and the theoretical behaviour expected for weakly elastically scattering objects ('weak phase objects').

In these conditions, the action of the electron microscope can be considered[2] as that of a linear system with transfer function $H(k)$ (k spatial frequency), with the projected potential of the object as 'input' and the image intensity recorded on the photographic plate as 'output'. The transfer function[3] describes the distorting and resolution limiting effects due to spherical aberration, defocusing, axial astigmatism, energy spread[4] and partial coherence [5,6]. Assuming an input with power spectrum $P(k)$ and additive signal-independent noise on the output side, the s/n ratio of the output becomes proportional to the signal variance[7].

$$\mathrm{var}(s) = \int |H(k)|^2 P(k) \mathrm{d}k \tag{2}$$

where the integration extends over the range of spatial frequencies within the bandlimit of the system.

Studies of carbon film

Pairs of electron micrographs of a thin (≈ 100 Å) carbon film were available, each pair taken with a different defocus but in otherwise identical conditions. Assuming a stable specimen, each pair contains the same signal portion distorted by the instrument in a characteristic way, with a superposed noise portion that is due to the photographic grain and electron noise and is by first approximation additive to the signal. For the experiment, a JEM 100B electron microscope was used at 100 kV, $\times 200{,}000$ magnification and large objective aperture (≈ 0.5 Å^{-1}). Defocus values and astigmatic focus difference (250 Å) were determined by optical diffraction using Thon's[8] method of evaluation, whereas the divergence of illumination (semiangle 2.7×10^{-4} rad) and the energy spread (1.8 eV) were obtained from bandlimit measurements (J.F., unpublished) using Young-type optical interference[9] for each micrograph pair. From these parameters, the transfer function can be constructed for any defocus value (Fig. 1). As power spectrum, the intensity curve for elastic scattering of thin amorphous carbon films as obtained by electron diffraction[10] was used. The overall behaviour of this function can be approximated in the interesting spatial frequency range by $P(k) \sim \exp(-|k|/\mathscr{H})$ with $\mathscr{H} = 0.18$ Å^{-1}. Since the highest bandlimit observed in the optical interference experiment was 0.38 Å^{-1}, the integration limit for the computation of the integral (2) was set at 0.4 Å^{-1}. For simplicity, the transfer functions were taken to be rotationally symmetric. As the astigmatism is very small, this approximation was expected to cause no appreciable error. The theoretical s/n ratio so obtained is shown in Fig. 2 as a function of defocus. For comparison, the case of a 'white' object spectrum $P(k) =$ constant was also considered (Fig. 3).

Broadly, the s/n ratio increases with increasing defocus and shows an oscillatory behaviour at positive defocus (that is, underfocus), with the maxima lying just above the defocus positions $[(2n+1)C_s\lambda]^{1/2}$; $n = 0, 1, 2...$ (C_s spherical aberration constant, λ electron wavelength) where large transfer intervals

(Reprinted from Nature, Vol. 256, No. 5516, pp. 376–379, July 31, 1975)

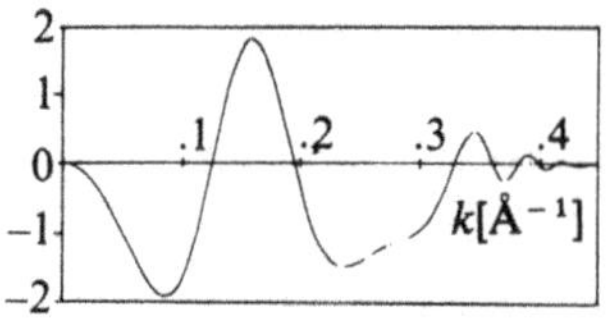

Fig. 1 Electron microscopic phase contrast transfer function for defocus $\Delta z = 1{,}860$ Å, spherical aberration constant $C_s = 1.8$ mm, electron wavelength 0.037 Å, chromatic aberration constant $C_c = 1.4$ mm, energy spread halfwidth $\Delta E = 1.8$ eV, and Gaussian source distribution with halfwidth 5.4×10^{-4} rad. The transfer function is computed from the coherent transfer function by using the envelope representations[5,6].

appear in the transfer function. For carbon scattering, the absolute maximum is assumed at $n = 0$ ('Scherzer focus'), whereas it appears at higher defocus values for 'white' scattering. Even if one ignores the oscillations, the curve is remarkably asymmetric about $\Delta z = 0$. Computations with different illumination and energy spread parameters show that this behaviour is typical for imaging conditions where the resolution is limited by illumination divergence rather than energy spread.

The electron micrographs were copied from plates (Ilford EM4) on to fine grain film (Ilford N4E.50). This process is linear within 4% in the interesting density range, as checked by calibration, and does not affect the s/n ratio appreciably. The images were scanned using a Joyce-Loebl Scandig densitometer, with a sampling distance of 25 μm, corresponding to optimum sampling[11] of data with a bandlimit at 0.4 Å^{-1}. The digital data were recorded on magnetic tape and processed using the IMPROC image processing software system[12]. The images were aligned using cross correlation[13], and the orientation checked by rotational correlation techniques[14]. For the computation of the cross-correlation coefficients, matrices of size 128×128 were selected from the scanned data. Since a sizeable portion of the full cross-correlation matrix had to be examined in the context of another study, the computation was done by using the Fourier relation, by forming the conjugate product of the respective discrete transforms, inverse transformation, and appropriate normalisation. As a check on the assumption of stationarity, the cross-correlation coefficient was computed for different areas of the scanned image field, and was found reproducible within 5%. The s/n ratio α was calculated using formula (1). The result was corrected for the decrease of the s/n ratio due to the densitometer. Detailed analysis leads to a correction formula of the form

$$1 + 1/\alpha_{true} = (1 + 1/\alpha)/(1 + 1/\alpha_d)$$

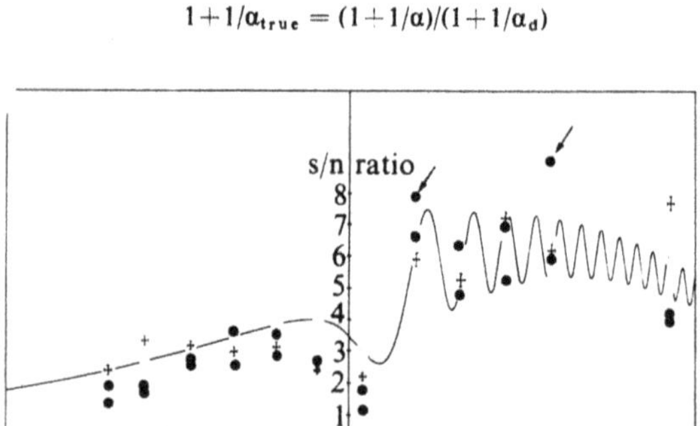

Fig. 2 Theoretical s/n ratio of electron micrographs of carbon as a function of defocus, and experimental s/n ratios α and $\tilde{\alpha}$ of a defocus series obtained from cross correlation (error bars) and from variance measurements (dots), respectively. The theoretical curve was scaled to fit the α measurements at $\Delta z = 790$, 1,320, 1,860 and 2,390 Å optimally. The sections of the theoretical curve which lie within the horizontal error margins of the α measurements are marked by dotting.

where the s/n ratio of densitometer measurement, α_d, is obtained from the cross correlation coefficient of two scans of the same image. For our measurements, α_d was found to be between 50 and 90. The resulting α_{true} values are plotted in Fig. 2, with the horizontal bars indicating the inaccuracy of defocus determination, estimated at ±50 Å, and the vertical bars the error limits due to the α_d range and the fluctuation of r when measured in different areas. The scaling factor for the theoretical curve was chosen in such a way that optimum fit occurs for the measurements at $\Delta z = 790$, 1,320, 1,860 and 2,390 Å.

Interpretation

Theoretically, we expect some deviation of the measured values from the theoretical curve because the latter does not account for the change of noise variance with the signal variance that can be expected for a Poisson process. The defocus dependence of the experimental values is nevertheless in excellent agreement with the predicted behaviour: the measured s/n values are low and follow a flat curve at negative defocus, they then jump to a high value between the Gaussian and the Scherzer focus, and follow the predicted variations at the sampled defocus positions. (Corroborating the detailed structure of the theoretical curve at positive defocus by experimental data would be a formidable task with the method used, involving the evaluation of dozens

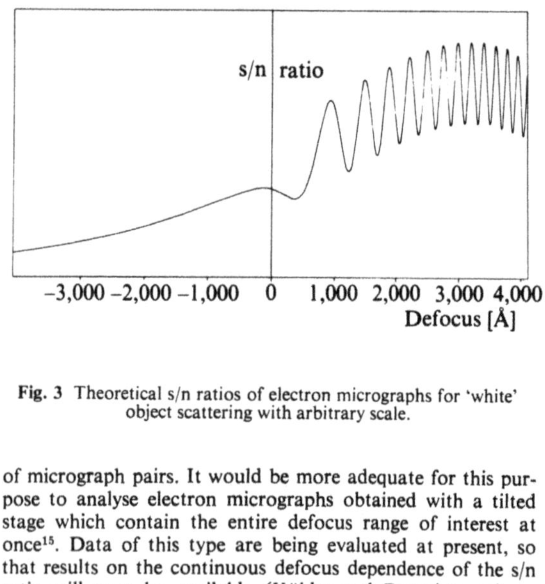

Fig. 3 Theoretical s/n ratios of electron micrographs for 'white' object scattering with arbitrary scale.

of micrograph pairs. It would be more adequate for this purpose to analyse electron micrographs obtained with a tilted stage which contain the entire defocus range of interest at once[15]. Data of this type are being evaluated at present, so that results on the continuous defocus dependence of the s/n ratio will soon be available (Kübler and Downing, private communication).) As an independent check of the experimental values, we have computed the variances of all image pairs var(i) ,and the variance of an image obtained with free electron exposure in the same optical density range, var(n). From these variances we obtain again with the assumption of additive superposition and uncorrelated noise,

$$\tilde{\alpha} = (\text{var(i)} - \text{var(n)})/\text{var(n)}$$

(Strictly speaking, these s/n estimates should be closer to the theoretical curve than the ones obtained by correlation, since they are obtained with the assumption of signal-independent noise variance.) We found good agreement between α and $\tilde{\alpha}$ (dots in Fig. 2) in the range of small α, and general agreement for large α. The two measurements marked by arrows belong to images whose histograms show a sharp peak on top of a Gaussian appearance. The origin of this artefact, which is missing in the corresponding twin image histograms, is unknown but it is thought that it is responsible for the large deviation of $\tilde{\alpha}$. On the other hand, no reason could be found for deviations at $\Delta z = 1{,}860$ Å and 3,900 Å. These discrepancies will be investigated further.

Several conclusions can be drawn from the results of our study: (1) the satisfactory agreement between experimental and theoretical defocus dependence suggests that the linear transfer theory gives an adequate description of contrast formation in the conditions prevailing in the experiment, which are typical for electron microscopy of thin biological specimens; (2) in particular, since the fit of the theoretical curve to the experimental data was achieved by using a scaling factor only, it can be concluded that all those scattering processes ignored in our treatment, which could produce an image with defocus-dependent variance, make only a negligible contribution to the image contrast, such as inelastic scattering[16]; (3) with regard to efforts to retrieve the original object function by restoration techniques, it is important to know that the amount of information is maximised at distinct defocus values. For our experimental conditions, the first four maxima have approximately the same height. On the other hand, it is known from the study of partially coherent transfer functions[17] that with increasing defocus the relative portion of high resolution information passed to the image increases. It may therefore be preferable, for the purpose of subsequent processing, to use defocus values larger than the Scherzer focus in the experiment, even though the electron micrographs may become less intelligible for direct interpretation; (4) as a practical conclusion, our measurements show that for the normal partially coherent illumination used, the gain in s/n ratio lies between two and three if one goes from overfocus to underfocus; (5) in the underfocus range, the positions of maxima and minima are fixed for a given instrument and may therefore be used as absolute focus references, possibly in connection with an objective current wobbler and on-line image readout, complementing methods recently proposed for controlled focusing[18].

The method of measuring the s/n ratio by cross correlation is not restricted to phase contrast electron microscopy, or electron microscopy for that matter; it could be used quite generally, as a performance criterion of electron optical and light optical systems as well as new imaging and recording techniques.

This work was supported by the SRC. We thank Dr E. D. Boyes for the electron micrographs, and Drs P. W. Hawkes, W. O. Saxton and H. Rose for comments.

Received May 6; accepted June 12, 1975.

1 Bershad, N. J., and Rockmore, A. J., *IEEE Trans.*, **IT, 20,** 112-113 (1974).
2 Lenz, F., in *Electron Microcopy in Material Science*, (edit. by Valdre, U.), (Academic, London, 1971).
3 cf. Hanszen, K.-J., *Adv. opt. Elect. Micr.* **4**, 1-84 (1971).
4 Hanszen, K.-J., and Trepte, L., *Optik*, **32**, 519-538 (1971).
5 Hanszen, K.-J., and Trepte, L., *Optik*, **33**, 182-198 (1971).
6 Frank, J., *Optik*, **38**, 519-536 (1973).
7 Goodman, J. W., *Introduction to Fourier Optics*, (McGraw-Hill, New York, 1968).
8 Thon, F., *Z. Naturforsch.*, **21a**, 476-478 (1966).
9 Frank, J., Bussler, P. H., Langer, R., and Hoppe, W., *Ber. Bunsenges. phys. Chem.*, **74**, 1105-1115 (1970).
10 Brünger, W., and Menz, W., *Z. Phys.* **184**, 271-285 (1965).
11 Rosenfeld, A., in *Picture Processing by Computer*, (edit. by Reinholdt, W.), (Academic, London and New York, 1969).
12 Saxton, W. O., *Comp. Graph. Im. Proc.*, **3**, 266-276 (1974).
13 Langer, R., Frank, J., Feltynowski, A., and Hoppe, W., *Ber. Bunsenges. phys. Chem.*, **74**, 1120-1126 (1970).
14 Saxton, W. O., *Eighth Int. Congr. Electron Microscopy Canberra.* **1**, 314-315 (1974).
15 Krakow, W., Downing, K. H., and Siegel, B. M., *Optik*, **40**, 1-13 (1974).
16 Misell, D. L., *J. Phys. D.*, **6**, L1-L5 (1973).
17 Frank, J., *Optik* (in the press).
18 Frank, J., *J. Phys. E.* (in the press).

Ultramicroscopy 2 (1977) 219–227
© North-Holland Publishing Company

MOTIF DETECTION IN QUANTUM NOISE-LIMITED ELECTRON MICROGRAPHS BY CROSS-CORRELATION *

W.O. SAXTON
Cavendish Laboratory, Free School Lane, Cambridge, England

J. FRANK
Division of Laboratories and Research, New York State Department of Health, Empire State Plaza, Albany, NY 12201, USA

Received 3 January 1977

The use of cross-correlation to detect randomly positioned low-dose realizations of a motif in an image field is quantitatively evaluated. For both the bright- and dark-field cases, we derive theoretical expressions for the minimum dose allowing detection in terms of motif size, resolution, and contrast. Model computations on simulated low-dose images of a spherical virus particle give results that agree with our theoretical formulas and demonstrate the feasibility of this approach.

1. Introduction

Our investigation was motivated by the success of low-dose electron microscopy of biological specimens exhibiting two-dimensional periodic order [1,2], in which a high-quality image of the repeating unit is reconstructed by spatial averaging or Fourier filtering. The high sensitivity of the technique allows the investigation of unstained specimens, provided these can be supported by a suitable embedding medium.

In a previous note [3] Frank discussed the possibility of extending this approach to nonperiodic arrangements of identical particles. The main complication introduced by the nonperiodic arrangement lies in the fact that the positions of the repeating units are unknown. It was proposed that these positions could be found by using generalized matched-filter detection techniques. However, this note did not give a quantitative estimate of the detection sensitivity in terms of dose, particle size, and contrast. The present paper attempts to answer this question for the two different modes of electron microscopic imaging. A short treatment of the bright-field case has been given previously [4].

* This paper is dedicated to Professor Otto Scherzer on the occasion of his 68th birthday.

2. Theory

2.1. Philosophy of the "self-detection" approach

Since the cross-correlation function plays a central role in the following analysis, we shall start by defining it briefly [5,6].

The discrete two-dimensional cross-correlation function (CCF) of two images $f(\boldsymbol{r}_i)$ and $g(\boldsymbol{r}_i)$ measured at regular sampling points $\boldsymbol{r}_i$, $i = 1 \ldots N$, is defined by

$$\rho(\boldsymbol{r}_k) = \sum_{i=1}^{N} f(\boldsymbol{r}_i)\, g(\boldsymbol{r}_i + \boldsymbol{r}_k)\,, \tag{1}$$

where $\boldsymbol{r}_k$ is an arbitrary vector contained in the sampling grid. The autocorrelation (ACF) is similarly obtained as the cross-correlation of an image with itself, i.e., by replacing g in this formula by f. This function normally displays a sharp peak at the origin. If the two images contain a common motif at different locations, the CCF will contain the ACF of the motif in an offset position, with the offset vector corresponding to the vector by which the motif appears to be shifted (fig. 1). The search for the relative position of the two realizations of the motif is thus reduced to the search for a sharp peak in the CCF. Since the width of this peak is of the order of the resolution distance,

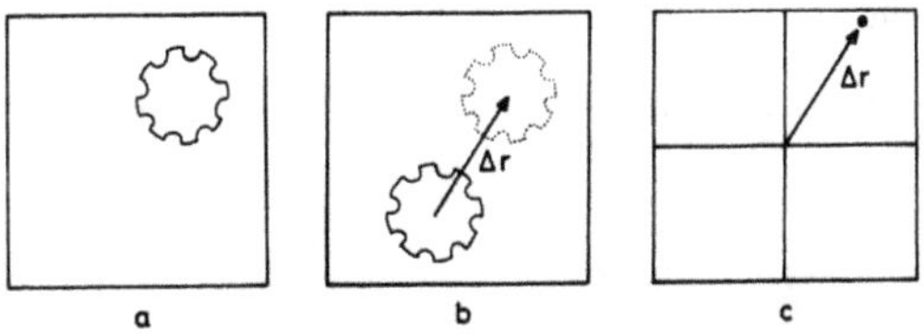

Fig. 1. Schematic diagram explaining the principle of particle alignment using the cross-correlation function. Panels a and b show the same particle at different positions; Δr is the difference vector between these positions. The CCF of a and b will show a sharp peak displaced by Δr with respect to the CCF origin (panel c)..

the alignment by using cross-correlation has a corresponding accuracy. Due to the symmetry of the peak, the accuracy may actually be better by a factor of 3 to 5.

We use the term *self-detection* for the mutual recognition of two arbitrarily positioned noisy motifs via CCF. Like the Fourier transform technique for detecting and filtering periodic arrangements of noisy motifs, this technique will give positional information without requiring any reference information. In contrast to this, the usual matched-filtering method [7] employs a noise-free reference image (template) which must be known beforehand.

Even though the self-detection approach gives us more freedom in principle, this is not necessarily true in practice. It is obvious from the definition of the CCF that the peak size and therefore the detectability of the motif increases in proportion to the relative area it occupies in the image. To achieve optimum conditions for detection, one must use as a reference an image portion which is just large enough to contain the entire motif. The first step, therefore, is to find the *approximate* position of at least one of the particles in the electron micrograph. In the case of negatively stained specimens, and possibly in some cases of unstained specimens with sufficient contrast, this can be achieved by using a micrograph of the same field with overcritical exposure, which will show the outline of the damaged particle at low resolution. An alternative solution to this problem – the use of randomly selected image portions – is discussed in section 2.9.

For a pair of "boxed-in" noisy motifs brought into alignment, the conditions under which the cross-correlation peak is detectable will be explored in the following mathematical derivation. The result of this analysis can be applied without modification to the practically important case where one such image section, used as a reference, is cross-correlated with a much larger image field.

2.2. Methodology and notation

The formal problem treated is the limitations imposed by quantum noise, for which the recorded image signal is governed by a Poisson distribution. Other sources of noise (e.g., photographic, structural noise, preparation artifacts and natural variation) could be treated by appropriate modifications, so far unexplored. We shall consider a statistical average CCF, taken over an ensemble of pairs of noisy digitized image arrays f_i and g_i with N points each, and shall regard a peak as detectable if its (ensemble average) height above the background is at least thrice the (ensemble average) standard deviation of the height. The noise is assumed to be uncorrelated from point to point. Angle brackets ($\langle\,\rangle$) denote spatial (array) averages, and $E(\,)$ and var ( ) denote ensemble averages (see appendix).

2.3. Simple derivation of the detection criterion for bright-field microscopy of weak objects

We let c_i denote the contrast at a given image point i, and p the number of electrons recorded per picture element *, so that $E(f_i) = p(1 + c_i)$ and $\mathrm{var}(f_i) = p$. The image g_i has similar properties, and f_i and g_i are statistically independent. We subtract the mean signal p from the data and from the cross-sum

$$\rho = \sum_i (f_i - p)(g_i - p) \tag{2}$$

which is the peak coefficient of the cross-correlation array.

If we denote the r.m.s. image contrast $\langle c_i^2 \rangle$ by c^2, application of the standard formulas given in the Appendix shows the mean $E(\rho)$ to be Np^2c^2 and the variance $\mathrm{var}(\rho)$ to be $Np^2(pc^2 + 1)$. Under the low-dose conditions of interest pc^2 is negligible compared with 1, so that $\mathrm{var}(\rho) = Np^2$. The average background height of the cross-correlation array is of course zero. The requirement that the peak height be thrice its

* A factor of approximately 0.7 can be assumed to account for the efficiency of photographic recording [8].

standard deviation thus gives the detection criterion

$$Np^2c^2 \geqslant 3p\sqrt{N} \quad \text{or} \quad p \geqslant \frac{3}{c^2\sqrt{N}} \tag{3}$$

showing that *the minimum dose for CCF self-detection is inversely proportional to the linear dimensions of the motif* [4].

At this point a comparison with Rose's [9] visual detection criterion is useful [4]. The recorded object dose per unit picture area is $\overline{p} = p/d^2$ where d is the sampling distance. The number of picture elements N can be expressed in terms of the object diameter D and the sampling distance as $N \approx D^2/d^2$. Our criterion for a significant cross-correlation peak becomes *

$$\overline{p} \geqslant \frac{3}{c^2Dd} . \tag{4}$$

On the other hand, the minimum dose for visual detection of details of size d and contrast c in an image field was found [8] to be

$$\overline{p} \geqslant \frac{25}{c^2d^2} \tag{5}$$

where the factor in the numerator was found by visual experiments. Comparison between the two expressions shows that cross-correlation self-detection requires a factor of about 8 D/d fewer quanta per unit area than direct visual recognition; indeed, a dose reduction as high as 500 can be achieved for a high resolution picture (d = 3 Å) of an assembly of large (D = 200 Å) particles (cf table 1). The improvement occurs because the self-detection process depends on contexts and not simply on individual picture details.

2.4. The general case

More generally, and especially in dark-field images, the noise statistics vary from point to point.

The recorded images are again denoted by f_i and g_i with the properties that $E(f_i) = p_i$, $\text{var}(f_i) = p_i$, etc., f_i and g_i being independent; p_i is the ideal number of electrons incident on the picture element i, and p $(=\langle p_i \rangle)$ is as before the mean number of electrons incident on the image per picture element. The means $\langle f_i \rangle$ and $\langle g_i \rangle$ subtracted from the image data before

* In [4] the factor 9 in the numerator of the corresponding expression is in error.

Table 1
Dose-reduction factors achievable by cross-correlation for various resolution distances d[Å] and particle diameters D[Å].

$d \backslash D$	50	100	200
3	133	266	532
10	40	80	160
20	20	40	80

formation of the cross-sum are here approximated by p; retention of the exact values would introduce only terms smaller by a factor of N than those derived below. We thus have again

$$\rho = \sum_i (f_i - p)(g_i - p) \tag{2}$$

from which the mean follows by application of the standard formulas, as

$$E(\rho) = \sum_i (p_i - p)^2 = Np^2(\beta - 1) \tag{6}$$

if $\beta = \langle p_i^2 \rangle / p^2$, the ratio of the mean square to the squared mean for the ideal image p_i. The variance can be similarly obtained as

$$\begin{aligned} \text{var}(\rho) &= \sum_i (2(p_i - p)^2 p_i + p_i^2) \\ &= Np^2((2\gamma + 2 - 4\beta)p + \beta) , \end{aligned} \tag{7}$$

where $\langle p_i^3 \rangle / p^3$ has been denoted by γ.

The detection criterion now becomes

$$N^2p^4(\beta - 1)^2 \geqslant 9Np^2((2\gamma + 2 - 4\beta)p + \beta) \tag{8}$$

which may be solved as a quadratic equation in p to give the criterion

$$p \geqslant q/N(1 + (1 + N/N_0)^{1/2}) , \tag{9}$$

where

$$q = 9(\gamma + 1 - 2\beta)/(\beta - 1)^2$$

and

$$N_0 = 9(\gamma + 1 - 2\beta)^2/(\beta(\beta - 1)^2) .$$

The general solution thus depends on the shape of the ideal image histogram and not simply, as in the bright-field weak-object case, on the contrast level and number of elements in the array. However, considerable simplification is possible when reasonable assumptions are made. Table 2 gives the values of q

Table 2
The histogram parameters q, N_0 and q_1 defined in the text for various histogram shapes.

Histogram shape	q	N_0	q_1
One-sided gaussian	28	18	7
Rectangular	27	7	10
Triangular	13	25	3
Exponential	27	41	4
Virus image *	15	53	2

* i.e., the virus model image used in the numerical test, squared after background removal.

and N_0 for various histogram shapes, and shows that N_0 is usually negligible compared with N. This approximation allows us to rephrase the detection criterion as

$$p \geqslant q_1/\sqrt{N} \tag{10}$$

with $q_1 = 3\sqrt{\beta}/(\beta - 1)$, so that the third moment of the ideal image histogram is no longer relevant. Values of q_1 for the various histogram shapes range from 2 to 10 (table 2).

For very small motifs, when N is not large compared with N_0, the necessary dose is substantially higher than $q_1/\sqrt{N}$.

2.5. Comparison of bright-field and dark-field criteria for weak objects

We can now compare the minimum object exposures permitting motif self-detection under bright- and dark-field illumination conditions. The bright-field image contrast was previously designated c_i, implying that the real part of the scattered wave at the image is a fraction $\frac{1}{2}c_i$ of the primary wave and carries a fraction $\frac{1}{4}c_i^2$ of the electron flux. The imaginary part may be assumed to carry a similar amount, so that an *object* dose of p electrons per picture element gives rise to a dark-field image $\frac{1}{2}pc_i^2$, with a spatial mean $\frac{1}{2}pc^2$ where c is as before the r.m.s. bright-field image contrast. Requiring that this exceed $q_1/\sqrt{N}$ gives a criterion

$$p \geqslant \frac{k}{c^2\sqrt{N}}, \tag{11}$$

where k ranges from 4 to 20. We conclude that *the object exposure necessary is at least as high, and possibly six times as high under dark-field conditions as it is under bright.*

2.6. Derivation of the bright-field weak-object criterion from the general case

If the ideal image histogram is narrowly distributed about a high mean value, so that $p_i = p(1 + c_i)$ with $|c_i| \ll 1$, it is easy to show that β closely approximates $1 + c^2$ and γ, $1 + 3c^2$. N_0 then becomes 9 and can always be neglected compared with N; q_1 becomes $3/c^2$; and the criterion

$$p \geqslant \frac{3}{c^2\sqrt{N}} \tag{3}$$

obtained before follows at once.

2.7. Motifs with a background shift

Images made solely by phase contrast inevitably have a zero mean; however, in many practical situations a particle image will stand up from the background even after subtraction of the background average. In these circumstances the value of c used should be that *within* the particle outline, not that for the local background shift. The reason is easily seen by regarding the ideal image as a superposition of two signals (fig. 2). It is the sharp peak on top of the roof shaped background ACF which must be detected if we are to reconstruct the interior of the particle.

2.8. Matched filtering with an ideal motif

Criteria appropriate to motif detection when a noise free reference motif p_i is available are easily obtained by applying the above argument to a cross-sum

$$\rho = \sum_i (f_i - p)(p_i - p). \tag{12}$$

The general result is

$$p \geqslant q/N \tag{13}$$

which is much lower for large N than the dose necessary for detection when a noisy motif is used, though the improvement is only two-fold when N is small. The criterion becomes $p \geqslant 9/c^2N$ in the bright-field weak-object case. Cross-correlation detection using an ideal reference motif is thus more sensitive than self-detection of two noisy repeats of the motif by a factor of $N/3$.

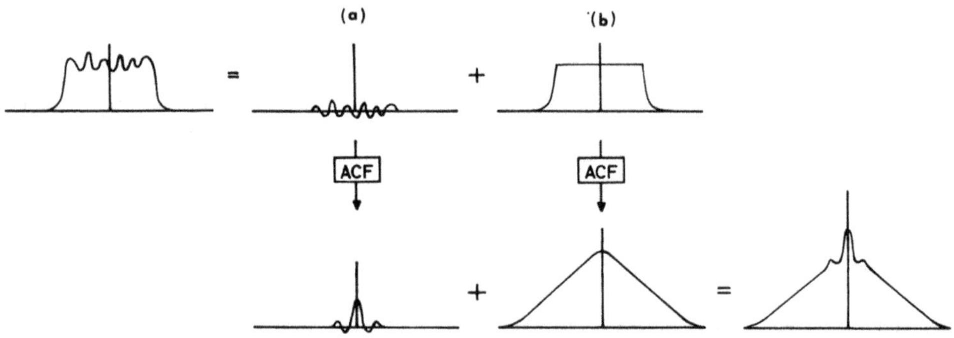

Fig. 2. The particle image (top left) may be considered the sum of two images: (a) the fine structure within the particle outline; and (b) the local background shift (particle shape function) which distinguishes the particle from the image background. The ACF of a sum of uncorrelated functions is the sum of the ACFs of these functions. Consequently we observe the ACF of the fine structure sitting on top of the roof shaped ACF of the particle function.

2.9. A note on the use of the ACF to locate particles

One of the methods suggested [3] to locate and superpose particles involves use of the ACF of the whole image. Our present analysis makes it clear that this method is in both respects less effective than the self-detection approach.

As for locating particles, it was proposed in [3] that the mutual displacement (Patterson) vectors between the particles in an image could be identified from the positions of peaks in the image ACF, but no estimate was given of the detectability of these peaks.

Let us assume a bright-field image containing M particles, with each particle occupying n image points and the full picture array N points. With the same notations as used previously, each cross-correlation peak that marks the overlap of two different particles in the picture ACF will have

$$E(\rho) = np^2c^2 \tag{14}$$

and

$$\mathrm{var}(\rho) = Np^2 \tag{15}$$

since the detection signal derives from n points only, while the background noise derives from all N points. Detection therefore requires that

$$n^2p^4c^4 \geqslant 9Np^2 \tag{16}$$

or that the minimum dose

$$p \geqslant \frac{3\sqrt{N}}{c^2 n} \tag{17}$$

which is larger than the minimum dose for the self detection method by a factor of N/n. The fact that N/n is equal to or (preferably) larger than M makes the autocorrelation method of localization clearly inferior to the other methods discussed in [3] and in this paper. Even if we cannot spot a particle by usual examinations which was the contingency providing in [3] the motivation for the autocorrelation approach, or by using a micrograph with overcritical dose, we can use many randomly chosen portions of the electron micrograph as references in turn until one of them gives significant cross-correlation peaks.

It was also suggested in [3] that once the Patterson peaks are found in the ACF, an effective noise reduction could be achieved by shifting the image along each of the Patterson vectors and superposing the images obtained after each shift. However, at the reconstruction site this procedure yields M superposed signals and $M(M-1)$ superposed independent noise functions. When they are summed, the signal rises by a factor of M, but adding the noise variances implies that the noise rises by $(M(M-1))^{1/2}$, or roughly the same factor. Thus the superposition of the whole image does not effect any appreciable noise reduction.

3. Numerical test for the detection criterion in the bright-field case

3.1. Methods

To demonstrate the feasibility of the outlined technique, we have constructed a model image from a cowpea chlorotic mottle virus (CCMV) picture (see fig. 2 in [10]).

CCMV is a small (250 Å) icosahedral RNA plant virus containing 32 morphological units (20 hexamers and 12 pentamers, making a total of 180 individual units). Its total mass is 4.6×10^6 daltons, of which 25% is RNA. The picture under consideration is the result of spatial averaging over 12 different negatively stained particles.

The model is a 128-point-square array containing four repeats of a CCMV particle at arbitrary locations with the same orientation (fig. 3a). For later comparison with the CCFs, the ACF of the particle image is also shown (fig. 3b).

The particle has a diameter of 27 sampling units, each corresponding to 9 Å on the object scale. For easier display of the CCF, we use here a zero mean image of the particle. (In the realistic case of images with nonzero mean, the CCF peak sits on top of a "roof", cf fig. 2.) The r.m.s. contrast of the particle was normalized to a value slightly higher than 5% as a rough estimate of the contrast expected for unstained preparations.

All image processing was done using the IMPROC Computer Software System [11] on the IBM 370/165 of the University of Cambridge. This program system allows arbitrary sequences of operations on images to be defined with a minimum of programming effort in a high-level language.

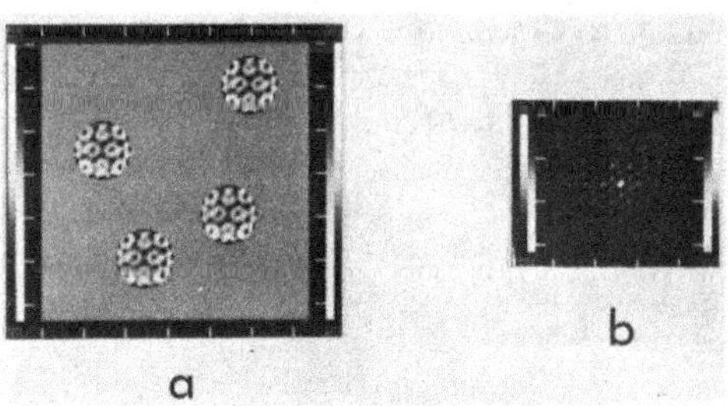

Fig. 3. (a) Model image created by placing a virus particle image at four arbitrary positions. (b) Auto-correlation function of a single virus particle.

Poisson noise-limited versions of the model were made (for details of this simulation see ref. [12]) at average exposures of 150, 120, 90, 60 and 30 electrons/picture element (fig. 4). With an efficiency factor of 0.7, these exposures correspond to object doses of 2.6, 2.1, 1.6, 1.15 and 0.6 electrons/Å^2 and are thus in the range of the critical exposures found by Unwin and Henderson for catalase and the purple membrane protein [1]. Similar noisy versions were made of a single particle contained in a 32-point square image which was masked off outside a circle of diameter 30 units (i.e., somewhat larger than the particle, accounting for the uncertainty of the boundary definition in the real situation) and embedded in a 128-point-square array, which served as the cross-correlation reference. Each model image is displayed here with the same apparent contrast: the upper and lower limits used are 0.8 and 1.3 times the image mean.

At each dose the CCF was computed. Because of the masking we are effectively cross-correlating circular arrays with diameters of 30 units, or 707 points in all. The resulting CCFs are displayed with their negative parts suppressed, and the prints are somewhat underexposed to make the peaks clear. We also include perspective displays which show more clearly the sizes of the peaks relative to the competing background peaks, this time with the lower threshold set at 40% of the peak height.

3.2. Results and discussion

The positions of the virus particles are marked by four sharp peaks in the CCF down to a dose level of 60 electrons/picture element (1.15 electrons/Å^2) but are no longer clear at 30 electrons/picture element. The theoretical expression $p \geqslant 3/c^2\sqrt{N}$ with $c = 0.05$ and $N = 707$ gives $p \geqslant 45$, so the agreement is excellent. The result is impressive since the eye cannot discern the particle boundaries at dose levels below 120 electrons/picture element.

As expected from the theory the ACF of the noisfree motif appears in the four shifted positions. The ACF pattern in fig. 3b can be clearly recognized in the original CCFs, if not in the reproductions, down to a dose level of 90 electrons/Å^2.

To reduce the dose further for a given contrast

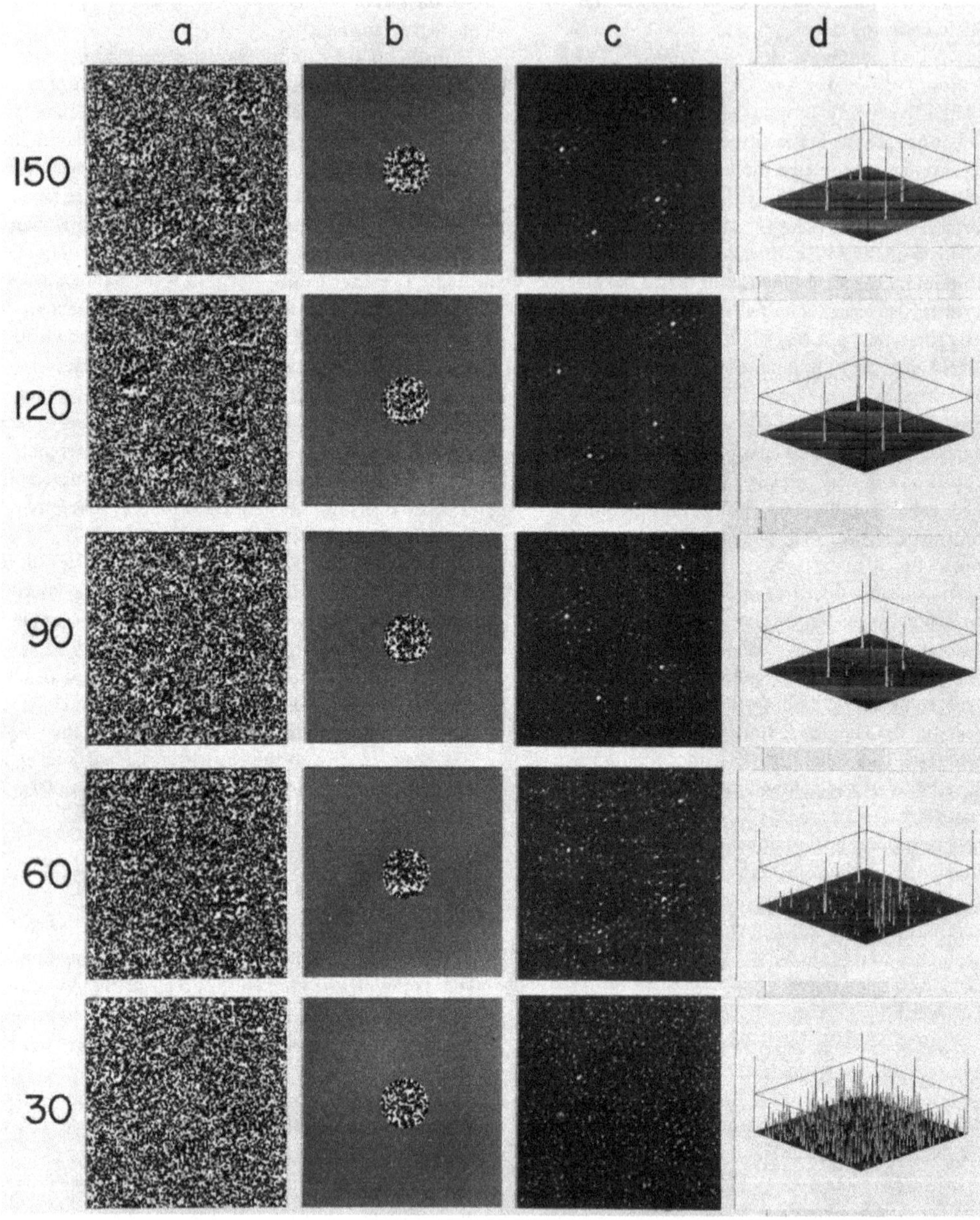

Fig. 4. Column (a) low-dose versions of the model image; (b) low-dose versions of a single particle; (c) cross-correlation functions of these two arrays; and (d) perspective displays of the cross-correlation functions. The numbers on the left indicate the exposures used (electrons per picture element).

and particle size, one must make the sampling grid coarser, so that the same area is sampled by fewer points; this effectively reduces the accuracy of alignment. In our case, a dose level of 0.6 electrons/Å^2 would be achieved by doubling the sampling distance to 18 Å. Three terms – the particle diameter D, its contrast level c, and the critical dose p_c for its destruction – together determine the finest image sampling at which undamaged particles remain detectable in an image. They thus define the ultimate resolution that can be achieved by the superposition of many images. Rephrasing eq. (4) (section 2.3) gives

$$d = 3/c^2 D p_c \,. \qquad (21)$$

For example, a 300-Å-diameter particle with 4% contrast tolerating a dose of 1.4 electron/Å^2 (corresponding to a recorded dose of 1 electron/Å^2) could be reconstructed with a resolution better than 7 Å, but if the contrast were only 2%, the resolution would only be 25 Å.

It is appropriate here to note the liminations of our model. We have assumed that the four particles have identical structures, but in practice some structural variation is likely, due to beam-induced damage and preparation artifacts. This will be increasingly serious for unstained specimens as the object exposure increases above 1 electron/Å^2. We have also ignored contrast due to the specimen-support film. Since the film structure differs for each particle, it can be treated as noise for our present purposes, and preliminary estimates indicate that it is unimportant compared to the Poisson noise at low exposure. Finally, we have ignored photographic noise, which seems likely to be a more substantial problem, though detection systems with a single-electron detection capability could be used to circumvent it. The cumulative effect of these factors will be to make the minimum dose permitting detection somewhat higher than has been estimated here.

A serious complicating factor discounted so far is the random orientation likely for particles in the image. This could necessitate repeated cross-correlation at many different angles; but more realistically, orientation-independent detection methods [13] might be used according to a scheme proposed in [3]. The sensitivity of these methods will be the subject of future studies.

4. Conclusion

Both our theoretical results and our model computations show that high-resolution localization of randomly positioned particles can be achieved under low-dose conditions comparable with those used in nondestructive electron microscopy [1]. This implies the possibility of securing high-resolution pictures of the radiation-sensitive particles by superposing many low-dose images.

If a reference (noise-free) particle is not known a priori, the best approach is found to be a self-detection version of matched filtering: cross-correlation of the entire image with a small portion containing one noisy particle. In the resulting CCF the site of each repeat of the particle is marked by a sharp peak. For discrimination of the peak against the background, the dose must exceed a certain minimum. Detection is possible at lower doses in bright-field than in dark-field conditions.

The particle diameter and contrast and the critical dose for its destruction by the electron beam determine the resolution of images ultimately attainable by this method. We estimate that for the low contrast expected in unstained specimen preparations, this method will only be of practical importance if the particle is larger than 100 Å.

Random orientation in the particles remains a complication for the present and is being studied further.

Acknowledgement

One of the authors (W.O. Saxton) gratefully acknowledges support by the U.K. Science Research Council. We would like to thank Dr. R.W. Horne for his kindness in making available the CCMV picture. We also wish to thank Dr. W. Goldfarb for critical comments.

Appendix

Standard formulae used in main text

x and y are random variables,
$E(\,)$ denotes the expectation value, or mean.
(1) $\mathrm{var}(x) = E(x^2) - E^2(x)$;

(2) $\text{cov}(x, y) = E(xy) - E(x)E(y)$ (= 0 if x and y are independent);
(3) $E(x + y) = E(x) + E(y)$;
(4) $\text{var}(x + y) = \text{var}(x) + \text{var}(y) + 2\ \text{cov}(x, y)$;
(5) If x and y are independent,
$\text{var}(xy) = E^2(x)\ \text{var}(y) + \text{var}(x)\ E^2(y) + \text{var}(x)\ \text{var}(y)$.

References

[1] P.N.T. Unwin and R. Henderson, J. Mol. Biol. 94 (1975) 425.
[2] R. Henderson and P.N.T. Unwin, Nature 257 (1975) 28.
[3] J. Frank, Ultramicroscopy 1 (1975) 159.
[4] J. Frank, Proc. Sixth Europ. Cong. El. Micr., Jerusalem (1976) 273.
[5] J.W. Goodman, Introduction to Fourier Optics (McGraw-Hill, 1968).
[6] J. Frank, Computer Processing of Electron Micrographs, in: Advanced Techniques in Biological Electron Microscopy, ed. J.K. Koehler (Springer Verlag, Heidelberg, 1973) p. 215.
[7] A. Van der Lugt, IEEE IT-10 (1964) 139.
[8] J.F. Hamilton and J.C. Marchant, 57 (1967) 232.
[9] A. Rose, Adv. Electron, 1 (1948) 131.
[10] R.W. Horne, J.M. Hobart and I. Pasquali-Ronchetti, J. Ultrastr. Res. 53 (1975) 319.
[11] W.O. Saxton, Comp. Graph. Im. Proc. 3 (1974) 266.
[12] J. Frank, J. Phys. E: Sci. Instr. 8 (1975) 582.
[13] W.O. Saxton, Proc. Eigth Int. Cong. El. Micr., Canberra I (1974) 314.

Ultramicroscopy 3 (1978) 283–290
© North-Holland Publishing Company

RECONSTRUCTION OF GLUTAMINE SYNTHETASE USING COMPUTER AVERAGING

J. FRANK and W. GOLDFARB
Division of Laboratories and Research, New York State Department of Health, Albany, NY 12201, USA

D. EISENBERG and T.S. BAKER *
Molecular Biology Institute, University of California-Los Angeles. Los Angeles, CA 90024, USA

Received 26 June 1978

The axial projection of the glutamine synthetase molecule has been reconstructed from electron micrographs of a stained preparation by using a new method of correlation search and averaging. The average over 50 individual molecules appears as a radial pattern with sixfold symmetry. The handedness evident in the average is attributed to nonuniformity of the negative stain.

1. Introduction

In recent years new techniques for interpreting computer images have been introduced into structural research with the electron microscope. One such technique is image averaging, now widely used to suppress substrate noise [1] and increasingly considered as a means of retrieving high resolution information from low-exposure micrographs of radiation-sensitive specimens [2]. This coherent filtering approach relies on the fact that in crystalline arrangements the repeated units are separated by integer combinations of the same basic translation vectors and that they have an identical orientation.

A large number of biological molecules, however, do not form crystals but exist as individual particles. For such specimens we are making use of a new method of averaging which does not require that the object be regularly arranged. This method employs a computer to search out the relative translation vector and orientation angle of each particle with respect to an arbitrary reference particle. We make use of correlation functions, the tool of classical approaches to pattern recognition, where only rigid body movement of the patterns is allowed.

According to electron microscopic studies by Valentine et al. [3] of glutamine synthetase (GS; EC 6.3.1.2), each molecule contains 12 subunits arranged in two eclipsed hexamers and has twofold axes perpendicular to its axis of sixfold symmetry. The maximum diameter of the hexamer face is 140 Å. Frey et al. [4] investigated seven-stranded cables formed by GS molecules under certain ionic conditions. A low-resolution three-dimensional reconstruction of these cables confirmed Valentine et al.'s data and characterized the subunits as oblate ellipsiods of revolution around axes parallel to the helical axis, with diameters of 63 Å when viewed down these axes. Arguments for the dihedral symmetry of the molecules were given by Kabsch et al. [5].

2. Electron microscopy

A sample of GS from *Bacillus caldolyticus* (0.1–0.5 mg/ml) was applied by the adhesion method to a carbon-stabilized, parloidion-covered electron microscope grid for 30 s, washed with 2–3 drops of double-distilled H_2O, stained with 2–3 drops of 1% uranyl acetate, blotted dry with filter paper, and allowed to air-dry. Electron microscopy was done without pre-

* Present address: MRC Laboratory of Molecular Biology University Postgraduate Medical School, Hills Road, Cambridge, CB22QH, England.

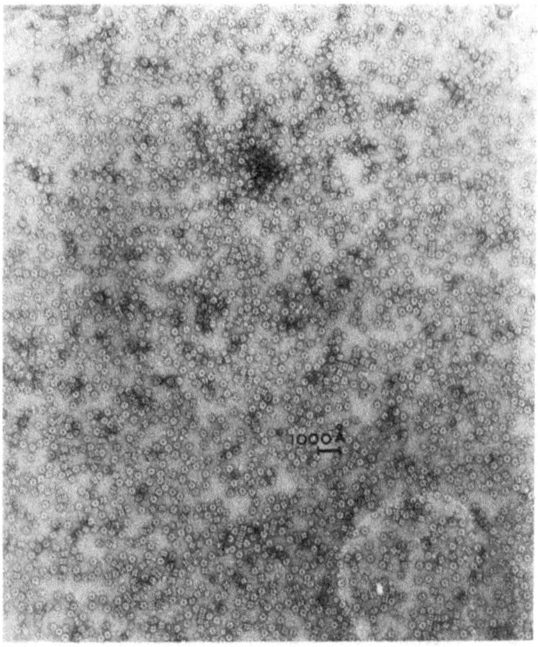

Fig. 1. Electron micrograph of glutamine synthetase from Bacillus caldolyticus stained with uranyl acetate.

cautions to limit radiation damage, using a Siemens 1 Å electron microscope at 80 kV and a nominal magnification of 46,000X. Particle size measurement determined the magnification value as 51 000X (see section 5). Optical diffraction showed a continuous transfer interval out to 0.05 $Å^{-1}$, which is sufficient to resolve the finest details to be expected in a highly irradiated, stained object. No attempt was made, therefore, to correct for instrument aberrations.

The electron micrograph (fig. 1) typically shows a large number of particles with apparent hexagonal symmetry, some lying on their side presenting a side view of the two hexamers, others stacked in long cylinders.

3. Image averaging

3.1. Principle

Particles visible in an electron micrograph contain noise from four different sources: (i) the random structure of the carbon support film, (ii) structural variations among the particles due to preparation artifacts and beam damage, (iii) granularity of the stain, and (iv) photographic noise.

Provided these noise components are random and not correlated to the observed structure, they can be virtually eliminated by averaging over a large number of repeats of the particle. The principle of this

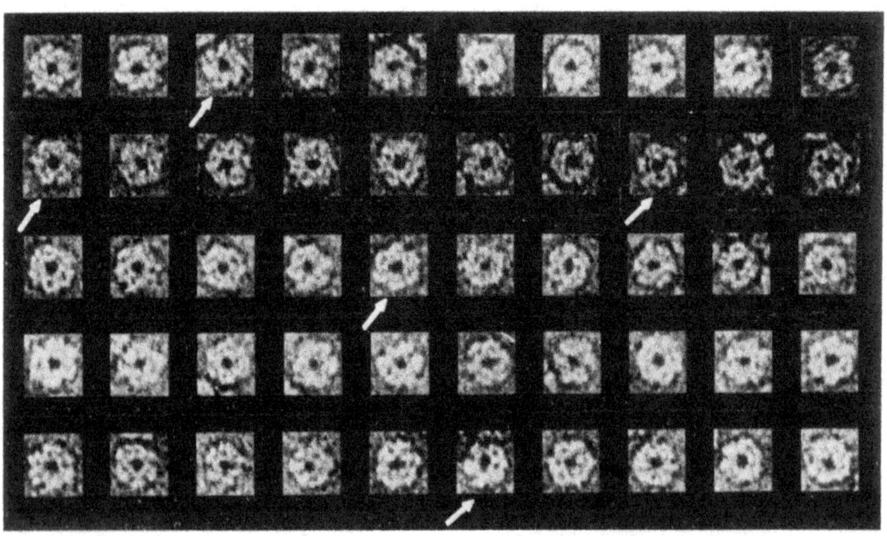

Fig. 2. Gallery of selected particles. The sidelength of the image arrays is 200 Å. Particles marked with arrows show the "windmill" features of the reconstruction (see section 4).

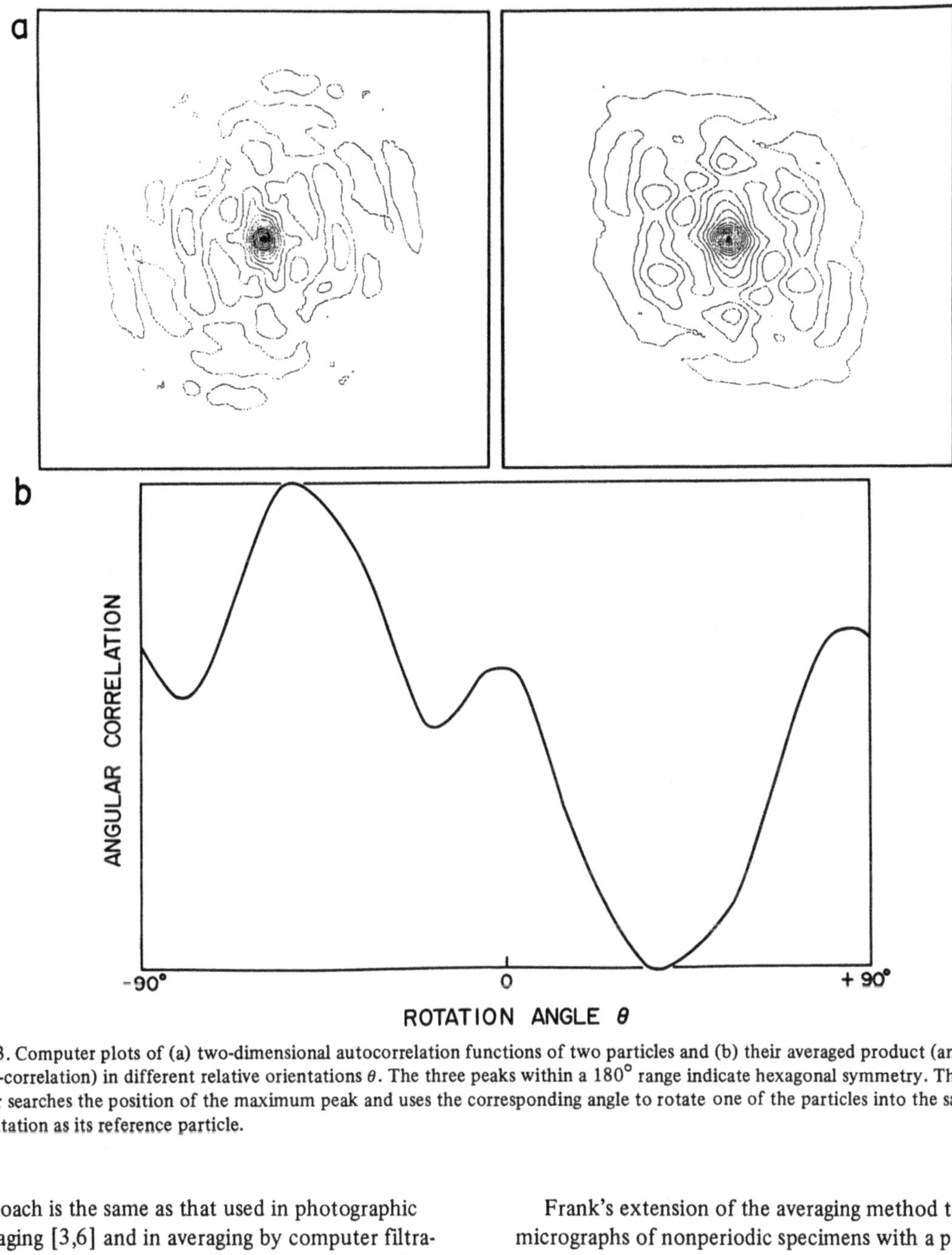

Fig. 3. Computer plots of (a) two-dimensional autocorrelation functions of two particles and (b) their averaged product (angular cross-correlation) in different relative orientations θ. The three peaks within a 180° range indicate hexagonal symmetry. The computer searches the position of the maximum peak and uses the corresponding angle to rotate one of the particles into the same orientation as its reference particle.

approach is the same as that used in photographic averaging [3,6] and in averaging by computer filtration of electron micrographs of stained (e.g. [1]) and unstained [2] crystalline objects. In the latter case, averaging was used to reduce the specimen dose to a level (0.5–1.0 e/Å^2) where no appreciable damage occurs.

Frank's extension of the averaging method to micrographs of nonperiodic specimens with a preferential attachment to the supporting film has been outlined previously [7,8]. Computer search techniques were proposed to determine translation and orientation in the plane of attachment. A computer study [9] explored the performance of the translation search under low-dose conditions, which is the ultimate goal of the present work. Early results of averaging studies using low-dose techniques have recently been presented [10,20].

Our approach is similar to that of Ottensmeyer et al. [11] except that (i) visual alignment is replaced by computer methods to establish the relative translation and orientation parameters and (ii) montage and optical filtration are replaced by a simple summation in the computer. We believe that for noisy images, computer alignment is crucial to assure reliable, artifact-free reconstruction.

3.2. Computer methods

Each reconstruction was done with 50 particles. Small areas large enough to include a GS molecule were scanned with a Perkin Elmer PDS 1010A flatbed microdensitometer. A sampling distance of 16 μm was used, corresponding to 3.14 Å at 51 000X magnification (see section 5). The particles contained in 64-by-64 image arrays (fig. 2) were brought into exact register with a reference particle by computer procedures that locate the rotation angle between two arbitrarily oriented and positioned two-dimensional motifs [12,13] and the translation vector between two motifs which have the same orientation but different translations [12–14].

For the orientation search the computer calculates the two-dimensional autocorrelation functions of the image arrays. (fig. 3a) and determines the angle at which their cross-product is maximized (fig. 3b). The size of the cross-product at the matching angle is a measure of the similarity of the autocorrelation functions or of the reproducibility of intramolecular vectors as they appear in projection. Autocorrelation functions are used because they make the orientation search independent from the translation search. A similar approach is employed in the so-called molecule-fitting methods in *x*-ray crystallography.

After the particles have been brought into matching orientation, using the angles found in the orientation search, the translation search is performed. Two image arrays will contain the same motif at different positions r_1 and $r_1 + \Delta r$. The two-dimensional cross-correlation function of these arrays will contain the autocorrelation function of the motif in the off-origin position Δr and will thus show the direction and size

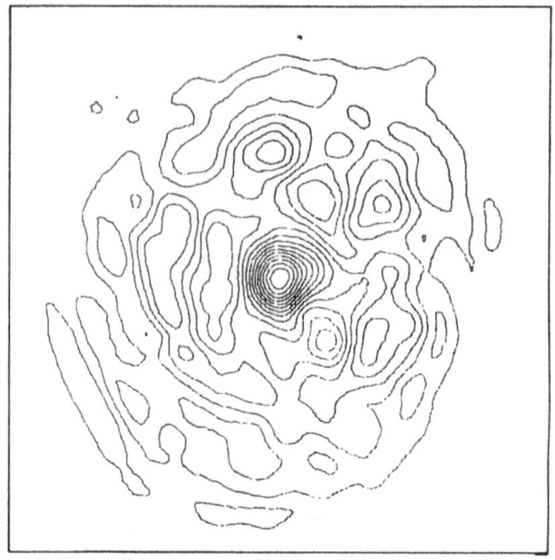

Fig. 4. Two-dimensional cross-correlation function of two GS particles that have been brought into matching orientations. The position of the peak with respect to the center indicates the relative shift between the two particles (from [19], courtesy of Springer-Verlag, Heidelberg).

of the relative translation vector (fig. 4). The size of the cross-correlation peak, when suitably normalized, is a measure of the similarity of the particles or of the reproducibility of their structure as shown in projection.

All computations were done on a PDP 11/45 under the RSX 11D executive. The software was the newly developed image-processing system SPIDER [15] in which 60 (and potentially more) operations, such as Fourier transformation, cross-correlation, rotation, and specific contrast-enhancement algorithms, can be specified in a simple command language. Through a DO loop command, any set of operations can be applied to a whole sequence of image files. The entire reconstruction – from locating the randomly positioned particles to visualizing the six-fold, rotationally averaged result – can thus be performed in a single step requiring no interaction by the user.

4. Results

Our approach is based on the assumption that the stained GS particles are short cylinders lying flat on the supporting film so that an axial projection is visible at zero tilting angle, provided the warping of the film is negligible.

For any mass-density distribution within the cylinder, there are two possible axial projections, and they are mirror images of each other. An electron micrograph of an assembly of such cylinders will show two different motifs related by mirror symmetry. Only one motif will be visible if the axial projection is centrosymmetric and thus identical to its mirror image, or if the axial projection is noncentrosymmetric but the cylinder always faces the supporting grid with the same side.

Earlier studies on the structure of GS from Escherichia coli [3–5], as well as the appearance of the particles in the electron micrographs, suggest that only one motif is present. There is no clustering of the values for angular correlation or for the normalized translational cross-correlation of the particles with respect to the selected reference particle. If two noncentrosymmetric motifs related by mirror symmetry were present, two distinct clusters would be expected in the latter values: a cluster at high correlation values indicating particles facing the same way as the reference particle, and a cluster at low correlation values indicating particles facing the opposite way.

The average obtained from 50 particles (fig. 2) after alignment (figs. 5a,b) does not exhibit the expected centrosymmetry. Instead, a "windmill" pattern with cyclic symmetry is visible, with six elliptic low-density regions at the periphery of the spiral arms. This pattern is not an accidental result for the

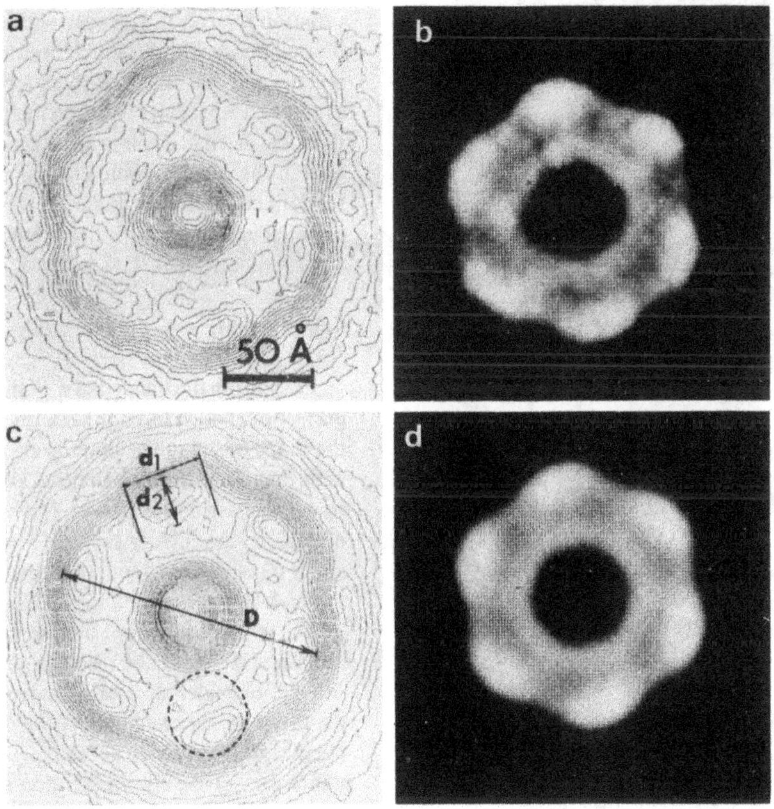

Fig. 5. Average of 50 particles: (a) contour representation; (b) enhanced gray-level display of stain-excluding region; (c) sixfold, rotationally averaged reconstruction in contour representation; and (d) same as c, with enhanced gray-level display of stain-excluding region. D, largest diameter scaled to 140 Å; d_1, d_2, minor and major axes of the elliptic region. The dotted line marks the approximately circular subunit region.

particular set of particles chosen, as could be shown by averaging over arbitrary subsets of the particles or by averaging over a different set of 50 particles. On closer inspection, handedness is clearly visible in a number of particles (fig. 2, arrows). The result of rotational averaging using the sixfold cyclic symmetry is shown in fig. 5c,d. The subunit of the rotationally averaged particle is thus the average of 300 individual subunits, with a 17-fold-enhanced signal-to-noise ratio.

To center the averaged particle for rotational averaging, we made use of the molecules' centrosymmetry. In this procedure the image containing a centrosymmetric motif is rotated 180°, and the cross-correlation function of original and rotated images is computed. The peak indicates the relative displacement between the motif in the original image and its 180°-rotated counterpart, which would be identical in the absence of noise. A shift of the image by half of this displacement vector then brings the motif into the center of the image.

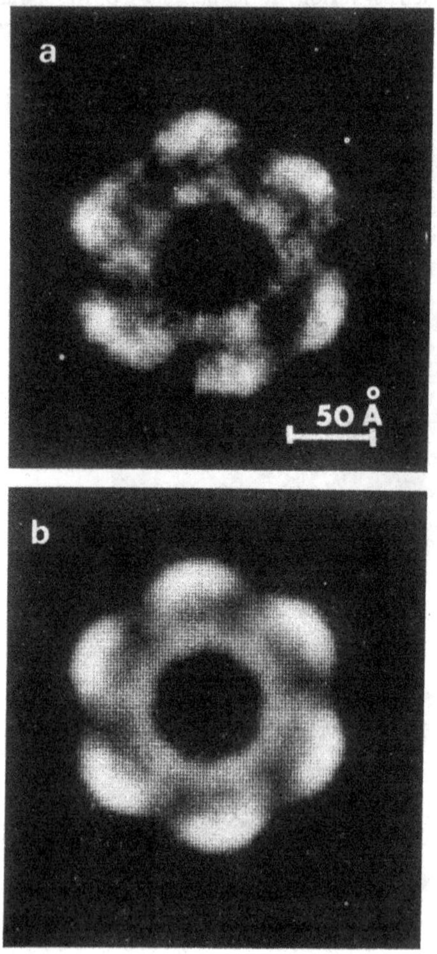

Fig. 6. Average of 15 particles with the highest correlation (a) before and (b) after sixfold rotational averaging.

To reduce the subjective element introduced by the selection of an arbitrary reference particle, we used the average of the first run as reference for a refinement run of the reconstruction. However, the contours of the stain-excluding region did not change significantly. This indicates that the reference particle was representative of a large subset of the particles.

Another concern is the extent of structural variation within the set used for reconstruction. If this variation is large, the reconstruction will produce a blurred image without much useful information. We investigated one aspect of this problem by selecting a subset of 15 particles with the highest cross-correlation coefficients. Their average showed enhanced contrast in the stain-excluding region and somewhat better-defined subunit boundaries. Otherwise it was in essential agreement with the overall average. (fig. 6).

5. Discussion

Except for the handedness of the stained molecule projection, which has not been previously observed, the shape of the molecule projection (fig. 7) agrees with that obtained by Valentine et al. for E coli GS. In our reconstruction however, the elliptic outside boundaries are slightly skewed, forming an angle of about 6° with the direction normal to the radius of the molecule.

The most pronounced feature in the stain-excluding region of the reconstruction is a low-density elliptic region at a radial distance of 61 Å, with the major axis skewed against the direction normal to the molecule radius by about 6°. The major and minor axes of the elliptical region are 42.5 and 24 Å long, respectively. These dimensions are based on Valentine et al.'s measurement of 140 Å for the outer diameter of the E. coli GS, since we expect the variation of the molecule size between species to be smal-

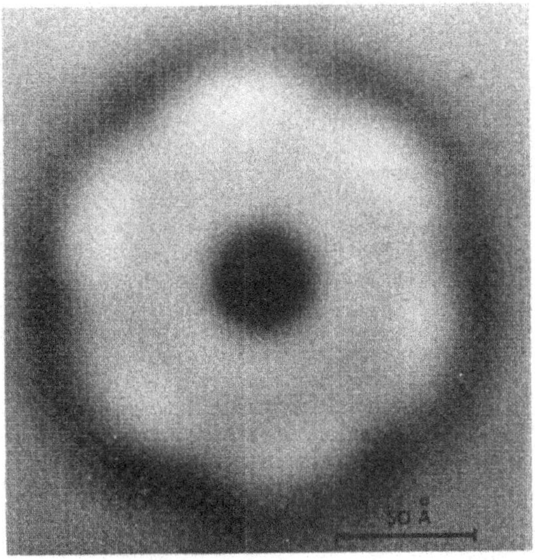

Fig. 7. Gray-level display of reconstructed particle, including highcontrast stain regions.

ler than the error of magnification calibration. With this scaling the magnification works out as 51 000X. The spiral arm and elliptic region form a roughly circular area with a diameter of 42.5 Å.

The most puzzling result is the evidence of handedness in the reconstructed projection. How can this finding be reconciled with the results of earlier studies which indicated dihedral point-group symmetry? We must assume that the stain distribution is consistently different on top of the particle (away from the carbon film) than on the bottom (the side attached to the carbon film). The contour map then contains some information on the structure of the single hexamer, but this is difficult to decipher without additional experiments using different stains and nonaxial projections of the particle.

Nonuniform stain distribution is well known in micrographs of virus particles. which are much larger than the GS molecules. Virus particles with a predominant sidedness in stain can often be picked out visually from others with more uniform staining on the same grid. When analyzed by optical diffraction, images of viruses and virus-related structures with helical symmetry give different sets of spots arising from the two sides of the particle [16]. This analysis can be used to verify the distinction between particles uniformly stained and those more strongly stained on one side.

Uniformity of stain distribution may also be strongly affected by the type of negative stain used. Nagington et al. [17] found that ammonium molybdate provided single-sided images of the spiral tubular thread structure in the wall of the Orf virus, while sodium phosphotungstic acid stain of the same virus showed discrete images of the top and bottom structures.

An alternative explanation may be a difference in shrinkage behavior between the top and bottom of the particle, if the bottom part is stabilized by the carbon film [18]. Because of the asymmetric structure of the subunits, which can be inferred from chemical data [3], such a shrinkage differential could distort the original dihedral point-group symmetry.

We have reconstructed GS particles from E. coli stained with ammonium molybdate, uranyl acetate, and phosphotungstic acid, as well as with uranyl formate. The results are very similar, with handedness evident in the reconstructions.

6. Conclusions

The full potential of the new computer averaging method will become available with the adoption of low-dose techniques and the replacement of stain by glass-like embedding media such as glucose. Preliminary averaging studies of subminimum-dose electron micrographs of stained GS [10,20] have confirmed the feasibility of the alignment procedure under these conditions.

The goal of the GS project is a high-resolution three-dimensional low-dose reconstruction of the molecule. Here the random azimuthal orientations of the particles will prove to be an advantage.

A single low-dose micrograph of a maximally tilted specimen will supply all the Fourier information contained in a cone up to that tilt angle. The missing part of the Fourier space can be recovered from a micrograph of long, stacked cylinders [4] (fig. 1).

Another application will be reconstruction from low-dose images of crystals with small patches of coherent order whose boundaries can be recognized in the high-dose reference image. Each such patch can be thought of as a particle and can be treated by the same alignment and averaging procedures used in this investigation.

Acknowledgements

We would like to thank Dr. F. Wedler, Rensselaer Polytechnic Institute, for a gift of glutamine synthetase and Dr. M. Kessel of this Division for valuable discussions.

References

[1] U. Aebi, P.R. Smith, J. Dubochet, C. Henry and E. Kellenberger, J. Supramol. Str. 1 (1973) 498.
[2] P.N.T. Unwin and R. Henderson, J. Mol. Biol. 94 (1975) 425.
[3] R.C. Valentine, B.M. Shapiro and E.R. Stadtmann, Biochemistry 7 (1968) 2143.
[4] T. Frey, D. Eisenberg and F.A. Eiserling, Proc. Nat. Acad. Sci. USA 72 (1975) 3402.
[5] W. Kabsch, H. Kabsch and D. Eisenberg, J. Mol. Biol. 100 (1976) 283.
[6] R. Markham, S. Frey and G.J. Hills, Virology 20 (1963) 88.
[7] J. Frank, Ultramicroscopy 1 (1975) 159.
[8] J. Frank, Ann. New York Acad. Sci. 306 (1978) 112.
[9] W.O. Saxton and J. Frank, Ultramicroscopy 2 (1977) 219.
[10] J. Frank, W. Goldfarb, M. Kessel, D. Eisenberg and T.S. Baker, Biophys. J. 21 (1978) 89a (Abstr).
[11] F.P. Ottensmeyer, J.W. Andrew, D.P. Bazett-Jones, A.S.K. Chan and J. Hewitt, J. Microscopy (1977) 259.
[12] R. Langer, J. Frank, A. Feltynowski and W. Hoppe, Ber. Bunsenges. Phys. Chem. 74 (1970) 1120.
[13] W.O. Saxton, Eighth Int. Congr. Electron Microscopy Canberra vol. I (1974) 314.
[14] J. Frank, in Advanced Techniques in Biological Electron Microscopy, ed. J. Koehler (Springer-Verlag, Berlin, 1973) p. 215.
[15] J. Frank and B. Shimkin, 9th Int. Cong. Electron Microscopy Toronto Vol. I (1978) p. 210.
[16] A. Klug and D. DeRosier, Nature 212 (1966) 29.
[17] J. Nagington, A.A. Newton and R.W. Horne, Virology 23 (1964) 461.
[18] M.F. Moody, Phil. Trans. Roy. Soc. Lond. B261 (1971) 181.
[19] J. Frank, The role of correlation techniques in computer image processing, in Computer Processing of Electron Microscope Images, ed., P.W. Hawkes (Springer-Verlag, Heidelberg, 1978) fig. 5.3, in press.
[20] J. Frank, W. Goldfarb and M. Kessel, 9th Int. Congr. Electron Microscopy, Toronto, Vol. II (1978) p. 8.

Ultramicroscopy 4 (1979) 247
© North-Holland Publishing Company

LETTER TO THE EDITOR

ADDENDUM TO "RECONSTRUCTION OF GLUTAMINE SYNTHETASE USING COMPUTER AVERAGING"

J. FRANK and W. GOLDFARB
Division of Laboratories and Research, New York State Department of Health, Albany, NY 12201, USA

D. EISENBERG and T.S. BAKER *
Molecular Biology Institute, University of California–Los Angeles, Los Angeles, CA 90024, USA

Received 9 January 1979

The article by Frank et al. [1] contains an erroneous term. The term occurs in statements that describe the symmetry of the axial projection of a cylindrical object. The term "centrosymmetric" was wrongly used to describe the presence of mirror symmetry in electron micrographs and in computed averages. Although no results or conclusions in the article are affected by the error, we feel obliged to correct it. This can be done by replacing the three paragraphs on page 287 in ref. [1] by the following:

"For any mass distribution within a cylinder, there are in general two axial projections, and they are mirror images of one another. Therefore an electron micrograph of a field of such cylinders will in general show two projections, related by mirror symmetry. However, only one projection occurs under either of the following conditions:

"(1) the cylinder faces the supporting grid with the same end, or

"(2) the cylinder mass distribution has at least one two-fold axis or symmetry perpendicular to the direction of projection. This would be the case for a stained glutamine synthetase molecule of symmetry 622 [2,3]. When condition (2) holds, the projection has a mirror line of symmetry.

"The average projection obtained from 50 glutamine synthetase molecules (fig. 2) after alignment (figs. 5a, b) does not exhibit a line of mirror symmetry. Instead, a 'windmill' pattern with cyclic symmetry is visible, with six elliptical low-density regions at the periphery of the spiral arms."

References

[1] J. Frank, W. Goldfarb, D. Eisenberg and T.S. Baker, Ultramicroscopy 3 (1978) 283.
[2] W. Kabsch, H. Kabsch and D. Eisenberg, J. Mol. Biol. 100 (1976) 283.
[3] T.G. Frey, D.S. Eisenberg and I.A. Eiserling, Proc. Natl. Acad. Sci. USA 72 (1975) 3402.

* Present address: MRC Laboratory of Molecular Biology, University Postgraduate Medical School, Hills Road, Cambridge, CB2 2QB, England.

Ultramicroscopy 6 (1981) 343–358
North-Holland Publishing Company

SPIDER—A MODULAR SOFTWARE SYSTEM FOR ELECTRON IMAGE PROCESSING

Joachim FRANK, Brian SHIMKIN * and Helen DOWSE **

Division of Laboratories and Research, New York State Department of Health, Albany, New York 12201, USA

Received 13 April 1981; revised 20 May 1981

The image-processing system SPIDER has been designed to operate on a minicomputer in a multiuser environment. SPIDER, which can be run either interactive or batch mode, makes a wide range of operations (including contrast enhancement, Fourier filtration, correlation averaging, and three-dimensional reconstruction) available for analysis of electron micrographs. The command language supports a hierarchical calling structure, branching commands, and DO-loops similar to those of FORTRAN.

1. Introduction

1.1. Electron image processing

Image processing has become increasingly important as a tool for interpretation and enhancement of electron micrographs. This is evident from recent reviews, which cover a wide range of applications [1–5]. The increased availability of minicomputers has now put the many image-processing schemes proposed over the years within the reach of electron microscopy groups.

Following Smith's scheme [6], the applications can be roughly grouped into three categories: restoration, image enhancement, and three-dimensional reconstruction.

Restoration is the attempt to eliminate distortions introduced by the objective lens of the electron microscope. Certain approximations in bright field electron microscopy of weakly scattering objects lead to a linear system description of image formation [7,8], which allows the object's potential distribution to be computed from the experimental data [9–11].

Image enhancement is a summary term denoting point-for-point operations, or operations involving local neighbors of image points, which aid in visual interpretation of images. Quantitative methods of image averaging aimed at enhancing the signal-to-noise ratio may also be included in this category. The importance of these methods in specimen-preserving high-resolution electron microscopy has been widely recognized since Unwin and Henderson's study of the purple membrane protein [12]. More recently, averaging methods for single particles have been developed [13–15], extending low-dose investigation to a larger class of biological specimens. Multivariate statistical techniques promise to be a powerful tool in structural analysis of single molecules and crystals [16,17].

Three-dimensional reconstruction is reconstruction, by various procedures, of an object's mass distribution in three dimensions from a series of micrographs showing the object in different views [4,18,19]. In the special case of an object having high symmetry, a single micrograph may be sufficient [18].

To represent an image recorded on a photographic film or plate in digital form, one must scan the film or plate on a computer-controlled digital microdensitometer, where optical density values are read on a raster, converted into digital form, and written onto a storage medium (tape, disk). In the reverse operation, often combined with the scanning function in the same instrument, digitally recorded images are displayed as a fine raster of points on a photographic film.

The increased flexibility of digital processing can best be realized when the system is designed in a modular way (fig. 1). Mathematical operations on images or sets of images are broken down into the simplest steps involving an entire image. These steps are standardized to permit their use in more than one context. In each step an input image (existing as a file on a mass storage medium) is processed, and an output file is created with the same format; the output file can then

* Present address: Gerber Systems Technology, Post Office Box 905, South Windsor, Connecticut 06074, USA.

** Present address: General Electric Company, 1 River Road, Schenectady, New York 12345, USA.

0304-3991/81/0000–0000/$02.50 © 1981 North-Holland

be used as input for a subsequent operation. List and display operations make it possible to check and debug each processing step separately. More involved schemes can be built from these elementary operations by simply

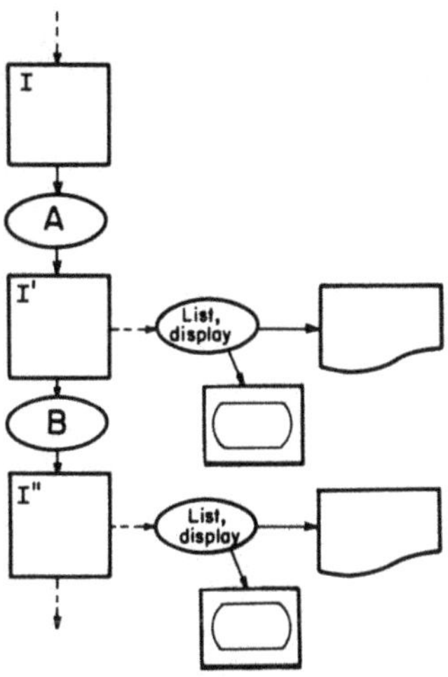

Fig. 1. Scheme of the data flow in a modular image processing system. Operation A creates image I′ from input image I. At the end of operation A, I′ resides along with I on a mass storage medium and can be listed and displayed before being subjected to step B.

stringing them together in a virtually infinite number of ways. The first comprehensive image-processing system with this design philosophy, VICAR [20], was developed at the Jet Propulsion Laboratory in Pasadena in the 1960s.

If a fast graytone display device is available, the decision on which processing step to choose next in a series of operations can be based on the result of the immediately preceding step, making the system interactive.

1.2. Existing modular software systems

The need for a modular design in electron image-processing systems has been recognized by a number of groups [6,21–30] and has been answered in various ways, depending on the particular computer installations of the originating laboratories. Of the software systems listed in table 1, IMPROC [27], MDPP [6], and EM [29] are written for large machines, while SEMPER [25], SPIDER [21], MIRAGE [22], PIC [28], and IMAGIC [30] are designed for minicomputers with 64 Kb storage or less.

Some correspondences exist between SPIDER and SEMPER [15,26] in the command language structure and in the system design. However, SEMPER is distinguished by freer rules of format and register assignments, while SPIDER has greater flexibility in the dynamic features of the language.

One difference between these two systems is that SPIDER supports a full dialogue with the user by printing out solicitation messages on the terminal, whereas SEMPER only returns results or error messages. Our experience is that solicitation messages in operations (as well as user-created solicitation messages in procedures; see section 3.6) are an invaluable aid in processing with a modular software system. With so many operations available, it is very difficult to memorize all of the various input sequences required. In contrast, the solicitation messages help even inexperienced users in interactive SPIDER sessions to comprehend and use the processing system within a short time. Interactive sessions also provide training in extended use of the language in procedure and batch command files.

The most comprehensive program package for three-dimensional reconstruction and various two-dimensional filtrations, which was developed at the Medical Research Council in Cambridge [31], is not a modular system in the sense described here. Rather it is a collection of programs that lacks the cohesiveness and versatility of other systems listed in table 1.

Development of a multipurpose software system is a very laborious task, often a byproduct of research activity. The successful design and implementation of such a system may therefore precede its publication by several years. For instance, EM existed in its basic form in 1970, 10 years before it was first described in the literature [29].

1.3. SPIDER hardware configuration and design considerations

SPIDER (System for Processing of Image Data in Electron microscopy and Related fields) was developed at the New York State Department of Health as a general-user facility for a wide range of applications associated with high-voltage and conventional-voltage microscopes. Among these applications are enhancement of low-contrast electron micrographs [32], computer filtration [33], single-molecule averaging [13–15], three-dimensional reconstruction [34], and evaluation of electron diffraction information.

Table 1
Modular image-processing systems in electron microscopy [a)]

Name	Ref.	Origin	Machine [b)]	(1)	(2)	(3)	(4)	(5)	(6)	(7)	(8)	(9)	(10)	(11)	(12)	(13)
EM	[29]	Max-Planck-Inst. Biochemie, Martinsried, FRG	IBM 360 Siemens 4004 VAX-11/780	x		x	x	x	x				x	x	x	
IMAGIC	[30]	Biochem. Lab. Univ. Groningen, Netherlands	NORD 10	x	x	x	x	x	x	x				x	x	x
IMPROC	[27]	Cavendish Lab., Cambridge, UK	IBM 370/165	x	x	x	x	x						x	x	
MDPP	[6]	Biozentrum Univ., Basel, Switzerland	IBM 360/370	x	x	x	x	x	x	x		x		x		
MIRAGE	[22]	CENG, Grenoble, France	Intertechnique Pherimat S	x	x	x	x	x				P		x	x	
OIDIPS	[23]	Dept. Appl. Phys., Osaka Univ., Japan	CEC-555 Interdata 7/32				x	x						x		
PIC	[28]	Natl. Inst. Health, Bethesda, MD, USA	PDP-11/70		x	x	x	x						x	x	
SEMPER	[25]	Cavendish Lab., Cambridge, UK	PDP-8/E IBM 370/165 ... [c)]	x	x	x	x	x			x		x	x	x	
SPIDER	[21]	New York State Dept. of Health, Albany, NY, USA	PDP-11/45 VAX-11/780	x	x		x	x	x		x	x	P		x	x

(1) Extended language=capability to handle DO-loops and procedures (macros) on the command language level.
(2) Plane lattice averaging.
(3) Rotational filtration.
(4) Image statistics.
(5) Correlation alignment.
(6) Handling of 3D info=three-dimensional file storage, general modular procedures for extraction sections from real space or Fourier space volumes.
(7) Helical reconstruction.
(8) ART reconstruction.
(9) Tilted lattice reconstruction.
(10) Tilted particle reconstruction.
(11) Image restoration.
(12) Interactive display.
(13) Multivariate statistics.

P=in preparation.

a) The table presents the results of a survey sent to the authors of the systems listed.
b) Where more than one installation is listed, the first is the one for which the system was originally developed.
c) Ten more machines listed by the author.

The hardware consists of a Digital Equipment Corporation (DEC) PDP-11/45 computer, two 1600-bpi tape units, two 176-Mb disk units, a VERSATEC printer/plotter, a line printer, a Princeton Electronics Product (PEP) 801 graytone storage display system, and a Perkin Elmer PDS 1010A flatbed microdensitometer. The microdensitometer is run by a DEC PDP-11/05 computer operating independently of the PDP-11/45 and equipped with a tape unit and a film-writing option.

Some of the design requirements underlying the SPIDER system were very similar to those spelled out by Smith [6]: that the system be easily accessible to users having limited experience with computers; that all operations be available in a single job; that options for display, listing, and permanent storage of images be available; that installation-dependent features be avoided, where possible; and that the system be capable of processing any number of images with arbitrary sizes (within limits) and formats. Additional restrictions were imposed by our reliance on a small computer in a multiuser environment.

Averaging of single molecules by correlation methods [13–15] requires the processing of a large number of images that are realizations of the same molecule projection. Other applications, such as three-dimensional reconstruction from projections [34,35], also involve a series of files that must be subjected to essentially identical operations. These applications demand an efficient mechanism for defining repeated operations or sets of operations over a series of images. DO-loops that can be nested up to three deep were created in SPIDER for this purpose.

Finally, there were requirements for a hierarchical command structure and for a general mechanism to transfer values from one operation to another and from one run of the program system to another.

All of these requirements have been realized in a way that may be of general interest in the design of modular software systems.

2. System design

2.1. General

The SPIDER system consists of a master task DRIVER and a set of slave tasks, which are run under the DEC RSX11M operating system in a multiuser environment (fig. 2). The slave tasks are started by DRIVER and return control to DRIVER when finished. Each slave task performs a subset of the 120 operations now available. A *session* is defined as a set of operations between the start of DRIVER and an end ("EN") command. Each session is distinguished from other parallel sessions by a unique *project code*. The user starts a session by executing DRIVER and specifying the project code and the data library he wishes to access. The image-processing commands are entered either directly on a terminal (interactive operation) or via a sequential command file previously prepared by using the text editor (batch operation). *Procedure files*, a special class of command files which allow run-time replacements, can be invoked interchangeably with the basic commands in either interactive or batch operation.

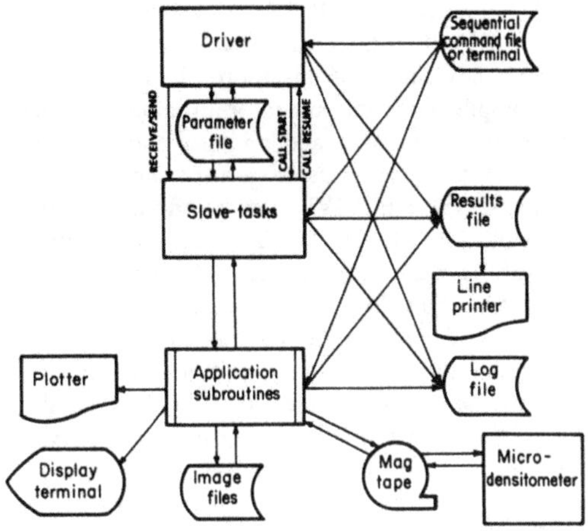

Fig. 2. Schematic representation of SPIDER procesing system [21]. DRIVER and the slave tasks communicate with each other by three means: (1) START/RESUME directives, (2) SEND/RECEIVE directives to pass the name of the current working area and project extension to the slave task, and (3) a parameter containing all information pertinent to the session and the current processing stage. The control sequence for the processing session is either contained in the sequential command (batch mode) or entered directly from the terminal (interactive mode). In the latter case, a log file stores all user input for later reference. (Reproduced with kind permission of the Microscopical Society of Canada.)

Upon encountering an operation command (e.g. "RT" for rotate), DRIVER activates the appropriate slave task and suspends itself. Once activated, the slave task performs the operation, soliciting any information required (names of input files, values of processing parameters, etc.) from the user. The slave task then asks for the next operation command and continues processing. When it encounters a command not contained among its operations, the slave task reactivates

DRIVER, passes on to DRIVER the current input line, and exits. DRIVER then continues as above.

The main advantages of this approach are minimization of compile and link time for program changes, minimization of core use, and an open-ended system design. DRIVER communicates with the slave tasks by means of a SEND/RECEIVE message and a sequential parameter file. The SEND/RECEIVE message contains the name of the user's working area and the name of the parameter file. This file, as well as other temporary system-generated files to be mentioned later, has a name unique to the session, so that two or more SPIDER sessions can be run simultaneously. Any information needed by the slave task (the file extension specifying the data library, the current command-file name, file pointers, etc.) is written to the parameter file by DRIVER and read by the slave task upon activation. Before exiting, the slave task updates the parameter file. Communication between DRIVER and the slave tasks, as well as the activation, suspension, and exiting of the various tasks, is transparent to the user.

DRIVER is the only task that retains all information pertinent to the session. It handles all global switches and system-related operations (DO-loops, procedures, etc.), while the slave tasks handle all operations involving data processing (e.g. Fourier transformation, masking, and rotation of images).

2.2. Coding

With a few expectations, all main programs and subroutines are written in FORTRAN IV. Subroutines for system-related functions (such as read/write accesses to the disk, communication with DRIVER, reading of parameter values, opening of files, and updating of statistical information) are standardized, facilitating the implementation of new operations into existing slave tasks or the creation of a new slave task.

2.3. Registers

Up to 100 storage spaces are available for storage and transfer of important values during the SPIDER session. The corresponding array of floating point numbers is passed from DRIVER to the slave tasks and vice versa as part of the parameter file. The storage spaces are called *registers* and are invoked in the command language by the symbols X0...X99, which may take the place of any integer or floating point number expected in the command stream. Registers are thus an important tool for exchange of information between operations within a session.

2.4. Global switches

Global switches are implemented in the system to invoke different modes of operation for the entire ses sion:

(a) Print output can be spooled to the line printer on completion of each operation or on termination of the session.

(b) A trace switch allows the user to follow the progress of a batch run from messages printed on the terminal.

(c) Two modes of error response in the batch operation are available. In one mode an error immediately terminates the session. In the other mode the processing is allowed to go on, skipping to the next intelligible command.

(d) Read-only access to data in other working areas is available on request.

(e) Two versions of each slave task may be activated, depending on the state of a switch. This feature enables testing of modified versions of a task without disrupting the use of the existing system. It also allows efficient use of the available memory in test runs, if one set of the tasks is built with a minimum of buffer space sufficient for small images.

2.5. Image format

Images are stored one record per line as direct-access files on disk. Each density reading is stored as a 4-byte word. Both Perkin Elmer 1010A tapes and Optronics tapes can be read by the system. SPIDER accesses the PDP-11 file directory through special subroutines.

The image file consists of the image data and a certain number of records needed to accommodate additional space for the Fourier transform, a 128-point histogram, and the SPIDER image-processing label. The historgram and part of the image label contain statistical information on the image that, once acquired, is available to any subsequent operation. The status of the statistical information is recorded by flags, which are interrogated in each operation to decide whether the information is already stored or whether it needs to be computed.

Thirteen file formats are used and distinguished in the SPIDER system (table 2). The different types of random-access files are distinguished by label flags, which are checked in each access for consistency with the operation. The format for storage of the Fourier data is consistent with the optimized fast Fourier transformation program by Fraser [36].

In the SPIDER system all file names have the standard form ⟨ABC⟩⟨LMN⟩. ⟨EXT⟩, where ⟨ABC⟩ are

Table 2
File formats used in the SPIDER system

Random access, unformatted	Sequential, formatted
Real 2-D	Document
Real 2-D polar	Procedure
Real 3-D	Batch
Fourier 2-D	Log [a)]
Fourier 2-D polar	Results [a)]
Fourier 3-D	
Nonimage	
parameter [a)]	

a) Created in each sessions.

three letters specifying a file series and ⟨LMN⟩ is a three-digit number between 001 and 999, specifying the number of the file in the series. ⟨EXT⟩ is the same file extension for the entire session and is used to restrict the access of the image procesing to a data library. Within the session only the first six letters of the file need to be specified, e.g. PIC008 or RES855.

To facilitate the processing of selected files from a file series, a variable file name specification has been created. With PIC00I, for example, the sequence "00I" is replaced by a three-digit number according to the value assigned to the index I in a command-level DO-loop. With RECX10 the three last characters denote a SPIDER register, causing a file name to be constructed from the pre-fix "REC" and a three-digit number stored in the register X10.

2.6. System–user communication

The system–user communication sequence can be summarized as: solicitation–user response–verification–log entry.

The solicitation line specifies the input that is next expected in the command sequence. In the interactive mode this line has an important mnemonic function. In the batch mode the solicitation line is printed out along with the input line to make the processing sequence intelligible.

The user response comes from the terminal or, in the batch mode, from the batch command file. The user input is then echoed in a verifying message to ensure the information accepted by the system corresponds to the input intended by the user. In addition, the opening of any file is accompanied by an explicit verification statement, which includes the complete file name, title, data and time of creation, data type and dimensions, and file disposition.

In the interactive mode the user input is copied into a sequential log file. It can be called up later to recreate the same results in a batch run. Changes of dimensions and parameters can easily be made by editing the log file after the session that created it.

All user input in interactive or batch operation is handled by special subroutines that combine the functions of printing the solicitation message, reading, echoing, and log-keeping. The input of parameter values is either explicit in free format or by reference to the contents of registers filled by previous computations.

3. Command language

At the beginning of the session and upon completing each operation, SPIDER solicits the next command by printing the message "OPERATION:". All possible user responses form a language. The syntactic rules of the language are set by rules of format (e.g., operation commands must be entered left-justified) and by the order of input parameters expected for a given operation. In the interactive mode the solicitation messages serve to enforce the rules of the language. In the batch mode the user must set up an error-free command text with the help of the user's manual and protocols of previous interactive sessions kept in a log file. (For handling of errors, see section 2.4.)

The various types of commands are listed in table 3.

3.1. Basic commands

There are over 100 basic commands relating to a variety of operations (see appendix 1). These commands can be roughly grouped into seven categories (examples are given in parentheses):

Table 3
Command types in SPIDER

Type of command	Example
Basic command	RT
Arithmetic interrogation	SIN(2−1./X10)+X20
Arithmetic assignment	X15=(2**X3)*150.6
Register interrogation	X10
Batch command	B43
Procedure command	PR1
Branching commands	
DO-loops	DO LB1 I=1, 10
labels	LB3
Conditional jumps	IF(X50.GT.4)GOTO LB3
Termination and return	EN, RE

file management (copy, file-information, delete, rename);
image movement and editing (mask, shift, rotate, interpolate, insert);
display (graytone display, contour, profile, list);
contrast enhancement (density stretching, histogram equalization, local averging);
Fourier operations (fast Fourier transform, filtration, cross- and autocorrelation);
three-dimensional reconstruction (project, back-project, stack, unstack);
interface with multivariate statistical analysis programs.

```
RUN DRIVER
(Spider V3 (06/10/80) on 09-Dec-80 at 19:07:20)
.enter project/data code:LMN/DAT
lmn/dat
.operation:  MO
   mo
.output file:  MOD001/TEST PICTURE
mod001.dat
.enter dims (nsam,nrow):  45,50
   45  50
db:[200,050]mod001.dat/test picture
(r)  45  50 created on 09-Dec-80 at 19:12:20 m
.(t)est/(s)ine/(c)irc/(w)ed/(r)an/(g)auss:  W
   w
.operation:  PD
   pd
.input1 file:  MOD001
   mod001.dat
db:[200,050]mod001.dat/test picture
(r)  45  50 created on 09-Dec-80 at 19:12:20 o
.output file:  PAD001/TEST PICTURE PADDED
   pad001.dat
.enter dims(nsam,nrow): 64,64
   64   64
db:[200,050]pad001.dat/test picture padded
(r)  64  64 created on 09-Dec-80 at 19:15:02 n
.average?(y/n), (c) circular option:  Y
   y
.top left coos:  10,7
     10    7
.operation: FT
      ft
.input file:  PAD001
  pad001.dat
db:[200,050]pad001.dat/test picture padded
(r)   64  64 created on 09-Dec-80 at 19:15:02 o
.operation:  PW
   pw
.input file:  PAD001
    pad001.dat
db:[200,050]pow001.dat/test picture padded
(r)  64  64 created on 09-Dec-80 at 19:24:51 n
.operation:
```

Fig. 3. Example of computer/user dialogue in interactive session. The SPIDER system is run with a project code LMN and data code DAT. The computer messages are in lowercase letters; the user input is in capital letters. In this example a model image MOD001 with dimensions 45, 50 is created ("MO") which contains a density wedge. It is padded ("PD") into a 64×64 image PAD001, with the image average as background. The padded image is subsequently Fourier-transformed ("FT"); this operation overwrites the image file with the Fourier transform. Finally the operation "PW" computes the modulus of the Fourier transform and stores it into POW001. Each open access to a file results in a two-line statement showing the file title, disposition, format and creation data. The first access to the file PAD001 after Fourier transformation shows it to have Fourier format, denoted "(F)".

A letter after the creation date indicates whether the file existed before (o) or is being created (n) or modified (m).

Each operation, when invoked by one of the basic commands, will solicit all necessary information. Normally this comprises the names of input and output files, option specification, and input parameter values pertinent to the operation. In the following examples lower-case letters denote messages printed by the computer, and capitals denote user responses:

·operation: SH
·input file: PIC001
·output file: OUT003
·shift components (x,y): 5,6
·operation:

(Here we have left out the verifying messages.) An example of a complete record of user/system interaction, such as would be recorded on a hardcopy terminal, is shown in fig. 3.

Basic commands may have options or outputs arguments. For instance, 'TRO' invokes the Optronics format option of the tape-read command. PK X10,X11,X12 is equivalent to the instruction "peak search, and put the x, y coordinates and the value of the largest peak into the registers X10, X11, and X12 respectively".

3.2. *Arithmetic interrogation (pocket calculator)*

Any arithmetic expression involving the five basic operations (+, −, /, *, **) and numbers or SPIDER registers can be evaluated. In addition, the functions SIN, COS, EXP, LOG, SQRT, and PAD can be used. The PAD function returns, for a given argument, the next larger number that can be represented by a power of 2. For instance, PAD(5) = 8, PAD(45) = 64. With this function images can be automatically padded for operations involving radix-2 Fourier transformations (e.g. ref. [37]).

3.3. *Arithmetic assignment*

Arithmetic expressions appearing on the right-hand side of an assignment statement are evaluated as above, and the result is put into the register appearing on the left-hand side. The assignment statement is thus exactly equivalent to a FORTRAN arithmetic assignment.

Register interrogation

If a register is specified in the command position, its contents are printed on the terminal or line printer.

3.5. *Batch command*

A batch command has the form B⟨NM⟩, where ⟨NM⟩ is a two-digit number. Such a command switches the SPIDER input command stream from the terminal to a sequential file named B⟨NM⟩.⟨PRJ⟩, where ⟨PRJ⟩ is the project code, thus terminating the interactive mode of operation. The batch file can contain any sequence of commands and parameter values as they would be entered by the user in the interactive session. The log file, which records each input line in an interactive session, can be used as a batch·command file to reproduce the processing sequence.

3.6. Procedure command

The full flexibility of the processing system has been achieved by creating a procedure calling structure. A procedure may be regarded as a batch command sequence in which certain input lines have been left unspecified until execution time. Any file name or input line where one or more numerical constants are expected can be replaced by a substitution line with the general format ?⟨character string⟩?, where ⟨character string⟩ is the desired execution-time solicitation message. Such a line causes DRIVER to fetch the next-scheduled input from the next-higher level of the calling hierarchy, i.e., from the terminal or from a calling batch or procedure command stream.

In the interactive mode the character string has the function of a solicitation message. The sequence of user-system interaction in the execution of a procedure is therefore identical to the sequence followed in the execution of a basic command. In a listing of the procedure all substitution strings stand out from the rest of the code, making the purpose of the procedure immediately evident. For example:

```
RT
PIC001
PIC002
50.
SH
PIC002
SHI002
-3,8
```

This is a command string to rotate the image stored in PIC001 by 50° and store the result in a file named PIC002. Subsequently PIC002 is to be shifted by a vector with the components −3,8, and the result is to be stored in SHI002.

If such a command string is to be executed many times with different source and destination file names and different values of the parameters (rotation angle, shift vector components), it may be worthwhile to make the command string into a procedure. If the intermediate image file resulting from the rotation is of no interest, the procedure would be written in the following form:

RT1.GLS

```
RT
?INPUT FILE?
SCR001
?ROTATION ANGLE?
SH
SCR001
?OUTPUT FILE?
?SHIFT VECTOR (X,Y)?
RE
```

where GLS is the project code assigned to this session and stored in a file named RT1.GLS. "RT1" is a freely chosen name of the procedure with the general naming convention ⟨AB⟩⟨N⟩, where ⟨AB⟩ is any two letters of the alphabet and ⟨N⟩ is any digit, including 0. Any image name or parameter value in the command string that was chosen to be a variable of the procedure is replaced by a solicitation line in the procedure sequence. A return command "RE" terminates the procedure, passing control to the next-higher level of the command input stream.

When called interactively, the command RT1 will now generate the following dialogue:

```
.operation: RT1
?input file? PIC001
?rotation angle? 50.
?output file? SHI002
?shift vector (x,y)? -3,8
```

which, for the arguments chosen, leads to the same processing sequence as the initial batch command sequence.

In the batch mode, similarly, the command RT1 can always be used to effect a rotation and subsequent shifting of an image. The rules for the sequence of the input lines and the types of value input follow directly from the sequence of the solicitation lines and types of parameters solicited in the procedure. (Note that all text after a semicolon is interpreted as a command.) For example:

B01. GLS

```
;B01     A BATCH COMMAND SEQUENCE TO
;        DEMONSTRATE
;        A CALL TO RT1
MO       ; CREATE MODEL PICTURE: DENSITY
;        WEDGE
PIC001   ; NAME OF OUTPUT FILE
32, 32   ; DIMENSION OF OUTPUT FILE
W        ; MAKE IT A WEDGE
RT1      ; CALL PROCEDURE RT1
PIC001   ; INPUT FILE TO PROCEDURE
50.      ; ROTATION ANGLE
SHI001   ; OUTPUT FILE FROM PROCEDURE
-3,8     ; SHIFT VECTOR COMPONENTS
PR       ; USE OVERPRINTING FOR DISPLAY
;        OF RESULT
SHI002   ; INPUT TO OVERPRINTING
N        ; NO CONTOURING
EN P     ; END SESSION AND SPOOL RESULT
;        TO LINE PRINTER
```

3.7. Branching commands

The branching commands in SPIDER essentially duplicate the functions of DO-loop statements, labels, and logical IF statements in FORTRAN. In contrast to the pseudo DO-loop in other systems, where the repetition is achieved by creating multiple copies of the original command stream [22], the DO-loop is here realized by execution-time substitutions in the reading routines. Up to threefold nesting of DO-loops is allowed; but this restriction, similar to the restriction of the number of procedure nesting levels, is arbitrary. Its sole purpose is to limit the storage area used by internal table of DRIVER to a practical size.

Branching commands are meaningful only if they appear in a batch or procedure command sequence.

3.8. Communication between operations within a session

Within a session important parameter values can be transferred from one operation to another by use of the 100 registers, denoted X0...X99. Any register can take the place of any floating-point number or integer in arithmetic expressions or input argument positions. In addition, it can appear in an output argument position, such as OR X10, where the register accepts a value computed in the operation. File names can also be generalized by the use of registers; for example, the input PICX10 is replaced PIC004 if X10 has been previously assigned the value 4.

The first 10 registers are reserved for system- and image-related quantities and are updated during each operation. Image dimensions and values of statistical parameters can thus be accessed and used in a dynamic design of the batch or procedure command sequence.

Each register, due to a special feature of DRIVER, is local to the batch or procedure in which it appears. The use of a particular register, say X20, in a given procedure does not interfere with the value stored in the register X20 used in a calling batch or procedure. For example:

B01.GLS

```
.
.
.
X10 = 5
PR1
X10
.
.
.
EN
```

PR1.GLS

```
X10 = 1
.
.
.
RE
```

Here the register X10 is local to procedure PR1, and will therefore have no effect on the value of X10 in the calling batch sequence B01: the value of X10 printed out in response to the interrogation is 5.

Transfer of values from one level of the calling hierarchy to another is achieved through a special command "RR" (read register). Execution of the sequence

B01.GLS

```
.
.
X10 = 5
PR1
X10
.
.
EN
```

PR1.GLS

```
RR X20
?ENTER HIGHEST REFLECTION ORDER?
.
.
.
RE
```

will transfer the value of register X10 into local register X20 in PR1 in response to the user-created solicitation message "ENTER HIGHEST REFLECTION ORDER".

Transfer of register values between one procedure and another can be achieved by using the command "SR", which has options for saving (S) and unsaving

(U) the values active in all 100 registers. For example:

```
B02.GLS        PR1.GLS         PR2.GLS

.              .               .
.              .               .
.              X10 = 200       SR U
PR1            X15 = 81.5      . .
.              .               .
.              .               RE
X10 = 5        SR S
.              .
.              .
PR2            RE
.
.
EN
```

Here the commands SR S and SR U cause the values of X10 and X15 to be transferred from PR1 to PR2 along with the contents of all other registers. Because of the local character of registers the transferred value of X10 is not changed by the assignment X10 = 5 in the calling sequence. Since the "SR" unsave call in PR2 affects only the local register contents, the value assigned to X10 in the calling sequence remains the same after the PR2 call.

Another means of communication, both within a session and between sessions, is the document file, which will be described in the following section.

3.9. Communication between sessions

All register contents vanish when a session is terminated. A general vehicle for communication of register contents between sessions has been created in the form of a document file. This is a sequential, formatted file organized into keyed document records. These records may or may not coincide with the physical 80-byte records used, depending on the number of registers transferred. The general saving or updating command is

SD ⟨key⟩, X⟨N1⟩, X⟨N2⟩, X⟨N3⟩,...,X⟨NJ⟩
⟨document file name⟩

where ⟨key⟩ is an integer number or a register containing an integer (with the meaning of a particle number, a DO-loop index, etc.) and X⟨N1⟩ ... X⟨NJ⟩ is an arbitrary sequence of J registers.

An unsave command of the form

UD ⟨key⟩, X⟨M1⟩, X⟨M2⟩, X⟨M3⟩,...,X⟨MI⟩
⟨document file name⟩

will pick out the last entry under the number ⟨key⟩ and transfer the contents of the document record into the arbitrary sequence of I registers X⟨M1⟩, X⟨M2⟩, X⟨M3⟩,...,X⟨MI⟩ with $I \leq J$. The transfer is determined, not by the register numbers themselves, but the sequence of the register numbers in the SD and UD commands. For example:

X10 = 1
X11 = 5.2
SD 5,X10,X11
DCM001

UD 5,X50,X20
DCM001

This sequence will transfer the values 1 and 5.2 from X10 and X11 into X50 and X20 respectively.

A typical application of the document file is the documentation of rotations, shift vector components and correlation coefficients calculated in the alignment of a series of particle. By the use of the document file any of these quantities may be applied in a subsequent batch run to a different set of files, or they may be used to recreate the set of aligned files from the raw data.

Finally a "list document" option allows the user to tabulate the entire document file using any desired headings. For example, the command

LD 'particle number', 'Angle', 'X shift', 'Y shift', 'corr'
DCM001

will cause the first four entries in each document record to be listed with the key numbers in ascending order and with the column headings 'Angle', 'X shift', etc., applied.

4. Example

An example of extended use of the language is the procedure RA0 (fig. 4), which performs *N*-fold rotational averaging of an uncentered particle contained in an image. The particle is assumed to be centrosymmetric. In this example a batch file B01 is prepared to subject 10 particles, contained in the image series PAR001-PAR010, to six fold rotational averaging. The batch commands instruct SPIDER to apply RA0 with $N = 6$ to 10 particles. RA0 brings the particle into the image center, generates $N-1$ symmetry-related positions by rotating the particle and adds these to the centered input image.

The method of centering a centrosymmetric motif makes use of cross-correlation [13,37]. First the shift vector between the motif as it appears in the 180°-rotated image is determined. This shift vector has the same direction, and is twice as long, as the vector needed to bring the particle into the image center. The procedure CT1 which is called at the beginning of RA0 is based on this principle. The image is rotated by 180° and cross-correlated with its 180°-rotated version. Both images are padded into a larger array as preparation for cross-correlation to prevent artifacts in the cross-correlation function due to circular overlap [39]. The shift vector found in the search of the correlation peak ("PK") is then halved and applied to the original image. The output file of CT1, TMP001 thus contains an image with the particle in centered position. Through the use of system registers containing image dimensions, procedures can be written in such a way that they are applicable to images of any size.

Another feature of the command language, the use of symbolic references to previously entered input lines, is evident from RA0. In this procedure P1 and P3 are used to invoke previously entered file names.

Another example, the procedures used to align particles that have random orientations and positions, was reported earlier [21].

a

```
; B01   DEMONSTRATE USAGE OF PROCEDURE RA0
DO LB1 I=1,10    ; DO-LOOP OVER PARTICLE SERIES
RA0     ; CALL GENERAL PROCEDURE FOR ROTATIONAL AVERAGING
PAR001  ; INPUT FILE FOR ROTATIONAL AVERAGING
(6.)    ; MAKE IT SIX-FOLD
RAV001  ; OUTPUT OF ROTATIONAL AVERAGING
LB1     ; END OF DO-LOOP
EN      ; END BATCH RUN
```

b

```
; RA0  PROCEDURE TO PERFORM AUTOMATIC CENTERING AND N-FOLD ROTATIONAL
; AVERAGING OF AN IMAGE CONTAINING A CENTRO-SYMMETRIC MOTIF.
FI      ; FILE INFORMATION TO INTRODUCE FILE TO BE AVERAGED
?IMAGE TO BE ROTATIONALLY AVERAGED?     ; INPUT TO FILE INFO = INPUT TO RA0
X11=X1  ; STORE NUMBER OF SAMPLES OF INPUT IMAGE IN REGISTER X11
X12=X2  ; STORE NUMBER OF ROWS OF INPUT IMAGE IN REGISTER X12
CT1     ; PROCEDURE CALL: CENTER MOTIF USING CROSS-CORRELATION
P1      ; INPUT TO CT1 = FILENAME ENTERED IN RESPONSE TO 1ST QUERY
TMP001  ; TEMPORARY FILE TO STORE CENTERED OUTPUT IMAGE
RR X20  ; READ SYMMETRY FACTOR INTO REGISTER X20
?SYMMETRY FACTOR?        ; VALUE OF SYMMETRY FACTOR = INPUT TO RA0
FI      ; FILE INFORMATION USED TO READ IN NAME OF OUTPUT FILE
?ROTATIONALLY AVERAGED IMAGE?   ; OUTPUT FILE OF RA0
IF(X20.EQ.1.)GOTO LB2   ; ONE-FOLD AVERAGING MEANS TAKE NO ACTION
IF(X20.LE.0.)GOTO LB3   ; GENERATE ERROR MESSAGE IF FACTOR .LE.0
; FROM HERE ON NORMAL SYMMETRY FACTORS GREATER THAN 1
BL      ; CREATE ZERO BACKGROUND FILE WITH THE DESIRED OUTPUT FILENAME
P3      ; OUTPUT OF BL = FILENAME ENTERED IN RESPONSE TO 3RD QUERY
X11,X12 ; DIMENSIONS OF FILE TO BE CREATED = DIMENSIONS OF INPUT IMAGE
;         THAT WERE PREVIOUSLY STORED
DO LB1 I=2,X20  ; DO-LOOP TO GENERATE AND ADD X20-1 ROTATED, SYMMETRY-
;                 RELATED IMAGES
X30=(I-1)*360./X20      ; CALCULATE ROTATION ANGLE
RT      ; ROTATE CENTERED FILE
TMP001  ; INPUT TO ROTATE = OUTPUT OF PROCEDURE CALL CT1 ABOVE
ROT001  ; OUTPUT OF ROTATE GOES INTO TEMPORARY FILE
X30     ; ROTATION ANGLE AS CALCULATED BEFORE
AD      ; ADD OUTPUT OF ROTATE TO PREVIOUS AVERAGE
P3      ; IMAGE TO BE ADDED TO = RESPONSE TO THIRD QUERY
ROT001  ; IMAGE TO BE ADDED = OUTPUT OF ROTATE
*       ; NO FURTHER IMAGES TO BE ADDED THIS STEP
LB1     ; END DO-LOOP
DE      ; DELETE TEMPORARY FILES
ROT001  ; NAME OF FILE TO BE DELETED
DE
TMP001
RE      ; NORMAL RETURN
LB2     ; JUMP HERE FOR TRIVIAL CASE OF ONE-FOLD SYMMETRY
CP      ; IN CASE OF ONE-FOLD AVERAGING, SIMPLY COPY INTO OUTPUT OF RA0
TMP001  ; OUTPUT OF CT1 = INPUT TO COPY
P3      ; OUTPUT OF COPY = FILENAME ENTERED IN RESPONSE TO 3RD QUERY
RE      ; RETURN FROM ONE-FOLD
LB3     ; JUMP HERE FOR ILLEGAL SYMMETRIZATION FACTORS
CK      ; *** ERROR *** SYMMETRIZATION FACTOR ZERO OR NEGATIVE
RE      ; RETURN FROM ERROR SECTION
```

c

```
; CT1.DOC  PROCEDURE TO CENTER A CENTRO-SYMMETRIC MOTIF
RT      ; ROTATE INPUT IMAGE BY 180 DEGREES TO CREATE SYMMETRY-RELATED IMAGE
?IMAGE TO BE CENTERED? ; INPUT TO ROTATION
INV001  ; OUTPUT OF ROTATION
(180.)  ; ROTATION ANGLE
X21=PAD(2*X1)    ; COMPUTE DIMENSIONS OF PADDED ARRAY. MAKE THEM TWICE THE
X22=PAD(2*X2)    ; DIMENSIONS OF IMAGE, AND ROUND UP TO NEXT INTEGER THAT IS A
;                  POWER OF TWO
PD      ; PAD INPUT IMAGE AS PREPARATION OF CROSS-CORRELATION
P1      ; INPUT TO PADDING = FILENAME ENTERED IN RESPONSE TO 1ST QUERY
PAD001  ; OUTPUT OF PADDING GOES INTO TEMPORARY FILE
X21,X22 ; DIMENSION OF PADDED ARRAY AS CALCULATED BEFORE
Y       ; YES, USE AVERAGE OF IMAGE FOR PADDING BACKGROUND
(1,1)   ; PAD INTO UPPER LEFT CORNER (NOTE THAT PADDING COORDINATES ARE
;         IRRELEVANT FOR CROSS-CORRELATION RESULT, PROVIDED THE SAME ONES
;         ARE USED FOR BOTH IMAGES)
PD      ; PAD 180-DEGREES ROTATED IMAGE
INV001  ; INPUT TO PADING = OUTPUT OF ROTATE
PAD002  ; OUTPUT OF PADDING GOES INTO TEMPORARY FILE
X21,X22 ; DIMENSIONS OF PADDED ARRAY AS ABOVE
Y       ; YES, USE AVERAGE OF IMAGE FOR PADDING BACKGROUND
(1,1)   ; USE SAME PADDING COORDINATES AS ABOVE
CC      ; CROSS-CORRELATE PADDED INPUT WITH PADDED, 180 DEGR. ROTATED IMAGE
PAD001  ; USE PADDED ORIGINAL AS FIRST INPUT TO CC; TO BE OVERWRITTEN BY
;         CROSS-CORRELATION FUNCTION
PAD002  ; USE PADDED, 180-DEGR ROTATED IMAGE AS SECOND INPUT TO CC; TO BE
;         OVERWRITTEN BY ITS FOURIER TRANSFORM
N       ; NO FILTRATION OF CONJUGATE FOURIER CROSS-PRODUCT
PK X11,X12      ; SEARCH THE CROSS-CORRELATION FUNCTION FOR PEAK, STORE
;       ; POSITIONS IN REGISTERS X11,X12 FOR LATER USE
PAD001  ; INPUT TO PEAK-SEARCH = CROSS-CORRELATION FUNCTION FROM PREVIOUS STEP
(3)     ; LIST OF 3 HIGHEST PEAKS ON PRINTOUT
X11=-X11/2      ; CALCULATE NEGATIVE HALF OF PEAK SHIFT VECTOR. THIS IS THE
X12=-X12/2      ; VECTOR BY WHICH THE IMAGE HAS TO BE SHIFTED TO BE CENTERED
SH      ; SHIFT ORIGINAL IMAGE
P1      ; INPUT TO SHIFT = FILENAME ENTERED IN RESPONSE TO 1ST QUERY
?CENTERED FILE? ; OUTPUT OF SHIFT = OUTPUT OF THIS PROCEDURE
X11,X12 ; USE VECTOR COMPONENTS COMPUTED ABOVE TO CENTER IMAGE.
DE      ; DELETE ALL TEMPORARY FILES
PAD001
DE
PAD002
DE
INV001
RE      ; RETURN
```

Fig. 4. Example of a three level calling structure in command language. Ten images are stored in PAR001,..., PAR010, each containing a particle projection in uncentered position. To be computed is the sixfold, rotationally symmetrized average of the centered particle. The user need only set up B01 (a), a batch file utilizing a DO-loop over the particle series, and a call to RA0 (b; general rotational symmetrization), specifying 6 as the symmetry count. The job of centering a centrosymmetric motif is delegated to the procedure CT1 (c) which is called by RA0. Both RA0 and CT1 are part of a standard procedure library and make use of existing basic operations such as cross-correlation, rotation, and shifting. Since no compilation and linking are involved, the development of the procedures takes no more than a few minutes.

5. Documentation

The system documentation consists essentially of four parts: a *user introduction* explaining the main concepts; a *command manual*, which outlines the system/user dialogue for each operation; a *cross-reference table*, which lists the commands and the relevant sections in the user introduction for a large number of keywords; and an alphabetical *list of subroutines* making up the package, along with brief explanations of their functions. All four parts of the documentation are updated each time an operation is changed or added to the existing package.

In addition there is documentation of the SPIDER procedures. User-built procedures are normally very specific to a project and have no practical value beyond the lifetime of the project. Only a few procedures that are generally useful are adopted into the system's procedure library. The system/user dialogue for such procedures is explained in the same way as the dialogue for basic commands in the command manual. Documentation of the remaining procedures is the user's responsibility.

6. Transportability

SPIDER was designed initially for the DEC RSX11D operating system and then converted to run under RSX11M version 3.2. The system layout into several tasks will be advantageous for any minicomputer system with multi-user environment. The main features affecting the transportability are that (1) initiation of slave tasks by a resident master task is supported; (2) SEND/RECEIVE communication allows the master task to initialize important parameters of the slave task; (3) file names are created and dynamically changed at various points in the program system according to the DEC file-11 naming convention; and (4) file information (existence of file, number of records and record length) can be interrogated in FORTRAN routines.

These functions are normally supported by most minicomputer configurations, and for these the implementation of the SPIDER system should offer no particular problem. The most drastic changes concern the system-specific file conventions.

7. Conclusions

Our software system resulted from an attempt to realize a large variety of image-processing operations on a small computer and to construct a control language that allows branching, iterating and procedure-nesting on several levels.

The price for the flexibility of the system is overhead time for searching of commands and swapping of slave tasks. However, this price may not be too high, considering the time and expenditures involved in the creation of rapidly ageing, project-oriented programs.

A separate paper will be devoted to applications of the SPIDER system in the main areas of electron image analysis.

Acknowledgements

Many colleagues and students have contributed to the gradual growth of SPIDER. We wish to acknowledge the help of Timothy Bilash, William Goldfarb, Richard Green, Robert Marshal, Richard Pelavin, and Vicky Riffle. Special thanks are due to Dr. William Moyer for many suggestions and assistance.

We thank Adriana Verschoor for stylistic corrections and comments. We are grateful to Dr. P.R. Smith for suggesting the survey in this form, and to all authors of other software systems for sending us detailed information.

Appendix 1. Table of SPIDER commands

(Commands created for interfacing multivariate statistical analysis programs are not included in this list.)

Operation	Short description	Function
AC	Autocorrelation	Auto-correlate an image using Fourier method
AD	Add	Add two or more images point-for-point
AF	Angular Fourier	Fourier transform in azimuthal direction
AI	Angular interpolation	Convert cartesian into polar representation
AR	Arithmetic operation	Perform point-for-point arithmetic operation on image
BC	Box convolution	Form local average and mix with original image

Appendix 1, Table of SPIDER commands (continued)

Operation	Short description	Function
BL	Blank	Create image with constant background
BP	Back projection	Back-project in two or three dimensions
CC	Cross-correlation	Cross-correlate two images using Fourier method
CE	Contrast enhancement	Stretch density scale/histogram-equalize
CF	Construct Fourier	Construct a Fourier transform from set of amplitudes and phases
CH	Correlation histogram	Plot histogram of correlation values or other values stored in document file
CN	Convolution	Convolute two images with each other using Fourier method
CO	Contour	Contour plot
CP	Copy	Copy
CR	Cross-reference	Extract Fourier from existing projection-Fourier stack
CS	Central slice	Obtain 2-D section of 3-D volume in arbitrary direction
CT	Concatenate	Concatenate two or more images
DC	Determine common line	Determine tilt axis and tilt angle from marker coordinates for back-projection
DE	Delete	Delete file
DF	Density foldover	Obtain display with bit-clipping
DO	DO-loop	Start of DO-loop
DU	Dust	Reset image points that are off the mean by 3 standard deviations or more
ED	Edge enhancement	Enhance edges in image by using recursive filtering
EF	Extract Fourier	Extract 2-D Fourier from 3-D Fourier
EN	End	End SPIDER session
EX	Exit	End SPIDER session and save LOG file
FC	File contour	Contour image by bit-clipping and superpose on image
FF	Fourier filter	Apply low- or high-pass filter function to Fourier transform
FI	File information	Show statistical attributes of image
FL	Fourier list	Print selected portions of Fourier transform
FP	Fourier interpolation	Interpolate by padding Fourier transform
FS	Find statistics	Compute statistical attributes of image
FT	Fourier transform	Compute Fourier transform or inverse
GF	General filter	Mask Fourier transform on reciprocal lattice
GP	Generate projections	Extract 1-D projection lines from image series for back projection
GS	Gray scale	Display image using Versatec halftone software
HI	Histogram	Compute and display histogram of image
IF	Logical IF	Conditional jump depending on arithmetic comparisons
IN	Insert	Insert an imago into a larger one
IP	Interpolate	Interpolate into arbitrary rectangular format using bilinear interpolation
LB#	Label	Destination label for conditional jumps and DO-loops
LD	List document	List contents of document file ordered by key
LI	List file	List any file by rows
MA	Mask	Apply circular mask to image
MD	Mode	Select global processing mode
ME	Menu	Display menu of commands
MO	Model	Create model image
MR	Mirror	Create mirror-related version of image

Appendix 1, Table of SPIDER commands (continued)

Operation	Short description	Function
MU	Multiply	Multiply two images point-for-point
OR	Orientation	Find orientation between two images or auto-correlation functions
PA	Patch	Add small image onto large image at arbitrary position
PD	Pad	Pad image with average or background constant to make it larger
PF	Profile	Plot profile of a selected image row
PH	Phase Fourier stack	Apply phase shift to projection-Fourier stack
PJ	projection	Compute 1-D or 2-D projection of 2-D or 3-D volume
PK	Peak search	Search positions of N highest peaks
PO	Poem	(Operation to celebrate the 100th command)
PR	Print	Display image using overprinting
PS	Pick slice	Pick slice from 3-D volume
PW	Power spectrum	Compute modulus of Fourier transform
RA	Ramp	Determine least-squares density wedge of image and subtract
RC	Real space convolution	Convolute image with arbitrary rectangular array
RD	reduce transform	Create reduced Fourier transform from amplitudes and phases of reflections
RE	Return	Return from procedure to next-higher level of command language
RF	Rotational filter	Filter angular Fourier
RN	Rename	Rename file
RO	Rotational average	Compute rotationally averaged profile
RR	Read register	Read number into register
RT	Rotate	Rotate image
SC	Scale Fourier stack	Scale projection-Fourier stack
SD	Save document	Store register contents in document file
SF	Stack Fourier	Edit projection-Fourier stack (add, delete, insert)
SH	Shift	Shift image
SI	Stack interpolation	Interpolate projection-Fourier stack into 3-D Fourier
SK	Stack 2-D slices	Create 3-D volume by stacking images representing slices
SL	Slice	Slice a 3-D volume in arbitrary direction
SQ	Square	Square image point-for-point
SR	Save registers	Save/unsave registers temporarily
SS	Serial section	Align images of serial section according to marker positions
ST	Set label	Edit statistical and protection label of file
SU	Subtract	Subtract two images point-for-point
SZ	Squeeze	Shear image to conform with arbitrary unit vector angle
TA	Tilt angle	Refine tilt angle using positions of reflections
TF	Tilted transfer function	Generate a transform function vs. defocus display
TI	Tape information	List contents of tape from microdensimeter
TM	Time	Print wall clock time
TP	Three-D plot	Make perspective plot of an image
TR	Tape read	Read image from tape to disk
TT	Title	Change title of image
TV	TV display	Display image on halftone display system
TW	Tape write	Write image onto tape in microdensitometer format
UD	Unsave document	Read registers from document file
WI	Window	Window out portion of image

Appendix 1, Table of SPIDER commands (continued)

Operation	Short description	Function
WT	TV Window	Window image interactively/index reciprocal lattice interactively
WU	Wurzel (square root)	Compute point-for-point square root of image
WV	Window averaging	Window out and sum portions from image according to vectors stored in document file

References

[1] P.W. Hawkes, Computer Graph. Image Processing 8 (1978) 406.

[2] D.L. Misell, Image Analysis, Enhancement and Interpretation (North-Holland, Amsterdam, 1978).

[3] W.O. Saxton, Computer Techniques For Image Processing, in: Advances in Electronics and Electron Physics, Suppl. 10, Ed. L. Marton (Academic Press, New York, 1978).

[4] R.A. Crowther and A. Klug, Ann. Rev. Biochem. 44 (1975) 161.

[5] J. Frank, J. Microsc. 117 (1979) 25.

[6] P.R. Smith, Ultramicroscopy 3 (1978) 153.

[7] P.W. Hawkes, in: Computer Processing of Electron Microscope Images, Ed. P.W. Hawkes (Springer, Berlin, 1980).

[8] J. Frank, in: Advanced Techniques in Biological Electron Microscopy, Ed. J.K. Koehler (Springer, Berlin, 1973).

[9] J. Frank, Biophys. J. 12 (1972) 484.

[10] O. Kübler, M. Hahn and J. Seredynski, Optik 51 (1978) 171, 235.

[11] R.E. Burge, T.C. Dainty and R.F. Scott, Ultramicroscopy 2 (1977) 169.

[12] P.N.T. Unwin and R. Henderson, J. Mol. biol. 94 (1975) 425.

[13] J. Frank, W. Goldfarb, D. Eisenberg and T.S. Baker, Ultramicroscopy 3 (1978) 283.

[14] H.P. Zingsheim, D.-Ch. Neugebauer, F.J. Barrantes and J. Frank, Proc. Natl. Acad. Sci. USA 77 (1980) 952.

[15] J. Frank and W. Goldfarb, in: Electron Microscopy in Molecular Dimensions; State of the Art and Strategies for the Future, Ed. W. Baumeister and W. Vogell (Springer, Berlin, 1980) p. 261.

[16] J. Frank and M. van Heel, in: Pattern Recognition in Practice, Eds. E.S. Gelsema and L.N. Kanal (North-Holland, Amsterdam, 1980) p. 235.

[17] M. van Heel and J. Frank, Ultramicroscopy 6 (1981) 187.

[18] D.. DeRosier and A. Klug, Nature 217 (1968) 130.

[19] J.E. Mellema, in: Computer Processing of Electron Microscope Images, Ed. P.W. Hawkes (Springer, Berlin, 1980) p. 89.

[20] F.C. Billingsley, Advances in Optical and Electron Microscopy, Vol. 4, Eds. Barer and V.E. Coslett (Academic Press, London, 1971) p. 127.

[21] J. Frank and B. Shimkin, in: Proc. 9th Intern. Congr. on Electron Microscopy, Ed. J.M. Sturgess (Microscopical Soc. Canada, Toronto, Ontario, 1978) Vol. I, p. 210.

[22] R.H. Wade, A. Brisson and L. Tranqui, J. Microsc. Spectrosc. Electron. 5 (1980) 699.

[23] S. Kawata, Y. Ichioka and T. Suzuki, J. Phys. E (Sci. Instr.) 11 (1978) 1191.

[24] S. Kawata, Y. Ichioka and T. Suzuki, Optik 52 (1978) 235.

[25] W.O. Saxton, T.J. Pitt and M. Horner, Ultramicroscopy 4 (1979) 343.

[26] M. Horner, in: Developments in Electron Microscopy and Analysis, Ed. J.A. Venables (Academic Press, London, 1976) p. 209.

[27] W.O. Saxton, Computer Graph. Image Processing 3 (1974) 266.

[28] B.L. Trus and A.C. Steven, Ultramicroscopy 6 (1981) 383.

[29] R. Hegerl, in: Electron Microscopy 1980, Eds. P. Brederoo and W. de Priester (7th European Congr. on Electron Microscopy Foundation, Leiden, 1980) Vol. II, p. 700.

[30] M. van Heel and W. Keegstra, Ultramicroscopy, to be published.

[31] D.J. DeRosier and P.B. Moore, J. Mol. Biol. 52 (1970) 355.

[32] W. Goldfarb and J. Frank, in: Proc. 9th Intern. Congr. on Electron Microscopy, Ed. J.M. Sturgess (Microscopical Soc. Canada, Toronto, Ontario, 1978) Vol. I, p. 22.

[33] W. Goldfarb, J. Frank, J.C. Hsung, C.H. Kim and T.E. King, in: Cytochrome Oxidase, Eds. T. King, Y. Orri, B. Chance and K. Okunuki (Elsevier–North Holland–Biomedical Press, Amsterdam, 1978) p. 161.

[34] J. Frank, J.N. Turner, M. Marko, K. Asmus and D.F. Parsons, in: Proc. 38th Ann. Meeting EMSA, San Francisco, 1980, Ed. G.W. Bailey (Claitor, Baton Rouge, LA) p. 46.

[35] R. Gordon, R. Bender and G.T. Herman, J. Theoret. Biol. 29 (1970) 471.

[36] D. Fraser, ACM Trans. Math. Softw. 5 (1979) 500.

[37] J. Frank, in: Computer processing of Electron Microscope Images, Ed. P.W. Hawkes (Springer, Berlin, 1980) p. 187.

Computer Averaging of Electron Micrographs of 40*S* Ribosomal Subunits

Abstract. *An enhanced lateral view of the 40S ribosomal subunit of HeLa cells has been obtained by computer averaging of single particles visualized in the electron microscope. Application of crystallographic criteria to independent averages shows that the reproducibility of the result is comparable to that obtained for thin, stained protein crystals by conventional Fourier filtration methods.*

Ribosomes are ribonucleoprotein particles that process the genetic information coded in messenger RNA (mRNA) for protein synthesis (*1*). Knowledge of their conformation is important for understanding their function in the assembly of polypeptide chains. Visualization by high-resolution electron microscopy offers obvious advantages for the study of the morphology of ribosomes. However, the large amount of noise in electron micrographs precludes detailed analysis, unless some method of image averaging (*2*, *3*) is employed. Studies of crystalline ribosome aggregates (*4*) have failed to give high-resolution structural information, because of the residual disorder in the arrangement of the ribosomes. We have used the computer to align and average large numbers of individual images corresponding to a single view of the ribosomal particle (*3*) for structural analysis of the small (40*S*) ribosomal subunit from HeLa cells [preliminary account in (*5*)].

Specimens for conventional transmission electron microscopy were prepared (*6*). A 0.5 percent aqueous uranyl acetate solution (not adjusted for *p*H) was used as contrasting stain. The grids were examined in an electron microscope (JEM 100B) operated at 80 kV, with a magnification of 70,000. The subunits (Fig. 1a) appear as elongate particles, approximately 250 by 130 Å, and display a strong preference for orientations referred to as right (R) and left (L) lateral views (*7*). The most characteristic structural feature is a beaklike protrusion (Fig. 1a, arrow).

Small areas containing particles in the L lateral view were selected from several electron micrographs and scanned into 128 by 128 arrays at 7.14 Å resolution. Each array was displayed on a computer-linked halftone monitor for interactive selection of a 64 by 64 window containing the particle (*8*).

A mask with a radius of 25 units (corresponding to 178 Å) was applied to each image to reduce the influence of noise in the further processing. Seventy-seven such masked arrays were aligned (Fig. 1b) by using auto- and cross-correlation functions (*3*, *9*), with one particle selected as reference. The reference particle was low-pass filtered (see Fig. 2b, legend) to a resolution of 1/28 $Å^{-1}$ to eliminate high-resolution portions of the noise. After the first alignment pass, the alignment was refined in two cycles, with the average of the previous cycle used as reference each time.

The first pass aligned the particles to within a few degrees of the reference. This approximate positioning allowed use of the direct method of orientation search (*10*) in the subsequent refinement cycles, in which the image, rather than its autocorrelation function, was subjected to the rotational cross-correlation. The combination of autocorrelation (*9*) and direct (*10*) methods was found to give a higher reproducible resolution in the final average than was given by the autocorrelation method alone.

The average of 77 aligned images (Fig. 2, a and b) shows the 40*S* subunit to have a complex morphology. It appears to be divided by stain incursions into three major regions, each characterized by three prominent minima in stain density.

Several different methods were used to ascertain the validity of the result: comparison of independent averages, analysis of the variance associated with

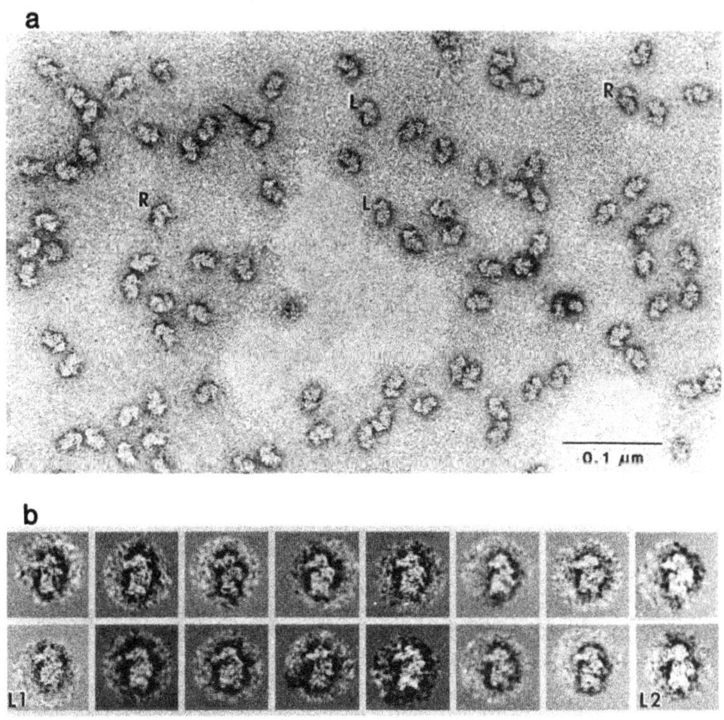

Fig. 1. (a) Electron micrograph showing left-oriented (L) and right-oriented (R) lateral views of 40*S* ribosomal subunits of HeLa cells. The particles in L views were selected for processing. Arrow points to the characteristic beaklike protrusion. (b) Gallery showing 16 of the 77 particles after alignment. L1 and L2 were used as the initial reference particles in the alignments.

0036-8075/81/1218-1353$01.00/0 Copyright © 1981 AAAS

the averaging of each image point, and quantitative comparison of the Fourier transforms of the averages. We concluded that the observed stain density minima are significant, and their locations vary by less than 6 Å. The overall reproducibility of the averaged projection of subunit and surrounding stain was found to be 1/32 $Å^{-1}$.

The images were arbitrarily split into two sets of 38 and 39, which were averaged separately (Fig. 2, c and d). The independent averages agree with one another and with the total average (Fig. 2a) in all major features. The positions of each of the nine density minimà differ by not more than 6 Å. The difference image (Fig. 2e) reveals that the main density differences are in regions where the stain penetrates the particle.

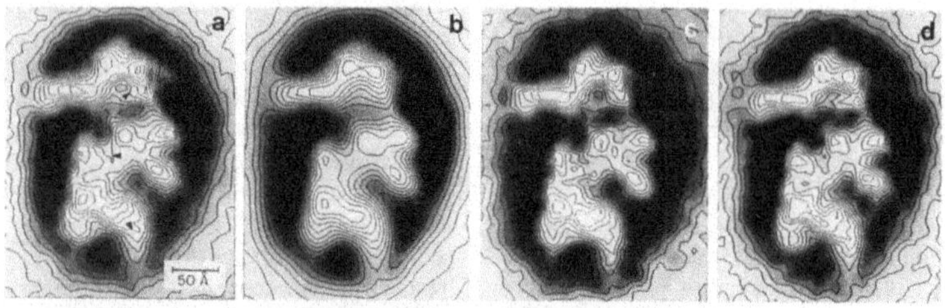

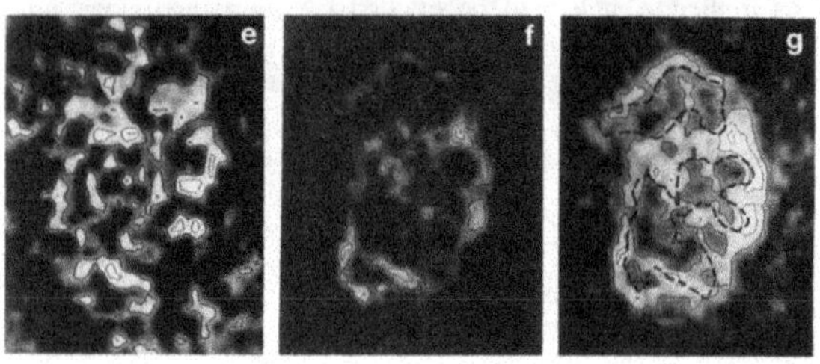

Fig. 2. (a) Total average obtained from all 77 particles, displayed at 1/14 $Å^{-1}$ resolution. Arrows indicate the contour level used to define the particle outline in (g). The contour increment corresponds to an optical density difference (Δd) of 0.01 on the electron micrographs. (b) Total average from (a) displayed with a limiting resolution of 1/32 $Å^{-1}$. For low-pass filtration the Fourier transform was multiplied by a Gaussian filter function that falls to $1/e$ at the limiting resolution. The contouring increment is Δd = 0.0075. (c and d) Averages obtained from independent sets of 38 and 39 particles, displayed at 1/14 $Å^{-1}$ resolution. The initial reference particle was L1 (Fig. 1b). The contour increment was the same as in (a). (e) Difference image of the averages in (c) and (d), displayed with the same contour increment. (f) Display of variance for total average image in (a), contoured with a 0.005 increment. (g) Display of standard deviation for (a), contoured with Δd = 0.01. The dashed line delimits the particle outline as defined by a selected contour level in (a).

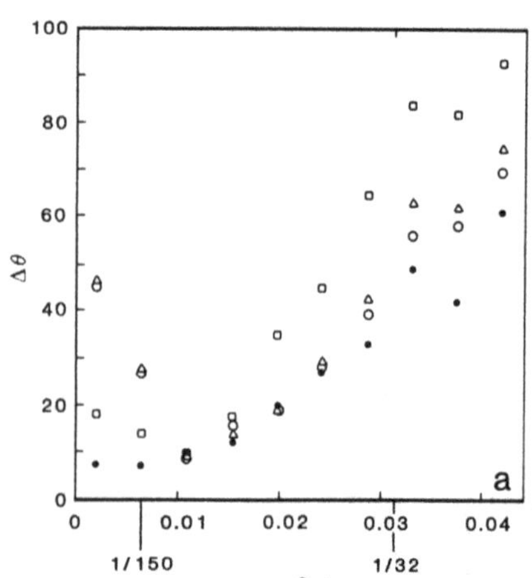

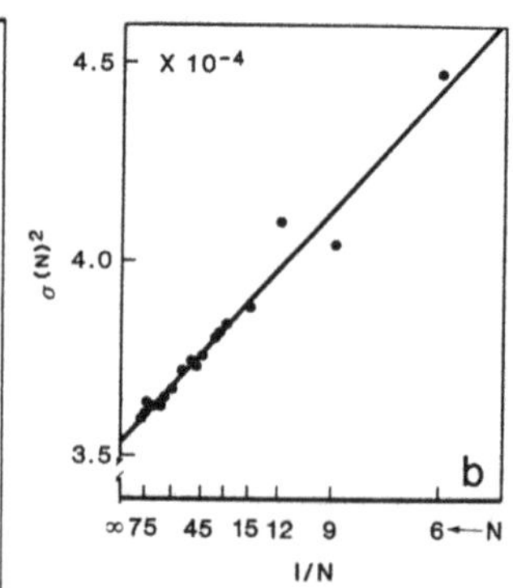

Fig. 3. (a) Differential plot of the phase residual obtained by comparing the Fourier transform coefficients of independent averages before refinement (□), after the first refinement (△), and after the second refinement (○). Corresponding averages of identical particle subsets, obtained with different initial reference particles (L1 and L2 in Fig. 1b), were also compared (●). Summation according to Eq. 3 in Fourier space is carried out over concentric ring zones $s = s_1 \pm \Delta s$ with $\Delta s = 0.0022$ $Å^{-1}$, and the results are plotted as a function of s_1. (b) Variance of the average image as a function of $1/N$, where N is the number of particles averaged. The plot shows the expected linear behavior for large N and enables the gain in s/n to be determined.

To test the significance of features observed in the average, we computed the variance of N realizations of each image element, where N is the number of images, according to (*11*)

$$v_{ik} = \frac{1}{N-1} \sum_{n=1}^{N} [d_{ik}^{(n)}]^2 - \bar{d}_{ik}^{\,2} N \quad (1)$$

Here $d_{ik}^{(n)} = D_{ik}^{(n)} - \bar{D}^{(n)}$ is an array representing the nth image on a sampling grid indexed by (i,k) obtained from the original optical density measurements $D_{ik}^{(n)}$ by floating; that is, by subtracting

$$\bar{D}^{(n)} = \frac{1}{M} \sum_{i,k} D_{ik}^{(n)}$$

(where M is the total number of image points). In Eq. 1, the average image obtained by averaging N particles is denoted by

$$\bar{d}_{ik} = \frac{1}{N} \sum_{n=1}^{N} d_{ik}^{(n)} \quad (2)$$

The resulting variance image (*11*) shows high values in those areas of the image field where the images differ most (Fig. 2f). A standard deviation image (Fig. 2g) was obtained by taking the square root, point by point, of the variance image. At the resolution of scanning representation (1/14 $Å^{-1}$), the standard deviation is high (up to three contour levels) only along the borders of the stain-excluding regions (enclosed by dashed lines in Fig. 2g). Within these regions, the standard deviation of the average does not exceed one contour interval, whereas the nine strong stain density minima stand out by two or three contour intervals from the local background.

For a quantitative assessment of overall reproducibility [see (*12*)], the phase residual [for example (*13*)] was computed from the Fourier transforms F_1 and F_2 of the independent averages (Fig. 2, c and d). The phase residual is defined as

$$\Delta\theta = \left\{ \frac{\Sigma(|F_1| + |F_2|)\,(\delta\theta^2)}{\Sigma(|F_1| + |F_2|)} \right\}^{1/2} \quad (3)$$

where $\delta\theta$ is the phase difference between F_1 and F_2, and the summations extend over a certain domain of the reciprocal space. Use of this criterion is possible because the averages to be compared are aligned with respect to one another, assuring the same phase origin for the corresponding transforms.

We computed the phase residuals for annuli of equal width in reciprocal space and plotted their values as a function of radius (Fig. 3a). We found such a differential plot was best suited for determining the resolution limit of agreement in reciprocal space and for assessing the relative merits of different alignment strategies. Above 1/150 Å^{-1}, the phase residual decreases substantially in the refinement cycles, whereas it shows the opposite behavior at and below 1/150 Å^{-1}. This behavior indicates that in going from the first to the second alignment cycle, accuracy in the alignment of strongly variable, low-resolution features of the outer stain layer is traded off for an increased accuracy in the alignment of high-resolution features.

By allowing a maximum phase residual of 45°, we obtain from Fig. 3a a limiting resolution of 1/32 Å^{-1}. Indeed, the total average of Fig. 2a when limited to this resolution (Fig. 2b) still shows the prominent features observed in the unfiltered average.

The phase residual was also used to check the dependence of the result on the choice of the initial reference. We repeated the entire three-cycle alignment procedure with a different particle as reference and determined the phase residual of averages over the same particle subset (Fig. 2c) obtained with the two different reference particles. The values of the phase residual (Fig. 3a) indicate that the agreement is closer than the agreement between averages over different subsets obtained with the same reference particle. This result is to be expected for the limited number of particles used.

A plot of the variance as a function of the number of images N averaged allows the gain in signal-to-noise ratio (s/n) to be determined (*14*). The variance of the average image, $\sigma^{(N)^2}$, is expected to decrease according to

$$\sigma^{(N)^2} = \sigma^2 + \frac{1}{N}\sigma_n^{\,2} \qquad (4)$$

where σ^2 is the variance of the signal common to the image set and $\sigma_n^{\,2}$ the variance of the noise, modeled as Gaussian and additive. Plotting $\sigma^{(N)^2}$ as a function of $1/N$ shows an approximately linear dependence above $N = 6$ (Fig. 3b). From the slope and intersection of the straight line with the variance axis ($N = \infty$), we could calculate the average initial s/n, $\sigma^2/\sigma_n^{\,2} = 0.55$, and the residual s/n in the average of all particles (Fig. 2a), $(N\sigma^2)/\sigma_n^{\,2} = 42.4$.

The reproducibility of the density distribution obtained by single-particle averaging approaches that of maps obtained by Fourier averaging of micrographs of ordered protein arrays [see, for example (*15*)]. In this range of accuracy, rounding errors and errors of interpolation in the digital analysis must be seriously considered.

The approach of single-particle averaging rests on the assumption that all of the images are comparable. However, interparticle variations, which will tend to blur the averaged image, arise from several sources. The strength of the outer stain layer varies relative to the mean density of the stain-excluding regions. The effective tilt angle may also vary. The lateral view analyzed may, in fact, comprise a small range of tilted views of the 40*S* subunit, with the tilt axis likely parallel to the particle's long axis.

Greater homogeneity of the image set to be averaged—and thus finer resolvable detail—can be obtained through application of multivariate statistical analysis (*16*). Even without this refinement, the realization of a quantitatively reproducible, averaged projection is a major step toward a high-resolution, three-dimensional model of the ribosome structure.

JOACHIM FRANK
ADRIANA VERSCHOOR
Division of Laboratories and Research, New York State Department of Health, Albany 12201
MILOSLAV BOUBLIK
Roche Institute of Molecular Biology, Nutley, New Jersey 07110

References and Notes

1. G. Chambliss, G. R. Craven, J. Davies, K. Davis, L. Kahan, M. Nomura, Eds., *Ribosomes* (University Park Press, Baltimore, 1980).
2. R. Markham, D. Hitchborn, G. Hills, S. Frey, *Virology* **22**, 342 (1964); F. P. Ottensmeyer, J. W. Andrew, D. P. Bazett-Jones, A. S. K. Chan, J. Hewitt, *J. Microsc.* **109**, 259 (1977); A. Klug and D. J. DeRosier, *Nature (London)* **212**, 29 (1966); H. P. Erickson and A. Klug, *Ber. Bunsenges. Phys. Chem.* **74**, 1129 (1970); G. Lutsch, K.-P. Pleissner, G. Wangermann, F. Noll, *Acta Biol. Med. Germ.* **36**, K-59 (1977).
3. J. Frank, W. Goldfarb, D. Eisenberg, T. S. Baker, *Ultramicroscopy* **3**, 283 (1978).
4. P. N. T. Unwin and C. Taddei, *J. Mol. Biol.* **114**, 491 (1977); W. Kühlbrandt and P. N. T. Unwin, in *Electron Microscopy at Molecular Dimensions*, W. Baumeister and W. Vogell, Eds. (Springer-Verlag, Berlin, 1980), pp. 108–116.
5. M. Boublik, J. Frank, B. Tesche, A. Verschoor, *Proc. Electron Microsc. Soc. Am.* **39**, 43 (1981).
6. M. Boublik, W. Hellmann, A. K. Kleinschmidt, *Cytobiologie* **14**, 293 (1977).
7. M. Boublik and W. Hellmann, *Proc. Natl. Acad. Sci. U.S.A.* **75**, 2829 (1978).
8. J. Frank, *Proc. 9th Int. Congr. Electron Microsc.* **3**, 87 (1978); ____ and B. Shimkin, *ibid.* **1**, 210 (1978); ____, H. Dowse, *Ultramicroscopy* **6**, 343 (1981).
9. J. Frank, *Ultramicroscopy* **1**, 159 (1975).
10. M. Steinkilberg and H. J. Schramm, *Hoppe-Seyler's Z. Physiol. Chem.* **361**, 1363 (1980).
11. J. L. Carrascosa and A. Steven, *Micron* **9**, 199 (1978).
12. J. Frank, *J. Microsc.* **117**, 25 (1979).
13. P. N. T. Unwin and A. Klug, *J. Mol. Biol.* **87**, 641 (1974).
14. W. Haenicke and H. P. Zingsheim have used this plot in a study of the reproducibility of acetylcholine receptor averages obtained by the same method of averaging.
15. T. S. Baker and L. A. Amos, *J. Mol. Biol.* **123**, 89 (1978).
16. M. van Heel and J. Frank, *Ultramicroscopy* **6**, 187 (1981).
17. We thank P. N. T. Unwin for a critical reading of the manuscript and D. DeRosier for discussions.

1 June 1981; revised 23 September 1981

 0036-8075/81/1218-1355$01.00/0 Copyright © 1981 AAAS

Ultramicroscopy 6 (1981) 187–194
North-Holland Publishing Company

USE OF MULTIVARIATE STATISTICS IN ANALYSING THE IMAGES OF BIOLOGICAL MACROMOLECULES

Marin VAN HEEL
Biochemisch Laboratorium der Rijksuniversiteit Groningen, Nijenborgh 16, 9747 AG Groningen, The Netherlands

and

Joachim FRANK
Division of Laboratories and Research, New York State Department of Health, Albany, New York 12201, USA

Received 5 March 1981

We have developed a new technique of analysis that allows automatic classification of molecule images according to subtle differences. Computer alignment and multivariate statistical methods were used to analyze electron micrographic images of horseshoe crab hemocyanin half-molecules. The molecule projections fell into four distinct classes related to four different positions of the molecule on the grid. Averages obtained for each images subset are interpreted in terms of a three-dimensional model arrangement for the four subunits forming the half-molecule.

1. Introduction

In an electron micrograph a population of chemically identical macro-molecules exhibit a variety of shapes. Even after low-pass filtration, limiting the image resolution to 20 Å, one observes variations due to differences in the angle of view, to genuine structural differences, or to systematic variations in stain distribution. These factors limit the effective resolution of average images obtained by computer alignment and superposition methods [1].

In a large population of particles the various structure-affecting factors tend to create subpopulations (subsets) of particles which appear more similar to each other than to members of any other subpopulation.

If an average image could be obtained for each subset, it would be expected to show a higher significant resolution than the average over the total population of images. Such subset averages would reflect the molecule in particular structural states, conditions of stain diffusion, angles of view, etc. Thus, once a method for defining and isolating homogeneous subsets in a population of particles is found, the different states can be separately studied and related to one another.

We applied the method of correspondence analysis [2], as suggested by J.P. Bretaudière of the New York State Department of Health, to a population of hemocyanin particles represented by computer-aligned electron micrographic images. By this approach we were able to order the particles into subsets according to two distinct structure-related features without introducing subjective criteria. Averages formed for each subset could be related to different positions of the molecule on the support film. A preliminary account of the results has been presented elsewhere [3,4].

Hemocyanins are the copper-containing respiratory proteins of certain molluscs and arthropods. The native hemocyanin molecule of the horseshoe crab *Limulus polyphemus* consists of 48 subunits in an assembly of eight hexamers [5] and has a molecular weight of 3.3×10^6. The particles in our experiment were half-molecules, each consisting of four hexamers, produced by the first dissociation step of the whole molecule [5,6].

0 304-3991/81/0000–0000/$02.50 © North-Holland

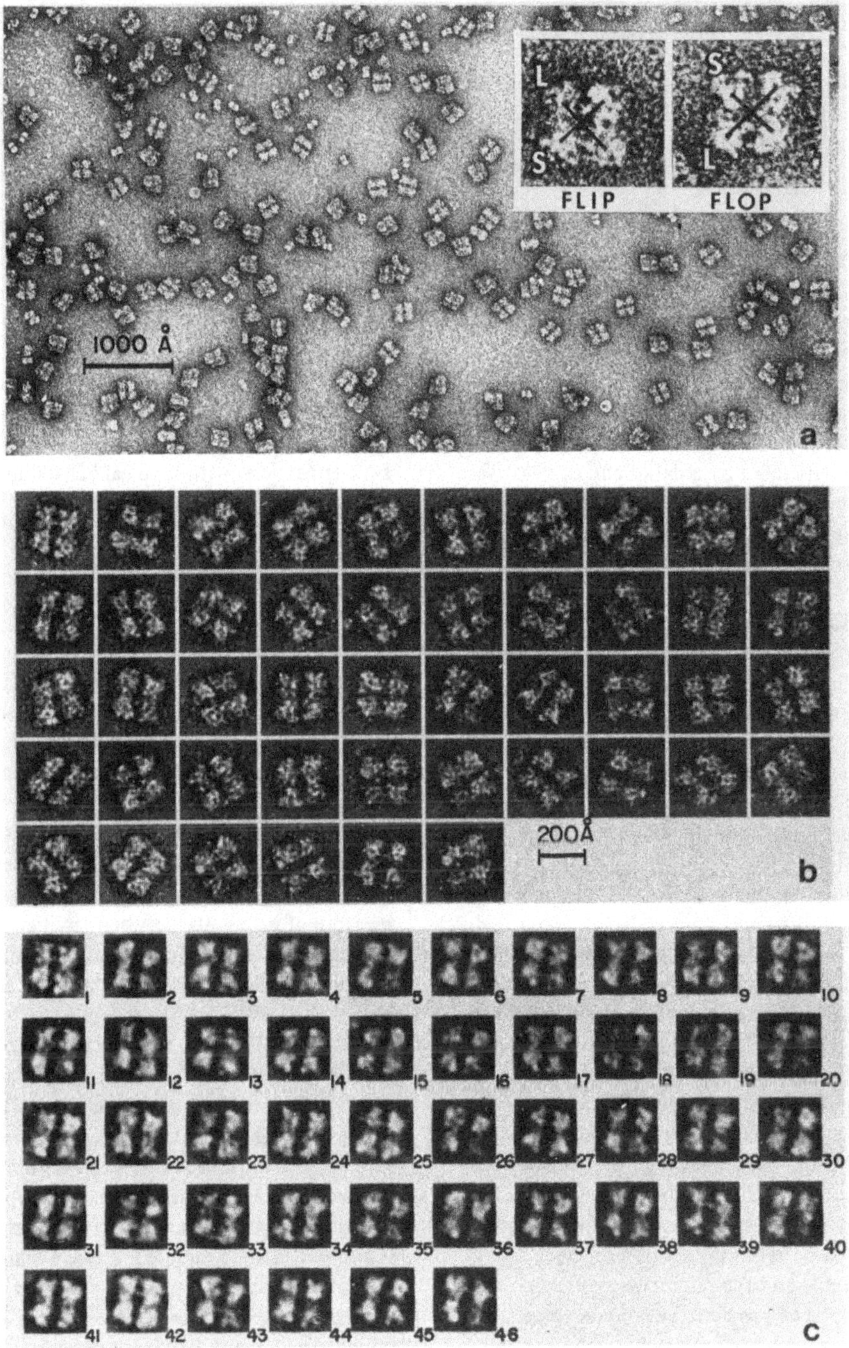

Fig. 1. (a) Electron micrograph of hemocyanin half-molecules from *Limulus polyphemus*. Inset shows examples of the two different views of the molecule apparent in the micrograph: flip and flop, or face-up and face-down. A short (S) and a long (L) axis joining diagonally opposed sub-units are superimposed for reference in the text. Note that all images in this paper represent the molecules as seen through the supporting carbon film. (b) Digitized and windowed molecule projections before computer alignment. The images are represented by 64 × 64 arrays. (c) Selected molecules from (b) after alignment and low-pass filtering to $1/34$ Å^{-1}. The images are now represented by 32 × 32 arrays.

2. Electron microscopy and image processing

Limulus hemolymph was diluted 1 : 200 (vol/vol) in 0.05M Tris-HCl + 0.01M EDTA, pH 6.5 (dissociating conditions). After 20 min specimens were prepared with the spray-droplet technique using 1% uranylacetate [5].

The electron micrograph (fig. 1a) was taken with a Philips EM 200 electron microscope at a linear magnification of 23,500 and 80 kV acceleration voltage. Most of the molecules in the electron micrograph (fig. 1a) appear as roughly rectangular projections. Each rectangular projection consists of four substructures which can be identified with the hexamers. These substructures are arranged as two double hexamers that are separated by a dark band of negative stain and connected by two narrow bridges. The electron micrograph also shows completely different molecular assemblies which were not included in the analysis.

Closer inspection reveals that the rectangular projections do not form a homogeneous set: The arrangement of the double hexamer is skewed by about 8°, with the direction of skewing equally divided among the molecules. The two types of projections are related by mirror symmetry and can be interpreted as corresponding to two positions – face-up and face-down or flip and flop – of the molecule on the carbon grid (fig. 1a insets).

For the computer analysis, the micrograph was first digitized with a flatbed microdensitometer and processed with the SPIDER program system [7] on a minicomputer. Forty-six molecules were selected interactively [8] with a computer-linked graytone system (fig. 1b). The molecules were aligned and centered by correlation methods [1]. The alignment of a molecule with respect to a reference is done in two steps: First the relative orientation is found by searching for an optimum match between the auto-correlation functions of molecule and reference, and in the second step, the relative translation is found by computing the cross-correlation function of the correctly oriented molecule and the reference.

Use of a flip-type projection as reference produced an excellent alignment of both types of projections. Fig. 1c shows the aligned images after low-pass filtration, represented at this stage as 32 X 32 arrays.

However, the method of orientational alignment using auto-correlation functions fails to distinguish between positions related by 180° rotation in the plane of the image. For molecules lacking centrosymmetry, an additional test [9] is necessary to allow this distinction to be made.

Because of the existence in our study of two different types of projections and the presence of noise, this test does not provide a conclusive answer for a sizable fraction of the molecules; the 180° orientational test is performed relative to the reference, which does not necessarily belong to the same subset as the molecule to be oriented.

We found that through the application of multivariate statistical methods of analysis, the flip-flop selection, the 180° rotation test, as well as the exclusion of particles with atypical behaviour could all be accomplished at the same time.

3. Statistical analysis

In multivariate statistics, correspondence analysis [2,10] belongs to the eigenvector methods of ordination. It is used to identify common trends and clusters in a set of experimental observations (cases) on which a set of measurements (variates) have been obtained.

Unlike principal components analysis, which is based on a Euclidian distance measure, correspondence analysis makes use of χ^2 distances [10] as a measure of proximity between any two elements, it is thus independent of the absolute magnitudes of the measurements and sensitive to relative measurements only.

In our application the experimental observations are the 32 X 32 elements of an array that represents the digitized image of a molecule projection as realized in 46 independent sets of measurements. All measurements are arranged in a 1024 X 46 matrix (input matrix). The analysis is done in two steps: First, the χ^2 distances between any two rows or columns of the input matrix are computed. These distances form a symmetric matrix. In the next step, the eigenvalues and eigenvectors of the distance matrix are determined. Each image may be represented by an expansion of these eigenvectors or factors, which are ordered by their relative importance. Only a few of the most significant factors carry meaningful information about major distinctions and trends present in the image data set. Thus, a projection of the data vectors representing the images into a map of the first two factors is often sufficient to reveal the presence of significant differences among the images.

The computer program for correspondence analysis (AFC) implemented on the New York State Department of Health PDP 11/45 is a modified version of Benzécri's program [11]. Special software had to be written to allow images stored in SPIDER [7] format to be analyzed.

4. Results of the statistical analysis

A map of the first two factors shows the images separated into four clusters (fig. 2A). A comparison with fig. 1c shows that molecule images lying closely together on the factorial map exhibit similar features. Examples are the image pairs (26,36), (28,37) and (38,44). That is, a clustering of images on the factorial map indicates a division into subsets of images that have certain features in common.

The images in the four clusters appear to be divided according to the two skewing positions, flip and flop (lower and upper pair of clusters in fig. 1c) respectively. Within each flip and flop subset, the images appear to be further divided according to the position of the subunit with the weakest staining in the upper left or the lower right of the image.

Before we draw any conclusions from this analysis, we have to ask: Is the grouping into the four clusters meaningful in terms of the molecule's structure or is it merely a result of stain variation? If the images were sorted according to accidental variations of stain among the four subunits, then the appearance of clustering within each flip and flop cluster pair would have no special significance, and the separate averaging of the corresponding images would only reinforce features artificially selected.

However, what we observe instead is that in both flip and flop groups, two out of four morphologically identical subunits vary strongly in staining while the other two exhibit little variation. Total averages of both flip and flop groups show the upper left and lower right subunit to have a systematically weaker staining pattern than the other two subunits. These observations indicate that the division into four groups of molecule projections is not due to noise but to structure-related differences, and that averages formed for each group are meaningful, showing structurally relevant features of the molecule projection.

Before discussing a tentative model of the subunit arrangement that would be consistent with the averages and with the statistical behaviour of the images, we will first interpret the factorial map in terms of two observable properties of the molecule projection. Averages obtained from the images belonging to the four clusters show either of two roughly mirror-related shapes (fig. 2A, insets, and fig. 2B). Within each flip and flop set, the two clusters contain projections that are related by 180° rotation as borne out by the corresponding average images (FLIP1, FLIP2 and FLOP1, FLOP2 in insets and in fig. 2B). The most prominent difference between 180° rotation-related averages is due to a stain imbalance

Fig. 2. (A) Correspondence analysis projection map showing the positions of the molecule images (numbers correspond to image numbers in fig. 1c) in the coordinate system spanned by the two most significant eigenvectors of the expansion. The two factors represent 17.6% and 9.4% respectively of the total information contained in the image set. Due to present program limitations, only 33 images can be analyzed at a time. The map is thus a combination of two maps, one for the particles 1–33, one for the particles 12–46. In the latter group, images 39 and 40 were omitted because they assumed extreme positions in a trial run. The images separate into four clusters. Average images (see insets) for the four clusters, obtained by averaging the aligned, unfiltered molecule images, reveal the nature of the separations. In the direction indicated as flip-flop direction, the images are segregated according to the direction of skewing of the geometric arrangement of the four substructures. The differing appearances of the flip and flop averages we attribute to two different positions of the molecule on the grid. The second separation, in the direction indicated on the map as rocking direction, results from the unequal distribution of stain in diagonally opposed substructures. We attribute this effect to instability of the molecule on the support film. In the flip inset averages, the substructures on the long diagonal axis of the molecule are affected by the stain imbalance, while in the flop averages the substructures on the short diagonal axis are affected. This behaviour would be expected if the long and the short axes were not coplanar. (B) Comparison of averages related by 180° rotation in the plane of the specimen support (FLIP1 versus FLIP2/180 and FLOP1 versus FLOP2/180), and comparison of mirror-related versions (indicated by /m). The four images on the left-hand side show a comparison of flip averages with mirror-inverted flip averages. The axis of mirror inversions is shown in the projections labeled as FLIP1.

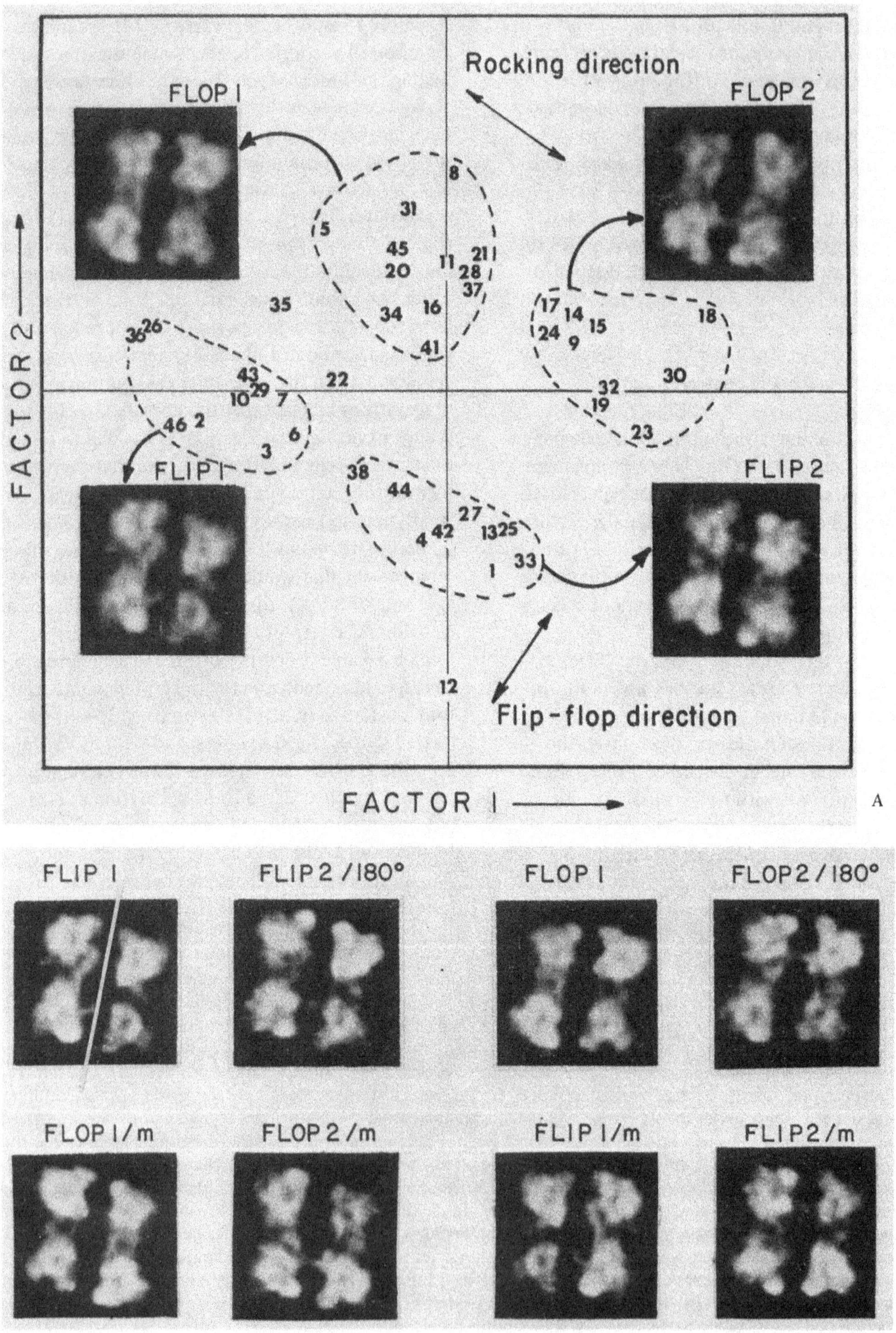
Rocking direction
FLOP I
FLOP 2
FLIP I
FLIP 2
Flip-flop direction
FACTOR 2
FACTOR I
A
FLIP I
FLIP 2 /180°
FLOP I
FLOP2 /180°
FLOP I/m
FLOP 2 /m
FLIP I/m
FLIP2/m
B

between two diagonally opposed subunits.

Two projections (39 and 40) were eliminated in a first run of the analysis because they assumed extreme positions on the factorial map. Three projections (12, 22 and 35) were not included in the averages because they fell outside the clusters and show an atypical behaviour compared with all other molecules: projection 12 is more strongly skewed than other projections in the gallery (fig. 1c), and both projections 22 and 35 are unusual in that they show no skewing at all. Their behaviour helps us to understand the way in which the images are ordered on the map: these latter two projections assume positions half-way between the flip and the flop cluster pairs; hence, the data separation in the direction of a line joining the two cluster pairs (flip-flop direction) appears to be associated with the skewedness of the molecule projections. Indeed, projections 12 and 8, which have extreme positions on the map, are strongly skewed.

Within each pair of clusters in fig. 2A, the molecule projections appear to be ordered, in the direction marked as "rocking direction", according to the degree of stain imbalance (see fig. 3). Due to the skewed arrangement of the substructures, a long and a short axis can be distinguished in the molecule (insets, fig. 1a). In the flip projection, the stain imbalance affects the substructures at the ends of the long axis; in the flop projection it affects the substructures at the ends of the short axis. Because of this effect, there is no mirror relationship between any flip average and any flop average (see fig. 2B).

The two clusters in each pair are not separated as much as the flip and flop clustering pairs; intermediate positions on the map are seen to be correlated with a closely balanced stain distribution, e.g. projections 37 and 38 (figs. 1c and 2A). Only such balanced projections appear to be related by mirror reflections. Examples for extreme positions within each cluster pair, and a corresponding extreme behaviour in terms of the stain imbalance, are projections 26 and 33.

Correspondence analysis separates the total infor-

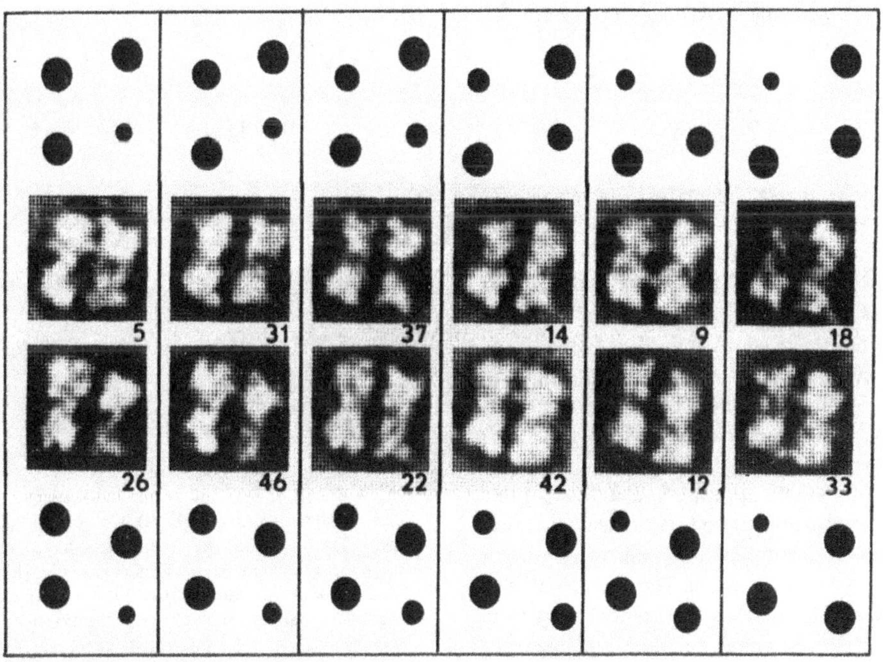

Fig. 3. Gallery of selected particles ordered by their positions in the rocking direction in fig. 2A. Upper and lower rows show flop and flip particles respectively. From left to right the distribution of stain between diagonally opposed substructures varies between two extremes. This behaviour is shown schematically in the accompanying drawings (the degree of skewing is exaggerated).

mation contained in the image set into basic components and ordinates these components according to their relative importance. This does not necessarily mean that the factorial axes coincide with the directions of the "logical" data separation on the factorial map (fig. 2A). As the FLIP1 group and the FLOP2 group differ both in the skewing *and* in stain imbalance, the χ^2 distance between these groups is larger than the distance between FLIP1 and FLIP2 or between FLIP1 and FLOP2. Therefore the FLIP1-to-FLOP2 direction coincides with the first factorial axis. The next strongest separation occurs in the FLIP2-to-FLOP1 direction, which coincides with the second factorial axis.

Additional distinctions among the molecule projections not related to skewedness and stain imbalance are found in higher-order factorial maps, which are being further investigated.

To summarize the results, the major separations found in the statistical analysis of the image set correspond to visually recognizable distinctions. This is confirmed by inspecting images assuming extreme positions on the correspondence analysis map, and images obtained by averaging over subsets that appear separated on the map.

5. Interpretation

The image set analyzed contains only two views of the half-molecule, which are obviously not sufficient to derive a three-dimensional model. However, the systematic stain behaviour revealed by the statistical analysis gives us some clues about the relative arrangement of the hexamers which are not available from the visual analysis of the micrograph. These clues can be obtained by analyzing the statistical distribution of the images on the factorial map, as well as by inspecting the subset averages.

Flip and flop groups appear completely separated, as would be expected for two totally disjunct classes formed by the realization of a non-mirror symmetric projection and its mirror image. The absence of transitions between flip and flop appearance (apart from projections 22 and 35 which could be explained by assuming small tilts of the molecule) supports the interpretation in terms of face-up and face-down positions of the molecule.

We do not believe the stain imbalance to be a random artifact because it consistently affects two different pairs of substructures, depending on the face-up or face-down position of the half-molecule. Furthermore dissociation experiments [5] have shown no morphological difference between the double-hexamers forming the half-molecule.

To reconcile these observations, we believe that the stain variation is a consequence of the three-dimensional arrangement of the hexamers. Unequal distribution of stain between two of the four hexamers would occur if they were not situated in the same plane, but rather pairwise in two different planes, as shown schematically in fig. 4. This would allow the half-molecule to assume different "rocking positions" in a continuous range, with the two extreme positions most likely to be encountered,

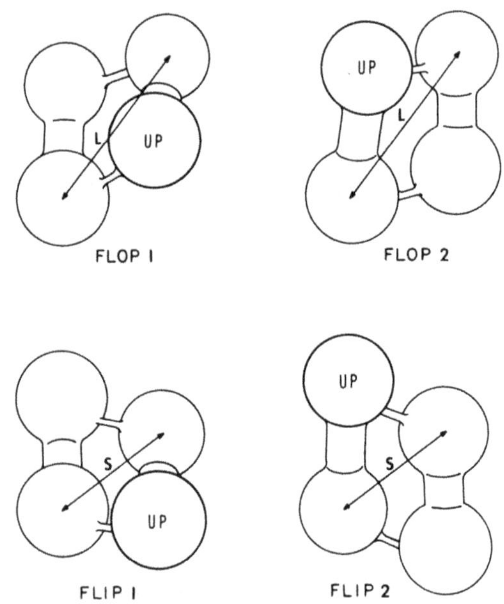

Fig. 4. Model of the hemocyanin half-molecule showing the hexamers (represented by balls) in a non-coplanar arrangement. The binding of the double hexamers is represented by thick bridges. The face-up and face-down positions of the model produce the flip and the flop views. In either position the model is able to rock between the extremes depicted; the rocking axis is the short diagonal (S) for the flip position, and the long diagonal (L) for the flop position. The model explains the appearance of the stain imbalance on different pairs of hexamers, depending on the face-up/face-down position of the molecule.

leading to identical, 180° rotated projections. Consistent with this interpretation is the fact that the clusters of projections related by 180° rotation show a considerable spread on the factorial map in the direction marked as "rocking direction". In a recent paper, Lamy et al. [12] found immunological evidence that a comparable molecule (scorpion hemocyanin, also consisting of four hexamers) has two different faces. Their findings are consistent with our model of the hexamer arrangement.

6. Conclusions

Our study has revealed the potential of a new approach, which combines computer alignment with multivariate statistical analysis to separate molecule images into groups on the basis of structure-related differences in appearance.

In our application these differences are most likely associated with different positions of the hemocyanin molecule on the support film. The separation allows partial averages to be computed, representing the typical, noise-free appearances of the different image subpopulations found.

Other possible applications of the new technique may lie in the localization of chemical markers and reaction sites through analysis of a mixed (treated/untreated) molecule population, and the extraction of high-resolution information from micrographs of thin, disordered crystals.

Acknowledgements

We thank W.G. Schutter and E.F.J. van Bruggen for providing the micrograph, J.P. Bretaudière and J.B. Kazmierczak for computer implementation of the AFC program, K. Azmus for scanning and display of images, and C. Manella and M.M.C. Bijlholt for commenting on the manuscript. One of the authors (M.v.H.) acknowledges a travel grant from the Nederlandse Vereniging voor Elektronenmicroscopie.

References

[1] J. Frank, W. Goldfarb, D. Eisenberg and T.S. Baker, Ultramicroscopy 3 (1978) 283.
[2] J.P. Benzécri, in Methodologies of Pattern Recognition, Ed. S. Watanabe (Academic Press, New York, 1969).
[3] J. Frank and M.G. van Heel, in: Proc. 7th European Congr. on Electron Microscopy, The Hague, Eds. P. Brederoo and W. de Priester (Leiden, 1980) Vol. 2, pp. 690–691.
[4] M.G. van Heel and J. Frank, in: Proc. 7th European Congr. on Electron Microscopy, The Hague, Eds. P. Brederoo and W. de Priester (Leiden, 1980) Vol. 2, pp. 692–693.
[5] W.G. Schutter, E.F.J. van Bruggen, J. Bonaventura and B. Sullivan, in: Structure and Function of Haemocyanin, Ed. J.V. Bannister (Springer, Berlin, 1977) pp. 13–21.
[6] M.M.C. Bijlholt, E.F.J. van Bruggen and J. Bonaventura, European J. Biochem. 95 (1979) 399.
[7] J. Frank and B. Shimkin, in: Proc. 9th Intern. Congr. on Electron Microscopy, Toronto, Ed. J.M. Sturgess (Imperial Press, Mississauga, Ontario, 1978) Vol. 1, pp. 210–211.
[8] J. Frank, in: Proc. 9th Intern. Congr. on Electron Microscopy, Toronto, Ed. J.M. Sturgess (Imperial Press, Mississauga, Ontario, 1978) Vol. 3, pp. 87–93.
[9] H.P. Zingsheim, D.-Ch. Neugebauer, F.J. Barrantes and J. Frank, Proc. Natl. Acad. Sci. USA 77 (1980) pp. 952–956.
[10] M.O. Hill, Appl. Statistics 23 (1974) 340.
[11] J.P. Benzécri, L'Analyse des Données, Vol. II, L'Analyse des Correspondences (Dunod, Paris, 1976) 2nd ed.
[12] J. Lamy, M.M.C. Bijlholt, P.Y. Sizaret, J. Lamy and E.F.J. van Bruggen, Biochemistry, in press.

APPENDIX

Correspondence Analysis of Aligned Images of Biological Particles

J. Frank

Division of Laboratories and Research
New York State Department of Health
Albany, NY 12201, U.S.A.

AND

M. van Heel

Biochemisch Laboratorium
Reijksuniversiteit te Groningen
Groningen, The Netherlands

The purpose of our analysis is to distinguish classes among the set of images, where each class is defined as a subset containing images more similar to each other than to images of any other subset. Differences in structure among the particles analyzed are reflected in differences in their projections. Conversely, if the angle of view is the same for all particles, then the existence of distinct classes in the particle images can be taken to indicate the presence of structural differences.

Let the images (numbered by i) of the particles in a given view be represented by an array of measurements (running index j) on a sampling grid, t_{ij}. Provided that two images t_{i_1j}, t_{i_2j} are aligned, the Euclidean distance

$$E_{i_1i_2} = \sum_j (t_{i_1j} - t_{i_2j})^2$$

is a possible choice for a measure of their similarity. Thus, a table containing the $n(n-1)/2$ Euclidean distances between every pair of images would contain sufficient information to group n images into classes. Classical principal component analysis, which is based on the Euclidean distance, could be used to achieve this. However, it is more appropriate to use an analysis technique that is invariant to changes in scaling, and that retains the symmetry between images and image elements in the initial data table t_{ij}. In our outline of correspondence analysis (Benzecri, 1969), which has these useful properties, we will closely follow the presentation by Lebart *et al.* (1977) and Nakache *et al.* (1977), but always with the application to image analysis (Frank & van Heel, 1980; van Heel & Frank, 1980*a*,*b*,1981) in mind.

We will call I the set of n images, represented by the rows of a matrix T, and J the set of p image elements, represented by the columns of T. Correspondence analysis deals with the analysis of I in the p-dimensional space R^p and, simultaneously, with the analysis of J in the n-dimensional space R^n.

Reprinted from *J. Mol. Biol.* (1982) **161**, 134–137

(a) *Analysis of* I *in* R^p

Normalized image densities y_{ij} are obtained from the original density measurements by setting:

$$y_{ij} = f_{ij}/f_{i\cdot},$$

where

$$f_{ij} = t_{ij}/\sum_{ij} t^{ij}$$

is the relative density of image element t_{ij}, and

$$f_{i\cdot} = \sum_{j} f_{ij}$$

is a scaling factor proportional to the mean density of image i. The normalized image densities may be arranged into a matrix Y with order (n, p).

We now consider the *image profiles*:

$$\mathbf{y}_i = \{y_{ij};\ j = 1, 2, \ldots, p\},$$

which are vectors in R^p. For later use, we also introduce the *image-element profiles*:

$$\mathbf{z}_j = \{z_{ij};\ i = 1, 2, \ldots, n\};$$

$$z_{ij} = f_{ij}/f_{\cdot j};\ f_{\cdot j} = \sum_{i} f_{ij}\ .$$

The *chi-square distance* between two image profiles $\mathbf{y}_{i_1}$ and $\mathbf{y}_{i_2}$ is defined by:

$$d^2(i_1, i_2) = \sum_{j} \frac{1}{f_{\cdot j}} (\mathbf{y}_{i_1} - \mathbf{y}_{i_2})^2,$$

where the weighting by the inverse of $f_{\cdot j}$ is used to avoid excessive contributions from image elements with a high average density. The chi-square distance has the property of distributional equivalence: if image elements are added that have the same image element profile, the distance between two image profiles remains unchanged.

We denote with M the positive definite matrix defining the metric in R^p. For the chi-square distance metric introduced above, M is a (p, p) diagonal matrix containing $1/f_{\cdot j}$ as diagonal elements. Furthermore, we introduce a diagonal "adjustment" matrix N with order (n, n), whose diagonal elements are the weights $f_{i\cdot}$. These are used to weight the contribution of each image profile to the squared-projection sum, to be computed below, according to its original, unnormalized distribution. Generally, a unitary vector $\mathbf{u}$ of R^p satisfies the orthonormalization relation $\mathbf{u}'M\mathbf{u} = 1$.

We would like to find a direction $\mathbf{u}$ in R^p for which the sum of weighted squares of projections of all image profiles is maximum. The n projections of image profiles on axis $\mathbf{u}$ are the n rows of $\hat{v} = YM\mathbf{u}$. Hence, the quantity to be maximized is:

$$\hat{v}N\hat{v} = \mathbf{u}'My'NYM\mathbf{u},$$

with the constraint $\mathbf{u}'M\mathbf{u} = 1$. This leads to the eigenvector equation:

$$Y'NYM\mathbf{u} = \lambda\mathbf{u}.$$

The solution of our problem is the eigenvector $\mathbf{u}_1$ associated with the largest eigenvalue λ_1. $\mathbf{u}_1$ is called the first factorial axis. Other solutions of the eigenvector equation $\mathbf{u}_k$ associated with eigenvalues λ_k, $k > 1$ present other directions of interimage variance in R^p with decreasing importance.

The projection operator on the axis $\mathbf{u}_k$, defined by $\phi_k = M\mathbf{u}_k$, is called the kth factor. Multiplying the eigenvector equation above from the left with M leads, after substitution of $\phi = M\mathbf{u}$, to the eigenvector equation for the factor:

$$MY'NY\phi = \lambda\phi.$$

Substitution of the chi-square metric M and the adjustment matrix N by unitary matrices would lead to the eigenvector equation of principal component analysis.

In general, a small number of eigenvectors, ordered by decreasing eigenvalues, is sufficient to represent major differences and trends present in the data table t_{ij}. To detect such differences, we project the image profiles into a low-dimensional subspace of the eigenvector expansion. Most convenient conceptually are two-dimensional factor maps. The co-ordinate of an image profile $\mathbf{y}_i$ with respect to the factor axis r is given by the projection formula

$$\Psi_{ir} = \frac{1}{\sqrt{\lambda_r}} \sum_{j \in J} y_{ij}\phi_{jr},$$

where ϕ_{jr} is the jth component of factor r, and λ_r is the corresponding eigenvalue.

For an analysis of the results, it is often sufficient to display the map of the first *versus* the second factor where a large portion of the interimage variance is concentrated. Clustering of the image profiles on this map indicates the presence of classes of images with different features. Averaging of images with the purpose of eliminating the noise should be carried out only over images within a class, since inter-class averaging is bound to blur out the significant structural differences that gave rise to the statistical separation.

There are essentially two ways to identify the image features that give rise to the data separation on the map: (1) by direct comparison of subset averages formed (e.g. by visual comparison, or by computing the difference image); (2) by making use of the simultaneous representation of image profiles and image element profiles, a feature unique to correspondence analysis.

(b) *Analysis of* J *in* R^n, *and the simultaneous representation*

Similarly, the set of image elements may be analyzed in R^n, leading to a representation of J by the same set of orthonormalized eigenvectors ϕ_r. The co-ordinates of image profiles and image element profiles with respect to the same factorial axis r are related by:

$$\Psi_{ir} = \beta^2 \sum_{j \in J} y_{ij}\phi_{jr}$$

$$\phi_{jr} = \beta^2 \sum_{i \in I} z_{ij}\Psi_{ir}.$$

These equations, known as *transition formulas*, make it possible to represent both image profiles and image element profiles in the same subspace. An exact simultaneous representation would require $\beta^2 = 1$, a situation impossible to achieve theoretically. It can be shown, however (Lebart *et al.*, 1977), that the choice $\beta^2 = 1/\lambda_r$ leads to an optimum simultaneous representation.

The simultaneous representation has an important consequence for the interpretation of the factor map: image elements whose profiles are important for the separation of an image cluster will be found in the same domain of the map; and *vice versa*: the position of a cluster of image elements on the factor map is "explained" by the images lying in the same domain of the map. This property suggests the following method of analysis (van Heel & Frank, 1980*a*,*b*). To characterize the distinguishing properties of a cluster of image profiles that occupies a domain A on the factor map, we identify all image elements J_A that project into A. We now form an image m_j with the same raster as the original set of t_{ij}, and the following properties:

$$m_j = 1 \quad \text{for } j \in J_A$$
$$m_j = 0 \quad \text{elsewhere.}$$

Such an image, called an *importance mask* (van Heel & Frank, 1980*a*,*b*), shows at a glance the location of those image regions that are most important in defining the cluster.

We will also make use of a related method of representation, the *importance image*, to understand the significance of the data separation occurring along a given factor axis. Weights are assigned to all image elements whose profiles have positive co-ordinates along the axis investigated, where the weights are made proportional to the co-ordinates. Thus, image elements sitting in an extreme position in the halfplane of the map will have a large weight, and an image formed by all weighted image elements will highlight all those image areas that are responsible for the separation of images in the direction of the selected axis. This procedure is repeated for the negative co-ordinates, and a pair of complementary importance images is obtained.

REFERENCES

Benzecri, J. P. (1969). In *Methodologies of Pattern Recognition* (Watanabe, S., ed.), pp. 35–74, Academic Press, New York.

Frank, J. & van Heel, M. (1980). In *Proc. 7th Eur. Congr. Elec. Microsc.* (Brederoo, P. & de Priester, W., eds), vol. 2, pp. 690–691, the Seventh European Congress on Electron Microscopy Foundation, Leiden.

Lebart, L., Morineau, A. & Tabard, N. (1977). *Techniques de la description statistique*, Dunod, Paris.

Nakache, J. P., Lebart, L. & Lorente, P. (1977). In *Medinfo 77* (Shires/Wolf, eds), p. 105, North-Holland, Amsterdam.

van Heel, M. & Frank, J. (1980*a*). In *Pattern Recognition in Practice* (Gelsema, E. S. & Kanal, L. N., eds), pp. 235–243, North-Holland, Amsterdam.

van Heel, M. & Frank, J. (1980*b*). In *Proc. 7th Eur. Congr. Elec. Microsc.* (Brederoo, P. & de Priester, W., eds), vol. 2, pp. 692–693, the Seventh European Congress on Electron Microscopy Foundation, Leiden.

van Heel, M. & Frank, J. (1981). *Ultramicroscopy*, **6**, 187–194.

Edited by A. Klug

Journal of Microscopy, Vol. 144, *Pt* 1, *October* 1986, *pp.* 1–14.
Received 4 *April* 1986; *accepted* 17 *May* 1986

Reconstitution of molecule images analysed by correspondence analysis: a tool for structural interpretation

by JEAN-PIERRE BRETAUDIERE *and* JOACHIM FRANK*, *Department of Pathology and Laboratory Medicine, The University of Texas Medical School, P.O.B.* 20708, *Houston, Texas* 77225, *and* **Wadsworth Center for Laboratories and Research, New York State Department of Health, Albany, New York* 12201, *U.S.A.*

KEY WORDS. Eigenimage, matrix analysis, image synthesis.

SUMMARY

Correspondence analysis is gaining increasing importance in the analysis of electron micrographs of macromolecules. Partial or complete reconstitution of images from their factorial representations is introduced as a useful tool in interpreting variational patterns and tracing their physical origin. The value of image reconstitution is demonstrated with two examples, one using a set of model images and the other a set of images of a haemocyanin molecule assembly product.

INTRODUCTION

Within a few years, correspondence analysis has become a powerful tool among the battery of computer programs used for analysis of electron microscopic images (Van Heel & Frank, 1981; Frank & Van Heel, 1982a). It has been utilized for separating labelled and unlabelled molecules in electron micrographs of a mixed preparation (Frank *et al.*, 1986), for separating molecule views or orientations (Van Heel & Frank, 1981; Verschoor *et al.*, 1985) and for classifying images of ribosomes (Frank *et al.*, 1982) or virus capsid fragments (J. Van Oostrum & R. Burnett, personal communication) according to the same level of staining.

Briefly correspondence analysis is an eigenvector method of ordination which describes the dispersion or shape of a cloud of p features in the n-dimensional space of objects or of a cloud of n objects in the p-dimensional space of features (Benzecri, 1969; Lebart *et al.*, 1977). A set of orthogonal linear coordinates replaces the original data set, creating a space of significantly lower dimensionality. Correspondence analysis is distinguished from other eigenvector ordination methods (e.g. Principal Component Analysis) in that it allows features and objects to be represented in the same space in a meaningful fashion.

The advantages of using such a method in analysing a multidimensional data set are the elimination of noise and the ordination of the information. The selection of the most significant eigenvectors—associated with the highest eigenvalues—results in the elimination of less significant or insignificant information represented by remaining eigenvectors. In particular, random noise always present in the data is discarded. The selection of the eigenvectors with the highest

© 1986 The Royal Microscopical Society

eigenvalues leads, moreover, to a reduction of the data volume from $n \times p$ to $(n+p) \times k$, the $(n+p)$ elements being characterized by their k most significant factorial coordinates ($k \leq \min(n, p)$; often, $k=8$ is found sufficient). Effectively, this method gives a visual representation of the data elements from their projections on planes determined by any two significant factorial axes. In many cases, however, visual inspection of 2D maps is not sufficient, and needs to be augmented by an objective method of classification (Benzecri, 1980; Van Heel *et al.*, 1982; Van Heel, 1985) or a method of 2D data representation of data with higher dimensionality (Radermacher & Frank, 1985). The fact that all eigenvectors are orthogonal to one another implies that the information expressed by a particular eigenvector is exclusive of that expressed by any other eigenvector; thus, an ordination of the information is achieved.

In the application to multiple images of single macromolecules developed by Frank & Van Heel (1982a; Van Heel & Frank, 1981) the 'objects' are macromolecules as they appear in electron micrographs, and the 'features' being analysed are sets of optical density measurements, each set representing multiple measurements of a point (pixel) in the molecule's absolute coordination system. To achieve this spatial correspondence of measurements, the scanned images are subjected to a correlation alignment procedure (Frank *et al.*, 1978, 1981). A mask is used to eliminate any pixels that lie outside the molecule boundary (Frank *et al.*, 1982). The result of the analysis is a grouping of images or pixels. It is the positions of pixels relative to those of the images that provides information on the meaning of factorial axes and the reasons for the observed grouping of the images in factorial space (Frank & Van Heel, 1982a; Frank *et al.*, 1982).

The interpretation of correspondence analysis results usually involves three considerations: the *statistical significance* of the factorial axes, their *meaning*, and their *relevance* for the structural problem being investigated. Most of the rules that have been given for the assessment of the *significance* are rather empirical and based on usage. For instance, the examination of the histogram of eigenvalues ranked in decreasing order is helpful, since it sometimes shows a clear gap between a group of prominent factorial axes associated with the largest eigenvalues and those axes associated with the subsequent eigenvalues. Such a concentration of power in a few factorial axes may justify the elimination of all higher factors in the interpretation of the data set. Unfortunately, it frequently happens that no such gap is observed, and no clear rules are available as to where to break off the series of factors.

Lebart *et al.* (1977) demonstrated by simulation that there is a relationship between the size of the original data matrix, the ranking of a factorial axis, and the significance of the associated eigenvalue. Significance charts have been published by these authors. However, these charts must be used with caution, since they only represent guidelines to avoid gross misinterpretation. It is our experience in image processing applications that the number of eigenvalues passing Lebart *et al.*'s significance test is well in excess of the eight factors normally considered.

The *meaning* of a factorial axis pertains to the existence of a relationship between the variational pattern associated with this axis and the physical variation of the macromolecule giving rise to it. It is often easy to identify this relationship for the eigenvectors having the greatest eigenvalues. Usually, the interpretation process continues through the eigenvectors having successively small eigenvalues until a point is reached where a meaningful relationship can no longer be established. This identification process relies on the availability of extraneous information, e.g. tilt experiments, knowledge of morphological constraints, and existing structural models derived from other experiments.

Finally, the *relevance* of an axis, given it is statistically significant and has a *traceable* physical origin, depends on the goal of the study. For example, in a study aimed at elucidating the quaternary structure of a macromolecular assembly, the exact pattern of variation in the amount of peripheral stain surrounding the particle may be totally irrelevant, although this variation may constitute the largest fraction of interparticle variance (see Frank *et al.*, 1982).

The aim of this paper is to provide an objective aid for interpreting the information associated with any eigenvector by reconstituting totally or partially the original data using the associated factorial coordinates and eigenvalues. Variational patterns are thereby presented in their purest form.

Theory of image reconstitution

Let T be a matrix of n rows (the images) and p columns (the pixels) having as general term t_{ij} with $i=1, \ldots, n$ and $j=1, \ldots, p$.*

From this matrix, a scaled matrix F is derived by

$$f_{ij}=\frac{t_{ij}}{t_{..}} \tag{1a}$$

with

$$t_{..}=\sum_{i=1}^{n}\sum_{j=1}^{p} t_{ij}. \tag{1b}$$

The columns and rows of this new matrix are termed pixel profiles and image profiles, respectively. It is the scaling of the original data in such a fashion which permits the use of the chi-square distance. The choice of this distance, as opposed to the usual euclidean distance, has the distinctive advantage of assuring the distributional equivalence property (Lebart *et al.*, 1977).

For later use, we also introduce the marginal weights, row-wise and column-wise, which play an important role in the formulation of the theory:

$$f_{i.}=\sum_{j=1}^{n} f_{ij} \quad \text{and} \quad f_{.j}=\sum_{i=1}^{p} f_{ij}. \tag{1c}$$

Correspondence analysis of the new matrix, having element f_{ij}, leads to a new coordinate system whose axes point in the directions of main dispersion of the data cloud. The magnitudes of these components of dispersion are quantified in the eigenvalues λ_α. In this new system, each image profile $\{f_{ij}; j=1\ldots p\}$ is represented by a set of factorial coordinates $\{\psi_{\alpha i}; \alpha=1, \ldots, k\}$ and, similarly, each pixel profile $\{f_{ij}; i=1\ldots n\}$ by a set $\{\phi_{\alpha j}; \alpha=1\ldots k\}$.

It can be demonstrated (Greenacre, 1984) that the original data matrix can be retrieved, exactly or approximately, using these factorial coordinates. The general term t_{ij} of matrix T can be obtained using the formula:

$$t_{ij}=f_{ij}\cdot t_{..}$$

with

$$f_{ij}=f_{i.}f_{.j}\left\{1+\sum_{\alpha=1}^{K}\sqrt{\lambda_\alpha}\,\psi_{\alpha i}\,\phi_{\alpha j}\right\} \tag{2}$$

where the summation extends over all $K=\min(n, p)$ (exact retrieval) or over only a subset of the most significant factors $\alpha_s<K$ (approximate retrieval). Since the term 'reconstruction' is commonly reserved for the retrieval of three-dimensional information from two-dimensional projections, we use the term 'reconstitution' for the retrieval, fully or approximately, of an image from its eigenvector representation (Greenacre, 1984).

The equation for reconstitution (Eq. 2) can be formally rewritten in the following way:

$$f_{ij}=f_{i.}\sum_{\alpha=0}^{k}\sqrt{\lambda_\alpha}\,\psi_{\alpha i}\,\phi_{\alpha j} \tag{3}$$

* In this presentation we follow the notation of Lebart *et al.* (1977); see also Appendix in Bretaudiere *et al.* (1981).

by setting

$$\lambda_0=1 \quad \text{and} \quad \begin{array}{l}\phi_{0j}\equiv 1,\\ \psi_{0i}\equiv 1,\end{array}$$

which may be regarded as the eigenvalue and eigenvectors of 0th order. From Eq. (3) it can be concluded that each image

$$\{f_{ij};\ j=1\ldots p\}$$

is a linear combination of basic images, or *eigenimages* (Van Heel, 1985)

$$e_{\alpha j}=f_{.j}\ \phi_{\alpha j};\ j=1\ldots p \tag{4}$$

where $e_{0j}=f_{.j}$; $j=1\ldots p$ is the average image.

The eigenimage relating to factor α is equal to the average image multiplied, point for point, with the pixel coordinates along that factorial direction.

The coefficients of the linear combination are

$$C_{\alpha i} \stackrel{def}{=} f_{i.}\sqrt{\lambda_\alpha}\ \psi_{\alpha i} \tag{5}$$

so that the final form of the linear combination becomes

$$f_{ij}^{(\alpha_s)}=\sum_{\alpha=0}^{\alpha_s} C_{\alpha i}\ e_{\alpha j};\ j=1\ldots p. \tag{6}$$

These images we call 'images reconstituted up to eigenfactor α_s'. The form of Eq. (6) may be compared with that of a Fourier expansion

$$f_j^{(k_{max})}=\sum_{k-0}^{k_{max}} F_k \exp(2\pi ikj/N);\ \ j=1\ldots N \tag{7}$$

where a single function (or, in the 2D case, a single image) is built up by summing a set of up to N exponential orthonormalized functions, which are in this case complex. The greater the number of orders k_{max} that are included in expansion (7), the more closely $\{f_j^{(k_{max})}\}$ approximates the original function $\{f_j\}$.

Similarly, any ith image of the set $\{f_{ij}\}$ may be built up from elementary eigenimages. Low-order eigenimages represent the most important variational patterns common to all images. We may see the use of a small subset of lowest-order eigenimages in the reconstitution as a filtration akin to Fourier low-pass filtration. Eigenimages representing irrelevant features, such as the variation of the stain level surrounding the molecule (Frank *et al.*, 1982) may be simply left out (see Radermacher & Frank, 1985; Verschoor *et al.*, 1984).

Next we observe that the $C_{\alpha i}$, and underlying coordinates $\Psi_{\alpha i}$, need not relate to existing experimental images. Within the multidimensional volume enclosing the data cloud, we can formally reconstitute an image that is not experimentally available, by using any image coordinates $\psi_{\alpha i}$, $\alpha=1\ldots\alpha_s$. The marginal weight $f_{i.}$ needed for the reconstitution may be chosen to be

$$\bar{f}_{i.}=1/n \tag{8}$$

The possibility of reconstituting hypothetical images is intriguing, since it allows experimental images having a small probability to be 'predicted' by extrapolation from a surrounding continuum of images occurring with high frequency. Such a prediction could of course only be based on the knowledge that the associated physical phenomenon possesses a continuous variation, such as the appearance of projections of a three-dimensional object tilted in a continuous manner (Frank & Van Heel, 1982b; Verschoor *et al.*, 1984; Van Heel, 1984).

Since the variations among the realizations of a pixel $\{f_{ij};\ i=1\ldots n\}$ are often small compared with the (spatial) variations among averaged pixels, it is useful for a characterization of features expressed by factorial axes to look at individual eigenimages, or linear combinations of these, omitting the term e_{0j}, the average image. The contributions of individual eigenimages in the sum (Eq. 7) can then be much better ascertained. To avoid confusion, we will denote partially reconstituted images by exact reference to their factorial constituents, e.g. a (1,2)-reconstituted image is built up by combining eigenimages relating to axes 1 and 2, but omitting the average image.

Because of the normalization of profiles (Eq. 1) the pixel vectors are grouped irrespective of their averaged (or marginal) values. To provide a visual display of the importance of each pixel vector in defining a factorial axis, *importance* images were previously introduced by Van Heel & Frank (1981); Frank & Van Heel (1982a). Accordingly, the importance image expressing the αth factor is simply

$$g_{\alpha j}=e_{\alpha j}/f_{.j}=\phi_{\alpha j};\ j=1\ldots p. \tag{9}$$

Therefore the importance image is a representation of the values of the eigenimage relative to those of the average image. Since in a halftone representation of such an image it is difficult to identify the zero level, importance images are usually presented in pairs

$$g^{+}_{\alpha j}=\begin{cases} g_{\alpha j} & \text{for } g_{\alpha j}>0 \\ 0 & \text{elsewhere} \end{cases} \qquad g^{-}_{\alpha j}=\begin{cases} 0 & \text{for } g_{\alpha j}>0 \\ -g_{\alpha j} & \text{elsewhere} \end{cases} \tag{10}$$

RESULTS

Two examples are given: a synthetic model image set which entirely avoids the difficulties of attributing a physical meaning and judging the relevance, and a well-characterized set of molecule images which allows us to introduce these aspects of interpretation.

1. *Synthetic example*

In order to illustrate the utilization of the various types of image reconstitution, a cartoon has been used as example (Frank, 1982). Eight configurations of a face were synthesized by combining three features varying in a binary fashion: round or oval facial shape, right- or left-looking eyes, round or broad mouth (Fig. 1, row a). Nine realizations of each configuration were generated by adding gaussian noise with an amplitude that exceeded the signal amplitude by a factor of 2. The resulting seventy-two images show very blurry patterns making their features almost indistinguishable (Fig. 1, row b). The histogram of the first eight eigenvalues (Fig. 2) from a correspondence analysis of these images indicates that the first three factorial axes stand out from the rest. The assumption that the grouping along these three axes follows the grouping of facial features is confirmed by the analysis of factor maps 1/2, 2/3 and 1/3 (Fig. 3). These maps indicate a separation into eight clusters, each of which comprises faces originating from the same noise-free prototype.

We will demonstrate that image reconstitution facilitates the interpretation of the information carried by each eigenvector, and the re-synthesis of images (experimental or extrapolated) from a selected set of eigenvectors.

The meaning of the information carried by each eigenvector is determined by reconstituting its associated eigenimage. Images reconstituted by using a single factor α are distinguished only by the size of a multiplier, $C_{\alpha i}$. Those for which $C_{\alpha i}$ has the same sign are indistinguishable in a display. It is therefore practical to represent eigenimages $\{e_{\alpha j};\ j=1\ldots p\}$ as pairs $\pm e_{\alpha j}$ to typify the distinguishing features of images falling on the positive or negative side of the origin of axis α, respectively. Figure 4 (columns a and b) shows the pairs of eigenimages for each of the axes 1,

Fig. 1. (a) Eight model images created by 'binary' variation of three features: the shape of the face (oval *v.* round), the case of the eyes (left *v.* right), and the shape of the mouth (round *v.* broad). (b) Examples of noisy realizations (s/n ratio=0·5) of each image in (a). c, d, e: Stepwise reconstitution of the complete image by addition of eigenimages for factors 1, 1–2 and 1–3, respectively, to the average image.

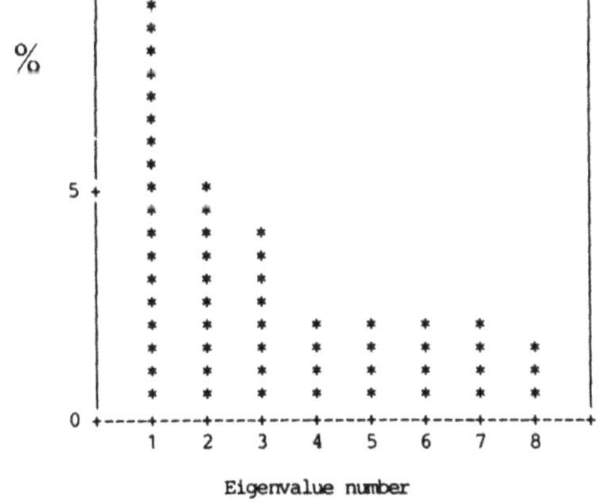

Fig. 2. Portion of the histogram of eigenvalues resulting from application of correspondence analysis to seventy-two (=8×9) noisy realizations of the model images. Note that the first three eigenvalues stand out from the rest of the histogram, suggesting that three factors are sufficient to express the systematic image variations.

Reconstitution of molecule images 7

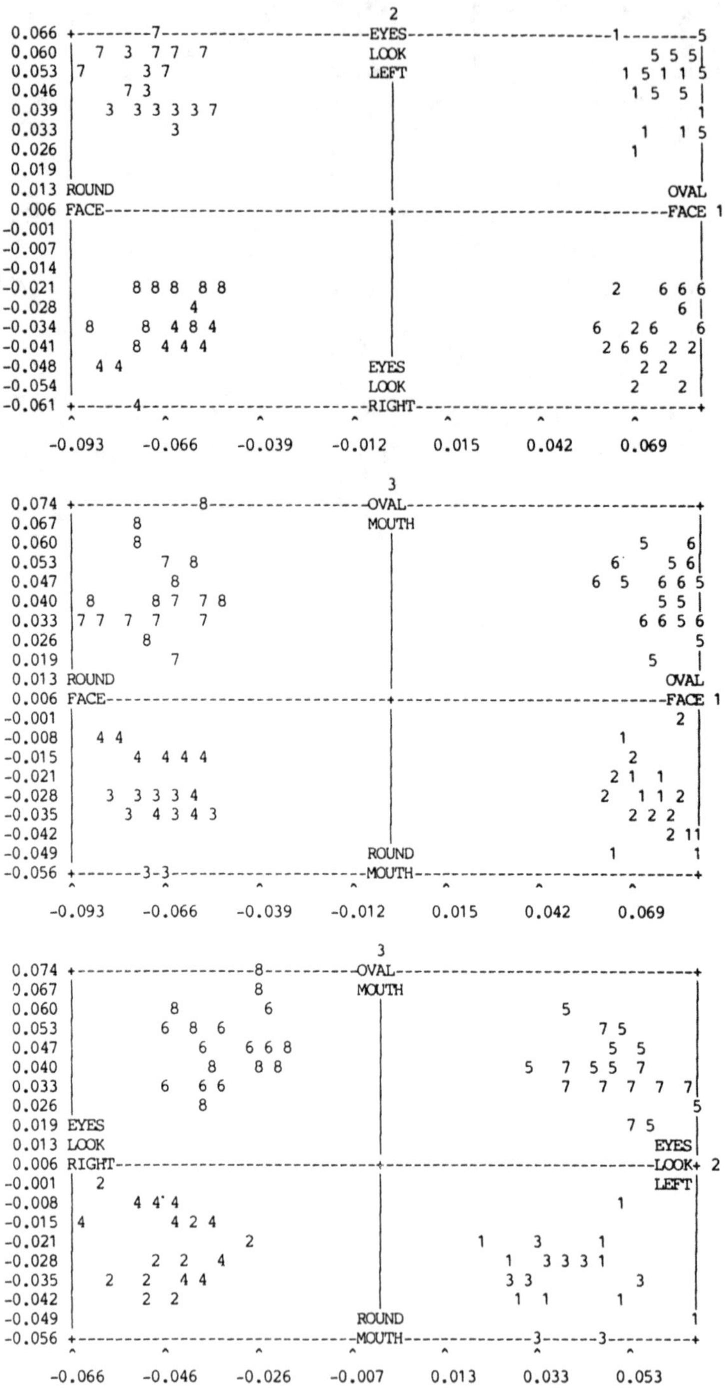

Fig. 3. Three factor maps showing the space spanned by the three predominant factors projected from three different directions. The separation along each factor is related to one of the independent feature variations, leading to eight clusters that each contain images originating from a single prototype.

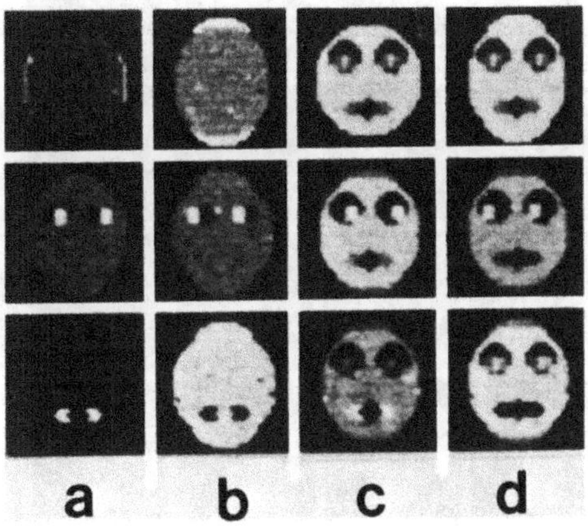

Fig. 4. Columns a, b (rows 1–3): eigenimage and its negative version for factor axes 1, 2 and 3, respectively. columns c, d (rows 1–3): six versions of the complete images, reconstituted at the two extremes of factor axes 1, 2 and 3, respectively. Note that the features expressed by the two axes not selected always appear averaged; e.g. the (0,1)-reconstituted image in row 1, column d, has a face with defined oval shape, but the features relating to axes 2 and 3, eyes and mouth, appear blurred.

2 and 3 (rows 1–3). The information explained by axis 1 pertains to the difference in the shape of the face, i.e. oval versus round. It also indicates that the rounder faces will be positioned toward the negative end of the axis and, conversely, the more oval faces toward the positive end. The information explained by axis 2 pertains to the direction in which the eyes look. The faces with the eyes shifted left and right will be found toward the negative and positive end of this axis, respectively.

Finally, the information explained by axis 3 pertains to the shape of the mouth. The faces with round and broad mouths are found toward the negative and positive end of this axis, respectively. It is interesting to note that the round portion of the mouth is common to both configurations—indeed, the only differentiating features are the corners on each side of the mouth. This explains why the round mouth appears (in Fig. 4, third row, column b) with the general pixel value of the face while the corners of the mouth show the variation.

It is important, before the groupings of the more complex halftone images analysed in electron microscopy can be interpreted (see below), to comprehend precisely how the variational patterns are expressed for the cartoon used in our example. Since image elements are independently analysed, without reference to their spatial interrelationships, all lateral changes of features or mass movements (e.g. those of the eyes) are in fact realized as composites of anticorrelated, spatially separated, in-place changes of mass. Thus, the movement of the pupil from left to right is seen as the disappearance of the pupil on the left and its correlated appearance on the right hand side of the eyeball. To understand the eigenimage associated with the eye movement (Fig. 4, second row, column a), one must make reference to the average image which has blurred pupils centered in each eyeball: from this average image, a right-focusing pupil is generated by subtracting mass on the left and adding it on the right. The eigenimage therefore shows a pair of anticorrelated masses in each eye, which are separated in horizontal direction.

The original images (Fig. 1a) may be retrieved from their noisy versions (Fig. 1b) by reconstitution up to factor 3. This is illustrated in Figs. 1(c)–1(e) where Eq. (6) was used for stepwise addition of the components from factors 1, 2 and 3 to the average image. The partially reconstituted images comprising factors (0,1), (0,1,2), and (0,1,2,3) in each column show

Fig. 5. Gallery of side views of the haemocyanin quarter-molecule from *Androctonus australis* (Lamy *et al.*, 1985) after image alignment and masking.

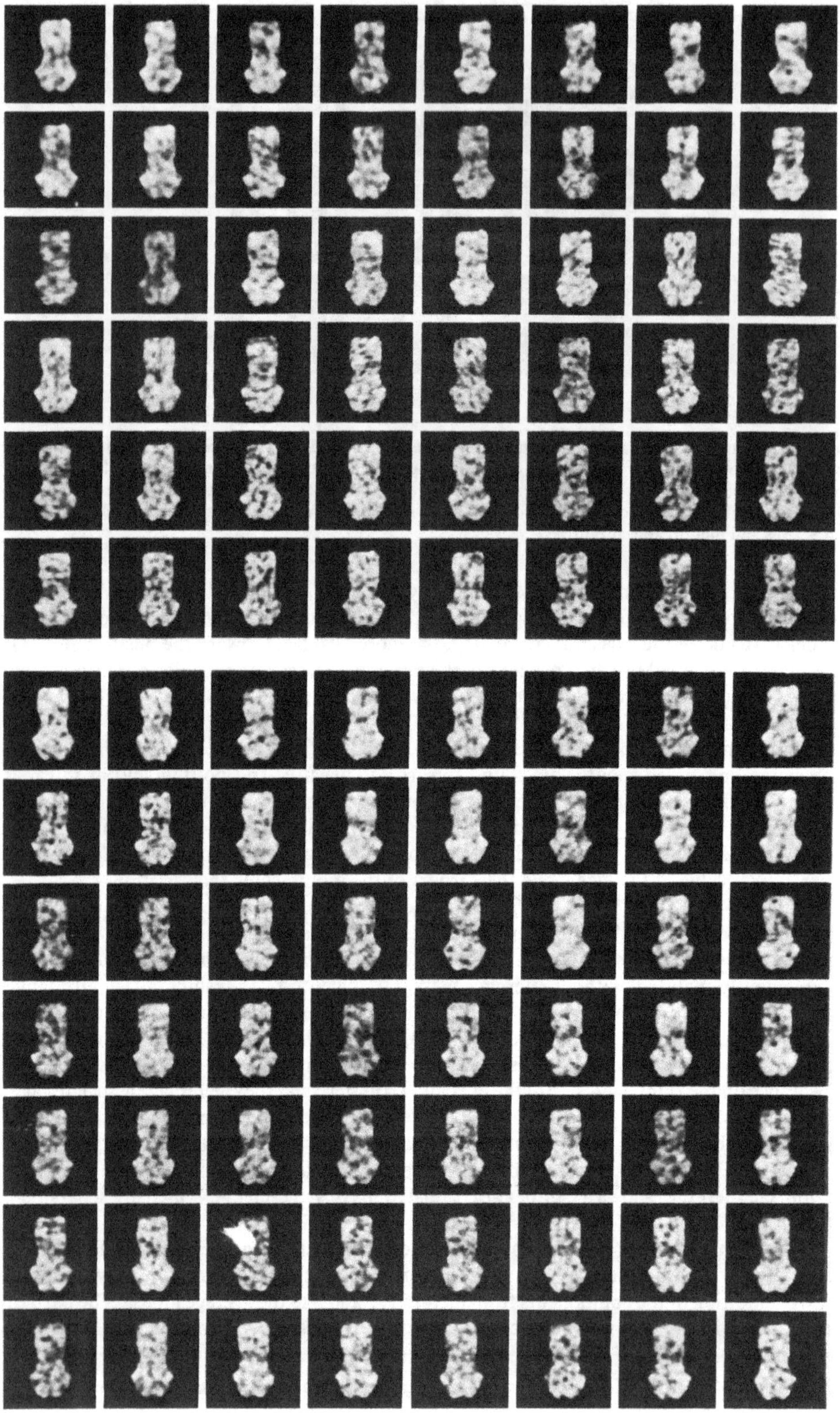

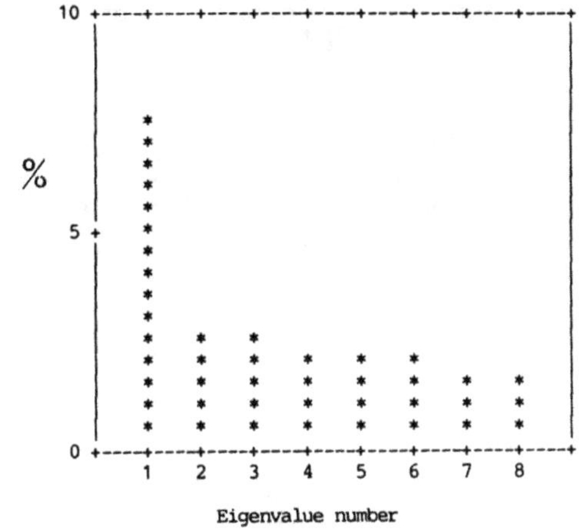

Fig. 6. Histogram of eigenvalues obtained by correspondence analysis of the image set in Fig. 5.

increasing similarity with their prototypes, expressing first the shape of the face, then the cast of the eyes, and finally the shape of the mouth. After the third component has been added, virtually noise-free versions of the original images are retrieved.

2. *Application to a macromolecular structure: haemocyanin*

Valuable information on the architecture of the haemocyanin molecule and its dissociation products has been gained from electron microscopy and multivariate classification (Van Heel & Frank, 1981; Bijlholt *et al.*, 1982; Van Heel & Keegstra, 1981; Frank *et al.*, 1982; Frank, 1982; Lamy *et al.*, 1982, 1985). We use here an example that exhibits a relatively simple statistical behaviour: the dodecameric quarter-molecule of a scorpion haemocyanin (Lamy *et al.*, 1982, 1985) which was first analysed with these techniques by van Heel & Keegstra (1981).

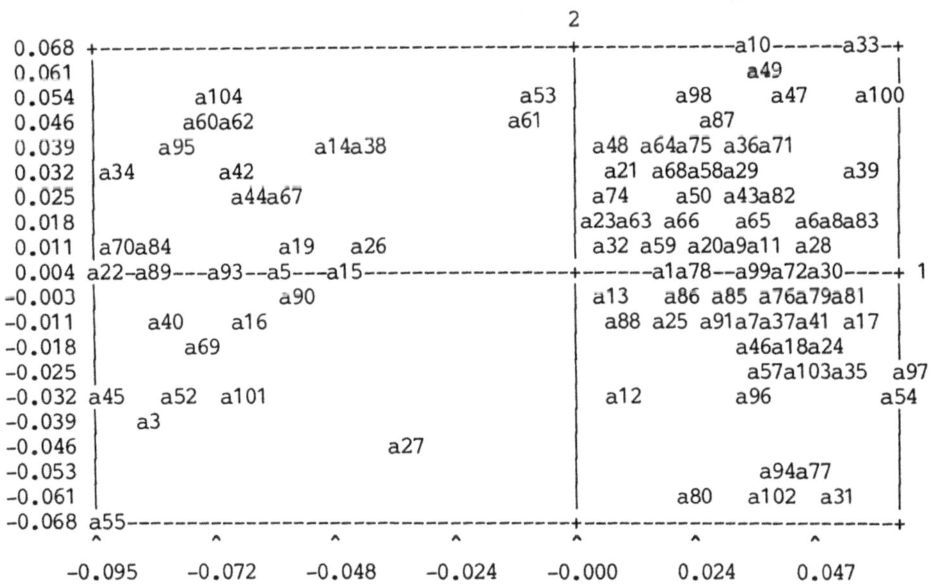

Fig. 7. Map of factor 1 *v.* 2 obtained in the correspondence analysis.

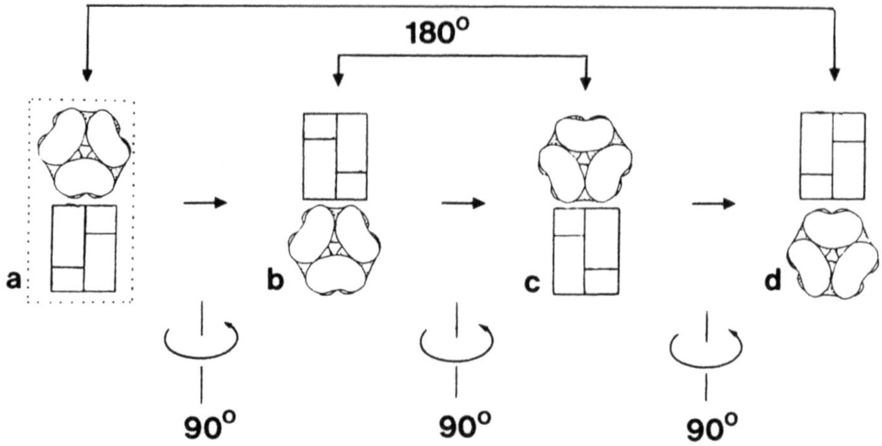

Fig. 8. Haemocyanin quarter-molecule in four stable positions (a–d) related by 90° rotations. The quarter-molecule is composed of twelve subunits arranged in two hexamers. The two hexamers lie side-by-side, related by 90° rotation around an axis perpendicular to their axis of three-fold symmetry. As a consequence, the four stable side views of the molecule fragment depicted here are pairwise identical, leading to two distinguishable mirror-related classes of views in the image analysis. (From Frank (1984), reproduced with permission by Elsevier Science Publishers B.V.).

Correspondence analysis of the set of images (Fig. 5) separates two classes unambiguously. The eigenvector histogram (Fig. 6) shows a wide separation between the first eigenvalue (7·6%) and the following ones (<3%) and the factor map shows the presence of two clusters along axis 1 (Fig. 7). The two side views of the molecule observed in the electron micrograph are related by mirror reflection; the molecule occurs in its 'flip' or 'flop' orientation. Due to the symmetry of the molecule, which is composed of two hexameric building blocks whose structures are indistinguishable at the electron microscopic resolution, the four stable orientations of the molecule related by 90° rotation produce the flip and the flop views alternately (Fig. 8).

The total average of all images, reproduced for later reference (Fig. 9a), shows rather blurred features, due to the superposition of two different structures. The main differences between the approximately mirror-related averages over the two clusters (Figs. 9b, 9c) are due to the asymmetry of mass distribution in the upper, rectangular-shaped region of the molecule projection: the masses A and B change positions from right to left and left to right, respectively. In the same way that the shift of the eye pupil in the previous example was expressed by a correlated subtraction and addition of mass at the source and the destination point, the shift of the two masses from Fig. 9(b) to Fig. 9(c) is expressed by two correlated subtractions and additions at the four mirror-related points A,A′,B,B′. Since the variations in the view associated with the flipping of the molecule are expressed by axis 1 in the correspondence analysis, the corresponding eigenimage (in its positive and negative versions; Figs. 9d and 9e) will most clearly reveal this shift of the masses.

Both the cluster averages (Figs. 9b, 9c) and the eigenimages (Figs. 9d, 9e) indicate that the precise mirror relationship does not actually hold. Evidence is provided by the following observations: (i) the sizes of the masses and their precise positions do not match; (ii) the lower, hexagonal region of the molecule projection shows a grouping of the six masses into three pairs, indicating a preference for visualization of the group of three subunits in contact with the support film; (iii) the statistical behaviour of the two sets of images is quite different: the molecule views of the type represented by Fig. 9(c) are seen more frequently, and they cluster more tightly than the type in Fig. 9(b). These effects may be due to preferential staining of the part of the molecule facing the support, and to differences in staining and adsorption of the

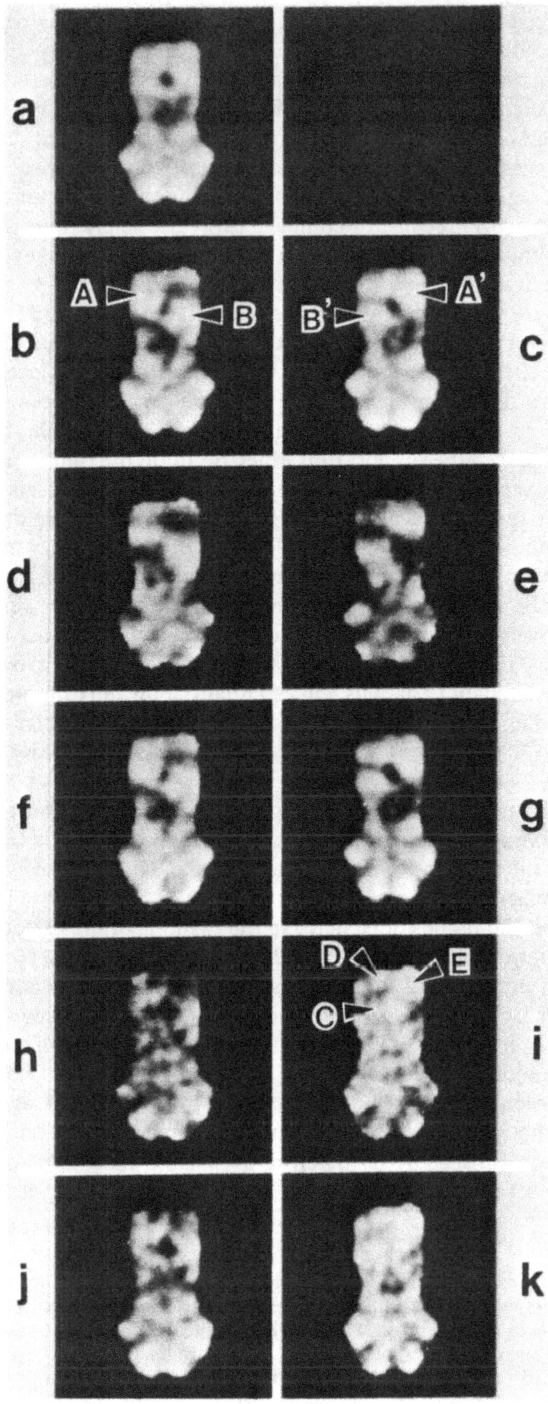
a
b
c
d
e
f
g
h
i
j
k
A
B
A'
B'
D
E
C

different subunits comprising the molecule. The same observation was made earlier by Van Heel & Keegstra (1981).

A pair of enhanced images of the molecule is obtained by using only axis 1 in a partial reconstitution (Figs. 9f, 9g), which differs from the eigenimages (Figs. 9d, 9e) by addition of the total average (Fig. 9a). The axis-1 coordinates of images lying closest to the cluster centres were used as coefficients $C_{\alpha i}$. It is instructive to compare these reconstitutions with the corresponding averages (Figs. 9b, 9c): some of the relevant features are more sharply defined in the reconstituted images. This is particularly noticeable in the comparison of Fig. 9(g) with Fig. 9(c). The cluster average contains higher-factor components that are unrelated to the flip/flop distinctions, while the (0,1)-reconstituted image expresses solely these distinctions.

The second and higher-order factors are related to complex changes in stain distribution, and do not necessarily carry clues to the molecule's architecture. As an example, factor 2 expresses the variations in the staining of three sites (marked C, D and E on the reconstituted images relating to factor 2, Figs. 9h–k). From the architecture of the molecule (see Fig. 8), it is quite clear that these sites are prone to staining variation as the overall level of staining changes: site C is part of the central protein-free tunnel of the hexamer, seen edge-on, which is only filled by stain when the molecule is immersed to well above 50% of its height. Sites D and E are the intersubunit gaps closest to the support film which occur on the right or left depending on the orientation of the molecule; both appear here simultaneously because factor 2 expresses the behaviour of *both* groups of molecule images. What we can infer from the eigenimages (Figs. 9h, 9i) and corresponding versions of the complete image (Figs. 9j, 9k) is that the invasion of the central channel by stain is correlated with strong staining of subunit gaps that are accessible from the side. This observation, although giving further support to existing theories about the mechanism of negative staining, does not significantly contribute to our understanding of the molecule's structure. Therefore, a representation of the molecule's two views that totally excludes such effects of stain variations will essentially show only the structure-related variation: such a representation is provided by the (0,1)-reconstitution (Figs. 9f, 9g).

CONCLUSIONS

The examples discussed demonstrate the power of correspondence analysis in separating classes of images in extreme noise situations, as well as the usefulness of reconstitution to guide the interpretation of the data, and to help to identify factors that are meaningful and relevant in the study of macromolecules.

A word of caution should be added here. In the example of the haemocyanin, as well as in the case of the synthetic image set, the individual factors found were seen to relate to strictly separate effects; e.g. mirroring of the mass distribution and movement of the eyes were each expressed by a single factor. There are, however, many situations where a physical effect may be expressed by two or more factors, in which case the interpretation of the results will be more involved than in the foregoing analysis.

Image reconstitution, in conclusion, is a means to precisely define the individual components of information that go into a representation of the important structure-related variations of the image set and its subsets. It is therefore a valuable tool both for the analysis of an experimental image set and for the re-synthesis of 'purified', intrinsically averaged images.

Fig. 9. (a) Total average of molecule images in Fig. 5. (b, c) Averages of molecule images falling into the left and right clusters in Fig. 7, respectively. A, B are two masses that change position from the flip (A, B) to the flop view (A′, B′) according to mirror reflection with respect to the long (vertical) axis of the molecule. (d, e) Positive and negative versions of the eigenimage relating to factor 1. (f, g) Two images (no. 85 and no. 93) representative of their classes, reconstituted as versions of the complete image, according to their positions along factor 1 axis only. (h, i) Positive and negative versions of the eigenimage relating to factor 2. In (i), the locations of strong stain variations expressed by factor 2 are marked as C, D and E. (j, k) Two extreme versions of the complete image, obtained by using extreme coordinates along both directions of factor 2.

This is in obvious analogy with Fourier filtration, which is similarly useful in re-synthesizing an image from significant components of structural information.

ACKNOWLEDGMENTS

We thank Ludovic Lebart and Alain Morineau, C.E.S.I.A., Paris, France, for assisting us with the formal description of reconstitution. We are grateful to Adriana Verschoor and Michael Radermacher for a discussion of this paper. This work was supported, in part (J.F.), by NIH grant 1R01 GM29169.

REFERENCES

Benzecri, J.P. (1969) Statistical analysis as a tool to make patterns emerge from data. In: *Methodologies of Pattern Recognition* (ed. by S. Watanabe). Academic Press, New York.

Benzecri, J.P. (1980) Introduction à la classification automatique d'aprés un example de données medicales. *Les Cahiers de l'Analyse des Données*, **5,** 311.

Bijlholt, M.M.C., Van Heel, M.G. & van Bruggen, E.F.J. (1982) Comparison of 4×6-meric hemocyanins from three different arthropods using computer alignment and correspondence analysis. *J. molec. Biol.* **161,** 139–153.

Bretaudiere, J.P., Dumont, G., Rej, R. & Bailly, M. (1981) Suitability of control materials. General principles and methods of investigation. *Clin. Chem.* **27,** 798–805.

Frank, J. (1982) Differential averaging of single molecule images using multivariate statistical classification. *Proc. 40th Ann. Meetg. EMSA* (ed. by G. W. Bailey), pp. 706–709. Claitor's Publishing Division, Baton Rouge.

Frank, J. & Van Heel, M. (1982a) Correspondence analysis of aligned images of biological particles, appendix to Frank, J., Verschoor, A. & Boublik, M. (1982) *J. molec. Biol.* **161,** 107–137.

Frank, J. & Van Heel, M. (1982b) Averaging techniques and correspondence analysis. *Proc. 10th Int. Congr. Electron Microsc. Hamburg*, Vol. 1, pp. 107–114.

Frank, J., Goldfarb, W., Eisenberg, D. & Baker, T.S. (1978) Reconstruction of glutamine synthetase using computer averaging. *Ultramicroscopy*, **3,** 283–290.

Frank, J., Verschoor, A. & Boublik, M. (1981) Computer averaging of electron micrographs of 40S ribosomal subunits. *Science*, **214,** 1352–1355.

Frank, J., Verschoor, A. & Boublik, M. (1982) Multivariate statistical analysis of ribosome electron micrographs (L and R lateral views of the 40S subunit from HeLa cells). *J. molec. Biol.* **161,** 107–137.

Frank, J., Verschoor, A. & Wagenknecht, T. (1985) Computer processing of electron microscopic images of single macromolecules. *New Methodologies in Studies of Protein Configuration* (ed. by T. T. Wu), pp. 36–89. Van Nostrand Reinhold, New York.

Greenacre, M.J. (1984) *Theory and Applications of Correspondence Analysis*. Academic Press, London.

Lamy, J., Sizaret, P.-Y., Frank, J., Verschoor, A., Feldmann, R. & Bonaventura, J. (1982) Architecture of *Limulus polyphemus* hemocyanin. *Biochemistry*, **21,** 6825–6833.

Lamy, J., Lamy, J., Billiald, P., Sizaret, P.-Y., Cave, G., Frank, J. & Motta, G. (1985) Approach to the direct intramolecular localization of antigenic determinants in *Androctonus australis* hemocyanin with monoclonal antibodies by molecular immunoelectron microscopy. *Biochemistry*, **24,** 5532–5542.

Lebart, L., Morineau, A. & Tabard, N. (1977) *Techniques de la Description Statistique*. Dunod, Paris.

Radermacher, M. & Frank, J. (1985) Use of nonlinear mapping in multivariate image analysis of molecule projections. *Ultramicroscopy*, **17,** 117–126.

Van Heel, M. (1984) Multivariate statistical classification of noisy images (randomly oriented biological macromolecules). *Ultramicroscopy*, **13,** 165–184.

Van Heel, M. (1985) Finding the characteristic views of macromolecules in extremely noisy electron micrographs. In: *Pattern Recognition in Practice II* (ed. by E. S. Gelsema and L. N. Kanal). North Holland Publ. Co., Amsterdam.

Van Heel, M. & Frank, J. (1981) Use of multivariate statistics in analyzing the images of biological macromolecules. *Ultramicroscopy*, **6,** 187–194.

Van Heel, M. & Keegstra, W. (1981) IMAGIC: a fast, flexible and friendly image analysis software system. *Ultramicroscopy*, **7,** 113–130.

Van Heel, M., Bretaudiere, J.P. & Frank, J. (1982) Classification and multi-reference alignment of images of macromolecules. *Proc. 10th Int. Cong. Electron Microsc. Hamburg*, 1982, Vol. I, pp. 563–564.

Verschoor, A., Frank, J., Radermacher, M., Wagenknecht, T. & Boublik, M. (1984) Three-dimensional reconstruction of the 30S ribosomal subunit from randomly oriented particles. *J. molec. Biol.* **178,** 677–698.

Verschoor, A., Frank, J., Wagenknecht, T. & Boublik, M. (1986) Computer-averaged views of the 70S monosome from *Escherichia coli*. *J. molec. Biol.* **187,** 581–590.

Journal of Microscopy, Vol. 146, *Pt* 2, *May* 1987, *pp.* 113–136.
Received 2 *July* 1986; *revised* 17 *November* 1986; *accepted* 10 *December* 1986

Three-dimensional reconstruction from a single-exposure, random conical tilt series applied to the 50S ribosomal subunit of *Escherichia coli*

by M. RADERMACHER, T. WAGENKNECHT, A. VERSCHOOR *and* J. FRANK, *Wadsworth Center for Laboratories and Research, New York State Department of Health, Empire State Plaza, Albany, NY* 12201, *U.S.A.*

KEY WORDS. Three-dimensional reconstruction, 50S ribosomal subunit, low-dose electron microscopy, image processing.

SUMMARY

We present a new reconstruction method that takes advantage of the fact that many biological macromolecular assemblies show a preferred orientation with respect to the plane of the specimen grid in the electron microscopic preparation. From one micrograph taken of such a specimen tilted by a large angle, a conical tilt series with random azimuthal angles can be extracted and used for a three-dimensional reconstruction. Our technique allows the determination of the molecular structure under low-dose conditions, which are not achievable with reconstruction methods that use conventional tilt series. The reconstruction method combines a number of existing image processing techniques with a newly developed weighted back-projection algorithm designed for three-dimensional reconstruction from projections taken with arbitrary projecting directions. The method is described as it was applied to the three-dimensional reconstruction of the structure of the 50S ribosomal subunit of *Escherichia coli* (*E. coli*).

INTRODUCTION

The three-dimensional structure of biomolecular assemblies can be determined from a series of projections taken in different directions in the electron microscope (DeRosier & Klug, 1968; Hoppe, 1969; Radon, 1917). The necessary projections may be obtained by tilting the object around one axis in small angular intervals. For a complete determination of the structure a tilt range of 180° would be necessary. In practice, the possible tilt angles extend from −60° to +60°. Outside of this angular range difficulties arise because the images of the molecules begin to be obscured by the bars of the specimen grid or by the specimen holder. Additionally, the thickness of the specimen traversed by the electron beam becomes so large that, at a typical accelerating voltage of 100 kV, the image no longer represents a projection of the object (e.g. Cohen *et al.*, 1984). The missing angular range of the available data caused by the limited tilt angle results in an elongation of the reconstructed three-dimensional image point in the direction perpendicular to the specimen plane, and in artefacts surrounding the structure (Hoppe *et al.*, 1976a). These effects can be reduced if data from a larger range of directions are available. This has led to the use of conical instead of single axis tilting geometries and the development of the appropriate reconstruction algorithms. In a conical tilt series the object is

© 1987 The Royal Microscopical Society

tilted by a large fixed angle (typically 45–60°) and then rotated in this tilted plane by small angular increments. The increased range of views that can be achieved in this geometry results in a reduction of the elongation of the image point and in a strong reduction of artefacts (e.g. Chiu *et al.*, 1979; Radermacher, 1980).

Many three-dimensional reconstructions have been done from actual tilt series, making use of up to forty-one images (Hegerl *et al.*, 1984) taken of the same specimen area. Although there has been an extensive discussion about the dose necessary for a three-dimensional reconstruction from a tilt series (e.g. Hoppe *et al.*, 1973; Saxberg & Saxton, 1981; van Heel, 1986), an accumulated dose as low as the theoretical minimum dose (on the order of 10×10^3 el/nm^2, Saxberg & Saxton 1981) has never been achieved with a conventional tilt series. When accumulated doses have been reported, they are in the range of 43×10^3 el/nm^2 (Hoppe *et al.*, 1976b) to 500×10^3 el/nm^2 (Hoppe & Grill, 1977; see also Knauer *et al.*, 1983, and Oettl *et al.*, 1983). When no figures are given (e.g. Dover *et al.*, 1981; Subirana *et al.*, 1985; Skoglund *et al.*, 1986), it can be assumed that normal-dose images (typically with 20×10^3 el/nm^2 per projection) were used and that the accumulated doses must therefore have been in or above the higher range of reported doses. The number of images needed for a reasonable resolution in three dimensions is already large for a single-axis tilt series (Crowther *et al.*, 1970) and increases about twofold for a conical tilt series (Radermacher, 1980; Radermacher & Hoppe, 1980). Biological specimens are extremely sensitive to damage by the electron beam (e.g. Scherzer, 1970; Glaeser, 1975; Cosslett, 1978), and the doses used in conventional tilt series are far above the limit at which substantial damage by the electron beam occurs even for negatively stained specimens (Williams & Fisher, 1970; Unwin, 1974). New approaches had to be found that allowed a three-dimensional reconstruction of single asymmetrical particles to be achieved without destroying the object in the course of data collection. To date it has been possible to minimize the electron dose, or at least the effects of radiation damage, in two situations. If the structure possesses internal symmetry, one image can provide all symmetry-related projections, thus reducing the number of exposures needed for the collection of projections over a sufficient angular range (e.g. Crowther *et al.*, 1970). If the specimen is a two-dimensional crystalline sheet, extremely low exposures are possible, because the image can be enhanced by averaging over the repeat of the array using either Fourier filtration (Erickson & Klug, 1970; Nathan, 1971) or correlation averaging techniques (Frank & Goldfarb, 1980; Saxton, 1980). As well as enhancing the signal-to-noise ratio, these techniques minimize the effects of radiation damage, which can be assumed to be statistically distributed over the area of the crystal and will thus be averaged out. However, since many biologically interesting macromolecular structures cannot be crystallized, this approach has been feasible only in a very limited number of applications.

The only way to collect a sufficient amount of data for a three-dimensional reconstruction of non-crystalline, asymmetric biomolecular assemblies under low-dose conditions is to combine images from a large number of particles. One such approach was tried by van Heel (1983) and Verschoor *et al.* (1984), who reconstructed the procaryotic 30S ribosomal subunit from single untilted micrographs. The 30S subunit lies with its long axis virtually parallel to the specimen plane and rotates around its long axis, inherently creating a single-axis tilt series. By assignment of angles to each image, which was done by a combination of multivariate statistical image analysis with interconversion studies (Wabl, 1973), the views of the 30S subunit could be used for a three-dimensional reconstruction. The disadvantage of this technique, however, lies in the limited accuracy of the angle assignment. Because of this limitation, no attempt has yet been made to apply the technique to low-dose data, although this is in principle feasible.

Our new technique also combines images of a large number of particles (several hundred) that can be extracted from a small number of micrographs, but it differs from the aforementioned technique in that it allows the exact measurement of all angles, leading to a much higher resolution in the three-dimensional reconstruction. The technique is applicable to the large group of biological particles that exhibit a preferred orientation in the plane of the specimen

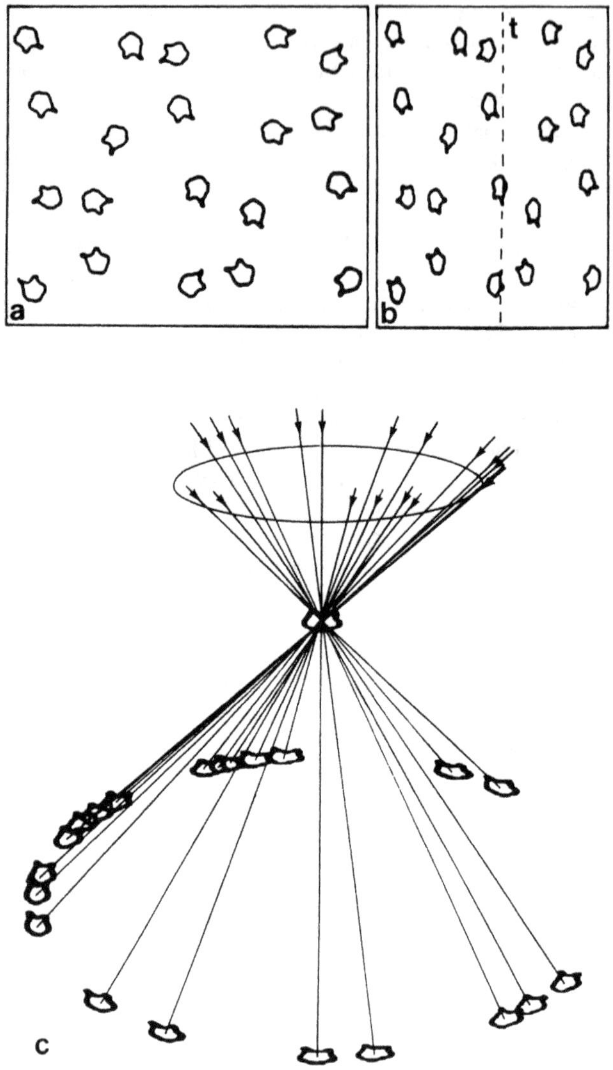

Fig. 1. Illustration of the basic principle of the reconstruction scheme: (a) view of a specimen with randomly oriented 50S particles lying flat in the plane of the specimen; (b) projection of the specimen in (a), tilted by 50°. The images which can be extracted from of the tilted image (b) form the conical tilt series shown in (c) equivalent to a tilt series of a single particle with random projection directions, all lying on the surface of a cone.

grid and random in-plane orientations (e.g. Frank *et al.*, 1978, 1981b; van Heel & Frank, 1981; Lamy *et al.*, 1982; Verschoor *et al.*, 1985). Indeed, at least subsets of particles exhibiting preferential orientations can be found in most cases, and throughout the history of electron microscopic structural research such typical views have been used to characterize the shape of molecules.

If such a specimen is tilted by one fixed angle, the set of particle images constitutes a conical tilt series with arbitrary azimuthal orientations (Fig. 1) (Frank *et al.*, 1978; Frank & Goldfarb, 1980). Only one exposure of the specimen area is necessary for collection of a conical tilt series that covers the complete 360° range. By combining data from several micrographs, an angular sampling can be achieved that is sufficiently fine for a high-resolution structure determination.

The dose that each particle receives can be kept as low as in two-dimensional low-dose electron microscopy. To measure the in-plane rotations of the particles, which are the azimuthal angles in the conical tilt series, a second exposure of the same specimen area is taken without tilt, but this later micrograph is not used in the reconstruction, so that the radiation damage encountered during this second exposure does not affect the final result. Thus our technique allows the radiation damage of the specimen to be minimized while achieving a sampling statistic sufficient for a three-dimensional reconstruction with high resolution and high signal-to-noise ratio.

There is an important difference between the method of data collection employed in our method and that employed in conventional tilt series where equal angular increments are used, namely, the randomness of the azimuthal angles. The reconstruction algorithms previously available were either developed for reconstructions from tilt series with regular angular increments or were very inefficient when dealing with a large number of input projections. We developed a new reconstruction algorithm that can be used for reconstructions from any arbitrary, irregular tilt geometry and is based on the principle of the weighted back-projection schemes previously developed for single-axis (Gilbert, 1972) and conical tilting geometries (Radermacher & Hoppe, 1980; Radermacher, 1980) with equal angular increments.

Three three-dimensional reconstructions of the procaryotic 50S ribosomal subunit have been reported to date. One reconstruction from a single-axis tilt series of single 50S particles from *E. coli* has been reported, but a high cumulative electron dose (160×10^3 el/nm^2) was used in this study (Oettl *et al.*, 1983; Hoppe, 1983). Three-dimensional reconstructions of the 50S ribosomal subunit done by crystallographic methods are restricted in resolution by the limited long-range order of the crystals. One such three-dimensional reconstruction of the 50S ribosomal subunit of *Bacillus stearothermophilus* has been done (Arad *et al.*, 1984), and a reconstruction of the 50S ribosomal subunit of *E. coli* by crystallographic methods is mentioned by Lake (1985).

The present paper describes all steps involved in carrying out a three-dimensional reconstruction from a single-exposure, random conical tilt series. The method has been applied to the three-dimensional reconstruction of the 50S ribosomal subunit of *E. coli*, and the individual steps are described in terms of this particular application, without loss of generality.

The reconstruction scheme starts with the determination of the direction of the tilt axis, which was done by triangulation. The method for determining the rotational alignment of particles was modified to deal with problems encountered with almost globular molecules. Large-scale organizational problems in the data entailed the establishment of a consistent bookkeeping system that ensures that each particle in the tilt series can be uniquely identified with its untilted counterpart throughout the procedure, even if some particles have already been excluded from further processing in early steps of the data analysis. By such an elimination of unsuitable particles in the early stages of the procedure, the processing time can be shortened significantly. All programs to carry out the reconstruction were implemented in the SPIDER image processing system (Frank *et al.*, 1981a), and a number of smaller modifications were made to the existing system so as to facilitate the application of the complete method to any suitable macromolecule. Extensive procedures were written using the command language of the SPIDER system, so that large parts of the reconstruction can now be carried out with minimal supervision.

OUTLINE OF THE METHOD

Following is a brief outline of the entire procedure. Subsequently each step is explained in more detail.

Pairs of micrographs are taken, one image of the specimen tilted by an angle ϑ between 45° and 60° (in our study 50°), and subsequently one of the non-tilted specimen (Fig. 2: 1). The particle images in the tilted view are used for the three-dimensional reconstruction, and the images in the untilted view are used to determine the azimuthal angles of the particles.

The direction of the tilt axis is determined from measurements on photographic prints of

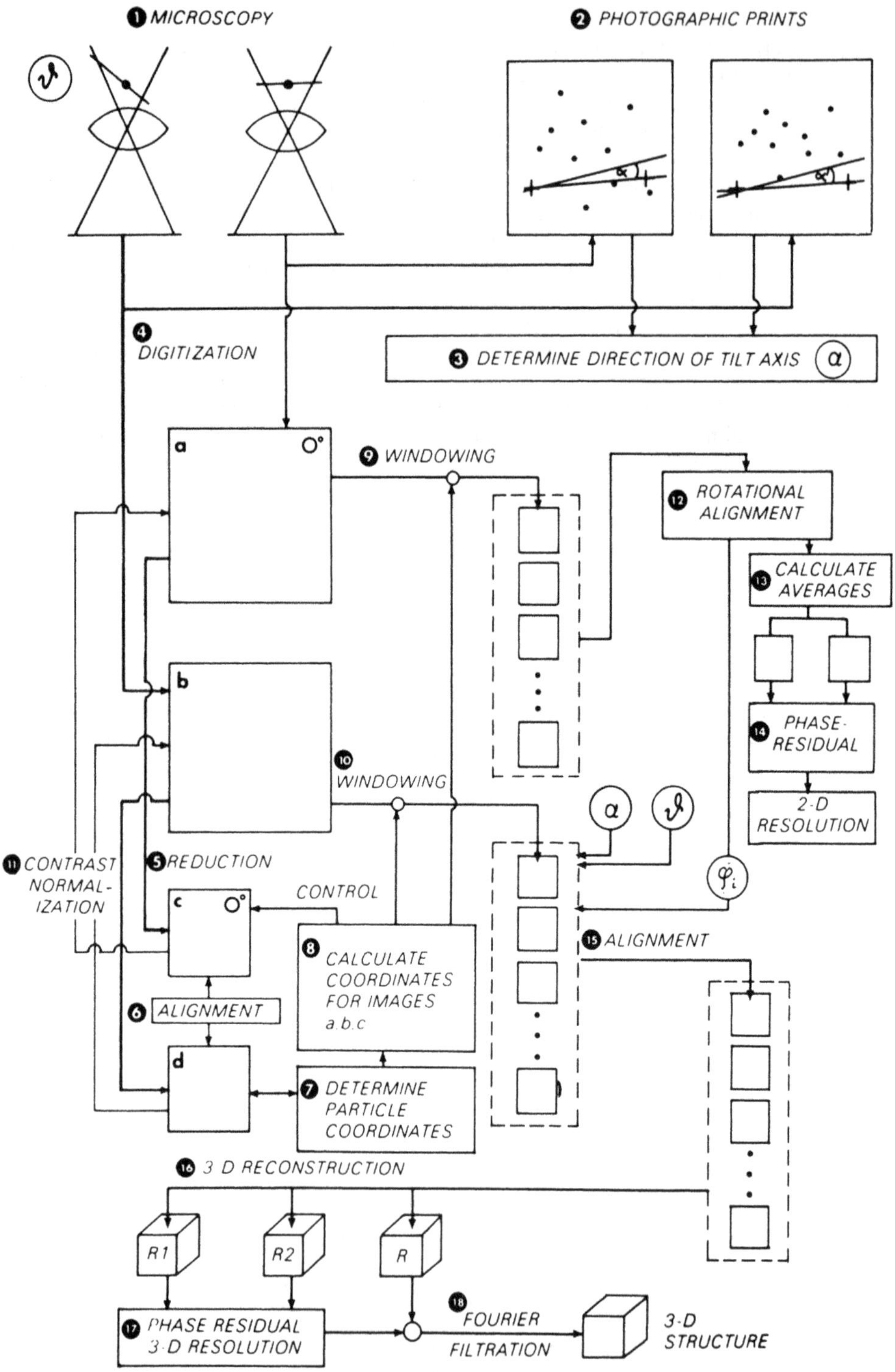

Fig. 2. Flow diagram outlining the complete reconstruction procedure. For explanations see text.

the micrographs relative to a line connecting two points in the image that are either marked with a sharp knife onto the emulsion or exist as clearly visible small features in the micrograph (Fig. 2: 2, 3). These marks are also used later to map the coordinates from the photographic prints onto the coordinates in the digitized images.

The micrographs of the tilted specimen show a wide range of defocus values varying perpendicular to the tilt axis. A portion of the specimen area usually lies in overfocus followed by a small strip where the specimen is in focus followed by an area of underfocus. To determine the different areas of defocus the micrographs are analysed in an optical diffractometer. Since the application of two-dimensional restoration techniques (Hoppe, 1971) was considered to be too laborious, only particles in the area of underfocus could be used for the three-dimensional reconstruction (Typke & Radermacher, 1982).

The parts of the micrographs that show the area of the specimen found to be in underfocus in the tilted image and that include the two marks (Fig. 2: 4) are digitized. Subsequently, a reduced version of each image is generated (Fig. 2: 5). Thus, four images are created representing the same area in the tilted and untilted specimen in both original and reduced versions. The reduced images are used in parallel to the large versions whenever operations involving the complete micrograph can be done on the small versions, and the results are later transferred to the large images. Such operations include alignment of large specimen areas, interactive particle selection, and visualization of the complete micrograph on a limited-sized (in our installation 640×512) raster display.

The reduced images of the tilted and untilted specimen are aligned by a cross-correlation procedure (Fig. 2: 6). The particle coordinates are measured interactively in one of the reduced images (Fig. 2: 7), which may be either the tilted or the untilted version. The coordinates of the particles are stored in lists, the so-called document files, used in the SPIDER image processing system.

Using the calculated direction of the tilt axis, the tilt angle known from the goniometer reading, and the location of the common origin determined from the alignment of the two reduced images, the particle coordinates in the second reduced image, as well as in both originally scanned micrographs, are calculated (Fig. 2: 8) and stored in three additional document files. Each particle image is numbered, and the same number is used as a key for the line in the document file which contains data associated with that specific particle. Care has been taken that the image number of a selected particle remains identical for its 0° image and its tilted counterpart throughout the procedure. If an image is to be excluded from further processing this is done by setting a flag in the corresponding line of the document file.

Using the four sets of coordinates, frames are drawn around the particles in the reduced images for visual control, and the single particle images are windowed out of both large micrographs (Fig. 2: 9, 10).

Since data from several micrographs are used, the contrast must be normalized. For this purpose frames are windowed out of the two reduced micrographs in areas without particles so that the influence of the stain surrounding the particles is avoided. The average optical density of these windows is calculated, and this value is used for normalization (Fig. 2: 11) using the formulas valid for phase-contrast images (e.g. Hoppe & Hegerl, 1980).

The images of the tilted particles are then stored for later processing. The particles from the 0° view are rotationally aligned (Fig. 2: 12), and two independent averages are calculated (Fig. 2: 13) to determine the limit of the resolution that can be expected under optimum conditions in a three-dimensional reconstruction (Fig. 2: 14). The angles ϕ_i found in the rotational alignment procedure are saved to be used as the azimuthal angles of the particles in the tilted view. These angles and the fixed tilt angle ϑ are stored in the headers of the tilted image files as well as in a document file. At this stage the tilted images constitute a conical tilt series in which all parameters are known.

The images of the tilted particles are subsequently aligned to a common origin (Fig. 2: 15) by a procedure using cross-correlations. The aligned tilted views are then processed with a weighted back-projection algorithm designed for arbitrary geometry, resulting in the three-dimensional reconstruction (Fig. 2: 16). Three reconstructions are calculated, one from the complete set of particle images and two from subsets that each cover the complete angular range but use only complementary halves of the complete particle set. The two independent reconstructions are used to determine the cross-resolution by phase-residual calculations (Fig. 2: 17), and the reconstruction from the complete set, representing the final result, is Fourier filtered to this resolution (Fig. 2: 18). This latter low-pass filtered reconstruction, which is represented both by halftone images of a series of slices through the three-dimensional volume and as surface displays, is the basis of the structural interpretation.

In the following sections the single steps are explained as they were applied to the three-dimensional reconstruction of the 50S ribosomal subunit.

APPLICATION OF THE METHOD TO THE 50S SUBUNIT

Electron micrographs show the 50S ribosomal subunit of *E. coli* in two preferred orientations, the 'crown' view and the 'kidney' views (Tischendorf *et al.*, 1974). In the crown view the orientation of the main part of the particle with respect to the specimen grid is well defined (Kiselev *et al.*, 1983; Verschoor *et al.*, 1985), and the analysis was based on this view. Five pairs of micrographs, the first one of each pair taken with the specimen grid tilted by 50° and the second taken without tilt, were used in the present analysis.

SPECIMEN PREPARATION AND MICROSCOPY

The ribosomes were isolated from *E. coli* strain MRE600 (Staehelin & Maglott, 1971), resolved into subunits by centrifugation through 10–30% sucrose gradients (Beckman SW41 rotor, 15·5 h at 25,000 rpm) using a buffer of 10 mM Tris, 1 mM magnesium acetate, 100 mM KCl, pH 7·5. For electron microscopy the subunits were diluted in 10 mM Tris, 10 mM NH_4Cl, 10 mM $MgCl_2$, 0·05 mM EDTA, pH 7·5 to OD_{260}=0·28 and applied to 300-mesh copper grids using a carbon sandwich technique (Tischendorf *et al.*, 1974), with 0·5% uranyl acetate as a negative stain.

Each pair of micrographs, at 50° and untilted, was recorded at an instrumental magnification of 49,000× at 100 kV under minimal-dose conditions (*c.* 1×10^3 el/nm^2; Fig. 3).

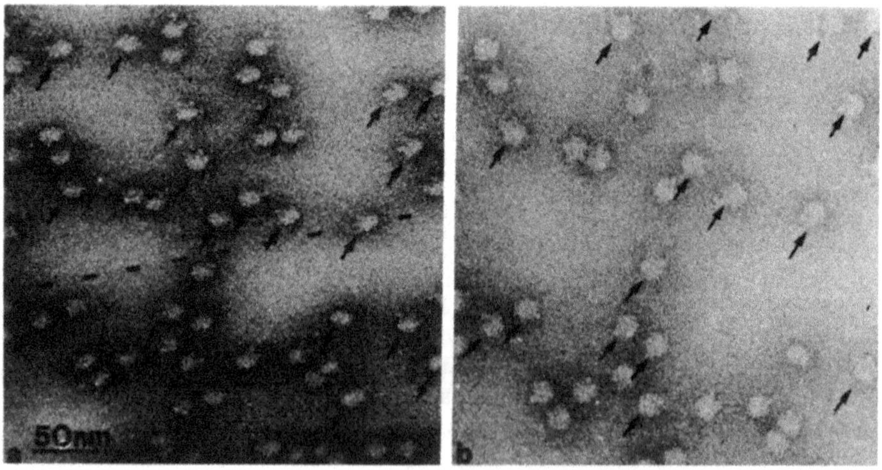

Fig. 3. Portion of the particle fields: (a) projection of the specimen tilted by 50°, the particles initially selected are marked at the lower left corner; (b) the same field without tilt, the particles finally used are marked. Only the area common to both micrographs and lying in the underfocus range in the tilted view (as determined by optical diffraction) was used for evaluation to avoid errors due to the sign change of the transfer function at low spatial frequencies. Average densities calculated in areas showing only carbon foil were used to normalize the contrast of the images.

The micrographs were analysed in an optical diffractometer to determine the defocus range. Unless two-dimensional restoration techniques are used (Hoppe, 1971), only the images of the tilted particles that lie in the underfocused range of the micrograph can be combined in a three-dimensional reconstruction (Typke & Radermacher, 1982). However, restoration techniques were not used because it seemed less time-consuming to ensure during microscopy that the major part of the image of the tilted specimen was in the underfocused range. All micrographs used contained an area with zero defocus. Thus, an upper limit of defocus contained in the image could be calculated using the tilt angle (50°), the magnification (49,000×), and the size of the micrograph (82 mm). With these values the maximum defocus is 1·3 μm. At this defocus the first zero of the transfer function is at 0·45 nm^{-1} (spherical abberation of the objective lens 2 mm, accelerating voltage 100 kV), which corresponds to a resolution of 2·2 nm. This resolution is better than the resolution that we expected in the three-dimensional reconstruction (see below), and thus a correction of the images for the transfer function was not deemed necessary.

The actual tilt angle of the specimen was calculated by comparing equivalent areas in the images of the tilted and untilted specimen. As the area decreases with the cosine of the tilt angle a slight deviation of the untilted image from the exact 0° position does not cause any major error in this calculation. The calculated tilt angle was within 0·3° of the angle obtained from the goniometer reading.

TILT AXIS DETERMINATION

The direction of the tilt axis was determined by triangulations among twenty-five to forty points defined by small features recognizable in both micrographs of each pair. First, two points M_1 and M_2, spaced rather far apart, were marked on the emulsion of each micrograph, thus defining a reference line r close to (within 20°) the visually estimated direction of the tilt axis. The micrographs were then printed using the same magnification for each pair, and recognizable features F_i were marked in the print of the untilted image and the tilted image. The coordinates of the two marks M_1, M_2 and of the selected points were subsequently measured with a sonar tablet. M_1, M_2 were also used to uniquely relate the coordinates measured with the sonar tablet to the coordinates in the digitized image. The direction of the tilt axis was determined by comparison of the lengths $l=\overline{F_iF_j}$ of corresponding lines in the tilted and untilted image (Fig. 4).

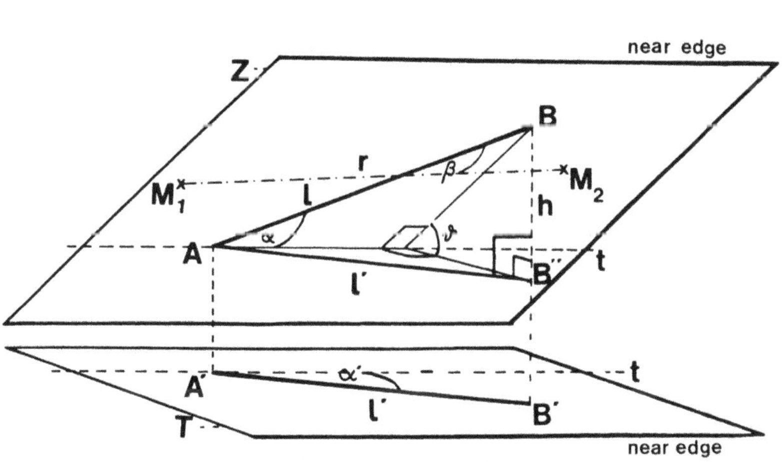

Fig. 4. Determination of tilt laxis. The tilt axis can be determined by comparison of corresponding lines in the image of the untilted specimen (Z) and the image of the tilted specimen (T). ϑ tilt angle. M_1 and M_2 points defining the reference line r. $l=\overline{AB}$ line connecting two selected points in the specimen and $l'=\overline{A'B'}$ corresponding line in the tilted image. α angle between l and the tilt axis t. β angle between the reference line r and l.

Let $l=\overline{AB}$ be the line connection two points in the specimen plane which projects onto $l'=\overline{A'B'}$ in the image of the tilted specimen. Let ϑ be the tilt angle, α between l and the tilt axis, α' the angle between l' and the tilt axis, and h the line $\overline{B''B}$, then

$$h=l \cdot \sin\alpha \, \sin\vartheta \tag{1}$$

and from the right triangle A, B, B″:

$$h^2+l'^2=l^2 \tag{2}$$

By combining (1) and (2) we obtain:

$$l^2-l'^2=l^2 \sin^2\alpha \, \sin^2\vartheta \tag{3}$$

and thus

$$\alpha=\pm\arcsin\left(\frac{\sqrt{l^2-l'^2}}{l \sin\vartheta}\right) \tag{4}$$

The angle α' can be calculated by combining (2) with:

$$h=l' \sin\alpha' \, \tan\vartheta \tag{5}$$

which results in the formula for α'

$$\alpha'=\pm\arcsin\left(\frac{\sqrt{l^2-l'^2)}}{l' \tan\vartheta}\right) \tag{6}$$

Formulas (4) and (6) determine the angles between l, l' and the direction of the tilt axis. If β is the angle between l and r (or β' the angle between l' and r'), the angles γ and γ' between the reference line and the tilt axis are:

$$\gamma=\alpha-\beta; \quad \text{and} \quad \gamma'=\alpha'-\beta' \tag{7}$$

The error of the calculated tilt angle from one measurement is

$$\Delta\alpha=\left|\frac{\partial\alpha}{\partial l}\right|\Delta l+\left|\frac{\partial\alpha}{\partial l'}\right|\Delta l' \tag{8}$$

$$=\frac{1}{\sqrt{\sin^2\vartheta-(l^2-l'^2)/l^2}}\left\{\frac{l'^2}{l^2\sqrt{l^2-l'^2}}\Delta l+\frac{l'}{l\sqrt{l^2-l'^2}}\Delta l'\right\} \tag{9}$$

As l' is a function of l, α, and ϑ, the error $\Delta\alpha$ also depends on these variables (Fig. 5).

To minimize the error, the arithmetic mean of all angles resulting from all possible connections between all marks was calculated. Lines l (l') that formed an angle of less than 20° or larger than 70° with the estimated tilt axis $\overline{M_1 M_2}$ were excluded from the calculations, as were lines shorter than 5 cm (estimated accuracy of the sonar tablet measurements, 1 mm). With these measures the accuracy of the tilt axis determination was estimated to be better than 1°, which was confirmed by repeated measurements.

DIGITIZATION

The micrographs were digitized with a spot size and scanning step of 25 μm, equivalent to a nominal resolution of 1.02 nm. The scanned areas typically had a width in the range of 1000–2000 pixels. The large images were reduced in size by box-convolution and subsequent undersampling to facilitate their display on a 640×512 raster display. The coordinates of the marks M_1, M_2 were measured in the digitized image to provide the reference for calculations involving the direction of the tilt axis.

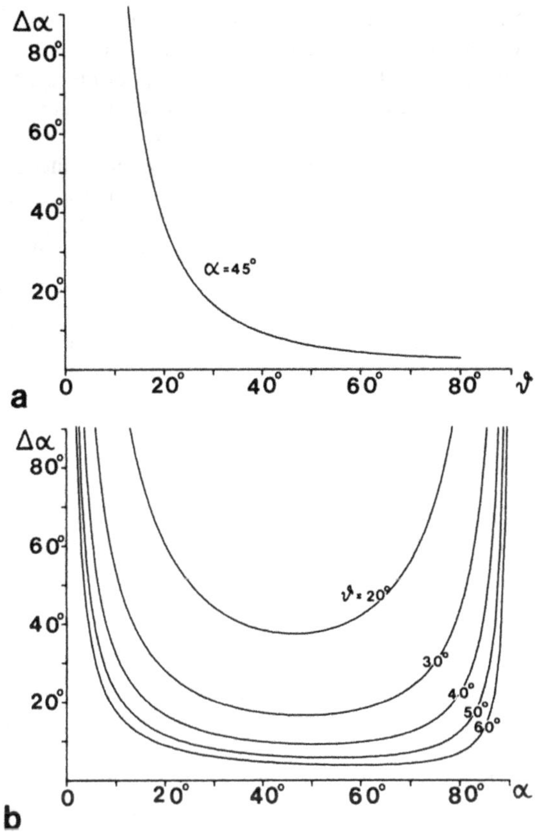

Fig. 5. Dependence of the error in the tilt axis determination from one measurement on (a) the tilt angle ϑ, (b) the angle α between l (see Fig. 3) and the tilt axis direction.

CONTRAST NORMALIZATION

Since images from separate micrographs which show slight differences in exposure, were used for the final reconstruction, the contrast had to be normalized. The formula used is:

$$D_i' = \frac{D_i - \overline{D}}{\overline{D}} \tag{10}$$

which gives the correct normalization for weak phase contrast images (e.g. Hoppe & Hegerl, 1980). D_i is the optical density of the ith pixel and $\overline{D}$ the average optical density of the image. As the weak phase approximation is not strictly valid for heavy-atom stained specimens (Typke & Radermacher, 1982), $\overline{D}$ was measured in areas where only the support foil was visible and no particles with their rather thick stain layer. In these areas, small images were interactively selected and their average density was used as $\overline{D}$ in (10).

PARTICLE SELECTION

For the selection of particles the relation between the coordinates in the tilted and the untilted images had to be established. In addition to the tilt axis direction and the tilt angle the origin must be defined. For this purpose two areas of size 256×256 pixels were windowed out of the reduced images at the same absolute pixel location. The centre of the area in the 0° view was chosen as the origin. Similarly as in the alignment procedure for single-axis tilt series (Guckenberger, 1982), the area windowed from the 50° micrograph was stretched by 1/cos(50°)

perpendicular to the tilt axis. The stretched image was then cross-correlated with the area taken out of the 0° image. The location of the maximum of the cross-correlation function determines the location of the origin in the tilted image. The corresponding origin coordinates in the originally scanned full-sized images are the product of the coordinates in the reduced images times the reduction factor.

The particle images may be interactively selected using the reduced version of either the tilted or the untilted micrograph (Fig. 3). Since at this stage the sole selection criterion is the sufficient separation of the particles in the tilted view, such that no adjacent particle projections overlap, the windowing is best done in the reduced tilted image, in which the chance for overlap is greater. We therefore used this method of selection for all micrographs except for the first pair, for which the 0° view was used. However, a disadvantage of using the tilted image for selection is that the recognition of particle shapes is more difficult than in the untilted image. If in other applications the ability to distinguish the shape of the particles, either visually or by automatic means, is used as a selection criterion, the untilted micrograph may be more suitable for the initial selection.

The corresponding coordinates of the particles in the untilted reduced image and in both originally scanned full-sized micrographs were calculated using the parameters from the alignment, the angle of the tilt axis and the reduction factor between the full-sized and reduced micrographs. With these coordinates the images of the single particles were subsequently windowed out of both originally scanned micrographs. The selection of true crown views was done in the course of the alignment procedure.

DETERMINATION OF THE AZIMUTHAL ANGLES

The in-plane orientations of the particles were determined from the untilted image using an orientational alignment procedure. The angles determined in the alignment are the azimuthal angles of the images in the random conical tilt series extracted from the tilted image. First the particles were centred in their frames, using a low-pass filtered image of a disc as a reference for cross-correlation alignment (cf. the method for automatic particle selection in Frank & Wagenknecht, 1984). The centred images were then rotationally aligned using a modification of the iterative direct alignment procedures described by Frank *et al.* (1978, 1981b) and Steinkilberg & Schramm (1980). In these procedures one-dimensional cross-correlation functions between the image of a reference particle and that of a centred particle are calculated along rings with selected radii. These angular cross-correlation functions are averaged and the rotation between the particles in the two images is determined from the location of the maximum in the resulting cross-correlation function. This procedure, applied to the images of the 50S ribosomal subunit, typically showed five peaks in the 360° range, reflecting a roughly

Table 1. Number of images originally selected versus number of images remaining after alignment. The additional four images were excluded from the combined set because of image defects.

Image pair number	No. of images: Originally selected	No. of images: Retained after alignment
1	221	112
2	155	80
3	117	87
4	93	69
5	216	145
Total	802	493
Excluded from combined 0° set:		3
Excluded from combined 50° set:		1
Remaining total:		489

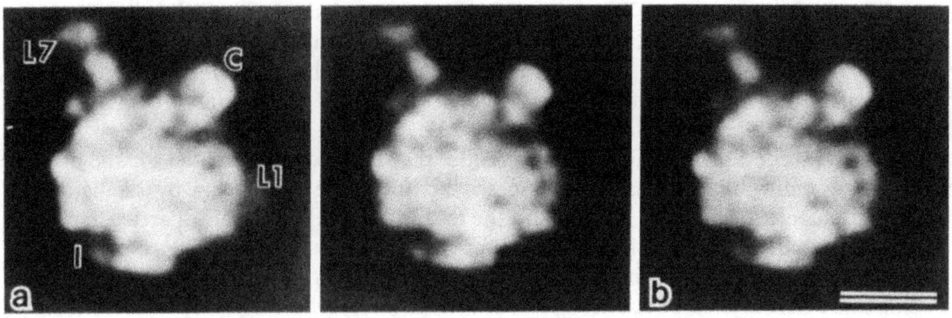

Fig. 6. Two-dimensional averages: (a) calculated from 245 images each; (b) calculated from all 490 images. The typical features visible in the two-dimensional projection are the L7/L12 stalk (L7), the L1 shoulder (L1), the central protuberance (C), and an indentation (I) at the bottom of the particle. The resolution of the averages was determined to be 2·2 nm from calculations of the phase residual between the independent averages shown in (a) (1·6 nm determined with the Fourier ring correlation criterion). This is the best resolution which can be expected in a three-dimensional reconstruction from this particle set. Averages in both (a) and (b) are displayed at this resolution. Scale bar 10 nm.

five-fold symmetric shape of the particle (see Fig. 6) caused by the L7/L12 stalk, the 'head', the L1 shoulder, and an indentation which divides the lower part of the particle. If the images were not precisely centred, then the highest peak found did not necessarily determine the correct rotation. To eliminate this ambiguity, the image was rotated into each of the three orientations corresponding to the three highest maxima in the angular cross-correlation function, and each of the three resultant images was two-dimensionally cross-correlated with the reference. The particle was then translationally realigned in the orientation showing the highest two-dimensional cross-correlation.

The procedure was carried out twice, first with a typical crown-view image as reference, and then with an average of aligned particles as a refined reference in a second pass. The alignment was checked visually, and misaligned particles were discarded. We found that the excluded images mainly showed distorted particles or particles not lying in the crown-view orientation. Table 1 shows the ratio between the number of images originally selected and the number of images finally aligned.

TWO-DIMENSIONAL AVERAGING

To check the consistency among the aligned particle images, two independent averages (Fig. 6), each containing the projections of 245 particles were calculated from the set of 490 aligned images. These averages showed a cross-resolution of 2·2 nm, determined by the 45° limit of the phase residual calculated between rings in the Fourier transform of the two averages (Frank *et al.*, 1981b). The resolution was also determined by the less stringent measure of Fourier ring correlation (Saxton & Baumeister, 1982; van Heel *et al.*, 1983; Radermacher *et al.*, 1986c), which resulted in a value of 1·6 nm. At the corresponding radius in Fourier space the phase residual has already reached a value of 87°. The resolution value of 2·2 nm, found in the two-dimensional average, is an upper bound of the resolution that can be expected under optimum conditions in a three-dimensional reconstruction from this particle set, provided that a sufficient number of projections is used.

ALIGNMENT OF THE TILTED IMAGES

After the azimuthal angles had been determined in the rotational alignment of the untilted images, the images of the tilted particles were sorted according to their azimuthal angles and aligned with respect to a common origin using a modification of the alignment procedures described by Radermacher (1980) and Guckenberger (1982). The alignment proceeds sequentially, each image being aligned to its neighbour. The images are first stretched by $1/\cos\vartheta$

perpendicular to the tilt axis. Thus, from the projections orthogonal to the electron beam inclined projections are created. Taking the conical geometry into consideration, each image is rotated by an angle equal to the difference between its azimuthal angle and that of its predecessor, such that both particles have the same orientation. Their relative displacement is subsequently determined by cross-correlation, and the original (unstretched and non-rotated) version of the second image in a pair is shifted by the shift vector calculated for the original coordinate system. The accumulation of errors in this procedure was checked by calculating the residual misalignment between the last and the first image in the conical tilt series.

The alignment and reconstruction procedures were first applied to the particle images extracted from the first pair of micrographs. The images of the tilted particles were precentred in a similar way as the untilted images by cross-correlation with the low-pass filtered image of an ellipse, which was created from an image of a circular disc by geometric compression corresponding to a 50° tilt. Subsequently the particles were aligned using the procedure described in the preceding paragraph. A three-dimensional reconstruction from this limited particle set was calculated with the method explained below. This reconstruction proved to have a resolution of better than 6 nm as determined by phase-residual calculations (Radermacher *et al.*, 1986a).

Since from the outset it could be assumed that the error accumulation in the set of 489 images (one of the tilted particle images was not used because of a defect in the image) would be considerable, the alignment of the complete set of images was done in several steps.

In a first attempt, projections were created from the low-resolution reconstruction corresponding to all directions found in the complete set of tilted images, and the original projections were aligned to the calculated images by cross-correlation. However, the reconstruction calculated from such aligned images showed a resolution of 4 nm, which was lower than we expected. Therefore the alignment procedure was replaced by a procedure consisting of three steps: The tilted images, aligned with the procedure described above, were averaged over 10° intervals thus creating a conical tilt series (A) of thirty-six images. This set of averages was then aligned by the serial alignment procedure for conical tilt series described above. The closing error of this alignment was less than 0·01 pixel (0·005 nm).

A second series (B) of seventy-two images was created by averaging the projections over 5° intervals. This series B was aligned relative to series A. In this step the alignment error proved to be about 1 pixel (0·5 nm). Finally, the series of original tilted images was aligned relative to series B, again resulting in an alignment with an accuracy of 1 pixel (0·5 nm). Since the resolution of the two-dimensional averages was 2·2 nm (4·4 pixels), which constitutes an upper limit for the resolution expected in our three-dimensional reconstruction, this alignment was considered to be sufficiently accurate.

We do not believe, however, that the alignment with respect to calculated projections from a preliminary low-resolution reconstruction is a necessary step, since the displacements between the precentred and the finally aligned projections were rather small (typically 2–3 pixels). Averaging of projections centred with the low-pass filtered image of an ellipse should be sufficient to create the initial image set of averaged projections.

Thus the alignment procedure we devised consists of six steps. First the tilted images are centred with a cross-correlation procedure using the strongly low-pass filtered image of a tilted disc. The centred tilted images are then averaged over 10° intervals. These averaged projections are aligned with each other in a consecutive series. Next the centred images are averaged over 5° intervals and this series is aligned relative to the 10° averaged series. Then the individual projections are aligned with the 5° averages.

THREE-DIMENSIONAL RECONSTRUCTION

With the tilt angle known from the experiment and the azimuthal angles determined in the two-dimensional alignment, each of the 489 tilted images has an assigned place in the conical tilt geometry. This projection series was used in the next step for the three-dimensional reconstruction of the ribosomal subunit.

A large number of algorithms are available for three-dimensional reconstruction from tilt series. Some are specialized for specific geometries, such as the Fourier-Bessel method for single-axis tilting with even angular increments (Cormack, 1963, 1964, 1973; Smith *et al.*, 1973) or helical symmetries (DeRosier & Klug, 1968), and the weighted back-projection method applicable to single-axis (Gilbert, 1972) and conical tilt series with equal angular increments (Radermacher & Hoppe, 1978). Other methods can be used for any geometry as most of the iterative direct reconstruction methods (e.g. Gordon *et al.*, 1970; Colsher, 1977) and Fourier interpolation methods (Hoppe *et al.*, 1968; Crowther *et al.*, 1970). Only the latter two classes of algorithms could be considerd in our application, as the set of images obtained with our method of data collection constitutes a tilt series with random and not equal angular increments. The Fourier interpolation methods, however, are mainly suitable for crystalline specimens, as in that situation the interpolation can be restricted to the interpolation along lattice lines. For non-crystalline, asymmetrical single particles the complete Fourier space would need to be interpolated, and the amount of calculations necessary with, e.g., a Whittaker Shannon interpolation (Whittaker, 1915; Shannon, 1949), becomes prohibitive. Similar arguments hold for the iterative methods available. With the large number of projections in our tilt series the computation of several iterations is far too time consuming.

The problems stated above led to the development of a new weighted back-projection algorithm that is suitable for any arbitrary geometry. In contrast to published schemes with analytically calculated weighting functions, the weighting function in our method is calculated numerically using the actual projection angles found in the experiment (Radermacher *et al.*, 1986a). A similar algorithm based on a different shape function has later been proposed by Harauz (Harauz & van Heel, 1986).

In the following we make use of the treatment in Radermacher & Hoppe (1978), Radermacher (1980) and Hoppe & Hegerl (1980). In a simple back-projection, three-dimensional back-projection bodies $\hat{p}_i(x_i', x_i', z_i')$ are created from the two-dimensional projections $p_i(x_i', y_i')$ of the object, x_i', y_i', z_i' are the coordinates in the coordinate system of the projection p_i with the projection angles ϑ_i, ϕ_i. These back-projection bodies consist of constant-valued rays in the direction z_i', perpendicular to the projection plane x_i', y_i' and with the value of the projection at the corresponding pixel, equivalent to a convolution of the two-dimensional projection $p_i(x_i', y_i')$ with the function

$$c_i(x_i', y_i', z_i') = \delta(x_i', y_i') \cdot g(z_i') \tag{11}$$

with $g(z_i') = 1$ for $-D/2 \leq z_i' \leq D/2$ and $g(z_i') = 0$ otherwise. (D diameter of the reconstruction volume.) The back-projection combines these back-projection bodies by summation in their proper geometrical orientations. The weighting is done by a multiplication of the three-dimensional Fourier transform of the back-projected object with the weighting function, which is equal to 1 divided by the Fourier transform of the three-dimensional point spread function. As the two-dimensional Fourier transform of a projection is a central section through the three-dimensional Fourier transform of the object the weighting function can be calculated in these two-dimensional sections and be applied to the corresponding projection.

In our method the weighting function is numerically calculated in the planes $x_i^{*\prime}, y_i^{*\prime}$ in Fourier space that correspond to the Fourier transforms of the projections p_i, by summation of the Fourier transforms of $c_i : C_i(x_i^{*\prime\prime}, y_i^{*\prime\prime}, z_i^{*\prime\prime})$. C_i are sinc functions originating in, and perpendicular to the planes $x_i^{*\prime\prime}, y_i^{*\prime\prime}$, which are the central sections corresponding to the Fourier transform of the projections p_i at angles ϑ_i, ϕ_i. The weighting function in the plane of the Fourier transform of a projection at angles ϑ_i, ϕ_i thus has the form

$$W_{\vartheta_i,\phi_i}(x_i^{*\prime}, y_i^{*\prime}) = 1 \Big/ \sum_j D \cdot \mathrm{sinc}[\pi D z_j^{*\prime\prime}(\vartheta_i, \phi_i)] \tag{12}$$

$z_j^{*\prime\prime}(\vartheta_i, \phi_i)$ are the z^*-coordinates of the point $(x_i^{*\prime}, y_i^{*\prime})$ in the coordinate systems $x_j^{*\prime\prime}, y_j^{*\prime\prime}, z_j^{*\prime\prime}$.

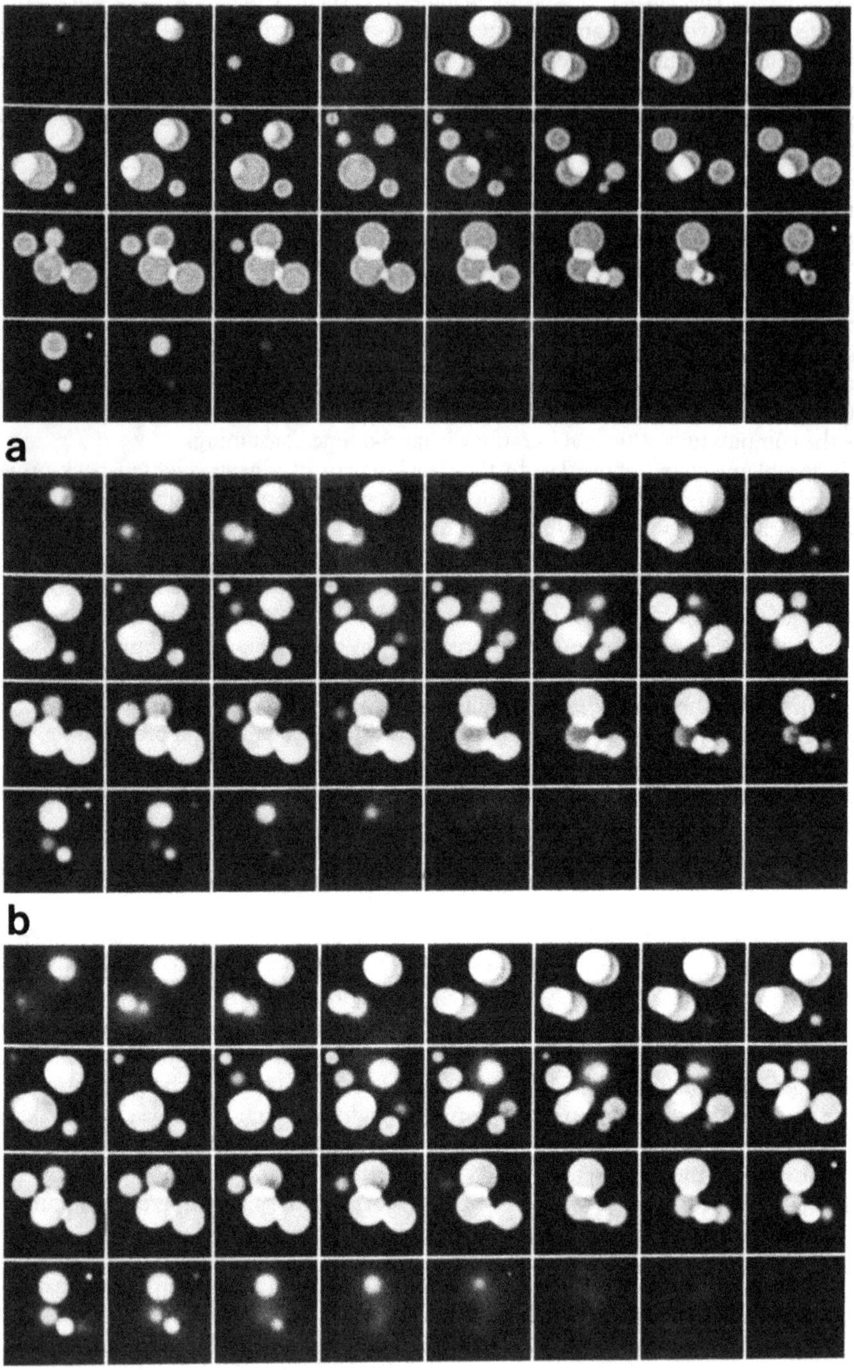

Fig. 7. Sections through a three-dimensional model reconstruction from seventy-two projections: (a) original object; (b) reconstructed with the weighting function described here; (c) reconstructed with the weighted back-projection scheme for conical tilting with even azimuthal increments.

This function $W_{\vartheta_i\phi_i}$ converges to the weighting functions of the r^*-weighted back-projection (Gilbert, 1972) and the weighting function for conical tilting with evenly spaced projections (Radermacher & Hoppe, 1978), if the same approximations are made as in the derivation of these functions ($D \rightarrow \infty$, ϕ continuous, ϑ constant). The values of $W_{\vartheta_i\phi_i}$ were limited to a value of 1·66 to limit enhancement of noise in areas where the Fourier transform of the object is not sufficiently sampled. As the weighted back-projection algorithm with the analytically calculated weight function for regular conical geometry, the new algorithm does not contain a resolution limitation. To limit artefacts the resolution has to be limited in a second step to a value that can be calculated for regular sampling (DeRosier & Klug, 1968; Radermacher, 1980). In our application, where irregular sampling is used, the resolution is determined through comparison of two independent reconstructions. The filtration may be done with a smooth filter function, applied either to the Fourier transforms of the projections before the reconstruction, or to the three-dimensional Fourier transform of the reconstructed object. We chose the latter method of filtration.

Model computations were done to check both the validity of the algorithm and its numerical

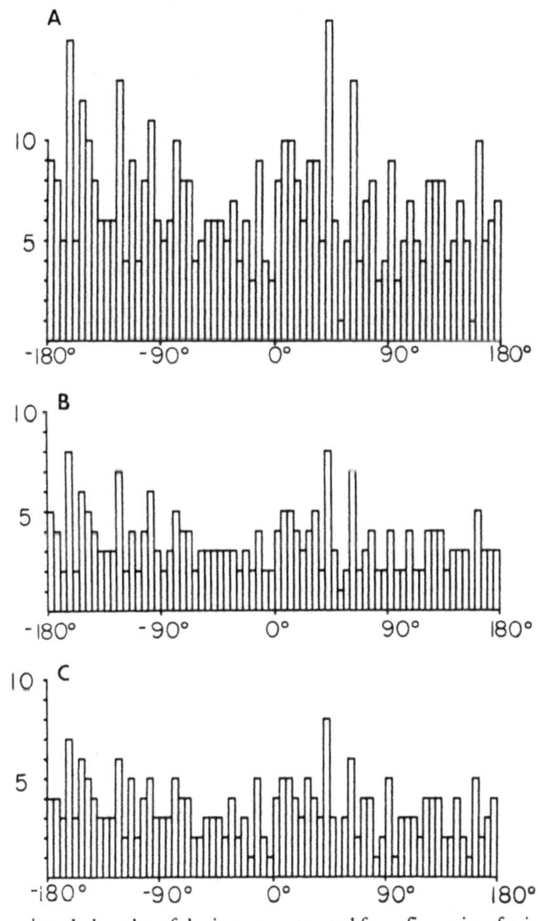

Fig. 8. Histogram of the azimuthal angles of the images extracted from five pairs of micrographs. (A) Histogram of the complete set; the entire range is covered over 5° intervals. (B), (C) Histograms of the two sets used for the two independent reconstructions. Only one 5° gap in each of the sets is present, and the comparison of the resolution to the resolution obtained from a conical tilt series with even angular increments of 5° still appears to be justified.

implementation. One of the test objects was a cube 64×64×64 containing twelve spheres with radii between 3 and 13 pixels. Seventy-two projections with equal and random angular spacings were calculated, and the object was reconstructed for either set and compared to the original. In the case of even angular spacing (Fig. 7) the reconstructions were compared to a reconstruction using the analytical weighted back-projection scheme (Radermacher & Hoppe, 1978). Visual inspection of the two reconstructions shows that the performance of the numerically weighted back-projection is of similar quality as that of the reconstruction with the analytical weighting scheme. To obtain a more objective measure, we calculated the normalized euclidean distances or discrepancies between the reconstructed model objects and the original object. The discrepancy is defined (Gilbert, 1972; Colsher, 1977) as:

$$\Delta = \left\{ \frac{\sum_i (a \cdot \tilde{D}_i - D_i - b)^2}{\sum_i (D_i - \bar{D})^2} \right\}^{1/2} \tag{13}$$

where $\tilde{D}_i$ are the voxels in the reconstructed volume, D_i the voxels in the original object and $\bar{D}$ the average of D_i. We introduced the constants a and b to compensate for scaling differences between the different reconstruction schemes. The resulting values were 0·48 for the analytical weighting scheme and 0·33 for the numerical weighting scheme. In a comparison of different iterative reconstruction schemes applied to a multiple sphere model reconstructed from twenty-four projections with a cone angle of 45°, Colsher (1977) finds values of the discrepancy in the range of 0·65–0·68. These values are considerably higher than our values, which can be attributed, to a large extent, to the lower number of projections and to the smaller cone angle used in Colsher's work. These figures are given here only to provide reference values.

The 489 projections of tilted particles were processed with this reconstruction algorithm and the structure of the 50S ribosomal subunit was calculated.

RESOLUTION

The resolution achievable in a three-dimensional reconstruction from the current image set under ideal conditions may be estimated with the formulas for conical tilting with even azimuthal angular increments (Radermacher, 1980):

$$N = \frac{2\pi D}{r} \sin \vartheta \quad \text{for } N \text{ even} \tag{14}$$

$$N = \frac{\pi D}{2r} \tan \vartheta \sqrt{16 \cos^2 \vartheta - r^2/D^2} \quad \text{for } N \text{ odd} \tag{15}$$

(N number of projections, D object diameter, r resolution, ϑ cone angle). Equations (14, 15) are derived from the condition that no sampling point on the surface of the resolution sphere in Fourier space is separated from its neighbours by more than $1/D$. The histogram of the angular distribution (Fig. 8) shows a complete coverage of the entire 360° range over 5° intervals. A 5° interval used in (14) ($N=72$, $\vartheta=50°$, $D=30$ nm) leads to a resolution of 2·0 nm. This value does not, however, take into account the limitations set by the quality of the specimen (negative stain, variations in particle shapes), electron optical limitations of the resolution (in our case better than 2·2 nm as determined from the diffractogram of the micrograph), and limitations set by the accuracy of the alignment and the tilt axis determination.

To determine the actual reproducible resolution, three reconstructions were calculated, one using the complete set of projections and two reconstructions each using half of the image set. The phase residuals for five corresponding slices through the centre of the latter two reconstructions were used to determine the resolution. As the resolution limit the Fourier

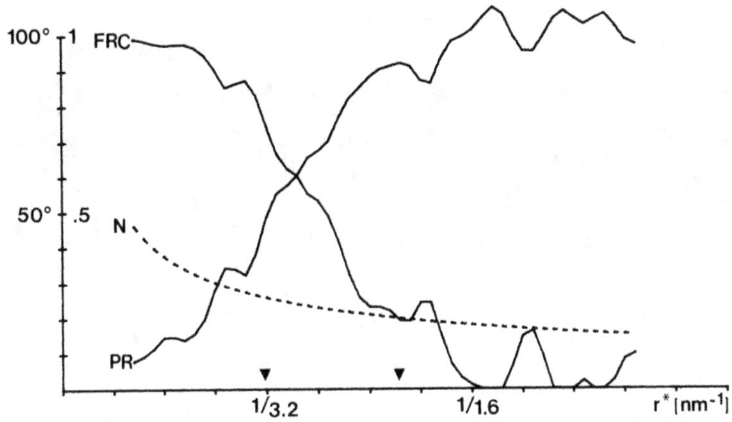

Fig. 9. The phase residual (PR) and Fourier ring correlation (FRC) as a function of the radius r^* in Fourier space; (*N*), the function $2/\sqrt{n}$, n number of coefficients used for the calculation of the Fourier ring correlation. The curves show the averaged functions calculated comparing five slices (marked in Fig. 10) near the centre of the two independent reconstructions. The resolutions found are marked on the r^* axis. The resolution values found with the phase residual criterion varied between 3·0 and 3·3 nm for the different slices, while the less stringent Fourier ring correlation criterion gave resolution values between 1·8 and 2·4 nm.

radius showing a phase residual between 45° and 50° was selected. The resolution found by this criterion was between 3·0 nm and 3·3 nm. For comparison the less stringent measure of Fourier ring-correlation (Saxton & Baumeister, 1982; van Heel *et al.*, 1983; Radermacher *et al.*, 1986c) was also calculated. This measure showed the resolution limit to be in the range of 1·8–2·4 nm (Fig. 9).

All three volumes were filtered to a resolution of 3·0 nm using the shape of the Fermi distribution as a filter function (Frank *et al.*, 1985):

$$F(r^*)=\frac{1}{1+\exp[-(r^*-r_0^*)/T]} \tag{16}$$

r^* radius in Fourier space, T temperature factor. The parameter T allows the smoothness of the falloff of this filter function to be varied over a wide range. For T approaching 0, $F(r^*)$ approaches the step function. For our filtration we used $r_0^*=10{\cdot}5$ and $T=2$, with a reconstruction volume of 64×64×64 voxels.

All three reconstructions were then displayed slice by slice (Fig. 10). The reconstruction calculated from the complete set of 489 projections was also represented as a surface display (Radermacher & Frank, 1984) (Fig. 11).

The resulting structure shows some major features found in earlier studies, specifically when compared with the results obtained by Oettl *et al.* (1983) and Hoppe (1983) from single-axis tilt data. There is a wealth of additional information contained in our reconstruction which is amenable to a detailed analysis and interpretation. The results will be discussed in a forthcoming publication (Radermacher *et al.*, 1986b, 1987).

POSSIBLE IMPROVEMENTS

The comparison of the resolution values found for the two-dimensional average and the three-dimensional reconstruction (2·2 nm versus 3 nm) and the value calculated with (14), 2·0 nm, suggests that even with the data set currently available a higher resolution might be possible. To improve the resolution, several steps of the procedure may be considered. One step at which improvements may be possible is the alignment of the tilted images. Instead of aligning the images with respect to one another, one could think of aligning them with respect

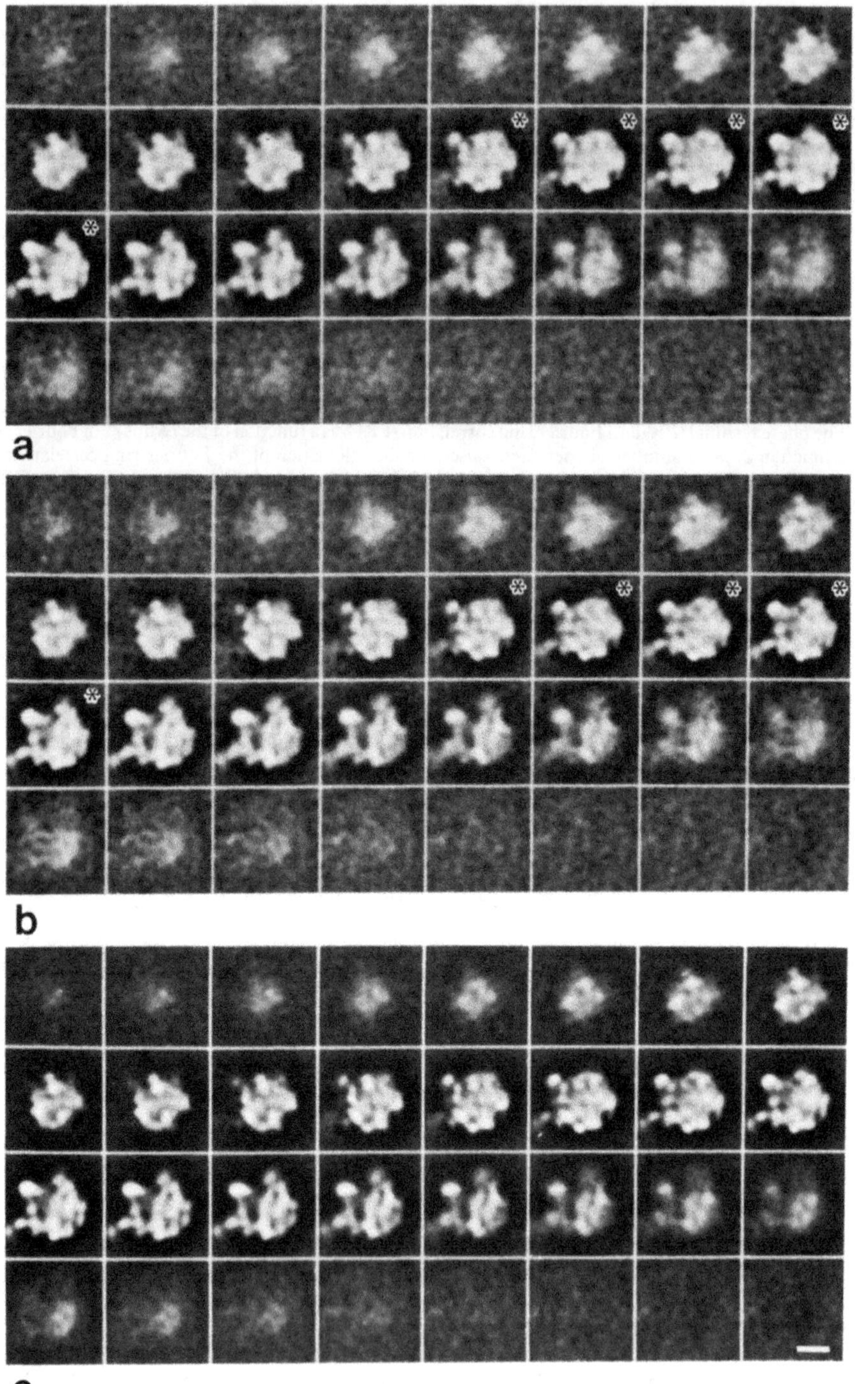

Fig. 10. Slices perpendicular to the Z-axis through the three reconstructions: (a), (b) the two reconstructions each calculated from half of the set and used for the determination of the resolution; (c) the reconstruction from the complete projection set. All three panels show representations of the structure limited to the resolution of 3 nm found by phase-residual analysis (Fig. 9). Scale bar 10 nm.

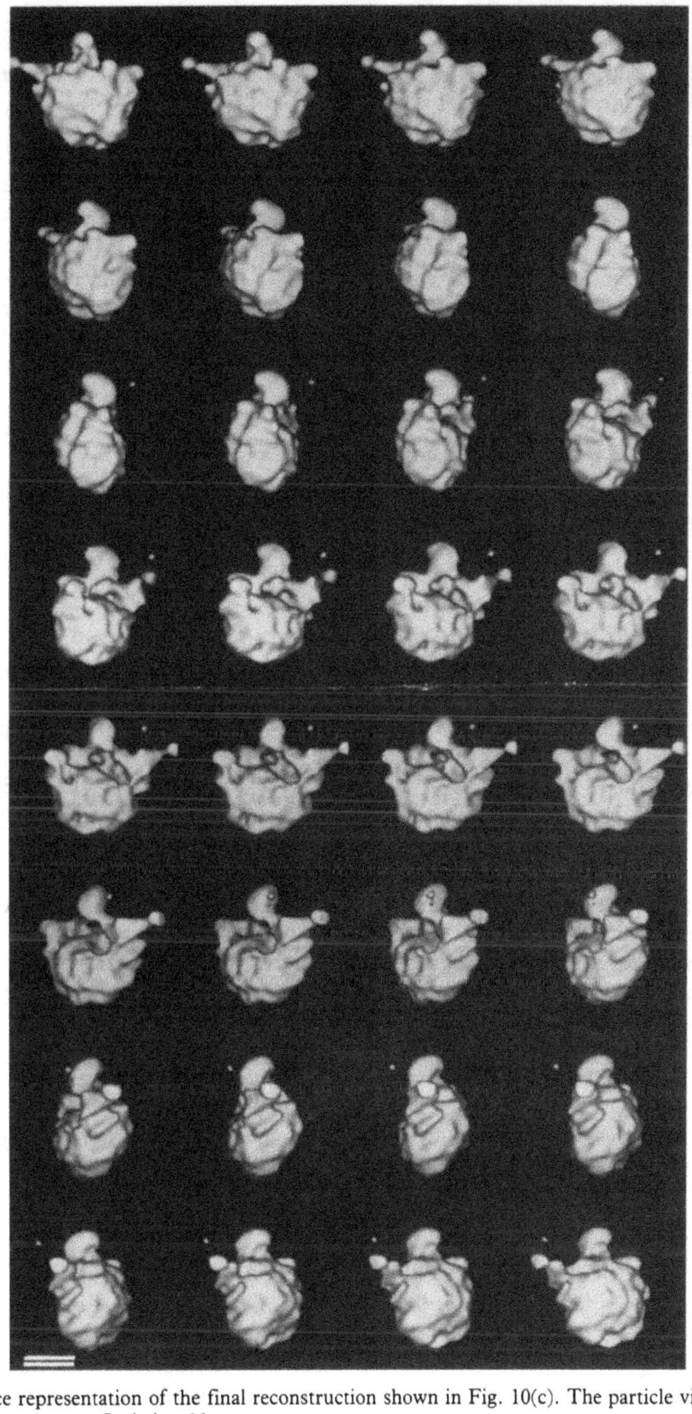

Fig. 11. Surface representation of the final reconstruction shown in Fig. 10(c). The particle views are shown rotated in 10° increments. Scale bar 10 nm.

to their 0° counterparts. Whether or not this kind of alignment results in a higher precision is currently under investigation.

A second change that could result in a higher resolution is a more stringent selection of particles. We applied correspondence analysis (van Heel & Frank, 1981; Frank & van Heel, 1982) to the current particle set, but the results were inconclusive. The first eight factors accounted for only about 8% of the total variations in the particle set. No clustering could be discerned, and the importance images (van Heel & Frank, 1981) failed to give any conclusive information.

We also tried to use the variances of the difference images between each single 0° image and the total average as a method of screening the particles. However, the resolution of the two-dimensional average did not improve by the exclusion of 10% of the images showing the highest variance.

We are investigating more intelligent multivariate statistical analysis schemes such as correspondence analysis using learning sets. Some methods of feature enhancement and extraction may prove valuable, yet most schemes found in the literature may not be applicable to low-dose electron micrographs because of their extremely low signal-to-noise ratios (typical values are on the order of 0·1).

CONCLUSION

A scheme for three-dimensional reconstruction from a single-exposure, random conical tilt series has been developed and successfully appied to the 50S ribosomal subunit, yielding a reconstruction with a resolution of 3 nm. The method is quite general and the range of possible applications reaches far beyond the one shown here.

The reconstruction technique essentially requires only one single exposure of each specimen area, thus facilitating the calculation of the structure under low-dose conditions. It may currently be the only method that will allow the calculation of a three-dimensional structure of asymmetrical single particles from unstained, frozen-hydrated preparations.

The structures of all single molecules that are amenable to two-dimensional image averaging techniques can now be analysed in three dimensions with relatively little additional effort. Our method is applicable to the large class of macromolecular assemblies that exhibit a preferred orientation in the electron microscopic preparation, e.g. the 40S ribosomal subunit of eukaryotes (Frank *et al.*, 1981c), the enzyme glutamine synthetase (Frank *et al.*, 1978), and the respiratory protein haemocyanin (Lamy *et al.*, 1982; van Heel & Frank, 1981). Our preliminary studies show that in frozen hydrated preparation the preferred orientation found in standard negative stain preparations can be maintained if the particles are allowed to adsorb to a carbon foil used as a substrate. There are indications that if a suspension of particles in ice is used, the water/air interface might serve as an attachment surface, though further investigations of this effect are necessary.

A vast field of applications to the study of ribosomal particles or, indeed, any other type of macromolecular assembly, would be immunoelectron microscopical studies, which have been done so far only in two dimensions by visual judgement with reference to the visually derived models (Tischendorf *et al.*, 1974). Our scheme would allow the localization of an epitope within the calculated three-dimensional structure.

ACKNOWLEDGMENTS

This work was supported by a grant of the National Institute of Health 1R01 GM 29169 and a grant of the National Science Foundation 8313405.

REFERENCES

Arad, T., Leonard, K., Wittmann, H.G. & Yonath, A. (1984) Three-dimensional reconstruction of *B. stearothermophilus* 50S ribosomal particles. *Proc. 8th Europ. Congr. Electron Microsc.* **2,** 1459–1460.

Chiu, M.Y., Barrett, H.H., Simpson, R.G., Chou, C., Arendt, J.W. & Gindi, G.R. (1979) Three-dimensional radiographic imaging with a restricted view angle. *J. Opt. Soc. Am.* **69,** 1323–1333.

Cohen, H.A., Schmid, M.F. & Chiu, W. (1984) Three-dimensional reconstructions at high resolution. *Ultramicroscopy,* **14,** 219–226.

Colsher, J.G. (1977) Iterative three-dimensional image reconstruction from tomographic projections. *Comp. Graph. Im. Proc.* **6,** 513–537.

Cormack, A.M. (1963) Representation of a function by its line integrals, with some radiological applications. *J. appl. Phys.* **34,** 2722–2727.

Cormack, A.M. (1964) Representation of a function by its line integrals, with some radiological applications. II. *J. appl. Phys.* **35,** 2908–2913.

Cormack, A.M. (1973) Reconstruction of densities from their projections, with applications in radiological physics. *Phys. Med. Biol.* **18,** 195–207.

Cosslett, V.E. (1978) Radiation damage in the high resolution electron microscopy of biological materials: a review. *J. Microsc.* **113,** 113–129.

Crowther, R.A., DeRosier, D.J. & Klug, A. (1970) The reconstruction of a three-dimensional structure from projections and its application to electron microscopy. *Proc. Roy. Soc.* A, **317,** 319–340.

DeRosier, D.J. & Klug, A. (1968) Reconstruction of three-dimensional images from electron micrographs. *Nature,* **217,** 130–134.

Dover, D.S., Elliot, A. & Kernaghan, A.K. (1981) Three-dimensional reconstruction from images of tilted specimens: the paramyosin filament. *J. Microsc.* **122,** 23–33.

Erickson, H.P. & Klug, A. (1970) The Fourier transform of an electron micrograph: effects of defocusing and aberrations, and implications for the use of underfocus contrast enhancement. *Ber. Bunseng. Phys. Chem.* **74,** 1129–1137.

Frank, J., Goldfarb, W., Eisenberg, D. & Baker, T.S. (1978) Reconstruction of glutamine synthetase using computer averaging. *Ultramicroscopy,* **3,** 283–290.

Frank, J. & Goldfarb, W. (1980) Methods for averaging of single molecules and lattice fragments. *Electron Microscopy at Molecular Dimensions* (ed. by W. Baumeister and W. Vogell), pp. 261–269. Springer, Berlin.

Frank, J., Shimkin, B. & Dowse, H. (1981a) SPIDER—A modular software system for electron image processing. *Ultramicroscopy,* **6,** 343–358.

Frank, J., Verschoor, A. & Boublik, M. (1981b) Computer averaging of 40S ribosomal subunits. *Science,* **214,** 1353–1355.

Frank, J. & van Heel, M. (1982) Correspondence analysis of aligned images of biological particles *J. molec. Biol.* **161,** 134–137.

Frank, J. & Wagenknecht, T. (1984) Automatic selection of molecular images from electron micrographs. *Ultramicroscopy,* **12,** 169–176.

Frank, J., Verschoor, A. & Wagenknecht, T. (1985) Computer processing of electron microscopic images of single macromolecules. *New Methodologies in Studies of Protein Configuration* (ed. by T. T. Wu), pp. 36–89. Van Nostrand Reinhold, New York.

Gilbert, P.F.C. (1972) The reconstruction of a three-dimensional structure from projections and its application to electron microscopy. II. Direct methods. *Proc. Roy. Soc.* B, **182,** 89–102.

Glaeser, R.M. (1975) Radiation damage and biological electron microscopy. *Physical Aspects of Electron Microscopy and Microbeam Analysis* (ed. by B. Siegel and D. R. Beaman), pp. 205–229. Wiley, New York.

Gordon, R., Bender, R. & Herman, G.T. (1970) Algebraic reconstruction techniques (ART) for three dimensional electron microscopy and X-ray photography. *J. theor. Biol.* **29,** 471–481.

Guckenberger, R. (1982) Determination of a common origin in the micrographs of tilt series in three-dimensional electron microscopy. *Ultramicroscopy,* **9,** 167–174.

Harauz, G. & van Heel, M. (1986) Exact filters for general three dimensional reconstruction. *Optik,* **73,** 146–156.

van Heel, M. & Frank, J. (1981) Use of multivariate statistics in analysing the images of biological macromolecules. *Ultramicroscopy,* **6,** 187–194

van Heel, M. (1983) Three-dimensional reconstruction of the 30S *E. coli* ribosomal subunit. *Proc. 41st Ann. Meeting EMSA* (ed. by G. W. Bailey), pp. 460–461. San Francisco Press, San Francisco.

van Heel, M., Keegstra, W., Schutter, W. & van Bruggen, E.F.J. (1983) Arthropod hemocyanin structures studied by image anaysis. *Structure and Function of Invertebrate Respiratory Proteins* (ed. by E. J. Wood), pp. 69–73. (Life Chemistry Reports Ser.: Supplement.) Harwood Academic, New York.

van Heel, M. (1986) Noise limited three-dimensional reconstructions. *Optik,* **73,** 83–86.

Hegerl, R., Hoppe, W., Knauer, V. & Typke, D. (1984) Some aspects of the 3D reconstruction of individual objects. *Proc. 8th Eur. Congr. Electron Microsc.* **2,** 1363–1373.

Hoppe, W., Langer, R., Knesch, G. & Poppe, Ch. (1968) Proteinkristallstrukturanalyse mit Elektronenstrahlen. *Naturwissenschaften,* **55,** 333–336.

Hoppe, W. (1969) Das Endlichkeitspostulat und das Interpolationstheorem der dreidimensionalen electronenmikroskopischen Analyse aperiodischer Strukturen. *Optik,* **29,** 617–621.

Hoppe, W. (1971) Use of zone correction plates and other techniques for structure determination of aperiodic objects at atomic resolution using a conventional electron microscope. *Phil. Trans. Roy. Soc.* B, **261,** 71–94.

Hoppe, W., Bussler, P., Feltynowski, A., Hunsmann, N. & Hirt, A. (1973) Some experience with computerized image reconstruction methods. *Image Processing and Computer-aided Design in Electron Optics* (ed. by P. W. Hawkes), pp. 92–126. Academic Press, London.

Hoppe, W., Schramm, H.J., Sturm, M., Hunsmann, N. & Gaßmann, J. (1976a) Three-dimensional electron microscopy of individual biological objects. Part I. Methods. *Z. Naturf.* **31a,** 645–655.

Hoppe, W., Schramm, H.J., Sturm, M., Hunsmann, N. & Gaßmann, J. (1976b) Three-dimensional electron microscopy of individual biological objects. Part III. Experimental results on yeast fatty acid synthetase. *Z. Naturf.* **31a,** 1380–1390.

Hoppe, W. & Grill, B. (1977) Prospects of three-dimensional high resolution electron microscopy of non-periodic structures. *Ultramicroscopy,* **2,** 153–168.

Hoppe, W. & Hegerl, R. (1980) Three-dimensional structure determination by electron microscopy. *Computer Processing of Electron Microscope Images,* (ed. by P. W. Hawkes), pp. 127–185. Topics in Current Physics 13). Springer, Berlin.

Hoppe, W. (1983) Elektronenbeugung mit dem Transmissions-Elektronenmikroskop als phasenbestimmendem Diffraktometer—von der Ortsfrequenzfilterung zur dreidimensionalen Strukturanalyse an Ribosomen. *Angew. Chem.* **95,** 465–494 (*Int. Ed. Engl.* **22,** 456–485).

Kiselev, N.A., Orlova, E.V., Stel'maschuk, V.Ya., Vasiliev, V.D., Selivanova, O.M., Kosykh, V.P., Pustovskikh, A.I. & Kirichuk, V.S. (1983) Strand-like structures and their three-dimensional organization in the large subunit of the Eschericia coli ribosome. *J. molec. Biol.* **169,** 345–350.

Knauer, V., Hegerl, R. & Hoppe, W. (1983) Three-dimensional reconstruction and averaging of 30S ribosomal subunits of *Escherichia coli* from electron micrographs. *J. molec. Biol.* **163,** 409–430.

Lake, J.A. (1985) Evolving ribosome structure: domains in archaebacteria, eubacteria, eocytes and eukaryotes. *Ann. Rev. Biochem.* **54,** 507–530.

Lamy, J., Sizaret, P., Frank, J., Verschoor, A., Feldmann, R. & Bonaventura, J. (1982) Architecture of *Limulus polyphemus* hemocyanin. *Biochemistry,* **21,** 6826–6833.

Liljas, A. (1982) Structural studies of ribosomes. *Prog. Biophys. molec. Biol.* **40,** 161–228.

Nathan, R. (1971) Image processing: enhancement procedures. *Advanc. opt. Electron Microsc.* **4,** 85–125.

Oettl, H., Hegerl, R. & Hoppe, W. (1983) Three-dimensional reconstruction and averaging of 50S ribosomal subunits of *Escherichia coli* from electron micrographs. *J. molec. Biol.* **163,** 431–450.

Radermacher, M. & Hoppe, W. (1978) 3-D reconstruction from conically tilted projections. *Proc. 9th int. Congr. Electron Microsc.* **1,** 218–219.

Radermacher, M. (1980) Dreidimensionale Rekonstruktion bei kegelförmiger Kippung im Elektronenmikroskop. Ph.D. thesis, Technische Universität München.

Radermacher, M. & Hoppe, W. (1980) Properties of 3-D reconstruction from projections by conical tilting compared to single axis tilting. *Proc. 8th Eur. Congr. Electron Microsc.* **1,** 132–133.

Radermacher, M. & Frank, J. (1984) Representation of three-dimensionally reconstructed objects in electron microscopy by surfaces of equal density. *J. Microsc.* **136,** 77–85.

Radermacher, M., Wagenknecht, T., Verschoor, A. & Frank, J. (1986a) A new reconstruction scheme applied to the 50S ribosomal subunit of *E. coli. J. Microsc.* **141,** RP1–RP2.

Radermacher, M., Wagenknecht, T., Verschoor, A. & Frank, J. (1986b) 3D reconstruction of the 50S ribosomal subunit of *Escherichia coli. Proc. 44th Ann. Meeting EMSA* (ed. by G. W. Bailey), pp. 40–41. San Francisco Press, San Francisco.

Radermacher, M., Frank, J. & Mannella, C.A. (1986c) Correlation averaging: Lattice separation and resolution assessment. *Proc 44th Ann. Meeting EMSA* (ed. by G. W. Bailey), pp. 140–143. San Francisco Press, San Francisco.

Radermacher, M., Wagenknecht, T., Verschoor, A. & Frank, J. (1987) Three-dimensionai structure of the large ribosomal subunit from *Escherichia coli. EMBO J.* **6** (in press).

Radon, J. (1917) Über die Bestimmung von Funktionen durch ihre Integral-werte längs gewisser Mannigfaltigkeiten. Berichte über die Verhandlungen der Königlich Sächsischen Gesselschaft der Wissenschaften zu Leipzig. *Math. Phys. Klasse,* **69,** 262–277.

Saxberg, B.E.H. & Saxton, W.O. (1981) Quantum noise in 2D projections and 3D reconstructions. *Ultramicroscopy,* **6,** 85–90.

Saxton, W.O. (1980) Matching and averaging over fragmented lattices. *Electron Microscopy at Molecular Dimensions* (ed. by W. Baumeister and W. Vogell), pp. 245–255. Springer, Heidelberg.

Saxton, W.O. & Baumeister, W. (1982) The correlation averaging of a regularly arranged bacterial envelope protein. *J. Microsc.* **127,** 127–138.

Scherzer, O. (1970) Die Strahlenschädigung der Objekte als Grenze für die hochauflösende Elektronenmikroskopie. *Ber. Bunseng.* **74,** 1154–1167.

Shannon, C.E. (1949) Communication in the presence of noise. *Proc. IRE,* **37,** 10–22.

Skoglund, U., Andersson, K., Strandberg, B. & Daneholt, B. (1986) Three-dimensional structure of a specific pre-messenger RNP particle established by electron microscope tomography. *Nature,* **319,** 560–564.

Smith, P.R., Peter, T.M. & Bates, R.H.T. (1973) Image reconstruction from a finite number of projections. *J. Phys.* A, **6,** 361–382.

Staehelin, T. & Maglott, D.R. (1971) Preparations of *Escherichia coli* ribosomal subunits active in polypeptide synthesis. *Meth. Enzymol.* **20,** 449–456.

Steinkilberg, M. & Schramm, J. (1980) Eine verbesserte Drehkorrelationsmethode für die Strukturbestimmung biologischer Makromoleküle durch Mittelung elektronenmikrosopischer Bilder. *Hoppe-Seyler's Z. physiol. Chem.* **361,** 1363–1369.

Subirana, J.A., Munoz-Guerra, S., Aymami, J., Radermacher, M. & Frank, J. (1985) The layered organization of nucleosomes in 30nm chromatin fibers. *Chromosoma,* **91,** 377–390.

Tischendorf, G.W., Zeichhardt, H. & Stöffler, G. (1974) Determination of the location of proteins L14, L17, L18, L19, L22, L23 on the surface of the 50S ribosomal subunit of *Escherichia coli* by immune electron microscopy. *Mol. gen. Genet.* **134,** 187–208.

Typke, D. & Radermacher, M. (1982) Determination of the phase of complex atomic scattering amplitudes from light-optical diffractograms of electron microscope images. *Ultramicroscopy,* **9,** 131–138.

Unwin, P.N.T. (1974) Electron microscopy of the stacked disk aggregate of tobacco mosaic virus protein. II. The influence of electron irradiation on the stain distribution. *J. molec. Biol.* **87,** 657–670.

Verschoor, A., Frank, J., Radermacher, M., Wagenknecht, T. & Boublik, M. (1984) Three-dimensional reconstruction of the 30S ribosomal subunit from randomly oriented particles. *J. molec. Biol.* **178,** 677–698.

Verschoor, A., Frank J. & Boublik, M. (1985) Investigation of the 50S ribosomal subunit by electron microscopy and image analysis. *J. Ultrastr. Res.* **92,** 180–189.

Wabl, M.R. (1973) Tilting experiments with negatively stained *E. coli* ribosomal subunits. An electron microscopical study. *Cytobiologie,* **7,** 1–9.

Whittaker, E.T. (1915) On the functions which are represented by the expansions of the interpolation-theory. *Proc. Roy. Soc. Edinb.* A, **35,** 181–194.

Williams, R.C. & Fisher, H.W. (1970) Electron microscopy of tobacco mosaic virus under conditions of minimal beam exposure. *J. molec. Biol.* **52,** 121–123.

J. Mol. Biol. (1990) **214**, 737–749

Three-dimensional Structure of the Mammalian Cytoplasmic Ribosome

Adriana Verschoor and Joachim Frank

Wadsworth Center for Laboratories and Research
New York State Department of Health
Box 509, Albany, NY 12201–0509, U.S.A.

(Received 28 June 1988; accepted 11 April 1990)

A three-dimensional reconstruction of the 80 S ribosome from rabbit reticulocytes has been calculated from low-dose electron micrographs of a negatively stained single-particle specimen. At 37 Å resolution, the precise orientations of the 40 S and 60 S subunits within the monosome can be discerned. The translational domain centered on the upper portion of the subunit/subunit interface is quite open, allowing considerable space between the subunits for interactions with the non-ribosomal macromolecules involved in protein synthesis. Further, the cytosolic side of the monosome is strikingly more open than the membrane-attachment side, suggesting a greater ease of communication with the cytoplasm, which would facilitate the inwards and outwards diffusion of a number of ligands. Although the 60 S subunit portion of the 80 S structure shows essentially all of the major morphological features identified for the eubacterial 50 S large subunit, it appears to possess a region of additional mass that evidently accounts for the more ellipsoidal form of the eukaryotic subunit.

1. Introduction

Because ribosomes from eukaryotes are larger and more complex than those from eubacteria, despite the striking interkingdom conservation of function, the determination of a structure for a eukaryotic ribosome at even a medium resolution should have important implications for investigation of structure–function relationships. Comparisons of two-dimensional projections from electron micrographs of ribosomes from eubacteria, methanogenic archaebacteria, and sulfur-dependent archaebacteria, and organellar and cytosolic ribosomes from lower and higher eukaryotes (e.g. see Lake *et al.*, 1982) suggest that a number of gross morphological features related to universal ribosomal functions are conserved. Other features that appear to be absent or vestigial in one taxon, yet well-developed in another, may be related to evolutionarily non-conserved aspects of ribosome function. An obvious example is the greater number of ribosomal proteins and RNAs and greater sizes of the RNAs in the (cytoplasmic) ribosome of a higher eukaryote as compared to that of a eubacterium.

Since the whole process of protein synthesis is more complex (or less streamlined) in the eukaryote, it is clear that additional functional sites must exist, and that even "universal" (conserved) functional domains may be modified to varying extents in size and morphology. Direct assignment of a particular function to a morphological feature can only be made from a specimen with a function-specific ligand bound. However, once a structure is available for the eukaryotic ribosome, approximate assignments for certain sites should be possible based on correlations with the eubacterial ribosome, for which a number of key functional sites have been mapped by various groups (e.g. see Stöffler & Stöffler-Meilicke, 1986).

Several three-dimensional (3D†) reconstructions of the eukaryotic ribosome have been published by Unwin and co-workers (Unwin, 1977; Kuhlbrandt & Unwin, 1982; Milligan & Unwin, 1986), who applied conventional Fourier methods to images of crystalline ribosome arrays from reptilian and avian sources. Since, however, the size and ordering (both long-range and short-range) of the crystals sufficed only for relatively low resolution (at best 55 to 60 Å: 1 Å = 0·1 nm) structural determinations, these reconstructions provide only limited information about the manner in which the large and small subunits are arranged in the ribosome, or about the detailed morphology of the overall monosome structure.

† Abbreviations used: 3D and 2D, 3-dimensional and 2-dimensional; LFF, left-featured frontal; Coran, correspondence analysis.

0022–2836/90/150737–13 $03.00/0

© 1990 Academic Press Limited

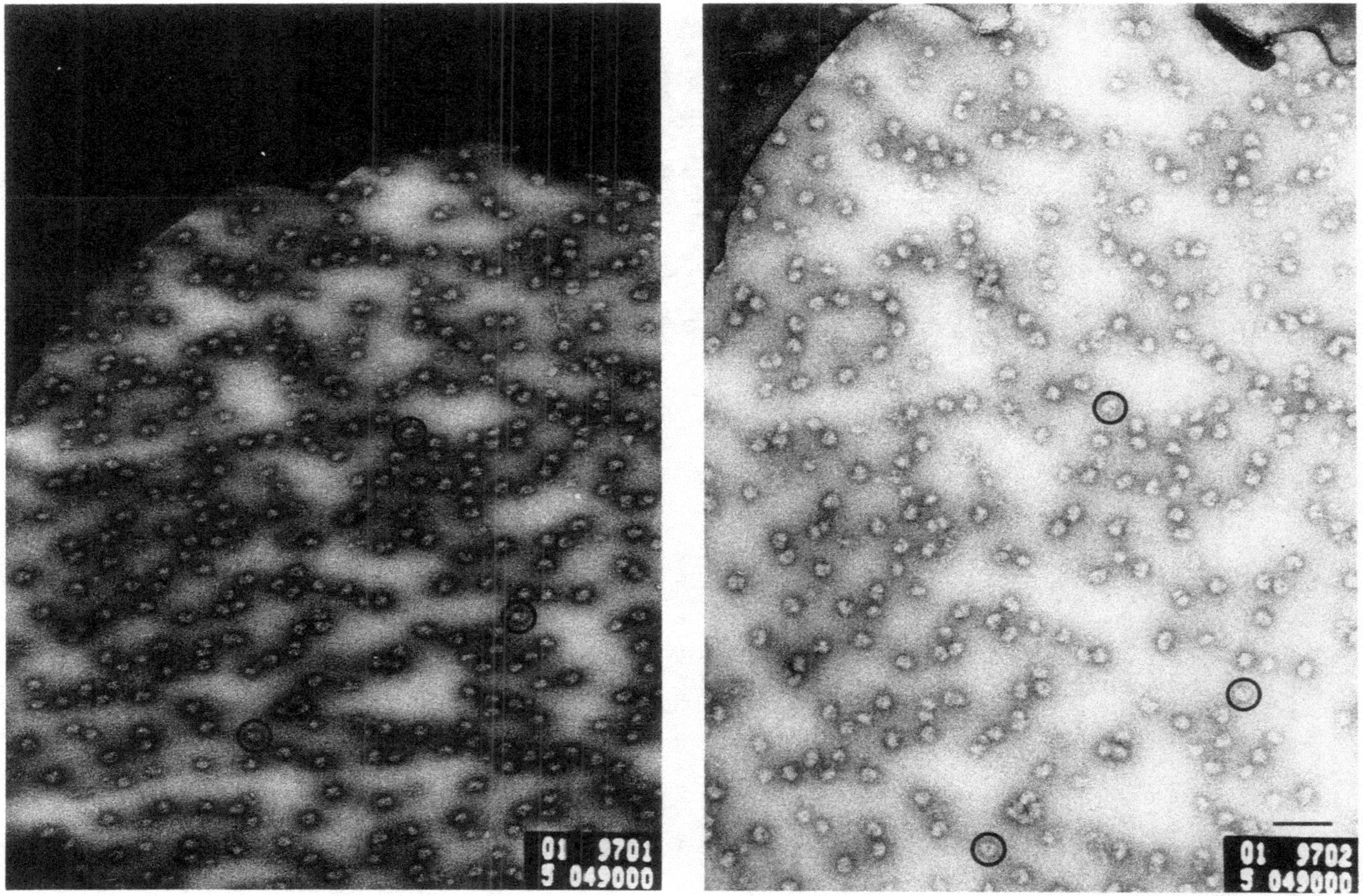

Figure 1. Representative field, shown at 50° tilt (left) and untilted (right), from a set of electron micrographs of 80 S ribosomes from rabbit reticulocytes. Left-featured frontal-view particle images (3 of which are circled) from the non-tilt micrographs are used for 2D averaging, resolution determination and statistical analysis (Coran), as well as for determination of the azimuthal angles of their tilt counterparts. The particle images at 50° tilt, taken from low-dose micrographs, are the actual input to the reconstruction algorithm. A reconstruction from 459 tilt images forming a random-conical tilt series was calculated. The tilt axis is slightly inclined to the horizontal (from upper right to lower left). The scale bar represents 1000 Å.

To gain higher-resolution 3D information, we have investigated a single-particle (i.e. non-crystalline) specimen of ribosomes from rabbit reticulocytes. A very effective low-irradiation technique for reconstruction of non-crystalline specimens has been developed, the single-exposure, random conical scheme (Radermacher *et al.*, 1987*a*). The method, which exploits a high-frequency occurrence within the specimen of a single stable orientation of the particles on the support, uses paired tilt and non-tilt micrographs to generate the equivalent of a conventional conical tilt series, as input to the reconstruction algorithm. One particular view of the 80 S ribosome, termed the left-featured frontal (LFF; Fig. 1; Nonomura *et al.*, 1971) view, is highly distinctive, despite its occurrence with only moderate frequency. We chose this view as the basis for the 3D analysis.

In the following, we describe the determination of a medium-resolution structure for the eukaryotic (cytoplasmic) ribosome, a structure that suggests a surprisingly large degree of interkingdom conservation of ribosome morphology. Despite the differences in specific features of the large and small subunits, most notably the numerous distinct protrusions or lobes on the eukaryotic 40 S subunit that are absent or vestigial on the eubacterial 30 S subunit, we can readily orient the new 80 S ribosome structure in such a way that direct comparisons to previously reconstructed eubacterial ribosomal particles (e.g. see Arad *et al.*, 1987; Radermacher *et al.*, 1987*a*,*b*; Wagenknecht *et al.*, 1987, 1989; Carazo *et al.*, 1988, 1989) can be made.

2. Materials and Methods

The 80 S ribosomes (a gift from T. Obrig) were isolated from rabbit reticulocytes according to Hardesty *et al.* (1971), and were run over a 10% to 30% (w/v) sucrose gradient before preparation for electron microscopy. Electron microscopic specimen preparation, recording protocols and image processing were performed as described (Radermacher *et al.*, 1987*a*,*b*). Briefly, low-dose (10 e/Å^2) micrographs were recorded at a magnification of 49,000 ×, on a Philips EM 420 microscope equipped with a tilt-rotation stage and low-dose kit, from double-carbon preparations (method modified from Tischendorf *et al.*, 1974) of ribosomes negatively contrasted with 0·5% (w/v) uranyl acetate. Micrographs were recorded in pairs for each specimen area, the first at a goniometer reading of 50° (tilt micrograph) and the second at 0° (non-tilt).

In our preparation of 80 S ribosomes (and according to our criteria of particle selection) 10 to 20% of the particles adsorbed to the specimen support in a distinctive orientation that was termed the left-featured frontal (LFF) orientation. A total of 510 particles presenting the LFF view (Fig. 1) were selected from 11 0° micrographs, and their counterparts were selected in parallel from the tilt micrographs through a modified (Radermacher, 1988) interactive selection program (Radermacher *et al.*, 1987*a*). After standardized alignment of the non-tilt images, the tilt images were individually aligned to a common origin by correlation with their non-tilt counterparts (Radermacher, 1988).

Correspondence analysis (Coran; van Heel & Frank, 1981; Frank & van Heel, 1982), a multivariate statistical technique, was used to identify variations within the set of LFF-view projections from the non-tilt micrographs. In this initial study, Coran was principally used to identify particles showing atypical characteristics in their non-tilt representations. By this screening for outliers, 7% of the total set of particles were eliminated from further analysis. A further 3% of the total set were discarded on the basis of image flaws in their tilt representations. The eigenvectors in the analysis were qualitatively assessed, and Factor 2, the eigenvector accounting for the second highest amount of interimage variance, was initially identified as relating to an apparent small-scale "rocking" of the LFF-orientation particles, around a subunit-heads to monosome-base axis (the vertical in Figs 2 and 3), with respect to the support. Although the magnitude of the variational range of the population of LFF particles did not appear large, it could not be quantified until some 3D information became available†. Accordingly, the set of particle images was not specifically restricted on the basis of Factor-2 values for the reconstruction.

3. Results

From the tilt counterparts of the set of 459 LFF images remaining after the exclusion of aberrant particles, a 3D reconstruction was calculated through the use of a weighted backprojection algorithm (Radermacher *et al.*, 1987*a*). Comparison of two independent half-set reconstructions, additionally calculated from complementary subsets of 229 and 230 images, gave a reproducible 3D resolution of 37 Å, according to the 45° differential phase residual criterion (Frank *et al.*, 1981; Radermacher *et al.*, 1987*a*). The total-set 3D reconstruction is shown in surface representation, rotated through 360° in 20° increments in Figure 2. It is shown in Figure 3 as a projection series, with a 10° rotational increment around the vertical (heads-to-base) axis.

The characteristic features of the two subunits can for the first time be recognized and located in three dimensions, and thus the precise subunit arrangement in the monosome is revealed (Fig. 2). In the 0° and 180° surface views in Figure 2, the two

† To derive an estimate of the putative rotation range, 3 refinement 3D reconstructions were subsequently computed from image subsets that were limited according to Factor-2 value. The result (Verschoor, 1989) was that the subset reconstructions showed no overall rotations with respect to one another; thus, the factor did not represent a rigid-body rotation of the particle. Rather, the systematic component of the variability related to the trend of the subunit/subunit interface, believed to result from some degree of initial rocking variability of the particles, with subsequent deformation to a common form with respect to the carbon layers. However, because of the approximately Gaussian distribution of the image population along the factor expressing this interface variation, the extreme-valued particles were not numerous, and thus their inclusion in the global reconstruction caused little perceptible degradation of the resultant structure.

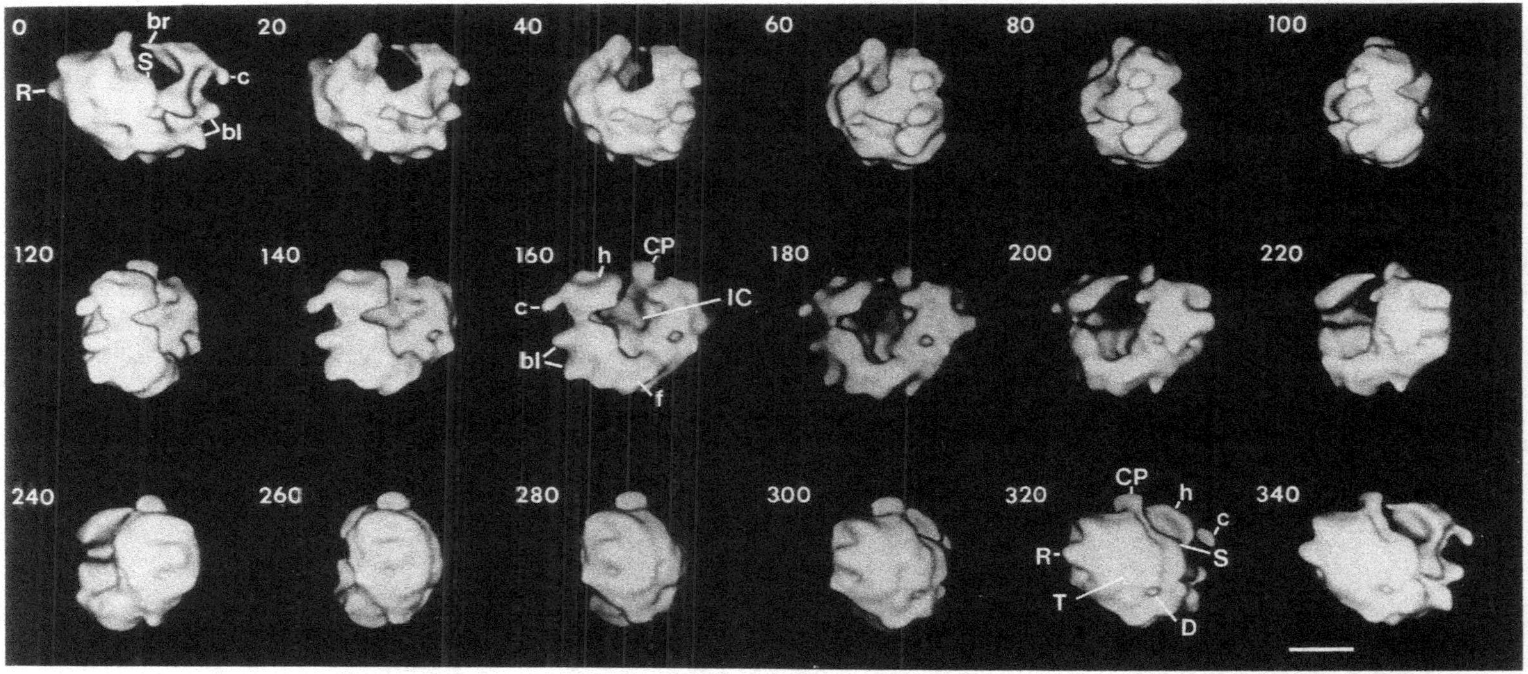

Figure 2. Surface views of the reconstructed 80 S ribosome, shown with a rotational increment of 20°. (The axis of rotation in the representation coincides with the axis around which the particle may have originally "rocked" by a small amount to give rise to the variability among the LFF views; see the footnote to p. 739.) The structure is shown at 37 Å resolution. The 180° view shows the surface that in the membrane-bound ribosome would be most freely accessible to translational ligands in the cytosol; note the large open area at the interface between the 40 S subunit (left-hand domain) and the 60 S subunit (right-hand domain). This 180° surface is also the side of the ribosome that adsorbed to the carbon film in the specimen preparation. The following 60 S subunit features are marked: CP, central protuberance; IC, interface canyon; R, stalk-base ridge; S, shoulder; D, exit-domain dimple; and T, putative exit-related trough. The following 40 S subunit features are marked: h, head; c, crest; br, bridge; bl, back lobes; and f, feet (or basal lobes). The scale bar represents 100 Å.

subunits appear roughly side by side. The long axis of the 40 S subunit is somewhat inclined to the vertical heads-to-base axis (corresponding roughly to the apparent original rocking axis of the LFF particle) of the monosome. As was previously inferred (Nonomura *et al.*, 1971; Lutsch *et al.*, 1972), the plane approximating the subunit/subunit interface surface in the frontal-view monosome is slightly oblique rather than normal to the plane of the support (see below).

The surface of the monosome that would lie tangential to the plane of the membrane, if the ribosome were membrane-bound†, represents the exit-domain region. It is on this convex and relatively featureless surface of the 80 S structure that the exit (Bernabeu *et al.*, 1983) and membrane-binding (e.g. see Unwin, 1977) sites have been mapped. The opposite side of the ribosome can be considered the cytosol-facing aspect of the ribosome. It is striking that almost the entire cytoplasmic aspect of the interface area is open, with a marked separation between the subunits except in the lowermost portion of the structure, where the bases of the subunits overlap.

The height (heads-to-base dimension) of the reconstructed ribosome is approximately 260 Å, and its maximum width is approximately 320 Å. These dimensions agree to within 15% with the dimensions obtained by X-ray scattering for the solvated ribosome (Damaschun *et al.*, 1974). The thickness of our reconstruction is roughly 210 Å, but no comparable measurement is available from the solution study. However, one crude determination of the particle thickness is available from the reconstruction of a single layer of a two-layer ribosome crystal in frozen-hydrated preparation (Milligan, 1985). The single crystalline layer was estimated to be 300 Å thick (Milligan, 1985). Because, however, the density envelope in the crystalline reconstruction did not have a sharp falloff, in part due to the poor resolution in a direction perpendicular to the plane of the crystal, we cannot make a reliable comparison between that reconstruction and ours, in terms of the density threshold at which the particle thickness was estimated. The comparison may, nevertheless, be of some use, in providing a rough upper limit for the possible amount of flattening in our structure.

(a) *Morphologies of the subunits in the reconstructed monosome*

(i) *The 40 S subunit*

The 40 S subunit in our reconstructed monosome is very roughly cylindrical, with a constricted neck separating the so-called head portion from the body portion of the subunit, and a marked curvature of the base (Fig. 2). Instead of the roughly conical shape, with a tapering base, that has been deduced for the eubacterial 30 S subunit (for a review, see Wittmann, 1986), the body of the 40 S subunit possesses basal lobes ("feet") that give it a broader, blunter base. It is in this basal region of the 40 S subunit that the association with the 60 S subunit is tightest; in the upper half of the interface region of the monosome there is no apparent contact of the two subunits (except for the bridge between the heads; see below). The basal lobes of the 40 S subunit are most clearly seen as a structure curving towards the 60 S subunit in the 140°, 160° and 180° surface views in Figure 2.

The upper body of the 40 S subunit also possesses two "back lobes" pointing more or less away from the direction of the interface with the 60 S subunit. The back lobes are seen in profile in the 160° and 340° surface views, and roughly end-on in the 60° and 80° views, in Figure 2. Lutsch *et al.* (1985) mapped (in 2 dimensions) the binding site for the largest of the initiation factors, eIF-3, to these back lobes, which are clearly the morphological equivalent of the platform feature of the eubacterial 30 S subunit.

The head of the 40 S subunit shows several distinctive protrusions. On the exterior aspect of the head is a well-defined "crest" feature (seen in profile in the 140° view, and end-on in the 80° view, Fig. 2), which appears as a more pronounced feature than was expected from our 2D studies of the isolated 40 S subunit (Frank *et al.*, 1982).

In contrast, the beak (also termed the archaebacterial bill; Lake *et al.*, 1982) is difficult to recognize in our reconstruction, although it has been considered one of the most characteristic features of the head in images of isolated 40 S subunits. Our previous 2D (Frank *et al.*, 1982) and 3D (Verschoor *et al.*, 1989) analysis of lateral-view 40 S subunits revealed this beak to be a substantial, asymmetrically attached protuberance, discernible on the "right-lateral" aspect of the head as a continuously tapering, extended structure, but meeting the head with a sharp break in profile on the "left-lateral" side (i.e. the side of the particle that adsorbed to the carbon film to give rise to a left-lateral view).

In our 80 S structure, we see only vestiges of a beak structure (Fig. 2, middle row). It appears to be directed towards the primary (Verschoor *et al.*, 1986) carbon, presumably an unfavorable direction for the preservation of an extended conformation for a thin, labile feature. The weak expression of this feature in the reconstruction volume undoubtedly represents a blurring of the mass due to its positional variability in the input projections.

Another feature extending from the head was unexpected, in that it had not been recognized from 2D and 3D studies of the isolated 40 S subunit (Frank *et al.*, 1982; Verschoor *et al.*, 1989). This feature, a thin, linear "bridge" structure (best seen in the 0° and 180° views in Fig. 2), partially or

† It should be noted that, although only a small proportion of reticulocyte ribosomes are membrane-bound, there is not conclusive evidence to suggest morphological differences between bound and non-bound ribosomes (unless possibly in the presence or absence of an actual membrane-attachment protrusion). There is thought to be free exchange between the 2 populations; see Burka & Schickling (1980).

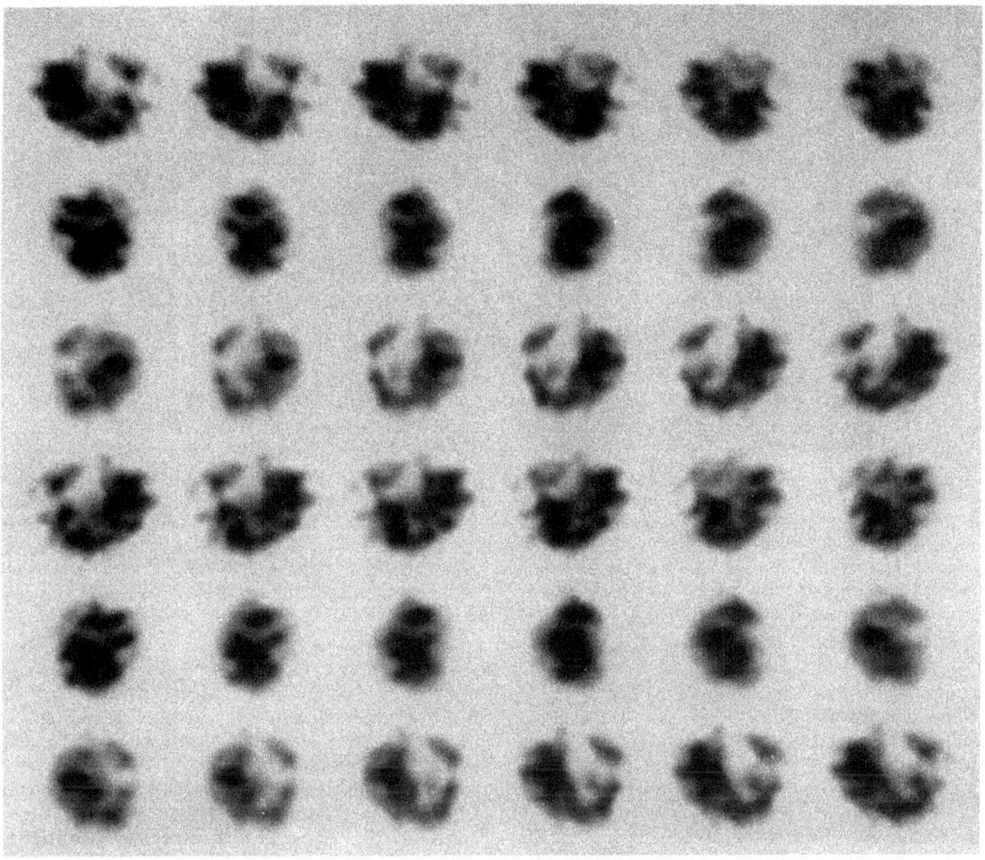

Figure 3. Thresholded projections of 3D reconstruction, shown with a 10° rotational increment around the vertical axis. These can be viewed in stereo, allowing the relative positions of density maxima (dark regions: contrast has been inverted) in the 2 subunits to be seen in differently overlapped positions. The 0° projection (row 1, 1st image) is, by definition, the LFF view. The 300° projection (row 6, 1st image) corresponds precisely to another characteristic view of the 80 S particle, termed the dorsal view (Lutsch *et al.*, 1972), which was not used in our reconstruction. Still a 3rd preferred view, the lateral view, is ambiguous: it could be simulated either by rotation around an orthogonal axis, which gives rise to end-on views (Verschoor, 1989), or it could be generated by another axial rotation as, for instance, the 90° projection (row 2, 4th image). These 2 very different orientations of the monosome give rise to quite similar projections, and the choice of which is the actual orientation seen in micrographs can not be made without additional information

completely (depending on the choice of threshold value in these surface representations) spans the intersubunit gap to join to the central protuberance of the 60 S subunit. This thin connection, which was already evident from 2D averaged images (data not shown) of the LFF-view monosome, thus appears to link the subunit heads at their point of closest approach.

(ii) *The 60 S subunit*

The 60 S subunit portion of the 80 S structure is, as noted earlier, easily distinguishable from the 40 S subunit portion, except at the base, where the two subunits are overlapped and in tight contact. As viewed from the back or exterior aspect (e.g. see 320° and 340° views in Fig. 2), the 60 S subunit has a different overall shape from the better-studied 50 S subunit (Radermacher *et al.*, 1987*a*,*b*). Rather than globular like the *Escherichia coli* structure, the 60 S subunit appears ellipsoidal. The major axis of the ellipsoid forms an angle of approximately 45° with the small, round central protuberance (which points vertically in the heads-to-base axial rotations; Figs 2 and 3).

The marked ellipticity of the subunit body, in contrast to the more equidimensional form of the *E. coli* 50 S subunit, may well be an indication of the fact that the mammalian large subunit contains areas of additional mass. However, in other views, the 60 S subunit form appears very similar to the 50 S subunit form; for example, if the monosome reconstruction is axially rotated by a small amount

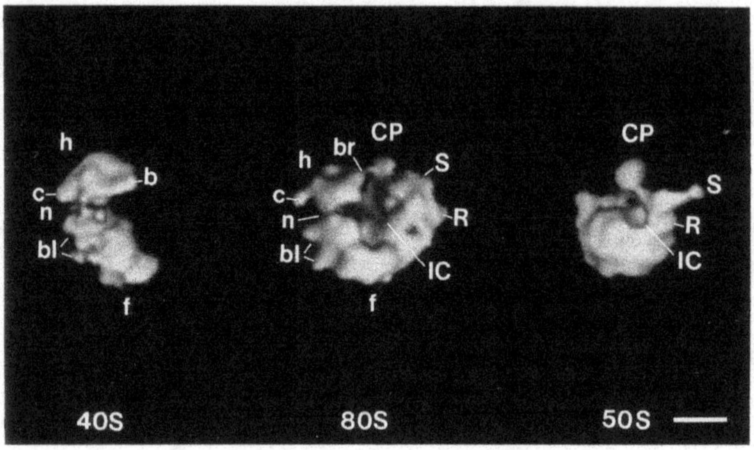

Figure 4. Comparison of the 80 S monosome with the 40 S small subunit from rabbit reticulocyte ribosomes (Verschoor *et al.*, 1989) and the *E. coli* 50 S large subunit (Radermacher *et al.*, 1987*a*) reconstructions. All 3 reconstructions are shown low-pass filtered to approximately the same resolution (37 to 38·5 Å), and the 2 subunit reconstructions have been re-oriented to provide crude positional matches with their counterparts in the 80 S monosome view. The following features of the 40 S subunit in the 40 S and 80 S reconstructions are labeled: h, head; n, neck; c, crest; b, beak, or br, bridge; bl, back lobes; and f, feet. Also labeled are the following large-subunit features in the 80 S and 50 S reconstructions; CP, central protuberance; IC, interface canyon; R, stalk-base ridge; and S, stalk (or putative stalk). The scale bar represents 100 Å.

so that the 60 S subunit portion is oriented with the "plane" of its interface side normal to the viewing direction (e.g. 300° view, Fig. 2). This rotation-dependent variation of the projected shape would suffice to explain the mixture of sub-round crown-type views and strongly elliptical views typically seen in micrographs of 60 S particles from higher eukaryotes (e.g. see Montesano & Glitz, 1988).

In the eubacterial 50 S subunit, three protuberances are grouped around one end of the structure. In an orientation analogous to the orientation of the 60 S subunit in the 320° and 340° views of the 80 S structure (Fig. 2), the central protuberance would be vertical, the L1 shoulder would be to the right of it, and the L7/L12 stalk would extend to the left of it. In our eukaryotic structure there is a low shoulder (marked S in Fig. 2) on the right, in good spatial and morphological agreement with the 50 S subunit L1 shoulder; however, to the left of the central protuberance we do not see an extended stalk structure. The region of the body below the central protuberance on this side does appear somewhat more massive than in the 50 S subunit, and one possibility is that this mass relates to the presence of the stalk-forming proteins in a folded-over orientation (see Discussion).

Aside from the central protuberance, a large "ridge" feature (marked R in Fig. 2) on the side of the body is perhaps the most distinctive protruding feature of the 60 S portion of our reconstruction. It appears to be a major structural as well as morphological element, comprising a densely packed bar of material running roughly horizontally through the middle of the subunit body (Fig. 3). Although it is a markedly larger and denser feature than the stalk-base ridge described for the 50 S subunit (Radermacher *et al.*, 1987*b*), we interpret it to be the morphological analog of that feature. In the *E. coli* 50 S subunit the stalk-base ridge, which occurs in the region of attachment of the L7/L12 stalk to the body of the 50 S subunit, persists in L7/L12-depleted 50 S subunits (Carazo *et al.*, 1988) and is thus not part of the stalk proper.

On the interface aspect of the 60 S subunit (which is difficult to study in the monosome structure because of the overlap of the 40 S subunit; e.g. Fig. 2, 140° to 160° views), the interface canyon feature (labeled IC) initially characterized on the eubacterial 50 S subunit (Radermacher *et al.*, 1987*a*,*b*) appears as a trough with moderately shelving sides. The trend and overall form of the canyon are very similar to what was described for *E. coli* in the 50 S subunit (Radermacher *et al.*, 1987*a*,*b*) and 70 S monosome (Wagenknecht *et al.*, 1987, 1989) reconstructions. The canyon trends diagonally across the front or interface side of the subunit, originating as a broad, shallow trough in the region between the shoulder and the central protuberance, and ending blindly as a more deeply scooped-out feature in the region of the stalk base near the other edge of the subunit body.

The back of the 60 S subunit in our reconstruction shows several concave surface features. One is a vertically trending, groove-like trough traversing the mid-back (Fig. 2, 280° to 340° views). Although this trough feature shows an apparent morphological similarity to the hypothetical exit "conduit" proposed by Ryabova *et al.* (1988), a mapping study would be required to demonstrate whether this trough indeed has an exit-related function.

To the right of the medial trough feature is a small but pronounced circular "dimple". This feature appears to be situated such that it would, in a membrane-bound ribosome, face the membrane almost directly (Fig. 2, 320° and 340° views; see also Milligan & Unwin, 1986). Again, however, we can not say whether this topographic feature has a function in either exit-related or membrane-binding processes.

A comparison of the 80 S structure, in an oblique orientation that reveals some features of the interface side of the 60 S subunit, with a view of the 50 S subunit reconstruction (Radermacher *et al.*, 1987*a*) in a similar orientation is shown in Figure 4. Despite the partial overlap of the 40 S subunit in the 80 S view, the match in such features as the central protuberance, the stalk-base ridge, and the interface canyon can be seen between the two large-subunit structures. Additionally shown in this Figure is a view of a reconstruction of the isolated (dissociated) 40 S subunit, also from rabbit reticulocytes and computed using the same random conical method (Verschoor *et al.*, 1989). The two determinations of the small subunit structure show a good agreement in the size, shape and placement of such features as the crest, neck, back lobes and feet.

4. Discussion

Our reconstruction of the 80 S ribosome allows us to examine in three dimensions the morphological features of the two complex shaped subunits that were previously known only from 2D studies. When viewed in projection, the reconstruction shows marked density variations, depending on the angle of view (Fig. 3). Because of a demonstrable agreement in the general pattern of interior density maxima, between a projection of this negative stain structure and an averaged 2D image of a tetrameric ribosome (Verschoor, 1989; Verschoor *et al.*, 1990), we conclude that the density variations contained in our reconstruction are meaningful in a broad sense.

The projection representations in Figure 3 allow comparison of projections computed from the reconstruction with experimental projections, i.e. particle images in an electron micrograph. One of the other (non-LFF) characteristic views of the 80 S particle, the so-called dorsal view (Lutsch *et al.*, 1972), which was recognized also by Boublik & Hellmann (1978), appears to be precisely matched by one of the axial rotations shown, namely the 120° projection (Fig. 3, row 3, 1st image). This dorsal view led to a certain amount of confusion in earlier attempts to infer the 3D structure from the characteristic projections. Lutsch *et al.* (1972) concluded that it was related to the frontal view by a "slight tilting", whereas the two projections of our structure that relate to the frontal and dorsal views are separated by an angle of 60°. Boublik & Hellmann (1978) concluded from interpretation of the dorsal type of view that the head-to-base axes of the two subunits must be orthogonal to one another in the monosome; again, a relationship not borne out by our reconstruction.

Another characteristic view recognized in micrographs, the so-called lateral view (Nonomura *et al.*, 1971; Lutsch *et al.*, 1972) is more equivocal; (at least) two very different orientations of the monosome could give rise to such an appearance, in which a smaller ellipsoidal mass appears to lie somewhat offset along one side of a larger blocky mass, forming a roughly triangular structure. The simplest interpretation is that the lateral view is generated by a rotation, again around the heads-to-base axis, of 90° away from the LFF-view orientation (and 30° from the dorsal view). This 90° projection (Fig. 3, row 2, 4th image) shows the subunits most completely overlapped and is the narrowest axial projection of the monosome.

An alternative explanation for the lateral view is that the monosome adsorbed more or less end-on to the primary carbon. The general shape of the particle in the lateral view is quite similar to that seen in a calculated "heads-on" rotation of the reconstruction, i.e. a view directly from the top (90° rotation around a horizontal axis, with respect to the convention used in Figs 2 and 3) rather than from the side.

Because, however, the individual lateral views in micrographs show so few distinctive features, we can not determine whether the lateral view is truly a single view, deriving from one or the other of the particle orientations suggested above, or whether it represents several unrelated but visually similar views. This serves to illustrate the ambiguities that have arisen in some immuno-electron microscopic studies of ribosomes, studies in which particle orientations, relationships among different views, and ultimately a 3D model for the structure, have been deduced mainly from visual appearances, without supporting tilt experiments.

(a) *Comparison to crystalline ribosome reconstructions*

A direct comparison of our structure with the reconstructions by Unwin's group from stained and unstained crystalline ribosomes (e.g. see Unwin, 1977; Kühlbrandt & Unwin, 1982; and, most particularly, Milligan, 1985; Milligan & Unwin, 1986) would be expected to be very informative. However, the resolution of the crystalline reconstructions has been too limited to reveal the presence of well-characterized morphological features of the subunits that we would wish to use as a basis for correlating and aligning the two structures. Even when our structure is low-pass filtered to 55 Å, the highest resolution quoted for the crystalline reconstructions, we still see a considerably more detailed morphology than can be discerned in published representations of the crystalline structures.

If we attempt to align our structure with the frozen-hydrated crystalline structure, by a best match of the low-resolution form ("envelope") of

Figure 5. Individual frontal-orientation 80 S ribosomes in frozen-hydrated preparation. The morphology may be compared to the 180° to 200° views of the 3D reconstruction in Fig. 2. Such features as the head (h) of the 40 S subunit, and the central protuberance (CP) and stalk (S) of the large subunit can be clearly recognized. The intersubunit gap, indicated by an arrow, represents a distinct separation between the upper portions of the 2 subunits. This demonstrates that, for our negatively stained 3D reconstruction, positive staining artefacts are insufficient to account for the existence of the gap feature. The scale bar represents 250 Å.

our reconstruction to the contour representation presented by Milligan & Unwin (1986), the result is unexpected. We obtain an orientation in which the situation of the subunits differs by a rotation of perhaps 90° from the assignment by Milligan & Unwin (1986). This marked discrepancy in the assignments for the locations and orientations of the two subunits of the ribosome must be resolved before any meaningful comparison can be made.

The crystalline 3D reconstruction was interpreted to show the 40 S subunits lying around the periphery of the tetramer motif (the "building block" for a 2D crystalline array), with the 60 S subunits towards the tetramer center. Following from these assignments, the elongate low-density feature in the crystalline structure that trended roughly normal to the presumed direction of the long axis of the small subunit was interpreted to represent a tunnel through the large subunit, connecting its (unresolved) interface aspect and the putative exit site, which had been very roughly mapped in 2D by Bernabeu *et al.* (1983).

In striking contrast to the existence of this apparent large tunnel feature, no interface gap between the upper portions of the subunits was discerned. In functional terms, this lack of intersubunit space to accommodate the non-ribosomal translational ligands, at least several of which are large, bulky molecules, was surprising. More recent reconstructions of prokaryotic monosomes in non-negatively stained preparations (e.g. see Arad *et al.*, 1987; Penczek *et al.*, 1990) have revealed intersubunit spaces evidently as large as that we describe for the eukaryotic structure.

For our non-crystalline 80 S structure, there is no ambiguity regarding the locations and orientations of the subunits within the envelope of the structure, because of the capability at our resolution of directly recognizing their characteristic morphological features. If four repeats of our structure were organized into a tetramer, the result would be a radial arrangement of alternating large and small subunits of the four frontal-orientation (with respect to the support or "membrane" plane) ribosomes. This arrangement gives rise to a pattern that agrees in detail with averaged images of actual tetramers, from another vertebrate species (Milligan, 1985), that we have analyzed (Verschoor, 1989; Verschoor *et al.*, 1990; see below). The bases of both the 40 S and the 60 S subunits are directed towards the tetramer center, with the subunit heads pointing out in the intratetramer link region. The isolated blob of dense material discerned in this region of the crystalline structure by Milligan and Unwin would represent the central protuberance of the 60 S subunit in our structure. The elongate low-density "tunnel" feature that Milligan and Unwin described as transecting the central dense core of the ribosome would presumably correspond in our structure to the intersubunit gap.

(b) *Possible specimen preparative effects on the structure*

One aspect of the reconstruction that we need to assess is the extent to which the specimen preparative techniques used might have caused alterations to the native particle structure. Among the principal effects to be considered in a uranyl acetate-stained single-particle ribosome preparation are, obviously, specimen collapse due to air drying and positive staining of rRNA.

(i) *Deformations*

Because our reconstruction was calculated from projections of an air-dried, negatively stained specimen, we assume that it shows a structure flattened to some extent in comparison to the native ribosome. The actual degree of flattening, however, is hard to estimate, in the absence of some reliable determination of the thickness of the solvated particle. (We note that actual mass loss or thinning of the specimen induced by irradiation should be comparatively minor at the electron dose used, at most about 10% (Berriman & Leonard, 1986).)

If a considerable degree of flattening can have occurred, one aspect of our random conical data collection scheme becomes particularly important; namely, that all of the particles contributing to the reconstruction will have suffered essentially the same deformations, since they lie in basically a single orientation with respect to the carbon support. This is crucial in ensuring consistency among the input projections to the reconstruction algorithm.

The main direction of the compression due to air-drying will trend normal to the plane of the specimen support. In the 180° surface view of the reconstruction (Fig. 2), this main direction of compression would be from back to front (i.e. directly towards the viewer). It can further be seen from Figure 2 that the axial rotation of the monosome in which the subunits lie most nearly side by side, i.e. with the least amount of overlap, is shown in the 200° view.

From this observation, we can state that a hypothetical interface plane, which we define as a vertical plane joining the points where the subunits meet one another on the two sides of the ribosome, and which thus approximately bisects the intersubunit gap, would be seen edge-on in the 200° view. Because this plane would intersect the back-to-front "compression" plane at an angle of only 20°, it appears that overlap-related shearing of the subunits should be minor.

We conclude that the original spatial relationship between the two subunits in the monosome should be minimally perturbed in our reconstruction from LFF-view particles†. The main effect of the deformation(s) is probably a certain reduction of the maximum width or thickness of the 60 S subunit. Since wrapping (Kellenberger *et al.*, 1982) of the two carbon films around the particle is likely to distribute the compressional forces to some extent, the actual particle morphology that we have reconstructed should represent a fairly smoothly deformed version of the native structure.

(ii) *Possible positive staining of RNA*

In the 80 S structure, as in other ribosome structures obtained from negatively stained specimens, the potential exists for artefacts to occur due to positive staining of exposed rRNA by the uranyl acetate. The actual occurrence of positive staining for any type of ribosomal particle would be dependent on a number of factors, including the experimental solution pH and the degree of exposure and the local organization and packing of the ribonucleoprotein.

Comparisons of the low-resolution 3D maps presented by Kühlbrandt & Unwin (1980, 1982) for crystalline ribosomes in uranyl acetate, aurothioglucose and glucose preparations naturally suggest the likelihood that a structure derived from a uranyl acetate-stained specimen will possess an undetermined amount of superficial positive staining, expressed as enhanced local stain accumulation. Without higher-resolution comparisons to unstained specimens, however, it is not possible to attempt any quantification of the surface effects of the stain on our structure. Certainly, no significant interior penetration of our structure by the stain appears to have occurred.

It is clear that the delineation of the characteristic morphological features of the subunits in our reconstruction is not a mere stain-induced exaggeration of slight topographical or compositional (ribonucleoprotein) variations, since all of the principal features can be recognized as well in unstained, frozen-hydrated images. In frozen-hydrated images of the isolated 40 S subunit (T. Wagenknecht & R. A. Milligan, unpublished results) we see the beak, back lobes and feet as sharply delineated and protuberant features.

Recently, micrographs of single-particle 80 S ribosomes in frozen-hydrated preparation were obtained in our laboratory (Verschoor *et al.*, 1990). Frontal-orientation particles revealed, among the more striking features, the head of the 40 S subunit, the pronounced subunit/subunit interface gap, the knob-like central protuberance of the 60 S subunit, and frequently traces of the stalk feature of the large subunit (see below). These raw frontal-view images (Fig. 5) can be matched directly to the approximately 180° to 200° views of the negative-stain 3D reconstruction (see Fig. 2).

We have, as noted earlier, obtained averaged images of frozen-hydrated tetramers of ribosomes lying in a frontal orientation (Verschoor, 1989); these show a projected topography of comparable intricacy to what was seen in 2D averages (not shown) over the isolated LFF 80 S ribosomes in the present study. Again, the extent of the agreement demonstrates that the details of the topography and the internal density variations seen in the negatively stained ribosome must more or less faithfully represent actual features of the ribosome structure.

We do not exclude the possibility that the large stain-defined space between the upper portions of the subunits in our reconstruction could have been enlarged to a certain degree by preferential stain accumulation around the near-surface rRNA along the interface aspects of both subunits. However, all of our supporting data from unstained specimens suggest that an appreciable intersubunit gap is a genuine attribute of the eukaryotic monosome, just as for the prokaryotic monosome.

(c) *Comparison with the* E. coli *50 S subunit*

Because images of the 60 S subunit have not been quantitatively studied even in two dimensions, the 3D structure of the subunit, and the extent to which it resembled the large subunit of the eubacterial

† It is because of this relatively "favorable" compression direction for our structure that a comparison with a negative stain reconstruction of the 70 S ribosome from *E. coli* (Wagenknecht *et al.*, 1987, 1989; Carazo *et al.*, 1989) is difficult. The 70 S reconstruction was based on a range of particle views in which the subunits were overlapped rather than side by side on the support, with the result that the compression has caused a complete loss of separation between the subunits. However, for a general comparison of features less seriously affected by this compression, see Frank *et al.* (1990).

ribosome, could only be conjectured until now. Further, while the morphology of the 40 S subunit appears quite constant among ribosomes from higher and lower eukaryotes, the morphology (Boublik *et al.*, 1982), and thus evidently the relative proportions of various preferred orientations, of the 60 S subunit seems to vary among species. In general, while in lower eukaryotes the form appears in electron micrographs to resemble that of the eubacterial 50 S subunit, micrographs of 60 S subunits from higher eukaryotes often reveal a larger and more oblate form for the particle, evidently due in large part to molecular weight increases of the 28 S RNA (Bielka, 1982) as compared to the eubacterial 23 S RNA. Such a difference in the form of the subunit body could indeed be recognized when the 60 S portion of our mammalian ribosome structure and the previously reconstructed 50 S subunit from the *E. coli* ribosome (Radermacher *et al.*, 1987*a*,*b*) were compared (see Results). Otherwise, however, there was a strong similarity in terms of the major morphological features that we recognize as characteristic for the large subunit (Fig. 4).

However, one immediate difference between the 60 S subunit morphology that we discerned in our 80 S reconstruction and the 50 S subunit morphology was the absence of a well-defined protruding stalk feature in the eukaryotic reconstruction, at least as visualized in a conventional surface representation (see Results). A stalk had likewise not been resolved in the crystalline 80 S ribosome reconstructions by Unwin and co-workers (Unwin, 1979; Kühlbrandt & Unwin, 1980; Milligan & Unwin, 1986).

The eukaryotic ribosome contains so-called P (phosphorylated) proteins structurally and functionally related to the stalk-forming L7/L12 proteins of the *E. coli* ribosome. For the two L7/L12 dimers in *E. coli*, labeling studies with monoclonal antibodies (Olson *et al.*, 1986) suggest that one dimer may fold over across the upper body of the subunit, while the other extends as the visible stalk. In eukaryotes, P1 and P2 appear, like L7/L12, to be present as dimers (Uchiumi *et al.*, 1987), and it is plausible that one such dimer forms the (rather labile) stalk that is visible in certain views of the 60 S subunit (Montesano & Glitz, 1988).

In our negatively stained 80 S LFF monosome data, a clearly recognizable stalk was not seen in the majority of the raw particle images. This could be due either to a stripping-off of the stalk from these particles, or a mere folding or repositioning of it, away from a fully extended conformation. Presumably, in the LFF orientation of the 80 S particle, an extended stalk would, if positioned analogously to the *E. coli* stalk, be brought against the carbon film at a large angle as the particle adsorbed, an obviously unfavorable geometry. Interestingly, frontal-orientation ribosomes in a frozen-hydrated preparation (Fig. 5) often showed traces of a variably extended feature in the region of the particle periphery from which the stalk would be expected to project.

In our negative-stain reconstruction, there are weak traces of an apparent extended stalk visible in a set of middle slices through the structure (Verschoor, 1989). Although these traces are too low in density to be visualized in the normal surface representations of the reconstruction, this weakness is likely to be due to positional variability of the stalk structure among the input projections. Such variability could, for instance, extend the stalk pattern in an average image into a planar fan of blurred-out protein density (Verschoor *et al.*, 1985).

The reconstruction also shows weak traces of another extended peripheral feature in the basal region of the ribosome. This basal feature, as well as the putative stalk feature, is seen in averaged images (Verschoor, 1989) of tetramers of frontal-orientation 80 S ribosomes, isolated from hypothermically treated chick embryos (Milligan, 1985). The stalk-analog feature is again very weak in the tetrameric ribosome images, but the basal protrusion is well defined. In the tetramers, the basal protrusions from each of the four ribosomes appear to join at the tetramer center; we have termed them "tether" structures (Verschoor, 1989).

It appears likely that this tether feature of the tetramers represents the same structure as the membrane-attachment structure seen in ribosomes in the crystalline arrays (which are simple aggregates of tetramers). The site of membrane attachment was first recognized by Unwin (1977) in 3D reconstructions from crystalline specimens: a thin, extended structure connected the lower back region of the 60 S subunit to the membrane. On purely morphological grounds, we might by extension conclude that the weak basal protrusion in the single-particle ribosomes also represents the same feature, but such a functional attribution would of course be totally speculative.

If we discount the lack of a strongly defined stalk structure in our reconstruction, the divergence of the overall body shape remains the principal morphological difference (at least at the present resolution) between eukaryotic and eubacterial large ribosomal subunits. The main mass increase appears to be somewhere on the back of the subunit, in the general vicinity of the ridge feature. However, we would consider a more precise localization unjustified, due to the likelihood of shearing (by the force exerted by the 2nd carbon layer during drying-down) of mass positioned away from the midline of the particle. Overall, however, the extent to which the major morphological features of the large subunit are conserved between a eubacterium and a higher eukaryote was unexpected. The implication is that much of the functional mapping and interpretation that has emerged from study of the *E. coli* ribosome can be transferred to the eukaryotic ribosome, as a first approximation. Subsequently, the actual mappings can be performed on the eukaryotic structure, and the degree of conservation of the (medium-resolution) morphology of the functional domains can be established. Mappings of the uniquely eukaryotic translational components may then reveal regions of morphological difference, related to the greater structural complexity of the eukaryotic ribosome.

(d) *Conclusions*

Because of the amount of morphological detail in the new structure that we can recognize and assign to locations in three dimensions, we can draw some very general conclusions about the two main functional domains of the eukaryotic ribosome, based on comparisons to the better-studied *E. coli* ribosome. It must, of course, be borne in mind in these morphological comparisons that the process of translation in eukaryotes, although believed to be organized in a manner broadly similar to that in eubacteria, involves additional reactions, sites, and components as, for example, the requirement of at least nine rather than three initiation factors (Bielka, 1982). Until mappings have been made of the major eukaryotic functional and ligand-binding sites, we can not determine the extent to which such sites are morphologically conserved. It appears that, in general, the interface surfaces of the subunits show a conservation of gross morphology, whereas features of the exterior surfaces appear to be elaborated in the eukaryotic structure; however, this may not hold true at a higher resolution.

The translational domain is centered on the upper-to-middle interface region. The comparative openness of this region could offer ease of entry, interaction, and exit for a number of non ribosomal components involved in translation. The exit domain of the eukaryotic ribosome, encompassing the lower exterior aspect of the large subunit, shares the same nascent chain-related exit functions as the exit domain of the eubacterial ribosome but, in addition, it contains the site of membrane attachment.

Clearly, much work needs to be done before the interactions that center on the two major functional domains of the eukaryotic ribosome come to be understood as fully as are the corresponding eubacterial interactions. Nevertheless, by recognizing that the eukaryotic and eubacterial ribosomes share a number of common morphological elements, we are now in a position where attempts to map major sites (e.g. the tRNA-binding sites, the peptidyl transferase center, factor-binding sites, and modifiable sites on the rRNAs) should be feasible.

We thank Tom Obrig for his generous gift of 80 S ribosomes, Terry Wagenknecht for much advice and assistance with the specimen preparation and microscopy as well as for a critical reading of the manuscript, and Michael Radermacher for invaluable advice during all phases of the image analysis and 3-dimensional reconstruction. This work was supported in part by grants from the National Institutes of Health (GM 29169) and the National Science Foundation (PCM-8313045).

References

Arad, T., Piefke, J., Weinstein, S., Gewitz, H.-S., Yonath, A. & Wittmann, H. G. (1987). *Biochimie,* **69**, 1001–1006.

Bernabeu, C., Tobin, E. M., Fowler, A., Zabin, I. & Lake, J. A. (1983). *J. Cell Biol.* **96**, 1471–1474.

Berriman, J. & Leonard, K. R. (1986). *Ultramicroscopy,* **19**, 349–366.

Bielka, H. (1982). Editor of *The Eukaryotic Ribosome,* Springer-Verlag, Berlin.

Boublik, M. & Hellmann, W. (1978). *Proc. Nat. Acad. Sci., U.S.A.* **75**, 2829–2833.

Boublik, M., Hellmann, W. & Jenkins, F. (1982). *Proc. 10th Int. Congr. Electron Microsc.* vol. 3, pp. 95–96, Deutsche Gesellschaft für Elektronenmikroskopie e.V., Frankfurt.

Burka, E. R. & Schickling, L. F. (1980). *Biochemistry,* **9**, 459–463.

Carazo, J. M., Wagenknecht, T., Radermacher, M., Mandiyan, V., Boublik, M. & Frank, J. (1988). *J. Mol. Biol.* **201**, 393–404.

Carazo, J. M., Wagenknecht, T. & Frank, J. (1989). *Biophys. J.* **55**, 465–477.

Damaschun, G., Muller, J. J., Bielka, H. & Bottger, M. (1974). *Acta Biol. Med. Germ.* **33**, 817–823.

Frank, J. & van Heel, M. (1982). *Proc. 10th Int. Congr. Electron Microsc.,* vol 1, pp. 107–114, Deutsche Gesellschaft für Elektronenmikroskopie e.V., Frankfurt.

Frank, J., Verschoor, A. & Boublik, M. (1981). *Science,* **214**, 1353–1355.

Frank, J., Verschoor, A. & Boublik, M. (1982). *J. Mol. Biol.* **161**, 107–137.

Frank, J., Verschoor, A., Radermacher, M. & Wagenknecht, T. (1990). In *The Ribosomes: Structure, Function and Evolution* (W. E. Hill *et al.*, eds), Amer. Soc. Microbiol. In the press.

Hardesty, B., McKeehan, W. & Culp, W. (1971). *Methods Enzymol.* **20**, 316–330.

Kellenberger, E., Haner, M. & Wurtz, M. (1982). *Ultramicroscopy,* **9**, 139–150.

Kühlbrandt, W. & Unwin, P. N. T. (1980). In *Electron Microscopy at Molecular Dimensions* (Baumeister, W. & Vogell, W., eds), pp. 108–116, Springer, Berlin.

Kühlbrandt, W. & Unwin, P. N. T. (1982). *J. Mol. Biol.* **156**, 431–448.

Lake, J. A., Henderson, E., Clark, M. W. & Matheson, A. T. (1982). *Proc. Nat. Acad. Sci., U.S.A.* **79**, 5948–5952.

Lutsch, G., Bielka, H., Wahn, K. & Stahl, J. (1972). *Acta Biol. Med. Germ.* **29**, 851–876.

Lutsch, G., Benndorf, R., Westermann, P., Behlke, J., Bommer, U.-A. & Bielka, H. (1985). *Biomed. Biochim. Acta,* **44**, K1–K7.

Milligan, R. A. (1985). Ph.D. thesis, Stanford University.

Milligan, R. A. & Unwin, P. N. T. (1986). *Nature (London),* **319**, 693–695.

Montesano, L. & Glitz, D. G. (1988). *J. Biol. Chem.* **263**, 4932–4938.

Nonomura, Y., Blobel, G. & Sabatini, D. (1971). *J. Mol. Biol.* **60**, 303–323.

Olson, H. M., Sommer, A., Tewari, D. S., Traut, R. R. & Glitz, D. G. (1986). *J. Biol. Chem.* **261**, 6924–6932.

Penczek, P., Srivastava, S. & Frank, J. (1990). In *Proc. XIIth Int. Congr. Electron Microsc.* San Francisco Press, San Francisco. In the press.

Radermacher, M. (1988). *J. Electron Microsc. Technique,* **9**, 359–394.

Radermacher, M., Wagenknecht, T., Verschoor, A. & Frank, J. (1987*a*). *J. Microsc.* **146**, 113–136.

Radermacher, M., Wagenknecht, T., Verschoor, A. & Frank, J. (1987*b*) *EMBO J.* **6**, 1107–1114.

Ryabova, L. A., Selivanova, O. M., Baranov, V. I., Vasiliev, V. D. & Spirin, A. S. (1988). *FEBS Letters*, **226**, 255–260.

Stöffler, G. & Stöffler-Meilicke, M. (1986). In *Structure, Function, and Genetics of Ribosomes* (Hardesty, B. & Kramer, G., eds), pp. 28–46, Springer-Verlag, New York.

Tischendorf, G. W., Zeichhardt, H. & Stöffler, G. (1974). *Mol. Gen. Genet.* **134**, 187–208.

Uchiumi, T., Wahba, A. J. & Traut, R. R. (1987). *Proc. Nat. Acad. Sci., U.S.A.* **84**, 5580–5584.

Unwin, P. N. T. (1977). *Nature (London)*, **269**, 118–122.

Unwin, P. N. T. (1979). *J. Mol. Biol.* **132**, 69–84.

van Heel, M. & Frank, J. (1981). *Ultramicroscopy*, **6**, 187–194.

Verschoor, A. (1989). Ph.D. thesis, State University of New York at Albany.

Verschoor, A., Frank, J. & Boublik, M. (1985). *J. Ultrastruct. Res.* **92**, 180–189.

Verschoor, A., Frank, J., Wagenknecht, T. & Boublik, M. (1986). *J. Mol. Biol.* **187**, 581–590.

Verschoor, A., Zhang, N.-Y., Wagenknecht, T., Obrig, T., Radermacher, M. & Frank, J. (1989). *J. Mol. Biol.* **209**, 115–126.

Verschoor, A., Milligan, R., Srivastava, S. & Frank, J. (1990). In *Proc. XIIth Int. Congr. Electron Microsc.* San Francisco Press, San Francisco. In the press.

Wagenknecht, T., Radermacher, M. & Frank, J. (1987). In *Fifth Conversation in Biomolecular Stereodynamics, Book of Abstracts* (Sarma, R. H., ed.), pp. 251–252, Institute of Biomolecular Stereodynamics, Albany.

Wagenknecht, T., Carazo, J. M., Radermacher, M. & Frank, J. (1989). *Biophys. J.* **55**, 465–477.

Wittmann, H. G. (1986). In *Structure, Function, and Genetics of Ribosomes* (Hardesty, B. & Kramer, G., eds), pp. 1–27, Springer-Verlag, New York.

Edited by A. Klug

Ultramicroscopy 40 (1992) 33–53
North-Holland

ultramicroscopy

Three-dimensional reconstruction of single particles embedded in ice

Pawel Penczek [a,b], Michael Radermacher [a] and Joachim Frank [a,c]

[a] Wadsworth Center, New York State Department of Health, P.O. Box 509, Albany, NY 12201-0509, USA
[b] Institute of Experimental Physics, Warsaw University, Warsaw, Poland
[c] Department of Biomedical Sciences, State University of New York at Albany, NY, USA

Received 17 July 1991; at Editorial Office 17 September 1991

Single particles embedded in ice pose new challenges for image processing because of the intrinsically low signal-to-noise ratio of such particles in electron micrographs. We have developed new techniques that address some of these problems and have applied these techniques to electron micrographs of the *Escherichia coli* ribosome. Data collection and reconstruction follow the protocol of the random-conical technique of Radermacher et al. [J. Microscopy 146 (1987) 113]. A reference-free alignment algorithm has been developed to overcome the propensity of reference-based algorithms to reinforce the reference motif in very noisy situations. In addition, an iterative 3D reconstruction method based on a chi-square minimization constraint has been developed and tested. This algorithm tends to reduce the effects of the missing angular range on the reconstruction, thereby facilitating the merging of random-conical data sets obtained from differently oriented particles.

Contents

0304-3991/92/$05.00 © 1992 – Elsevier Science Publishers B.V. All rights reserved

1. Introduction

1.1. Cryo-electron microscopy of single particles

The method of cryo-electron microscopy is currently revolutionizing electron microscopic structure research. This method allows biological macromolecules to be imaged in their fully hydrated state. To take full advantage of the new imaging technique, one has to use minimum dose (below 10 $e^-/Å^2$) and average over a large number of structural repeats. Specimens that are either crystalline or highly symmetric lend themselves most easily to low-dose averaging and application of crystallographic 3D reconstruction techniques [1]. Impressive and informative three-dimensional images have been obtained for viruses [2,3] and membrane proteins [4,5], to give but a few examples. As a consequence, a more accurate picture is beginning to emerge about the fidelity of the negative staining technique in rendering macromolecular structure.

In the current paper, we describe how techniques of averaging and 3D reconstruction can be extended to macromolecular assemblies embedded in vitreous ice and occurring in the form of "single particles". To a large extent, the methodology of single particle averaging and reconstruction can be adopted in an unchanged form, as developed for negatively stained specimens imaged at low dose [6]. However, due to the intrinsically low contrast and low signal-to-noise ratio found in micrographs of frozen-hydrated specimens, some of the image processing procedures used had to be revised or entirely rethought before they could be successfully applied. These procedures are described in the following, based on our experience with the 70S ribosome data set.

1.2. Test case: the ribosome

The ribosome is a highly complex particle, formed by association of proteins and rRNA, which is responsible for protein synthesis in all forms of life. While numerous steps in the translation cycle of the *Escherichia coli* (*E. coli*) ribosome have been well characterized, the precise way in which the ribosome interacts with mRNA, tRNA and the factors required for protein synthesis is as yet unknown. Knowledge of the three-dimensional structure of the ribosome is a prerequisite for studying function in terms of spatial macromolecular interactions.

Because of the slow progress in attempts to crystallize ribosomes, crystallographic studies have thus far failed to make important contributions to the understanding of ribosomal structure. The advantage of the approach outlined in the following is that it does not require crystallization and that it yields results within a relatively short period of time. Our reconstruction of the 70S *E. coli* ribosome, although limited in resolution to 1/(40 Å), is already able to give conclusive answers to questions regarding the relative orientation of the two subunits, as well as regarding topology, size and accessibility of the intersubunit space where the translation process is known to take place. In addition, the spatial distribution of ribosomal RNA in the interior has been inferred [7].

Before we can start describing the various steps of the image processing methodology, it is necessary to give a brief account of the "materials and methods" of ribosome imaging. Ribosomes were applied to grids covered with thick holey carbon film, whose holes were in turn covered with a second, thin, carbon film. The frozen-hydrated preparation was done as described previously [8] using GATAN cryo-transfer equipment and specimen stage on a PHILIPS EM420 electron microscope. The images were taken in pairs at 2 μm defocus, 100 kV voltage, and a dose of less than 10 $e^-/Å^2$. The first was taken with the specimen tilted by 50° while the second was taken without tilt. A total of 838 particles from two electron micrograph pairs were selected for image analysis, using an interactive selection program [9] within the SPIDER image processing system [10].

1.3. Strategy of reconstruction

In our approach to reconstruction [1,9], we make use of the multiple occurrence and preferred orientation of the particle to be studied. A specimen field containing particles in the preferred orientation is imaged twice: once with the specimen tilted to the maximum angle feasible, and once with the specimen in untilted position. It is then easy to see that the particle projections appearing in the tilted specimen image form a random conical projection set, because the particles assume random orientations with respect to an axis perpendicular to the specimen plane ("in-plane orientations"). With a sufficient number of particles, the azimuthal coverage is good enough to allow three-dimensional reconstruction.

The image of the specimen field in the untilted position is required for two reasons: (i) in this micrograph, all particles should look the same, so it is straightforward to find the in-plane orientations by correlation methods [11,6] and thereby the azimuthal placement of the corresponding tilted-specimen projection on the cone; (ii) after this alignment step, the untilted-specimen data set can be analyzed more closely for heterogeneity by using multivariate statistical analysis (MSA) [12,13]. Particles are accepted only if they fall within a certain variance margin of the main group; otherwise they are considered outliers. This step is supposed to narrow the orientational variation that always exists in a preferred view, as well as eliminate particles that may have been deformed or damaged in the course of specimen preparation. After the particles are screened in this way, their corresponding tilted-specimen projections are used for 3D reconstruction by weighted back-projection [9,1].

Although the conical data collection geometry reduces the range of missing angular information as compared to single-axis tilting, the residual artifacts still present a concern in the interpretation of the results. Our present strategy attempts to fill the angular gap by combining data sets derived from different preferred views. Indeed, the method described for a single group can easily be generalized if several groups of 0°-views exists, as suggested by MSA and classification: in this case, *several* reconstructions can be computed, one for each group, provided that the number of particles is sufficient for adequate azimuthal coverage. Such a strategy was first applied by Carazo et al. [14] in the study of the 50S ribosomal subunit of *E. coli* depleted of L7/L12 protein, where different conformations of the particle were found to coexist.

It is this latter strategy that we have adopted for analyzing the 70S particles, which occur in a number of different, distinct views. After reconstructions are obtained for each of the different views, we employ a 3D-orientation search and find the rotations that bring the reconstructions into best match. It can then be verified, by comparing the 3D-aligned reconstructions, that we are indeed dealing with the same structure, and that the differences in the 0°-appearance are largely attributable to changes in orientation of this structure. Ice-embedded particles, unlike particles prepared by negative staining and air-drying, can be expected to keep the same shape irrespective of orientation.

It is therefore legitimate to compute a "merged-data reconstruction" by using all data sets in a common coordinate system. If the differences in the orientations of the particle in the individual reconstructions are sufficiently large, as is the case for our data, then the merged-data reconstruction will no longer suffer from artifacts due to missing angular coverage, and will possess virtually isotropic resolution.

After this introduction into the overall strategy, each step will now be separately discussed in greater detail.

2. Methods

2.1. The alignment problem

The problem of aligning a set of single-particle projections has been extensively treated in the literature [11,15–17]. Usually it is assumed that a reference image is available, to which each particle is matched using the maximum of the cross-correlation function to determine the translation and rotation parameters. The disadvantage of this approach is that the resulting parameters are biased by the choice of the reference image. As a remedy, the alignment is normally repeated once or twice, using the average obtained in the previous step as new reference [11]. However, this measure fails when the images are extremely noisy: features of the original reference still tend to dominate the final average. The ribosome images are in fact very noisy: using the cross-correlation coefficient between two raw images and the method described by Frank and Al-Ali [18], we found a signal-to-noise ratio of 1.4 at a resolution of 1/(40 Å). In addition, certain steps in the preparation of a reference image, such as masking and filtration, require choices of parameters that are more or less intuitive. The so-called multi-reference alignment scheme of Van Heel [15] addresses the problem of aligning several groups of images simultaneously by iterating the steps of classification and of alignment within groups found. Again, the results have been found to critically depend upon user-made decisions [19]. From this brief outline of the problem it is clear that in order to obtain a result that does not depend on subjective decisions the use of a reference image must be entirely eliminated.

Recently, a reference-free alignment method has been suggested [20]. In this approach, certain rotation-, translation- and mirror-invariant functions (so-called double auto-correlation functions) are derived from the raw images. In this way MSA and classification can be done *prior* to the alignment. This interesting approach has the advantage that the large set of images can be initially split into more homogenous groups, thereby simplifying the choice of the reference images for the further analysis. The disadvantage of this method is that the "phases" of the input data are discarded in the process of calculating the double auto-correlation functions. The phase information (referring to the Fourier representation of an image) is largely responsible for defining the shape of an object; thus it is uncertain if this method will work for particles, such as ribosomes, whose 3D overall shape is spherical and whose projections differ only by "secondary" features. In addition, the abandonment of phases may have the consequence that entirely different particles are lumped into the same class. We present here a different approach that avoids the complications arising from these invariant functions.

In devising a reference-free alignment scheme, it is advantageous to use a concept of alignment as a feature of the entire image set. The first attempt to formulate the problem in these general terms has been made by Frank et al. [6]. They proposed a functional and defined the objective of alignment procedure as finding the minimum of this functional. Here we will follow this pathway: we propose a reference-free algorithm based on a related functional, and provide some analysis of its behavior.

2.2. Formal description of the problem

Let us assume that a set X of N images is available

$$X = \{x_i;\ i = 1, \ldots, N\}. \tag{1}$$

For the sake of simplicity we will treat images x_i as continuous functions of the variable u. The position of each image with respect to its initial position is described by three parameters: s_α^i denoting the rotation around the origin and s_x^i, s_y^i denoting shifts in x and y directions, respectively. Thus, the relative positions of all N images are entirely described by the set of $3N$ parameters:

$$P = \left\{s_\alpha^i,\ s_x^i,\ s_y^i;\ i = 1, \ldots, N\right\}. \tag{2}$$

Using expressions (1) and (2) we can define the alignment of the set of images in the following way:

Definition 1: The set of N images is called *aligned* if a set of $3N$ parameters P is found such that the expression

$$L_1\{X, P\} = \int_{-\infty}^{+\infty} \sum_{i=1}^{N} \left[x_i\left(u;\ s_\alpha^i,\ s_x^i,\ s_y^i\right) - \langle \boldsymbol{x} \rangle_i\right]^2 \mathrm{d}u, \tag{3}$$

where

$$\langle \boldsymbol{x} \rangle_i = \frac{1}{N-1} \sum_{\substack{j=1 \\ j \neq i}}^{N} x_j\left(u;\ s_\alpha^j,\ s_x^j,\ s_y^j\right) \tag{4}$$

is minimized.

After going through some algebra it can be shown that minimization of the functional given by eq. (3) is equivalent to maximization of the expression

$$L_1'\{X, P\} = \int_{-\infty}^{+\infty} \left[\sum_{i=1}^{N} x_i\left(u; s_\alpha^i, s_x^i, s_y^i\right) \right]^2 du, \tag{5}$$

or to the maximization of the following expression

$$L_1''\{X, P\} = \int_{-\infty}^{+\infty} \sum_{i=1}^{N-1} \sum_{j=i+1}^{N} x_i\left(u; s_\alpha^i, s_x^i, s_y^i\right) \times x_j\left(u; s_\alpha^j, s_x^j, s_y^j\right) du. \tag{6}$$

Maximization of eq. (6) is in turn equivalent to the minimization of

$$L_2\{X, P\} = \int_{-\infty}^{+\infty} \sum_{i=1}^{N-1} \sum_{j=i+1}^{N} \left[x_i\left(u; s_\alpha^i, s_x^i, s_y^i\right) - x_j\left(u; s_\alpha^j, s_x^j, s_y^j\right) \right]^2 du, \tag{7}$$

which can be recognized as the objective function of the paper by Frank et al. [6]. The form of eq. (7) suggests that the direct minimization of L_2 can be accomplished by bringing all pairs of images into register. Even if this was computationally possible, the very low SNR discussed previously makes the position parameters found between two raw images very uncertain and biased by the error induced by noise. At the same time, the form of eq. (3) suggests rather the usage of partial sums of images. Provided that reasonable first approximations of the parameter set P for the whole set of images can be found, such a sum should have much better SNR and the process should be robust. Thus, for the purpose of the algorithm analysis in our paper we find eqs. (3) and (5) more suitable than eq. (7).

2.3. The reference-free alignment algorithm

As was pointed out in the previous section, the direct solution of the alignment problem (definition 1) exceeds the capacities of current computers. However, since the images in EM are of rather limited resolution, an approximate solution should lead to satisfying results. The new "reference-free" algorithm which we propose here was designed to find such a solution. The description of this algorithm contains no specification of the method used for the position search between two images. In fact, the algorithm does not depend on the choice of such a method.

The algorithm consists of two parts (figs 1a, b): in the first, a "random approximation" of the global average is found and in the second, this average is iteratively refined until convergence is achieved.

The purpose of the first part (fig. 1a) is to find a good approximation of the global average without using a reference image. From eq. (3) we note that for two images the minimum of the functional can be always found (although the low SNR can result in incorrect position parameters). These two images positioned according to the minimum can be added, thereby providing the first approximation of the average with improved SNR compared with a single image. Then this average can be used to find the position of the third image, which again can be added to the sum of the two initial images. This procedure has the following features:

(a) it improves the SNR in each step;

(b) if the order in which images are included is random the resulting average will not be biased by the motif from any particular image;

(c) in each step the optimal decision for the choice of position parameters is made, which means that the procedure gives a good approximation of the initial alignment; the whole procedure, however, is not optimal since the inclusion of an additional image to a set of aligned images may in general require the change of the orientation of already positioned images to maintain the minimum of functional (3).

From point (c) it is clear that an additional procedure is needed in which the position parameters found are refined. Again, from eq. (3) we note that such refinement can be achieved by corrections of the position of each single image with respect to the sum of the rest.

The first part called "random approximation" of the global average contains the following steps:

(i) choose randomly two images x_i and x from the set $\boldsymbol{X}$ of N available images;

(ii) bring into mutual register x_i and x_j and initialize the average a_2 by setting

$$a_2 = \left(x_i \oplus x_j\right)/2,$$

where $\oplus$ denotes the algebraic sum of the im-

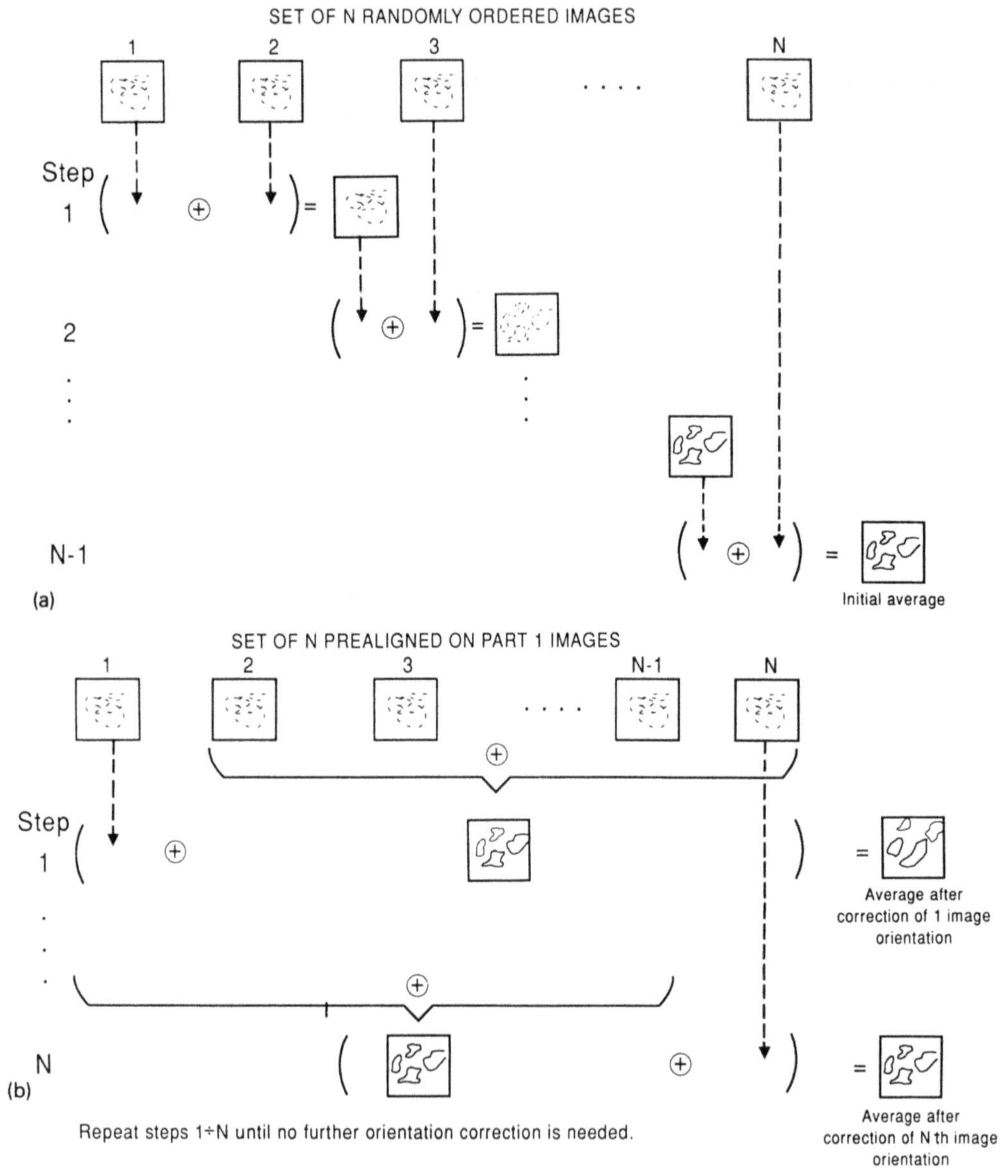

Fig. 1. (a) Block diagram of the first part of the alignment algorithm – the "random approximation" of the global average. (b) Block diagram of the second part of the alignment algorithm – the iterative refinement of the initial average.

ages using the position parameters found;

(iii) set a counter m to 3;

(iv) choose randomly an image x_k from the set of $N-m+1$ remaining images;

(v) bring into register x_k and a_m and update the average a_m by

$$a_m = \left(x_k \oplus (m-1)a_{m-1}\right)/m;$$

(vi) increase m by 1 and if $m=N$ then stop, else go to step (iv).

The second part of the algorithm (fig. 1b) is the iterative refinement of the average $A=a_N$. The steps are as follows:

(i) set a counter $m=1$;

(ii) remove the image x_m from the set and create subaverage A' by

$$A' = \left(NA \ominus x_m\right)/(N-1),$$

where $\ominus$ denotes the algebraic difference of the images using the current position parameters;

(iii) bring into register x_m and A' and update the new average A by

$$A = \left(x_m \oplus (N-1)A'\right)/N;$$

(iv) increase m by 1 and if $m \le N$ then go to step (ii);

(v) if in step (iii) any of the images changed its position then go to step (i), else stop.

It has to be noted that the exclusion of images in step (ii) of the second part is done in sequential order.

To show that the algorithm is correct we have to show that it converges and that the minimum of L_1 can be found. The convergence of the procedure can be concluded from the fact that due to the steps (ii) and (iii) in part two, the value of L_1 can only decrease or remain constant, and since L_1 is a non-negative function the algorithm ultimately has to terminate.

The necessary condition for the extremum of the functional L_1 is to have for all i and for o standing for α, x or y

$$\frac{\partial}{\partial s_o^i} L_1(s_o^i) = 0 \tag{8}$$

fulfilled, with s_o^i being the ith position parameter. To investigate this condition we will rather use the equivalent functional L_1'.

$$\begin{aligned}
\frac{\partial}{\partial s_o^k} L_1'(s_o^k) &= \frac{\partial}{\partial s_o^k} \int_{-\infty}^{+\infty} \left[\sum_{i=1}^{N} x_i\left(u - s_o^i\right) \right]^2 du \\
&= \frac{\partial}{\partial s_o^k} \int_{-\infty}^{+\infty} \left\{ \left[x_k\left(u - s_o^k\right)\right]^2 \right. \\
&\quad + \left[\sum_{\substack{i=1 \\ i \ne k}}^{N} x_i\left(u - s_o^i\right) \right]^2 \\
&\quad \left. + 2x_k\left(u - s_o^k\right) \left[\sum_{\substack{i=1 \\ i \ne k}}^{N} x_i\left(u - s_o^i\right) \right] \right\} du.
\end{aligned} \tag{9}$$

The first term on the right-hand side is constant, the second does not depend on s_o^k, and thus they both vanish in the differentiation. The third term can be rewritten as follows:

$$\frac{\partial}{\partial s_o^k} \int_{-\infty}^{+\infty} \left\{ x_k\left(u - s_o^k\right) \left[\sum_{\substack{i=1 \\ i \ne k}}^{N} x_i\left(u - s_o^i\right) \right] \right\} du = 0, \tag{10}$$

which can be recognized as the condition for the maximum of the cross-correlation between the kth image and the sum of all other images with respect to the position parameter s_o^k. For the rotation parameter s_α^k a similar result is obtained after changing the coordinate system into a polar one. Thus, if the position search method implemented uses the condition (10) to find the position parameters, then we can say that the algorithm will always terminate in a local minimum of the functional L_1. This is seen from the fact that in the step (v) the change of the position parameters was used as the condition for the termination of the algorithm. We can also conclude that in general the minimum found is not the global minimum since a further reduction of L_1 might be achieved by simultaneously changing the positions of three, four, or any higher number of images.

The main features of the algorithm described are as follows:

(a) a reference image is not being used, thus the problem of its selection and preparation is eliminated, and the result of the alignment no longer depends on the choice of any particular reference image;

(b) the assumption of similarity among the input images is not used, thus the algorithm is able to align even images that share only a small amount of common features;

(c) the algorithm is suboptimal, and thus the result depends on the initial estimate of the average. This initial estimate is calculated in the first part of the algorithm, which by itself is suboptimal, but nevertheless gives a good approximation of the global minimum and mitigates the bias caused by the choice of a particular sequence of images.

2.4. Implementation

For the practical implementation of the algorithm described in the previous section, the question of the choice of a position search method must be answered. For both parts of the algorithm such search is required to find three parameters (x and y shift components and rotation angle) which determine the position of a single image with respect to the partial average of the rest. Since the shift operation is not separable from rotation, the direct solution must entail a minimization procedure of the steepest-descent kind for the search of the optimal position. Unfortunately, such a search in real space is particularly sensitive to noise in the data and the procedure tends to terminate in local minima, resulting in incorrect position parameters. Another approach described in ref. [6] separates the shift from the rotational search and uses the cross-correlation functions (CCF) to determine the optimal parameters. In our case we made use of the latter method which means that in fact two programs were used: one applying the reference-free algorithm described above to find the shift alignment and a second to find the rotational alignment.

Both programs were implemented in a way that minimizes the time of calculations. The shift alignment program reads each image into the memory and calculates its Fourier transform using the mixed-radix FFT (fast Fourier transform) algorithm [21]. This makes possible the use of images with almost any size (except for dimensions involving large prime numbers), in particular with the size very close to the actual size of the particles. The limitation of the surrounding background noise achieved in this way improves the SNR of the CCFs calculated. All following operations (shifts, averaging, calculating of CCF) are performed in Fourier space on the transformed images stored in memory. No padding is done, which means that images are treated as circularly closed (i.e., the left side of the image borders the right and the top borders the bottom). This approach may lead to a certain bias caused by the possible circular overlap of the particles included in the field, but the fact that the problems of normalization of the CCF and uneven variance of noncircular CCF coefficients are avoided outweighs this disadvantage. An additional option allows checking for the 180° rotation of images (also readily implemented in Fourier space), which further speeds up the whole alignment process. The output of the program contains the shift parameters found for each image. An important feature in the design of this program is that it does not enforce any particular position of the final average.

The next step, the rotational alignment, requires all particles to be placed in such a way that their centers coincide with the origin of the coordinate system used for rotation. Since such centers are for obvious reasons difficult to define or found for the raw data, we decided to use instead the average obtained from the shift alignment step for the estimate of necessary centering parameters. This average has usually high SNR and can be easily centered using for example a low-pass-filtered disc as reference. The parameters found by the shift alignment program have to be subsequently corrected by this additional shift so that all particles will be moved to the central position.

The rotational alignment program calculates

the rotational cross-correlation function in polar coordinates:

$$c(\phi) = \int_{r_1}^{r_2} \int_0^{2\pi} f(r, \theta) g(r, \theta + \phi) |r| \, d\theta \, dr, \tag{11}$$

where r_1 and r_2 are the integration limits which depend on the size of the object and the particular region of particle the alignment process should be focussed on.

The discrete form of eq. (11) is

$$c(k \Delta\phi) = \sum_{l=n_1}^{n_2} \sum_{m=1}^{M} f(l \Delta r, m \Delta\theta) \times g(l \Delta r, m \Delta\theta + k \Delta\phi) \Delta\theta |l \Delta r|, \tag{12}$$

where n_1 and n_2 are "ring numbers" corresponding to the integral limits in eq. (11) and M is the number of sampling points in each ring. Eq. (12) is the "rotation function" suggested in refs. [22,6], except for the fact that we did not include a weighting function. The reason is that in our formulation of the alignment problem (3) we do not allow for parameters that would make the result of the alignment procedure dependent on choices made by the user. Since we do not single out any reference image, it would be difficult to enhance certain parts of the particles without knowing a priori what the result should look like.

The direct implementation of eq. (12) suggested in ref. [6] requires calculation of $n_2 - n_1 + 1$ one-dimensional CCF's for each ring, and then calculating the sum to find the maximum of final CCF function c. This is unnecessarily time-consuming as it uses highly oversampled rings at small radii.

In our current implementation we decided to vary the number of sampling points M over the rings, using $M = M_l$ sampling points for the lth ring, where M_l is the power-of-two number closest to $2\pi l$. Thus, the lth one-dimensional CCF can be written as

$$c_l(k \Delta\phi_l) = \sum_{m=1}^{M_l} f(l \Delta r, m \Delta\theta_l) \times g(l \Delta r, m \Delta\theta_l + k \Delta\phi_l) \Delta\theta_l. \tag{13}$$

By exploiting the Fourier convolution theorem we can write this equation as

$$c_l(k \Delta\phi_l) = \sum_{m=0}^{M_l - 1} F(l \Delta r, m \Delta\theta_l) \times G^*(l \Delta r, m \Delta\theta_l) \Delta\theta_l \times \exp[2\pi i(m \Delta\theta_l k \Delta\phi_l)]. \tag{14}$$

To equalize the lengths of all one-dimensional CCF's given by eq. (14) we pad with zeros each product of two functions F and G in Fourier space up to the length $M = 2 \max_l\{M_l\}$, an operation which simply amounts to performing a Fourier interpolation of the resulting real-space function. We also double the maximum length to increase the accuracy with which the maximum of the CCF can be located in real space. Next, we exchange the order of summation:

$$c(k \Delta\phi) = \sum_{m=0}^{M-1} \exp[2\pi i(m \Delta\theta_M k \Delta\phi)] \Delta\theta_M \times \sum_{l=n_1}^{n_2} F'(l \Delta r, m \Delta\theta_M) \times G'^*(l \Delta r, m \Delta\theta_M) |l \Delta r|. \tag{15}$$

The primes denote padded versions of corresponding nonprimed functions. In this way a substantial gain in speed and accuracy of calculations is achieved, since only one inverse FFT of length M is needed and the resulting CCF function is already the sum of properly interpolated versions of individual CCF's.

The actual implementation of the rotational alignment program is as follows: at first the limits of integration (in terms of inside and outside rings) must be decided. The outer radius has to correspond to the particle size. To avoid oversampling (and thus the relatively large error of interpolation) an inner radius equal to 5 was found to be sufficient for all practical purposes. With these limits set each image is read into the memory and its B-spline interpolation is calculated [23]. The choice of the order of the spline is left to the user (we found cubic splines to be sufficiently accurate for our data). This interpolation is used to resample the image in polar coor-

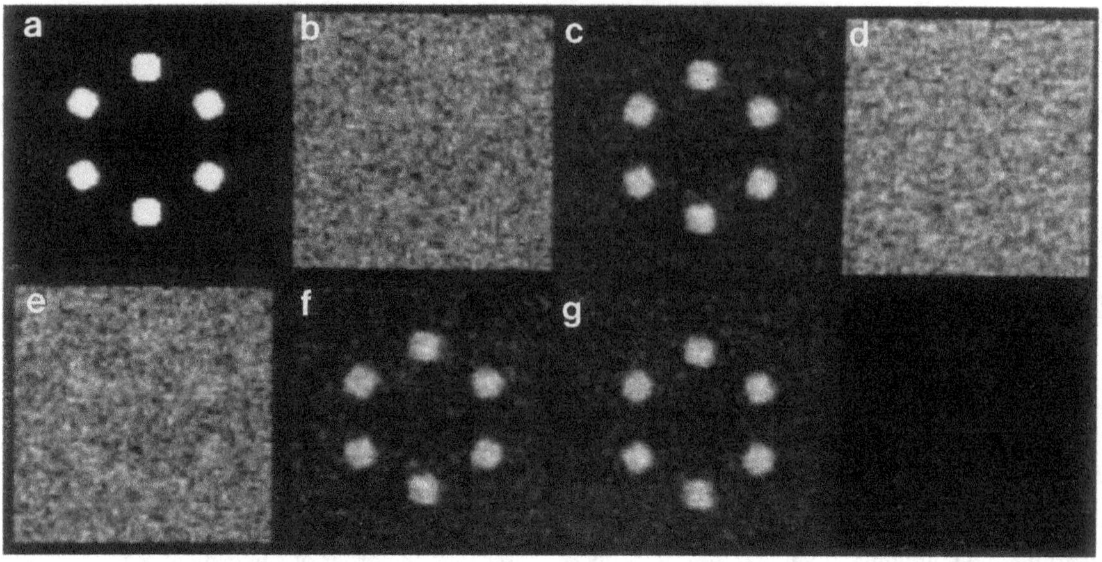

Fig. 2. Test of the reference-free alignment algorithm using model data. (a) Reference image used to test the standard alignment procedure. (b) Example of one of 128 64×64 test images containing Gaussian noise with the average equal to zero and variance equal to one. (c) Average of the series of 128 noise images aligned by the standard procedure. (d) Average of the series of 129 images (including the reference of the first test) after using the "reference-free" alignment program. (e) Modified version of the random image (b): the hexagonal reference image is added with a weight such that the resulting SNR is 0.75. (f) Average of the series of 128 modified noise images aligned by the standard procedure. (g) Average of the series of 129 modified images (including the reference of the first test) after using the "reference-free" alignment program.

dinates, with the number of points in each ring equal to the power-of-two number that is closest to $2\pi l$, where l is the ring radius. Then the Fourier transform of each ring is calculated and this initialization procedure is repeated for each image. The "reference-free" algorithm described in the previous section is implemented entirely in the Fourier domain using eq. (15) for the search of the optimum rotation parameter. Finally both programs – shift and rotational alignment – are used alternately until no further changes of image positions are encountered.

In order to test the performance of the new program we compared it with the approach based on the use of reference image (fig. 2). The first test was designed to demonstrate the bias caused by the choice of a particular reference pattern in the case of an image set containing pure noise. As reference image we used an image containing six discs with radius 3 pixels arranged in a hexagonal pattern (fig. 2a). Each disc was located 15 pixels from the image center. As image series to be aligned we generated 128 64×64 images containing Gaussian noise with zero average and unity variance (fig. 2b). To this set we applied the standard alignment procedure described in ref. [6]: the rotation search between auto-correlation functions (ACF's) of the reference image and each image of the set was used to determine the rotational angle, then the shift parameters were found using the padded version of the reference image as template. In each search the CCF was used to find the translation parameters. The average of the aligned series of 128 noise images is shown in fig. 2c. This average has a striking resemblance to the reference pattern despite the fact that only random noise was aligned. The differential phase residual analysis indicates 11 Fourier units as resolution for this set of images, which would lead us to believe – in a situation where the origin of the images is unknown – that the images contain a substantial amount of common low-resolution information. Obviously, the alignment algorithm allows the reference to pick out those features from the random patterns that tend to agree with it, a situation analogous to the result of quasi-optical filtration of a random pattern using a preconceived reciprocal lattice.

In the second test we applied our new reference-free alignment program to the same full set of 129 images (including the reference of the first test). We used five steps: shift, rotational, shift, rotational and final shift alignment. According to the full scheme of this approach, the centering of the average after each shift alignment is required. To keep the proximity to the previous test as close as possible we decided to use the hexagonal dot pattern image as centering reference in this step (although a disc would have been preferable as a "neutral shape" to avoid any bias).

The final average obtained after application of the steps described (fig. 2d) does not reveal any similarity whatsoever to the one particular image included in the set, *even when this image was forced to participate in the initialization*. The resolution test also did not indicate any significant degree of similarity.

Next two tests were designed to check the ability of the algorithm to align data in the presence of high noise. As input images we used modified versions of the random images from the previous test: we added the hexagonal reference image to each image containing Gaussian noise in such a way that the resulting SNR (defined as the ratio of the variance of the signal to the variance of the noise) was 0.75 (fig. 2e). Both alignment strategies were applied to this set in the same way as in the first two tests. The averages obtained are in figs. 2f and g. Since the resolution found was the same for both sets (9 Fourier units) we used the mean square error of the final shifts found for each series. These shifts should be equal zero, since we did not apply any initial transformation to the input series. The errors for the standard alignment procedure were 11.72 and 14.85 for x and y shift directions, respectively, and 2.54 and 2.93 for the "reference-free" program.

These results prove the ability of the "reference-free" alignment program to outperform the standard alignment procedure. In any practical case the results of the standard approach would probably be even worse, since the "perfect" reference image which we used is normally not available.

2.5. Application to ribosome data

The reference-free alignment program was applied to 838 images containing 70S ribosome data taken from two 0° micrographs. Each image had a size of 64×64 pixels with 5.55 Å pixel size. The five steps of alignment were used as in the tests described in the previous section. The resulting global average was not meaningful since the data contained different views of the particle. Correspondence analysis was therefore used to separate the whole data set into homogenous classes. The first six factors were found to contain most of the variability in the data and these factors were used in the hierarchical clustering program with complete linkage as merging criterion [24]. As a result of this analysis we obtained a linkage tree (fig. 3) which allowed us to investigate different possible partitions of the whole data set. We decided to use 21 groups. This particular partition produced a number of reasonably populous groups having a within-group resolution of 1/(33 Å) [#1] and averages with distinctly different features. The number of groups represented a compromise between two conflicting statistical trends: a lower number of classes resulted in structurally inhomogeneous groups, whereas a higher number of classes produced groups too small to have a reasonable within-group resolution. Only three groups were finally used in the further analysis. We decided to use those groups which had the highest number of images and at the same time represented distinct views. All three groups had a resolution of 1/(33 Å) according to the differential phase residual analysis [11], and the numbers of objects were 93 for the first group, 144 for the

[#1] By within-group resolution we understand the resolution found by the differential phase residual analysis of partial averages of images randomly drawn from a group found by cluster analysis.

second, and 66 for the third. The corresponding averages are shown in fig. 4.

The results obtained are interesting because our reference-free alignment program was actually used on the set of images containing distinct views of the same particle. From the test described in section 2.4 we already know that the resulting alignment should not be biased by any particular view. The results of the 70S data alignment show even more: among the set of aligned images containing different patterns, those which are similar to one another are automatically aligned. To check this conclusion we applied the alignment program separately to the groups of images extracted by the classification. There was no change in orientation of images (within 1.4°), a result which confirms the very robust performance of the program and indicates that the local minimum of the functional of eq. (3) found is close to the global minimum. From the fact that this functional is equivalent to eq. (7) it becomes clear that similar images must emerge in similar orientations.

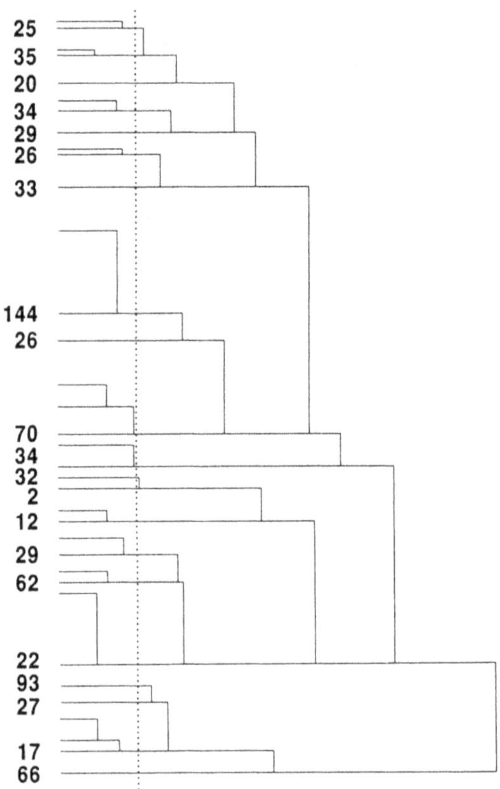

Fig. 3. The linkage tree obtained from the hierarchical ascendant classification using complete linkage as a merging criterion. A total of 838 images were clustered using the first six factors from correspondence analysis. The dashed line indicates the cutting level used.

2.6. *Alignment of tilted-particle images*

The method for the translational alignment of tilted particle images, which was described in refs. [9,25] and successfully used for the negatively stained particles, relies on the robust cross-correlation peak which can be found between the 0° image of a particle and its tilted counterpart. According to Guckenberger [26], for this procedure to work, the particle images must first be stretched by a factor $1/\cos(\theta)$ in the direction perpendicular to the tilt axis (θ being the tilt angle). Unfortunately, this method failed for our images of ice-embedded ribosomes. The lower SNR makes the cross-correlation function too noisy and the correct maximum becomes difficult to detect after cosine-stretching.

Thus the problem of alignment of tilted-particle images requires careful re-examination. However, since the resolution of the ribosome study was limited to 1/(40 Å) by electron-optical conditions, we were able to use a simplified procedure, which implicitly uses the fact that the particle has a near-spherical shape: by aligning the images directly by cross-correlation, we are effectively bringing the disk-shaped particle projections into register without reference to their internal features. The errors of this procedure are surprisingly small, as can be assessed from a comparison of reconstructions obtained from particles falling in different orientations (see below).

2.7. *Three-dimensional reconstruction algorithm*

The problem of three-dimensional reconstruction from projections in electron microscopy is particularly challenging. The main reason is the fact that due to the limited tilt range a certain amount of angular information is missing. Additional problems are posed by the high noise level

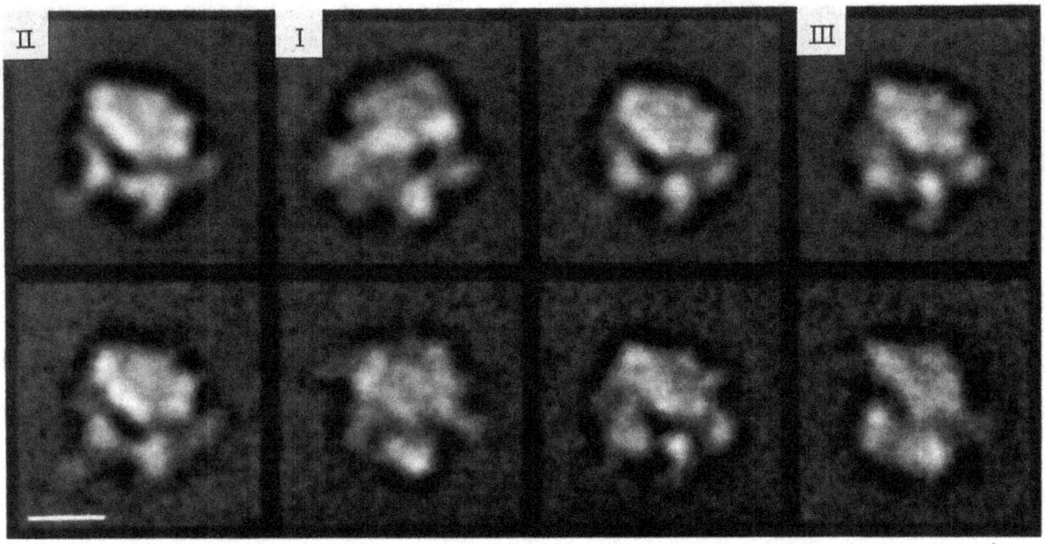

Fig. 4. Averages of the eight most populated classes determined by hierarchical ascendant classification (scale bar = 100 Å). Roman letters denote averages of classes used in the 3D reconstruction.

in the available data and by the errors in azimuthal angle assignment caused by imperfect alignment of 0° images and the error in assigning a common tilt angle to all particles. Special care must therefore be given to the choice of the 3D reconstruction method. A readily available and commonly used 3D tomographic reconstruction method is the *filtered back-projection* (FBP). Earlier FBP methods [26b,26c] call for a rather large number of equally spaced projections. However, this condition cannot be satisfied in the case of the random-conical data collection. The design of this scheme implies that data will not be available inside the missing cone and that it is very unlikely to obtain an even approximately equidistant distribution of projection angles. Moreover, it must be noted that in the case of limited angular information the Radon inversion formula does not hold [27], so the reconstructed object is only an approximation of the "true" 3D structure. At least a partial remedy to this problem is provided by approximated Radon inversion formulae [27,28]. A version of FBP with general filtration accounting for arbitrarily spaced projections was introduced to electron microscopy by Radermacher [25] and Independently by Harauz and van Heel [29]. Even if such filtering is applied, the resolution of the reconstructed object in the direction of the missing cone is deteriorated by a factor estimated to be 1.3 for a missing cone angle of 50° (e.g., ref. [9]). Our tests of FBP method applied to the reconstruction of a sphere from the projections with missing cone angle 50° revealed that the resulting 3D image is elongated by 30%. It is important to note that the elongation depends not only on the missing cone angle but also on the structure itself and on its orientation with respect to the missing cone direction.

As was discussed before, when certain angular information is missing, the reconstructed object will contain spurious details and will be distorted. Further, in applications of FBP methods some discrepancies between the projections of the reconstruction and the original data can be observed. In the following we introduce an iterative 3D reconstruction method which is designed to reduce these artifacts.

We treat the problem of computer tomography as an algebraic one (see Gordon et al. [30]). If we describe the projection operation by the matrix P

(non-square, in general), then the relation between the 3D structure $\boldsymbol{d}$ and the projections $\boldsymbol{g}$ can be written as

$$\boldsymbol{g} = P\boldsymbol{d}. \tag{16}$$

In this notation, the problem of tomographic reconstruction becomes the problem of finding the solution of a system of equations. Because of the presence of noise and the inconsistencies among the data, this system is in general ill-conditioned [31]. If we take into account the fact that a certain amount of angular information is missing we find that the this problem also becomes severely ill-posed [27]. In such a situation, only an approximate solution can be found: i.e., the least-squares estimate, which can be calculated by the singular value decomposition of the matrix P. If, however, the matrix P is sparse (i.e., most of its entries are equal to zero), which is obviously the case in the tomographic problem, a faster and preferable way is to use the conjugate gradient method [31]. This method is iterative, but the sparsity of matrix P reduces the number of required calculations by a large number.

Since the equation system (16) is ill-conditioned, the use of the straightforward least-squares solution would result in the amplification of errors contained in the data. Therefore, the solution has to be "stabilized", and the method of regularized least-squares is the obvious candidate [27]. In our approach we decided to use a least-squares regularizing (or cost) criterion that implies that the reconstructed object should have minimal variance. The agreement with the projection data is enforced by the concurrent minimization of the chi-square statistic describing the discrepancy between projected structure and known projections. The ratio between these two contradictory requirements is established by a weighting factor λ. In other words, this method can be regarded as an attempt to find a 3D structure that simultaneously minimizes the *variability* as measured by the sum of squares, and the *total squared residual error*, as measured by a chi-squared term, with the relative importance of the two being specified by the choice of the value for parameter λ.

Thus, the approach chosen can be mathematically formulated as follows:

Determine the three-dimensional structure $\boldsymbol{d}$ *that optimizes the cost functional*

$$J(\{\lambda, \boldsymbol{d}\}) = W + \lambda C, \tag{17}$$

where

$$C = \sum_{i=1}^{K} \frac{(f_i - g_i)^2}{\sigma_i^2} \tag{18}$$

is the chi-square error between the known projection data f *and the projected structure* $\boldsymbol{g} = P\boldsymbol{d}$, P *denoting the projection matrix,* σ_i *denoting the error of* f_i, K *the number of data points, and*

$$W = \sum_{i=1}^{M} d_i^2 \tag{19}$$

the cost criterion, with M *equal to the number of points in the 3D structure.* The parameter λ is chosen such that for the solution found, C is close to its expected value K.

Other possible cost criteria can be found in ref. [32]. In particular, different entropy measures are thought to yield the best solution, but they call for a rather complex minimization algorithm. The least-squares criterion chosen here is certainly the simplest one and for incorrectly estimated errors in eq. (18) the "overshoot" artifacts can corrupt the 3D structure found. In particular, the analytical partial derivatives with respect to d_i can be easily calculated and Powell's powerful conjugate gradient method for fast optimization can be used [23]. It is interesting to note that in the algebraic statement of the tomographic problem the missing cone question does not arise explicitly. Neither is it required to provide equally spaced projections. From algebraic considerations [33] it also follows that the number of projections required is smaller than the number predicted by the Crowther et al.'s formula [34].

All iterative methods using chi-square constraints require the error distribution to be defined or found from the experimental data. In our case, the limited number of data does not allow this error to be estimated from the variance of projections measured at the same angle (see ref.

[35]). We therefore decided to use the error estimated from the Poisson distribution of the electron counts in each pixel. This means that in our case $\sigma_i = d_i$, which further simplifies the algorithm. However, it must be noted that because neither conformational changes nor errors caused by the projection angles incorrectly assigned are taken into account, our approach results in an underestimate of the actual noise. This means that the actual resolution of the reconstructed object is lower than the resolution expected from the number of projections used.

The implementation of the method described requires the alternate usage of projection and back-projection operations. For both of these operations, the particular reconstruction geometry defining the limitation of summations has to be decided. The most natural one to choose is the spherical geometry. Thus, to minimize the memory requirements and the time of calculations, only the projection data contained within the circular mask are stored in the computer memory and the reconstruction is performed only within a sphere.

The choice of the spherical geometry obviously constitutes an implicit constraint on the reconstruction. The radius of the sphere limits the spatial extension of the 3D object which must be estimated from the projections.

A theoretical analysis of this reconstruction method leads to the conclusion that the results should be superior to those obtained by using FBP. The usage of a spherical mask and the minimization of the cost criterion W introduces a-priori information into the reconstruction algorithm which results in a reduction of the elongation and of the spurious details present in the reconstruction. To test the program we applied it to the reconstruction of a sphere from the projections with missing cone angle 50°. The data used were the same as for the test of the FBP method. The radius of the sphere was 20 pixels and the reconstruction was calculated within a sphere with a radius of 28 pixels. The elongation of the reconstructed sphere found with the new method was only 10%, compared with the 30% found for the FBP method.

Further comparisons were made on the actual reconstructions of the 70S ribosome calculated by both methods. Fig. 5 shows central slices of the structures calculated by the FBP method using the general weighting function and the iterative reconstruction, respectively. The latter apparently contains fewer artifacts and has better-defined features. It is also notably shorter in the direction of the missing cone (z-axis; vertical in figure).

The iterative reconstruction method was applied independently to all three experimental 70S ribosome data sets. To reduce the time of calculations we assumed that the azimuthal angles ϕ assigned to the projections by the alignment of the 0° images had only limited accuracy. We

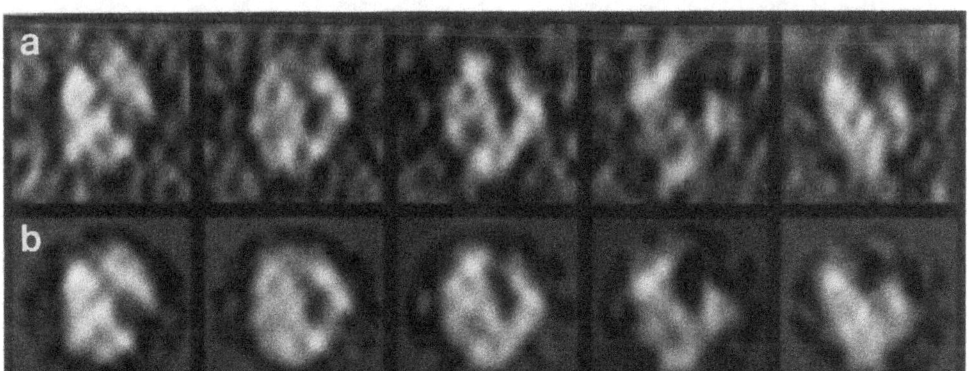

Fig. 5. Selected $X-Z$ slices (27, 30, 33, 36, 39) of the 70S ribosome structure calculated by the different reconstruction techniques. (a) FBP method with general weighting function. (b) The iterative reconstruction program.

assumed that this accuracy was no better than 0.5°, and at any rate a better accuracy was not needed, considering the intrinsic resolution limitations of this study. This assumption allowed us to merge projections that differed by no more than 0.5° in ϕ. Thus, the numbers of projections eventually used to calculate the 3D reconstructions for the three groups found by the classification of 0° images were reduced to 69, 93, and 66, respectively (the third group contained no merging projections).

3. Results

3.1. The individual reconstructions

Shaded surface representations of the structures obtained are shown in fig. 6a. Here the surface in each case was determined with the aid of a histogram of the 3D density (see below). The reconstructed particle is seen to lie in three different orientations, which are defined by the initial orientations of the three particle classes on the carbon grid, and – trivially – by the effect of the alignment algorithm on particles belonging to different classes. In order to compare the reconstructions, and obtain a final "merged" reconstruction, the relative orientations must be determined by using a suitable search program.

3.2. Determination of the relative three-dimensional orientations

The results from the correspondence analysis and classification enable us to compare three separate three-dimensional reconstructions of the same object, which differ by their orientations. The orientations can each be described by a set of three Eulerian angles ϕ, θ, ψ (see appendix), with θ having the meaning of the "rocking angle" (see ref. [12]).

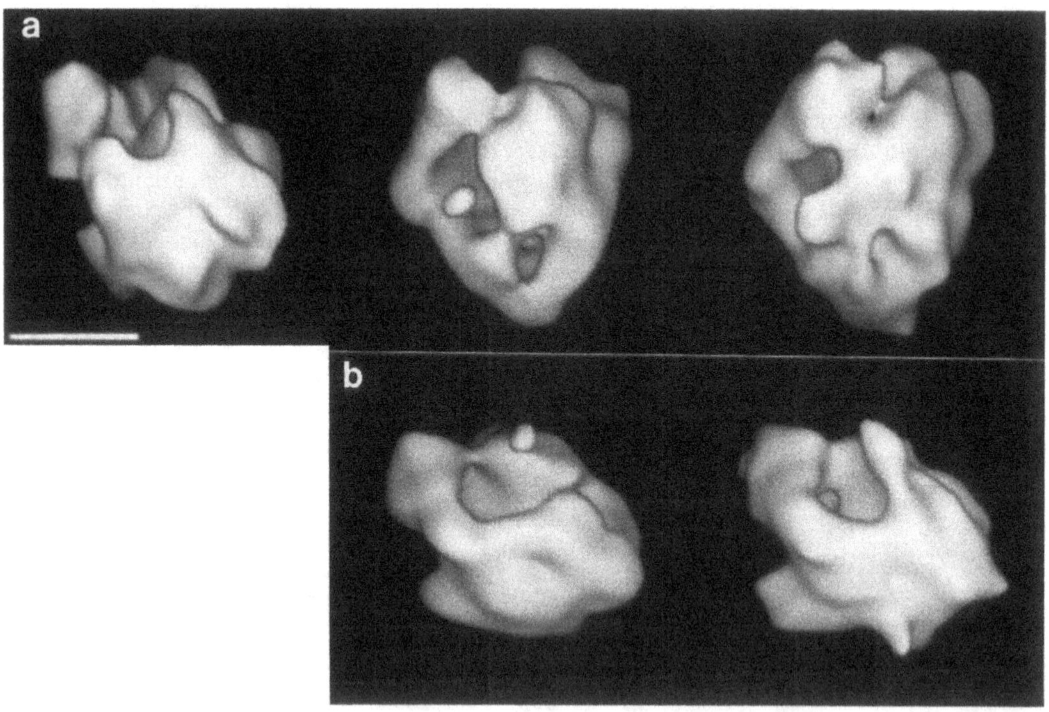

Fig. 6. Shaded versIons of the 70S ribosome structure calculated from the three independent data sets. The numbers of projections used were 69, 93, and 66, respectively. (a) Before orientational alignment. (b) After alignment of reconstructions II and III to I.

As a measure of the consistency between two differently oriented 3D structures we used the correlation coefficient. Thus, the problem of orientation search can be stated as the minimization of the functional given by the following expression:

$$1-\rho(\boldsymbol{d}_1, \boldsymbol{d}_2(\phi, \theta, \psi)), \tag{20}$$

where ρ is the correlation coefficient, $\boldsymbol{d}_1$ and $\boldsymbol{d}_2$ are the two three-dimensional structures, the first one to be thought of as a "reference" and the second with the orientation in terms of three Eulerian angles to be determined. To solve this problem we used a direct search in real space. The analytical forms of derivatives of eq. (20) cannot be calculated, thus we used the IMSL minimization procedure [23] based on a quasi-Newton method and a finite-difference gradient.

The disadvantage of this approach is that the minimization scheme does not guarantee reaching the global minimum (i.e., the best matching position). A guess of the initial Eulerian angles is required and, unless this guess is close to the optimal solution, the procedure will tend to terminate in one of the local minima. However, the average time of running the program is very short (of the order of minutes on the VAX 3100 workstation for 32^3 volumes), which makes it acceptable to use many trials with different starting points to determine the best solution. Our original data were contained in 64^3 volumes. Since the reliable information in the reconstructions is contained only in frequencies below 1/40 Å^{-1}, we were able to use scaled-down versions (from size 64 to size 32) of the three reconstruction volumes. The correlation coefficient was calculated within a sphere of radius 14 only, a measure which further reduced the time of calculations.

The final results, in terms of the three Eulerian angles, are listed in table 1. The consistency among these results can be verified in the following way: when any two of the three solutions are combined, the resulting Eulerian angle should match the remaining solution (*closure check*). If we use the set of angles between structures III and II and between II and I, the combined rotation is expressed by the angles $\phi = 88.0$, $\theta = 80.2$,

Table 1
Eulerian angles (ϕ, θ, ψ) and corresponding correlation coefficient (ρ) describing the 3D orientation found between three pairs of reconstructed 70S ribosomes

Rotated structure	Reference structure I	Reference structure II
II	$\phi = 93.0$ $\theta = 71.4$ $\psi = -39.6$ $\rho = 0.794$	
III	$\phi = 90.0$ $\theta = 79.8$ $\psi = -34.3$ $\rho = 0.748$	$\phi = 235.7$ $\theta = -10.5$ $\psi = 120.8$ $\rho = 0.86$

and $\psi = -33.7$, which match the rotation angles between structures III and I from table 1 quite closely. If we now use the set of angles between structures III and I and between II and I, the combined rotation leads to angles $\phi = 239.2$, $\theta = -9.8$, and $\psi = 119.1$, which again are in good agreement with the rotation angles between structures III and II shown in table 1.

The small size of the "rocking angle" found between reconstructions II and III, $\theta = 10.5°$, is consistent with the high degree of similarity of the corresponding 0°-cluster averages. Another confirmation is given by the high level of visual similarity among the three structures which can be observed after applying the rotations to structures II and III (fig. 6b). This illustration also helps visualizing the degree to which reconstructions of presumably identical structures may differ due to the missing-cone artifacts and other orientation-related errors.

3.3. The merged reconstruction

In the next step we applied the 3D orientation angles directly to the projections of groups 2 and 3. In this way we were able to calculate a "merged" 3D reconstruction (fig. 7) using the three sets of projections in a common coordinate system. We used the same iterative 3D reconstruction program as was used for the separate reconstructions. From the analysis of the θ-angles of each set of projections it follows that the missing cone is now virtually eliminated. The reliability of this merged reconstruction is also improved, compared with that of the individual reconstructions, since the total number of projections has now increased to 222.

The merged reconstruction shows the 30S and 50S subunits which are clearly recognizable by their characteristic shapes. The representation of the outer particle surface brings out only one aspect from a continuum of volumetric information; a whole set of interior surfaces or some type of volume rendering are obviously needed for an exhaustive representation. To decide on the appropriate density thresholds for the particle boundary and interior surfaces, we made use of a histogram of densities in the reconstructed volume [7], which records as a function of density d the number of volume elements (voxels) that have density values within density slice $(d,\ d+\delta d)$. This histogram (fig. 8) shows a sharp peak, due to the relatively uniform background density of ice, followed by a broader range of medium and higher densities that has been attributed to protein and RNA [7].

The diagonal position of the 30S subunit, relative to the axis of pseudo-mirror symmetry of the 50S subunit (which runs through the central protuberance), confirms visually derived models [36]. The two subunits are fused at two points involving extended contact regions at the top and bottom of the small subunit. This leaves a large gap between the subunits, termed the *interface gap*, where the binding of the t-RNAs and m-RNA and all steps of protein synthesis are thought to take place. Thus, when seen in the direction of

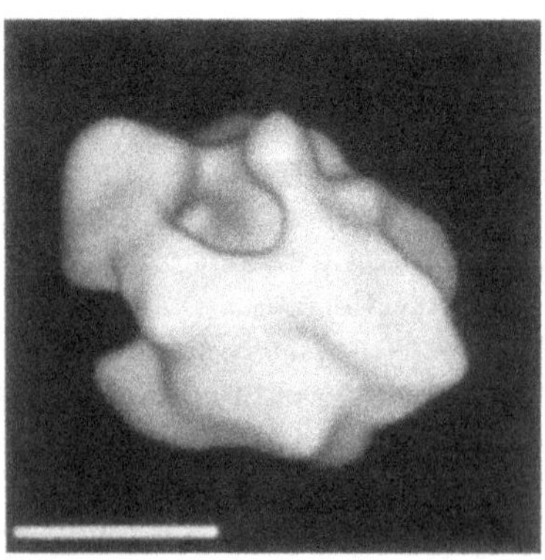

Fig. 7. The merged 3D reconstruction calculated from the three sets of projections in the same coordinate system.

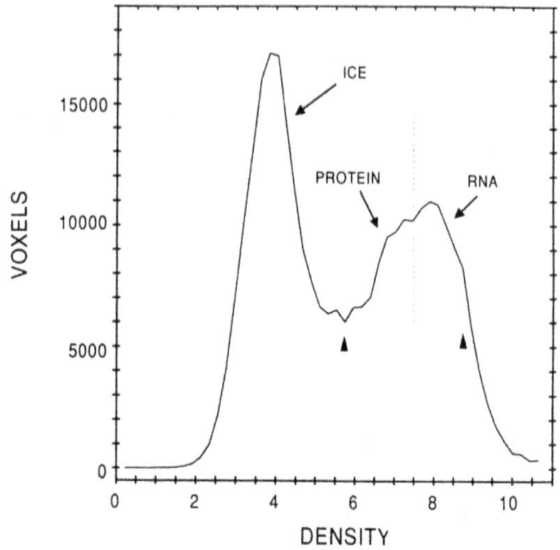

Fig. 8. Histogram of 3D density distribution. The most distinct feature is the "ice peak" which is separated by a sharp minimum from the particle portion of the data on the left. The position of this minimum was used to define the density threshold in our surface representations. (Reproduced from ref. [7] by copyright permission of the Rockefeller University Press.)

the gap (fig. 9), the ribosome has the approximate shape of a ring with a nonconcentric opening.

3.4. Discrimination between protein and ribosomal RNA

In the interior of the reconstructed ribosome mass, regions of higher density can be distinguished. In a paper describing the results in greater detail [7], arguments were put forth in favor of an interpretation of this higher density as RNA or RNA-rich. The volume estimate for the total ribosomal RNA (165 RNA component in the small subunit and 5S and 23S component in the large subunit) leads to an enclosing surface associated with the density threshold d_{RNA} on the histogram (fig. 8). Interestingly, these higher densities thus identified in the small and the large subunit are connected in the region of the small subunit "platform", the region where the 16S–23S RNA contact has in fact been established previously.

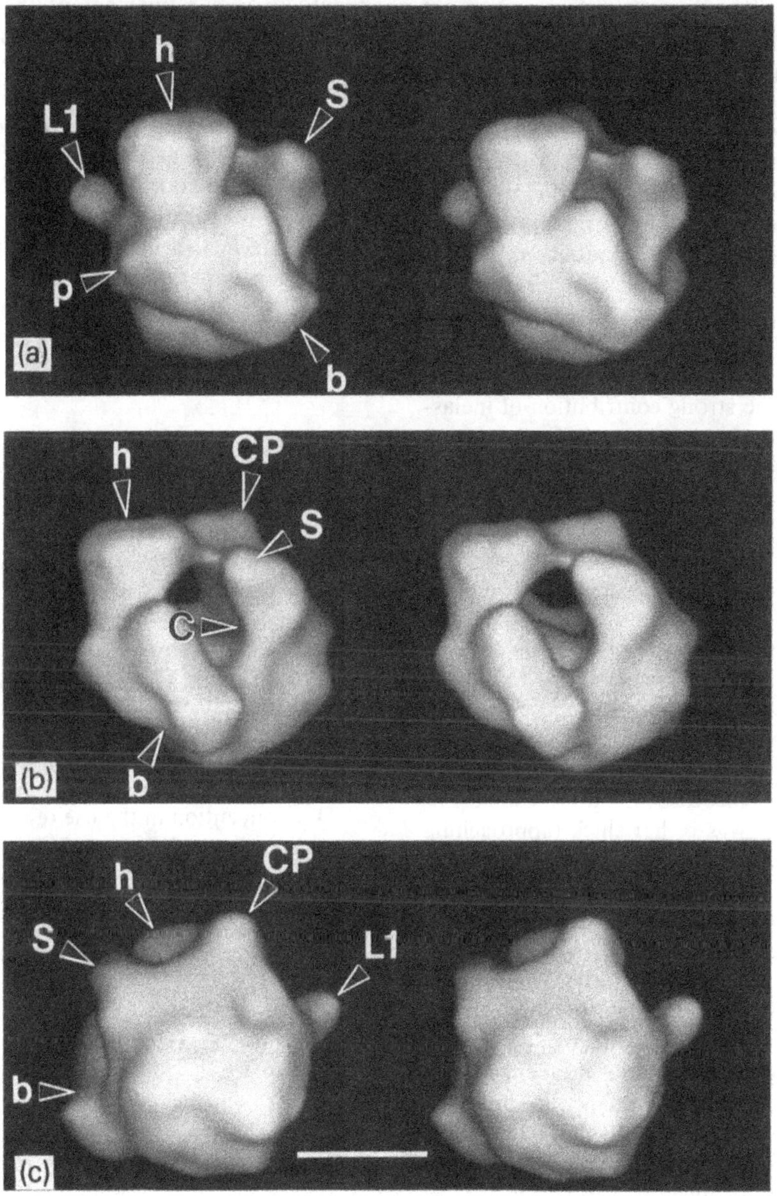

Fig. 9. Merged reconstruction, presented as stereo-pairs from three viewing directions: (a) ribosome oriented in such a way that the 30S subunit is in front, presenting Lake's asymmetric view; (b) view of the ribosome after 40° rotation from (a) around an axis lying vertical in the plane of the image; (c) view after 150° rotation from (a) around the same axis. Meaning of symbols: 30S subunit: p = platform, h = head, b = main body. 50S subunit: L1 = L1 ridge, CP = central protuberance, S = stalk base (the stalk is not visible in the merged reconstruction because of its high mobility).

4. Discussion

In this paper we present a number of new techniques that are designed to overcome the problems posed by low signal-to-noise ratios and distortions due to missing data. The close match among reconstructions of particles lying in different orientations testifies both to the superior performance of vitrified ice as an embedding medium (cf. reconstructions from negatively stained ribosomes [37]) and to the self-consistency of our reconstruction method.

The most obvious gap in the treatment thus far is the absence of a contrast transfer function correction. While the theory of contrast transfer under bright-field illumination conditions is well understood (see ref. [36]) and lends itself to corrections [38,39], the strong contribution of inelastic scattering and consequent multiple scattering events in ice pose special difficulties. Questions such as the spatial distribution of ribosomal RNA in the interior of the ribosomal subunits require a correction of the low spatial frequency range to receive a satisfactory answer. An increase in resolution obviously requires some efforts to extend the effective transfer interval towards high spatial frequencies. Both low and high spatial frequency bands are affected by inelastic scattering. The nature and intensity of these effects are dependent upon the thickness of the ice. We have reason to believe that in the case of our ribosome images, the ice layer was rather thick (approaching 1000 Å), because we observe a strong low spatial frequency component in the Fourier transform of the micrographs which cannot be attributed to elastic scattering. This component, which is believed to be due to the inelastic "dark-field" image (i.e., the result of the inelastically scattered wave interfering with itself), effectively prevents us from restoring the low spatial frequency range of the (elastic) bright-field image. In order to obtain interpretable micrographs that allow tracking of the image due to elastic scattering, future experiments should include special efforts to monitor and minimize the thickness of the ice (i.e., to match it more closely to the z-depth of the specimen). We plan to address the question of contrast transfer function correction in single-particle reconstruction in a separate paper [40].

5. Conclusions

The techniques of reference-free alignment and reconstruction presented in this paper can be regarded as a refinement of the single-particle reconstruction methodology. The new techniques have enabled us to extend the application of this methodology to particles embedded in ice. Great strides are now necessary to improve the resolution to a range that allows important functional questions to be addressed. Better control over the specimen preparation was mentioned as one of the factors in this improvement. The emergence of the new generation of digitally controlled microscopes and the addition of energy filtration gives us hope that significant progress will be achieved in the near future.

Acknowledgements

We are grateful to Witold Grochulski for valuable suggestions and to Jose-Maria Carazo for a critical reading of the manuscript. Special thanks also go to Suman Srivastava and Robert Grassucci for assistance and discussions.

This work was supported by grants NIH 1R01 GM 29169, 1S10 RR 03998, and NSF 8721474.

Appendix

Our convention in the use of Eulerian angles is according to the following definition of the rotation matrix:

$$R(\psi, \theta, \phi) = R_1(\psi) R_2(\theta) R_1(\phi)$$

$$= \begin{pmatrix} \cos\psi & \sin\psi & 0 \\ -\sin\psi & \cos\psi & 0 \\ 0 & 0 & 1 \end{pmatrix} \begin{pmatrix} \cos\theta & 0 & -\sin\theta \\ 0 & 1 & 0 \\ \sin\theta & 0 & \cos\theta \end{pmatrix} \times \begin{pmatrix} \cos\phi & \sin\phi & 0 \\ -\sin\phi & \cos\phi & 0 \\ 0 & 0 & 1 \end{pmatrix}.$$

References

[1] M. Radermacher, T. Wagenknecht, A. Verschoor and J. Frank, J. Microscopy 146 (1987) 113.

[2] T.-W. Jeng, R.A. Crowther, G. Stubbs and W. Chiu, J. Mol. Biol. 205 (1989) 251.

[3] M. Yeager, K.A. Dryden, N.H. Olson, H.B. Greenberg and T.S. Baker, J. Cell Biol. 110 (1990) 2133.
[4] J.M. Valpuesta, R. Henderson and T. Frey, J. Mol. Biol. 214 (1990) 237.
[5] P.N.T. Unwin and P.D. Ennis, Nature 307 (1984) 609.
[6] J. Frank, M. Radermacher, T. Wagenknecht and A. Verschoor, in: Methods in Enzymology, Eds. H.F. Noller and K. Moldave (Academic Press, San Diego, 1988) p. 3.
[7] J. Frank, P. Penczek, R. Grassucci and S. Srivastava, J. Cell Biol. 115 (1991) 597.
[8] T. Wagenknecht, R. Grassucci and J. Frank, J. Mol. Biol. 199 (1988) 137.
[9] M. Radermacher, J. Electron Microsc. Tech. 9 (1988) 359.
[10] J. Frank, B. Shimkin and H. Dowse, Ultramicroscopy 6 (1981) 343.
[11] J. Frank, A. Verschoor and M. Boublik, Science 214 (1981) 1353.
[12] M. van Heel and J. Frank, Ultramicroscopy 6 (1981) 187.
[13] J. Frank and M. van Heel, J. Mol. Biol. 161 (1982) 134.
[14] J.M. Carazo, T. Wagenknecht, M. Radermacher, V. Mandiyan, M. Boublik and J. Frank, J. Mol. Biol. 201 (1988) 393.
[15] G. Harauz, E. Boekema and M. van Heel, in: Methods in Enzymology, Eds. H.F. Noller and K. Moldave (Academic Press, San Diego, 1988) p. 35.
[16] J. Frank, Optik 63 (1982) 67.
[17] M. van Heel and M. Stöffler-Meilicke, EMBO J. 4 (1985) 2389.
[18] J. Frank and L. Al-Ali, Nature 256 (1975) 376.
[19] E.J. Boekema, J.A. Berden and M. van Heel, Biochem. Biophys. Acta 851 (1986) 353.
[20] M. Schatz and M. van Heel, Ultramicroscopy 32 (1990) 255.
[21] R.C. Singleton, IEEE Trans. Audio Electroacoust. AU-17 (1969) 93.
[22] J. Frank, W. Goldfarb, D. Eisenberg and T.S. Baker, Ultramicroscopy 3 (1978) 283.
[23] IMSL, User's Manual, Version 1.0, April 1987.
[24] J.E. Mezzich and H. Solomon, Taxonomy and Behavioral Science (London, Academic Press, 1980).
[25] M. Radermacher, T. Wagenknecht, A. Verschoor and J. Frank, J. Microscopy 141 (1986) RP1.
[26] (a) R. Guckenberger, Ultramicroscopy 9 (1982) 167;
(b) P.F.C. Gilbert, J. Theor. Biol. 36 (1972) 105;
(c) M. Radermacher and W. Hoppe, in: Proc. 9th Int. Congr. on Electron Microscopy, Toronto, 1978, Vol. 1, p. 218.
[27] A.K. Louis, F. Natterer, Proc. IEEE 71 (1983) 379.
[28] M.E. Davison and F.A. Grunbaum, Commun. Pure Appl. Math. 34 (1981) 77.
[29] G. Harauz and M. van Heel, Optik 73 (1986) 146.
[30] R. Gordon, B. Bender and G.T. Herman, J. Theor. Biol. 29 (1970) 471.
[31] W.H. Press, B.P. Flannery, S.A. Teukolsky and W.T. Vetterling, Numerical Recipes (Cambridge University Press, Cambridge, 1986).
[32] T.S. Durrani and C.E. Goutis, IEE Proc. 127E (1980) 161.
[33] M.B. Katz, Question of Uniqueness and Resolution in Reconstruction from Projections, Vol. 26 of Lecture Notes In Biomathematics (Springer, Berlin, 1978).
[34] R.A. Crowther, D.J. DeRosier and A. Klug, Proc. Roy. Soc. (London) A 317 (1970) 319.
[35] Weiping Liu, in: Proc. 49th Annu. EMSA Meeting 1991, Ed. G.W. Bailey (San Francisco Press, San Francisco, 1991) p. 542.
[36] P.W. Hawkes, in: Computer Processing of Electron Microscope Images, Ed. P.W. Hawkes (Springer, Berlin, 1980) p. 1.
[37] T. Wagenknecht, J.M. Carazo, M. Radermacher and J. Frank, Biophys. J. 55 (1989) 455.
[38] J. Frank, Biophys. J. 12 (1972) 484.
[39] C. Toyoshima and P.N.T. Unwin, Ultramicroscopy 25 (1988) 279.
[40] M. Pujari, P. Penczek and J. Frank, in preparation.

J. Mol. Biol. (1992) **226**, 301–304

Eukaryotic Initiation Factor 3 Does Not Prevent Association Through Physical Blockage of the Ribosomal Subunit–Subunit Interface

Suman Srivastava, Adriana Verschoor and Joachim Frank

Wadsworth Center for Laboratories and Research
N.Y.S. Dept. of Health
Albany, NY 12201-0509, U.S.A.

(Received 4 December 1991; accepted 3 April 1992)

The "native" 40 S ribosomal subunit, in which the protein eukaryotic initiation factor 3 is bound to the 40 S small ribosomal subunit, has been reconstructed to 48 Å resolution. Comparison with a previous three-dimensional reconstruction of the "derived" 40 S subunit lacking any non-ribosomal components reveals the attachment site and morphology of the factor. It is a large (~165 to 170 Å long), bilobed, elongate structure, attached to the back lobes of the 40 S subunit by two strand-like features. Significantly, the factor is oriented away from the 60 S-subunit–40 S-subunit interface surface of the 40 S particle, suggesting that its anti-association activity is not accomplished *via* simple physical blockage of that surface.

Keywords: eukaryotic initiation factor 3; 40 S ribosomal subunit; three-dimensional reconstruction; anti-association; eIF-3–40 S complex

In the process of eukaryotic protein synthesis, the large (60 S) and small (40 S) subunits of the 80 S ribosome are prevented from associating with one another when active translation is not occurring. This anti-association is accomplished through binding of eukaryotic initiation factor 3 (eIF-3†) to the small 40 S subunit, forming the so-called "native" 40 S subunit. eIF-3 is a multicomponent protein with at least three known functions. It is required for stabilization of the 40 S initiation complex, for binding of mRNA to the initiation complex, and for preventing the association of ribosomal subunits.

The 40 S–eIF-3 complex can be isolated from ribosomes as the native 40 S subunit (in distinction to the "derived" 40 S subunit, which contains no non-ribosomal components) by variation of the KCl concentration (Lutsch *et al.*, 1986). At low (e.g. 25 to 75 mM) concentrations of KCl, negative-stain electron microscopy reveals a 40 S particle having the same morphology and assuming the same characteristic orientations as the derived 40 S subunit, except that a large elliptical mass is attached to the "back lobes" (Frank *et al.*, 1981; Verschoor *et al.*, 1989) of the 40 S structure. At high (e.g. 200 mM) concentrations of KCl, the particles lack this additional mass and appear identical with derived 40 S subunits (Lutsch *et al.*, 1986).

Native 40 S subunits from rabbit reticulocyte ribosomes in a buffer containing 75 mM-KCl were examined by negative-stain electron microscopy, using the sandwiched carbon technique (Fig. 1). Initially, 1622 individual particle image pairs, selected from eleven pairs of digitized 0- and 50-deg. micrographs, were analyzed according to the protocols of the random-conical reconstruction scheme of Radermacher *et al.* (1987). A two-dimensional average (Fig. 2) was computed over a subset of 895 aligned 0-deg. images that showed the characteristic L-lateral-view 40 S subunit morphology (Frank *et al.*, 1981, 1982; Verschoor *et al.*, 1989), with the addition of an elongate mass attached at two points to the back lobes of the 40 S subunit; these back lobes are the counterpart of the platform of the eubacterial 30 S small ribosomal subunit. A three-dimensional reconstruction was calculated by weighted back projection over the corresponding set of 895 tilt images.

In the reconstructed complex visualized at a resolution of 48 Å (Fig. 3(a): 1 Å = 0·1 nm), the factor appears as a bilobed structure attached to the back lobes of the 40 S subunit, with its long axis parallel to the subunit's long axis. The 40 S subunit

† Abbreviation used: eIF-3, eukaryotic initiation factor 3.

0022-2836/92/140301-04 $03.00/0

© 1992 Academic Press Limited

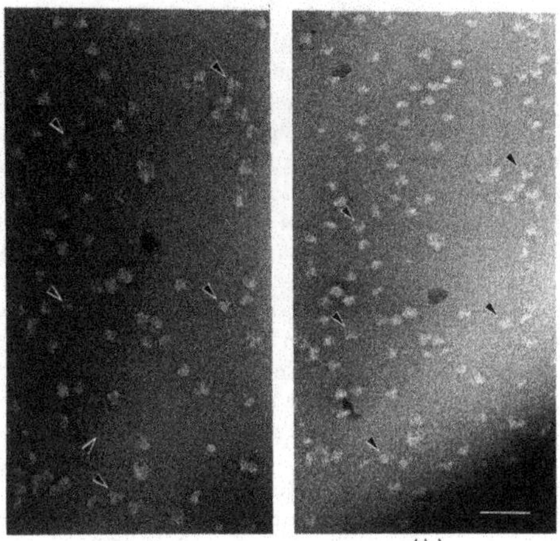

Figure 1. Field of native 40 S subunits from electron micrographs recorded at (a) 0-deg. and (b) 50-deg. tilts (tilt axis, broken line). Arrowheads indicate typical 40 S–eIF-3 complexes. Scale bar represents 100 nm. Native 40 S subunits were prepared from rabbit reticulocyte ribosomes (Lutsch *et al.*, 1986) and diluted into a buffer containing 75 mM-KCl. For electron microscopy, specimens were prepared by a sandwiched carbon technique using 0·5% (w/v) uranyl acetate as negative stain (Tischendorf *et al.*, 1974). Pairs of tilt (50 deg.) and non-tilt (0 deg.) micrographs were recorded from suitable regions of the grid, using low-dose techniques, on a Philips EM420 operating at 100 kV and at a magnification of 49,000. In a control preparation, the concentration of KCl was raised to 200 mM, and the specimen was examined as above (without tilting). SDS gel electrophoresis confirmed that the two specimens differed only by the presence in the native sample of bands corresponding to the pattern seen for purified eIF-3 subunits, as had been found by Lutsch *et al.* (1986).

Figure 2. Average images for derived and native 40 S subunits, and for control specimen in high concentrations of KCl. Scale bar represents 200 Å. (a) Average image from a previous study (Verschoor *et al.*, 1989) of derived 40 S subunits in the L-lateral view. Double-carbon preparation. (b) Average image computed over 895 aligned images of native 40 S subunits selected from double-carbon regions of specimen grids, and lying in the equivalent of the L-lateral orientation. (The lateral-view terminology was originally coined for derived 40 S subunits). (c) Average image computed over 23 aligned L-views from control micrograph, which was recorded from an evident single-carbon region of a double-carbon specimen grid. (See Frank *et al.* (1985) for a discussion of effects of incomplete stain embedding on single-layer images.) In projection, the morphology of the native 40 S subunit matches that of the derived subunit (Frank *et al.*, 1981, 1982; Verschoor *et al.*, 1989) and of subunits in the control preparation in high concentrations of KCl, except for the presence of the large elongate mass, identified with eIF-3, attached to the back lobes of the ribosomal particle. A total of 11 digitized micrograph pairs (11 fields recorded at 0- and 50-deg. tilts) were read into the computer, and particles were interactively selected from the fields on a graphics terminal. The non-tilt images were aligned in several cycles by means of correlation methods, using the SPIDER image-processing software (e.g., Frank *et al.*, 1988). The reference in each cycle was the cumulative total average of successfully aligned particle images from the previous cycle. From an initial set of 1622 pairs of individual images of the complex, elimination of non-L lateral views, misaligned images and images from non-double carbon regions of the specimen yielded a set of 895 suitable 0-deg. images. The corresponding tilt images were then centered and labeled by their azimuthal angles in the random-conical tilt geometry. Three-dimensional reconstruction using these 895 tilt projections was by a weighted back-projection algorithm (Radermacher *et al.*, 1987). Comparison of independent half-set reconstructions of 448 and 447 projections (not shown) gave a 45-deg. phase residual resolution estimate of 48 Å.

itself appears identical to the previously reconstructed derived subunit (Verschoor *et al.*, 1989). The upper lobe of the factor mass is globular and, when seen in slices (not shown) through the reconstruction, appears to have a low-density center and a higher-density rim or shell. The lower lobe of the structure is smaller, more irregular in shape, with a "bootlike" extension, and variable in size among the input images (A.V., S.S. & J.F., unpublished results) and thus is likely to be somewhat blurred in this "global" reconstruction. The factor is about 165 to 170 Å long, and about 90 Å in diameter (about 40 to 50 Å in diameter at its "waist"). The 40 S subunit in this reconstruction is about 300 Å in length; thus, the factor molecule is more than half the length of the particle to which it binds.

On the basis of the density threshold used for the visualization (Frank *et al.*, 1988) in Figure 3, an approximate volume of 710 nm^3 for the factor was calculated (A.V., S.S. & J.F., unpublished results). In comparison, a volume figure of 780 nm^3 was derived by Behlke *et al.* (1986) from the relative molecular mass (M_r = 650,000 to 660,000) and partial specific volume of the factor.

We note that, although various other initiation factors, in addition to eIF-3, have been reported to be present when the native 40 S complex is isolated (see references cited in Lutsch *et al.*, 1985), eIF-3 is the only one found in stoichiometric quantities. SDS gel electrophoresis of our specimen revealed no bands other than those identified with the polypeptides of the 40 S subunit proteins or eIF-3 (see legend to Fig. 1). It is possible, nevertheless, that a characteristic morphological–volumetric variation that we discern for the factor among individual particles in the population (A.V., S.S. & J.F., unpublished results) relates to the sub-stoichio-

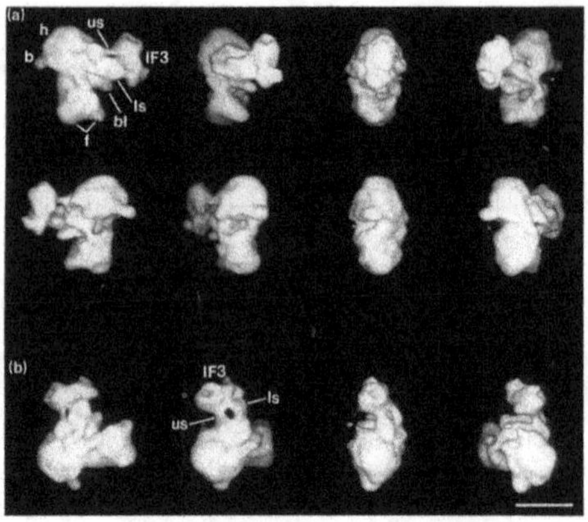

Figure 3. Surface series computed from 3-dimensional reconstruction of native 40 S subunit from 895 tilt projections. Rotations in both series are about a vertical axis, with a rotational increment of 45 deg. Scale bar represents 200 Å. (a) Rotational series around long axis of reconstructed native 40 S subunit. Labeled are the beak (b), head (h), back lobes (bl) and feet (f) of the 40 S subunit, the factor (IF3), and the thin upper attachment strand (us) and thick lower strand (ls). (b) Rotational series in which the long axis of the 40 S subunit has been brought to the horizontal by a 90-deg. rotation of the reconstruction. These rotations around a "head-on" view of the complex (3rd image) show the positioning of the 2 attachment strands on the interface-wards aspect (see Fig. 4) of the back lobes of the 40 S subunit.

metric presence of one or more of the smaller factors.

Perhaps the most interesting characteristic of the complex brought out by our reconstruction is the attachment of the factor to the subunit by two strand-like structures, a thin upper strand attaching to the upper back lobe, and a much thicker strand attaching to the lower back lobe (Fig. 3(a), column 1). The factor itself is symmetrically placed (Fig. 3(a), row 1, 3rd image) on the subunit's back lobes. However, the reconstruction shows the two strands to curve around the side of the back lobes (Fig. 3(b)) that faces the subunit–subunit interface in the monomeric ribosome.

What does our reconstruction reveal about how the factor might function in anti-association? The two alternative mechanisms generally proposed are: (1) steric hindrance, in which the presence of the factor on the 40 S subunit physically blocks the 60 S subunit from associating; or (2) induction of some type of conformational change in the 40 S subunit that alters its 60 S-subunit binding surface or even its binding affinity for the 60 S subunit. The conclusion reached by Nygard & Nilsson (1990) in a recent review was that eIF-3 is likely to function by a partial blockage of the subunit's interface with the 60 S subunit, as had originally been suggested by Emanuilov *et al.* (1978).

Initially, our results would seem to argue strongly against the first alternative. The factor does not directly block the interface surface. We can delineate the area of contact between the 40 S and 60 S subunits when they associate based on our previous reconstruction of the monomeric 80 S ribosome (Verschoor & Frank, 1990). From the positioning of the factor on the side of the subunit almost directly opposite the aspect that would come into contact with the 60 S subunit in the monomeric ribosome (Fig. 4), it is difficult to visualize in what way the factor could sterically hinder the 60 S subunit. On the basis of our reconstruction, the only

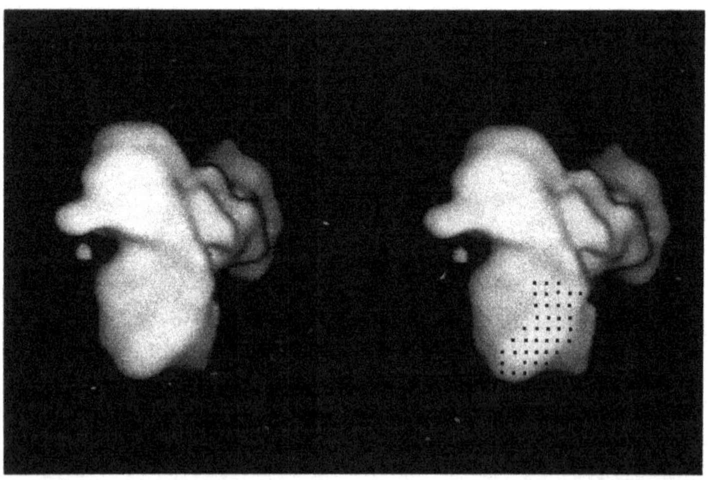

Figure 4. Surface of native 40 S subunit that would face the large 60 S subunit if the 2 subunits were associated in the 80 S monomeric ribosome. Stippling on right-hand member of stereo pair indicates the only portion of 40 S particle that is in tight contact with the 60 S subunit in the monomeric ribosome: the upper parts of both subunits "lean" away from each other. This information derives from our previous 3-dimensional reconstruction of the 80 S ribosome (Verschoor & Frank, 1990), in which the 2 subunits could be discerned as separate, except in this basal region.

obvious means by which the factor could block the interface and physically obstruct 60 S association is by the positioning of the two attachment strands, which appear to curve toward the interface surface (Fig. 3(b)).

The alternative explanation for the anti-association activity is that the binding of eIF-3 to the 40 S subunit produces a conformational change in some component or region of the 40 S subunit, thus preventing its binding of the 60 S subunit. This possibility is more difficult for us to confirm or reject, especially given the limited resolution of the present study. Certainly, conformational changes may be very small, yet still able to produce a profound effect. A quantitative comparison of the new native 40 S reconstruction with the derived-subunit reconstruction previously calculated (Verschoor *et al.*, 1989) shows only a single region of the 40 S subunit proper in which the two structures do not match closely (A.V., S.S. & J.F., unpublished results). It is possible, or even probable, however, that this density difference in the "throat" region of the subunit arises from staining differences between the two preparations. Repetition of the two reconstruction studies using frozen hydrated preparations would be required to eliminate this possibility.

We thank T. Wagenknecht for a critical reading of the manuscript. Partial support was provided by the National Institutes of Health (GM 29169).

References

Behlke, J., Bommer, U.-A., Lutsch, G., Henske, A. & Bielka, H. (1986). Structure of initiation factor eIF-3 from rat liver. *Eur. J. Biochem.* **157**, 523–530.

Emanuilov, I., Sabatini, D. D., Lake, J. & Freienstein, C. (1978). Localization of eukaryotic initiation factor 3 on native small ribosomal subunits. *Proc. Nat. Acad. Sci., U.S.A.* **75**, 1389–1893.

Frank, J., Verschoor, A. & Boublik, M. (1981). Computer averaging of electron micrographs of 40 S ribosomal subunits, *Science*, **214**, 1353–1355.

Frank, J., Verschoor, A. & Boublik, M. (1982). Multivariate statistical analysis of ribosome electron micrographs. *J. Mol. Biol.* **161**, 107–137.

Frank, J., Verschoor, A. & Wagenknecht, T. (1985). Computer processing of electron microscopic images of single macromolecules. In *New Methodologies In Studies of Protein Configuration* (Wu, T. T., ed.), pp. 36–89, Van Nostrand Reinhold, New York.

Frank, J., Radermacher, M., Wagenknecht, T. & Verschoor, A. (1988). Studying ribosome structure by electron microscopy and computer-image processing. *Methods Enzymol.* **164**, 3–35.

Lutsch, G., Benndorf, R., Westermann, P., Bommer, U.-A. & Bielka, H. (1985). Electron microscopic investigation on the location of rat liver ribosomal proteins. *Biomed. Biochim. Acta*, **44**, K1–K7.

Lutsch, G., Benndorf, R., Westermann, P., Bommer, U. A. & Bielka, H. (1986). Structure and location of initiation factor eIF-3 within native small ribosomal subunits from eukaryotes. *Eur. J. Cell. Biol.* **40**, 257–265.

Nygard, O. & Nilsson, L. (1990). Translational dynamics. *Eur. J. Biochem.* **191**, 1–17.

Radermacher, M., Wagenknecht, T., Verschoor, A. & Frank, J. (1987). Three-dimensional reconstruction from a single-exposure random conical tilt series applied to the 50 S ribosomal subunit of *E. coli*. *J. Microsc.* **146**, 113–136.

Tischendorf, G. W., Zeichhardt, H. & Stöffler, G. (1974). Determination of the location of proteins L14, L17, L18, L19, L22 and L23 on the surface of the 50 S ribosomal subunit of *E. coli* by immune electron microscopy. *Mol. Gen. Genet.* **134**, 187–208.

Verschoor, A. & Frank, J. (1990). Three-dimensional structure of mammalian cytoplasmic ribosome. *J. Mol. Biol.* **214**, 737–749.

Verschoor, A., Zhang, N. Y., Wagenknecht, T., Obrig, T., Radermacher, M. & Frank, J. (1989). Three-dimensional reconstruction of mammalian 40 S ribosomal subunit. *J. Mol. Biol.* **209**, 115–126.

Edited by D. J. De Rosier

ELSEVIER

Ultramicroscopy 53 (1994) 251–270

ultramicroscopy

The ribosome at improved resolution: new techniques for merging and orientation refinement in 3D cryo-electron microscopy of biological particles

Pawel A. Penczek [a], Robert A. Grassucci [a], Joachim Frank [a,b]

[a] *Wadsworth Center for Laboratories and Research, New York State Department of Health, P.O. Box 509, Albany, NY 12201-0509, USA*

[b] *Department of Biomedical Sciences, State University of Albany at New York, Albany, NY 12222, USA*

(Received 18 June 1993; in final form 7 October 1993)

Abstract

Cryo-electron microscopy of single biological particles opens up new possibilities for structure analysis: the particle can be reconstructed in its native shape and internal features are preserved. To take advantage of these possibilities we have developed new methods of data collection and image processing and we have applied them to the 70S *Escherichia coli* ribosome. A method of orientation search is proposed, which makes it possible to relate random-conical data sets to one another even if they are collected from low-tilt micrographs. A technique for 3D alignment of projections is described and applied to the single-micrograph 3D reconstruction.

1. Introduction

The random-conical reconstruction scheme [1–4] is now in routine use in several laboratories [5–8] (for a survey see Ref. [9]) as a tool to reconstruct 3D images of macromolecular assemblies. The method uses pairs of micrographs, of a tilted and untilted specimen, to determine the Eulerian angles required to calculate the 3D reconstruction. The fact that the angles are determined from comparable 0°-views makes this technique powerful, since those images are not used in the actual reconstruction, allowing the electron dose to be reduced. The progress in applying this technique to ice-embedded particles has been slow because of obstacles that are either technical (the difficulty of obtaining micrographs of sufficient quality for highly [50° or above] tilted specimen grids) or intrinsic to unstained specimens (low signal-to-noise ratio [SNR]). It has become clear that new image-processing tools must be developed to deal with these problems. Our previous paper [10] described two important advances: an algorithm for reference-free alignment of 0°-views was developed, and the problem of missing angular information was solved by merging several reconstructions that were each based on particles lying in different orientations.

The current work addresses some shortcomings of the merging method, which entails a pairwise real-space orientation search among individual reconstructions: it is restricted to high-tilt

0304-3991/94/$07.00 © 1994 Elsevier Science B.V. All rights reserved
SSDI 0304-3991(93)E0155-2

data, and the Eulerian angles found are unavoidably biased by the partial absence of angular information. To overcome these problems we developed a new method to orient entire random-conical projection sets with respect to each other. This method is no longer burdened by the missing angular information and allows calculation of a *low-tilt 3D reconstruction* by directly merging a number of projection sets corresponding to different 0°-views.

In addition we put forward a *3D projection alignment method* making it possible to refine the Eulerian angles assigned to the individual projections. Application of this method results in improved resolution of the structure. The method makes it also possible to calculate a 3D reconstruction from single, very-low-tilt or even 0°-tilt micrographs *without using tilt pairs*, employing a low-resolution 3D reconstruction as a reference instead.

2. Methods

2.1. Background

The problem of determining the relative orientations of 3D structures from their 2D projections has two known solutions: the method of moments and the so-called common-lines approach. The first method [11,12], with its high sensitivity to errors in the data, is of rather academic interest. The second, originally proposed by Crowther et al. [13], is routinely used in the 3D reconstruction of virus structures with high symmetry and has been extended to general non-symmetrical structures [11,14]. This approach is based on the Central-Section Theorem, which implies that in real space any two projections of a 3D object will share a common 1D projection. The equivalent Fourier space formulation states that the 2D Fourier transforms of any two 2D projections of a 3D object are central sections of the 3D Fourier transform of that object, and thus will share a common line in Fourier space. The known directions of three such common lines fix the orientation of three planes (either in real or in Fourier space) and are sufficient to determine the orientations of the three associated projections (up to an ambiguity of handedness of the 3D reconstruction). Thus, in principle, one would need only three 2D projections of an unknown 3D object to calculate the required Eulerian angles. Such an approach was described in Ref. [14], but no clues were given for how to deal with errors in the data and how to include a larger number of projections. Goncharov [11] suggested including new projections by calculating common lines with a number of pairs of already aligned projections and excluding those solutions that do not fulfill the requirements of geometric consistency. Nevertheless, no discussion of errors was given, nor was a refinement of the angles found proposed.

An interesting elaboration of the common-lines approach was described in Ref. [15]. The authors recognized that even under modest noise conditions the average error in the position of the common axis becomes unacceptably large. To increase the accuracy of the method they used a large number of projections and used quaternion mathematics [16] to find a consistent solution and minimize the errors. The method proved sufficiently robust in the test situation but has yet to be proved applicable to real data. Although the authors made an attempt to enhance the performance of the common-lines method by inclusion of multiple common-lines comparisons, they were unable to include all the possible correlations. Another drawback of the method is that the orientations of common lines and corresponding discrepancies are calculated only once, during the first step of the method. All refinements of the solution are done only by enforcing the consistency of the solution, without recalculation of the corresponding error in the space of common lines.

In what follows we will describe a method which uses the elements of common-lines methods and takes full advantage of the random-conical data collection scheme [3,4]. Instead of finding the orientations of individual projections it determines the orientation between two entire *sets* of projections with the orientations among projections staying fixed within the sets. Such an approach considerably increases the signal-to-noise ratio (SNR) in the comparison of the data.

2.2. The method of Orientation Search using Sets of Projections (OSSP)

In the framework of the random-conical 3D reconstruction of non-symmetrical particles the problem of orientation determination is automatically solved: the random-conical scheme of data collection provides all Eulerian angles required for the 3D reconstruction. Two of these angles are determined with high accuracy from the tilt geometry and the third one is found through the alignment of particles from the untilted-specimen micrograph. Provided that the structure occurs on the specimen grid in a preferred orientation, its 0° projections differ only by a rotation in the plane of the grid, and the corresponding angles can be found using an alignment procedure. The important advantage of this approach is that only very similar projections are compared and the resulting average has high SNR, which facilitates the determination of the missing third Eulerian angle with high accuracy. Reconstructions obtained in this way are, however, limited by the missing angular region, and efforts must be made to fill this region by using more than one preferred orientation. Therefore the problem of orientation determination resurfaces, but this time it can be solved by relating *entire* data sets to one another. We achieve this goal by finding the best match between the two corresponding sets of projections, assuming the geometry within each set (in terms of Eulerian angles of each projection) to be known. A preliminary account of this new approach was given in Ref. [17].

The random-conical data collection implies that the Fourier transforms of the projections form a set of planes in Fourier space tangential to the cone (Fig. 1). Thus, the problem of finding the orientation between two 3D structures can be formulated as the problem of finding the best matching orientations between two such sets of Fourier planes tangential to their respective cones.

To understand the derivation of the method it is sufficient to consider two Fourier planes belonging to two different cones. Such two planes intersect in 3D along a single line (except in the degenerate case in which they coincide), the *common* line. Along this line we wish to calculate the *discrepancy* $1-\rho_{12}$, where ρ_{12} is the cross-correlation coefficient. The direction of the common line in 3D space is determined by the orientations of the planes in the coordinate systems associated with their cones, given by rotation matrices R_1 and R_2, respectively, and by the relative orientation between the two sets of planes (or two cones) given by the rotation matrix R_T. In the coordinate systems associated with the planes, the direction of the intersection line is α_1 for the first plane and α_2 for the second plane. Thus, the problem can be stated as follows: find two unknown angles α_1 and α_2 given rotation matrices R_1, R_2 and R_T

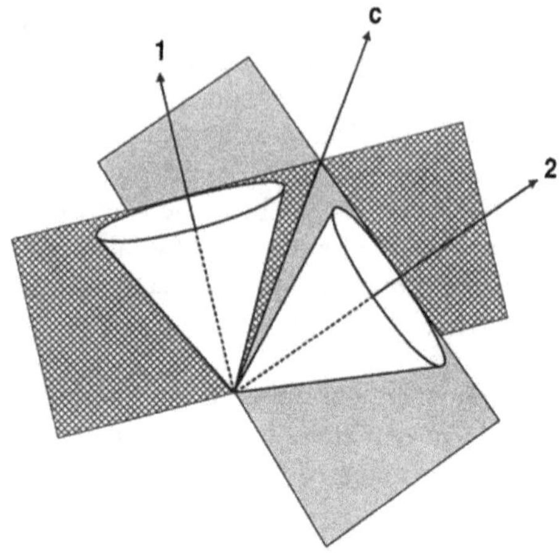

Fig. 1. Principle of the OSSP method. A pair of planes in Fourier space tangentlal to their corresponding cones in the random-conical data collection scheme intersect each other along the common line C.

The unitary vectors describing the orientation of the line on corresponding planes are defined as follows:

$$\boldsymbol{n}_k = \begin{pmatrix} \cos\alpha_k \\ \sin\alpha_k \\ 0 \end{pmatrix}, \quad k=1,2. \tag{1}$$

The orientation of these vectors in 3D space can be found using corresponding rotation matrices R_k describing the orientation of the planes in the coordinate system associated with their cones:

$$\boldsymbol{u}_k = R_k^{-1}\boldsymbol{n}_k. \tag{2}$$

The two vectors $\boldsymbol{u}_1$ and $\boldsymbol{u}_2$ are brought into coincidence by the rotation matrix R_T (which describes the orientation between the two cones):

$$\boldsymbol{u}_1 = R_T^{-1}\boldsymbol{u}_2, \tag{3}$$

or

$$R_1^{-1}\boldsymbol{n}_1 = R_T^{-1}R_2^{-1}\boldsymbol{n}_2. \tag{4}$$

To solve this system of equations for two unknown angles α_1 and α_2 it is convenient to introduce the product of three rotation matrices:

$$R_p = R_2 R_T R_1^{-1}, \tag{5}$$

and replace Eq. (4) by the simpler one:

$$R_p\boldsymbol{n}_1 = \boldsymbol{n}_2. \tag{6}$$

Since matrix R_p is a product of three rotation matrices it must have the form of a rotation matrix itself. Using our convention of Eulerian angles [10] we can write: (see Eq. (7) below). In Eq. (7) the subscript p has been omitted for simplicity. Using the last of the three equations given by (6) we obtain:

$$\cos\phi_p \cos\alpha_1 + \sin\phi_p \sin\alpha_1 = 0,$$

which is equivalent to

$$\cos(\alpha_1 - \phi_p) = 0,$$

from which it follows that

$$\alpha_1 = 90° + \phi_p. \tag{8}$$

To find angle α_2 we can multiply both sides of Eq. (6) by R_p^{-1} and obtain

$$\boldsymbol{n}_1 = R_p^{-1}\boldsymbol{n}_2, \tag{9}$$

which can be solved in a similar way as Eq. (6) leading to

$$\alpha_2 = 90° - \psi_p. \tag{10}$$

From Eq. (5) we know that the angles ϕ_p and ψ_p are functions of the Eulerian angles describing the orientation of both planes in their own system of coordinates as well as the parameter angles describing the relative orientation between the two structures. The explicit equations would be very complicated, but they are not needed since these angles can be easily retrieved from the elements of the rotation matrix R_p. Indeed, we can always find the numerical values of the elements r_{ij}^p the matrix R_p using Eq. (5), and from its elements calculate angles and ϕ_p and ψ_p. The procedure is as follows: since Eulerian angle θ_p is defined to be in the range $(0, \pi)$ we can find its value uniquely from

$$\theta_p = \arccos(r_{33}^p), \tag{11}$$

and the two remaining angles from

$$\phi_p = \text{arctane}(r_{23}^p, r_{13}^p),$$

$$\psi_p = \text{arctane}(r_{32}^p, -r_{31}^p), \tag{12}$$

where arctane is the extended arcus tangent function of two arguments, which gives the result in the whole $(0, 2\pi)$ range. In the case when $r_{33}^p = \pm 1$ the angle θ is equal to zero or π and the rotation matrix degenerates to a matrix representing in-plane rotation. In that case ψ_p can be set to zero, and the rotation angle becomes

$$\phi_p = \text{arctane}(r_{21}^p, r_{11}^p). \tag{13}$$

We define the *global discrepancy* between two *sets of planes* in the following way:

$$D(I,II;R_T) = \sum_{m_1=1}^{M_1}\sum_{m_2=1}^{M_2}(1-\rho(m_1,m_2)), \tag{14}$$

where M_1 and M_2 are the numbers of projections (or Fourier planes) in structures I and II, respec-

$$R_p(\psi_p,\theta_p,\phi_p) = R^a(\psi_p)R^b(\theta_p)R^a(\phi_p) = \begin{pmatrix} \cos\psi & \sin\psi & 0 \\ -\sin\psi & \cos\psi & 0 \\ 0 & 0 & 1 \end{pmatrix}\begin{pmatrix} \cos\theta & 0 & -\sin\theta \\ 0 & 1 & 0 \\ \sin\theta & 0 & \cos\theta \end{pmatrix}\begin{pmatrix} \cos\phi & \sin\phi & 0 \\ -\sin\phi & \cos\phi & 0 \\ 0 & 0 & 1 \end{pmatrix}$$

$$= \begin{pmatrix} \cos\phi\cos\theta\cos\psi - \sin\phi\sin\psi & \sin\phi\cos\theta\cos\psi + \cos\phi\sin\psi & -\sin\theta\cos\psi \\ -\cos\phi\cos\theta\sin\psi - \sin\phi\cos\psi & -\sin\phi\cos\theta\sin\psi + \cos\phi\cos\psi & \sin\theta\sin\psi \\ \cos\phi\sin\theta & \sin\phi\sin\theta & \cos\theta \end{pmatrix}. \tag{7}$$

tively, and $\rho(m_1,m_2)$ is the correlation coefficient calculated along the common line between planes belonging to the first and second structure. The directions of the line are given by Eqs. (8) and (10).

The best matching orientation is defined by the rotation matrix R_T for which the global discrepancy in Eq. (14) is minimized. To find this minimum, one of the standard minimization procedures may be used such as the procedure implemented in the IMSL package [18], which is based on a quasi-Newton method using finite-difference gradients. The evaluation of the global discrepancy D requires numerous calculations of correlation coefficients along common lines (for one evaluation there are $M_1 \times M_2$ such lines). Thus, in the actual implementation of the program the lines are precalculated with given angular accuracy. To facilitate the calculations we took advantage of the fact that a line in 3D Fourier space going through the origin of the coordinate system is equivalent to the Fourier transform of the 1D line projection calculated at the same angle. Thus, since we assume that the 2D projections of the 3D object are available we can avoid Fourier transforms altogether and use the following strategy instead: for every plane (i.e., projection) a circular mask is applied and line projections are calculated with angular step $\Delta\alpha$ in the range $(0, \pi)$. Line projections in the range $(\pi, 2\pi)$ are mirrored versions of the projections from the first set and for the sake of saving computer memory this symmetry is taken into account in the program. The line projections corresponding to all the planes become the input to the Orientation Search using Sets of Projections (OSSP) program. To avoid cumbersome interpolations during the calculations, the angles α_1 and α_2 describing orientations of intersection lines are rounded to coincide with the directions of precalculated lines.

2.3. Tests of the OSSP program

The OSSP program described in the previous section was tested on a simulated structure. A model (denoted S_1) was built in a cube of $64 \times 64 \times 64$ size as a sum of three 3D Gaussian functions (the design is similar to that described in Ref. [19]):

$$f(x,y,z) = \sum_{i=1}^{3} G_i(x,y,z), \tag{15}$$

where

$$G_i(\boldsymbol{x}) = A_i \exp\left\{-\tfrac{1}{2}(R_i\boldsymbol{x} + \boldsymbol{x}_{0i})^{\mathrm{T}} C_i^{-1}(R_i\boldsymbol{x} - \boldsymbol{x}_{0i})\right\}. \tag{16}$$

C_i is a convariance matrix:

$$C_i = \begin{bmatrix} \sigma_{xi} & 0 & 0 \\ 0 & \sigma_{yi} & 0 \\ 0 & 0 & \sigma_{zi} \end{bmatrix}, \tag{17}$$

and R_i is a rotation matrix. All the relevant parameters used for the simulation are listed in Table 1.

The model structure (denoted by S_1, shown in Fig. 2) described above was rotated using three sets of the Eulerian angles included in Table 2a simulating particles lying in different orientations (denoted S_2, S_3, S_4).

In the first set of tests we emulated the conical tilt geometry of data collection. The orientation search was applied to pairs of structures, always using structure S_1 as a reference. All model structures (S_1, S_2, S_3, S_4) were projected into 36 evenly distributed directions ($\Delta\phi = 10°$) with the tilt angle $\theta = 50°$. Projections of structure S_1 are

Table 1
Parameters used in simulation of the model shown in Fig. 2; the model is built of three Gaussian functions

i	A_i	x_{0i}	y_{0i}	z_{0i}	σ_{xi}	σ_{yi}	σ_{zi}	ψ_i	θ_i	ϕ_i
1	10	0.0	0.0	−5.2	9.5	5.5	6.5	0	65	0
2	10	1.8	−2.5	12.8	7.0	4.0	4.5	−30	35	0
3	10	−2.7	4.95	8.25	1.5	4.5	8.25	−37	54	162

shown in Fig. 3. Thus, every structure was represented by a set of 36 noise-free 2D projections. From these projections two new sets of projections were created by adding Gaussian noise of zero mean and standard deviation chosen in such a way that the resulting noisy projections had a signal-to-noise ratio equal to 1 for the first set and 0.5 for the second set. All 2D projections were projected onto sets of 180 lines ($\Delta\alpha = 1°$) using a circular mask with a radius of 24 pixels. These line projections were used as input to the OSSP program.

During each test we used a number of different initial angles and we chose the solution with the smallest global discrepancy value. Since there is always a possibility for the minimization procedure to be trapped in a local minimum the procedure should be started from several different sets of initial parameters to verify the solution and exclude any false answers. The results of the test runs for the noise-free as well as noisy data are listed in Table 2b.

To verify the OSSP program for low-tilt data we repeated the previous tests for projections calculated at 30° tilt (projections of structure S_1 are shown in Fig. 4). The results are listed in Table 2c. In both cases (50° as well as 30° tilt) the Eulerian angles found by the OSSP program match the original angles closely. The error increases with the the noise level, but in all cases the discrepancy is acceptable.

The original motivation for the design of the method was to facilitate the data collection for the random-conical 3D reconstruction. Nevertheless, there is nothing in this design that would prevent use of the program in other situations where the data-collection geometry is limited to a certain subset. An obvious example of a prospective application is the single-axis tilt 3D reconstruction. In this scheme the data are collected

Table 2
(a) Eulerian angles used to rotate model structure S_1;
(b) Relative orientation angles found from the set of 36 evenly distributed projections at the tilt angle 50° with structure S_1 used as a reference;
(c) Relative orientation angles found from the set of 36 evenly distributed projections at the tilt angle 30° with structure S_1 used as a reference;
(d) Relative orientation angles found from the set of nine single-axis tilt projections with structure S_1 used as a reference

		ψ	θ	ϕ
(a)	$S_1 \to S_2$	90.0	50.0	50.0
	$S_1 \to S_3$	−90.0	−110.0	40.0
	$S_1 \to S_4$	−90.0	40.0	90.0

		Calculated using noise-free projections			Calculated using noisy projections (SNR = 1.0)			Calculated using noisy projections (SNR = 0.5)		
		ψ	θ	ϕ	ψ	θ	ϕ	ψ	θ	ϕ
(b)	$S_1 \to S_2$	90.1	50.0	−50.0	91.0	48.7	−47.5	90.9	48.3	−50.3
	$S_1 \to S_3$	−89.7	−110.5	40.3	−90.0	−110.9	38.6	−88.9	−111.0	39.2
	$S_1 \to S_4$	−90.2	40.1	89.3	−90.3	39.3	90.0	−93.9	41.2	96.8
		ψ	θ	ϕ	ψ	θ	ϕ	ψ	θ	ϕ
(c)	$S_1 \to S_2$	89.7	49.3	−50.6	91.4	51.8	−50.3	93.6	50.0	−48.2
	$S_1 \to S_3$	−90.0	−110.4	40.0	−90.5	−108.7	33.5	−89.2	−109.9	38.7
	$S_1 \to S_4$	−90.3	40.0	90.1	−91.3	39.0	90.0	−93.1	39.5	92.8
		ψ	θ	ϕ	ψ	θ	ϕ	ψ	θ	ϕ
(d)	$S_1 \to S_2$	90.2	49.9	−50.0	90.9	49.0	−47.2	92.9	50.3	−52.4
	$S_1 \to S_3$	−89.2	−111.3	40.6	−88.8	−112.9	36.4	−86.8	−106.5	9.7
	$S_1 \to S_4$	−90.0	40.0	90.6	−91.8	39.7	88.2	−93.4	36.0	93.3

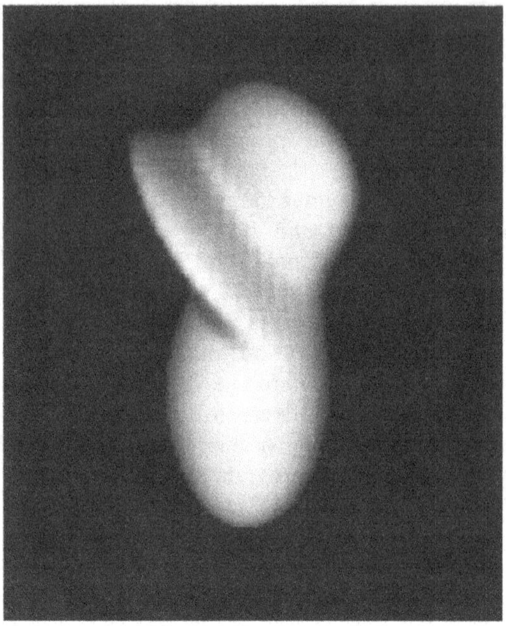

Fig. 2. Model structure built as a sum of three 3D Gaussian functions and used in the orientation search tests (z-axis is vertical to the picture plane).

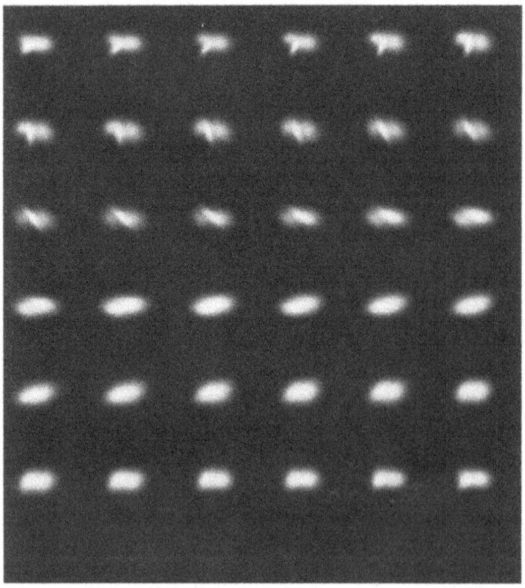

Fig. 3. Projections of the first test structure S_1 taken at a θ angle of 50° and ϕ angle in 10° increments.

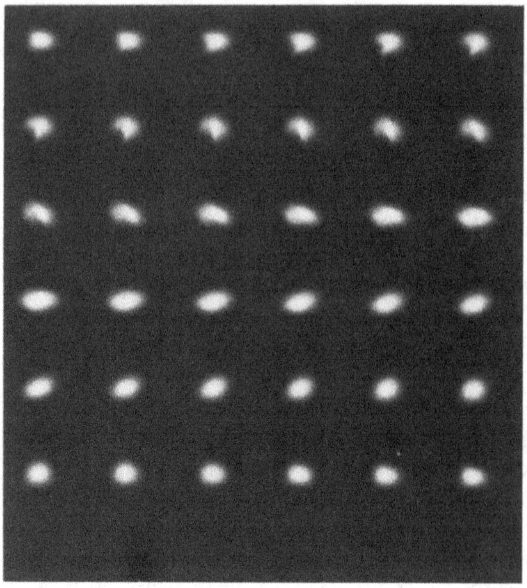

Fig. 4. Projections of the first test structure S_1 taken at a θ angle of 30° and ϕ angle in 10° increments.

using a series of tilts around the same axis, usually in the range of $-60°$ to $+60°$ [20–22]. To test the applicability of the OSSP method in this case we calculated for each structure S_1–S_4 a series of projections with angles $\psi = \phi = 0°$ and tilt angles $\theta = 0°$, $\pm 15°$, $\pm 30°$, $\pm 45°$, $\pm 60°$; thus each structure was represented by a set of nine 2D projections. Two additional data sets were created from the first one by adding Gaussian noise of zero mean and standard deviation chosen in such a way that the resulting noisy projections had SNR = 1 for the first set and 0.5 for the second set. All 2D projections were converted to line projections using the same parameter values as for the conical tilt case (mask radius equal to 24 pixels, $\Delta\alpha = 1°$). The tests were run in the same way as in the case of conical tilt geometry data and the results are listed in Table 2d.

The Eulerian angles found match the original angles closely for noise-free data and for data with SNR = 1. In the case of SNR = 0.5 the solutions for the structures S_2 and S_4 are acceptable, but not for the structure S_3, for which angle ϕ is entirely incorrect. In general, the angles found for the single-tilt axis case have larger errors than those found for both conical tilt cases. This can be easily explained by the fact that, although the SNR's for both tests were the same (1.0 and 0.5), the numbers of projections representing the structures were quite different. The number of 2D projections for the conical tilt case was 36 for each structure, thus there were $36^2 = 1296$ common lines contributing to the global discrepancy measure D. For the single-tilt axis the number of 2D projections was 9, thus the number of common lines was only $9^2 = 81$. Since the measure D is a sum of correlation coefficients between single lines it is obvious that an increase in the number of 2D projections and a resulting increase in the number of common lines should improve the overall performance of the OSSP program.

The designs of the model structure and the second set of tests (single-tilt axis orientation search) were similar to those described in Refs. [19,23]; thus it is possible to compare the two orientation search methods. The accuracy of both methods seems to be similar, but the OSSP approach certainly uses more accessible mathematical tools, and all the steps involved are easier to control, since all functions are real-valued (sinograms) and can be displayed.

2.4. Angular refinement

2.4.1. The need for refinement

The tests described in the previous section were designed to verify the ability of the program to find the orientation between two structures represented by sets of 2D projections collected in conical and single-axis tilt geometries. The underlying assumption was that the Eulerian angles describing the orientation of each 2D projection with respect to its own set of projections are known correctly. Unfortunately, in the case of EM application this is not true – the Eulerian angles are known only indirectly and with limited accuracy. To increase this accuracy one could think of collecting a much larger number of particles than are currently used and being more restrictive during the classification step, hoping to find truly homogenous groups to be used for the 3D reconstruction. This approach, which probably would lead to high-quality reconstructions, is

Table 3
(a) Eulerian angles used to calculate three 2D projections P_1–P_3 of the structure S_1;
(b) Relative orientation angles found for the single projections using 180 noise-free projections of the structure S_1 as a reference;
(c) Relative orientation angles found for the single projections using noise-free projections of the structure S_1 as a reference

		ψ			θ			ϕ		
(a)	P_1	40.0			48.0			−60.0		
	P_2	27.0			23.0			77.0		
	P_3	−70.0			33.0			55.0		
		Calculated using noise-free projections			Calculated using noisy projections (SNR = 1.0)			Calculated using noisy projections (SNR = 0.5)		
		ψ	θ	ϕ	ψ	θ	ϕ	ψ	θ	ϕ
(b)	$P_1 \rightarrow S_1$	39.5	47.6	−59.2	39.5	47.7	−90.7	42.3	47.8	−90.1
	$P_2 \rightarrow S_1$	−26.8	22.7	77.1	−38.2	18.8	88.6	−40.2	27.8	89.4
	$P_3 \rightarrow S_1$	−71.6	32.3	58.3	−68.6	32.1	56.3	−72.8	153.3	90.1
		ψ	θ	ϕ	ψ	θ	ϕ	ψ	θ	ϕ
(c)	$P_1 \rightarrow S_1$	38.6	48.0	−69.3	37.7	48.0	−71.8	34.9	42.0	−102.8
	$P_2 \rightarrow S_1$	−29.4	22.0	80.1	−25.7	20.0	76.0	−30.9	24.0	88.5
	$P_3 \rightarrow S_1$	−71.5	32.0	56.6	−68.1	32.0	52.8	−74.5	30.0	56.0

difficult to apply in practice in the case of ice-embedded specimens because of the low yield. Additional problems are: very high noise levels, intrinsic variability of the specimen, inaccuracies of the Eulerian angles found from the micrographs geometry and from alignment of 0° particles, to mention but a few. Thus, it seems more reasonable to assume that the initial Eulerian angles are known only approximately and to develop a procedure to refine them.

2.4.2. The use of common lines

In the previous section we described how the orientation between two sets of projections can be found. It seems that it should be straightforward to use the approach described to find the orientation of a single projection with respect to the 3D volume represented by a set of 2D projections using the discrepancy measure along common lines. To test such a possibility we created three arbitrarily oriented 2D projections P_1–P_3 of the structure S_1 (the corresponding Eulerian angles are listed in Table 3a). Structure S_1 was used as reference. To cover the 3D Fourier space approximately evenly, we calculated 90 2D projections with angles ψ and ϕ set to zero and $\theta = 0°, 2°, 4°, \ldots, 178°$ and additional 90 2D projections with angles $\psi = 0°$, $\phi = 90°$, and $\theta = 0°$, $2°, 4°, \ldots, 178°$. Every 2D projection was projected onto 180 lines ($\Delta\alpha = 1°$) using a circular mask with a radius of 24 pixels.

To determine the orientation we used the discrepancy measure defined by Eq. (14) with $M_2 = 1$ (i.e., there is only one projection in the second set) and the minimization program described in Section 2.1. After the first few trials it became clear that in this case the discrepancy measure has a very large number of local minima resulting in entirely incorrect solutions. It proved to be impossible to find the correct solution when using only a few of different starting points for the minimization procedure. In this situation an initial exhaustive search becomes necessary. Such a search requires a choice of Eulerian angles that cover the whole range of projection directions evenly.

An even distribution of projections can be found by the following reasoning: the direction of a normal to the projection plane is determined by two angles ϕ and θ; angle ψ is responsible for the rotation of the projection "in-plane". The corresponding normal vector is defined by:

$$\boldsymbol{n} = \begin{pmatrix} \cos\phi\sin\theta \\ \sin\phi\sin\theta \\ \cos\theta \end{pmatrix}, \tag{18}$$

while the requirement of even distribution of directions means that the norm of the derivative of this vector should be constant:

$$\|\mathrm{d}\boldsymbol{n}\| = (\sin\theta\,\mathrm{d}\phi)^2 + (\mathrm{d}\theta)^2 = \text{constant}, \tag{19}$$

from which it immediately follows that

$$\mathrm{d}\phi = \frac{\mathrm{d}\theta}{\sin\theta}. \tag{20}$$

Thus, for a chosen constant increment of the θ angle the increment of the ϕ angle should vary according to Eq. (20). The increment of the ψ angle has to be chosen equal to the increment of angle θ.

The tests proved that the exhaustive search with an angular increment of 10° followed by a search using the minimization procedure was sufficient to correctly identify the solution. The results are listed in Table 3b. For noise-free data the method performed remarkably well and all but one angle were found with an accuracy of better than one degree. The errors increased significantly for data with SNR = 1.0 and became unacceptable for data with SNR = 0.5.

2.4.3. Projection matching

Since the prospective method of angles correction needs to be robust, even in a very low SNR case, we decided to explore another approach. Instead of using the similarity measure in the space of 1D projections we can try to match the projection data *directly* to the 2D projections of the 3D structure. A similar method was originally described by Dengler [25] as a bootstrapping technique. He developed a refinement procedure for the single-axis tilt 3D reconstruction, which was used to correct two translation parameters and one angle (in-plane rotation) of every projection by comparison with the projection of the 3D structure in the corresponding direction.

In the more general case, the number of parameters to be corrected is larger since the orientation of every projection is described by three Eulerian angles. This problem was explored by van Heel [26] and Harauz and Ottensmeyer [27]. They described the method of refining orientation parameters of projections by matching them to the projections of a known 3D structure. In Harauz and Ottensmeyer's application of the method to real data [28] the initial Eulerian angles were assigned to the projection data by visual comparison with the calculated projections of the assumed model of the 3D structure. Van Heel [26] starts with random assignments of projection angles.

In what follows we will describe a computationally efficient method of exhaustive search of all three Eulerian angles that does not require knowledge of initial orientation parameters. Thus, instead of 2D projections calculated in the direction of original projections, we need a sufficient number of 2D projections covering the space evenly. This problem calls for the determination of three Eulerian angles, two of which (ϕ and θ) are responsible for the direction of the projection in 3D space, while the third one (ψ) describes the rotation of the projection "in-plane". Thus, it suffices to calculate the 2D projections of the 3D reference structure while varying only the first two angles (using Eq. (20) to determine the increment of angle ϕ) and to calculate the cross-correlation coefficient between the reference projection and projection data in polar coordinates. The second step determines the similarity and the third angle ψ at the same time. This calculation can be performed efficiently using 1D Fast Fourier Transforms (FFT) [10].

The computer program designed to determine the Eulerian angles of single projections includes the following steps (illustrated in Fig. 5):

(a) the increment in θ angle is chosen and, using Eq. (20), the set of Eulerian angles (ϕ, θ) is calculated;

(b) the 2D projections of the 3D reference structure are calculated according to the angles precalculated in step (a);

(c) all 2D reference projections are sampled in polar coordinates, and Fourier transforms of all the rings chosen are calculated and stored on disk;

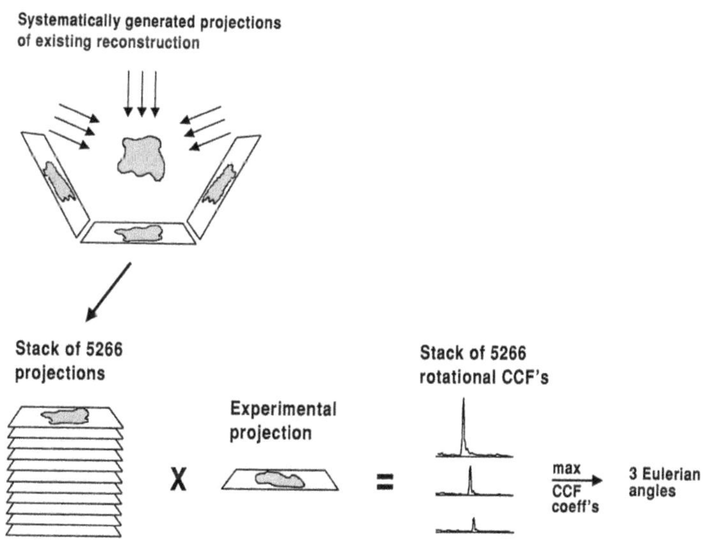

Fig. 5. Block diagram of the program designed to determine the Eulerian angles of single projections. In this example 5266 evenly distributed projections are used corresponding to an accuracy $\Delta\theta = 2.0°$.

Table 4
Number of quasi-uniformly distributed 2D projection directions for various increments in θ angle (the "mirror" projections are excluded)

Increment in θ angle	Number of 2D projections
10.0	229
5.0	869
2.5	3390
2.0	5266
1.5	9314
1.0	20847

(d) the projection data to be matched are read, sampled in polar coordinates, and the Fourier transforms of the data along rings are calculated;
(e) the 1D cross-correlation functions in polar coordinates between the projection data and all the reference projections are calculated;
(f) the projection direction (in terms of angles ϕ and θ) is determined by the largest cross-correlation coefficient, and angle ψ is determined by the position of the maximum in the corresponding cross-correlation function in polar coordinates.

Since the projection calculated in the direction (ϕ, θ) has its "mirrored" version in the direction ($180 + \phi$, $180 - \theta$), only half of the projections have to be actually calculated in step (a), and the range of θ can be restricted to (0, 90°). This redundancy is taken into account in step (e) where two CCF's are calculated: one using complex conjugation of the reference Fourier transforms, and the second without it. The latter step corresponds to an implicit mirroring of the reference image and reduces the amount of required memory by half.

The precalculations in steps (a)–(c) are done only once, no matter how many projections have to be matched. Together with the precalculations in step (d) they limit the amount of computer time required substantially. The calculations done during step (e) have to be repeated as many times as there are reference projections for every projection to be matched, but they actually amount only to a small number of multiplications of complex numbers and one inverse 1D FFT.

The method described above was tested using three projections P_1–P_3 of the structure S_1. The angular increment was chosen to be 2.0° which corresponds to 5266 2D reference projections of the structure (Table 4). The angles found (Table 3c) for all three cases (noise-free, SNR = 1.0, and SNR = 0.5) match the original angles, although the error increases with increasing noise level and considerable discrepancies are eventually noticed in the ϕ angle.

2.4.4. Comparison of test results

It is obvious that the first method using the discrepancy measure along common lines performs better for noise-free data. The accuracy of all three Eulerian angles is better than one degree, which corresponds to the accuracy with which line projections were calculated (1°). The second method has lower accuracy (particularly for angles θ and ϕ, which is understandable since both these angles are "quantized" by the angular increment chosen), but it is superior in the case of noisy data. The deviations from the original angles are noticeable, but never is the answer completely erroneous, as in the case of the first method.

The relatively poor performance of the common-lines method in the form tested here can be explained by the observation that for objects of finite size and limited resolution, two central sections of its 3D Fourier transform share more information than just a common line. In fact, such Fourier planes have finite "thickness" and the common region in Fourier space clearly depends on the angle between them. The common-lines approach neglects this fact and restricts the similarity measure just to the one-line region. The second method, using 2D projections, utilizes the Fourier information to the full extent and thus appears to be more robust in the very-high-noise case.

It has to be noted that the tests performed do not reflect the features of the EM data entirely. The noise which was added was Gaussian, while in reality there is a strong component of noise that can be classified as "salt and pepper" – imperfections of various kind, pieces of dust and so on. Moreover, in the case of the random conical data collection scheme, the projection data do not in reality correspond to the same objects: the projections actually come from different particles, whereas it is only assumed that these particles are identical. For these reasons the performance of the first method using the discrepancy along the common lines is even worse for real data than it was for the test case. In fact, the application of this method to the correction of the Eulerian angles proved to be unsuccessful, while the second method using 2D reference projections gave good results.

An additional factor to be considered is the time of calculations for both methods. As was earlier discussed, the application of the first method requires an exhaustive search of all possible orientations to avoid false solutions. A choice of 10° for the increment means that for every projection 15192 discrepancies have to be calculated. This requires 30 min on our current computer equipment (VAX 4000 model 60 workstation). Assuming a total number of 1000 projections to be matched we would have to spend 3 weeks to process all of them. The second method requires precalculation of a large number of 2D projections, which can be done with the speed of 1000 projections/hour, and then it takes only 3 min to match one projection to 5266 preprocessed reference projections. For 1000 projections the total computing time is just above 2 days, which is acceptable. The computation can be still made faster by using an intermediate search with lower accuracy.

2.5. 3D projection alignment method

The 3D projection alignment procedure proposed here combines both the orientation refinement method described in the previous section with the refinement of the translation parameters. Combined they constitute an iterative refinement procedure analogous to the 2D alignment procedure. Such a 3D alignment procedure uses the 3D structure obtained from the "merged" reconstruction as a reference and refines the orientation and translation parameters of the projections used. The reference structure is projected into directions of original projections and, using cross-correlation functions, the translation parameters are corrected. Next, using quasi-evenly distributed projections of the reference structure, the Eulerian angles are corrected (using the orientation refinement method described in the previous section). The corrected parameters are used to calculate a refined 3D reconstruction, which is subsequently used as a reference in the next step of the procedure. In each step the 3D reference structure is carefully centered, preferably locating its center of gravity at the origin of the coordinate system. The centering is done using the phase-approximation method, as described in Ref. [29].

Using the principle of the 3D projection alignment method it is very easy to include additional projection data in the process. The calculated cross-correlation coefficient between the projected structure and the projection data allows those projections to be removed that do not match the reference structure very well.

The same 3D projection alignment method makes it possible to calculate a high-resolution 3D reconstruction of a macromolecule using as a "nucleus" an already calculated reconstruction of the same structure and a new set of projection images collected from single micrographs. Such a "nucleus" is used as a reference in a first step of 3D projection alignment; thus the initial Eulerian angles are not needed and the procedure of collecting tilt pairs data is avoided altogether. Hence, the initial, tedious steps of data analysis (scanning, windowing, and so on) can be done much faster. In particular, it is relatively easy to use data obtained at lower defocus, enabling 3D reconstructions of higher resolution.

3. Results

3.1. Low-tilt 3D reconstruction of the 70S *E. coli* ribosome

The OSSP program was designed to facilitate the data-collection step in the framework of the random conical 3D reconstruction scheme. The applicability of the program was tested on a 70S *E. coli* ribosome data set. As was known from our recent 3D reconstruction [30], this particle can be found in a number of distinct views. If these views correspond to a wide range of 3D orientations the whole angular space should be covered by projections, and the missing cone problem should be eliminated even for relatively low-tilt data.

A total of six pairs of micrographs containing 70S ribosome particles (frozen-hydrated on a carbon-coated grid) were collected with tilt angles 29°, 36°, 32°, 23°, 31° and 34°. The defocus used was 2.0 μm for the 0° images and 1.5 μm for the tilted images. The negatives were digitized using a pixel size corresponding to 5.2 Å on the specimen scale. A set of 667 pairs of images was selected during the windowing step. All images were normalized using a histogram correction procedure described in Ref. [31] and any linear trends were removed ("ramp subtraction"). The untilted images were subjected to eight iterations of the reference-free alignment [10], multivariate statistical analysis, and classification using the hierarchical clustering program with complete linkage as merging criterion [24].

As a result we obtained 26 groups (Table 5). The averages of the first 25 groups are shown in Fig. 6 (group number 26 contained only three images and was omitted). From this set eight groups showing particles in possibly dissimilar orientations (numbered 1, 6, 11, 13, 15, 17, 21, and 25) were selected for the OSSP search. Since the number of images in each group was small (between 21 and 75) we used the following strategy to minimize the possible errors: for each group all tilted images were approximately centered using cross-correlation with a low-pass filtered disk and the 3D reconstruction was calculated. The volumes were projected into the directions of original projection and, using cross-correlation, the centering of the projection data was corrected. New 3D volumes were calculated and projected into 36 evenly spaced directions with a tilt angle of 35° (thus contained within the range of tilt of the original data). These projections were used as input to the OSSP program using a circular mask with a radius of 28 pixels and angular accuracy of 1°. One of the groups (number 13) was arbitrarily chosen as reference. All the Eulerian angles found are listed in Table 6. The results were verified using the concept of "closure error" [10] – for example, using groups

Table 5
Number of images in the groups obtained from the hierarchical clustering program (averages shown in Fig. 6) and corresponding within-group resolutions (according to the differential phase-residual analysis [32])

Group number	Number of images in the group	Resolution [1/Å]
1	38	39
2	7	60
3	12	60
4	25	44
5	17	44
6	31	60
7	11	60
8	15	39
9	2	–
10	19	51
11	48	60
12	47	60
13	32	51
14	21	51
15	21	44
16	13	60
17	57	35
18	52	51
19	27	51
20	25	51
21	75	51
22	9	60
23	11	51
24	15	51
25	33	44
26	3	–

6, 13, 17 and combining rotation between groups 6 and 17 and between 6 and 13, we obtain rotation angles $\psi = 54°$, $\theta = 111°$ and $\phi = 263°$ between groups 17 and 13, which match the rotation angles from Table 6 quite closely.

In the next step we applied the 3D orientation angles determined by the OSSP program directly to the projections of the seven groups listed in Table 6. Including group 13 (originally used as reference) the total number of projections was 335. We used this set to calculate a "merged" 3D reconstruction (Fig. 7). From an analysis of the θ angles of each set of projections it follows that the missing cone should be almost entirely eliminated. It is therefore justified to use this volume as an initial reference in an iterative procedure to refine the translation and orientation parameters of the projection data (3D projection alignment).

Table 6
Rotation (in terms of Eulerian angles) found by the OSSP program using group 13 as a reference

Group number	Orientation angles		
	ψ	θ	ϕ
1	44	113	−36
6	217	110	196
11	233	100	170
15	49	88	−84
17	57	97	267
21	48	93	267
25	107	106	55

The increment of the θ angle used was 2°. In each refinement step the 3D resolution of the reconstruction was estimated using the differential phase residual (DPR) criterion with a maximum of 45° phase discrepancy [32]: the set of projections was split into two equal subsets and two independent reconstructions were calculated and compared in the whole range of Fourier space over concentric shells using the DPR criterion.

The resolution of the first reconstruction (Fig. 7) was $1/55.5$ Å^{-1}. After one step of correction of rotation and translation parameters the resolution of the reconstruction (Fig. 8) improved to $1/47.5$ Å^{-1}. At that stage we decided to include into the process the remaining projections, increasing their total number to 667. After four additional steps of 3D projection alignment and removing 100 projections with smallest cross-correlation coefficients, the final 3D reconstruction was calculated using 567 projections. The result (Fig. 9) has a resolution of $1/39.9$ Å^{-1}. The reconstruction is free of missing cone artifacts, since the projections cover the whole range of angles. The distribution of projection directions (Fig. 10) is far from even, but the largest gap near $\theta = 0°$ spans no more than 20°.

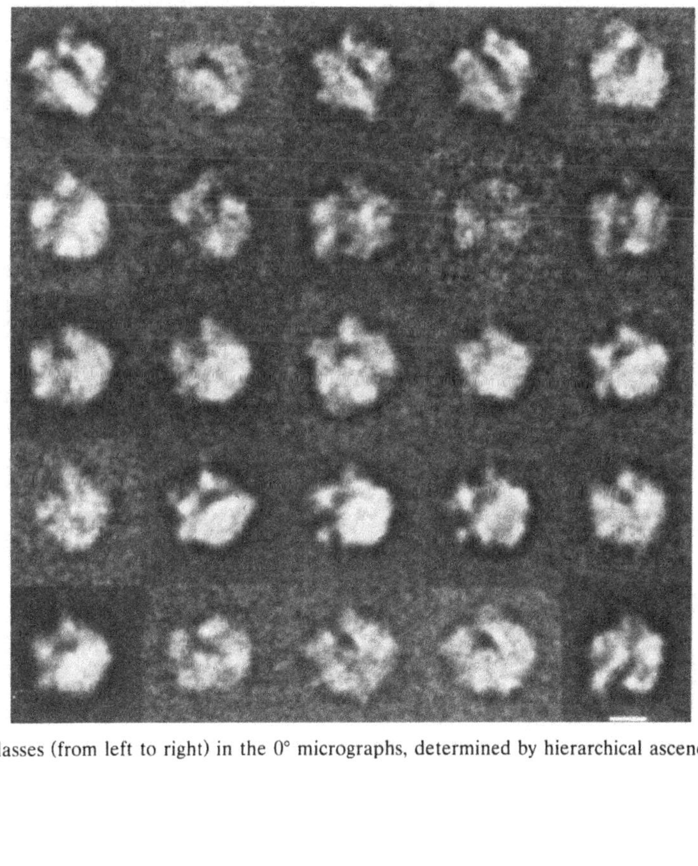

Fig. 6. Averages of 25 classes (from left to right) in the 0° micrographs, determined by hierarchical ascendant classification (scale bar = 100 Å).

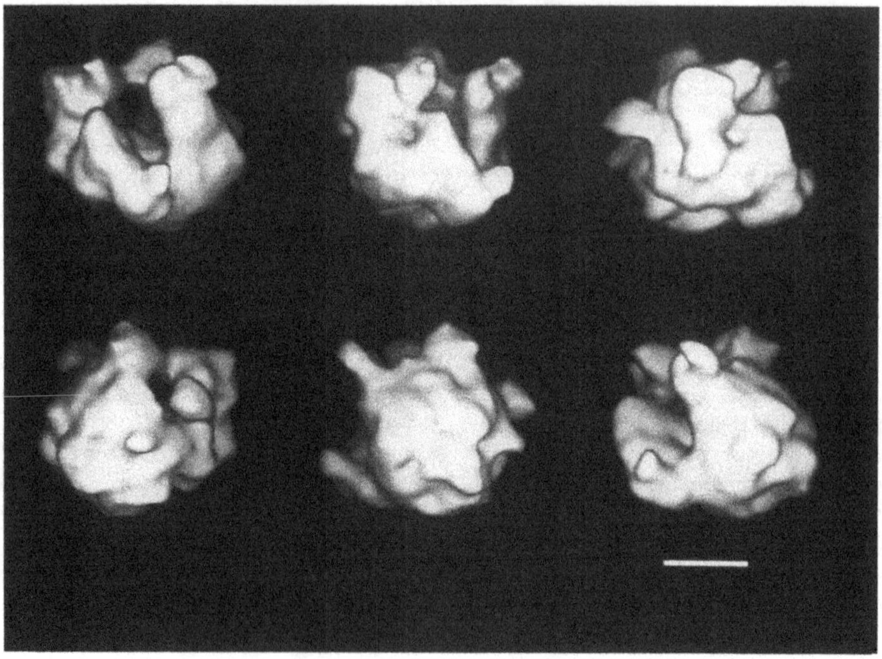

Fig. 7. "Merged" reconstruction of the 70S ribosome from 335 projections from seven classes (scale bar = 100 Å).

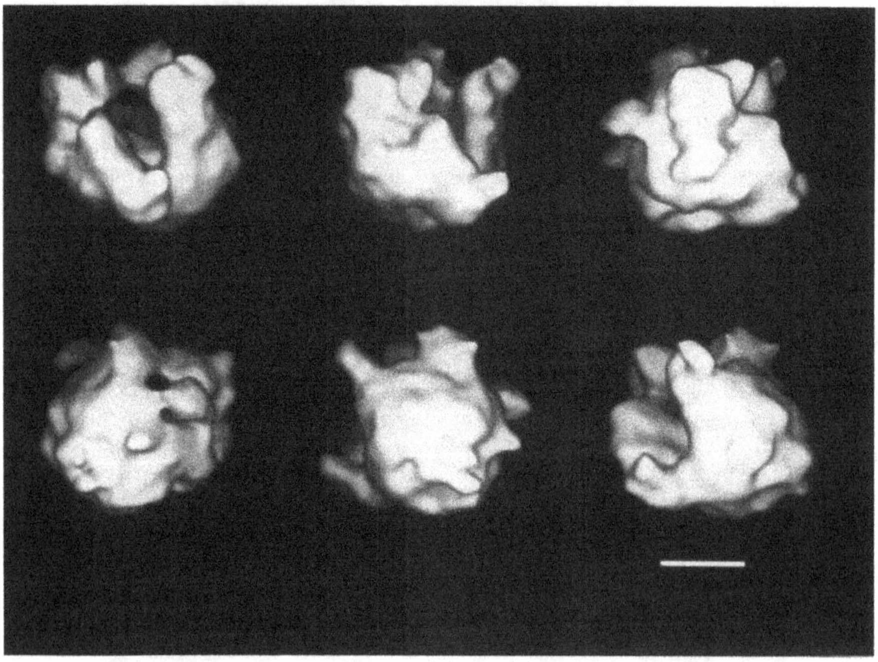

Fig. 8. "Merged" reconstruction of the 70S ribosome from 335 projections after one step of refinement of orientation parameters (scale bar = 100 Å).

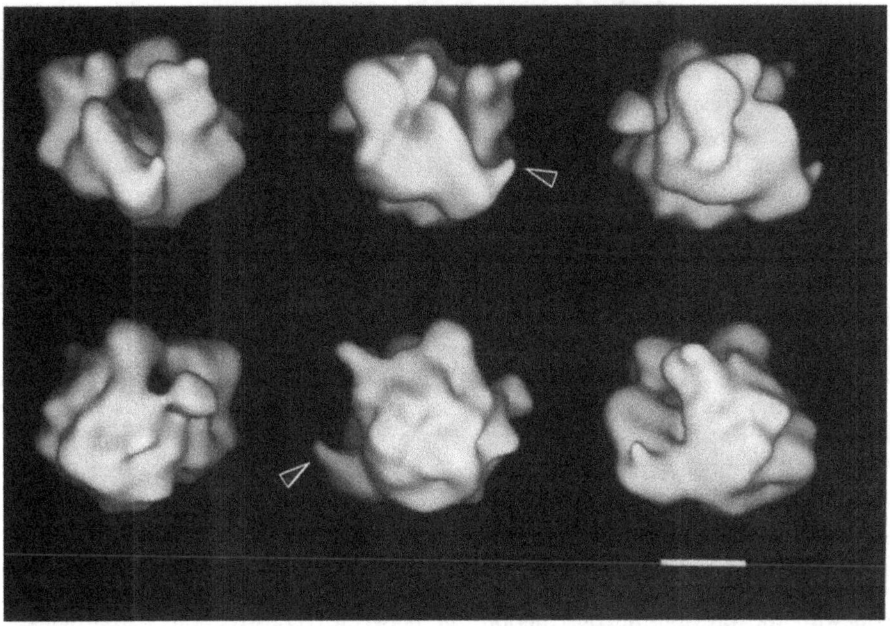

Fig. 9. Low-tilt reconstruction of the 70S ribosome from 567 projections after four steps of refinement of orientation parameters (scale bar = 100 Å). The structure was filtered to 1/39.9 Å^{-1} resolution. The arrows point to the extension in the "foot" region.

Both the unrefined, "merged", and the refined reconstruction agree in all essential details with the previous results [30]. There is one notable exception, however: a prominent extension of the 30S subunit in the "foot" region. This feature, which has not been previously observed, disappears when the structure is filtered to 1/55.5 Å^{-1} resolution (see Fig. 12 below). Such an extension is in fact observed in a particular view that occurred in several images of an older data set not used in the current analysis. It was also found after re-examination of the published data set (Fig. 4, second average in bottom row [10]). If we project the reconstruction into a direction such that this view is reproduced, an extension appears at the precise place where it was previously observed. We therefore conclude that the extension is a real feature of the specimen, but only observable in a fortuitous viewing direction.

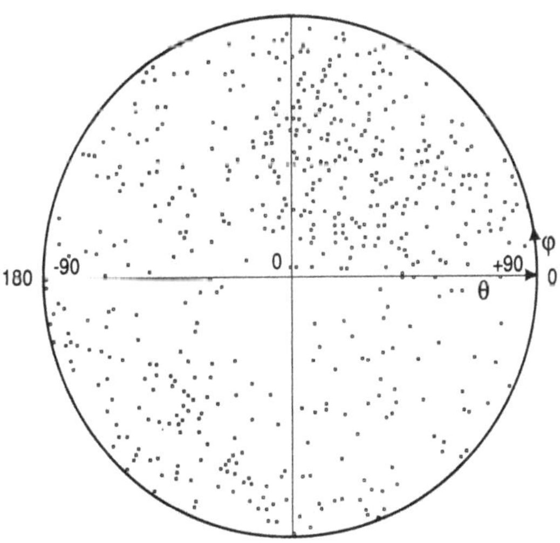

Fig. 10. Distribution of directions of 567 projections used for the final reconstruction of the 70S ribosome (shown in Fig. 9).

3.2. Single-micrograph 3D reconstruction of 70S *E. coli* ribosome

The method outlined in the previous section makes it possible to calculate the 3D reconstruction of a macromolecule using as reference an already calculated reconstruction of the same structure. To test this idea we collected 16 micrographs containing 70S ribosome data. The defocus used was 0.9 μm. The negatives were digitized using a pixel size corresponding to 5.2 Å on the specimen scale. Ten micrographs had a small tilt equal to 20° (as read from the goniometer on the microscope; thus the value is not accurate), and six were taken without tilt. The total number of particles selected was 942 (647 from tilted-specimen micrographs and 295 from untilted). All images were normalized using a histogram correction procedure [31] and subjected to the 3D projection alignment. In the first step the low-tilt 3D reconstruction of the 70S ribosome (described in the previous section) was used as a reference. Three more steps of alignment were used and the final reconstruction was calculated after exclusion of 200 projections with smallest cross-correlation coefficients. The resulting structure, shown in Fig. 11, has a resolution of 1/29.9 $Å^{-1}$.

Comparison with the previous reconstruction (Fig. 9) confirms a gain in resolution: the new, high-resolution 70S reconstruction has a large number of previously unnoticeable details.

3.3. Verification of the results

Both methods of 3D reconstruction described are not as straightforward as the method using exclusively the random-conical data collection scheme with 50° tilt. The latter, although producing a reconstruction marred by the missing cone, provides robust and easily verifiable results: it suffices to compare the projection along the z-axis of the reconstructed structure with the 2D average obtained from the alignment of untilted images to determine whether the method of analysis was self-consistent. Since the procedure is straightforward the overall shape of the particle has to be correct (even accounting for errors made in the assignment of θ angles to the projections and for distortions caused by the missing information). This simplifies further steps of the analysis; in particular, it is possible to verify visually and computationally the fidelity of match between reconstructions of particles lying in different orientations. In the case of the low-tilt reconstruction put forward here, advantage is taken of the random-conical data collection

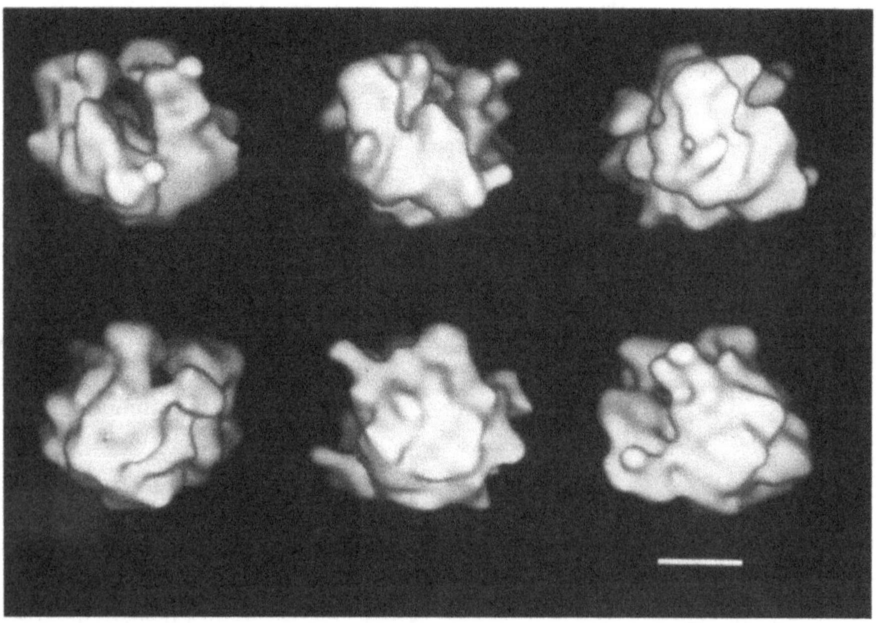

Fig. 11. Single-tilt reconstruction of the 70S ribosome from 742 projections using four steps of 3D projection alignment (scale bar = 100 Å). The structure was filtered to 1/29.9 $Å^{-1}$ resolution.

scheme, and a comparison with the 2D average is still possible. However, owing to the large missing cone the 3D reconstruction does not provide any clues about the overall shape of the particle. This shape becomes known only *after* projections exhibiting different views have been combined and a merged 3D reconstruction has been calculated. It is clear that in such a situation special care must be taken to verify all the steps which lead to the reconstruction.

Since in our case we already have a 3D reconstruction of the 70S ribosome obtained from 50°-tilt data [30] we were able to check to what resolution the new reconstructions are consistent with the known features. In all the following tests we measured the agreement by calculating the 3D cross-resolution in the full range of Fourier space (none of the reconstructions had a missing cone) between two volumes, using the DPR criterion. The agreement between our old 70S reconstruction and both new reconstructions extends to 1/55.5 Å^{-1}. It is worse than the previously reported resolution [30], but that value (1/45.0 Å^{-1}) merely described the agreement between the 0° 2D average and the projection of the reconstructed structure in the corresponding direction.

It should be noted that such a comparison tends to be optimistic, since simple geometrical considerations in Fourier space make it clear that only one plane of the 3D Fourier transform of the reconstruction is being compared. This plane is actually composed of single lines coming from individual 2D projections. The orientations of these lines are determined by ϕ and ψ angles only (both of which are known with high accuracy), but not by the θ angle. Since we expect to have considerable errors in the estimation of tilt angles, it becomes clear why the overall 3D resolution (estimated by comparison of two 3D subset reconstructions) will necessarily give worse results than the comparison of a 2D 0° projection with the 0° 2D average.

The cross-resolution between the two new 70S reconstructions was found to be 1/37.0 Å^{-1}. Since the first one (used as a reference to obtain the second one) had a resolution of 1/39.9 Å^{-1} and the second of 1/29.9 Å^{-1}, this result con-

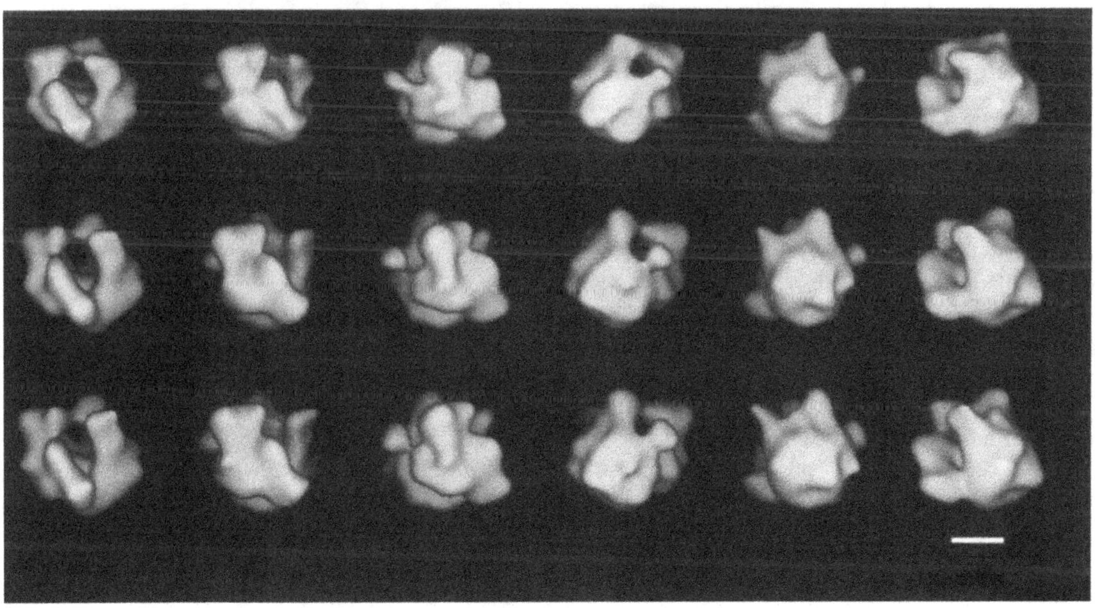

Fig. 12. Comparison of three reconstructions of the 70S ribosome (scale bar = 100 Å). (Top row) previous reconstruction [10]; (middle row) low-tilt reconstruction from Fig. 9; (bottom row) single tilt reconstruction from Fig. 11. All structures were filtered to 1/55.5 Å^{-1} resolution.

firms the validity of our 3D projection alignment procedure.

The agreement between all three reconstructions can be visually confirmed (Fig. 12). The agreement between the two new reconstructions is understandable, since the computation of the second volume was based on the first one, but the old reconstruction was obtained independently and using a different orientation search procedure. Nevertheless the similarity among all three reconstructions is very high, and all major features first reported in Ref. [30] are seen to be reproduced.

4. Discussion

In this paper we present two new major techniques designed to overcome the problems posed by the occurrence of multiple, non-homogenous 0°-views often encountered in ice-embedded specimen and by difficulties with obtaining high-tilt data. The approach of OSSP solves the problem of finding the Eulerian angles between two sets of low-tilt data, taking advantage of the information provided by the random-conical data collection scheme. The tests and verification of the results presented here prove that the errors in the determination of angles are within acceptable limits. The current implementation of the program uses one of the projection sets as reference with respect to which all other available sets are oriented, a procedure which can cause a certain bias. A possible solution would be to include all the projection sets in the orientation search program simultaneously, with an appropriate modification of the global discrepancy measure defined by Eq. (14).

The concept of the "bootstrapping" approach to the refinement of the 3D reconstruction described in the section "3D projection alignment" proved to be valuable. Its application led not only to an improvement of the resolution of low-tilt 3D reconstruction, but also to the high-resolution, low-defocus 3D reconstruction from the single-micrograph data. Interestingly, in the case of the low-tilt reconstruction, the final resolution of the 3D structure was better than the resolution of any of the 2D averages of the data sets initially used for the reconstruction. Possible explanations for this improvement are either that the classes were too small to produce averages with better resolution, or that they were not sufficiently homogenous. In either case this seems not to be the major obstacle, taking into account the overall strategy. The 3D projection alignment program is an important step in this strategy and it makes it possible to correct the Eulerian angles assigned to the projections after an initial, low-resolution 3D reconstruction has been calculated. In this way the classification of 2D, 0° images, which always contains a measure of ambiguity and subjective judgment [24], no longer determines the quality of the final 3D reconstruction. Nevertheless, as was already discussed in the previous sections, special care and verification of results are required to avoid systematic distortions in the 3D structure obtained.

The successful reconstruction of two 70S structures at different defocus values (1.5 and 0.9 μm) should make it possible to correct the solution for the contrast transfer function effect and obtain a more detailed description of rRNA distribution [Penczek and Frank, work in progress].

5. Conclusions

The techniques of orientation search using sets of projections and 3D projection alignment, together with the previously published reference-free alignment, constitute a new strategy of 3D reconstruction of particles embedded in ice. Taking advantage of the random-conical data collection scheme, this new strategy extends the applicability of the method to the multiple 0°-views specimens and makes it possible to refine the resolution of the 3D structure. In addition, using the 3D projection alignment program makes the 3D reconstruction from *single* micrographs possible. The method proposed should open the possibility of detecting structural changes or binding sites in a specimen: assuming that the "basic" structure of a particle is solved, it should be possible to use this structure as a reference in the alignment of projection views of the particle that has been modified.

6. Acknowledgements

This work was supported, in part, by grant 1 R01 GM 29169 to J.F.

7. References

[1] J. Frank, W. Goldfarb, D. Eisenberg and T.S. Baker, Ultramicroscopy 3 (1978) 283.
[2] J. Frank and W. Goldfarb, in: Electron Microscopy at Molecular Dimensions, Eds. W. Baumeister and W. Vogel (Springer, Berlin, 1980) p. 261.
[3] M. Radermacher, T. Wagenknecht, A. Verschoor and J. Frank, J. Microscopy 146 (1987) 113.
[4] M. Radermacher, J. Electron Microsc. Tech. 9 (1988) 359.
[5] N. Boisset, J.-C. Taveau, J. Lamy, T. Wagenknecht, M. Radermacher and J. Frank, J. Mol. Biol. 216 (1990) 743.
[6] D. Typke, K. Dierksen and W. Baumeister, in: Proc. 49th Annual EMSA Meeting, Ed. G.W. Bailey (San Francisco Press, San Francisco, 1991) p. 544.
[7] J.M. Carazo, I. Benavides, F.F. Rivera and E. Zapata, Ultramicroscopy 40 (1992) 13.
[8] J.P. Schroeter, T. Wagenknecht, S.J. Kolodziej, J.-P. Bretaudiere, D.K. Strickland and J.K. Stoops, FASEB J. 5 (1991) A452.
[9] J. Frank and M. Radermacher, Ultramicroscopy 46 (1992) 241.
[10] P. Penczek, M. Radermacher and J. Frank, Ultramicroscopy 40 (1992) 33.
[11] A.B. Goncharov, Preprint of Cybernetic Council Acad. Sci. (1986) [in Russian].
[12] D.B. Salzman, Comput. Vision Graph. Image Process.50 (1990) 129.
[13] R.A. Crowther, D.J. DeRosier and A. Klug, Proc. Roy. Soc. (London) A 317 (1970) 319.
[14] M. van Heel, Ultramicroscopy 21 (1987) 111.
[15] N.A. Farrow and F.P. Ottensmeyer, J. Opt. Soc. Am. A 9 (1992) 1749.
[16] G. Harauz, Ultramicroscopy 33 (1990) 209.
[17] J. Frank, P. Penczek and L. Weiping, Scanning Microsc. Suppl. 6 (1992) 11.
[18] IMSL, User's Manual, Version 1.0, April 1987.
[19] R.H. Vogel and S.W. Provencher, Ultramicroscopy 25 (1988) 223.
[20] J. Frank, Microsc. Microanal. Microstruct. 3 (1992) 45.
[21] W. Hoppe, J. Gassman, N. Hunsmann, J. Schramm and M. Sturm, Hoppe-Seylers Z. Physiol. Chem. 355 (1974) 1483.
[22] V. Knauer, R. Hegerl and W. Hoppe, J. Mol. Biol. 163 (1983) 409.
[23] S.W. Provencher and R.H. Vogel, Ultramicroscopy 25 (1988) 209.
[24] J. Frank, Q. Rev. Biophys. 23 (1990) 281.
[25] J. Dengler, Ultramicroscopy 30 (1989) 337.
[26] M. van Heel, in: Proc. 8th Eur. Congr. on Electron Microscopy, Budapest, 1984, Vol. 2, Eds. A. Csanády, P. Röhlich and D. Szabó, p. 1347.
[27] G. Harauz and F.P. Ottensmeyer, Ultramicroscopy 12 (1984) 309.
[28] G. Harauz and F.P. Ottensmeyer, Science 226 (1984) 936.
[29] B.M. Sadler and G.B. Giannakis, J. Opt. Soc. Am. A 9 (1992) 57.
[30] J. Frank, P. Penczek, R. Grassucci and S. Srivastava, J. Cell Biol. 115 (1991) 597.
[31] N. Boisset, P. Penczek, F. Pochon, J. Frank and J. Lamy, J. Struct. Biol. (1993) in press.
[32] J Frank, A. Verschoor and M. Boublik, Science 214 (1981) 1353.

Erratum

Equations (12) and (13) on page 254 should be as follows:

$$\phi_p = \text{arctane}(r^p_{32}\ ,\ r^p_{31}\ )$$

$$(12)$$

$$\psi_p = \text{arctane}(r^p_{23}\ ,-r^p_{13}\ )$$

$$\phi_p = \text{arctane}(r^p_{12}\ ,\ r^p_{11}\ ). \qquad (13)$$

Erratum

The paragraph containing equation (20) should read as follows:

"from which it immediately follows that for a chosen constant increment of the θ angle the increment of the ϕ angle should vary according to

$$d\phi = \frac{d\theta}{\sin\theta}. \qquad (20)$$

The increment of the ψ angle has to be chosen equal to the increment of angle θ."

A model of protein synthesis based on cryo-electron microscopy of the *E. coli* ribosome

Joachim Frank*†, Jun Zhu*†, Pawel Penczek*, Yanhong Li*‡, Suman Srivastava*, Adriana Verschoor*, Michael Radermacher*†, Robert Grassucci*, Ramani K. Lata* & Rajendra K. Agrawal*

* Wadsworth Center, New York State Department of Health, Empire State Plaza, PO Box 509, Albany, New York 12201-0509, USA
Departments of † Biomedical Sciences and ‡ Computer Science, State University of New York at Albany, 1400 Washington Avenue, Albany, New York 12222, USA

THE ribosome is formed by assembly of proteins and nucleic acids, and synthesizes proteins according to genetic instructions in all organisms. Many of the biochemical steps of this fundamental process are known, but a detailed understanding requires a well-defined structural model of the ribosome. Electron microscopy combined with image reconstruction of two-dimensional crystals[1-3] or single ribosomes[4] has been the most promising technique, but the resolution of the resulting models has been insufficient. Here we report a 25-Å reconstruction of the ribosome from *Escherichia coli*, obtained by combining 4,300 projections of ice-embedded single particles. Our new reconstruction reveals a channel in the small ribosomal subunit and a bifurcating tunnel in the large subunit which may constitute pathways for the incoming message and the nascent polypeptide chain, respectively. Based on these new findings, a three-dimensional model of the basic framework of protein synthesis is presented.

An energy-filtering electron microscope was used to visualize ribosomes prepared as frozen hydrated samples[5] (Fig. 1). Our reconstruction (Fig. 2*a–c*) shows the two ribosomal subunits as well defined separate masses of density that are joined by five bridges across the subunit–subunit interface. Morphological features established for the two subunits are easily recognized: the head (h) and platform (p) for the 30S subunit, and the L1 arm (L1), central protuberance (CP), stalk (St), and interface canyon (IC) (ref. 6) for the 50S subunit. A new spur-like feature (sp) appears at the base of the small subunit[7]. The subunits enclose a space ('intersubunit space') that accommodates mes-

LETTERS TO NATURE

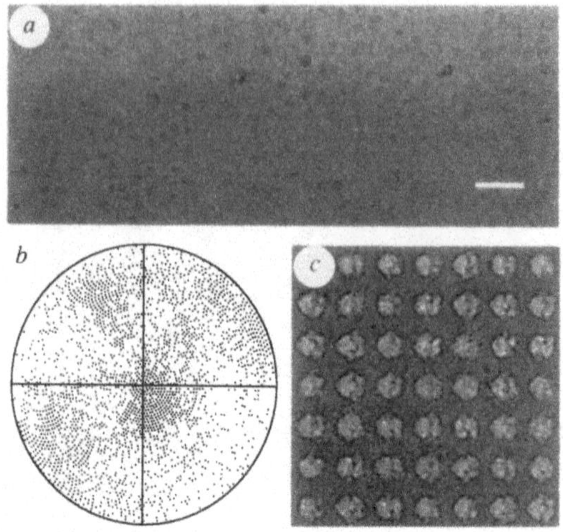

FIG. 1 Images of ice-embedded ribosomes and orientations of the 4,300 selected particles. *a*, Portion of a micrograph showing ribosomes in random orientations (scale bar, 1,000 Å). *b*, Orientations of the particles. Each particle is represented by a point in a (θ, ϕ) angular coordinate system. *c*, Averages of clusters found by classification.
METHODS. Inelastic scattering of the electrons in the ice causes a loss of contrast and makes it difficult to correct the electron microscope's contrast transfer function. The use of an energy-filtering electron microscope eliminates these problems entirely[21]. Micrographs (20) of untilted specimen grids were recorded under low-dose conditions, using different defocus settings (Δz, 2.0 μm and 2.5 μm) and an energy window of 14 eV. Particles (2,043 and 2,254, respectively) were picked and subjected to reference-free alignment[22] and classification[23]. The orientation of each projection was determined individually by matching it[7] to computed projections of the previous 29-Å model[7] that was obtained without energy filtering.

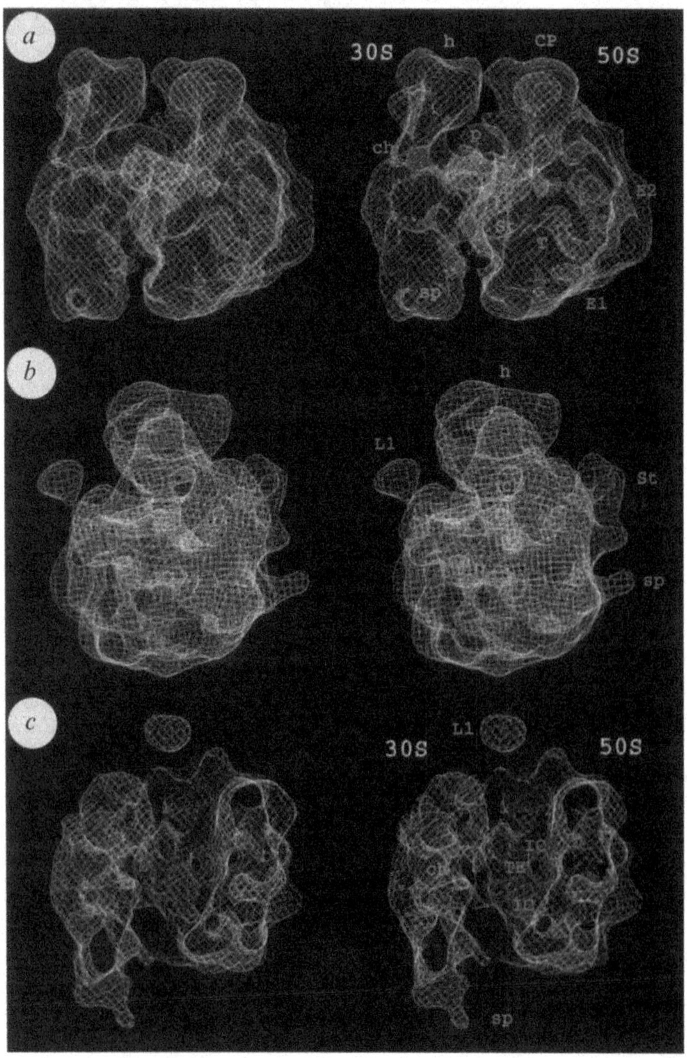

FIG. 2 Three-dimensional map of the *E. coli* ribosome presented from different viewing angles. Lower-case letters refer to features of the 30S subunit: h, head; p, platform; ch, channel; sp, spur. Upper-case letters refer to features of the 50S subunit: L1, L1 arm; CP, central protuberance; St, stalk; IC, interface canyon; T, tunnel; TE, tunnel entry; E1 and E2, exit sites of tunnel after branching. *a*, Side view showing part of the tunnel through the 50S subunit. The tunnel connects with the bottom of the interface canyon at higher contouring levels (see *c* and Fig. 3). b, View down the channel through the neck of the 30S subunit (generated from *a* by rotation by 90 around vertical axis, then by −15 around the horizontal axis). *c*, View down the tunnel through the 50S subunit (generated from *a* by rotation by 75 around horizontal axis and use of cutting plane to remove part of the ribosome). A higher contour level was used to visualize tunnel entry.
METHODS. The reconstruction was obtained by the simultaneous iterative reconstruction technique (SIRT)[22,24]. To verify that the solution is stable and unique, we wobbled the angles of the projections randomly by 15 , and refined against the resulting 45-Å low-resolution reconstruction using the 3D projection alignment procedure[7]. Two such refinements produced virtually identical results. In addition, an independent reconstruction from cluster averages using the method of common lines and a new minimization technique further confirmed the results (P.P., J.Z. and J.F., in preparation). The two sets of projections were combined in a new iterative 3D reconstruction procedure, in which the different contrast transfer functions (CTFs) of the electron microscope were explicitly taken into account (J.Z., P.P., R. Schröder and J.F., in preparation). The most important effect of the CTF correction is that it restores the correct weight of the low spatial frequencies, which carry information on the relative magnitudes of densities that are separated by large distances. Uncorrected reconstructions show several tunnels, but some of these are found to possess only slightly reduced density when the volume is CTF-corrected. Similarly, the identification of RNA-rich regions on the basis of their higher scattering density can be made with greater certainty on such a corrected volume.

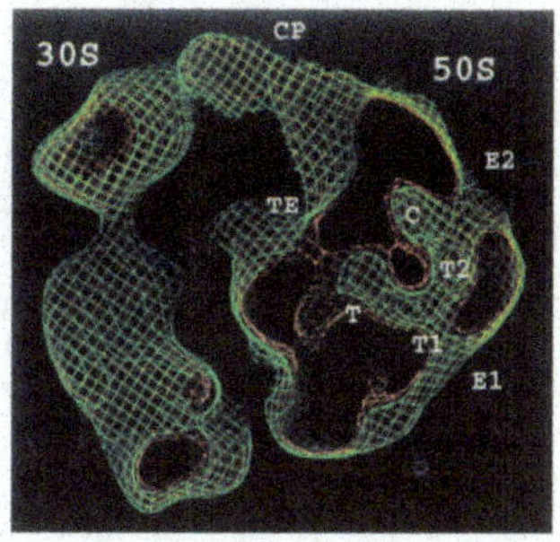

FIG. 3 Cross-section of the 3D map showing the bifurcating tunnel in the 50S subunit. Two different contour levels were used. For the normal contouring level (green), the primary tunnel (T) ends upwards in a cul-de-sac, probably because the upper segment has a diameter substantially smaller than the resolution. If present, such a thin tunnel segment will still be recognized as a line of low density. Indeed, the high contouring level (red) shows a connection of the tunnel with the bottom of the interface canyon (TE). (Other expansions of the tunnels, visible in green, can be explained by the removal of protein.) CP, central protuberance; T1 and T2, lower and upper tunnel branches, respectively; C, side compartment; E1 and E2, tunnel exit sites.

senger RNA, transfer RNAs and associated factors during protein synthesis.

The new three-dimensional density map contains several new features that allow tentative identification of the pathways of the incoming mRNA and the exiting nascent polypeptide, in turn suggesting the locations of the anticodon-binding region and peptidyltransferase centre: (1) the 'neck' portion of the small subunit is penetrated by a channel (ch in Fig. 2*a*, *b*) leading from the cytoplasmic side into the intersubunit space; (2) the cup-shaped platform of the small subunit (p in Fig. 2*a*) forms a well defined cleft with the main body, into which the channel leads, and (3) a tunnel (T), whose upper portion is seen at elevated contouring levels, starts at the bottom of the interface canyon (tunnel entry, TE, in Figs 2*c*, and 3) and bifurcates into two tunnels (T1, T2 in Fig. 3), which exit at different points (E1, E2) on the back of the 50S subunit. In addition, a compartment (C) is recognized (Fig. 3). A single, low-resolution tunnel through the 50S subunit was previously observed[1,2,8].

Our interpretation (Fig. 4) is based on the locations of the new features relative to sites previously mapped, and on their morphologies. The incoming mRNA threads through the channel in the small-subunit neck, makes a U-turn[9] and exits by a pathway circumscribed by the platform rim. Threading of the mRNA through the neck has been postulated[10] as a means of unfolding the mRNA into a single-stranded form. In fact, the channel is located in a region of the subunit that is thought to be traversed by a single-stranded segment of the 16S ribosomal RNA[11,12]. The suggested locations of entry and exit of the mRNA restrict the possible sites of binding of the aminoacyl and peptidyl tRNAs (A and P sites) to a region 20 Å wide in the cleft, to which a 16S ribosomal RNA region implicated in mRNA binding has been mapped[13].

In agreement with previous observations and conjectures[1-3,8,14,15], we interpret the large-subunit tunnel as the conduit through which the nascent polypeptide exits the ribosome. However, our results suggest the existence of not one but two

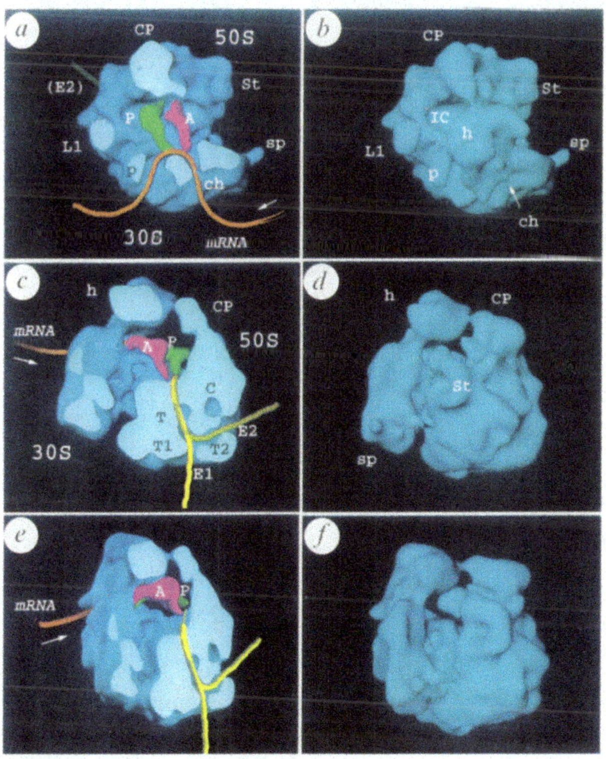

FIG. 4 Model (3D) of the basic framework of protein synthesis using a surface representation of the 25-Å reconstruction. The 12.5 Å map of tRNA–tRNA complex[19] (A site, pink; P site, green) was placed such that the anticodon ends face the cleft region immediately next to the channel entry. Aminoacyl ends were placed at two extreme positions related by rotation of the complex around mRNA axis within the range allowed by space exclusion. The loop-shaped mRNA pathway entering the channel from cytosol and encircling the back portion of neck is shown in orange. Two possible pathways of exit of the polypeptide chain are shown in olive green and yellow. *a*, Plane that sections the channel (ch) and the cleft that it leads into. L1, L1 arm; CP, central protuberance; St, stalk; p, platform; sp, spur. The tRNA complex (A and P) has been placed such that the anticodon ends are in the cleft, and aminoacyl ends at the tunnel entry as shown in *c*. Olive strand marked (E2) is a possible path of polypeptide chain exiting through the opening E2 that is marked in *c*. *b*, Corresponding top view of whole ribosome. IC, interface canyon; h, head of small subunit. *c*, Plane that sections the bifurcating tunnel (T, T1, T2). C, compartment contiguous with T2; E1 and E2, alternative exit sites. The tRNA complex is placed with aminoacyl ends at the lowest possible position. *d*, Corresponding side view of whole ribosome. *e*, Sectioning plane as in *c*, but the tRNA complex is placed with aminoacyl ends at the highest possible position limited by contact to the 'roof' formed by the small subunit head. The model has been rotated around the vertical axis, allowing a glancing view along the canyon wall[6] of the large subunit. The broken line marks the range of possible positions of the peptidyltransferase centre along that wall. *f*, Corresponding side view of the whole ribosome.

pathways, created by a bifurcation at the bottom of the primary tunnel T into a short lower tunnel, T1, and a long upper tunnel, T2. The short branch T1 (Fig. 3) emerges at the large-subunit base, within the 'exit domain' delineated by immunoelectron microscopy[14], whereas the long branch T2 emerges towards the middle of the subunit back, approximately 65 Å above the first.

The main pathway (T, T1 in Fig. 3) has a linear length of 85 Å, roughly equivalent to a 40-residue α-helical segment of polypeptide, the span found to be protected from proteases[16]. The exit of this tunnel, E1, lies in a domain of the large subunit believed to be tightly associated with the cytoplasmic membrane[17], forming part of the secretory pathway. The length of the branch T2, from the point of bifurcation to its exit, E2, is 50 Å. Towards E2 it becomes contiguous with a globular compartment, C (Figs 3 and 4*b*, *c*). For a membrane-bound ribosome, E2 would lead into the cytoplasm. However, the way in which two alternative pathways could be used in translocation (see, for example, ref. 18) is unknown.

The tunnel entry point (TE in Fig. 3) at the bottom of the interface canyon implies a location of the peptidyltransferase centre somewhere above this point. To restrict the range of possible locations of the peptidyltransferase centre we must dock A- and P-site tRNAs to the putative anticodon-binding region in the cleft, and further consider the spatial constraints imposed by the intersubunit space and the steric interactions between mRNA and A- and P-site tRNAs.

We have used the atomic model of the A-tRNA·P-tRNA complex[19] in S configuration[20] to produce a low-resolution image, as would be obtained by an electron microscope with 12.5 Å resolution. When the complex is docked onto the putative codon-binding site (Fig. 4*a*), it has freedom to rotate within the intersubunit space by 30° around the axis defined by the two codons. Two extreme positions are shown (Fig. 4*c*, *e*). In the first, the aminoacyl end of the P-site tRNA is proximal to the tunnel entry; in the other extreme, it is about 25 Å above this point. Either position would leave ample space towards the back of the A- and P-tRNA complex for an additional E-site tRNA. Detailed three-dimensional mapping experiments are needed to confirm our hypothesis and to allow refinement of our basic model of protein synthesis. □

Received 12 May; accepted 23 June 1995.

1. Yonath, A., Leonard, K. R. & Wittmann, H. G. *Science* **236,** 813–816 (1987).
2. Eisenstein, M. *et al. Biochimie* **73,** 879–886 (1991).
3. Yonath, A. & Wittmann, H. G. *Trends biochem. Sci.* **14,** 329–335 (1989).
4. Frank, J., Penczek, P., Grassucci, R. & Srivastava, S. *J. Cell Biol.* **115,** 597–605 (1991).
5. Lepault, J., Booy, F. P. & Dubochet, J. *J. Microsc.* **129,** 89–102 (1983).
6. Radermacher, M., Wagenknecht, T., Verschoor, A. & Frank, J. *EMBO J.* **6,** 1107–1114 (1987).
7. Penczek, P., Grassucci, R. & Frank, J. *Ultramicroscopy* **53,** 251–270 (1994).
8. Frank, J., Verschoor, A., Radermacher, M. & Wagenknecht, T. in *Ribosomes* (eds Hill, W. E. *et al.*) 107–113 (American Society for Microbiology, Washington DC, 1990).
9. Montesano-Roditis, L. & Glitz, D. G. *J. biol. Chem.* **269,** 6458–6470 (1994).
10. Lim, V. *et al. Nucleic Acids Res.* **20,** 2627–2637 (1992).
11. Shatsky, I., Bakin, A. V., Bogdanov, A. A. & Vasiliev, V. D. *Biochimie* **73,** 937–945 (1991).
12. Malhotra, A. & Harvey, S. C. *J. molec. Biol.* **240,** 308–340 (1994).
13. Oakes, M. I. & Lake, J. A. *J. molec. Biol.* **211,** 897–906 (1990).
14. Bernabeu, C. & Lake, J. A. *Proc. natn. Acad. Sci. U.S.A.* **79,** 3111–3115 (1982).
15. Avila-Sakar, A. J. *et al. J. molec. Biol.* **239,** 689–697 (1994).
16. Blobel, G. & Sabatini, D. D. *J. Cell Biol.* **45,** 130–145 (1970).
17. Crowley, K. S., Reinhart, G. D. & Johnson, A. E. *Cell* **73,** 1101–1115 (1993).
18. Kudlicki, W. *et al. J. biol. Chem.* **270,** 10650–10657 (1995).
19. Easterwood, T. R., Major, F., Malhotra, A. & Harvey, S. C. *Nucleic Acids Res.* **22,** 3779–3786 (1994).
20. Sundaralingam, M., Brenman, T., Yathindra, N. & Ichikawa, T. in *Structure and Conformation of Nucleic Acids and Protein–Nucleic Acid Interactions* (eds Sundaralingam, M. & Rao, S. T.) 101–115 (University Park, Baltimore, 1975).
21. Schröder, R. R., Hofmann, W. & Menetret, J.-F. *J. struct, Biol.* **105,** 28–34 (1990).
22. Penczek, P., Radermacher, M. & Frank, J. *Ultramicroscopy* **40,** 33–53 (1992).
23. Frank, J. *Q. Rev. Biophys.* **23,** 281–329 (1990).
24. Gilbert, P. *J. theor. Biol.* **36,** 105–117 (1972).

ACKNOWLEDGEMENTS J.F. and J.Z. thank K. Holmes and R. Schröder at the Max Planck Institute for Medical Research, Biophysics Department for hospitality during J.F.'s sabbatical visit supported by the Alexander-von-Humboldt-Stiftung and NIH, and for the use of the Zeiss EM 912 electron microscope. We acknowledge use of the Wadsworth Center's Research Computing Core. We thank R. Brimacombe for discussion of the tRNA fitting; S. Harvey for providing the atomic coordinates of the tRNA–tRNA model; A. Leith for assistance with the modelling; and T. Wagenknecht for discussion. This work was supported by grants from NIH and NSF to J.F.

Direct Visualization of A-, P-, and E-Site Transfer RNAs in the *Escherichia coli* Ribosome

Rajendra K. Agrawal, Pawel Penczek, Robert A. Grassucci, Yanhong Li, ArDean Leith, Knud H. Nierhaus, Joachim Frank*

Transfer RNA (tRNA) molecules play a crucial role in protein biosynthesis in all organisms. Their interactions with ribosomes mediate the translation of genetic messages into polypeptides. Three tRNAs bound to the *Escherichia coli* 70S ribosome were visualized directly with cryoelectron microscopy and three-dimensional reconstruction. The detailed arrangement of A- and P-site tRNAs inferred from this study allows localization of the sites for anticodon interaction and peptide bond formation on the ribosome.

In protein biosynthesis, the tRNAs, carrying covalently attached amino acids which are the building blocks of the polypeptide chain, are known to occupy successively three sites on the ribosome: one for the incoming aminoacyl tRNA (A site), one for the peptidyl tRNA (P site) (*1*), and the third for the exiting tRNA (E site) (*2, 3*). These binding sites account for the codon-anticodon interaction between mRNA and tRNA, the correct positioning of tRNA acceptor and donor arms during peptide bond formation, and the movement of mRNA relative to the ribosome. Studies conducted during the past two decades clearly place the anticodon and aminoacyl ends of A- and P-site tRNAs onto the 30S and 50S subunits of the *E. coli* ribosome, respectively. Although some attempts have been made to map the tRNA binding site on the 30S ribosomal subunit (*4*), no studies have succeeded in directly visualizing tRNA molecules bound to the 70S ribosome (comprising the 30S and 50S subunits). We have obtained a three-dimensional (3D) cryoelectron microscopy map of a poly(U) programmed *E. coli* 70S ribosome in which the sites are occupied by three deacylated $tRNA^{Phe}$ molecules, and we have computed a difference map by comparison with the previous cryoreconstruction of an unoccupied (naked) ribosome (*5*). All three tRNAs are clearly distinguishable and occur in an open configuration, with the A- and P-site tRNAs enclosing an angle of about 160°. In addition, a conformational change of the tRNA relative to that seen in its crystal structure (*6*) was observed. The physiological relevance of our findings is supported by the observation (*7, 8*) that the positions of the deacylated tRNAs in 70S ribosomes are close to those with aminoacyl or peptidyl tRNAs.

The difference map obtained by subtraction of the map of the naked 70S ribosome (*5*) from the new reconstruction of the $tRNA^{Phe}$-bound 70S ribosome (*9*) shows a mass of density with complex shape in the intersubunit space (Fig. 1 and Fig. 2, A and B), which can be readily interpreted (Fig. 1 and Fig. 2, C and D) as an arrangement of three tRNA molecules that appear to be partially fused because of resolution limitations. The locations of three elbow-shaped bulges observed in the difference mass, interpreted as elbow regions of the three tRNAs, were used as the main guide for fitting of the atomic models of tRNA. In accordance with the existing body of knowledge, we identify the tRNA molecule closest to the L7/L12 stalk [for identification of this and other morphological markers, see

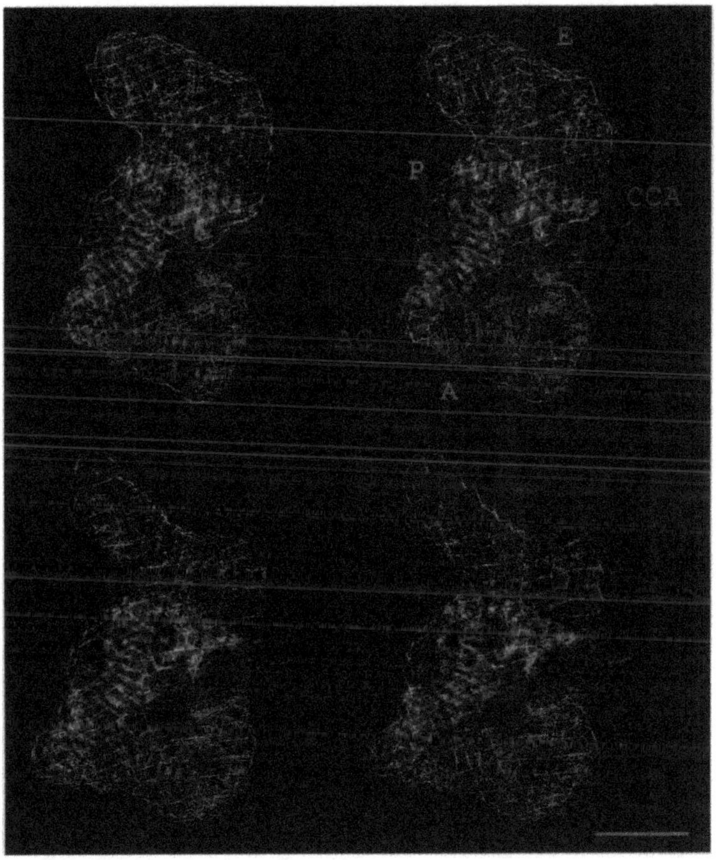

Fig. 1. Stereoview representation of the difference map (white contour) found in the intersubunit space of the ribosome (also see Fig. 2, A and B) and its interpretation in terms of the crystal structure of tRNA (6). A, A site (pink); P, P site (green); E, E site (brown) tRNA. AC, anticodon portion; CCA, aminoacyl ends of the A- and P-site tRNAs. (**Upper portion**) CTF-corrected difference map. (**Lower portion**) Difference map without CTF correction, showing that the fit extends to the anticodon ends when high-resolution features are enhanced. Scale bar, 25 Å.

R. K. Agrawal, P. Penczek, R. A. Grassucci, Y. Li, A. Leith, Wadsworth Center, New York State Department of Health, Empire State Plaza, Albany, NY 12201–0509, USA.
K. H. Nierhaus, Max-Planck-Institut für Molekulare Genetik, Ihnestrasse 73, D-14195 Berlin, Germany.
J. Frank, Wadsworth Center, New York State Department of Health, Empire State Plaza, Albany, NY 12201–0509, USA, and Department of Biomedical Sciences, State University of New York at Albany, 1400 Washington Avenue, Albany, New York 12222, USA.

*To whom correspondence should be addressed.

(5)] as the A site, the one next to it as the P site, and the one in the close vicinity of the L1 protein, farthest away from the A-site tRNA, as the E site. This interpretation immediately places the anticodon loops and CCA ends of A- and P-site tRNAs in close proximity to each other, respectively. With this positioning of A- and P-site tRNAs, the remaining mass tightly accommodates another tRNA molecule, interpreted as the E site. We believe that the reason the CCA ends are not visible in the difference mass is because they are single-stranded, with a thickness below the resolution limit (25 Å). Some of the remaining discrepancies observed in this fitting may be due to the mass of density contributed by the subsets of the ribosomal populations present in their hybrid (A/P and P/E) states (*10*). The absence of any detectable negative difference peaks in the immediate vicinity of the difference mass makes it unlikely that part of the mass is due to a conformational change of the ribosome. Additionally, the fitting of the atomic model of tRNA to our experimental data revealed a conformational change in its ribosome-bound state as compared with the crystal structure (6): The anticodon arm of the A-site tRNA was found to be bent in the middle (Fig. 2), out of the plane of the molecule, toward the side of the D loop. Similar bends in the anticodon arm of tRNAs have been observed in the structures of tRNA-tRNA synthetase complexes (*11*). However, accounting for this change would move the inferred position of the CCA end only slightly. The difference map also shows a separate, smaller mass that is possibly due to a conformational change in the region of the L7/L12 stalk of the 50S subunit.

The mutual arrangement of the A- and P-site tRNAs differs from both "R" and "S" configurations (*12*) in that their planes enclose an angle of about 160° [compare with (*5, 13, 14*)], whereas the position of the E-site tRNA appears to exclude codon-anticodon contact [compare with (*15*)]. However, the positions we have found satisfy most of the data from cross-linking (*13, 16, 17*) and electron microscopic mapping (*4, 18*). The anticodon ends of the A- and P-site tRNAs meet at the interface side of the small subunit's "neck" region, in close vicinity to the channel previously seen (*5*) but on the L7/L12 side of the channel rather than toward the platform [compare with (*5*)]. The elbow of the A-site tRNA is perched in the cavity formed by the intersubunit bridge, the base of the L7/L12 stalk, and the interface canyon. The position of the P-site tRNA is such that the anticodon stem-loop side of the elbow touches the 30S subunit head and the rim of the platform, and the CCA end points toward the entry of the tunnel situated at the bottom of the interface canyon, approximately as postulated by Frank and co-workers (*5*). This position agrees with chemical protection

Table 1. Comparison of our results with FRET (*20*) measurements. Dash indicates no data available.

A-site tRNA residue number	P-site tRNA residue number	Distance measurements (Å)	
		FRET	This study
37 (anticodon loop)	37	24 ± 4	15
37	16, 17 (D loop)	46 ± 12	43
16, 17	37	38 ± 10	36
16	16, 17	35 ± 9	44,48
17	16, 17	35 ± 9	51,55
8	8	26 ± 4	28
76 (CCA end)	76	—	12

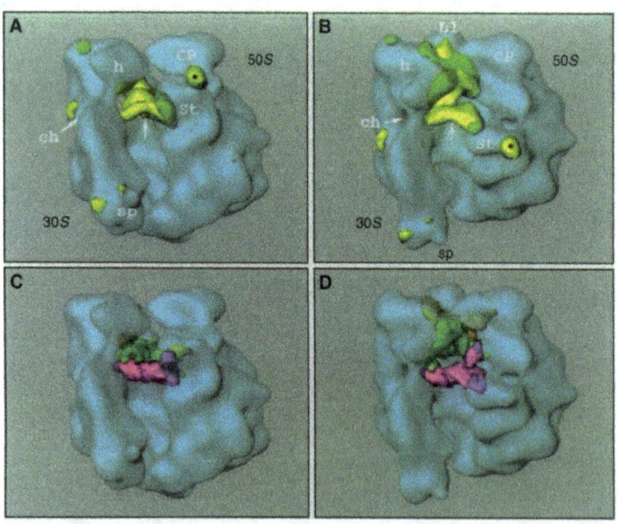

Fig. 2. CTF-corrected difference map (yellow, appearing green in regions of perspective overlap) superimposed on the ribosome (shown in blue as transparent map). Morphological features of the 50S subunit: CP, central protuberance; L1, L1 protein; St, L7/L12 stalk. Morphological features of the 30S subunit; h, head; ch, channel; sp, spur. (**A**), side view; (**B**), top view [generated from (A) by rotation of the map by 35° around the horizontal axis]. The arrow below the difference map points to a bend in the region interpreted as the anticodon arm of the A-site tRNA and the asterisk indicates the conformational change in the L7/L12 stalk region. (**C** and **D**) Interpretation of the difference map viewed as in (A and B) in terms of three tRNA molecules, each represented by the envelope of a 5 Å resolution map.

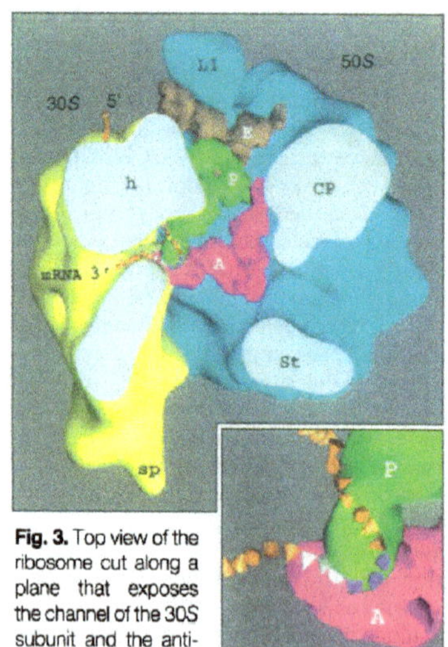

Fig. 3. Top view of the ribosome cut along a plane that exposes the channel of the 30S subunit and the anticodon binding region, and schematic outline of the inferred mRNA pathway. Morphological features are marked as in Figs. 1 and 2. Ribonucleotide bases are represented by cones whose orientations follow the pitch of 11 bases per 360°, except in the region of anticodon binding (white and purple) where an additional 70° turn is accommodated by a stretch of four bases (that is from the center of the A-site codon to the center of the P-site codon), as necessitated by the orientations of the A and P tRNAs. Inset shows a magnified view of the codon-anticodon interaction region.

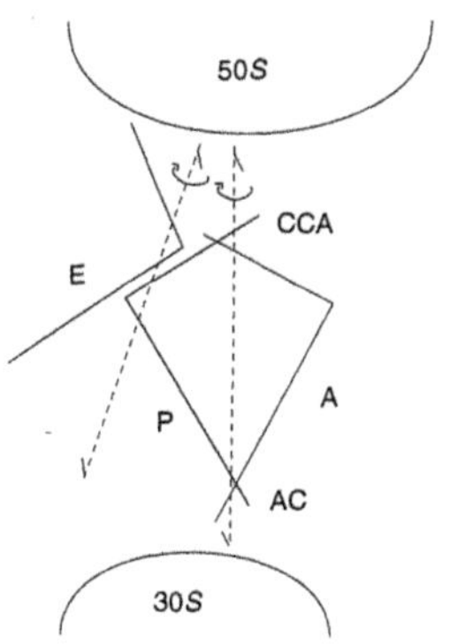

Fig. 4. Schematic diagram depicting the inferred movements of the tRNA from A to P site and from P to E site. A→P is a combination of a 160° rotation and an upward shift along the rotation axis by 9 Å. P→E is a combination of a 145° rotation and an upward shift by approximately 20 Å. The positions of the corresponding two left-handed screw axes are indicated by dashed lines. CCA, CCA ends; AC, anticodons of A- and P-site tRNAs. For both A- and E-site tRNAs, the D loop in the elbow region faces the viewer, whereas it faces away for the P-site tRNA.

and cross-linking data (*7, 8, 16*). We find the E-site tRNA in a topographically distinct position, with the D-loop side facing (and almost touching) the D-loop side of the aminoacyl arm of the P-site tRNA, such that the inner bend of the molecule faces the L1 protein. Its CCA end appears to occupy the space between the L1 protein and the central protuberance of the 50S subunit, while its anticodon arm spans the space between the L1 protein and the platform of the 30S subunit (Fig. 3), which is in good agreement with chemical protection (*7*) and protein cross-linking (*19*) data.

The arrangement obtained by fitting three copies of the atomic model of the tRNA (Fig. 1 and Fig. 2, C and D) into the difference map (Fig. 1 and Fig. 2, A and B) agrees with much of the fluorescence-resonance energy-transfer (FRET) data (*20*) (Table 1). Discrepancies occur in the distances between the D loops and residues 37 of A- and P-site tRNAs, which are the regions that would be most affected by the conformational change in the A-site tRNA.

The relative position between A- and P-site tRNAs can be described as a rotation by 160° around the anticodon-CCA axis and a simultaneous shift by approximately 9 Å. A similar staggered arrangement between A- and P- site tRNAs has been postulated by Lim and co-workers (*12*), albeit with a smaller angle between the molecules. Thus, when both sites are occupied, the 3′ end of the A-site tRNA anticodon is in close proximity (slightly below) to the 5′ end of the P-site tRNA anticodon. When the natural conformation of mRNA and the crystal structure of tRNA are considered, this arrangement would imply that to allow cogent hydrogen bonding to be formed for the two adjacent codons, the mRNA may undergo a twist of close to 70°. The resulting strain must be absorbed by the six consecutive base pairings, which include the wobble bases of the anticodons (Fig. 3). If the tRNA molecule is constrained to move along the A→P path outlined in Fig. 4, by an as yet unknown mechanism, then the advance of the mRNA by one codon (*21*) could be effected by a force that acts on the A-site tRNA. The L7/L12 stalk, whose base is now found in close proximity to the A-site tRNA elbow, has been long suspected to have an active role in translocation (*22*). Indeed, our data indicate a conformational change in that region of the map (Fig. 2, A and B) where the stalk seems to be extended away from the body of the ribosome. Experiments with a translationally defined system using aminoacyl and peptidyl tRNAs may further refine the data presented in this study.

REFERENCES AND NOTES

1. J. D. Watson, *Bull. Soc. Chim. Biol.* **46**, 1399 (1964).
2. H.-J. Rheinberger, H. Sternbach, K. H. Nierhaus, *Proc. Natl. Acad. Sci. U.S.A.* **76**, 5310 (1981).
3. R. A. Grajevskaja, Yu. V. Ivanov, E. M. Saminsky, *Eur. J. Biochem.* **128**, 47 (1982); S. V. Kirrilov, E. M. Makarov, Yu. P. Semenkov, *FEBS Lett.* **157**, 91 (1983); R. Lill, J. M. Robertson, W. Wintermeyer, *Biochemistry* **25**, 3245 (1984).
4. P. Gornicki, K. Nurse, W. Hellmann, M. Boublik, J. Ofengand, *J. Biol. Chem.* **259**, 10493 (1984); T. Wagenknecht, J. Frank, M. Boublik, K. Nurse, J. Ofengand, *J. Mol. Biol.* **203**, 753 (1988).
5. J. Frank *et al.*, *Nature* **376**, 441 (1995).
6. S.-H. Kim *et al.*, *Science* **185**, 435 (1974); J. L. Sussman and S.-H. Kim, *Biochem. Biophys. Res. Commun.* **68**, 89 (1976); J. L. Sussman, S. R. Holbrook, R. W. Warrant, G. M. Church, S.-H. Kim, *J. Mol. Biol.* **123**, 607 (1978).
7. D. Moazed and H. F. Noller, *Cell* **57**, 585 (1989).
8. ———, *J. Mol. Biol.* **211**, 135 (1990).
9. The *E. coli* tRNA^{Phe} was radiolabeled at the 5′ end, as described by Silberklang and co-workers (*23*), and its binding saturation curve was studied (*2*). A 32-fold molar excess of tRNA^{Phe} was required to saturate the poly(U) coded ribosome when all three sites were supposed to be occupied (*2, 3*). Activated 30*S* tight-couple subunits (50 pmol) were incubated with poly(U) (10 μg) in 10 μl of buffer containing 50 mM tris-HCl (pH 7.8), 15 mM magnesium acetate, 160 mM NH_4Cl, and 5 mM β-mercaptoethanol. Then an equimolar amount of the 50*S* tight-couple subunit was added (total volume 15 μl) and further incubated under the same conditions for 60 min. *Escherichia coli* tRNA^{Phe} (1.6 nmol) was added in the reaction mixture (total volume 25 μl) and the incubation was continued for another 30 min. The reaction mixture was subjected to centrifugation (10 to 30% sucrose gradient) in the same buffer to purify ribosomes from the unbound tRNAs, mRNAs, and ribosomal subunits. Cryogrids were prepared at a 0.85 A_{260}/ml concentration of the sample. Two sets of ribosome projections were obtained by cryoelectron microscopy (Philips EM 420): A total of 855 projections were obtained with 2.0-μm defocus and a total of 1636 projections were obtained with 2.5-μm defocus. The angles were determined with the use of a projection matching procedure (*24*) with the control (*5*). Two volumes were calculated, one for each set of projections at 2.0-μm and 2.5-μm defocus, respectively, and combined by means of a contrast transfer function (CTF) correction procedure (J. Zhu, P. Penczek, R. Schröder, J. Frank, in preparation). The difference map was formed by subtraction of the control reconstruction of the naked ribosome (*5*) from the new merged reconstruction of the ribosome-tRNA complex.
10. D. Moazed and H. F. Noller, *Nature* **342**, 142 (1989); H. F. Noller *et al.*, in *The Ribosome, Structure, Function and Evolution*, W. E. Hill *et al.*, Eds. (American Society for Microbiology, Washington, DC, 1990), pp. 73–92.
11. M. A. Rould, J. J. Perona, D. Söll, T. A. Steitz, *Science* **246**, 1135 (1989); M. Ruff *et al.*, *ibid.* **252**, 1682 (1991).
12. V. Lim *et al.*, *Nucleic Acids Res.* **20**, 2627 (1992).
13. R. Brimacombe, *Eur. J. Biochem.* **230**, 365 (1995).
14. H. Stark *et al.*, *Structure* **3**, 815 (1995).
15. K. H. Nierhaus, *Biochemistry* **29**, 4997 (1990).
16. R. A. Zimmermann, C. L. Thomas, J. Wower, in *The Ribosome, Structure, Function and Evolution*, W. E. Hill *et al.*, Eds. (American Society for Microbiology, Washington, DC, 1990), pp. 331–347.
17. J. Wower, K. V. Rosen, S. S. Hixon, R. A. Zimmermann, *Biochimie* **76**, 1235 (1994).
18. G. Stöffler and M. Stöffler-Meilicke, in *Structure, Function and Genetics of Ribosomes*, B. Hardesty and G. Krammer, Eds. (Springer-Verlag, New York, 1986), pp. 28–46; M. Oakes, E. Henderson, A. Scheinman, M. Clark, J. A. Lake, *ibid.*, pp. 47–67.
19. J. Wower, P. Scheffer, L. A. Sylvers, W. Wintermeyer, R. A. Zimmermann, *EMBO J.* **12**, 617 (1993).
20. A. E. Johnson, H. J. Adkins, E. A. Matthews, C. R. Cantor, *J. Mol. Biol.* **156**, 113 (1982); H. Paulsen, J. M. Robertson, W. Wintermeyer, *ibid.* **167**, 411 (1983).
21. D. Beyer, E. Skripkin, J. Wadzak, K. H. Nierhaus, *J. Biol. Chem.* **269**, 30713 (1994).
22. D. P. Burma, S. Srivastava, A. K. Srivastava, S. Mahanti, D. Dash, in *Structure, Function and Genetics of Ribosomes*, B. Hardesty and G. Krammer, Eds. (Springer-Verlag, New York, 1986), pp. 438–453; A. Liljas and A. T. Gudkov, *Biochimie* **69**, 1043 (1987); W. Moller, *ibid.* **73**, 1093 (1991); A. V. Oleinkov, B. Perroud, B. Wang, R. R. Traut, *J. Biol. Chem.* **268**, 917 (1993).
23. M. Silberklang, A. M. Gillum, U. L. RajBhandary, *Methods Enzymol.* **59**, 58 (1979).
24. P. Penczek, R. Grassucci, J. Frank, *Ultramicroscopy* **53**, 251 (1994).
25. Supported by NIH grants 1R01 GM29169 and P41 RR01219 and by NSF grant BIR 9219043. We thank S. Srivastava for helpful suggestions and discussions and T. Wagenknecht for a critical reading of the manuscript and suggestions. Coordinates of the complex formed by three copies of tRNA^{Phe} as determined by fitting have been deposited in the Brookhaven Protein Data Bank, with tracking number T7700. The 3D density maps of occupied and unoccupied ribosomes and their difference map will be made available on the World Wide Web site www.wadsworth.org.

2 November 1995; accepted 21 December 1995

Alignment of Conduits for the Nascent Polypeptide Chain in the Ribosome-Sec61 Complex

Roland Beckmann,* Doryen Bubeck, Robert Grassucci, Pawel Penczek, Adriana Verschoor, Günter Blobel, Joachim Frank

An oligomer of the Sec61 trimeric complex is thought to form the protein-conducting channel for protein transport across the endoplasmic reticulum. A purified yeast Sec61 complex bound to monomeric yeast ribosomes as an oligomer in a saturable fashion. Cryo–electron microscopy of the ribosome-Sec61 complex and a three-dimensional reconstruction showed that the Sec61 oligomer is attached to the large ribosomal subunit by a single connection. Moreover, a funnel-shaped pore in the Sec61 oligomer aligned with the exit of a tunnel traversing the large ribosomal subunit, strongly suggesting that both structures function together in the translocation of proteins across the endoplasmic reticulum membrane.

The existence of a protein-conducting channel (PCC) for protein transport across the endoplasmic reticulum (ER) was proposed in 1975 (*1*). Electrophysiological experiments in 1991 provided the first direct evidence for the existence of the PCC (*2*). Moreover, fluorescently labeled nascent chains in membrane-bound ribosomes remain in an aqueous environment sealed from the cytoplasm and accessible to fluorescence quenching from the lumen of the ER (*3*). An aqueous pore with a diameter of 40 to 60 Å during cotranslational translocation is suggested by similar experiments (*4*).

The Sec61 trimeric complex is a strong candidate for the PCC of the ER in yeast and mammalian cells (*5*, *6*). The Sec61 complex provides the principal binding site for ribosomes at the ER during protein translocation (*7*, *8*) and, together with other membrane proteins, is associated with ribosomes after solubilization of rough microsomes with digitonin (*6*). A two-dimensional map of the purified Sec61 complex obtained by electron microscopy has revealed a quasi-pentagonal, circular structure with a central depression (*9*).

The three-dimensional (3D) structure of monomeric ribosomes is currently known at various resolutions for *Escherichia coli* (*10*), wheat germ (*11*), and yeast (*12*). Among the structural features recognized is a tunnel that traverses the large ribosomal subunit and has been considered a candidate for the nascent chain conduit. Here, we present a 3D reconstruction of the ribosome-Sec61 complex.

For purification of the trimeric Sec61 complex (*13*) containing the Sec61α, Sec61β, and Sec61γ subunits (Sec61p, Sbh1p, and Sss1p), a heptameric complex (*14*) was isolated first with protein A–tagged Sec63 protein, followed by elution of the trimeric Sec61 complex with Triton X-100 (Fig. 1A). To determine whether the trimeric Sec61 complex could bind to ribosomes (*15*) in a membrane-free system, we incubated the purified Sec61 complex with ribosomes and analyzed the incubation mixture by sucrose density-gradient centrifugation (*16*). The Sec61 complex incubated without ribosomes remained in the top fraction of the gradient. In the presence of ribosomes, however, the Sec61 complex migrated with ribosomes as determined by immunoblotting (*16*) with antibodies to Sec61β (anti-Sec61β) (Fig. 1B) and Sec61α (anti-Sec61α) (*17*). In agreement with the known salt sensitivity of the Sec61-ribosome interaction, there was no binding at 1 M KOAc (OAc, acetate) (*17*). Incubation of a fixed amount of ribosomes with increasing amounts of Sec61 complex resulted in saturation of ribosome-binding sites (Fig. 1, C and D). On the basis of the amount of Sec61α and ribosomes, we estimate that, at saturation, two to four Sec61 trimers were bound per ribosome and that the dissociation constant K_d is about 10 nM.

The ribosome-Sec61 complex formed under saturating conditions was examined by cryo–electron microscopy (*18*). In the

R. Beckmann and G. Blobel, Howard Hughes Medical Institute, Laboratory of Cell Biology, Rockefeller University, 1230 York Avenue, New York, NY 10021, USA.
D. Bubeck, R. Grassucci, A. Verschoor, Wadsworth Center, New York State Department of Health, Empire State Plaza, Albany, NY 12201–0509, USA.
P. Penczek and J. Frank, Wadsworth Center, New York State Department of Health, and Department of Biomedical Sciences, State University of New York at Albany, Empire State Plaza, Albany, NY 12201–0509, USA.

*To whom correspondence should be addressed. E-mail: beckmar@rockvax.rockefeller.edu

electron micrographs (Fig. 2A), the ribosome-Sec61 complex appears in random orientations (Fig. 2B). This distribution allowed an artifact-free 3D reconstruction by means of a 3D projection alignment procedure (*19*) with an existing reconstruction of the ribosome from yeast (*12*) as a reference. In side views of the ribosome (marked by arrows in Fig. 2B), an ~100 Å–long ellipsoidal mass of density appears at the surface of the large subunit. This location on the ribosome is the same as the site where, in projection, the exiting polypeptide chain was located on both eubacterial and eukaryotic ribosomes by immuno–electron microscopy (*20*).

In the resulting reconstruction (Fig. 3), which has a resolution of 26 Å (*21*), the Sec61 complex appears as a slightly pentagonally shaped toroidal structure with an outer diameter of 95 Å, an inner diameter ranging from 15 to 35 Å depending on depth, and an overall thickness of 40 Å supported by a single stem attached to the base of the large ribosomal subunit. A single site of rigid attachment may facilitate lateral opening of the channel, at the opposite site, to allow the release of nascent transmembrane segments into the lipid bilayer. The surface of the Sec61 oligomer facing the ribosome is parallel to the surface of the large subunit. The distance between these surfaces ranges from 15 to 20 Å. Hence, the attachment of the Sec61 oligomer to the ribosome does not appear to form the tight seal that was implicated in nascent chain fluorescence-quenching experiments. Seal-forming proteins could be missing, or, more likely, the signal sequence may be necessary for seal formation (*2*, *3*, *8*, *22*).

The central pore of the Sec61 oligomer aligns precisely with an opening in the large ribosomal subunit that represents the exit of a tunnel (Fig. 3B). This tunnel runs from the interface canyon (*12*) to the lower portion of the large subunit (Fig. 3D). At the threshold level chosen (*23*), a small segment of the tunnel is blocked because of the limiting resolution, but it appeared to be open at increased threshold levels (*17*). A similar effect has been observed in the reconstruction of the 70S *E. coli* ribosome (*10*). These tunnels, seen in cryo–electron microscopy maps of the ribosome from *E. coli* (*10*) and *Saccharomyces cerevisiae* (*12*),

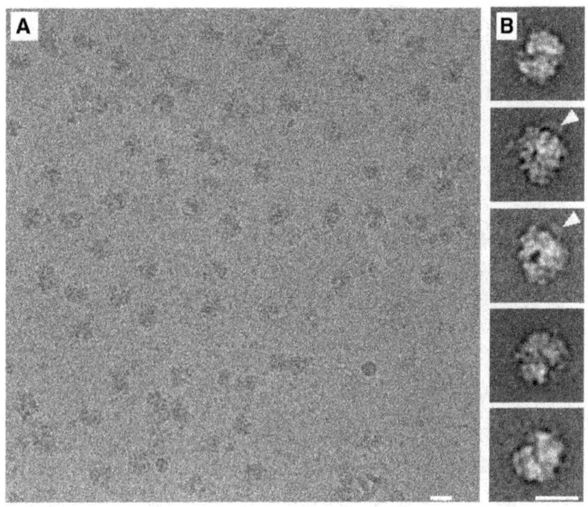

Fig. 2. (**A**) Cryo–electron micrograph showing a field of yeast ribosome-Sec61 complexes. Scale bar, 200 Å. (**B**) Averaged projections of the ribosome-Sec61 complex obtained by classification. Particles marked with arrowheads show the ribosome in side view, with the Sec61 complex visible as a 100 Å–long mass lying parallel to the ribosome surface. Scale bar, 200 Å.

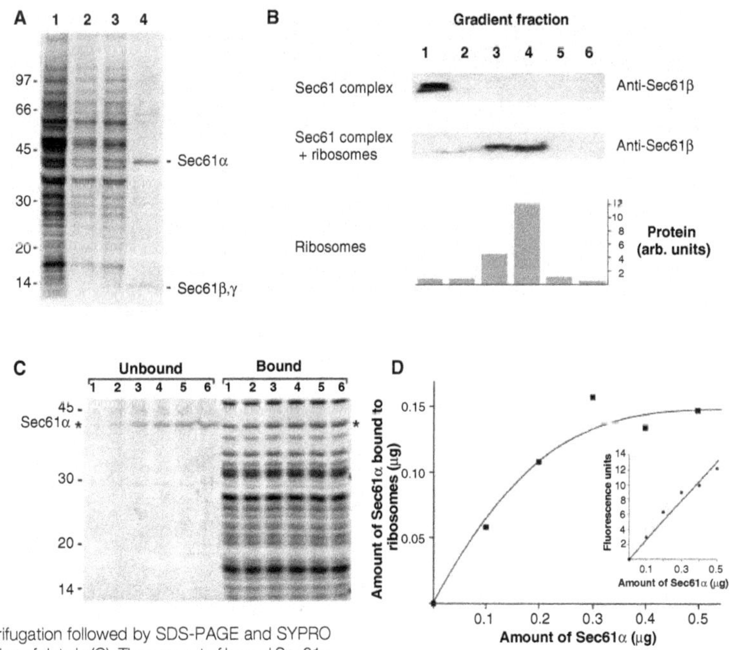

Fig. 1. Purification and binding of the trimeric Sec61 complex to ribosomes. (**A**) Purification of the Sec61 complex. Crude nuclear envelopes from *Sec63prA* cells were extracted with digitonin, and the digitonin extract was incubated with IgG-Sepharose. The trimeric Sec61 complex was eluted with Triton X-100 and analyzed by SDS-PAGE followed by Coomassie blue staining. Lane 1, crude nuclear envelopes (1/30,000); lane 2, digitonin extract (1/30,000); lane 3, IgG-Sepharose flow through (1/30,000); and lane 4, Triton X-100 eluate (1/10). Ten to 15 micrograms of Sec61α [by comparison with a bovine serum albumin standard] were typically purified from 80 g of packed cells. Molecular mass is given in kilodaltons at the left. (**B**) The Sec61 complex binds to ribosomes. The Sec61 complex (0.5 μg) was incubated with and without ribosomes (0.5 A_{260}), and the incubation mixture was analyzed by sucrose density-gradient centrifugation followed by SDS-PAGE analysis and immunoblotting of gradient fractions. Lanes 1 to 6 represent top to bottom fractions probed with a peptide antibody to Sec61β. For the ribosomes, the Coomassie-stained proteins were quantitated with NIH image (arb. units, arbitrary units). (**C**) Binding of the Sec61 complex to ribosomes is saturable. Increasing amounts of the Sec61 complex (0 μg of Sec61α in lane 1 to 0.5 μg of Sec61α in lane 6) were incubated with a fixed amount of ribosomes (0.125 A_{260}) and analyzed by density-gradient centrifugation followed by SDS-PAGE and SYPRO Red staining. Asterisks denote Sec61α. (**D**) Quantitation of data in (C). The amount of bound Sec61α was calculated according to the calibration shown in the inset.

have been proposed as exit pathways of the nascent polypeptide chain, although the evidence for that is indirect (*20*). The precise alignment between the pore of the Sec61 oligomer and the tunnel (Fig. 3, C and D) provides strong support for this hypothesis. Conversely, this structural arrangement also implies that the Sec61 oligomer indeed constitutes the PCC.

Detailed analysis of the Sec61 oligomer (Fig. 4, A to C) shows that the azimuthal distribution of mass is irregular, with the bulk of the mass lying on the side attached to the stem. Again, as speculated above, the thinner wall opposite the attachment stem might facilitate lateral opening of the channel. There is a further asymmetry in the Sec61 oligomer: the pore is funnel-shaped, with a diameter of 15 Å at the lumenal site of the ER, widening toward the ribosome, so that a small vestibule with a diameter of 35 Å is formed. Consistent with the estimate from kinetic data, the measured volume of the Sec61 oligomer would accommodate two trimeric Sec61 complexes. The reasons for the asymmetric appearance of the ribosome-bound Sec61 oligomer are presently not clear.

It is likely that the Sec 61 oligomer bound to the ribosome represents the PCC in its inactive and closed conformation. Future structural analyses of an in vitro–assembled complex containing a ribosome, a nascent chain, and the Sec61 complex will lead to information regarding the active state of the PCC.

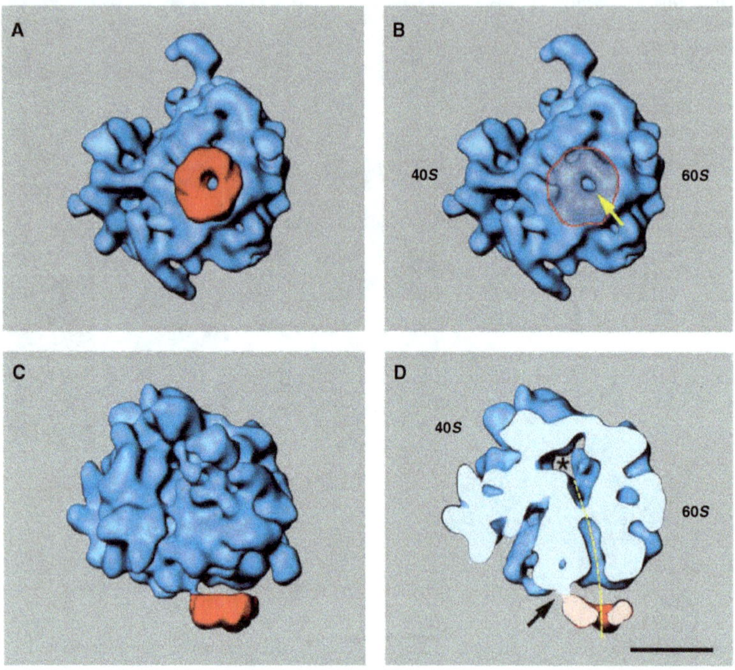

Fig. 3. Three-dimensional reconstruction of the ribosome-Sec61 complex in a surface representation. (**A**) Front view, with the Sec61 oligomer shown in red. (**B**) Front view, with the Sec61 oligomer shown as transparent, to demonstrate the alignment of the Sec61 oligomer pore with the tunnel exit of the large ribosomal subunit indicated with the yellow arrow. (**C**) Side view, with the Sec61 oligomer shown in red. (**D**) Ribosome-Sec61 complex lying in the same orientation as in (C), cut along a plane that cross sections the pore of the Sec61 oligomer and the ribosome tunnel. The arrow indicates the stem connecting the ribosome with the Sec61 oligomer; the space between the two ribosomal subunits is indicated by an asterisk. The ribosomal tunnel and its alignment with the Sec61 pore is indicated by a broken yellow line. Scale bar, 100 Å.

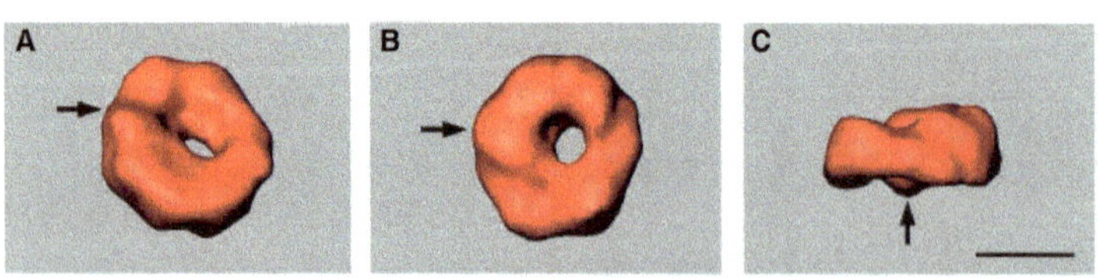

Fig. 4. Three closeup views of the Sec61 oligomer. (**A**) Surface facing the ribosome. There is a vestibule (diameter of 35 Å) formed by the funnel-like structure and the pore (diameter of 15 Å). (**B**) Surface facing away from the ribosome. (**C**) View of the side opposite the attachment site. The ribosome would be located underneath the channel. The wall opposite the attachment site is thinner and more irregular. Arrows indicate the attachment site. Scale bar, 50 Å.

REFERENCES AND NOTES

1. G. Blobel and B. Dobberstein, *J. Cell Biol.* **67**, 835 (1975).
2. S. M. Simon and G. Blobel, *Cell* **65**, 371 (1991); *ibid.* **69**, 677 (1992).
3. K. S. Crowley, G. D. Reinhart, A. E. Johnson, *ibid.* **73**, 1101 (1993); K. S. Crowley, S. Liao, V. E. Worrel, G. D. Reinhart, A. E. Johnson, *ibid.* **78**, 461 (1994).
4. B. D. Hamman, J.-C. Chen, E. E. Johnson, A. E. Johnson, *ibid.* **89**, 535 (1997).
5. R. J. Deshaies and R. Schekman, *J. Cell Biol.* **105**, 633 (1987); C. J. Stirling, J. Rothblatt, M. Hosobuchi, R. Deshaies, R. Schekman, *Mol. Biol. Cell* **3**, 129 (1992).
6. D. Görlich, S. Prehn, E. Hartmann, K. U. Kalies, T. A. Rapoport, *Cell* **71**, 489 (1992); D. Görlich and T. A. Rapoport, *ibid.* **75**, 615 (1993).
7. K. U. Kalies, D. Görlich, T. A. Rapoport, *J. Cell Biol.* **126**, 925 (1994).
8. B. Jungnickel and T. A. Rapoport, *Cell* **82**, 261 (1995).
9. D. Hanein *et al.*, *ibid.* **87**, 721 (1996).
10. J. Frank *et al.*, *Nature* **376**, 441 (1995); H. Stark *et al.*, *Structure* **3**, 815 (1995).
11. A. Verschoor, S. Srivastava, R. Grassucci, J. Frank, *J. Cell Biol.* **133**, 495 (1996).
12. A. Verschoor, J. R. Warner, S. Srivastava, R. A. Grassucci, J. Frank, *Nucleic Acids Res.*, in press.
13. A yeast strain (*Sec63prA*) was constructed in which the genomic copy of the gene encoding Sec63p was tagged by COOH-terminal, in-frame integration of a DNA fragment encoding for the immunoglobulin G (IgG)–binding domains of protein A. The DNA fragment encoding the protein A gene and adjacent *HIS3* and *URA3* markers was amplified by polymerase chain reaction with specific primers for *SEC63* and a template plasmid as described [J. D. Aitchison, M. P. Rout, M. Marelli, G. Blobel, R. Wozniak, *J. Cell Biol.* **131**, 1133 (1995)]. A crude nuclear pellet was prepared from a 36-liter culture as described (C. Strambio-de-Castillia, G. Blobel, M. P. Rout, *ibid.*, p. 19). Thirty milliliters of crude nuclei (one-fifth of the preparation) was extracted by incubation with deoxyribonuclease I (20 μg/ml) and Heparin (1 mg/ml) in buffer E [10 mM bis-tris-Cl (pH 6.5), 1 mM $MgCl_2$, 10 mM KOAc, 1 mM dithiothreitol (DTT), and 0.5 mM phenylmethylsulfonyl fluoride (PMSF)] for 20 min at 25°C and 40 min on ice. The crude nuclear envelopes were sedimented for 40 min at 145,000*g* at 4°C and extracted with 10 ml of buffer S [3% digitonin, 0.4 M sucrose, 10 mM triethanolamine-OAc (pH 7.5), 750 mM KOAc, 1.5 mM $Mg(OAc)_2$, 0.5 mM EDTA, and 1 mM DTT] for 30 min on ice. After pelleting by centrifugation of insoluble material for 35 min at 145,000*g* at 4°C, the extract was diluted with 1 volume of buffer D [0.4 M sucrose, triethanolamine-OAc (pH 7.5), 1.5 mM $Mg(OAc)_2$, 1 mM PMSF, and 1 mM DTT] and incubated overnight with 1 ml of IgG-Sepharose (Cappel, Durham, NC) at 4°C. The column was washed with 10 volumes of buffer W [1% digitonin, 100 mM KOAc, Azolectin (0.5 mg/ml), 10 mM triethanolamine-OAc (pH 7.5), 10% glycerol, 3 mM $Mg(OAc)_2$, 0.5 mM DTT, and 1 mM $CaCl_2$], and the trimeric Sec61 complex was eluted with buffer T [1% Triton X-100, 200 mM KOAc, 10 mM triethanolamine-OAc (pH 7.5), 10% glycerol, 3 mM $Mg(OAc)_2$, 0.5 mM DTT, and 1 mM $CaCl_2$]. The

identity of the proteins was confirmed by specific antibodies to Sec61α (Sec61p) and Sec61β (Sbh1p) (*17*).

14. S. Panzner, L. Dreier, E. Hartmann, S. Kostka, T. A. Rapoport, *Cell* **81**, 561 (1995)
15. For the purification of ribosomes, the yeast strain DF5 was grown in 3.5 liters of yeast extract, peptone, and dextrose medium. At an optical density of 600 nm (OD_{600}) of 1.0, the cells were washed with water and incubated for 15 min at 25°C in 100 mM tris-SO_4 (pH 9.4) and 10 mM DTT. After homogenization by French press in buffer A [50 mM triethanolamine-OAc (pH 7.5), 50 mM KOAc, 5 mM $MgCl_2$, 1 mM DTT, and 0.5 mM PMSF], the homogenate was centrifuged for 30 min at 100,000*g* at 4°C. The supernatant was layered over a continuous 10 to 40% sucrose gradient in buffer A. After centrifugation for 4.5 hours at 200,000*g* at 4°C, the monomeric ribosomes were pooled according to the A_{254} profile. The ribosomes were pelleted by centrifugation for 4.5 hours at 145,000*g* at 4°C, resuspended in water, and frozen in liquid N_2.
16. Purified Sec61 complex and ribosomes were incubated for 30 min on ice in a buffer containing 0.5% Triton X-100, 100 mM KOAc, 5 mM triethanolamine-OAc (pH 7.5), 5% glycerol, 1.5 mM $Mg(OAc)_2$, 0.5 mM DTT, and 0.5 mM $CaCl_2$. To separate unbound Sec61 complex from ribosome-bound Sec61 complex, we carried out gradient centrifugation using a 10 to 50% sucrose step gradient in 0.5% Triton X-100, 100 mM KOAc, 10 mM triethanolamine-OAc, 1 mM DTT, and 3 mM $Mg(OAc)_2$. After centrifugation for 60 min at 240,000*g* at 4°C, six fractions were collected manually. For saturation assays, the first two fractions were pooled as the unbound fraction and the following two as the bound fraction. Fractions were analyzed by SDS–polyacrylamide gel electrophoresis (PAGE) and stained with SYPRO Red. The amount of protein was quantitated with the STORM system (red fluorescence) and NIH image. For immunoblotting, proteins were precipitated, separated on 10 to 20% SDS-PAGE gradient gels, transferred to a nitrocellulose membrane, incubated consecutively with anti-Sec61α or anti-Sec61β and horseradish peroxidase–conjugated donkey antibodies to rabbit, and detected by ECL as described (Amersham).
17. R. Beckmann *et al.*, data not shown.
18. Incubation to form the ribosome Sec61 complex was performed as described (*16*), and the mixture was diluted with 4 volumes of water immediately before it was applied to the grid. Grids for cryo–electron microscopy were prepared as described [T. Wagenknecht, R. Grassucci, J. Frank, *J. Mol. Biol.* **199**, 137 (1988); J. Dubochet *et al.*, *Q. Rev. Biophys.* **21**, 129 (1988)]. Micrographs were recorded under low-dose conditions on a Philips EM 420, with 1.5-μm defocus and magnification of 52,200 ± 2% as checked by a tobacco mosaic virus standard.
19. P. A. Penczek, R. Grassucci, J. Frank, *Ultramicroscopy* **53**, 251 (1994).
20. C. Bernabeu, E. M. Tobin, A. Fowler, I. Jabin, J. A. Lake, *J. Cell Biol.* **96**, 1471 (1983).
21. Micrographs were checked for drift, astigmatism, and presence of Thon rings by optical diffraction. Scanning was done with a step size of 25 μm corresponding to 4.78 Å on the object scale, on a Perkin-Elmer PDS 1010 A microdensitometer. Particles were selected by an automated selection procedure that differed from the one previously described [K. R. Lata, P. Penczek, J. Frank, *Ultramicroscopy* **58**, 381 (1995)] in that the particle candidates were compared directly with the reference set of 87 quasi–evenly spaced projections (*19*) of an existing reconstruction of the ribosome from yeast (*12*). A total of 13,178 particles were picked. The reconstruction was done with two independent approaches to obtain the orientations of the projections. In the first approach, an existing reconstruction of the ribosome from yeast (*12*) was used as a reference in the 3D projection alignment procedure (*19*). In the second approach, an initial reconstruction was obtained with the simultaneous minimization technique [P. Penczek, J. Zhu, J. Frank, *Ultramicroscopy* **63**, 205 (1996)]. In both cases, four steps of the 3D projection alignment procedure (*19*) were applied with a 2° angular interval. In each step, the refined 3D structure was calculated with 70% of the best matching particles (on the basis of the value of the cross-correlation coefficient). Both reconstructions proved to be indistinguishable within the measured resolution range. The final resolution, estimated with the Fourier shell correlation with a cutoff value at 0.5 [B. Böttcher, S. A. Wynne, R. A. Crowther, *Nature* **386**, 88 (1997)], was 26 Å.
22. R. Gilmore and G. Blobel, *Cell* **42**, 497 (1985).
23. In trying to gauge the correct threshold value, we were led by two criteria: (i) We observed the structure as the threshold was increased. There is normally a "plateau," a range of threshold values within which the appearance (or the volume encompassed) varies only slightly. (ii) Three-dimensional connectivity must not be violated, which means that, in this case, we could not choose a threshold, within the plateau defined above, that makes the connecting rod disappear. Thus, the plateau was further narrowed. These criteria were applied separately in the preparation of the 3D representations of the ribosome and the channel. Because of the residual uncertainty in the molecular boundaries, the measurements for pore size and the distance between channel and ribosome have an uncertainty of 25%.

24. We thank A. Fischer for the purification of the Sec61 complex; members of the Blobel lab and R. Agrawal for discussions; S. Darst and A. Malhotra for discussions, support with the electron microscopy, and assistance with the image processing; A. Heagle for help with the illustrations; and the National Center for Supercomputer Applications, University of Illinois at Urbana-Champaign, for computing support. Supported by grants from NIH (1R01 GM29169) and NSF (BIR 9219043) (to J.F.) and a fellowship of the Deutsche Foschungsgemeinschaft (to R.B.).

26 September 1997; accepted 17 November 1997

Scanning Microscopy Vol. 11, 1997 (Pages 147-154) 0891-7035/97$5.00+.25
Scanning Microscopy International, Chicago (AMF O'Hare), IL 60666 USA

THREE DIMENSIONAL RECONSTRUCTION WITH CONTRAST TRANSFER COMPENSATION FROM DEFOCUS SERIES

P.A. Penczek[1,2]*, J. Zhu[1], R. Schröder[3] and J. Frank[1,2]

[1]Wadsworth Center, New York State Department of Health, Albany, NY
[2]Department of Biomedical Sciences, School of Public Health, State University of New York at Albany, NY
[3]Department of Biophysics, Max-Planck Institute for Medical Research, Heidelberg, Germany

Abstract

Cryo-electron microscopy provides the means to quantitatively study macromolecules in their native state. However, the original mass distribution of the macromolecule is distorted by the contrast transfer function (CTF) of the electron microscope. In addition, the zeros of the CTF put a practical limit on the resolution that can be achieved. Substantial improvement to the quality of the results can be accomplished by collecting the data using a series of defocus settings. Such data sets can be combined and the resolution can be extended beyond the first zero of the CTF. This procedure can be applied either at the stage of raw data, or more effectively at the stage of reconstructed volumes which have a high signal-to-noise ratio as a result of averaging over many projections. A method of three-dimensional (3D) reconstruction that combines an algebraic, iterative 3D reconstruction technique with CTF correction is proposed. The potential to incorporate *a priori* knowledge into the reconstruction process is discussed. This approach was used to obtain a 3D reconstruction of the *E. coli* 70S ribosome from energy-filtered cryo-images.

Key Words: Cryo-electron microscopy, energy-filtered cryo-images, three-dimensional reconstruction, contrast transfer function correction.

*Address for correspondence:
P.A. Penczek
Wadsworth Center
New York State Department of Health
Empire State Plaza, P.O. Box 509
Albany, NY 12201-0509

Telephone number: (518) 474-4235
FAX number: (518) 486-2191
E-mail: pawel@wadsworth.org

Introduction

Single particle reconstruction methods [21] make it possible to obtain three-dimensional density maps of macromolecular complexes that are not amenable to crystallization. When combined with cryo techniques [8, 16, 29, 33] and recent advances in image processing methods [17, 18], these methods yield structural maps that can be quantitatively interpreted [9]. Even though substantial improvement in the resolution of three-dimensional results has been achieved, further progress is hampered by the image/contrast distortions introduced by the electron microscope (EM). These effects can be described within a linear, weak-phase approximation by the contrast transfer function (CTF) of the microscope. Additional distortions are caused by the inelastically scattered electrons, which contribute a high and slowly varying background to the image. Moreover, if multiple scattering events occur, the inelastically scattered electrons can give rise to elastic bright field images, acting as "internal sources", and the image we observe can be thought of as a superposition of a large number of bright field images, each originating from an inelastically scattered electron having different angular incidence and energy [22]. The conventional linear contrast transfer function cannot account for these inelastic and multiple scattering effects. However, zero-loss energy filtering in the EM eliminates virtually all inelastically scattered electrons and thus reduces non-linear effects to a minimum. In ice, elastic scattering from nucleic acids is predicted to be more than twice as strong as from protein, whereas the amount of inelastic scattering is similar for the two types of molecules. Thus, energy filtration should also greatly increase the contrast between RNA and protein [26], a desired effect in the study of the protein-nucleic acid complexes such as the ribosome. Finally, by combining energy-filtered cryo-EM with CTF correction one can, in principle, obtain absolute densities of biological molecules that agree very well with those from X-ray structures [14, 25].

Linear Theory of the Contrast Transfer Function

The contrast transfer theory was formulated in the

seventies [4, 7, 11, 12, 15]. It can be demonstrated [29] that under the weak-phase approximation the relationship between the object and the bright field image can be described by the linear transfer theory. In Fourier space,

$$I(\boldsymbol{k}) = H(k)\Phi(\boldsymbol{k}) \qquad (1)$$

where $\Phi(\boldsymbol{k})$ is the Fourier transform of the object (linearly related to the Coulomb potential distribution), $I(\boldsymbol{k})$ is the Fourier transform of the observed image, $\boldsymbol{k}$ is the spatial frequency vector, and k $|\boldsymbol{k}|$ θ/λ, where θ is the scattering vector and λ is the electron wavelength. $H(k)$ is the CTF, with the explicit form

$$H(k) = \sin\gamma(k) - W\cos\gamma(k) \qquad (2)$$

The term W (here assumed to be constant in the spatial frequency range considered) is called *amplitude contrast ratio*, and $\gamma(k)$ is the phase shift produced by the lens aberrations and defocusing. $\gamma(k)$ is rotationally symmetric assuming that the astigmatism is negligible. The contrast transfer functions for our choice of defocus values are shown in Figure 1. Since the amplitude contrast is different for different atomic species, Equation 2 is strictly correct only for a single species of atoms [6, 10]. The use of single value for W implies that W will be to some extent specimen-dependent.

The presence of additional effects in the image formation process requires expansion of the CTF model given by Equation 2. These effects include: finite source size [7], energy spread [12, 31], resolution-limiting effects of the film [3], scanner [14], and drift [5, 13]. All these effects have a frequency-limiting impact on the EM image which can be approximately described by so called *envelope functions*. To demonstrate the effect of envelope functions, the same CTFs of Figure 1 are again shown for the conditions of our experiment (Fig. 2; for details see the section on Application to Energy-Filtered Images).

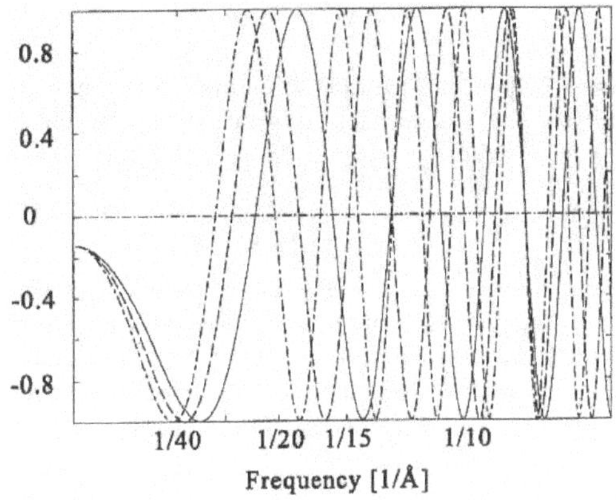

Figure 1. Theoretical contrast transfer function (eqn. 2) for three defocus values: 1.5 μm (_____), 2.0 μm (----), and 2.5 μm (-·-·-).

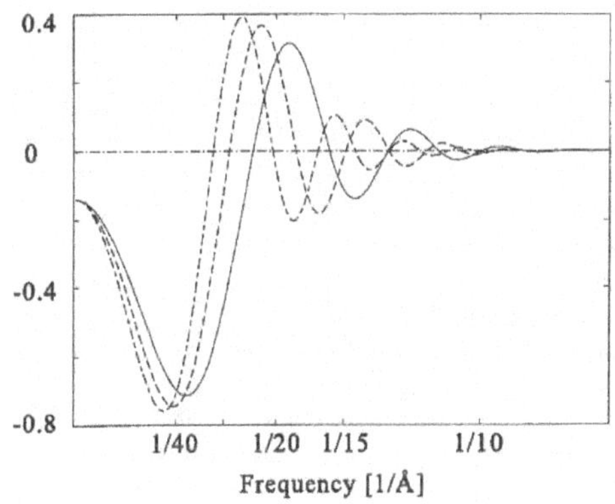

Figure 2. Theoretical contrast transfer function with envelope functions effects for three defocus values: 1.5 μm (_____), 2.0 μm (----), and 2.5 μm (-·-·-).

Methods of Correction for the CTF

The goal of the CTF correction (for a review of some of the earlier literature, see [28]) is to recover the original values of the amplitudes of the Fourier transform of the image, within the framework of the linear approximation.

According to Equation 1, Fourier transform of the observed image is the product of the CTF and the Fourier transform of the object. Thus, it is tempting to divide the Fourier transform by the CTF (estimated or known) to recover the exact values of the object:

$$\Phi(\boldsymbol{k}) \cong I(\boldsymbol{k}) / H(k) \qquad (3)$$

Unfortunately, this simple solution rarely yields desired results. In practice, the measured image is corrupted by noise, which will be enhanced upon division by the CTF in regions where the CTF is small. The CTF of the EM is low in the low-spatial frequency region, but also tends to zero at high spatial frequencies. Most problematic are zero crossings in the range between these extremes where division is impossible. Thus, a better approach is to take advantage of the fact that in most practical cases the noise is not correlated with the original image. Using this assumption one would seek a filter function $F(\boldsymbol{k})$ that minimizes the expected mean squared deviation of the restored object $\tilde{\Phi}(\boldsymbol{k}) = F(k)I(\boldsymbol{k})$ from the original object [19]:

3-D reconstruction with contrast transfer compensation

$$E[\,|\,\Phi(\boldsymbol{k}) - F(k)I(\boldsymbol{k})|^2] \tag{4}$$

where $E[.]$ denotes the expectation over an ensemble of images. The solution to this problem is given by so-called *Wiener filter* [24]:

$$F(k) = \frac{H^*(k)}{|\,H(k)\,|^2 + 1/SNR} \tag{5}$$

Here SNR denotes the signal-to-noise ratio of the original image and the noise (in the general case it can be frequency dependent and given by the ratio of the respective power spectra). In the noiseless case ($SNR \to \infty$) the Wiener filter approximates the pseudo-inverse filter [1] defined by:

$$F_{ps}(k) = \begin{cases} \dfrac{1}{H(k)} & H(k) \neq 0 \\ & \text{for} \\ 0 & H(k) = 0 \end{cases} \tag{6}$$

Thus, it also approximates the simple inversion, as given by Equation 3. For the other extreme case, namely $SNR \to 0$, the application of the Wiener filter will approximately correspond to the multiplication by the CTF. In intermediate cases, depending on the value of the *SNR* used, the Wiener filter will in general amplify frequencies in those regions where CTF has moderate to large values and suppress frequencies in regions where the CTF has small values. In addition, as follows from Equation 5, wherever the CTF has zero value the filtered image will have its Fourier amplitude set to zero. Thus, the Wiener filtration does not make any attempts to recover the information in the missing frequency regions, a property that limits its appeal as a tool for the CTF correction for EM data. Moreover, the Wiener filters are often criticized for excessive suppression of the measurement noise, while performing only minor corrections [2].

The iterative methods for the CTF corrections [2] attempt to find an approximation $\tilde{\phi}(\boldsymbol{r})$ of an original image such that the squared residual error is minimized over the appropriate region of support:

$$|i(\boldsymbol{r}) - h(r) * \tilde{\phi}(\boldsymbol{r})|^2 \to min \tag{7}$$

Equation 7 is given in real space and * denotes convolution. The solution is found in a recursive manner starting from the initial approximation $\tilde{\phi}^{(i)}(\boldsymbol{r})$ (usually a blank image). The advantage of the iterative methods is that additional constraints can be incorporated into the process, either linear, in form of regularization, or non-linear, for example, as a positivity constraint. In some cases these additional constraints, often referred to as *a priori* knowledge on the image, can help to recover information suppressed by the zeros of the CTF.

The presence of zeros in the CTF of the electron microscope and the dependence of the locations of these zeros on the defocus value suggest that the CTF correction can be augmented by collecting a defocus series data set. Such data, if combined in Fourier space, should for all the practical purposes cover the whole range of frequencies. Assuming that L defocus values Δz_l have been used, both the Wiener filter method and iterative methods can be used to combine the data sets and correct for the CTF effects. The Wiener filter for the *l'th* data set has the form [10, 27]:

$$F(k) = \frac{SNR_l H_l^*(k)}{\sum_{l=1}^{L} SNR_l |\, H_l(k)\,|^2 + 1} \tag{8}$$

and the Fourier transform of the CTF corrected object is given by

$$\Phi(k) = \sum_{l=1}^{L} F_l(k) I_l(\boldsymbol{k}) \tag{9}$$

As in the case of a single defocus data set (Eqn. 5), the spatial frequency regions corresponding to the zeros of the CTFs are set to zero in the respective data sets. Thus, it becomes evident that the knowledge about the precise settings of the defocus during data collection is essential for this method to succeed.

The iterative method can be easily extended to the defocus case series by having Equation 7 modified to

$$\sum_{l=1}^{L} w_l |i(\boldsymbol{r}) - h_l(r) * \tilde{\phi}_l(\boldsymbol{r})|^2 \to min \tag{10}$$

where $w_l > 0$ defines the relative importance of the *l'th* defocus data set. As in the case of a single defocus data set (Eqn. 7) the solution is found in a recursive manner starting from the appropriate approximations $\tilde{\phi}_l^{(0)}(x)$ and the *a priori* knowledge about the original image can be easily incorporated.

In the field of electron microscopy the data are available in the form of projections of the biological macromolecule. Thus, in principle, the CTF correction should be applied directly to the 2D data (i.e., original micrographs). Unfortunately, due to the very low signal-to-noise ratio in cryo-EM images this method is not likely to succeed. Instead, the CTF correction must be applied after all the orientation parameters were found and a 3D reconstruction could be calculated. At this point, due to the heavy oversampling in Fourier space, the structure has much higher SNR, and errors in individual projections are less likely to dominate the result. Since our measured data are 2D projections of the structure, the best solution is to make CTF correction an integral part of the 3D reconstruction procedure. In this way the final volume is obtained in one single step and any accumulation of errors is avoided.

To design the method of three-dimensional reconstruction with CTF compensation from defocus series, we have to modify Equation 10. Both 3D reconstruction and CTF application can be treated as algebraic problems. In real space, the effect of the CTF can be described by a point spread function, represented by a square matrix, and the projection operation by a non-square matrix P. Moreover, since the CTF function was assumed to be rotationally symmetric, it can be applied to the 3D object instead of 2D projections. Thus, denoting the unknown, CTF artifacts-free object by g and using matrix notation, we obtain from equation 10

$$(1-a)\sum_{l=1}^{L} w_l |d_l - P_l \hat{H}_l g|^2 + a|Bg|^2 \rightarrow \min \tag{11}$$

where d is a vector containing all the available 2D projections, matrix B is a discrete approximation of the Laplacian and a is a Langrange multiplier. The value of a determines the relative smoothness of the solution g. This additional term will regularize the solution [20] and prevent adverse amplification of noise. The solution of Equation 11 is found using the steepest descent method [20]:

$$g^{(n+1)} = g^{(n)} + \lambda\{(1-a)[\sum_{l=1}^{L}(\hat{H}_l P_l^T d_l - \sum_{l=1}^{L}(\hat{H}_l^T P_l^T P_l(\hat{H}_l g^{(n)})] - a B^T B g^{(n)}\} \tag{12}$$

The first sum can be pre-calculated and stored as a 3D volume in the computer memory. Thus, the input projections d_l have to be read only once and are never again accessed during the course of the iterations. In addition, the product B^TB is the Laplacian of the 3D volume $g^{(n)}$, which can be calculated more efficiently without actually creating the matrix B^TB. The point spread function is space-invariant, so H_l has a block-Toeplitz structure [2], and H_l is equal to H_l^T. Because of the large size of the matrix it is more convenient to use the CTF H_l in Fourier space and to modify the Fourier transform of the volume instead of using matrix multiplication or real space convolution.

Application to Energy-Filtered Images of the 70S *E. coli* Ribosome

The specimen was prepared following the procedure of Wagenknecht *et al.* [32]. Molybdenum 400-mesh grids were used to minimize thermal effects. The specimens were examined in a Zeiss/LEO EM912 (Oberkochen, Germany) transmission electron microscope operated at 120 kV. The microscope is equipped with a Köhler illumination system and an Omega energy filter. All the microscopy was done at a temperature below -170°C using an Oxford cryo-holder CT3500 (Oxford Instruments, Oxford, UK) and temperature controlling unit. A 90 μm object aperture was used in all experiments. Three images were taken at zero degree tilt of the same specimen area using the defocus settings 1.5 μm, 2.0 μm and 2.5 μm at a magnification of 52000x, which was calibrated using catalase crystals. The dose for each exposure was 5e$^-$/Å^2, thus the total dose accumulation reached 15e$^-$/Å^2 for the data collected at 2.5 μm defocus. The width of the zero-loss energy filter was 14 eV. Images were recorded on Kodak (Rochester, NY) SO163 films and processed in the developer D19 at full strength for 12 minutes. The negatives were scanned using the Perkin Elmer (Norwalk, CT) flatbed PDS 1010A microdensitometer with a step size of 20 μm. The corresponding pixel size on the specimen is 3.8Å.

The parameters of the CTFs (defocus, source size, energy spread, amplitude contrast ratio) were determined using the method described by Zhu *et al.* [34]. Briefly, the power spectrum of a large field is computed and rotationally averaged. The resulting profile is background corrected and fitted on the basis of CTF theory and analytical expressions for the envelope terms. The amplitude contrast ratio was determined as 0.14. After windowing the particles were subjected to the 3D projection alignment procedure [18] with an angle increment of 2°. A previous reconstruction of the 70S ribosome [18] was used as initial reference for a 2.5 μm-defocus group (2254 images). The resulting reconstruction was used as a reference for a 2.0 μm-defocus group (2043 images) and the result was in turn used as a reference for a 1.5 μm-defocus group (1539 images) (Figs. 3a and 3b). Finally, after all the necessary parameters were established, images from all three groups were used in a 3D reconstruction with the CTF compensation program (Figs. 3c and 3d), as described in the section on "Methods of Correction for the CTF". To test the influence of the non-linear constraints on the 3D reconstruction process we decided to use the most intuitive one, namely a positivity constraint. Since the CTF-corrected volume should be linearly related to the Coulomb potential distribution within the object it is reasonable to assume that the density values should be positive. The 3D reconstruction calculated by enforcing positivity of the volume is shown in Figures 3e and 3f.

The gain in resolution after the CTF correction and merging of three defocus groups is easy to notice in the surface representations (Fig. 3d *versus* Fig. 3b), but is particularly striking in the appearance of the central slice

3-D reconstruction with contrast transfer compensation

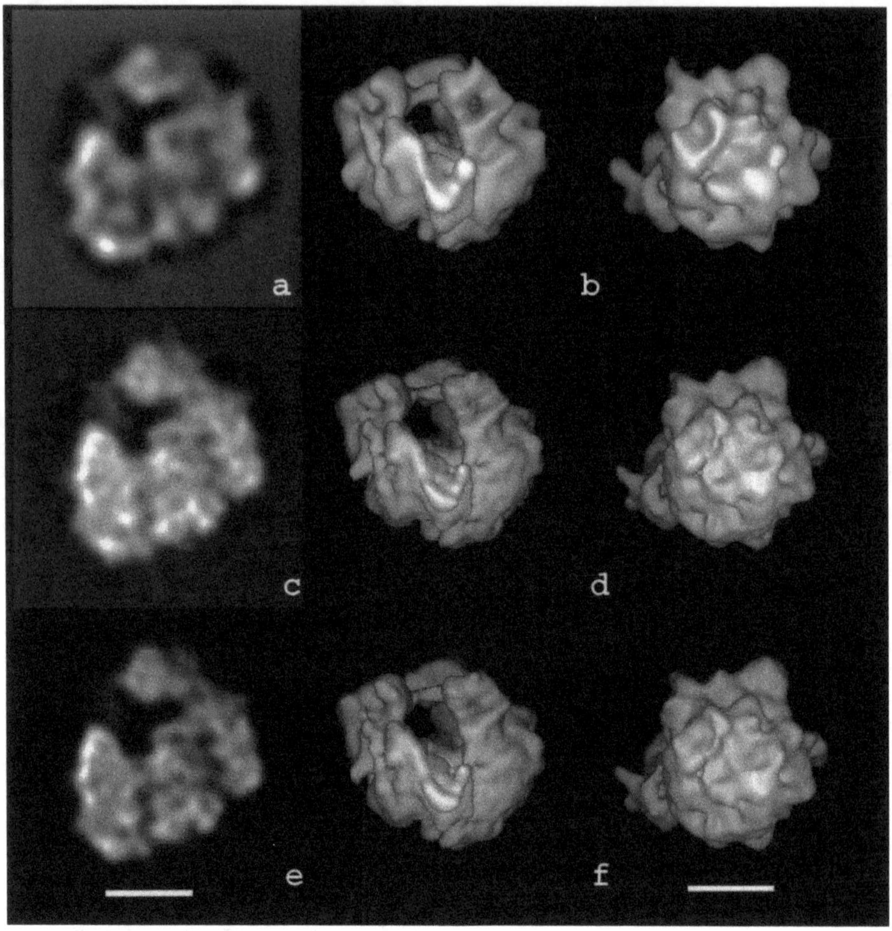

Figure 3. 3D reconstruction of the 70S *E. coli* ribosome. (**a**) Central slice of the 1.5 μm data set reconstruction. (**b**) Surface representations of the 1.5 μm data set reconstruction. (**c**) Central slice of the CTF-corrected reconstruction using three data sets with defocus settings 1.5 μm, 2.0 μm and 2.5 μm, respectively. (**d**) Surface representations of the CTF-corrected reconstruction using three data sets with defocus settings 1.5 μm, 2.0 μm and 2.5 μm, respectively. (**e**) Central slice of the CTF-corrected reconstruction using the positivity constraint and three data sets with defocus settings 1.5 μm, 2.0 μm and 2.5 μm, respectively. (**f**) Surface representations of the CTF-corrected reconstruction using three data sets with defocus settings 1.5 μm, 2.0 μm and 2.5 μm, respectively. Scale bars = 100 Å.

(Fig. 3c *versus* Fig. 3a). The number of high-frequency internal details increases dramatically and broad maxima have become better localized. The addition of the positivity constraint to the 3D reconstruction and CTF correction process does not change the surface representation in any significant way (Fig. 3f *versus* Fig. 3d). The main difference can be noticed in the central slice (Fig. 3e) where the artifacts surrounding the structure are all but eliminated.

The analysis of resolution was done using Fourier Shell Correlation (FSC) curves (FSC is a 3D version of Fourier Ring Correlation measure [23]). For each 3D reconstruction all the input projections were divided randomly into two equal subsets, for each subset a 3D reconstruction was calculated, and these two volumes were compared in Fourier space using FSC curves (Fig. 4). The curve for 1.5 μm-defocus data set has a first minimum at approximately 1/25

P.A. Penczek et al.

Å^{-1}, which is the position of the first zero of the corresponding CTF (Fig. 2). The curve for CTF-corrected reconstruction using three data sets with defocus settings 1.5 µm, 2.0 µm and 2.5 µm has a minimum at lower frequency and then rises at 1/25 Å^{-1}, which proves that the gap in frequency range caused by the zero of the CTF was successfully filled by the information originating from two other data sets. The FSC curve for the CTF-corrected reconstruction decreases quickly at about 1/19 Å^{-1}. In this region CTFs for both 2.0 µm and 2.5 µm data sets intersect zero. The CTF for 1.5 µm data set has a maximum in this region, but as it is clear from its FSC curve that SNR is insufficient to provide significant contributions to the merged reconstruction. Thus, due to the particular choice of defocus values there was no phase information recovered past the spatial frequency of 1/20 Å^{-1}. The addition of the positivity constraint modified the FSC significantly. In the whole frequency range considered the FSC has higher values and extends to almost 1/15 Å^{-1}. Nevertheless, the visual analysis of the corresponding structures and, most significantly, the central slices (Figs. 3c and 3e) leads to the conclusion that the gain in resolution is mostly due to the elimination of artifacts in the background, not due to the increase in details of the structure.

The influence of CTF correction on the reconstructed structures is best illustrated on the histogram of density (Fig. 5). The 70S ribosome should have two main density components, namely protein and RNA. In addition, the ice surrounding the structure and filling its cavities and tunnels should have a major contribution. This can be noticed on the corresponding histograms: both 1.5 µm defocus data set and CTF-corrected volume have large maximum in their histogram of mass around the zero value (in the course of the 3D reconstruction process the density of ice was arbitrarily set to zero). The main difference is the position of the second maximum, which in the case of the 1.5 µm reconstruction is hardly separated from the "ice maximum", while after CTF correction the gap between the two maxima becomes wider. The histogram of density of the third reconstruction has a shape similar that of the CTF-corrected: two maxima are separated, but the negative values are eliminated.

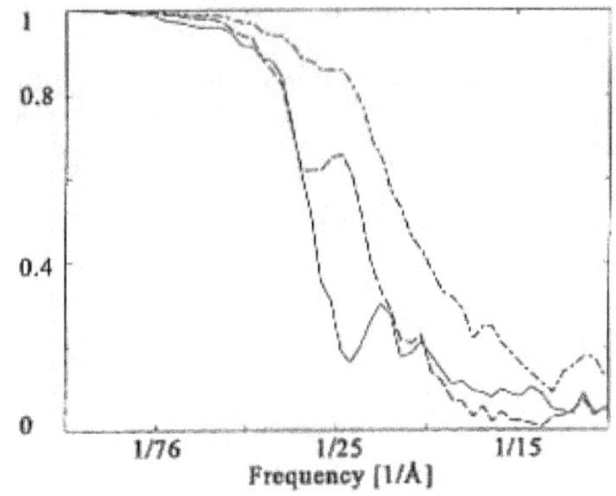

Figure 4. Resolution estimations using Fourier Shell Correlation curves for 1.5 µm (_______), CTF-corrected (-----), and CTF-corrected with positivity constraints (-·-·-) 3D reconstruction of the 70S *E. coli* ribosome.

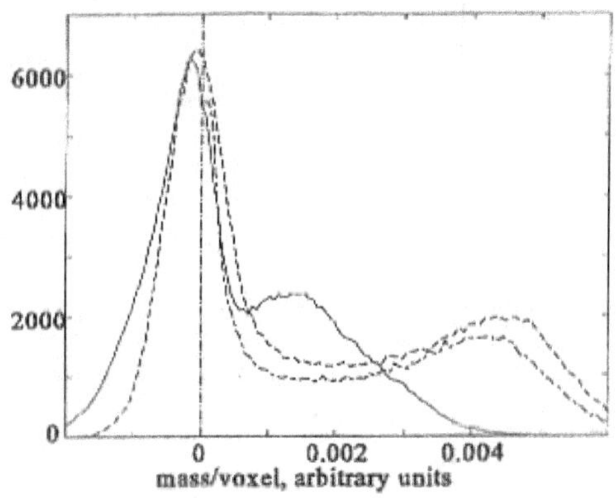

Figure 5. Histogram of mass distribution for 1.5 µm (_______), CTF-corrected (-----), and CTF-corrected with positivity constraints (-·-·-) 3D reconstruction of the 70S *E. coli* ribosome.

Conclusions

The application of the 3D reconstruction with CTF compensation procedure to the energy filtered EM images of 70S *E. coli* ribosome proves that this strategy leads to the significant increase in the resolution (as measured by FSC curves). Due to the availability of the reference volume and application of the 3D projection alignment procedure the results could be obtained in a relatively short time. The FSC curves agree with the resolution limits set by the parameters of CTFs, as estimated from the micrographs using a newly designed retrieval procedure. The particular choice of the defocus settings in the collected defocus series (1.5 µm, 2.0 µm and 2.5 µm) proved to be the most important factor in the limiting the resolution of the merged, CTF-corrected reconstruction. Further experiments with other choices of defocus settings are necessary to establish the practical limit to the resolution that can be obtained using energy filtered EM data.

Acknowledgements

This work was supported, in part, by grant NIH 1R01 GM 29169. We thank Bob Grassucci for assistance with electron microscopy.

References

1. Andrews HC, Hunt BR (1977) Digital Image Restoration. Prentice-Hall, Englewood Cliffs, NJ.

2. Biemond J, Lagendijk RL, Mersereau RM (1990) Iterative methods for image deblurring. Proc IEEE **78**: 856-883.

3. Downing KH, Grano DA (1982) Analysis of photographic emulsions for electron microscopy of two-dimensional crystalline specimen. Ultramicroscopy **7**: 381-404.

4. Erickson HP, Klug A (1970) The Fourier transform of an electron micrograph: effects of defocusing and aberrations, and implications for the use of underfocus contrast enhancement. Ber Bunsenges Phys Chem **74**: 1129-1137.

5. Frank J (1969) Nachweis von Objektbewegungen im lichtoptischen Diffraktogramm von elektronenmikroskopischen Aufnahmen (Demonstration of movement of objects in the light optical diffractogram of electron microscopic images). Optik **30**: 171-180.

6. Frank J (1972) A study on heavy/light atom discrimination in bright field electron microscopy using the computer. Biophys J **12**: 484-511.

7. Frank J (1973) The envelope of electron microscopic transfer functions for partially coherent illumination. Optik **38**: 519-536.

8. Frank J, Penczek P, Grassucci R, Srivastava S (1991) Three-dimensional reconstruction of the 70S *Escherichia coli* ribosome in ice - the distribution of ribosomal RNA. J Cell Biol **115**: 597-605.

9. Frank J, Zhu J, Penczek P, Li Y, Srivastava S, Verschoor A, Grassucci R, Lata RK, Agrawal RK (1995) A model of protein synthesis based on cryo-electron microscopy of the *E. coli* ribosome. Nature **376**: 441-444.

10. Frank J, Penczek P (1995) On the correction of the contrast function in biological electron microscopy. Optik **98**: 125-129.

11. Hanszen KH, Trepte L (1971) The contrast transfer function of the electron microscope with partial coherent illumination. Part A: the ring condensor. Optik **33**: 166-181.

12. Hanszen KH, Trepte L (1971) The contrast transfer function of the electron microscope with partial coherent illumination. Part B: disk shape source. Optik **33**: 182-198.

13. Henderson R, Glaeser RM (1985) Quantitative analysis of image contrast in electron micrographs of beam-sensitive crystals. Ultramicroscopy **16**: 139-150.

14. Langmore JP, Smith MF (1992) Quantitative energy-filtered electron microscopy of biological molecules in ice. Ultramicroscopy **46**: 349-373.

15. Lenz F (1971) Transfer of image information in the electron microscope. In: Electron Microscopy in Material Science. Valdrè U (ed). Academic Press, New York. pp 540-568.

16. Lepault J, Booy FP, Dubochet J (1983) Electron microscopy of frozen biological suspensions. J Microsc **129**: 89-102.

17. Penczek P, Radermacher M, Frank J (1992) Three-dimensional reconstruction of single particles embedded in ice. Ultramicroscopy **40**: 33-53.

18. Penczek PA, Grassucci RA, Frank J (1994) The ribosome at improved resolution: new techniques for merging and orientation refinement in 3D cryo-electron microscopy of biological particles. Ultramicroscopy **53**: 251-270.

19. Pratt WK (1991) Digital Image Processing. John Wiley & Sons, Inc., New York.

20. Press WH, Teukolsky SA, Vetterling WT, Flannery BP (1992) Numerical Recipes. Cambridge University Press, Cambridge.

21. Radermacher M, Wagenknecht T, Verschoor A, Frank J (1987) Three-dimensional reconstruction from a single-exposure, random conical tilt series applied to the 50S ribosomal subunit of *Escherichia coli*. J Microsc **146**: 113-136.

22. Reimer L, Ross-Messemer M (1990) Contrast in the electron spectroscopic imaging mode of a TEM. J Microsc **159**: 143-160.

23. Saxton WO, Baumeister W (1982) The correlation averaging of a regularly arranged bacterial envelope protein. J Microsc **127**: 127-138.

24. Schiske P (1973) Image processing using additional statistical information about the object. In: Image Processing and Computer-Aided Design in Electron Optics. Hawkes PW (ed). Academic Press, London. pp 82-90.

25. Schröder RR, Manstein DJ, Jahn W, Holden H, Rayment I, Holmes KC, Spudich JA (1993) Three-dimensional atomic model of F-actin decorated with *Dictyostelium* myosin S1. Nature **364**: 171-174.

26. Smith MF, Langmore JP (1992) Quantitation of molecular densities by cryo-electron microscopy. Determination of the radial density distribution of tobacco mosaic virus. J Mol Biol **226**: 763-774.

27. Trussel HJ (1984) A priori knowledge in algebraic reconstruction methods. In: Image Reconstruction in Incomplete Observations (Huang TS, ed). Jai Press, London.

28. Typke D, Hegerl R, Kleinz J (1992) Image restoration for biological objects using external TEM control and electronic image recording. Ultramicroscopy **46**: 157-173.

29. van Heel M (1987) Angular reconstitution: a posteriori assignment of projection directions for 3D reconstruction. Ultramicroscopy **21**: 111-124.

30. Wade RH (1992) A brief look at imaging and contrast transfer. Ultramicroscopy **46**: 145-156.

31. Wade RH, Frank J (1977) Electron microscope transfer functions for partially coherent axial illumination and chromatic defocus spread. Optik **49**: 81-92.

32. Wagenknecht T, Grassucci R, Frank J (1988) Electron microscopy and computer image averaging of ice-embedded large ribosomal subunits from *Escherichia coli.* J Mol Biol **199**: 137-147.

33. Wagenknecht T (1992) Three-dimensional reconstruction of noncrystalline macromolecular assemblies. In: Electron Tomography. Frank J (ed). Plenum, New York. pp. 359-389.

34. Zhu J, Penczek PA, Schröder RR, Frank J (1997) Three-dimensional reconstruction with CTF correction of the 70S *Escherichia coli* ribosome using energy-filtered cryo-electron microscopy. J Struct Biol **118**: 197-219.

Proc. Natl. Acad. Sci. USA
Vol. 95, pp. 6134–6138, May 1998
Cell Biology

Visualization of elongation factor G on the *Escherichia coli* 70S ribosome: The mechanism of translocation

RAJENDRA K. AGRAWAL*, PAWEL PENCZEK*†, ROBERT A. GRASSUCCI*, AND JOACHIM FRANK*†‡

*Wadsworth Center, New York State Department of Health and †Department of Biomedical Sciences, State University of New York at Albany, Empire State Plaza, Albany, NY 12201-0509

Communicated by Peter B. Moore, Yale University, New Haven, CT, March 30, 1998 (received for review March 11, 1998)

ABSTRACT **During protein synthesis, elongation factor G (EF-G) binds to the ribosome and promotes the step of translocation, a process in which tRNA moves from the A to the P site of the ribosome and the mRNA is advanced by one codon. By using three-dimensional cryo-electron microscopy, we have visualized EF-G in a ribosome–EF-G–GDP–fusidic acid complex. Fitting the crystal structure of EF-G–GDP into the cryo density map reveals a large conformational change mainly associated with domain IV, the domain that mimics the shape of the anticodon arm of the tRNA in the structurally homologous ternary complex of Phe-tRNA^{Phe}, EF-Tu, and a GTP analog. The tip portion of this domain is found in a position that overlaps the anticodon arm of the A-site tRNA, whose position in the ribosome is known from a study of the pretranslocational complex, implying that EF-G displaces the A-site tRNA to the P site by physical interaction with the anticodon arm.**

The elongation cycle of prokaryotic protein synthesis on the ribosome involves two elongation factors (EFs), EF-Tu and EF-G. EF-Tu delivers aminoacyl tRNA to the A site of the elongating ribosome, as part of the aminoacyl-tRNA–EF-Tu–GTP ternary complex, and EF-G promotes the translocation step, in which the A- and P-site tRNAs move to the P and E sites, respectively, and mRNA is advanced by one codon. Both EF-related events are accompanied by GTP hydrolysis. Furthermore, the intrinsic GTPase activity of both EF-Tu–GTP and EF-G–GTP is stimulated by the empty ribosome (1, 2) and by the tRNA-bound ribosome (3, 4). Whether the energy liberated in GTP hydrolysis is used directly for tRNA binding and translocation or for the release of the factors is not well understood. However, in a recent study, it has been proposed that it is the hydrolysis of GTP by EF-G that drives the translocation step (5).

In the present study, we present the three-dimensional (3D) binding position of the EF-G on the *Escherichia coli* 70S ribosome. Earlier immunoelectron microscopic (6) and chemical cross-linking (7) studies suggested that EF-G interacts with both subunits (30S and 50S) of the ribosome. These studies also suggested the tentative binding position of the EF-G on the ribosome. We have obtained a 3D cryo-electron microscopy map of the ribosome–EF-G–GDP–fusidic acid complex at 20-Å resolution and inferred the precise location of EF-G on the ribosome. To stablilize the ribosome–EF-G complex for cryo-electron microscopy, we used fusidic acid, an inhibitor of the translational elongation cycle that binds to the ribosome–EF-G–GDP complex and prevents the dissociation of EF-G from the ribosome (8). Because of its high occupancy, EF-G is directly visible in our 3D cryo map. A detailed comparison of the cryo electron microscopy mass attributable to EF-G with the crystal structure of EF-G–GDP (9, 10) reveals the 3D interaction sites between the ribosome and the various domains of the EF-G. Most interesting is the observation that the position of domain IV of the EF-G overlaps with the anticodon arm of the A-site tRNA (11, 12).

MATERIALS AND METHODS

Tight-coupled 70S ribosomes from *E. coli* MRE 600 and EF-G from its 100,000 × *g* supernatant were isolated as described by Burma *et al.* (13), except that the EF-G preparation was subjected to one additional step of purification by size-exclusion chromatography on Sephacryl S-200 (Sigma). The homogeneity of EF-G preparation was checked by PAGE. Both ribosomes and EF-G preparations were dialyzed in buffer containing 20 mM Tris·HCl (pH 7.5), 6 mM $MgCl_2$, 150 mM NH_4Cl, 2 mM spermidine, 0.4 mM spermine, and 5 mM 2-mercaptoethanol. The ribosome–EF-G–GDP–fusidic acid complex was prepared in 100 μl of the same buffer by incubating activated 70S ribosomes (0.032 μM) with EF-G (0.8 μM) and GTP (15 μM) at 37°C for 7 min and continuing the reaction for another 5 min after the addition of fusidic acid (0.5 mM). The complex was immediately applied to the electron microscope grid for cryo grid preparation and electron microscopy.

Micrographs were recorded as two defocus groups, at 1.67 μm and 1.00 μm, by using the low-dose protocol (10 $e^-/Å^2$) on a Philips EM420, equipped with low-dose kit and a GATAN (model 626) cryo transfer holder, at a magnification of ×52,200 (±2%) as checked by a tobacco mosaic virus standard. Micrographs were checked for drift, astigmatism, and the presence of Thon rings by optical diffraction and scanned on a Perkin–Elmer PDS 1010A microdensitometer with a step size of 25 μm corresponding to 4.78 Å on the object scale. For 3D reconstruction, a total of 3,237 ribosomes in the 1.67-μm group and 4,326 ribosomes in the 1.00-μm group were selected from a total of 25 micrographs by an automated selection procedure (14), and particle candidates were directly compared with the reference set of 87 quasi-evenly spaced projections (15) of an existing best-resolution reconstruction. For each defocus group, one step of the 3D projection alignment procedure (15) was applied and the merged contrast transfer function (CTF)-corrected structure was computed as described (16). Starting each time with the CTF-modified reconstruction of the previous step, this refinement was repeated five times with a 2.0° angular step. The final resolution was 20 Å as estimated with the Fourier shell correlation with a cutoff value of 0.5 (17). The occupancy of EF-G was estimated by comparing the average densities within L1 protein regions with the region corresponding to EF-G. By using as a reference the known density of the surrounding ice, the EF-G occupancy was calculated as ≈96%. The region corresponding to EF-G was

The publication costs of this article were defrayed in part by page charge payment. This article must therefore be hereby marked "*advertisement*" in accordance with 18 U.S.C. §1734 solely to indicate this fact.

© 1998 by The National Academy of Sciences 0027-8424/98/956134-5$2.00/0
PNAS is available online at http://www.pnas.org.

Abbreviations: EF, elongation factor; 3D, three-dimensional.

‡To whom reprint requests should be addressed. e-mail: joachim@wadsworth.org.

separated by using the envelope of the 70S ribosome without EF-G.

To check the reproducibility of the reconstruction, we made an attempt to compute two maximally dissimilar structures derived from the same data set. We divided the available set of ribosome images into halves based on the values of the cross-correlation coefficient between the experimental images and projections of the reconstruction. For the two subsets we calculated the corresponding 3D volumes, which served as seeds in an iterative procedure in which the ribosome projections were dynamically reassigned to one of two possible classes based on their similarity to the updated volumes. After reaching a stable projection assignment the two final volumes were compared and it was verified that the density mass interpreted as domain IV occupied the same position, within an angstrom. The bias in these two reconstructions revealed itself in a number of minor differences in the overall density distribution, but none have any impact on the conclusions about the domain's positions.

RESULTS AND DISCUSSION

Because of the high (>95%) occupancy of the EF-G, it is directly visible in our 3D cryo map (Fig. 1). Five sites of interaction with the ribosome are observed: three on the 50S subunit, all located on the base of the L7/L12 stalk, and two sites on the 30S subunit, at the junction of the platform and neck and in a region on the lower body. Most notable is the binding position of an arm-like structure in the neck region, at the origin of the platform, near the anticodon binding region of the A-site tRNA (11, 12).

Comparison of the x-ray structure of the EF-G–GDP complex of *Thermus thermophilus* (9, 10), which consists of five domains (numbered I–V: domain I is presented as two subdomains, the GTP binding core G domain and an insert G′ domain; see ref. 10), to the density mass attributable to EF-G shows a close match in overall size and shape (Fig. 2) and reveals a significant conformational change of EF-G in the ribosome-bound state. Domain IV, in conjunction with domains III and V, represents a substructure that mimics the tRNA portion of the ternary complex (19). By itself, domain IV with its elongate shape mimics the anticodon arm of tRNA (19).

We took three approaches in an attempt to obtain the best fit of the x-ray structure (9) into the EF-G density map: (*i*) by aligning the x-ray structure into the density map to achieve a best overall fit (because the EF-G visible in the density map is elongated compared with its x-ray structure, this approach gives a poor fit, with a cross-correlation value of 0.72); (*ii*) by using domains I and II of the x-ray structure as the main guide, which leads to serious misalignment in the arm-like region associated with domain IV (Fig. 2*A*) and decreases the cross-correlation value to 0.67; and (*iii*) by keeping domains I and II at the same position, as obtained from the second approach, but moving an entire substructure formed by domains III–V (Fig. 2*B*). This third approach yields a cross-correlation value of 0.81 and thus provides the best explanation of the density map. Although the structure of domain III is only partially defined by x-ray crystallography (see ref. 20), the structure obtained by modeling (9) fits very well into our EF-G density map (Fig. 2*B*).

Sequence analysis of EF-G from fusidic acid-resistant mutants (21) shows that amino acid residues responsible for fusidic acid binding are present in the GTP binding core G domain, as well as in domains III and V. The variant of the EF-G–GDP structure obtained by the joint fitting of domains III–V into the cryo map (Fig. 2*B*) may represent a transient conformation of the ribosome-bound factor that follows GTP hydrolysis and is prevented from assuming the conformation of the unbound factor by the binding of fusidic acid. It is also

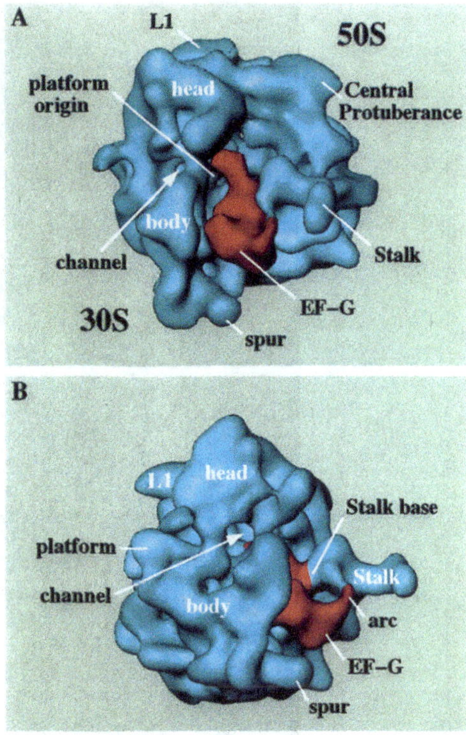

FIG. 1. The 3D cryo electron microscope maps of *E. coli* 70S ribosome complexed with EF-G–GDP–fusidic acid. (*A*) Side view showing the binding position of EF-G (red) in the intersubunit space. The 30S subunit is on the left and 50S subunit is on the right. EF-G is directly visible and seen to interact with both 30S and 50S subunits (6, 7). The tip of the elongated portion of EF-G is situated in the neck region of 30S subunit, where the anticodon loop region of A site tRNA binds (11, 12). (*B*) The 30S side view, showing the arc-like structure formed between the EF-G and a protusion at the base of the L7/L12 stalk. Note the clearly defined well-extended L7/L12 stalk. In this view, the tip (domain IV) of EF-G can be seen through the 30S channel (18) where the anticodon of A site tRNA in a pretranslocational complex binds (ref 12; R.K.A., P.P., R.A.G., N. Burkhardt, R. Jünemann, K. H. Nierhaus, and J.F., unpublished results; also see Fig. 4). Landmarks: L1, protein L1 of the 50S subunit; arc, arc-like connection between EF-G and an extension from the stalk base (see text).

possible that the crystal structure, because of involvement of crystal packing forces, represents a nonphysiological conformation of EF-G (9). The structure of the GTP form of EF-G and the way it might be affected by GTP hydrolysis in the ribosome-bound state are unknown. However, a conformational change corresponding to that seen between the GTP and GDP forms of EF-Tu (22) does not appear to take place in the case of unbound EF-G (20).

The fit between the x-ray structure and the density map attributable to EF-G (Fig. 2*B*) shows that the tip of domain IV makes contact with the inner edge of the origin of the platform in the immediate vicinity of the proposed mRNA channel (18) and in close proximity to the head of the 30S subunit. This is the region where the anticodon arm of the A site tRNA is situated (11, 12). EF-G is known to form a cross-link with the S19 protein (7), which is mapped in the head region of the 30S subunit (23), and it is likely that this cross-link involves domain IV. Domain II also makes contact with the 30S subunit (Figs. 3*A* and 4), in the vicinity of proteins S4 and S12 (23), of which the S12 protein is known to form a cross-link with EF-G (7, 25).

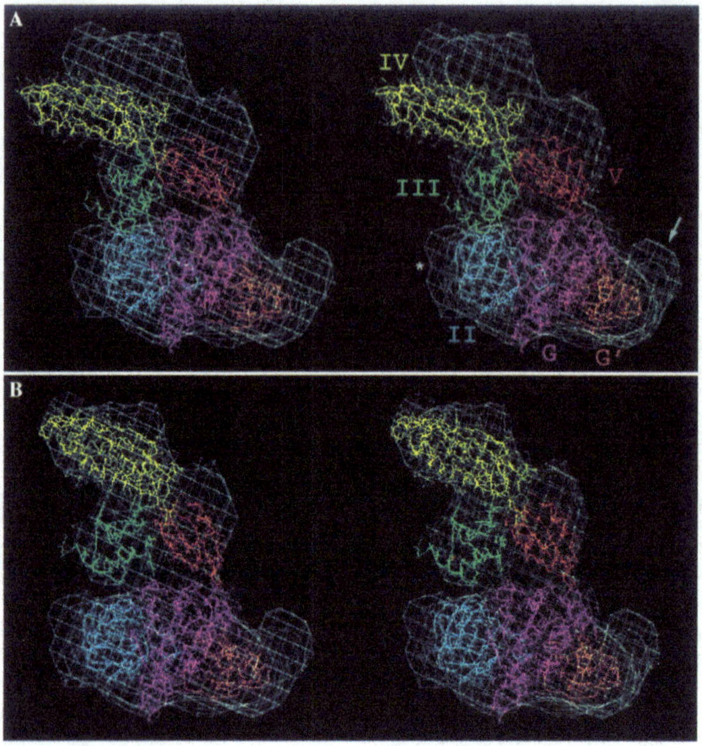

FIG. 2. Stereo view showing the fitting of the crystal structure (9) of EF-G–GDP into the extra mass of density corresponding to EF-G. The density map corresponding to EF-G is shown in white wire-mesh form. The various domains of the crystal structure are shown in different colors. (*A*) Fitting using domains I and II as the main guide to align the whole crystal structure into the density map (see text). (*B*) Separate fitting of the substructure formed by domains III–V (see text). To obtain an optimum fit, this structure was shifted by approximately 10 Å toward the central protuberance of the 50S subunit and rotated by about 10°, around a pivoting point in the contact region between domains II and III, toward the L1 side, whereas domains I and II were left in the same position as in *A*. The arrow points to the portion of the density map that is partially occupied by the G′ domain and involved in making the arc-like connection with the L7/L12 stalk (see also Figs. 1*B* and 3*B*). The asterisk shows a partially unoccupied region that probably comes from the ribosome because of a conformational change at the site of interaction with domain II.

Domain III is in close proximity to a region below the channel and the origin of the platform. Domain II of both EF-G and EF-Tu in the ternary complex appears to make contact with the same region of the 30S subunit (see ref. 24).

The EF-G contact with the 50S subunit, in the region of the L7/L12 stalk base (see Fig. 1*A*), is made through its domains I (core G and insert G′) and V (Figs. 3*A* and 4). The core G domain of EF-G is a structural homolog of the G domain of EF-Tu, which is a conserved feature of the GTPase superfamily (see refs. 9, 10, and 19), whereas the insert portion of domain I of EF-G has no counterpart in EF-Tu (9, 10, 19). The fitting of the x-ray structure reveals that the GDP-binding region of the G domain (9) touches a point at the lower part of the stalk base, whereas the G′ domain connects with a protrusion closer to the stalk, along a thin arc-like structure (Figs. 1*B* and 3*B*). A similar connection seen in the ternary complex was interpreted as a bridge between the C-terminal end of the L7/L12 stalk and the G domain of EF-Tu (24). However, because the stalk is composed of four copies of L7/L12 protein (25), and in our map it is a well-extended separate feature, whose C-terminal domains are located at its tip (25), the arc connection probably involves the C-terminal domain of one of the folded L7/L12 molecules (26).

Domain V, which is important for ribosome binding of EF-G (27), touches and fits into a structurally complementary groove region at the base of the stalk (Figs. 3 and 4). In addition to protein L7/L12, EF-G is known to form cross-links with proteins L6, L14, and L31 (7, 25) present at the base of the stalk. Because the stalk is highly mobile (25, 28) and implicated in both factor binding and translocation (25, 28, 29), its appearance as a well-delineated structure at the moment frozen by fusidic acid binding could indicate that its movement is precisely controlled and choreographed in the course of the translocation process.

The 2,660 loop, one of the universally conserved nucleotide regions of 23S rRNA of the 50S subunit, is the binding site of ribotoxins, α-sarcin and ricin, which abolish all EF-dependent functions of both prokaryotic (30) and eukaryotic (31) ribosomes. Several lines of evidence indicate that this 23S RNA region, along with nucleotides 1,052–1,112 of 23S RNA, of which nucleotides 1,054–1,081 have been cross-linked to EF-G (32), are present near the base of the L7/L12 stalk. Both EF-G and EF-Tu protect some common nucleotides in the 2,660-loop region from chemical modification, whereas EF-G protects additional nucleotides at positions 1,067 (the binding site of the antibiotic thiostrepton) and 1,069 of 23S RNA (33). Protein L11, which is present at the base of the L7/L12 stalk, binds to the region of positions 1,052–1,112 (34) and the pentameric protein complex L10-$(L7/L12)_2$ binds to up- and downstream neighboring regions, positions 1,113–1,124 and 1,028–1,051, of the 23S RNA (35). These results imply that (*i*) the 2,660 region of the 23S RNA is the site where the structurally homologous G domains of both EF-Tu and EF-G interact and (*ii*) the 1,052–1,112 region of the 23S RNA is

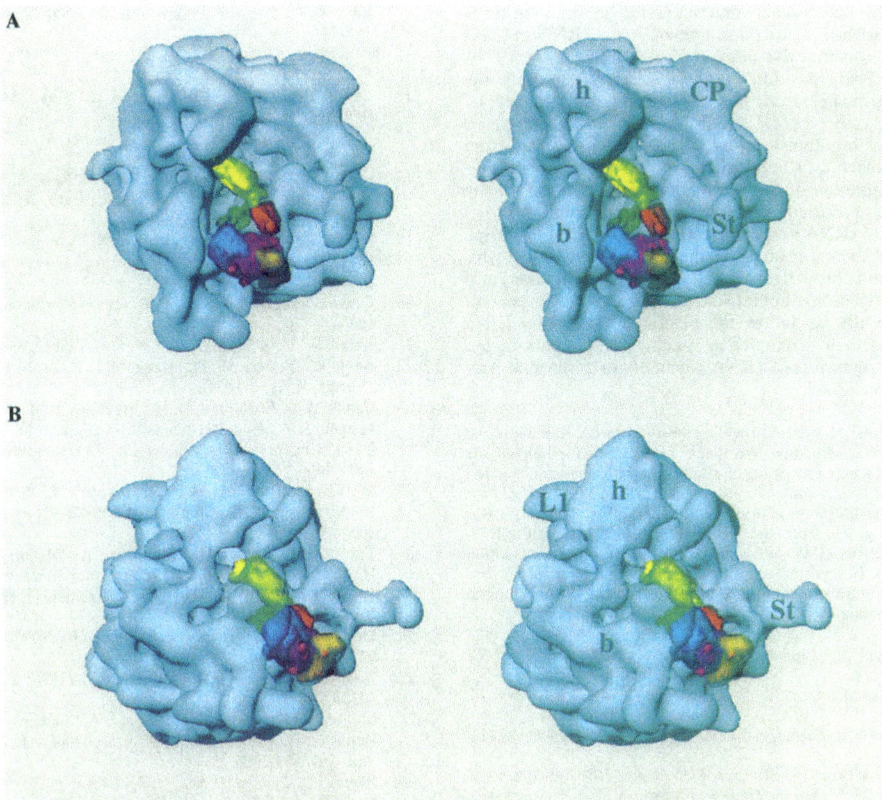

FIG. 3. Stereo view of a low-resolution version of the EF-G x-ray structure shown in Fig. 2*B*, superimposed onto the 3D map of the 70S–EF-G–GDP–fusidic acid complex (shown as a transparent blue surface). Color coding for different domains of the x-ray structure is the same as in Fig. 2: domain I, magenta and brown (core G domain, magenta; insert G′ domain, brown); domain II, blue; domain III, green; domain IV, yellow; and domain V, red. (*A*) In the intersubunit face view, the 30S subunit is on the left and the 50S subunit on the right. (*B*) Same as *a* but presented from the 30S-solvent side to show that the G′ domain is involved in the formation of the arc-like structure with the base of the L7/L12 stalk. Because only the lower part of the arc-like structure is occupied by the x-ray structure, the upper part is probably contributed by the C-terminal domain of one of the folded L7/L12 molecule (see text), which becomes pronounced only upon factor binding (see also ref. 24). Landmarks have been abbreviated to avoid visual complexity of the stereo picture. For 30S subunit: h, head; b, body. For 50S subunit: CP, central protuberance; St, L7/L12 Stalk; L1, L1 protein.

located at a point closer to the stalk, probably where domain V of EF-G interacts with the 50S subunit.

The multiple interactions between the L7/L12 stalk base and domains I and V (Figs. 3 and 4) suggest a complex mechanism of translocation in which these domains may act in concert with the flexible stalk. The stalk may act as an intermediary (see refs. 25, 28, and 29) and effect the movement of domain IV through domain V, which fits into a groove-like

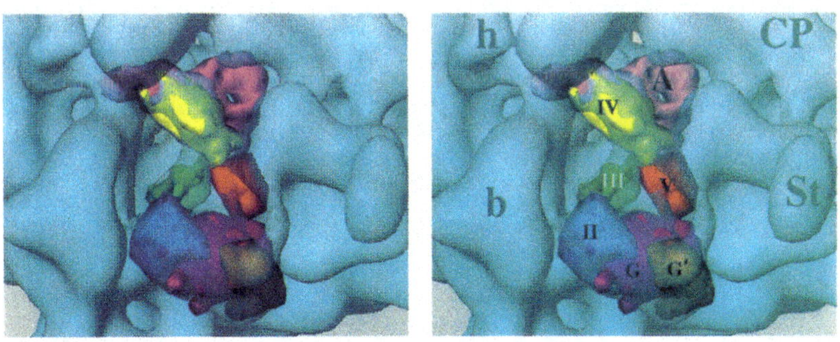

FIG. 4. Same as Fig. 3*A* but showing a magnified view of the site of interaction of different domains of EF-G with the ribosome. Particularly important to note is the partial overlap of domain IV with the A-site tRNA (pink), as obtained from a cryo map of a pretranslocational complex (R.K.A., P.P., R.A.G., N. Burkhardt, R. Jünemann, K. H. Nierhaus, and J.F., unpublished results). Also, note the proximity of the anticodon end of the A site tRNA to the tip of the domain IV. Various domains are labeled with roman numerals and the A-site tRNA is marked as A. Other landmarks are as in Fig. 3.

structure at its base. Moreover, the overlap between the tip of domain IV with the anticodon arm of A-site tRNA (11, 12) (Fig. 4) suggests that this domain displaces the A-site tRNA toward the P-site position by physical interaction with the anticodon arm, i.e., from the L7/L12-stalk side to the L1 protein side (see Figs. *1B* and *3C* for structural reference), thereby effecting translocation, as is also suggested from recent nonstructural studies (5, 36).

After completion of this study, the EF-G position on the ribosome was reported, as derived by determining the proximal regions of rRNA to specific amino acids of EF-G by using the hydroxyl radical probing method (37). Even though this study does not throw light on the conformational change of EF-G in the ribosome-bound state, it is in striking agreement with our results, as far as the positions of various EF-G domains and their proximity to specific components of the ribosomal structure (i.e., rRNA segments or ribosomal proteins) are concerned.

We thank Peter Moore for providing the all-atom coordinates of the EF-G–GDP x-ray structure. We thank Yiwei Chen for help in the initial stage of image processing and Amy Heagle for assistance in the preparation of illustrations. We also thank Terence Wagenknecht for critical reading of the manuscript and suggestions. This work was supported by grants from the National Institutes of Health (R37-GM29169 and R01-GM55440) and the National Science Foundation (BIR9219043). Computing support was kindly provided by the National Center for Supercomputer Applications, University of Illinois at Urbana–Champaign.

1. Nishizuka, Y. & Lipmann, F. (1966) *Proc. Natl. Acad. Sci. USA* **55,** 212–219.
2. Kawakita, M., Arai, K. & Kaziro, Y. (1974) *J. Biochem.* **76,** 801–809.
3. Chinali, G. & Parmeggiani, A. (1982) *Eur. J. Biochem.* **125,** 415–421.
4. Voigt, J. & Nagel, K. (1993) *J. Biol. Chem.* **268,** 100–106.
5. Rodnina, M. V., Savelsbergh, A., Katunin, V. I. & Wintermeyer, W. (1997) *Nature (London)* **385,** 37–41.
6. Girshovich, A. S., Kurtskhalia, T. V., Ovchinikov, Y. A. & Vasiliev, V. D. (1981) *FEBS Lett.* **130,** 54–59.
7. Sköld, S. E. (1982) *Eur. J. Biochem.* **127,** 225–229.
8. Willie, G. R., Richman, N., Godtfredson, W. O. & Bodley, J. W. (1975) *Biochemistry* **14,** 1713–1718.
9. Czworkowski, J., Wang, J., Steitz, T. A. & Moore, P. B. (1994) *EMBO J.* **13,** 3661–3668.
10. Al-karadaghi, S., Ævarsson, A., Garber, M., Zheltonosova, J. & Liljas, A. (1996) *Structure* **4,** 555–565.
11. Agrawal, R. K., Penczek, P., Grassucci, R. A., Li, Y., Leith, A., Nierhaus, K. H. & Frank, J. (1996) *Science* **271,** 1000–1002.
12. Stark, H., Orlova, E. V., Rinke-Appel, J., Jünke, N., Mueller, F., Rodnina, M., Wintermeyer, W., Brimacombe, R. & van Heel, M. (1997) *Cell* **88,** 19–28.
13. Burma, D. P., Srivastava, A. K., Srivastava, S. & Dash, D. (1985) *J. Biol. Chem.* **260,** 10517–10525.
14. Lata, K. R., Penczek, P. & Frank, J. (1995) *Ultramicroscopy* **58,** 381–391.
15. Penczek, P., Grassucci, R. A. & Frank, J. (1994) *Ultramicroscopy* **53,** 251–270.
16. Zhu, J., Penczek, P. A., Schröder, R. & Frank, J. (1997) *J. Struct. Biol.* **118,** 197–219.
17. Böttcher, B., Wynne, S. A. & Crowther, R. A. (1997) *Nature (London)* **386,** 88–91.
18. Frank, J., Zhu, J., Penczek, P., Li, Y., Srivastava, S., Verschoor, A., Radermacher, M., Grassucci, R., Lata, R. K. & Agrawal, R. K. (1995) *Nature (London)* **376,** 441–444.
19. Nissen, P., Kjeldgaard, M., Thirup, S., Polekhina, G., Reshetnikova, L., Clark, B. F. C. & Nyborg, J. (1995) *Science* **270,** 1464–1472.
20. Czworkowski, J. & Moore, P. B. (1997) *Biochemistry* **36,** 10327–10334.
21. Johanson, U. & Hughes, D. (1994) *Gene* **143,** 55–59.
22. Abel, K., Yoder, M. D., Hilgenfeld, R. & Jurnak, F. (1996) *Structure* **4,** 1153–1159.
23. Capel, M. S., Engleman, D. M., Freeborn, B. R., Kjeldgaard, M., Langer, J. A., Ramakrishnan, V., Schindler, D. G., Schneider, D. K., Schoenborn, B. P., Sillers, I. -Y., *et al.* (1987) *Science* **238,** 1403–1406.
24. Stark, H., Rodnina, M., Rinke-Appel, J., Brimacombe, R., Wintermeyer, W. & van Heel, M. (1997) *Nature (London)* **389,** 403–406.
25. Traut, R. R., Tewari, D. S., Sommer, A., Govino, G. R., Olson, H. M. & Glitz, D. G. (1986) in *Structure Function and Genetics of Ribosomes*, eds. Hardesty, B. & Kramer, G. (Springer, New York), pp. 286–308.
26. Dey, D., Oleinikov, A. V. & Traut, R. R. (1995) *Biochimie* **77,** 925–930.
27. Hou, Y., Yaskowiak, E. S. & March, P. E. (1994) *J. Bacteriol.* **176,** 7038–7044.
28. Möller, W., Schrier, P. I., Maassen, J. A., Zantema, A., Schop, E., Reinalda, H., Cremers, A. F. M. & Mellema, J. E. (1983) *J. Mol. Biol.* **163,** 553–573.
29. Burma, D. P., Srivastava, S., Srivastava, A. K., Mahanti, S. & Dash, D. (1986) in *Structure Function and Genetics of Ribosomes*, eds. Hardesty, B. & Kramer, G. (Springer, New York), pp. 438–453.
30. Hausner, T.-P., Atmadja, J. & Nierhaus, K. H. (1987) *Biochimie* **169,** 911–923.
31. Wool, I. G., Gluck, A. & Endo, Y. (1992) *Trends Biochem. Sci.* **17,** 266–269.
32. Sköld, S. E. (1983) *Nucleic Acids Res.* **11,** 4923–4932.
33. Moazed, D., Robertson, J. M. & Noller, H. F. (1988) *Nature (London)* **334,** 362–364.
34. Schmidt, F. J., Thompson, J., Lee, K., Dijk, J. & Cundliffe, E. (1981) *J. Biol. Chem.* **266,** 12301–12305.
35. Beauclerk, A. A. D., Cundliffe, E. & Dijk, J. (1984) *J. Biol. Chem.* **259,** 6559–6563.
36. Borowski, C., Rodnina, M. V. & Wintermeyer, W. (1996) *Proc. Natl. Acad. Sci. USA* **93,** 4202–4206.
37. Wilson, K. S. & Noller, H. F. (1998) *Cell* **92,** 131–139.

Cell, Vol. 93, 1105–1115, June 26, 1998, Copyright ©1998 by Cell Press

A 9 Å Resolution X-Ray Crystallographic Map of the Large Ribosomal Subunit

Nenad Ban,[1] Betty Freeborn,[2] Poul Nissen,[1] Pawel Penczek,[4,5] Robert A. Grassucci,[4] Robert Sweet,[6] Joachim Frank,[4,5] Peter B. Moore,[1,2] and Thomas A. Steitz[1,2,3,7]
[1] Department of Molecular Biophysics and Biochemistry
[2] Department of Chemistry
[3] Howard Hughes Medical Institute
Yale University
New Haven, Connecticut 06520-8114
[4] Wadsworth Center
New York State Department of Health
Albany, New York 12201-0509
[5] Department of Biomedical Sciences
State University of New York
Albany, New York 12201-0509
[6] Department of Biology
Brookhaven National Laboratory
Upton, New York 11973

Summary

The 50S subunit of the ribosome catalyzes the peptidyl-transferase reaction of protein synthesis. We have generated X-ray crystallographic electron density maps of the large ribosomal subunit from *Haloarcula marismortui* at various resolutions up to 9 Å using data from crystals that diffract to 3 Å. Positioning a 20 Å resolution EM image of these particles in the crystal lattice produced phases accurate enough to locate the bound heavy atoms in three derivatives using difference Fourier maps, thus demonstrating the correctness of the EM model and its placement in the unit cell. At 20 Å resolution, the X-ray map is similar to the EM map; however, at 9 Å it reveals long, continuous, but branched features whose shape, diameter, and right-handed twist are consistent with segments of double-helical RNA that crisscross the subunit.

Introduction

In all organisms, messenger-directed protein synthesis is catalyzed by ribonucleoprotein particles called ribosomes. Bacterial ribosomes sediment at about 70S and consist of two nonequivalent subunits: a small, 30S subunit and a large, 50S subunit. The latter has a molecular mass of about 1.5×10^6 Da and catalyzes peptide bond formation, possibly functioning as a ribozyme (Noller et al., 1992). The 50S subunit contains a 2900 nt RNA and a 120 nt RNA and about 33 different proteins (Wittmann-Liebold, 1986). If the structure of this macromolecular assembly and its substrate complexes were solved at atomic resolution, a significant increase in the understanding of the mechanism of protein synthesis would result, as well as valuable insights into RNA structure and RNA–protein interactions.

[7] To whom correspondence should be addressed.

All of the current information on the ribosome structure derives from physical and chemical methods capable of producing relatively low-resolution information. In the 1970s, the shape and quaternary organization of ribosomes were illuminated by electron microscopy (EM) (Oakes et al., 1986; Stoeffler and Stoeffler-Meilicke, 1986), and by the end of the 1980s, the positions of all of the proteins in the 30S subunit and many of the proteins in the 50S subunit had been determined by neutron scattering (Capel et al., 1987; May et al., 1992). Recent advances in single-particle cryo-EM have led to three-dimensional reconstructions of the *Escherichia coli* 70S ribosome and its complexes with tRNAs and elongation factors at resolutions between 15 and 25 Å (Frank et al., 1995a, 1995b; Stark et al., 1995, 1997a, 1997b; Agrawal et al., 1996; Zhu et al., 1997; Malhotra et al., 1998). Detailed models of the conformation of 16S RNA in the small subunit have been proposed based on this kind of structural information and RNA footprinting data (Malhotra and Harvey, 1994; Noller et al., 1995; Mueller and Brimacombe, 1997a, 1997b; Mueller et al., 1997). While these models integrate much of the structural and functional information available on the ribosome, studies of other, less complicated enzymes leave little doubt that only high-resolution crystal structures of the ribosome will yield accurate models that are sufficiently detailed so that the mechanisms of processes such as substrate recognition, peptide bond formation, and translocation can be fully understood.

Ada Yonath and coworkers made a seminal contribution to the quest for a high-resolution structure of the ribosome by obtaining the first crystals of ribosomes and of ribosomal subunits, and by discovering crystals of the *Haloarcula marismortui* large subunit that diffract beyond 3 Å resolution (Yonath et al., 1980; Shevack et al., 1985; van Bohlen et al., 1991; Franceschi et al., 1993). This accomplishment established that, in principle, an atomic level structure of the ribosome can be determined using X-ray crystallography. However, the published electron density maps of the *H. marismortui* 50S subunit, which were based on X-ray diffraction data from crystals derivatized by heavy atom cluster compounds, do not show the continuous density features expected (Schlunzen et al., 1995). For this reason, we decided to pursue a strategy that depends on using EM maps to phase the X-ray diffraction data at low resolution in order to assist in the location of heavy atoms, as Harrison and coworkers did in their crystallographic studies of tomato bushy stunt virus more than 20 years ago (Jack et al., 1975).

Although many high-resolution crystal structures of spherical viruses, which are larger than the ribosome, have been determined during the last two decades, the structure of the ribosome presents a more formidable problem because it combines large size with lack of symmetry. The high symmetry of viruses helped their structure determination in two ways: the asymmetric unit of a virus crystal is only a small fraction of the entire particle; furthermore, averaging of the identical subunits in the asymmetric unit results in stunning map improvement even if the initial heavy atom phased map is poor

(Rossmann, 1995). However, averaging cannot be used when there is only a single 1.5 million molecular weight particle in the asymmetric unit. In order to obtain measurable diffraction differences from heavy atoms bound to such a large asymmetric structure, a very large number of single heavy atoms must be bound. As an approach to this problem, heavy atom cluster compounds, which scatter like super heavy atoms at very low resolution but become decreasingly effective at modestly higher resolutions, have been used to increase the size of intensity differences greatly (Ladenstein et al., 1987; O'Halloran et al., 1987; Thygesen et al., 1996; Knablein et al., 1997; Neuefeind et al., 1997). Whatever heavy atom compounds are used, the essential first step in solving the crystal structure of the ribosome is to locate correctly and model bound heavy atoms in several derivatized ribosome crystals, which we have now successfully achieved.

We present here X-ray crystallographically determined electron density maps of the *H. marismortui* large ribosomal subunit calculated at resolutions between 20 Å and 9 Å; they begin to unveil the molecular architecture that underlies this large, ribonucleoprotein complex. The X-ray maps were calculated using phases determined by multiple isomorphous replacement and anomalous scattering (MIRAS) from three heavy atom derivatives. Important to the success of this study was the molecular replacement of a 20 Å resolution EM map into the crystal cell, which provided initial phases for low-resolution reflections. Using these EM-derived phases, it was possible to locate the high-occupancy sites in heavy atom-derivatized crystals in difference Fourier maps and thus to initiate the process of heavy atom refinement. The X-ray map calculated at 20 Å resolution shares many features with the 20 Å resolution EM map, but the region believed to contain ribosomal protein L1 is oriented differently. Although protein cannot be clearly distinguished from RNA at 9 Å resolution, the map contains many continuously connected and branched densities, which are about 20 Å in diameter in many places, consistent with their representing an RNA double helix. Further, a right-handed twist arising from duplex backbone can be seen. Thus, we believe that a critical step has been taken on the path leading to an electron density map of the large ribosomal subunit that can be interpreted in atomic detail.

Results

Electron Microscopic Images

Two electron microscopic reconstructions of large ribosomal subunits were used in the course of this work. The 50S subunit reconstruction used initially was abstracted from a 23 Å resolution map of the intact 70S ribosome from *E. coli* (Frank et al., 1995a) using a lower resolution map of the *E. coli* 30S ribosomal subunit (Lata et al., 1996) as a guide to determine the location of the boundary between the 50S and 30S subunits. Subsequently, a reconstruction of isolated large ribosomal subunits from *H. marismortui* was done (see Experimental Procedures) because it was deemed more appropriate for these crystallographic studies. That reconstruction was derived from 13,170 images of individual particles taken at two levels of defocus under ionic conditions similar to those used for crystal growth. *H. marismortui* image orientations were determined initially by comparison with images computed from the *E. coli* 50S reconstruction just described, and then refined. Figure 1 shows the dependence of the Fourier shell correlation coefficient (FSC) of this reconstruction as a function of resolution (Böttcher et al., 1997). In an ideal reconstruction, the FSC would be 1.0 out to some limiting value and would then fall to 0. In this case, it begins to fall below 1.0 at about 50 Å resolution, reaches 0.5 at 20 Å resolution, and is nonzero to nearly 12 Å. Thus, while the nominal resolution of this map is 20 Å, it clearly contains useful information to higher resolution (see below).

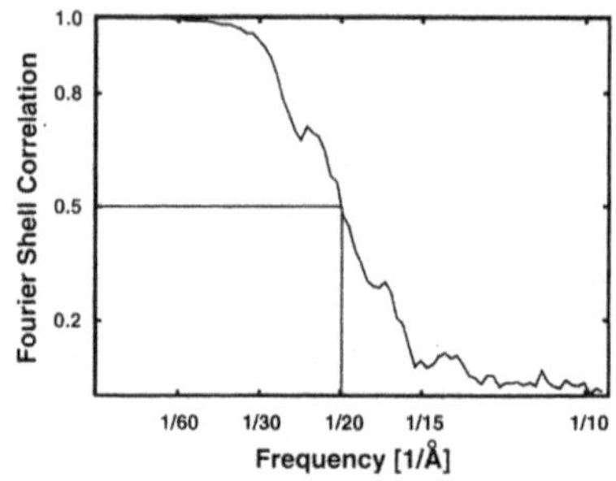

Figure 1. Resolution Estimation of EM Reconstruction of the HM 50S Subunit

Nominal resolution is 20 Å for a FSC cut-off level 0.5 (Böttcher et al., 1997). FSC: $\Sigma_R[F_1(R)\,F_2(R)]/(\Sigma_R|F_1(R)|^2\Sigma_R|F_2(R)|^2)^{1/2}$, where R is the shell radius in Fourier space; F1 and F2 are Fourier transforms of 3D structures calculated from two randomly divided subsets of the available EM projection data. The summation is carried over shells in Fourier space.

Molecular Replacement

Using modified forms of the molecular replacement procedures implemented by the computer programs X-PLOR and CNS, EM reconstructions of the large ribosomal subunit were positioned in the crystal unit cell with good packing. The crystals of *Haloarcula marismortui* large ribosomal subunits (HM 50S) used in this study were grown following procedures described by Yonath and coworkers (van Bohlen et al., 1991). They are orthorhombic, space group $C222_1$, and with unit cell dimensions of: a = 210, b = 300, c = 570 Å. Although these crystals diffract to better than 3.5 Å resolution, the data used for this study were measured to 7.0 Å resolution at Brookhaven National Laboratory (Table 1).

After rigid body refinement, the best molecular replacement solution found for EC 50S had an R factor of 0.45 and a correlation coefficient of 0.43 for reflections between 60 and 30 Å resolution. The resulting unit cell packing shows no unacceptable contacts or overlaps between subunits. Subsequently, when the HM 50S EM reconstruction became available, the rotation and translation searches were repeated and the same packing solution was obtained (Figure 2). Encouragingly, the signal-to-noise ratio was better in all steps (Table 2); the R factor was lower (0.39) and the correlation coefficient

Table 1. Summary of Molecular Replacement Statistics

Direct Rotation Search (25.0–80.0 Å)			
	Correct Solution (RF)	False Peak (RF)	S/N
H. marismortui	0.18	0.15	1.2
E. coli	0.20	0.22	0.92
PC Refinement of the Rotation Function Solutions (25.0–80.0 Å)			
H. marismortui	0.14	0.074	1.9
E. coli	0.14	0.087	1.6
Translation Search (25.0–80.0 Å)			
	Correct Solution (E2E2)	False Peak (E2E2)	S/N
H. marismortui	0.48	0.36	1.33
E. coli	0.44	0.35	1.26
Final Statistics (60.0–30.0 Å; 360 reflections)			
	R factor	Correlation Coefficient	
H. marismortui	0.39	0.51	
E. coli	0.45	0.43	

E2E2, Patterson correlation coefficient, where E is a normalized structure factor as defined in X-PLOR; S/N, signal to noise; RF, Rotation function value as defined in X-PLOR; R factor, $\Sigma_{hkl}(||Fobs_{hkl}| - |Fcalc_{hkl}||)/|Fobs_{hkl}|$, where $|Fobs_{hkl}|$ and $|Fcalc_{hkl}|$ are the observed and calculated structure factor amplitudes.

was higher (0.51). Both of these packing studies were done using step function representations of EM maps (see Experimental Procedures) similar to the procedures described by Podjarni and Urzhumtsev (1997). Even better phases resulted when continuous EM-determined electron density was positioned in the unit cell. The W18 difference Fourier peak increased from 13 σ to 14 σ (see below) while the R factor decreased from 44% to 41% for the 1322 reflections between 100 and 20 Å resolution.

HM 50S subunits pack in this C-centered, orthorhombic cell in tight pairs that are related by a crystallographic 2-fold axis parallel to the b direction. The L7/L12 regions of subunits related by that dyad axis interact with each other at the interface between the two subunits. There are only weak contacts between molecules related by the crystallographic 2-fold axis along the a axis of the unit cell.

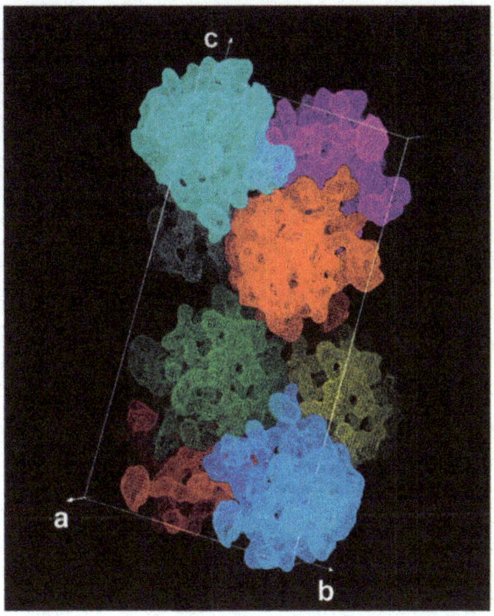

Figure 2. The Packing of *H. marismortui* 50S Subunits in the Orthorhombic Unit Cell Viewed Approximately down Its a Axis

The eight subunits related by crystallographic symmetry are shown in different colors. Less extensive crystallographic contacts are formed between subunits colored green and orange, while green/red and orange/lavender subunit pairs have extensive contact surfaces.

Heavy Atom Derivatives

The positions of the major heavy atom binding sites in the three different derivatives were found using difference Patterson maps as well as difference electron density maps, calculated with EM-derived phases. The first derivative obtained was prepared by soaking crystals in solutions containing an 18-atom tungsten cluster compound (W18). The position of the highest occupancy W18 binding site was evident in the three Harker sections of a 14 Å resolution difference Patterson map, which contained correlated peaks at about twice the noise level. Recently, data were recollected from this derivative with the X-ray wavelength set to the tungsten absorption edge, and the combined isomorphous and anomalous difference Patterson map calculated contained 7 σ peaks in the same positions (Figure 3a).

We were able to verify simultaneously both the correctness of the molecular replacement solution shown in Figure 2 as well as our interpretations of derivative difference Patterson maps by calculating difference electron density maps, using EM-derived phases. For example, a difference Fourier map of the W18 derivative calculated using X-ray amplitudes between 35 and 14 Å resolution and phases derived from the EC 50S reconstruction contained a 6 σ peak at the expected position, whereas a similar map calculated using phases derived from the HM 50S reconstruction showed a 13 σ peak (Figure 3b). The dramatic improvement in the signal-to-noise ratio obtained using the HM 50S reconstruction shows that it is a significantly more accurate representation of the subunits in these crystals.

To examine the resolution limit of phasing provided by EM reconstructions, difference isomorphous Fourier maps were calculated at various resolution ranges and the heights of the heavy atom peaks compared. The peak in Figure 3b remained the strongest peak in a difference map calculated using data between 18 and 14 Å, suggesting that the EM map is still generating useful phase information at this resolution, which is significantly higher than its nominal 20 Å resolution, consistent with Figure 1.

A second heavy atom derivative containing a bound W11 compound produced difference Patterson maps in which one major site could be identified, and several additional sites were found in difference Fourier maps. The major sites of a third derivative, prepared using a

Table 2. Data Collection and Phasing Statistics

Data Collection

	Native	W18	W11	Ta
Soaking time		6 days	2 days	5 days
Cluster concentration (mM)		1.0	1.0	1.0
Wavelength (Å)	1.21	1.21	1.21	1.25
Resolution	100–7.2	100–7.0	100–7.0	100–7.0
Observations				
Total	98,000	204,087	191,540	15,1534
Unique	26,324	50,576	57,103	56,172
Completeness	98.3	92.7	99.2	94.2
R_{sym}	11.1	4.5	5.2	5.5
No. Sites		7	12	10
R_{iso}		23.9	27.3	19.5
$I/\sigma I$ (highest resolution bin)	11.1 (2.8)	31.7 (13.8)	24.4 (7.7)	21.6 (9.4)

MIRAS Phasing Statistics

	Resolution Shells (Å)—~6400 Reflections per Bin				
	90.0–14.2	14.2–11.3	11.3–9.9	9.9–9.0	Total (25,540 reflections)
W18					
Phasing power	1.28	1.13	0.77	0.56	1.10
R_{cullis} (centric)	0.67	0.78	0.87	0.91	0.72
W11					
Phasing power	1.4	1.2	0.85	0.63	1.15
R_{cullis} (centric)	0.65	0.78	0.83	0.88	0.70
Ta					
Phasing power	0.83	0.84	0.64	0.58	0.75
R_{cullis} (centric)	0.79	0.86	0.92	0.91	0.83
Mean figure of merit	0.82	0.71	0.54	0.43	0.63

R_{iso}: $\Sigma|F_{PH}-F_P|/\Sigma F_{PH}$, where F_{PH} and F_P are the derivative and the native structure factor amplitudes, respectively. R_{sym}: $\Sigma\Sigma_i|I_{(h)}-I_{(h)i}|/\Sigma\Sigma_i I_{(h)i}$, where $I_{(h)}$ is the mean intensity. Phasing power: rms isomorphous difference divided by the rms residual lack of closure. R_{cullis}: $\Sigma(||F_{PH}-F_P|-|F_{H(calc)}||)/\Sigma|F_{PH}-F_P|$, where F_{PH} is the structure factor of the derivative and F_P is that of the native data. The summation is valid only for centric reflections.

tantalum cluster compound, were identified by a combination of anomalous and isomorphous difference Fourier maps.

The positions and occupancies of bound clusters were refined as rigid bodies by maximum likelihood methods to minimize the lack of closure, and phases were calculated using multiple isomorphous replacement and anomalous scattering (MIRAS). Using these phases, more accurate difference Fourier maps were computed that revealed the locations of numerous minor cluster binding sites in all three derivatives. Using the positions of the major and minor heavy atom binding sites, difference Patterson maps were computed that match the major features in the corresponding experimental difference Patterson maps (Figure 3a). The phasing power of each of the three derivatives and the average figure of merit examined as a function of resolution suggest that the quality of the MIRAS phasing is excellent to 12.5 Å but drops significantly beyond 9 Å. For this reason, a 9 Å resolution electron density map was calculated using MIRAS phasing from all three derivatives.

The Structure of the 50S Subunit

The X-ray crystallographically derived electron density map of the *H. marismortui* large ribosomal subunit resembles the electron microscopic reconstruction of the same particle. Comparison of a surface rendering of the HM 50S EM 20 Å resolution reconstruction (Figure 4a) with a similar rendering of a 20 Å resolution map derived from X-ray data alone (Figure 4b) shows that both structures have the same overall shape and density features. The resolution of the X-ray map appears substantially higher, however, presumably because the amplitudes of its high-resolution components are larger. Nevertheless, most of the structural characteristics of the large ribosomal subunit previously visualized by EM reconstruction (Frank et al., 1995a) and identified biochemically by immunoelectron microscopy (Oakes et al., 1986) are easily recognized (Figures 4b and 4c). The most obvious differences involve the region of the 50S subunit known to include ribosomal protein L1 and a region where the 50S subunit contacts the small subunit. The X-ray map suggests that the positions of these protrusions are shifted in the crystal, which may mean that these domains are capable of adopting several different conformations.

Comparison of sections through the crystallographic unit cell taken from maps computed using X-ray intensities and either EM-derived phases (Figure 5b) or X-ray-derived phases (Figure 5a) show some similar features of continuous electron density. Since these two maps have only the X-ray amplitudes in common, we conclude that both the EM and the X-ray-derived phases are basically correct.

X-ray electron density maps calculated at 14 Å (Figures 6a and 6b) and 9 Å resolution (Figures 4c, 4d, and 6c) show long, continuous rods of electron density that

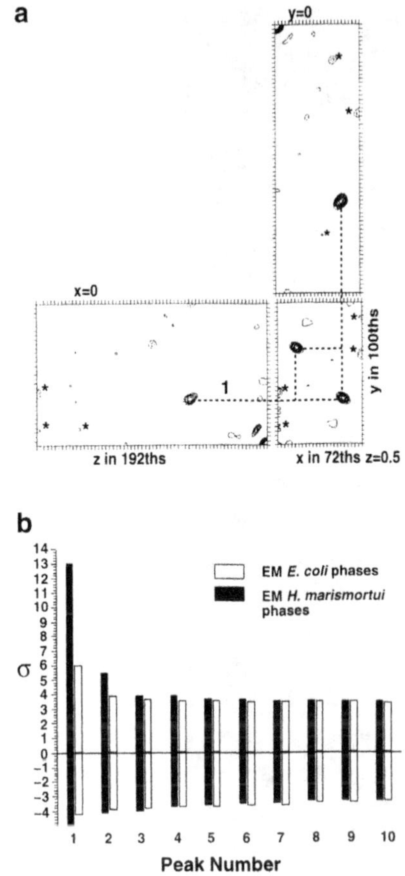

Figure 3. Heavy Atom Locations Determined by Difference Patterson and Difference Fourier Maps

(a) The three Harker sections from a combined isomorphous-anomalous difference Patterson map of the W18 derivatized crystals of the *Haloarcula marismortui* 50S ribosomal subunit with the predicted positions indicated. The map is contoured at 2 σ with 1 σ increments. The number 1 and the stars indicate the Patterson peaks expected from the major and minor sites, respectively.

(b) Histograms of the largest positive and negative isomorphous peaks identified in the 14 Å isomorphous difference Fourier maps of the HM 50S subunit crystals using phases derived from the HM 50S reconstruction or the EC 50S reconstruction (Frank et al., 1995a). The phases were weighted according to the SigmaA scheme for the data between 35 and 14 Å resolution.

are bent, branched in places, about 20 Å in diameter, and have distinct, right-handed spiral features expected from double-stranded RNA (Figure 6c). These density features are likely to represent the branched, double-helical RNA that constitutes about two-thirds of the subunit's mass. There are regions where the presumed RNA density is well separated from other features and regions of more extensive, unresolved density that could correspond to more tightly packed RNA helices and/or clusters of protein subunits with RNA (Figures 6a and 6b). Aside from suggesting where protein L1 might be, we have made no attempt to assign density to protein separately from RNA, since the resolution is insufficient. Among the features visible on the side facing the 30S subunit, several protrusions are seen whose locations suggest that they furnish some of the intersubunit bridges observed in the cryo-EM map of the *E. coli* 70S ribosome (Frank et al., 1995a, 1995b).

Discussion

We conclude that cryo-EM reconstructions of single particles can be used to phase the X-ray diffraction amplitudes of crystals of macromolecular assemblies in order to locate bound heavy atoms in derivative crystals. It is clear that the EM maps obtained from *H. marismortui* are excellent representations of the 50S subunit in the crystals studied here. Further, since the 20 Å resolution maps generated using either EM or X-ray-derived phases are similar, both methods must be producing correct structures at this resolution. In contrast, our all-X-ray map does not resemble the previously published X-ray electron density map of the *H. marismortui* large ribosomal subunit, which had a nominal resolution at 7 Å, nor is the packing of subunits in the unit cell described here the same as that deduced in that study from a solvent flattened map (Schlunzen et al., 1995).

The EM reconstruction of the large subunit from *H. marismortui* is a significantly better model for the subunit in these crystals than the EM map of the 50S subunit carved out of the reconstruction of the 70S ribosome from *E. coli*, as judged by molecular replacement statistics and by the increase in size of the difference Fourier peak for the W18 cluster. There are several possible explanations for the differences between these two EM maps. Perhaps the most likely is that the structures of the eubacterial and archaebacterial ribosomes are significantly different at low resolution, as the sequence differences between their RNAs and proteins might imply (Wittmann-Liebold et al., 1990; Woese, 1996). It is also possible that there are conformational differences between isolated 50S subunits and 50S subunits complexed with small subunits. Alternatively, it could be that we failed to distinguish density arising from the 50S subunit from density belonging to 30S subunits correctly when abstracting the 50S map from maps of the whole *E. coli* ribosome.

In order to help identify electron density features in the 9 Å resolution HM 50S map, it has been compared with a 9 Å resolution map of 5S RNA fragment 1 calculated from its atomic coordinates (Correll et al., 1997). Many of the strands in the *H. marismortui* 50S map have distinct, right-handed spiral ridges that resemble the ridges in the 5S fragment map that arise from the duplex backbone. Presumably, the spiral ridges in the HM 50S map also represent the sugar phosphate backbone. In both maps, the spacings between the backbone ridges vary from that of A-form RNA due to the presence of non-Watson-Crick base pairs. The right-handedness of the strands in the HM 50S map proves that the hand of the ribosome was indeed correctly determined by earlier EM studies (Leonard and Lake, 1979) and by the method of single-particle reconstruction (Radermacher et al., 1987).

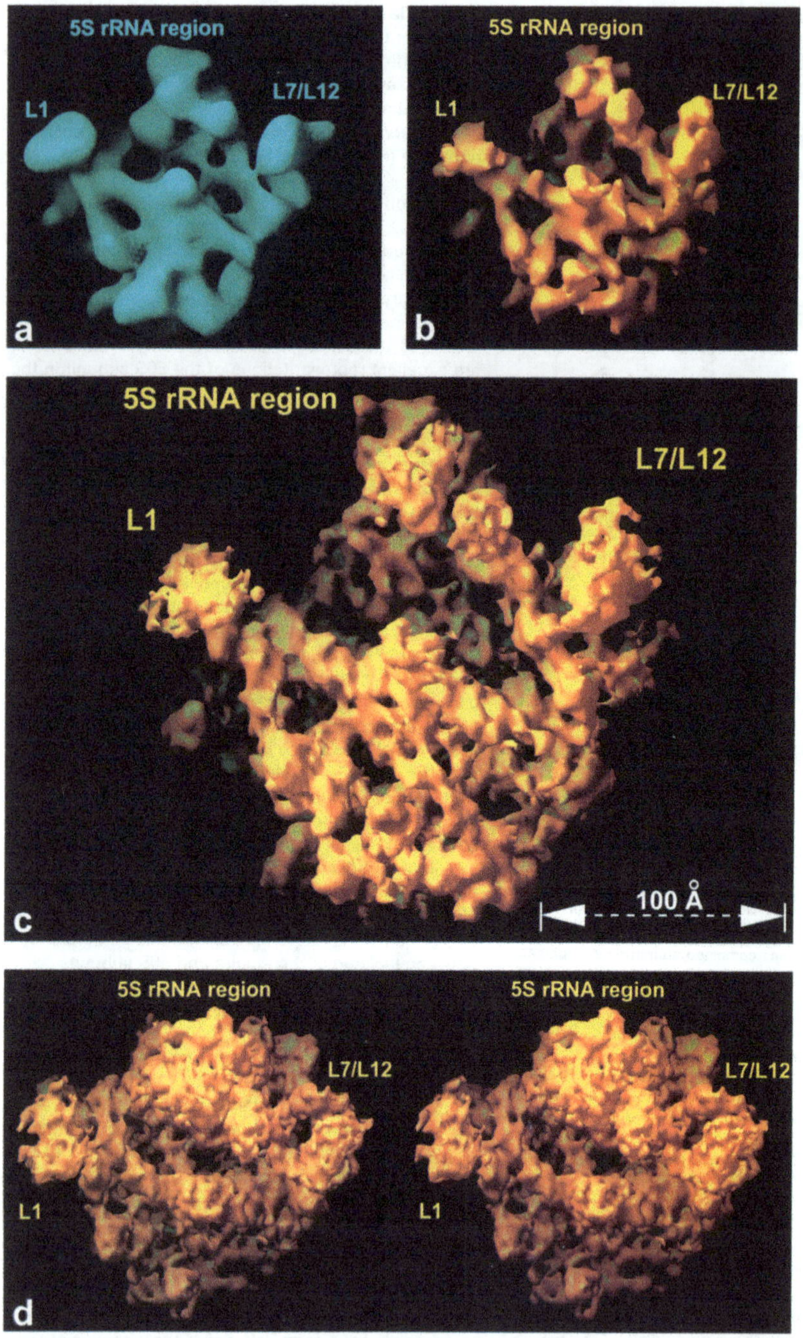

Figure 4. Comparison of the 20 Å Resolution EM Reconstruction with Both 20 Å and 12 Å Resolution X-Ray Maps of the HM 50S Subunit

(a) A surface rendering of the EM reconstruction of the 50S subunit at about 20 Å resolution. This structure was used in the molecular replacement study. Previously identified features such as the position of ribosomal protein L1, the L7/L12 region, and the 5S ribosomal RNA in the central protuberance are labeled. The large subunit is shown in the crown view from the side that interacts with the 30S subunit.

(b) A surface rendition of a 20 Å resolution "all X-ray" electron density map calculated using phases derived by MIRAS using three heavyatom derivatives. The volume occupied by the map contoured at this level is about 85% of the expected volume of the large ribosomal subunit calculated from molecular mass. The subunit is shown in the same view as in (a).

Crystal Structure of the 50S Ribosomal Subunit
1111

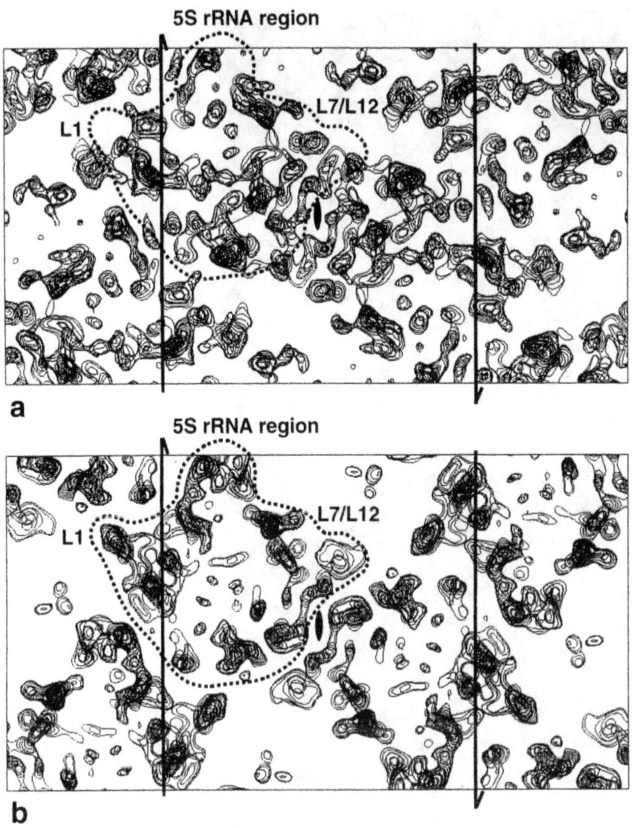

Figure 5. Comparison of HM 50S Electron Density Maps Calculated Using X-Ray- or EM-Derived Phases and X-Ray Amplitudes

(a) An approximately 20 Å thick superposition of sections through the HM 50S electron density map, calculated at 20 Å resolution using MIRAS X-ray phasing derived from three derivatives (Table 2) with no solvent flattening. The dotted line represents the 50S subunit envelope derived from the EM map positions by molecular replacement. The map shows a clear solvent and subunit boundary, and many of the features, such as the continuous density near the central 2-fold axis, are visible in both the X-ray and EM-phased maps. The slice through the unit cell was chosen to portray prominent features, such as the L1 region, central protuberance, and the L7/L12 region.

(b) For comparison, the superposition of the same sections of a map calculated using HM 50S EM-derived phases and observed X-ray diffraction amplitudes is shown.

Since rods of electron density, presumed to be RNA, crisscross the entire large subunit, it is obvious that the general architectural principle of this protein–nucleic acid assembly is distinctly different from those of nucleosomes or viruses. RNA and protein are intermingled, rather than the nucleic acid being wrapped around a protein core or encased in a protein shell. Present models of the small ribosomal subunit based on neutron scattering, chemical modification, and cryo-EM likewise place the proteins within a forest of RNA helices (Malhotra and Harvey, 1994; Noller et al., 1995; Mueller and Brimacombe, 1997a, 1997b; Mueller et al., 1997). It appears (Figures 6a and 6b) that the 50S ribosome structure is formed by struts of RNA rods whose branching cross-links the struts. Proteins may serve to stabilize further the branched RNA scaffold.

Rotation of the map shown in Figure 4c by about 50 degrees around its horizontal axis reveals a tunnel (Figure 4d). This tunnel aligns with tunnels observed in cryo-EM maps of large ribosomal subunits from *H. marismortui* (this study), *E. coli* (Frank et al., 1995a; Malhotra et al., 1998), and yeast (Beckmann et al., 1997; Verschoor et al., 1998), when the subunits themselves are aligned. A similar observation was made by Dube et al. (1998), who compared ribosomes from rat liver and *E. coli*. The tunnel, presumably identical with that found by Yonath et al. (1987) in the 70S ribosome and by Milligan and Unwin (1986) in the 80S, is thought to conduct the nascent polypeptide chain from the peptidyltransferase center to a point where it exits the ribosome (reviewed by Eisenstein et al., 1994). In agreement with this hypothesis, the acceptor end of the P-site-bound-fMet-tRNA is found in the immediate vicinity (~15 Å) of the tunnel entrance (Malhotra et al., 1998), and the tunnel exit, on the back of the large subunit, was found to align with the central pore of the Sec61 channel in yeast (Beckmann et al., 1997). According to this interpretation, the peptidyltransferase center must be in the immediate neighborhood of the tunnel entrance shown in the X-ray-derived map in Figure 4d.

(c) An X-ray-phased 9 Å resolution map oriented as in (a).

(d) The same map as in (c) rotated by about 50° about a horizontal axis to show a tunnel that lies at the back of the peptidyltransferase cleft in the approximate center of the particle. All maps are displayed by RIBBONS, a crystallographic molecular graphics program (Carson, 1991).

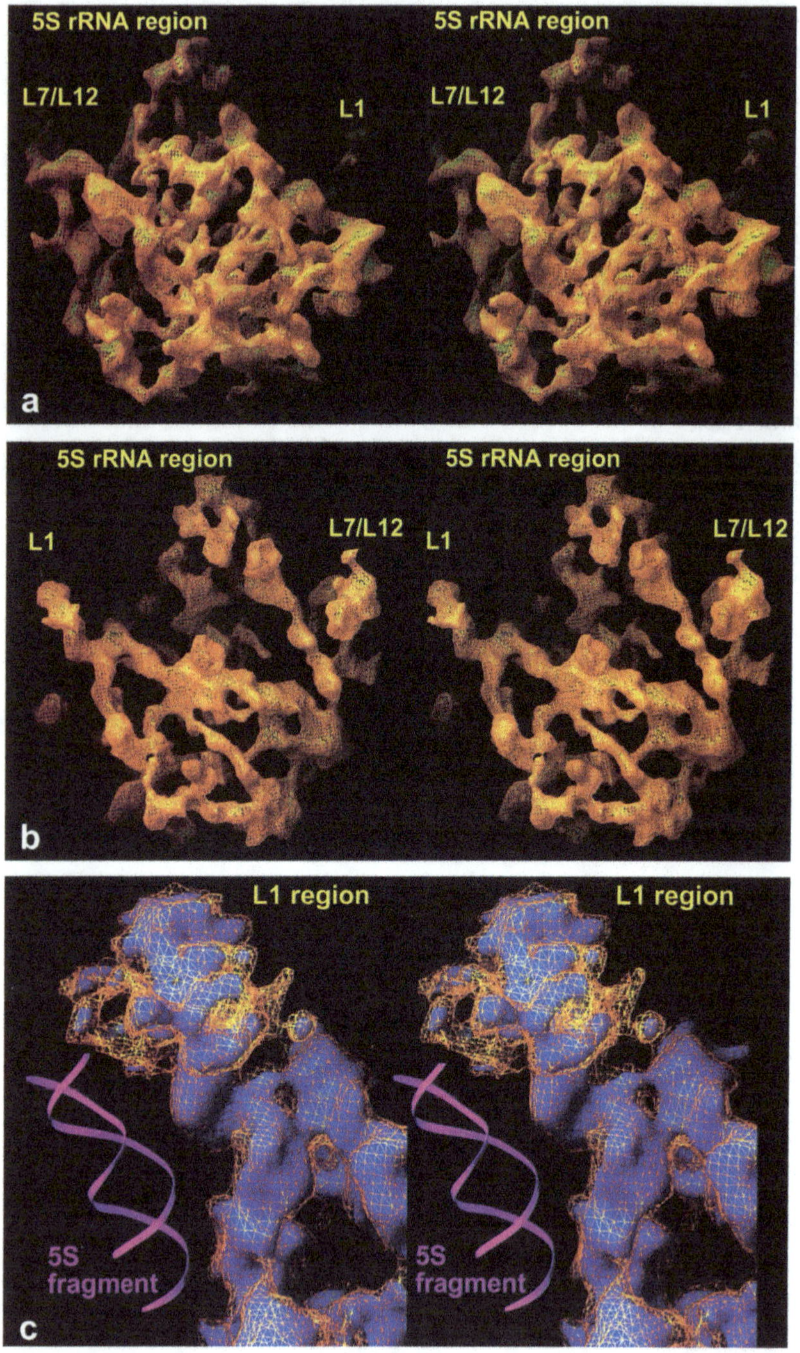

Figure 6. X-Ray Maps at 14 and 9 Å Resolution

(a) The MIRAS map in the crown view at 14 Å resolution shown in stereo from the side opposite the 30S subunit. Previously identified features such as ribosomal protein L1, L7/L12 region, and 5S ribosomal RNA in the central protuberance are labeled. To show the underlying strut-like architecture, the map is contoured at a high-density level and all density not connected has been removed.

(b) The same MIRAS map shown in the crown view.

This X-ray crystallographic map of a ribosomal subunit shows electron density features expected from other structural studies. Although the quality of its phasing and the resulting electron density diminishes at resolutions higher than 9–10 Å, it constitutes an important beachhead for launching an attack on the structure of the ribosome at higher resolution. These 9 Å resolution phases should make it possible to find the heavy atom sites in derivatives containing large numbers of single heavy atoms by difference electron density maps. Additionally, improved modeling of the heavy atom cluster compounds may improve the phasing quality at higher resolution. Thus, we anticipate obtaining higher resolution insights into the ribosome's structure and function in the near future.

Abbreviations

Abbreviations are as follows: HM50S, *Haloarcula marismortui* large ribosomal subunit; EC50S, *Escherichia coli* large ribosomal subunit; EM, electron microscopy SIR(AS) single isomorphous replacement (anomalous scattering); MIR(AS), multiple isomorphous replacement (anomalous scattering); W18, heavy atom cluster $(AsW_9O_{33})_2(PhSn)_4$; W11, heavy atom cluster $Cs_5(PW_{11}O_{39}\{Rh_2(CH_3COO)_2\})$; Ta, heavy atom cluster $Ta_6Br_{12}^{+2}$; and L7/L12, the region of the ribosome where proteins L7 and L12 are located.

Experimental Procedures

Cells and Ribosomes

H. marismortui (ATCC 43049) was grown on a slightly modified version of ATCC culture medium 1230, which was supplemented with 4.3 g of yeast extract, 5.1 g of Tris, and 3.4 g of glucose per liter. Bacteria were grown at 37°C to an OD_{550nm} between 1.0 and 2.2. They were harvested by centrifugation and stored at −80°C. Cells were ruptured using a French press. Ribosomes were prepared from lysates by centrifugation, and subunits were isolated on sucrose gradients (Shevack et al., 1985).

Electron Microscopy

Grids were prepared for cryomicroscopy using standard methods (Dubochet et al., 1988; Wagenknecht et al., 1988). The sample was diluted to an optical density at 260 nm of 1.2 immediately before preparation of cryo grids to prevent aggregation, and it was applied onto grids and flash-frozen for cryo-electron microscopy. Micrographs were recorded using low-dose protocols on a Philips EM420 microscope at a magnification of 52,200× ± 2%, which was established using a tobacco mosaic virus as a standard.

Image Processing

Micrographs were checked for drift, astigmatism, and presence of Thon rings by optical diffraction. Scanning was done with a step size of 25 μm, which corresponds to 4.79 Å on the ribosome, using a Perkin Elmer PDS 1010A microdensitometer. A total of 40 micrographs were used, grouped into two defocus settings: 1.25 and 1.80 μm. HM 50S subunit images were automatically selected from micrographs by comparing them with a reference set of 87 quasi-evenly spaced projection images (Penczek et al., 1994) of the 50S subunit portion of a 70S *E. coli* ribosome reconstruction (Frank et al., 1995a). This 50S image was carved out of the 70S map using a lower resolution reconstruction of the 30S subunit from *E. coli* (Lata et al., 1996). The number of particles selected was 6150 for 1.25 μm group and 7020 for 2.80 μm group. For each defocus group, one cycle of a 3D projection alignment procedure (Penczek et al., 1994) was applied, and a merged, contrast transfer function (CTF)–corrected reconstruction was computed as described earlier (Zhu et al., 1997). The resulting reconstruction was subjected to five cycles of refinement. Each cycle started with the appropriately CTF-modified reconstruction of the previous step, and the step size was 2°. The resolution of the reconstruction was calculated to be approximately 20 Å (Figure 5).

Preparation of Crystals

Crystals of *H. marismortui* 50S subunit were grown in high-salt buffers using PEG 6000 as a precipitant, as described by Yonath and coworkers (van Bohlen et al., 1991). They were grown in hanging drops by vapor diffusion at room temperature. The thin plates that resulted had maximum dimensions of 0.5 × 0.5 × 0.1 mm and were harvested after ~2 weeks. Crystals were stabilized by gradual transfer into a solution containing ethylene glycol in addition to other components of the crystallization buffer and flash frozen in liquid propane. Heavy atom derivatives were prepared by soaking crystals in stabilization solutions containing the heavy atom compound of interest at a concentration of 1 mM for 1–6 days. In most cases, the diffraction properties of derivatized crystals did not change upon soaking in heavy atom solutions.

X-Ray Data Collection and Processing

All data were collected at the National Synchrotron Light Source (Brookhaven) from crystals frozen at 100 K, using beamlines X12c and X12b. For each heavy atom derivative, anomalous diffraction data were collected at the wavelength where peak anomalous scattering is observed. Diffraction data were recorded using either a 300 mm MAR imaging plate, or a Brandeis 2 × 2 CCD detector (W. C. Phillips, M. Stanton, D. O'Mara, C. Ingersoll, H. Qian, A. Stewart, and R. M. S., unpublished data) or a Quantum 4 CCD detector (Area Detector Systems Corporation). The beam size was 200 × 200 μm. The crystals were aligned along the long axis of the unit cell (570 Å) so that 1.5° oscillations could be used to collect reflections out to 7.0 Å resolution at the edge of the detector. The crystal-to-detector distances varied between 450.0 mm and 900.0 mm depending on crystal quality and beam divergence, and they were chosen so that maximum resolution data could be collected while avoiding overlapping of spots. Data sets were processed by using DENZO and SCALEPACK (Otwinowski, 1993).

Computational Techniques

The initial molecular replacement solution was obtained using an envelope derived from an EM reconstruction of the *E. coli* 50S subunit (see above). The map was stored as an array of densities, $\rho(x,y,z)$, on an arbitrary scale, sampled along all three directions at equal intervals, which were ~1/3 of the calculated maximum resolution of the map. The map was reduced to a three-dimensional step function by assigning unit weight to all points of the array for which $\rho(x,y,z)$ was greater than some minimum value and setting all other points to zero. For the purposes of further computation, it was represented as a collection of atoms within the envelope. Direct rotational searches, based on the Patterson correlation coefficient (Huber, 1985; Brünger, 1990, 1997), were performed using X-PLOR (Brünger, 1992) and CNS (Brünger et al., 1998). In these searches, the linear scale factor of the reconstruction (the sampling interval) was varied, as was the threshold at which the envelope was defined. These two parameters are not completely independent because calculated structure factors depend primarily on the outer envelope, as noted previously (Jack et al., 1975). The orientations determined this way were subjected to Patterson correlation coefficient refinement using data between 100 and 30 Å. The results that yielded

(c) A stereo close-up of a 9 Å resolution density map showing the region that includes ribosomal protein L1; a backbone ribbon model of part of the 5S RNA is shown adjacent for scale. The upper part of the density contains ribosomal protein L1, and the rest may represent rRNA that interacts with L1. The yellow cage is at a lower contour level than the solid blue. The figure was generated with RIBBONS (Carson, 1991).

relatively high-correlation coefficients were used for translation function searches (Fujinaga and Read, 1987).

MIRAS Phasing

The heavy atom positions for the major binding sites were determined from combination difference anomalous and isomorphous Patterson maps at various resolution ranges. Solutions were cross-checked using anomalous and isomorphous difference Fourier maps, based on EM-derived phases. Refinement of the heavy atom cluster compound parameters was performed by maximum likelihood minimization of the lack of closure. All atoms in each of the clusters were included. Anomalous scattering was assumed to result only from heavy atoms in the cluster. All aspects of the isomorphous phasing including difference Patterson, difference Fourier map calculation, and heavy atom cluster rigid body refinement against lack of closure were done using the CNS program package (Brünger et al., 1998). Temperature factors were maintained at 40 $Å^2$ for all atoms in each of the clusters, while occupancies were refined as a group, and positions of atoms in clusters were refined as rigid bodies.

Acknowledgments

We thank Lu Min, Szilvia Szép, and Malcolm Capel (Brookhaven) for help with the X-ray data collection; Michael Pope and Gerta Sazani (George Washington University), Nevenka Brničević (Ruđer Bošković Institute, Croatia), and Duward F. Shriver and Nicolas Prokopuk (Northwestern University) for gifts of heavy atom cluster compounds; Axel Brünger for advice on molecular replacement procedures and heavy atom cluster refinement; Jimin Wang for helpful suggestions; and Doryen Bubeck and Rajendra Agrawal for assistance with image processing of electron micrographs and discussions. This work was supported by NIH grants GM-22778 to T. A. S., GM-54216 to P. B. M., and GM-29169 to J. F. N. B. was supported by a postdoctoral fellowship from the Cancer Research Fund of the Damon Runyon–Walter Winchell Foundation and P. N. by a grant from the Danish Research Council. Computing support for J. F. was provided by the National Center for Supercomputer Applications, University of Illinois at Urbana-Champaign.

Received April 1, 1998; revised May 6, 1998.

References

Agrawal, R.K., Penczek, P., Grassucci, R.A., Li, Y., Leith, A., Nierhaus, K.H., and Frank, J. (1996). Direct visualization of A-, P-, and E-site transfer RNAs in the *Escherichia coli* ribosome. Science *271*, 1000–1002.

Beckmann, R., Bubeck, D., Grassucci, R., Penczek, P., Verschoor, A., Blobel, G., and Frank, J. (1997). Alignment of conduits for the nascent polypeptide chain in the ribosome–Sec61 complex. Science *278*, 2123–2126.

Böttcher, B., Wynne, S.A., and Crowther, R.A. (1997). Determination of the fold of the core protein of hepatitis B virus by electron cryomicroscopy. Nature *386*, 88–91.

Brünger, A.T. (1990). Extension of molecular replacement: a new search strategy based on Patterson correlation refinement. Acta Crystallogr. *A46*, 46–57.

Brünger, A. (1992). X-PLOR Version 3.1. System for X-Ray Crystallography and NMR (New Haven, CT: Yale University Press).

Brünger, A.T. (1997). Patterson correlation searches and refinement. Methods Enzymol. *276*, 558–580.

Brünger, A.T., Adams, P.D., Clore, G.M., DeLano, W.L., Gros, P., Grosse-Kunstleve, R.W., Jiang, J.-S., Kuszewski, J., Nilges, M., Pannu, N.S., et al. (1998). Crystallography and NMR system (CNS): a new software suite for macromolecular structure determination. Acta Crystallogr., in press.

Capel, M.S., Engelman, D.M., Freeborn, B.R., Kjeldgaard, M., Langer, J.A., Ramakrishnan, V., Schindler, D.G., Schneider, D.K., Schoenborn, B.P., Sillers, I.-Y., et al. (1987). A complete mapping of the proteins in the small ribosomal subunit of *E. coli*. Science *238*, 1403–1406.

Carson, M. (1991). Ribbons 2.0. J. Appl. Crystallogr. *13*, 458–961.

Correll, C.C., Freeborn, B., Moore, P.B., and Steitz, T.A. Metals, motifs, and recognition in the crystal structure of a 5S rRNA domain. (1997). Cell *91*, 705–712.

Dube, P., Wieske, M., Stark, H., Schatz, M., Stahl, J., Zemlin, F., Lutsch, G., and van Heel, M. (1998). The 80S rat liver ribosome at 25 Å resolution by electron cryomicroscopy and angular reconstitution. Structure *6*, 389–399.

Dubochet, J., Adrian, M., Chang, J.-J., Homo, J.-C., Lepault, J., McDowall, A.W., and Schultz, P. (1988). Cryo-electron microscopy of vitrified specimens. Q. Rev. Biophys. *21*, 129–228.

Eisenstein, M., Hardesty, B., Odom, O.W., Kudlicki, W., Kramer, G., Arad, T., Franceschi, F., and Yonath, A. (1994). Modeling and experimental study of the progression of nascent proteins in ribosomes. In Biophysical Methods in Molecular Biology, G. Pifat, ed. (Rehovot, Israel: Balaban Press), pp. 213–246.

Franceschi, F., Weistein, S., Evers, U., Arndt, E., Jahn, W., Hansen, H.A.S., von Bohlen, K., Berkovitch-Yellin, Z., Eisenstein, M., Agmon, I., et al. (1993). Towards atomic resolution of prokaryotic ribosomes: crystallographic, genetic and biochemical studies. In The Translational Apparatus, K.H. Nierhaus, F. Franceschi, A.R. Subramanian, V.A. Erdmann, and B. Wittmann-Liebold, eds. (New York: Plenum Press), pp. 397–410.

Frank, J., Zhu, J., Penczek, P., Li, Y., Srivastava, S., Verschoor, A., Radermacher, M., Grassucci, R., Lata, R.K., and Agrawal, R.K. (1995a). A model for protein synthesis based on cryo-electron microscopy of the *E. coli* ribosome. Nature *376*, 441–444.

Frank, J., Verschoor, A., Li, Y., Zhu, J., Lata, R.K., Radermacher, J., Penczek, P., Grassucci, R., Agrawal, R.K., and Srivastava, S. (1995b). A model of the translational apparatus based on a three-dimensional reconstruction of the *Escherichia coli* ribosome. Biochem. Cell Biol. *73*, 757–765.

Fujinaga, M., and Read, R.J. (1987). Experiences with a new translation-function program. J. Appl. Crystallogr. *20*, 517–521.

Huber, R. (1985). In Molecular Replacement, Proceedings of the Daresbury Study Weekend, Daresbury, February, P.A. Mochin, ed. (Daresbury, United Kingdom: Science and Engineering Research Council, The Librarian, Daresbury Laboratory), pp. 58–61.

Jack, A., Harrison, S.C., and Crowther, R.A. (1975). Structure of tomato bushy stunt virus II. Comparison of results obtained by electron microscopy and X-ray diffraction. J. Mol. Biol. *97*, 163–172.

Knablein, J., Neuefeind, T., Schneider, F., Bergner, A., Messerschmidt, A., Lowe, J., Steipe, B., and Huber, R. (1997). $Ta_6Br^{(2+)}{}_{12}$, a tool for phase determination of large biological assemblies by X-ray crystallography. J. Mol. Biol. *270*, 1–7.

Ladenstein, R., Bacher, A., and Huber, R. (1987). Some observations of a correlation between the symmetry of large heavy-atom complexes and their binding sites on proteins. J. Mol. Biol. *195*, 751–753.

Lata, K.R., Agrawal, R.K., Penczek, P., Grassucci, R., Zhu, J., and Frank, J. (1996). Three-dimensional reconstruction of the *Escherichia coli* 30S ribosomal subunit in ice. J. Mol. Biol. *262*, 43–52.

Leonard, K.R., and Lake, J.A. (1979). Ribosome structure: hand determination by electron microscopy of 30S subunit. J. Mol. Biol. *128*, 155–163.

Malhotra, A., and Harvey, S.C. (1994). A quantitative model of the *Escherichia coli* 16S RNA in the 30S ribosomal subunit. J. Mol. Biol. *240*, 308–340.

Malhotra, A., Penczek, P., Agrawal, R.K., Gabashvili, I.S., Grassucci, R.A., Jünemann, R., Burkhardt, N., Nierhaus, K.H., and Frank, J. (1998). *Escherichia coli* 70S ribosome at 15 Å resolution by cryo-electron microscopy: localization of fMet-$tRNA_f^{Met}$ and fitting of L1 protein. J. Mol. Biol., in press.

May, R.P., Nowotny, V., Nowotny, P., Voss, H., and Nierhaus, K.H. (1992). Inter-protein distances within the large subunit from *Escherichia coli* ribosomes. EMBO J. *11*, 373–378.

Milligan, R.A., and Unwin, P.N.T. (1986). Location of exit channel for nascent protein in 80S ribosome. Nature *319*, 693–695.

Mueller, F., and Brimacombe, R. (1997a). A new model for the three-dimensional folding of *Escherichia coli* 16S ribosomal RNA. I. Fitting the RNA to a 3D electron microscopic map at 20 Å. J. Mol. Biol. *271*, 524–544.

Mueller, F., and Brimacombe, R. (1997b). A new model for the three-dimensional folding of *Escherichia coli* 16S ribosomal RNA. II. The RNA–protein interaction data. J. Mol. Biol. *271*, 545–565.

Mueller, F., Stark, H., van Heel, M., Rinke-Appel, J., and Brimacombe, R. (1997). A new model for the three-dimensional folding of *Escherichia coli* 16S ribosomal RNA. III. The topography of the functional center. J. Mol. Biol. *271*, 566–587.

Neuefeind, T., Bergner, A., Schneider, F., Messerschmidt, A., and Knablein, J. (1997). The suitability of Ta6Br12(2+) for phasing in protein crystallography. Biol. Chem. *378*, 219–221.

Noller, H.F., Hoffarth, V., and Zimniak, L. (1992). Unusual resistance of peptidyl transferase to protein extraction procedures. Science *256*, 1416–1419.

Noller, H.F., Green, R., Heilek, G., Hoffarth, V., Huttenhofer, A., Joseph, S., Lee, I., Lieberman, K., Mankin, A., Merryman, C., et al. (1995). Structure and function of ribosomal RNA. Biochem. Cell Biol. *73*, 997–1009.

Oakes, M., Henderson, E., Scheinman, A., Clark, M., and Lake, J.A. (1986). Ribosome structure, function, and evolution: mapping ribosomal RNA, proteins, and functional sites in three dimensions. In Structure, Function and Genetics of Ribosomes, B. Hardesty and G. Kramer, eds. (New York: Springer-Verlag), pp. 47–67.

O'Halloran, T.V., Lippard, S.J., Richmond, T.J., and Klug, A. (1987). Multiple heavy-atom reagents for macromolecular X-ray structure determination. Application to the nucleosome core particle. J. Mol. Biol. *194*, 705–712.

Otwinowski, Z. (1993). DENZO. In Data Collection and Processing, L. Sawyer, N. Isaacs, and D. Bailey, eds. (Warrington, England: SERC Daresbury Laboratory), pp 56–62.

Penczek, P., Grassucci, R.A., and Frank. J. (1994). The ribosome at improved resolution: new techniques for merging and orientation refinement in 3D cryo-electron microscopy of biological particles. Ultramicroscopy *53*, 251–270.

Podjarni, A.D., and Urzhumtsev, A.G. (1997). Low resolution Phasing. Methods Enzymol. *276*, 641–658.

Radermacher, M., Wagenknecht, T., Verschoor, A., and Frank, J. (1987). Three-dimensional reconstruction from a single-exposure, random conical tilt series applied to the 50S ribosomal subunit. J. Microsc. *146*, 113–136.

Rossmann, M.G. (1995). Ab initio phase determination and phase extension using non-crystallographic symmetry. Curr. Opin. Struct. Biol. *5*, 650–655.

Schlunzen, F., Hansen, H.A.S., Thygesen, J., Bennett, W.S., Volkmann, N., Levin, I., Harms, J., Bartles, H., Zaytzev-Bashan, A., Berkovitch-Yellin, Z., et al. (1995). A milestone in ribosomal crystallography: the construction of preliminary electron density maps at intermediate resolution. Biochem. Cell Biol. *73*, 739–749.

Shevack, A., Gewitz, H.S., Hennemann, B., Yonath, A., and Wittmann, H.G. (1985). Characterization and crystallization of ribosomal particles from *Haloarcula marismortui*. FEBS Lett. *184*, 68–71.

Stark, H., Mueller, F., Orlova, E.V., Schatz, M., Dube, P., Erdemir, T., Zemlin, F., Brimacombe, R., and van Heel, M. (1995). The 70S *Escherichia coli* ribosome at 23 Å resolution: fitting the ribosomal RNA. Structure *3*, 815–821.

Stark, H., Orlova, E.V., Rinke-Appel, J., Junke, N., Mueller, F., Rodnina, M., Wintermeyer, W., Brimacombe, R., and van Heel, M. (1997a). Arrangment of tRNAs in pre- and posttranslocational ribosomes revealed by electron cryomicroscopy. Cell *88*, 19–28.

Stark, H., Rodnin, M.V., Rinke-Appel, J., Brimacombe, R., Wintermeyer, W., and van Heel, M. (1997b). Visualization of elongation factor Tu on the *Escherichia coli* ribosome. Nature *389*, 403–406.

Stoeffler, G., and Stoeffler-Meilicke, M. (1986). Immuno electron microscopy on *Escherichia coli* ribosomes. In Structure, Function, and Genetics of Ribosomes, B. Hardesty and G. Kramer, eds. (New York: Springer-Verlag), pp. 28–46.

Thygesen, J., Weinstein, S., Franceschi, F., and Yonath, A. (1996). The suitability of multi-metal clusters for phasing in crystallography of large macromolecular assemblies. Structure *4*, 513–518.

van Bohlen, K., Makowski, I., Hansen, H.A.S., Bartels, H., Berkovitch-Yellin, Z., Zaytzev-Bushan, A., Meyer, S., Paulke, C., Franscheschi, F., and Yonath, A. (1991). Characterization and preliminary attempts for derivatization of crystals of large ribosomal subunits from *Haloarcula marismortui* diffracting to 3 Å resolution. J. Mol. Biol. *222*, 11–15.

Verschoor, A., Warner, J.R., Srivastava, S., Grassucci, R.A., and Frank, J. (1998). Three-dimensional structure of the yeast ribosome. Nucleic Acids Res. *26*, 655–661.

Wagenknecht, T., Grassucci, R., and Frank. J. (1988). Electron microscopy and computer image averaging of ice-embedded large ribosomal subunits from *E. coli*. J. Mol. Biol. *199*, 137–147.

Wittmann-Liebold, B. (1986). Ribosomal proteins: their structure and evolution. In Structure, Function, and Genetics of Ribosomes, B. Hardesty and G. Kramer, eds. (New York: Springer-Verlag), pp. 326–361.

Wittmann-Liebold, B., Kopke, A.K.E., Arndt, E., Kromer, W., Hatakeyama, T., and Wittmann, H.G. (1990). Sequence comparison and evolution of ribosomal proteins and their genes. In The Ribosome: Structure, Function, and Genetics, W.E. Hill, A. Dahlberg, R.A. Garrett, P.B. Moore, D. Schlessigner, and J.R. Warner, eds. (Washington, D.C.: American Society for Microbiology), pp. 598–616.

Woese, C.R. (1996). The world of ribosomal RNA. In Ribosomal RNA: Structure, Evolution, Processing, and Function in Protein Biosynthesis, R.A. Zimmermann and A. E. Dahlberg, eds. (Boca Raton: CRC Press, Inc.), pp. 23–48.

Yonath, A., Mussig, J., Tesche, B., Lorenz, S., Erdmann, V.A., and Wittmann, H.G. (1980). Crystallization of the large ribosomal subunits from *Bacillus stearothermophilus*. Biochem. Int. *1*, 428–435.

Yonath, A., Leonard, K.R., and Wittmann, H.G. (1987). A tunnel in the large ribosomal subunit revealed by three-dimensional reconstruction. Science *23*, 813–816.

Zhu, J., Penczek, P.A., Schroeder, R., and Frank, J. (1997). Three-dimensional reconstruction with contrast transfer function correction from energy-filtered cryoelectron micrographs: procedure and application to the 70S *Escherichia coli* ribosome. J. Struct. Biol. *118*, 197–219.

Cell, Vol. 100, 537–549, March 3, 2000, Copyright ©2000 by Cell Press

Solution Structure of the *E. coli* 70S Ribosome at 11.5 Å Resolution

Irene S. Gabashvili,† Rajendra K. Agrawal,†‡ Christian M. T. Spahn,*† Robert A. Grassucci,*† Dmitri I. Svergun,§‖ Joachim Frank,*†‡# and Pawel Penczek†‡
*Howard Hughes Medical Institute
Health Research, Inc.
†Wadsworth Center
Empire State Plaza
Albany, New York 12201-0509
‡Department of Biomedical Sciences
State University of New York at Albany
Empire State Plaza
Albany, New York 12201-0509
§European Molecular Biology Laboratory
c/o DESY
Notkestraße 85
D-22603 Hamburg
Germany
‖Institute of Crystallography
Russian Academy of Sciences
Leninsky pr. 59
117333 Moscow
Russia

Summary

Over 73,000 projections of the *E. coli* ribosome bound with formyl-methionyl initiator $tRNA_f^{Met}$ were used to obtain an 11.5 Å cryo-electron microscopy map of the complex. This map allows identification of RNA helices, peripheral proteins, and intersubunit bridges. Comparison of double-stranded RNA regions and positions of proteins identified in both cryo-EM and X-ray maps indicates good overall agreement but points to rearrangements of ribosomal components required for the subunit association. Fitting of known components of the 50S stalk base region into the map defines the architecture of the GTPase-associated center and reveals a major change in the orientation of the α-sarcin-ricin loop. Analysis of the bridging connections between the subunits provides insight into the dynamic signaling mechanism between the ribosomal subunits.

Introduction

The *Escherichia coli* 70S ribosome is a complex macromolecular assembly that consists of three rRNA molecules and over 50 proteins. To date, cryo-electron microscopy (cryo-EM) has been the only viable approach to the study of its structure (Frank et al., 1995b; Stark et al., 1995; Malhotra et al., 1998). The importance of cryo-EM studies lies in the fact that they allow ribosomes to be readily visualized in many different functional states without the need for crystallization (Agrawal et al., 1996, 1998, 1999a, 1999c; Stark et al., 1997a, 1997b).

Crystals suitable for X-ray crystallographic studies have been obtained only for ribosomes of organisms living in extreme environments (i.e., halophilic archeon: Ban et al., 1998, 1999; and thermophilic bacteria: Cate et al., 1999; Clemons et al., 1999; Harms et al., 1999; Schlünzen et al., 1999). Comparison of structures resulting from these studies, now in the resolution range of 5–8 Å, with those of mesophilic *E. coli* will be extremely useful in shining light on the process of molecular adaptation. At the same time, comparison between X-ray and cryo-EM maps helps in identifying flexible domains that are distorted on account of the packing in the crystal (Penczek et al., 1999). The best-defined 3D motifs, of double-stranded A form rRNA, are 17–20 Å in diameter with major grooves that can reach 13.5 Å in depth. Thus, resolutions in the 13 to 10 Å range are sufficient for detection and interpretation of such structural features and elements richly presented on the surface and at the subunit interface (Ban et al., 1999; Cate et al., 1999; Clemons et al., 1999).

Here, we present an 11.5 Å cryo-EM map of the *E. coli* 70S ribosome initiation-like complex with fMet-$tRNA_f^{Met}$ at the P site. Guided by recent placements of proteins and rRNA elements in X-ray maps, we were able to identify some of the ribosomal components and the nature of connections between the ribosomal subunits. The map also elucidates the architecture of functionally important domains. Structural details seen at this resolution, especially when put into the context of cryo-EM mappings of EF-G (Agrawal et al., 1998, 1999a, 2000), begin to reveal the identity and functional interplay of the connections between the regions associated with decoding, peptidyl transfer, and GTPase activity.

Results

Cryo-Electron Microscopy and Three-Dimensional Reconstruction

The cryo-EM density map was obtained from a total of 73,523 projections of the fMet-$tRNA_f^{Met}$·ribosome complex in a defocus range of 0.73 to 4.34 μm (see Experimental Procedures). The Fourier shell correlation curve (Figure 1) of the final, contrast-transfer function-corrected reconstruction indicates a resolution of 11.5 Å for a 0.5 cutoff (Böttcher et al., 1998). Although some useful information extends to 8.4 Å (3 σ criterion; Orlova et al., 1997), its inclusion into the final map is not justified as the signal-to-noise ratio falls below 1 (Malhotra et al., 1998). In the reconstruction, Fourier amplitudes at higher spatial frequencies are always underrepresented due to charging, instrument instabilities, specimen drift, and partial coherence. To compensate for these effects, we obtained scattering intensities using X-ray solution scattering measurements for *E. coli* ribosomes in the range up to 1/8 $Å^{-1}$, and applied a correction to the Fourier amplitudes up to the 1/11.5 $Å^{-1}$ cutoff point (Figure 1). The correction corresponds approximately to

To whom correspondence should be addressed (e-mail: joachim@wadsworth.org).

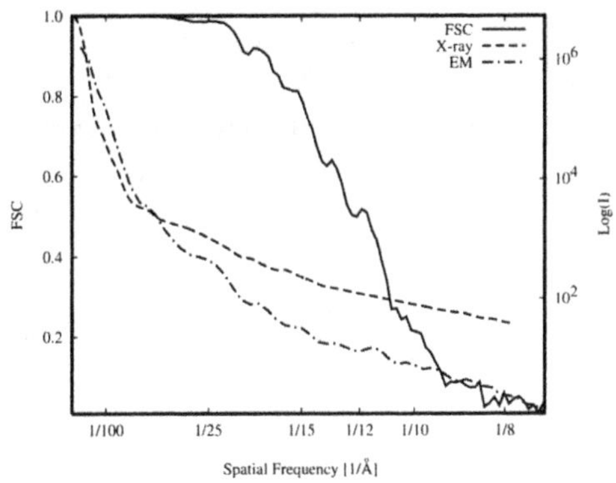

Figure 1. Assessment of Resolution, and Basis for Fourier Amplitude Correction

The Fourier shell correlation (FSC) curve (solid line) falls below 0.5 at 11.5 Å. Averaged radial falloff of Fourier intensities obtained from cryo-EM reconstruction (dashed/dotted line) versus measured using low-angle solution scattering of *E. coli* ribosomes (dashed line). The ratio of the two latter curves was used as a basis for correcting the EM Fourier amplitudes to obtain the final density map.

a temperature factor with $B = -310$ Å^2 (definition as in Glaeser and Downing, 1992). Application of this relatively small factor is justified in the range of spatial frequencies where the Fourier transform is statistically well defined.

For surface representations, threshold values were selected using the molecular volume as a guide. The protein level was initially chosen to encompass the total volume of 2.6×10^6 Å^3, based on the chemical molecular weight (Wittmann et al., 1982). We used a slightly enlarged volume of 2.68×10^6 Å^3 to retain continuity of some structural features. The RNA level was chosen to yield a protein/RNA ratio of 0.72:1.0. Around this threshold level, the appearance of the strands changed little when the level was varied.

Structure of Initiator-tRNA

The quality of a 3D cryo-EM reconstruction can be assessed when a component can be identified that is known at atomic resolution, by comparing its appearance with that of a low-resolution map computed from atomic coordinates. For comparison with the tRNA portion of the cryo-EM map, we used the X-ray coordinates of the yeast initiator tRNA (Basavappa and Sigler, 1991) and *E. coli* fMet-tRNA$_f^{Met}$ interacting with the methionyl initiator tRNA$_f^{Met}$ transformylase (Schmitt et al., 1998; inset in Figure 2, right). The structure corresponding to fMet-tRNA$_f^{Met}$ reveals details expected at 10 Å resolution (inset in Figure 2, left), proving the reliability of the other fine structural detail observed in the density map (Figure 2). Moreover, the overall shape and geometry of the tRNA in the cryo-EM map show specific structural features of *E. coli* fMet-tRNA$_f^{Met}$ (Schmitt et al., 1998), as opposed to the initiator tRNA of yeast. The acceptor and anticodon ends of the tRNA are seen as spirally distributed densities with distinguishable major and minor grooves. Both T and D loops are readily recognized. Connections of the tRNA with the 70S ribosome observed earlier (Malhotra et al., 1998) have become much more distinct.

Secondary-Structure Elements Visible in the Cryo-EM Map

The 3D map of the *E. coli* 70S ribosome at 11.5 Å resolution (Figure 2) shows numerous strands of relatively high densities with well-defined geometry such as right-handed twists and deep grooves characteristic of the A form conformation of rRNA. At this level of resolution it is not possible to unambiguously differentiate between RNA and proteins or to identify their recognition and binding determinants. Some of the helices, interpretable as rRNA, appear relatively isolated in the ribosome and run a long way before merging with other ribosomal secondary structure elements. A number of such structures are located between the ribosomal subunits, strongly resembling the corresponding elements of the crystallographic maps at ~8–5 Å.

30S Subunit

The interface side of the 30S subunit mainly consists of high-density material. Separate helical strands form the frames of body and shoulder (Figures 2a–2c) and define the skeleton of the platform (Figure 3, bottom). Helical elements also form the spur and two other strands intersecting with it at the bottom of the subunit (Figures 2b–2d). Separate strands connecting the head with the body are clearly seen. They surround a wide, curved core, identified as the neck, the only head-to-body connection visible in the isolated 30S subunit (Lata et al., 1996; Gabashvili et al., 1999a).

Comparison of our structure with the X-ray map of the *T. thermophilus* 30S subunit (Clemons et al., 1999) allows us to identify a number of structural elements whose appearance is virtually identical in both maps although their relative positions change (see below). Similarities are seen especially well when the cryo-EM map is displayed at a higher threshold for which the volume is estimated to enclose mainly RNA mass. The conformation of the long vertically running helix, identified as helix 44 of 16S RNA in the X-ray maps (Cate et al., 1999; Clemons et al., 1999), appears practically the same, except that it is slightly longer, consistent with the 5 nucleotide difference in secondary structure, and that it shows a slight kink close to the 1430/1470 bulge where bridge B5b (see section on Intersubunit Bridges below) connects. Two structures coming up from the shoulder, looking similar in both X-ray (Cate et al., 1999; Clemons et al., 1999) and cryo-EM maps (marked as * and # in Figure 3c), are clearly seen as differently positioned in all three maps. While only one of these structures (*) participates in the head-to-shoulder contact in the cryo-EM map of the 70S ribosome and the X-ray map of the isolated 30S subunit (Clemons et al., 1999), both appear to interact with the head in the X-ray map of the 70S ribosome (Cate et al., 1999). The truncated appearance of a major head domain in the X-ray map of the subunit (Clemons et al., 1999) could be due to disorder, while the enlarged form of the shoulder structure contacting the head (marked # in Figure 3c; Cate et al., 1999) could be due to difficulties in finding the correct molecular boundary (see Schlünzen et al., 1999).

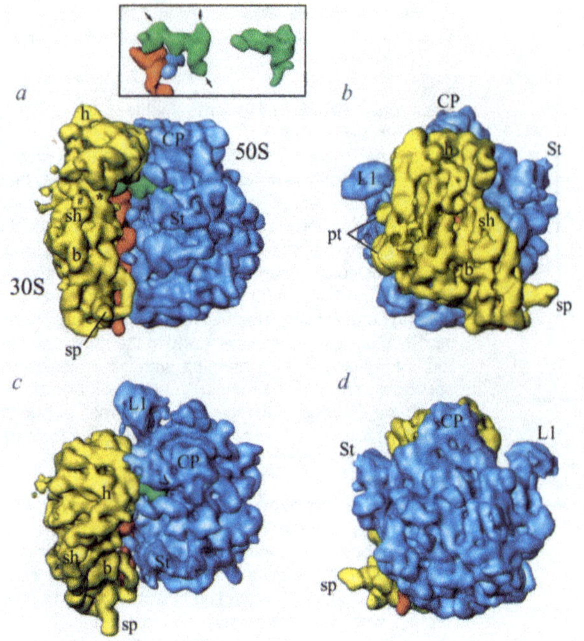

Figure 2. The 11.5 Å Resolution Cryo-EM Map of the *E. coli* 70S Ribosome

The cryo-EM map is shown in (a) subunit–subunit side view, (b) 30S subunit solvent-side view, (c) top view, and (d) the 50S subunit solvent-side view. Domains and features of the 30S subunit (yellow, with helix 44 shown in red) and 50S subunit (blue) are as follows: b, body; h, head; pt, platform; sp, spur; sh, shoulder; St, stalk base; L1, protein L1; and CP, central protuberance. * and # mark sites of the shoulder contacting the head in Cate et al., 1999. V marks the v-shaped structure possibly representing helices of the 5S rRNA. Inset (upper left) displays cryo-EM map of the fMet-tRNA$_f^{Met}$ (green), with arrows pointing to sites interacting with ribosomal components (left), and the expected appearance of fMet-tRNA$_f^{Met}$ (Schmitt et al., 1998) at ~10 Å resolution (right).

50S Subunit

Among easily distinguishable RNA helices is the one bearing the L1 protein (Figure 2, Figure 3, top and center), partially truncated in the X-ray map (Ban et al., 1999). Other well-defined helices can be seen on the solvent side, next to the central protuberance (CP) of the subunit (Figures 2a and 2c). Helical-appearing structural elements of the 50S subunit making contact with helix 44 of the 16S RNA (Figure 3, top and bottom) and others located at the interface (Figure 3, center) are identified as RNA helices in the interpretation of the 5 Å X-ray map (Ban et al., 1999). Variations in the 23S RNA secondary structure among the species are recognizable in some regions of the subunit. It is clearly seen in the case of helix 63 (localized by Spahn et al., 1999), known to be 22 (*H. marismortui*) or 16 (*T. thermophilus*) nucleotides longer in *E. coli*. There are some notable differences in the conformations between elements with close secondary structure, e.g., the helices of the CP (Figure 2): two strands connected in a V-shaped arrangement (marked V in Figure 2c) are aligned perpendicularly to the interface, interacting with the head of the 30S subunit (also see Cate et al., 1999). In the X-ray map of the isolated 50S subunit (Ban et al., 1999) these helices point upwards, toward the top of the CP. The most striking differences are seen in the intersubunit region where the cryo-EM map contains helical strands pointing away from the 50S, toward the 30S, participating in the formation of intersubunit bridges (Figure 3, center: B1a, B2a, B3 in the nomenclature introduced below). In the X-ray structure of the 50S subunit from *H. marismortui* (Ban et al., 1999), the corresponding strands are bent to run close to the intersubunit surface, as though confined by the crystal packing.

Intersubunit Bridges

Several intersubunit bridges were described earlier in low-resolution maps of the *E. coli* ribosome (Frank et al., 1995a; Lata et al., 1996). These bridges are sharply resolved in the present 11.5 Å map, in some cases into several thinner bridges (Figure 3). To characterize them, we use the previously introduced nomenclature B1–B6 (Frank et al., 1995a; Lata et al., 1996), further elaborated by Cate and coworkers (1999).

The earlier-identified bridge B1 between the head of the 30S subunit and the CP of the 50S subunit is now seen to comprise three minor connections, named B1a, B1b, and B1c. B1b and B1c spring from the CP, where the 5S rRNA is located, while B1a is closer to the L7/L12 stalk base and is clearly associated with a 23S RNA helix.

Another bridge previously identified, B2, involving an extended contact between the platform of the 30S subunit and the rim of the interface canyon of the 50S subunit (Frank et al., 1995a; Lata et al., 1996), is now resolved into five separate connections, B2a–B2e. The position of B2a is at the point where helix 44 of the 16S RNA merges with the platform. This upper portion of the helix is directly involved in the formation of the strongest bridging connection between the subunits. Bridges B2b–B2e are associated with other helical structures of the platform. B2d and B2e evidently belong to the same helix positioned at the upper edge of the platform. B2a, B3, B5a, and B5b are formed by four strands of density projecting off the central interface region of the 50S subunit, interacting with four areas of closest approach along h44 of 16S RNA on the interface side of the 30S subunit (Figure 3). B4 was already very well resolved in a lower-resolution map (Frank et al., 1995a).

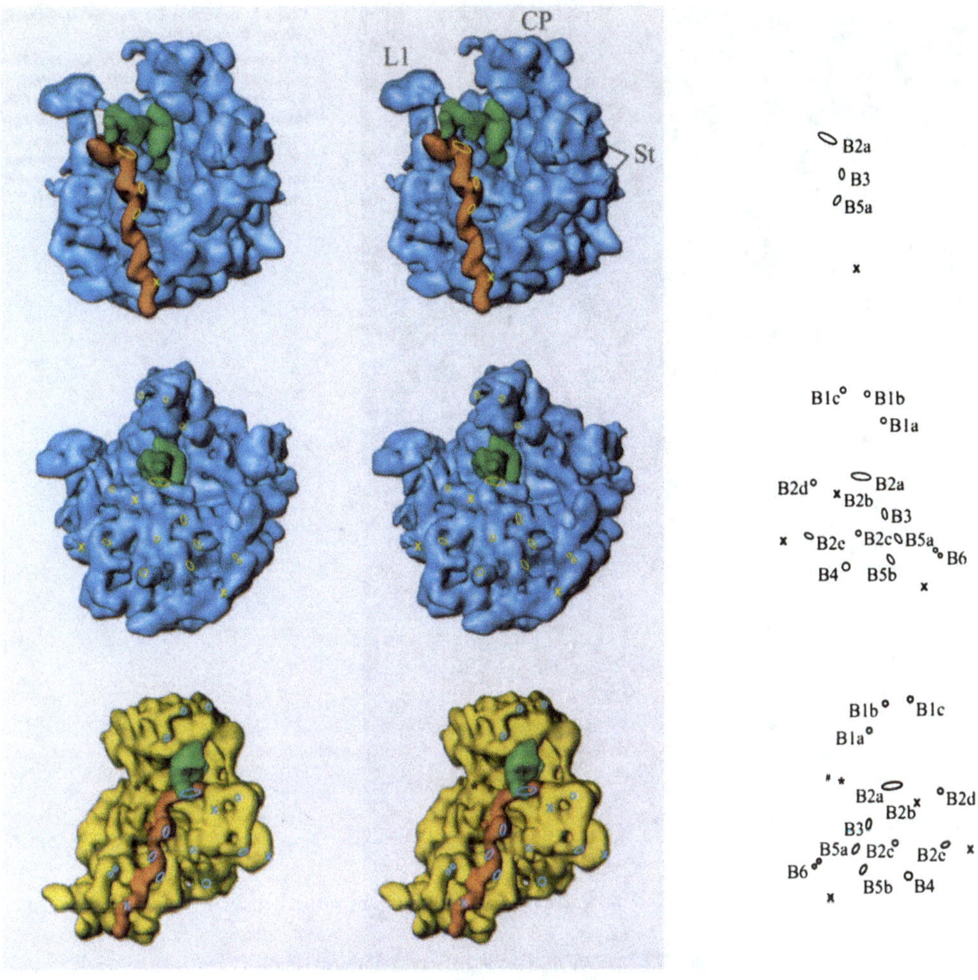

Figure 3. Stereo Views of Large and Small Subunit after Separation of the Ribosome Map

Stereo views of large (top and center panels) and small (bottom panel) subunits, as would be seen from the intersubunit space, obtained by separation of the 70S ribosome map. The portion of the map identified as helix 44 of 16S RNA, which is involved in the association of the subunits, is displayed in red in top and bottom panels. The density attributed to initiator tRNA is shown in green. The arrangement of intersubunit bridges (open circles) is shown on the right of all three panels as mirror-related footprints that can be superimposed on the corresponding subunit views. X marks regions of close contact that do not form a bridge. Other landmarks are the same as introduced in Figure 2.

Finally, a low-density contact in the lower part of the 30S subunit earlier identified as B6 (Frank et al., 1995a) is now seen to involve two different structures. This bridge is absent in the *T. thermophilus* 70S ribosome, though Cate and coworkers (1999) apply this label to another, apparently unrelated structure at a different location.

Placement of Known X-Ray Structures of Ribosomal Components

Each element (protein, RNA helix, or RNP fragment) was docked into the cryo-EM map independently using both Iris Explorer and O (Jones et al., 1991), with initial guidance provided by the placements of Clemons and coworkers (1999) for the 30S subunit, and of Ban and coworkers (1999) and Agrawal and coworkers (2000) for the 50S subunit.

Proteins and rRNA Fragments of the 30S Subunit

All helices of the 16S RNA central domain traced in the X-ray map of the isolated 30S subunit (Clemons et al., 1999) were identified in the cryo-EM map. Helices 19 and 27 (numbering by Müller and Brimacombe, 1997) are continued by a long strand apparently including helices 2 and 28, which form the neck of the subunit. The three-way junction of helices 20, 21, and 22, interacting

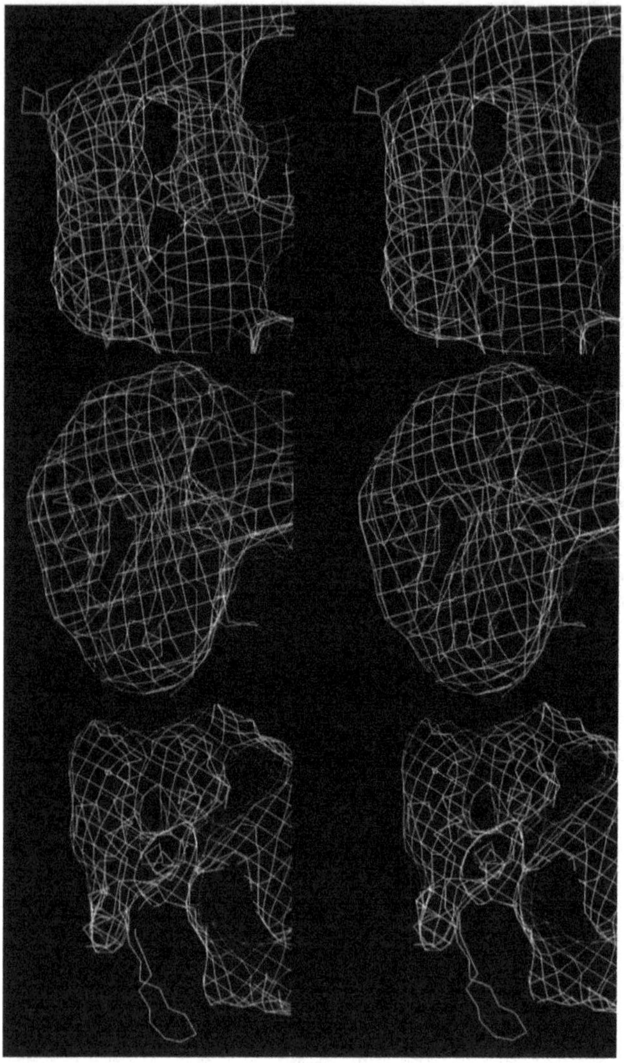

Figure 4. Examples for Fittings of Proteins into the Cryo-EM Density Map

Fittings of protein crystal structures (carbon backbones) into the cryo-EM density (white, wire-mesh), displayed in stereo. Top, S5 (Ramakrishnan and White, 1992); center, S6 (Lindahl et al., 1994); bottom, S7 (Hosaka et al., 1997; Wimberly et al., 1997). These protein structures are displayed on the 30S subunit map in Figure 5 in corresponding colors.

with protein S15, was found to lie in the vicinity of bridge B4. Relative positions of h44, h27, and P site tRNA modeled by Cate and coworkers (1999) showed a very good agreement with the positions of the corresponding elements in the cryo-EM map.

The following proteins were placed: S4, S5, S6, S7, S8, and S15 (Figure 5). In the case of structures identified by Clemons and coworkers (1999) as S17 and S20, the interior boundaries are poorly defined in our map, precluding an accurate fitting. Stereo pairs showing the fittings of three proteins (S5, S6, S7) are shown in Figure 4. Except for the apparently flexible loop region of S7, the overall shape and features of these proteins, all located on the surface of the 30S subunit, are as would be expected at 11.5 Å resolution.

Proteins and rRNA Fragments of the 50S Subunit

Components of the 50S Stalk Base Region. A bilobed region of the large subunit commonly referred to as the stalk base is formed, in major part, by two functionally important components: (1) the thiostrepton-binding L11–23S RNA (58 nt) fragment (upper lobe in Figures 6 and 7) and (2) L6 and the α-sarcin-ricin loop (lower lobe) (see Ban et al., 1999; Agrawal et al., 2000).

Areas of contact between elongation factor G and this region were determined by cryo-EM of several EF-G·ribosome complexes (Agrawal et al., 1998, 1999a). Fitting of the crystal structure of the EF-G domains (Agrawal et al., 1998, 1999c) along with hydroxyl-radical probing (Wilson and Noller, 1998) helped in the placement of ribosomal components interacting with EF-G

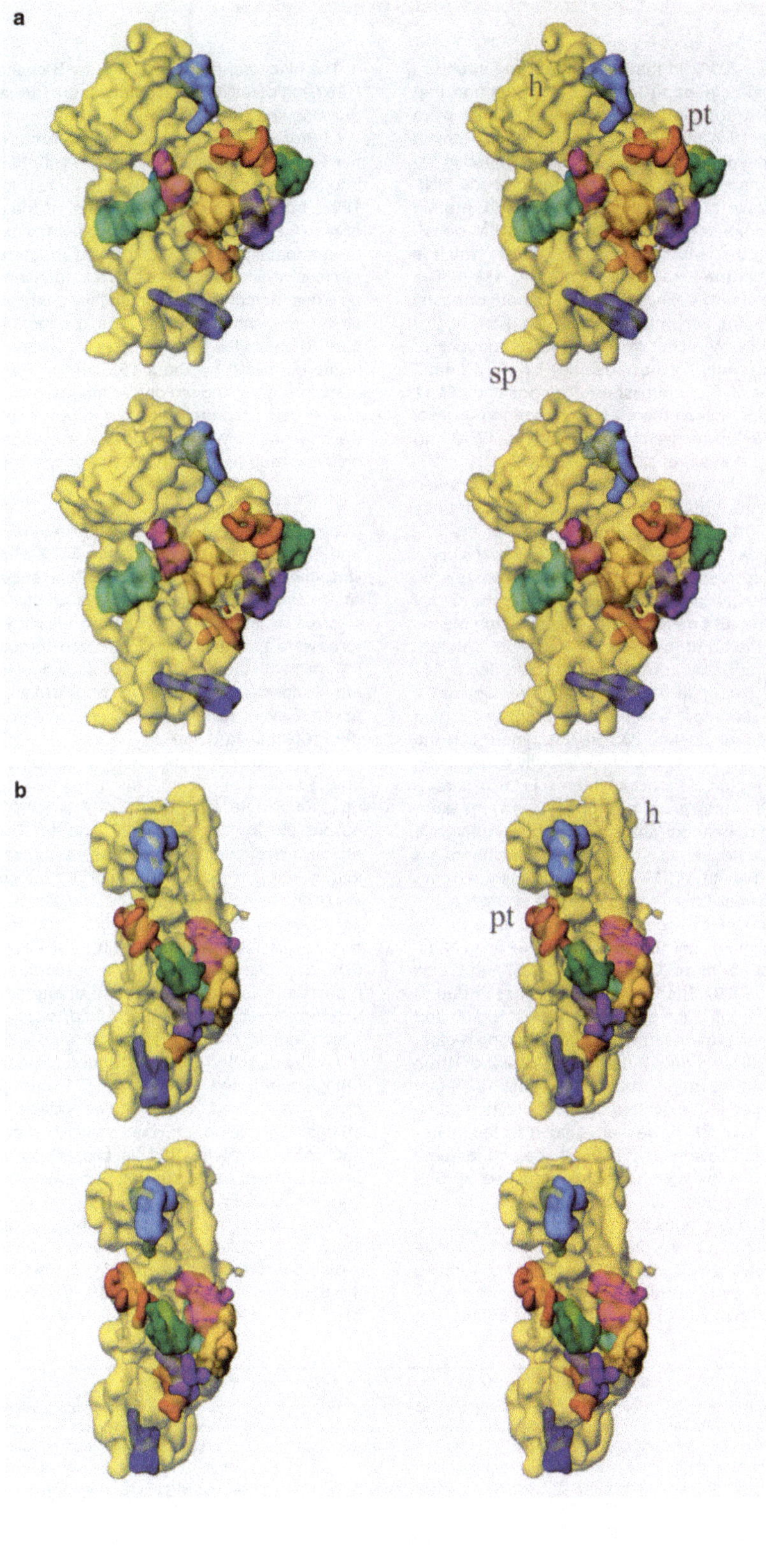
a
h
pt
sp
b
h
pt

(Agrawal et al., 2000). In that work, the X-ray structure of the α-sarcin-ricin loop (Correll et al., 1998) and that of an L11-RNA complex (Wimberly et al., 1999) were fitted into the 15 Å map of the ribosome (Malhotra et al., 1998). Those fittings have been further refined in the new 11.5 Å map (Figure 6a). The 58-nucleotide RNA fragment (shown in blue) along with the L11 protein (green) are completely covered by the cryo-EM densities, including the N-terminal domain of L11, which is not observed in the X-ray map (Ban et al., 1999). The cryo-EM map clearly defines the shape of both domains of L11 (Figure 6a), which was directly localized in this part of the stalk by recent cryo-EM reconstructions of an L11-lacking mutant of the ribosome (R. K. A., J. Linde, K. H. Nierhaus, J. F., unpublished). The position of L11 reveals that the observed arc-like connection (Agrawal et al., 1998, 1999a) involves the G′ domain of EF-G and the N-terminal domain of L11 (see Figure 7).

Among various other EF-G domains, the G domain and domain V also make direct contacts with the base of the stalk. Fitting of the X-ray structure of the EF-G·GDP complex (Czworkowski et al., 1994) into the cryo-EM density corresponding to EF-G (Agrawal et al., 1999a) revealed an unoccupied region near the G domain in the cryo-EM density. This region, which makes contact with the tip of the α-sarcin-ricin loop structure (see below), probably corresponds to disordered 24–amino acid residues in the X-ray structure. Domain V of EF-G makes contact with the groove-shaped region between the 58-nucleotide 23S RNA fragment and the N-terminal domain of L11 protein (Wimberly et al., 1999).

A 23S RNA region of particular interest is the α-sarcin-ricin loop (SRL), because of its crucial roles in the elongation factor–dependent activities of both prokaryotic and eukaryotic ribosomes. Its immediate neighbors are L6 and L14 (Ban et al., 1999). All three elements are located in the stalk base region, with SRL sharply delineated on account of its higher-density S turn profile (yellow in Figure 6a). In the fitting of SRL we used X-ray structures both from rat (Correll et al., 1998) and *E. coli* (Correll et al., 1999). The latter gives a better fit due to the absence of one base pair in the flexible region and a changed arrangement between stem and loop region. Thus, the small structural differences observed in the X-ray maps of SRL fragments (Correll et al., 1999) can also be identified in the context of the ribosome matrix.

In locating L6 and L14, we were guided by the placement of Ban and coworkers (1999). Shape and features of the L6 protein, which was slightly modified by Ban and coworkers (1999), are recognizable in the surrounding stalk base densities (green in Figure 6a). L6 (aa 103) is found in close proximity to a minor groove of SRL, near base 2667, with the long axis of L6 running approximately perpendicular to SRL. On the other side, SRL is flanked by L14 but forms no direct contact with it. The placement of this protein by Ban and coworkers (1999) indicates its involvement in the formation of bridges B5a and B6.

L1 Protein. Previously, the X-ray structure of this protein from *T. thermophilus* (Nikonov et al., 1996) was fitted into the 15 Å map of the 70S ribosome (Malhotra et al., 1998), based on shape constraints. At the current level of resolution, fitting of X-ray coordinates can be guided by subdomain details of the protein structure. Moreover, the relative orientation of the two L1 domains appears to be different from the orientation they possess in the X-ray structure, which depicts the protein without RNA interaction. To obtain the best fit (Figure 7), we therefore had to rotate domain II by about 15 degrees, using the interdomain-connecting regions as hinges. It was previously shown that L1 protein has a different conformation in the ribosomal crystal structure when compared with the cryo-EM map (Ban et al., 1998; Penczek et al., 1999).

Comparison of Placements of Various rRNA Helices and Proteins in Cryo-EM versus 5 Å X-Ray Maps

Obtaining a Common Frame of Reference

At the time of preparation of this article, none of the electron densities of the recently published X-ray structures were available. Thus, in order to compare the cryo-EM map with the X-ray maps of individual subunits, the atomic coordinates of proteins and RNA elements fitted to the relevant density maps were used (available from the Protein Data Bank).

In the case of 30S, the docking (see previous section on Placement of Known Structures) was done for proteins S4, S5, S6, S7, S8, S15, S17, and S20, as well as helices 20, 21, 22, 23, 24, 27, and 44. The full set of atomic coordinates obtained was treated as a rigid body and overlapped with the model of Clemons and coworkers (1999) using a least-squares transformation, including changes of scale, translation, and rotation. After outliers (S4, S6, h23) and fittings of high uncertainty (S17, S20) were removed, the least-squares fitting was repeated. To evaluate the extent of any discrepancies, a root-mean-square deviation (rmsd) was calculated between atomic coordinates of the model derived from the X-ray map and the transformed coordinates of the fittings performed on the cryo-EM map. This overall rmsd was 4.5 and 5.4 Å with two independent sets of fittings. In addition, an rmsd value was calculated for each of the proteins and helices, which was decomposed into two parts, attributed to translation and rotation, respectively.

In the case of the 50S subunit, the analysis described above was applied to the following four elements: the α-sarcin-ricin loop (SRL), protein L6, the RNA portion of the thiostrepton-binding L11/23S RNA fragment (Wimberly et al., 1999), and the C-terminal domain of L11

Figure 5. Positions of Fitted Proteins and 16S RNA Helix 23 in Cryo-EM and X-Ray Map of the 30S Subunit

Semitransparent stereo representations of the 30S subunit cryo-EM density map, showing the positions of some of the proteins and helix 23, as fitted into the cryo-EM map (top panels) and into the X-ray map (Clemons et al., 1999) (bottom panels). The relative positions between coordinates derived from fittings into cryo-EM and X-ray maps were obtained by an overall least-squares fit (see text). (a) view of the subunit from the interface side, (b) from the side of the platform (i.e., from the right in [a]). Color coding: h23, red; S4, blue-green; S5, magenta; S6, green; S7, light blue; S8, yellow; S15, violet; S17, orange; S20, dark blue. Landmarks: h, head; pt, platform; sp, spur. (Models of S17 and S20 were derived from C-α coordinates, all others from all-atoms coordinates.)

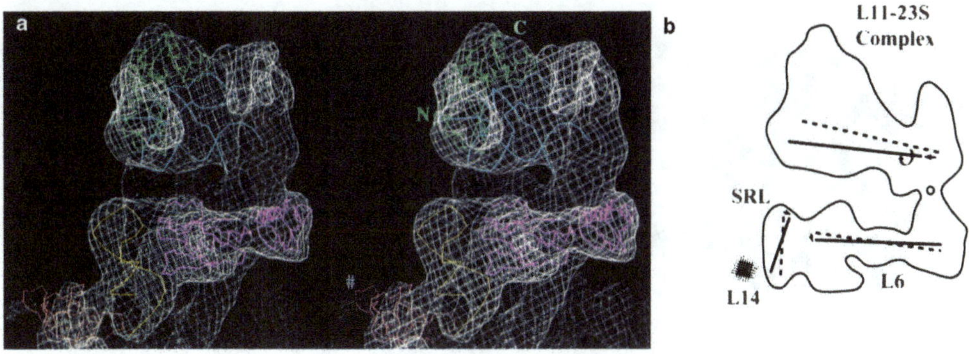

Figure 6. Architecture of the GTPase-Associated Center, and Its Rearrangement

(a) Closeup on the stalk base region of the 50S subunit, along with fittings of PDB coordinates of three components, displayed in stereo. These fittings were obtained using all-atom coordinates; however, only backbones are shown for clarity. Yellow, α-sarcin-ricin loop (SRL; Correll et al., 1999); magenta, protein L6; green and blue, L11 (green)/23S RNA fragment (blue) (Wimberly et al., 1999). Orange, L14, whose placement was adopted from Ban and coworkers (1999). # points to the proximity of mass corresponding to bridge B5a and a portion of L14. N and C indicate N- and C-terminal domains of protein L11, respectively.

(b) Schematic diagram depicting the changes in the positions of SRL, L6 and L11-23S complex from X-ray fittings (*H. marismortui*, Ban et al., 1999; dashed line) to those obtained for the 11.5 Å cryo-EM map (*E. coli*). The outline shown is the boundary of the two lobes of the stalk base density map as it appears in the plane in which the three fitted components lie. The lines indicate, respectively, the position of the helix axis for SRL, the approximate position of the long axis for L6, and a line joining the stem containing the 5′ and 3′ ends of the RNA (58 nt) fragment (on the right) with the center of L11 (on the left). Circle represents the peripheral hinge point between the two lobes of the stalk base.

(Conn et al., 1999), using the placements supplied by Ban and coworkers (1999) for comparison. In this case, the overall rmsd was 3.4 Å.

30S Subunit

Fittings of various proteins and RNA fragments into the 30S portion of the cryo-EM map are presented in Figure 5. For comparison, the fittings of the same components into the X-ray map (Clemons et al., 1999) are also shown in the framework of our map (bottom panels in Figures 5a and 5b). It can be seen in the side view (Figure 5b) that the general direction of changes in position, going from the X-ray fittings to our fittings, is inward. S4, situated on the shoulder, a portion of which forms the border of the mRNA channel, moves in the direction of the channel toward the 50S subunit. S6, which is located on the lower lobe of the bilobed platform, moves by ~6 Å away from the 50S subunit, while helix 23, which forms the rim of the upper lobe, moves by ~10 Å in the same direction. The different sizes of these movements define a twist of the underlying platform structure, which curls toward the head of the 30S subunit. Helix 23 is instrumental in the association of the subunits and evidently participates in the formation of bridges B2d and B2e. The same change in platform morphology was observed when we compared the density maps of the isolated and the 50S-bound 30S (Lata et al., 1996; Agrawal et al., 1999b).

The shift of S7 (head portion) by >5 Å toward the 50S subunit is probably associated with a slight movement of the 30S head in that direction, again in agreement with a conformational change upon subunit association (Agrawal et al., 1999b). Finally, S20 (modeled by Clemons et al., 1999), which could not be fitted because of the lack of internal features in this region of our map, can nevertheless be recognized by the shape of its boundary. Its placement in the X-ray map is clearly in conflict with our map since it protrudes from it by almost 10 Å. Thus S20, located close to the lower end of h44, indicates a major rearrangement in that part of the two maps. The main thrust of the rearrangement is an inward movement of S20 from its more protruding position in the X-ray map, roughly following a diagonal direction.

50S Subunit: Region of GTPase-Associated Center

One of the most interesting changes between the cryo-EM and X-ray structure of the 50S subunit, from a functional point of view, takes place in the constellation formed by the α-sarcin-ricin loop (SRL), proteins L6, and L14. The most substantial change when going from the X-ray to the cryo-EM map is a rotation of the helix axis of SRL by 17° in the plane of Figure 6a around a point located on the upper half of the helix axis. In addition, we observe a minor rotation (~6°) of L6 in that same plane (Figure 6b). The movement of SRL brings its 3′ and 5′ ends closer to protein L14. When comparing the relative positions of the L11/23S rRNA fragment with respect to the tip of SRL in the two maps, we find a shift, by 8 Å, in the direction of L11 (horizontal in Figure 6) accompanied by a rotation of 10.5°, effectively narrowing the distance between the two structures by ~10 Å as we go from the X-ray to the cryo-EM map. Since these two components of the 50S subunit are sitting in two distinct lobes of the stalk base, their relative movement is associated with a relative twisting movement of the lobes around a peripheral flex point (marked with a circle in Figure 6b).

Discussion

Assignments of Interface Elements Forming the Bridges

Available biochemical information and existing rRNA models, as well as recent placements in crystallographic maps (Ban et al., 1999; Clemons et al., 1999) and analysis of bridging connections in *T. thermophilus* (Cate et al.,

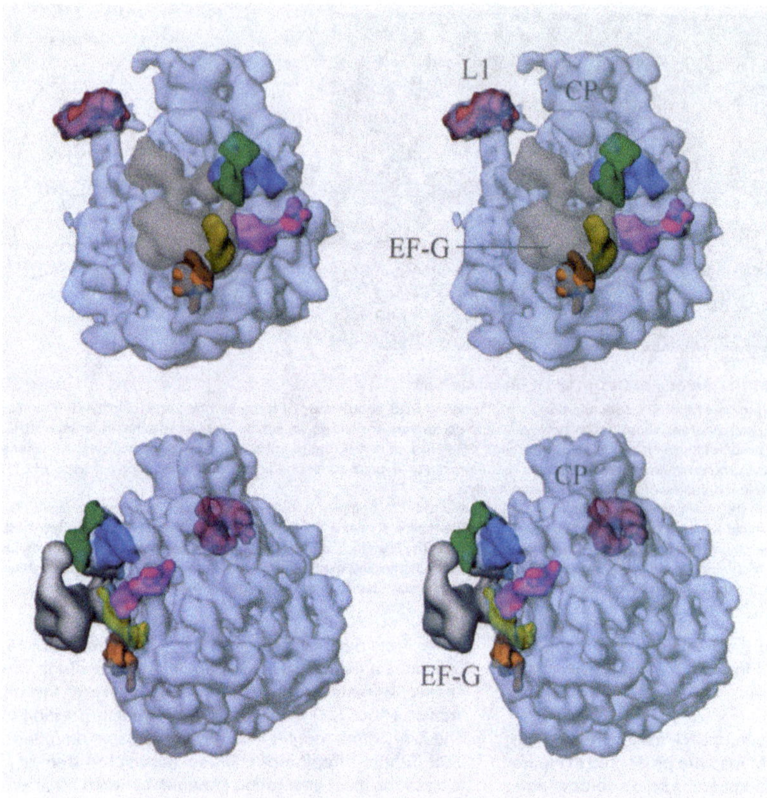

Figure 7. Positions of Fitted Proteins, RNA Fragments, and Elongation Factor G in the 50S Subunit

Semitransparent stereo representation of the 50S subunit cryo-EM density map (light blue), showing the fitted positions of proteins and RNA fragments at the GTPase-associated center, as well as L1. Color coding: red, L1; magenta, L6; orange, L14; green, L11; blue, 58 nt 23S RNA fragment (Wimberly et al., 1999); yellow, α-sarcin-ricin loop; semi-transparent gray, density corresponding to EF-G in the ribosome·$(tRNA)_2$·EF-G·GDP·fusidic acid complex (Agrawal et al., 1999a). EF-G is seen to contact the tip of the α-sarcin-ricin loop with its GTP-binding domain, and the N-terminal domain of L11 with the tip of its G domain (Agrawal et al., 2000). Landmark: CP, central protuberance.

1999) help to identify structural elements forming the intersubunit bridges in *E. coli* ribosome (Figure 3).

Several bridging connections of the ribosome have been unambiguously identified in the recent paper of Cate and coworkers (1999). Bridge B4 has been shown (Culver et al., 1999) to be formed by protein S15 and helix 34 of the 23S RNA. The participation of helices 44 of the 16S RNA and 69 of 23S RNA in the construction of B2a, and of helices 23, 24, and 27 of the 16S RNA in forming B2b (h24), B2d (h23.2), B2e (h23.1), and B2c (h27) is also known (see Cate et al., 1999). The corresponding contacting structures from the 23S RNA are possibly h68 and h71. We believe that bridges B2b and B2d are formed by a single helix, h68, as, on the one hand, it is the closest neighbor of h69 (B2a) and, on the other hand, it has been cross-linked both with protein L1, localized in the vicinity of this bridge, and with tRNAs located close to this part of the interface canyon (Osswald et al., 1990, 1995). Some of the B2 bridges with the 30S subunit platform could involve portions of the large proteins L2 (B2e) and L3 (B2c) of the 50S subunit and S11 (B2d) of the small subunit. L2 and S11 have been cross-linked to each other (Lambert and Traut, 1981), as well as to the bridge-forming (h23 of 16S RNA) or neighboring (h64 of 23S RNA) helices (Osswald et al., 1990). These three proteins are known to play pivotal functional roles in protein synthesis that require a high degree of coordination: L2 is associated with the peptidyl-transferase activity (Wittmann-Liebold, et al., 1995), S11 has been implicated in tRNA selection (Fanning et al., 1978), and L3 in controlling maintenance of the mRNA reading frame in particular cases (Peltz et al., 1999).

Participation of L2 in bridge B2e is supported by the placement of L2 by Ban and coworkers (1999). This bridge is absent in the 70S map of *T. thermophilus* (Cate et al., 1999), although in *E. coli* we do observe a very strong connection of this 50S subunit region with the 16S RNA helix h23. This may be one of the species-related differences between mesophilic and thermophilic 70S ribosomes.

According to protection data (Merryman et al., 1999),

helix 38 of 23S RNA is the most likely candidate for bridge B1a. This structure was identified earlier as long bridge (Malhotra et al., 1998), or A site finger (Stark et al., 1997a) in a truncated form.

Bridge B6 is located near the variable loop of h44 of the 16S RNA. The participating structural elements of the 50S subunit are seen in the cryo-EM map as belonging to two different but neighboring structures. The lower strand is in the immediate neighborhood of helix 63 of the 23S RNA (identified by genetic tagging and cryo-EM; Spahn et al., 1999). Thus, it may involve neighboring nucleotides in the RNA secondary structure (shown to cross-link with proteins L19 and L14). The upper structure participating in the formation of this bridge from the 50S side may represent protein L14 (Ban et al., 1999). Bridge B6 is absent and replaced by another bridge in *T. thermophilus* located on h44, apparently providing a restraint for the movement of this helix at the subunit–subunit interface.

Dynamic Nature of Connections between the Subunits

Comparison of the cryo-EM map with the recent X-ray maps of the extremophile ribosomes (Ban et al., 1999; Cate et al., 1999; Clemons et al., 1999), along with recent results showing rearrangement of bridges in the *E. coli* ribosome (Gabashvili et al., 1999b), indicates movements in these bridging elements in the ribosomal interface. The most interesting region in this respect is the penultimate loop, i.e., h44 of the 16S RNAs 3′ minor domain, connected to the 50S subunit in four regions (bridges B2a, B3, and B5a,b). The appearance of these long connections suggests their role in the dynamics of the ribosome, as small changes in the conformation of the contacting 23S RNA helices would result in relatively big displacements of h44 and components of the body of the 30S subunit connected with it. It is highly likely that these structures play a crucial role in the association and cooperative functioning of the ribosomal subunits. The upper portion of h44 participates in the mRNA decoding and controls fidelity of the codon–anticodon interactions along with the mRNA translocation. Of particular importance in this respect is h27, located between B3 and B2c (Clemons et al., 1999; Cate et al., 1999), which switches conformation, thereby modulating translational accuracy (Lodmell and Dahlberg, 1997). Indeed, a recent cryo-EM study of ribosomes from mutants with h27 locked in alternative conformations indicates large rearrangements affecting the position of h44 as well as more distant parts of the ribosome (Gabashvili et al., 1999b). B1b and B1c were seen to change their architecture in this transition from one proofreading state to the other. Comparison with the X-ray structure of the unbound 30S subunit (Clemons et al., 1999) indicates a shift of helix 44 upon subunit–subunit association. Presumably, this helix serves as a binding sensor, promoting the 30S–50S subunit association and initiating structural rearrangements of the decoding region.

The absence of intersubunit cross-links and regions protected by subunit–subunit association in the head of the 30S subunit is in accordance with the observed relative mobility of this domain (Gabashvili et al., 1999b; Agrawal et al., 2000).

Interpretation of Conformational Changes

The changes in positions of proteins fitted into the 30S part of the 70S cryo-EM map from *E. coli* compared to those reported for the isolated 30S subunit from *T. thermophilus* (Clemons et al., 1999) can be explained on the basis of an architectural rearrangement of the subunit apparently required for its association with the 50S subunit. These changes involve a curling-in of the platform toward the head of the 30S subunit (thereby creating a channel for the upstream region of mRNA; see Frank et al., 2000b) and a shift of the head toward the 50S subunit (Lata et al., 1996; Agrawal et al., 1999b). It is likely that the nature of this rearrangement is the same for *T. thermophilus*, as far as it can be judged from a visual comparison with published X-ray maps (Cate et al., 1999; Clemons et al., 1999).

To assess the significance of the observed changes in the region of the GTPase-associated center, it is necessary to recall all the aspects in which the cryo-EM map differs from the X-ray map: (1) species (*H. marismortui* versus *E. coli*): it is unlikely that difference of this size will occur in one of the most conserved, functionally important regions of the ribosome; (2) crystal packing (crystal versus solution structure): the size and nature of the SRL movement argue against such an explanation; however, crystal packing forces cannot be ruled out as an explanation for the rearrangement of the peripherally located lobes of the stalk base that gives rise to the change in the distance between SRL and L11/23S rRNA; (3) free versus 30S-bound 50S subunit; (4) tRNA-bound versus unbound ribosome. The two last factors are likely to be the most important in explaining the observed conformational rearrangement. Since the cryo-EM structure shows the ribosome in a defined state (initiation-like), it is quite possible that the X-ray map of the isolated large subunit (Ban et al., 1999) could represent another conformational state along the elongation cycle.

The effect of the SRL rotation on L6 is that it changes the geometry of the contact zone between the two structures. Since L6 and the L11/23S rRNA structure are interconnected and strategically located in the region of the stalk base, their movements are likely to produce changes in the orientation of the L7/L12 stalk, which may in turn cause relatively large changes in the position of the stalk tip. On the side opposite L6, the decrease in distance between SRL and protein L14 brought about by the rotation of SRL is noteworthy because of the participation of L14 in forming two bridges spanning the two subunits, B5a and B6. Thus, the formation of the contact between SRL and L14 could be part of the dynamic signaling that is thought to coordinate the actions at the decoding region of the 30S subunit with that of the centers of the 50S subunits GTPase and peptidyltransferase activities.

Experimental Procedures

Microscopy and Image Acquisition

Initiation-like fMet-tRNA$_f^{Met}$-ribosome complexes were prepared for cryo-microscopy as described by Malhotra and coworkers (1998). Micrographs were recorded using low-dose protocols (10e$^-$/Å^2) on a Philips CM200 C1 field emission gun (FEG) electron microscope at 200 kV and a nominal magnification of 51,200×. They were checked for drift, astigmatism, and presence of Thon rings by optical diffraction, and scanned on a Perkin Elmer PSD 1010A microdensitometer using a step size of 15 μm, corresponding to 2.93 Å on the object scale.

Image Processing
Ribosomes were selected by a quasi-automated selection procedure (Lata et al., 1995) and by direct comparison of the particle candidates with the reference set of 83 projections (Penczek et al., 1994) of the existing map of the fMet-$tRNA_f^{Met}$ complex (Malhotra et al., 1998). A subset of particles was chosen based on the size of the cross-correlation coefficient and was further reduced following visual inspection to a final set of 54,479.

The CTF parameters were initially calculated based on the power spectrum estimated from each micrograph using the method of averaged overlapping periodograms (Zhu et al., 1997). All particles were sorted according to defoci derived from CTF analysis and placed into 23 groups. The defoci ranged from 2.17 to 4.34 μm and the number of particles per group ranged from 974 to 3831. The same analysis was repeated for the previously published set of fMet-$tRNA_f^{Met}$-ribosome particles (Malhotra et al., 1998), from which 19,044 best-quality particles were selected. They were placed in additional eight defocus groups with defoci ranging from 0.73 to 2.32 μm and the number of particles per defocus group ranging from 860 to 3300. Since this data set was collected on a different microscope (Philips EM 420 at 52,200× magnification), the pixel size was adjusted and additional normalization of the data set was performed.

The total set of 73,523 particles was subjected to 3D projection alignment, as described in Frank et al., 2000a. The refinement was repeated eight times with the angular step gradually decreasing from the initial 2.0° to 0.2° for the final step. During the refinement an additional verification and adjustment of CTF parameters were performed. After CTF correction (Frank and Penczek, 1995), the estimation of resolution was done by randomly splitting the data set into two groups, calculating separate volumes, and comparing them in Fourier space using the Fourier shell correlation (van Heel, 1986).

All steps of image processing were performed using the SPIDER image processing system (Frank et al., 1996). Computations for refinement and 3D reconstruction were done on the Origin 2000 (Silicon Graphics, Mountain View, CA).

Correction of the Fourier Amplitudes
Synchrotron radiation X-ray scattering data were collected using standard procedures on the X33 camera of the EMBL on the storage ring DORIS III of DESY and multiwire proportional chambers with delay line readout (Koch and Bordas, 1983; Boulin et al., 1988). The samples were measured at 70S ribosome concentrations of 5 and 20 mg/ml in the same buffer as used for EM studies. Scattering data were recorded at sample-detector distances of 3.9 and 1.4 m covering the momentum transfer ranges $0.01 < s < 1.7\ nm^{-1}$ and $0.03 < s < 5.4\ nm^{-1}$, respectively ($s = 4\pi \sin\theta/\lambda$, where $\lambda = 0.15$ nm is the X-ray wavelength and 2θ the scattering angle). The data were normalized and processed using the program SAPOKO (D. I. S. and M. Koch, unpublished), and the processed curves at the two sample-detector distances were merged to yield a composite scattering curve. The latter was further processed using the indirect Fourier transform program GNOM (Svergun et al., 1988) to compute the distance distribution function in the range $0 < r < 27$ nm. A sine Fourier transformation of this function yielded the final smoothed, extrapolated scattering intensity up to $1/7.85\ Å^{-1}$. Comparison between the final X-ray scattering intensity distribution and the power spectrum of the cryo-EM map (Figure 1) shows a falloff toward higher spatial frequencies. Fourier amplitudes of the density map were multiplied by the square root of the ratio of these curves before application of a low-pass filter to limit the resolution to 11.5 Å.

Visualization and Interpretation of 3D Maps
Extraction of 3D elements from the final volumes was done using SPIDER with the help of Tinkerbell (Li et al., 1997) and IRIS Explorer (Numerical Algorithms Group, Inc., Downers Grove, IL). Visualization of the density map and docking of X-ray crystal structures was done using IRIS Explorer and O (Jones et al., 1991). For docking in Explorer, the atomic coordinate files were converted into electron density maps by computing averaged densities within volume elements scale-matched to those of the EM map (2.93 Å). Surface representations of these computed density maps were then used. Docking in O was done using a surface representation of the EM map and direct visualization of atomic models imported from the PDB.

Acknowledgments

We thank Knud Nierhaus for providing MF-RNA and fMet-$tRNA_f^{Met}$ and David Stokes from New York University for making a Philips CM200 FEG electron microscope available for our studies. We are grateful to Venki Ramakrishnan for providing all-atom coordinates of the L11-RNA fragment complex. We also thank Arun Malhotra for helpful suggestions in the initial phase of the work, Ardean Leith for writing software for examination of contact surfaces, Gregor Blaha for preparing specimens for solution scattering, and Amy Heagle for preparing the illustrations. Some of the computing support was provided by NCSA, University of Illinois at Urbana-Champaign. This work was supported by grants NIH R37 GM29169, R01 GM55440, and NSF BIR 9219043 (to J. F.).

Received December 13, 1999; revised February 10, 2000.

References

Agrawal, R.K., Penczek, P., Grassucci, R.A., Li, Y., Leith, A., Nierhaus, K.H., and Frank, J. (1996). Direct visualization of A-, P-, and E-site transfer RNAs in the *Escherichia coli* ribosome. Science *271*, 1000–1002.

Agrawal, R.K., Penczek, P., Grassucci, R.A., and Frank, J. (1998). Visualization of elongation factor G on the *Escherichia coli* 70S ribosome: the mechanism of translocation. Proc. Natl. Acad. Sci. USA *95*, 6134–6138.

Agrawal, R.K., Heagle, A.B., Penczek, P., Grassucci, R.A., and Frank, J. (1999a). EF-G-dependent GTP hydrolysis induces translocation accompanied by large conformational changes in the 70S ribosome. Nat. Struct. Biol. *6*, 643–647.

Agrawal, R.K., Lata, K.R., and Frank, J. (1999b). Conformational variability in *E. coli* 70S ribosome as revealed by 3D cryo-electron microscopy. Intl. J. Biochem. Cell Biol. *31*, 243–254.

Agrawal, R.K., Penczek, P., Grassucci, R.A., Burkhardt, N., Nierhaus, K.H., and Frank, J. (1999c). Effect of buffer conditions on the position of tRNAs on the 70S ribosome as visualized by cryo-electron microscopy. J. Biol. Chem. *274*, 8723–8729.

Agrawal, R.K., Heagle, A.B., and Frank, J. (2000). Studies of elongation factor G-dependent tRNA translocation by three-dimensional cryo-electron microscopy. In The Ribosome: Structure, Function, Antibiotics, and Cellular Interactions, R.A. Garrett, S.R. Douthwaite, A. Liljas, A.T. Matheson, P.B. Moore, and H.F. Noller, eds. (Washington, D.C.: ASM Press), pp. 53–62.

Ban, N., Freeborn, B., Nissen, P., Penczek, P., Grassucci, R.A., Sweet, R., Frank, J., Moore, P.B., and Steitz, T.A. (1998). A 9 Å resolution X-ray crystallographic map of the large ribosomal subunit. Cell *93*, 1105–1115.

Ban, N., Nissen, P., Hansen, J.C., Capel, M., Moore, P.B., and Steitz, T.A. (1999). Placement of protein and RNA structures into a 5Å-resolution map of the 50S ribosomal subunit. Nature *400*, 841–847.

Basavappa, R., and Sigler, P.B. (1991). The 3 Å crystal structure of yeast initiator tRNA: functional implications in initiator/elongator discrimination. EMBO J. *10*, 3105–3111.

Böttcher, B., Schwarz, L., and Graber, P. (1998). Direct indication for the existence of a double stalk in CF_0F_1. J. Mol. Biol. *281*, 757–762.

Boulin, C.J., Kempf, R., Gabriel, A., and Koch, M.H.J. (1988). Data acquisition systems for linear and area X-ray detectors using delay line readout. Nucl. Instrum. Methods *A269*, 312–320.

Cate, J.H., Yusupov, M.M., Yusupova, G.Z., Earnest, T.N., and Noller, H.F. (1999). X-ray crystal structures of 70S ribosome functional complexes. Science *285*, 2095–2104.

Clemons, W.M., Jr., Davies, C., White, S.W., and Ramakrishnan,

V. (1998). Conformational variability of the N-terminal helix in the structure of ribosomal protein S15. Structure *6*, 429–438.

Clemons, W.M., Jr., May, J.L.C., Wimberly, B.T., McCutcheon, J.P., Capel, M., and Ramakrishnan, V. (1999). Structure of a bacterial 30S ribosomal subunit at 5.5 Å resolution. Nature *400*, 833–840.

Conn, G.L., Draper, D.E., Lattman, E.E., Gittis, A.G. (1999). Crystal structure of a conserved ribosomal protein-RNA complex. Science *284*, 1171.

Correll, C.C., Munishkin, A., Chan, Y.L., Ren, Z., Wool, I.G., and Steitz, T.A. (1998). Crystal structure of the ribosomal RNA domain essential for binding elongation factors. Proc. Natl. Acad. Sci. USA *95*, 13436–13441.

Correll, C.C., Wool, I.G., and Munishkin, A. (1999). The two faces of the *Escherichia coli* 23S rRNA sarcin/ricin domain: the structure at 1.11 Å resolution. J. Mol. Biol. *292*, 275–287.

Culver, G.M., Cate, J.H., Yusupova, G.Z., Yusupov, M.M., and Noller, H.F. (1999). Identification of an RNA-protein bridge spanning the ribosomal subunit interface. Science *285*, 2133–2135.

Czworkowski, J., Wang, J., Steitz, T.A., and Moore, P.B. (1994). The crystal structure of elongation factor G complexed with GDP, at 2.7 Å resolution. EMBO J. *13*, 3661–3668.

Fanning, T.G., Cantrell, M., Shih, C.Y., and Craven, G.R. (1978). Evidence that proteins S1, S11 and S21 directly participate in the binding of transfer RNA to the 30S ribosome. Nucleic Acids Res. *5*, 933–950.

Frank, J., and Penczek, P. (1995). On the correction of the contrast transfer function in biological electron microscopy. Optik *98*, 125–129.

Frank, J., Verschoor, A., Li, Y., Zhu, J., Lata, R.K., Radermacher, R., Penczek, P., Grassucci, R., Agrawal, R.K., and Srivastava, S. (1995a). A model of the translational apparatus based on a three-dimensional reconstruction of the *Escherichia coli* ribosome. Biochem. Cell Biol. *73*, 757–765.

Frank, J., Zhu, J., Penczek, P., Li, Y., Srivastava, S., Verschoor, A., Radermacher, M., Grassucci, R., Lata, R.K., and Agrawal, R.K. (1995b). A model of protein synthesis based on cryo-electron microscopy of the *E. coli* ribosome. Nature *376*, 441–444.

Frank, J., Radermacher, M., Penczek, P., Zhu, J., Li, Y., Ladjadj, M., and Leith, A. (1996). SPIDER and WEB: processing and visualization of images in 3D electron microscopy and related fields. J. Struct. Biol. *116*, 190–199.

Frank, J., Penczek, P., Agrawal, R.K., Grassucci, R.A., and Heagle, A.B. (2000a). Three-dimensional cryo-electron microscopy of ribosomes. Methods Enzymol. *317*, 276–291.

Frank, J., Penczek, P., Grassucci, R.A., Heagle, A.B., Spahn, C.M.T., and Agrawal, R.K. (2000b). Cryo-electron microscopy of the translational apparatus: experimental evidence for the paths of mRNA, tRNA, and the polypeptide chain. In The Ribosome: Structure, Function, Antibiotics, and Cellular Interactions, R.A. Garrett, S.R. Douthwaite, A. Liljas, A.T. Matheson, P.B. Moore, and H.F. Noller, eds. (Washington D.C.: ASM Press), pp. 45–51.

Gabashvili, I.S., Agrawal, R.K., Grassucci, R.A., and Frank, J. (1999a). Structure and structural variations of the *E. coli* 30S ribosomal subunit as revealed by three-dimensional cryo-electron microscopy. J. Mol. Biol. *5*, 1285–1291.

Gabashvili, I.S., Agrawal, R.K., Grassucci, R.A., Squires, C.L., Dahlberg, A.E., and Frank, J. (1999b). Major rearrangements in the 70S ribosomal 3D structure caused by a conformational switch in 16S ribosomal RNA. EMBO J. *18*, 6501–6507.

Glaeser, R.M., and Downing, K.H. (1992). Assessment of resolution in biological electron crystallography. Ultramicroscopy *47*, 256–265.

Harms, J., Tocilj, A., Levin, I., Agmon, I., Stark, H., Kölln, I., van Heel, M., Cuff, M., Schlünzen, F., Bashan, A., Franceschi, F., and Yonath, A. (1999). Elucidating the medium-resolution structure of ribosomal particles: an interplay between electron cryo-microscopy and X-ray crystallography. Structure *7*, 931–941.

Hosaka, H., Nakagawa, A., Tanaka, I., Harada, N., Sano, K., Kimura, M., Yao, M., and Wakatsuki, S. (1997). Ribosomal protein S7: a new RNA-binding motif with structural similarities to a DNA architectural factor. Structure *5*, 1199–1208.

Jones, T.A., Zhou, J.Y., Cowan, S.W., and Kjeldgaard, M. (1991). Improved methods for building protein models in electron density maps and the location of errors in these models. Acta Crystallogr. A *A47*, 110–119.

Koch, M.H.J., and Bordas, J. (1983). X-ray diffraction and scattering on disordered systems using synchroton radiation. Nucl. Instrum. Methods *208*, 461–469.

Lambert, J.M., and Traut, R.R. (1981). The subunit interface of the *Escherichia coli* ribosome. Identification of proteins at the interface between the 30 S and 50 S subunits by crosslinking with 2-iminothiolane. J. Mol. Biol. *149*, 451–476.

Lata, K.R., Penczek, P., and Frank, J. (1995). Automatic particle picking from electron micrographs. Ultramicroscopy *58*, 381–391.

Lata, K.R., Agrawal, R.K., Penczek, P., Grassucci, R., Zhu, J., and Frank, J. (1996). Three-dimensional reconstruction of the *Escherichia coli* 30 S ribosomal subunit in ice. J. Mol. Biol. *262*, 43–52.

Li, Y., Leith, A., and Frank, J. (1997). Tinkerbell—a tool for interactive segmentation of 3D data. J. Struct. Biol. *120*, 266–275.

Lindahl, M., Svensson, L.A., Liljas, A., Sedelnikova, S.E., Eliseikina, I.A., Fomenkova, N.P., Nevskaya, N., Nikonov, S.V., Garber, M.B., Muranova, T.A. (1994). Crystal structure of the ribosomal protein S6 from *Thermus thermophilus*. EMBO J. *13*, 1249–1254.

Lodmell, J.S., and Dahlberg, A.E. (1997). A conformational switch in *Escherichia coli* 16S ribosomal RNA during decoding of messenger RNA. Science *277*, 1262–1267.

Malhotra, A., Penczek, P., Agrawal, R.K., Gabashvili, I.S., Grassucci, R.A., Jünemann, R., Burkhardt, N., Nierhaus, K.H., and Frank, J. (1998). *Escherichia coli* 70 S ribosome at 15 Å resolution by cryo-electron microscopy: localization of fMet-$tRNA_f^{Met}$ and fitting of L1 protein. J. Mol. Biol. *280*, 103–116.

Müller, F., and Brimacombe, R. (1997). A new model for the three-dimensional folding of *Escherichia coli* 16S ribosomal RNA. I. Fitting the RNA to a 3D electron microscopic map at 20Å. J. Mol. Biol. *271*, 524–544.

Nikonov, S., Nevskaya, N., Nliseikina, I., Fomenkova, N., Nikulin, A., Ossina, N., Garber, M., Jonsson, B.H., Briand, C., al-Karadaghi, S., et al. (1996). Crystal structure of the RNA binding ribosomal protein L1 from *Thermus thermophilus*. EMBO J. *15*, 1350–1359.

Orlova, E.V., Dube, P., Harris, J.R., Beckman, E., Zemlin, F., Markl, J., and van Heel, M. (1997). Structure of keyhole limpet hemocyanin type 1 (KLH1) at 15 Å resolution by electron cryomicroscopy and angular reconstitution. J. Mol. Biol. *271*, 417–437.

Osswald, M., Greuer, B., and Brimacombe, R. (1990). Localization of a series of RNA-protein cross-link sites in the 23S and 5S ribosomal RNA from *Escherichia coli*, induced by treatment of 50S subunits with three different bifunctional reagents. Nucleic Acids Res. *18*, 6755–6760.

Osswald, M., Doring, T., and Brimacombe, R. (1995). The ribosomal neighbourhood of the central fold of tRNA: cross links from position 47 of tRNA located at the A, P or E site. Nucleic Acids Res. *23*, 4635–4641.

Peltz, S.W., Hammell, A.B., Cui, Y., Yasenchak, J., Puljanowski, L., and Dinman, J.D. (1999). Ribosomal protein L3 mutants alter translational fidelity and promote rapid loss of the yeast killer virus. Mol. Cell. Biol. *19*, 384–391.

Penczek, P.A., Grassucci, R.A., and Frank, J. (1994). The ribosome at improved resolution: new techniques for merging and orientation refinement in 3D cryo-electron microscopy of biological particles. Ultramicroscopy *53*, 251–270.

Penczek, P., Ban, N., Grassucci, R, Agrawal, R.K., and Frank, J. (1999). *Haloarcula marismortui* 50S subunit—complementarity of EM and X-ray crystallographic information. J. Struct. Biol. *128*, 44–50.

Ramakrishnan, V., and White, S.W. (1992). The structure of ribosomal protein S5 reveals sites of interaction with 16S rRNA. Nature *358*, 768–771.

Schlünzen, F., Kolln, I., Janell, D., Gluhmann, M., Levin, I., Bashan, A., Harms, J., Bartels, H., Auerbach, T., Pioletti, M., et al. (1999). The identification of selected components in electron density maps of prokaryotic ribosomes at 7 angstrom resolution. J. Synchrotron Radiat. *6*, 928–941.

Schmitt, E., Panvert, M., Blanquet, S., and Mechulam, Y. (1998). Crystal structure of methionyl-tRNA$_f^{Met}$ transformylase complexed with the initiator formylmethionyl-tRNA$_f^{Met}$. EMBO J. *17*, 6819–6826.

Spahn, C.M.T., Grassucci, R.A., Penczek, P., and Frank, J. (1999). Direct three-dimensional localization and positive identification of RNA helices within the ribosome by means of genetic tagging and cryo-electron microscopy. Structure *7*, 1567–1573.

Stark, H., Müller, F., Orlova, E.V., Schatz, M., Dube, P., Erdemir, T., Zemlin, F., Brimacombe, R., and van Heel, M. (1995). The 70S *Escherichia coli* ribosome at 23 Å resolution: fitting the ribosomal RNA. Structure *3*, 815–821.

Stark, H., Orlova, E.V., Rinke-Appel, J., Junke, N., Müller, F., Rodnina, M., Wintermeyer, W., Brimacombe, R., and van Heel, M. (1997a). Arrangement of tRNAs in pre- and posttranslocational ribosomes revealed by electron cryomicroscopy. Cell *88*, 19–28.

Stark, H., Rodnina, M.V., Rinke-Appel, J., Brimacombe, R., Wintermeyer, W., and van Heel, M. (1997b). Visualization of elongation factor Tu on the *Escherichia coli* ribosome. Nature *389*, 403–406.

Svergun, D.I., Semenyuk, A.V., and Feigin, L.A. (1988). Small-angle scattering data treatment by the regularization method. Acta Crystallogr. *A44*, 244–250.

van Heel, M. (1986). Noise-limited three-dimensional reconstructions. Optik *73*, 83–86.

Wilson, K.S., and Noller, H.F. (1998). Mapping the position of translational elongation factor EF-G in the ribosome by directed hydroxyl radical probing. Cell *92*, 131–139.

Wimberly, B.T., White, S.W., and Ramakrishnan, V. (1997). The structure of ribosomal protein s7 at 1.9-angstrom resolution reveals a beta-hairpin motif that binds double-stranded nucleic acids. Structure *5*, 1187–1198.

Wimberly, B.T., Guymon, R., McCutcheon, J.P., White, S.W., and Ramakrishnan, V.R. (1999). A detailed view of a ribosomal active site: the structure of the L11-RNA complex. Cell *97*, 491–502.

Wittmann, H.G., Mussig, J., Piefke, J., Gewitz, H.S., Rheinberger, H.J., and Yonath, A. (1982). Crystallization of *Escherichia coli* ribosomes. FEBS Lett. *146*, 217–220.

Wittmann-Liebold, B., Uhlein, M., Urlaub, H., Müller, E.C., Otto, A., and Bischoff, F.R. (1995). Structural and functional implications in the eubacterial ribosome as revealed by protein-rRNA and antibiotic contact sites. Biochem. Cell Biol. *73*, 1187–1197.

Zhu, J., Penczek, P.A., Schröder, R., and Frank, J. (1997). Three-dimensional reconstruction with contrast transfer function correction from energy-filtered cryoelectron micrographs: procedure and application to the 70S *Escherichia coli* ribosome. J. Struct. Biol. *118*, 197–219.

Protein Data Bank ID Code

Coordinates of all ribosomal proteins (C-α backbones) and RNA fragments (P atoms) displayed in the figures have been deposited in the Protein Data Bank (ID code 1EG0), along with the fMet-tRNA$_f^{Met}$ (Schmitt et al., 1998) that was fitted into the cryo-EM density.

A ratchet-like inter-subunit reorganization of the ribosome during translocation

Joachim Frank*†‡ & Rajendra Kumar Agrawal†‡

* *Howard Hughes Medical Institute, † Health Research Incorporated at the Wadsworth Center, and ‡ Department of Biomedical Sciences, State University of New York at Albany, Empire State Plaza, PO Box 509, Albany, New York 12201-0509, USA*

The ribosome is a macromolecular assembly that is responsible for protein biosynthesis following genetic instructions in all organisms. It is composed of two unequal subunits: the smaller subunit binds messenger RNA and the anticodon end of transfer RNAs, and helps to decode the mRNA; and the larger subunit interacts with the amino-acid-carrying end of tRNAs and catalyses the formation of the peptide bonds. After peptide-bond formation, elongation factor G (EF-G) binds to the ribosome, triggering the translocation of peptidyl-tRNA from its aminoacyl site to the peptidyl site, and movement of mRNA by one codon[1]. Here we analyse three-dimensional cryo-electron microscopy maps of the *Escherichia coli* 70S ribosome in various functional states, and show that both EF-G binding and subsequent GTP hydrolysis lead to ratchet-like rotations of the small 30S subunit relative to the large 50S subunit. Furthermore, our finding indicates a two-step mechanism of translocation: first, relative rotation of the subunits and opening of the mRNA channel following binding of GTP to EF-G; and second, advance of the mRNA/(tRNA)$_2$ complex in the direction of the rotation of the 30S subunit, following GTP hydrolysis.

We analysed the amplitude-corrected three-dimensional cryo-electron microscopy (3D cryo-EM) density maps of five complexes presenting the 70S ribosome in different functional states: (1) an initiation-state complex[2], bound with a mRNA and fMet–tRNA$_f^{Met}$ at the peptidyl (P) site (the 'control'); (2) the pre-translocational state complex[3], bound with mRNA, aminoacyl (A)- and P-site tRNAs, and EF-G in the presence of a nonhydrolysable GTP analogue, GMPP(CH$_2$)P (the 'GTP form'); (3) as (2), but without mRNA and tRNAs[3]; (4) the post-translocational state complex[3], bound with mRNA, P and exit (E)-site tRNAs, and EF-G in the presence of GTP and fusidic acid (the 'GDP form'); and (5) as (4), but without mRNA and tRNAs[3,4]. The comparison of these refined maps shows that, upon binding of EF-G/GMPP(CH$_2$)P to the ribosome, the 30S subunit rotates with respect to the 50S subunit by an angle of about 6 ($\pm$ 0.7) degrees, resulting in a maximum displacement, at the periphery of the ribosome, of 19 ($\pm$ 3) Å. The rotation is counterclockwise when seen from the solvent side of the 30S subunit (compare Fig. 1a and b). Following EF-G-dependent GTP hydrolysis, in the presence of fusidic acid the 30S subunit rotates back (Fig. 1c) by an angle of about 3 ($\pm$ 0.7) degrees. The conformational rearrangement is marked if one compares the distances between the L1-protein region of the 50S subunit and the platform and head regions of the 30S subunit (Fig. 1a and b). The positions of beak and spur of the 30S subunit are also strongly affected. Figure 1 shows maps of 70S/EF-G complexes lacking mRNA and tRNAs. The lower occupancy of EF-G/GMPP(CH$_2$)P in the complex containing tRNA at the A site[3,5] makes its presentation confusing, however the rotations of the subunit in the maps with or without mRNA and tRNAs are of the same size.

The rotation is accompanied by many conformational changes of the two subunits. Among these is a bifurcation of the stalk of the 50S subunit in the GTP state[3,5] (Fig. 1b) and its reversal to a single, elongated form in the GDP state[3–5] (Fig. 1c), showing an arc-like connection between the G′ domain of EF-G and the stalk base.

Going from the control to the GTP state, we also see a transition to a less compact structure[6], such that the radius of gyration increases[7] by 1.1 Å), and a complex rearrangement in the multi-stranded mass near the mRNA channel on the 30S subunit[8–11]. Significantly, both the entrance channel (conducting the down-stream 3′ end of the mRNA) and the exit channel (conducting the upstream 5′ end of the mRNA) are affected (Fig. 2). The entrance channel is defined by a contact between the 30S subunit shoulder and head[12,13] involving the 530 loop, which is dynamic during decoding[14], and the 1050/1200 regions of the 16S rRNA[15], respectively. The entrance channel is much wider in both GTP (an increase in cross-sectional area by ~68%) and GDP (by ~118%) states than in the control (Fig. 3). Thus, the opening at the entrance site is wider (by ~30%) in the GDP (Figs 2b and 3c) than in the GTP (Figs 2a and 3b) state. The shape and position of the exit channel, formed between the platform and the head[11], are known to be variable[12,13] The behaviour of this channel is the reverse of that of the entrance channel: it is wider in the GTP (Fig. 2c) than in the GDP state (Fig. 2d). Thus, a dynamic clamping mechanism apparently exists that alternatively secures the mRNA in the entrance channel or leaves it free to move following EF-G/GTP binding and GTP hydrolysis.

We find clear evidence in our cryo-maps that the positions of the tRNAs with respect to the 30S subunit are unchanged in the GTP state (Fig. 3b; ref. 16), indicating that translocation of the tRNA from the A to the P site has not taken place[3]. However, a contact between the head and the anticodon arm of the A-site tRNA (formed before EF-G binding) is severed (arrow, Fig. 3b), which would be required as a preparation for the translocation of tRNA. In addition, the subunit head flattens from a more convex shape and gains in length on the platform side. After GTP hydrolysis (Fig. 3c), the A site is vacated and the P and E sites are occupied. *In vitro*, the E site is always occupied by deacylated tRNA[17], and the corresponding mass is visible in Fig. 3b and c. However, the density corresponding to the E-site tRNA is higher after GTP hydrolysis (not shown). The conformation of the upper half of the helix 44 of the 16S RNA[18–20], which is clearly distinguishable in all maps (Fig. 3) and the tip of which forms the decoding region, is affected by both EF-G binding and GTP hydrolysis.

These observations allow us to sketch out the following sequence of events (Fig. 4): EF-G binding alone triggers a relative movement of the 30S versus the 50S subunit accompanied by a widening of the mRNA channel, without changing the position of the tRNA ends relative to the 30S or 50S subunit, although inevitably leading to a skewed tRNA orientation. Subsequent GTP hydrolysis triggers conformational changes that result in three coupled events: (1) translocation of the mRNA/$(tRNA)_2$ complex with respect to the small subunit, which requires the mRNA to be free to move; (2) release of EF-G; and (3) reverse rotation toward the original positions of the subunits, a movement that is arrested in an intermediate position owing to the presence of fusidic acid[3,4].

Thus, the translocation process entails two relative rotations of the subunits in opposite directions, which are rapid on the timescale of the elongation cycle, and can only be observed by trapping intermediate states. The observation of a rotation preceding the actual translocation calls for a more detailed terminology in describing ribosome states (Fig. 4): 'pre', before EF-G binding; 'pre′', EF-G/GTP bound and subunits rotated, but before GTP hydrolysis; 'post′', with EF-G/GDP bound but immediately after GTP hydrolysis; 'post^{fus}', EF-G/GDP bound in the presence of fusidic acid; and 'post', after release of EF-G and the subunits have

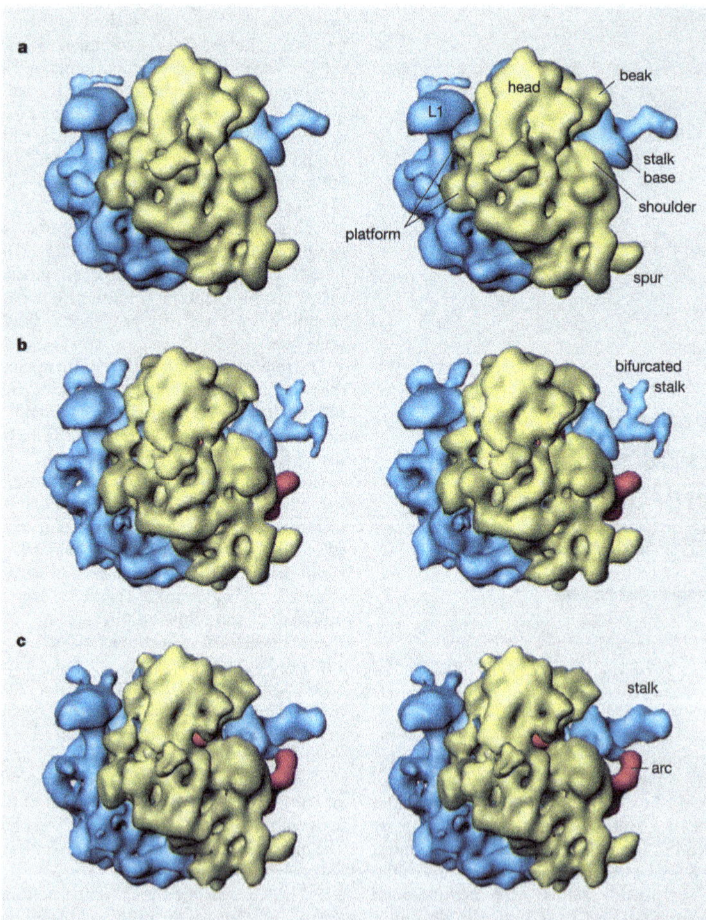

Figure 1 Stereo representation of the rotational movement of the 30S subunit (yellow) with respect to 50S subunit (blue) upon EF-G (red) binding. **a**, 70S/fMet–$tRNA_f^{Met}$ complex; **b**, 70S/EF-G/GMPP(CH_2)P complex; and **c**, 70S/EF-G/GDP/fusidic acid complex. The three-dimensional maps of the 70S complexes are shown from the 30S solvent side. The masses corresponding to EF-G in **b** and **c** are only partially seen in this view. The movement of the 30S subunit is anticlockwise from **a** to **b** and clockwise from **b** to **c**. Landmarks: L1, protein L1; arc, arc-like connection between EF-G and stalk base.

rotated into their original relative position. The orientation between the subunits is the same in the pre and post states[17,21]. The same relative orientations of the subunits in both states would be required for the binding of EF-G and the highly similar EF-Tu/Phe-tRNA/GMPPNP ternary complex[22] to overlapping binding sites on the ribosome[3–5,23]. A recent cryo-EM study of EF-G binding, in the presence of an antibiotic thiostrepton[24], does not show the ratchet-like movement, presumably because thiostrepton, by binding to the GTPase-associated centre[25], prevents EF-G from binding in a way required to induce the dramatic conformational change of the ribosome. Furthermore, our findings do not support earlier models[26,27] according to which translocation occurs simultaneously with the relative movement between the subunits. We also note that an earlier ratchet model[28] deals exclusively with the relative movement of tRNAs and mRNA.

Cryo-EM[8,18] and X-ray[19] maps of the 70S ribosome show a number of bridges connecting the subunits, which could have either passive (registration, elastic storage of energy) or active (transmission of force, signalling) roles in the control of the subunit movements. The bridges most strongly affected by the inter-subunit rotation, B1a, B1b and B1c, link the 30S subunit head with the central protuberance of the 50S subunit[18,19], whose conformation changes dramatically (Fig. 3). Our results, obtained by direct visualization, explain previous experimental findings[6,7,29,30], which were taken to indicate the existence of different relative constellations of two ribosomal subunits during translation. □

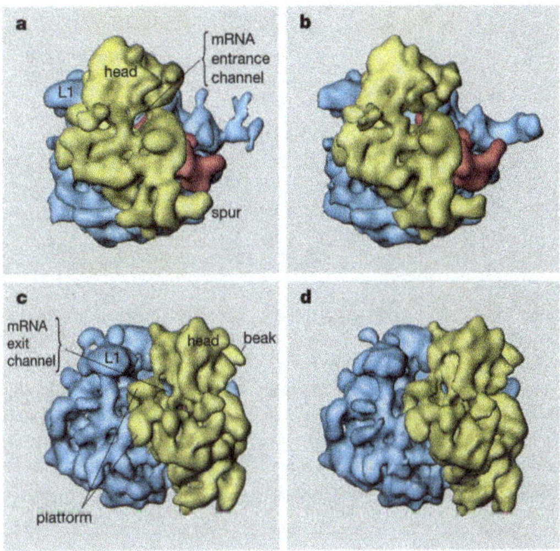

Figure 2 Effect of EF-G-dependent GTP hydrolysis on the dynamics of the mRNA channel. **a**, **c**, 70S/EF-G/GMPP(CH$_2$)P complex; and **b**, **d**, 70S/EF-G/GDP/fusidic acid complex. **a**, **b**, Maps of 70S ribosome from the channel entrance side, on the solvent side near the shoulder, in an orientation different to that in Fig. 1 to highlight the channel opening. **c**, **d**, Maps from the channel exit side, or platform side of the 30S subunit.

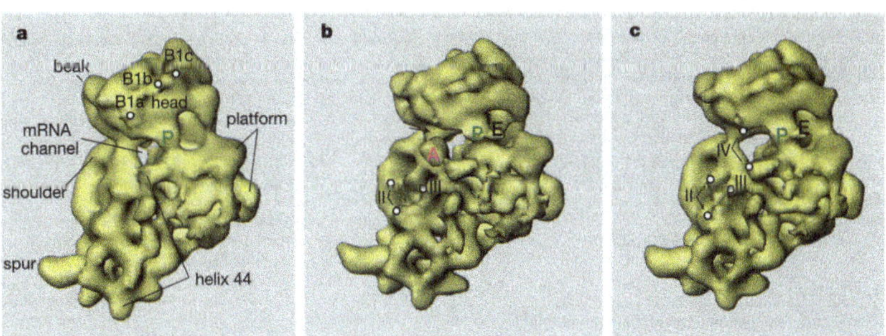

Figure 3 Effect of rotation on the tRNA position before and after GTP hydrolysis. The 30S subunit from various complexes is shown from the subunit interface side, that is, rotated by about 180 degrees around the vertical axis of the view in Fig. 1. Subsequently, the subunit is slightly rotated around the horizontal axis to show the mRNA entrance channel. The 50S subunit, EF-G and portions of tRNAs have been removed computationally. Only portions of the tRNAs making contact with the 30S subunit are retained. **a**, 70S/fMet–tRNA$_f^{Met}$ complex, showing the P site filled. Upper (decoding site) and lower ends of helix 44 of 16S RNA[18–20] are marked. **b**, 70S/(tRNA)$_2$/EF-G/GMPP(CH$_2$)P complex, showing both A and P sites filled, even though the 30S subunit has already rotated with respect to the 50S subunit. In this panel, mRNA entrance channel is partially obscured by the density of the A-site tRNA. **c**, 70S/(tRNA)$_2$/EF-G/GDP/fusidic acid complex, showing the A site vacated but the P site occupied. Landmarks: B1a, B1b and B1c, areas of the 30S subunit head involved in formation of bridges[18,19] with the central protuberance of the 50S subunit in the control; A, portion of the A-site tRNA contacting the subunit; P, portion of the P-site tRNA contacting the head and platform of the subunit; E, anticodon portion of the E-site tRNA. II, III, and IV indicate the areas of the 30S subunit that make contact with EF-G domains II and III in the GTP state (**b**), and II, III, and IV in the GDP state[3–5] (**c**).

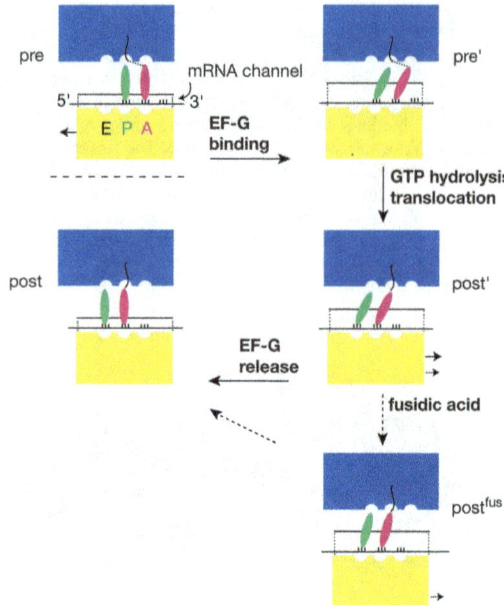

Figure 4 Diagram showing three steps of the translocation process. 'pre' state: tRNAs are in A (magenta) and P (green) sites on both 30S (yellow) and 50S (blue) subunits; the peptide bond has already been formed and the elongated peptidyl moiety is now on the A-site tRNA; the subunits are in the original orientations; both entrance and exit channels for mRNA, indicated by its 3′ and 5′ ends, are narrow. 'pre′' state, following EF-G binding: 30S subunit has rotated with respect to the 50S subunit, without changing the positions of the tRNAs relative to either subunit; mRNA channel has widened enabling mRNA advance in the next step. 'post′' state, following GTP hydrolysis (transient state, not experimentally observed): tRNA and associated mRNA have advanced from A and P sites to P and E sites (which might involve transitional, A/P and P/E hybrid states[29]); relative subunit orientations are unchanged. 'post^{fus}' state, assumed when fusidic acid is present: small subunit has partially rotated back, complete reversal is blocked by the presence of EF-G/GDP which is not released. 'post' state, assumed after EF-G is released under normal conditions: small subunit is fully rotated back into original orientation; tRNAs are in P and E sites; mRNA channel is tight again, securing mRNA in its new position.

Methods

Five separate 3D cryo-EM reconstructions of ribosomal complexes[2,3] with resolutions varying from 15 Å to 18 Å have been analysed. Preparation of complexes, electron microscopy and image processing were described in detail previously[2–4] (70S/fMet–$tRNA_f^{Met}$ complex[2]; 70S/EF-G/GMPP(CH_2)P complex[3]; 70S/(tRNA)$_2$/EF-G/GMPP(CH_2)P complex[3]; 70S/EF-G/GDP/fusidic acid complex[3,4]; and 70S/(tRNA)$_2$/EF-G/GDP/fusidic acid complex[3]). For the quantitative comparison, all density maps were individually filtered such that their radial Fourier amplitude profile matched the profile obtained by X-ray solution scattering of the 70S ribosome[18]. Finally, all maps were low-pass filtered to a common resolution of 18 Å, as determined by the Fourier shell correlation using a threshold cut-off of 0.5 (ref. 2). The final common resolution of these maps would be 13 Å if measured and filtered by the less conservative 3-σ criterion used by others in the field[23,24] (see ref. 2 for a discussion of methods of calculating resolution of cryo-EM maps). Radii of gyration of all three-dimensioal maps were calculated using the SPIDER software[11]. An animated sequence alternating between the control and the 70S/EF-G/GMPP(CH_2)P complex while rotating around a vertical axis was computed using the 3D modelling/animation software (Lightwave 3D, version 5.5, NewTek Inc., San Antonio, Texas). The animation is available as Supplementary Information.

Received 30 March; accepted 15 June 2000.

1. Wilson, K. S. & Noller, H. F. Molecular movement inside the translational engine. *Cell* **92,** 337–349 (1998).
2. Malhotra, A. *et al. Escherichia coli* 70 S ribosome at 15 Å resolution by cryo-electron microscopy: localization of fMet-$tRNA_f^{Met}$ and fitting of L1 protein. *J. Mol. Biol.* **280,** 103–116 (1998).
3. Agrawal, R. K., Heagle, A. B., Penczek, P., Grassucci, R. A. & Frank, J. EF-G-dependent GTP hydrolysis induces translocation accompanied by large conformational changes in the 70S ribosome. *Nature Struct. Biol.* **6,** 643–647 (1999).
4. Agrawal, R. K., Penczek, P., Grassucci, R. A. & Frank, J. Visualization of elongation factor G on the *Escherichia coli* 70S ribosome: the mechanism of translocation. *Proc. Natl Acad. Sci. USA.* **95,** 6134–6138 (1998).
5. Agrawal, R. K., Heagle, A. B. & Frank, J. in *The Ribosome: Structure, Function, Antibiotics, and Cellular Interactions* (eds Garrett, R. A. *et al.*) 53–62 (ASM, Washington DC, 2000).
6. Spirin, A. S., Baranov, V. I., Polubesov, G. S., Serdyuk, I. N. & May, R. P. Translocation makes the ribosome less compact. *J. Mol. Biol.* **194,** 119–128 (1987).
7. Serdyuk, I. *et al.* Structural dynamics of translating ribosomes. *Biochimie* **74,** 299–306 (1992).
8. Frank, J. *et al.* A model of protein synthesis based on cryo-electron microscopy of the *E. coli* ribosome. *Nature* **376,** 441–444 (1995).
9. Agrawal, R. K. *et al.* Direct visualization of A-, P-, and E-site transfer RNAs in the *Escherichia coli* ribosome. *Science* **271,** 1000–1002 (1996).
10. Mueller, F., Stark, H., van Heel, M., Rinke-Appel, J. & Brimacombe, R. A new model for the three-dimensional folding of *Escherichia coli* 16 S ribosomal RNA. III. The topography of the functional centre. *J. Mol. Biol.* **271,** 566–587 (1997).
11. Frank, J. *et al.* in *The Ribosome: Structure, Function, Antibiotics, and Cellular Interactions* (eds Garrett, R. A. *et al.*) 45-51 (ASM, Washington DC, 2000).
12. Lata, K. R. *et al.* Three-dimensional reconstruction of the *Escherichia coli* 30 S ribosomal subunit in ice. *J. Mol. Biol.* **262,** 43–52 (1996).
13. Agrawal, R. K., Lata, R. K. & Frank, J. Conformational variability in *Escherichia coli* 70S ribosome as revealed by 3D cryo-electron microscopy. *Int. J. Biochem. Cell Biol.* **31,** 243–254 (1999).
14. Powers, T. & Noller, H. F. The 530 loop of 16S rRNA: a signal to EF-Tu? *Trends Genet.* **10,** 27–31 (1994).
15. Newcomb, L. F. & Noller, H. F. Directed hydroxyl radical probing of 16S rRNA in the ribosome: spatial proximity of RNA elements of the 3′ and 5′ domains. *RNA* **5,** 849–855 (1999).
16. Rodnina, M. V., Savelsbergh, A., Katunin, V. I. & Wintermeyer, W. Hydrolysis of GTP by elongation factor G drives tRNA movement on the ribosome. *Nature* **385,** 37–41 (1997).
17. Agrawal, R. K. *et al.* Visualization of tRNA movements on the *E. coli* ribosome during the elongation cycle. *J. Cell. Biol.* (in the press).
18. Gabashvili, I. S. *et al.* Solution structure of the *E. coli* 70S ribosome at 11.5 Å resolution. *Cell* **100,** 537–549 (2000).
19. Cate, J. H., Yusupov, M. M., Yusupova, G. Z., Earnest, T. N. & Noller, H. F. X-ray crystal structures of 70S ribosome functional complexes. *Science* **285,** 2095–2104 (1999).
20. Clemons, W. M. *et al.* Structure of a bacterial 30S ribosomal subunit at 5.5 Å resolution. *Nature* **400,** 833–840 (1999).
21. Stark, H. *et al.* Arrangement of tRNAs in pre- and posttranslocational ribosomes revealed by electron cryomicroscopy. *Cell* **88,** 19–28 (1997).
22. Nissen, P. *et al.* Crystal structure of the ternary complex of Phe-$tRNA^{Phe}$, EF-Tu, and a GTP analog. *Science* **270,** 1464–1472 (1995).
23. Stark, H. *et al.* Visualization of elongation factor Tu on the *Escherichia coli* ribosome. *Nature* **389,** 403–406 (1997).
24. Stark, H., Rodnina, M. V., Wieden, H. -J., van Heel, M. & Wintermeyer, W. Large scale movement of elongation factor G and extensive conformational change of the ribosome during translocation. *Cell* **100,** 301–309 (2000).

letters to nature

25. Porse, B. T., Leviev, I., Mankin, A. S. & Garrett, R. A. The antibiotic thiostrepton inhibits a functional transition within protein L11 at the ribosomal GTPase center. *J. Mol. Biol.* **276,** 391–404 (1998).
26. Bretscher, M. S. Translocation in protein synthesis: A hybrid structure model. *Nature* **218,** 675–677 (1968).
27. Spirin, A. S. A model of the functioning ribosome: locking and unlocking of the ribosome subparticles. *Cold Spring Harbor Symp. Quant. Biol.* **34,** 197–207 (1969).
28. Woese, C. Molecular mechanics of translation: a reciprocating ratchet mechanism. *Nature* **226,** 817–820 (1970).
29. Moazed, D. & Noller, H. F. Intermediate states in the movement of transfer RNA in the ribosome. *Nature* **342,** 142–148 (1989).
30. Agrawal, R. K. & Burma, D. P. Sites of ribosomal RNAs involved in the subunit association of tight and loose couple ribosomes. *J. Biol. Chem.* **271,** 21285–21291 (1996).
31. Frank, J. *et al.* SPIDER and WEB: processing and visualization of images in 3D electron microscopy and related fields. *J. Struct. Biol.* **116,** 190–199 (1996).

Supplementary information is available on *Nature*'s World-Wide Web site (http://nature.com).

Acknowledgements

This work was supported by grants from the National Institutes of Health. We thank A. Heagle for preparing the illustrations and animated sequence, and I. Gabashvili and P. Penczek for help with image processing.

Correspondence and requests for materials should be addressed to J.F.

Cell, Vol. 107, 361–372, November 2, 2001, Copyright ©2001 by Cell Press

Architecture of the Protein-Conducting Channel Associated with the Translating 80S Ribosome

Roland Beckmann,[1,7,8] Christian M.T. Spahn,[2,3] Narayanan Eswar,[4] Jürgen Helmers,[1] Pawel A. Penczek,[5] Andrej Sali,[4] Joachim Frank,[2,3,6] and Günter Blobel[1]
[1]Laboratory of Cell Biology
Howard Hughes Medical Institute
The Rockefeller University
1230 York Avenue
New York, New York 10021
[2]Howard Hughes Medical Institute, Health Research Inc.
[3]Wadsworth Center
Empire State Plaza
Albany, New York 12201
[4]Laboratory of Molecular Biophysics
Pels Family Center for Biochemistry and Structural Biology
The Rockefeller University
New York, New York 10021
[5]Department of Biochemistry and Molecular Biology
University of Texas–Houston Medical School
6431 Fannin
Houston, Texas 77030
[6]Department of Biomedical Science
State University of New York at Albany
Albany, New York 12222
[7]Institut für Biochemie der Charité
Humboldt Universität zu Berlin
Monbijoustr. 2,
10117 Berlin
Germany

Summary

In vitro assembled yeast ribosome-nascent chain complexes (RNCs) containing a signal sequence in the nascent chain were immunopurified and reconstituted with the purified protein-conducting channel (PCC) of yeast endoplasmic reticulum, the Sec61 complex. A cryo-EM reconstruction of the RNC-Sec61 complex at 15.4 Å resolution shows a tRNA in the P site. Distinct rRNA elements and proteins of the large ribosomal subunit form four connections with the PCC across a gap of about 10–20 Å. Binding of the PCC influences the position of the highly dynamic rRNA expansion segment 27. The RNC-bound Sec61 complex has a compact appearance and was estimated to be a trimer. We propose a binary model of cotranslational translocation entailing only two basic functional states of the translating ribosome-channel complex.

Introduction

A large number of mRNAs code for secretory and membrane proteins, which are translocated across, or inserted into, the membrane of the eukaryotic endoplasmic reticulum (ER) or the bacterial plasma membrane. In the case of cotranslational protein translocation, the translating ribosome is targeted in a signal sequence-dependent manner to the membrane, and associates with the protein-conducting channel (PCC) (Matlack et al., 1998). The so-called translocon is formed by association of the PCC with additional membrane protein complexes, which interact with the translocating polypeptide, such as the signal peptidase complex (SPC) and the oligosaccharyl transferase (OST) (Johnson and van Waes, 1999).

The PCC can specifically bind and insert a signal sequence, which leads to a high salt-resistant interaction with the ribosome (Görlich and Rapoport, 1993; Jungnickel and Rapoport, 1995; Kalies et al., 1994; Mothes et al., 1998) and to accessibility of an aqueous environment for the translocating peptide in the channel (Crowley et al., 1994; Gilmore and Blobel, 1985; Simon and Blobel, 1991). Hydrophobic domains of translocating nascent polypeptides have access to the hydrophobic core of the lipid bilayer and can partition from the channel into the membrane by lateral opening of the PCC (Borel and Simon, 1996; Heinrich et al., 2000; Martoglio et al., 1995). The cytosolic domains of membrane proteins are translated and released from the ribosome-channel complex (Mothes et al., 1997). The structural basis of these different activities of the PCC is unknown.

Importantly, during the different phases of polypeptide translocation or insertion, the ion permeability barrier of the membrane must be maintained. Based on fluorescence quenching studies using probes in the translocating nascent chain, it is believed that the necessary seal is provided by a tight junction between the translocon and the ribosome (Crowley et al., 1994). A diameter of 15 Å has been estimated for the aqueous pore in the inactive channel (plugged on the lumenal side by a Hsp70-like protein called BiP) by quenching experiments (Hamman et al., 1998), and of up to 60 Å for the pore of the active channel (Hamman et al., 1997). This creates a continuous space from the ER lumen through the channel into the tunnel of the large ribosomal subunit. For the release of cytosolic loops or transmembrane domains of nascent integral membrane proteins, the seal between ribosome and channel must be opened, which would lead to ion conductance. However, communication between the ribosome and the PCC has been suggested to coordinate the release of cytosolic loops with vertical and lateral gating of the channel such that an ion barrier is maintained at all times (Liao et al., 1997).

The PCC is formed in eukaryotic cells by the trimeric Sec61 complex, consisting of the Sec61α, β, and γ subunits (Sec61p, Sbh1p, and Sss1p in yeast) with, altogether, 12 transmembrane helices (Görlich and Rapoport, 1993; Wilkinson et al., 1996). Two-dimensional (2D) electron microscopy (EM) maps revealed that three to four Sec61 complexes could form ring-like structures of 100 Å diameter, with a central indentation or pore (Hanein et al., 1996). Similar oligomeric structures are

[8]Correspondence: roland.beckmann@charite.de

Cell
362

formed by the bacterial homolog, the SecYEG complex (Manting et al., 2000; Meyer et al., 1999). The first cryo-EM three-dimensional (3D) structure of the ribosome-Sec61 complex showed a donut-shaped Sec61 complex associated with the large ribosomal subunit (Beckmann et al., 1997). The channel is positioned such that its central pore is aligned with the exit of the tunnel in the large ribosomal subunit representing the conduit for the nascent chain. Surprisingly, a gap between the channel and the ribosome was observed, but it was attributed to the absence of a signal sequence, which was assumed to induce a tight seal formation (Beckmann et al., 1997). A recent cryo-EM study of mammalian ribosome-Sec61 complexes reported a similar spatial arrangement, with several connections forming the ribosome-channel junction, but still leaving a gap (Ménétret et al., 2000). This gap was observed with both purified Sec61 complexes and with less defined native translocons, and even under conditions where a signal sequence was assumed present. This led to a challenge of the current view of seal formation between ribosome and channel. It is difficult to understand, however, that these maps of presumably translating ribosomes show no tRNA density and that no conformational difference between active and inactive channels could be detected (Ménétret et al., 2000).

Taken together, it is not clear on what structural basis the ribosome-PCC complex can fulfill its different functions in cotranslational translocation without compromising the ion permeability barrier of the membrane. Here, we present a cryo-EM reconstruction of a complex containing the translating yeast 80S ribosome, a peptidyl-tRNA, a nascent polypeptide chain including a signal sequence, and the Sec61 complex at a resolution of 15.4 Å. The fact that the tRNA is visible with high occupancy gives us confidence that the complex indeed represents a translating ribosome. Based on homology modeling, we provide a molecular analysis of the complex and propose a mechanistic model for cotranslational translocation.

Results and Discussion

Programming, Purification, and Reconstitution of RNC-Channel Complexes

We chose a strategy to first isolate a homogenous population of ribosome-nascent chain complexes (RNCs) carrying a signal sequence, and then, in a second step, to use these RNCs for reconstitution of the RNC-channel complex in a membrane-free system for cryo-EM and 3D reconstruction. As a nascent chain, we used the first 120 amino acids of the yeast vacuolar type II membrane protein dipeptidylaminopeptidase B (DAP2), which is translocated cotranslationally in vivo (Ng et al., 1996). The transmembrane domain of this protein in positions 30 to 45 serves as an uncleaved signal sequence, which inserts into the membrane in a loop, resulting in a cytosolic orientation of the N terminus.

The ribosomes were programmed in a yeast cell-free translation system with a truncated synthetic mRNA lacking a stop codon. Under optimized conditions, this construct yielded 15%–20% stalled ribosome-nascent chain complexes, with the nascent chains still bound to tRNA in the P site of the ribosome (data not shown). To immunopurify RNCs based on the presence of the nascent chain, we translated mRNA encoding for the 120 N-terminal residues with an additional N-terminal HA-tag (9 amino acids). Translation of the truncated mRNA of both the tagged and the untagged construct (called HA-DP120 and DP120) resulted in the appearance of two major bands on the autoradiography after SDS-PAGE (Figure 1a). The lower band represented the translated polypeptide; the larger product, migrating at about 40 kDa, represented the polypeptide still bound to tRNA (peptidyl-tRNA). A smear reaching from the higher to the lower band is a result of ongoing hydrolysis of the peptidyl-tRNA bond in the basic environment of the gel during the run (Figure 1a). Treatment with puromycin after translation led to a single band below 25 kDa and disappearance of the smear (data not shown). The translation products of the HA-tagged construct ran slightly higher than the untagged ones and were recognized by a monospecific antibody (Figure 1a, lower panel).

A crude ribosome fraction was isolated by spinning the translation reaction through a high salt/high sucrose cushion, which also resulted in dissociation of nonribosomal proteins from RNCs. The mixture of empty ribosomes and RNCs was then incubated with a biotinylated anti-HA antibody and streptavidin-coupled magnetic beads. After washing the beads with high salt and detergent containing buffers, the bound RNCs were eluted by incubation with an excess of corresponding HA-peptide and spun through a sucrose cushion again. For the tagged construct, this procedure resulted in a highly enriched RNC fraction (Figure 1b) with small amounts of detectable ribosomal proteins but a strong ^{35}S-signal due to the presence of labeled nascent chains. Most importantly, however, the purification procedure did not lead to any detectable background when performed with the untagged control construct (DP120, Figure 1b). We scaled up the procedure to purify sufficiently large amounts (0.25 OD_{260}, ~5 pmol) for further studies. The isolated RNCs showed the characteristic pattern of ribosomal proteins (Figure 1c) and were stable for at least 9 hr on ice without significant hydrolysis or dissociation of the peptidyl-tRNA (data not shown). Thus, we describe a method for the isolation of a homogenous fraction of programmed (80S) ribosomes, which (1) carry a chosen nascent chain kept in place by the P site tRNA, (2) are virtually free of empty ribosomes and additional factors, (3) are stable enough for further experiments, and (4) can be isolated in quantities sufficient for biochemical and structural studies.

Next, we attempted to reconstitute the active, i.e., translating and translocating, ribosome-channel complex using the RNC preparation and purified Sec61 complex in a membrane-free system. The length of the nascent chain (HA-DP120) was chosen such that the signal sequence and at least 40 additional amino acids between the signal sequence and the C terminus should be fully emerged from the tunnel of the large ribosomal subunit (Blobel and Sabatini, 1970). This configuration is known to facilitate a productive interaction of the signal sequence, with the channel leading to the insertion of the nascent chain as a loop (Jungnickel and Rapoport, 1995; Mothes et al., 1998; Shaw et al., 1988).

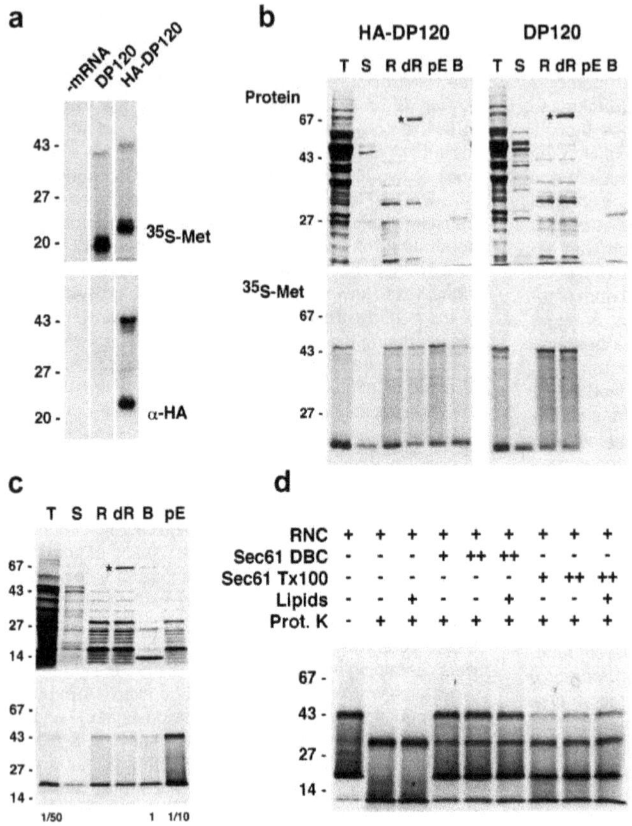

Figure 1. Immunopurification of RNCs and Reconstitution of the RNC-Sec61 Complex

(a) Truncated mRNA coding for the first 120 amino acids of dipeptidylpeptidase B with (HA-DP120) or without (DP120) N-terminal HA-tag was translated in a yeast cell-free system in the presence of ^{35}S-labeled methionine. The translation reaction was subjected to SDS-PAGE, blotted onto nitrocellulose, and either autoradiographed (top) or probed using a monoclonal anti-HA antibody (bottom).

(b) Translation reactions with tagged and untagged nascent chains were subjected to an analytical immunopurification procedure. Aliquots of fractions were applied to SDS-PAGE, blotted onto nitrocellulose, amido black stained (Protein), and autoradiographed to detect nascent polypeptide chains (^{35}S-Met). T, total; S, supernatant; R, crude ribosomes; dR, depleted crude ribosomes after antibody and magnetic bead incubation; pE, eluate using excess HA-peptide; B, magnetic beads after elution. Note the enrichment of HA-DP120 RNCs (pE) and the background-free control eluate (pE for DP120). Asterisks mark bovine serum albumin, which is part of the antibody preparation.

(c) Same as shown for HA-DP120 in (b) but upscaled for preparative purpose. Note the characteristic pattern of ribosomal proteins in the peptide eluate (pE).

(d) Immunopurified RNCs carrying the HA-DP120 chain were reconstituted with different amounts of solubilized Sec61 complex in DeoxyBigChap (DBC) or Triton X-100 detergent in the presence or absence of phospholipids. Subsequently, the samples were subjected to protease digestion by incubation with proteinase K and analyzed by SDS-PAGE followed by autoradiography.

The RNCs were incubated with increasing amounts of solubilized Sec61 complex under different conditions and, subsequently, nascent chain insertion was monitored by protease protection assays. After digestion with proteinase K of RNCs alone, the nascent chain signal shifted completely to a lower molecular weight (regardless of the presence or absence of lipids), demonstrating that only a small part, buried in the ribosomal tunnel (30–35 aa), was protected (Figure 1d). However, incubation of the RNCs with the Sec61 complex resulted in protection of the full-length chain with different degrees of efficiency; most efficient was an excess of Sec61 complex in the detergent DeoxyBigChap (DBC), leading to a fraction of approximately 75%–85% protected chains. Higher concentrations of the Sec61 complex or the presence of phospholipids did not significantly change the degree of protection. In contrast, incubation with the Sec61 complex in Triton X-100 led only to 15%–25% protected chains (Figure 1d). Thus, by using RNCs with the solubilized Sec61 complex in DeoxyBigChap, we successfully reconstituted an active RNC-channel complex with fully inserted nascent chain. This result agrees well with the results of previous experiments demonstrating (1) that the channel interacts with the ribosome in detergent solution (Beckmann et al., 1997; Prinz et al., 2000), (2) that direct and productive interaction of RNCs with the channel can take place without involvement of the SRP/SR targeting system (Jungnickel and Rapoport, 1995), and (3) that the mammalian Sec61 complex, as well as the heptameric Sec complex from yeast, display translocation activity in detergent solution (Matlack et al., 1997; Mothes et al., 1998).

Reconstruction and Overall Structure of the RNC-Channel Complex

The RNC-channel complex reconstituted in DBC was used for structure determination by cryo-EM and single particle reconstruction. As a control, empty ribosomes were isolated from a translation system lacking mRNA and reconstituted under the same conditions with the Sec61 complex. Figure 2 shows the structures of the active and inactive complexes determined at a resolution of 15.4 Å and 18.9 Å, respectively, based on the Fourier shell correlation criterion using a cutoff at 0.5 (corresponding to 10.9 Å and 13.7 Å using a cutoff at 3σ). The lower resolution of the nontranslating ribosome is most likely a result of higher conformational variability, due to a less defined functional state of the ribosome. The 15.4 Å (10.9 Å) reconstruction represents the highest resolution achieved so far for an eukaryotic 80S ribosome. It was further processed to separate rRNA from

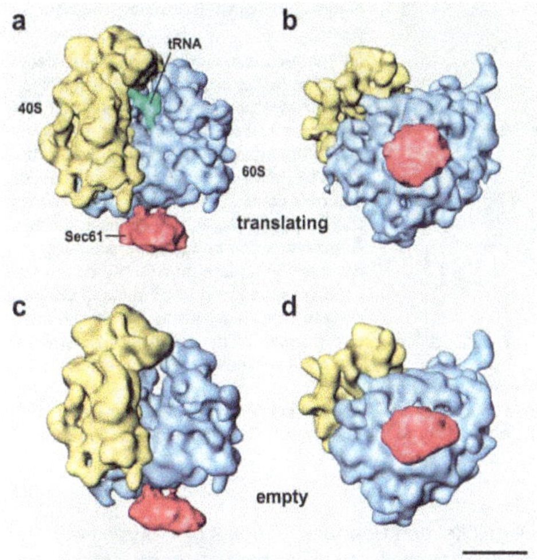

Figure 2. Cryo-EM Structures of the Translating and the Empty Ribosome-Sec61 Complex

(a) Reconstruction of the translating and translocating RNC-Sec61 complex in DBC at 15.4 Å resolution. Note the presence of tRNA density in the P site and the Sec 61 channel with the shape of a compact disk in the active complex. Color coding: yellow, small ribosomal subunit (40S); blue, large ribosomal subunit (60S); red, Sec61 complex; green, P site tRNA.

(b) Same as (a) but rotated upward by 90°.

(c) Reconstruction of the empty complex in DBC at 18.9 Å resolution. Note the absence of tRNA density and the elongated shape of the channel. Color coding as described for (a).

(d) Same as (c) but rotated upward by 90°. Bar = 100 Å.

protein densities (Spahn et al., 2000) and interpreted by docking of rRNA and protein structure models (see accompanying paper by Spahn et al., 2001 [this issue of *Cell*]).

Both structures show the same overall appearance typical for the yeast 80S ribosome, with the small (40S) and large (60S) ribosomal subunits clearly recognizable (Figure 2) (Beckmann et al., 1997; Gomez-Lorenzo et al., 2000; Morgan et al., 2000). Extra density, representing the Sec61 complex, is visible in both reconstructions at the exit site of the large ribosomal subunit tunnel, which is believed to function as a conduit for the nascent chain (Beckmann et al., 1997; Bernabeu et al., 1983; Nissen et al., 2000). The shape of the PCC in its active and inactive state is similar to the shape we observed before (Beckmann et al., 1997), with a gap of at least 15 Å between the channel surface and the ribosome, regardless of the presence of a signal sequence. However, in both states, additional connections between the PCC and the ribosome are visible and, interestingly, the PCC now appears as a compact disk with a central indentation rather than a toroidal mass with a central pore. The inactive PCC is somewhat flattened in the direction perpendicular to the plane of the membrane and elongated in the plane of the membrane when compared to the active PCC. Furthermore, additional density is visible in the intersubunit space of the programmed ribosome, unambiguously recognizable as tRNA.

To quantify the occupancy of RNCs with channel and to obtain additional confirmation of the contour levels chosen in this study, we sorted the translating complex particles computationally, dependent on the presence of density in the channel region (C.M.T.S., P.A.P., J.F., unpublished data). The particle distribution between the two resulting populations indicated that about two-thirds of the ribosomes were carrying the channel, which agrees approximately with the degree of nascent chain protection in the protease protection assay (Figure 1d). The reconstruction of the two subsets yielded two structures, both with clearly visible tRNA densities. As expected, only one of the structures had a channel density. The shape of this density is essentially identical to the structure shown, but the relative density of the channel is stronger (data not shown).

Presence of tRNA in the Programmed Ribosome

The tRNA density is shown with the same contour level as the ribosome. It is located in the P site of the ribosome and corresponds to an occupancy of at least 80%–90% of the ribosomes (Figure 2; see also Spahn et al., 2001). Difference maps, using programmed and empty ribosomes, did not reveal any significant occupancy of the A site or E site with tRNA. tRNAs occupying A or E site were probably removed during the wash step using high salt concentrations. This result demonstrates that the translation of truncated mRNA indeed leads to a stalled peptidyl-tRNA, with the vast majority tightly bound to the P site in a manner resistant to high salt concentrations. Furthermore, these results confirm the efficiency of the new protocol and the quality of the resulting preparation. The high occupancy of ribosomes with peptidyl-tRNA serves as proof for the presence of the signal sequence-containing nascent chain, which is small (~14 kDa) and too elongated to be visualized at this resolution. Similarly, cryo-EM visualization of 70S ribosomes charged with tRNA in vitro (Agrawal et al., 1996, 2000) or of translating dimeric ribosomes formed in vivo (Stark et al., 1997) also led to clearly identifiable tRNA densities in the intersubunit space. Therefore, the presence of the tRNA density in our map served as a prerequisite to permit the interpretation of the structures based on different functional states.

Movement of the Expansion Segment 27

Close to the channel attachment site, the large subunit bears a rod-like density 150 Å in length with a twist typical for helical rRNA. The density originates at the subunit interface and was identified as the main helix of expansion segment 27 (ES27) of 25S rRNA (Spahn et al., 2001), one of the rRNA insertions characteristic for 80S ribosomes (Gerbi, 1996). In several reconstructions of different nontranslating yeast ribosome complexes (Beckmann et al., 1997; Gomez-Lorenzo et al., 2000; Morgan et al., 2000; and not shown), this helix can be found close to the L1 arm ("L1 position"), but also, following a counterclockwise rotation by 90°, reaching over the tunnel exit site ("exit position"). A distance of approximately 15 Å between the tip of the helix and the tunnel exit leaves sufficient space for the nascent chain to emerge from the ribosome. The preference for one or the other position varies in these reconstructions, and

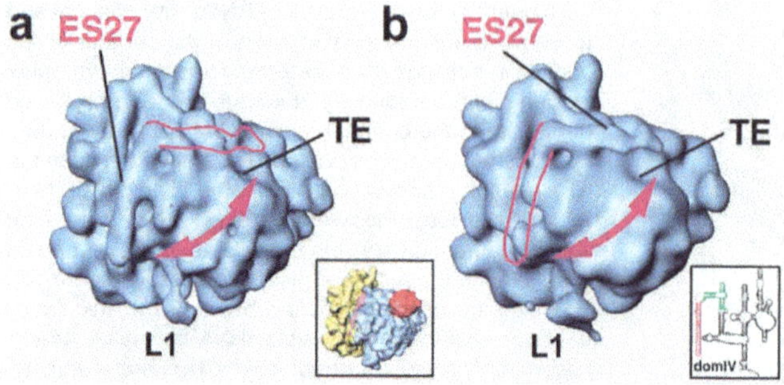

Figure 3. Dynamic Expansion Segment 27 (ES27)

(a) The reconstruction of a ribosome-Sec61 complex in Triton X-100 (Beckmann et al., 1997) shown without the channel density. TE marks the tunnel exit and L1 the L1 protrusion. Note the expansion segment 27 (ES27) in the L1 position, distant from the tunnel exit. The pink contour shows the exit position of ES27 as observed in (b). Left inset: the 15.4 Å reconstruction in the same orientation as the maps in (a) and (b) with the color code used in Figure 2 and ES27 in the L1 position colored pink. Right inset: Secondary structure diagram of domain IV of *S. cerevisiae* 25S rRNA with colored ES27 (green and pink) and the moving helix labeled pink (diagram from R.R. Gutell, http://www.rna.icmb.utexas.edu).

(b) A reconstruction of the ribosome without attached Sec61 complex at a resolution of 24.8 Å. Note that in this structure no density is visible in the L1 position. Instead, the ES27 is in the exit position with its tip in close proximity to the tunnel exit. The pink contour shows the L1 position as observed in (a).

the density is often present in both positions. Therefore, the ES27 helix is a highly dynamic structure with two main conformations in functional equilibrium (Figure 3). Notably, in reconstructions of ribosome-channel complexes, the ES27 helix was located exclusively in the L1 position, thereby not interfering with the channel in the exit position (Figure 3). Thus, binding of the channel to the ribosome must coincide with a conformational change in the ribosomal periphery, shifting the equilibrium for the ES27 helix entirely to the L1 position.

We suggest that this rotating rRNA structure may play a role in coordinating access of nonribosomal factors, such as chaperones, modifying enzymes, SRP, or the translocon to the tunnel exit site and thereby to the emerging nascent chain. In addition, the segment may interact directly with the nascent chain or with nascent chain-interacting proteins. In any case, the function of ES27 is essential and conserved. It has been demonstrated in *Tetrahymena* that deletion of this insertion is lethal but can be complemented with a corresponding insert from other species (Sweeney et al., 1994). After the L1 protuberance (Gomez-Lorenzo et al., 2000), the ES27 helix is the second example of an extremely dynamic rRNA segment in the periphery of the 80S ribosome.

The Ribosome-Channel Connection

The connection between ribosome and channel is not circumferentially sealed, but leaves a lateral opening of ~15 Å (Figures 2, 4, and 5) in both reconstructions. This finding is similar to what has been observed at lower resolution in several ribosome-channel reconstructions from yeast as well as from mammalian complexes (Beckmann et al., 1997; Ménétret et al., 2000). There are no significant differences in the dimensions of these gaps when comparing active and inactive complexes (Figures 2 and 4). Even at very low contour levels, the ribosome-channel connection is never completely sealed regardless of the functional state (not shown). This observation agrees well with protease protection experiments showing that emerging nascent chains are accessible to added protease when they are either too short to interact productively with the channel or form extended cytosolic loops (Hegde and Lingappa, 1996; Jungnickel and Rapoport, 1995). It also agrees with the finding that nascent chains such as preprolactin 86mer and the substrate used in this study are protected from digestion. These nascent chains form a loop reaching directly from the tunnel exit into the channel, with the space between the channel surface and ribosome being too small for the protease to enter (Jungnickel and Rapoport, 1995; Shaw et al., 1988).

However, with a lateral opening as large as 15 Å, it is unlikely that the ribosome-channel connection can function as an ion-tight seal maintaining the ion permeability barrier of the ER membrane, as suggested based on fluorescence quenching experiments (Crowley et al., 1994; Hamman et al., 1997). It is possible, however, that it is not the PCC itself, but lipids or additional proteins that lead to the observed seal formation in native mammalian membranes. On the other hand, a seal formed by the ribosome-membrane junction may be ineffective in light of the existence of additional tunnels in the large ribosomal subunit (T2 and T3, Figure 4), which are also present in bacterial and archaebacterial ribosomes (Ban et al., 2000; Frank et al., 1995; Gabashvili et al., 2001). Since these tunnels connect the conduit for the nascent chain with the cytosolic environment, they may allow ion flow between the cytosol and the ER lumen, even if a seal-forming junction between ribosome and membrane is present (Figure 4).

Docking of rRNA and protein models into separated rRNA and protein densities (see Spahn et al., 2001) allowed the analysis of the PCC attachment sites at the molecular level (Table 1). In both functional states, the channel is attached to the ribosome with the same connections, one of which we observed before (connection 1) at lower resolution (Beckmann et al., 1997). Connection 1 is formed by a protrusion consisting of the helix 59 of rRNA domain III and the N-terminal domain of the ribosomal protein rpL19 (Table 1). However, three additional connections are now visible (Figures 4 and 5). Connection 2 most likely also involves rRNA domain III (helix 53 and helix 50) and the N-terminal domain of rpL19 as well as rpL25. Connection 3 is in closest proximity to the tunnel exit engaging rRNA helix 24 of domain I and, in addition, an extended loop of rpL26. Connection 4 is the most substantial one, involving two

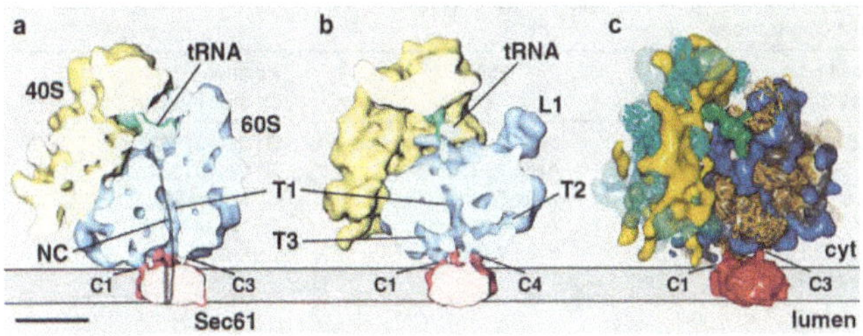

Figure 4. Connections, Gaps, and Tunnels in the RNC-Sec61 Structure

(a) The 15.4 Å reconstruction of the RNC-Sec61 complex as shown in Figure 2a but cut perpendicularly to the plane of the membrane along the tunnel in the large ribosomal subunit. Color code as in Figure 2. C1 to C4 mark the connections between the ribosome and the channel. T1 marks the vertical tunnel, which functions as conduit for the nascent chain. The expected path of the nascent chain is indicated (NC). Note the gap between the ribosome and the channel. Cyt, cytosolic side; lumen, ER lumen. Bar = 100 Å.

(b) As in (a) but rotated by 60° around the z axis. T2 and T3 indicate the position of two horizontal tunnels connecting the main vertical tunnel (T1) with the cytosolic environment. Note the gap between the ribosome and the channel.

(c) Structure of the RNC-Sec61 complex after separating rRNA and protein densities, and docking of homology protein structure models into their densities. Color code: red, Sec61 complex; green, P site tRNA; blue, 25S/5S rRNA density; transparent orange, 60S protein density; orange, backbone of 60S protein models; yellow, 18S rRNA density; transparent turquoise, 40S protein density; turquoise, backbone of 40S protein models. Note the gap between the ribosome and the channel, and the presence of rRNA as well as protein near the channel connections.

ribosomal proteins, rpL25 and rpL35, and the tip of rRNA helix 7 of domain I (Figures 4 and 5). It is interesting to find that ribosomal proteins are most likely involved in all ribosome-channel connections. The isolated rRNA of the large ribosomal subunit was shown to be sufficient to bind to the channel with high affinity, even across different species (Prinz et al., 2000). However, in contrast to other parts of the ribosomal periphery, the entire region near the tunnel exit, including rRNA as well as proteins, appears to be relatively conserved. At a resolution between 25–30 Å, the mammalian ribosome has also a very similar appearance in this region (Morgan et al., 2000), and four ribosome-channel connections have been observed in similar positions (Ménétret et al.,

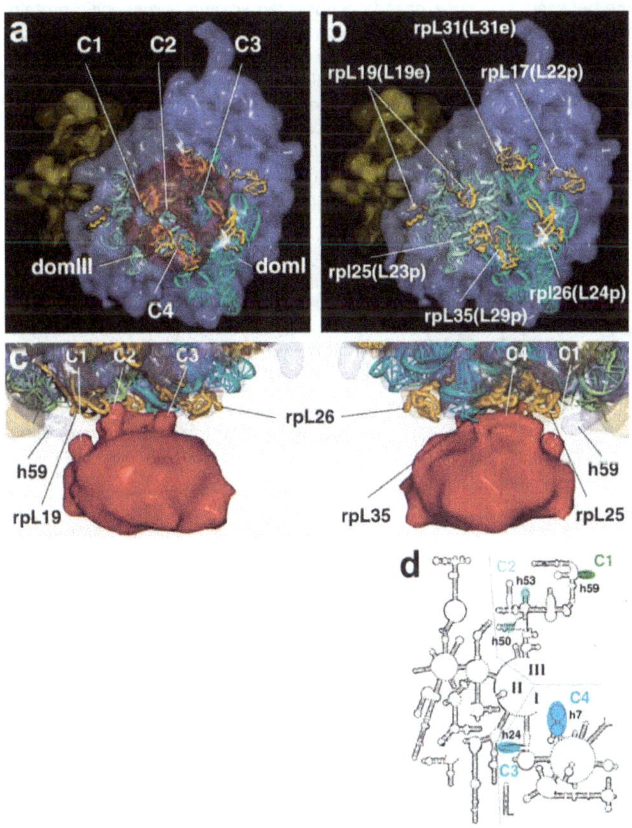

Figure 5. Molecular Analysis the RNC-Sec61 Connections

(a) The RNC-Sec61 complex is shown in the same orientation and color code as in Figure 2b with transparent densities. Docked models of the 25S rRNA domains (turquoise, domI; green, domIII) involved in channel attachment and the backbones of homology models (orange) of the ribosomal proteins near the tunnel exit are shown. The positions of the four connections (C1–C4) are indicated by holes in the Sec61 density. Note the particularly strong participation of ribosomal proteins in connection 4.

(b) Same as in (a) but without the Sec61 density. Proteins are labeled according to the yeast nomenclature with the corresponding ribosomal protein family name given in parenthesis.

(c) Close up of the RNC-Sec61 junction in a similar orientation as in Figure 2a (left side) and rotated 180° around the z axis (right side). The separated 25S rRNA and the 60S protein densities are shown in transparent blue and transparent orange, respectively. Docked models of the 25S rRNA domain I (turquoise), domain III (green), and proteins (orange) are shown as ribbons. The positions of the four connections (C1–C4) are indicated. Density for helix 59 of rRNA domain III is marked by h59. Participation in connection 1, rRNA domain III and rpL19 (L19e); connection 2, rRNA domain III and rpL19 (L19e); connection 3, rRNA domain I and rpL26 (L24p); connection 4, rRNA domain I, rpL25 (L23p) and rpL35 (L29p). For details see Table 1.

(d) Secondary structure diagram of the yeast 25S rRNA domains I, II, and III (diagram from R.R. Gutell, http://www.rna.icmb.utexas.edu). The regions involved in connection are labeled accordingly (for details see Table 1).

Table 1. Molecular Analysis of the Ribosome-Channel Connections

Connection	Protein	Position (yeast models)	25S rRNA	Position (archea model)
1	rpL19 (L19e)	P25–T48	helix 59 (domIII)	G1627–G1634
2	rpL19 (L19e)	P25–T48	helix 53 (domIII)	G1498–C1507
	rpL25 (L23p)	S68–I80	helix 50 (domIII)	C1421–A1424
3	rpL26 (L24p)	T86–P96	helix 24 (domI)	G504–G487
4	rpL25 (L23p)	S68–I80, A127–A135	helix 7 (domI)	G81–G94
	rpL35 (L29p)	A2–K5, K25–K49		

2000). This region is also similar comparing the yeast ribosome with the archeabacterial 50S subunit (Nissen et al., 2000), an important difference being a shorter rRNA helix 59 in *H. marismortui*. This helix in the bacterial *E. coli* ribosome (Gabashvili et al., 2000) is comparable in length to the yeast ribosome, but a rpL19-like protein is missing (L19e in *Archea*). Thus, despite the fact that there are some differences, the overall spatial arrangement of the ribosome-channel interaction is conserved in bacterial, archeabacterial, and eukaryotic cells and involves both large subunit RNA and proteins.

Structure and Function of the Protein-Conducting Channel

The dimensions of the PCC are similar to findings of earlier EM studies (Beckmann et al., 1997; Hanein et al., 1996; Ménétret et al., 2000). The distance between cytosolic and lumenal surface is approximately 48 Å, hence easily spanning the membrane. The diameter of the translocating channel is between 85 and 95 Å (Figure 6). The channel associated with the empty ribosome appears slightly flattened and elongated toward the back of the large ribosomal subunit with a diameter of approximately 100 Å. This elongation is lost at higher contour levels, where translocating and nontranslocating channels appear more similar (data not shown). Although the shape of the channel displays structural details, the slightly pentagonal shape, which was described in previous studies of the Sec61 complex (Beckmann et al., 1997; Hanein et al., 1996) and in one study of its bacterial homolog, the SecYE complex (Meyer et al., 1999), is not recognizable.

For estimating the Sec61 oligomer stoichiometry, we used the shape of the channel in the following way: each Sec61 trimer is believed to have 12 membrane-spanning helices plus two amphipathic helices (Johnson and van Waes, 1999; Wilkinson et al., 1996). Assuming membrane-specific helix packing, we determined the number of helices of known membrane protein structures that could fit into the channel density. This approach is reliable even in the presence of flexible and thereby invisible domains outside the membrane, in contrast to calculations based on observed volume and specific protein density. Assuming less dense packing in the central region, the outline of the channel offers space for 35 helices, for example provided by five rhodopsin molecules (Palczewski et al., 2000) (Figures 6b and 6c). This helix packing corresponds to an average area between 180 and 190 Å²/helix, which is in good agreement with experimental data (184 Å²/helix for rhodopsin and 186 Å²/helix for bacteriorhodopsin; for review see Collinson et al., 2001). This result argues strongly in favor

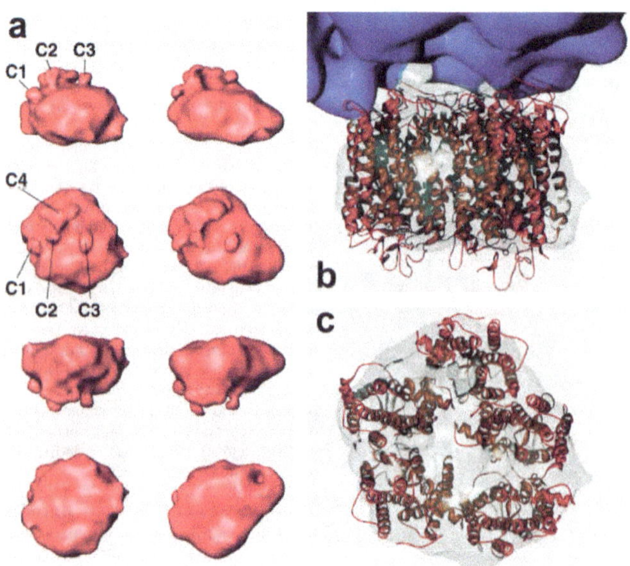

Figure 6. Structure and Stoichiometry of the Protein-Conducting Channel

(a) Isolated density of the Sec61 complex in the translocating state (left column) and the inactive state (right column). The view on top is the same as in Figures 2a and 2c. The views below are rotated 90°, 180°, and 270° toward the viewer. The four connections are marked by C1 through C4. Note the central indentation at the surface facing the ER lumen and the ribosome. The channel attached to the empty ribosome appears elongated and flattened compared to the one engaged in translocation.

(b) Closeup of the RNC-Sec61 junction in the same orientation as in (a), with the large ribosomal subunit shown in blue and the Sec61 complex in transparent green. In order to estimate the stoichiometry of the Sec61 oligomer, the transmembrane helices of five rhodopsin molecules (red ribbons) were fitted into the channel density.

(c) Top view of (b), but only the channel density with fitted membrane domains. The number of approximately 35 helices fitting into the density suggests a trimer of Sec61 trimers (36 helices).

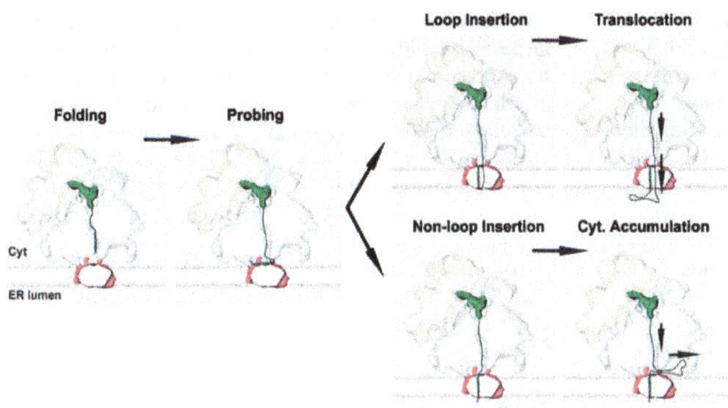

Figure 7. Binary Model of Cotranslational Protein Translocation

The model is based on the finding that a gap exists between the RNC and the PCC, and that the translating PCC has a compact conformation. Hence, the PCC can provide a seal to maintain the ion permeability barrier of the ER membrane:

(1) The tunnel in the large ribosomal subunit facilitates folding of α-helical segments.

(2) The emerging segment is probed by the PCC before insertion. Hydrophobicity, helicity, and the nature of the flanking regions (i.e., positive charges) determine if and in what orientation the segment is inserted.

(3) Insertion can occur in two different orientations with the channel opening just wide enough to accommodate the inserted nascent polypeptide.

(4) In case of *loop insertion*, the following nascent chain is guided through the membrane, and translocation is possible. In case of *non-loop insertion*, the following nascent chain cannot translocate, accumulating on the cytosolic side of the membrane. Because of the sealed channel and the existence of the gap between ribosome and PCC, such cytosolic domains can easily exit into the cytosol at any time without compromising the ion permeability barrier.

Nascent chain translocation after loop insertion and cytosolic accumulation after non-loop insertion are the only two principally different functional states of the RNC-PCC complex. A simple secretory protein would experience only one loop insertion of the signal sequence after targeting by SRP and translocate. For a polytopic membrane protein, the states would alternate with every new hydrophobic transmembrane domain. This model may be expanded by including additional regulation (i.e., pause transfer sequences [Chuck and Lingappa, 1992]) and exceptions.

of three Sec61 trimers forming the isolated active PCC. For the homologous SecYEG complex, it was reported that a slightly higher average area of 199 Å^2 per helix is occupied in the plane of the membrane (Collinson et al., 2001). If this area is similar for the Sec61 complex, the observed part of the density map could encompass approximately 33 transmembrane helices, still supporting the estimation of three Sec61 trimers per channel. The lack of a 3-fold rotational symmetry may be a feature of the channel itself, or could be induced by the asymmetric interaction with the ribosome.

Remarkably, in both functional states, the channel appears compact and has only a small indentation instead of a central pore (Figure 6). This indicates that the interaction with the signal sequence does not necessarily lead to a widely open channel conformation. It appears compact in both maps, even at higher contour levels, a result which is different from our previous observation with the empty ribosome-channel complex in Triton X-100 detergent solution (Beckmann et al., 1997). The open conformation in Triton X-100 may be induced by this strong detergent and may explain the accessibility of the nascent chain to protease under these conditions (Figure 1d). The slightly flattened and more elongated shape of the inactive channel could indicate that gating by the signal sequence leads to a small rotation of the membrane helices toward a position more perpendicular to the plane of the membrane. This iris-like movement would be reminiscent of the conformational change suggested for the gating of gap junctions (Unwin and Zampighi, 1980). Our results do not support conclusions from fluorescence quenching experiments with native mammalian ER membranes suggesting a pore of 9–15 Å diameter in the inactive translocon (Hamman et al., 1998), and of 40–60 Å diameter in the translocating one (Hamman et al., 1997). Yet, in contrast to the purified channels used by us, the channels in native membranes are known to be associated with additional proteins in different oligomeric states (Ménétret et al., 2000; Wang and Dobberstein, 1999). In a recent cryo-EM study of mainly mammalian ribosome-channel and mixed ribosome-translocon complexes (Ménétret et al., 2000), the channels are shown with open conformations. However, the contour levels, chosen in that study based on a molecular weight/specific density calculation, appear to be very high. At slightly lower, more realistic contour levels, closed conformations of the channels are visible, which is in accord with our results. Taken together, in all studies of the eukaryotic PCC, using either 2D (Hanein et al., 1996) or 3D reconstructions (Beckmann et al., 1997; Ménétret et al., 2000), a central pore or at least an indentation has been observed, which is juxtaposed to the nascent chain tunnel exit of the ribosome. This result is a strong indication that the central region in the channel is less dense and very flexible, in accordance with its function in protein conductance across the membrane.

Binary Model of Cotranslational Translocation and Membrane Protein Insertion

We observed the translocating channel in a compact conformation, and a gap between the channel and the ribosome. The compact appearance indicates that gating of the channel by the signal sequence can lead to an opening just large enough to be completely occupied by the inserted nascent polypeptide chain (Figures 3 and 7) without allowing ion conductance. Based on these findings, we propose a binary model, describing how the ribosome-PCC complex could function in cotranslational translocation and membrane protein insertion (Figure 7). First, we suggest that the conserved tunnel in the large ribosomal subunit represents an important functional domain of the ribosome, allowing exclusively the folding of α-helical secondary structure. The tunnel dimensions are such that α-helix folding is supported, but β sheet formation is sterically unlikely, because the average tunnel diameter of 15 Å is too small. A preferred sequence of folding events starting with α-helix formation would provide folding seeds along the sequence and considerably reduce the difficulty of folding. In the translocation context, it has been shown that helicity as well as hydrophobicity of a signal sequence are necessary for its productive interaction with the signal recognition particle (SRP) and with the PCC (Keenan et al., 2001; Plath et al., 1998). Probing of the nascent chain for hydrophobic helices by the channel would lead to the insertion and capture of any sufficiently hydrophobic segment in such a way that the inserted polypeptide is tightly accommodated. Depending upon its orientation in the channel, the hydrophobic segment can expose its C-terminal end either to the cytosolic or to the lumenal side, in the latter case resulting in loop formation of the remaining nascent chain in the channel (Figure 7). Importantly, insertion would not lead to further opening of the pore to any predefined size. As a result, the following nascent chain will either accumulate on the cytosolic side or, only if it has been coinserted in a loop, translocate across the membrane. This scenario would have two major implications: (1) After ribosome targeting to the PCC, further events are almost exclusively dependent on the presence and nature of hydrophobic segments and their interaction with the channel, especially their preferred orientation for channel insertion. (2) The channel itself would have the property of a seal, preventing ion flow across the membrane independent of the presence of the ribosome. It would require that the pore accessible to hydrated ions in the presence of a translocating chain is not larger than 6 Å in diameter (Johnson and van Waes, 1999). How could the channel provide such flexibility? One possibility is that the overall arrangement of transmembrane helices and subunits can be adjusted accordingly. At the same time, as observed for SRP (Keenan et al., 2001), flexible amino acid side chains could behave like bristles to provide the appropriate environment and accommodate the large number of different translocating polypeptide chains.

Conclusions

Using immunopurified RNCs for a molecular analysis of the RNC-PCC complex, we find conserved rRNA segments as well as proteins involved in forming the ribosome-channel junction. We observe the translocating channel in a compact conformation and a gap between the channel and the ribosome. This observation provides the basis for a binary model of cotranslational translocation. The model needs to postulate only two principally different functional states of the RNC-PCC complex to explain the translocation of secretory proteins or insertion of almost all kinds of membrane proteins: (1) Translation preceded by a loop insertion resulting in translocation across the membrane (e.g., secretory proteins), and (2) translation preceded by the channel insertion without loop resulting in accumulation in the cytosol (e.g., cytosolic domains of membrane proteins or signal anchor proteins). For polytopic membrane proteins, these two states would alternate. Thus, the sealing properties of the translocating channel and the gap between the channel and the ribosome, as others and we observe it, would significantly reduce the regulatory and spatial requirements for the RNC-PCC complex in processes such as release of cytosolic domains and partitioning of transmembrane domains into the bilayer. Furthermore, seal-like behavior of the Sec61 complex may also maintain the permeability barrier of the membrane in posttranslational or retrograde protein translocation.

The spatial arrangement of the RNC-PCC complex containing nascent chains of polytopic integral membrane proteins or associated with additional translocon components will be addressed in future studies.

Experimental Procedures

Purification of Ribosome-Nascent Chain Complexes

For the generation of RNCs we used a cell-free yeast translation system, which was programmed with truncated synthetic mRNA coding for the 120 N-terminal amino acids of the type II membrane protein dipeptidylpeptidase B (DPAP-B). An additional N-terminal HA-tag (YPYDVPDYA) was used for the immunopurification of the stalled RNCs.

Capped mRNA was synthesized using the SP6 polymerase-based Message Machine kit from Ambion with DNA fragments as templates. The DNA fragments were generated by PCR, with the full-length gene as a template and extended primers. The 5′-primer added a SP6 site, followed by a high initiation efficiency 5′-untranslated region, an optimized AUG context and, in case of the tagged construct, the additional 9 amino acids of the HA-tag (5′-ATTTAGG TGACACTATAGAAACCAAACAAAACAAATAAAACAAAAACACAAT GTCTTACCCATACGATGTTCCAGATTACGCTGAAGGTGGCGAAGA AGAAGTTG-3′). The 3′-primer determined the length of the nascent chain (5′-ATCGTAGACAGATTTAACAACGTA-3′ for HA-DP-120). The amounts of mRNA in translation reactions were optimized for each batch of yeast extract.

The yeast translation extracts were prepared and used as previously described (Waters and Blobel, 1986). We used a haploid strain derived from DF5α. Gel filtration of the extract was carried out in buffer A (20 mM HEPES [pH 7.5], 100 mM KOAc, 2 mM $Mg(OAc)_2$, 2 mM DTT, 0.5 mM PMSF, and 10% w/v glycerol). The extract was treated with nuclease S7, frozen in liquid nitrogen, and stored at −80°C.

The 100 μl translation reaction contained 33% yeast extract (∼100 OD_{260}/ml), 20 mM HEPES (pH 7.5), 150 mM KOAc, 2.5 mM $Mg(OAc)_2$, 2 mM DTT, 20 mM creatine phosphate, 0.1 mg/ml creatine kinase, 1 mM ATP, 0.5 mM GTP, 0.1 mg/ml yeast tRNA, 0.2 mM amino acids, 3.3% glycerol, 0.8 U/μl RNAsin (Promega), and ^{35}S-methionine (NEN). Reactions were incubated for 70 to 90 min at 17°C and terminated by addition of 2 μl of 10 mg/ml cycloheximide. Aliquots were analyzed by SDS-PAGE. After subsequent semidry blotting onto nitrocellulose membranes, proteins were visualized with amido black. The membranes were dried and exposed to X-ray film or probed with α-HA antibody followed by ECL detection (Amersham).

Analytical immunopurification of RNCs was performed by using a 100 μl reaction as starting material for each type of nascent chain. The purification was done as described below for the preparative purification except that the amounts of reagents and volumes were 5- to 8-fold smaller.

Preparative immunopurification was performed with 8 × 100 μl reactions. After addition of cycloheximide, two reactions were combined and spun through 800 μl of a high salt sucrose cushion (50 mM Tris.Cl [pH 7.0], 500 mM KOAc, 25 mM $Mg(OAc)_2$, 2 mM DTT, 1 M sucrose, 10 μg/ml cycloheximide, 0.1% Nikkol, and 0.1% pill/ml [1 pill complete protease mix/ml H_2O, Boehringer Mannheim]) at 355,000 g_{av} for 45 min. Each of the resulting pellets was resuspended in 200 μl of ice-cold buffer B (32.5 mM Tris.Cl [pH 7.0], 125 mM KOAc, 26.25 mM $Mg(OAc)_2$, 1.5 mM DTT, 250 mM sucrose, 100 μg/ml cycloheximide, 0.1% Nikkol, 0.1% pill/ml, and 0.4 U/μl RNAsin). 4 μg of a biotinylated anti-HA antibody (clone 12CA5 from Boehringer Mannheim) was added, followed by incubation for 60 min on ice and 30 min at RT. Each suspension was then added to 3 mg of equilibrated (2 × 200 μl buffer B) streptavidin-coupled magnetic beads (Dynabeads M-280 Strepavidin from Dynal) and incubated for 30 min at RT. The beads were washed with 2 × 400 μl of ice cold buffer B and 2 × 400 μl ice cold buffer C (50 mM Tris.Cl [pH 7.0], 500 mM KOAc, 25 mM $Mg(OAc)_2$, 1 mM DTT, 250 mM sucrose, 100 μg/ml cycloheximide, 0.1% Nikkol, 0.1% pill/ml, and 0.4 U/μl RNAsin). To elute the bound RNCs, the beads were incubated with 200 μl of buffer C including 1 mg/ml HA peptide (Boehringer Mannheim) for 30 min on ice and 15 min at RT. After rinsing of the beads with 100 μl buffer C, eluates were combined in two fractions of 600 μl and each spun through a 400 μl high salt sucrose cushion as described above. The resulting pellets were slowly resuspended in 50–100 μl of buffer G (20 mM Tris.Cl [pH 7.0], 50 mM KOAc, 10 mM $Mg(OAc)_2$, 1 mM DTT, 125 mM sucrose, 100 μg/ml cycloheximide, 0.05% Nikkol, 0.5% pill/ml, and 0.2 U/μl RNAsin), shock frozen, and stored at −80°C. The concentration of the final fraction was between 2.5 and 5 OD_{260}/ml corresponding to an overall yield of about 5–10 pmol. Aliquots of the various purification steps were precipitated with 7% trichloroacetic acid in the presence of deoxycholate for analysis by SDS-PAGE.

Reconstitution of RNC-Sec61 Complexes

The Sec61 complex was purified as previously described (Beckmann et al., 1997). Exchange of detergent from Triton X-100 to DeoxyBigChap (DBC, Calbiochem) was done essentially as described by Gorlich and Rapoport (1993) by binding of the complex to SP-Sepharose (Pharmacia) followed by equilibration and elution in 0.2% DBC. Before reconstitution with RNCs or empty ribosomes, the Sec61 complex in DBC and Triton X-100 was dialyzed at a molecular cutoff of 3.5 kDa against 20 mM Tris.Cl (pH 7.0), 100 mM KOAc, 5 mM $Mg(OAc)_2$, and 1 mM DTT.

RNC-Sec61 complexes were reconstituted by incubating 0.25 pmol of RNCs in a volume of 25 μl with 4–40 fold molar excess of Sec61 in 0.2% DBC or 1% Triton X-100. The incubation was done in the presence and absence of lipids (0.4 mg/ml of phosphatidylcholine and phosphatidylethanolamine, 4:1) for 10 min on ice and 15 min at RT in 30 mM Tris.Cl (pH 7.0), 100 mM KOAc, 10 mM $Mg(OAc)_2$, 1.5 mM DTT, 12.5 mM sucrose, 8% glycerol, 20 μg/ml cycloheximide, and 0.005% Nikkol. Protease protection was assayed by subsequent addition of 2.7 μl of 0.1 mg/ml proteinase K solution and incubation on ice for 15 min. The reaction was stopped by addition of 2.5 μl 0.1 M PMSF, precipitated with trichloroacetic acid, and analyzed by SDS-PAGE followed by autoradiography.

For cryo-EM, 1 pmol RNCs was reconstituted with a 10-fold excess of DBC-Sec61 complex in a volume of 20 μl under the described conditions without lipids. Empty ribosome-Sec61 complexes were generated by a mock translation reaction in the absence of mRNA and the presence of 0.5 mM puromycin. The ribosomes were spun through a high salt sucrose cushion as described above, resuspended in buffer G, and reconstituted with the DBC-Sec61 complex for cryo-EM as described for RNCs.

Electron Microscopy and Image Processing

Cryo-EM grids were prepared as described (Wagenknecht et al., 1988). Micrographs were recorded in a defocus range between 1.7 μm and 6.0 μm on a Philips CM200 and a Philips F20 (FEI/Philips, Eindhoven) at a magnification of about 52,000× and scanned on a HiScan drum scanner (Eurocore, Saint-Denis) with a pixel size of 4.68 Å on the object scale.

The data were analyzed using the SPIDER system. After automated particle selection and visual verification, the data sets were subdivided into defocus groups and refined CTF-corrected 3D reconstructions were calculated as described (Gabashvili et al., 2000). The resolution was estimated by the Fourier shell correlation with a cutoff value of 0.5. The falloff of Fourier amplitudes toward higher spatial frequencies was corrected as described previously (Gabashvili et al., 2000), using the X-ray solution scattering intensity distribution of 70S ribosomes from *E. coli*.

For the translating complex, a subset of 22,788 particles in 30 defocus groups was used for the 3D reconstruction at a final resolution of 15.4 Å. In an attempt to estimate the occupancy of RNCs with channel, this dataset was sorted according to the presence of density in the channel position, resulting in two subsets consisting of 17,934 (+channel) and 10,492 particles (−channel), which were used for independent reconstructions (not shown). For the non-translating complex, a subset of 18,519 particles in 23 defocus groups was used for the 3D reconstruction at a final resolution of 18.9 Å.

The data for the map shown in Figure 3b of the empty ribosome (purified as described (Beckmann et al., 1997) from the strain *Egd2aprA*) were collected on a Philips CM12, and a set of 14,028 particles in 10 defocus groups between 0.8 μm and 2.7 μm was used for the 3D reconstruction at a final resolution of 24.8 Å.

Separation of rRNA and protein density of the ribosomal subunits (Spahn et al., 2000), as well as calculation and docking of RNA and protein structure models, is described in the accompanying paper by Spahn et al., 2001. Visualization and interpretation of the maps and the docked models was done using SPIDER, IRIS Explorer (Numerical Algorithms Group, Inc., Downers Grove, IL), O (Jones et al., 1991), Ribbons (Carson, 1991), and POV-Ray.

Acknowledgments

We thank Sean Campbell for excellent technical assistance, Bob Grassucci, Mike Lewis and David Stokes for support with the electron microscopy, members of the Blobel lab for discussions, Jan Giesebrecht for generation of raytraced images, and Michael Watters for assistance with part of the illustrations. This work was supported by a grant of the VolkswagenStiftung (to R.B.) and grants from NIH (R37 GM29169 to J.F. and R01 GM60635 to P.A.P.) and NSF (BIR 9219043 to J.F.).

Received August 8, 2001; revised September 21, 2001.

References

Agrawal, R.K., Penczek, P., Grassucci, R.A., Li, Y., Leith, A., Nierhaus, K.H., and Frank, J. (1996). Direct visualization of A-, P-, and E-site transfer RNAs in the Escherichia coli ribosome. Science *271*, 1000–1002.

Agrawal, R.K., Spahn, C.M., Penczek, P., Grassucci, R.A., Nierhaus, K.H., and Frank, J. (2000). Visualization of tRNA movements on the Escherichia coli 70S ribosome during the elongation cycle. J. Cell Biol. *150*, 447–460.

Ban, N., Nissen, P., Hansen, J., Moore, P.B., and Steitz, T.A. (2000). The complete atomic structure of the large ribosomal subunit at 2.4 A Resolution. Science *289*, 905–920.

Beckmann, R., Bubeck, D., Grassucci, R., Penczek, P., Verschoor, A., Blobel, G., and Frank, J. (1997). Alignment of conduits for the nascent polypeptide chain in the ribosome-Sec61 complex. Science *278*, 2123–2126.

Bernabeu, C., Tobin, E.M., Fowler, A., Zabin, I., and Lake, J.A. (1983). Nascent polypeptide chains exit the ribosome in the same relative position in both eucaryotes and procaryotes. J. Cell Biol. *96*, 1471–1474.

Blobel, G., and Sabatini, D.D. (1970). Controlled proteolysis of nascent polypeptides in rat liver cell fractions. I. Location of the polypeptides within ribosomes. J. Cell Biol. *45*, 130–145.

Borel, A.C., and Simon, S.M. (1996). Biogenesis of polytopic membrane proteins: membrane segments assemble within translocation channels prior to membrane integration. Cell *85*, 379–389.

Carson, M. (1991). Ribbons 2.0. Appl. Cryst. *24*, 103–106.

Chuck, S.L., and Lingappa, V.R. (1992). Pause transfer: a topogenic sequence in apolipoprotein B mediates stopping and restarting of translocation. Cell *68*, 9–21.

Collinson, I., Breyton, C., Duong, F., Tziatzios, C., Schubert, D., Or, E., Rapoport, T., and Kuhlbrandt, W. (2001). Projection structure and oligomeric properties of a bacterial core protein translocase. EMBO J. *20*, 2462–2471.

Crowley, K.S., Liao, S., Worrell, V.E., Reinhart, G.D., and Johnson, A.E. (1994). Secretory proteins move through the endoplasmic reticulum membrane via an aqueous, gated pore. Cell *78*, 461–471.

Frank, J., Zhu, J., Penczek, P., Li, Y., Srivastava, S., Verschoor, A.,

Active Ribosome-Channel Complex
371

Radermacher, M., Grassucci, R., Lata, R.K., and Agrawal, R.K. (1995). A model of protein synthesis based on cryo-electron microscopy of the E. coli ribosome. Nature *376*, 441–444.

Gabashvili, I.S., Agrawal, R.K., Spahn, C.M., Grassucci, R.A., Svergun, D.I., Frank, J., and Penczek, P. (2000). Solution structure of the E. coli 70S ribosome at 11.5 A resolution. Cell *100*, 537–549.

Gabashvili, I.S., Gregory, S.T., Valle, M., Grassucci, R., Worbs, M., Wahl, M.C., Dahlberg, A.E., and Frank, J. (2001). Role of ribosomal proteins L4 and L22 in gating mechanisms of the polypeptide tunnel system. Mol. Cell *8*, 181–188.

Gerbi, S.A. (1996). Expansion segment: regions of variable size that interrupt the universal core secondary structure of ribosomal RNA. In Ribosomal RNA: Structure, Evolution, Processing, and Function in Protein Synthesis. R.A. Zimmermann and A.E. Dahlberg, eds. (New York: CRC Press), pp. 71–87.

Gilmore, R., and Blobel, G. (1985). Translocation of secretory proteins across the microsomal membrane occurs through an environment accessible to aqueous perturbants. Cell *42*, 497–505.

Gomez-Lorenzo, M.G., Spahn, C.M., Agrawal, R.K., Grassucci, R.A., Penczek, P., Chakraburtty, K., Ballesta, J.P., Lavandera, J.L., Garcia-Bustos, J.F., and Frank, J. (2000). Three-dimensional cryo-electron microscopy localization of EF2 in the Saccharomyces cerevisiae 80S ribosome at 17.5 A resolution. EMBO J. *19*, 2710–2718.

Görlich, D., and Rapoport, T.A. (1993). Protein translocation into proteoliposomes reconstituted from purified components of the endoplasmic reticulum membrane. Cell *75*, 615–630.

Hamman, B.D., Chen, J.C., Johnson, E.E., and Johnson, A.E. (1997). The aqueous pore through the translocon has a diameter of 40–60 Å during cotranslational protein translocation at the ER membrane. Cell *89*, 535–544.

Hamman, B.D., Hendershot, L.M., and Johnson, A.E. (1998). BiP maintains the permeability barrier of the ER membrane by sealing the lumenal end of the translocon pore before and early in translocation. Cell *92*, 747–758.

Hanein, D., Matlack, K.E., Jungnickel, B., Plath, K., Kalies, K.U., Miller, K.R., Rapoport, T.A., and Akey, C.W. (1996). Oligomeric rings of the Sec61p complex induced by ligands required for protein translocation. Cell *87*, 721–732.

Hegde, R.S., and Lingappa, V.R. (1996). Sequence-specific alteration of the ribosome-membrane junction exposes nascent secretory proteins to the cytosol. Cell *85*, 217–228.

Heinrich, S.U., Mothes, W., Brunner, J., and Rapoport, T.A. (2000). The Sec61p complex mediates the integration of a membrane protein by allowing lipid partitioning of the transmembrane domain. Cell *102*, 233–244.

Johnson, A.E., and van Waes, M.A. (1999). The translocon: a dynamic gateway at the ER membrane. Annu. Rev. Cell Dev. Biol. *15*, 799–842.

Jones, T.A., Zhou, J.Y., Cowan, S.W., and Kjeldgaard, M. (1991). Improved methods for building protein models in electron density maps and the location of errors in these models. Acta Crystallogr. A *A47*, 110–119.

Jungnickel, B., and Rapoport, T.A. (1995). A posttargeting signal sequence recognition event in the endoplasmic reticulum membrane. Cell *82*, 261–270.

Kalies, K.U., Gorlich, D., and Rapoport, T.A. (1994). Binding of ribosomes to the rough endoplasmic reticulum mediated by the Sec61p-complex. J. Cell Biol. *126*, 925–934.

Keenan, R.J., Freymann, D.M., Stroud, R.M., and Walter, P. (2001). The Signal Recognition Particle. Annu. Rev. Biochem. *70*, 755–775.

Liao, S., Lin, J., Do, H., and Johnson, A.E. (1997). Both lumenal and cytosolic gating of the aqueous ER translocon pore are regulated from inside the ribosome during membrane protein integration. Cell *90*, 31–41.

Manting, E.H., van Der Does, C., Remigy, H., Engel, A., and Driessen, A.J. (2000). SecYEG assembles into a tetramer to form the active protein translocation channel. EMBO J. *19*, 852–861.

Martoglio, B., Hofmann, M.W., Brunner, J., and Dobberstein, B. (1995). The protein-conducting channel in the membrane of the endoplasmic reticulum is open laterally toward the lipid bilayer. Cell *81*, 207–214.

Matlack, K.E., Plath, K., Misselwitz, B., and Rapoport, T.A. (1997). Protein transport by purified yeast Sec complex and Kar2p without membranes. Science *277*, 938–941.

Matlack, K.E., Mothes, W., and Rapoport, T.A. (1998). Protein translocation: tunnel vision. Cell *92*, 381–390.

Ménétret, J., Neuhof, A., Morgan, D.G., Plath, K., Radermacher, M., Rapoport, T.A., and Akey, C.W. (2000). The structure of ribosome-channel complexes engaged in protein translocation. Mol. Cell *6*, 1219–1232.

Meyer, T.H., Menetret, J.F., Breitling, R., Miller, K.R., Akey, C.W., and Rapoport, T.A. (1999). The bacterial SecY/E translocation complex forms channel-like structures similar to those of the eukaryotic Sec61p complex. J. Mol. Biol. *285*, 1789–1800.

Morgan, D.G., Menetret, J.F., Radermacher, M., Neuhof, A., Akey, I.V., Rapoport, T.A., and Akey, C.W. (2000). A comparison of the yeast and rabbit 80 S ribosome reveals the topology of the nascent chain exit tunnel, inter-subunit bridges and mammalian rRNA expansion segments. J. Mol. Biol. *301*, 301–321.

Mothes, W., Heinrich, S.U., Graf, R., Nilsson, I., von Heijne, G., Brunner, J., and Rapoport, T.A. (1997). Molecular mechanism of membrane protein integration into the endoplasmic reticulum. Cell *89*, 523–533.

Mothes, W., Jungnickel, B., Brunner, J., and Rapoport, T.A. (1998). Signal sequence recognition in cotranslational translocation by protein components of the endoplasmic reticulum membrane. J. Cell Biol. *142*, 355–364.

Ng, D.T., Brown, J.D., and Walter, P. (1996). Signal sequences specify the targeting route to the endoplasmic reticulum membrane. J. Cell Biol. *134*, 269–278.

Nissen, P., Hansen, J., Ban, N., Moore, P.B., and Steitz, T.A. (2000). The structural basis of ribosome activity in peptide bond synthesis. Science *289*, 920–930.

Palczewski, K., Kumasaka, T., Hori, T., Behnke, C.A., Motoshima, H., Fox, B.A., Le Trong, I., Teller, D.C., Okada, T., Stenkamp, R.E., et al. (2000). Crystal structure of rhodopsin: A G protein-coupled receptor. Science *289*, 739–745.

Plath, K., Mothes, W., Wilkinson, B.M., Stirling, C.J., and Rapoport, T.A. (1998). Signal sequence recognition in posttranslational protein transport across the yeast ER membrane. Cell *94*, 795–807.

Prinz, A., Behrens, C., Rapoport, T.A., Hartmann, E., and Kalies, K.U. (2000). Evolutionarily conserved binding of ribosomes to the translocation channel via the large ribosomal RNA. EMBO J. *19*, 1900–1906.

Shaw, A.S., Rottier, P.J., and Rose, J.K. (1988). Evidence for the loop model of signal-sequence insertion into the endoplasmic reticulum. Proc. Natl. Acad. Sci. USA *85*, 7592–7596.

Simon, S.M., and Blobel, G. (1991). A protein-conducting channel in the endoplasmic reticulum. Cell *65*, 371–380.

Spahn, C.M., Penczek, P.A., Leith, A., and Frank, J. (2000). A method for differentiating proteins from nucleic acids in intermediate-resolution density maps: cryo-electron microscopy defines the quaternary structure of the Escherichia coli 70S ribosome. Struct. Fold. Des. *8*, 937–948.

Spahn, C.M., Beckmann, R., Eswar, N., Penczek, P.A., Sali, A., Blobel, G., and Frank, J. (2001). Structure of the 80S ribosome from *Sacharomyces cerevisiae*—tRNA-ribosome and subunit-subunit interactions. Cell *107*, this issue, 373–386.

Stark, H., Orlova, E.V., Rinke-Appel, J., Junke, N., Mueller, F., Rodnina, M., Wintermeyer, W., Brimacombe, R., and van Heel, M. (1997). Arrangement of tRNAs in pre- and posttranslocational ribosomes revealed by electron cryomicroscopy. Cell *88*, 19–28.

Sweeney, R., Chen, L., and Yao, M.C. (1994). An rRNA variable region has an evolutionarily conserved essential role despite sequence divergence. Mol. Cell. Biol. *14*, 4203–4215.

Unwin, P.N., and Zampighi, G. (1980). Structure of the junction between communicating cells. Nature *283*, 545–549.

Wagenknecht, T., Grassucci, R., and Frank, J. (1988). Electron microscopy and computer image averaging of ice-embedded large ribosomal subunits from Escherichia coli. J. Mol. Biol. *199*, 137–147.

Wang, L., and Dobberstein, B. (1999). Oligomeric complexes involved in translocation of proteins across the membrane of the endoplasmic reticulum. FEBS Lett. *457*, 316–322.

Waters, M.G., and Blobel, G. (1986). Secretory protein translocation in a yeast cell-free system can occur posttranslationally and requires ATP hydrolysis. J. Cell Biol. *102*, 1543–1550.

Wilkinson, B.M., Critchley, A.J., and Stirling, C.J. (1996). Determination of the transmembrane topology of yeast Sec61p, an essential component of the endoplasmic reticulum translocation complex. J. Biol. Chem. *271*, 25590–25597.

Cell, Vol. 107, 373–386, November 2, 2001, Copyright ©2001 by Cell Press

Structure of the 80S Ribosome from *Saccharomyces cerevisiae*—tRNA-Ribosome and Subunit-Subunit Interactions

Christian M.T. Spahn,[1,2,9] Roland Beckmann,[4,5,9] Narayanan Eswar,[6] Pawel A. Penczek,[7] Andrej Sali,[6] Günter Blobel,[4] and Joachim Frank[1,2,3,8]

[1]Howard Hughes Medical Institute, Health Research Inc.

[2]Wadsworth Center
Empire State Plaza
Albany, New York 12201

[3]Department of Biomedical Science
State University of New York at Albany
Albany, New York 12222

[4]Laboratory of Cell Biology
Howard Hughes Medical Institute
The Rockefeller University
1230 York Avenue
New York, New York 10021

[5]Institut für Biochemie der Charité
Humboldt Universität zu Berlin
Monbijoustr. 2, 10117
Berlin
Germany

[6]Laboratory of Molecular Biophysics
Pels Family Center for Biochemistry and Structural Biology
The Rockefeller University
New York, New York

[7]University of Texas—Houston Medical School
6431 Fannin
Houston, Texas 77030

Summary

A cryo-EM reconstruction of the translating yeast 80S ribosome was analyzed. Computationally separated rRNA and protein densities were used for docking of appropriately modified rRNA models and homology models of yeast ribosomal proteins. The core of the ribosome shows a remarkable degree of conservation. However, some significant differences in functionally important regions and dramatic changes in the periphery due to expansion segments and additional ribosomal proteins are evident. As in the bacterial ribosome, bridges between the subunits are mainly formed by RNA contacts. Four new bridges are present at the periphery. The position of the P site tRNA coincides precisely with its prokaryotic counterpart, with mainly rRNA contributing to its molecular environment. This analysis presents an exhaustive inventory of an eukaryotic ribosome at the molecular level.

Introduction

The translation of the genetic message is carried out by ribosomes, large macromolecular machines that consist of ribosomal RNA (rRNA) and 50–80 ribosomal proteins (Green and Noller, 1997). Recently, the structures of the small 30S subunit from the thermophile bacterium *Thermus thermophilus* (Schlünzen et al., 2000; Wimberly et al., 2000) and the large 50S subunit from the halophile archaebacterium *Haloarcula marismortui* (Ban et al., 2000) were solved by X-ray crystallography. These landmark structures indicate that ribosomes are RNA-based machines, in agreement with biochemical evidence (Green and Noller, 1997; Noller, 1991). The 23S rRNA of the large subunit is responsible for the catalytic activity of the ribosome, the peptidyl transferase activity (Nissen et al., 2000), while the 16S rRNA of the small subunit is used almost exclusively for the decoding process, the selection of the cognate tRNA (Carter et al., 2000). The atomic models of the ribosomal subunits, together with structural investigations of functional ribosomal complexes by X-ray crystallography (Cate et al., 1999; Yusupov et al., 2001; Carter et al., 2001; Ogle et al., 2001; Pioletti et al., 2001) and 3D cryo-EM (Agrawal and Frank, 1999; Agrawal et al., 2000; Frank and Agrawal, 2000; Gabashvili et al., 2000; Stark et al., 1997a, 1997b, 2000), will eventually lead to an understanding of the structure and dynamics of protein biosynthesis.

Most information about the ribosome has been obtained in prokaryotic systems. In contrast, our knowledge about eukaryotic ribosomes is sparse. Because of the evolutionary conservation of rRNA and ribosomal proteins, it can be anticipated that the ribosomal subunits share a similar spatial arrangement and that the fundamental mechanism of protein biosynthesis is the same. However, the actual degree of similarity of the whole ribosomes is not known, and significant differences are known to exist. Due to insertion elements in the rRNA molecules (Gerbi, 1996) and the presence of 20 to 30 additional proteins (Wittmann-Liebold, 1986; Wool et al., 1990; Planta and Mager, 1998), eukaryotic ribosomes are larger than their prokaryotic counterparts (Figure 1). There are functional differences, especially in the mechanism of initiation, which is far more complex in eukaryotes (Sachs et al., 1997). There are also differences in the elongation phase; e.g., yeast has a third elongation factor, EF3 (Triana-Alonso et al., 1995), and several antibiotics are specific for one class of ribosomes or the other (Spahn and Prescott, 1996). These and other differences in function are likely to be reflected by differences in structure.

Furthermore, the ribosome has to play a role in other biological processes in the cell, such as protein transport (Beckmann et al., 1997; Menetret et al., 2001; Beckmann et al., 2001) and protein folding (Hardesty et al., 2000), and is subjected in eukaryotes to a transport process through the nuclear pore complex (Ho et al., 2000). Binding and recognition events in these processes do not necessarily take place at those ribosomal sites that are highly conserved among bacterial and eukaryotic ribosomes. Therefore, structural understanding of protein biosynthesis and related processes in eukaryotes will depend on the availability of structural information about the 80S ribosome.

We present here a molecular analysis of the translat-

[8]Correspondence: joachim@wadsworth.org

[9]These authors contributed equally to this work.

Cell
374

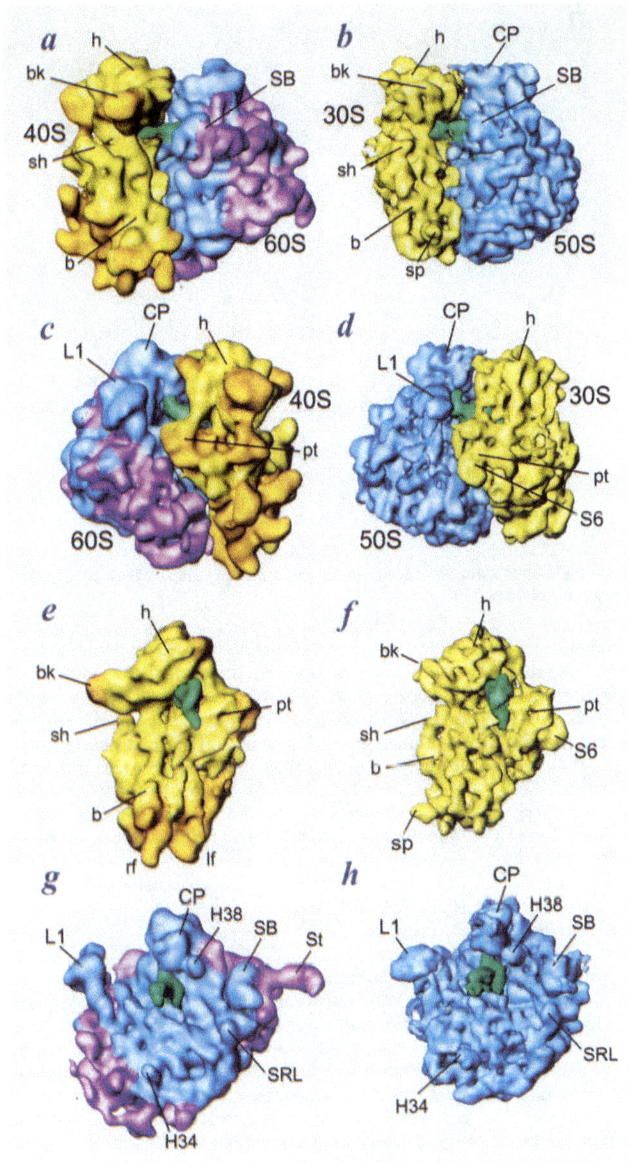

Figure 1. Comparison of the 80S Ribosome from *S. cerevisiae* with the *E. coli* 70S Ribosome

The cryo-EM map of the yeast 80S ribosome (a, c, e, and g) is shown together with the cryo-EM map of the *E. coli* 70S ribosome (b, d, f, and h; Gabashvili et al., 2000). The ribosomes are shown from the L7/L12 side (a and b) and the L1 side (c and d). The computationally isolated small subunits (e and f) and large subunits (g and h) are shown from the interface sides. The small subunits are in yellow, the large subunits in blue, and the P site-bound tRNA in green. Additional parts of the eukaryotic 80S ribosome that are due to expansion segments in the rRNAs (Figure 2) and nonhomologous proteins are shown in gold (40S subunit) and purple (60S subunit). Landmarks for the 40S subunit: b, body; bk, beak; h, head; lf, left foot; rf, right foot; pt, platform; sh, shoulder; sp, spur. Landmarks for the 60S subunit: CP, central protuberance; L1, L1 protuberance; SB, stalk base; St, L7/L12 stalk; H34, helix 34; H38, helix 38; SRL, sarcin-ricin loop.

ing 80S ribosome from yeast by combining a cryo-EM map at approximately 15 Å resolution (Figure 1, see Beckmann et al., 2001) with molecular rRNA and protein models. This analysis reveals the positions of all major rRNA expansion elements and of additional proteins. The model of the 80S ribosome derived describes the intersubunit bridges and the contacts between the 80S ribosome and the P site-bound peptidyl-tRNA at the molecular level. In the accompanying paper, this model is used to derive molecular information about the interaction of the 80S ribosome with the protein-conducting channel Sec61 (Beckmann et al., 2001 [this issue of *Cell*]).

Results and Discussion

Docking of Atomic Models into the Cryo-EM Map

In the absence of an atomic resolution map of a macromolecular complex, docking of atomic models of its components into a lower-resolution cryo-EM map is a

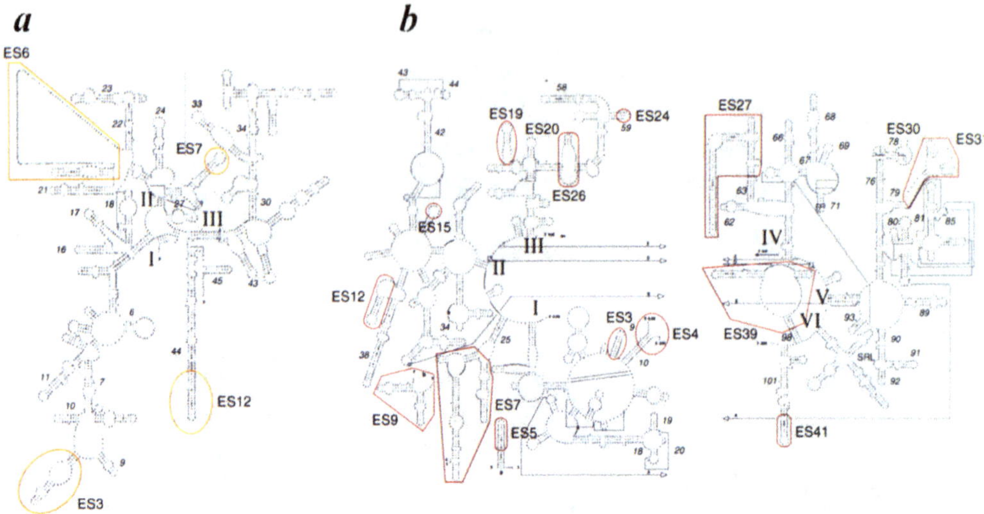

Figure 2. rRNA Secondary Structure Diagrams

Secondary structure diagrams of the 18S rRNA (a) and the 5.8/25S rRNA (b) of *S. cerevisiae* (http://www.rna.icmb.utexas.edu). Expansion segments in the rRNAs are indicated, using the nomenclature of Gerbi (1996).

powerful tool to obtain molecular level information about at least part of the macromolecular assembly. It is estimated that atomic models of known substructures can be positioned in a cryo-EM map with accuracy exceeding the resolution of the map several-fold (Rossmann, 2000).

To facilitate the docking, the map of the 80S ribosome was computationally separated (Spahn et al., 2000) into rRNA and protein maps (Figures 3–5). rRNA models of the recently solved crystal structures of the 30S subunit from *T. thermophilus* (Wimberly et al., 2000) and the 50S subunit from *H. marismortui* (Ban et al., 2000) were fitted into the resulting maps for the small subunit (SSU) rRNA (Figures 3c–3f) and large subunit (LSU) rRNA of yeast (Figure 5), respectively. Where necessary, the X-ray models were modified by moving nonfitting parts (e.g., helices) as rigid bodies relative to the rest of the model. Homology models were calculated for those yeast ribosomal proteins that were found to be related to a bacterial or an archaebacterial protein of known 3D structure. In total, homology models were obtained for 43 yeast ribosomal proteins (15 for the 40S subunit [Figures 3a and 3b], 28 for the 60S subunit [Figure 4]) and positioned in the protein map of the 80S ribosome. In calculating the homology models, it was advantageous that the X-ray structure of the 50S subunit was from an archaebacterial species, since the protein composition of eukaryotic ribosomes is more closely related to archaebacterial ribosomes than to bacterial ones.

Accuracy of the RNA-protein separation method is not absolute; e.g., flexible RNA can be assigned to the protein map, and a small protein cluster to the rRNA map. However, the overall agreement of the separated maps with the atomic models is very good. There is generally excellent agreement between the shape of the protein densities and the homology models, showing that the observed sequence similarities are indeed reflected in conserved tertiary structures (Figures 3 and 4). Prediction of extended protein tails (Ban et al., 2000; Schlünzen et al., 2000; Wimberly et al., 2000) or smaller parts of a protein that intimately interact with rRNA is, of course, beyond the resolution limit of the current cryo-EM map of the yeast 80S ribosome. However, in the case of evolutionary conservation of such protein segments, their positions within the 80S ribosome can be inferred from the atomic structures of the prokaryotic ribosomal subunits, as can the positions of the three proteins (rpS13, rpL19, and rpL42) whose density was assigned to the RNA partition.

The 40S Subunit versus the 30S Subunit

The 40S ribosomal subunit shows the classical division into head, body, and platform (Verschoor et al., 1998; Dube et al., 1998). However, it is not easy to align the overall shape of the 40S subunit from yeast with the bacterial 30S subunit. Due to extra density in the 40S subunit and a change in orientation of some elements, the 40S subunit is longer than its bacterial counterpart and shows a strong subdivision at the bottom part of the body into a "left foot" and a "right foot" (Figure 1). Extra density is mainly present (1) at the bottom of the 40S subunit, responsible for the increase in length of the subunit, (2) below the platform, forming the "left foot" and the "back lobe" of the 40S subunit, (3) at the "beak," and (4) at the solvent side of the head, building the "head lobe."

The 40S ribosomal subunit of *S. cerevisiae* is composed of the 1798 nucleotide (nt)-long 18S rRNA and 32 ribosomal proteins. The 18S rRNA is 256 nt longer than the 16S rRNA of *E. coli*, and the yeast 40S subunit contains 11 more proteins than the 30S subunit from *E. coli*. Consequently, both insertion elements in the 18S rRNA (Figure 2a) and additional ribosomal proteins (Supplemental Table S1) are responsible for the extra mass of the small subunit in eukaryotes.

The Structure of the 18S rRNA

The X-ray model of the 16S rRNA is in good agreement with the map of the 18S rRNA in the region that corresponds to the common core of the SSU rRNA, showing that the phylogenetic conservation of secondary structure is indeed reflected in the conservation of the tertiary structure. However, some differences due to conformational changes are apparent. An unexpected difference is in the conformation of the 18S rRNA helices that make up the shoulder of the 40S subunit (i.e., helices 16 and 17; Figure 3). In bacterial 30S subunits, helix 16 is folded toward helix 18 (the 530 pseudoknot structure), and helix 17 runs parallel to the long axis of the subunit toward the bottom of the body domain (Wimberly et al., 2000; Schlünzen et al., 2000; Spahn et al., 2000). In the yeast 40S subunit, helix 16 is rotated toward the back side of the subunit where it sticks out into the solvent (Figure 3). Helix 17 is shorter in yeast, containing more irregular base pairs. It appears to make a sharp turn below the position of protein rpS9 (S4p; Figure 3b) and folds toward the apical loop of helix 21 (Figure 3d). Helix 16 in a mammalian 40S subunit has the same conformation as in the yeast 40S; it becomes broadly fused with the head domain upon binding of the HCV IRES (Spahn et al., 2001). Therefore, the difference in the shoulder region between bacteria and eukaryotes might be important for the distinct mechanisms of translation initiation.

Another difference between prokaryotic and eukaryotic 40S subunits is in the position and shape of the beak. The corresponding rRNA element, helix 33, is slightly longer in yeast than helix 33 of 16S rRNA, but it does not contain helix 33a, so that the overall rRNA element is shorter (Figure 2a). The curvature of the beak is less pronounced in yeast and its length is increased (Figures 1e and 1f) due to the presence of an additional protein (Figure 3, see below). Finally, there are differences between yeast 18S rRNA and *T. thermophilus* 16S rRNA in the position of the lower part of the body. In yeast it is located more toward the side of the shoulder, a difference which also affects the lower part of helix 44. A movement of components of the lower body in this direction has also been observed when the structure of the vacant *T. thermophilus* 30S subunit was compared with the 50S-bound 30S subunit of *E. coli* (Gabashvili et al., 2000). Therefore, this conformational change might be at least in part due to subunit association.

A straightforward localization of an insertion element is possible wherever unaccounted density is present in the 18S rRNA map and the comparison of the secondary structure diagrams reveals an insertion (Figures 3e and 3f). Helix 44 is clearly resolved, and its extension (ES12) builds the "right foot" of the 40S subunit (cf. Dube et al., 1998). ES6, the helix 21 insertion, emerges at the solvent side of the platform, appearing to branch into two irregular helices (Figure 3f). The smaller of these helices builds one of the "back lobes" and runs toward the shoulder, parallel to helix 21. The larger branch runs toward the bottom, constituting the "left foot" (Figures 3e and 3f). At its lower end, it appears to interact with expansion segment ES3. This structure creates a network of interactions connecting the platform of the 40S subunit with the lower part of the body, a feature that is specific to eukaryotic ribosomes.

The Distribution of Proteins in the 40S Subunit

Fifteen of the SSU proteins of yeast have homologous proteins in bacteria (Supplemental Table S1), while seventeen have no homologous counterparts. Density in the 40S protein map that is not accounted for by homology models predicts the locations of these proteins or of additional domains in the homologous proteins (Figure 3). There are eight regions with additional protein density, most of which are present at the solvent side of the 40S subunit (Figure 3b). Two such clusters of protein density are in the head of the 40S subunit, where they build the headlobe, located above protein rpS0 (S2p), and part of the beak, replacing helix 33a of the 16S rRNA (Figure 3). The protein density at the top of the platform surrounding rpS14 (S11p) is enlarged, leading to a stronger contact with rpS5 (S7p) (Figure 3). This head-platform interaction defines part of the mRNA exit channel. Also, the region around protein rpS0 (S2p) is enlarged, with additional protein density being present at the branch site of the helix 21 insertion (ES6) in 18S rRNA. Another large cluster of protein density is located below helix 21 of 18S rRNA where it "replaces" the bacterial protein S16 (Figure 3b). The presence of this protein cluster might be responsible, to some extent, for the difference in the path of helix 17 of the SSU rRNA (Figures 3c and 3f, see above). Finally, two protein clusters are at the 60S side of the 40S subunit at the bottom of the "left" and "right foot" (Figure 3a). The protein cluster at the "left foot" is at a similar position as the bacterial S20p protein, which has no homologous protein in eukaryotes. A more detailed analysis is unfortunately not possible at this time, because data on the quaternary structure of eukaryotic ribosomes by IEM, cross-linking, or footprinting experiments are scarce.

Four bacterial SSU proteins do not have detectably homologous eukaryotic proteins, but two of these proteins, S16p and S20p, might have analogous counterparts (see above). Interestingly, however, there is no protein density corresponding to proteins S6p and S18p, leaving the 18S rRNA (the junction between helices 22 and 23; Figure 2a) accessible at the solvent side of the platform (Figures 1 and 3). This difference might have functional implications, since the IRES element of HCV binds the 40S subunit in this region (Spahn et al., 2001).

The 60S Subunit versus the 50S Subunit

The 60S ribosomal subunit of *S. cerevisiae* is built from 25S rRNA (3392 nt), 5.8 S rRNA (158 nt), 5S rRNA (121 nt), and 45 ribosomal proteins. Since the 5.8S rRNA is homologous to the 5′-region of the 23S rRNA, the LSU rRNA (Figure 2b) in yeast is 646 nt longer than the bacterial 23S rRNA from *E. coli* (2904) and 505 nt longer than the archaebacterial 23S rRNA from *H. marismortui* (3045 nt). The yeast 60S subunit has 12 proteins more than the *E. coli* 50S subunit (33 different proteins) and 14 proteins more than the *H. marismortui* 50S subunit (31 proteins).

As in the case of the 40S subunit, the atomic model of the *H. marismortui* 23S rRNA (Ban et al., 2000) could

80S Ribosome Structure
377

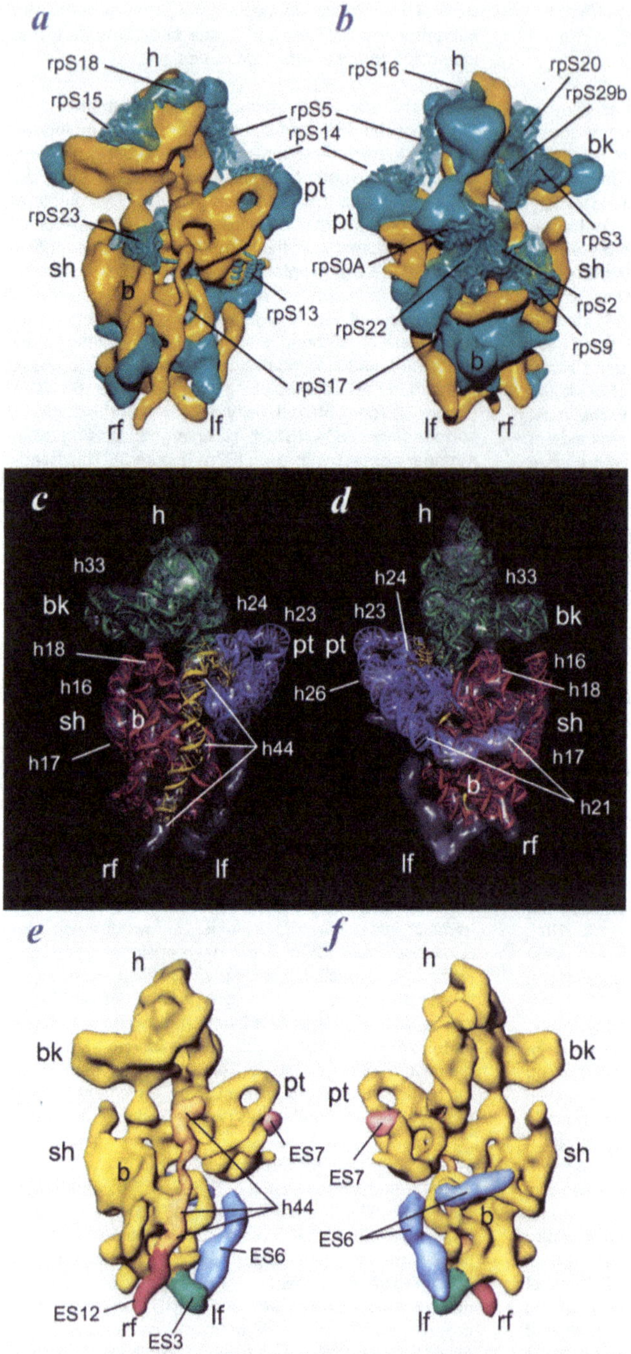

Figure 3. RNA versus Protein Separation for the 40S Subunit

The 40S subunit is shown from the intersubunit side (a, c, and e) and the solvent side (b, d, and f).

(a and b) The RNA partition is in yellow at high contour level; the protein partition is in turquoise. Wherever homology models could be docked into the map, the protein partition is shown with transparency. Therefore, solid parts of the protein partition predict the position of additional proteins without homologous counterparts in prokaryotes. The fitted homologous proteins are shown as ribbon models (Carson, 1991) and are designated with their name.

(c and d) The RNA map is shown transparently superposed on the ribbons model for the common conserved core of the 18S rRNA. The domains of the ribbons model for the SSU rRNA are color coded: domain I, red; domain II, blue; domain III major, green; domain III minor, yellow.

(e and f) The common core of the 18S rRNA is shown in yellow with helix 44 and the expansion segments (defined in Figure 2a) highlighted in different colors. Several helices of the 18S rRNA are annotated with their numbers. Landmarks of the 40S subunit are as in Figure 1.

Cell
378

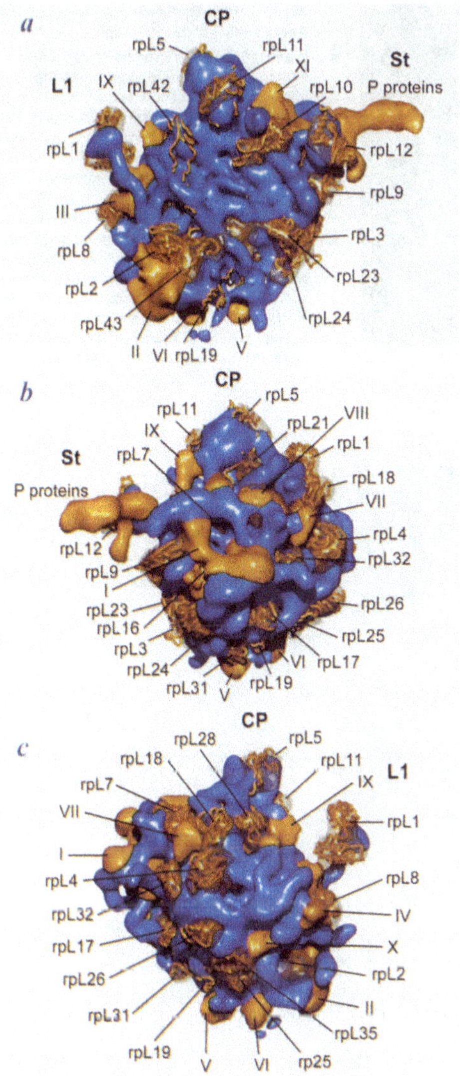

Figure 4. RNA and Protein Separation for the 60S Subunit and Docking of Homology Models for the Proteins

The 60S subunit is shown from the intersubunit side, in the classical crown view (a), from the bottom of the L7/L12 side (b), and from the bottom of the L1 side (c). The RNA partition is shown in blue at a high contour level; the protein partition is in orange. Wherever homology models could be docked into the map, the protein partition is shown transparently. Therefore, solid parts of the protein partition predict the position of additional proteins with no homologous counterpart in prokaryotes. These clusters of density are labeled with a Roman numeral, and the neighboring components are summarized in Supplemental Table S3. The fitted homologous proteins are shown as ribbon models and are designated with their name.

be fitted well into the RNA map of the 60S subunit, again indicating a conserved tertiary core of the 25S rRNA (Figure 5). The X-ray structure of the 50S subunit is disordered in the L1 and L11 regions, so the 25S rRNA model (Figure 5) was supplemented with the L1 region of domain IV of 23S rRNA from *T. thermophilus* (Yusupov et al., 2001) and L11 binding RNA from *T. maritima* (Wimberly et al., 1999), respectively. However, the position of some elements had to be readjusted to achieve a satisfying agreement with the EM density. The underlying conformational changes concern (i) domain VI of 25S rRNA, (ii) helices 7, 18, 19, 20, and 24 of domain I around the exit site of the nascent peptide, (iii) helix 25, (iv) the sarcin-ricin loop (SRL) with helices 89 and 91, (v) helix 42 that leads to the stalk base, and (vi) 5S rRNA. These conformational changes might be species-specific as in the case of helix 25 (see below), but, since these changes are relatively small and located in functionally important regions, they might be caused by subunit association or the active functional state of the 80S ribosome. Compared to the recent X-ray structure of the *T. thermophilus* 70S ribosome (Yusupov et al., 2001), large differences can be seen at the L1- and L11-bearing protuberances. The L11-bearing helices 43 and 44 are shifted by approximately 15 Å toward the central protuberance in *T. thermophilus*; the L1-bearing helix 78, by even more than 30 Å. Movements of both protuberances by a similar magnitude have been observed previously in the yeast ribosome in response to EF2 binding (Gomez-Lorenzo et al., 2000). Both protuberances are apparently highly dynamic, in line with their emerging roles in facilitating the entering and exiting of the tRNAs.

A pronounced species-related conformational change can be observed in the LSU rRNA when the 80S ribosome from *S. cerevisiae* is compared to the 70S ribosome from *E. coli* (Gabashvili et al., 2000), which also carries a tRNA in the P site. Helix 58 of domain III of the LSU rRNA occupies the same position in the archaebacterium *H. marismortui* (Ban et al., 2000) and in yeast, and is located at the bottom of the subunit below the L1 protuberance (Figure 5a). In contrast, in the *E. coli* 70S ribosome and also in the recent X-ray structure of the *T. thermophilus* 70S ribosome (Yusupov et al., 2001), this helix is folded upwards such that it runs along protein L2 toward the apical loops of helices 10 and 79 of 23S rRNA. The apical loop of helix 58 is shifted by more than 50 Å in the bacterial 70S ribosome when compared to its position in the archaebacterial 50S subunit or the eukaryotic 80S ribosome.

Expansion Segments in the 5.8S/25S rRNA

The yeast LSU rRNA contains several expansion segments (Gerbi, 1996; Gutell et al., 2001) that were identified as extra density in the separated RNA map, unaccounted for by the X-ray model of the *H. marismortui* 23S rRNA. Expansion segments are distributed over the whole 5.8S/25S rRNA and are present in all 6 domains (Figure 2b). The expansion segments are located at the surface, roughly at two opposite sides of the 60S subunit (Figure 5). One region is behind the central protuberance and the stalk region, when looking at the 60S subunit in the crown view; the other region extends from the bottom of the 60S to the base of the L1 protuberance. The expansion segments are not mere solvent-exposed protuberances, but are frequently involved in tertiary and quaternary contacts, providing additional interdomain interactions. Two expansion segments even form

80S Ribosome Structure
379

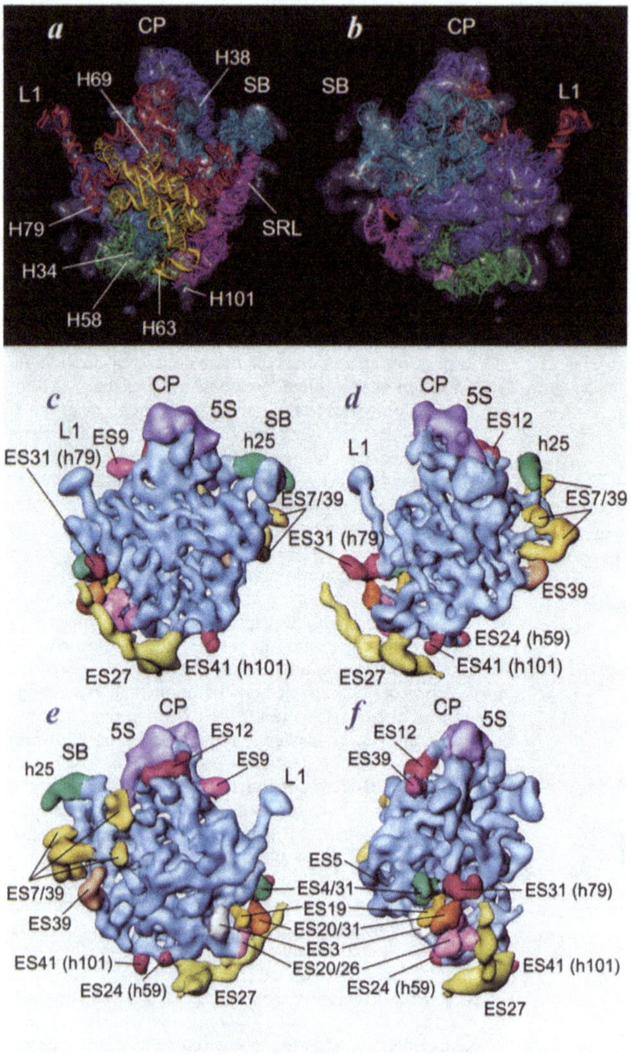

Figure 5. High-Threshold Rendering of the RNA Partition of the 60S Subunit

(a and b) The RNA map is shown transparently superposed with the ribbons model for the common conserved core of the 5.8S/25S and 5S rRNAs. The domains of the ribbons model for the LSU rRNA are color coded: domain I, blue; domain II, cyan; domain III, green; domain IV, yellow; domain V, red; domain VI, magenta; 5S rRNA, blue.

(c–f) The common core of the LSU rRNA is shown in blue, with the expansion segments (defined in Figure 2b) highlighted in different colors. The RNA map is shown from the intersubunit side (a and c), the solvent side (b and e), the L7/L12 side (d), and the L1 side (f). Several helices of the LSU rRNA are annotated with their number. Landmarks are as in Figure 1.

additional bridges with the 40S subunit (Figure 6, see below).

Expansion segments in 25S rRNA domains II and VI (Figure 2b) are responsible for the additional RNA density behind the stalk region (Figure 5). Domain II of 25S rRNA contains a major expansion segment in helix 25, ES7. However, it cannot be completely resolved, since it is broadly fused with another major expansion segment, ES39, and it has a fragmented appearance, apparently because some part of the RNA was assigned to the protein partition in the density separation, indicating flexibility (see Spahn et al., 2000). ES39 is an addition to helix 98 in domain VI (Figure 2b). Helix 98 is missing in *H. marismortui* but has been identified by genetic tagging in *E. coli* ribosomes (Spahn et al., 1999). The beginning of ES39 in 25S rRNA is resolved (Figure 5) before it becomes fused with ES7, and its orientation is similar to that of helix 98 in *E. coli*.

Two smaller insertions into domain II of 25S rRNA, ES9 and ES12, are located at the back of the central protuberance (Figure 5e). ES12, an insertion into helix 38, appears to be of mechanistic interest because of the prominent role of helix 38: it is involved in the formation of a bridge with the 40S subunit and in forming a contact with the elbow of the A site-bound tRNA (A site finger). ES12 runs at the back of the central protuberance toward the L1 side of the ribosome (Figures 5e and 5f). It might act as a strut that could, together with 5S rRNA, be involved in the coupling of conformational changes of ribosomal components that interact with the elbow regions of A and P site-bound tRNAs. At its apical loop, it makes contact with a eukaryote-specific sequence of the 5S rRNA binding protein rpL5 (L18p). rpL5 (L18p) and rpL11 (L5p) interact directly with each other as well as indirectly via 5S rRNA. rpL11 (L5p), similarly to helix 38 in the A site, interacts with the elbow of the

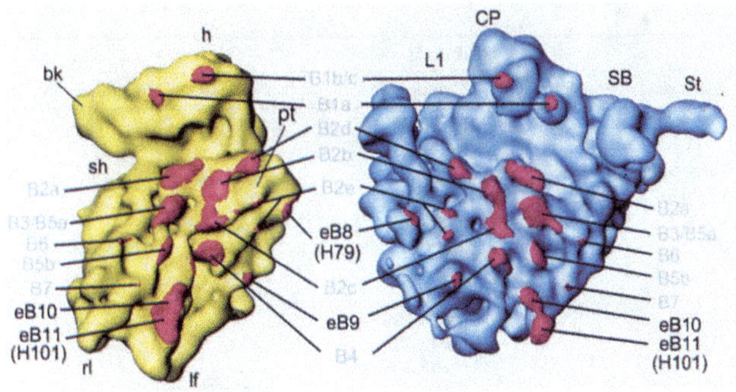

Figure 6. Position of the Intersubunit Bridges
The 40S subunit in yellow (a) and the 60S subunit in blue (b) are shown from their intersubunit sides. The intersubunit bridges are shown in red and are annotated with their numbers. Bridges also observed in prokaryotic ribosomes are numbered B1–B7. Additional intersubunit connections in the yeast ribosome are named eB8–eB11 (see also Table 1).

P site-bound tRNA and is also involved in the formation of a bridge with the 40S subunit (Figures 4–6).

Insertions into domains I, II, and V of 5.8S/25S rRNA appear to form a new network of strands below the L1 stalk (Figure 5). This additional cluster of density is formed by ES4, ES5, ES19, ES20, ES26, and ES31. Interestingly, due to the additional rRNA elements, helix 79 of 25S rRNA undergoes tertiary contacts via ES31 with domains I and III. Helix 79, in turn, is stacked on the protein L1-bearing helix 76, an arrangement which might have implications for the large movement of the L1 stalk observed in the yeast ribosome (Gomez-Lorenzo et al., 2000). Furthermore, ES31 is involved in the formation of an intersubunit bridge (Figure 6, see below). Adjacent to this region, the cluster of expansion segments around ES31 forms another connection between domains I and III of 5.8S/25S rRNA. Helix 9, which is absent in the *H. marismortui* 23S rRNA, appears to interact with helix 54 (Figure 5). An equivalent connection can be observed in the cryo-EM map of the *E. coli* 70S ribosome.

The third major insertion into yeast 25S rRNA, ES27 in helix 63, is located at the L1 side, toward the bottom of the 60S subunit (Figure 5). Helix 63 runs at the bottom of the intersubunit face of the 60S subunit from the L7 side (L7/L12 is replaced in eukaryotes by the acidic P proteins) to the L1 side. This feature is consistent with the location of helix 63 in *E. coli* (Spahn et al., 1999), which is larger than its *H. marismortui* counterpart. At the position of the apical loop of helix 63 in *E. coli*, the additional RNA density emerges from the 25S rRNA (Figure 5). In agreement with the predicted secondary structure, ES27 consists of a more convoluted RNA structure, from which a long rod emerges that runs up at the L1 side of the 60S subunit (Figure 5). This rod has clear helical features, and a part of it was also named "yeast spine" in a lower-resolution cryo-EM structure of the yeast 80S ribosome (Morgan et al., 2000). In our present structure, the density of this feature is lower than expected for RNA. In contrast, the same feature is strong when ribosomes for cryo-EM are purified without subjecting them to an in vitro translation system (Gomez-Lorenzo et al., 2000). This behavior can be partially explained by a large conformational change occurring at this element, discussed in the accompanying paper (Beckmann et al., 2001).

Two more expansion segments are located at the bottom of the 60S subunit. ES24, an extension of helix 59 of 25S rRNA, appears to be involved in the binding of Sec61 (Beckmann et al., 2001). ES41, an extension of helix 101 of 25S rRNA (Figure 2b), makes an intersubunit bridge with the 40S subunit (Figure 6, see below).

The Protein Distribution in the 60S Subunit

Homology models of 28 yeast LSU proteins were calculated and docked into the cryo-EM map (Supplemental Table S2). Density clusters in the 60S subunit protein map that are not accounted for by homology models predict the positions of the 17 proteins for which the exact position is still unknown, as well as additional domains in the homologous proteins (Figure 4). There are 12 regions with additional protein density at the surface of the 60S subunit, most of which are present at the solvent side.

The large protein cluster building the extended stalk can be identified as the P0/P1/P2 complex, the homolog of the bacterial $L10(L7/L12)_4$ complex. As noted earlier, the extended stalk in yeast has a different orientation compared to the extended stalk in bacteria (Gomez-Lorenzo et al., 2000). Interestingly, the P0/P1/P2 density has a sidelobe that is located close to the N-terminal domain of rpL12 (L11p). Exactly at this location, a contact occurs between the stalk-base region and elongation factor EF2 (Gomez-Lorenzo et al., 2000), a contact that has never been observed in any of the 70S•EF-G complexes from *E. coli* (Frank and Agrawal, 2000). A direct and unique interaction of the P proteins with EF2 would give a structural explanation for their ability to discriminate between EF2 and EF-G (Uchiumi et al., 1999).

Apart from the cluster corresponding to the P proteins, the lack of data about the quaternary structure of eukaryotic ribosomes by IEM, cross-linking, or footprinting experiments does not allow further analysis. An overview of the unidentified protein densities and their neighborhood in the 60S subunit is presented in Supplemental Table S3. The vast majority of these clusters appear to contact more than one component, and the additional ribosomal proteins in yeast might have structural tasks, e.g., stabilizing the tertiary fold of the LSU rRNA, or they might interact with components of the eukaryotic environment.

Intersubunit Bridges between the 40S and the 60S Subunit

Protein synthesis requires communication between the small and the large subunit. For example, signals from

Table 1. Bridges between the 40S and 60S Ribosomal Subunits

		40S		60S	
Bridge	Type	Component	Approximate Position	Component	Approximate Position
B1a	p-R	rpS15 (S19p)		25S h38	889
B1b/c	p-p	rpS18 (S13p)		rpL11 (L5p)	
B2a	R-R	18S h44	1408/1493 min	25S h69	1910-1920 min
B2b	R-R	18S h24	784/800	25S h68	1847
	R-R	18S h45	1515	25S	1939-1941
B2c	R-R	18S h27	899	25S h62	1693
				25S h67	1832
B2d	R-R	18S h23	702	25S h68	1848/1895 min
B2e	R-p	18S h22	671	rpL2 (L2p)	
	R-p	18S h23	713	rpL2 (L2p)	
				rpL43 (L37ae)	
B3	R-R	18S h44	1420/1481 min	25S h71	1948/1960 min
B4	R-R	18S h11	247	25S h34	711/720 min
	R-R	18S h20	583/761 min	25S h34	716 lo
	p-R	rpS13 (S15)		25S h34	716 lo
B5a	R-p	18S h44	1422 bb	rpL23 (L14p)	
B5b	R-R	18S h44	1428/1472 min	25S h62	1689/1704 min
B6	R-p	18S h14	343	rpL23 (L14p)	
B7	R-p	18S h44	1446	rpL24 (L24e)	
eB8	p-R	rpSx		25S h79	Expansion
eB9	p-p	rpSx		RpLx	
	R-p	18S h21	Expansion		
eB10	R-R	18S h11	272	25S h63	1713/1747 min
	p-R	rpSx			
eB11	R-R	18S h9	187	25S h101	Expansion
	p-R	rpSx			

bb, backbone; lo, loop; min, minor groove

the decoding center on the SSU to the peptidyl transferase center on the LSU can be transmitted either by ligands that contact both regions, e.g., bound tRNAs, or by the intersubunit bridges that connect the subunits. The structure of the intersubunit bridges and their dynamics are therefore of particular importance for an understanding of the mechanism of protein synthesis.

Six bridges were observed earlier in low-resolution maps of the *E. coli* 70S ribosome (Frank et al., 1995; Lata et al., 1996). The nomenclature B1–B6 to describe them was further elaborated in the context of the analysis of a 7.8 Å resolution X-ray map of the *T. thermophilus* 70S ribosome (Cate et al., 1999) and a 11.5 Å resolution cryo-EM map of the *E. coli* 70S ribosome (Gabashvili et al., 2000). A first attempt has been made to describe the bridges in a mammalian 80S ribosome at approximately 25 Å resolution (Morgan et al., 2000). The present 15.4 Å resolution cryo-EM map of the *S. cerevisiae* 80S ribosome allows us to describe the intersubunit bridges in eukaryotic ribosome at the level of detail that is comparable to that achieved for the intersubunit bridges in bacterial ribosomes (Figure 6). The docking of X-ray models of the rRNAs and homology models of the ribosomal proteins has allowed us to go even a step further and determine the molecular identity of most of the ribosomal components involved in subunit-subunit interaction, as has been done in the recent 5.5 Å resolution X-ray map of the *T. thermophilus* 70S ribosome (Yusupov et al., 2001).

All the bridges that have been discovered in the bacterial 70S ribosomes have a corresponding bridge in the yeast 80S ribosome. This remarkable evolutionary conservation demonstrates the importance of these bridges. Most of the intersubunit contacts involve RNA-RNA interactions (see also Mitchell et al., 1992; Merryman et al., 1999a, 1999b; Yusupov et al., 2001), although various RNA-protein interactions and one protein-protein contact can be identified (Table 1). Several of the RNA interactions clearly involve minor groove sites. Interestingly, in addition to the conserved set of bridges, four new bridges can be identified in the yeast 80S ribosome that might be specific for eukaryotic ribosomes (Figure 6). We therefore extend the current B1–B7 nomenclature for bacterial ribosomes and name the additional intersubunit contacts eB8–eB11.

Two bridges connect the head of the 40S subunit with the central protuberance of the 60S subunit. The apical loop of helix 38 of 25S rRNA (A site finger) interacts with rpS15 (S19p) to form bridge B1a. The 30S part of B1a in *T. thermophilus* has been identified as S13 and not S19 (Yusupov et al., 2001). Since rpS18 (S13p) appears to have a somehow changed position in yeast compared to *T. thermophilus* and also lacks the C-terminal extension that contacts the P site-bound tRNA, this potential difference might be species-related, although higher resolution of the yeast ribosome will be required to solve this question unequivocally. B1b, the sole bridge that does not involve RNA, is formed by an interaction of rpS18 (S13p) with rpL11 (L5p) (Figure 6 and Table 1).

Most of the bridges that connect the main part of the 60S subunit with the body/platform domains of the 40S subunit are lying approximately on three parallel lines that run from the top to the bottom of the subunits. The bridges on the line closest to the shoulder of the 40S or the stalk region of the 60S subunit are B2a, B3, B5a, B5b, eB10, and eB11 (Figure 6). B2a, B3, and B5b are all formed by minor groove-minor groove interactions between helix 44 of 18S rRNA and helices of domain IV

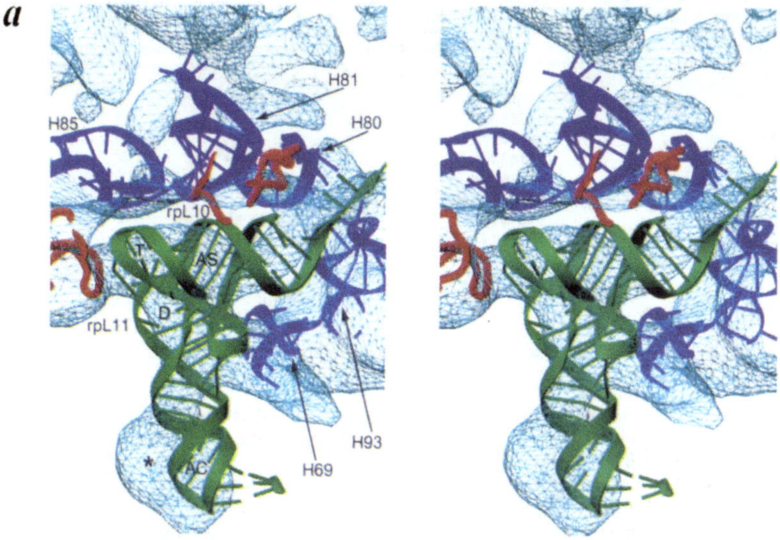

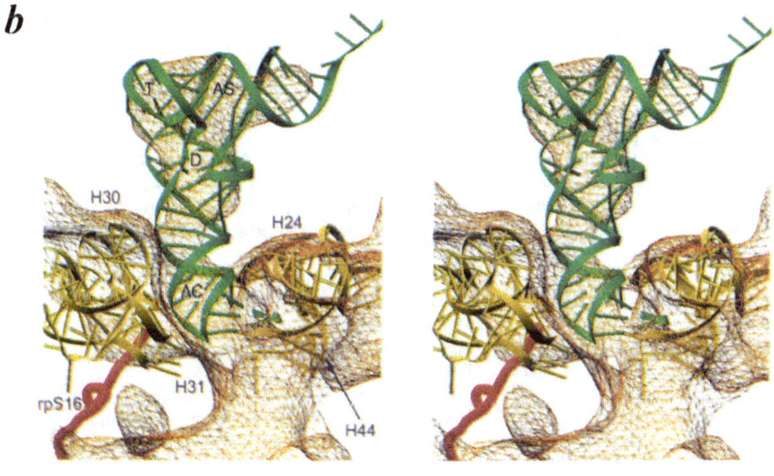

Figure 7. Molecular Environment of the P Site-Bound tRNA in the Yeast 80S Ribosome

Stereo pictures show the tRNA density together with surrounding density of the 60S subunit in blue (a) and 40S subunit in yellow (b). Fitted atomic models, corresponding to the P site-bound tRNA and its molecular environment (Table 2), are shown in the ribbon representation. The ribosomal components are designated, as are the domains of the tRNA: AC, anticodon loop; AS, acceptor stem; D, D loop; T, T loop. The asterix (a) indicates density that belongs to the 40S subunit and is broadly fused to the tRNA density.

of 25S rRNA. B5a is due to an RNA-protein contact between helix 44 of 18S rRNA and rpL23 (L14p). Further down the hypothetical line that connects these bridges are the additional bridges eB10 and eB11, which are among the strongest intersubunit connections in the yeast 80S ribosome. eB10 involves an RNA-RNA contact between 18S rRNA helix 11 and 25S rRNA helix 63. The participation of an unknown eukaryote-specific protein in this bridge might explain why it is not formed in bacterial ribosomes. eB11 is formed by an RNA-RNA interaction between 18S rRNA helix 9 and the expansion of helix 101 of 25S rRNA. This bridge also involves an unknown protein of the 40S subunit.

Helix 44 of 16S rRNA is bent toward the shoulder side of the 40S subunit to become the "left foot" at its apical domain. The kink is approximately at the position of the bridge B5b (Figure 6). Further down, B7 is yet another bridge that involves helix 44 of 18S rRNA. Its partner on the 60S subunit is rpL24 (L24e). B7 has been previously observed in the *T. thermophilus* 70S ribosome, where

Table 2. Contacts between the P Site Bound Peptidyl-tRNA and the 80S Ribosome from Yeast

Ribosomal Subunit	Approximate tRNA Position	Ribosomal Component	Approximate Ribosome Position
40S	29/30	18S h30	1229/1230
	32	rpS16 (S9p)	Tyr141/Arg142
	34	18S h31	966
	39	18S h24	790
	41/42	18S h43	1338/1339
	mRNA codon	18S h44	1400/1402
60S	2	rpL10 (L10e)	Lys101
	3/4	25S h80	2285/2286
	11-13	25S h69	1908-1910
	14	25S h69	1924
	51/52/63	rpL10 (L10e)	Arg24
	56	rpL11 (L5p)	Tyr51
	71/72	25S h93	2594
	73	25S h93	2602

it has been named B6 (Cate et al., 1999). rpL24 has no homologous protein in bacteria, and therefore, this bridge is formed by an analogous protein in *T. thermophilus*. Indeed, this contact has been attributed to protein L19p in the X-ray structure of the *T. thermophilus* 70S ribosome at improved resolution (Yusupov et al., 2001).

Bridges B2b, B2c, and B4 are lying on the second and central line (Figure 6). B2b and B2c are formed by RNA-RNA interactions that again involve domain IV of 25S rRNA. B2b involves helix 24 of 18S rRNA and the last helix at the 3′ end, helix 45. B2c encompasses helix 27, the switch helix of 18S rRNA. B4 has been previously identified in *T. thermophilus* as an RNA-protein contact between S15 and helix 34 of 23S rRNA (Culver et al., 1999). In yeast, this bridge is also formed by helix 34 of 25S rRNA and rpS13 (S15p), but it appears to involve additional RNA-RNA contacts between helix 34 of 25S rRNA and helices 11 and 10 of 18S rRNA (Table I).

The third line of bridges on the L1 side of the 60S subunit comprises B2d, B2e, and eB9 (Figure 6). B2d is a contact between helix 23 of 18S rRNA in the platform of the 40S subunit and the minor groove of helix 68 of domain IV of 25S rRNA. B2e was first observed in the *E. coli* 70S ribosome (Gabashvili et al., 2000) and was named B8 in the *T. thermophilus* 70S ribosome (Yusupov et al., 2001). B2e is an RNA-protein contact that involves rpL8 (L2p). The presence of protein L2 in *E. coli* has been shown to be crucial for the ability of the 50S subunit to associate with the 30S subunit (Diedrich et al., 2000). In yeast, rpL43 (L37ae) also appears to be involved in this connection, and the 40S counterparts are helices 22 and 23 of 18S rRNA. eB9 is built by components that are not present in bacterial ribosomes (Figure 6). It appears to make use of a protein-protein contact between unknown proteins, or an RNA-protein contact between an unknown LSU protein and the helix 21 insertion in 18S rRNA.

Two more bridges are present at the outside of the intersubunit face of the subunits (Figure 6). B6 is formed by a contact between 18S rRNA helix 14 and protein rpL23 (L14p). eB8, located below the L1 protuberance, is formed by an unknown 40S protein and ES31, an insertion into helix 79 of 25S rRNA (see also Figure 5). Since helix 79 is stacked onto the L1-bearing helix 76, bridge eB8 could be important in controlling the large conformational change that affects the position of the L1 protuberance in the yeast 80S ribosome.

tRNA-Ribosome Interaction at the P Site

During protein synthesis, a tRNA has to interact with the ribosome in at least two regions. The anticodon loop of the tRNA and the mRNA codon have to interact with the ribosomal decoding center on the SSU. The 3′-CCA end of the tRNA has to contact the peptidyltransferase center located on the LSU. tRNA-ribosome interactions in these two regions are well documented for bacterial ribosomes by various biochemical techniques, e.g., footprinting, cross-linking and genetic studies (for review see Green and Noller, 1997; Brimacombe, 1995) and have been directly observed by cryo-EM (Stark et al., 1997a; Agrawal et al., 2000; Gabashvili et al., 2000) and X-ray crystallography (Cate et al., 1999; Yusupov et al., 2001; Nissen et al., 2000; Carter et al., 2000). In addition, tRNA-ribosome interactions at the elbow region of the tRNA have been suggested by phosphorothioate footprints that occur in the T loop and D loop of the tRNA upon ribosome binding (Dabrowski et al., 1995). Indeed, contacts between the tRNA and the ribosome at the tRNA elbow have been visualized by cryo-EM in various functional complexes of the *E. coli* 70S ribosome (Agrawal et al., 2000; Gabashvili et al., 2000; Malhotra et al., 1998). However, no information is available yet on the structural basis of the tRNA-ribosome interaction in eukaryotic cells.

The 15.4 Å resolution cryo-EM map of the translating 80S ribosome from yeast clearly displays an L-shaped density in the intersubunit space that corresponds to the P site-bound peptidyl-tRNA in its entirety (Figure 7). Only the single-stranded 3′-CCA end is not resolved. The density shows indications of major and minor grooves of the tRNA molecule and allows an accurate docking of the atomic structure of the tRNA into the density map. The X-ray structure of the isolated $tRNA^{Phe}$ from yeast (Hingerty et al., 1978) and the structure of the tRNA derived from a tRNA•70S complex from *T. thermophilus* (Cate et al., 1999; Yusupov et al., 2001) were used for the docking. The latter tRNA model has a somewhat changed conformation when compared to the isolated tRNA, implying a conformational change of the tRNA upon binding to the ribosomal P site. This tRNA model appears to fit better into the yeast cryo-

EM map. The conformational change of the tRNA that is implicit in the structure of the 70S-bound *T. thermophilus* tRNA (Cate et al., 1999; Yusupov et al., 2001) might therefore also occur in yeast. Moreover, a conformational change of the tRNA upon binding to the ribosome has been suggested by phosphorothiate footprinting experiments (Dabrowski et al., 1995).

The tRNA density is fused with the ribosome density at several discrete positions. These apparent contacts involve all four domains of the tRNA (Figure 7). The docking of models for the rRNAs, the ribosomal proteins, and the tRNA allows the parts of the molecules involved in these contacts to be determined, and thereby strong candidates for actual molecular interactions to be identified (Table 2). Only a smaller part of the tRNA interacts with the 40S subunit of yeast (see also Nierhaus et al., 2000). The anticodon stem-loop is positioned between the head and body/platform of the 40S subunit (Figure 1e). On the body/platform side, the P site codon appears to interact with the top of helix 44 of 18S rRNA and tRNA with the apical loop of helix 24 in the platform. From the opposite side, the tRNA is clamped down by 40S components belonging to the head. Helices 30, 31, and 43 of 18S rRNA, as well as an extended part of rpS16 (S9p), are likely to be involved in these interactions (Figure 7 and Table 2), which in yeast appear to be very similar to those described for a bacterial system (Carter et al., 2000; Yusupov et al., 2001).

A second protein-tRNA contact has been described for *T. thermophilus* (Carter et al., 2000; Yusupov et al., 2001). However, the long tail of S13 that interacts with the tRNA in *T. thermophilus* does not have a corresponding sequence for rpS18. Moreover, rpS18 (S13p) in yeast appears to have a somewhat changed position compared to the position of S13 in the *T. thermophilus* 30S subunit. Therefore, the tRNA contact with the S9p protein family is evolutionary conserved, in contrast to the tRNA contact with the S13p protein family.

Three of the four tRNA domains interact with the large ribosomal subunit (Table 2). The D loop makes a contact with helix 69 of 25S rRNA and the T loop interacts with rpL11 (L5p). The same interactions have been proposed for the *E. coli* ribosome (Spahn et al., 2000) and the *T. thermophilus* ribosome (Yusupov et al., 2001). Interestingly, both components of the LSU are involved in the formation of bridges with the SSU, a fact which might have implications for the dynamics of translation, i.e., specifically the mechanism of translocation. The 3′-CCA end of the tRNA is not resolved in the yeast cryo-EM map, but the end of the tRNA acceptor stem appears to contact the 60S subunit. This interaction might involve helices 80 (the apical loop of helix 80 participates in a base-pairing interaction with the 3′-CCA end of the tRNA [Samaha et al., 1995; Nissen et al., 2000]) and 93 of 25S rRNA, as well as protein rpL10 (L10e). At a lower density threshold (not shown), an additonal contact can be observed that involves the TΨC stem of the tRNA and protein rpL10 (L10e). We note that in *T. thermophilus*, no interaction between the P site-bound tRNA and L16p (Yusupov et al., 2001), which occupies the same position as rpL10 (L10e) in yeast, has been reported, and it remains to be seen if this weaker contact reflects an actual interaction.

Conclusions

We have presented here an interpretation of a cryo-EM map of an eukaryotic 80S ribosome in terms of atomic RNA and protein models. The common core of the eukaryotic ribosome agrees very well with X-ray structures of the bacterial and archaebacterial subunits and reinforces the notion that the fundamental mechanism of protein synthesis is highly conserved throughout all kingdoms. This view is supported by the finding that the intersubunit bridges and the ribosome-tRNA interactions are highly conserved, as well. Nevertheless, a few unexpected differences exist even in this highly conserved region. The most pronounced differences in appearance between eukaryotic and bacterial ribosomes are due to expansion segments in the rRNAs and additional proteins that are present mainly at the surface of the solvent sides of the 40S and 60S subunits. Even though these elements are far away from the primary functional centers, the additional tertiary and quaternary interactions could be important for ribosome dynamics, as also suggested by the emergence of four new intersubunit bridges. Moreover, the additional components are likely to interact with other cellular components, outside of the immediate process of protein synthesis. Our model of the yeast 80S ribosome provides the basis for molecular-level interpretations of the interactions of the eukaryotic ribosomes within the context of the eukaryotic environment.

Experimental Procedures

The cryo-EM map of the 80S ribosome (Beckmann et al., 2001) was computationally separated into rRNA and protein maps using a recently developed method that takes into account the differences in the density distribution of RNA and proteins, as well as the molecular masses and contiguity constraints (Spahn et al., 2000).

Homology models for the yeast ribosomal proteins were constructed using the large-scale protein structure modeling pipeline as implemented in ModPipe (Sanchez and Sali, 1998) and deposited into a relational database for homology models, ModBase ((Sanchez et al., 2000); http://guitar.rockefeller.edu/modbase). Structural templates used to calculate the models consisted of all the individual chains from structures in PDB (as of September 2000), clustered such that the sequences of no two chains from any two clusters were more than 95% identical. In addition, the structures of the small subunit from *T. thermophilus* (PDB code: 1FJF) and the large subunit from *H. morismortui* (PDB code: 1FKF) were considered as separate sets of templates.

Sequence structure matches were established by aligning the psi-blast generated profile (Altschul et al., 1997) of the yeast sequences against each of the template sequences and also by scanning the yeast sequences against a database of psi-blast generated profiles of the template sequences using IMPALA (Schaffer et al., 1999). Significant alignments covering distinct regions of the yeast sequences were chosen for modeling. Models were calculated for each one of the sequence-structure matches using MODELLER (Sali and Blundell, 1993). The resulting models were then evaluated using a scheme based on statistical potentials that uses the compactness of the model, the sequence identity of the sequence-structure match, and energy z-scores to the assess the quality of the model.

Models for docking into the EM-map were selected based on the e-value of the alignment, the quality of the final model, and the template structure on which they were based. The selected models represent alignments with sequence identities in the range 20%–56% (with an average of 32%) and e-values better than 10^{-4}. The coverage of the models (fraction of the yeast ribosomal sequence modeled) ranges between 34%–99% (with an average of 75%).

Docking of atomic models into the cryo-EM density map was done manually using O (Jones et al., 1991). The figures were prepared using IRIS Explorer (Numerical Algorithms Group, Inc., Downers Grove, IL), POV-Ray, and Ribbons (Carson, 1991).

Acknowledgments

We are grateful to Michael Watters and Jan Giesebrecht for assistance with the illustrations and to Dr. Ursula Pieper for her help with ModBase. A.S. is an Irma T. Hirschl Trust Career Scientist. This work was supported by a grant of the Volkswagen Stiftung (to R.B.), a Merck Genome Research Award (to A.S.), and grants from NIH (GM 54762 to A.S., R37 GM29169 and P41 RR01219 to J.F., and R01 GM60635 to P.A.P.) and NSF (BIR 9219043 to J.F.).

Received August 8, 2001; revised September 21, 2001.

References

Agrawal, R.K., and Frank, J. (1999). Structural studies of the translational apparatus. Curr. Opin. Struct. Biol. *9*, 215–221.

Agrawal, R.K., Spahn, C.M.T., Penczek, P., Grassucci, R.A., Nierhaus, K.H., and Frank, J. (2000). Visualization of tRNA movements on the *Escherichia coli* 70S ribosome during the elongation cycle. J. Cell Biol. *150*, 447–459.

Altschul, S.F., Madden, T.L., Schaffer, A.A., Zhang, J., Zhang, Z., Miller, W., and Lipman, D.J. (1997). Gapped BLAST and PSI-BLAST: a new generation of protein database search programs. Nucleic Acids Res. *25*, 3389–3402.

Ban, N., Nissen, P., Hansen, J., Moore, P.B., and Steitz, T.A. (2000). The complete atomic structure of the large ribosomal subunit at 2.4 Å resolution. Science *289*, 905–920.

Beckmann, R., Bubeck, D., Grassucci, R., Penczek, P., Verschoor, A., Blobel, G., and Frank, J. (1997). Alignment of conduits for the nascent polypeptide chain in the ribosome-Sec61 complex. Science *278*, 2123–2126.

Beckmann, R., Spahn, C.M.T., Eswar, N., Helmers, J., Penczek, P.A., Sali, A., Frank, J., and Blobel, G. (2001). Architecture of the protein-conducting channel associated with the translating 80S ribosome. Cell *107*, this issue, 361–372.

Brimacombe, R. (1995). The structure of ribosomal RNA: a three-dimensional jigsaw puzzle. Eur. J. Biochem. *230*, 365–383.

Carson, M. (1991). Ribbons 2.0. J. Appl. Cryst. *24*, 103–106.

Carter, A.P., Clemons, W.M., Jr., Brodersen, D.E., Morgan-Warren, R.J., Wimberly, B.T., and Ramakrishnan, V. (2000). Functional insights from the structure of the 30S ribosomal subunit and its interactions with antibiotics. Nature *407*, 340–348.

Carter, A.P., Clemons, W.M., Jr., Brodersen, D.E., Morgan-Warren, R.J., Hartsch, T., Wimberly, B.T., and Ramakrishnan, V. (2001). Crystal structure of an initiation factor bound to the 30S ribosomal subunit. Science *291*, 498–501.

Cate, J.H., Yusupov, M.M., Yusupova, G.Z., Earnest, T.N., and Noller, H.F. (1999). X-ray crystal structures of 70S ribosome functional complexes. Science *285*, 2095–2104.

Culver, G.M., Cate, J.H., Yusupova, G.Z., Yusupov, M.M., and Noller, H.F. (1999). Identification of an RNA-protein bridge spanning the ribosomal subunit interface. Science *285*, 2133–2135.

Dabrowski, M., Spahn, C.M., and Nierhaus, K.H. (1995). Interaction of tRNAs with the ribosome at the A and P sites. EMBO J. *14*, 4872–4882.

Diedrich, G., Spahn, C.M.T., Stelzl, U., Schafer, M.A., Wooten, T., Bochariov, D.E., Cooperman, B.S., Traut, R.R., and Nierhaus, K.H. (2000). Ribosomal protein L2 is involved in the association of the ribosomal subunits, tRNA binding to A and P sites and peptidyl transfer. EMBO J. *19*, 5241–5250.

Dube, P., Bacher, G., Stark, H., Müller, F., Zemlin, F., van Heel, M., and Brimacombe, R. (1998). Correlation of the expansion segments in mammalian rRNA with the fine structure of the 80 S ribosome; a cryoelectron microscopic reconstruction of the rabbit reticulocyte ribosome at 21 Å resolution. J. Mol. Biol. *279*, 403–421.

Frank, J., and Agrawal, R.K. (2000). A ratchet-like inter-subunit reorganization of the ribosome during translocation. Nature *406*, 318–322.

Frank, J., Zhu, J., Penczek, P., Li, Y., Srivastava, S., Verschoor, A., Radermacher, M., Grassucci, R., Lata, R.K., and Agrawal, R.K. (1995). A model of protein synthesis based on cryo-electron microscopy of the *E. coli* ribosome. Nature *376*, 441–444.

Gabashvili, I.S., Agrawal, R.K., Spahn, C.M.T., Grassucci, R.A., Frank, J., and Penczek, P. (2000). Solution structure of the *E. coli* 70S ribosome at 11.5 Å resolution. Cell *100*, 537–549.

Gerbi, S.A. (1996). Expansion segment: regions of variable size that interrupt the universal core secondary structure of ribosomal RNA. In Ribosomal RNA: Structure, Evolution, Processing, and Function in Protein Synthesis. R.A. Zimmermann and A.E. Dahlberg, eds. (New York: CRC Press), pp. 71–87.

Gomez-Lorenzo, M.G., Spahn, C.M.T., Agrawal, R.K., Grassucci, R.A., Penczek, P., Chakraburtty, K., Ballesta, J.P.G., Lavandera, J.L., Garcia-Bustos, J.F., and Frank, J. (2000). Three-dimensional cryo-electron microscopy localization of EF2 in the *Saccharomyces cerevisiae* 80S ribosome at 17.5 Å resolution. EMBO J. *19*, 2710–2718.

Green, R., and Noller, H.F. (1997). Ribosomes and translation. Annu. Rev. Biochem. *66*, 679–716.

Gutell, R.R., Subashchandran, S., Schnare, M., Du, Y., Lin, N., Madabusi, L., Müller, K., Pande, N., Shang, Z., Date, S., et al. (2001). Comparative Sequence Analysis and the Prediction of RNA Structure (http://www.rna.icmb.utexas.edu).

Hardesty, B., Kramer, G., Tsalkova, T., Ramachandiran, V., McIntosh, B., and Brod, D. (2000). Folding of nascent peptides on ribosomes. In The Ribosome: Structure, Function, Antibiotics, and Cellular Interactions, R.A. Garrett, S.R. Douthwaite, A. Liljas, A.T. Matheson, P.B. Moore, and H.F. Noller, eds. (Washington, D.C.: ASM Press), pp. 287–300.

Hingerty, B., Brown, R.S., and Jack, A. (1978). Further refinement of the structure of yeast $tRNA^{Phe}$. J. Mol. Biol. *124*, 523–534.

Ho, J.H., Kallstrom, G., and Johnson, A.W. (2000). Nmd3p is a Crm1p-dependent adapter protein for nuclear export of the large ribosomal subunit. J. Cell Biol. *151*, 1057–1066.

Jones, T.A., Zhou, J.Y., Cowan, S.W., and Kjeldgaard, M. (1991). Improved methods for building protein models in electron density maps and the location of errors in these models. Acta Crystallogr. A *A47*, 110–119.

Lata, K.R., Agrawal, R.K., Penczek, P., Grassucci, R., Zhu, J., and Frank, J. (1996). Three-dimensional reconstruction of the *Escherichia coli* 30 S ribosomal subunit in ice. J. Mol. Biol. *262*, 43–52.

Malhotra, A., Penczek, P., Agrawal, R.K., Gabashvili, I.S., Grassucci, R.A., Jünemann, R., Burkhardt, N., Nierhaus, K.H., and Frank, J. (1998). *Escherichia coli* 70 S ribosome at 15 Å resolution by cryo-electron microscopy: localization of fMet-$tRNA_f^{Met}$ and fitting of L1 protein. J. Mol. Biol. *280*, 103–116.

Menetret, J., Neuhof, A., Morgan, D.G., Plath, K., Radermacher, M., Rapaport, T.A., and Akey, C.W. (2001). The structure of ribosome-channel complexes engaged in protein translocation. Mol. Cell *6*, 1219–1232.

Merryman, C., Moazed, D., Daubresse, G., and Noller, H.F. (1999a). Nucleotides in 23S rRNA protected by the association of 30S and 50S ribosomal subunits. J. Mol. Biol. *285*, 107–113.

Merryman, C., Moazed, D., McWhirter, J., and Noller, H.F. (1999b). Nucleotides in 16S rRNA protected by the association of 30S and 50S ribosomal subunits. J. Mol. Biol. *285*, 97–105.

Mitchell, P., Osswald, M., and Brimacombe, R. (1992). Identification of intermolecular RNA cross-links at the subunit interface of the *Escherichia coli* ribosome. Biochemistry *31*, 3004–3011.

Morgan, D.G., Ménétret, J.F., Radermacher, M., Neuhof, A., Akey, I.V., Rapoport, T.A., and Akey, C.W. (2000). A comparison of the yeast and rabbit 80S ribosome reveals the topology of the nascent chain exit tunnel, inter-subunit bridges and mammalian rRNA expansion segments. J. Mol. Biol. *301*, 301–321.

Nierhaus, K.H., Spahn, C., Burkhardt, N., Dabrowski, M., Diedrich, G., Einfeldt, E., Kamp, D., Marquez, V., Patzke, S., Schafer, M.A., et al. (2000). Ribosomal elongation cycle. In The Ribosome: Structure,

Function, Antibiotics, and Cellular Interactions. R.A. Garrett, S.R. Douthwaite, A. Liljas, A.T. Matheson, P.B. Moore, and H.F. Noller, eds. (Washington, D.C.: ASM Press), pp. 319–336.

Nissen, P., Hansen, J., Ban, N., Moore, P.B., and Steitz, T.A. (2000). The structural basis of ribosome activity in peptide bond synthesis. Science *289*, 920–930.

Noller, H.F. (1991). Ribosomal RNA and translation. Annu. Rev. Biochem. *60*, 191–227.

Ogle, J.M., Brodersen, D.E., Clemons, W.M., Jr., Tarry, M.J., Carter, A.P., and Ramakrishnan, V. (2001). Recognition of cognate transfer RNA by the 30S ribosomal subunit. Science *292*, 897–902.

Pioletti, M., Schlünzen, F., Harms, J., Zarivach, R., Gluhmann, M., Avila, H., Bashan, A., Bartels, H., Auerbach, T., Jacobi, C., et al. (2001). Crystal structures of complexes of the small ribosomal subunit with tetracycline, edeine and IF3. EMBO J. *20*, 1829–1839.

Planta, R.J., and Mager, W.H. (1998). The list of cytoplasmic ribosomal proteins of *Saccharomyces cerevisiae*. Yeast *14*, 471–477.

Rossmann, M.G. (2000). Fitting atomic models into electron-microscopy maps. Acta Crystallogr. D *D56*, 1341–1349.

Sachs, A.B., Sarnow, P., and Hentze, M.W. (1997). Starting at the beginning, middle, and end: translation initiation in eukaryotes. Cell *89*, 831–838.

Sali, A., and Blundell, T.L. (1993). Comparative protein modelling by satisfaction of spatial restraints. J. Mol. Biol. *234*, 779–815.

Samaha, R.R., Green, R., and Noller, H.F. (1995). A base pair between tRNA and 23S rRNA in the peptidyl transferase centre of the ribosome. Nature *377*, 309–314.

Sanchez, R., and Sali, A. (1998). Large-scale protein structure modeling of the *Saccharomyces cerevisiae* genome. Proc. Natl. Acad. Sci. USA *95*, 13597–13602.

Sanchez, R., Pieper, U., Mirkovic, N., de Bakker, P.I., Wittenstein, E., and Sali, A. (2000). MODBASE, a database of annotated comparative protein structure models. Nucleic Acids Res. *28*, 250–253.

Schaffer, A.A., Wolf, Y.I., Ponting, C.P., Koonin, E.V., Aravind, L., and Altschul, S.F. (1999). IMPALA: matching a protein sequence against a collection of PSI-BLAST-constructed position-specific score matrices. Bioinformatics *15*, 1000–1011.

Schlünzen, F., Tocilj, A., Zarivach, R., Harms, J., Glühmann, M., Janell, D., Bashan, A., Bartels, H., Agmon, I., Franceschi, F., and Yonath, A. (2000). Structure of functionally activated small ribosomal subunit at 3.3 Å resolution. Cell *102*, 615–623.

Spahn, C.M.T., and Prescott, C.D. (1996). Throwing a spanner in the works: antibiotics and the translation apparatus. J. Mol. Med. *74*, 423–429.

Spahn, C.M.T., Grassucci, R.A., Penczek, P., and Frank, J. (1999). Direct three-dimensional localization and positive identification of RNA helices within the ribosome by means of genetic tagging and cryo-electron microscopy. Structure *7*, 1567–1573.

Spahn, C.M.T., Penczek, P., Leith, A., and Frank, J. (2000). A method for differentiating proteins from nucleic acids in intermediate-resolution density maps: cryo-electron microscopy defines the quaternary structure of the *Escherichia coli* 70S ribosome. Structure Fold. Des. *8*, 937–948.

Spahn, C.M.T., Kieft, J.S., Grassucci, R.A., Penczek, P., Zhou, K., Doudna, J.A., and Frank, J. (2001). Hepatitis C virus IRES RNA-induced changes in the conformation of the 40S ribosomal subunit. Science *291*, 1959–1962.

Stark, H., Orlova, E.V., Rinke-Appel, J., Junke, N., Müller, F., Rodnina, M., Wintermeyer, W., Brimacombe, R., and van Heel, M. (1997a). Arrangement of tRNAs in pre- and posttranslocational ribosomes revealed by electron cryomicroscopy. Cell *88*, 19–28.

Stark, H., Rodnina, M.V., Rinke-Appel, J., Brimacombe, R., Wintermeyer, W., and van Heel, M. (1997b). Visualization of elongation factor Tu on the *Escherichia coli* ribosome. Nature *389*, 403–406.

Stark, H., Rodnina, M.V., Wieden, H.J., van Heel, M., and Wintermeyer, W. (2000). Large-scale movement of elongation factor G and extensive conformational change of the ribosome during translocation. Cell *100*, 301–309.

Triana-Alonso, F.J., Chakraburtty, K., and Nierhaus, K.H. (1995). The elongation factor 3 unique in higher fungi and essential for protein biosynthesis is an E site factor. J. Biol. Chem. *270*, 20473–20478.

Uchiumi, T., Hori, K., Nomura, T., and Hachimori, A. (1999). Replacement of L7/L12.L10 protein complex in *Escherichia coli* ribosomes with the eukaryotic counterpart changes the specificity of elongation factor binding. J. Biol. Chem. *274*, 27578–27582.

Verschoor, A., Warner, J.R., Srivastava, S., Grassucci, R.A., and Frank, J. (1998). Three-dimensional structure of the yeast ribosome. Nucleic Acids Res. *26*, 655–661.

Wimberly, B.T., Guymon, R., McCutcheon, J.P., White, S.W., and Ramakrishnan, V.R. (1999). A detailed view of a ribosomal active site: The structure of the L11-RNA complex. Cell *97*, 491–502.

Wimberly, B.T., Brodersen, D.E., Clemons, W,M., Jr., Morgan-Warren, R.J., Carter, A.P., von Rhein, C., Hartsch, T., and Ramakrishnan, V. (2000). Structure of the 30S ribosomal subunit. Nature *407*, 327–339.

Wittmann-Liebold, B. (1986). Ribosomal proteins: their structure and evolution. In Structure, Function, and Genetics of Ribosomes, B. Hardesty and G. Kramer, eds. (New York: Springer-Verlag), pp. 326–361.

Wool, I.G., Endo, Y., Chan, Y.L., and Glück, A. (1990). Structure, function, and evolution of mammalian ribosomes. In The Ribosome: Structure, Function, & Evolution. W.E. Hill, A.E. Dahlberg, R.A. Garrett, P.B. Moore, D. Schlessinger, and J.R.Warner, eds. (Washington, D.C.: ASM Press), pp. 203–214.

Yusupov, M.M., Yusupova, G.Z., Baucom, A., Lieberman, K., Earnest, T.N., Cate, J.H., and Noller, H.F. (2001). Crystal structure of the ribosome at 5.5 Å resolution. Science *292*, 883–896.

Accession Numbers

The atomic coordinates of the modeled yeast 80S structure have been deposited in the RCSB Protein Data Bank with accession codes 1K5X, 1K5Y, and 1K5Z.

Hepatitis C Virus IRES RNA–Induced Changes in the Conformation of the 40S Ribosomal Subunit

Christian M. T. Spahn,[1,2]* Jeffrey S. Kieft,[3]* Robert A. Grassucci,[1,2] Pawel A. Penczek,[2]† Kaihong Zhou,[3] Jennifer A. Doudna,[3]‡ Joachim Frank[1,2,4]‡

Initiation of protein synthesis in eukaryotes requires recruitment of the 40S ribosomal subunit to the messenger RNA (mRNA). In most cases, this depends on recognition of a modified nucleotide cap on the 5′ end of the mRNA. However, an alternate pathway uses a structured RNA element in the 5′ untranslated region of the messenger or viral RNA called an internal ribosomal entry site (IRES). Here, we present a cryo-electron microscopy map of the hepatitis C virus (HCV) IRES bound to the 40S ribosomal subunit at about 20 Å resolution. IRES binding induces a pronounced conformational change in the 40S subunit and closes the mRNA binding cleft, suggesting a mechanism for IRES-mediated positioning of mRNA in the ribosomal decoding center.

Initiation of protein synthesis in eukaryotes requires the ordered assembly of ribosomal preinitiation complexes. The ribosome dynamics that accompany this process are not understood, although recent progress in determining the structure of ribosomes and functional ribosomal complexes by x-ray crystallography (*1–4*) and cryo–electron microscopy (cryo-EM) (*5–8*) has revealed changes and variations in ribosome structure that occur during the elongation stage of protein synthesis in prokaryotes. Canonical eukaryotic cap-dependent translation initiation involves a complex pathway during which a large number of factors interact with the ribosome and one another (*9*) (Fig. 1A). An alternate pathway, called internal initiation of translation, is a cap-independent, end-independent mechanism of recruiting, positioning, and activating the eukaryotic protein synthesis machinery, driven by structured RNA sequences called internal ribosome entry sites (IRESs) located in the mRNA 5′ untranslated region (UTR) (*10*) (Fig. 1A). These sequences were first identified in many viral RNAs, but the number of known eukaryotic mRNAs that use an IRES mechanism is growing.

In hepatitis C virus (HCV), a human pathogen and world-wide health threat, the minimal sequence and secondary structure requirements of HCV IRES–driven translation have been defined, the presence and architecture of an IRES RNA tertiary fold has been described, and the identity and binding sites of necessary cofactors have been determined (*11*) (Fig. 1B). The HCV IRES forms an extended structure that binds the 40*S* ribosomal subunit and eukaryotic initiation factor 3 (eIF3) through multiple intermolecular contacts, resulting in precise placement of the start codon in the ribosome decoding site (*12–14*). To address the molecular mechanism by which the HCV IRES RNA recruits the eukaryotic translation machinery, we have generated three-dimensional reconstructions of HCV IRES RNA bound to the rabbit 40*S* ribosomal subunit. Reconstructions at 20 Å resolution produced by cryo-EM in combination with the single-particle approach (*15*) reveal that HCV IRES RNA actively manipulates the translational machinery through extensive and specific intermolecular interactions.

The structure of the rabbit 40*S* ribosomal subunit obtained as a control displays the classical division into body, head, and platform domains (Fig. 2, A to C). It represents the first reconstruction of a eukaryotic ribosomal subunit from zero-tilt images and has a greatly improved resolution compared with a previous random-conical tilt reconstruction (*16*). Although the resolution is lower than currently obtained with 70*S* (*5*, *17*) or 80*S* ribosomes, the overall structure of the rabbit 40*S* subunit compares well with the 40*S* part of a reconstruction of the yeast 80*S* ribosome (*18*). Density attributable to the HCV IRES RNA appears directly in the map of the IRES RNA–40*S* subunit complex and can be seen in the difference map between the complex and the vacant 40*S* subunit as an extended structure (Fig. 2, D and F). The cross-sectional diameter and overall shape of the IRES density is consistent with A-form helices extending from regions of denser packing or more convoluted structure (Fig. 1, C and D). Previous small-angle x-ray scattering experiments suggested that the unbound, folded HCV IRES RNA maintained some conformational flexibility (*14*). The new results show that the IRES assumes a single conformation when bound to the 40*S* subunit.

The IRES RNA is bound to the solvent side of the 40*S* subunit in a position that is consistent with the proposed path of the mRNA through the subunit (*19*). The 5′ end of the single-stranded coding RNA is expected to exit the ribosome through a channel that is built by the neck, the head, and the platform domains (*20*). Two segments of the IRES RNA density extend from this exit channel, one toward the head, and the other to the back of the platform (*21*). There are no contacts with the intersubunit face of the 40*S* subunit, consistent with the observation that 80*S* ribosomes can be assembled without removal of the HCV IRES RNA (*12*).

To assign the HCV IRES density to RNA structural domains, we used a truncated HCV IRES construct in which domain II is deleted; this mutant binds the 40*S* subunit with nearly wild-type affinity (Fig. 1B) (*13*). IRES RNA density missing from the structure of this truncated IRES–40*S* subunit complex (Fig. 2, G to I) unambiguously identifies domain II (Fig. 2, D to F, and Fig. 3). From this assignment, we deduce that the globular density that contacts the back of the platform contains domain IIId/e/f stem-loops and junctions, whereas the remaining density extending away from the 40*S* subunit surface corresponds to IIIb, IIIa, and IIIc (Fig. 3). Domain IIIb is not involved in 40*S* subunit binding, but is known to make specific contacts with eIF3 (*13*, *22*). An earlier reconstruction of eIF3 bound to the 40*S* subunit (*23*) shows that this factor interacts with the 40*S* subunit in the region where domain IIIb of the HCV IRES RNA is seen to extend from the ribosome surface (Fig. 3). Thus, even in the absence of eIF3, the IRES RNA domain IIIb is properly positioned to bind this factor.

HCV IRES domain II is a particularly interesting feature of the bound RNA structure. This domain loops away from the 40*S* subunit surface, making contact only at the apical half of the domain. The sequence of

[1]Howard Hughes Medical Institute, Health Research Inc. at the [2]Wadsworth Center, Empire State Plaza, Albany, New York 12201–0509, USA. [3]Department of Molecular Biophysics and Biochemistry and Howard Hughes Medical Institute, Yale University, New Haven, CT 06520–8114, USA. [4]Department of Biomedical Sciences, State University of New York at Albany, Albany, NY 12222, USA.

*These authors contributed equally to this work.
†Present address: Department of Biochemistry, University of Texas, Health Science Center, Houston, TX 77030, USA.
‡To whom correspondence should be addressed. E-mail: doudna@csb.yale.edu and joachim@wadsworth.org

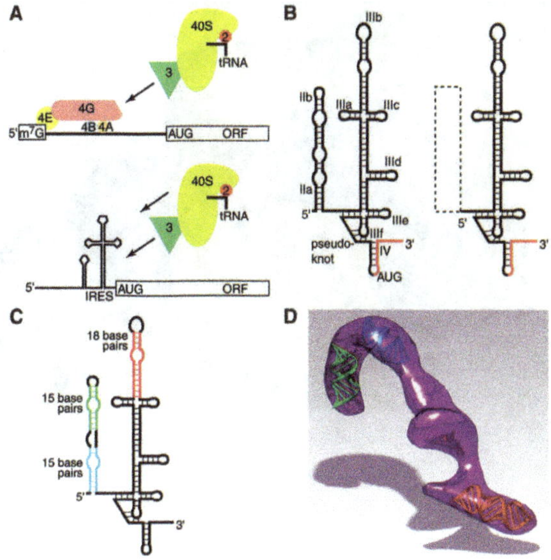

Fig. 1. Hepatitis C virus ribosome recruitment strategy, IRES secondary structure, and modeling of example A-form RNA helices into HCV IRES density. **(A)** Comparison of canonical eukaryotic cap-dependent translation initiation (top) and the internal initiation mechanism used by hepatitis C virus (bottom). Cap-dependent translation requires recognition of a modified nucleotide at the 5′ terminus of the mRNA by the eIF4F complex, which includes eIF-4G, eIF-4E, eIF-4B, and eIF-4A. The 43*S* particle (40*S* subunit/eIF3/Met-tRNA/eIF2/GTP) binds to the mRNA/eIF4F complex primarily through interactions between eIF3 and eIF4G. In contrast, HCV IRES–driven translation initiation occurs through direct, specific recognition of the 40*S* subunit and eIF3 by the IRES RNA tertiary structure, eliminating the need for the cap-structure and cap-binding factors. **(B)** Schematic secondary structures of the HCV IRES RNA constructs used in this study. The full-length construct (left) consists of nucleotides 40 to 372 of HCV IRES genotype 1b, and is fully active in initiation translation. Thirty nucleotides of the coding RNA are included in this construct and are shown in red. The deletion mutant (right) contains nucleotides 119 to 372, has diminished translation initiation activity, but binds the 40*S* subunit with wild-type activity. **(C)** Locations within the secondary structure and sizes of RNA helices that were fit into the HCV IRES density (purple). For the purpose of modeling, structure throughout internal loops was approximated by continuous base stacking into helices of 15 base pairs (cyan), 15 base pairs (green), and 18 base pairs (red). **(D)** Placement of helices into a surface representing the 40*S* bound IRES RNA density reveals a good fit both in terms of helical diameter and lengths. A-form RNA helices shown in (D) were generated in NAMOT2 and SPOCK; images were created with RIBBONS (*47*) and POVRAY.

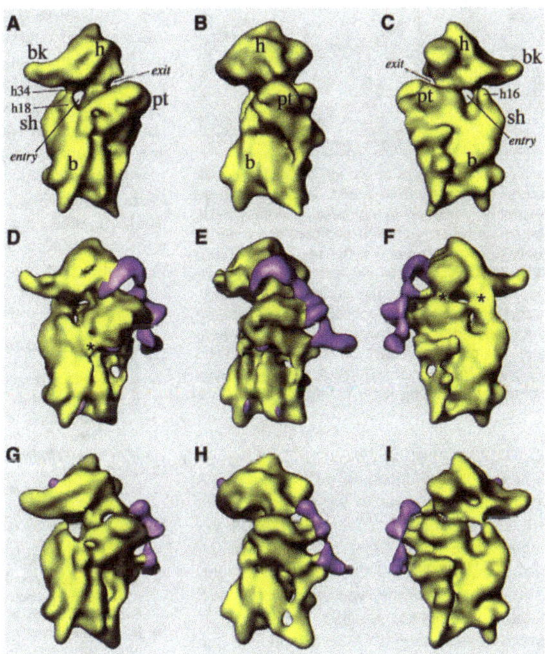

Fig. 2. Surface representation of the vacant 40*S* ribosomal subunit from rabbit reticulocytes (**A** to **C**), the 40*S* subunit in complex with the HCV IRES RNA (**D** to **F**), and the 40*S* subunit in complex with the truncated HCV IRES RNA, where domain II has been deleted (**G** to **I**). The cryo-EM maps are shown in yellow. The difference maps corresponding to the full-length HCV IRES RNA (D to F) or the deletion mutant (G to I) are superposed and shown in purple. The views are from the 60*S* side of the 40*S* subunit (A, D, and G), the platform side (B, E, and H), and the solvent side (C, F, and I). Some landmarks of the 40*S* subunit are indicated: b, body; bk, beak; h, head; pt, platform; sh, shoulder. "Entry" and "exit" label the proposed entry and exit path, respectively, of the mRNA (*26*, *31*). h16, h18, and h34 indicate the positions of helices 16, 18, and 34, respectively, of the 18*S* rRNA as identified by comparison with a cryo-EM map of the 80*S* ribosome from yeast (*28*). Some conformational changes in the IRES RNA–40*S* subunit complex referred to in the text are indicated with asterisks (D and F).

the IIb terminal loop and adjacent helix are found in other viral IRESs, suggesting this contact may be an important determinant not only of HCV IRES function but also of other, similar, viral IRESs (*24*). The contact between domain II and the 40*S* subunit occurs in a region of the head and the edge of the platform. Thus, domain II partially overlaps with the E site, the binding site occupied by deacylated tRNA before it exits the ribosome (*4*, *25*). This contact is the only HCV IRES–40*S* subunit interaction that approaches the active site of the ribosome, and it lies near the 60*S* subunit binding face. The binding of domain II closes the mRNA binding cleft from the 60*S* side, restricting the movement of coding RNA within the cleft. Single-stranded RNA is not likely to be resolved at the present resolution and is difficult to detect even in higher-resolution cryo-EM maps of the 70*S* ribosome (*5*, *26*). However, a small extension of density pointing into the mRNA binding cleft, which grows when the contour level of the difference map is lowered (not shown), might be part of the single-stranded coding RNA. This observation suggests that the domain II terminal loop may contact the single-stranded coding part of the viral RNA. Although the presence of noise in the difference map at such low contour level (not shown) prevents a definite conclusion, it is tempting to speculate that part of the function of domain II is to place the single-stranded coding part of the RNA into the decoding center.

Binding of the HCV IRES RNA to the 40*S* subunit induces a conformational change in the 40*S* subunit, the first reported example of a structured viral RNA that binds to, and actively manipulates, the structure of a cellular machine. The structural change involves a change in the orientation of the head relative to the body, which can be recognized in a different position of the beak part of the head domain in the IRES RNA–40*S* subunit complex (Fig. 2, D to F) compared with the vacant 40*S* subunit (Fig. 2, A to C) (*27*). The platform domain also changes shape, and several more localized changes can be observed (marked with asterisks in D and F of

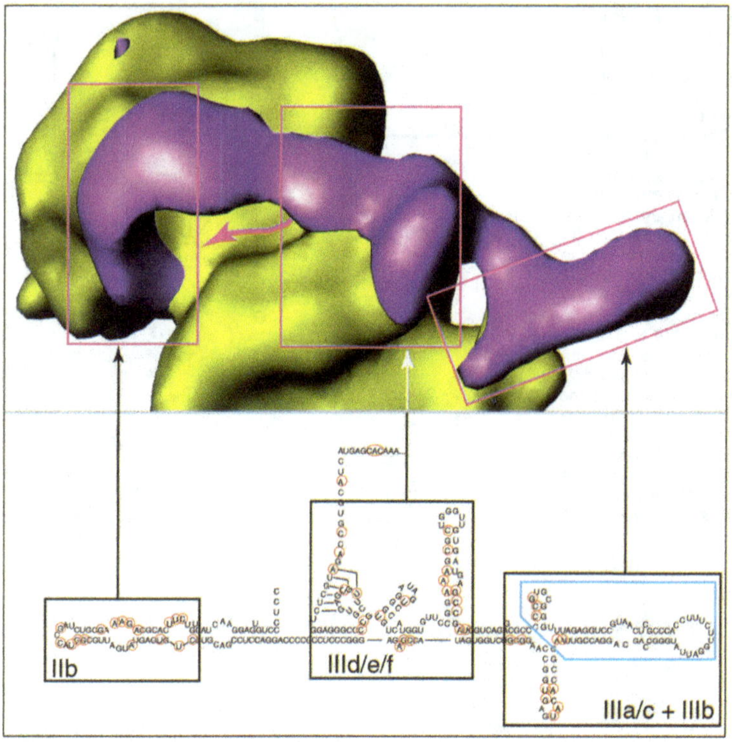

Fig. 3. Correlation of chemical probing data with the cryo-EM structure of the HCV IRES RNA–40S subunit complex, and assignment of IRES RNA structural domains within the structure. The three IRES RNA tertiary structural elements that were previously identified by chemical and enzymatic probing are boxed within the secondary structure. The secondary structure is depicted in a manner that better reflects its organization on the 40S subunit surface. Arrows link the tertiary structure domains to the corresponding elements within the IRES RNA cryo-EM structure. The strategy used to assign density to these domains is described in the text. Within the secondary structure, nucleotides that were previously reported to be inaccessible to solvent upon 40S subunit binding are circled in red (*13*). The location of the eIF3 binding site on domain IIIb and IIIa is boxed in blue; this domain extends away from the 40S subunit surface. The protection pattern sites agree with the locations of intermolecular contacts observed in the cryo-EM representations. A magenta arrow pointing to the 3′ end delineates the probable location of the coding RNA in the mRNA binding groove.

Fig. 2). The rotation of the head and a slight rearrangement of the shoulder region lead to a broad fusion of the shoulder with the underside of the head. In a detailed analysis of a higher-resolution cryo-EM map of the yeast 80*S* ribosome, the rod-like structure that extends from the top of the 40*S* subunit shoulder toward the solvent side and fuses with the head upon HCV IRES binding (Fig. 2F, asterisk on right) was identified as helix 16 of 18*S* ribosomal RNA (rRNA) (*28*). Thus, the position of helix 16 differs from that in prokaryotes, where it is folded toward helix 18 of 16*S* rRNA (the 530 pseudoknot structure). Helix 18, in turn, is involved in a noncovalent junction with helix 34 of the head domain (*2*, *3*, *29*, *30*). This junction or latch has been proposed to be part of the mRNA entry channel and to clamp around the incoming strand of the mRNA (*31*). The fusing of the helix 16 density with the underside of the head effectively alters the shape of the entry site of the mRNA, apparently clamping down on the solvent side (compare C and F of Fig. 2). However, the latch between helices 18 and 34, which can be observed in the vacant 40*S* subunit, is opened, possibly helping the 3′ end of the coding RNA to thread into the mRNA entry channel. In this respect, the HCV IRES might work analogously to the clamp-loading machinery in DNA replication (*32*).

The conformational change induced by the binding of full-length HCV IRES RNA is not observed when the domain II deletion mutant is bound to the 40*S* subunit and, therefore, must be a consequence of domain II's interaction at or near the E site (Fig. 2, G to I). Although domain II does not contribute to IRES affinity for the 40*S* subunit, and isolated domain II does not bind to the 40*S* subunit (*33*), removal of domain II is deleterious to the ability of the IRES to initiate translation (*34*).

The HCV IRES RNA-induced structural change in the 40*S* ribosomal subunit could have significance in activating, or promoting initiation of, translation without the help of certain canonical initiation factors. Domain II may hold the coding RNA in position in the decoding site until the rest of the translation machinery has assembled and elongation has begun (*35*). In addition, HCV IRES binding could capture a 40*S* subunit conformation that is favorable for continued preinitiation complex assembly (*36*).

Previous models of HCV IRES action have focused on the binding of the 40*S* subunit to an IRES RNA scaffold in a mechanism reminiscent of the role of the Shine-Dalgarno sequence in prokaryotes. Our findings call for an expansion of these models such that IRES RNA actively manipulates the host cell's translation machinery to orchestrate the assembly of the preinitiation complex. This raises the interesting questions of whether other viral or eukaryotic IRES RNAs use similar mechanisms to alter the conformation of the translational machinery, and whether these observed structural changes may be common and required steps in cap- and end-dependent pathways of eukaryotic translation initiation. A deeper understanding of the conformational changes and intermolecular signals that occur in IRES-driven translation initiation may offer insights into the events that are common features of all types of translation initiation in eukaryotes.

References and Notes

1. N. Ban, P. Nissen, J. Hansen, P. B. Moore, T. A. Steitz, *Science* **289**, 905 (2000).
2. B. T. Wimberly *et al.*, *Nature* **407**, 327 (2000).
3. F. Schlunzen *et al.*, *Cell* **102**, 615 (2000).
4. J. H. Cate, M. M. Yusupov, G. Zh. Yusupova, T. N. Earnest, H. F. Noller, *Science* **285**, 2095 (1999).
5. I. S. Gabashvili *et al.*, *Cell* **100**, 537 (2000).
6. J. Frank, R. K. Agrawal, *Nature* **406**, 318 (2000).
7. H. Stark *et al.*, *Nature* **389**, 403 (1997).
8. R. K. Agrawal, J. Frank, *Curr. Opin. Struct. Biol.* **9**, 215 (1999).
9. W. C. Merrick, J. W. B. Hershey, *Translational Control*, J. W. B. Hershey, M. B. Mathews, N. Sonenberg, Eds. (Cold Spring Harbor Laboratory Press, Cold Spring Harbor, NY, 1996), p. 31.
10. R. J. Jackson, *Translational Control*, J. W. B. Hershey, M. B. Mathews, N. Sonenberg, Eds. (Cold Spring Harbor Laboratory Press, Cold Spring Harbor, NY, 1996), p. 71.
11. R. C. Rijnbrand, S. M. Lemon, *Curr. Top. Microbiol. Immunol.* **242**, 85 (2000).
12. T. V. Pestova, I. N. Shatsky, S. P. Fletcher, R. J. Jackson, C. U. T. Hellen, *Genes Dev.* **12**, 67 (1998).
13. J. S. Kieft, K. Zhou, R. Jubin, J. A. Doudna, *RNA* **7**, 194 (2001).
14. J. S. Kieft *et al.*, *J. Mol. Biol.* **292**, 513 (1999).
15. The 40S subunits from rabbit reticulocytes and HCV IRES RNA were prepared as described previously (*13*). To generate IRES RNA–40S subunit complexes, RNA was annealed by heating to 75°C for 1 min, then cooled to room temperature. RNA was then added to a tube containing folding/binding buffer [20 mM

tris-HCl, 100 mM potassium acetate, 200 mM KCl, 2.5 mM $MgCl_2$, 1 mM dithiothreitol (DTT)], followed by addition of stoichiometric amounts of 40S subunit and incubation at 37°C for 15 min. The final complex concentration was about 500 nM. The samples were snap-frozen with liquid nitrogen and stored at −80°C until use.
Samples of 40S subunits (final concentration about 50 nM) were applied to cryo-EM grids and shock-frozen in liquid ethane (*37, 38*). Micrographs were recorded under low-dose conditions in a defocus range between 1.7 μm and 5.7 μm on a Philips F20 (FEI/Philips) at a magnification of 50.000 ± 2%. From the micrographs, 69 were selected for the vacant 40S subunit, 74 for the IRES-40S complex, and 61 for the IRESΔdII-40S complex. The micrographs were scanned on a Highscan drum scanner (Eurocore/Saint-Denis, France) with a resolution of 1069 dpi corresponding to a pixel size of 4.78 Å on the object scale. The data were analyzed using the SPIDER system (*39*). After automated particle preselection, manual verification, and selection by cross-correlation, 18,801 particles were chosen for the vacant 40S subunit, 20,939 particles for the IRES-40S complex, and 13,613 particles for the IRESΔdII-40S complex. The data sets were subdivided each into 16 to 23 defocus groups, and a refined, three-dimensional reconstruction that was corrected for the contrast transfer function was calculated for each of the 40S samples (*40*). The final resolution was estimated by the Fourier shell correlation with a cutoff value of 0.5. It is 22.7 Å for the vacant 40S subunit, 19.8 Å for the IRES-40S complex, and 21.9 Å for the IRESΔdII-40S complex. According to the more lenient (*17*) 3σ criterion (*41*), the resolution values would be 15.9 Å, 14.5 Å, and 15.4 Å, respectively.

16. S. Srivastava, A. Verschoor, M. Radermacher, R. Grassucci, J. Frank, *J. Mol. Biol.* **245**, 461 (1995).
17. A. Malhotra *et al.*, *J. Mol. Biol.* **280**, 103 (1998).
18. M. G. Gomez-Lorenzo *et al.*, *EMBO J.* **19**, 2710 (2000).
19. J. Frank *et al.*, *Nature* **376**, 441 (1995).
20. J. Frank *et al.*, *Biochem. Cell Biol.* **73**, 757 (1995).
21. To our knowledge, the ribosomal region at the back of the platform has not been previously implicated in molecular recognition of a ribosomal ligand. As does the protein conducting channel Sec61 (translocon) (*42, 43*), the HCV IRES recognizes the evolutionarily more divergent outer surfaces of the ribosome. Thus, the surfaces available for specific molecular recognition extend beyond the highly conserved binding sites for tRNAs and elongation factors. The fact that the HCV IRES binds to the more variable regions of the 40S subunit surface may explain its ability to distinguish the 40S subunit of rabbit from that of wheat germ (*12*).
22. D. V. Sizova, V. G. Kolupaeva, T. V. Pestova, I. N. Shatsky, C. U. Hellen, *J. Virol.* **72**, 4775 (1998).
23. S. Srivastava, A. Verschoor, J. Frank, *J. Mol. Biol.* **226**, 301 (1992).
24. M. Honda, M. R. Beard, L. H. Ping, S. M. Lemon, *J. Virol.* **73**, 1165 (1999).
25. R. K. Agrawal *et al.*, *J. Cell Biol.* **150**, 447 (2000).
26. J. Frank *et al.*, *The Ribosome: Structure Function Antibiotics and Cellular Interactions*, R. A. Garrett *et al.*, Eds. (ASM Press, Washington, DC, 2000), p. 45.
27. An animated sequence showing the conformational change is available as supplementary material on *Science* Online at www.sciencemag.org/cgi/content/full/291/5510/1959/DC1.
28. C. M. Spahn, R. Beckmann, G. Blobel, J. Frank, unpublished observations.
29. C. M. T. Spahn, P. Penczek, A. Leith, J. Frank, *Structure Fold. Des.* **8**, 937 (2000).
30. A. P. Carter *et al.*, *Nature* **407**, 340 (2000).
31. K. R. Lata *et al.*, *J. Mol. Biol.* **262**, 43 (1996).
32. S. Waga, B. Stillman, *Annu. Rev. Biochem.* **67**, 721 (1998).
33. J. S. Kieft, K. Zhou, J. A. Doudna, unpublished observations.
34. Moreover, deletion of domain II strongly reduces an ultraviolet radiation-induced cross-link between the IRES RNA and ribosomal protein S9 (the homolog of bacterial protein S4) (*44*). Electron microscopy of antibody-labeled 40S subunits located S9 at the back of the shoulder, below helix 16, near the mRNA binding cleft entry site (*45*), in agreement with the placement of bacterial S4 within the 30S subunit (*2, 3, 46*). This position is quite far away from the structured part of the HCV IRES RNA in our density map, suggesting that the IRES RNA-S9 cross-link occurs to the coding part of the IRES RNA just 3′ of the start codon (*12*). Thus, the domain II–induced changes in the ribosome structure around the coding RNA, as well as the direct restriction of movement of the single-stranded coding part of the RNA by the terminal loop of domain II (see above), may explain the differing efficiencies of the S9-RNA cross-link between IRES RNA with and without domain II.
35. This would be consistent with toe-printing experiments that show a decrease in the intensity of the stop associated with mRNA bound in the cleft when domain II is removed (*44*). Removal of domain II also causes a slight change in the position of IRES RNA domain IIIb relative to the surface of the 40S subunit, although binding data show that the overall IRES RNA affinity for the 40S subunit remains unchanged and that there are no changes in ribosome structure in this region. The significance of this observation is presently not clear.
36. It is difficult to imagine domain II remaining in place during elongation, as this would restrict both coding RNA and ribosome movement; some mechanism must exist to release this contact. Perhaps once elongation has begun, tRNA moving into the E site during the translocation step might displace domain II, effectively releasing the ribosome from its IRES-induced conformation, opening up the entry site for incoming coding RNA, and allowing unrestricted ribosome movement during elongation. Alternatively, subunit association could move the 40S subunit out of its IRES-induced conformation, thus ejecting domain II from its contact point.
37. T. Wagenknecht, R. Grassucci, J. Frank, *J. Mol. Biol.* **199**, 137 (1988).
38. J. Dubochet *et al.*, *Q. Rev. Biophys.* **21**, 129 (1988).
39. J. Frank *et al.*, *J. Struct. Biol.* **116**, 190 (1996).
40. J. Frank, P. Penczek, R. K. Agrawal, R. A. Grassucci, A. B. Heagle, *Methods Enzymol.* **317**, 276 (1999).
41. E. V. Orlova *et al.*, *J. Mol. Biol.* **271**, 417 (1997).
42. R. Beckmann *et al.*, *Science* **278**, 2123 (1997).
43. J. F. Ménétret *et al.*, *Mol. Cell* **6**, 1219 (2000).
44. V. G. Kolupaeva, T. V. Pestova, C. U. Hellen, *J. Virol.* **74**, 6242 (2000).
45. G. Lutsch, H. Bielka, G. Enzmann, F. Noll, *Biomed. Biochim. Acta* **42**, 705 (1983).
46. W. M. Clemons Jr. *et al.*, *Nature* **400**, 833 (1999).
47. M. Carson, *J. Appl. Crystallogr.* **24**, 103 (1991).
48. This work was funded, in part, by grants from NIH (R37 GM29169 to J.F. and GM60635 to P.A.P) and NSF (DBI 9871347 to J.F.). We thank Y. Chen for assistance with the illustrations and R. T. Batey, P. Adams, and P. Masters for critically reading the manuscript.

18 December 2000; accepted 25 January 2001

The EMBO Journal Vol. 21 No. 13 pp. 3557–3567, 2002

Cryo-EM reveals an active role for aminoacyl-tRNA in the accommodation process

Mikel Valle[1], Jayati Sengupta[2], Neil K.Swami[3], Robert A.Grassucci[1], Nils Burkhardt[4], Knud H.Nierhaus[4], Rajendra K.Agrawal[2,5] and Joachim Frank[1,5,6]

[1]Howard Hughes Medical Institute, Health Research, Inc., at the [2]Wadsworth Center, New York State Department of Health, Empire State Plaza, Albany, NY 12201-0509, [3]Columbia High School, 962 Luther Road, East Greenbush, NY 12061, [5]Department of Biomedical Sciences, State University of New York at Albany, Empire State Plaza, Albany, NY 12201-0509, USA and [4]Max-Planck-Institut für Molekulare Genetik, Ihnestrasse 73, D-14195 Berlin, Germany

[6]Corresponding author
e-mail: joachim@wadsworth.org

During the elongation cycle of protein biosynthesis, the specific amino acid coded for by the mRNA is delivered by a complex that is comprised of the cognate aminoacyl-tRNA, elongation factor Tu and GTP. As this ternary complex binds to the ribosome, the anticodon end of the tRNA reaches the decoding center in the 30S subunit. Here we present the cryo-electron microscopy (EM) study of an *Escherichia coli* 70S ribosome-bound ternary complex stalled with an antibiotic, kirromycin. In the cryo-EM map the anticodon arm of the tRNA presents a new conformation that appears to facilitate the initial codon–anticodon interaction. Furthermore, the elbow region of the tRNA is seen to contact the GTPase-associated center on the 50S subunit of the ribosome, suggesting an active role of the tRNA in the transmission of the signal prompting the GTP hydrolysis upon codon recognition.

Keywords: decoding/electron microscopy/elongation factor Tu/ribosome/tRNA

Introduction

In protein biosynthesis the decoding process is the fundamental step that assures the accuracy of the translation from mRNA into polypeptide. The codon in the mRNA has to be correctly recognized by the anticodon of the cognate tRNA, leading to its incorporation, while incorrect matches must lead to rejection. Elongation factor Tu (EF-Tu) plays a crucial role in this function of delivering the aminoacyl-tRNA (aa-tRNA) as part of the ternary complex (aa-tRNA·EF-Tu·GTP). When this ternary complex binds to the ribosome, the tRNA is bound in the A/T state, as its anticodon is able to interact with the mRNA in the A-site while its acceptor end is still bound to the factor (Moazed and Noller, 1989). In this way, the codon–anticodon match can be tested before the new amino acid is incorporated. When a cognate tRNA is present, EF-Tu catalyzes the hydrolysis of GTP and changes its conformation (Berchtold *et al.*, 1993; Kjeldgaard *et al.*, 1993; Polekhina *et al.*, 1996), thereby reducing its affinity for the ribosome and the tRNA (Dell *et al.*, 1990). The remaining EF-Tu·GDP binary complex leaves the ribosome, so that the tRNA is free to move its CCA end towards the peptidyltransferase center (PTC), allowing the polypeptide to be elongated by the new amino acid residue (Kaziro, 1978). The transition of tRNA from the A/T site to A-site is known as accommodation.

Several important questions have remained unanswered. First, how the anticodon of the aa-tRNA recognizes the codon in the mRNA while it is delivered by EF-Tu at a very different angle compared with the orientation assumed in the final A-site. Secondly, how this codon–anticodon recognition acts as a signal to trigger the GTP hydrolysis and the release of EF-Tu. And finally, what mechanism is employed that makes possible a tRNA rotation towards the PTC while maintaining the codon–anticodon pairing. As a signal for GTP hydrolysis, one possibility is that the cognate tRNA–mRNA interaction in the decoding center promotes conformational changes in the ribosome (Geigenmüller and Nierhaus, 1990; Powers and Noller, 1994; Lodmell and Dahlberg, 1997) that are communicated through the inter-subunit bridges to the GTPase-associated center of the 50S subunit (Rodnina and Wintermeyer, 2001). Another possibility that has been discussed is that a conformational change in the tRNA following codon–anticodon pairing (Moras *et al.*, 1985, 1986) might act as the signal (Rodnina *et al.*, 1994; Yarus and Smith, 1995).

The ribosome-bound EF-Tu ternary complex has previously been visualized by cryo-electron microscopy (EM) techniques in the presence of the antibiotic kirromycin (Stark *et al.*, 1997; Agrawal *et al.*, 2000b). Kirromycin is known to block the conformational changes in EF-Tu that follow GTP hydrolysis (Wolf *et al.*, 1977; Parmeggiani and Stewart, 1985). As a result of this inhibition, the factor remains bound to the ribosome in a conformation that is thought to be an intermediate between the GTP and the GDP states. This previous study suggested that EF-Tu binds to the base of L7/L12 stalk of the 50S subunit, close to the universally conserved α-sarcin–ricin loop (SRL) of the 23S rRNA, and showed the anticodon loop of the tRNA interacting with the decoding region. One of the puzzles posed by the results of previous cryo-EM studies is that the orientation of the tRNA deduced does not facilitate codon–anticodon interaction.

In the present work we have employed cryo-EM to study a complex of 70S ribosomes with P-site tRNA and EF-Tu ternary complex carrying cognate aa-tRNA and GDP in the presence of kirromycin. The first three-dimensional (3D) reconstruction of the whole set of single particles (see Agrawal *et al.*, 2000b) showed low (~50%)

© European Molecular Biology Organization

density in the region of the ternary complex. Computation of a difference map revealed the overall binding position of the complex, but a detailed interpretation of binding sites was not possible. We have now applied a supervised classification strategy based on the similarity of these images with distinct 3D references. One of the resultant 3D maps shows the ternary complex with much higher definition than seen in the earlier reports (Stark *et al.*, 1997; Agrawal *et al.*, 2000b). Most importantly, EF-Tu and the tRNA are now distinctly recognizable, and the tRNA is seen to interact with the base of the L7/L12 stalk and with protein S12. Detailed fitting, supported by quantitative measures, reveals that the tRNA structure is different from the reported X-ray structures and from the structure deduced in the former cryo-EM work (Stark *et al.*, 1997) due to a change of orientation in the anticodon arm, towards an orientation that facilitates codon–anticodon interaction. Furthermore, the T-loop side of the tRNA is seen to interact with the 58-nucleotide segment of the 23S rRNA that forms a complex with protein L11 (Wimberly *et al.*, 1999), a region known as the GTPase-associated center (GAC). The results bring new insights in understanding the incorporation of tRNA into the ribosome and strongly implicate tRNA as an active player in this process.

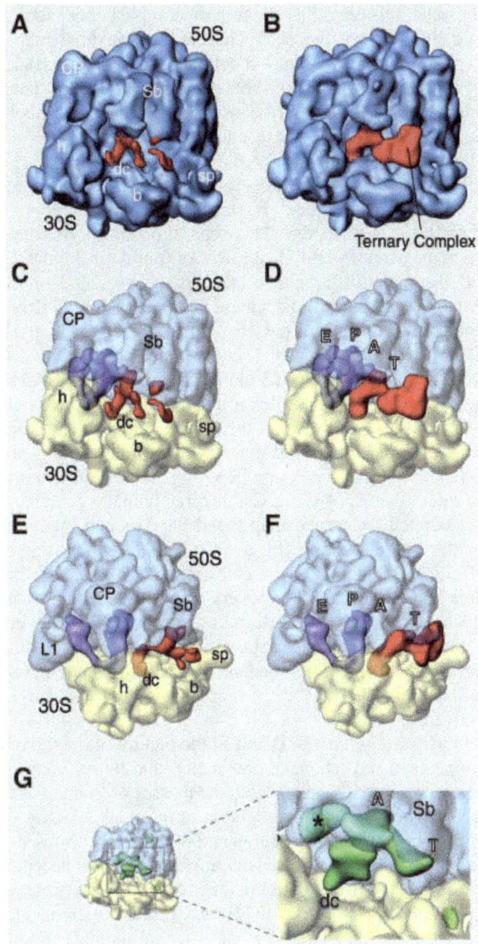

Fig. 1. Cryo-EM maps resulting from the classification, and their analysis. Map 1 (**A**, **C** and **E**) and map 2 (**B**, **D** and **F**) are depicted in two orientations related by a 90° rotation around a vertical axis in the plane [(A–D) show the side views, (E) and (F) the top views of the 70S ribosome). Side views are rotated in the plane by 90° from their conventional presentation (see Supplementary figure S1) to maintain continuity in the tRNA positions with (E) and (F)]. In (C)–(F) the ribosomal subunits are represented as semi-transparent surface (blue for the 50S and yellow for the 30S subunit), so that the positions of P- and E-site tRNAs (purple) and the mass attributable to the ternary complex (red) can be seen more clearly. (**G**) Representation of the difference map (see Materials and methods) calculated for the cryo-EM map 1, shown in green. Lobe of mass labeled with an asterisk is attributed to a conformational change in the L1 region. The ribosome is shown in same orientation as in (C). Landmarks are as follows: CP, central protuberance; Sb, L7/L12 stalk base; sp, spur; b, body; h, head; dc, decoding center; L1, stalk of L1 protein. Labels in tRNA positions: T, T-site within the ternary complex; A, aminoacyl site; P, peptidyl site; E, exit site.

Results

3D reconstruction and analysis of the cryo-EM maps

A set of 22 905 projections was selected by a previously described quasi-automated method (Lata *et al.*, 1995). By image processing techniques (see Materials and methods), a first map was calculated. The map showed the presence of extra density located in the inter-subunit space, which we attributed to the ternary complex. However, the extra density was fragmented, and could only be seen by rendering the volume at lowered density threshold. This initial poor representation of the putative ternary complex can be understood either as the result of averaging a population of ribosomes with low occupancy of the factor, or as a consequence of merging particles where the complex exists in different conformations, or as a combination of both effects. Evidence that low occupancy is a major factor comes from the filter-binding assay (see Materials and methods), which estimated ~50% occupancy for the ternary complex.

In order to separate the heterogeneous particle set into more homogeneous subsets, we have used a supervised classification method (details described in the Supplementary data, available at *The EMBO Journal* Online), where the images are classified according to their similarity to distinct references (see Frank, 1996). Briefly, our strategy for separation makes use of the resemblance that the ternary complex and EF-G share in their architecture and their overall binding position on the ribosome, and the fact that the projection of ribosomes carrying the ternary complex would show more similarity with an EF-G bound ribosome than with a ribosome with an empty A-site. After the selection, new 3D reconstructions were computed. Application of the same method of classification to a control set of empty ribosomes did not show a mass at the A-site in the reconstruction from the subset having highest similarity with the EF-G bound reference (see Supplementary data).

The final 3D cryo-maps are depicted in Figure 1A (map 1) and B (map 2). Final map 1 (Figure 1A) was reconstructed from particles more similar to an empty ribosome. It is clear that the ternary complex occupancy is poor. In contrast, the occupancy is large in final map 2 (Figure 1B), a reconstruction obtained from those particles most similar to an EF-G-bound ribosome. In this latter map (Figure 1B) the extra density (in red) is a continuous mass with a volume large enough to accommodate the whole ternary complex.

We isolated the densities corresponding to tRNAs and tRNA–EF-Tu complex bound to the ribosome (see Materials and methods). This separation shows that both the maps carry P- and E-site tRNAs (purple in Figure 1), and the main difference relates to the presence of the ternary complex (red). [It should be noted that in all tRNA binding experiments (e.g. Cate *et al.*, 1999; Agrawal *et al.*, 2000a), a mass of density corresponding to the E-site tRNA is always observed.] In map 1 (Figure 1A, C and E) the putative ternary complex is under-represented and only few parts of the tRNA and the EF-Tu can be seen. Some density clearly appears at the expected binding site of the anticodon loop of the A-site tRNA on the 30S subunit, but the rest of the tRNA is not visible directly in the volume. In a calculated difference map (see Materials and methods) for this reconstruction (in green in Figure 1G) no major difference in the ribosome conformation is detected, but there is a large mass that covers the position of an A-site tRNA reaching from the decoding site in the 30S subunit to the PTC in the 50S subunit. The region that should correspond to the CCA end of the tRNA extends below the base of L7/L12 stalk, toward the region that EF-Tu binds to.

In map 2 (Figure 1B, D and F) the picture of the ternary complex (in red) changes drastically, and its full mass is now represented, together with the masses of P- and E-site tRNAs (purple). The anticodon arm of the tRNA is situated in the decoding region and the acceptor arm is still bound to the factor. No density attributable to an acceptor arm connected to the A-site in the 50S subunit is visible. In this reconstruction, both tRNA and protein interact with the base of the L7/L12 stalk, where contact points important for the activity of the elongation factors are located. The interactions between the ternary complex and the ribosome have been studied using this map.

Views of the mass attributable to the ternary complex from the solvent (Figure 2A) and from side of the inter-subunit space (Figure 2B) reveal a structure distinct enough to permit visual identification of tRNA and the domains of the protein factor. To understand the nature of the elements within the complex, the X-ray structure of the Phe-tRNA^{Phe}·EF-Tu·GTP analog complex from *Thermus aquaticus* (Nissen *et al.*, 1995) was filtered to the same resolution and displayed in similar orientation (Figure 2C and D). The tRNA and EF-Tu were colored separately for the purpose of clarity. We note that the structures come from different organisms (EF-Tu from *T.aquaticus* and from *Escherichia coli* share 68% sequence identity), the complexes have different guanine nucleotides bound, and while our cryo-EM map shows the ternary complex bound to the ribosome, the X-ray structure represents the complex in the close packing of a 3D crystal. Nevertheless, the overall mass distribution and the shape of the ribosome-bound ternary complex in our cryo-EM map and in the ternary complex crystal structure are strikingly similar.

The result of the rigid-body fitting of the X-ray structure form of the ternary complex into the cryo-EM map is presented in Figure 2E and F. Clearly, the domains of the elongation factor have not undergone a large rearrangement in the interaction with the ribosome, but the tRNA has. In the cryo-EM map, the acceptor arm of the tRNA that is held by EF-Tu apparently maintains its binding with the protein as in the crystal structure, but the elbow of the tRNA has moved significantly from its position in the crystal structure. However, in both versions of the tRNA, the anticodon loops are in the same position due to an apparent kink in the anticodon arm. In this way, the T and D loops forming the elbow of the tRNA in our cryo-EM map move toward the base of the L7/L12 stalk of the ribosome, while the two ends of the L-shaped structure remain in the same place, changes that imply a large internal reorganization of the tRNA structure.

Docking of tRNA and EF-Tu X-ray structures

The *E.coli* EF-Tu has been divided into three domains: domain I, the guanine nucleotide binding domain that accounts for the first 204 amino acids of the N-terminus; domain II, residues 205–298; and domain III, the C-terminal domain that covers positions 299–393. In the X-ray structure of *T.aquaticus* EF-Tu, the constellation of these domains changes significantly from the GTP (Figure 2G; from Kjeldgaard *et al.*, 1993) to the GDP (Figure 2I; from Polekhina *et al.*, 1996) states. It is clear that the architecture of EF-Tu within the kirromycin-stalled ternary complex (Figure 2H) resembles more closely the EF-Tu·GTP state in both, binary (Figure 2G) and ternary complexes (Figure 2C and D) than the GDP state (Figure 2I). This similarity with the GTP state is revealed by computation of cross-correlation coefficients between the crystal structures and the cryo-EM density (Table I). For *E.coli*, the X-ray structure was calculated in its GDP state (Song *et al.*, 1999), and in order to obtain an acceptable fit, domain I needed to be moved significantly from its original position, close to domain III, to the position where the contacts between these two domains of the protein are minimized and enough room is created to accommodate the tRNA acceptor arm in the space available between them (Figure 3). Those changes produce a much better fit (Table I) and establish a GTP-state architecture.

In the case of the tRNA, the Phe-tRNA^{Phe} within the ternary complex from *T.aquaticus* in the presence of a non-hydrolyzable GTP analog (Nissen *et al.*, 1995) was used. It was not possible to get a full adjustment. The best docking was achieved by using the putative T and D loops in the elbow region of the L-shaped molecule as the main guide. The cryo-EM mass describes the above-mentioned change in the shape of the anticodon arm ('kinked' appearance; the position of the kink is indicated in Figure 3A by a dotted line), and this distortion causes the anticodon loop of the atomic coordinates of the Phe-tRNA^{Phe} (Nissen *et al.*, 1995) to lie partially outside the density mass (Figure 3A).

Inside domain I of EF-Tu there is a segment of the sequence that overlaps with the fitted tRNA structure in the CCA end (Figure 3B). This part of the protein corresponds to a single turn of an α-helix known as the switch-I region (depicted in cyan in the figures), a putative

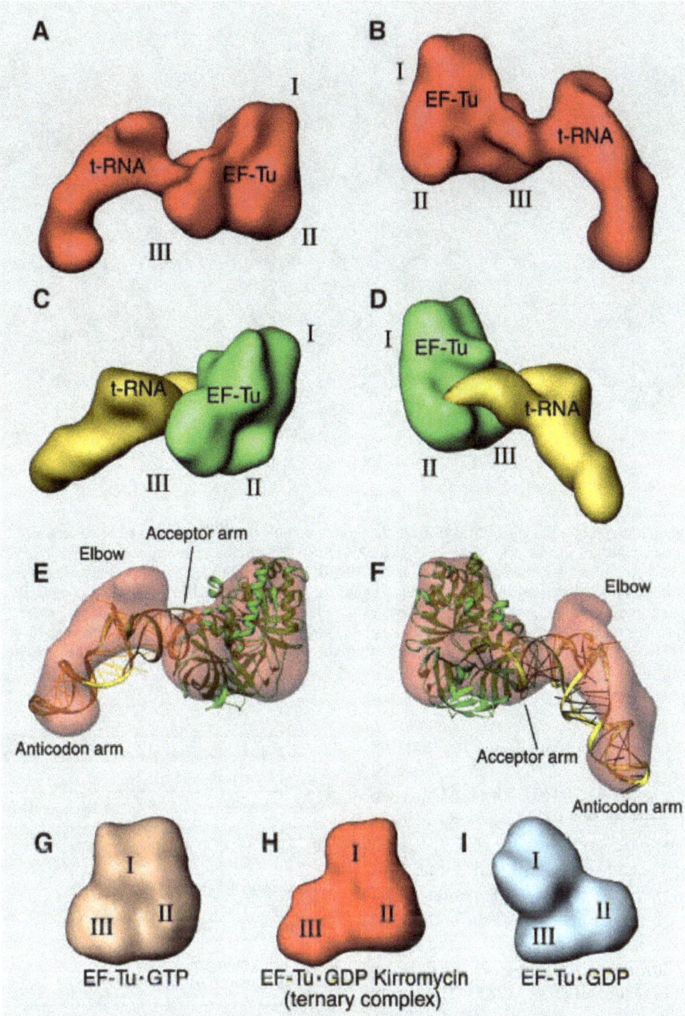

Fig. 2. Comparison of cryo-EM density and X-ray structure of the ternary complex. (**A** and **B**) Ternary complex density, isolated from the cryo-EM map: (A) as seen from the solvent side; (B) as seen from the intersubunit space side. (**C** and **D**) Equivalent orientations of the X-ray crystal structure from *T.aquaticus* ternary complex (Nissen *et al.*, 1995), filtered to the resolution of the cryo-EM map. EF-Tu is shown in green, aa-tRNA in yellow. (**E** and **F**) Ribbons representation of the fitting of the ternary complex crystal structure into the cryo-EM density. (**G–I**) EF-Tu from *T.aquaticus* in the GTP (Kjeldgaard *et al.*, 1993) and GDP (Polekhina *et al.*, 1996) states are shown next to the EF-Tu·GDP·kirromycin inferred from the cryo-EM map at the same resolution.

ribosomal interaction site that undergoes a dramatic conversion in conformation to a β structure in the transition from the GTP to the GDP state (Abel *et al.*, 1996). Obviously, either the position of this part of the protein or the position of the tRNA must be changed in order to avoid the overlap.

Interaction of the ternary complex with the ribosome

Fitting of X-ray structures into the cryo-EM map reveals three of the most conserved regions in the ribosome that are present in the immediate vicinity of the ternary complex: the decoding site on the 30S subunit (see Ogle *et al.*, 2001); the L11–rRNA complex (Wimberly *et al.*, 1999); and the SRL (Wool *et al.*, 1992) at the base of the stalk of the large subunit. These ribosomal components were fitted in both subunits to study their interactions with Phe-tRNA^{Phe} and with EF-Tu. The results of this analysis are shown in Figures 4 and 5. According to the positions found, some of the interactions can now be described.

Turning first to the 50S subunit, the T loop of the aa-tRNA closely interacts with the region of the 58-nucleotide rRNA segment that binds to the L11 protein (Figure 4A and B). This portion of the ribosome is known

Aminoacyl-tRNA in the accommodation process

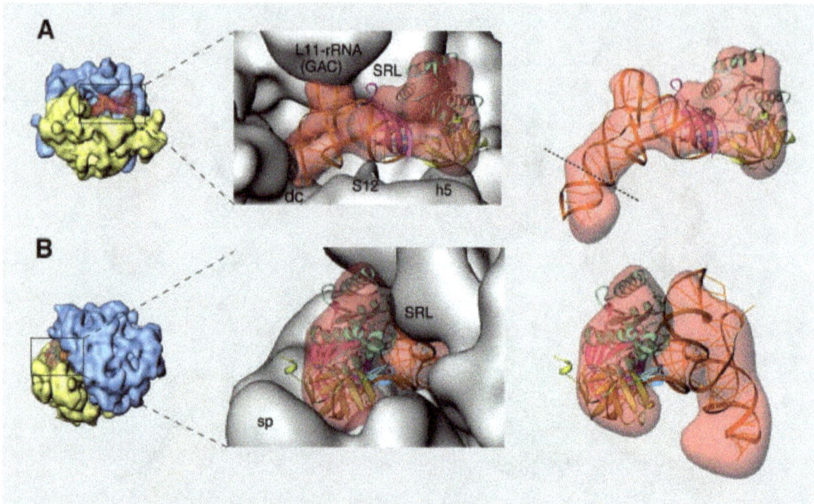

Fig. 3. Docking of EF-Tu and aa-tRNA into the cryo-EM density of the ternary complex: (**A**) as seen from the solvent side; (**B**) as seen from the inter-subunit side. (Views are similar, but not identical to those in Figure 2A and B.) Fitted atomic coordinates of *E.coli* EF-Tu (Song *et al.*, 1999) and Phe-tRNA^{Phe} [(from the ternary complex of *T.aquaticus* (Nissen *et al.*, 1995)] are shown inside the semi-transparent cryo-EM density of the ternary complex. The domains of EF-Tu were fitted independently (domain I shown in green, domain II in yellow and domain III pink). The switch-I region within domain I is highlighted in cyan. The dotted line in (A) indicates the place of the kink in the anticodon arm of the tRNA. Orientations of the ribosome are shown as thumbnails on the left. SRL, α-sarcin–ricin loop; L11–rRNA (GAC), protein L11 and the segment of 58 nucleotides of the 23S rRNA, also known as the GTPase-associated center; S12, protein S12 from the small subunit; h5, helix 5 from the 16S rRNA; dc, decoding center in the A-site of the 30S subunit.

as the GTPase-associated center (GAC) due to the fact that antibiotics that bind to it strongly affect the activity of EF-G and EF-Tu (Vazquez *et al.*, 1979). The GAC is seen to interact with the T arm of the tRNA through a loop, around positions 1060–1075 in the 23S rRNA, which protrudes between the C- and the N-terminal domains of L11. In this loop, a single transversion in the residue 1067 impairs the function of the elongation factors (Saarma *et al.*, 1997). Residues around position 2660 of the rRNA (within the SRL) were defined as binding site for EF-Tu by footprinting experiments (Moazed *et al.*, 1988). The fitting leaves a distance between the SRL and EF-Tu large enough (10–15 Å) to be uncertain about the existence of a real binding contact. However, the effector loop region inside domain I of EF-Tu switches from an α-helical to a β-sheet structure and moves within the protein in the transition between GTP and GDP states (Abel *et al.*, 1996). In Figure 4C and D, those two different positions for the effector loop are represented. When the coordinates of the EF-Tu·GTP form are used (Figure 4C), this switch-I region interacts with the SRL and avoids the earlier mentioned steric conflict with the CCA tip of the tRNA that the GDP state shows (Figure 4D).

In the 30S subunit the ternary complex forms three contacts, one with domain II of EF-Tu and two with the aa-tRNA. Domain II contacts helix 5 of 16S rRNA; the acceptor arm of the tRNA interacts with the protein S12; and the anticodon loop reaches the decoding center. In Figure 4A and B it can be seen that protein S12 extends into a loop that goes close to the decoding center. This loop includes the conserved PNSA (Pro-Asn-Ser-Ala) amino acid sequence at positions 48–51. The mass in the region of

Table I. Correlation between the fitted X-ray structures and the cryo-EM density

X-ray structure	Rigid body fitting	Flexible fitting
T.aquaticus EF-Tu·GDP	0.51	–
T.aquaticus EF-Tu·GTP	0.78	–
E.coli EF-Tu·GDP	0.52	0.81
Phe-tRNA^{Phe}·EF-Tu·GTP	0.69	0.83
Phe-tRNA^{Phe}	0.77	0.81[a]

[a]Value when the anticodon tip of the A-site tRNA was used for the fitting.

the anticodon tip of the tRNA is unexpectedly wide, but the additional mass could be explained as the result of a conformational change of this loop of S12 such that it reaches the density in immediate vicinity to the tRNA (labeled with an asterisk in Figure 4A).

Particular interest is focused on the codon–anticodon interaction and in the dynamic transition of the tRNA from the T- to the A-site. The rigid-body fitting of the tRNA structure inside the cryo-EM map leaves the anticodon loop partially outside the density (Figure 5A). Evidently, the tRNA crystal structure is different from the structure of the tRNA in the transition state observed here, and hence cannot be used to study the interactions in the region. However, it is important for understanding the process of tRNA accommodation that the anticodon portion of the tRNA interacting with the mRNA in the A-site (Agrawal

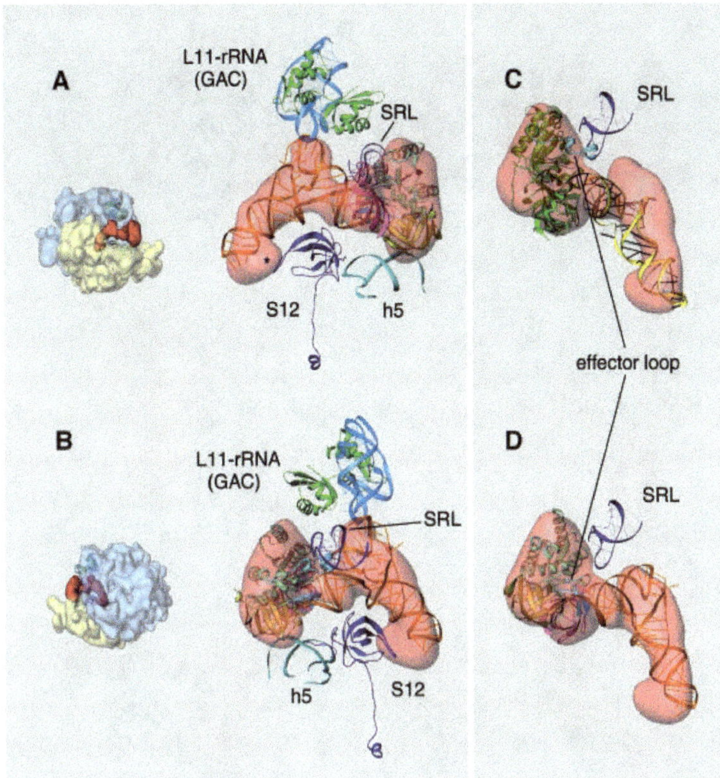

Fig. 4. Interaction of EF-Tu and aa-tRNA with the ribosome. (**A** and **B**) Ribbons representation of the docked EF-Tu and aa-tRNA within the ternary complex. (**C** and **D**) Focus on the interaction between the α-sarcin–ricin loop (SRL) and the effector loop within domain I of EF-Tu (cyan). In (C) the coordinates of the whole ternary complex from *T.aquaticus* with a GTP analog (Nissen *et al.*, 1995) were used for the fitting, while in (D) the crystal structure of EF-Tu from *E.coli* bound to GDP (Song *et al.*, 1999) was used. Orientation of the ribosomes for (A) and (B) are shown as thumbnails on the left. Labeling is the same as in Figure 5.

et al., 2000a; Yusupova *et al.*, 2001) partially overlaps the cryo-EM density of the tRNA up to the observed kink, and that the correlation coefficient is improved in the docking (Table I). The position of this A-site tRNA was deduced by the alignment of the P- and E-site tRNAs from the X-ray structure (Yusupova *et al.*, 2001) with those from our density map. Figure 5C shows a merged representation of both tRNAs where the kink in the anticodon arm separates an A-site anticodon loop (red) from the rest of the tRNA still residing in the T-site (gold).

We have used the accommodated position of the tRNA to represent the interactions in the decoding center. In Figure 5B the density identified with the anticodon is broadened on both sides. The extra mass not filled with the tRNA structure might be understood to be a result of contributions from the tight interactions of the anticodon end with the surrounding ribosomal elements (Figure 5D). This way, the parts of helix 69 from 23S rRNA and helix 44 from 16S rRNA, which interact tightly with the codon of the mRNA (Figure 5E), and as well as a putative contribution due to a conformational change of S12 (Figure 5F), would be displayed as a compact mass at the resolution of our map.

Discussion

Ternary complex in the presence of kirromycin

The biggest problem that arose in our sample was its heterogeneity. Taking into account the occupancy of the ternary complex that the filter binding assays determined, it was reasonable to expect a significant fraction of the population of ribosomes to be without the factor. The classification produced two 3D reconstructions with different features. One of the maps, which showed very low occupancy, has to be interpreted with caution because the set of individual particles that it comes from represents, in all likelihood, a mixture of factor-free and factor-bound ribosomes. However, it shows an interesting presence of some densities in the calculated difference map that we attribute to the mass of tRNAs in the process of moving to the A-site from the initial position where they were bound to EF-Tu. This interpretation requires the assumption that kirromycin fails to block the conformational change of EF-Tu at least in a fraction of the ribosomes.

The good representation of the ternary complex in the high-occupancy map suggests the existence of a uniform subset of particles, and the inferred structure for the

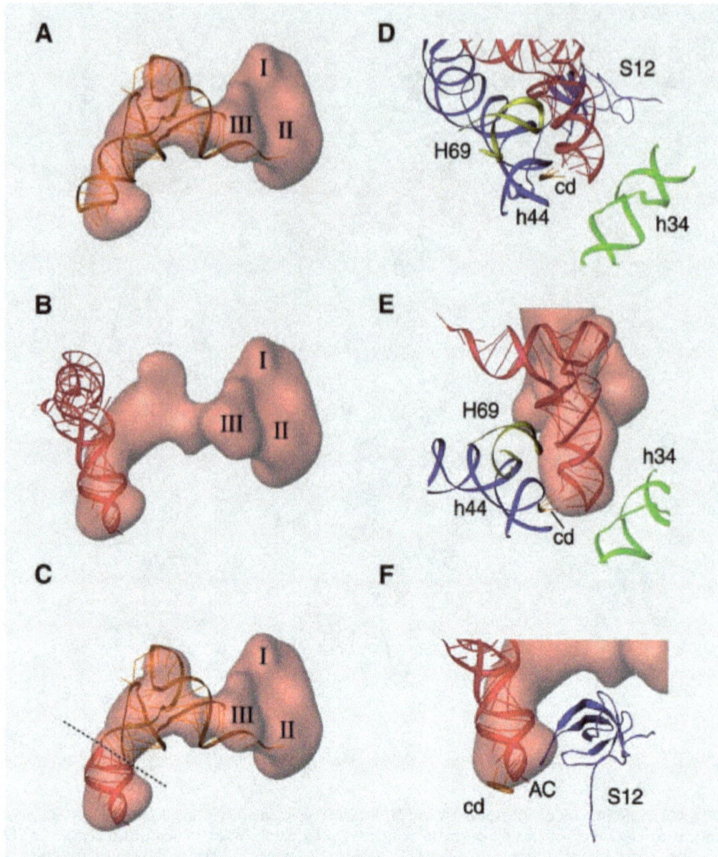

Fig. 5. Interaction in the decoding center and accommodation of the aa-tRNA. (**A–C**) Semi-transparent representation of the ternary complex density from the cryo-EM map showing the fitted tRNA (gold) and the A-site tRNA (red) with corresponding mRNA codon (Yusupova *et al.*, 2001). (C) A tRNA construct in which the anticodon position, up to the kink, is adopted from (B) and the rest of the tRNA from (A). (**D–F**) Interaction in the anticodon loop of the tRNA in the decoding site. H69, helix 69 from 23S rRNA; h44, helix 44 from 16S rRNA; h34, helix 34 from 16S rRNA; cd, A-site codon in the mRNA; AC, anticodon loop; S12, protein S12 in the 30S subunit.

ternary complex has to be understood as the most abundant conformation of the complex in the sample, i.e. the conformation corresponding to the bottleneck in the inhibition of the decoding by the antibiotic. The kirromycin binding site in EF-Tu is formed by residues of domains I and III (Abdulkarim *et al.*, 1994; Mesters *et al.*, 1994), situated at a crucial place to stall the rearrangements of the interdomain positions that the change from GTP to GDP forms requires (Berchtold *et al.*, 1993; Kjeldaarg *et al.*, 1993; Polekhina *et al.*, 1996; Song *et al.*, 1999). Evidently, as our fitting shows, the structure that EF-Tu·GDP·kirromycin displays in the cryo-EM map is closer to a GTP constellation of domains. Furthermore, we have seen that the switch-I region (termed effector loop) inside the domain G of EF-Tu cannot be fitted properly into the isolated density when the coordinates from the *E.coli* factor in its GDP form are used. The effector loop is known to move and modify the secondary structure upon the change of the guanine nucleotide, and it is considered a putative interaction region with the ribosome (Song *et al.*, 1999). By using the X-ray structure from the whole ternary complex with a GTP analog (Nissen *et al.*, 1995) it is seen that the effector loop contacts the SRL from the ribosome as in the case of EF-G (Wriggers *et al.*, 2000), and avoids a clash with the docked tRNA. Altogether, the data indicate that the ternary complex density in this work represents a stalled structure where the factor has been caught in an intermediate conformation, which is similar to the GTP form of the X-ray structure.

Interaction of the ternary complex with proteins and rRNA in the entrance channel of the ribosome

Previously, the only ribosome-binding region defined for EF-Tu was the above-mentioned SRL (Hausner *et al.*, 1987; Moazed *et al.*, 1988) interaction that in our map, as noted above, appears to include the switch-I region of domain I of EF-Tu. In the same work, chemical probes revealed that EF-G binds to the GAC in domain II of 23S rRNA, but no interaction from EF-Tu was detected in this region. On the other hand, the rRNA segment that folds with the L11 protein is known to have a direct influence on the GTPase activity of both factors when they are bound to the ribosome: the GAC is the binding site for antibiotics such as thiostrepton and micrococcin that affect the GTPase activity not only of EF-G, but also of EF-Tu (Vazquez, 1979). Even a single change of the nucleotide in position 1067 of the 23S rRNA (located within the GAC) was found to impair the function of both elongation factors (Saarma *et al.*, 1997). In our cryo-EM map there is a clear interaction between the ternary complex and the GAC, but the interaction involves the tRNA instead of EF-Tu. More precisely, it is the T-loop region of the tRNA, the part of the nucleic acid that contacts this region. In the experiments probing 23S rRNA protection by the ternary complex against chemical modification (Moazed *et al.*, 1988), a tRNA–GAC interaction was not observed, but neither did P- or A-site tRNAs give any effect in the 23S rRNA labeling pattern. The discrepancy of the results seems to be related to the reduced sensitivity of the technique for tRNA–rRNA interactions rather than to a real difference in the sample.

When comparing the overall structure of the ternary complex with that of EF-G, one is struck by their similarity (Nissen *et al.*, 2000) and finds that individual domains of EF-G are mimicking parts of the ternary complex. In the case of the tRNA within the ternary complex, the domains in EF-G that build up a more or less compact structure are domains III, IV and V. Domain IV goes into the decoding site in the 30S subunit, the same place the anticodon of the tRNA binds to. Domain V of EF-G is the substitute for the T region of the tRNA. It is this domain that binds to the GAC in the case of EF-G (Agrawal *et al.*, 1998, 1999, 2001). According to our present findings, the molecular mimicry goes beyond this, and now includes the fact that the interaction of protein S12 with the acceptor arm of the tRNA inside the ternary complex closely parallels its interaction with domain III of EF-G (Agrawal *et al.*, 1998). In our fitting the protein is also close to the anticodon loop of the tRNA in the decoding site.

In the previous cryo-EM reconstruction of a ribosome–ternary complex (Stark *et al.*, 1997), where the ternary complex density was seen without requiring any classification, a connection between EF-Tu and the base of the L7/L12 stalk close to the L11–rRNA complex was shown at the same place where the arc-like connection was observed in the EF-G·GDP·ribosome complex (Agrawal *et al.*, 1998, 1999, 2000b). In our new map this connection cannot be seen. An interpretation of the binding in this region was made that implicated L7/L12 in the interaction (Stark *et al.*, 1997). However, this and other differences in fine features can be caused by the difference in resolution of the reconstructions [using the same criteria as defined in Orlova *et al.* (1997), the resolution of our map is 11 Å, as compared with the 18 Å resolution reported by those authors].

An active role for the tRNA

The incorporation of tRNAs into the A-site in the ribosome by EF-Tu follows a sequence of closely related steps: initial binding of the ternary complex to the ribosome; codon recognition; GTP hydrolysis; release of the EF-Tu·GDP accompanied by the accommodation of the tRNA in the A-site; and formation of a new peptide (Rodnina *et al.*, 1994). Two main questions that remained unanswered have been how the codon–anticodon interaction triggers the GTP hydrolysis by EF-Tu and how the conformational changes after this event facilitate both the exit of the factor and the accommodation of the tRNA. In the present work, the ribosome-bound ternary complex was blocked with kirromycin, so that the structure visualized in our cryo-EM map represents an intermediate state after the GTP hydrolysis and before the completion of the accommodation process. The antibiotic kirromycin is known to stimulate the GTP hydrolysis by EF-Tu (Wolf *et al.*, 1977) and we cannot distinguish between a GTPase stimulation by the antibiotic from an activation due the codon–anticodon recognition. Nevertheless, the cognate aa-tRNA shows a clear interaction with the decoding center, and we assume that the anticodon-pairing signal has been triggered.

The elbow in the L shape of the tRNA that comprises the D- and T-loop regions moves significantly toward the ribosome, and the T loop contacts the GAC of the 23S rRNA. Evidence of this movement, together with the evidence that cognate codon recognition is not sufficient for the GTPase activation and that an intact tRNA is required for the transmission of the signal (Piepenburg *et al.*, 2000), points directly to the role of tRNA as the communicator that translates a cognate interaction in the decoding center into a signal for GTPase activation in the EF-Tu interaction with the ribosome. It is necessary to point out that in the previous cryo-EM work (Stark *et al.*, 1997) a change in the relative position between the tRNA and EF-Tu was described, but no changes were detected in the structure of the aa-tRNA, nor did the study detect any interaction between the tRNA and the GAC.

In dynamic studies of the interaction of the ternary complex with the ribosome, Rodnina *et al.* (1994) found that the D loop of the tRNA, labeled with proflavin, changed its structure after codon–anticodon recognition and before GTP hydrolysis. These changes were deduced from a rapid increase of fluorescence followed by a slow decrease, reflecting two different levels of exposure of the label. In that work, the same experiment in the presence of kirromycin revealed an initial rapid increase of the signal, while the slow decrease was completely suppressed. Thus, the change that the tRNA structure undergoes upon codon recognition is not reversed when the complex is bound to the antibiotic. We conjecture that the structural change in the tRNA deduced in the dynamic experiments (Rodnina *et al.*, 1994) is captured by the cryo-EM map we have obtained, where the tRNA shows a distorted structure and reaches the GAC in the 50S subunit, apparently sending the signal for the GTP hydrolysis.

The idea of the tRNA having an active role in translation is not new and has been postulated in several works by Woese (1970) (and more recently Woese, 2001), a concept that is at odds with the currently prevailing view of tRNAs as mere adaptors during the process.

Flexibility of tRNA: codon recognition and accommodation

By proposing that the relative movement of parts of the tRNA acts as signal after the codon–anticodon recognition we assume that the tRNA possesses flexibility. Indeed, in our data, the tRNA appears distorted and we have been unable to fit any atomic structure of tRNA from the CCA end all the way up to the anticodon loop without leaving significant parts of the nucleic acid outside the map. The change in the structure of the tRNA occurs during the interaction with the ribosome following cognate tRNA–mRNA recognition (Rodnina *et al.*, 1995). The question is, can the codon–anticodon interaction promote structural changes in the aa-tRNA? Interesting results were introduced with the crystal structure of a yeast $tRNA^{Asp}$ (Dumas *et al.*, 1985). In the 3D crystals the tRNA molecules were interacting via their GUC anticodon sequence with a slight mismatch at the uridine position, and the structure was considered a model for the effects of a codon–anticodon binding on the tRNA (Moras *et al.*, 1986). The comparison with the yeast $tRNA^{Phe}$ structure revealed a more rigid anticodon loop and a gain in flexibility of the T and D loops in $tRNA^{Asp}$ versus $tRNA^{Phe}$. Studies with tRNA mutants have shown that tRNAs with altered D and T arm interactions have different degrees of stability on the ribosome and favor near-cognate recognition (reviewed in Yarus and Smith, 1995). These authors proposed a general 'waggle' theory that attributes some role to the flexibility and structural reorganization of the elbow region in the tRNA during the decoding process. Thus, there are data that suggest conformational changes in the D and T arms of the tRNA after the pairing of the anticodon loop with the complementary RNA sequence, and indicate that the disruption of this region has direct effects on the accuracy of the decoding process. The present work brings direct evidence of a structural modification in the tRNA within the ternary complex after the codon–anticodon pairing.

In the delivery of the aa-tRNA to the ribosome it moves from the A/T position to the position in the A-site. The tRNA structure inferred from the density map shows a better fit in the anticodon loop if we use the position of a full A-site tRNA interacting with the mRNA (Yusupova *et al.*, 2001) than the tRNA structure within the ternary complex crystal structure (Nissen *et al.*, 1995). The kink that the anticodon arm shows represents a modification from the known structures of tRNAs that could explain the phase mismatch between an anticodon loop already accommodated and an acceptor arm still residing in the T site. The kirromycin has stalled the complex at the stage where the tRNA has moved toward the ribosome to contact the GAC. This interaction could trigger the GTPase activity of EF-Tu, and the subsequent liberation of the acceptor arm of the tRNA from the binding with the factor could allow the acceptor arm to swing and reach the PTC. Furthermore, the contacts of the tRNA with two different parts of protein S12 would connect the movement of the acceptor arm with the previously described reorganization of the decoding site, particularly the helix 44 of 16S rRNA (VanLoock *et al.*, 2000; see also Carter *et al.*, 2001), in a way that both changes could be synchronized. In any event, the flexibility of the anticodon arm of tRNA would have an interesting role as a mechanism to absorb the temporary discrepancy between the positions of acceptor and anticodon arms of the tRNA in the transition toward the A-site position.

A 5′-stacked anticodon loop?

After the submission of the present work, Simonson and Lake (2002) presented the 'transorientation hypothesis', where they revive the idea that the tRNA might flip between a 5′-stacked and a 3′-stacked anticodon loop structure (Woese, 1970), and that these two conformations might facilitate both the codon recognition and the accommodation process. All the X-ray-determined structures for tRNAs are in the 3′-stacked form, and no direct structural evidence for the existence of the 5′-stacked form has been reported. Nevertheless, the authors of this hypothesis constructed an energy-minimized model for a tRNA with a 5′-stacked anticodon loop. We have used this model to see if its conformation is compatible with our cryo-EM density (details shown in Supplementary data). By using the anticodon of this modeled tRNA we can fill the whole density of our cryo-EM map satisfactorily. However, in the best fit, the codon and the anticodon run in the same 3′→5′direction, showing a parallel orientation rather than the antiparallel one required for base pairing (S.M.Stagg and S.C.Harvey, personal communication). In order to accomplish the codon recognition, the whole tRNA would have to be rotated by about 180° around the anticodon arm. Thus, given the antiparallel base pairing requirements, and the known orientation of the mRNA, our map is not compatible with such a tRNA structure.

Materials and methods

Preparation of 70S·aa-tRNA·EF-Tu·GDP·kirromycin complex

Escherichia coli ribosomes (10 pmol) were incubated with 15 pmol of MF-mRNA (Triana-Alonso *et al.*, 1995) and 14 pmol of [^{35}S]fMet-$tRNA_f^{Met}$ (575 d.p.m./pmol) at 37°C for 10 min in 30 μl of a buffer containing 20 mM HEPES–KOH pH 7.5, 6 mM $MgCl_2$, 150 mM NH_4Cl, 4 mM 2-mercaptoethanol, 0.05 mM spermine and 2 mM spermidine according to Blaha *et al.* (2000). Three microliters of a 100 mM kirromycin solution were added to the reaction mixture and the incubation was continued for another 5 min. Separately, another incubation with 14 pmol of [^{14}C]Phe-$tRNA^{Phe}$ (996 d.p.m./pmol), 20 pmol of EF-Tu, and 8 μl of a 2.5 mM solution of [α-^{32}P]GTP (150 d.p.m./pmol) was carried out for 10 min at the same temperature. The samples with the 70S ribosomes and with the ternary complex were mixed and another 10 min incubation at the same temperature was performed before the cryo-grids were prepared. The occupancy of individual components on the ribosome was measured by filter-binding assays in parallel incubation mixtures, and was found to be ~60% for both, the P-site fMet-$tRNA_f^{Met}$ and the A-site Phe-$tRNA^{Phe}$. The radioactivity contributed by the α-^{32}P moiety of the GDP served to calculate a ~50% occupancy for the elongation factor.

Electron microscopy and image processing

The grids for cryo-EM were prepared following standard procedures (Wagenknecht *et al.*, 1988). Micrographs were taken following low-dose procedures on a Philips EM420 at 120 kV and a magnification of ×52 000. The micrographs were digitized with a step size of 14 μm on a Zeiss/Imaging scanner (Z/I Imaging Corporation, Huntsville, AL), corresponding to 2.69 Å on the object scale. The 3D reconstructions were calculated using 3D projection alignment procedures (Penczek *et al.*, 1994; Frank *et al.*, 2000). The volumes were CTF-corrected in defocus groups, with an average of ~1000 individual images per group (Frank and Penczek, 1995) and the resolution was estimated by the Fourier shell correlation with a 0.5 threshold (Malhotra *et al.*, 1998). X-ray solution scattering data for the *E.coli* ribosome were used to correct the Fourier amplitudes of the maps (Gabashvili *et al.*, 2000). Difference maps were calculated by subtracting an *E.coli* 70S ribosome reference volume that had both P- and E-sites occupied with tRNAs (M.Valle, unpublished). Before the subtraction, the maps were normalized taking into account their average density and their variability (σ) and in the rendering only significant values (>2σ) were displayed. The isolation of tRNAs and ternary complex within the volumes was performed by masking out the ribosomal subunits and selecting the remaining masses by clustering procedures.

Fitting of atomic structures

After docking of X-ray structures into the cryo-EM volume using program O (Jones *et al.*, 1991); the visualization of the density maps was performed in IRIS Explorer (Numerical Algorithms Groups, Inc., Downers Grove, IL). EF-Tu domains were taken from the *E.coli* crystal structure (PDB code 1EFC, from Song *et al.*, 1999) and aa-tRNA or the whole ternary complex (PDB code 1TTT, from Nissen *et al.*, 1995) as the initial coordinates for fitting. The EF-TU·GTP (PDB code 1EFT, from Kjeldgaard *et al.*, 1993) and EF-TU·GDP (PDB code 1TUI, from Polekhina *et al.*, 1996) were also used. The coordinates of the ribosomal proteins and rRNAs were taken from published atomic structures (PDB code 1FFK for the 50S subunit, from Ban *et al.*, 2000; PDB code 1FJF for the 30S, from Wimberly *et al.*, 2000). For the final rendering of fitted structures, the Ribbons package (Carson, 1997) was used. For computation of the cross-correlation coefficient, the fitted PDB coordinates were converted into electron densities, which were low-pass filtered to the resolution of the cryo-EM map.

Coordinates

The atomic coordinates of *E.coli* EF-Tu (GDP form; Song *et al.*, 1999) and yeast Phe-tRNA^{Phe} (Nissen *et al.*, 1995), independently fitted into the cryo-EM map, have been deposited in the Protein Data Bank (http://www.rcsb.org) with the accession code 1LS2. In addition, the anticodon loop region of the tRNA, fitted separately into the cryo-EM map, has been deposited with the accession code 1LU3.

Supplementary data

A description of the supervised classification method is provided as Supplementary data, available at *The EMBO Journal* Online.

Acknowledgements

We thank Christian Spahn for helpful suggestions. We are grateful to Michael Watters and Yu Chen for the preparation of the illustrations. The work was supported by National Institutes of Health Grants R37 GM29169, R01 GM55440, P41 RR01219 and NSF BIR 9219043 (to J.F.), and R01 GM61576 (to R.K.A.).

References

Abdulkarim,F., Liljas,L. and Hughes,D. (1994) Mutations to kirromycin resistance occur in the interface of domains I and II of EF-Tu–GTP. *FEBS Lett.*, **352**, 118–122.

Abel,K.M., Yoder,M.D., Hilgenfeld,R. and Jurnak,F. (1996) An α to β conformational switch in EF-Tu. *Structure*, **4**, 1153–1159.

Agrawal,R.K., Penzeck,P., Grassucci,R.A. and Frank,J. (1998) Visualization of elongation factor G on the *Escherichia coli* 70S ribosome: the mechanism of translocation. *Proc. Natl Acad. Sci. USA*, **95**, 6134–6138.

Agrawal,R.K., Heagle,A.B., Penzeck,P., Grassucci,R.A. and Frank,J. (1999) EF-G-dependent GTP hydrolysis induces translocation accompanied by large conformational changes in the 70S ribosome. *Nat. Struct. Biol.*, **6**, 643–647.

Agrawal,R.K., Spahn,C.M.T., Penzeck,P., Grassucci,R.A., Nierhaus, K.H. and Frank,J. (2000a) Visualization of tRNA movements on the *Escherichia coli* ribosome during the elongation cycle. *J. Cell Biol.*, **150**, 447–459.

Agrawal,R.K., Heagle,A.B. and Frank,J. (2000b) Studies of elongation factor G dependent tRNA translocation by three-dimensional cryo-electron microscopy. In Garrett,R.A., Douthwaite,S.R., Liljas,A., Matheson,T., Moore,P.B. and Noller,H.F. (eds), *The Ribosome: Structure, Function, Antibiotics and Cellular Interactions.* ASM Press, Washington, DC, pp. 53–62.

Agrawal,R.K., Linde,J., Sengupta,J., Nierhaus,K.H. and Frank,J. (2001) Localization of L11 protein on the ribosome and elucidation of its involvement in EF-G-dependent translocation. *J. Mol. Biol.*, **311**, 777–787.

Ban,N., Nissen,P., Hansen,J., Moore,P.B. and Steitz,T.A. (2000) The complete atomic structure of the large ribosomal subunit at 2.4Å resolution. *Science*, **289**, 905–920.

Berchtold,H., Reshetnikova,L., Reiser,C.O., Schirmer,N.K., Sprinzl,M. and Hilgenfeld,R. (1993) Crystal structure of active elongation factor Tu reveals major domain rearrangements. *Nature*, **365**, 126–132.

Blaha,G., Stelzl,U., Spahn,C.M.T., Agrawal,R.K., Frank,J. and Nierhaus,K.H. (2000) Preparation of functional ribosomal complexes and the effect of buffer conditions on tRNA positions observed by cryoelectron microscopy. *Methods Enzymol.*, **317**, 292–309.

Carson,M. (1997) Ribbons. *Methods Enzymol.*, **277**, 493–505.

Carter,A.P., Clemons,W.M., Brodersen,D.E., Morgan-Warren,R.J., Hartsch,T., Wimberly,B.T. and Ramakrishnan,V. (2001) Crystal structure of an initiation factor bound to the 30S ribosomal subunit. *Science*, **291**, 498–501.

Cate,J.H., Yusupov,M.M., Yusupova,G.Z., Earnest,T.N. and Noller,H.F. (1999) X-ray structures of 70S ribosome functional complexes. *Science*, **285**, 2095–2104.

Dell,V.A., Miller,D.L. and Johnson,A.E. (1990) Effects of nucleotide- and aurodox-induced changes in elongation factor Tu conformation upon its interaction with aminoacyl-tRNA. A fluorescence study. *Biochemistry*, **29**, 1757–1763.

Dumas,P., Ebel,J.P., Giegé,R., Moras,D., Thierry,J.C. and Westhof,E. (1985) Crystal structure of yeast tRNA^{Asp}: atomic coordinates. *Biochimie*, **67**, 597–606.

Dvorak,D., Kidson,C. and Winzor,D.J. (1978) Conformational changes in tRNA: consequences of aminoacylation and codon–anticodon recognition. *FEBS Lett.*, **90**, 187–188.

Frank,J. (1990) Classification of macromolecular assemblies studied as 'single particles'. *Rev. Biophys.*, **23**, 281–329.

Frank,J. (1996) *Three-dimensional Electron Microscopy of Macromolecular Assemblies.* Academic Press, San Diego, CA.

Frank,J. and Penczek,P. (1995) On the correction of the contrast transfer function in biological electron microscopy. *Optik*, **98**, 125–129.

Frank,J., Penczek,P., Agrawal,R.K., Grassucci,R.A. and Heagle,A.B. (2000) Three-dimensional cryo-electron microscopy of ribosomes. *Methods Enzymol.*, **317**, 276–291.

Gabashvili,I.S., Agrawal,R.K., Spahn,C.M.T., Grassucci,R.A., Svergun, D.I., Frank,J. and Penczek,P. (2000) Solution structure of *E. coli* 70S ribosome at 11.5 Å resolution. *Cell*, **100**, 537–549.

Geigenmüller,U. and Nierhaus,K.H. (1990) Significance of the third tRNA binding site, the E site, on *E.coli* ribosomes for the accuracy of translation: an occupied E site prevents the binding of non-cognate aminoacyl-tranfer RNA to the A site. *EMBO J.*, **9**, 4527–4533.

Hausner,T.P., Atmadja,J. and Nierhaus,K.H. (1987) Evidence that the G2661 region of 23S rRNA is located at the ribosomal binding sites of both elongation factors. *Biochimie*, **69**, 911–923.

Jones,T.A., Zhou,J.Y., Cowan,S.W. and Kjeldgaard,M. (1991) Improved methods for binding protein models in electron density maps and the location of errors in these models. *Acta Crystallogr. A*, **47**, 110 119.

Kaziro,Y. (1978) The role of guanosine-5′-triphosphate in polypeptide chain elongation. *Biochim. Biophys. Acta*, **505**, 95–127.

Kjeldgaard,M., Nissen,P., Thirup,S. and Nyborg,J. (1993) The crystal structure of elongation factor EF-Tu from *Thermus aquaticus* in the GTP conformation. *Structure*, **1**, 35–50.

Lata,K.R., Penczek,P. and Frank,J. (1995) Automatic particle picking from electron micrographs. *Ultramicroscopy*, **58**, 381–391.

Lodmell,J.S. and Dahlberg,A.E. (1997) A conformational switch in *Escherichia coli* 16S ribosomal RNA during the decoding of messenger RNA. *Science*, **277**, 1262–1267.

Malhotra,A., Penczek,P., Agrawal,R.K., Gabashvili,I.S., Grassucci,R.A., Burhardt,N., Jünemann,R., Nierhaus,K.H. and Frank,J. (1998) *E. coli* 70S ribosome at 15Å resolution by cryo-electron microscopy: localization of fMet-tRNA$_f^{Met}$ and fitting of L1 protein. *J. Mol. Biol.*, **280**, 103–116.

Mesters,J., Zeef,L., Hilgenfeld,R., Degraaf,J., Kraal,B. and Bosch,L. (1994) The structural and functional basis for the kirromycin resistance of mutant EF-Tu species in *E.coli*. *EMBO J.*, **13**, 4877–4885.

Moazed,D. and Noller,H.F. (1989) Intermediate states in the movement of transfer RNA in the ribosome. *Nature*, **342**, 142–148.

Moazed,D., Robertson,J.M. and Noller,H.F. (1988) Interaction of elongation factors EF-G and EF-Tu with a conserved loop in 23S rRNA. *Nature*, **334**, 362–364.

Moras,D., Dock,A.C., Dumas,P., Westhof,E., Romby,P., Ebel,J.P. and Giegé,R. (1985) The structure of yeast tRNA (Asp). A model for tRNA interacting with messenger RNA. *J. Biomol. Struct. Dyn.*, **3**, 479–493.

Moras,D., Dock,A.C., Dumas,P., Westhof,E., Romby,P., Ebel,J.P. and Giegé,R. (1986) Anticodon–anticodon interaction induces conformational changes in tRNA: yeast tRNA^{Asp}, a model for tRNA-mRNA recognition. *Proc. Natl Acad. Sci. USA*, **83**, 932–936.

Nissen,P., Kjeldgaard,M., Thirup,S., Polekhina,G., Reshetnikova,L., Clark,B.F.C. and Nyborg,J. (1995) Crystal structure of the ternary complex of the Phe-tRNA^{Phe}, EF-TU and a GTP analog. *Science*, **270**, 1464–1472.

Nissen,P., Kjeldaarg,M. and Nyborg,J. (2000) Molecular mimicry. *EMBO J.*, **19**, 489–495.

Ogle,J.M., Brodersen,D.E., Clemons,W.M., Tarry,M.J., Carter,A.P. and Ramakrishnan,V. (2001) Recognition of cognate transfer RNA by the 30S ribosomal subunit. *Science*, **292**, 868–869.

Orlova,E.V., Dube,P., Harris,J.R., Beckman,E., Zemlin,F., Marki,J. and van Heel,M. (1997) Structure of keyhole limpet hemocyanin type 1 (KLH1) at 15 Å resolution by electron cryomicroscopy and angular reconstitution. *J. Mol. Biol.*, **271**, 417–437.

Parmeggiani,A. and Stewart,G.W. (1985) Mechanism of action of kirromycin-like antibiotics. *Annu. Rev. Microbiol.*, **39**, 557–577.

Penczek,P., Grassucci,R.A. and Frank,J. (1994) The ribosome at improved resolution: new techniques for merging and orientation refinement in 3D cryo-electron microscopy of biological particles. *Ultramicroscopy*, **53**, 251–270.

Piepenburg,O., Pape,T., Pleiss,J.A., Wintermeyer,W., Uhlenbeck,O.C. and Rodnina,M.V. (2000) Intact aminoacyl-tRNA is required to trigger GTP hydrolysis by elongation factor Tu on the ribosome. *Biochemistry*, **39**, 1734–1738.

Polekhina,G., Thirup,S., Kjeldgaard,M., Nissen,P., Lippman,C. and Nyborg,J. (1996) Helix unwinding in the effector region of elongation factor EF-Tu–GDP. *Structure*, **4**, 1141–1151.

Powers,T. and Noller,H.F. (1994) The 530 loop of 16S rRNA: a signal to EF-Tu? *Trends Genet.*, **10**, 27–31.

Rodnina,M.V. and Wintermeyer,W. (2001) Fidelity of aminoacyl-tRNA selection on the ribosome: kinetic and structural mechanism. *Annu. Rev. Biochem.*, **70**, 415–435.

Rodnina,M.V., Fricke,R. and Wintermeyer,W. (1994) Transient conformational states of aminoacyl-tRNA during ribosome binding catalyzed by elongation factor Tu. *Biochemistry*, **33**, 12267–12275.

Rodnina,M.V., Fricke,R., Kohn,L. and Wintermeyer,W. (1995) Codon-dependent conformational change of elongation factor Tu preceding GTP hydrolysis on the ribosome. *EMBO J.*, **14**, 2613–2619.

Saarma,U., Remme,J., Ehrenberg,M. and Bilgin,N. (1997) An A to U transversion at position 1067 of 23S rRNA from *Escherichia coli* impairs EF-Tu and EF-G function. *J. Mol. Biol.*, **272**, 327–335.

Schwarz,U., Möller,A. and Gassen,H.G. (1978) Induced structural transitions in tRNA. *FEBS Lett.*, **90**, 189.

Simonson,A.B. and Lake,J.A. (2002) The transorientation hypothesis for codon recognition during protein synthesis. *Nature*, **416**, 281–285.

Song,H., Parsons,M.R., Rowsel,S., Leonard,G. and Phillips,S.E.V. (1999) Crystal structure of intact elongation factor EF-Tu from *Escherichia coli* in GDP conformation at 2.05Å resolution. *J. Mol. Biol.*, **285**, 1245–1256.

Stark,H., Rodnina,M.V., Rinke-Appel,J., Brimacombe,R., Wintermeyer,W. and van Heel,M. (1997) Visualization of elongation factor Tu on *Escherichia coli* ribosome. *Nature*, **389**, 403–406.

Stark,H., Rodnina,M.V., Wieden,H., van Heel,M. and Wintermeyer,W. (2000) Large scale movement of elongation factor G and extensive conformational change of the ribosome during translocation. *Cell*, **100**, 301–309.

Triana-Alonso,F.J., Dabrowski,M., Wadzack,J. and Nierhaus,K.H. (1995) Self-coded 3′-extension of run-off transcripts produces aberrant products during *in vitro* transcription with T7 RNA polymerase. *J. Biol. Chem.*, **270**, 6298–6307.

VanLoock,M.S., Agrawal,R.K., Gabashvili,I.S., Qi,L., Frank,J. and Harvey,S.C. (2000) Movement of the decoding region of the 16S ribosomal RNA accompanies tRNA translocation. *J. Mol. Biol.*, **304**, 507–515.

Vazquez,D. (1979) *Inhibitors of Protein Synthesis*. Springer Verlag, New York, NY.

Wagenknecht,T., Grassucci,R.A. and Frank,J. (1988) Electron microscopy and computer image averaging of ice-embedded large ribosomal subunits from *Escherichia coli*. *J. Mol. Biol.*, **199**, 137–145.

Wimberly,B.T., Guymon,R., McCutcheon,J.P., White,S.W. and Ramakrishnan,V. (1999) A detailed view of a ribosomal active site: the structure of the L11–RNA complex. *Cell*, **97**, 491–502.

Wimberly,B., Brodersen,D.E., Clemons,W.M., Morgan-Warren,R.J., Carter,A.P., von Rhein,C., Hartsch,T. and Ramakrishnam,V. (2000) Structure of the 30S ribosomal subunit. *Nature*, **407**, 327–339.

Woese,C.R. (1970) Molecular mechanics of translation: a reciprocating ratchet mechanism. *Nature*, **226**, 817–820.

Woese,C.R. (2001) Translation: in retrospect and prospect. *RNA*, **7**, 1055–1067.

Wolf,H., Chinali,G. and Parmeggiani,A. (1977) Mechanism of the inhibition of protein synthesis by kirromycin. *Eur. J. Biochem.*, **75**, 67–75.

Wool,I.G., Gluck,A. and Endo,Y. (1992) Ribotoxin recognition of ribosomal-RNA and a proposal for the mechanism of translocation. *Trends Biochem. Sci.*, **17**, 266–269.

Wriggers,W., Agrawal,R.K., Drew,D.L., McCammon,A. and Frank,J. (2000) Domains motions of EF-G bound to the ribosome: insights from a hand-shaking between multi-resolution structures. *Biophys. J.*, **79**, 1670–1678.

Yarus,M. and Smith,D. (1995) tRNA on the ribosome: a waggle theory. In Söll,D. and RajBahandary,U. (eds), *tRNA: Structure, Biosynthesis and Function*. American Society for Microbiology, Washington, DC, pp. 443–468.

Yusupova,G.Z., Yusupov,M.M., Cate,J.H.D. and Noller,H.F. (2001) The path of the messenger RNA through the ribosome. *Cell*, **106**, 233–241.

Received February 25, 2002; revised April 16, 2002; accepted May 3, 2002

Supplementary Material

1. Single particle classification

The overall strategy is depicted in figure 1. All 22,905 particles were compared to a set of 83 views of two references. For the individual images the corresponding 83 correlation coefficients are calculated and the maximum coefficient is taken to define the best- matching reference view (Penczek et al., 1994). In this selection step, two volumes were used, a 70S ribosome-fMet-tRNA$_f^{Met}$ complex (Gabashvili et al., 2000) (fig. 1B) and a 13-Å resolution map of the 70S ribosome bearing Elongation Factor G (fig. 1C) in the presence of GDP and fusidic acid (M. Valle, unpublished results).

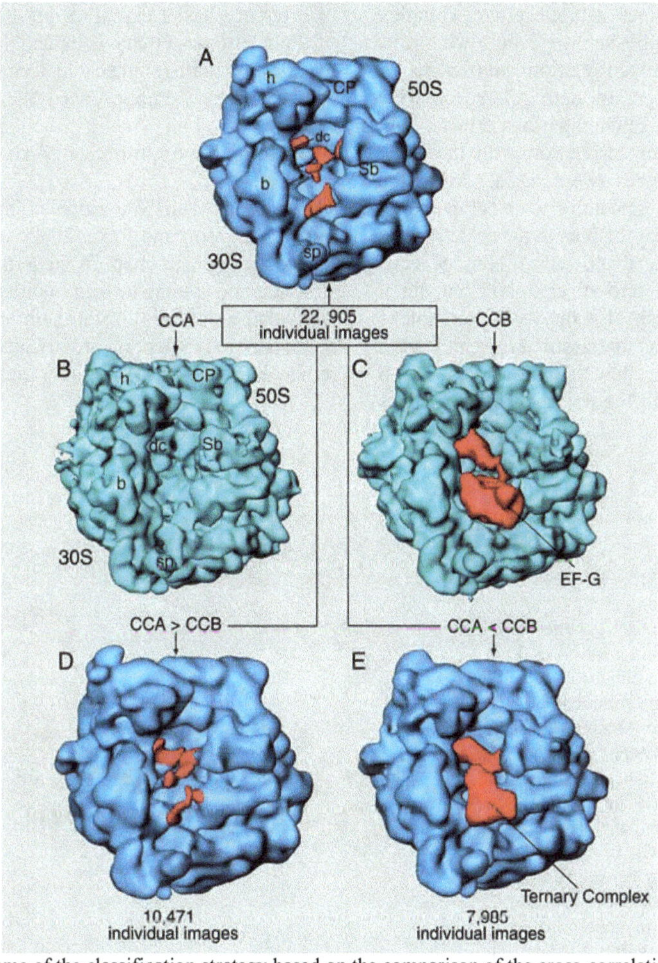

Fig. 1. General scheme of the classification strategy based on the comparison of the cross-correlation coefficients (CC) of the individual particles with two different references (CCA = with reference A, *versus* CCB = with reference B). (A) Cryo-EM map of the 70S ribosome-ternary complex using 22,905 particles. (B) Reference A. Cryo-EM map of 70S ribosome·fMet-tRNA$_f^{Met}$ complex (Gabashvili et al., 2000). (C) Reference B. Cryo-EM map of 70S ribosome·EF-G·GDP·fusidic acid complex (M. Valle, unpublished results). (D) Cryo-EM map #1, computed from 10,471 individual particles, with a resolution of 16.8 Å. (E) Cryo-EM map #2, computed from 7,985 individual particles, with a resolution of 16.8 Å.

Landmarks are as follows: CP, central protuberance; Sb, L7/L12 stalk base; sp, spur; b, body; h, head; dc, decoding center.

In the ribosome, Elongation Factor G (EF-G) binds to the same regions as EF-Tu, and both show similar mass distributions (Agrawal.,et al., 1998; 2000a; 2000b; Stark et al., 1997; 2000). Indeed, the structure of the unbound EF-G closely resembles that of the EF-Tu·tRNA·GTP complex (Nissen et al., 2000). Our strategy for separation made use of this resemblance: the projection of ribosomes carrying the ternary complex would show more similarity (i.e., higher correlation) with the EF-G bound ribosome (reference in fig. 1C) than with the one that has no factor bound (reference in fig. 1B). Conversely, it can be expected that ribosomes without ternary complex will behave in the opposite way. Following this strategy, the normalized correlation coefficients of the particles with respect to both references were compared. In order to exclude spurious effects, only particles with maximum correlation with the same view for both references were used. After the selection, new 3-D reconstructions were computed.

It is obvious that the bias in the selection has allowed us to separate a sub-set of homogeneous particles that carry the ternary complex, but did it induce an artificial density for the ternary complex? The use of reference bias in classification can lead to incorrect results when low signal-to-noise ratio images are employed. In the present work two facts rule out the possibility of a "made up" density for the ternary complex. First, despite the overall similarity between EF-G and ternary complex, it is obvious that the isolated density is different in detail from the original EF-G density that was present in one of the references. This is an indication of "real" signal for the ternary complex coming from the experimental images rather that an induced alignment of noise. Second, the same classification procedure was carried out with a control set of images from a 70S ribosome bearing a P-site tRNA but no tRNA at the A site (unpublished results). We applied the same procedure of reference-based classification as used in the case of the ribosome-ternary complex. The two final 3-D maps were undistinguishable from each other, and showed a clearly empty A site (data not shown).

2. Docking of a hypothetical 5'-stacked tRNA.

The 5'-stacked tRNA modeled by Simonson and Lake (Simonson and Lake, 2002) was docked inside the isolated density map for the ternary complex. The structure needs to be broken in order to fit the kinked anticodon arm (fig. 2c). Furthermore, in the best fit, the codon and anticodon are oriented parallel (fig 2e) rather than anti-parallel as required by base-pairing.

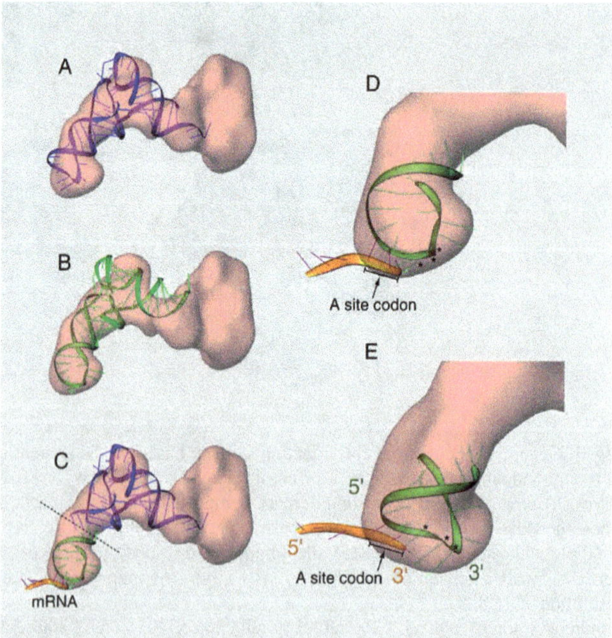

Fig. 2. Fittings of the tRNA model for the 5'-stacked tRNA (Simonson and Lake, 2002; pdb code 1KS1) either following the acceptor arm (A) or the antidocon loop (B) as main guide for the docking. Panel (C) shows a hybrid between (A) and (B). Panels (D) and (E) focus on the codon-anticodon interaction. The orientation of the anticodon end in panel (D) is as in panels (A) through (C). In order to show the co-linear proximity between codon and anticodon, the region has been rotated (E) from its orientation in panel (D) by 50 degrees around a vertical axis. The anticodon in the tRNA is labeled with asterisks in panels (D) and (E).

References

Agrawal, R. K., Penzeck, P., Grassucci, R. A., and Frank, J. (1998) Visualization of elongation factor G on the *Escherichia coli* 70S ribosome: the mechanism of translocation. *Proc. Natl. Acad. Sci.* **95**, 6134-6138.

Agrawal, R. K., Spahn, C. M. T., Penzeck, P., Grassucci, R. A., Nierhaus, K. H., and Frank, J. (2000a) Visualization of tRNA movements on the *Escherichia coli* ribosome during the elongation cycle. *J. Cell Biol.* **150**, 447-459.

Agrawal, R. K., Heagle, A. B., and Frank, J. (2000b) Studies of elongation factor G-dependent tRNA translocation by three-dimensional cryo-electron microscopy. In *The Ribosome: Structure, Function, Antibiotics and Cellular Interactions*. ASM Press, Washington, DC, pp. 53-62.

Gabashvili, I. S., Agrawal, R. K., Spahn, C. M. T., Grassucci, R. A., Svergun, D. I., Frank, J., and Penczek, P. (2000) Solution structure of *E. coli* 70S ribosome at 11.5Å resolution. *Cell* **100**, 537-549.

Nissen, P., Kjeldaarg, M., and Nyborg, J. (2000) Molecular mimicry. *EMBO J.* **19**, 489-495.

Penczek, P., Grassucci, R. A., and Frank, J. (1994) The ribosome at improved resolution: new techniques for merging and orientation refinement in 3D cryo-electron microscopy of biological particles. *Ultramicroscopy* **53**, 251-270.

Simonson, A.B. and Lake, J.A. (2002) The transorientation hypothesis for codon recognition during protein synthesis. *Nature*, **416**, 281-285.

Stark, H., Rodnina, M. V., Rinke-Appel, J., Brimacombe, R., Wintermeyer, W., and van Heel, M. (1997) Visualization of elongation factor Tu on *Escherichia coli* ribosome. *Nature* **389**, 403-406.

Stark, H, Rodnina, M. V., Wieden, H., van Heel, M., and Wintermeyer, W. (2000) Large scale movement of elongation factor G and extensive conformational change of the ribosome during translocation. *Cell* **100**, 301-309.

Cell, Vol. 113, 789–801, June 13, 2003, Copyright ©2003 by Cell Press

Study of the Structural Dynamics of the *E. coli* 70S Ribosome Using Real-Space Refinement

Haixiao Gao,[1,7] Jayati Sengupta,[2,7] Mikel Valle,[1] Andrei Korostelev,[4] Narayanan Eswar,[5] Scott M. Stagg,[6] Patrick Van Roey,[2] Rajendra K. Agrawal,[2,3] Stephen C. Harvey,[6] Andrej Sali,[5] Michael S. Chapman,[4] and Joachim Frank[1,2,3,*]
[1]Howard Hughes Medical Institute
Health Research, Inc.
[2]Wadsworth Center
Empire State Plaza
Albany, New York 12201
[3]Department of Biomedical Sciences
State University of New York at Albany
Empire State Plaza
Albany, New York 12201
[4]Kasha Laboratory
Institute of Molecular Biophysics
and Department of Chemistry and Biochemistry
Florida State University
Tallahassee, Florida 32306
[5]Departments of Biopharmaceutical Sciences and Pharmaceutical Chemistry
and California Institute for Quantitative Biomedical Research
University of California, San Francisco
San Francisco, California 94143
[6]School of Biology
Georgia Institute of Technology
Atlanta, Georgia 30332

Summary

Cryo-EM density maps showing the 70S ribosome of *E. coli* in two different functional states related by a ratchet-like motion were analyzed using real-space refinement. Comparison of the two resulting atomic models shows that the ribosome changes from a compact structure to a looser one, coupled with the rearrangement of many of the proteins. Furthermore, in contrast to the unchanged inter-subunit bridges formed wholly by RNA, the bridges involving proteins undergo large conformational changes following the ratchet-like motion, suggesting an important role of ribosomal proteins in facilitating the dynamics of translation.

Introduction

The ribosome is a highly organized dynamical macromolecule responsible for protein biosynthesis. The dynamic features can be captured by trapping functionally meaningful states and imaging the structure by cryo-electron microscopy (cryo-EM). A recent cryo-EM study of the ribosome (Frank and Agrawal, 2000) found a large conformational reorganization during the elongation cycle of protein synthesis. Following the binding of elongation factor G (EF-G), the small and large subunits undergo a relative ratchet-like motion, accompanied by many local conformational changes during tRNA and mRNA translocation. The question raised by these findings is the nature of the molecular mechanisms underlying the inter-subunit motion. Because of the limited resolution of the cryo-EM maps, currently in the range of 10 Å, the link between the observed global movement of the subunits and their local molecular reorganization is still unknown.

Recently, several atomic structures of ribosomal subunits from the archaean *Haloarcula marismortui* and the bacteria *Thermus thermophilus* and *Deinococcus radiodurans* have been determined by X-ray crystallography at resolutions around 2–3 Å (Ban et al., 2000; Wimberly et al., 2000; Schluenzen et al., 2000; Ogle et al., 2001; Harms et al., 2001). With the aid of the atomic structures of the *T. thermophilus* 30S and *H. marismortui* 50S subunits, a model for the RNA and protein backbone of *T. thermophilus* 70S ribosome was subsequently constructed based on a 5.5 Å X-ray map (Yusupov et al., 2001). The availability of these atomic crystal structures makes it possible to further investigate the ribosomal dynamics by combining structural data from both cryo-EM and X-ray crystallography.

Methods of interpretation of cryo-EM maps of large complexes employing docking and fitting of X-ray structures have received much attention recently (Volkmann and Hanein, 1999; Roseman, 2000; Rossmann, 2000; Wriggers and Birmanns, 2001). In the absence of crystal structures showing the macromolecular complex in different dynamic states, docking techniques still make it possible to construct approximate atomic models. In the present work, a stereochemically restrained real-space refinement technique (Tronrud et al., 1987; Chapman, 1995; Chen et al., 2001) was employed for docking the atomic structures of the 70S ribosome components, including comparative models of proteins, rRNA and tRNA, into the experimental density maps representing different functional states that were obtained by cryo-EM and single-particle reconstruction (Frank, 1996).

Our aim in this work was to understand the local molecular reorganization underlying the "ratchet-like" inter-subunit motion, which is triggered by the binding of EF-G in the presence of a nonhydrolyzable GTP analog ("EF-G·GTP bound state"). The two states that were analyzed are an initiation-like state, with tRNA at the P site, and the EF-G·GTP bound state. These states are not consecutive, but are separated by states associated with decoding and accommodation. Nevertheless, there are indications that the initiation-like complex has a similar conformation as the ribosome prior to EF-G binding. Real-space refinement was used to build atomic resolution models based on the cryo-EM maps of these two states. Since most of the cryo-EM studies and biochemical observations so far were made with the *E. coli* 70S ribosome for which no atomic structure is available, the *E. coli* ribosomal RNA structure was modeled from the crystal structures of the *T. thermophilus* 30S (Ogle et al., 2001) and *H. marismortui* 50S subunits (Ban et al.,

*Correspondence: joachim@wadsworth.org
[7]These authors contributed equally to this work.

2000), accounting for the species-dependent variations. In addition, ribosomal proteins from *E. coli* were modeled by comparative protein structure prediction based on their sequence similarity to structurally defined homologs from other species (Martí-Renom et al., 2000).

The building of the models was guided mainly by the experimental density maps obtained from cryo-EM. The atomic structures of RNA and protein were cut into pieces of presumed stability and fitted into the experimental density map while observing stereo-chemical constraints. In the light of the atomic models that represent the two states of the ratchet-like movement, we analyzed the local molecular conformational rearrangements of RNA and proteins. Our study shows that the most extensive molecular rearrangements are seen in the ribosomal proteins. Specifically, proteins S13, S19, L2, L5, and L14 are seen to be involved in the control of the relative movement of the subunits through their inter-subunit bridges.

Results

Structural Modeling

The three-dimensional cryo-EM density maps of *Escherichia coli* 70S ribosome complexes studied by real-space refinement are: (1) an 11.5 Å map (Gabashvili et al., 2000) of the initiation-like complex with fMet-tRNA$_f^{Met}$ at the P site (Malhotra et al., 1998) and (2) a 12.3 Å map of the EF-G·GTP bound complex, i.e., a ribosome bound with EF-G in the presence of a nonhydrolyzable GTP analog. This latter map is a higher-resolution version (M.V., unpublished data) of the earlier 18 Å map of this complex (Agrawal et al., 1999a; Frank and Agrawal, 2000).

The *E. coli* 70S ribosome consists of two subunits, which are designated as the 30S subunit, comprising 16S rRNA (1542 nucleotides) and 21 proteins, and the 50S subunit, comprising 23S rRNA (2904 nucleotides), 5S rRNA (120 nucleotides), and 36 proteins. The models of *E. coli* 23S rRNA and 5S rRNA were generated from the crystal structure of *H. marismortui* (PDB code 1FFK), while the model of *E. coli* 16S rRNA was generated from the crystal structure of *T. thermophilus* (PDB code 1IDL). The molecular modeling package Insight II (Accelrys Inc.) was used to model the sequences of the ribosomal RNA of the *E. coli* based on the crystal structures and also to model the helices 38, 43, 44, 69, and 76 of 23S rRNA, which were unresolved in the *H. marismortui* 50S structure.

All of the *E. coli* ribosomal protein structure models were calculated by ModPipe, a completely automated large-scale protein structure-modeling pipeline (Sanchez and Sali, 1998; Pieper et al., 2002). The crystal structures of the proteins from the small subunit of *T. thermophilus* (PDB code 1FJG) were chosen as the structural templates to model 19 proteins of the 30S small subunit (S2–S20) of *E. coli*. For proteins of the 50S subunit, 29 out of all the 36 *E. coli* proteins were modeled based on the 50S crystal structure of *H. marismortui* (PDB code 1JJ2; L2, L3, L4, L6, L14, L15, L16, L18, L24, L29, and L30), the 50S crystal structure of *D. radiodurans* (1LNR; L5, L11, L13, L17, L19, L20, L21, L22, L23, L27, L31, L32, L33, L34, L35, and L36), and the 70S crystal structure of *T. thermophilus* (1GIY; L9, L25), respectively. Model building began by generating a multiple sequence profile for each one of the *E. coli* sequences and the potential template structures. Next, the sequence-structure matches for model building were obtained by aligning the profile of each *E. coli* sequence against the sequences of the template structures using PSI-Blast (Altschul et al., 1997), and also by aligning the profiles of the template structures against the *E. coli* sequences using IMPALA (Schaffer et al., 1999). Alignments that had a significant e-value and covered distinct regions of the *E. coli* sequences were then chosen for modeling. Models were calculated for each one of the sequence-structure matches using MODELLER (Sali and Blundell, 1993). The resulting models were then evaluated using statistical potentials of model compactness, the sequence identity of the sequence-structure match, and energy z-scores to assess the quality of the model. The final selection of models for docking was determined from the e-value of the alignment, the quality of the final model, and the quality of the template structure. The characteristics of each protein model (sequence identity, e-value of the alignment, and model score) are listed in the Supplemental Data available at http://www.cell.com/cgi/content/full/113/6/789/DC1. Proteins L1 and L7/L12 were not included because their corresponding cryo-EM densities are not well defined, or are fragmented due to their high flexibilities. Additionally, several proteins, including L21, L32 and L35 in the 50S subunit, were excluded from the final model as the qualities of their homology models are very low. L34 was excluded because its features cannot be distinguished in the cryo-EM maps.

The starting model was built by manually docking the RNA and protein structures, each as a rigid body, into the cryo-EM density maps using the interactive program O (Jones et al., 1991). The initial position of each RNA structure and protein was taken from the position of the corresponding crystal structure template.

The program RSRef (Chapman, 1995), a real-space refinement module for the TNT program (Tronrud et al., 1987), was employed for automatically and simultaneously refining both the stereochemistry and the fit of the atomic structures to the density map. Since the resolutions of the experimental density maps are not suitable for refinement of independent atoms, a multi-rigid-body refinement was employed. The progressive cutting into smaller rigid units of RNA and protein was based on the statistics of both the stereochemistry and the fit to the density map in each round of refinement. Initially, the RNA was cut into domains and each protein molecule was treated as a single rigid body. Real-space refinement produced significant overall improvement, but a large number of the poor van der Waals contacts could not be resolved, and some local areas, such as the spur and beak regions of the 30S subunit, showed substantial conflicts with the density map. Inspection showed that the regions of poor fit to the experimental density map and an excessive number of close contacts were predominantly regions with known flexibility, species-dependent structural variations, or experimental evidence of disorder. The RNA and proteins were therefore further cut into components/blocks in those poorly fitted regions. The integrity of the double-helical regions was maintained with cuts only applied in single-stranded and loop regions. Most of the proteins were retained as single rigid units. Exceptions included S2, S7, S13 in the 30S subunit and L2, L3, L5, L6, L9, L11, L18, and L24 in the 50S subunit, which were cut into their major domains. In addition, for some proteins, namely S7, S9, S11, S12, L3, L4, L15, L17, L20, L23, L27, and L32, unstructured flexible extensions that are attached to RNA were removed to avoid severe van der Waals overlap. Typically, proteins moved by up to 10 Å during the refinement. The final assignment of RNA rigid bodies is shown in Figure 1, including 43 pieces of 16S rRNA, 62 pieces of 23S rRNA, and 4 pieces of 5S rRNA.

The 70S Ribosomal Models in Two States

An *E. coli* 70S ribosome model for the initiation-like state was generated by fitting a total of 165 fragments of RNA and proteins into the 11.5 Å EM density map using the real-space refinement algorithm. As shown in Figure 2, all of the 44 proteins, 23S rRNA, 16S rRNA, and 5S rRNA fit very well into the density map. The overall appearance of the model is very similar to the crystal structures from other species. A total of 19 proteins in the 30S subunit were fitted, most of which are near the solvent surface, except for S12, S13, and S19. In the 50S subunit, among all the 25 fitted proteins, L2, L5, L14, and L19 are located on the inter-subunit surface, forming bridges with the 30S subunit.

Similarly, a model of the *E. coli* 70S ribosome in the EF-G·GTP bound state was built from the same set of coordinates used to model the ribosome in the initiation-like state. Since protein S2 is fragmented in the cryo-EM map of the EF-G·GTP bound state and part of the mass of protein L9 is missing, a total of 162 rigid fragments of RNA and proteins were subjected to the refinement in this case.

For a more detailed appraisal of the two fitted structures of the initiation-like and the EF-G·GTP bound state, Figure 3 depicts the fittings in the maps of the 30S subunit and the 50S subunit separately. In the 30S subunit in both states (Figures 3A and 3B), the whole 16S rRNA is seen to be satisfactorily placed in the experimental maps, especially in the regions of the spur, beak, and shoulder, which are known to be the most different from the crystal structures of the 30S subunit (Ogle et al., 2001). For many of the proteins that are located near the surface and have distinct shapes, such as S2, S6, S11, and S15, the fitting was particularly straightforward and reliable. Similarly, in the case of the 50S subunit (Figures 3C and 3D), the overall models are also in excellent agreement with the density maps.

The quality of the model was assessed with three quantitative indicators: crosscorrelation coefficient, real-space R factor, and number of bad van der Waals contacts. Shown in Table 1 is a summary of the fitting statistics of the models in the initiation-like and the EF-G·GTP bound states before and after the real-space refinement. As indicated by the correlation coefficient and R factor, the real-space refinement resulted in substantial improvements in the fitting of the density maps, and the numbers of van der Waals overlaps are also reduced to a modest range (1000–2000, compared to a total 120,000 atoms). Furthermore, to assess the local fitting quality, the correlation coefficients were computed for each of the ribosomal RNA fragments and proteins (Supplemental Data available at http://www.cell.com/cgi/content/full/113/6/789/DC1).

All three RNA structures (16S rRNA, 23S rRNA, and 5S rRNA) achieved high correlation coefficients in both states: 0.74–0.80 for the initiation-like state and 0.68–0.69 for the EF-G·GTP bound state. For the proteins, with few exceptions, the correlation coefficients of the proteins are above 0.6, indicating satisfactory fitting. Among the small-subunit proteins, the correlation coefficient of S18 was the poorest. Interestingly, S18 is located near S1, which is known to have very weak association with the ribosome (Laughrea and Moore, 1977). As S1 directly interacts with S18 (Sengupta et al., 2001), it is conceivable that the detachment of S1, either partially or fully, might introduce conformational changes that affect S18. In addition, those proteins that show high mobility and undergo significant rearrangement, such as S7, S13, and S19 in the 30S subunit and L2, L9, and L13 in the 50S subunit, usually appear to have lower correlation coefficients compared to less mobile proteins. The lower correlation coefficient can be easily explained as a result of the averaging over different conformations exhibited in the sample.

Inter-Subunit Bridges

The inter-subunit bridges are important for maintaining the overall architecture of the ribosome (Frank et al., 1995a; Cate et al., 1999; Gabashvili et al., 2000), but can also be expected to play a role in the dynamics of translation (Frank and Agrawal, 2000, 2001). An earlier cryo-EM study (Gabashvili et al., 2000) identified a total of 13 inter-subunit connections based on the separation of the EM density map of the 11.5 Å initiation-like complex. Recently, Yusupov and coworkers (2001), using their 5.5 Å crystal structure, identified more than 30 individual inter-subunit interactions spread among 12 bridges in the *T. thermophilus* 70S ribosome.

In our newly generated *E. coli* model in the initiation-like state, all of the earlier identified bridges are interpreted as discrete, sometimes multiple atomic contacts. Table 2 lists the bridges and their constituent components from both subunits in the initiation-like complex, following the nomenclature of Frank and coworkers (1995a) and expanded in subsequent studies (Cate et al., 1999; Gabashvili et al., 2000; Yusupov et al., 2001). For each bridge identified in the *T. thermophilus* ribosome, there exists a counterpart in the *E. coli* structure. Furthermore, many bridges in *E. coli* are virtually identical to the bridges found in the *T. thermophilus*, in terms of their locations and the RNA helices and proteins involved. The high similarity suggests that the inter-subunit bridge regions are highly conserved across species, and that their atomic makeup is essential for function. The same observation was made in a comparison of the inter-subunit bridges between bacterial and yeast ribosomes (Spahn et al., 2001).

More than 80 percent of the bridges in the *E. coli* involve contacts with helices 23, 24 (central domain) and 44 (3′ minor domain) from the 30S subunit side (Figure 4A), and helices 64 and 67–71 (domain IV) from the 50S subunit side (Figure 4B). One of the bridge regions that

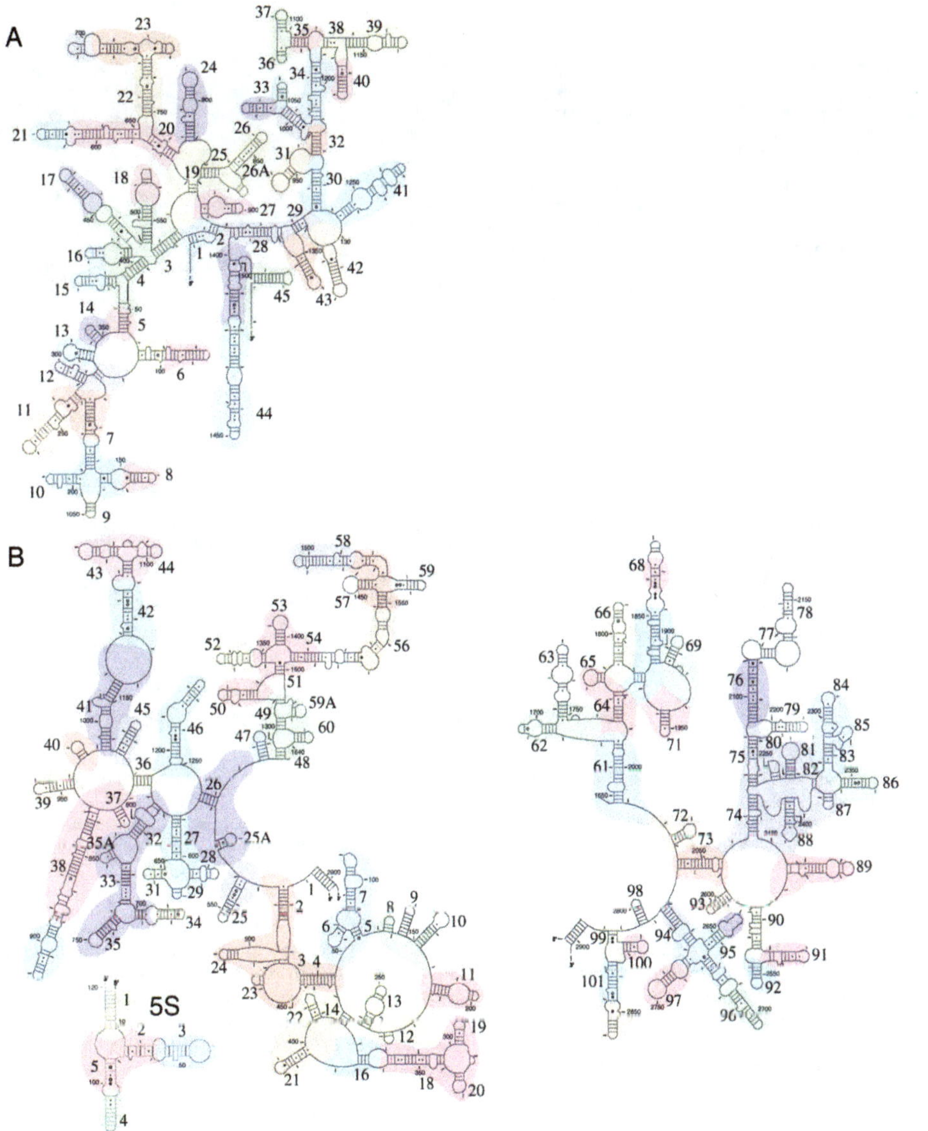

Figure 1. The Rigid Body Assignment of Ribosomal RNA Secondary Structure Elements
The rigid bodies are shaded in different colors on the secondary structure maps of (A) 16S rRNA, and (B) 23S rRNA and 5S rRNA, respectively. Uncolored regions are not included in the model (see details in the Supplemental Data available at http://www.cell.com/cgi/content/full/113/6/789/DC1).

differed between the previous *E. coli* and *T. thermophilus* structures is at the connection between S13 and L5, which is depicted as two distinct bridges, B1b and B1c, in *E. coli*, but only one connection was identified in *T. thermophilus* as B1b. However, inspection of our new model in the initiation-like state indicates that there are no significant differences between *E. coli* and *T. thermophilus* in the S13 residues that participate in this bridge. In addition, several newly identified bridging contacts include protein S19 (B1a) and helix 22 (B7b) of the 30S subunit, protein L19 (B8), and helices 66 (B2c) and 71 (B5) of the 50S subunit.

In order to investigate all the conformational changes in the bridge regions and the underlying mechanism, the bridges in the EF-G·GTP bound complex were also identified (Table 2), based on the fitted model of the 70S ribosome in this state. A comparison of the two sets of bridges in Table 2 indicates that the changes take place

Structural Dynamics of the 70S Ribosome
793

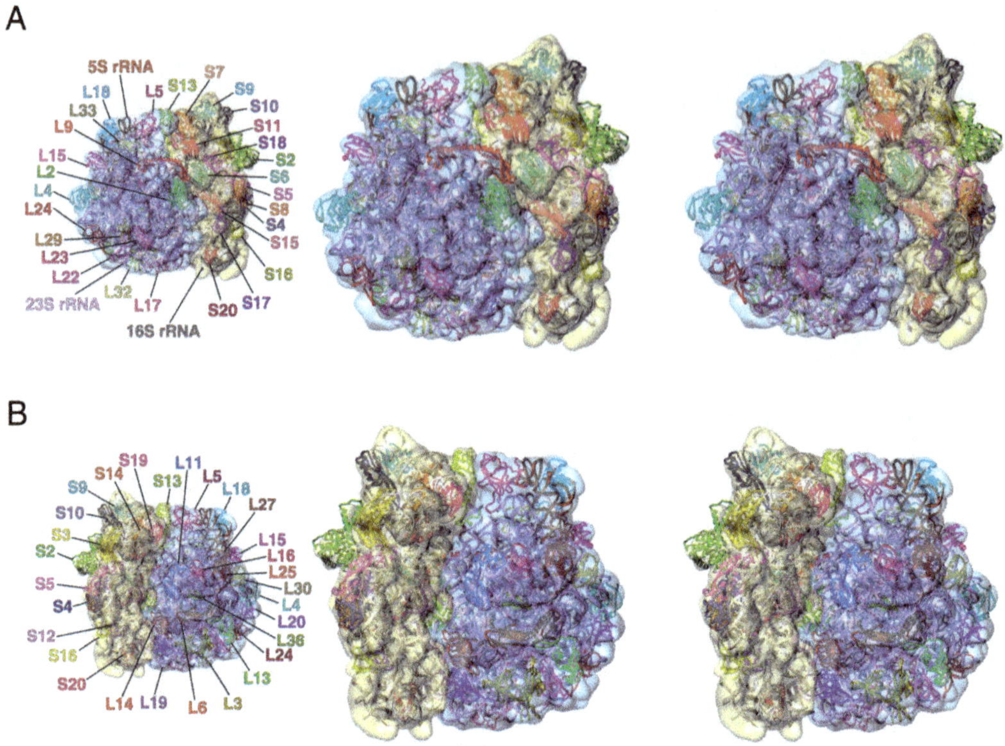

Figure 2. Stereo Views of the Overall Fitting of the Initiation-Like State of the 70S Ribosome Complex

(A) Side view, with the 30S subunit on the right and the 50S subunit on the left.

(B) View obtained by rotating the ribosome by 180° from (A) around an axis running vertical in the image plane. The ribosome is shown by semitransparent contours in yellow (30S) and blue (50S). The models of 16S rRNA (silver), 23S rRNA (purple), 5S rRNA (orange) and proteins (in varied colors) are displayed by Ribbons (Carson, 1991).

mainly in bridges involving protein-protein or protein-RNA interactions as in B1a, B1b, and B7b. B1a and B1b were earlier identified as being strongly affected by the ratchet motion (Frank and Agrawal, 2000). A detailed discussion of these bridges follows later.

Atomic Interpretation of the Ratchet-Like Movement
Superimposition of the experimental density maps of the initiation-like state and the EF-G·GTP bound state immediately reveals the ratchet-like rotation of the 30S subunit relative to the 50S subunit. The observed relative inter-subunit movement is very similar to that from the earlier study (Frank and Agrawal, 2000), except that the overall rotation angle is smaller than previously reported (4° instead of 6.5°). In order to quantitatively analyze the local movement of the ribosomal RNA for each residue of the RNA structure, we calculated its displacement from the initiation-like to the EF-G·GTP bound state on the basis of the two fitted models. We computed a displacement map (Figure 5), in which each residue was assigned a color according to its absolute displacement. Going from the blue to the red regions, the value of the local displacement is increasing, implying an increase in the size of the movement. Overall, the movement range of 16S rRNA is between 0–10 Å (Figure 5A). The large-scale rotation of the 30S subunit is represented mainly by a movement of the head (4°), the shoulder (3°), and spur region (8°). The regions showing largest movements (red areas in Figure 5A) are the head domain of the 16S rRNA, particularly helices 39, 41 and 42, and the spur (helix 6). Interestingly, the mobility of the head has been predicted both from the 16S secondary structure (Yusupov et al., 2001) and from neutron scattering (Serdyuk et al., 1992). The region showing the smallest movements (colored in blue in Figure 5A) contains the geometric center of the rotation comprising helices 2, 7, 14, 19, 21, 26, 27, and 44. Most of this relatively stable 30S subunit region participates in the formation of the centrally located inter-subunit bridges, which anchor it to the 50S subunit. In general, going from the inner region of the 16S RNA to the peripheral region, the displacement of the local structure gradually increases as indicated by the change of the assigned color from blue to yellow and then red.

The overall rotation of the 16S rRNA with respect to the 23S rRNA can be characterized in an orthogonal coordinate system by three rotational angles, around the x-axis (normal to the subunit interface), the y-axis (head to bottom), and the z-axis (shoulder to platform) with the values of 3.3°, 0.5°, and 0.7°, respectively. The

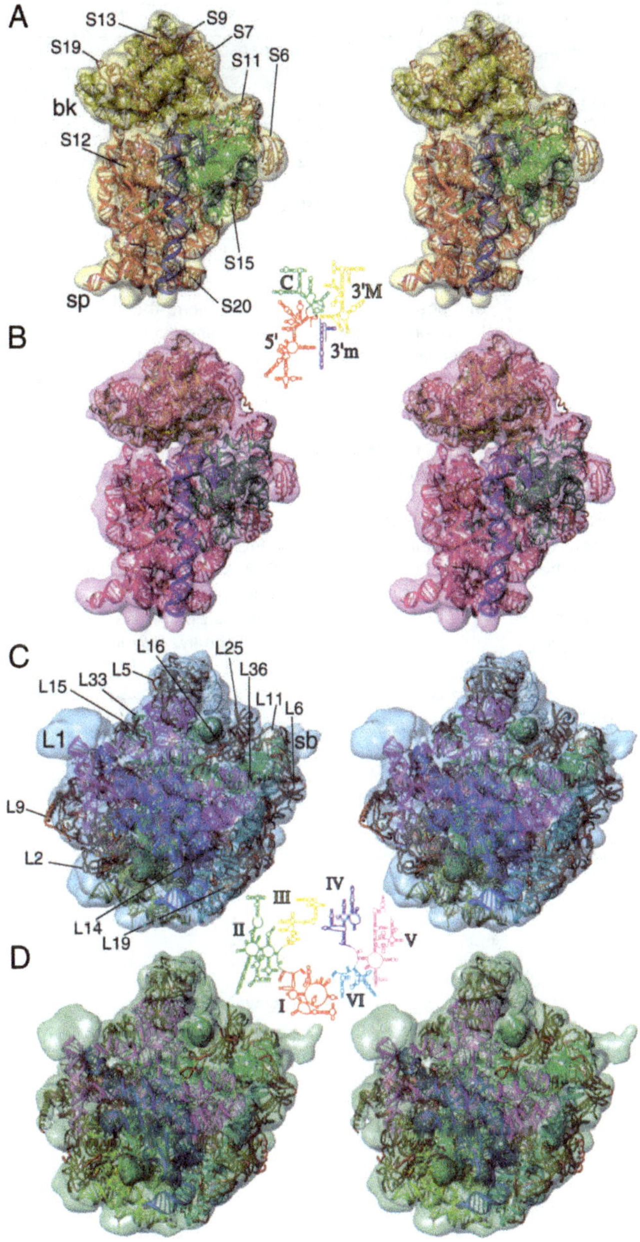

Figure 3. Stereo Views of the Fittings of the Initiation-Like Complex and the EF-G·GTP Bound Complex

The fittings of the initiation-like state complex and the EF-G·GTP bound complex are presented in pairs.

(A and B) Inter-subunit view of the 30S subunit and (C and D) inter-subunit view of the 50S subunit. The density maps are shown as semitransparent contours (30S initiation-like, yellow; 30S EF-G·GTP bound, pink; 50S initiation-like, blue; 50S EF-G·GTP bound, green). The RNA structures are colored according to domains. For 16S rRNA, these are 5′ (red), central (green), 3′ major (yellow) and 3′ minor domain (blue). For 23S rRNA and 5S rRNA, these are domain I (red), II (green), III (yellow), IV (blue), V (pink), VI (cyan) and 5S (silver). All the proteins are in orange. Landmarks: bk, beak; sp, spur; L1, protein L1; sb, L11 stalk base.

30S subunit regions that undergo substantial reorganizations in the course of the ratchet-like movement are located at the mRNA entrance and exit channels, and in the head region that forms the only protein-protein inter-subunit bridge. A comparison between the fitted models of the two states shows that these local rearrangements are mostly related to the movement of proteins.

The path of mRNA in the 30S ribosome was proposed and mapped by cryo-EM (Frank et al., 1995b; Frank et al., 2000) and X-ray crystallography (Yusupova et al., 2001). The mRNA entrance channel is formed mainly by proteins S3, S4, and S5. With protein S3 in the lower region of the head, S4 near the shoulder, and S5 at the top center of the body, these three proteins define a triangular opening between head and shoulder of the 30S subunit. The relative movements between S3, S4, and S5, from the initiation-like state to the EF-G·GTP

Table 1. Fitting Statistics of *E. coli* 70S Ribosome Models in the Two States

	Initiation-Like State	EF-G-GTP Bound State
Map resolution (Å)	11.5	12.3
Initial correlation coefficient	0.53	0.37
Final correlation coefficient	0.71	0.67
Initial R factor	0.29	0.32
Final R factor	0.23	0.24
Initial van der Waals close contacts	>10,000	>10,000
Final van der Waals close contacts	~1900	~1200

bound state (Figure 6A), result in an opening up of the mRNA entrance channel (Frank and Agrawal, 2000). Similarly, the mRNA exit channel, which is located between head and platform, and is surrounded by S7, S11, S18 and helices 23, 24, 28, and 45 of the 16S rRNA, becomes wider in the EF-G·GTP bound state than in the initiation-like state, clearly due to the relative movements between proteins and part of the head region of 16S RNA (Figure 6B). S13, the protein bridging the head of the 30S subunit with the central protuberance of the 50S subunit, stands out among all the proteins of the 30S small subunit by displaying the largest movement. The displacement of the N-terminal region of S13 is about 12 Å between the two states. In the initiation-like state, the C-terminal of S13 is close to the location of the P-site tRNA, connecting with the tip of helix 38 of 23S RNA of the 50S subunit. In the EF-G·GTP bound state, it departs from the previous position by an ~15 Å movement, thereby breaking the connection with the protruding tip of helix 38. Another region that undergoes a large conformational rearrangement comprises S6, S18, and S11, which form a rather compact cluster in the initiation-like state while changing into a looser one in the EF-G·GTP bound state.

S12, the only protein close to the decoding center (Brodersen et al., 2002), is also the only known protein in the 30S subunit that interacts with EF-G (Agrawal et al., 1998). S12 is therefore expected to have an important role during translocation. From the initiation-like to the EF-G·GTP bound state, S12 is observed to undergo a rotation by about 19 degrees toward the inter-subunit space, possibly positioning the protein such that it is able to contact EF-G. Interestingly, two independent

Table 2. Bridges between the 30S and 50S Ribosomal Subunits[a]

Bridge	Type	30S Component	RNA or Protein Positions	50S Component	RNA or Protein Positions	*E.coli* Nomenclature[b]
B1a	p-R	S13	102–115	H38	881–883[c]	(B1a)
	p-R	S19	83–84	H38	884–885,891–892	
B1b	p-p	S13	2–8	L5	110–114,130–136	(B1c)
	p-p	S13	56–67	L5	110–114,130–136	(B1b)
B2a	R-R	h44	1409–1410,1495–1496	H69	1912–1913	(B2a)
B2b	R-R	h24	783–786,791–792	H67	1836–1838,	(B2b)
				H69	1922–1923,1928–1929	
		h45	1514–1516	H67,H71	1833–1834,1932–1933	
B2c	R-R	h24	770–774	H66,H67	1793–1794,1830–1833	(B2c)
		h27	900–901	H67	1832	
B3	R-R	h44	1483–1486	H71	1948–1949,1960–1962	(B3)
B4	R-R	h20	762	H34	715–716	(B4)
	p-R	S15	59–63,86–87	H34	714–716	
B5	R-R	h44	1418–1419	H64	1718–1719	(B3a,b)
	R-R	h44	1420–1421	H71	1950–1951	
	R-p	h44	1421–1423	L14	45–50	
	R-R	h44	1473–1476	H62	1689–1690,1702–1703	
				H64	1988–1989	
B6	R-R	h44	1429–1431	H62	1704–1705	
	R-p	h44	1432–1433	L19	107–110	
	R-p	h44	1463–1465	L19	108–113	
B7a	R-R	h23	698–699,701–703	H68	1847–1848,1896–1897	(B2d)
B7b	R-p	h23	712–714	L2	240–244	(B2e)
	R-p	h24	773–776	L2	226–227,247–251	
	R-p	h22	669–671	L2	235–238	
B8	R-p	h14	339–340,345–346	L14	13,97–98	(B6)
	R-p	h14	345–346	L19	39–43	

R, RNA; p, protein; h, helices in 30S; H, helices in 50S.

[a] In the EF-G.GTP bound state, B1a and B7b are broken, and resides involved in B1b change into:

	Type	30S Component	RNA or Protein Positions	50S Component	RNA or Protein Positions	
	p-p	S13	17–22	L5	141–145	
	p-p	S13	64–70	L5	110–113,132–135	

[b] Gabashvili et al, 2000.

[c] The missing residues 886–890 from the model are very likely involved in B1a.

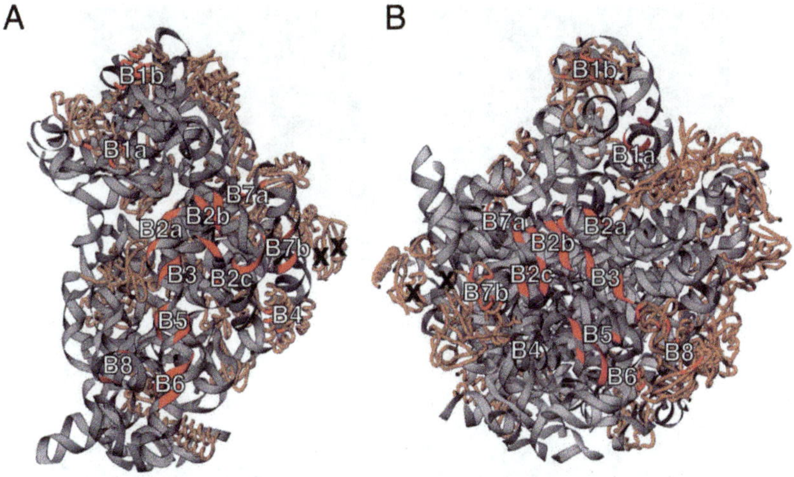

Figure 4. Inter-Subunit Bridges of *E. coli* 70S Ribosome in the Initiation-Like State

Inter-subunit view of 30S subunit (A) and 50S subunit (B), with the bridges numbered and shaded in red. X marks regions of close contact between proteins S6 and L2/L9. RNA and proteins are colored in gray and golden, respectively.

cryo-EM studies of the *E. coli* 70S ribosome bound aa-tRNA·EF-Tu·GTP complex (Valle et al., 2002; Stark et al., 2002) both reported a significant movement of S12.

All the proteins of the 30S subunit are found in both the initiation-like and EF-G·GTP bound states, except for S2. The mass attributable to S2 was partially missing from the EF-G·GTP bound state but fully presented in the initiation-like state. S2 is located at the hinge between the head and body of the 30S subunit (Brodersen et al., 2002), two regions that undergo a large relative movement. This displacement suggests that the partial absence of S2 in the map of the EF-G·GTP bound state could be associated with the local instability of the region to which it is attached.

In the case of the 50S subunit, the superimposition of the two EM density maps indicates an absence of large-scale movements. However, many local movements can be deduced from the comparison of the fitted models in the two states. As shown in Figure 5B, most of the regions in 23S rRNA and 5S rRNA appear to be less mobile than average, in contrast to the behavior of 16S rRNA, with movements generally less than 3 Å, with the exception of 5S rRNA and helices 43, 44, 76, and 86 of 23S rRNA.

Helix 76 of domain V of the 23S rRNA belongs to the so-called L1 stalk along with helices 77, 78, and protein L1. The mobility of the L1 stalk has been inferred from its different locations in different cryo-EM (Gomez-Lorenzo et al., 2000; Valle et al., submitted) and crystal structures (Harms et al., 2001; Yusupov et al., 2001). It

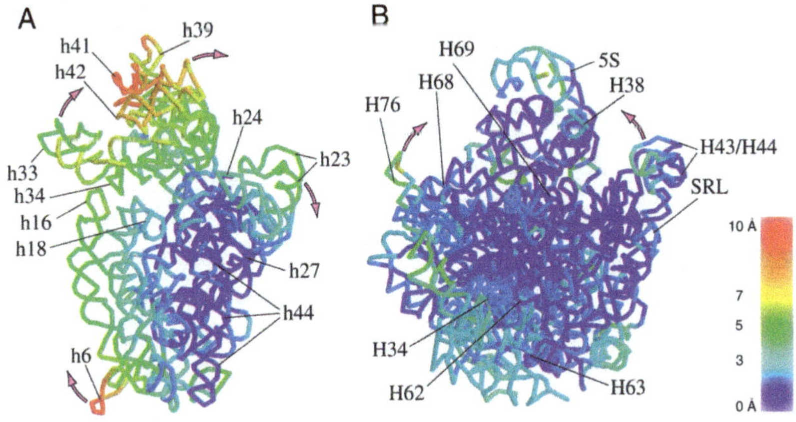

Figure 5. Displacement of RNA Structure between the Initiation-Like and the EF-G·GTP Bound States

(A) 16S rRNA, (B) 23S, and 5S rRNA. Each residue was assigned a certain color according to its displacement calculated from the models. The color bar represents the displacement value assigned to each color.

Structural Dynamics of the 70S Ribosome
797

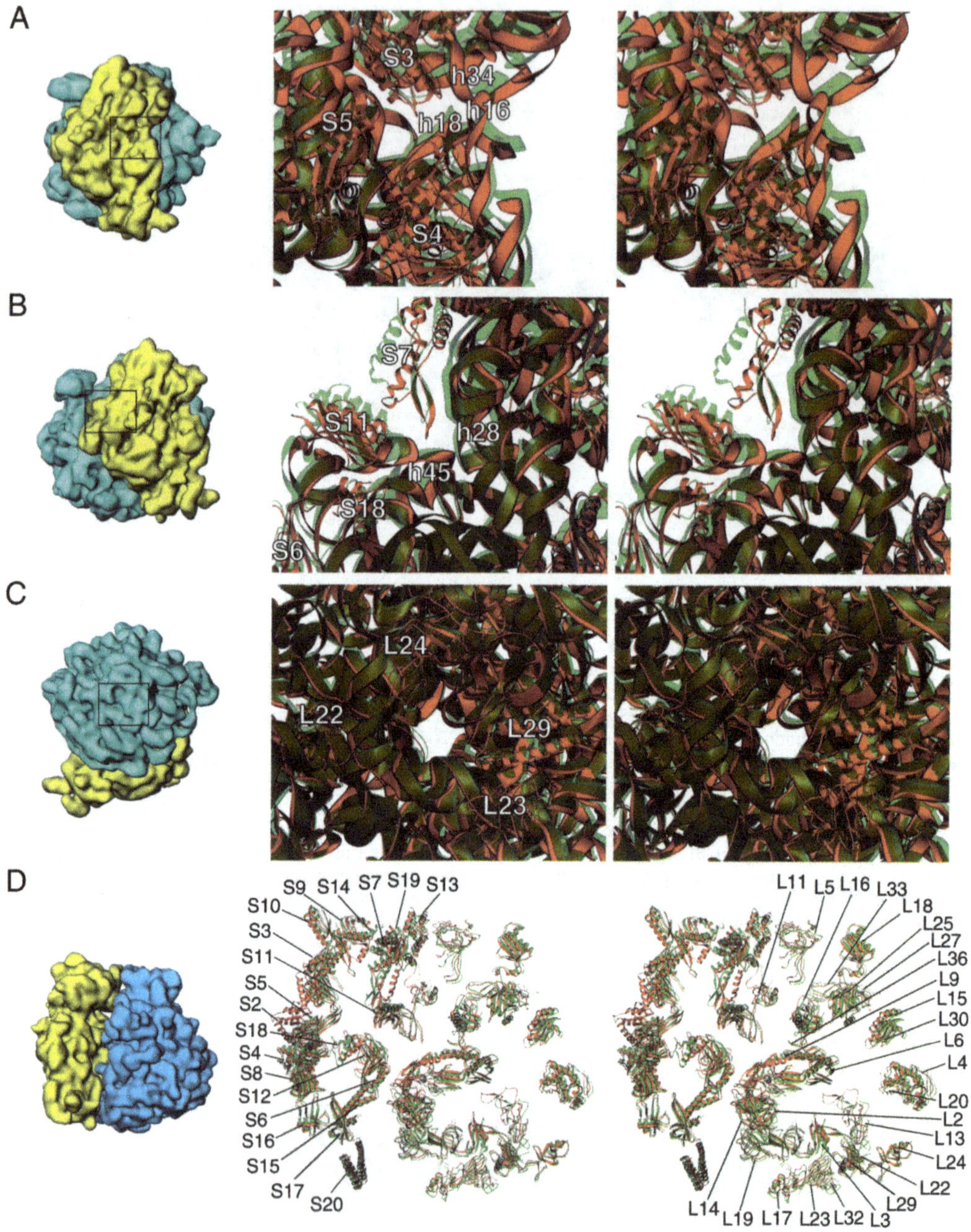

Figure 6. Stereo Views of the Protein Conformational Changes between the Initiation-Like and the EF-G·GTP Bound States

The proteins in the initiation-like and the EF-G·GTP bound states are colored in red and green, respectively.

(A) mRNA entrance channel.

(B) mRNA exit channel.

(C) Polypeptide exit tunnel.

(D) 70S proteins.

has been proposed to facilitate the release of the E-site tRNA (Agrawal et al., 1999b; Gomez-Lorenzo et al., 2000; Yusupov et al., 2001; Harms et al., 2001) and to be actively involved in the translocational movement of tRNA from the P site to the E site (Valle et al., submitted). Despite the absence of helices 77, 78, and protein L1 from our current model, a comparison of the positions of helix 76 in the two states indicates a pronounced movement as it swings by 15° toward the central protuberance (CP), in effect blocking the inter-subunit space, when going from the initiation-like to the EF-G·GTP bound state.

Forming the RNA part of the L7/L12 stalk base, helices 43 and 44 constitute another flexible region in the 50S subunit, as observed from both X-ray crystallography (Harms et al., 2001) and a cryo-EM study of EF-G-dependent translocation (Agrawal et al., 2001). From the initiation-like to the EF-G·GTP bound state, there is a movement of helices 43 and 44 toward helix 38 coupled to a 3–4 Å movement of the L11 N-terminal domain toward the G domain of EF-G (Agrawal et al., 2001). The C-terminal domain of L11, being tightly attached to helices 43 and 44, rides along with them. These movements are associated with the interaction of this region with domain V of EF-G in the EF-G·GTP bound state. In addition, the protruding tip of helix 38 shows significant movements between the two states, accompanied by the changes of the inter-subunit bridge B1a.

In the case of the 50S subunit proteins, from the initiation-like to the EF-G·GTP bound state, many proteins near the solvent surface follow an outward movement, especially those proteins that lie close to the different branches of the polypeptide exit tunnel (Frank et al., 1995b; Nissen et al., 2000; Gabashvili et al., 2001), such as L22, L23, L24, and L29 (Figure 6C).

One intriguing phenomenon emerging from this study concerns the 50S subunit proteins close to the subunit interface, in particular L5, L2, and L14. From the initiation-like to the EF-G·GTP bound state, these proteins undergo significant local movements and rearrangements. L5 is the only protein in the 50S subunit that forms a protein-protein bridge (B1b), with protein S13 of the 30S subunit. The overall movement of L5 between the initiation-like and the EF-G·GTP bound states can be characterized as a 13° rotation, apparently related to the large movement of protein S13 as the subunits rotate relative to each other. This large movement of L5 thus helps to maintain the connection between L5 and S13. Similarly, proteins L2 and L14, which are involved in B5, B7b, and B8, show high mobility. As discussed shortly, the movements of these proteins can be correlated with the crucial dynamic role of the inter-subunit bridges during translocation.

Discussion

Following the first report of the ratchet-like movement of 70S ribosome during EF-G-dependent translocation (Frank and Agrawal, 2000), similar motions have been observed in other functional ribosomal complexes (Valle et al., submitted; Spahn et al., submitted). A recent normal mode analysis of the 70S crystal structure of *T. thermophilus* (Tama et al., submitted) indicated a similar inter-subunit movement, in striking agreement with that observed experimentally, implying that this motion might be an intrinsic behavior that follows from the overall shape of the ribosome.

Due to the limitations in resolution and quality of the experimental density maps, and the fact that no structure data are available for some proteins and RNA components, the real-space refinement protocol was chosen as the best approach for exploiting the currently available structural information to the maximum extent. RSRef, the real-space refinement method selected to build models of the ribosome in different functional states, produced satisfactory results as the new models account for the experimental observations. To give an example for the pronounced sensitivity of the quality indicators (local correlation coefficient and R factor) to small positional changes, a 2 Å shift reduces the local correlation coefficient from 0.73 to 0.66 and degrades the R factor from 0.24 to 0.26, for a typical protein (L6; see Supplemental Data available at http://www.cell.com/cgi/content/full/113/6/789/DC1). As a rule, the fitting quality of the initiation-like state exceeds that of the EF-G·GTP bound state, evidently since the EF-G·GTP bound state is characterized by a looser, less stable conformation.

During the preparation of this paper, a homology model of the *E. coli* 30S ribosomal subunit based on the crystal structure of the *T. thermophilus* 30S ribosomal subunit was published (Tung et al., 2002). The crosscorrelation coefficient between the 16S rRNA of this homology model and the 11.5 Å EM density map is 0.56. Our real-space-refined model for the initiation-like state gives a substantially better value of 0.80. This discrepancy is most likely due to the lack of a restraint to the cryo-EM data in the modeling of Tung and coworkers.

The Dynamics of the Inter-Subunit Bridges

Based on examination of the inter-subunit bridges identified in the two states, the behavior of the bridges subjected to the ratchet-like inter-subunit movement can be classified into two types: (1) bridges that are rather stable and almost unchanged and (2) bridges whose molecular contacts are changed or broken as they absorb large displacements far away from the geometrical axis of rotation. The first type includes all the RNA-RNA bridges, such as B2a-c, B3, B5, and B7a. As the large-scale relative motion between the 30S and 50S subunit takes place, these bridges remain stable and show only a small (<3Å) local rearrangement. In contrast, most of the bridges involving proteins behave in quite a different way. A case in point is B1b, the only bridge formed by proteins, connecting S13 and L5. As both S13 and L5 undergo large movements during the large-scale inter-subunit movement, their residues involved in the B1b contacts apparently switch from one set to another (Table 2), likely playing the important role of regulating and stabilizing the ribosome as it goes from one state to the other. Other examples are the bridges that are well connected in the initiation-like state and subsequently broken in the EF-G·GTP bound state, including B1a and B7b. B1a is located between S13, S19 of the 30S subunit and helix 38 of the 50S subunit, and B7b is between helices 22–24 of the 30S subunit and L2 of the 50S subunit. The two bridges are both formed by RNA-protein interactions, and the disruptions are due to the protein side of the bridges undergoing large movements, while the RNA side is quite static. Interestingly, the two bridges (B5 and B8) that involve protein L14 show no significant change in terms both of the connection distance and the participating residues, although L14 itself undergoes a 14° rotation. Closer inspection reveals that the movement of the bridge is small because the bridge interaction region on protein L14 is close to the protein's rotation center.

Our results show that the inter-subunit RNA-RNA bridges in the central part of the 70S ribosome remain virtually unchanged, evidently maintaining the stability of the 70S assembly while allowing the functionally important ratchet motion to occur in response to EF-G·GTP binding. It is likely that due to the stability of the subunit connection formed by the central bridges, the ratchet motion imposes a strain on the RNA framework of the subunits that forces it to return to the "ground state" of the control.

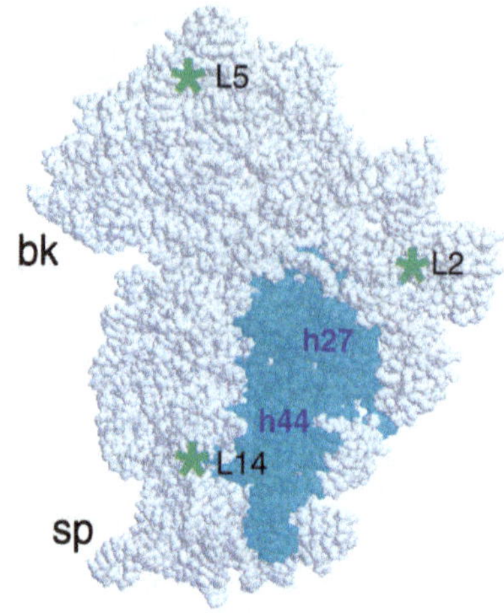

Figure 7. Diagrammatic Presentation of Interaction between the 30S Subunit and Proteins L2, L5, and L14

The interacting regions of L2, L5, and L14 on the 30S subunit are marked with stars. The geometrically rotational center region of the 30S subunit is colored in cyan. Landmarks: bk, beak; sp, spur.

The Role of Ribosomal Proteins

The importance of RNA in the translation process has been stressed since the functional core of the ribosome including the decoding center and the peptidyl-transferase center were observed to be rich in RNA in the crystal structures (Ban et al., 2000, Wimberly et al., 2000). Therefore, the role of the ribosomal proteins is now generally regarded as ancillary, mainly important for assembly of the complete 70S structure and for maintaining its stability (Ban et al., 2000; Moore and Steitz, 2002). In the analysis of the ratchet-like movement, we found that many ribosomal proteins participate directly and actively in the large conformational changes (Figure 6D), especially in important functional regions, such as the mRNA entrance and exit channel, polypeptide exit tunnel, and some of the inter-subunit bridges, in contrast to the low mobility of most of the RNA regions. From the initiation-like to the EF-G·GTP bound state, the opening up of the mRNA channel in the 30S subunit apparently facilitates the passing of the mRNA (Frank and Agrawal, 2000). An overall outward movement of the proteins in the 50S subunit is likely associated with the smaller compactness of the ribosome in the EF-G·GTP bound state.

Since the protein structure always has much more freedom than RNA, a possible explanation of the protein behaviors is that the large changes in the protein are induced by small changes in the RNA. One typical example is the switch of the residues 885 and 888 in helix 27 of 16S rRNA, which induces a large rearrangement of RNA and proteins (Gabashvili et al., 1999). While the RNA structure of the 50S subunit in most parts shows little change, especially in the central bridging regions, three proteins residing on the periphery near the inter-subunit surface, L2, L5, and L14, show considerable movement. Since these proteins are involved in bridge formation with the 30S subunit, it is conceivable that they have a pivotal role in modulating the ratchet-like movement of 30S subunit relative to the 50S subunit. Strikingly, a close check shows that all three proteins interact with functionally important RNA regions of the 50S subunit that display prominent movement. L2 is located near the base of the L1 stalk, and the observed movement of L2 from the initiation-like to the EF-G·GTP bound state is in concert with the large "swing" motion of the L1 stalk (helix 76) mentioned in the preceding section, implying that both movements are coupled. Protein L2 was proven to be functionally important in the peptidyl-transferase activity (Uhlein et al., 1998; Cooperman et al., 1995). Turning to L14, it is the only protein that interacts directly with the sarcin-ricin loop (SRL), which is in immediate vicinity to the GTPase-associated center. From the initiation-like to the EF-G·GTP bound state, L14 performs a 14 degree rotation while SRL moves by less than 3 Å, suggesting that the large movement of L14 is likely induced by the small changes of SRL. Similarly, L5 interacts directly with 5S rRNA, which is one of the RNA regions that undergo substantial changes during the ratchet-like motion. The large movement of L5 is therefore likely correlated with the changes in the 5S rRNA. More strikingly, a superimposition of the 30S subunit and the above three proteins reveals that these proteins lie close to the periphery of the 30S subunit and form a large triangle, whose center is close to the rotational center of the 30S subunit (Figure 7). Thus, the positions of L2, L5, and L14, and their considerable movements between the initiation-like and the EF-G·GTP bound states suggest that these proteins are optimally placed for modulating the rotation of the 30S subunit.

Experimental Procedures

Cryo-EM

The sample preparation, electron microscopy and image processing of the initiation-like complex were described in detail previously (Gabashvili et al., 2000). The 12.3 Å EF-G·GTP bound complex was obtained from a total of 50,854 projections of the EFG·GMPP(CH2)·

ribosome complex as described earlier (Agrawal et al., 1999a). The micrographs were recorded on a Philips FEI (Eindhoven, The Netherlands) Tecnai F20 field emission gun electron microscope at a magnification of 50,000 and at 200 kV with an electron dose of ~20e⁻/Å². After a check for drift, astigmatism, and the presence of Thon rings, micrographs were scanned on a Zeiss-Imaging scanner (Z/I Imaging Corporation, Huntsville, AL) with a sampling step corresponding to a pixel size of 2.82 Å on the object scale. The complete 3D reconstruction was done in SPIDER (Frank et al, 1996) following standard procedures. The final resolution was determined by using the Fourier shell correlation curve with a 0.5 cut off.

Both volumes were amplitude-corrected by using low-angle X-ray scattering data. Prior to the application of the real-space refinement, the EF-G fragments in the GTP state volume were removed. The two volumes were brought to the same scale, by changing the pixel size of the initiation-like volume from 2.93 to 2.82 Å, and then aligned with each other.

RSRef and TNT

The molecule was placed into a unit cell of dimensions 366.6 × 366.6 × 366.6 Å, calculated from the dimension (130 × 130 × 130 pixels) and pixel size (2.82 Å) of the cryo-EM map. The high-resolution limits were set to the resolutions of the cryo-EM density maps. The low-resolution limit was set to 100 Å for both volumes. As an all-atom structure was used in the computation, over 120,000 atoms were included in the refinement. Additional parameters included: (1) a refinement cut off radius (3.5 Å), limiting the distance between atoms and grid points used for scaling and derivative calculation and (2) a calculation cut off radius (7.5 Å) beyond which the contribution of an atom to the electron density is assumed to be zero. Computations for the real-space refinement were done on an Origin 2000 (Silicon Graphics, Mountain View, CA).

After real-space refinement, fitted coordinates were subjected to few steps of energy minimization (Discover, Accelrys Inc.) to relieve unfavorable steric interactions, without deviating from the fitted backbone atomic positions.

Acknowledgments

We thank Michael Watters and Yu Chen for preparing the illustrations. This work was supported by NIH R37 GM 29169 (J.F.), NIH R01 GM 55440 (J.F.), NIH P01 GM 64676 (T.A. Cross/MSC), NSF DBI-98-08098 (M.S.C.), NIH R01 GM 54762 (A.S.), NIH P50 GM62529, Sun Academic Equipment Grant. EDUD-7824-020257-US (A.S.), NIH R01 GM 53827(SCH), and NIH R01 GM 61576 (R.K.A.).

Received: March 10, 2003
Revised: April 17, 2003
Accepted: May 8, 2003
Published: June 12, 2003

References

Agrawal, R.K., Penczek, P., Grassucci, R.A., and Frank, J. (1998). Visualization of elongation factor G on the Escherichia coli 70S ribosome: the mechanism of translocation. Proc. Natl. Acad. Sci. USA *95*, 6134–6138.

Agrawal, R.K., Heagle, A.B., Penczek, P., Grassucci, R.A., and Frank, J. (1999a). EF-G-dependent GTP hydrolysis induces translocation accompanied by large conformational changes in the 70S ribosome. Nat. Struct. Biol. *6*, 643–647.

Agrawal, R.K., Penczek, P., Grassucci, R.A., Burkhardt, N., Nierhaus, K.H., and Frank, J. (1999b). Effect of buffer conditions on the position of tRNA on the 70S ribosome as visualized by cryoelectron microscopy. J. Biol. Chem. *274*, 8723–8729.

Agrawal, R.K., Linde, J., Sengupta, J., Nierhaus, K.H., and Frank, J. (2001). Localization of L11 protein on the ribosome and elucidation of its involvement in EF-G-dependent translocation. J. Mol. Biol. *311*, 777–787.

Altschul, S.F., Madden, T.L., Schaffer, A.A., Zhang, J., Zhang, Z., Miller, W., and Lipman, D.J. (1997). Gapped BLAST and PSI-BLAST: a new generation of protein database search programs. Nucleic Acids Res. *25*, 3389–3402.

Ban, N., Nissen, P., Hansen, J., Moore, P.B., and Steitz, T.A. (2000). The complete atomic structure of the large ribosomal subunit at 2.4 Å resolution. Science *289*, 905–920.

Brodersen, D.E., Clemons, W.M., Carter, A.P., Wimberly, B.T., and Ramakrishnan, V. (2002). Crystal structure of the 30S ribosomal subunit from Thermus thermophilus: structure of the proteins and their interactions with 16S RNA. J. Mol. Biol. *316*, 725–768.

Carson, M. (1991). Ribbons 2.0. J. Appl. Crystallog. *24*, 958–961.

Cate, J.H., Yusupov, M.M., Yusupova, G.Z., Earnest, T.N., and Noller, H.F. (1999). X-ray crystal structures of 70S ribosome functional complexes. Science *285*, 2095–2104.

Chapman, M.S. (1995). Restrained real-space macromolecular atomic refinement using a new resolution-dependent electron density function. Acta Crystallogr. *A51*, 69–80.

Chen, L.F., Blanc, E., Chapman, M.S., and Taylor, K.A. (2001). Real space refinement of acto-myosin structure from sectioned muscle. J. Struct. Biol. *133*, 221–232.

Cooperman, B.S., Wooten, T., Romero, D.P., and Traut, R.R. (1995). Histidine 229 in protein L2 is apparently essential for 50S peptidyl transferase activity. Biochem. Cell Biol. *73*, 1087–1094.

Frank, J., Verschoor, A., Li, Y., Zhu, J., Lata, R.K., Radermacher, M., Penczek, P., Grassucci, R., Agrawal, R.K., and Srivastava, S. (1995a). A model of the translational apparatus based on a three-dimensional reconstruction of the Escherichia coli ribosome. Biochem. Cell Biol. *73*, 757–765.

Frank, J., Zhu, J., Penczek, P., Li, Y., Srivastava, S., Verschoor, A., Radermacher, M., Grassucci, R., Lata, R.K., and Agrawal, R.K. (1995b). A model of protein synthesis based on cryo-electron microscopy of the E. coli ribosome. Nature *376*, 441–444.

Frank, J. (1996). Three-Dimensional Electron Microscopy of Macromolecular Assemblies (San Diego, CA: Academic Press).

Frank, J., Radermacher, M., Penczek, P., Zhu, J., Li, Y., Ladjadj, M., and Leith, A. (1996). SPIDER and WEB: processing and visualization of images in 3D electron microscopy and related fields. J. Struct. Biol. *116*, 190–199.

Frank, J., and Agrawal, R.K. (2000). A ratchet-like inter-subunit reorganization of the ribosome during translocation. Nature *406*, 318–322.

Frank, J., and Agrawal, R.K. (2001). Ratchet-like movements between the two ribosomal subunits: their implications in elongation factor recognition and tRNA translocation. Cold Spring Harbor Symposia on Quantitative Biology. (Cold Spring, NY: Cold Spring Harbor Laboratory Press), pp. 67–75.

Frank, J., Penczek, P., Grassucci, R.A., Heagle, A.B., Spahn, C.M.T., and Agrawal, R.K. (2000). Cryo-electron microscopy of the translational apparatus: experimental evidence for the paths of mRNA, tRNA, and the polypeptide chain. In The Ribosome: Structure, Function, Antibiotics and Cellular Interactions, R.A. Garrett, S.R. Douthwaite, A. Liljas, A.T. Matheson, P.B. Moore, and H.F. Noller, eds. (Washington, D.C.: ASM Press), pp. 45–51.

Gabashvili, I.S., Agrawal, R., Grassucci, R., Squires, C.L., Dahlberg, A.E., and Frank, J. (1999). Major rearrangements in the 70S ribosomal 3D structure caused by a conformational switch in 16S ribosomal RNA. EMBO J. *18*, 6501–6507.

Gabashvili, I.S., Agrawal, R.K., Spahn, C.M.T., Grassucci, R.A., Svergun, D.I., Frank, J., and Penczek, P. (2000). Solution structure of the E. coli 70S ribosome at 11.5 Å resolution. Cell *100*, 537–549.

Gabashvili, I.S., Gregory, S.T., Valle, M., Grassucci, R., Worbs, M., Wahl, M.C., Dahlberg, A.E., and Frank, J. (2001). The polypeptide tunnel system in the ribosome and its gating in erythromycin resistance mutants of L4 and L22. Mol. Cell *8*, 181–188.

Gomez-Lorenzo, M.G., Spahn, C.M.T., Agrawal, R.K., Grassucci, R.A., Penczek, P., Chakraburtty, K., Ballesta, J.P., Lavandera, J.L., Garcia-Bustos, J.F., and Frank, J. (2000). Three-dimensional cryo-electron microscopy localization of EF2 in the Saccharomyces cerevisiae 80S ribosome at 17.5 Å resolution. EMBO J. *19*, 2710–2718.

Harms, J., Schluenzen, F., Zarivach, R., Bashan, A., Gat, S., Agmon,

I., Bartels, H., Franceschi, F., and Yonath, A. (2001). High resolution structure of the large ribosomal subunit from a mesophilic eubacterium. Cell *107*, 679–688.

Jones, T.A., Zou, J.Y., Cowan, S.W., and Kjeldgaard, M. (1991). Improved methods for building protein models in electron density maps and the location of errors in these models. Acta Crystallogr. *A47*, 110–119.

Laughrea, M., and Moore, P.B. (1977). Physical properties of ribosomal protein S1 and its interaction with the 30S ribosomal subunit of Escherichia coli. J. Mol. Biol. *112*, 399–421.

Malhotra, A., Penczek, P., Agrawal, R.K., Gabashvili, I.S., Grassucci, R.A., Junemann, R., Burkhardt, N., Nierhaus, K.H., and Frank, J. (1998). Escherichia coli 70 S ribosome at 15 Å resolution by cryo-electron microscopy: localization of fMet-tRNAfMet and fitting of L1 protein. J. Mol. Biol. *280*, 103–116.

Martí-Renom, M.A., Stuart, A.C., Fiser, A., Sanchez, R., Melo, F., and Sali, A. (2000). Comparative protein structure modeling of genes and genomes. Annu. Rev. Biophys. Biomol. Struct. *29*, 291–325.

Moore, P.B., and Steitz, T.A. (2002). The involvement of RNA in ribosome function. Nature *418*, 229–235.

Nissen, P., Hansen, J., Ban, N., Moore, P.B., and Steitz, T.A. (2000). The structural basis of ribosome activity in peptide bond synthesis. Science *289*, 920–930.

Ogle, J.M., Brodersen, D.E., Clemons, W.M., Jr., Tarry, M.J., Carter, A.P., and Ramakrishnan, V. (2001). Recognition of cognate transfer RNA by the 30S ribosomal subunit. Science *292*, 897–902.

Pieper, U., Eswar, N., Stuart, A.C., Ilyin, V.A., and Sali, A. (2002). MODBASE, a database of annotated comparative protein structure models. Nucleic Acids Res. *30*, 255–259.

Roseman, A.M. (2000). Docking structures of domains into maps from cryo-electron microscopy using local correlation. Acta Crystallogr. *D56*, 1332–1340.

Rossmann, M.G. (2000). Fitting atomic models into electron-microscopy maps. Acta Crystallogr. *D56*, 1341–1349.

Sali, A., and Blundell, T.L. (1993). Comparative protein modeling by satisfaction of spatial restraints. J. Mol. Biol. *234*, 779–815.

Sanchez, R., and Sali, A. (1998). Large-scale protein structure modeling of the Saccharomyces cerevisiae genome. Proc. Natl. Acad. Sci. USA *95*, 13597–13602.

Schaffer, A.A., Wolf, Y.I., Ponting, C.P., Koonin, E.V., Aravind, L., and Altschul, S.F. (1999). IMPALA: matching a protein sequence against a collection of PSI-BLAST-constructed position-specific score matrices. Bioinformatics *15*, 1000–1011.

Schluenzen, F., Tocilj, A., Zarivach, R., Harms, J., Gluehmann, M., Janell, D., Bashan, A., Bartels, H., Agmon, I., Franceschi, F., and Yonath, A. (2000). Structure of functionally activated small ribosomal subunit at 3.3 Å resolution. Cell *102*, 615–623.

Sengupta, J., Agrawal, R.K., and Frank, J. (2001). Visualization of protein S1 within the 30S ribosomal subunit and its interaction with messenger RNA. Proc. Natl. Acad. Sci. USA *98*, 11991–11996.

Serdyuk, I., Baranov, V., Tsalkova, T., Gulyamova, D., Pavlov, M., Spirin, A., and May, R. (1992). Structural dynamics of translating ribosomes. Biochimie *74*, 299–306.

Spahn, C.M.T., Beckmann, R., Eswar, N., Penczek, P.A., Sali, A., Blobel, G., and Frank, J. (2001). Structure of the 80S ribosome from *Saccharomyces cerevisiae*--tRNA-ribosome and subunit-subunit interactions. Cell *107*, 373–386.

Stark, H., Rodnina, M.V., Wieden, H.J., Zemlin, F., Wintermeyer, W., and van Heel, M. (2002). Ribosome interactions of aminoacyl-tRNA and elongation factor Tu in the codon-recognition complex. Nat. Struct. Biol. *9*, 849–854.

Tronrud, D.E., Ten Eyck, L.F., and Matthews, B.W. (1987). An efficient general-purpose least-squares refinement program for macromolecular structures. Acta Crystallogr. *A43*, 489–501.

Tung, C.S., Joseph, S., and Sanbonmatsu, K.Y. (2002). All-atom homology model of the Escherichia coli 30S ribosomal subunit. Nat. Struct. Biol. *9*, 750–755.

Uhlein, M., Weglohner, W., Urlaub, H., and Wittmann-Liebold, B. (1998). Functional implications of ribosomal protein L2 in protein biosynthesis as shown by in vivo replacement studies. Biochem. J. *331*, 423–430.

Valle, M., Sengupta, J., Swami, N.K., Grassucci, R.A., Burkhardt, N., Nierhaus, K.H., Agrawal, R.K., and Frank, J. (2002). Cryo-EM reveals an active role for aminoacyl-tRNA in the accommodation process. EMBO J. *21*, 3557–3567.

Volkmann, N., and Hanein, D. (1999). Quantitative fitting of atomic models into observed densities derived by electron microscopy. J. Struct. Biol. *125*, 176–184.

Wimberly, B.T., Brodersen, D.E., Clemons, W.M., Morgan-Warren, R.J., Carter, A.P., Vonrhein, C., Hartsch, T., and Ramakrishnan, V. (2000). Structure of the 30S ribosomal subunit. Nature *407*, 327–339.

Wriggers, W., and Birmanns, S. (2001). Using Situs for flexible and rigid-body fitting of multi resolution single-molecule data. J. Struct. Biol. *133*, 193–202.

Yusupov, M.M., Yusupova, G.Z., Baucom, A., Lieberman, K., Earnest, T.N., Cate, J.H., and Noller, H.F. (2001). Crystal structure of the ribosome at 5.5 Å resolution. Science *292*, 883–896.

Yusupova, G.Z., Yusupov, M.M., Cate, J.H., and Noller, H.F. (2001). The path of messenger RNA through the ribosome. Cell *106*, 233–241.

Accession Numbers

Atomic models have been deposited in the Protein Data Bank under the following ID codes: 1P6G (30S, EF-G·GTP bound state); 1P85 (50S, EF-G·GTP bound state); 1P87 (30S, initiation-like state); and 1P86 (50S, initiation-like state).

A cryo-electron microscopic study of ribosome-bound termination factor RF2

Urmila B. S. Rawat*†‡, Andrey V. Zavialov‡§, Jayati Sengupta†, Mikel Valle*†, Robert A. Grassucci*†, Jamie Linde†, Bente Vestergaard||, Måns Ehrenberg§ & Joachim Frank*†¶

* *Howard Hughes Medical Institute, Health Research, Inc., Empire State Plaza, Albany NY 12201-0509, USA*
† *Wadsworth Center, Empire State Plaza, Albany, New York 12201-0509, USA*
§ *Department of Cell and Molecular Biology, BMC, Uppsala University, Box 596, S-75124 Uppsala, Sweden*
|| *Institute of Molecular and Structural Biology, University of Aarhus, DK-8000 Aarhus C, Denmark*
¶ *Department of Biomedical Sciences, State University of New York at Albany, New York 12222, USA*
‡ *These authors contributed equally to this work.*

Protein synthesis takes place on the ribosome, where genetic information carried by messenger RNA is translated into a sequence of amino acids. This process is terminated when a stop codon moves into the ribosomal decoding centre (DC) and is recognized by a class-1 release factor (RF). RFs have a conserved GGQ amino-acid motif, which is crucial for peptide release and is believed to interact directly with the peptidyl-transferase centre (PTC) of the 50S ribosomal subunit[1,2]. Another conserved motif of RFs (SPF in RF2) has been proposed to interact directly with stop codons in the DC of the 30S subunit[3]. The distance between the DC and PTC is ~73 Å. However, in the X-ray structure of RF2, SPF and GGQ are only 23 Å apart[4], indicating that they cannot be at DC and PTC simultaneously. Here we show that RF2 is in an open conformation when bound to the ribosome, allowing GGQ to reach the PTC while still allowing SPF–stop-codon interaction. The results indicate new interpretations of accuracy in termination, and have implications for how the presence of a stop codon in the DC is signalled to PTC.

A highly efficient *in vitro* system for bacterial protein synthesis was used to generate a release complex (RC) containing ribosomes with a UAA stop codon in the A site and fMet-Phe-Thr-Ile–tRNA^{Ile} in the P site[5]. To study the interaction of the RC with RF2 before and after peptide release, we obtained two types of complex: (1) by reacting wild-type RF2 (RF2(wt)) with RC, in which case the peptide was rapidly removed; (2) by reacting a mutant RF2 (RF2(GAQ)) with RC, in which case the peptide release by RF2 was inhibited. RF2(GAQ) was derived by changing the second glycine residue in the conserved GGQ motif to alanine. This mutation slows down peptide release drastically, but leaves other interactions between factor and ribosome unchanged[2].

Cryo-electron microscopy (cryo-EM) maps depicting the RC bound to RF2(GAQ) (Fig. 1a) and to RF2(wt) (Fig. 1b) were obtained at resolutions of 10.9 Å and 12.8 Å, respectively. The cryo-maps (Fig. 1a and b) revealed the presence of an electron density attributable to RF2. We see RF2 as a mass in the intersubunit space extending from the DC on the 30S subunit to the PTC on the 50S subunit. There is also extra density in the region of the L7/L12 stalk base of the large subunit that contains the GTPase-associated centre (GAC)[6]. Apart from a small change in the density directed towards the PTC (Fig. 1c and d), the open, tri-lobed shape of RF2 and its position on the ribosome are the same for the mutant (Fig. 1c) and wild-type (Fig. 1d) factors.

The open, tri-lobed shape of the RF2 cryo-EM density before (Fig. 1c) and after (Fig. 1d) peptide release is markedly different from the compact shape of the X-ray structure of RF2 (ref. 4;

Fig. 2a). The fitting of RF2 into the corresponding cryo-EM density was virtually impossible without altering the crystal structure (cross-correlation coefficient (CCC) = 0.3). Instead, we fitted the RF2 structure by allowing domains I and III and superdomain II–IV to rotate as rigid bodies around flexible regions in movements characterized by low-energy barriers[7]. Domains I and III were rotated by ~35° and ~75°, respectively, relative to superdomain II–IV. Protein domain movements in the range ~22° to ~150° have previously been observed during protein–ligand interactions[7]. We obtained the best fit (CCC = 0.73) by fitting domains I and III into the density directed towards the GAC and the PTC on the 50S subunit, respectively, while placing superdomain II–IV in the thick density strand reaching the DC on the 30S subunit (Fig. 1c and d). The flexible loop region at the tip of domain III containing the GGQ motif was fitted separately.

Domain I of RF2 lies in close proximity to the L7/L12 stalk base, which thus allows its interaction with the amino-terminal domain of protein L11 and with the associated 23S ribosomal RNA segment (Fig. 3a). This ribosomal component forms part of the GAC and is involved in the binding of antibiotics that impede ribosomal G-factor-dependent processes[8]. Interestingly, L11 (ref. 9) and the L7/L12 proteins[10] have been implicated in high-affinity RF–ribosome interactions. It is significant in this regard that binding of RF2(GAQ) to the RC induces a movement in the L11 N-terminal domain by ~7 Å, causing it to curl towards the PTC, the only major change in ribosome conformation compared with the RF2-free RC control. This movement is less pronounced for RF2(wt). Indeed, earlier genetic studies suggested correlated changes at the GAC and the PTC during the termination process[11].

RF2 domain III, with the conserved GGQ motif at its tip, is directed towards the PTC on the ribosome, and its distal end lies close to helices 89 and 93 of the 23S rRNA (Fig. 3b), as well as to the CCA end of the P-site tRNA. This is consistent with the proposal that GGQ participates in peptidyl-tRNA hydrolysis[12,13]. Helix 93 contains nucleotide A2602, situated between the CCA ends of the A- and P-site tRNAs[14]. Mutations of A2602 abolish RF-dependent hydrolysis of the ester bond in peptidyl-tRNA (A. Mankin, personal communication). The only conformational difference observed in domain III between RF2(GAQ) and RF2(wt), at the PTC, is a change in the shape of the finger-like density (Fig. 1c and d) that presumably represents the flexible GGQ-loop region. The change might be attributed to the mutation or to a conformational change in RF2 caused by peptide release. The peptide release mediated by RF2(wt) did not alter the position of the P-site or E-site tRNA.

The RF2 superdomain II–IV is directed towards the DC, and comes very close to the stop codon. Helices 34 and 44 of the 16S rRNA and protein S12, known to stabilize codon–anticodon interactions in the DC[8], are in close proximity to this domain (Fig. 3c). In addition, domain II is near helix 69 of the 23S rRNA, which participates in several intersubunit contacts[15]. The SPF motif, a loop region at the tip of domain II, is seen close to the A-site codon (Fig. 3c). This finding strongly supports the earlier conclusion, based on genetic data, that SPF is vital in stop codon recognition[3]. Direct RF–stop-codon interaction has been suggested by other lines of evidence: (1) RFs and suppressor tRNA compete for the stop codon[16], (2) cross-linking data[17] and (3) hydroxyl radical probing[18].

Because both RFs and tRNAs act at the DC and PTC, RFs have been proposed to mimic tRNAs structurally[19]. However, superposition of the two cryo-EM structures of mutant (Fig. 1c) and wild-type (Fig. 1d) RF2 with the X-ray structure of tRNA (aligning parts of the molecules interacting with the DC and PTC separately) shows that the width of the factor is greater, by ~10 Å, than that of tRNA, whereas its length along the superdomain II–IV is shorter by ~23 Å than the anticodon stem. Accordingly, the overall shape of the ribosome-bound structure of RF2 bears little resemblance to the

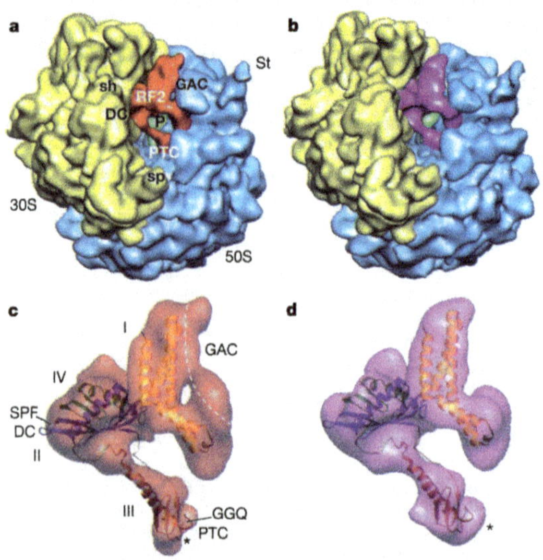

Figure 1 Interaction of RC with RF2. Three-dimensional maps of the RC with RF2(GAQ) mutant (**a**) and RF2(wt) (**b**). Yellow, 30S subunit; blue, 50S subunit. **c**, **d**, Fitting of modified X-ray structure of RF2 into cryo-density of RF2(GAQ) (**c**) and RF2(wt) (**d**). Ribosome features: sh, shoulder; sp, spur; St, L7/L12 stalk; P, P-site tRNA; DC, decoding centre; PTC, peptidyl-transferase centre; GAC, GTPase-associated centre. Colour coding for RF2: orange, RF2(GAQ); magenta, RF2(wt); gold, domain I; purple, domain II; red, domain III; green, domain IV; magenta, SPF; grey, GGQ. Flexible regions between RF2 domains: cyan, I–II; pale red, II–III; pale green, III–IV. Asterisks in **c** and **d** mark conformational changes in domain III. Density to the right of the dotted line in **c** is due to a conformational change of the ribosome.

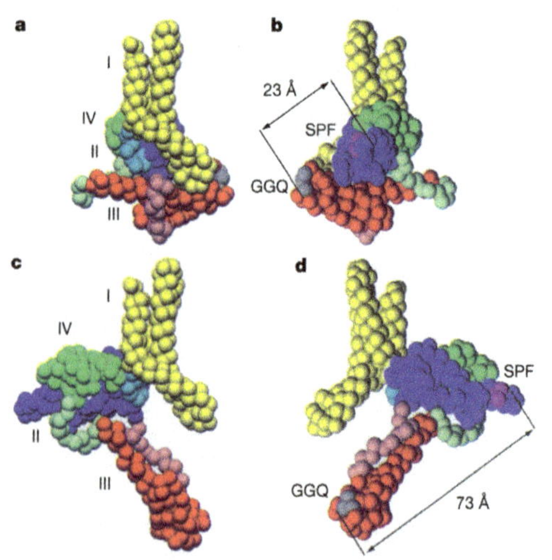

Figure 2 CPK (Corey–Pauling–Koltun) representation of RF2 domain constellations in the two conformations. **a**, **b**, X-ray structure; **c**, **d**, cryo-EM structure. The X-ray and cryo-EM structures of RF2 are shown in orientations such that domain I is aligned in each. Colour coding for RF2: gold, domain I; purple, domain II; red, domain III; green, domain IV; magenta, SPF; grey, GGQ. Flexible region between RF2 domains: cyan, I–II; pale red, II–III; pale green, III–IV.

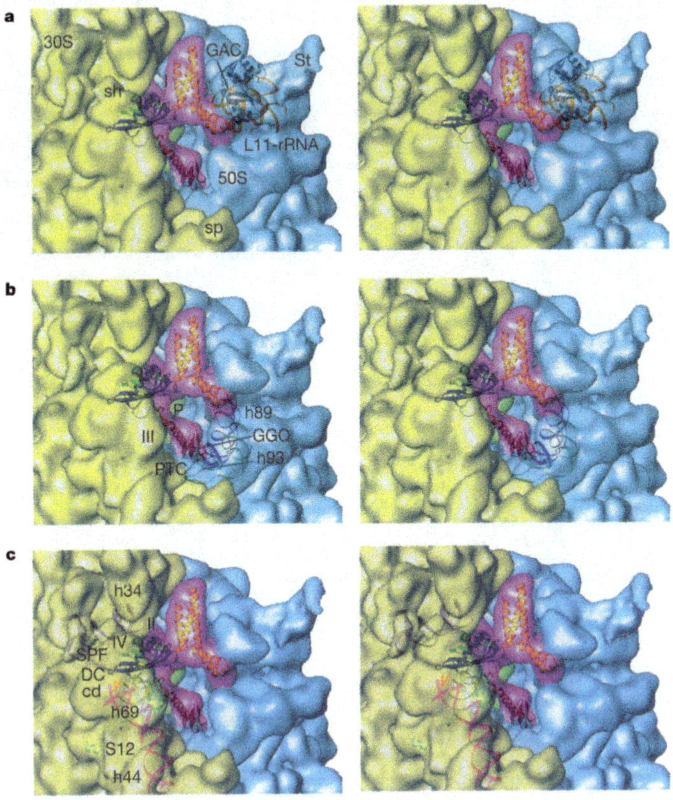

Figure 3 Stereo views showing interaction between functional centres on the RC and RF2 domains. **a**, GTPase-associated centre (GAC): RF2 domain I, protein L11 with 58-nucleotide RNA (L11-rRNA). **b**, Peptidyl-transferase centre (PTC): RF2 domain III with GGQ, helices 89, 93 and P-site tRNA (P). **c**, Decoding centre (DC): RF2 domain II with SPF, domain IV, helices 44, 34 and 69, protein S12 and A-site codon (cd). Other ribosome features: sh, shoulder; sp, spur; St, L7/L12 stalk. The colour coding for RF2 and the orientation of the complex is the same as in Fig. 1b, with the 30S subunit coloured yellow, the 50S subunit blue, and RF2(wt) magenta; all are semi-transparent.

shape of tRNA. In contrast, our data support the notion of a functional mimicry between RF2 and tRNA, in which SPF and the anticodon of tRNA interact with the mRNA, whereas GGQ and CCA interact with the PTC.

The present study demonstrates that RF2 has an open conformation in its post-termination complex with the ribosome (Fig. 2c and d) that is markedly different from its closed X-ray structure (Fig. 2a and b). In the absence of physical measurements, the conformation of RF2 in solution is unknown. However, the X-ray structures of RF2 determined independently from crystals with different cell parameters are virtually identical[4], attesting to the stability of this conformation. Furthermore, opening of the closed structure exposes conserved hydrophobic residues at the interface of domain III with superdomain II–IV, which increases the water-accessible surface by 15%. This suggests that a closed rather than an open structure is present also in solution. If so, an attractive new mechanism for the action of RFs could be proposed: RFs would enter the decoding site of the ribosome in a conformation that forms a relatively weak complex with the ribosome. When the RF encounters a sense codon in the DC it dissociates with high probability, but when it encounters its cognate stop codon it attains a fully extended structure that has high affinity for the DC (Figs 1 and 2c, d). Biochemical[2] and cryo-EM data demonstrate that the extended structures of RF2(GAQ) (Fig. 1a) and RF2(wt) (Fig. 1b) are very strongly bound to the ribosome both before and after the removal of peptide. The biological importance of strong binding of RFs for ester bond hydrolysis is emphasized by the fact that bacteria have invested in a G protein, RF3, that removes class 1 RFs after peptide release and that they consume one GTP molecule at each termination event[2,5].

According to our hypothesis, the presence of a stop codon in the DC is 'signalled' to the PTC by a conformational change of RF2, which increases the distance between the SPF and GGQ and allows the GGQ domain to reach into the PTC in the 50S subunit (Fig. 3b). This proposal has the advantage that the induction of hydrolysis of the ester bond in peptidyl-tRNA is performed by the extended, strongly bound structure of RF2 (Fig. 1a and b), and the precise codon discrimination, which requires a low affinity of RF2 for the ribosome, is performed by a compact, weakly binding form of RF2. □

Methods

RCs were generated as described[1]. Biochemical experiments were performed to study the interactions between RF2(wt) or RF2(GAQ) and RC (complex A). RF2(wt) binding was obtained when complex A (0.13 μM) was incubated with RF2(wt) (8 μM) for 2 min at 37 °C, in polymix buffer[20] (complex B). Before application to the cryo-grid, an aliquot of complex B was diluted into the same buffer containing 1 μM RF2(wt), to a final ribosome concentration of 32 nM. In parallel, another aliquot of complex B was withdrawn and subjected to scintillation counting of the [^{14}C]Ile tetrapeptide released from the P-site tRNA, on the interaction of RF2(wt) with complex A. This yielded a value of ~80% for the occupancy of RF2(wt) in complex B. In conditions similar to those described above for RF2(wt), binding of the RF2(GAQ) mutant to complex A was obtained (complex C). The occupancy of RF2(GAQ) factor was assumed to be the same as for the wild type, because the factors have similar dissociation constant (K_d) values of 0.7 and 1.8 nM, respectively. Complex A served as the control for complex C, whereas the control for complex B was obtained by incubation of complex A (0.13 μM) with puromycin (0.4 mM) in the same polymix buffer for 2 min at 37 °C (complex D). Immediately after the incubation of each complex, the samples were stored on ice and then applied to the cryo-grids at a concentration of 32 nM in a cold room at 4 °C. The micrographs were recorded on a Philips FEI (Eindhoven, The Netherlands) Tecnai F20 field emission gun electron microscope at a magnification of 50,000 × (±2%) and at 200 kV with an electron dose of ~20e$^-$ Å^{-2}. Micrographs having Thon rings that showed an absence of astigmatism and drift were selected and scanned on a Zeiss-Imaging scanner (Z/I Imaging Corporation, Huntsville, Alabama, USA) with a sampling step corresponding to a pixel size of 2.82 Å on the object scale. After an automatic particle picking procedure followed by manual and cross-correlation-based selection, totals of 14,965, 18,199, 18,958 and 13,157 ribosome particles were selected for complexes A, B, C and D, respectively. Each data set was subdivided into defocus groups and analysed with the SPIDER system[21] to obtain refined three-dimensional cryo-density maps. The final resolutions of the CTF-corrected volumes were estimated by using the Fourier shell correlation coefficient with a cut-off value of 0.5, and resolutions of 11.25, 12.8, 10.9 and 14.1 Å were obtained for complexes A, B, C and D, respectively. When assessed by the 3σ criterion[22], the resolutions were in the range 7.4–9.5 Å. The final maps were amplitude-corrected by using low-angle X-ray scattering data[23], and displayed by using IRIS Explorer (Numerical Algorithms Group Inc., Downers Grove, Illinois, USA). SPIDER was used for the computational isolation of the RF2 cryo-EM density. Docking of the X-ray structure of RF2 (pdb code 1GQE), ribosomal proteins and rRNA into the corresponding cryo-densities was performed with O (ref. 24). Individual domains of the RF2 X-ray structure were fitted into the density mass attributable to RF2, and the resulting coordinates were energy-minimized using Insight II to relieve strain from unfavourable steric interactions. Coordinates for the ribosomal proteins and rRNAs were taken from the published atomic structures for the 50S subunit[25] (pdb code 1FFK) and the 30S subunit[26] (pdb code 1FJF). Atomic coordinates for the A-site codon were obtained from ref. 27 (pdb code 1JGO). The figures were generated using IRIS Explorer, Ribbons[28] (University of Alabama at Birmingham) and Insight II LS (Accelrys Inc., San Diego, California, USA).

Received 4 July; accepted 4 October 2002; doi:10.1038/nature01224.

1. Kisselev, L. L. & Buckingham, R. H. Translational termination comes of age. *Trends Biochem. Sci.* **25**, 561–566 (2000).
2. Zavialov, A. V., Mora, L., Buckingham, R. M. & Ehrenberg, M. Release of peptide promoted by the GGQ motif of class 1 release factors regulates the GTPase activity of RF3. *Mol. Cell* **10**, 789–798 (2002).
3. Ito, K., Ebihara, K. & Nakamura, Y. A tripeptide anticodon deciphers stop codons in messenger RNA. *Nature* **403**, 680–684 (2000).
4. Vestergaard, B. *et al.* Bacterial polypeptide release factor RF2 is structurally distinct from eukaryotic eRF1. *Mol. Cell* **8**, 1375–1382 (2001).
5. Zavialov, A. V., Buckingham, R. H. & Ehrenberg, M. A posttermination ribosomal complex is the guanine nucleotide exchange factor for peptide release factor RF3. *Cell* **107**, 115–124 (2001).
6. Wimberly, B. T., Guymon, R., McCutcheon, J. P., White, S. W. & Ramakrishnan, V. R. A detailed view of the ribosomal active site: the structure of the L11-RNA complex. *Cell* **97**, 491–502 (1999).
7. Gerstein, M., Lesk, A. M. & Chothia, C. Structural mechanisms for domain movements in proteins. *Biochemistry* **33**, 6739–6749 (1994).
8. Ramakrishnan, V. Ribosome structure and the mechanism of translation. *Cell* **108**, 557–572 (2002).
9. Tate, W. P., Dognin, M. J., Noah, M., Stöffler-Meilicke, M. & Stöffler, G. The NH_2 terminal domain of *Escherichia coli* ribosomal protein L11. *J. Biol. Chem.* **259**, 7317–7324 (1984).
10. Brot, N., Tate, W. P., Caskey, C. T. & Weissbach, H. Requirement for ribosomal proteins L7 and L12 in peptide-chain termination. *Proc. Natl Acad. Sci. USA* **71**, 89–92 (1974).
11. Arkov, A. L. & Murgola, E. J. Ribosomal RNAs in translation termination: facts and hypothesis. *Biochemistry (Moscow)* **64**, 1354–1359 (1999).
12. Seit-Nebi, A., Frolova, L., Justesen, J. & Kisselev, L. Class-1 translation termination factors: invariant GGQ minidomain is essential for release activity and ribosomal binding but not for stop codon recognition. *Nucleic Acids Res.* **29**, 3982–3987 (2001).
13. Song, H. *et al.* The crystal structure of human eukaryotic release factor eRF1—mechanism of stop codon recognition and peptidyl-tRNA hydrolysis. *Cell* **100**, 311–321 (2000).
14. Nissen, P., Hansen, J., Ban, N., Moore, P. B. & Steitz, T. A. The structural basis of ribosome activity in peptide bond synthesis. *Science* **289**, 920–930 (2000).
15. Yusupov, M. M. *et al.* Crystal structure of the ribosome at 5.5 Å resolution. *Science* **4**, 897–902 (2001).
16. Drugeon, G. *et al.* Eukaryotic release factor 1 (eRF1) abolishes readthrough and competes with suppressor tRNAs at all the three termination codons in messenger RNA. *Nucleic Acids Res.* **25**, 2254–2258 (1997).
17. Tate, W. P., Greuer, B. & Brimacombe, R. Codon recognition in polypeptide chain termination: site directed cross linking of termination codon to *Escherichia coli* release factor 2. *Nucleic Acids Res.* **18**, 6537–6544 (1990).
18. Wilson, K. S., Ito, K., Noller, H. F. & Nakamura, Y. Functional sites of interaction between release factor RF1 and the ribosome. *Nature Struct. Biol.* **7**, 866–870 (2000).
19. Ito, K., Ebihara, K., Uno, M. & Nakamura, Y. Conserved motifs in prokaryotic and eukaryotic polypeptide release factors: tRNA-protein mimicry hypothesis. *Proc. Natl Acad. Sci. USA* **93**, 5443–5448 (1996).
20. Jelenc, P. C. & Kurland, C. G. Nucleoside triphosphate regeneration decreases the frequency of translation errors. *Proc. Natl Acad. Sci. USA* **76**, 3174–3178 (1979).
21. Frank, J. *et al.* SPIDER and WEB: processing and visualization of images in 3D electron microscopy and related fields. *J. Struct. Biol.* **116**, 190–199 (1996).
22. Orlova, E. V. *et al.* Structure of keyhole limpet hemocyanin type 1 (KLH1) at 15 Å resolution by electron cryomicroscopy and angular reconstitution. *J. Mol. Biol.* **271**, 417–437 (1997).
23. Gabashvili, I. S. *et al.* Solution structure of the *E. coli* 70S ribosome at 11.5 Å resolution. *Cell* **100**, 537–549 (2000).
24. Jones, T. A., Zhou, J. Y., Cowan, S. W. & Kjeldgaard, M. Improved methods for building protein models in electron density maps and the location of errors in these models. *Acta Crystallogr. A* **47**, 110–119 (1991).
25. Nissen, P., Hansen, J., Ban, N., Moore, P. B. & Steitz, T. A. The complete atomic structure of the large ribosomal subunit at 2.4 Å resolution. *Science* **289**, 905–920 (2000).
26. Wimberly, B. T. *et al.* Structure of the 30S ribosomal subunit. *Nature* **407**, 327–339 (2000).
27. Yusupova, G. Z., Yusupov, M. M., Cate, J. H. & Noller, H. F. The path of messenger RNA through the ribosome. *Cell* **106**, 233–241 (2001).
28. Carson, M. Ribbons 2.0. *J. Appl. Cryst.* **24**, 103–106 (1991).

Acknowledgements We thank M. Kjeldgaard (University of Aarhaus, Denmark) for pre-releasing the RF2 coordinates, T. Tenson (Tartu University, Estonia) for a critical reading of the manuscript and helpful suggestions, and M. Watters for assistance with the illustrations. This work was supported by the Howard Hughes Medical Institute and by grants from the National Institute of Health (to J.F.) and by the Swedish Foundation for Strategic Research and the Swedish Research Council (to A.V.Z. and M.E.).

Competing interests statement The authors declare that they have no competing financial interests.

Correspondence and requests for materials should be addressed to J.F. (e-mail: joachim@wadsworth.org). The coordinates of the modified RF2 structure, obtained by fitting individual domains of RF2 (pdb code 1GQE) into the cryo-EM density of RF2(wt), have been deposited in the Protein Data Bank with accession number 1MI6. The coordinates of the ribosomal components (1G1X, 1G1Y) interacting with the RF2 domains at the GAC, PTC and DC aligned to the cryo-EM map of the 70S *Escherichia coli* ribosome have also been deposited with accession number 1MVR. The cryo-EM maps of the RC (EMD-1006) and its interaction with the RF2(GAQ) mutant (EMD-1008) and the wild type (EMD-1007) have been deposited at the 3D-EM database, EMBL-European Bioinformatics Institute, Cambridge, UK.

REPORTS

Visualizing tmRNA Entry into a Stalled Ribosome

Mikel Valle,[1]* Reynald Gillet,[2]* Sukhjit Kaur,[1] Anke Henne,[3] V. Ramakrishnan,[2]† Joachim Frank[1,4]†

Bacterial ribosomes stalled on defective messenger RNAs (mRNAs) are rescued by tmRNA, an ~300-nucleotide-long molecule that functions as both transfer RNA (tRNA) and mRNA. Translation then switches from the defective message to a short open reading frame on tmRNA that tags the defective nascent peptide chain for degradation. However, the mechanism by which tmRNA can enter and move through the ribosome is unknown. We present a cryo–electron microscopy study at ~13 to 15 angstroms of the entry of tmRNA into the ribosome. The structure reveals how tmRNA could move through the ribosome despite its complicated topology and also suggests roles for proteins S1 and SmpB in the function of tmRNA.

During the normal course of protein synthesis, a problem occurs if the ribosome reaches the 3′ end of a defective or degraded mRNA before it encounters a stop codon. This situation has two possible consequences: the ribosome can stall, and the incomplete polypeptide made as a result may be toxic to the cell. In bacteria, both these problems are solved simultaneously by the intervention of an RNA molecule called 10Sa RNA, SsrA, or most commonly, tmRNA, because it incorporates within a single molecule the functions of both tRNA and mRNA (*1–3*). The tmRNA molecule is ~260 to 430 nucleotides long, depending on bacterial species, and contains both a tRNA-like domain (TLD) that can be charged with alanine at its 3′ CCA end and an internal stretch of RNA that contains a short open reading frame (ORF). The molecule first

[1]Howard Hughes Medical Institute, Wadsworth Center, Health Research, Inc., Empire State Plaza, Albany, NY 12201–0509, USA. [2]MRC Laboratory of Molecular Biology, Hills Road, Cambridge CB2 2QH, UK. [3]Göttingen Genomics Laboratory, Institute of Microbiology and Genetics, Grisebachstrasse 8, 37077 Göttingen, Germany. [4]Department of Biomedical Sciences, State University of New York at Albany, Empire State Plaza, Albany, New York 12201–0509, USA.

*These authors contributed equally to this work.
†To whom correspondence should be addressed. E-mail: ramak@mrc-lmb.cam.ac.uk (V.R.); joachim@wadsworth.org (J.F.)

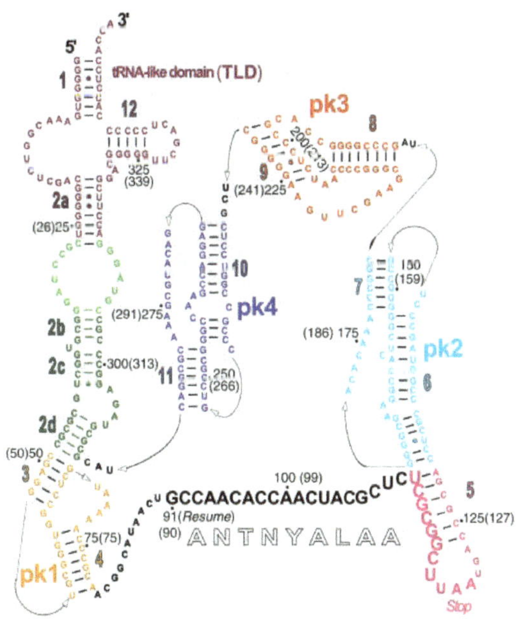

tmRNA (Thermus thermophilus)

Fig. 1. Secondary structure diagram of tmRNA. Base pairs are linked by lines, whereas GU pairs are represented by dots. The TLD, the four pseudoknots (PK1 to PK4), and helices (numbered from 1 to 12) are shown in the colors that are used to represent these modules in subsequent figures. The nucleotides within the ORF are shown in a larger font. *T. thermophilus* sequence numbers are shown, with the corresponding *E. coli* numbering in parentheses. The figure is adapted from the diagram for *E. coli* on the tmRDB Web site (*29, 41*).

REPORTS

binds in the A site of the stalled ribosome and allows its alanine to be added to the nascent polypeptide chain through its TLD. After peptidyl transfer, the TLD must translocate to the ribosomal P site. The template for translation is then switched from the original defective mRNA to the ORF of tmRNA, a process termed trans-translation. Normal termination occurs when the stop codon at the end of the ORF is reached. The result is that the stalled ribosome is rescued and the defective protein is tagged for degradation at its C-terminus (*4*). Apart from acting on defective mRNA, tmRNA has also been implicated in acting on real stop codons or internal mRNA sites (*5*, *6*). However, a recent study shows that the bacterial toxin relE cleaves rare codons on mRNA in the A site (*7*), so these cases of tmRNA rescue may also involve truncated mRNAs. Although tmRNA does not have to decode mRNA in the usual way, it has been shown that EF-Tu (elongation factor Tu) and GTP (guanosine triphosphate) bind tmRNA to form a ternary complex as they do with tRNAs (*8*, *9*). However, it has also been reported that EF-Tu cross-links to two different sites on tmRNA (*10*). Apart from EF-Tu, at least two other proteins have been implicated in tmRNA function. A small protein, SmpB, has been shown to be required for tmRNA-mediated peptide-tagging activity (*11*), and ribosomal protein S1, which is known to bind single-stranded RNA, has been proposed to be required for tmRNA binding to the ribosome (*12*).

The secondary structure of tmRNA is well established (*13–15*). In addition to the TLD and ORF, the molecule typically contains four pseudoknots, PK1 to PK4 (Fig. 1). Because of its function, tmRNA lacks an anticodon stem-loop. This is replaced by an irregular helix, H2, that is broken by an internal loop, followed by a disrupted helix that is connected to the first pseudoknot, PK1. The internal ORF follows PK1 and is partially located within helix H5. It encodes a short hydrophobic peptide that is a suitable substrate for multiple proteases (*16*). The three other pseudoknots, PK2 to PK4, connect the ORF back to the TLD. In some species, tmRNA is made in two pieces that fold together to produce a similar overall structure (*17*).

We have determined the structure of a complex of tmRNA with the ribosome by cryo-electron microscopy (cryo-EM). Kirromycin (kir) allows GTP hydrolysis by EF-Tu but stalls the EF-Tu•GDP complex with tRNA on the ribosome before the accommodation of the tRNA into the peptidyl transferase site (*18–21*). To visualize the analogous state for tmRNA, we formed a complex of the 70S ribosome from *Thermus thermophilus* with initiator tRNA and an mRNA that contains a Shine-Dalgarno sequence and ends with a start codon in the P site. This complex, with no codon in the A site, was reacted with alanylated tmRNA, SmpB, EF-Tu, and GTP in the presence of kirromycin (*22*). A cryo-EM map at ~13 Å resolution was obtained by single-particle reconstruction (Fig. 2, A and B) (*22*). Density is clearly present for the 70S ribosome, as it is for the P-site tRNA, whereas the E site is empty. By comparison of the 70S•tmRNA•EF-Tu•GDP•SmpB•kir complex with control 70S, a density attributable to EF-Tu, SmpB, and tmRNA (in red in Fig. 2, A and B) is observed at the entrance of the ribosomal inter-subunit space, between the base of the L7/L12 stalk in the 50S subunit and the decoding site in the 30S subunit. The mass protrudes prominently along the beak of the 30S subunit and forms an arc at the solvent side that reaches the vicinity of the mRNA channel entrance (*23*, *24*).

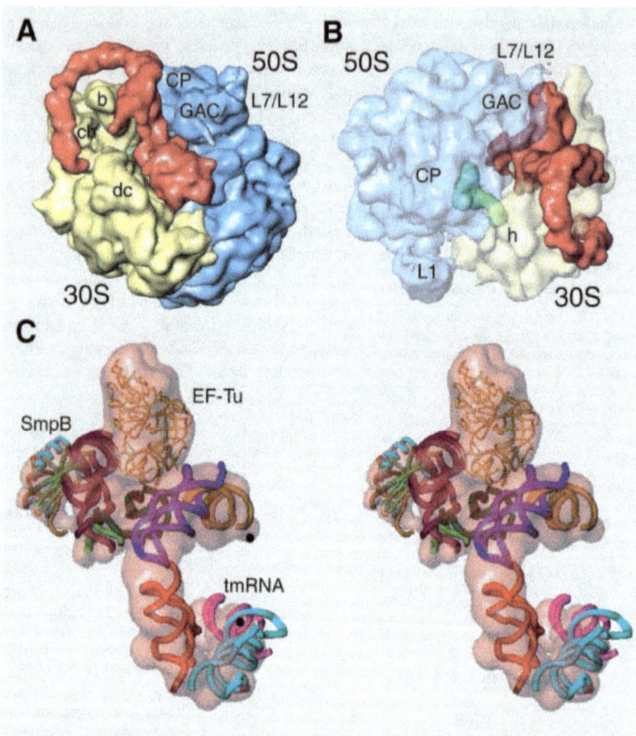

Fig. 2. Model for tmRNA, EF-Tu, and SmpB in the cryo-EM map of 70S ribosome in complex with tmRNA. Cryo-EM map obtained for 70S•tRNA•EF-Tu•tmRNA•SmpB in the presence of GTP and kirromycin (**A** and **B**). The 50S subunit is depicted in blue, the 30S subunit in yellow, and the P-site tRNA in green. Semitransparent ribosomal subunits in (B) show the relative position between EF-Tu•tmRNA•SmpB and P-site tRNA. Landmarks on the 50S subunit: L1, stalk of protein L1; CP, central protuberance; L7/L12, stalk of the proteins L7/L12; and GAC, GTPase-associated center. Landmarks on the 30S subunit: h, head; b, beak; dc, decoding center; and ch, entrance of mRNA channel. The density attributable to the EF-Tu•tmRNA•SmpB complex is colored in red and depicted semitransparent in the stereo pair of (**C**) (orientation as in panel 2B). The docked atomic coordinates are shown in a ribbons representation generated with the programs RIBBONS or Insight. The fitting of the coordinates for EF-Tu•GTP analog [Protein Data Bank (PDB) code 1TTT] (*25*) and the 3D model for tmRNA (*29*) did not account for the whole density. The coordinates for SmpB (PDB code 1K8H) (*26*) were satisfactorily docked in the unexplained region. The ribbon colors for elements of tmRNA model are the same as in the other figures: TLD, burgundy; loop within helix H2, light green; helix H2b-d, dark green; PK1, orange; PK4, dark blue; PK3, red; PK2, light blue; and helix H5, magenta.

Along with the secondary structure of tmRNA, the previously known structures of the ternary complex of EF-Tu (*25*), SmpB (*26*), the 30S subunit and 70S ribosome-tRNA complex from *T. thermophilus* (*27*, *28*), and the cryo-EM structure of the ternary complex bound to the *Escherichia coli* ribosome (*20*, *21*) allowed us to interpret the cryo-EM density map. In addition, a three-dimensional model in the tmRNA database (tmRDB) (*29*) was useful in providing model structures of the helices and pseudoknots suitable for fitting into the cryo-EM map as modules. The resulting assignments of these elements within the cryo-EM density map are depicted in the model shown in Fig. 2C. Because a mass shaped like EF-Tu occurs in a location virtually identical to that in the cryo-EM structure of the ternary complex bound to the ribosome (*19–21*), EF-Tu could readily be placed. This immediately allowed the placement of the TLD into the density, because its contacts with EF-Tu are similar to those of tRNA. However, in tmRNA an increased angle of ~110° between the acceptor and "anticodon" arms (rather than the ~90° of canonical tRNAs) was predicted by transient electric birefringence data (*30*). This increased angle is in good agreement with the density obtained in the present work. Although no prior evidence on the role of kirromycin on EF-Tu bound to tmRNA existed, we reasoned that its effect on EF-Tu would be the same as with canonical tRNAs, and this effect indeed is observed. As expected, EF-Tu•GDP is bound to tmRNA by interacting with its acceptor arm, as with tRNAs. This result suggests that, even though tmRNA does not participate in normal decoding of mRNA, GTP hydrolysis by EF-Tu is essential for its function, just as it is for tRNAs. Cross-linking experiments have suggested unusual modes of interaction of EF-Tu with tmRNA, including a second binding site for EF-Tu (*10*). These interactions are not supported by our model for EF-Tu binding, which is very similar to that of tRNA. However, it is possible that such interactions could occur outside the ribosome.

With the position of the TLD defined, and taking into account the connectivity in both tmRNA's secondary structure and the cryo-EM density, we could assign all the elements in the tmRNA molecule inside the isolated density (Fig. 2C). Furthermore, after the placement of EF-Tu and tmRNA in the three-dimensional (3D) map, additional density in the vicinity of the acceptor arm of the TLD was still unexplained. This density was assigned to SmpB on the basis of its size, shape, and previous biochemical data showing that the protein interacts with the TLD (*31*, *32*). The NMR structure of SmpB from *Aquifex aeolicus* (*26*) was positioned into the density in a preferred orientation (Fig. 2C). The position of the docked protein agrees with previous footprinting data as well as mutational analysis, which suggests that the protein interacts with the elbow and the lower portion of TLD through a contact surface that includes the C-terminus of the protein (*31*, *32*). In our interpretation of the cryo-EM map, the protein bridges the TLD and helices H69, H71, and H89 of the 23S rRNA (Fig. 3A), thus enhancing the binding of tmRNA to the ribosome. In the delivery of canonical tRNAs, the interaction of H69 is established with the tRNA itself (*20*, *21*); thus, SmpB seems to complement the TLD to facilitate this contact. The close interactions of SmpB with the 50S subunit and the TLD suggest that SmpB may facilitate GTPase (guanosine triphosphatase) activation by EF-Tu even in the absence of a normal decoding signal from codon-anticodon interactions. The proposal that two or three copies of SmpB may interact at the same time with tmRNA (*33*) is not supported by the current maps, but its occurrence cannot be excluded during other stages of trans-translation.

In our model, the TLD interacts with the GTPase-associated center (GAC) in the 50S subunit (Fig. 3A) and with protein S12 in the 30S subunit (Fig. 3B), in a way similar to that of tRNA during the elongation cycle (*20*, *21*). In agreement with a previous suggestion (*30*), the increased angle of ~110° between the acceptor and "anticodon" arms of the TLD places a highly conserved internal loop between helices 2a and 2b deep into the decoding site. It is bound to the vicinity of the tip of helices h44 and h38 from the 16S rRNA of the 30S subunit, with a resulting sharp turn in tmRNA's structure (Figs. 2C and 3B). It is possible that, independent of a codon, interactions of this region of the tmRNA with the 30S subunit trigger conformational changes in the decoding site and the GTP hydrolysis by EF-Tu, leading to the movement of the TLD into the peptidyl-transferase site. The protein SmpB may also play a role in this process. A direct interaction between H69 of the 50S subunit and elongator tRNA (*20*, *21*) is mediated through SmpB in the case of tmRNA. Because H69 forms an intersubunit bridge with h44 on the 30S subunit close to the decoding site, this interaction may be important for transmission of the decoding signal to EF-Tu. The rest of helix 2 of tmRNA is redirected along the 30S subunit beak, ending in the fork between the pseudoknots PK1 and PK4 near the top of the beak.

The single-stranded RNA that includes part of the ORF is not visible, but it presumably connects PK1 to helix 5, approximately in the regions labeled with black dots in Fig. 2C. However, it is apparent that in this state, the resume codon of the ORF cannot yet have entered the decoding site in the 30S subunit and will do so only after accommodation of the TLD into the peptidyl-transferase site. The 3′ end of the ORF forms part of the double-stranded helix 5. This part is close to the entrance of the mRNA channel in the 30S subunit (*23*, *24*). However, the tmRNA ternary complex analyzed here is representative of the initial selection step for the TLD. The movements that result from accommodation of the TLD, peptidyl transfer, and translocation of the TLD into the P site will need to pull PK1 and its downstream sequence toward the decoding site to place the resume codon in the correct position for decoding in the A site. It is very likely that in this position, interactions between the highly conserved PK1 downstream region and ribosomal elements are required for the correct positioning of the resume codon in the decoding site and resumption of translation from the ORF.

The remaining string of pseudoknots PK4-PK2 and helix 5 form the aforementioned arc that wraps around the beak of the 30S subunit

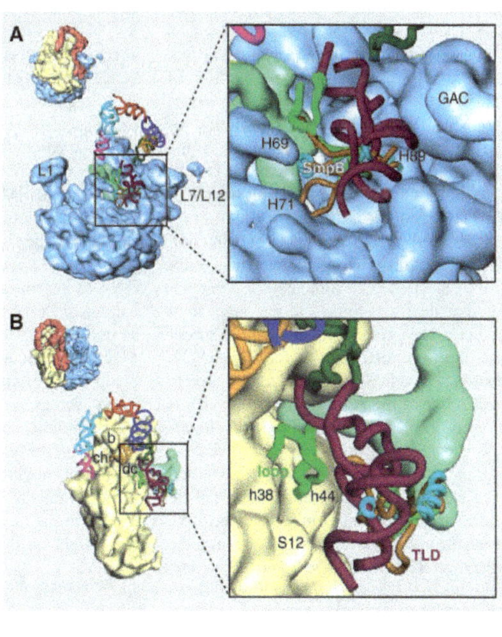

Fig. 3. Interactions of SmpB•tmRNA with the 70S ribosome. Two different ribosomal orientations are used to illustrate the interactions of SmpB and tmRNA with the 50S subunit (**A**) and the 30S subunit (**B**). The orientations are defined by ribosomal thumbnails on the left. Landmarks on the 50S subunit (A): L1, L1 stalk; L7/L12, stalk of proteins L7/L12; GAC, GTPase-associated center; H69, H71, and H89, helices of the 23S rRNA. Labeling on the 30S subunit (B): ch, entrance of the mRNA channel; dc, decoding site; b, beak; loop, RNA loop within helix H2 of tmRNA; TLD, tRNA-like domain of tmRNA; S12, ribosomal protein S12; h44 and h38, the position of those helixes from 16S rRNA. Color coding for the tmRNA is as in Fig. 2C. For clarity, EF-Tu is not shown.

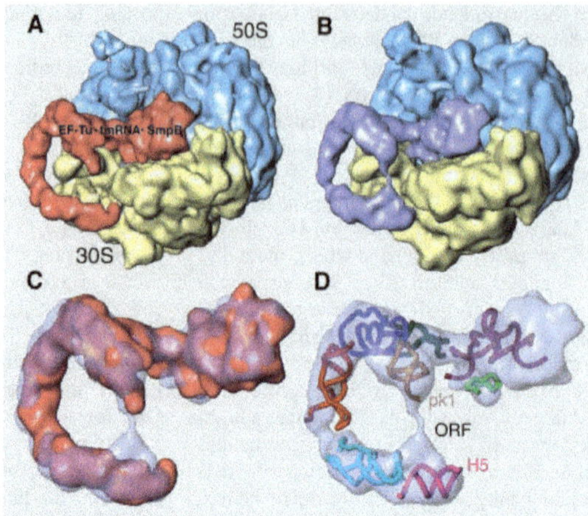

Fig. 4. Comparison of maps of tmRNA complexes with and without S1. Cryo-EM maps for the 70S•tRNA•EF-Tu•tmRNA•SmpB complex in the presence (**A**) and absence (**B**) of ribosomal protein S1. Different colors for the density attributable to EF-Tu•tmRNA•SmpB are used to allow a comparison in (**C**), where the purple semitransparent map from (B) is depicted together with the solid red from (A). In the absence of S1 (B) an additional connection is seen between PK1 and helix H5 of the tmRNA (**D**) that reveals the presence of the folded ORF.

(Figs. 2C and 3B). Those elements are not required for tmRNA function and can be replaced by single-stranded RNA, which suggests that they are involved in stabilizing tmRNA rather than participating directly in its function (*34*). Our model provides a structural explanation of this finding, because the dispensable pseudoknots PK2 to PK4 form the arc outside the head of the 30S subunit without forming extensive contacts with it. Such an absence of extensive contacts could be required to allow movement of the arc while the ORF and TLD progress inside the ribosome during trans-translation. Nevertheless, the existence of the pseudoknots suggests that they play a role in the presentation of the tmRNA to the ribosome and in helping to orient the ORF and TLD. These pseudoknots are positioned such that, despite their large size, they would not interfere with the subsequent binding of elongation factors and tRNAs. Our model supports the idea that the unwinding of RNA from helix 5 allows movement of the ORF through the ribosome without requiring a large change in the conformation of the rest of tmRNA.

Ribosomal protein S1 is thought to be associated with tmRNA and to be required for tmRNA function (*12*). Although the 70S complex described was formed in the presence of S1, the protein could not be localized in our map, either on tmRNA or on the 30S subunit. Ribosomal protein S1 was localized in *E. coli* in the shoulder of the 30S subunit (*35*), which is far away from the position that tmRNA displays in our map. However, it is known that the association of S1 with the ribosome is weak and reversible (*36*). A separate reconstruction of the identical tmRNA complex in the absence of S1 reveals interesting features (Fig. 4). First, the ternary complex of tmRNA can stably bind ribosomes in the absence of S1 (Fig. 4B). However, in this case, an extension of density from PK1 is now visible, suggesting that a portion of the single-stranded region including the 5′-end of the ORF is more structured in the absence of S1. By contrast, in the presence of S1 (Fig. 4A) this element cannot be seen, probably because it is unfolded. This difference suggests that S1, without being bound to the ribosome, may play a role in unwinding part of the single-stranded region, thus allowing correct presentation of the ORF. This role would be consistent with prior biochemical evidence showing that S1 can disrupt helical regions in mRNAs (*37*). Previous UV-induced cross-linking experiments showed the importance of nucleotide U85 from *E. coli* tmRNA in the binding of S1 (*12*). This nucleotide is situated in the vicinity of the 86 to 88 span, a major determinant for the correct positioning of the resume codon at nucleotides 90 to 92 (*38*, *39*). Similarly, nucleotides in, and upstream of, the resume codon become more accessible to structural probes when S1 binds (*40*). Regardless of the exact role of S1, it is clear that a stable complex of tmRNA with the ribosome can be formed in the absence of S1, contrary to previous suggestions (*12*).

The cryo-EM structure presented here addresses a number of interesting questions regarding the functioning of tmRNA. It shows that EF-Tu brings aminoacylated tmRNA to the ribosome in a manner similar to canonical tRNAs. The protein SmpB bridges tmRNA and the 50S subunit. A mimicry of codon-anticodon base pairing between distal parts of tmRNA is shown to be unlikely; rather, an internal loop between helices 2a and 2b is situated close to the decoding site. Protein S1 is potentially involved in unwinding a portion of the ORF. In addition, the cryo-EM structure reveals the locations of the pseudoknots and sheds some light on their individual roles.

References and Notes

1. A. W. Karzai, E. D. Roche, R. T. Sauer, *Nature Struct. Biol.* **7**, 449 (2000).
2. R. Gillet, B. Felden, *Mol. Microbiol.* **42**, 879 (2001).
3. J. H. Withey, D. I. Friedman, *Curr. Opin. Microbiol.* **5**, 154 (2002).
4. K. C. Keiler, P. R. Waller, R. T. Sauer, *Science* **271**, 990 (1996).
5. E. D. Roche, R. T. Sauer, *EMBO J.* **18**, 4579 (1999).
6. ———, *J. Biol. Chem.* **276**, 28509 (2001).
7. K. Pedersen *et al.*, *Cell* **112**, 113 (2003).
8. J. Rudinger-Thirion, R. Giege, B. Felden, *RNA* **5**, 989 (1999).
9. S. Barends, J. Wower, B. Kraal, *Biochemistry* **39**, 2652 (2000).
10. M. I. Zvereva *et al.*, *J. Biol. Chem.* **276**, 47702 (2001).
11. A. W. Karzai, M. M. Susskind, R. T. Sauer, *EMBO J.* **18**, 3793 (1999).
12. I. K. Wower, C. W. Zwieb, S. A. Guven, J. Wower, *EMBO J.* **19**, 6612 (2000).
13. K. P. Williams, D. P. Bartel, *RNA* **2**, 1306 (1996).
14. C. Zwieb, I. Wower, J. Wower, *Nucleic Acids Res.* **27**, 2063 (1999).
15. B. Felden *et al.*, *RNA* **3**, 89 (1997).
16. S. Gottesman, E. Roche, Y. Zhou, R. T. Sauer, *Genes Dev.* **12**, 1338 (1998).
17. K. C. Keiler, L. Shapiro, K. P. Williams, *Proc. Natl. Acad. Sci. U.S.A.* **97**, 7778 (2000).
18. M. V. Rodnina, R. Fricke, L. Kuhn, W. Wintermeyer, *EMBO J.* **14**, 2613 (1995).
19. H. Stark *et al.*, *Nature* **389**, 403 (1997).
20. M. Valle *et al.*, *EMBO J* **21**, 3557 (2002).
21. H. Stark *et al.*, *Nature Struct. Biol.* **9**, 849 (2002).
22. See Supporting Online Material.
23. J. Frank *et al.*, *Biochem. Cell Biol.* **73**, 757 (1995).
24. G. Z. Yusupova, M. M. Yusupov, J. H. Cate, H. F. Noller, *Cell* **106**, 233 (2001).
25. P. Nissen *et al.*, *Science* **270**, 1464 (1995).
26. G. Dong, J. Nowakowski, D. W. Hoffman, *EMBO J.* **21**, 1845 (2002).
27. B. T. Wimberly *et al.*, *Nature* **407**, 327 (2000).
28. M. M. Yusupov *et al.*, *Science* **292**, 883 (2001).
29. B. Knudsen, J. Wower, C. Zwieb, J. Gorodkin, *Nucleic Acids Res.* **29**, 171 (2001).
30. S. M. Stagg, A. A. Frazer-Abel, P. J. Hagerman, S. C. Harvey, *J. Mol. Biol.* **309**, 727 (2001).
31. S. Barends, A. W. Karzai, R. T. Sauer, J. Wower, B. Kraal, *J. Mol. Biol.* **314**, 9 (2001).
32. K. Hanawa-Suetsugu, M. Takagi, H. Inokuchi, H. Himeno, A. Muto, *Nucleic Acids Res.* **30**, 1620 (2002).
33. J. Wower, C. W. Zwieb, D. W. Hoffman, I. K. Wower, *Biochemistry* **41**, 8826 (2002).
34. N. Nameki, T. Tadaki, H. Himeno, A. Muto, *FEBS Lett.* **470**, 345 (2000).
35. J. Sengupta, R. K. Agrawal, J. Frank, *Proc. Natl. Acad. Sci. U.S.A.* **98**, 11991 (2001).
36. M. Laughrea, P. B. Moore, *J. Mol. Biol.* **112**, 399 (1977).
37. D. G. Bear *et al.*, *Proc. Natl. Acad. Sci. U.S.A.* **73**, 1824 (1976).
38. K. P. Williams, K. A. Martindale, D. P. Bartel, *EMBO J.* **18**, 5423 (1999).
39. S. Lee, M. Ishii, T. Tadaki, A. Muto, H. Himeno, *RNA* **7**, 999 (2001).
40. V. Bordeau, B. Felden, *Biochimie* **84**, 723 (2002).
41. The tmRDB Web site, http://psyche.uthct.edu/dbs/tmRDB/tmRDB.html.
42. We thank T. Ueda for a clone of AlaRS, and M. Tarry, F. Murphy, and D. Brodersen for purified 70S ribosomes, EF-Tu, and S1, respectively. We thank R. Grassucci for taking micrographs of a control, M. Watters for assistance with the illustrations, and A. Verschoor for helpful comments on the manuscript. V.R. was supported by grants from the Medical Research Council (UK), GM 44973 from NIH, and the Human Frontier Science Program. R.G. was the recipient of an EMBO postdoctoral fellowship. J.F. was supported by Howard Hughes Medical Institute and grants GM29169 and GM55440 from NIH.

Supporting Online Material

www.sciencemag.org/cgi/content/full/300/5616/127/DC1
Materials and Methods
Fig. S1
References

20 December 2002; accepted 13 February 2003

ARTICLES

nature
structural biology

Incorporation of aminoacyl-tRNA into the ribosome as seen by cryo-electron microscopy

Mikel Valle[1,7], Andrey Zavialov[2], Wen Li[3], Scott M Stagg[4], Jayati Sengupta[3], Rikke C Nielsen[5], Poul Nissen[5], Stephen C Harvey[4], Måns Ehrenberg[2] & Joachim Frank[1,6]

Aminoacyl-tRNAs (aa-tRNAs) are delivered to the ribosome as part of the ternary complex of aa-tRNA, elongation factor Tu (EF-Tu) and GTP. Here, we present a cryo-electron microscopy (cryo-EM) study, at a resolution of ~9 Å, showing that during the incorporation of the aa-tRNA into the 70S ribosome of *Escherichia coli*, the flexibility of aa-tRNA allows the initial codon recognition and its accommodation into the ribosomal A site. In addition, a conformational change observed in the GTPase-associated center (GAC) of the ribosomal 50S subunit may provide the mechanism by which the ribosome promotes a relative movement of the aa-tRNA with respect to EF-Tu. This relative rearrangement seems to facilitate codon recognition by the incoming aa-tRNA, and to provide the codon-anticodon recognition-dependent signal for the GTPase activity of EF-Tu. From these new findings we propose a mechanism that can explain the sequence of events during the decoding of mRNA on the ribosome.

In translation, aa-tRNA, EF-Tu and GTP form a ternary complex that binds to the ribosome, allowing codon-anticodon pairing in the 30S ribosomal subunit. Cognate codon recognition triggers fast GTP hydrolysis on EF-Tu[1], and the ensuing conformational changes of the factor[2,3] lead to its dissociation. This allows the aa-tRNA[4] to accommodate into the ribosomal intersubunit cavity, where it reaches the peptidyl-transferase center (PTC) in the ribosomal 50S subunit[5]. The orientation of the incoming aa-tRNA differs substantially from the orientation of the tRNA residing in the ribosomal A site[6–9]. Thus, for the initial codon-anticodon pairing to form, either the binding of the ternary complex to the ribosome must be different from that in the complexes of 70S ribosome, aa-tRNA and EF-Tu described[6–9], or the process requires an incoming aa-tRNA to be able to recognize the A site codon in a sterically less favorable approach. Recent cryo-EM structures[8,9] have suggested that the anticodon arm of the aa-tRNA in the A/T site, while interacting with EF-Tu, has a conformation that differs from those of the known structures for tRNAs, and that this modified tRNA structure might allow a codon recognition in a local, steric configuration similar to that seen for the A-site tRNA. However, the limited resolution of those studies (~15–20 Å) did not provide enough definition to model the postulated structural change in the A/T-site tRNA, and the appearance of the anticodon arm of the tRNA was qualitatively described as "kinked"[8] or "distorted"[9].

During ternary complex recognition, the interaction of EF-Tu with the ribosome greatly stimulates the factor's GTPase activity[10], and the ensuing hydrolysis of the nucleotide separates the two steps for the selection of cognate tRNAs[11]: 'initial selection' before GTP hydrolysis and 'proofreading'[12,13] after the hydrolysis. The kinetics of the different binding steps have been characterized in detail[11,14]. In the initial selection, codon-anticodon interaction leads to structural changes in the aa-tRNA[15]. These are followed by a local conformational change in the vicinity of the GTP binding site of EF-Tu[16] that ultimately leads to hydrolysis of the GTP molecule on the factor. Both the GTP binding region in EF-Tu and the GAC in the ribosome are distant from the decoding center in the 30S subunit. Nevertheless, rapid triggering of the GTP hydrolysis on EF-Tu requires that the codon in the A site is recognized by the cognate anticodon[15] of an intact aa-tRNA[17]. Unanswered questions thus far are (i) how the codon recognition by the aa-tRNA affects the active center of EF-Tu to induce the hydrolysis of GTP, and (ii) what role the GAC plays in this process.

In the present work we have addressed these questions by analyzing cryo-EM maps depicting the *E. coli* ribosome in two states of tRNA incorporation, characterized as A/T-site and A-site binding states. This analysis reveals that the decoding process is facilitated by a dynamic interplay between the ribosome and the tRNA in which the tRNA acts like a molecular spring.

[1]Howard Hughes Medical Institute, Health Research, Inc. at the Wadsworth Center, Empire State Plaza, Albany, New York 12201-0509, USA. [2]Department of Cell and Molecular Biology, Biomedical Center, Box 596, S-751 24 Uppsala, Sweden. [3]Wadsworth Center, Empire State Plaza, Albany, New York 12201-0509, USA. [4]School of Biology, Georgia Institute of Technology, 310 Ferst Drive, Atlanta, Georgia 30332, USA. [5]Department of Molecular Biology, University of Aarhus, Gustav Wieds vej 10C, DK-8000 Aarhus C, Denmark. [6]Department of Biomedical Sciences, State University of New York at Albany, Empire State Plaza, Albany, New York 12201-0509, USA. [7]Present address: Centro Nacional de Biotecnología, CSIC, Universidad Autónoma de Madrid, Cantoblanco 28049, Madrid, Spain. Correspondence and requests for materials should be addressed to J.F. (joachim@wadsworth.org).

Published online 19 October 2003; doi:10.1038/nsb1003

ARTICLES

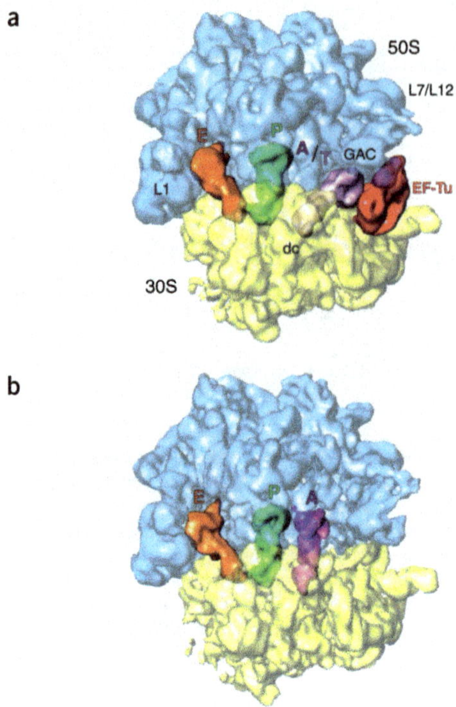

Figure 1 Cryo-EM maps showing the incorporation of the aa-tRNA into the ribosomal A site. (**a**) 70S–fMet-tRNA^{fMet}–Phe-tRNA^{Phe}–EF-Tu–GDP–kir complex at a resolution of 9 Å. The stalled ternary complex is bound to the ribosome below the L7/L12 stalk of the 50S subunit. EF-Tu is red, and A/T-site tRNA pale pink. (**b**) The map at a resolution of 9 Å of the 70S–tRNA^{fMet}–MF-tRNA^{Phe} complex, with A site dipeptidyl-tRNA in magenta. In both panels the ribosomal subunits are rendered semitransparent to illustrate the densities for tRNAs in the ribosomal intersubunit space. The sites for the tRNAs are labeled as follows: A/T, A/T site for the aa-tRNA bound to EF-Tu, reaching the decoding site in the A site of the 30S ribosomal subunit; A, aminoacyl site; P, peptidyl site; E, exit site. The P-site tRNA is green, the E-site tRNA orange. L7/L12, stalk formed by proteins L7/L12; GAC, GTPase-associated center; L1, stalk of the ribosomal protein L1; dc, decoding center.

RESULTS

We have obtained cryo-EM maps at ~9-Å resolution for ribosomes, originally with fMet-tRNA^{fMet} in the P site, that were allowed to react with the ternary Phe-tRNA^{Phe}–EF-Tu–GTP complex either in the presence or in the absence of kirromycin (kir) (see Methods). The antibiotic kir binds to the ternary complex and effectively prevents dissociation of EF-Tu from the ribosome after GTP hydrolysis on the factor[18,19]. In the presence of the antibiotic, the ribosome contained a stalled Phe-tRNA^{Phe}–EF-Tu–GDP–kir complex. In its absence, the ribosome contained fMet-Phe-tRNA^{Phe} in the A site and deacylated tRNA^{fMet} in the P site. In the map for the 70S–fMet-tRNA^{fMet}–Phe-tRNA^{Phe}–EF-Tu–GDP–kir (**Fig. 1a**), the stalled ternary complex is clearly visible. The overall position of the aa-tRNA–EF-Tu–GDP–kir complex is as described before[6,8,9], but the densities for Phe-tRNA^{Phe} (pale pink) and EF-Tu (red) can be discerned as separate as a result of the improved resolution. When the complex is made in the absence of kir, the fMet-Phe-tRNA^{Phe} appears in the A site (magenta) after accommodation and peptide bond formation (**Fig. 1b**). Because of the high affinity of uncharged tRNAs for the E site, this region is seen occupied by deacylated tRNAs in both maps.

The decoding site

Because the present maps represent direct structural data on the steps of decoding and accommodation, we analyzed the conformation of the decoding site in the 30S ribosomal subunit both with aa-tRNA in the A/T site (**Fig. 2a**) and, after accommodation as peptidyl-tRNA in the A site. By comparing the present maps with a control map, which had fMet-tRNA^{fMet} in the P site and a vacant A site, we detected in both cases an additional mass in the A-site codon region (asterisk in **Fig. 2a**). This can be rationalized as the presence of density from the helix fragment formed by codon and anticodon in the cognate mRNA-tRNA pairing. There is also an empty space in SH44 (ref. 20; indicated with an arrow in **Fig. 2a**), in comparison with the control. (For brevity, rRNA helices are denoted as 'SH' for the 16S rRNA from the small 30S subunit, or 'LH' for the 23S rRNA from the large 50S subunit.)

These results invite a comparison with results from X-ray crystallography. The crystal structure of the 30S subunit from *Thermus thermophilus* in complex with mRNA and a fragment of the cognate tRNA composing the anticodon stem loop (ASL)[21] served to characterize the changes in the structure of the decoding center upon codon-anticodon recognition. Introduction of the cognate ASL fragment induced a flipping-out of the universally conserved bases A1492 and A1493 from their original positions inside an internal loop of SH44. These two bases are essential for viability and A-site tRNA binding. When they interact with the minor groove of the first two base pairs of the codon-anticodon helix[21], they vacate the positions they originally occupied in SH44. We have used the atomic coordinates from that work to interpret the structure of the decoding site as seen in our cryo-EM maps. To this end, the crystal structure of the 30S subunit was docked as a rigid body inside the density of the small subunit of our cryo-EM map of the ribosome with a stalled ternary complex. Inspection of the decoding center (**Fig. 2b**) reveals that the low-density region within SH44 in the cryo-EM density map (**Fig. 2a**) matches the internal loop in the tip of SH44 that becomes vacant when A1492 and A1493 flip out as the result of a cognate codon-anticodon interaction. This means that changes in the crystal structure of the 30S subunit resulting from interactions between a tRNA analog and mRNA are reproduced in the more realistic setting of the present study.

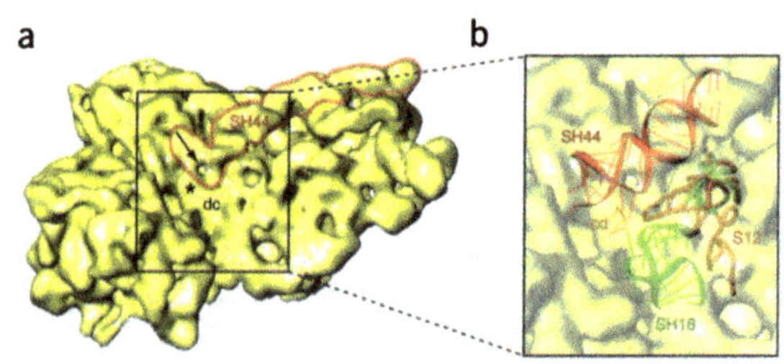

Figure 2 Topology of the decoding site. (**a**) 30S subunit portion isolated from the 70S–fMet-tRNA^{fMet}–Phe-tRNA^{Phe}–EF-Tu–GDP–kir complex, seen from the intersubunit space. The architecture of the decoding site (dc) shows a density in the region of the A-site codon (labeled with an asterisk) and a low-density region (indicated by an arrow) within SH44.
(**b**) Ribbons representation of the docked atomic coordinates for SH44, ribosomal protein S12 (S12), SH18 and the A site codon (cd) from Ogle and co-workers[21] within the density region extracted from **a**.

ARTICLES

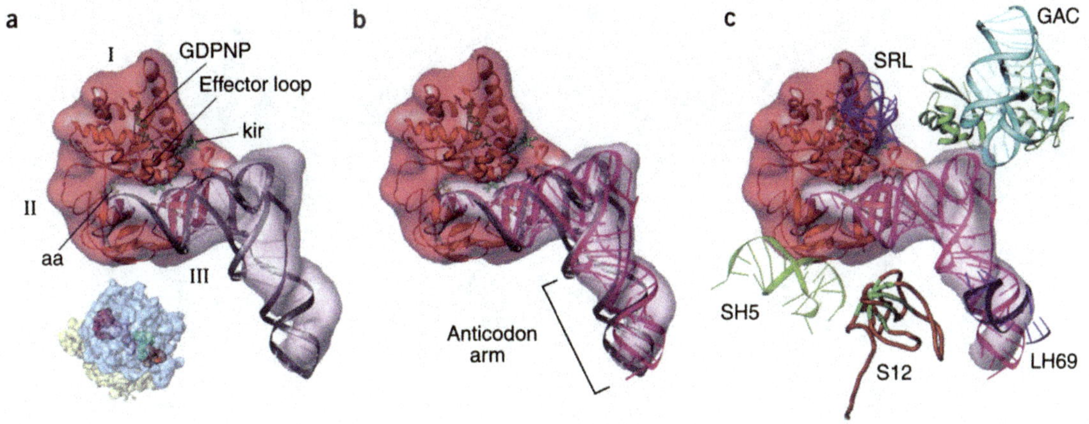

Figure 3 Ternary complex stalled with kirromycin. (**a**) Atomic coordinates for the Phe-tRNA^{Phe}–EF-Tu-GDPNP–kir from R.C.N. and co-workers (unpublished data), docked in the cryo-EM density for the stalled ternary complex isolated from the map of the 70S–fMet-tRNA^{fMet}–Phe-tRNA^{Phe}–EF-Tu-GDP–kir. The fitting was done using the coordinates as a rigid body and looking for an optimal docking for EF-Tu. The domains of EF-Tu are labeled I, II and III. The positions of the GTP analog from the crystal structure (GDPNP), the antibiotic kirromycin (kir), the aminoacyl moiety (aa) and the effector loop within domain I of EF-Tu are also labeled. The thumbnail indicates the orientation of the complex within the context of the ribosome. (**b**) After allowing a relative movement between aa-tRNA and EF-Tu, the new position of the aa-tRNA is magenta, versus the position of the aa-tRNA in the rigid body fitting of the ternary complex (gray). The coordinates for the aa-tRNA do not match the features of the cryo-EM density in the anticodon arm. (**c**) The panel shows the relevant ribosomal elements after the docking of the corresponding atomic coordinates inside the cryo-EM map for the 70S–fMet-tRNA^{fMet}–Phe-tRNA^{Phe}–EF-Tu-GDP–kir complex. The densities for EF-Tu and aa-tRNA in the A site are depicted as semitransparent. GAC, GTPase-associated center; LH69, helix 69 of large subunit; SH5, helix 5 of small subunit; SRL, sarcin-ricin loop; S12, ribosomal protein S12.

Arrangement of EF-Tu and aa-tRNA on the ribosome

The resolution of the cryo-EM map for the 70S–fMet-tRNA^{fMet}–Phe-tRNA^{Phe}–EF-Tu-GDP–kir complex depicted in **Figure 1b** is substantially better (~9 Å) than in former reports[6–9]. The improved definition has been used to refine the description of the structural relationship among EF-Tu, aa-tRNA and the ribosome. Recently, the structure for the Phe-tRNA^{Phe}–EF-Tu–GDPNP–kir complex was solved (R.C.N. *et al.*, unpublished data), and we used those coordinates for the docking of the ternary complex inside our cryo-EM densities for the aa-tRNA residing in the A/T site and for EF-Tu (**Fig. 3a**). The fitting was initially done with aa-tRNA–EF-Tu taken as a rigid body, but a relative movement between EF-Tu and aa-tRNA was found to improve the fit to the map (**Fig. 3b**).

The atomic structure of EF-Tu fits well into the cryo-EM map. The docking of the factor did not require any relative movement among the three domains of EF-Tu, indicating that the architecture of EF-Tu–GDP, stalled by kir in the ternary complex with Phe-tRNA^{Phe} on the ribosome, is very similar to the crystal structure of EF-Tu–GDPNP in complex with Phe-tRNA^{Phe} and kir. It is only the 'effector loop' region in domain I that is outside of the cryo-EM density for EF-Tu (**Fig. 3a**). The secondary structure and relative position of this region change in the transition from EF-Tu–GTP to EF-Tu–GDP[22,23]. The effector loop is very close to the guanine nucleotide binding pocket (GDPNP in the crystal structure) and contains some of the residues postulated to have a central role in the GTPase activity of EF-Tu[24]. Our failure to detect a density for this region indicates that the effector loop is flexible in the stalled ternary complex and that its density disappears as a result of averaging. Similarly, the crystal structure of free EF-Tu–GDP–aurodox (aurodox is a methylated derivative of kir) shows disorder in the effector loop[25]. The local conformational changes in the GTP binding region of EF-Tu that are associated with GTP hydrolysis on the ribosome are to some extent stabilized in the presence of kir[16]. Thus, the EF-Tu–GDP–kir structure that we see may correspond to the one that is activated for the hydrolysis of the nucleotide, and it is the conformation that immediately follows GTP hydrolysis, yet precedes the conformational change to the free GDP-bound structure and hence EF-Tu release. Additionally, the residues Arg58 and Ile60 in EF-Tu (*E. coli* numbering for aminoacyl residues is used throughout), postulated to participate in the mechanism of GTP hydrolysis by EF-Tu[24], are both within the effector loop. Thus, the movement of the flexible effector loop might be the change in EF-Tu that is responsible for the activation of its GTPase activity during its interaction with the ribosome.

Available crystal structures for the ribosomal subunits[21,26] were used as a source for atomic coordinates that were fitted into the cryo-EM map for the 70S–Met-tRNA^{fMet}–Phe-tRNA^{Phe}EF-Tu–GDP–kir complex. The resulting atomic model allowed us to study the interactions between the ternary complex and several ribosomal elements that are represented in **Figure 3c**. EF-Tu interacts with the sarcin-ricin loop (SRL) around positions 2661–2663 of the 23S rRNA, as described earlier[27,28]. This binding is probably established through residues within the effector loop of EF-Tu, but the putative flexibility of this region would prohibit a precise definition of the contact. Domain II of EF-Tu is bound to SH5 around the base pair formed by nucleotides 57 and 356 of 16S rRNA. The amino acid in EF-Tu that is closest to SH5 is Gly222 (*E. coli* numbering), and mutations in Gly222 inhibit the ribosome-dependent GTPase activity of the factor[29,30].

Apart from the codon-anticodon pairing in the decoding site of the ribosomal 30S subunit[21], three additional interactions between the tRNA and the ribosome can be defined. In the acceptor arm of the aa-tRNA, near nucleotide 69, there is a contact with amino acids 78–80 of protein S12. This is the only interaction that we observe between the tRNA and S12, and in our interpretation of the cryo-EM map, the previously postulated interaction of S12 with the D stem of the tRNA[9] is

ARTICLES

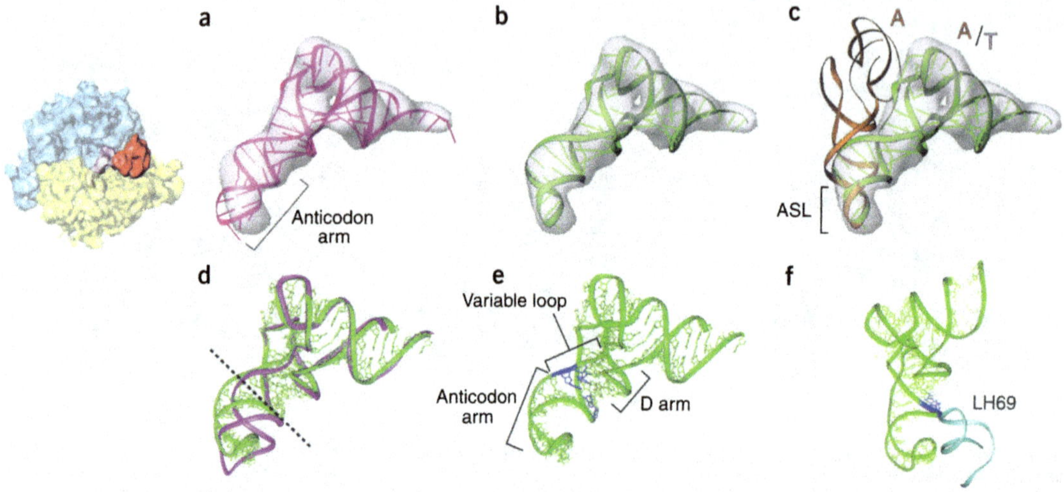

Figure 4 Model of A/T site tRNA and its interaction with ribosomal LH69. (**a**) Rendering of the fitted tRNA^{Phe} inside the cryo-EM density for the A/T-site tRNA. Evidently, the anticodon arm of the X-ray structure does not fit the features of the density for the A/T site tRNA. (**b**) Modeled structure for the A/T-tRNA is represented in green ribbons, showing an evident improvement in the docking of the coordinates within the density. (**c**) Comparison of the modeled coordinates for the A/T-site tRNA (green) and the coordinates for the A-site tRNA (gold). (The docking of the A-site tRNA in our cryo-EM density did not require any modification in the anticodon arm of the X-ray structure.) The ASLs of both tRNAs are similarly oriented. (**d**) Modeled structure for the A/T-site tRNA (green), shown with the ribbons representation of the phosphate backbone using the coordinates for the tRNA before modeling (magenta). The dashed line indicates the location of the 'kink' in the modeled tRNA structure that changes the relative angle between D and anticodon stems. (**e**) Representation of the modeled A/T-site tRNA with nucleotides 44, 45 and 26, which define the putative hinge region in the tRNA, highlighted in dark blue. (**f**) Ribbons representation of the modeled A/T-site tRNA (green) and the fitted position of LH69. The interaction between these two elements involves nucleotides 24–25 of the tRNA (dark blue).

absent. Nucleotide 56 within the T arm, situated in the elbow region of the tRNA within the conserved TΨCG sequence (position 56 is the C), is close enough to interact with residue 1067 from LH43 in the GAC. The interaction between this highly conserved region in the tRNA and the fragment of the 23S rRNA is clearly discerned, but again, in contrast to previous claims based on lower-resolution cryo-EM results[9], the tRNA is not seen to contact protein L11 situated on the same lobe of the GAC. Finally, LH69 establishes a clear interaction (around the nucleotide in position 1914) with the D arm of the aa-tRNA (approximately with residues 24–25 of the aa-tRNA).

A flexible tRNA

We have shown that the atomic structure of the ternary complex cannot be docked to the ribosome in its entirety as a rigid body, suggesting the existence of relative movements between EF-Tu and aa-tRNA, which must be partly stabilized by kir binding. However, even if relative movements between rigid EF-Tu and aa-tRNA molecules are allowed, the ASL will necessarily remain outside the cryo-EM map for the A/T site tRNA (**Fig. 4a**). Fortunately, the good definition of the helical regions of tRNA at 9-Å resolution in the present map allowed us to model a distinct conformation of the anticodon arm (see Methods) that fits the obtained density quite well (**Fig. 4b**) and puts the ASL in a position and orientation that closely match those seen in the A site tRNA (**Fig. 4c**). Most notably, this novel tRNA conformation could allow a similar codon-anticodon interaction for a tRNA in the A/T and A sites.

A comparison of the structure of the aa-tRNA in our model with the atomic coordinates from X-ray crystallography (**Fig. 4d**) reveals that in the modeled coordinates the anticodon arm is reoriented by a bending at the junction of the anticodon and D stems that defines the 'kink'[8] in the tRNA (dashed line in **Fig. 4d**). In the highly conserved architecture of tRNAs, the connection between the D and the anticodon stems is a complex structure that involves tertiary interactions between residues from the D stem and the variable region of the tRNA. This D–anticodon stem junction was understood as a hinge in one of the first reports describing the atomic structure of a tRNA[31]. In all the available crystal structures for tRNAs, the nucleotide at position 26, which separates the D and the anticodon stems, is an unpaired base that is partially intercalated with residues 44 and 45 of the variable loop (nucleotides 26, 44 and 45 are dark blue in **Fig. 4e**), and this particular conformation was associated with the relative tilt between the D and anticodon stems[31]. The exact details of the tRNA deformation cannot be determined at atomic resolution from the available data, but it clearly occurs near this putative hinge. In our model the deformation leading to the movement of the ASL relative to the acceptor–T stem is distributed smoothly over the four base pairs immediately surrounding the junction between the D and anticodon stems. Thus the model preserves the local features of the tRNA crystal structure for the lower part of the ASL, for the upper part of the D stem and for the rest of the molecule. We expect the changes in the tertiary structure to increase the electrostatic repulsion and the torsion angle energy of the phosphate backbone.

Reorientation of the anticodon stem loop

Our results show that the aa-tRNA is flexible enough to adopt a conformation that allows optimal codon-anticodon interaction to develop while the tRNA is in the A/T site. There is no indication that the mRNA codon in the A site can change its orientation. However, our present resolution does not allow an unambiguous determination of the structure of the codon-anticodon interaction, so we cannot rule out the existence of some rearrangements in the decoding site. In any case, the observed reorientation of the anticodon arm of the aa-tRNA seems to be an essential mechanism during this cognate codon recognition by the tRNA while it resides in the A/T position. In the crystal structure of the aa-tRNA–EF-Tu–GDP–kir complex (R.C.N. *et al.*, unpublished results), the atoms in the anticodon arm have the highest temperature factors, suggesting that this region is indeed flexible. Increased temperature factors for the ASL relative to the rest of the molecule have been known from X-ray crystallographic studies on tRNA[32], and flexibility in this region was predicted by early molecular dynamics simulations[33].

The structural change in the aa-tRNA may be induced by several variables: (i) interaction with ribosomal elements (ii) conformational changes in the decoding center upon codon recognition or (iii) GTP hydrolysis by EF-Tu. However, the conformational changes within the aa-tRNA are known to precede GTP hydrolysis by EF-Tu[15]. The remaining problem is thus to determine whether the aa-tRNA changes its conformation before, or in response to, codon-anticodon pairing. Because the codon recognition by the aa-tRNA in the A/T position requires a structural distortion of the tRNA, it is natural to assume that the aa-tRNA deformation precedes the codon-anticodon interaction, and that the energetic cost of this deformation is compensated by cognate tRNA-mRNA hydrogen bonding and tRNA-ribosome interactions. The tRNA deformation, which could be part of the mechanism that triggers the GTPase activity of EF-Tu[15], is thus seen to be an essential component of the recognition of the cognate codon.

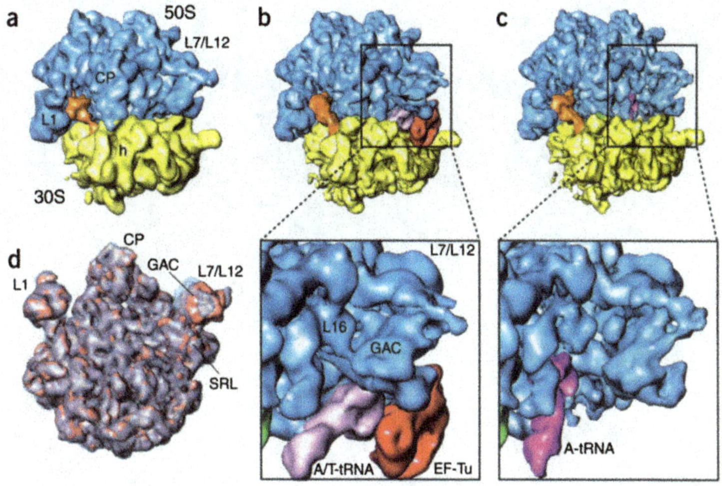

Figure 5 Conformational change of the GAC of the 50S subunit upon interaction of the ternary complex with the ribosome. (**a**) Cryo-EM map for the 70S–fMet-tRNA^{fMet} complex, showing a vacant A site. (**b**) Cryo-EM map for the 70S–fMet-tRNA^{fMet}–Phe-tRNA^{Phe}–EF-Tu–GDP–kir complex, with the A/T-site tRNA colored in pale pink and EF-Tu in red. (**c**) Cryo-EM map for the 70S–tRNA^{fMet}–MP-tRNA^{Phe}, with the A site tRNA in magenta. Insets in **b** and **c** focus on the conformational change of the GAC, and the 30S subunit is not rendered. The view from the intersubunit space allows the movement of the GAC to be observed. Ribosomal subunits are blue (50S subunit) and yellow (30S subunit). The density that corresponds to tRNAs in the E site is shown in orange. (**d**) Superposition of the 50S ribosomal subunits from the ribosome bound with the stalled ternary complex (semitransparent blue) and from the ribosome bearing a dipeptidyl-tRNA in the A site (solid red). L7/L12, stalk of proteins L7/L12 from the 50S ribosomal subunit; GAC, GTPase-associated center; SRL, sarcin-ricin loop; L16, ribosomal protein L16; CP, central protuberance; L1, stalk of ribosomal protein L1; h, head of the 30S subunit.

Ribosomal LH69 interacts with A/T-site tRNA close to the proposed hinge between the D and anticodon stems (**Fig. 4f**). This contact involves residues around position 1914 from LH69 (ref. 26) and nucleotides 24–25 from the D stem of the aa-tRNA (residues 24–26 are dark blue in **Fig. 4f**) in the aforementioned putative hinge of the tRNA structure, whose tertiary interactions we expect to change when compared with the canonical atomic structure of tRNAs. These observations suggest that the interaction of LH69 with the D arm in this critical hinge region facilitates the structural distortion of the tRNA that produces the correct orientation of the ASL. If so, one would expect that mutations either in LH69 or in the hinge region of the tRNA could impair the decoding process. Indeed, there are mutations in residues 1914 and 1916 in H69 that impair tRNA selection, enhance frame shifting and increase the readthrough of all three stop codons[34]. A mutation in nucleotide 24 of tRNA^{Trp}, far away from the anticodon, enhances its readthrough of UGA stop codons[35]. Other mutations in the D arm sequence of several tRNAs that modify the tertiary interactions in this region enhance misreading of several codons[36]. This could indicate that a perturbation of the LH69-tRNA interaction enhances misreading by distorting the orientation of the ASL, and that modifications in any of those interacting regions might cause different orientations of the ASL or changes in the flexibility of the tRNA, which would eventually lead to noncanonical decoding.

Conformational change of the GTPase-associated center

Apart from the ribosomal cryo-EM maps for the stalled ternary complex and the accommodated A site tRNA, we have obtained a cryo-EM map for a 70S ribosome bearing an initiator fMet-tRNA^{fMet} in the P site and a vacant A site (**Fig. 5a**), which serves as a control. Thus, the three maps shown in **Figure 5a–c** can be understood as consecutive snapshots of the ribosome during ternary complex binding and accommodation of aa-tRNA in the A site. The binding of the ternary complex promotes a large conformational change in the base of the L7/L12 stalk (indicated with an arrow in **Fig. 5b**), which affects the two lobes very differently. The base of the L7/L12 stalk is composed of two main lobes. One is formed by protein L6 and LH95, and is known as the SRL[26]. The SRL contacts the G domain near the GTP binding site of both EF-Tu and EF-G and is thought to stimulate the GTPase activity of these factors when they interact with the ribosome[27,37,38]. The other lobe contains protein L11 and a 58-nucleotide fragment of the 23S rRNA comprising LH43 and LH44 (refs. 26,39). This L11-rRNA complex interacts with EF-G–GTP and aa-tRNA–EF-Tu–GTP, in a region far away from their guanine nucleotide binding sites[8,9,27]. Despite the distance, the binding of antibiotics to the L11-rRNA region[40] and a mutation at position 1067 in LH43 (ref. 41) interfere with the GTPase activity of EF-Tu and EF-G. From these findings, the L11–rRNA complex is also known as the GTPase-associated center (GAC). (There is no uniform use of this term in the literature; it has been used either with the present definition or with reference to both the L11–58-nt rRNA and the SRL, which form two proximate, distinct morphological features of the 50S subunit.)

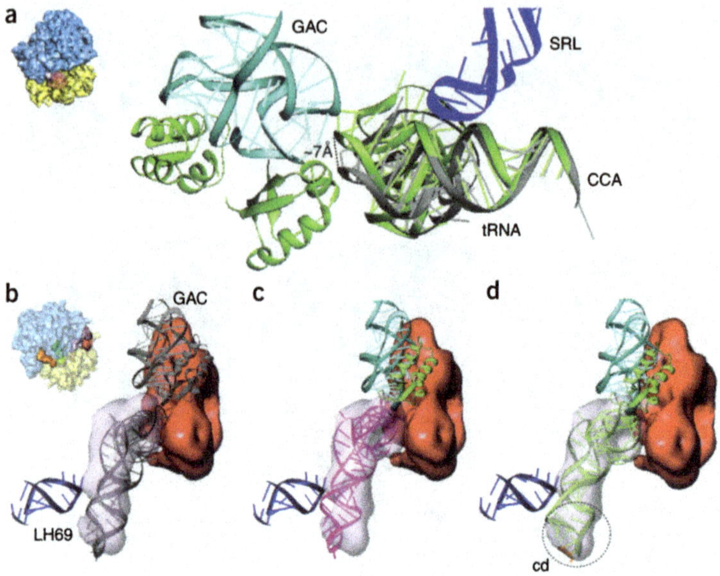

Figure 6 Coupling of the conformational change of the GAC with the movement of aa-tRNA. (**a**) Comparison of the aa-tRNA position defined inside the cryo-EM density for the ternary complex (green) versus the position of the aa-tRNA in the atomic coordinates of Phe-tRNA^{Phe}–EF-Tu–GDPNP–kir (gray). The ribbons representation of the GAC and the SRL serves as a frame of reference to make it apparent that the change in the aa-tRNA position follows the conformational change of the GAC while the aa-tRNA maintains its position in the region of the SRL. The GAC position shown correspond to the fitting of the atomic coordinates in the 'closed' GAC from the 70S–fMet-tRNA^{fMet}–Phe-tRNA^{Phe}–EF-Tu–GDP–kir complex. The ribosomal orientation is depicted in the thumbnail. (**b–d**) The postulated coupling between the movement of the GAC and the approach of the aa-tRNA to LH69. In a hypothetical initial binding, the 'open' GAC would interact with the aa-tRNA delivered in the ternary complex by EF-Tu (**b**). The transition of the GAC to the closed conformation (**c**) brings the aa-tRNA in contact with LH69. The aa-tRNA–LH69 interaction facilitates a distortion in the aa-tRNA (**d**) that reorients the anticodon arm (in the region encircled) and allows a codon (cd)-anticodon recognition.

When the ternary complex is present on the ribosome, the lobe carrying the GAC region moves toward the body of the 50S subunit, approaching ribosomal protein L16 (inset in **Fig. 5b**). When EF-Tu leaves the ribosome and fMet-Phe-tRNA^{Phe} assumes its position in the A site, the GAC lobe moves back to its original position (inset in **Fig. 5c**). The SRL lobe, in contrast, maintains its position during the entire process (**Fig. 5d**). We will refer to the two conformations of the GAC lobe as 'open', before and after the ribosomal interaction with the ternary complex (**Fig. 5a,c**), and as 'closed' when the ribosome interacts with the stalled ternary complex (**Fig. 5b**).

DISCUSSION

The accuracy in translation is based on discrimination between cognate and near-cognate tRNAs in two steps, which are separated by hydrolysis of GTP on EF-Tu[11–13]. The initial selection of ternary complexes is accompanied by local structural changes in the decoding site of the 30S subunit that are triggered by codon-anticodon interactions[21]. When the ternary complex contains an aa-tRNA cognate to the codon in the A site, GTP on EF-Tu is rapidly hydrolyzed and the factor dissociates from the ribosome. When, in contrast, the ternary complex contains a noncognate aa-tRNA, the ternary complex preferentially dissociates before hydrolysis of GTP. After dissociation of EF-Tu in the GDP form, the aa-tRNA moves from the A/T to the A site; near-cognate aa-tRNAs are frequently rejected in the proofreading step[11–13], whereas cognate aa-tRNAs with high probability become stably bound in the A site. All these steps have been characterized biochemically and are understood as sequential, stochastic changes of the system that with high probability lead to the final acceptance of a cognate tRNA or the dissociation of a near- or noncognate tRNA. However, the structural basis that can explain the sequence of events remains unknown. Here we propose a model that explains mechanistically how aa-tRNAs are delivered to and selected by the ribosome following the temporal sequence of the events.

The first selection involves the codon recognition by the aa-tRNA, and the triggering of the GTP hydrolysis by EF-Tu. In this cryo-EM study we have observed two conformational changes in addition to the change of the tRNA structure reported before; one change affects the position of the GAC lobe in the ribosomal 50S subunit, and another change alters the relative position between aa-tRNA and EF-Tu as a cognate ternary complex binds to the ribosome. These observations suggest that the two conformational changes might be linked, because the relative movement between aa-tRNA and EF-Tu could be rationalized if it was mirrored by corresponding relative movements of their ribosomal binding sites. It is not known how aa-tRNA and EF-Tu interact with the ribosome in the initial binding that precedes the cognate codon-anticodon interaction. However, all the available crystal structures for the ternary complex display the same relationship between aa-tRNA and EF-Tu, and we assume therefore that those atomic coordinates accurately describe the free ternary complex. Thus, given the two different relative positions between aa-tRNA and EF-Tu, and two conformations for the GAC, we can compare the movements and inspect how closely they match each other. For this purpose, we have kept EF-Tu bound to the immobile SRL in a fixed position. In **Figure 6a** it can be seen that the aa-tRNA defined in our cryo-EM map (green) has moved from its position in the crystal structure of the ternary complex (gray) by ~7 Å in the elbow region. This distance was measured between the phosphates of nucleotide 56 (the residue that is closest to the GAC) from the tRNAs in the two positions. After docking of the GAC into our cryo-EM maps, we see that nucleotide 1067 in LH43, the region that interacts with the aa-tRNA, has also moved by ~7 Å in the transition from the open to the closed conformation. The movement between the two positions of the tRNA corresponds to a rotation of the acceptor arm that follows the conformational change of the GAC while maintaining its position in the vicinity of the SRL, where EF-Tu is bound (**Fig. 6a**). Obviously, this movement must modify the interface between the aa-tRNA and EF-Tu. However, it must be kept in mind that the GTPase activity of EF-Tu is greatly stimulated when a cognate ternary complex interacts with the ribosome[1], and we interpret this very movement of the aa-tRNA with respect to EF-Tu as part of the mechanism that activates GTP hydrolysis on the factor.

The hypothesis that a coupled movement of GAC–tRNA triggers GTP hydrolysis on EF-Tu receives further support from the observation that a mutation at position 1067 in the 23S rRNA, situated within the GAC and in contact with the aa-tRNA in the A/T site, reduces the GTPase activity of EF-Tu[42]. Additionally, the aforementioned effect of a mutation in G222 within EF-Tu on the GTP hydrolysis by the factor[29,30] can now be explained, because an impaired interaction between EF-Tu and the 30S subunit (established through this particular residue) would leave the factor without a necessary anchor point on the ribosome to maintain a stable position while the aa-tRNA moves following the movement of the GAC.

It is remarkable that the GAC, a part of the 50S subunit, seems to have a pivotal role in the decoding of the mRNA, which resides on the small 30S subunit. Another element of the 50S subunit, LH69, seems to have an important function during this process as well. It is the only ribosomal element that interacts with the aa-tRNA close to the region where the distortion of the incoming aa-tRNA structure is seen in the present work.

We have sought an integrated model that (i) would fit the observed interplay of interactions and conformational changes (ii) would explain the sequence of events during the delivery of aa-tRNA to the ribosome by EF-Tu and (iii) would give both the GAC and LH69 a prominent place. In our proposed model, the ternary complex, presumably in a conformation as defined by its crystal structures, approaches a ribosome that has its GAC lobe in the open conformation (**Fig. 6b**). Two of the contacts between the ternary complex and the ribosome that are seen in the kirromycin-stalled complex are maintained in this hypothetical constellation. These are the contact between EF-Tu and SRL and the contact between residue 1067 from the GAC and the elbow region of the aa-tRNA. The only contact with 23S rRNA that is absent, because of the leveraging of tRNA produced by the different orientation of the GAC lobe, is the contact of the tRNA D loop with nucleotide 1914 of LH69. We postulate that this contact, once established, increases the intrinsic flexibility of the tRNA structure and facilitates the reorientation of the ASL that is required for codon-anticodon contact. Thus, the conformational change of the GAC to the closed state, which leads to the crucial tRNA D loop contact with LH69 (**Fig. 6c**), may be seen as the start of a causal chain leading to codon-anticodon pairing (**Fig. 6d**). For the energy cost of tRNA deformation to be compensated, a cognate codon-anticodon pairing would be essential. A cognate tRNA will stabilize the closed GAC lobe conformation and conserve the relative movement between aa-tRNA and EF-Tu in the ternary complex, leading to rapid hydrolysis of GTP on EF-Tu. When, in contrast, the aa-tRNA is noncognate, its interaction with the codon will fail to stabilize the closed GAC lobe conformation and the structural rearrangement of the ternary complex, leading to slow GTP hydrolysis on EF-Tu[11].

The second selection of the specific aa-tRNA takes place after the hydrolysis of GTP and the dissociation of EF-Tu. The tRNA conformation in the A/T site that we observe is energetically unfavorable, which would explain why it has not been observed before in the structure of the isolated molecule. We suggest that after dissociation of EF-Tu and liberation of the acceptor arm from its interaction with the factor, the aa-tRNA will resume its energetically more favorable structure while maintaining the same codon-anticodon interaction. This means that the ASL will not change its position or orientation, so that the tension in the tRNA structure is resolved by a movement of the acceptor arm toward the A site in the 50S subunit (see **Supplementary Fig. 1** and **Supplementary Videos** 1 and 2 online). According to our model, the tRNA can be understood as a molecular spring, whose flexibility permits (i) codon recognition in the presence of EF-Tu and (ii) rapid movement of the CCA end of tRNA toward the A site on the 50S subunit. It is clear that if during the movement of the aa-tRNA from the A/T to the A site, the interaction between codon and anticodon is not strong enough, the aa-tRNA will dissociate from the ribosome. Thus, in addition to explaining the steps of codon-anticodon interaction and accommodation of a cognate tRNA, our model would provide a ready explanation for the structural basis of proofreading of near-cognate tRNAs.

METHODS

Preparation of ribosomal complexes. All the ribosomal samples were prepared using an *in vitro* translation system in polymix buffer[43]. For the 70S–tRNA^{fMet}–MF-tRNA^{Phe} complex, an initiation mix (200 µl), containing 4.4 µM ribosomes, 1.5 µM each IF1, IF2 and IF3, 12.5 µM MFstop-mRNA, 14.1 µM fMet-tRNA^{fMet}, energy regeneration system (50 ng µl^{-1} PK, 1 mM GTP and 2 mM PEP), was incubated at 37 °C for 25 min. An elongation mix (200 µl), containing 21.2 µM [^{14}C]Phe, 20 µM [^{14}C]Ile, 21.8 µM tRNA^{Phe}, 29 µM EF-Tu, 12.5 µM EF-Ts, 90 U of Phe-tRNA synthetase, an energy regeneration system (50 ng µl^{-1} PK, 1 mM GTP, 2 mM PEP, 8 ng µl^{-1} MK, 1 mM ATP) was incubated separately at 37 °C for 15 min. Peptide elongation was started by adding together the initiation and elongation mixtures. After 15 s incubation at 37 °C, the mixture was diluted with ice-cold polymix buffer to 1 µM ribosomal concentration and frozen in liquid nitrogen. The occupancy of A site, which was determined as the amount of peptide that could be released by RF1 in a stop codon–dependent reaction after fMet-Phe-tRNA^{Phe} translocation with addition of EF-G, was >75% of ribosome. The release of peptide was detected by filtration of the mixture through nitrocellulose. No detectable A site tRNA was translocated without EF-G. In addition, HPLC analysis showed that >90% of the A site–bound tRNA carried the dipeptide. To prepare the 70S–fMet-tRNA^{fMet}–Phe-tRNA^{Phe}–EF-Tu–GDP–kir complex, 0.25 mM kirromycin was added into the elongation mixture. No detectable Phe-tRNA^{Phe} was translocated when EF-G was added to this complex. To prepare the control ribosomal complex with fMet-tRNA^{fMet} in the P site, a mix (400 µl) containing 2.8 µM *E. coli* ribosomes, 0.75 µM each IF1, IF2 and IF3, 5.6 µM M stop-mRNA, 6.3 µM fMet-tRNA^{fMet}, energy regeneration system (100 ng µl^{-1} PK, 1 mM GTP and 6 mM PEP) was incubated at 37 °C for 20 min. Then the complex was pelleted in the airfuge (Beckman Coulter) for 15 min at 100,000 rpm (199,000*g*), dissolved in 600 µl of polymix buffer and frozen in liquid nitrogen. To prepare samples for cryo-EM, an aliquot of the ribosomal complex was thawed on ice and diluted with ice-cold polymix buffer to 32 nM concentration of ribosome. More than 70% of the ribosomal P site was occupied with fMet-tRNA^{fMet}, as detected by scintillation counting of [^{3}H]fMet.

Cryo-electron microscopy and image processing. The solutions containing ribosomal complexes were diluted in the same polymix buffer to a final concentration of 32 nM. Grids for cryo-EM were prepared following standard procedures. Micrographs were collected on a Philips Tecnai F20 at 200 kV and a magnification of 50,000× (± 2%) in low-dose conditions. After scanning the micrographs on a Zeiss Imaging scanner (Z/I Imaging Corporation) with a step size of 14 µm, the final pixel size corresponded to 2.82 Å on the object scale. The image processing included a 3D projection alignment procedure with correction of the contrast transfer function and enhancement of the high-resolution Fourier amplitudes based on X-ray solution scattering data. The total numbers of particles used in the final cryo-EM maps were 75,996 for the 70S–fMet-tRNA^{fMet}–Phe-tRNA^{Phe}–EF-Tu–GDP–kir sample, 52,181 for the 70S–tRNA^{fMet}–MF-tRNA^{Phe} complex and 19,865 individual particles for the 70S–fMet-tRNA^{fMet} control. The resolution for the maps was estimated to be 9 Å, 9 Å and 11 Å, respectively, using a cutoff of 0.15 in the Fourier shell correlation, equivalent to a 0.5 figure of merit in X-ray crystallography[44], and the maps were low-pass filtered accordingly. The cutoffs of 0.5 and 3 σ in the FSC corresponded to resolutions of ~10.25 Å and ~7.5 Å in the cases of the 70S–fMet-tRNA^{fMet}–Phe-tRNA^{Phe}–EF-Tu–GDP–kir and 70S–tRNA^{fMet}–MF-tRNA^{Phe}, and 14 Å and 9 Å for the control. The density thresholds that defined the rendered boundaries were chosen to display the approximate dimensions of published ribosomal atomic structures, and the final volumes represent ~120–140% the calculated volume for the ribosome. The segmentation of the 3D maps was carried out by cluster analysis using control maps of ligand-free ribosomes.

Docking of atomic structures and modeling the A/T site tRNA. The docking of X-ray structures into the cryo-EM maps was made manually using O[45], and the visualization was carried out in IRIS Explorer, RIBBONS and Insight II LS

ARTICLES

(Accelrys). The estimated accuracy of the fittings within the density maps is ±~1 Å, as checked by cross-correlation traces. The coordinates of the ribosomal proteins and rRNAs were taken from published atomic structures (PDB entry 1FFK for the 50S subunit from Ban and co-workers[26]; 1IBM for the 30S subunit from Ogle and co-workers[21]), and the docking was done using the coordinates as rigid bodies. Coordinates from R.C.N. and co-workers (PDB entry 1OB2; unpublished data) were used for the docking of EF-Tu and aa-tRNA in the cryo-EM density for the ternary complex. The model for the A/T site tRNA was constructed by interpolating between the tRNA positions for two independent fittings of the tRNA into the cryo-EM density. In one position, model A, the tRNA (PDB entry 1OB2) was fitted such that the T/acceptor stem came to lie in its corresponding density, but the anticodon stem was outside the density. In the second position, model B, the anticodon was fitted to the density while the T/acceptor stem was allowed to stick out. To create a model in which both domains fit in the density, we did a linear interpolation in which the T/acceptor stem was held fixed for model A, and the anticodon was held fixed in model B. The positions for the remaining atoms were determined by interpolating between the two models in a smooth and continuous fashion. The resulting model was subjected to an energy minimization to convergence using steepest descent minimization and the AMBER force field[46] in the Insight II package.

Coordinates. For the A/T state, the following coordinates were deposited in the Protein Data Bank: A/T-site tRNA (modeled) (1QZA); EF-Tu (R.C.N. *et al.*, unpublished data) (1OB2) fitted into ribosome environment (1QZD); GAC in closed position (1R2X). For the accommodated state, the following coordinates were deposited in the Protein Data Bank: A-site tRNA (1QZB) and GAC in open position (1R2W). The ribosomal environment is represented by the elements SH44, protein S12, LH69 and SRL (jointly deposited as 1QZC). The cryo-EM density maps of the ribosomal complexes in the A/T (AMD-1055) and the accommodated state (EMD-1056) have been deposited in the 3D-EM database, EMBL-European Bioinformation Institute.

Note: Supplementary information is available on the Nature Structural Biology website.

ACKNOWLEDGMENTS
We thank M. Yarus for a discussion and helpful suggestions and M. Watters for assistance with the illustrations. This work was supported by the Howard Hughes Medical Institute (J.F.), by grants from the US National Institutes of Health (J.F. and S.C.H.) and the US National Science Foundation (J.F.), by the Swedish Foundation for Strategic Research and the Swedish Research Council (A.V.Z. and M.E.), by an Ole Romer research grant from the Danish Research Council and the EMBO Young Investigator Program (P.N.) and by a scholarship from Novo Nordisk/Novozymes (R.N.).

COMPETING INTERESTS STATEMENT
The authors declare that they have no competing financial interests.

Received 7 July; accepted 5 September 2003
Published online at http://www.nature.com/naturestructuralbiology/

1. Rodnina, M.V., Pape, T., Fricke, R., Kuhn, L. & Wintermeyer, W. Initial binding of the elongation factor Tu·GTP·aminoacyl-tRNA complex preceding codon recognition on the ribosome. *J. Biol. Chem.* **271**, 646–652 (1996).
2. Berchtold, H. *et al.* Crystal structure of active elongation factor Tu reveals major domain rearrangements. *Nature* **365**, 126–132 (1993).
3. Kjeldgaard, M., Nissen, P., Thirup, S. & Nyborg, J. The crystal structure of elongation factor EF-Tu from *Thermus aquaticus* in the GTP conformation. *Structure* **1**, 35–50 (1993).
4. Dell, V.A., Miller, D.L. & Johnson, A.E. Effects of nucleotide- and aurodox-induced changes in elongation factor Tu conformation upon its interaction with aminoacyl-tRNA. A fluorescence study. *Biochemistry* **29**, 1757–1763 (1990).
5. Kaziro, Y. The role of guanosine-5′-triphosphate in polypeptide chain elongation. *Biochim. Biophys. Acta* **505**, 95–127 (1978).
6. Stark, H. *et al.* Visualization of elongation factor Tu on *Escherichia coli* ribosome. *Nature* **389**, 403–406 (1997).
7. Agrawal, R.K. *et al.* Visualization of tRNA movements on the *Escherichia coli* ribosome during the elongation cycle. *J. Cell Biol.* **150**, 447–459 (2000).
8. Valle, M. *et al.* Cryo-EM reveals an active role for aminoacyl-tRNA in the accommodation process. *EMBO J.* **21**, 3557–3567 (2002).
9. Stark, H. *et al.* Ribosome interactions of aminoacyl-tRNA and elongation factor Tu in the codon-recognition complex. *Nat. Struct. Biol.* **9**, 849–854 (2002).
10. Mohr, D., Wintermeyer, W. & Rodnina, M.V. GTPase activation of elongation factors Tu and G on the ribosome. *Biochemistry* **41**, 12520–12528 (2002).
11. Rodnina, M.V. & Wintermeyer, W. Fidelity of aminoacyl-tRNA selection on the ribosome: kinetic and structural mechanism. *Annu. Rev. Biochem.* **70**, 415–435 (2001).
12. Thompson, R.C. & Stone, P.J. Proofreading of the codon-anticodon interaction on ribosomes. *Proc. Natl. Acad. Sci. USA* **74**, 198–202 (1977).
13. Ruusala, T., Ehrenberg, M. & Kurland, C.G. Is there proofreading during polypeptide synthesis? *EMBO J.* **1**, 415–435 (1982).
14. Pape, T., Wintermeyer, W. & Rodnina, M.V. Complete kinetic mechanism of elongation factor Tu-dependent binding of aminoacyl-tRNA to the A site of *E. coli* ribosome. *EMBO J.* **17**, 7490–7497 (1998).
15. Rodnina, M.V., Fricke, R. & Wintermeyer, W. Transient conformational states of aminoacyl-tRNA during ribosome binding catalyzed by elongation factor Tu. *Biochemistry* **33**, 12267–12275 (1994).
16. Rodnina, M.V., Fricke, R., Kohn, L. & Wintermeyer, W. Codon-dependent conformational change of elongation factor Tu preceding GTP hydrolysis on the ribosome. *EMBO J.* **14**, 2613–2619 (1995).
17. Piepenburg, O. *et al.* Intact aminoacyl-tRNA is required to trigger GTP hydrolysis by elongation factor Tu on the ribosome. *Biochemistry* **39**, 1734–1738 (2000).
18. Wolf, H., Chinali, G. & Parmeggiani, A. Mechanism of the inhibition of protein synthesis by kirromycin. *Eur. J. Biochem.* **75**, 67–75 (1977).
19. Parmeggiani, A. & Stewart, G.W. Mechanism of action of kirromycin-like antibiotics. *Annu. Rev. Microbiol.* **39**, 557–577 (1985).
20. Wimberly, B. *et al.* Structure of the 30S ribosomal subunit. *Nature* **407**, 327–339 (2000).
21. Ogle, J.M. *et al.* Recognition of cognate transfer RNA by the 30S ribosomal subunit. *Science* **292**, 897–902 (2001).
22. Abel, K.M., Yoder, M.D., Hilgenfeld, R. & Jurnak, F. An α to β conformational switch in EF-Tu. *Structure* **4**, 1153–1159 (1996).
23. Polekhina, G. *et al.* Helix unwinding in the effector region of elongation factor EF-Tu-GDP. *Structure* **4**, 1141–1151 (1996).
24. Hilgenfeld, R., Mesters, J. & Hogg, T. Insights into the GTPase mechanism of EF-Tu from structural studies. In *The Ribosome: Structure, Function, Antibiotics, and Cellular Interactions* (eds. Garret, R.A. *et al.*) 347–357 (ASM Press, Washington, DC, 2000).
25. Vogeley, L., Palm, G.J., Mesters, J.R. & Hilgenfeld, R. Conformational change of elongation factor Tu (EF-Tu) induced by antibiotic binding. *J. Biol. Chem.* **276**, 17149–17155 (2001).
26. Ban, N., Nissen, P., Hansen, J., Moore, P.B. & Steitz, T.A. The complete atomic structure of the large ribosomal subunit at 2.4 Å resolution. *Science* **289**, 905–920 (2000).
27. Moazed, D., Robertson, J.M. & Noller, H.F. Interaction of elongation factors EF-G and EF-Tu with a conserved loop in 23S rRNA. *Nature* **334**, 362–364 (1988).
28. Hausner, T.P., Atmadja, J. & Nierhaus, K.H. Evidence that the G2661 region of 23S rRNA is located at the ribosomal binding sites of both elongation factors. *Biochimie* **69**, 911–923 (1987).
29. Swart, G.W.M. & Parmeggiani, A. Effects of the mutation glycine-222→aspartic acid on the functions of elongation factor Tu. *Biochemistry* **26**, 2047–2054 (1987).
30. Vorstenbosch, E., Pape, T., Rodnina, M.V., Kraal, B. & Wintermeyer, W. The G222 mutation in elongation factor Tu inhibits the codon-induced conformational changes leading to GTPase activation on the ribosome. *EMBO J.* **15**, 6766–6774 (1996).
31. Robertus, J.D. *et al.* Structure of yeast phenylananine tRNA at 3 Å resolution. *Nature* **250**, 546–551 (1974).
32. Sussman, J.L., Holbrook, S.R., Warrant, R.W., Church, G.M. & Kim, S.H. Crystal structure of yeast phenylalanine transfer RNA. I. Crystallographic refinement. *Science* **123**, 607–630 (1978).
33. Harvey, S.C., Prabhakaran, M., Mao, B. & McCammon, J.A. Phenylalanine transfer RNA: molecular dynamics simulations. *Science* **223**, 1189–1191 (1984).
34. O'Connor, M. & Dahlberg, A.E. The involvement of two distinct regions of 23S ribosomal RNA in tRNA selection. *J. Mol. Biol.* **254**, 838–847 (1995).
35. Hirsh, D. Tryptophan transfer RNA as the UGA suppressor. *J. Mol. Biol.* **58**, 439–458 (1971).
36. Yarus, M., & Smith, D. tRNA on the ribosome: a Waggle theory. In ***tRNA**: Structure, Biosynthesis, and Function* (eds. Söll, D. & RajBahandary, U.) 443–468 (American Society for Microbiology, Washington, DC, 1995).
37. Hausner, T.P., Atmadja, J. & Nierhaus, K.H. Evidence that the G2661 region of 23S rRNA is located at the ribosomal binding sites of both elongation factors. *Biochimie* **69**, 911–923 (1987).
38. Wriggers, W., Agrawal, R.K., Drew, D.L., McCammon, A. & Frank, J. Domains motions of EF-G bound to the ribosome: insights from a hand-shaking between multi-resolution structures. *Biophys. J.* **79**, 1670–1678 (2000).
39. Wimberly, B.T., Guymon, R., McCutcheon, J.P., White, S.W. & Ramakrishnan, V. A detailed view of a ribosomal active site: the structure of the L11-RNA complex. *Cell* **97**, 491–502 (1999).
40. Vazquez, D. *Inhibitors of Protein Synthesis* (Springer, New York, 1979).
41. Saarma, U., Remme, J., Ehrenberg, M. & Bilgin, N. An A to U transversion at position 1067 of 23S rRNA from *Escherichia coli* impairs EF-Tu and EF-G function. *J. Mol. Biol.* **272**, 327–335 (1997).
42. O'Connor, M. & Dahlberg, A.E. The involvement of two distinct regions of 23S ribosomal RNA in tRNA selection. *J. Mol. Biol.* **254**, 838–847 (1995).
43. Zavialov, A.V., Buckingham, R.H. & Ehrenberg, M. A posttermination ribosomal complex is the guanine exchange factor for peptide release factor RF3. *Cell* **107**, 1–20. (2001).
44. Rosenthal, P. & Henderson, R. Optimal determination of particle orientation, absolute hand, and contrast loss in single particle electron cryomicroscopy. *J. Mol. Biol.*, in press.
45. Jones, T.A., Zhou, J.Y., Cowan, S.W. & Kjeldgaard, M. Improved methods for building protein models in electron density maps and the location of errors in these models. *Acta Crystallogr.* A**47**, 110–119 (1991).
46. Cornell, W.D. *et al.* A second generation force field for the simulation of proteins, nucleic acids, and organic molecules. *J. Am. Chem. Soc.* **117**, 5179–5197 (1995).

Cell, Vol. 114, 123–134, July 11, 2003, Copyright ©2003 by Cell Press

Locking and Unlocking of Ribosomal Motions

Mikel Valle,[1,2] Andrey Zavialov,[4] Jayati Sengupta,[2] Urmila Rawat,[1,2] Måns Ehrenberg,[4] and Joachim Frank[1,2,3,*]
[1]Howard Hughes Medical Institute
[2]Health Research Incorporated at the Wadsworth Center
[3]Department of Biomedical Sciences
State University of New York, Albany
Empire State Plaza
Albany, New York 12201
[4]Department of Cell and Molecular Biology
Biomedical Center
S-75124 Uppsala
Sweden

Summary

During the ribosomal translocation, the binding of elongation factor G (EF-G) to the pretranslocational ribosome leads to a ratchet-like rotation of the 30S subunit relative to the 50S subunit in the direction of the mRNA movement. By means of cryo-electron microscopy we observe that this rotation is accompanied by a 20 Å movement of the L1 stalk of the 50S subunit, implying that this region is involved in the translocation of deacylated tRNAs from the P to the E site. These ribosomal motions can occur only when the P-site tRNA is deacylated. Prior to peptidyl-transfer to the A-site tRNA or peptide removal, the presence of the charged P-site tRNA locks the ribosome and prohibits both of these motions.

Introduction

During the protein elongation cycle, the ribosome alternates between two conformational states: the pretranslocational ("pre") state, where the ribosome has a deacylated tRNA in the P site and a peptidyl-tRNA in the A site; and the posttranslocational ("post") state, where the ribosome has a peptidyl-tRNA in the P site and an empty A site. The transition of the ribosome from the pre- to the poststate is catalyzed by EF-G·GTP in a reaction stimulated by hydrolysis of GTP to GDP and inorganic phosphate (Ishitsuka et al., 1970; Rodnina et al., 1997). In this transition, the mRNA·(tRNAs) complex moves through the inter-subunit space of the ribosome by one codon, and the A- and P-site tRNAs are translocated to the P and E sites, respectively. Even in the absence of GTP hydrolysis, the interaction between EF-G and the ribosome induces a ratchet-like rotation of the small (30S) subunit relative to the large (50S) subunit (Frank and Agrawal, 2000).

The pre- and poststate ribosomes carry out different functional tasks, and major conformational rearrangements could therefore be expected. A postribosome must be stable and have a structurally well-defined A site that allows for rapid and accurate codon recognition by the anticodon of incoming cognate or noncognate aminoacyl-tRNAs, in complex with elongation factor Tu (EF-Tu) and GTP (Stark et al., 1997a, 2002; Agrawal et al., 2000a; Valle et al., 2002). In contrast, the prestate must allow for the large conformational changes in the ribosome that bring about the translocation event (Agrawal et al., 1999; Frank and Agrawal, 2000). However, despite their different agendas, the ribosomal pre- and poststates have very similar morphologies, at least as judged from cryo-electron microscopy (cryo-EM) maps at resolutions in the 15–20 Å range (Stark et al., 1997b; Agrawal et al., 2000b).

Recent biochemical experiments have revealed that the presence or absence of a nascent peptide on the P-site bound tRNA controls the mode of interaction between the ribosome and three translation factors; initiation factor 2 (IF2), EF-G, and peptide release factor 3 (RF3) (Zavialov and Ehrenberg, 2003 [this issue of *Cell*]; Zavialov et al., 2001, 2002). Normally, EF-G·GTP binds to the ribosome bearing a deacylated P-site tRNA and a peptidyl-tRNA in the A site, i.e., a prestate ribosome. However, EF-G·GTP also binds with high affinity to a postribosome, obtained by removing the peptide from the peptidyl-tRNA originally in the P site by the antibiotic puromycin (Zavialov and Ehrenberg, 2003 [this issue of *Cell*]). Puromycin mimics the aminoacyl-CCA end of the A-site tRNA in the peptidyl-transferase reaction; it accepts the polypeptide from the P-site tRNA and dissociates from a ribosome with a deacylated P-site tRNA and an empty A site (Traut and Monro, 1964). This observation suggests that the removal of the peptide from the P-site tRNA by peptidyl-transfer allows the ribosome to switch into a conformation similar to that induced by EF-G·GTP binding to the prestate, even in the absence of a peptidyl-tRNA in the A-site tRNA (Zavialov and Ehrenberg, 2003 [this issue of *Cell*]).

Using cryo-EM, we have studied ribosomal complexes with a vacant A site and bearing either a tetrapeptidyl-tRNA or, after treatment of the poststate ribosome with puromycin, a deacylated tRNA in the P site. As in the case of the pre- and poststates (Stark et al., 1997b; Agrawal et al., 2000b), no large structural difference is observed. When the study was extended to several 70S ribosome·EF-G complexes, it became clear that as long as the P-site tRNA is still charged with the tetrapeptide, no relative rotation between subunits is induced in the presence of EF-G. In contrast, puromycin treatment and subsequent removal of the peptide from the P-site tRNA allows a high affinity interaction with EF-G·GTP and produces the ratchet-like subunit rotation. Thus, the presence or absence of a peptide bound to the P-site tRNA appears decisive in determining the ribosome's disposition for conformational change. In the presence of peptide, the ribosome is locked into a state that prohibits the conformational change (i.e., the ratchet rotation) required for translocation, while its absence leaves the ribosome in a labile state that allows this rotation to occur upon EF-G binding.

Interestingly, during the ribosomal ratchet rotation,

*Correspondence: joachim@wadsworth.org

the deacylated P-site tRNA changes its position and moves to a hybrid P/E position (Bretscher, 1968; Moazed and Noller, 1989), a possible translocation intermediate. In addition, the L1 stalk of the 50S subunit moves toward the inter-subunit space, where it interacts with the observed P/E tRNA. In line with previous suggestions and observations (Gomez-Lorenzo et al., 2000; Harms et al., 2001), this conformational change suggests that the L1 stalk plays a pivotal role in the translocation process.

These results, together with other findings (Zavialov and Ehrenberg, 2003 [this issue of *Cell*]; Zavialov et al., 2002), support the notion that the P-site tRNA—apart from mRNA the only nonribosomal element always present on the ribosome during protein synthesis—is a major determinant of the ribosomal behavior, and that its state (peptide or no peptide) controls the sequence of interactions between the ribosome and several translation factors. This regulation includes the locking and unlocking of characteristic motions intrinsic to the ribosomal architecture.

Results and Discussion

Unlocking of Ribosomal Motions

The presence versus absence of a peptide or just a single amino acid on the P-site bound tRNA strongly affects the affinity of the bacterial ribosome to the three translation factors IF-2, EF-G, and RF3 (Zavialov and Ehrenberg, 2003 [this issue of *Cell*]). The most likely explanation is that there are certain structural changes in the ribosome that regulate the action and GTPase activity of these essential G proteins. Cryo-EM and three-dimensional (3D) reconstruction techniques were applied to several *E. coli* 70S complexes (see Experimental Procedures). The first analysis was made with 70S ribosomes, containing either the tetrapeptidyl-tRNA MFTI-tRNA^{Ile} (Figure 1A) or its deacylated counterpart tRNA^{Ile} (Figure 1B) in the P site. The former complex is a poststate ribosome formed in an EF-G driven translocation event. The latter complex, obtained by transfer of the tetrapeptide from the P-site tRNA to the antibiotic puromycin, can be recycled by RRF and EF-G (Zavialov and Ehrenberg, 2003 [this issue of *Cell*]). It is functionally and structurally indistinguishable from a posttermination ribosome where the ester bond in peptidyl-tRNA has been hydrolyzed by the action of a class I peptide release factor (Rawat et al., 2003) that has been subsequently removed by the action of the class II peptide release factor RF3 (Zavialov et al., 2001, 2002). Comparison of these complexes before (Figure 1A) and after (Figure 1B) puromycin treatment shows that the overall ribosomal architecture is very similar in the two cases. Not even the position of the P-site tRNA (in green) changes when its peptide link with the peptidyl-transferase center in the 50S subunit (Figure 1A) is removed by the action of puromycin (Figure 1B). It can therefore be concluded that puromycin treatment per se does not induce a structural change in the ribosome of sufficient magnitude to be recognized at the resolution (10–13 Å) obtained in our study.

The structural study includes the interaction of these two ribosomal complexes with EF-G in the presence of either GDP and fusidic acid (fus) or GDPNP. The resulting 3D maps are shown in Figures 1C–1E. Ribosomes with MFTI-tRNA^{Ile} in the P site (Figure 1A) have low affinity to EF-G·GDPNP (Zavialov and Ehrenberg, 2003 [this issue of *Cell*]) and, as expected, no such complex could be seen by cryo-EM. However, the use of GDP instead GTP allowed us to obtain a 70S·EF-G·GDP·fus complex with peptidyl-tRNA in the P site. The 3D map for the 70S·MFTI-tRNA^{Ile}·EF-G·GDP·fus complex shows EF-G (in red) bound below the base of the L7/L12 stalk as in previous cryo-EM studies (Agrawal et al., 1998, 1999, 2000a; Stark et al., 2000). Fus is an antibiotic that stalls the ribosome in complex with EF-G·GDP in a state proposed to precede the conformational changes that release EF-G from the ribosome after translocation (reviewed in Rodnina et al., 2001). Previous experiments showed that binding of EF-G·GTP to a prestate ribosome (peptidyl-tRNA in the A site and deacylated tRNA in the P site) in the presence of fus promotes a "ratchet-like" rotation of the 30S relative to the 50S subunit that stalls the ribosome in a long-lived complex with EF-G·GDP (Frank and Agrawal, 2000). In the complex depicted in Figure 1C, in contrast, no such conformational change can be observed after binding of EF-G·GDP together with fus, although EF-G in its GDP form becomes stably bound to the ribosome in the presence of the drug; despite its strong interaction with the factor, the ribosomal subunits remain in the relative position they had before the interaction with EF-G (Figure 1A). Accordingly, the ribosome appears to be locked in a state that prohibits any structural rearrangement upon the binding of the factor.

The scenario changes when the binding reaction is made in the presence of puromycin, which removes the peptide from peptidyl-tRNA. In this case, the interaction of EF-G·GDP·fus with the 70S ribosome promotes several major changes (Figure 1D). The tRNA in the P site has moved to what appears to be a P/E hybrid site (Figure 1D) as postulated by Moazed and Noller (1989), very different from the previous P or E site positions (Figures 1A, 1B, and 1C). The anticodon of the tRNA is still in the P site of the 30S subunit, while its CCA end is in the E site of the 50S subunit. The tRNA previously in the E site has dissociated from the ribosome, and a more detailed comparison with the EF-G free complexes (Figures 1A and 1B) reveals that the previously described ratchet-like motion (Frank and Agrawal, 2000) has occurred. Here, however, this movement has taken place in the absence of a peptidyl-tRNA at the A site. When EF-G binds to the ribosome (Figure 1D), the 30S subunit rotates counterclockwise with respect to the 50S subunit. Furthermore, the L1 stalk of the 50S subunit moves in the direction opposite to the movement of the 30S subunit head, toward the inter-subunit cavity (compare Figures 1B and 1D), where it interacts with the P/E-tRNA.

Removal of the peptide from the tRNA at the P site of a poststate ribosome enhances its affinity to EF-G in the GTP form and induces EF-G dependent hydrolysis of GTP (Zavialov and Ehrenberg, 2003 [this issue of *Cell*]). The 3D map of such a ribosome in complex with EF-G and the nonhydrolyzable GTP analog GDPNP is depicted in Figure 1E. Both the rotation between the subunits and the movement of the L1 stalk that were seen as a result of EF-G·GDP·fus binding after puromy-

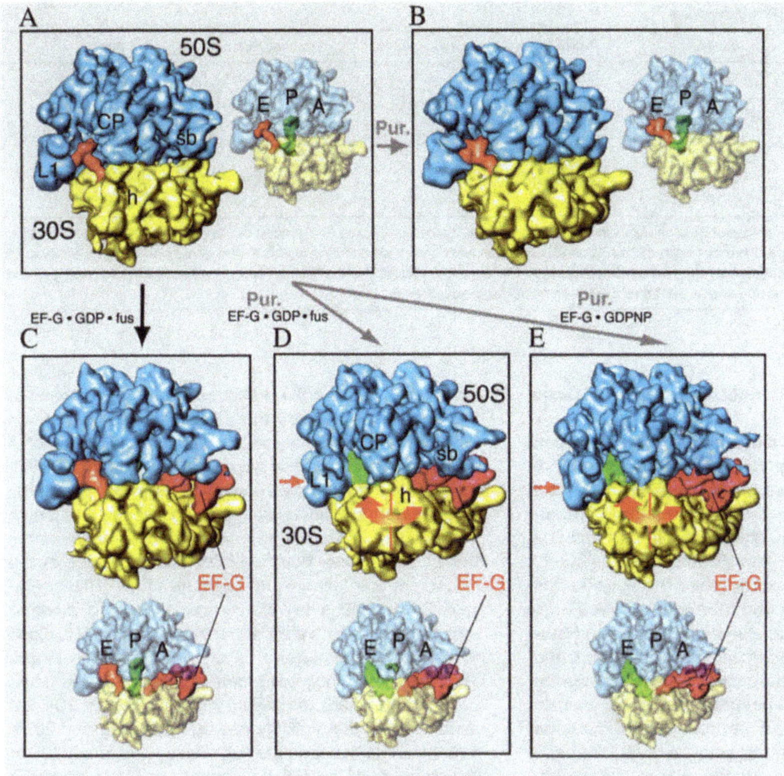

Figure 1. Unlocking of Ribosomal Conformational Changes

Solid and semitransparent representations of ribosomal 3D maps. The complexes are: 70S·MFTI-tRNA^{Ile} (A); 70S·MFTI-tRNA^{Ile}+puromycin (B); 70S·MFTI-tRNA^{Ile}·EF-G·GDP·fusidic acid (C); 70S·MFTI-tRNA^{Ile}·EF-G·GDP·fus+puromycin (D); and 70S·MFTI-tRNA^{Ile}·EF-G·GDPNP+puromycin (E). In the semi-transparent views the A, P, and E sites are labeled. EF-G is depicted in red; P-site tRNAs in green; E-site tRNAs in orange; and the P/E hybrids in an intermediate color (between the green of the P-site tRNA and the orange of the E-site tRNA). The conformational changes in the ribosome are indicated by dashed arrows in (D) and (E). The red arrows in (D) and (E) indicate the direction of the conformational changes.

Landmarks: CP, central protuberance; sb, base of the L7/L12 stalk; h, head; L1, L1 stalk of the 50S subunit.

cin treatment (Figure 1D) do occur also here and, similarly, only a P/E-tRNA is present in the ribosome (Figure 1E). Since EF-G in the GTP conformation cannot bind to postribosomal complexes (Figure 1A; Zavialov and Ehrenberg, 2003 [this issue of *Cell*]), we conclude that binding of EF-G in the GTP form requires an unlocked ribosome, i.e., a ribosomal state that is able to switch to a conformation with high affinity to EF-G·GDPNP. With this consideration in mind, we will refer with the term “locked” to ribosomes bearing a charged P-site tRNA, while ribosomes bearing a decylated tRNA in the P site will be described as “unlocked”.

The relative movement between ribosomal subunits during the 70S·EF-G interaction has been analyzed in several complexes in previous work, and some of the previous and present cryo-EM maps are summarized in Table 1. In the first complex, EF-G and a GTP analog were bound to an empty 70S ribosome, containing neither tRNA nor mRNA (Agrawal et al., 1999; Frank and Agrawal, 2000). The binding of the factor did induce the relative rotation between subunits, seen here in the presence of both tRNA and mRNA (Figures 1D and 1E). From this it can be concluded that the conformational change is an intrinsic property of the two-subunit design of the ribosome itself, and that the rearrangement is catalyzed by EF-G. However, the presence of tRNAs alters this behavior. Specifically, the state of the P-site tRNA regulates the ratchet rotation upon the interaction of the ribosome with EF-G. A charged tRNA at the P site (poststate ribosome in Table 1) does not allow a high-affinity interaction between the ribosome and EF-G·GTP, and even when the factor binds in its GDP form with fus, the subunit rotation does not take place. As shown by the use of puromycin with those complexes (poststate ribosome and puromycin in Table 1), when this tRNA is deacylated, high-affinity interactions become allowed for both EF-G·GDPNP and EF-G·GDP·fus, since now the ratchet-like rotation of the subunits can

Table 1. Observations in 70S·EF-G Complexes

Ribosomal Complex	EF-G	P-site tRNA	A-site tRNA	Ratchet Rotation	L1 Stalk
Empty 70S[a]	GTP analog	-	-	+	open/closed
Prestate[a]	GTP fus	deacylated	charged	+	open/closed
Poststate	GTP analog No binding	charged	-	-	open
Poststate	GDP fus	charged	-	-	open
Poststate and puromycin	GTP analog	deacylated	-	+	closed
Poststate and puromycin	GDP fus	deacylated	-	+	closed

The table shows a summary of conformational changes in the ribosome in different complexes analyzed in our group. The ratchet-like rotation between ribosomal subunits and the movement of the L1 stalk correlate with deacylated-tRNAs in the P-site or with empty ribosomes. L1 open/closed indicates a superposition of the two conformations in the cryo-EM maps. The complexes with ([a]) were described in Frank and Agrawal, 2000; the rest of the complexes in the table are taken from the present work.

take place, presenting the correct binding constellations for these ligands.

It is clear that the 70S ribosome becomes unlocked by the removal of the peptide from the peptidyl-tRNA in the P site, but this event per se does not lead to a detectable change in the ribosome structure. One attractive explanation for the unlocking event is that it is linked to the ability of the P-site tRNA to enter the hybrid P/E site, which can only occur when the peptide has been cleaved off the CCA end of the tRNA. The P-site tRNA interacts with several elements on the inner surface of the central protuberance in the 50S subunit, and its anticodon stem is buried between the head and the body of the 30S subunit. Among the previously characterized contacts are those with protein L5 in the large subunit and protein S13 in the small subunit (Yusupov et al., 2001). In the ratchet rotation motion, the central protuberance and the head of the 30S subunit experience the maximal relative displacement. Another, and perhaps complementary explanation, is that as long as the P-site tRNA is connected to the peptidyl-transferase center (PTC) by its peptide, it might act as a stabilizing link between these ribosomal parts, thereby preventing their relative movement. Severing the tRNA from the peptide would, following this explanation interrupt its connection to the PTC, thus enabling the movement.

Movement of the L1 Stalk

The L1 stalk of the large subunit is composed of helices H76–78 from 23S rRNA and protein L1. It interacts with the elbow region of the E-site tRNA, and it has been suggested that dissociation of this tRNA requires the L1 stalk to move (Gomez-Lorenzo et al., 2000; Yusupov et al., 2001). Recently, the crystal structure of the 50S subunit from *Deinococcus radiodurans* (Harms et al., 2001) revealed that the entire L1 stalk is tilted by 30° compared to its orientation in the crystal structure of the 70S ribosome from *Thermus thermophilus* (Yusupov et al., 2001). This observation led Harms and coworkers (2001) to postulate, in line with earlier suggestions based on cryo-EM of the eukaryotic ribosome (Gomez-Lorenzo et al., 2000), that the ribosome structure alternates between an "open" conformation, where the exit path of the E-site tRNA is free (as observed in the structure of *D. radiodurans*), and a "closed" conformation, in which the exit path is blocked (as in *T. thermophilus*).

In the present work, binding of EF-G to 70S ribosome complexes made in the presence of puromycin leads to a tRNA that is held by the L1 stalk in a P/E position (Figures 1D and 1E), in a similar way that the E-site tRNA is held in the factor-free ribosome (Figures 1A and 1B) or in the locked complex (Figure 1C). The contact of the L1 stalk with the tRNA residing in the P/E site and E site, respectively, is apparently maintained by a movement of the L1 stalk, away from the inter-subunit space of the ribosome, that follows the trajectory of the tRNAs. Figures 2A and 2B show clearly that there is a pivoting movement of the L1 stalk at the base of helix H76, close to the ribosomal proteins L2 and L9 (labeled in Figure 2B). The size of this movement can be measured when both positions are displayed together (Figure 2C): the distal part of the stalk moves by approximately 20 Å, and the angle between the two positions is about 20°. In the structure of the 70S ribosome from *T. thermophilus* (Yusupov et al., 2001) the E-site tRNA was held by the L1 stalk in the closed position (as in Figure 2B), but its position is different from the E site that we deduce in the present work, where the tRNA interacts with the L1 stalk in the open position (as in Figure 2A). Those different E-site tRNA positions, together with the P/E tRNA, could be part of a pathway of the tRNA moving from the P site to the final position, from where it exits the ribosome (Figures 2D–2G). It must be noted that the tRNAs in Figures 2E–2G, in the P/E, E_Y (from Yusupov et al., 2001) and E states, are seen to have a close interaction with protein L1 and helix H79 of the 23S rRNA in the L1 stalk.

In our complexes, the L1 stalk movement and the subunit rotation occur only when ribosomes with a deacylated P-site tRNA interact with EF-G (Table 1). Similar changes in the L1 region were detected in the *Saccharomyces cerevisiae* ribosome (Gomez-Lorenzo et al., 2000) and were correlated with the presence or absence of tRNAs and elongation factor 2 (EF2, the eukaryotic homolog of EF-G). In the *E. coli* ribosome, upon its interaction with EF-G, the L1 stalk appeared as a bilobed structure (Agrawal et al., 1999; Frank and Agrawal, 2000), which was explained as a mixture of conformations corresponding to the closed and the open states (Table 1).

Coupling of L1 stalk movement and tRNA translocation, directly visualized in our reconstructions, was suggested by several findings. A P-site tRNA analog, comprising only the anticodon arm and therefore unable to interact with the L1 stalk, inhibits translocation (Joseph

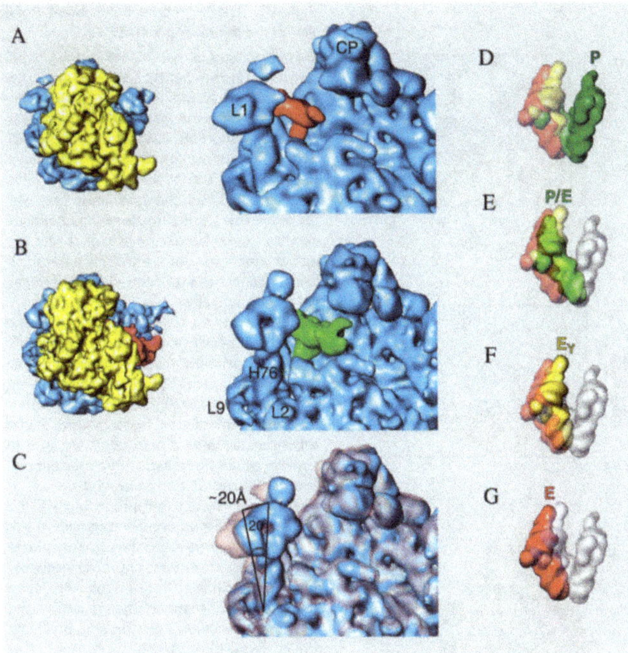

Figure 2. Coupling of the L1 Stalk Motion with the Movement of tRNAs

Rendering of the 70S·MFTI-tRNA^{Ile}+puromycin complexes before (A) and after (B) the interaction with EF-G·GDPNP (a nonhydrolyzable GTP analog). The L1 stalk region of the 50S subunit can be seen open (A) or closed (B) upon the interaction with EF-G. In both images a view of the 50S subunit from the inter-subunit cavity is presented, while the thumbnails show the orientation of the ribosome. The conformational change of the L1 stalk can be observed in (C) with both conformations of the 50S subunit represented together, with the one from (A) shown in semi-transparent pink, and the one from (B) in solid blue.

(D–G) show the experimentally observed positions of different tRNAs, as follow: P, P-site tRNA; P/E, P/E hybrid observed in this work; E_Y, E-site tRNA in the crystal structure from Yusupov et al., 2001; E, E-site tRNA shown in this work.

Labels and landmarks: L9, ribosomal protein L9; L2, protein L2; H76, helix 76 from the 23S rRNA; L1, L1 protuberance or stalk in the 50S subunit.

and Noller, 1998). Ribosomes lacking protein L1 have a reduced rate of protein synthesis (Subramanian and Dabbs, 1980) and modifications of the 3′-end of the P-site tRNA adversely affect the GTPase activity of EF-G and reduce the translocation rate by up to 40-fold. This suggested a cooperative function between the E site and the EF-G binding site mediated by conformational changes of the ribosome (Lill et al., 1989).

Together with these previous findings, our data suggest that upon the binding of EF-G, and combined with the ratchet-like rotation, the L1 stalk moves closer to the P site region, takes hold of the tRNA (in the closed conformation), then, upon transition to the open state, pulls it toward the exit of the ribosomal inter-subunit space. The P/E tRNA position defined in the present work could represent a translocation intermediate in the transition of the tRNA from the P to the E site. Therefore, the L1 stalk appears to participate in the translocation process itself, rather than being limited to that of a gate for the E-site tRNA.

Molecular Aspects of the Ratchet-like Rotation

The relative rotational motions of the two subunits during the interaction with EF-G were suggested to be fundamental for the translocation process (Frank and Agrawal, 2000). However, these motions are still not well understood since it has proved difficult to obtain sufficiently stable and uniform complexes of reaction intermediates to allow for cryo-EM analysis of their structures. Although we could improve the resolution to the 12–13 Å range of several previously analyzed 70S ribosome·EF-G complexes (as those described in Frank and Agrawal, 2000), undefined regions remained in the final 3D maps (data not shown). This suggested that the a priori biochemical uniformity of the samples did not translate into a homogeneous structural display among the individual ribosomes.

One puzzle was the low occupancy of the EF-G·GTP form (with a noncleavable GTP analog) in the ribosomal complexes containing both P- and A-site tRNAs (Agrawal et al., 1999). The results of the present experiments now suggest this low occupancy to be the inevitable consequence of a general locking/unlocking phenomenon. During normal translocation mediated by EF-G, the peptidyl-tRNA in the A site is moved to the P site. Since, however, a peptidyl-tRNA in the P site is incompatible with binding of EF-G in the GTP form (Zavialov and Ehrenberg, 2003 [this issue of *Cell*]) because the ribosome is locked (this work), no such stable complexes can be expected. In the complexes presented here, the absence of a peptidyl-tRNA in the A site prevents the normally occurring EF-G dependent transition to a post-ribosome with peptidyl-tRNA in the P site. Therefore, these complexes allow characterization of the EF-G dependent ratchet-like rotation between the ribosomal subunits.

Figure 3 presents a detailed comparison between the puromycin-treated 70S·MFTI-tRNA^{Ile} complex before (Figures 3A and 3B) and after (Figures 3C and 3D) interaction with EF-G·GDPNP. The largest relative movement between the 50S and the 30S subunits occurs at the connection between the central protuberance of the 50S subunit and the head of the 30S subunit. In Figures 3B and 3D some of the structural elements present in this region are shown, fitted into the densities of both cryo-

Cell
128

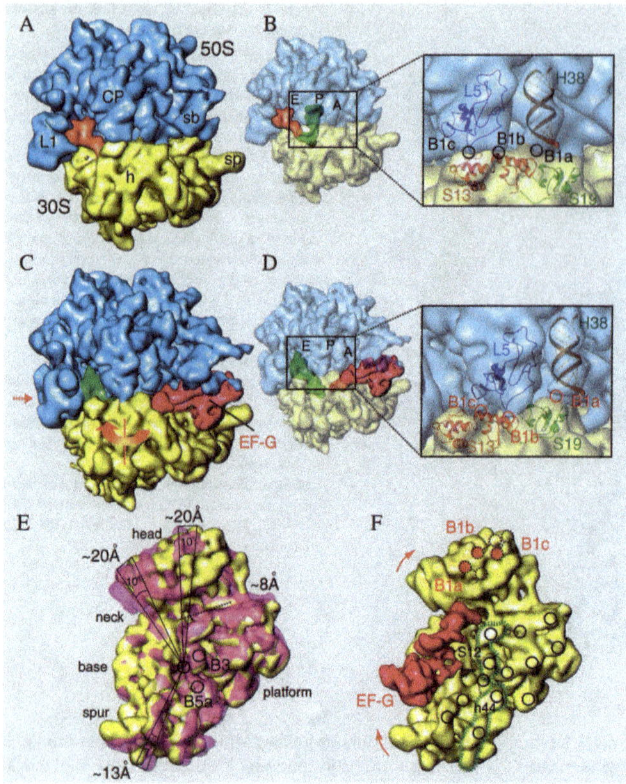

Figure 3. Relative Ribosomal Subunit Rotation in the Interaction with EF-G

Cryo-EM maps of the 70S·MFTI-tRNA^{Ile}+puromycin samples before (A and B) and after (C and D) the addition of EF-G·GDPNP. The ratchet-like rotation of the 30S subunit with respect to the 50S subunit is clear in (C), where the connection between the central protuberance (CP) of the 50S and the head (h) of the 30S has changed compared with (A). In (B) and (D), the ribosomal subunits are semitransparent and the insets show the elements that form the central protuberance-head connections of inter-subunit bridges (B1a, B1b, and B1c).

The rotation of the 30S subunit seen from the inter-subunit cavity is depicted in (E) and (F). In (E), the position of the small subunit before the interaction with EF-G (semitransparent pink) is compared with its position upon ribosome·EF-G interaction (solid yellow). If the subunit is taken as a rigid body, the axis of rotation would be between the inter-subunit bridges B3 and B5a on helix 44 (E).

(F) shows EF-G (solid red) interacting with the 30S subunit of the complex depicted in (C). The inter-subunit bridges that do not change in the ratchet movement are represented by empty black circles. Those bridges that move during the conformational change are marked by filled red circles. Their original positions are labeled with dashed black circles.

EM maps. There are three bridges in this region (Gabashvili et al., 2000), which change significantly when EF-G binds to the ribosome. In two of the bridges, B1b and B1c, the contacts with protein L5 move to different regions of protein S13. In the third bridge, B1a, the connection that the helix H38 of the 50S subunit forms with S13 (in the factor-free ribosome, Figure 3B) is lost and protein S19 becomes the new contact point on the 30S subunit side (in the ribosome·EF-G·GDPNP complex, Figure 3D). It is obvious that proteins L5 and S13 play a central role in this motion. The fact that the P-site tRNA interacts with both proteins (Yusupov et al., 2001) supports the idea of a mechanical coupling between this tRNA and the ribosomal subunit rotation, which can explain the locking and unlocking of the conformational change. The other bridges are conserved and display no changes at the resolution of the maps.

The alternate bridge connections define two extreme positions in the reorganization of the ribosome. In Figure 3E, we compare the positions of the 30S subunit in the two complexes when viewed from the inter-subunit space. The described changes in the inter-subunit bridges that connect the head of the 30S subunit with the central protuberance of the 50S subunit allow the large displacements in the head and the spur regions associated with the subunit rotation (Figure 3F). The axis of the rotation passes through a point between the bridges B3 and B5a (Gabashvili et al., 2000; Yusupov et al., 2001) in helix h44 of the 16S rRNA, above the region of helix h27. This observation is compatible with the recent study of the ratchet-like rotation motion carried out by real-space refinement (Gao et al., 2003), where helix h27 of the 16S rRNA, known to switch between two conformations (Lodmell and Dahlberg, 1997), has been identified as the axis of the ratchet-like rotation in the 30S ribosomal subunit. The rotation of the subunit head around this axis (as measured between the different positions of the fitted proteins S13 and S19) is about 10°, and the lateral movement of the tip of h44 parallel to the mRNA pathway is about 8 Å in agreement with a previous estimate (Agrawal et al., 1999; VanLoock et al., 2000). The observed conformational changes can be explained by the interaction of EF-G with the 30S subunit (Figure 3F): the factor binds in the neck region, acting as a molecular wedge between the head and the base, pushing the head of the 30S in the direction of the subunit rotation. EF-G interacts directly with h44, promoting the movement of this helix toward the P site. In the rotation of h44 the pivotal point is close to protein S12, another region that EF-G binds to.

In the ribosomal recycling step, carried out by EF-G and ribosomal recycling factor RRF (Karimi et al., 1999; Hirokawa et al., 2002), there are initially a deacylated-tRNA in the P site and a stop codon in the A site (see

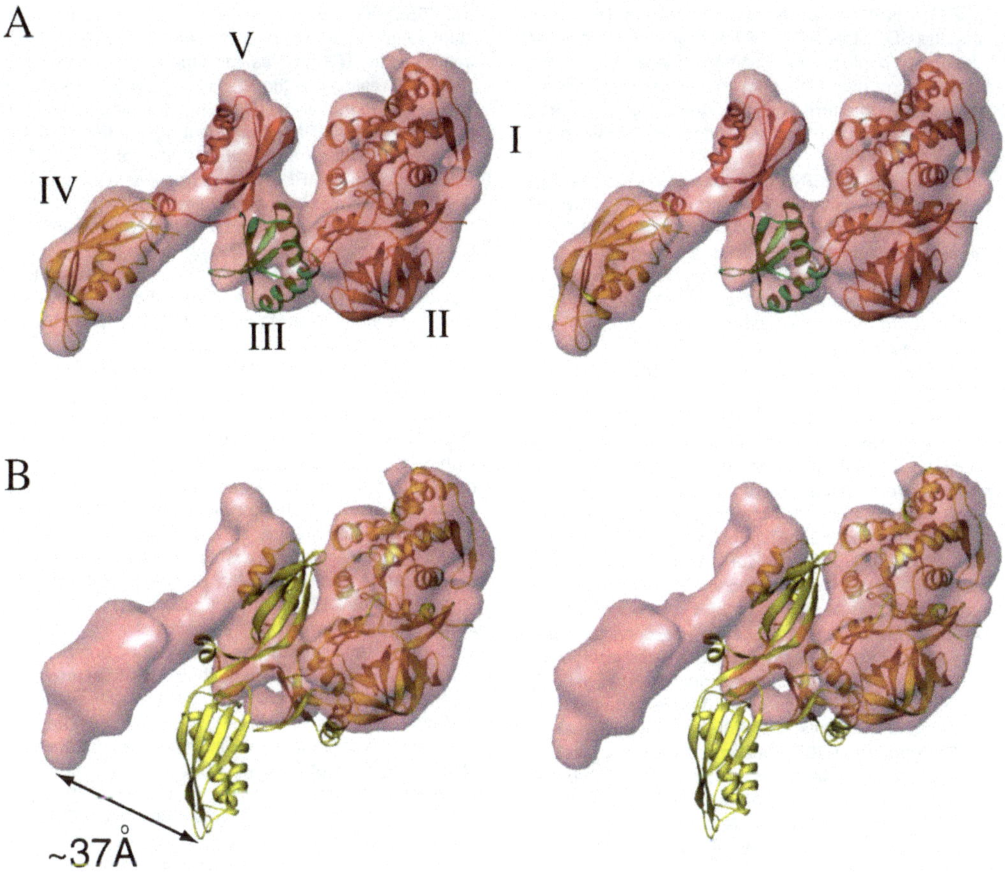

Figure 4. Docking of EF-G

Stereo views of the fitted EF-G·GDP structure in the cryo-EM density for EF-G·GDPNP (semitransparent red). In (A) relative movements between domains were allowed, while in (B) the EF-G·GDP coordinates were used as a rigid body and only domains I and II were docked.

Figure 1B). Binding of EF-G·GTP to such a ribosomal complex presumably causes the same conformational change as seen in Figures 3E–3F. In the presence of RRF, which binds in the inter-subunit space of the ribosome and overlaps with the CCA end of the P-site tRNA (Lancaster et al., 2002), the ratchet-like movement could extensively disrupt the inter-subunit bridges. This, we suggest, is the structural basis for ribosome splitting into subunits by RRF and EF-G (Karimi et al., 1999).

Figure 4 presents the density of EF-G in complex with the noncleavable GTP analog GDPNP and the result of the fitting of the X-ray structure of a mutant factor (the only structure with the complete coordinates for domain III for EF-G) in complex with GDP (Laurberg et al., 2000). The docking was carried out manually in O (Jones et al., 1991) using the domains of the protein as rigid bodies. The good representation of the factor and the detailed resemblance with the local features of the X-ray structure (Figure 4A) reflect the high occupancy and the uniformity of the sample, and allow a high degree of accuracy in the docking with an error estimate of ±1 Å (see Supplemental Data available at http://www.cell.com/cgi/content/full/114/1/123/DC1). Domains I and II did not require any relative movement compared to their position in the EF-G·GDP structure and they were docked as one piece (Figure 4B). In the case of domains III, IV, and V the conformational change between the two forms is clear: in the cryo-EM density of the complex in the GTP state, those three domains shift and rotate moving the tip of domain IV by approximately 37 Å from its position in the crystal. Comparing the position of domain IV of our docked EF-G with the coordinates of the wild-type EF-G in complex with GDP (Czworkowski et al., 1994), this distance is around 33 Å. This domain forms the aforementioned contact point of the factor with h44 and the head of the 30S subunit, close to the decoding center. The relative flexibility of these domains is important for the coupling of GTP hydrolysis on EF-G with ribosomal translocation (Peske et al., 2000; Rodnina et al., 2001). We propose that after the GTP hydrolysis on EF-G, a transition of the factor from the GTP (Figure 4A) to the GDP state (Figure 4B) is required to move the 30S subunit back to the position displayed before the ratchet-like rotation. This back movement could be accompanied by the mRNA shift of one codon in relation to the 30S subunit that occurs during every translocation step.

In the case of the EF-G·GDP·fus complexes, the isolated cryo-EM densities for the factor suggest an overall structure similar to the EF-G·GTP state described in Figure 4A (data not shown). However, the maps did not provide a detailed definition of the local features for EF-G domains; the masses were discontinuous and did not sum up to the whole mass of the factor. The structural definition was particularly poor in the region of domains III, IV, and V. This indicates, firstly, that EF-G·GDP·fus has a structure closer to the GTP than to the GDP conformation of EF-G and, secondly, that the presence of fus and GDP makes the factor flexible enough to adapt to the different ribosome conformations that are available before and after unlocking.

In all the structural characterizations of the 70S ribosome·EF-G interaction performed in our group (Agrawal et al., 1998, 1999, 2000a, 2000b), as well as EF2 complexes with the eukaryotic ribosome (Gomez-Lorenzo et al., 2000), the arrangement of the domains of the elongation factors and their binding sites on the ribosome are very similar. In contrast, Stark and coworkers (2000) postulate drastic conformational changes of EF-G combined with migrations of the binding sites on the two ribosomal subunits. This latter model is based on cryo-EM studies of 70S ribosome·EF-G complexes in the presence of the antibiotic thiostrepton, which binds to the L11 protein region at the base of the L7/L12 stalk of the 50S subunit (Highland et al., 1975), one of the binding sites for EF-G. Therefore, the interference by the antibiotic could have led to an aberrant structure of the elongation factor. Furthermore, the conformational changes they postulate for the translocation factor are difficult to reconcile with the space that is available below the L7/L12 stalk (Ramakrishnan, 2002), which makes difficult to assess the validity of their interpretation.

Among the binding sites for EF-G, the stalk formed by proteins L7/L12 and its base have been seen in close interaction with the factor. Particularly important is the putative interaction of the stalk that triggers the GTPase activity of EF-G (Savelsbergh et al., 2000). The stalk is known to be very flexible, but it is apparently stabilized in some ribosomal binding states and thus could be morphologically described (Agrawal et al., 1998; Stark et al., 2000). However, in the current cryo-EM maps, no such stabilization is seen, as the stalk appears reduced by conformational averaging, so no conclusions can be drawn about this region of the map.

Implications for the Translocation Process

An A-site tRNA, and particularly its anticodon, is required for EF-G dependent translocation (Joseph and Noller, 1998). Its absence in the complexes presented here blocks the transition to the poststate, thereby preserving strong binding of EF-G in the GTP form and, at the same time, preventing mRNA movement. When, in the 70S·EF-G complex in the presence of puromycin (sample in Figure 1E), we substituted the GDPNP (GTP analog) with GTP, the resulting map clearly showed a tRNA still residing in the P/P site, while EF-G was not visible due to its turnover. This result indicates that the blockage of the mRNA movement due to the vacant A site prevented the translocation (data not shown). The P/E position of the tRNA that we observe in the 70S·EF-G·GDPNP complex can be understood as an effect of this blockage. Previously, Moazed and Noller (1989) followed the movements of tRNAs through the ribosomal cavity using chemical footprinting. From that work, it was believed that immediately after the transfer of the peptide from the P site to the A-site tRNA, these tRNAs move to the P/E and A/P hybrid sites, respectively. The concept of tRNAs that are translocated in two steps and the existence of intermediate hybrid positions were predicted before (Bretscher, 1968), but in the work by Moazed and Noller (1989) this first transition was seen without the participation of EF-G. Furthermore, a treatment with puromycin of an aminoacylated tRNA in the P site (a ribosomal complex with a vacant A site) led to a P/E tRNA intermediate, as well (Moazed and Noller, 1989). This latter complex is essentially equivalent to our poststate ribosome in the presence of puromycin (shown in Figure 1B), but in our case the tRNA remains in the P/P site after peptidyl-transfer.

To further investigate this discrepancy between results obtained with different methods, we have analyzed a prestate complex containing a deacylated P-site tRNA and a peptidyl fMet-Phe-$tRNA^{Phe}$ in the A site after the peptide transfer reaction. The complex was obtained by short incubation of initiated ribosomes that carry an fMet-$tRNA^{fMet}$ in the P site with a Phe-$tRNA^{Phe}$ in the presence of EF-Tu and GTP. No antibiotics or GTP analogs were used that could inhibit peptidyl-transfer (90% of the A-site tRNAs were bearing the fMet-Phe dipeptide as judged by HPLC analysis). The final 3D map is shown in Figure 5. As in most of the ribosomal complexes visualized in our lab, the E site is occupied by tRNA, which we attribute to free deacylated tRNAs that have high affinity for the ribosomal exit site. Despite the presence of a peptide in the A-site tRNA and the deacylation of the P-site tRNA, the deacylated tRNA has stayed in the P/P position. In the case of the A-site tRNA, we cannot rule out an A/P hybrid state because the 3D map does not allow an unambiguous tracing of the ssRNA at its CCA end, and this flexible region could reach close to the P site on the 50S subunit. However, the occupation of this region by the P/P tRNA suggests that fMet-Phe-$tRNA^{Phe}$ is still in the A/A configuration. This indicates that, in the high-efficiency mRNA translation system we have used, there is no movement of tRNAs to hybrid positions after peptide bond formation and prior to the interaction of EF-G with the ribosome.

A spontaneous translocation of tRNAs in the 50S subunit after the peptide bond formation would result in P/E and A/P hybrid states for the tRNAs. The postulated driving force in the mechanism is the difference in affinities that the 50S A, P, and E sites have for the 3′-end of tRNAs (Noller et al., 2002). The specific affinity of the 50S P site for the peptidyl moiety and the high affinity of the 50S E site for deacylated tRNAs (Lill et al., 1989) could move the acceptor stems of A- and P-site tRNAs

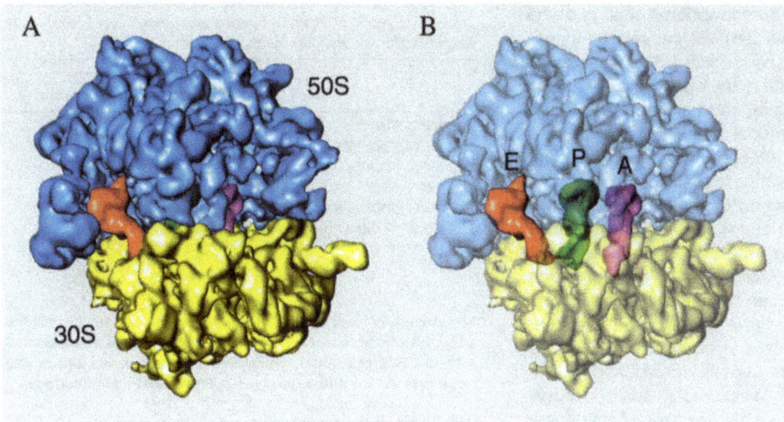

Figure 5. A Pretranslocational Complex

Rendering of the cryo-EM map for 70S ribosome bearing deacylated tRNA^{fMet} in the P site (green) and dipeptidyl fMet-Phe-tRNA^{Phe} in the A site (magenta). The E site is occupied by deacylated tRNA (orange) available in the solution.

(A) shows the segmented map in solid, while in (B) a semitransparent representation of the ribosomal subunits allows the visualization of the tRNAs in A, P, and E sites.

to the E and P sites before the interaction of EF-G with the ribosome (Noller et al., 2002). However, recent results using enzymatically active 50S subunits from *Haloarcula marismortui* presented evidence that this spontaneous movement in the 50S subunit does not immediately follow the peptidyl-transferase reaction (Schmeing et al., 2002). In the crystal structure of the 50S subunit the CC-puromycin used as an A-site tRNA analog remained in the 50S A site after accepting the peptide from the P site CCA analog, and this deacylated CCA was found in the 50S P site. Despite the fact that the peptidyl-transferase reaction was fully accomplished in the crystals, the P and A site substrates did not move from their original position, and it is reasonable to think that those analogs could have been moved easier than complete tRNA, which have multiple interactions with both ribosomal subunits. Nevertheless, the results in that work (Schmeing et al., 2002) were still understood as fitting into a model of spontaneous movement of tRNAs in the 50S subunit, and the structure was presented as a "pretranslocational intermediate." Our cryo-EM study supports a different interpretation, namely that the movement of tRNAs to their hybrid sites occurs only through the action of EF-G, and agrees with the X-ray structure of the 50S subunit after the peptide bond formation (Schmeing et al., 2002).

In addition, the transition of the tRNAs to A/P and P/E hybrid states due to the ribosomal interaction with EF-G would explain why the preribosomes are not puromycin-reactive (Moazed and Noller, 1989) until the binding of the elongation factor (Borowski et al., 1996). This lack of reactivity against the antibiotic would be the consequence of the fact that, before the binding of EF-G to the prestate ribosome, the 3′ end of the peptidyl-tRNA in the A site does not move to the P site in the 50S subunit spontaneously, and therefore no substrate is yet available to be accepted by puromycin. Furthermore, it has been shown that the P-site tRNA does not progress to the P/E state immediately after the peptide bond formation (Wower et al., 2000).

There is an integrated model for the protein elongation cycle; the α-ε model (reviewed in Marquez et al., 2002). This postulates a reciprocal coupling between the A and E sites that limits the number of tRNAs that can be simultaneously present on the ribosome (at the A and P sites, or P and E sites) during protein synthesis to maximally two. Our cryo-EM map of the pretranslocation complex does not support this model, since Figure 5 clearly shows three tRNAs, A-, P-, and E-site tRNAs, bound to the ribosome just after peptidyl-transfer in an authentic elongation cycle. In the α-ε model the translocation of tRNAs is accomplished by movable ribosomal domains, and the tRNA-ribosome interactions are conserved. We observe that the ratchet-like rotation is accompanied by the movement of the L1 stalk, but the combination of both motions still requires tRNA transients (as the visualized P/E hybrid position) that interact differently with the ribosomal subunits.

Accordingly, we propose a model for the translocation mechanism in which the peptidyl-transfer reaction leads to a ribosome with a deacylated tRNA in the P/P site and a peptidyl-tRNA in the A/A site (Figure 5). The ribosome has been unlocked in this step so that it now can switch to a conformation with high affinity to EF-G in the GTP form. The switch involves a relative movement of the small subunit and a conformational change of the L1 stalk, and the combination of both motions results in an L1 stalk-tRNA interaction. During the switch, peptidyl-tRNA in the A/A site moves to the A/P site and deacylated tRNA in the P/P site moves to the P/E site. After the hydrolysis of GTP and dissociation of inorganic phosphate, EF-G adopts the GDP conformation (Rodnina et al., 2001), and the conformational change of the factor (compare Figures 4A and 4B) causes the movement of the 30S subunit to be reversed. Simultaneously, the transition of the L1 stalk to the open form (i.e., away from the inter-subunit space) carries the former P-site tRNA to the E site. In our model, the 30S subunit moves twice with respect to the 50S subunit. The required displacement of the mRNA relative to the 30S subunit could take place as a consequence of the first (ratchet-like rotation) or the second (reversion to the original relative position between subunits) conformational change, or as a result of both. The regulation during the elongation cycle by the P-site tRNA can be summarized in terms of effecting an alternation between the pre- (unlocked) and post- (locked) ribosome structure. A deacylated tRNA in the P site (prestate ribosome, unlocked) allows the incoming EF-G·GTP to promote conformational changes in the ribosome for the purpose of translocation. A peptidyl-tRNA in the P site (poststate ribosome, locked) will display a strong resistance to relative movements between the ribosomal subunits, defining a stable architecture, required for the rapid and accurate reading of the following codon in the mRNA.

In order to show the ribosomal motions described in the present work, we have provided a movie as Supplemental Data (see Supplemental Data). In a superimposition of the 70S·MFTI-tRNA^{Ile}+puromycin maps after (described in Figure 1B) and before (Figure 1E) interaction with EF-G·GDPNP, the conformational changes in the ribosome can be seen from different orientations in a dynamic representation.

Experimental Procedures

Preparation of Ribosomal Complexes

Release complexes (RC) carrying MFTI tetrapeptidyl-tRNA in the P site and UAA stop codon in the A site were prepared using in vitro translation according to Zavialov and coworkers (2001). In the puromycin-treated samples, 0.13 μM RC were incubated with 0.4 mM puromycin. In the samples containing EF-G and guanine nucleotides 4 μM EF-G and 0.4 mM GDPNP or GDP were included in the incubation. In the case of GDP, 0.4 mM fusidic acid was present. After 5 min of incubation at 37°C, the reaction mixture (50 μl) was diluted to 32 nM concentration of RC and was stored on ice. Cryogrids were prepared in a cold room (at 4°C). The occupancy of EF-G in the complexes was about 80%, which was determined as the amount of [^{3}H]GDPNP or [^{3}H]GDP bound to EF-G-ribosomal complexes (here and below, Zavialov and Ehrenberg, 2003 [this issue of *Cell*]).

To prepare a complex with fMet-Phe-tRNA^{Phe} in the A site, EF-G was excluded from the elongation mixture (Zavialov et al., 2001). An initiation mix (200 μl), containing 4.4 μM ribosomes, 1.5 μM of IF1, IF2 and IF3, 12.5 μM MFstop-mRNA, 14.1 μM fMet-tRNA^{fMet}, and energy regeneration system (50 ng/μl PK, 1 mM GTP and 2 mM PEP) was incubated at 37°C for 25 min. An elongation mix (200 μl), containing 21.2 μM [^{14}C] Phe, 21.8 μM tRNA^{Phe}, 29 μM EF-Tu, 12.5 μM EF-Ts, 90 U of Phe-tRNA synthetase, an energy regeneration system (50 ng/μl PK, 1 mM GTP, 2 mM PEP, 8 ng/μl MK, and 1 mM ATP) was incubated separately at 37°C for 15 min. Peptide elongation was started by adding together the initiation and elongation mixtures. After 15 s incubation at 37°C, the mixture was diluted with ice-cold polymix buffer to 1 μM ribosomal concentration and frozen in liquid nitrogen. To prepare samples for cryo-EM, an aliquot of the mixture was thawed on ice and diluted with ice-cold polymix buffer to 32 nM concentration of ribosome. The occupancy of the A site, which was determined as the amount of peptide that could be released by RF1 in a stop codon dependent reaction after fMet-Phe-tRNA^{Phe} translocation with addition of EF-G (Zavialov and Ehrenberg, 2003 [this issue of *Cell*]), was more than 75% of ribosome.

Table 2. Number of Particle Images and Resolution Achieved for Each Ribosomal Complex Reconstructed

	Number of Images	Resolution (Å)
70S·MFTI-tRNA^{Ile}	14,965	11.2
70S·MFTI-tRNA^{Ile}·EF-G·GDP·Fus.	26,547	13.1
70S·MFTI-tRNA^{Ile}+Pur.	20,892	12.8
70S·MFTI-tRNA^{Ile}·EF-G·GDP·Fus.+Pur.	41,161	11.7
70S·MFTI-tRNA^{Ile}·EF-G·GDPNP+Pur	38,858	10.8
70S·tRNA^{fMet}· fMet-Phe-tRNA^{Phe}	52,181	10.0

The release of peptide was detected by filtration of the mixture through nitrocellulose. No detectable A-site tRNA was translocated without EF-G. In addition, using HPLC analysis it was shown that more than 90% of the A-site bound tRNA carried the dipeptide.

Cryo-Electron Microscopy and Image Processing

The solutions containing ribosomal complexes were diluted in the same polymix buffer to a final concentration of 32 nM. Cryogrids were prepared following standard procedures, and micrographs were taken in low-dose conditions on a Philips Tecnai F20 at 200 kV and a magnification of 49,700×. The micrographs were digitized with a step size of 14 μm on a Zeiss/Imaging scanner (Z/I Imaging Corporation, Huntsville, AL), corresponding to 2.82 Å on the object scale. The three-dimensional reconstructions were calculated using the 3D projection alignment procedure (see Frank et al., 2000). The resulting 3D maps were CTF-corrected and X-ray solution scattering data were used to correct the Fourier amplitudes (Gabashvili et al., 2000). The resolution of the final 3D maps was estimated by the 0.5 cutoff in the Fourier shell correlation. The final number of individual images and the resolution achieved is listed in Table 2.

In the rendering of the cryo-EM maps, the density cutoffs were set for the display of envelopes that represent ~120%–140% of the a priori estimated volume for the 70S ribosome complexes.

Docking of Atomic Structures

The docking of X-ray structures into the cryo-EM maps was made using O (Jones et al., 1991) and the visualization was performed in IRIS Explorer. EF-G domains were taken from the *Thermus thermophilus* H573A mutant crystal structure in complex with GDP (PDB code: 1FNM from Laurberg et al., 2000). The coordinates of the ribosomal proteins and rRNAs were taken from published atomic structures (PDB code 1FFK for the 50S subunit from Ban et al., 2000; 1FJF for the 30S subunit from Wimberly et al., 2000) and the docking was performed using the coordinates as rigid bodies.

Acknowledgments

We thank Michael Watters for assistance with the illustrations. This work was supported by HHMI and NIH grants R37 GM29169 and R01 GM55440 (to J.F.), and by the Swedish Research Council and the Swedish Foundation for Strategic Research (to M.E.).

Received: January 7, 2003
Revised: May 27, 2003
Accepted: June 6, 2003
Published: July 10, 2003

References

Agrawal, R.K., Penzeck, P., Grassucci, R.A., and Frank, J. (1998). Visualization of elongation factor G on the *Escherichia coli* 70S ribosome: the mechanism of translocation. Proc. Natl. Acad. Sci. USA *95*, 6134–6138.

Agrawal, R.K., Heagle, A.B., Penzeck, P., Grassucci, R.A., and Frank, J. (1999). EF-G-dependent GTP hydrolysis induces translocation accompanied by large conformational changes in the 70S ribosome. Nat. Struct. Biol. *6*, 643–647.

Agrawal, R.K., Heagle, A.B., and Frank, J. (2000a). Studies of elongation factor G-dependent tRNA translocation by three-dimensional cryoelectron microscopy. In The Ribosome: Structure, Function, Antibiotics and Cellular Interactions. Garret, R.A., Douthwaite, S.R., Liljas, A., Matheson, T., Moore, P.B., and Noller, H.F., eds. (Washington, D.C.: ASM Press), pp 53–62.

Agrawal, R.K., Spahn, C.M.T., Penczek, P., Grassucci, R.A., Nierhaus, K.H., and Frank, J. (2000b). Visualization of tRNA movements on the *Escherichia coli* 70S ribosome during elongation cycle. J. Cell Biol. *150*, 447–460.

Ban, N., Nissen, P., Hansen, J., Moore, P.B., and Steitz, T.A. (2000). The complete atomic structure of the large ribosomal subunit at 2.4 Å resolution. Science *289*, 905–920.

Borowski, C., Rodnina, M.V., and Wintermeyer, W. (1996). Truncated elongation factor G lacking the G domain promotes translocation of the 3′ end but not of the anticodon domain of peptidyl-tRNA. Proc. Natl. Acad. Sci. USA *93*, 4202–4206.

Bretscher, M.S. (1968). Translocation in protein synthesis: a hybrid structure model. Nature *218*, 675–677.

Czworkowski, J., Wang, J., Steitz, T.A., and Moore, P.B. (1994). The crystal structure of elongation factor G complexed with GDP, at 2.7 Å resolution. EMBO J. *13*, 3661–3668.

Frank, J., and Agrawal, R.K. (2000). A ratchet-like inter-subunit reorganization of the ribosome during translocation. Nature *406*, 318–322.

Frank, J., Penczek, P., Agrawal, R.K., Grassucci, R.A., and Heagle, A.B. (2000). Three-dimensional cryo-electron microscopy of ribosomes. Methods Enzymol. *317*, 276–291.

Gabashvili, I.S., Agrawal, R.K., Spahn, C.M.T., Grassuci, R.A., Svergun, D.I., Frank, J., and Penczek, P. (2000). Solution structure of the *E. coli* 70S ribosome at 11.5 Å of resolution. Cell *100*, 537–549.

Gao, H., Sengupta, J., Valle, M., Korostelev, A., Eswar, N., Stagg, S.M., Van

Roey, P., Agrawal, R.K., Harvey, S.C., Sali, A., et al. (2003). Study of the structural dynamics of the *E. coli* 70S ribosome using real-space refinement. Cell *113*, 789–801.

Gomez-Lorenzo, M.G., Spahn, C.M.T., Agrawal, R.K., Grassucci, R.A., Penczek, P., Chakraburtty, K., Ballesta, J.P.G., Lavandera, J.L., Garcia-Bustos, J.F., and Frank, J. (2000). Three-dimensional cryo-electron microscopy localization of EF2 in the *Saccharomyces cerevisiae* 80S ribosome at 17.5 Å resolution. EMBO J. *19*, 2710–2718.

Harms, J., Schluenzen, F., Zarivach, R., Bashan, A., Gat, S., Agmon, I., Bartels, H., Franceschi, F., and Yonath, A. (2001). High resolution structure of the large ribosomal subunit from mesophilic eubacterium. Cell *107*, 679–688.

Highland, J.H., Howard, G.A., Ochsner, E., Hasenbank, R., Gordon, J., and Stoffler, G. (1975). Identification of a ribosomal protein necessary for thiostrepton binding to *Escherichia coli* ribosomes. J. Biol. Chem. *250*, 1141–1145.

Hirokawa, G., Kiel, M.C., Muto, A., Selmer, M., Raj, V.S., Liljas, A., Igarashi, K., Kaji, H., and Kaji, A. (2002). Post-termination complex disassembly by ribosome recycling factor, a functional tRNA mimic. EMBO J. *21*, 2272–2281.

Ishitsuka, H., Kuriki, Y., and Kaji, A. (1970). Release of transfer ribonucleic acid from ribosomes. A G factor and guanosine triphosphate-dependent reaction. J. Biol. Chem. *245*, 3346–3351.

Jones, T.A., Zhou, J.Y., Cowan, S.W., and Kjeldgaard, M. (1991). Improved methods for binding protein models in electron density maps and the location of errors in these models. Acta Crystallogr. A *47*, 110–119.

Joseph, S., and Noller, H. (1998). EF-G catalyzed translocation of anticodon stem-loop analogs of transfer RNA in the ribosome. EMBO J. *17*, 3478–3483.

Karimi, R., Pavlov, M.Y., Buckingham, R.H., and Ehrenberg, M. (1999). Novel roles for classical factors at the interface between translation termination and initiation. Mol. Cell *3*, 601–609.

Lancaster, L., Kiel, M.C., Kaji, A., and Noller, H.F. (2002). Orientation of ribosome recycling factor in the ribosome from directed hydroxyl radical probing. Cell *111*, 129–140.

Laurberg, M., Kristensen, O., Martemyanov, K., Gudkov, A.T., Nagaev, I., Hughes, D., and Liljas, A. (2000). Structure of a mutant EF-G reveals domain III and possibly the fusidic acid binding site. J. Mol. Biol. *303*, 593–603.

Lill, R., Robertson, J.M., and Wintermeyer, W. (1989). Binding of the 3′ terminus of tRNA to 23S rRNA in the ribosomal exit site actively promotes translocation. EMBO J. *8*, 3933–3938.

Lodmell, J.S., and Dahlberg, A.E. (1997). A conformational switch in *Escherichia coli* 16S ribosomal RNA during decoding of messenger RNA. Science *277*, 1262–1267.

Marquez, V., Wilson, D.N., and Nierhaus, K.H. (2002). Functions and interplay of the tRNA-binding sites of the ribosome. Biochem. Soc. Trans. *30*, 133–140.

Moazed, D., and Noller, H.F. (1989). Intermediate states in the movement of transfer RNA in the ribosome. Nature *342*, 142–148.

Noller, H.F., Yusupov, M.M., Yusupova, G.Z., Baucom, A., and Cate, J.H.D. (2002). Translocation of tRNA during protein synthesis. FEBS Lett. *514*, 11–16.

Peske, F., Matassova, N.B., Savelsbergh, A., Rodnina, M.V., and Wintermeyer, W. (2000). Conformational restricted elongation factor G retains GTPase activity but is inactive in translocation on the ribosome. Mol. Cell *6*, 501–505.

Ramakrishnan, V. (2002). Ribosome structure and the mechanism of translocation. Cell *108*, 557–572.

Rawat, U.B.S., Zavialov, A.V., Sengupta, J., Valle, M., Grassucci, R.A., Linde, J., Vestergaard, V., Ehrenberg, M., and Frank, J. (2003). A cryo-electron microscopic study of ribosome-bound termination factor RF2. Nature *421*, 87–90.

Rodnina, M.V., Savelsberg, A., Katunin, V.I., and Wintermeyer, W. (1997). Hydrolysis of GTP by elongation factor G drives tRNA movement on the ribosome. Nature *385*, 37–41.

Rodnina, M.V., Semenkov, Y.P., Savelsbergh, A., Katunin, V.I., Peske, F., Wilden, B., and Wintermeyer, W. (2001). Mechanism of tRNA translocation on the ribosome. Mol. Biol. *35*, 559–568.

Savelsbergh, A., Mohr, D., Wilden, D., Wintermeyer, W., and Rodnina, M.V. (2000). Stimulation of the GTPase activity of translation elongation factor G by ribosomal protein L7/L12. J. Biol. Chem. *275*, 890–894.

Schmeing, T.M., Seila, A.C., Hansen, J.L., Freeborn, B., Soukup, J.K., Scaringe, S.A., Strobel, S.A., Moore, P.B., and Steitz, T.A. (2002). A pre-translocational intermediate in protein synthesis observed in crystals of enzymatically active 50S subunits. Nat. Struct. Biol. *9*, 225–230.

Stark, H., Rodnina, M.V., Rinke-Appel, J., Brimacombe, R., Wintermeyer, W., and van Heel, M. (1997a). Visualization of elongation factor Tu on *Escherichia coli* ribosome. Nature *389*, 403–406.

Stark, H., Orlova, E.V., Rinke-Appel, J., Jünke, N., Mueller, F., Rodnina, M., Wintermeyer, W., Brimacome, R., and van Heel, M. (1997b). Arrangement of tRNAs in pre- and posttranslocational ribosomes revealed by electron cryomicroscopy. Cell *88*, 19–28.

Stark, H., Rodnina, M.V., Wieden, H., van Heel, M., and Wintermeyer, W. (2000). Large-scale movement of elongation factor G and extensive conformational change of the ribosome during translocation. Cell *100*, 301–309.

Stark, H., Rodnina, M.V., Wieden, H., Zemlin, F., Wintermeyer, W., and van Heel, M. (2002). Ribosome interactions of aminoacyl-tRNA and elongation factor Tu in the codon-recognition complex. Nat. Struct. Biol. 9, 849–854.

Subramanian, A.R., and Dabbs, E.R. (1980). Functional studies on ribosomes lacking protein L1 from mutant *E. coli*. Eur. J. Biochem. *112*, 425–430.

Traut, R.R., and Monro, R.E. (1964). The puromycin reaction and its relation to protein synthesis. J. Mol. Biol. *10*, 63–72.

Valle, M., Sengupta, J., Swami, N.K., Grassucci, R.A., Burkhardt, N., Nierhaus, K.H., Agrawal, R.K., and Frank, J. (2002). Cryo-EM reveals an active role for aminoacyl-tRNA in the accommodation process. EMBO J. *21*, 3557–3567.

VanLoock, M.S., Agrawal, R.K., Gabashvili, I.S., Qi, L., Frank, J., and Harvey, S.C. (2000). Movement of the decoding region of the 16S ribosomal RNA accompanies tRNA translocation. J. Mol. Biol. *304*, 507–515.

Wimberly, B.T., Brodersen, D.E., Clemons, W.M., Jr., Morgan-Warren, R.J., Carter, A.P., Vonrhein, C., Hartsch, T., and Ramakrishnan, V. (2000). Structure of the 30S ribosomal subunit. Nature *407*, 327–339.

Wower, J., Kirillov, S.V., Wower, I.K., Guven, S., Hixson, S.S., and Zimmermann, R.A. (2000). Transit of tRNA through the *Escherichia coli* ribosome. J. Biol. Chem. *275*, 37887–37894.

Yusupov, M.Y., Yusupova, G.Z., Baucom, A., Lieberman, K., Earnest, T.N., Cate, J.H.D., and Noller, H.F. (2001). Crystal structure of the ribosome at 5.5 Å resolution. Science *292*, 883–896.

Zavialov, A.V., and Ehrenberg, M. (2003). Peptidyl-tRNA regulates the GTPase activity of translation factors. Cell *114*, this issue, 113–122.

Zavialov, A.V., Buckingham, R.H., and Ehrenberg, M. (2001). A post-termination ribosomal complex is the guanine exchange factor for peptide release factor RF3. Cell *107*, 115–124.

Zavialov, A.V., Mora, L., Buckingham, R.H., and Ehrenberg, M. (2002). Release of peptide promoted by the GGQ motif of class 1 release factors regulates the GTPase activity of RF3. Mol. Cell *10*, 789–798.

Accession Numbers

Atomic models have been deposited in the Protein Data Bank under the following ID codes: 1PN6 (EF-G); 1PN8 (S12 and L11 proteins, and E-site tRNA from the 70S structure of Yusupov et al. 2001), fitted separately into the cryo-EM map, with the tRNA fitted into the P/E position; and 1PN7 (S12 and L11 proteins, and P-site tRNA from the 70S structure of Yusupov et al., after global fitting of the 70S map).

LETTER TO THE EDITOR

A twisted tRNA intermediate sets the threshold for decoding

MICHAEL YARUS,[1] MIKEL VALLE,[2] and JOACHIM FRANK[2,3]
[1]Department of Molecular, Cellular and Developmental Biology, University of Colorado, Boulder, Colorado 80309-0347, USA
[2]Howard Hughes Medical Institute, Health Research, Inc., Albany, New York 12201-0509, USA
[3]Department of Biomedical Sciences, State University of New York, Albany, Albany, New York 12201-0509, USA

ABSTRACT

Putting together consistent cryo-EM structure, transient kinetic and mutant tRNA suppressor data, it appears that a deformed or waggling aminoacyl-tRNA is the critical transitional structure examined by the ribosome during decoding at the A site. The unusual conformation may be required for effective proofreading of the codon–anticodon complex.

Once information gets into peptides, you cannot get it back out. A corollary to this famous adage is that the precision of translation itself must be determined before the peptide bond forms. New findings unite very diverse experiments to confirm that transfer RNA (tRNA) deformation accompanies and helps to guide those early steps of translation that must be accurate—from the binding of an aminoacyl-tRNA up to its incorporation at the ribosome's A site.

The newest information is structural (Valle et al. 2002), from a cryo-EM reconstruction of *Escherichia coli* 70S ribosomes carrying bound phe-tRNA:EF-Tu:GTP in the presence of kirromycin. The latter antibiotic allows hydrolysis of GTP to GDP but then jams further progress by bridging domains of EF-Tu that otherwise close like scissors. One therefore gets a view of EF-Tu and tRNA that represents the structure before the incoming aminoacyl end enters the peptidyl transferase center. As will be evident in Figure 1a, the entire factor–tRNA structure can be seen (red), but the anticodon arm of the free tRNA or EF-Tu:aa-tRNA structure cannot be fit to the observed electron density (Fig. 1b). Instead, the anticodon arm must be bent. In addition, the D and TΨ loops of the tRNA have moved, apparently to contact the GTPase-associated center (GAC) of the 23S rRNA (around nt 1060–1075 in *E. coli*). This alteration suggests that the tRNA structure may help to signal the need for GTPase activity from the decoding center to the GTPase domain of EF-Tu (see below). Colored beads on the purple tRNA structure highlight the nucleotides discussed next.

Rapid kinetic analysis agrees with this picture: During EF-Tu:aa-tRNA association with ribosomes, a change in the conformation of the D-loop (blue beads in Fig. 1b) is the first detectable event (Pape et al. 1998). This altered tRNA shape is then resolved when the tRNA is released by EF-Tu:GDP and enters the peptidyl transferase center. The rates at which tRNAs transit these steps are dramatically different for tRNAs that match their codons or that have one mispair or that do not match at all (for review, see Rodnina and Wintermeyer 2001). Therefore these in vitro kinetics suggested that D-loop shape changes during codon recognition, GTPase activity, and accommodation to the peptidyl transferase center are essential for accurate decoding (Rodnina and Wintermeyer 2001). Note that dissociation before peptide transfer also appears influential in the coding decision.

Similar ideas arose from directed mutation of suppressor tRNAs, particularly from tRNA mutations that enable unusual coding at the first and third (wobble) codon positions (Schultz and Yarus 1994). These mutations taken together (green beads, Fig. 1b) define a conjoined D-loop–anticodon arm tRNA domain whose conformation is apparently crucial (Yarus and Smith 1995) to acceptance or rejection by the ribosome. For example, a combination of D-arm and anticodon arm mutations made it possible to read a normally forbidden first position wobble pair with about 50% efficiency (Schultz and Yarus 1994). This suggested that altered tRNA conformations are more permissive, and in particular that the D-arm–anticodon arm domain must "waggle" during coding decisions (Yarus and Smith 1995).

These conformational shifts can be explained by the logic of proofreading (Yarus and Smith 1995). The codon–anticodon match must first be tested while the aminoacyl arm

Reprint requests to: Michael Yarus, Department of Molecular, Cellular and Developmental Biology, University of Colorado, Boulder, CO 80309-0347, USA; e-mail: yarus@stripe.colorado.edu.
Article and publication are at http://www.rnajournal.org/cgi/doi/10.1261/rna.2184703.

 RNA (2003), 9:384–385. Published by Cold Spring Harbor Laboratory Press. Copyright © 2003 RNA Society.

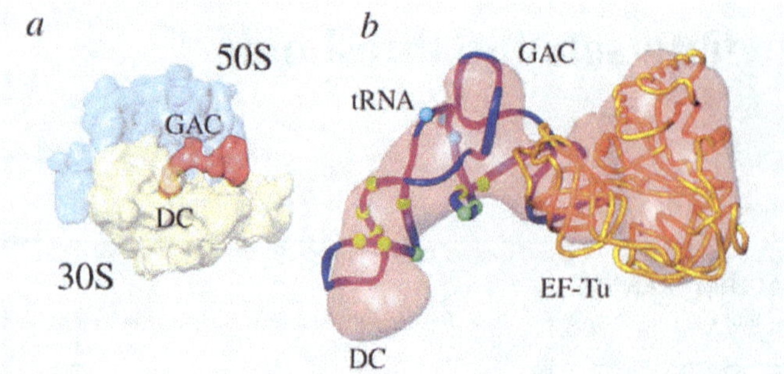

FIGURE 1. (*a*) Phe-tRNA bound to EF-Tu: GDP (red) on the 70S ribosome (50S subunit, blue, 30S subunit, yellow). The complex has been stalled with kirromycin just after GTP hydrolysis but before EF-Tu rearrangements. (*b*) Ribbons representation of the Phe-tRNA (purple) and EF-Tu (orange) inside the semitransparent cryo-EM density of the ternary complex. The beads on the tRNA label mutations that alter the accuracy of translation (dark and light green) and sites of reporter fluors used in kinetic experiments (blue). (DC) decoding center; (GAC) GTPase-associated center.

is outside the peptidyl transferase center, on EF-Tu. The codon–anticodon match is tested again after GTP hydrolysis drops the aminoacyl arm of a tRNA into the A site, before the peptide bond is formed. Throughout these displacements of the tRNA's aminoacyl arm, the anticodon must be in the same place, bound to the codon (at DC in Fig. 1b). Accordingly, we have cryo-EM structures that clearly indicate deformation of the aminoacyl-tRNA anticodon arm and D/TΨ loop structures immediately after recognition of the tRNA's codon–anticodon match (that is, just after GTP hydrolysis by EF-Tu). We have kinetics that imply that D-loop deformation is likely to be a universal early step during successful encounters with the ribosomal decoding site in vitro (before EF-Tu departs). Finally, we have genetic data indicating that easier deformation of the proximal anticodon arm and D-arm lowers the threshold for biologically acceptable coding in the steady state within growing cells. Separately and especially taken together, these findings show that ribosomes deform incoming tRNAs while discriminating among them. The ribosome distorts the tRNA in order to proofread, and may cleverly exploit successful distortion to help communicate a codon–anticodon match to EF-Tu GTPase (Rodnina and Wintermeyer 2001).

ACKNOWLEDGMENTS

The publication costs of this article were defrayed in part by payment of page charges. This article must therefore be hereby marked "advertisement" in accordance with 18 USC section 1734 solely to indicate this fact.

Received November 21, 2002; accepted January 9, 2003.

REFERENCES

Pape, T., Wintermeyer, W. and Rodnina, M.V. 1998. Complete kinetic mechanism of elongation factor Tu-dependent binding of aminoacyl-tRNA to the A site of the *E. coli* ribosome. *EMBO J.* **17:** 7490–7947.

Rodnina, M.V. and Wintermeyer, W. 2001. Fidelity of aminoacyl-tRNA selection on the ribosome: Kinetic and structural mechanisms. *Annu. Rev. Biochem.* **70:** 415–435.

Schultz, D.W. and Yarus, M. 1994. tRNA structure and ribosomal function II. Interaction between anticodon helix and other tRNA mutations. *J. Mol. Biol.* **235:** 1395–1405.

Valle, M., Sengupta, J., Swami, N.K., Grassucci, R.A., Burkhardt, N., Nierhaus, K.H., Agrawal, R.K., and Frank, J. 2002. Cryo-EM reveals an active role for aminoacyl-tRNA in the accommodation process. *EMBO J.* **13:** 3557–3567.

Yarus, M. and Smith, D. 1995. tRNA on the ribosome: A waggle theory. In *transfer RNA* (eds. D. Soll and U. RajBhandary), pp. 443–469. American Society of Microbiology Press, Washington, DC.

The EMBO Journal (2004) 23, 1008–1019 | © 2004 European Molecular Biology Organization | All Rights Reserved 0261-4189/04
www.embojournal.org

THE EMBO JOURNAL

Domain movements of elongation factor eEF2 and the eukaryotic 80S ribosome facilitate tRNA translocation

Christian MT Spahn[1,2,3,*], Maria G Gomez-Lorenzo[2,4], Robert A Grassucci[1,2], Rene Jørgensen[5], Gregers R Andersen[5], Roland Beckmann[6], Pawel A Penczek[7], Juan PG Ballesta[8] and Joachim Frank[1,2,9,*]

[1]Wadsworth Center, Health Research Inc., Howard Hughes Medical Institute, Albany, NY, USA, [2]Wadsworth Center, Albany, NY, USA, [3]Institut für Medizinische Physik und Biophysik der Charité, Humboldt Universität zu Berlin, Berlin, Germany, [4]Centro Nacional de Biotecnología-CSIC, Campus Universidad Autonoma, Madrid, Spain, [5]Department of Molecular Biology, Aarhus University, Århus, Denmark, [6]Institut für Biochemie der Charité, Humboldt Universität zu Berlin, Berlin, Germany, [7]University of Texas – Houston Medical School, Houston, TX, USA, [8]Centro de Biología Molecular 'Severo Ochoa', CSIC and UAM de Madrid, Madrid, Spain and [9]Department of Biomedical Science, State University of New York at Albany, USA

An 11.7-Å-resolution cryo-EM map of the yeast 80S·eEF2 complex in the presence of the antibiotic sordarin was interpreted in molecular terms, revealing large conformational changes within eEF2 and the 80S ribosome, including a rearrangement of the functionally important ribosomal intersubunit bridges. Sordarin positions domain III of eEF2 so that it can interact with the sarcin–ricin loop of 25S rRNA and protein rpS23 (S12p). This particular conformation explains the inhibitory action of sordarin and suggests that eEF2 is stalled on the 80S ribosome in a conformation that has similarities with the GTPase activation state. A ratchet-like subunit rearrangement (RSR) occurs in the 80S·eEF2·sordarin complex that, in contrast to *Escherichia coli* 70S ribosomes, is also present in vacant 80S ribosomes. A model is suggested, according to which the RSR is part of a mechanism for moving the tRNAs during the translocation reaction.

The EMBO Journal (2004) **23**, 1008–1019. doi:10.1038/sj.emboj.7600102; Published online 19 February 2004
Subject Categories: structural biology
Keywords: cryo-EM; eEF2; 80S ribosome; sordarin; tRNA translocation

Introduction

Proteins in all living cells are synthesized by ribosomes, which are large, RNA-based macromolecular machines (Ban *et al*, 2000; Wimberly *et al*, 2000). During protein synthesis, the macromolecular ligands of the ribosome, that is, mRNA, tRNAs, as well as the nascent peptide chain, move through the ribosome in a precise and controlled manner. These movements of the ligands are accompanied and facilitated by corresponding movements in the ribosome itself (Spahn and Nierhaus, 1998; Frank and Agrawal, 2000; Noller *et al*, 2002). In line with the paradigm of the ribosome as a molecular machine, structural investigations of ribosomes in defined functional states by cryo-EM and X-ray crystallography are providing increasing evidence for movements of ribosomal parts and are thereby providing a first direct look into the dynamic behavior of the ribosome (for reviews, see Ramakrishnan and Moore, 2001; Ramakrishnan, 2002; Yonath, 2002).

The translocation reaction is accompanied by large-scale movements. During this step, the ribosome changes from the pre-translocational (PRE) to the post-translocational (POST) state as the A- and P-site bound tRNAs move to the P and E sites, respectively. The translocation step is catalyzed by elongation factor EF-G in prokaryotes and eEF2 in eukaryotes. In the classical view of translocation, EF-G/eEF2 acts by a GTPase switch mechanism similar to a regulatory G protein (Kaziro, 1978). However, this view was challenged by recent time-resolved pre-steady-state experiments (Rodnina *et al*, 1997). These experiments indicated that GTP hydrolysis by EF-G precedes tRNA translocation and that the release of inorganic phosphate is coupled with the tRNA translocation step, suggesting that the chemical energy of GTP hydrolysis is used to perform mechanical work on the ribosome. However, it should be noted that mere binding of EF-G without GTP hydrolysis plays a major role in the catalysis of tRNA translocation and contributes more to the acceleration of translocation than GTP hydrolysis (1000-fold versus 50-fold acceleration, respectively; Rodnina *et al*, 2001).

The molecular mechanism of translocation and its catalysis by EF-G/eEF2 is largely unknown. However, binding of prokaryotic EF-G induces large-scale conformational changes in the ribosome that might be related to tRNA movement (Frank and Agrawal, 2000). Much less information is available about eukaryotic ribosomes, but an earlier cryo-EM structure of eEF2 bound to the yeast 80S ribosome at 17-Å resolution showed overall similarity to the prokaryotic system as well as some significant differences (Gomez-Lorenzo *et al*, 2000). We present here a highly improved cryo-EM map of the yeast eEF2·80S complex in the presence of the antibiotic sordarin at 11.7-Å resolution. Docking of the recent eEF2·sordarin X-ray structure (Jørgensen *et al*, 2003) and of a molecular model for the yeast 80S ribosome (Beckmann *et al*, 2001; Spahn *et al*, 2001a) reveals large conformational changes within eEF2 and the 80S ribosome. Interpreted in molecular terms, these indicate how the ribosome might actively facilitate tRNA translocation. In addition, the mechanism for the inhibitory action of the antifungal drug sordarin is suggested.

*Corresponding authors. J Frank, Wadsworth Center, Health Research Inc., Howard Hughes Medical Institute, Empire State Plaza, Albany, NY 12201-0509, USA. Tel.: +1 518 474 7002; Fax: +1 518 486 2191; E-mail: joachim@wadsworth.org; CMT Spahn, Institut für Medizinische Physik und Biophysik, Charité, Ziegelstr. 5-8, 10117 Berlin, Germany. Tel.: +49 30 450524131; Fax: +49 30 450524931; E-mail: christian.spahn@charite.de

Received: 23 June 2003; accepted: 8 January 2004; published online: 19 February 2004

©2004 European Molecular Biology Organization

Results and discussion

Cryo-EM reconstruction of the ribosomal eEF2·80S complex and docking of atomic models

The reconstruction of the 80S·eEF2·sordarin complex (Figure 1) at 11.7-Å resolution (0.5 threshold; Supplementary Figure S1) was interpreted in molecular terms by docking the recently determined eEF2·sordarin (eEF2-sor) X-ray structure (Jørgensen *et al*, 2003) and atomic models of ribosomal components into the density map. The docking of ribosomal component models was facilitated by a previous atomic model of the POST 80S ribosome based on a 15.4-Å cryo-EM map (Spahn *et al*, 2001a). For this purpose, the cryo-EM map of the 80S·eEF2·sordarin complex was brought into the same orientation as the map of the POST 80S ribosome (Spahn *et al*, 2001a) by first aligning the 60S subunit parts of both maps. After this alignment, it became obvious that the 40S subunits of the two maps differ in position and conformation (see below). Therefore, a cross-correlation-based 3D orientation search was performed (Spahn *et al*, 2001b) to find the alignment parameters for the head and body/platform domains of the 40S part of the 80S·eEF2·sordarin map relative to the map of the POST 80S ribosome (Table I). Different alignment parameters for body/platform domains and the head domain were obtained and subsequently used to transform the corresponding parts of our previous atomic model of the translating 80S ribosome (Spahn *et al*, 2001a). The fitting was further improved by moving nonfitting parts (e.g., RNA helices) as rigid bodies relative to the rest of the model in order to account for local conformational changes in the 40S and 60S subunits. It is estimated that atomic models of known substructures can be positioned in a cryo-EM map with an accuracy exceeding the resolution of the map several-fold (Rossmann, 2000). Indeed, estimation of the accuracy of docking for eEF2 by correlation traces (Valle *et al*, 2003b) indicated a positional accuracy of better than ±2 Å (Supplementary Figure S2). The resulting atomic model allows an interpretation of the interaction between eEF2 and the 80S ribosome and an analysis of conformational changes within the yeast 80S ribosome at the molecular level.

Large-scale conformational change in eEF2 upon binding to the 80S ribosome

The quality of the cryo-EM map allowed eEF2-sor to be docked unambiguously (Figure 2A and B). However, it was necessary to separate the eEF2 structure into two blocks corresponding to domains I/G′/II and III–V and to fit these separately as rigid bodies. The resulting model for ribosome-bound eEF2 (eEF2-cryo) is distinct from both eEF2-sor and apo-eEF2, the nucleotide-free form of eEF2 (Jørgensen *et al*, 2003). The rotation of domains III–V relative to domains I/G′/II in eEF2-cryo is somewhere intermediate between the two X-ray structures, and closer to that for apo-eEF2 (Figure 2A).

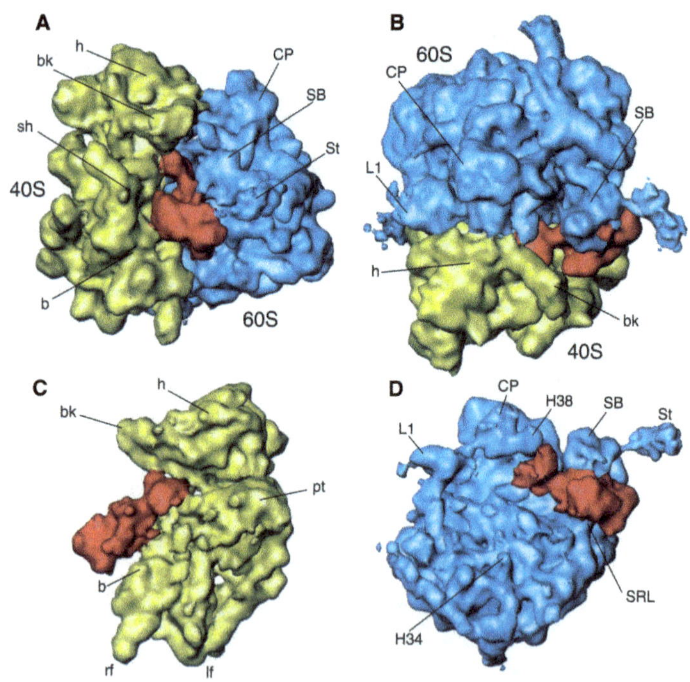

Figure 1 A 11.7-Å-resolution cryo-EM map of the yeast 80S·eEF2·sordarin complex. The cryo-EM map is shown (**A**) from the side; (**B**) from the top ; (**C**) from the 60S side, with 60S removed; and (**D**) from the 40S side, with 40S removed. The ribosomal 40S subunit is painted yellow, the 60S subunit blue and eEF2 red. Landmarks for the 40S subunit: b, body; bk, beak; h, head; lf, left foot; rf, right foot; pt, platform; sh, shoulder; sp, spur. Landmarks for the 60S subunit: CP, central protuberance; L1, L1 protuberance; SB, stalk base; St, stalk; H34, helix 34; H38, helix 38; SRL, sarcin–ricin loop.

©2004 European Molecular Biology Organization

Table I Orientation search and alignment between different parts of the 80S ribosome

	Initial alignment at 60S subunits Angles: $\alpha=0°$ $\beta=0°$ $\gamma=0°$	Alignment at complete 40S subunit Angles: $\alpha=5°$ $\beta=-4°$ $\gamma=1°$	Alignment at 40S body/platform Angles: $\alpha=5°$ $\beta=-3°$ $\gamma=1°$	Alignment at 40S head Angles: $\alpha=4°$ $\beta=-14°$ $\gamma=7°$
		Crosscorrelation coefficient *Angles α/β/γ (in case of optimal alignment)*		
60S subunit	**0.89** **0/0/0**	0.76	n.d.	n.d.
40S subunit body/platform	0.83	0.92	**0.93** **5/−3/1**	0.59
40S subunit head	0.74	0.82	0.80	**0.92** **4/−14/7**

The corresponding ribosomal parts of the POST 80S ribosome (Spahn *et al*, 2001a) and of the 80S·eEF2·sordarin complex are compared in different orientations with the crosscorrelation coefficient. The relative rotational orientation is given in an orthogonal coordinate system, where the *x*-axis is parallel to the anticodon stem helix of the P-site bound tRNA, the *y*-axis is parallel to the acceptor stem helix, and the *z*-axis points from the A-site region toward the E-site region (see Spahn *et al*, 2001b). The cartoon shows the outline of the 40S subunit from the 60S site and the outline of the P-site bound tRNA, together with the axes representing the coordinate system. The angles α, β and γ describe rotations around the *x*-, *y*- and *z*-axis, respectively. n.d., not determined. Optimal alignment for a certain ribosomal region is indicated by bold numbers and by the angles in the table. The angles describe the rotational rearrangement from the conformation of the POST 80S ribosome to the conformation of the 80S·eEF2·sordarin complex.

The conformation of apo-eEF2 is thought to be closely related to both the eEF2·GDP and eEF2·GTP forms and should, therefore, represent the conformation of eEF2 in solution (Jørgensen *et al*, 2003). This leads us to conclude that the conformational change from apo-eEF2 to eEF2-cryo (Figure 2A) occurs upon ribosome binding and that it represents a structural transition that occurs during the translocation reaction. This conformational change is overall similar to the one that has been observed in the *Escherichia coli* system (Agrawal *et al*, 1999). However, the conformational change in the yeast system is more pronounced, and the tip of domain IV moves by some 25–30 Å between apo-eEF2 and eEF2-cryo. Additionally, a movement of domain III of eEF2 relative to domains IV and V that is present in eEF2-sor but not in apo-eEF2 (Jørgensen *et al*, 2003) persists in the structure of eEF2-cryo. The conformational change from apo-eEF2 to eEF2-cryo involves a rotation of domains III, IV and V relative to domains I and II and an additional rearrangement of domain III.

Molecular interactions between eEF2 and the 80S ribosome

The interaction between eEF2·sordarin and the yeast 80S ribosome involves both ribosomal subunits and all five domains of the factor (Figures 1, 2C and D). The α-sarcin-ricin loop (SRL, H95 of 25S rRNA; we will designate rRNA helices of the large subunit (LSU) by Hmn, where mn is the helix number, and helices of the small subunit by hmn), the so-called GTPase-associated center (GAC; rpL12, H43, H44), and the P proteins (P0, P1α, P1β, P2α, P2β; Ballesta and Remacha, 1996) are contact elements of the large ribosomal subunit (Table II). The latter two elements form the stalk base and stalk, respectively, of the 60S ribosomal subunit. An additional contact element on the 60S ribosomal subunit is the intersubunit bridge B2a. With respect to the 40S subunit, eEF2 interacts with the rRNA (h5, 15, 33, 34, 44) in the head and body domains, as well as the ribosomal protein rpS23 (S12p).

The highly conserved SRL of the LSU rRNA is an essential element for the binding and function of several translation factors, forming a strong interaction with the GTP-binding face of domain I (Figure 2D, Table II). The minor-groove side of the SRL stem is adjacent to His27-Asp29 of eEF2. This eEF2 sequence is part of the highly conserved G1 motif, which comprises the phosphate-binding loop (P-loop) (Jørgensen *et al*, 2003). A second tentative contact involves the tetraloop of the SRL and the G3 motif of eEF2 around Asp110. In addition, the interaction of domain I of eEF2 with the ribosome appears to involve rpL9 (L6p). Interestingly, the position of domain III of eEF2·sordarin appears to be such that the conserved sequence around Leu536 interacts with the tetraloop of the SRL (Figure 2D, Table II). An equivalent interaction between the SRL and domain III of EF-G has not yet been described within the bacterial system.

A second ribosomal determinant of translation factor interaction is the GAC in combination with the stalk proteins. These elements undergo a very complex interaction with eEF2 that is to some extent different from the corresponding interaction with EF-G in bacteria (Gomez-Lorenzo *et al*, 2000). In a separate study, we have identified an interaction between the P proteins and the α-helix D of domain I of eEF2 (Gomez *et al*, 2004 in preparation). The apical loops of H43 and H44 as well as rpL12 interact with domain V of eEF2 (Figure 2D, Table II). Domain V of eEF2 is further contacted by the apical loop of H89, which from its location could transmit conformational signals into the peptidyltransferase center.

The tip of domain IV of eEF2 interacts with both the 40S and 60S subunit, in the region of the decoding center. It contacts the top part of h44 and the apical loop of H69, which form the intersubunit bridge B2a (Figure 2C, Table II). rpS23 (S12p), located adjacent to the decoding center, interacts with domain III of eEF2. As domain III also interacts with

©2004 European Molecular Biology Organization

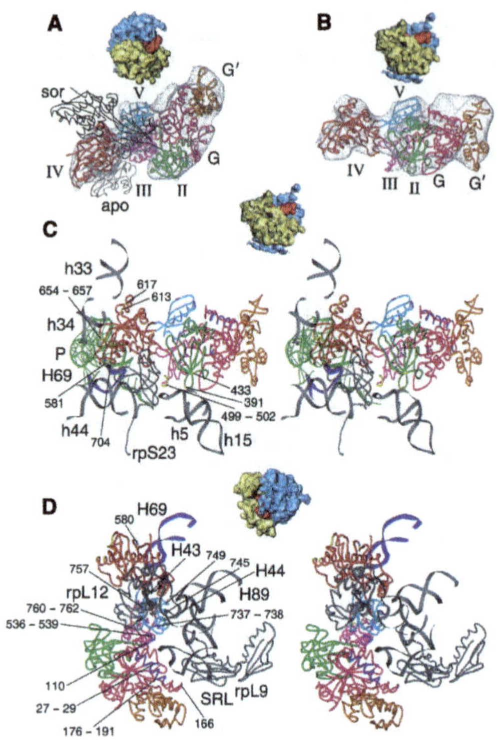

Figure 2 Docking of the X-ray model for eEF2 into the cryo-EM density and interactions between eEF2 and the 80S ribosome. Fitting of eEF2 (colored ribbon representation) into the corresponding cryo-EM density (**A, B**) and ribosomal environment of eEF2 (**C, D**). Thumbnails are included as an orientation aid. In (A), the X-ray structures of eEF2 · sordarin (black, thin ribbon, designated *sor*) and the apo form of eEF2 (gray, thin ribbon, designated *apo*) are shown superposed onto domains I and II of the 80S bound eEF2, in order to show the conformational changes. The domains of ribosome-bound eEF2 are color-coded: G (pink), G′ (orange), II (green), III (purple), IV (red), V (cyan). Fitted atomic models of eEF2 and ribosomal components (gray ribbons) in the neighborhood of the factor in ribbon representation are displayed in stereo (C, D). The upper panel (C) focuses on the interactions between the 40S subunit and eEF2, and the lower panel (D) shows a close-up on the stalk base region of the 60S subunit. Residues of eEF2 that possibly interact with the 40S subunit are highlighted in yellow, residues that might be in contact with the 60S subunit are in blue and their numbers are indicated. (C) includes the position of the P-site bound tRNA (in green) that was derived from the POST 80S complex (Spahn *et al*, 2001a), in order to show the neighborhood of the tip of domain IV of eEF2 to the tRNA anticodon:codon complex.

the SRL (see above), this contact could be part of an information relay system between the decoding center and the SRL. A prominent difference from EF-G is the prong-like appearance of domain IV of eEF2 (Gomez-Lorenzo *et al*, 2000), attributed to an insertion in domain IV and a C-terminal addition (Jørgensen *et al*, 2003). These changes lead to a yeast-specific interaction between domain IV of eEF2 and h33 in the head of the 40S subunit. The 40S subunit also interacts with domain II of eEF2. This interaction with h5 and h15 at the shoulder of the 40S subunit (Figure 2C, Table II) appears to be similar to the corresponding interaction in the *E. coli* system (Frank and Agrawal, 2001).

In line with our previous suggestion (Gomez-Lorenzo *et al*, 2000), an interaction between the tip of domain IV of eEF2 and the P-site bound tRNA is possible according to the more recent maps, with an experimentally determined model of the P-site bound peptidyl-tRNA (Spahn *et al*, 2001a) and the model of ribosome-bound eEF2 · sordarin. In a superposition of both models, His694-Ile698 of eEF2 proves to be close enough to interact with the codon–anticodon duplex between P-site bound tRNA and mRNA (Figure 2C). Moreover, the adjacent residue His699 is post-translationally modified to diphthamide and ADP ribosylation of diphthamide by bacterial toxins that inactivate the factor. Mutations in eEF2 that prevent diphthamide formation impair factor function (Foley *et al*, 1995). Taken together, this suggests an additional function for the tip of domain IV of eEF2. It is known that the accuracy of codon–anticodon interaction in the A site is enhanced by A1492 and A1493 of SSU rRNA (*E. coli* numbering), which interact with the minor groove of the codon–anticodon base pairs (Ogle *et al*, 2002). These A-minor interactions have to be disrupted for the tRNA to move from the A to the P site. The tip of domain IV of eEF2 could take over the stabilization of codon–anticodon pairing during the transition phase. In this way, eEF2 would ensure that the mRNA follows the movement of the tRNAs, thereby reducing the possibility of frameshifts.

A ratchet-like subunit rearrangement takes place in the yeast 80S ribosome

The binding of EF-G to the *E. coli* 70S ribosomes induces a ratchet-like subunit rearrangement (RSR) within the 70S ribosome (Frank and Agrawal, 2000). The RSR presumably corresponds to a transition state of translocation and is accompanied by a movement of a deacylated tRNA in the P site to a P/E hybrid site (Valle *et al*, 2003b; Zavialov and Ehrenberg, 2003). Previously, we have not detected an RSR-type conformational change upon eEF2 binding, when the 80S · eEF2 · sordarin map at 17.5-Å resolution was compared to the map of the vacant 80S ribosome from yeast (Gomez-Lorenzo *et al*, 2000). However, when we tried to dock the atomic model of the translating yeast 80S ribosome (Spahn *et al*, 2001a) into the new 80S · eEF2 · sordarin map at 11.7-Å resolution, it became obvious that a pronounced RSR is present in the 80S · eEF2 · sordarin complex compared to the translating yeast 80S ribosome (Figure 3; Supplementary data: animation, Supplementary Figure S3).

A comparison with the older maps of the vacant 80S ribosome and the 80S · eEF2 · sordarin map at 17.5-Å resolution (Gomez-Lorenzo *et al*, 2000) shows that the reason why the RSR has not been observed previously is because the RSR state is already present in the vacant yeast 80S ribosome. The translating yeast 80S ribosome contains a peptidyl-tRNA in the P site (Beckmann *et al*, 2001; Spahn *et al*, 2001a) and is therefore in the POST state. The subunit organization of the yeast POST 80S ribosome is similar to the translating or vacant bacterial 70S ribosome, consistent with the hypothesis that the peptidyl moiety in the P site locks the prokaryotic ribosome (Valle *et al*, 2003b; Zavialov and Ehrenberg, 2003). The unexpected presence of the RSR in the vacant yeast ribosome shows that the conformational properties of ribosomes of yeast are different from those of *E. coli*. The yeast

©2004 European Molecular Biology Organization

Table II Contacts between eEF2 and the 80S ribosome from yeast

Domain of eEF2	eEF2 position[a]	Ribosomal subunit	rRNA helix[b] or ribosomal protein	rRNA or protein position[a,c]
I	H27–D29 (G1 motif, P-loop)	60S	H95 (SRL)	2656/2663 min
I	D110	60S	H95 (SRL)	2660–2662
I	E166	60S	rpL9 (L6p)	H96, G119
I	Q176–T191	60S	P proteins	
II	K391	40S	h5	359
II	R433	40S	h15	368
III	N499–P502	40S	rpS23 (S12p)	N99
III	L536, E538, E539	60S	H95 (SRL)	2660–2662
IV	P580–K582	60S	H69	1911–1913 lo
IV	N581, Q704	40S	H44	1492, 1493
IV (inset)	K613, R617	40S	H33	1044
IV	Q654–H657	40S	H34	1208
V	E737, Q738	60S	H44	1095
V	E757	60S	H43	1067
V	S745, K749	60S	H89	2473
V	R760–G762	60S	rpL12 (L11p)	K24–K29

[a]Amino acids and nucleotides are given that are closest to the observed contact.
[b]18S rRNA helices are designated with hmn, where mn is the helix number and 25S rRNA helices with Hmn.
[c]rRNA nucleotide numbers are according to *E. coli* numbering. lo: loop; min: minor groove.

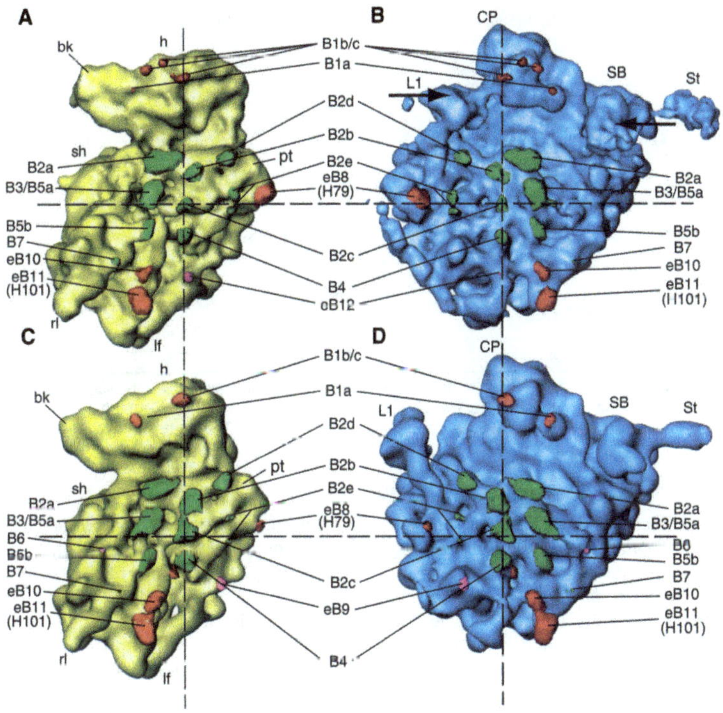

Figure 3 Comparison of the yeast 80S ribosome in two different conformations, and positions of the intersubunit bridges. The 80S · eEF2 · sordarin complex (**A, B**) is compared to the POST 80S ribosome (Spahn *et al*, 2001a) (**C, D**). The two maps were computationally aligned at their respective 60S subunits (see text). The 40S subunits in yellow (A, C) and the 60S subunits in blue (B, D) are shown from their intersubunit sides. Intersubunit bridges are color-coded. Bridges that are preserved in both structures are painted green, those that are formed by different components are painted red, and those that are specific for one of the conformations are painted pink. The dashed registration lines intersect at bridge b2c, the center of rotation for the RSR. Arrows (B) indicate the inward movements of the L1 protuberance and the stalk region.

ribosome can stably adopt the RSR conformation without being stabilized by eEF2.

Normal-mode analysis is a computational approach to predict and explore global conformational changes of macromolecular assemblies. One of the motions derived from a normal-mode analysis of the X-ray structure of the *Thermus thermophilus* 70S ribosome (Yusupov *et al*, 2001) shows striking similarities to the RSR (Tama *et al*, 2003). The

©2004 European Molecular Biology Organization

Table III Intersubunit bridges

Bridge	60S component[a]	40S component[a]	Difference between eEF2·80S and POST 80S[b]
B1a	H38 (886–888)	rpS15/S19p (12) H33 (1033)	Different 40S component
B1b/c	rpL11/L5p (91, 92, 170) rpL11/L5p (84) rpL11/L5p (34,35)	rpS15/S19p (70) rpSx or rpS15/S19p (N-term) rpS18/S13p 111–118	Different 40S component
B2a	H69 (1910–1920)	H44 (1408/1493)	3.5–4 Å
B2b	H68 (1847) 25S (1939–1941)	H24 (784) H45 (1515)	2.5 Å
B2c	H67 (1832)	H27 (899) H24 (771)	0.5–1 Å
B2d	H68 (1848/1895)	H23 (698–703)	5–5.5 Å
B2e	rpL2 (136) rpL43/L37ae (C-term)	H23 (712/713) H22 (671)	4 Å 3 Å
B3	H71 (1948/1960)	H44 (1418/1482)	2–3 Å
B4	H34 (716)	H20 (580/761) rpS13/S15p	1.5 Å; no interaction between minor groove of 25S H34 and 18S H11
B5a	rpL23/L14p	H44 (1422)	3.5 Å
B5b	H62 (1689/1704)	H44 (1428/1472)	3–3.5 Å
B6	rpL23/L14p (132)	H14	No bridge
B7	rpL24/L24e (47)	H44 (1446)	6 Å
eB8	H79 (exp)	rpSx	Different part of rpSx; stronger
eB9	rpLx	rpSx/h21 (exp)	No bridge
eB10	H63 (1713/1747) H101	H11 (272) rpSx	Different ribosome position
eB11	H101 (exp)	H9 (187) rpSx	Local change
eB12	rpL19/L19e (142)	n.d.	New bridge

[a]Amino acids and nucleotides are given that are closest to the observed contact; rRNA nucleotide numbers are according to *E. coli* numbering.
[b]If a distance is given, the corresponding bridge appears to be formed by the same components in both states. The distance was inferred geometrically from the rigid body movement of the 40S subunit associated with the RSR. The actual relative movement might be counteracted by local changes.

center of the rotational movement predicted by the normal-mode analysis of the bacterial ribosome was found to be h27, the same center as obtained in our experimental analysis of the RSR in the yeast 80S ribosome (see below). Therefore, the mechanical properties of the ribosome that gives rise to the RSR appear to be evolutionary conserved. This, however, leads to the question of why the conformations of the vacant 70S ribosome and the vacant 80S ribosome are different, that is, the latter shows the RSR and the former does not.

One possibility is that the two ribosomal conformations exist in an equilibrium, which is shifted toward one state in prokaryotes and toward the other in eukaryotes. In eukaryotic ribosomes, however, there might be a second factor that influences the dynamic behavior. Most of the evolutionary conserved intersubunit bridges, especially the centrally located RNA–RNA bridges, do not have to be broken for the conformation to switch (Figure 3, Table III). However, molecular interactions do have to be temporarily broken and re-formed between some components for the eukaryotic-specific outer bridges (see below), which requires a larger amount of activation energy for the transition to occur. Therefore, the vacant 80S ribosome might be kinetically trapped in the RSR conformation, and the transition between the two conformations might be impeded without proper catalysis.

Local conformational changes in the 40S subunit

Part of the RSR of the 80S·eEF2·sordarin complex is a rotational movement of the head domain of the 40S subunit relative to the body/platform domains. This rotation has to occur around the neck region, which covalently connects the two parts of the 40S subunit (Supplementary Figure S3) and affects the noncovalent interactions between the head and the body/platform, which must be disrupted or accommodated by more local changes upon 40S subunit rearrangement.

Such a noncovalent head–body interaction ('latch') that is part of the mRNA entry channel is formed between h18 in the body with h34 and rpS3 in the head (Schlünzen *et al*, 2000; Wimberly *et al*, 2000; Spahn *et al*, 2001a). In the 80S·eEF2·sordarin structure, the rotation of the head

©2004 European Molecular Biology Organization

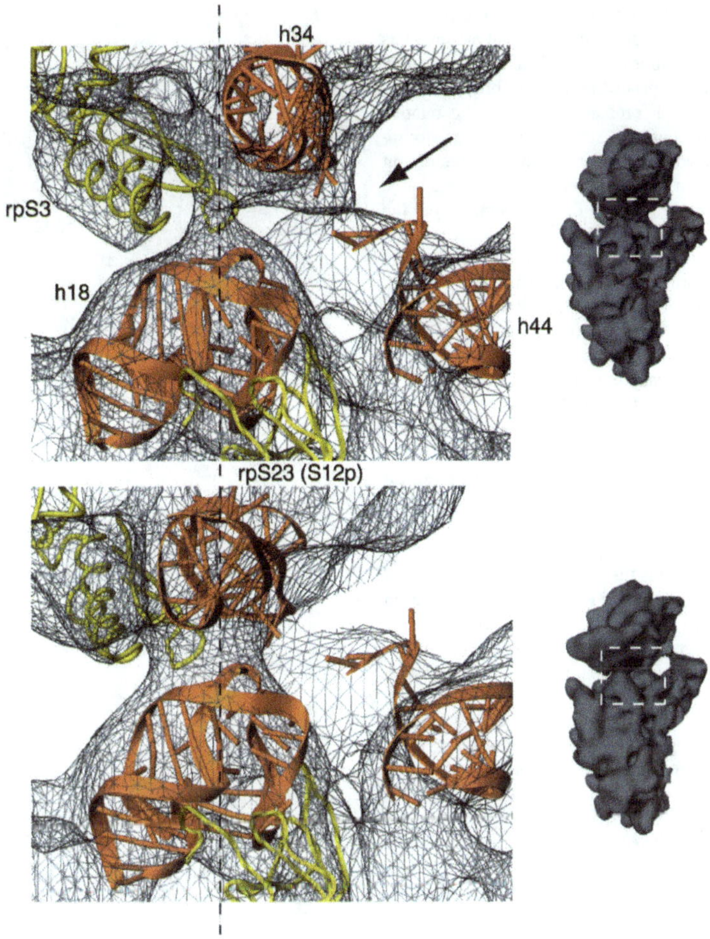

Figure 4 Rearrangement of the latch region of the 40S subunit. Comparison of the latch region in the 80S · eEF2 · sordarin complex (upper panel) with the POST 80S ribosome (lower panel) (Spahn *et al*, 2001a). The corresponding cryo-EM densities are shown as a gray wire-mesh. Docked models for h18, h34 and the top of h44 are shown as orange ribbons, and models for rpS3 and rpS23 (S12p) as yellow ribbons. The dashed line goes through the center of the latch. Small inset on the right: the 40S subunits as an orientation aid. The white, dashed box indicates the latch region. The arrow in the upper panel highlights an additional connection between the head and the body of the 40S subunit that is present in the 80S · eEF2 · sordarin complex.

moves h34 inward by about 15 Å (Figure 4). This movement breaks the contact between the minor groove of h34 with h18. Instead, the interaction of h18 is predominantly with rpS3. This change of interactions seems to be facilitated by a movement of protein rpS3 itself. Interestingly, the region of h34 that forms the latch in the POST 80S ribosome is engaged in an interaction with domain IV of eEF2 (Table II). The insert in domain IV interacts with the nearby h33 and might have evolved to control and stabilize the opening of the latch during translocation.

An independent movement of the head domain of the 30S subunit relative to the body/platform domains is also part of the RSR in *E. coli* (Frank and Agrawal, 2000). Some movements of proteins S3, S4 and S5, involved in forming the mRNA entry channel, as well as a movement of h34 in the range of 5 Å have been reported (Gao *et al*, 2003). However, the magnitude of the head movement we observe in yeast ($\sim$15 Å) and of the concurring change in the latch interaction is unprecedented as yet in prokaryotes.

Another strong network of interactions between the head and body of the 40S subunit is present at the solvent side of the 40S subunit. It involves protein rpS0A (S2p) and h36. In general, the relative movement of the head in the 80S · eEF2 · sordarin complex appears to be accommodated by local conformational changes, which include a movement of the upper subdomain of the 3′ major domain of 18S rRNA (Supplementary Figure S3). This subdomain makes only a few packing contacts with the other parts of the 16S RNA in *T. thermophilus* (Wimberly *et al*, 2000). The absence of rigid contacts might be important to allow flexibility of this subdomain, which appears necessary to facilitate the head movement.

 ©2004 European Molecular Biology Organization

Local conformational changes in the 60S subunit

Localized changes in the 60S subunit are present mainly in its three protuberances. A conformational change takes place in the central protuberance, producing a more broadened appearance (see Figure 3). It probably facilitates the relative movement of the two ribosomal subunits, since this ribosomal region is involved in the formation of intersubunit bridges with the head of the 40S subunit (see below). Large movements can be observed at the L1 protuberance and the stalk base, which comprises the GAC.

As observed previously (Gomez-Lorenzo *et al*, 2000), the binding of eEF2 · sordarin induces a movement of the GAC toward the central protuberance (Figure 3, Supplementary Figure S4). The inward rotation around a hinge region within H42 results in an ~15 Å movement at the apical loops of H43 and H44, where the interaction with eEF2 takes place. A similar but smaller movement of the GAC was also observed in the *E. coli* 70S ribosome upon binding of EF-G (Frank and Agrawal, 2001), aa-tRNA · EF-Tu · GTP ternary complex in the presence of kirromycin (Valle *et al*, 2003a) and RF2 (Rawat *et al*, 2003). The movement of the GAC might be therefore part of a general mechanism of loading translation factors into the ribosome's factor binding site.

Another dynamic region is the L1 protuberance, the top of which is formed by rpL1, H77 and H78, while the connection to the body of the 60S subunit is formed by H76. The L1 protuberance has been observed in different positions (Harms *et al*, 2001; Valle *et al*, 2003b; Yusupov *et al*, 2001), which in the yeast ribosome span about 50 Å (Gomez-Lorenzo *et al*, 2000). Furthermore, its flexibility can be inferred from various degrees of disorder in the X-ray structures of the 50S subunit and 70S ribosomes, making an atomic interpretation of this region difficult (Ban *et al*, 2000; Harms *et al*, 2001; Yusupov *et al*, 2001). However, the X-ray structure of the isolated L1 protuberance at atomic resolution has been presented recently (Nikulin *et al*, 2003).

In the present map of the 80S · eEF2 · sordarin complex, the L1 protuberance is essentially in an 'in-position', close to the central protuberance (Figure 3), whereas in the cryo-EM map of the POST 80S ribosome (Spahn *et al*, 2001a) the L1 protuberance is mostly in an 'out-position'. Even though the cause for the change in L1 position is as yet unknown, it is not related to eEF2 binding, because in vacant 80S ribosomes (Gomez-Lorenzo *et al*, 2000), the L1 protuberance is in the 'in-position' as well. The L1 movement is likely to participate in the active release of the E-site bound tRNA (Gomez-Lorenzo *et al*, 2000; Harms *et al*, 2001; Valle *et al*, 2003b; Yusupov *et al*, 2001).

The conformational change at the L1 protuberance from its 'out-position' to the 'in-position' in the 80S · eEF2 · sordarin complex appears to be rather complex (Figure 5). It entails two rotations around separate hinge points. One rotation is around a hinge point close to the H75/H76/H79 junction, and the other is around a point located within the rod-like H76. While H76 is a regular RNA double helix in prokaryotes, it contains a bulged-out nucleotide in eukaryotes exactly in the region of the second hinge. Furthermore, H76 in the 'in-position' appears to make contact with the body of the 60S subunit at the position of this unpaired nucleotide. It is likely that the extra nucleotide provides flexibility for the second hinge movement, which would explain why much larger movements of the L1 protuberance are observed in yeast. In addition to the rotations, a third conformational change of the L1 protuberance takes place at the top of the 'mushroom', resulting in a different domain arrangement within rpL1 (Figure 5). We note that the L1 protein can be crystallized in conformations with different arrangements of the N- and C-terminal domains (Nevskaya *et al*, 2000). A conformational change in rpL1 seems to be required to allow an interaction of rpL1 with the central protuberance that takes place in the 'in-position' (Figures 1, 3 and 5). This interaction in turn might stabilize the L1 protuberance in the 'in-position'.

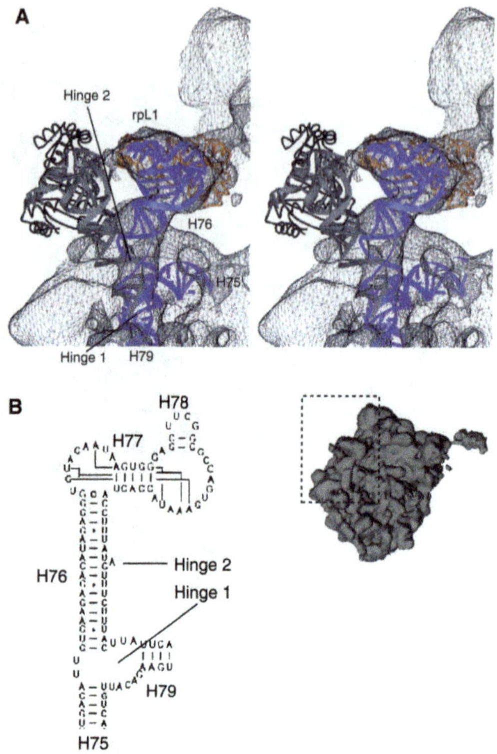

Figure 5 Movement of the L1 protuberance. (**A**) Close-up on the L1 protuberance. Elements of 25S rRNA and 60S ribosomal proteins fitted into the 80S · eEF2 · sordarin complex are shown as blue and orange ribbons, respectively. The thumbnail of the 60S subunit is included as an orientation aid. The cryo-EM density of the 80S · eEF2 · sordarin complex is shown as a gray wire-mesh. Docked models for the L1 protuberance in the outer position (Spahn *et al*, 2001a) are superposed as gray (25S rRNA) and black (rpL1) ribbons. Arrows indicate the hinge 1 and 2 regions of the conformational change. (**B**) Part of the secondary structure diagram of the 25S rRNA from *Saccharomyces cerevisiae* (http://www.rna.icmb.utexas.edu) showing the RNA corresponding to the L1 protuberance. Helices are indicated with their number. A eukaryotic-specific bulged-out nucleotide that is located at the hinge 2 region is marked by an arrow.

Dynamics of the ribosomal intersubunit bridges

The RSR of the subunits has to be regulated and made possible by conformational changes of the intersubunit bridges. The dynamic behavior of the bridges, thought to take part in the communication between the small and the large subunit, is of great functional importance. Therefore, we analyzed the intersubunit bridges of the

©2004 European Molecular Biology Organization

80S·eEF2·sordarin complex and compared the bridging regions between the subunits with the corresponding regions in our previous map of the POST 80S ribosome (Spahn *et al*, 2001a) (Figure 3, Table III). The RSR can be approximately described by rotational rigid-body movements of the body/platform and head domains of the 40S subunit relative to the 60S subunit. Dependent on the distance to the center of the rotation, components of both subunits that are engaged in intersubunit bridges move relative to each other, unless the movements brought about by the subunit rotation are counteracted by local changes and independent movements of the structural components forming the bridges.

Most of the central bridges are formed by RNA–RNA contacts, whereas RNA–protein and protein–protein contacts are located more toward the periphery (Spahn *et al*, 2001a; Yusupov *et al*, 2001). All central RNA–RNA intersubunit bridges (B2a, B2b, B2c, B2d, B3 and B5) are preserved in the 80S·eEF2·sordarin complex. This is very similar to the behavior observed for the bridges of the prokaryotic 70S ribosome (Gao *et al*, 2003; Valle *et al*, 2003b). As inferred from the two atomic models for the POST 80S ribosome (Spahn *et al*, 2001a) and the 80S·eEF2·sordarin complex, the RSR moves the corresponding RNA components relative to each other by distances ranging from 2 to 5.5 Å (Table III). Although small movements of bridge components cannot be directly observed at our resolution, it is likely that local conformational changes occur which preserve the connections despite the overall movement of the subunits. This possibility is in excellent agreement with the observed flexibility of bridge components in the X-ray structures of the ribosomal subunits (for a review, see Yonath, 2002). There is practically no relative movement at B2c, which involves the prominent switch helix h27 (Lodmell and Dahlberg, 1997). B2c can therefore be regarded as the center of the rotation of the 40S subunit relative to the 60S subunit (Figure 3; see similar observations for the 70S ribosome: Gao *et al*, 2003; Tama *et al*, 2003).

Bridges B2e and B4 involve RNA–protein contacts and are located relatively close to B2c, that is, to the center of rotation. B2e, which has been reported to be broken in *E. coli* ribosomes due to the RSR (Gao *et al*, 2003), can nevertheless be observed in both states of the yeast 80S ribosome (Table III). However, there are changes in B4. The apical loop of H34 interacts with h20 and rpS13 (S15p) in both states of the ribosome, but the second interaction, which involves the minor groove of H34 in the POST yeast 80S ribosome, appears to be absent in the 80S·eEF2·sordarin complex (Figure 3). H34 belongs to domain II of 25S rRNA, which also encompasses the GAC. Conformational signals induced at the GAC by the binding of ligands such as eEF2 might be transmitted to H34 and thereby control the subunit rearrangement.

Another candidate for an element important for controlling and regulating the intersubunit arrangement is protein rpL23 (L14p), since it is located directly at the foot of the SRL. In the POST 80S ribosome, rpL23 (L14p) participates in the formation of two bridges, B5a and B6 (Spahn *et al*, 2001a). Moreover, rpL23 (L14p) interacts with rpL24 (L24e), which in turn participates in B7. In support of a more active role of rpL23 (L14p) in controlling subunit rearrangement, B6 is not formed in the 80S·eEF2·sordarin complex, whereas B5a can be observed only at a low contour level (Figure 3, Table III). The L14 bridges are not changed in response to the RSR in *E. coli* ribosomes, but the observed conformational change of L14 (Gao *et al*, 2003) is in line with an active role of this protein.

In addition to those intersubunit bridges that are conserved among prokaryotes and eukaryotes, the eukaryotic 80S ribosome has extra bridges at the periphery (Spahn *et al*, 2001a). Bridge eB9 is disrupted in the 80S·eEF2·sordarin complex (Figure 3, Table III). However, other eukaryotic-specific bridges are present in both RSR-related conformations of the yeast 80S ribosome, apparently because intersubunit contacts are alternately formed between different parts of the subunits. This is the case for eB8. In the two states, different sites of a 40S protein density of unknown identity interact with the expansion of H79 of the 60S subunit. Different pairs of interactions for the two 80S conformations also exist for eB10, whereas the subunit movement appears to be accommodated by a local change in the case of eB11 (Table III). The new bridge eB12 involves rpL19 (L19e) and might be specific for the conformation of the RSR.

Intersubunit bridges formed by the head of the 40S subunit represent a special class. The rotation of the head, which takes place in addition to the RSR of the 40S subunit, results in a movement by about 15 Å relative to the 60S subunit when the POST 80S ribosome and the 80S·eEF2·sordarin complex are compared (see Figure 3). This movement leads to changed intersubunit bridges in the 80S·eEF2·sordarin complex, similar to the changes that have been observed for the *E. coli* 70S ribosome (Valle *et al*, 2003b). H38, the so-called A-site finger, continues to participate in bridge B1a, but the attachment site on the 40S subunit is different and appears to involve the N-terminus of rpS15 (S19p) and h33 (Table III). Moreover, the bridge can be observed only at a low contour level of the cryo-EM density map.

Protein rpL11 (L5p) of the central protuberance is also engaged in the formation of intersubunit bridges in both states. Again, bridge B1b/c is formed with differing proteins of the 40S subunit. A single-bridge interaction in the POST 80S ribosome is replaced by three interactions in the 80S·eEF2·sordarin complex that are with rpS15 (S19p) and rpS18 (S13p) (Figure 3, Table III). The set of different interactions at bridge B1b/c between the head of the 40S subunit and the central protuberance might be facilitated by concomitant conformational changes of the head proteins rpS15 (S19p) and rpS18 (S13p) and of the central protuberance.

Interaction between the 80S ribosome and eEF2 and the mechanism of sordarin

The interaction between the ribosome and the factor EF-G/eEF2 is normally transient, but the addition of nonhydrolyzable GTP analogs or certain drugs locks the system in a defined state. Fusidic acid inhibits translocation in both prokaryotic and eukaryotic systems by preventing the dissociation of EF-G/eEF2·GDP from the POST ribosome (for a review, see Spahn and Prescott, 1996). In contrast, sordarin, which has been used in this study, is highly specific for eEF2 from fungi. Similar to fusidic acid, sordarin prevents the dissociation of eEF2 from the ribosome (Justice *et al*, 1998), and therefore allows direct visualization of eEF2 bound to the ribosome by cryo-EM (Gomez-Lorenzo *et al*, 2000, and this work). However, the actual modes of inhibition of the translocation reaction by fusidic acid and sordarin appear to be different (Dominguez *et al*, 1999). The binding site for

©2004 European Molecular Biology Organization

sordarin on eEF2 is located between domains III, IV and V (Capa *et al*, 1998; Jørgensen *et al*, 2003), whereas the fusidic acid binding site has been suggested to be located between the G domain and domains II and III (Laurberg *et al*, 2000).

Key to the inhibitory action of sordarin appears to be the particular conformation of domain III of eEF2 that is present in eEF2-sor (Jørgensen *et al*, 2003) and eEF2-cryo, which allows domain III to interact with the SRL (Figure 2). The remarkable increase in affinity of eEF2 for sordarin upon binding to the ribosome (Dominguez and Martin, 1998) is compatible with the possibility that the observed position of eEF2 domain III is also adopted on the ribosome, even in the absence of sordarin. In this case, part of the energy cost for the domain rearrangement of eEF2, which allows productive sordarin interaction with domains III, IV and V, would be paid by the energy released in the eEF2–ribosome interaction, explaining the increased sordarin affinity. Sordarin in turn might prevent domain III from moving away from the SRL, and thereby prevent the dissociation of eEF2 from the ribosome. The particular conformation of domain III of eEF2 has not yet been detected in the prokaryotic system. However, the importance of domain III for factor binding has been demonstrated for bacterial EF-G by a domain III deletion mutant (Martemyanov and Gudkov, 2000). If domain III is deleted, GTPase activity is impaired, but not the interaction of the ribosome with EF-G·GDP. As sordarin at high concentrations can stimulate the uncoupled GTPase activity when eEF2 is in excess over ribosomes (Dominguez *et al*, 1999), the particular conformation and ribosome interaction of domain III of eEF2 might therefore be important for GTPase activation.

Mechanism of tRNA translocation

In the course of the tRNA movement, the interactions of the tRNAs with the ribosome at the A and P sites have to be

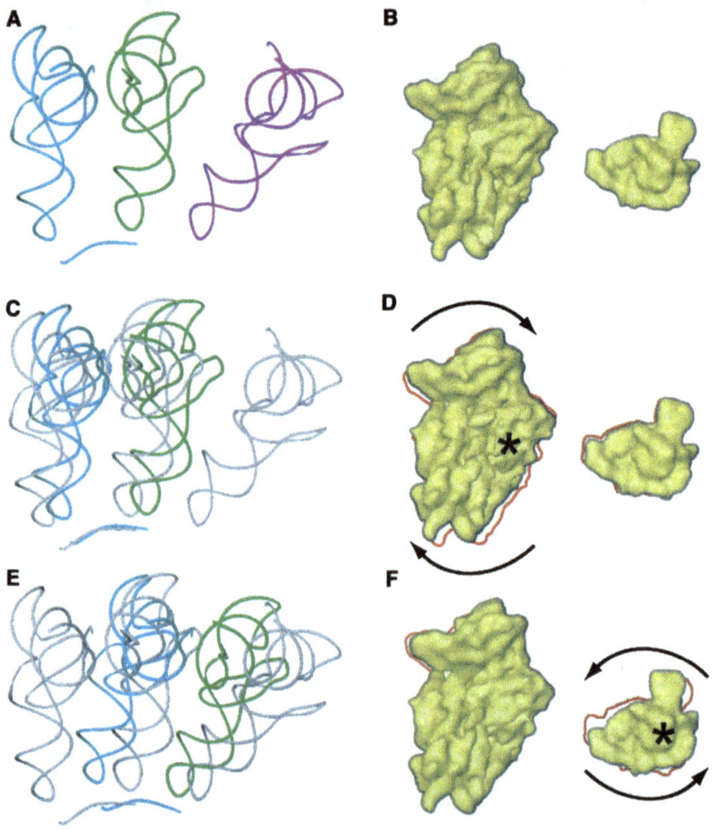

Figure 6 Model for tRNA movement during the translocation reaction. The coordinate transformations of the RSR, when applied to the A- and P-site bound tRNAs (**A**), result in hypothetical intermediates of the tRNAs during translocation (**C, E**) and a plausible trajectory for the path of the tRNAs from the A to the P and from the P to the E site. The corresponding state of the 40S subunit and its movements are shown in the cartoon in (**B, D, F**). The 40S subunit is shown on the left from the intersubunit side, and the head domain on the right from the top of the subunit. The crystallographically determined positions of the tRNAs in A (cyan), P (green) and E (purple) sites and an mRNA fragment (cyan) (Yusupov *et al*, 2001) are shown in (A) and are included in (C, E) as references painted in gray. (C) Shows the tRNAs in A (cyan) and P (green) sites after the coordinate transformation of the RSR for the body/platform domains of the 40S subunit (D) has been applied. The rotation of the 40S subunit is indicated by arrows, and the rotation axis at h27 of 18S rRNA by a star (D). The outline of the untransformed 40S (B) is included in (D) by the red line. The tRNA positions are further transformed (E) according to the additional movement of the head domain (F). This rotation occurs around a rotation axis through the neck of the 40S subunit (indicated by a star in (F)). The red line in (F) indicates the outline of the 40S subunit in (D).

©2004 European Molecular Biology Organization

broken and new interactions have to be formed at the P and E sites, respectively. The conformational changes that the ribosome undergoes during tRNA translocation are likely to occur for the purpose of facilitating the movement of tRNA, especially since the movement of the 30S subunit during the RSR is in the expected direction of tRNA movement from A to P and P to E sites (Frank and Agrawal, 2000). Furthermore, kinetic experiments have indicated that a ribosomal conformational change precedes tRNA translocation (Savelsbergh *et al*, 2003). Indeed, as observed very recently, the RSR in *E. coli* 70S ribosomes is accompanied by the transition of a deacylated tRNA from the P site to a P/E hybrid site (Valle *et al*, 2003b). In an analogous way, the RSR could be involved in moving the A-site bound peptidyl-tRNA into an A/P hybrid site, directly linking the RSR to tRNA movement (Valle *et al*, 2003b; Zavialov and Ehrenberg, 2003). However, in contrast to the original hybrid-site model (Moazed and Noller, 1989; Noller *et al*, 2002), the formation of hybrid sites requires EF-G binding and does not occur spontaneously upon the peptidyl-transferase reaction.

Similar to the *E. coli* system (Valle *et al*, 2003b; Zavialov and Ehrenberg, 2003), the RSR in yeast could be directly coupled to tRNA movement. Part of the RSR in yeast is a rotational movement of the head domain of the 40S subunit that has not been observed in this form in the prokaryotic system. An intriguing possibility is that the rotation of the small subunit might occur first, prior to the additional movement of the head domain. As the tRNAs are tightly bound between body/platform and head domains, this first rotation would be forced on the tRNAs (Figure 6) and in turn break the tRNA interactions at the elbow with the large ribosomal subunit. The 3′-CCA end of the tRNAs is flexible and could move independently. It is therefore possible that this part of the RSR would lead to hybrid site formation, as suggested for the *E. coli* 70S ribosome.

The additional movement of the head domain that would follow next could participate in the translocation reaction also in a direct manner (Figure 6). As a result of the curvature of the head domain, the rotational movement around the neck region results in a translational movement of ribosomal residues bound to the tRNAs in the direction of the trajectory of the tRNAs. In the *T. thermophilus* A site, both h34 and the 965 loop of 16S rRNA in the head domain interact with the tRNA anticodon stem (Yusupov *et al*, 2001). In the 80S·eEF2·sordarin complex, the homologous residues are closer to the P site than to the A site. The same holds for head residues that interact with the P-site tRNA (Spahn *et al*, 2001a; Yusupov *et al*, 2001). In the 80S·eEF2·sordarin complex, these residues have moved close to the E-site region.

If the transformation parameters for the movement of the head domain from the POST 80S ribosome (Spahn *et al*, 2001a) to the 80S·eEF2·sordarin complex are applied to tRNAs located in A and P sites, the transformed tRNA positions are indeed close to the known P and E sites (Figure 6E and F). (We postulate in analogy to the *E. coli* system that in yeast the conformation of the PRE ribosome is similar to that of the POST ribosome.) It is therefore tempting to speculate that the movement of the 40S head domain occurs together with the tRNAs, such that the contacts between these moieties remain intact. The 40S head would act as a mechanical conveyer for transporting the anticodon stem loops of the tRNAs. As domain IV of eEF2 occupies the A site, it could prevent the tRNAs from moving backward when the RSR is reversed in completion of translocation. This would explain why at least an anticodon stem loop of the tRNA has to be present in the A site for translocation to occur (Joseph and Noller, 1998), as it would constitute a minimum target for interaction with domain IV. This model is reminiscent of the concept of a movable ribosomal domain (α–ε model; Dabrowski *et al*, 1998; Spahn and Nierhaus, 1998), except that in our new model the translocation reaction includes the resetting of the moveable head domain.

Supplementary data

Supplementary data are available at *The EMBO Journal* Online.

Acknowledgements

We are grateful to Dr Sean Connell for help and discussion and to Michael Watters for assistance with the illustrations. This work was supported by HHMI and grants R37 GM29169 and R01 GM55440 from NIH and DBI 9871347 from NSF (to JF), R01 GM60635 (to PP), a grant of the Volkswagen Stiftung (to CS) and by grant PM99-0108 (MCyT) and Fundación Ramón Areces (to JB). The coordinates for the docked eEF2 and the ribosomal components have been deposited in the Protein Data Bank with accession numbers: 1S1H for the 40S subunit and eEF2; 1S1I for the 60S subunit. The cryo EM map has been deposited at the 3D-EM database, EMBL-European Bioinformatics Institute, Cambridge, UK (accession number EMD-1067).

References

Agrawal RK, Heagle AB, Penczek P, Grassucci RA, Frank J (1999) EF-G-dependent GTP hydrolysis induces translocation accompanied by large conformational changes in the 70S ribosome. *Nat Struct Biol* **6:** 643–647

Ballesta JP, Remacha M (1996) The large ribosomal subunit stalk as a regulatory element of the eukaryotic translational machinery. *Prog Nucleic Acid Res Mol Biol* **55:** 157–193

Ban N, Nissen P, Hansen J, Moore PB, Steitz TA (2000) The complete atomic structure of the large ribosomal subunit at 2.4 Å resolution. *Science* **289:** 905–920

Beckmann R, Spahn CMT, Eswar N, Helmers J, Penczek PA, Sali A, Frank J, Blobel G (2001) Architecture of the protein-conducting channel associated with the translating 80S ribosome. *Cell* **107:** 361–372

Brunger AT, Adams PD, Clore GM, DeLano WL, Gros P, Grosse-Kunstleve RW, Jiang JS, Kuszewski J, Nilges M, Pannu NS, Read RJ, Rice LM, Simonson T, Warren GL (1998) Crystallography & NMR system: a new software suite for macromolecular structure determination. *Acta Crystallogr D Biol Crystallogr* **54** (Part 5): 905–921

Capa L, Mendoza A, Lavandera JL, Gomez de las Heras F, Garcia-Bustos JF (1998) Translation elongation factor 2 is part of the target for a new family of antifungals. *Antimicrob Agents Chemother* **42:** 2694–2699

Carson M (1991) Ribbons 2.0. *J Appl Cryst* **24:** 103–106

Dabrowski M, Spahn CM, Schafer MA, Patzke S, Nierhaus KH (1998) Protection patterns of tRNAs do not change during ribosomal translocation. *J Biol Chem* **273:** 32793–32800

©2004 European Molecular Biology Organization

Dominguez JM, Gomez-Lorenzo MG, Martin JJ (1999) Sordarin inhibits fungal protein synthesis by blocking translocation differently to fusidic acid. *J Biol Chem* **274:** 22423–22427
Dominguez JM, Martin JJ (1998) Identification of elongation factor 2 as the essential protein targeted by sordarins in *Candida albicans*. *Antimicrob Agents Chemother* **42:** 2279–2283
Foley BT, Moehring JM, Moehring TJ (1995) Mutations in the elongation factor 2 gene which confer resistance to diphtheria toxin and *Pseudomonas* exotoxin A. Genetic and biochemical analyses. *J Biol Chem* **270:** 23218–23225
Frank J, Agrawal RK (2000) A ratchet-like inter-subunit reorganization of the ribosome during translocation. *Nature* **406:** 318–322
Frank J, Agrawal RK (2001) Ratchet-like movements between the two ribosomal subunits: their implications in elongation factor recognition and tRNA translocation. *Cold Spring Harb Symp Quant Biol* **66:** 67–75
Frank J, Radermacher M, Penczek P, Zhu J, Li Y, Ladjadj Leith A (1996) SPIDER and WEB: processing and visualization of images in 3D electron microscopy and related fields. *J Struct Biol* **116:** 190–199
Gabashvili IS, Agrawal RK, Spahn CMT, Grassucci RA, Frank J, Penczek P (2000) Solution structure of the *E. coli* 70S ribosome at 11.5 Å resolution. *Cell* **100:** 537–549
Gao H, Sengupta J, Valle M, Korostelev A, Eswar N, Stagg SM, Van Roey P, Agrawal RK, Harvey SC, Sali A, Chapman MS, Frank J (2003) Study of the structural dynamics of the *E. coli* 70S ribosome using real-space refinement. *Cell* **113:** 789–801
Gomez-Lorenzo MG, Spahn CMT, Agrawal RK, Grassucci RA, Penczek P, Chakraburtty K, Ballesta JPG, Lavandera JL, Garcia-Bustos JF, Frank J (2000) Three-dimensional cryo-electron microscopy localization of EF2 in the *Saccharomyces cerevisiae* 80S ribosome at 17.5 Å resolution. *EMBO J* **19:** 2710–2718
Gomez-Lorenzo MG, Spahn CMT, Linde J, Penczek PA, Ballesta JPG, Frank J (2004) Structural studies of the *Saccharomyces cerevisiae* ribosomal stalk, in preparation
Harms J, Schluenzen F, Zarivach R, Bashan A, Gat S, Agmon I, Bartels H, Franceschi F, Yonath A (2001) High resolution structure of the large ribosomal subunit from a mesophilic eubacterium. *Cell* **107:** 679–688
Jones TA, Zhou JY, Cowan SW, Kjeldgaard M (1991) Improved methods for building protein models in electron density maps and the location of errors in these models. *Acta Crystallogr A Found Crystallogr* **A47:** 110–119
Jørgensen R, Ortiz PA, Carr-Schmid A, Nissen P, Kinzy TG, Andersen GR (2003) Two crystal structures demonstrate large conformational changes in the eukaryotic ribosomal translocase. *Nat Struct Biol* **10:** 379–385
Joseph S, Noller HF (1998) EF-G-catalyzed translocation of anticodon stem-loop analogs of transfer RNA in the ribosome. *EMBO J* **17:** 3478–3483
Justice MC, Hsu MJ, Tse B, Ku T, Balkovec J, Schmatz D, Nielsen J (1998) Elongation factor 2 as a novel target for selective inhibition of fungal protein synthesis. *J Biol Chem* **273:** 3148–3151
Kaziro Y (1978) The role of guanosine 5′-triphosphate in polypeptide chain elongation. *Biochim Biophys Acta* **505:** 95–127
Laurberg M, Kristensen O, Martemyanov K, Gudkov AT, Nagaev I, Hughes D, Liljas A (2000) Structure of a mutant EF-G reveals domain III and possibly the fusidic acid binding site. *J Mol Biol* **303:** 593–603
Lodmell JS, Dahlberg AE (1997) A conformational switch in *Escherichia coli* 16S ribosomal RNA during decoding of messenger RNA. *Science* **277:** 1262–1267
Martemyanov KA, Gudkov AT (2000) Domain III of elongation factor G from *T. thermophilus* is essential for induction of GTP hydrolysis on the ribosome. *J Biol Chem* **275:** 35820–35824
Moazed D, Noller HF (1989) Intermediate states in the movement of transfer RNA in the ribosome. *Nature* **342:** 142–148
Nevskaya N, Tischenko S, Fedorov R, Al-Karadaghi S, Liljas A, Kraft A, Piendl W, Garber M, Nikonov S (2000) Archaeal ribosomal protein L1: the structure provides new insights into RNA binding of the L1 protein family. *Struct Fold Des* **8:** 363–371
Nikulin A, Eliseikina I, Tishchenko S, Nevskaya N, Davydova N, Platonova O, Piendl W, Selmer M, Liljas A, Drygin D, Zimmermann R, Garber M, Nikonov S (2003) Structure of the L1 protuberance in the ribosome. *Nat Struct Biol* **10:** 104–108
Noller HF, Yusupov MM, Yusupova GZ, Baucom A, Cate JH (2002) Translocation of tRNA during protein synthesis. *FEBS Lett* **514:** 11–16
Ogle JM, Murphy FV, Tarry MJ, Ramakrishnan V (2002) Selection of tRNA by the ribosome requires a transition from an open to a closed form. *Cell* **111:** 721–732
Ramakrishnan V (2002) Ribosome structure and the mechanism of translation. *Cell* **108:** 557–572
Ramakrishnan V, Moore PB (2001) Atomic structures at last: the ribosome in 2000. *Curr Opin Struct Biol* **11:** 144–154
Rawat UB, Zavialov AV, Sengupta J, Valle M, Grassucci RA, Linde J, Vestergaard B, Ehrenberg M, Frank J (2003) A cryo-electron microscopic study of ribosome-bound termination factor RF2. *Nature* **421:** 87–90
Rodnina M, Savelsbergh A, Katunin VI, Wintermeyer W (1997) Hydrolysis of GTP by elongation factor G drives tRNA movement on the ribosome. *Nature* **385:** 37–41
Rodnina MV, Semenkov Iu P, Savelsbergh A, Katunin VI, Peske F, Wilden B, Wintermeyer W (2001) Mechanism of tRNA translocation on the ribosome]. *Mol Biol (Mosk)* **35:** 655–665
Rossmann MG (2000) Fitting atomic models into electron-microscopy maps. *Acta Crystallogr D Biol Crystallog* **D56:** 1341–1349
Saad A, Ludtke SJ, Jakana J, Rixon FJ, Tsuruta H, Chiu W (2001) Fourier amplitude decay of electron cryomicroscopic images of single particles and effects on structure determination. *J Struct Biol* **133:** 32–42
Savelsbergh A, Katunin VI, Mohr D, Peske F, Rodnina MV, Wintermeyer W (2003) An elongation factor G-induced ribosome rearrangement precedes tRNA–mRNA translocation. *Mol Cell* **11:** 1517–1523
Schlünzen F, Tocilj A, Zarivach R, Harms J, Glühmann M, Janell D, Bashan A, Bartels H, Agmon I, Franceschi F, Yonath A (2000) Structure of functionally activated small ribosomal subunit at 3.3 Å resolution. *Cell* **102:** 615–623
Spahn CMT, Beckmann R, Eswar N, Penczek PA, Sali A, Blobel G, Frank J (2001a) Structure of the 80S ribosome from *Saccharomyces cerevisiae*—tRNA–ribosome and subunit–subunit interactions. *Cell* **107:** 373–386
Spahn CMT, Blaha G, Agrawal RK, Penczek P, Grassucci RA, Trieber CA, Connell SR, Taylor DE, Nierhaus KH, Frank J (2001b) Localization of the tetracycline resistance protein Tet(O) on the ribosome and the inhibition mechanism of tetracycline. *Mol Cell* **7:** 1037–1045
Spahn CMT, Nierhaus KH (1998) Models of the elongation cycle: an evaluation. *Biol Chem* **379:** 753–772
Spahn CMT, Prescott CD (1996) Throwing a spanner in the works: antibiotics and the translation apparatus. *J Mol Med* **74:** 423–429
Tama F, Valle M, Frank J, Brooks 3rd CL (2003) Dynamic reorganization of the functionally active ribosome explored by normal mode analysis and cryo-electron microscopy. *Proc Natl Acad Sci USA* **100:** 9319–9323
Valle M, Zavialov A, Li W, Stagg SM, Sengupta J, Nielsen RC, Nissen P, Harvey SC, Ehrenberg M, Frank J (2003a) Incorporation of aminoacyl-tRNA into the ribosome as seen by cryo-electron microscopy. *Nat Struct Biol* **10:** 899–906
Valle M, Zavialov A, Sengupta J, Rawat U, Ehrenberg M, Frank J (2003b) Locking and unlocking of ribosomal motions. *Cell* **114:** 123–134
Wimberly BT, Brodersen DE, Clemons Jr WM, Morgan-Warren RJ, Carter AP, Vonrhein C, Hartsch T, Ramakrishnan V (2000) Structure of the 30S ribosomal subunit. *Nature* **407:** 327–339
Yonath A (2002) The search and its outcome: high-resolution structures of ribosomal particles from mesophilic, thermophilic, and halophilic bacteria at various functional states. *Annu Rev Biophys Biomol Struct* **31:** 257–273
Yusupov MM, Yusupova GZ, Baucom A, Lieberman K, Earnest TN, Cate JH, Noller HF (2001) Crystal structure of the ribosome at 5.5 resolution. *Science* **292:** 883–896
Zavialov AV, Ehrenberg M (2003) Peptidyl-tRNA regulates the GTPase activity of translation factors. *Cell* **114:** 113–122

©2004 European Molecular Biology Organization

Cell, Vol. 121, 703–712, June 3, 2005, Copyright ©2005 by Elsevier Inc. DOI 10.1016/j.cell.2005.03.023

The Cryo-EM Structure of a Translation Initiation Complex from *Escherichia coli*

Gregory S. Allen,[1] Andrey Zavialov,[2] Richard Gursky,[1] Måns Ehrenberg,[2] and Joachim Frank[1,3,*]
[1]Howard Hughes Medical Institute
Health Research, Inc. at the Wadsworth Center
Empire State Plaza
Albany, New York 12201
[2]Department of Cell and Molecular Biology
Biomedical Center
Box 596
S-751 24 Uppsala
Sweden
[3]Department of Biomedical Sciences
The State University of New York at Albany
Empire State Plaza
Albany, New York 12201

Summary

The 70S ribosome and its complement of factors required for initiation of translation in *E. coli* were purified separately and reassembled in vitro with GDPNP, producing a stable initiation complex (IC) stalled after 70S assembly. We have obtained a cryo-EM reconstruction of the IC showing IF2•GDPNP at the intersubunit cleft of the 70S ribosome. IF2•GDPNP contacts the 30S and 50S subunits as well as fMet-tRNA^{fMet}. IF2 here adopts a conformation radically different from that seen in the recent crystal structure of IF2. The C-terminal domain of IF2 binds to the single-stranded portion of fMet-tRNA^{fMet}, thereby forcing the tRNA into a novel orientation at the P site. The GTP binding domain of IF2 binds to the GTPase-associated center of the 50S subunit in a manner similar to EF-G and EF-Tu. Additionally, we present evidence for the localization of IF1, IF3, one C-terminal domain of L7/L12, and the N-terminal domain of IF2 in the initiation complex.

Introduction

Initiation of protein synthesis in eubacteria, during which the 50S and 30S ribosomal subunits are joined to form an elongation-competent 70S ribosome, requires a 30S preinitiation complex. In this preinitiation complex, the 30S subunit of the ribosome is associated with mRNA, initiator fMet-tRNA^{fMet}, and initiation factors 1 (IF1), 2 (IF2), and 3 (IF3) (Lockwood et al., 1971; Gualerzi and Pon, 1990). IF1 and IF2 have homologs in most organisms (Gualerzi and Pon, 1990; Lee et al., 1999), while IF3 lacks a eukaryotic homolog.

The Shine-Dalgarno sequence of mRNA, upstream of the AUG start codon, binds specifically to the 3′-terminal sequence of the 16S rRNA (Shine and Dalgarno, 1974). In this way, the mRNA becomes anchored to the 30S subunit, and the AUG start codon is positioned near the P site of the 30S subunit.

IF1 and IF3 are important for accelerating the rate of assembly of the preinitiation complex (Kay and Grunberg-Manago, 1972; Weiel and Hershey, 1982; Wintermeyer and Gualerzi, 1983; Pon and Gualerzi, 1984; Zucker and Hershey, 1986; Gualerzi and Pon, 1990), for prevention of initiation of protein synthesis with elongator tRNAs (Gualerzi et al., 1971; Canonaco et al., 1986; Hartz et al., 1989; Hartz et al., 1990; Gualerzi and Pon, 1990), and for formation of naked 70S ribosome complexes that lack tRNA in the P site (Grunberg-Manago et al., 1975). An important role for the G protein IF2 is to promote association of the ribosomal subunits (Grunberg-Manago et al., 1975) when an initiator tRNA is present on the 30S subunit (Lockwood et al., 1971; Antoun et al., 2003; Antoun et al., 2004). Although IF2 binds 30S subunits in the absence of GTP (Weiel and Hershey, 1982), rapid subunit association requires that IF2 be on the 30S subunit in the GTP form (Antoun et al., 2003; Antoun et al., 2004). The C terminus of IF2 recognizes both the 3′ single-stranded acceptor stem and the formylated methionine at the extreme 3′ end of fMet-tRNA^{fMet} (Guenneugues et al., 2000; Petersen et al., 1979; Petersen et al., 1981; Wakao et al., 1989). When IF2 binds to the 30S subunit together with initiator tRNA, its affinity to GTP is radically enhanced (Antoun et al., 2003), suggestive of cooperative binding of IF2 and fMet-tRNA^{fMet} to the 30S subunit (Groner and Revel, 1973; Wakao et al., 1990), and GTP binding is accompanied by a conformational change of IF2. Subsequent release of IF2 from the 70S initiation complex requires GTP hydrolysis on the factor (Grunberg-Manago et al., 1972; Luchin et al., 1999; Antoun et al., 2003; Antoun et al., 2004). The resulting 70S postinitiation complex (PIC) is the end product of initiation of protein synthesis in eubacteria.

While both X-ray and cryo-EM structures exist for two of the initiation factors bound to the 30S subunit, IF1•30S (Carter et al., 2001) and IF3•30S (McCutcheon et al., 1999; Pioletti et al., 2001), no structure of IF2 in complex with either the 30S subunit or the 70S ribosome has been reported. Roll-Mecak and coworkers solved the crystal structure of IF2 from *Methanobacterium thermoautotrophicum* bound to GTP; in this structure, the factor resembles a "chalice"; the N-terminal GTPase domain forms part of the cup portion of the chalice, while a long helix (H12) and the C-terminal fMet-tRNA^{fMet} binding domain form the stem and base, respectively (Roll-Mecak et al., 2000). Using various truncation mutants of IF2 in binding reactions with 30S or 50S subunits, which were then purified by pelleting of the resulting complex through a sucrose cushion, Moreno and coworkers delineated the specific domains of IF2 responsible for binding to the ribosomal subunits (Moreno et al., 1998; Moreno et al., 1999). Accordingly, the GTPase domain and domain III of IF2 interact with the 50S subunit, while portions of domain II, as well as the N-terminal domain of IF2, interact with the 30S subunit. Marzi and coauthors, using various nucleolytic

*Correspondence: joachim@wadsworth.org

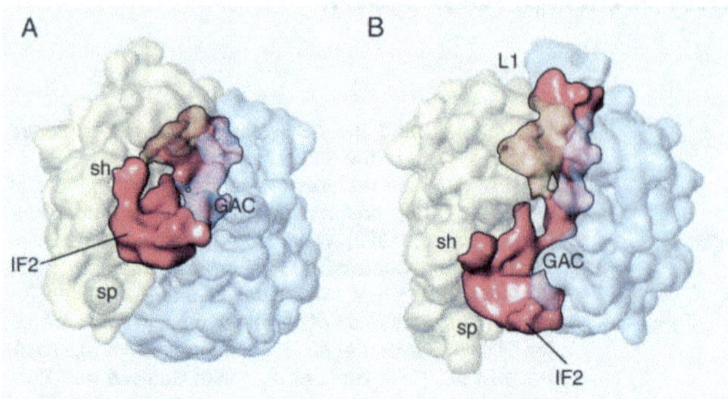

Figure 1. Cryo-EM Structure of the GDPNP-Stalled *E. coli* 70S Ribosome IC with IF2

The ligand density (red with black outline) was obtained by subtraction of the PIC density from the IC density.

(A) In this view, the N-terminal domains of IF2 are in the foreground. These interact with the 50S subunit (blue) at the GTPase-associated center (GAC) and with the 30S subunit (yellow) over a large surface on the intersubunit side of the shoulder (sh).

(B) The initiation complex has, with respect to (A), been rotated by ~45° around a horizontal axis and ~30° around the long axis of the ligand density. Additional labeled features also indicate the spur of the 30S (sp) subunit and the L1 stalk (L1).

reagents tethered to cysteines in IF2 from *Bacillus stearothermophilus*, demonstrated cleavages at specific sites on 16S and 23S rRNA from *Escherichia coli*. This information, in combination with the above-mentioned results, was used to model how IF2 binds to the 70S ribosome (Marzi et al., 2003).

We have formed a stable intermediate in the initiation pathway by substituting the nonhydrolyzable GTP analog GDPNP for GTP in our highly active in vitro system for protein synthesis (Zavialov et al., 2001). This ribosomal complex was obtained by mixing *E. coli* ribosomes with IF1, IF2, GDPNP, IF3, mRNA, and initiator tRNA. The rates of IF2-dependent subunit association with GTP and GDPNP are comparable (Antoun et al., 2003; Antoun et al., 2004). The conformation of IF2 in this complex provides the structural basis for our model of IF2-promoted association of subunits in the presence of initiator tRNA and GTP.

Our cryo-EM map of this stable intermediate shows IF2 in the intersubunit space, in contact with the 50S subunit, the 30S subunit, and initiator tRNA. The crystal structures of the GTP form of IF2 (PDB code 1G7T) and yeast $tRNA^{Phe}$ (PDB code 1EHZ), when fitted into the difference volume (i.e., IC − PIC), reveal the mutual interactions of these ligands and their contacts with the ribosome. However, our cryo-EM structure of IF2 within the 70S complex differs significantly from the crystal structure of free IF2 in its GTP form; cryo-EM in concert with X-ray crystallography can supply a more complete view of the relation between the structure and function of IF2 during initiation of protein synthesis.

Results

Reconstruction of the IC Volume

70S initiation complexes (ICs) with IF2 were formed in vitro for cryo-EM from individually prepared 70S *E. coli* ribosomes, IF1, IF2, IF3, fMet-$tRNA^{fMet}$, and mRNA together with the noncleavable GTP analog GDPNP. In the initial reconstruction from 55,270 particles, we made use of a previously determined cryo-EM reconstruction of the PIC containing mRNA and fMet-$tRNA^{fMet}$ in the P site (Valle et al., 2003a). Reconstruction and further refinement of the IC, using the PIC as a reference, yielded an 11.6 Å resolution (0.5 FSC) density map containing the 70S ribosome plus scattered densities, attributable to IF2, in the intersubunit space near the GTPase-associated center (GAC); the indistinctness of the mass suggested low occupancy of IF2 among the selected particles.

To circumvent this problem, we applied a method of supervised classification (Gao et al., 2004) to the same particle set, using reconstructions of the PIC and an EF-G bound 70S ribosome (Valle et al., 2003b) as alternative references. To reconstruct a density map of the IC with 100% IF2 occupancy, we computed the cross-correlation of each particle for both references and then selected that subset of particles with a higher crosscorrelation for the EF-G bound reference. This subset of particles was used in a reference-based alignment, with the EF-G bound 70S reconstruction as the initial reference. The resulting reconstruction of the IC was then iteratively refined to a resolution of 13.8 Å (0.5 FSC) (Supplemental Data) or 8.6 Å, as determined by the 3σ criterion. In this IC map, the ligand portion appears as a solid mass, with density comparable to that of the ribosomal proteins (Figure 1). We then fitted the coordinates of yeast $tRNA^{Phe}$ (Shi and Moore, 2000) and IF2 bound to GTP (see Experimental Procedures) into the ligand mass isolated from the new density map and refined the positions of these models as rigid bodies using RSRef (Chen et al., 2003; Gao et al., 2003; see Experimental Procedures).

Structural Model of IF2 in the IC

As an aid to interpretation of the IC volume, we isolated the ligand mass using the PIC as a guide (Figure 2). The shape of this mass was significantly different from the overall shape of the crystal structure of IF2•GTP (Figure 3). However, we found that straightforward changes in the relative orientations of the domains of IF2 could render its crystal form (Roll-Mecak et al., 2000; see Experimental Procedures) into a shape similar to its envelope in the difference map (Figure 2). Nevertheless, the portion of the map clearly attributable to the GTP binding domain (G domain) of IF2 is significantly larger than the G domain of the reoriented crystal structure. Therefore, to achieve an objective placement of the G domain within this region of the map, we applied a rapid motif-search algorithm (RAMOS/MOTIF in the SPIDER package; Rath et al., 2003) to the IC mass, using only the G domain coordinates of the crystal structure as a search motif. The highest peak found by the motif

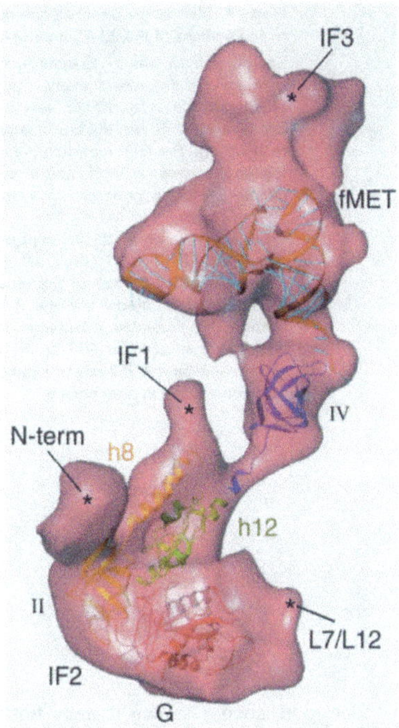

Figure 2. Interpretation of the IC-PIC Difference Map

The difference map is in approximately the same orientation as in Figure 1B. IF2 and fMet-tRNA^{fMet} (fMet) have been fitted to the cryo-EM density. Note that domain names and colors for this IF2 model follow the conventions of Roll-Mecak et al. (2000). Details of the interaction between IF2 domain IV and fMet-tRNA^{fMet} are apparent; our model suggests that domain IV of IF2 interacts only with single-stranded portions of the tRNA. The feature denoted N-term has no counterpart in the crystal structure of IF2: in the preparation of our sample we used the βIF2 protein, which contains an additional 150 amino acids N-terminal of the G domain. Density proximal to the C terminus of helix 8 (H8) of IF2 (domain II) is designated IF1, since this is the location of IF1 in the crystal structure (Carter et al., 2001). The difference density also may include a CTD of L7/L12 and IF3. Also note that domain III is not visible in this view.

search within this region of the map corresponded to a placement of the G domain of IF2 (Figure 2) in the same site on the ribosome as that occupied by the G domains of both ribosome bound EF-G (Figure 4A) and EF-Tu (Figure 4B) (Valle et al., 2003a). We found it striking that each G domain was independently fitted to different cryo-EM density maps yet they all concurred as to their placement and orientation relative to the 70S ribosome. The remainder of IF2 was fitted to the density as described in Experimental Procedures. We note that the resulting model of IF2 is in general agreement with the recently proposed model based on rRNA-cleavage footprinting data (Marzi et al., 2003).

One of the conformational changes in IF2 that we posit, based on comparison of the X-ray structure to our cryo-EM model, is a movement of helix H8 in domain II, which packs against the remainder of domain II and links it to domain III in the X-ray structure; movement of helix H8 is necessary and sufficient to fill a portion of the difference density map connecting the previously fitted domains II and III. At the C-terminal end of the reoriented helix H8 is a globular density in our map that we identify as IF1 for two reasons. First, binding studies have shown that IF1 interacts with domain II of IF2 (Moreno et al., 1999). Second, this placement of IF1 agrees with the factor's position in the crystal complex with the 30S subunit (see Experimental Procedures; Carter et al., 2001).

Figure 5 displays further details of the difference volume and its interactions with the ribosomal subunits. As expected, the G domain of IF2 contacts the 50S subunit in the vicinity of the GTPase-associated center (GAC) (Figure 5A); it approaches the 23S rRNA near bases 2650–2660 and protein L6 near residues 90–94. Domain II of IF2 covers a large area of the 30S subunit (Figure 5B), including 16S rRNA bases 55–58, 357–360, and 366–369. Domain III of IF2 interacts extensively with the surface of L14, while domain IV contacts a large surface of L16 as well as 23S rRNA bases 2550–2556 and 2601–2602. In all, IF2 renders inaccessible 3900 Å^2 of 70S surface area that was formerly solvent accessible: 2600 Å^2 on the 50S subunit and 1300 Å^2 on the 30S subunit (see Experimental Procedures).

30S Subunit Rotation

In Figure 6 we compare the 30S subunits from the IC and PIC density maps. When the two maps are superimposed so as to produce maximal crosscorrelation, their 50S subunit regions are virtually indistinguishable (not shown), with the exception of the head of the L1 stalk, while the 30S subunit portions of the 50S-aligned

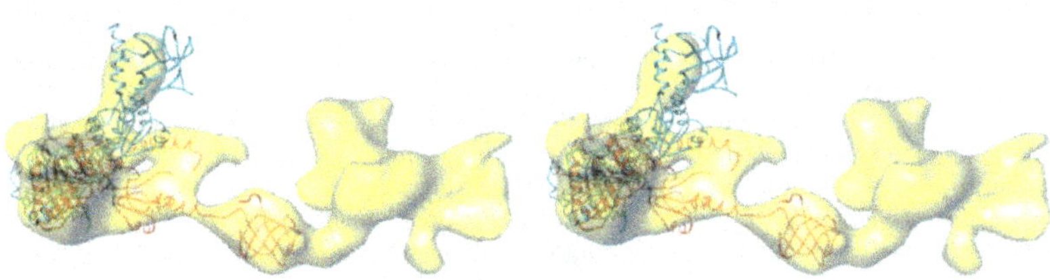

Figure 3. Stereoview Comparison of the Fit of Our IF2 model and the X-Ray Structure to the IC-PIC Difference Map

The crystal structure of IF2 from *M. thermoautotrophicum* (blue; PDB code 1G7T), with its G domain superimposed upon the G domain of our model of IF2 (red). The difference map to which our model was fitted is shown in yellow (see Figure 2). It is seen that our model of IF2 in the IC adopts a conformation quite different from the crystal structure of IF2.

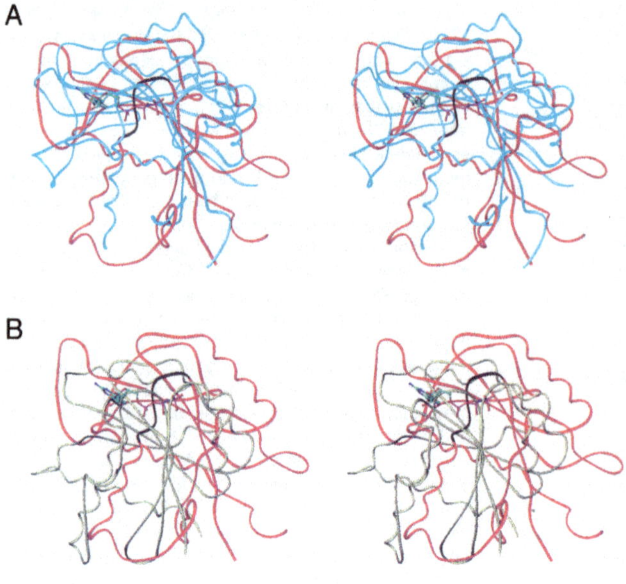

Figure 4. Stereoview Comparisons of G Domain Positions of IF2, EF-G, and EF-Tu

Each G domain was independently fitted to different cryo-EM density maps.

(A) EF-G (Valle et al., 2003b) versus IF2. G domains of EF-G and IF2 are blue and red, respectively. The GTP molecule is bound at the loop, between a strand and a helix, that defines the highly conserved P loop (black portion of IF2).

(B) EF-Tu (Valle et al., 2003b) versus IF2. G domains of EF-Tu and IF2 are green and red, respectively (the P loop of IF2 is colored black). The GTP model and the P loop illustrate the predicted orientation of, and binding-site location for, GTP on IF2's G domain in both panels; IF2 was brought into the same orientation in both panels.

maps differ significantly. The difference map shows a mass between the head of the L1 stalk and the fMet-tRNA (effectively occluding the E site) that may be partially due to motion of the stalk head but probably indicates the presence of another factor, the possible identity of which is discussed below. Evidently, the 30S subunit of the IC undergoes a general counterclockwise rotation (by ~4°) with respect to the 30S subunit of the PIC. The rotation is reminiscent of the conformational changes ("ratcheting") that were observed when EF-G•GDPNP was bound to the release complex in the presence of puromycin (Valle et al., 2003b), as well as changes seen earlier that were related to EF-G-mediated translocation (Frank and Agrawal, 2000).

Novel Hybrid P Site

Our placement of the C-terminal domain of IF2 in the IC volume sterically precludes classical P site binding of initiator tRNA (data not shown). To better compare the positions of the P site and P/E site tRNAs with the IC tRNA, we superimposed models of the three tRNAs on the basis of maximization of the correlation coefficient between the 30S portions of the relevant density maps (see Experimental Procedures); we found that the anticodon loops of all three maps superimpose. Remarkably, the IF2-fMet-tRNA^{fMet} interaction in the IC map exposes a novel site for initiator tRNA, intermediate between the classical P site and the P/E site, which we name the P/I site (i.e., P site on the 30S subunit, new position "I" on the 50S subunit) (Figure 7). Upon hydrolysis of GTP in the IC and subsequent dissociation of the bound initiation factors, the PIC is formed, with a classical P site fMet-tRNA^{fMet}. Clearly, then, the P/I site tRNA must move relative to the 30S subunit following hydrolysis of GTP. The motion required is a simple rotation in the plane of the tRNA by 20° about the anticodon loop; such a motion thereby translates the acyl moiety through a distance of 28 Å (Figure 7B).

Identification of Additional Factors and Domains

The ligand mass also reveals several regions of density not accounted for by the placement of IF2 and fMet-tRNA^{fMet} (Figure 2). One region of the density map for which the interpretation seems obvious lies near the G domain and domain II. This density extends across the surface of the 30S subunit toward the shoulder (sh) and interacts with 16S bases 37–38, 413–420, and 497–498. We assign this density to the N-terminal domain of *E. coli* βIF2, which does not exist in IF2 from *M. thermoautotrophicum* and therefore is absent from its crystal structure (Roll-Mecak et al., 2000). This location of the N-terminal domain of *E. coli* βIF2 agrees with crosslinking data for the complex of *B. stearothermophilus* IF2 with 16S rRNA, which demonstrated that the N-terminal domain of *B. stearothermophilus* IF2 is near guanine 423 of the 16S rRNA (Marzi et al., 2003); G423 of our 30S model is indeed within 20 Å of the density denoted N-term in Figure 2.

Another density region of unknown identity exists between the GAC and the G domain of IF2. Of course, our model includes L11 and other portions of the 50S subunit adjacent to the density in question, but these do not account for the unknown density. However, L7/L12 has been crosslinked to IF2 (Heimark et al., 1976), and the C-terminal domain (CTD) of L7/L12 has been strongly crosslinked to L11 (Dey et al., 1998). Furthermore, a cryo-EM structure of a 70S ribosome containing nanogold-labeled L7/L12 proteins showed the CTD to be next to L11 (Montesano-Roditis et al., 2001). Therefore, we suggest that the density in question is the CTD of L7/L12. Indeed, the recent NMR structure

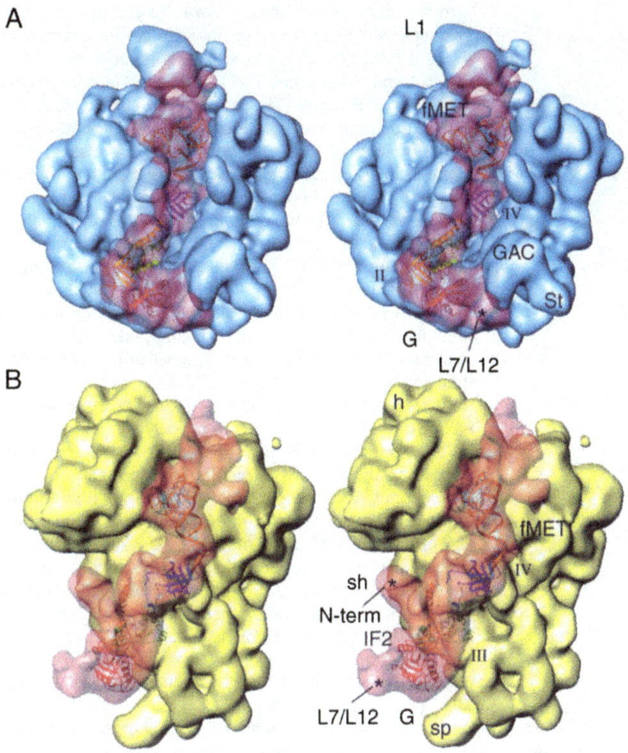

Figure 5. Difference Map-Ribosomal Subunit Contacts in the IC

(A) View of the IC-PIC difference map and its contacts with the intersubunit surface of the 50S subunit. With reference to Figure 1B, the 30S subunit has now been removed, and the 50S subunit has been rotated by ~90° around a vertical axis. St, stalk; L1, L1 stalk; G, II, and IV, G domain and domains II and IV of IF2; L7/L12, CTD of one monomer of L7/L12; N-term, N-terminal domain of βIF2; fMet, fMet-tRNA^{fMet}.

(B) View of the difference map and its interactions with the intersubunit surface of the 30S ribosomal subunit. With reference to Figure 1B, the 50S subunit has been removed, and the 30S subunit has been rotated by ~90° around a vertical axis. We clearly see the interaction of fMet-tRNA^{fMet} with domain IV of IF2. sh, shoulder; sp, spur; III, domain III of IF2.

of the CTD of L7/L12 (Bocharov et al., 2004) fits the density well, as shown in Figure S3.

Even when these tentative assignments have been made, a region of difference density remains unidentified—specifically, density on the side of the initiator tRNA distal to IF2 (Figure 2). Due to its proximity to the E site, we attempted to fit a tRNA to this unassigned density, but without success. However, we noticed that the difference density is strikingly similar in position and shape to a recent model of IF3 bound to the 30S subunit (Dallas and Noller, 2001), and we have therefore tentatively fitted this region of density with the atomic structures of IF3 (Figure S3).

Discussion

We successfully reconstituted the *E. coli* 70S initiation complex, stalled by GDPNP, using separately purified factors and ribosomes. Translation assays performed in parallel with complex assembly and using components from the same lots of purified factors and ribosomes gave high activity (data not shown). Therefore, we are confident that the cryo-EM structure of the stalled *E. coli* 70S initiation complex that we report here represents the true solution structure of the complex. As expected, IF2 and fMet-tRNA^{fMet} were located in the intersubunit space. Owing to the interaction with IF2, fMet-tRNA^{fMet} is forced into a novel hybrid P site on the 30S subunit. We have also located the N-terminal domain of *E. coli* IF2 in the long "thumb" of density extending across the surface of the 30S subunit in our map of the IC.

We also found that IF2 buries 2600 Å^2 of the 50S subunit's surface area. Burial of surfaces of comparable size commonly occurs at dimerization interfaces; indeed, IF2's role as the catalyst of 50S subunit association is analogous to that of a partner in dimerization. A survey of published binding affinities for dimerization interactions shows that such interactions have a high ΔG (<−10 kcal/mole), which corresponds to binding affinities on the order of $K_D \sim$ nM or better (Brooijmans et al., 2002). Therefore, we hypothesize that the burial of IF2 surface area probably accounts for the mechanism of subunit association as well as the relatively high stability of the 70S intermediate containing IF2•GDPNP, here called the IC (Antoun et al., 2003; Antoun et al., 2004).

The observation that IF2's affinity for GTP on the 30S subunit is greatly enhanced by the presence of initiator tRNA suggests the occurrence of a GTP-dependent conformational change of the factor; this change is a prerequisite for rapid subunit association (Antoun et al., 2003; Antoun et al., 2004). Furthermore, the GDP and GTP forms of IF2 in the crystal are significantly different from one another, suggesting that GTP hydrolysis on IF2 in the 70S initiation intermediate reduces the area of the IF2•50S subunit interface, which in turn leads to decreased affinity and rapid dissociation of the factor.

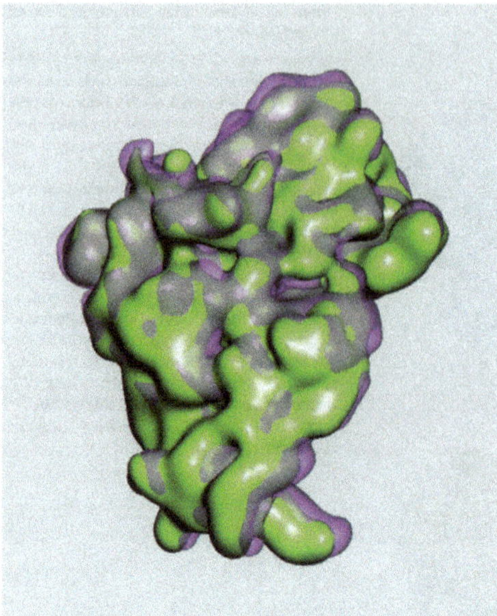

Figure 6. Rotation of the 30S Ribosomal Subunit between the IC and PIC Complexes

The IC and PIC (Valle et al., 2003b) maps were filtered at 13.8 Å and aligned with respect to their 50S subunits only. The 30S subunit from the PIC is green, while that from the IC is violet. Most of the differences between the two complete ribosome density maps appear in the 30S subunit region. There is a general counterclockwise rotation of the 30S subunit in the IC by ~4° with respect to the 30S subunit in the PIC.

In the absence of a structure for the 30S preinitiation complex, it is not known whether the conformations of IF2 and fMet-tRNA^{fMet} change upon association of the 50S subunit followed by 70S formation. It is clear, however, that GTP hydrolysis on IF2 in the 70S IC must result in a structural rearrangement of the ribosome to allow for fMet-tRNA^{fMet} movement into the classic P/P site. Since IF2 recognizes structural elements of the initiator tRNA, it could provide an accuracy step in initiation, ensuring that the IC is formed with only initiator tRNA (Huang et al., 1997; Shin et al., 2002). If all noninitiator tRNAs bind to IF2, albeit at a much lower affinity than does the initiator tRNA (Mayer et al., 2003), and if any tRNA can occupy the P site, a means for excluding noninitiator tRNAs must exist. We suggest that a higher standard free energy exists for any tRNA at the P/I site than for that tRNA at the P site; this increase in standard free energy raises the energy barrier too high for elongator tRNAs, which would already be destabilized by the noncognate interaction with the AUG start codon. Accordingly, P/I site binding and formation of the 70S IC could be strongly inhibited for elongator tRNAs but could be allowed for fMet-tRNA^{fMet}. Thus, the IC with IF2 in the GTP form may provide a checkpoint at which initiation can be aborted when the cognate initiator tRNA has been substituted by an elongator tRNA. Abortion may be effected either by redissociation of ribosomal subunits or else by dissociation of the elongator tRNA from the IC, followed by recycling of the 70S ribosome (minus tRNA) by the action of the RRF and EF-G (Karimi et al., 1999).

It is clear from the results of our placement of the G domain of IF2, using the motif search on the IC volume, that the G domains of EF-Tu, EF-G, and IF2 all occupy the same site on the ribosome, in roughly equivalent orientations (Figure 3). Furthermore, although these G domains have significant differences in both sequence

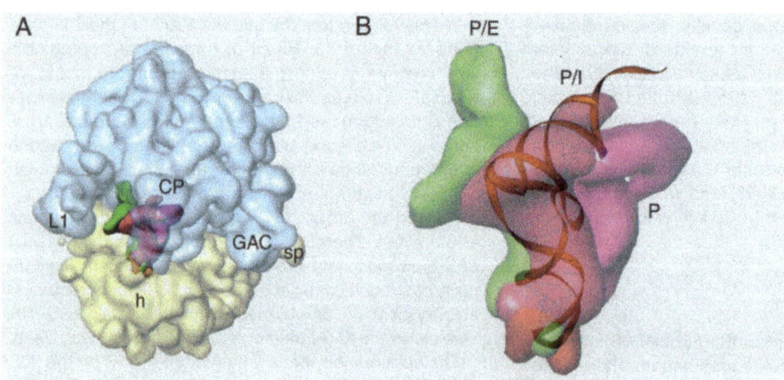

Figure 7. Comparison of tRNA in the P/I Site of the Initiation Complex with Its P Site and P/E Site Positions

(A) This view of the IC density map optimally shows the various P site tRNAs. The P site on the 30S subunit (yellow) is visible due to transparent rendering of both subunits (50S subunit in blue); the central protuberance (CP) is the chief structure here. Canonical P site (purple); P/E hybrid site (green); P/I site discovered in this work (red); h, head; sp, spur; L1, L1 stalk; GAC, GTPase center.

(B) Close-up of the P site tRNA variants. In going from the P site to the P/I site, the initiator tRNA rotates by ~20° about an axis through the anticodon, normal to the plane of the page; its acyl moiety moves by about 28 Å. By comparison, the P-to-P/E transition involves a rotation of ~35° about the long axis of the tRNA, as well as a rotation of ~35° about an axis through the anticodon, normal to the plane of the page; in this case, the acyl moiety moves by about 68 Å.

and structure, the number of features that they have in common is striking (Roll-Mecak et al., 2000). These observations suggest that there exists a general mechanism by which the ribosome effects GTP hydrolysis on these factors. However general this mechanism might be for the ribosome, it probably does not conform to the canonical GTPase mechanism, which normally involves intrusion of activity partners into the GTP binding pocket so as to provide a conserved side chain, such as lysine or arginine, that is directly involved in hydrolysis (Vetter and Wittinghofer, 2001). Yet, in the case of the ribosome-factor interactions noted above, there are no ribosomal protein residues proximal to the binding site of the factors. The factor's P loop, which is the site of GTP binding, could only interact with the sarcin-ricin loop (SRL) of the 23S rRNA, approximately 8 Å distant. Thus, the SRL has been identified as the GTPase-activating factor (Blanchard et al., 2004). Possibly the phosphate backbone of the SRL coordinates a metal ion, which could foster hydrolysis (Beaudry et al., 1979). However, previous structural and crosslinking data, taken together with our data, suggest that the C-terminal domain of L7/L12 intercalates between L11 and the G domain of IF2. Additionally, L7/L12 is known to be required for rapid hydrolysis of GTP on EF-Tu and EF-G (Mohr et al., 2002). Therefore, we suggest that the L7/L12 C-terminal domain induces conformational changes in the G domain, and these changes then stimulate the GTPase activity of IF2 (Mohr et al., 2002; Bocharov et al., 2004).

We have previously shown that ribosome-dependent GTP hydrolysis on ribosome release factor RF3 and EF-G requires a deacylated tRNA in the P site (Zavialov and Ehrenberg, 2003), allowing for the tRNA's movement into a hybrid P/E site (Valle et al., 2003b). When a noninitiator tRNA occupies the P site, the same is true for IF2; i.e., there is strong GTP hydrolysis when the tRNA is deacylated, but not otherwise. In contrast, when the P site is occupied by fMet-tRNA^{fMet}, there is strong idling GTPase activity by IF2 (Zavialov and Ehrenberg, 2003). The explanation for this behavior, we suggest, is that GTP hydrolysis on RF3, EF-G, and IF2 requires that, in each case, the CCA end of the P site tRNA move toward the E site. When the P site tRNA lacks peptide, this is easily achieved through hybrid P/E site binding of the tRNA. When the P site tRNA has a peptide, movement of the tRNA's CCA end toward the E site is effectively prevented by a high standard free-energy barrier so that the GTPase activity of all three factors is inhibited. It is only when the P site tRNA is fMet-tRNA^{fMet} that IF2 in the GTP form, with its strong affinities for both the initiator tRNA and the 50S subunit, can drive the ribosome into the intermediate IC, in which hydrolysis of IF2 is induced. This interpretation lends strong support to the above-suggested accuracy function of the IC: if an aminoacyl tRNA rather than an initiator tRNA is present in the preinitiation complex, then the IC cannot be formed, and initiation will be aborted.

Finally, we propose that IF3 is present in the IC. However, the authors who initially proposed the manner in which IF3 binds to the 30S subunit point out that their model predicts a clash between the C-terminal domain of IF3 and helix 69 of the 50S subunit (Dallas and Noller, 2001). We note that helix 69 is visible in both the IC map and the PIC map. When the two density maps are aligned so as to maximize the crosscorrelation, both helices 69 overlay precisely—that is, helix 69 does not move at all. Also, in the PIC map, helix 69 contacts fMet-tRNA, which is in the P site, but in the IC map, fMet-tRNA has been pushed out of the P site by IF2, thus making space between helix 69 and fMet. The space created between helix 69 and fMet also corresponds to a mass of density in the difference map (Figure S3); the volume associated with the mass in the difference map is approximately the right size for the C-terminal domain (CTD) of IF3. Biochemical data (Gualerzi and Pon, 1990; Pon and Gualerzi, 1986) suggest that 70S ribosome formation from the 50S subunit and a preinitiation 30S complex requires either that IF3 leave the 30S subunit before the 50S subunit can associate or else that IF3 dissociate from the 70S ribosome as a result of subunit association. In either scenario, there should therefore be no IF3 in the 70S initiation complex. It is, however, possible that IF3 is released from the 70S ribosome upon hydrolysis of GTP on IF2, in which case the presence of IF3 in the IC with IF2•GDPNP would be consistent with the second scenario of Gualerzi and collaborators (Gualerzi and Pon, 1990). Indeed, results of recent experiments using kinetic techniques suggest that IF3 remains in the IC complex until P_i is released (Grigoriadou et al., 2004). Early experiments (Goss et al., 1982) also suggested that 70S ribosomes can complex with IF3. Our present results lead us to conclude that IF3 is present in the IC (GDPNP).

Experimental Procedures

Preparation of Initiation Complexes

For preparation of the complex, a 300 μl mixture containing 2.7 μM *E. coli* ribosomes, 6.7 μM *E. coli* IF1, 6.7 μM *E. coli* β form of IF2 (Nyengaard et al., 1991), 4 μM *E. coli* IF3, 8.1 μM mRNA encoding for MFTI tetrapeptide (Zavialov et al., 2001), 7.6 μM *E. coli* fMet-tRNA^{fMet}, and 0.6 mM GDPNP was incubated at 37°C for 25 min. Then the mixture was diluted to 1 μM ribosome concentration with 0.5 mM GDPNP and 2 μM IF1, and it was aliquoted and frozen with liquid nitrogen. Prior to the cryo-EM analysis, an aliquot containing the complex was further diluted to 32 nM ribosome concentration with a dilution mixture containing 4 μM IF1, 2 μM βIF2, and 0.5 mM GDPNP. The incorporation of [^{3}H]fMet-tRNA^{fMet} into the initiation complex detected by nitrocellulose filtration was greater than 75%. Translation assays performed in parallel with the purified initiation complex, using βIF2, IF1, and GDPNP, showed that the initiated ribosome is nearly saturated with βIF2•GDPNP (Zavialov and Ehrenberg, 2003).

Cryo-EM and Image Processing

Thin carbon was floated onto Quantifoil grids (Quantifoil Micro Tools GMBH, Jena, Germany). A 5 μl aliquot was placed on each grid. The grids were then blotted, plunge frozen in liquid ethane, and stored in liquid nitrogen until data collection. Data were collected on a Tecnai F20 (FEI Company, Hillsboro, Oregon) at 200 kV and a magnification of 50,000 (±2%), under low-dose conditions. Micrographs were scanned at 14 μm resolution, giving 2.82 Å per pixel. One hundred and fourteen micrographs were each assigned to one of 21 defocus groups ranging from 0.93 to 3.93 μm underfocus. Particles (55,270) could easily be identified on the micrographs as 70S ribosomes. Each of the 70S particle images was initially aligned by crosscorrelation to 83 equispaced projections of a 3D reference, here the PIC, yielding the particle orientation as a set of Eulerian angles. The aligned particle images were backprojected to yield a 3D reconstruction of the IC. Alignment and orientation parameters of the particles were iteratively refined with decreasing angular increments using standard methods. The SPIDER/WEB suite of programs (Frank et al., 1996) was used for all image processing. The resolution was estimated by a Fourier shell correlation cutoff of 0.5 to be 11.6 Å.

At this point it was clear that a significant number of the particles were 70S ribosomes from which one or more of the factors had dissociated. A method of supervised classification, also implemented in SPIDER, was used to separate images according to ligand occupancy. In this method, each image is compared with two reference density maps, here the PIC and an EF-G bound 70S ribosome (Valle et al., 2003b), and is then assigned to one of two classes, depending on the size of its crosscorrelation coefficient with respect to the two references (Gao et al., 2004). Figure S1 shows a histogram plot of the difference between the crosscorrelation values of each image with respect to both references; the class of images having a larger crosscorrelation coefficient with respect to the EF-G bound ribosome (particles plotted to left of zero, i.e., ligand bound form) was then used to reconstruct the IC volume. The initial image set contained 55,270 particles, which supervised classification pared to 20,283 particles for the IC volume. Following supervised classification, the resolution was estimated by a Fourier shell correlation cutoff of 0.5 to be 13.8 Å and by the 3σ criterion to be 8.6 Å.

Docking and Modeling of IF2, IF1, and P/I Site tRNA

The ModPipe server provided an excellent *E. coli* IF2 model (Pieper et al., 2004) based on the X-ray crystal structure of IF2 from *M. thermoautotrophicum* (PDB code 1G7T, Roll-Mecak et al., 2000). Since the region of the IC map corresponding to the G domain probably contains density for other molecules and domains (i.e., L7/L12 and domain III), we initially placed the G domain of IF2 using the rapid motif-search procedure implemented in SPIDER (Rath et al., 2003). The IC density map was searched systematically for crosscorrelation maxima using a 3D volume constructed from the coordinates of the G domain. Using the visualization program O (Jones et al., 1991), domain II was placed into the difference density using the location of the G domain and the crystal-structure conformation of IF2 as a reference (Roll-Mecak et al., 2000). Additionally, the cleavage patterns of rRNA by two mutant IF2s with tethered nucleases in domain II provided further constraints on the conformation of domain II (Marzi et al., 2003). Domain III could only be fitted by eye since no constraints were available and since the apparent conformation would necessitate large movements from the crystal-structure form. However, helix 12 was easily recognizable in the final refined density map. As helix 12 connects domain III with domain IV, a domain which is globular and interacts with fMet-tRNA^{fMet}, we found it straightforward to fit domain IV and the initiator tRNA. To improve the fit to the IC map, the hand-fitted models of IF2 and fMet-tRNA^{fMet} were subjected to rigid-body refinement using a real-space refinement program (RSRef, Gao et al., 2003). To calculate the surface area buried by IF2, the software suite CNS was applied using a 1.4 Å probe radius (Brunger et al., 1998).

We fitted the IF1 model by overlaying the IF1:30S crystal-structure model (Carter et al., 2001) on the 30S subunit coordinates obtained by RSRef applied to the IC map; the resulting IF1 model nicely filled the extra EM density found at the end of helix 8 of our model of IF2. Note that the domain numbering (i.e., G–IV) follows the crystal-structure convention (Roll-Mecak et al., 2000), whereas the residue numbering adheres to the βIF2 sequence (i.e., the first residue of the crystal structure is residue 224 of βIF2).

The P site tRNAs were superimposed on the basis of the translational shifts that maximized the crosscorrelation of the respective 30S subunits with which they were associated. In order to superimpose the three fMet-tRNAs, as in Figure 7, we performed real-space refinement with the coordinates of the P site tRNA from Valle and coworkers (Valle et al., 2003a). As expected, this procedure brought all three anticodons into virtually perfect alignment.

The models were manipulated by hand in O (Jones et al., 1991). Figures were prepared in RIBBONS (Carson, 1997) and IRIS Explorer 5.

Supplemental Data

Supplemental Data include three figures and are available with this article online at http://www.cell.com/cgi/content/full/121/5/703/DC1/.

Acknowledgments

We thank Michael Watters for assistance with the illustrations, and we thank Adriana Verschoor, Jayati Sengupta, Ayman Antoun, and Michael Pavlov for comments on the manuscript. This work was supported by HHMI and NIH grants R37 GM29169 and R01 GM 55440 (J.F.) and by the Swedish Research Council and the Swedish Foundation for Strategic Research (M.E.).

Received: December 21, 2004
Revised: March 14, 2005
Accepted: March 22, 2005
Published: June 2, 2005

References

Antoun, A., Pavlov, M.Y., Andersson, K., Tenson, T., and Ehrenberg, M. (2003). The roles of initiation factor 2 and guanosine triphosphate in initiation of protein synthesis. EMBO J. *22*, 5593–5601.

Antoun, A., Pavlov, M.Y., Tenson, T., and Ehrenberg, M.M. (2004). Ribosome formation from subunits studied by stopped-flow and Rayleigh light scattering. Biol. Proced. Online *6*, 35–54.

Beaudry, P., Sander, G., Grunberg-Manago, M., and Douzou, P. (1979). Cation-induced regulatory mechanism of GTPase activity dependent on polypeptide initiation factor 2. Biochemistry *18*, 202–207.

Blanchard, S.C., Gonzalez, R.L., Kim, H.D., Chu, S., and Puglisi, J.D. (2004). tRNA selection and kinetic proofreading in translation. Nat. Struct. Mol. Biol. *11*, 1008–1014.

Bocharov, E.V., Sobol, A.G., Pavlov, K.V., Korzhnev, D.M., Jaravine, V.A., Gudkov, A.T., and Arseniev, A.S. (2004). From structure and dynamics of protein L7/L12 to molecular switching in ribosome. J. Biol. Chem. *279*, 17697–17706.

Brooijmans, N., Sharp, K.A., and Kuntz, I.D. (2002). Stability of macromolecular complexes. Proteins *48*, 645–653.

Brunger, A.T., Adams, P.D., Clore, G.M., DeLano, W.L., Gros, P., Grosse-Kunstleve, R.W., Jiang, J.S., Kuszewski, J., Nilges, M., Pannu, N.S., et al. (1998). Crystallography & NMR system: A new software suite for macromolecular structure determination. Acta Crystallogr. D Biol. Crystallogr. *54*, 905–921.

Canonaco, M.A., Calogero, R.A., and Gualerzi, C.O. (1986). Mechanism of translational initiation in prokaryotes. Evidence for a direct effect of IF2 on the activity of the 30 S ribosomal subunit. FEBS Lett. *207*, 198–204.

Carson, M. (1997). Ribbons. Methods Enzymol. *277*, 493–505.

Carter, A.P., Clemons, W.M., Jr., Brodersen, D.E., Morgan-Warren, R.J., Hartsch, T., Wimberly, B.T., and Ramakrishnan, V. (2001). Crystal structure of an initiation factor bound to the 30S ribosomal subunit. Science *291*, 498–501.

Chen, J.Z., Furst, J., Chapman, M.S., and Grigorieff, N. (2003). Low-resolution structure refinement in electron microscopy. J. Struct. Biol. *144*, 144–151.

Dallas, A., and Noller, H.F. (2001). Interaction of translation initiation factor 3 with the 30S ribosomal subunit. Mol. Cell *8*, 855–864.

Dey, D., Bochkariov, D.E., Jokhadze, G.G., and Traut, R.R. (1998). Cross-linking of selected residues in the N- and C-terminal domains of Escherichia coli protein L7/L12 to other ribosomal proteins and the effect of elongation factor Tu. J. Biol. Chem. *273*, 1670–1676.

Frank, J., and Agrawal, R.K. (2000). A ratchet-like inter-subunit reorganization of the ribosome during translocation. Nature *406*, 318–322.

Frank, J., Radermacher, M., Penczek, P., Zhu, J., Li, Y., Ladjadj, M., and Leith, A. (1996). SPIDER and WEB: processing and visualization of images in 3D electron microscopy and related fields. J. Struct. Biol. *116*, 190–199.

Gao, H., Sengupta, J., Valle, M., Korostelev, A., Eswar, N., Stagg, S.M., Van Roey, P., Agrawal, R.K., Harvey, S.C., Sali, A., et al. (2003). Study of the structural dynamics of the E coli 70S ribosome using real-space refinement. Cell *113*, 789–801.

Gao, H., Valle, M., Ehrenberg, M., and Frank, J. (2004). Dynamics of EF-G interaction with the ribosome explored by classification of a heterogeneous cryo-EM dataset. J. Struct. Biol. *147*, 283–290.

Goss, D.J., Parkhurst, L.J., and Wahba, A.J. (1982). Kinetic studies on the interaction of chain initiation factor 3 with 70 S Escherichia coli ribosomes and subunits. J. Biol. Chem. *257*, 10119–10127.

Grigoriadou, C., Marzi, S., Seo, H.S., Dongli, P., Kirillov, S., Gualerzi, C.O., and Cooperman, B.S. (2004). IF2, IF3 and thiostrepton interaction during initiation complex formation. In Translational Control, A. Hinnebusch, T. Pestova, and J. Richter, eds. (Cold Spring Harbor, New York: Cold Spring Harbor Laboratory Press), pp. 407.

Groner, Y., and Revel, M. (1973). Escherichia coli initiation factors: isolation of an IF2-fMet-transfer RNA complex. J. Mol. Biol. *74*, 407–410.

Grunberg-Manago, M., Dondon, J., and Graffe, M. (1972). Inhibition by thiostrepton of the IF-2-dependent ribosomal GTPase. FEBS Lett. *22*, 217–221.

Grunberg-Manago, M., Dessen, P., Pantaloni, D., Godefroy-Colburn, T., Wolfe, A.D., and Dondon, J. (1975). Light-scattering studies showing the effect of initiation factors on the reversible dissociation of Escherichia coli ribosomes. J. Mol. Biol. *94*, 461–478.

Gualerzi, C., Pon, C.L., and Kaji, A. (1971). Initiation factor dependent release of aminoacyl-tRNAs from complexes of 30S ribosomal subunits, synthetic polynucleotide and aminoacyl tRNA. Biochem. Biophys. Res. Commun. *45*, 1312–1319.

Gualerzi, C.O., and Pon, C.L. (1990). Initiation of mRNA translation in prokaryotes. Biochemistry *29*, 5881–5889.

Guenneugues, M., Caserta, E., Brandi, L., Spurio, R., Meunier, S., Pon, C.L., Boelens, R., and Gualerzi, C.O. (2000). Mapping the fMet-tRNA(f)(Met) binding site of initiation factor IF2. EMBO J. *19*, 5233–5240.

Hartz, D., McPheeters, D.S., and Gold, L. (1989). Selection of the initiator tRNA by Escherichia coli initiation factors. Genes Dev. *3*, 1899–1912.

Hartz, D., Binkley, J., Hollingsworth, T., and Gold, L. (1990). Domains of initiator tRNA and initiation codon crucial for initiator tRNA selection by Escherichia coli IF3. Genes Dev. *4*, 1790–1800.

Heimark, R.L., Hersehy, J.W., and Traut, R.R. (1976). Cross-linking of initiation factor IF2 to proteins L7/L12 in 70 S ribosomes of Escherichia coli. J. Biol. Chem. *251*, 779–784.

Huang, H.K., Yoon, H., Hannig, E.M., and Donahue, T.F. (1997). GTP hydrolysis controls stringent selection of the AUG start codon during translation initiation in Saccharomyces cerevisiae. Genes Dev. *11*, 2396–2413.

Jones, T.A., Zou, J.Y., and Cowan, S.W. (1991). Improved methods for building protein models in electron density maps and the location of errors in these models. Acta Crystallogr. A *47*, 110–119.

Karimi, R., Pavlov, M.Y., Buckingham, R.H., and Ehrenberg, M. (1999). Novel roles for classical factors at the interface between translation termination and initiation. Mol. Cell *3*, 601–609.

Kay, A.C., and Grunberg-Manago, M. (1972). The mechanism of action of initiation factor IF1: non-analogy with elongation factor EFTs. Biochim. Biophys. Acta *277*, 225–230.

Lee, J.H., Choi, S.K., Roll-Mecak, A., Burley, S.K., and Dever, T.E. (1999). Universal conservation in translation initiation revealed by human and archaeal homologs of bacterial translation initiation factor IF2. Proc. Natl. Acad. Sci. USA *96*, 4342–4347.

Lockwood, A.H., Chakraborty, P.R., and Maitra, U. (1971). A complex between initiation factor IF2, guanosine triphosphate, and fMet-tRNA: an intermediate in initiation complex formation. Proc. Natl. Acad. Sci. USA *68*, 3122–3126.

Luchin, S., Putzer, H., Hershey, J.W., Cenatiempo, Y., Grunberg-Manago, M., and Laalami, S. (1999). In vitro study of two dominant inhibitory GTPase mutants of Escherichia coli translation initiation factor IF2. Direct evidence that GTP hydrolysis is necessary for factor recycling. J. Biol. Chem. *274*, 6074–6079.

Marzi, S., Knight, W., Brandi, L., Caserta, E., Soboleva, N., Hill, W.E., Gualerzi, C.O., and Lodmell, J.S. (2003). Ribosomal localization of translation initiation factor IF2. RNA *9*, 958–969.

Mayer, C., Kohrer, C., Kenny, E., Prusko, C., and RajBhandary, U.L. (2003). Anticodon sequence mutants of Escherichia coli initiator tRNA: effects of overproduction of aminoacyl-tRNA synthetases, methionyl-tRNA formyltransferase, and initiation factor 2 on activity in initiation. Biochemistry *42*, 4787–4799.

McCutcheon, J.P., Agrawal, R.K., Philips, S.M., Grassucci, R.A., Gerchman, S.E., Clemons, W.M., Jr., Ramakrishnan, V., and Frank, J. (1999). Location of translational initiation factor IF3 on the small ribosomal subunit. Proc. Natl. Acad. Sci. USA *96*, 4301–4306.

Mohr, D., Wintermeyer, W., and Rodnina, M.V. (2002). GTPase activation of elongation factors Tu and G on the ribosome. Biochemistry *41*, 12520–12528.

Montesano-Roditis, L., Glitz, D.G., Traut, R.R., and Stewart, P.L. (2001). Cryo-electron microscopic localization of protein L7/L12 within the Escherichia coli 70 S ribosome by difference mapping and Nanogold labeling. J. Biol. Chem. *276*, 14117–14123.

Moreno, J.M., Kildsgaard, J., Siwanowicz, I., Mortensen, K.K., and Sperling-Petersen, H.U. (1998). Binding of *Escherichia coli* initiation factor IF2 to 30S ribosomal subunits: a functional role for the N-terminus of the factor. Biochem. Biophys. Res. Commun. *252*, 465–471.

Moreno, J.M., Drskjotersen, L., Kristensen, J.E., Mortensen, K.K., and Sperling-Petersen, H.U. (1999). Characterization of the domains of E. coli initiation factor IF2 responsible for recognition of the ribosome. FEBS Lett. *455*, 130–134.

Nyengaard, N.R., Mortensen, K.K., Lassen, S.F., Hershey, J.W., and Sperling-Petersen, H.U. (1991). Tandem translation of E. coli initiation factor IF2 beta: purification and characterization in vitro of two active forms. Biochem. Biophys. Res. Commun. *181*, 1572–1579.

Petersen, H.U., Roll, T., Grunberg-Manago, M., and Clark, B.F. (1979). Specific interaction of initiation factor IF2 of E. coli with formylmethionyl-tRNA f Met. Biochem. Biophys. Res. Commun. *91*, 1068–1074.

Petersen, H.U., Kruse, T.A., Worm-Leonhard, H., Siboska, G.E., Clark, B.F., Boutorin, A.S., Remy, P., Ebel, J.P., Dondon, J., and Grunberg-Manago, M. (1981). Study of the interaction of Escherichia coli initiation factor IF2 with formylmethionyl-tRNAMetf by partial digestion with cobra venom ribonuclease. FEBS Lett. *128*, 161–165.

Pieper, U., Eswar, N., Braberg, H., Madhusudhan, M.S., Davis, F.P., Stuart, A.C., Mirkovic, N., Rossi, A., Marti-Renom, M.A., Fiser, A., et al. (2004). MODBASE, a database of annotated comparative protein structure models, and associated resources. Nucleic Acids Res. *32* (Database issue), D217–D222.

Pioletti, M., Schlunzen, F., Harms, J., Zarivach, R., Gluhmann, M., Avila, H., Bashan, A., Bartels, H., Auerbach, T., Jacobi, C., et al. (2001). Crystal structures of complexes of the small ribosomal subunit with tetracycline, edeine and IF3. EMBO J. *20*, 1829–1839.

Pon, C.L., and Gualerzi, C.O. (1984). Mechanism of protein biosynthesis in prokaryotic cells. Effect of initiation factor IF1 on the initial rate of 30 S initiation complex formation. FEBS Lett. *175*, 203–207.

Pon, C.L., and Gualerzi, C.O. (1986). Mechanism of translational initiation in prokaryotes. IF3 is released from ribosomes during and not before 70 S initiation complex formation. FEBS Lett. *195*, 215–219.

Rath, B.K., Hegerl, R., Leith, A., Shaikh, T.R., Wagenknecht, T., and Frank, J. (2003). Fast 3D motif search of EM density maps using a locally normalized cross-correlation function. J. Struct. Biol. *144*, 95–103.

Roll-Mecak, A., Cao, C., Dever, T.E., and Burley, S.K. (2000). X-Ray structures of the universal translation initiation factor IF2/eIF5B: conformational changes on GDP and GTP binding. Cell *103*, 781–792.

Shi, H., and Moore, P.B. (2000). The crystal structure of yeast phenylalanine tRNA at 1.93 A resolution: a classic structure revisited. RNA *6*, 1091–1105.

Shin, B.S., Maag, D., Roll-Mecak, A., Arefin, M.S., Burley, S.K., Lorsch, J.R., and Dever, T.E. (2002). Uncoupling of initiation factor eIF5B/IF2 GTPase and translational activities by mutations that lower ribosome affinity. Cell *111*, 1015–1025.

Shine, J., and Dalgarno, L. (1974). The 3′-terminal sequence of Escherichia coli 16S ribosomal RNA: complementarity to nonsense triplets and ribosome binding sites. Proc. Natl. Acad. Sci. USA *71*, 1342–1346.

Valle, M., Zavialov, A., Li, W., Stagg, S.M., Sengupta, J., Nielsen, R.C., Nissen, P., Harvey, S.C., Ehrenberg, M., and Frank, J. (2003a). Incorporation of aminoacyl-tRNA into the ribosome as seen by cryo-electron microscopy. Nat. Struct. Biol. *10*, 899–906.

Valle, M., Zavialov, A., Sengupta, J., Rawat, U., Ehrenberg, M., and Frank, J. (2003b). Locking and unlocking of ribosomal motions. Cell *114*, 123–134.

Vetter, I.R., and Wittinghofer, A. (2001). The guanine nucleotide-binding switch in three dimensions. Science *294*, 1299–1304.

Wakao, H., Romby, P., Westhof, E., Laalami, S., Grunberg-Manago, M., Ebel, J.P., Ehresmann, C., and Ehresmann, B. (1989). The solution structure of the Escherichia coli initiator tRNA and its interactions with initiation factor 2 and the ribosomal 30 S subunit. J. Biol. Chem. *264*, 20363–20371.

Wakao, H., Romby, P., Laalami, S., Ebel, J.P., Ehresmann, C., and Ehresmann, B. (1990). Binding of initiation factor 2 and initiator tRNA to the Escherichia coli 30S ribosomal subunit induces allosteric transitions in 16S rRNA. Biochemistry *29*, 8144–8151.

Weiel, J., and Hershey, J.W. (1982). The binding of fluorescein-labeled protein synthesis initiation factor 2 to Escherichia coli 30 S ribosomal subunits determined by fluorescence polarization. J. Biol. Chem. *257*, 1215–1220.

Wintermeyer, W., and Gualerzi, C. (1983). Effect of Escherichia coli initiation factors on the kinetics of N-Acphe-tRNAPhe binding to 30S ribosomal subunits. A fluorescence stopped-flow study. Biochemistry *22*, 690–694.

Zavialov, A.V., and Ehrenberg, M. (2003). Peptidyl-tRNA regulates the GTPase activity of translation factors. Cell *114*, 113–122.

Zavialov, A.V., Buckingham, R.H., and Ehrenberg, M. (2001). A post-termination ribosomal complex is the guanine nucleotide exchange factor for peptide release factor RF3. Cell *107*, 115–124.

Zucker, F.H., and Hershey, J.W. (1986). Binding of Escherichia coli protein synthesis initiation factor IF1 to 30S ribosomal subunits measured by fluorescence polarization. Biochemistry *25*, 3682–3690.

Accession Numbers

The cryo-EM maps have been deposited in the 3D-EM database, EMBL-European Bioinformatics Institute, Cambridge: EMD-3523 (70S·IF2·GDPNP·IF1·IF3·mRNA·fMet-tRNA^{fMet} initiation complex) and EMD-3666 (IC-PIC difference map). Atomic models have been deposited in the Protein Data Bank under the following ID codes: IZ01 (IF2·GDPNP, IF1, and tRNA fitted to the IC map) and IZ03 (P site and P/E site tRNA used for comparison with P/I).

The structure of the 80S ribosome from *Trypanosoma cruzi* reveals unique rRNA components

Haixiao Gao*, Maximiliano Juri Ayub†, Mariano J. Levin†, and Joachim Frank*‡§

*Howard Hughes Medical Institute, Health Research, Inc., at the Wadsworth Center, Empire State Plaza, Albany, NY 12201-0509; ‡Department of Biomedical Sciences, State University of New York, Empire State Plaza, Albany, NY 12201-0509; and †Laboratory of Molecular Biology of Chagas Disease (LaBMECh), Instituto de Ingenieria Genetica y Biologia Molecular, National Research Council (CONICET), 1428 Buenos Aires, Argentina

Edited by Thomas A. Steitz, Yale University, New Haven, CT, and approved June 8, 2005 (received for review March 22, 2005)

We present analysis, by cryo-electron microscopy and single-particle reconstruction, of the structure of the 80S ribosome from *Trypanosoma cruzi*, the kinetoplastid protozoan pathogen that causes Chagas disease. The density map of the *T. cruzi* 80S ribosome shows the phylogenetically conserved eukaryotic rRNA core structure, together with distinctive structural features in both the small and large subunits. Remarkably, a previously undescribed helical structure appears in the small subunit in the vicinity of the mRNA exit channel. We propose that this rRNA structure likely participates in the recruitment of ribosome onto the 5′ end of mRNA, in facilitating and modulating the initiation of translation that is unique to the trypanosomes.

eukaryote | translation | initiation

The trypanosomatidae family encompasses a group of flagellated protozoans causing a series of severe diseases: *Trypanosoma cruzi*, *Trypanosoma brucei*, and species of *Leishmania*. They have evolved very differently from bacteria, yeast, and animal cells, and have developed unique cellular and genetic pathways. These protozoans are also among those few lower eukaryotic organisms that possess unusually structured mRNAs, as products of a special transsplicing RNA editing mechanism in posttranscription (1–3). As the result of transsplicing, every trypanosome mRNA has 39 identical nucleotides at the 5′ terminus, the so-called spliced leader (SL). Besides the universally conserved 7-methyl guanosine cap, which is linked to the first nucleotide via a 5′-5′ triphosphate bridge, the first four nucleotides of the 39-nt SL are all methylation-modified. This unusually modified cap, known as the cap-4 structure (4–7), is the most highly modified cap structure among all eukaryotic cells.

The functionality of the specially constructed mRNAs in the trypanosome and the reason for their occurrence in only a few organisms are as yet unknown. Counting on the important role of the mRNA 5′ end during eukaryotic translation initiation (8), the SL sequence and its associated cap structure are expected to play a significant part in translation initiation, and possibly to be involved in a unique mechanism of initiation. Indeed, mRNAs processed by transsplicing were found to be translated with higher efficiency than the regularly constructed mRNA, on account of the synergistic cooperation between SL sequence and its associated cap structure (9, 10). There is increasing evidence in biochemical studies for direct interactions between ribosome and SL sequence/cap structure through complementary base pairing (11–13). In kinetoplastid *Leishmania*, the cap-4 structure and the first two thirds of the SL sequence have been found to be critical for the association of mRNAs with polysomes (13).

Here, we have used cryo-EM to obtain the structure of the *T. cruzi* 80S ribosome. In comparison with ribosomal structures from other species, the density map of *T. cruzi* 80S ribosome shows the presence of unusual structural components, related to large expansion segments in the rRNA secondary structure. We postulate that these unusual ribosomal components are involved in the unique mechanism of translation initiation in trypanosomes.

Materials and Methods

Purification of the *T. cruzi* 80S Ribosome. Ribosomes were purified from epimastigotes of *T. cruzi* according to the protocol described by Gomez *et al.* (14). All ribosome purification steps were performed at 4°C in the presence of protease inhibitors. The parasites were disrupted in SKS (0.25 M sucrose/5 mM KCl) supplemented with 0.25% of deoxycholate. The homogenate was centrifuged at 1,000 × *g* for 10 min. The supernatant (SN) was repeatedly centrifuged at 12,000 × *g* for 20 min until no pellet was detected. The SN was centrifuged at 105,000 × *g* for 2 h, the pellet was resuspended in buffer I (20 mM Tris·HCl, pH 7.5/100 mM $MgCl_2$/500 mM ammonium acetate/5 mM 2-mercaptoethanol). Afterward, the suspension was clarified several times by centrifugation at 12,000 × *g* for 10 min until no pellet was evident. The SN was loaded on a two-layer discontinuous sucrose gradient formed by a bottom 40% (wt/vol) sucrose solution and a top 20% (wt/vol) sucrose layer. The sample was centrifuged at 105,000 × *g* for 18 h. The translucent pellet was resuspended in buffer II (10 mM Tris·HCl, pH 7.5/12.5 mM $MgCl_2$/80 mM KCl/5 mM 2-mercaptoethanol) and diluted in buffer II. Samples were frozen and stored at −80°C.

Cryo-EM and Image Processing. Buffer solutions containing *T. cruzi* 80S ribosome with final concentration of 32 nM were applied to cryo-EM grids at 4°C following the standard procedure using the Vitrobot (FEI, Eindhoven, The Netherlands). A total of 173 micrographs were recorded in a defocus range between 1.0 and 3.3 μm on a Philips Tecnai F20 electron microscope at 200 kV with a calibrated magnification of 49,650, under low-dose conditions (18 $e^-/Å^2$). The micrographs were digitized on a Zeiss Imaging scanner (Z/I Imaging, Huntsville, AL) with a step size of 14 μm, corresponding to a pixel size of 2.82 Å on the object scale. After evaluation of drift, astigmatism, and the presence of Thon rings in the power spectrum of each micrograph, 161 good micrographs were selected and divided into 31 defocus groups. Ribosomal particles were selected from these micrographs through three steps: preliminary automated selection, manual verification, and selection based on the size of the cross-correlation coefficient with a template. A total of 53,505 particles were obtained and subjected to the single particle reconstruction. To start, two separate reconstructions were performed on a small dataset of *T. cruzi* 80S ribosome, based on two different reference volumes (80S ribosome from *Thermomyces lanuginosus* and *Saccharomyces cerevisiae*) for projection matching. The two reconstructed volumes were compared and found to be highly consistent in terms of the new structural features.

This paper was submitted directly (Track II) to the PNAS office.

Freely available online through the PNAS open access option.

Abbreviations: SL, spliced leader; CP, central protuberance; ES, expansion segment; cryo-EM, cryo-electron microscopy.

§To whom correspondence should be addressed. E-mail: joachim@wadsworth.org.

© 2005 by The National Academy of Sciences of the USA

 www.pnas.org/cgi/doi/10.1073/pnas.0500926102

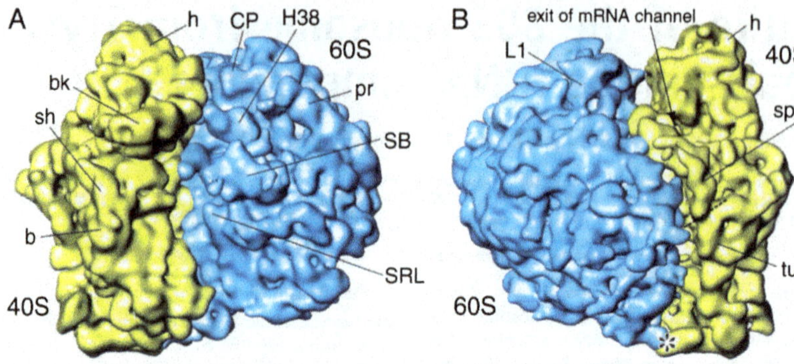

Fig. 1. The 12-Å resolution cryo-EM density map of the 80S ribosome from *T. cruzi*. The density map is shown in two side views of the 80S ribosome. The 40S subunit is in yellow and the 60S subunit in blue. Landmarks for the 40S subunit: h, head; bk, beak; sh, shoulder; b, body; tu, turret; landmarks for the 60S subunit: CP, central protuberance; H38, helix 38; SB, stalk base; pr, prong; SRL, sarcin–ricin loop; L1, L1 stalk. The dashed line defines the spire of the turret. The asterisk marks the previously undescribed connection between the small and large ribosomal subunits.

One of the newly reconstructed volumes was then used as reference volume applied to the entire data set for the final reconstruction. The final resolution of the CTF-corrected volume was estimated by the Fourier Shell Correlation criterion with a cutoff value of 0.5. The falloff of the Fourier amplitudes toward higher spatial frequencies was corrected as described (15), using the x-ray solution scattering intensity distribution of the *Escherichia coli* 70S ribosome. RNA and protein components were computationally separated using a method (16) based on the differences in the density distribution of RNA and proteins, taking into account the molecular masses and RNA contiguity constraints.

All of the steps of image processing were performed by using the SPIDER package (17). The atomic RNA core structure was fitted as rigid body into the EM density map, except that the L1 stalk region was manually docked as a separated rigid body using O (18). The figures were prepared by using IRIS EXPLORER (Numerical Algorithms Group, Downers Grove, IL) and the CHIMERA package (19).

Results

Using the cryo-EM technique and the single-particle reconstruction method (20), a density map of the *T. cruzi* 80S ribosome with a resolution of 12 Å was reconstructed from purified ribosomal particles. Considering the relatively low total number of particles used in the reconstruction, such resolution indicates that the ribosomal particles in the preparation are highly homogenous. The overall structure of the *T. cruzi* 80S ribosome exhibits well defined small (40S) and large (60S) subunits, in keeping with the feature fundamentally and evolutionally conserved among eukaryotic ribosomes (Fig. 1). Landmark characteristics of the typical ribosome structure are clearly identified in the density map. Compared with the eukaryotic 80S ribosome from yeast, both the small and large ribosomal subunits from *T. cruzi* are larger (1.3 times in volume), mostly on account of the size increments of the ribosomal RNA molecules, with the *T. cruzi* rRNAs [18S rRNA: 2,315 nt (GenBank accession no. AF245382); 28S rRNA: 4,151 nt (www.genedb.org)] being one-fifth larger than the yeast rRNA (18S rRNA: 1,798 nt; 25S rRNA: 3,392 nt), in terms of the total number of nucleotides.

Superposition of the density map of the *T. cruzi* 80S ribosome and the model of the rRNA core structure built on the basis of the cryo-EM map for yeast (21) produces a good fit of the morphological features, indicating that this microorganism possesses the phylogenetically conserved eukaryotic rRNA core structure. As shown in Fig. 2, the spatial ribosomal configuration formed by rRNA core structure is preserved as indicated by the well fitted known RNA components, such as helix 44 from the small subunit and helix 69 from the large subunit.

In terms of the ribosomal proteins, sequence analysis using BLAST search indicates that almost all yeast ribosomal proteins have counterparts in *T. cruzi* (except for L41e and S31), with sequence identities varying between 30% and 70%. Surprisingly, comparison of the density maps of the 40S ribosomal subunits from *T. cruzi* and yeast reveals that RACK1, a large scaffold protein in the head of the 40S subunit is absent from the *T. cruzi* ribosome (Fig. 3). Recently identified as a ribosomal protein by mass spectroscopy (22) and cryo-EM (23), RACK1 is a constituent of all eukaryotic ribosomes visualized by cryo-EM thus far, and apparently plays a direct role in the regulation of eukaryotic translation initiation through the recruitment of protein kinase C (23). Because data mining indicates that a homolog of the RACK1 gene is present in the *T. cruzi* genome, the absence of RACK1 from the *T. cruzi* 80S ribosome revealed by our study might be a reflection of an earlier stage of eukaryotic evolution.

Although the *T. cruzi* 80S ribosome possesses common ribosomal features, it exhibits many distinctive structural features in both the small and large subunits. Compared with prokaryotic and other eukaryotic ribosomes, the *T. cruzi* ribosomal 40S subunit appears expanded, due to the addition of a large piece of density adjacent to the platform region (Fig. 3). As indicated in the secondary structure of the *T. cruzi* 18S rRNA (Fig. 4), this extra density must be attributed to two large expansion segments in domain II of the 18S rRNA, ES6 and ES7, designated as insertions of helices 21 and 26. These are the two largest expansion segments in the *T. cruzi* 18S rRNA, involving 504 and 147 nucleotides, respectively.

Part of ES6/ES7 makes up a large helical structure (which we term the "turret"), located at the most lateral side of the 40S subunit (Fig. 4). The turret measures 205 Å in length, thus forming the longest helical structure ever observed in a ribosome. The upper end of the turret appears as a sharp, freestanding spiral of 50 Å in length, which we term "spire," apposed to the exit of the mRNA channel. The distance between the spire and the mRNA exit is ≈130 Å. The lower portion of the turret extends all of the way to the bottom of the 40S subunit. At its lower end, it bends by almost 90° and forms a bridge with the 60S subunit (Fig. 1). This is a unique type of connection between the small and large subunits, as compared with all other ribosomal

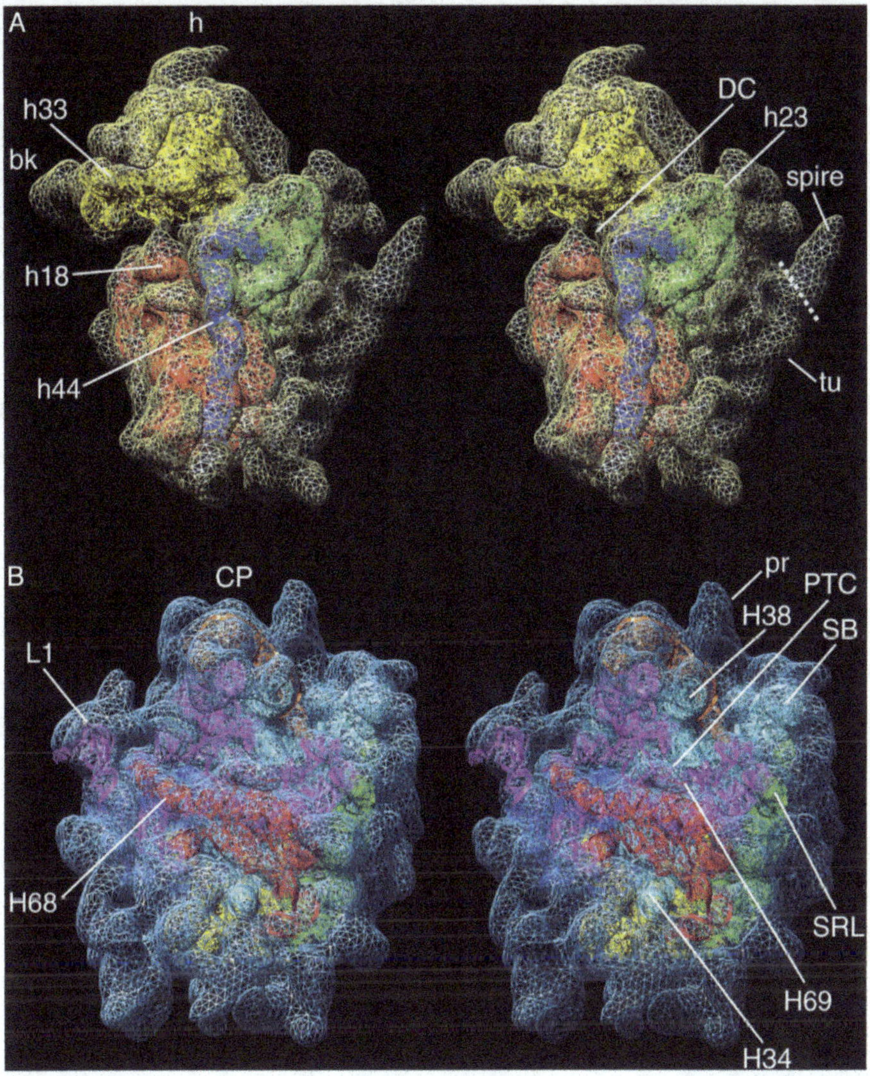

Fig. 2. Stereoviews of the conserved rRNA core structure in 40S (*A*) and 60S (*B*) ribosomal subunits from *T. cruzi*. The density maps of the 40S and 60S ribosomal subunits are represented in wire mesh. The atomic structures were directly adopted from the rRNA core structure from yeast (Protein Data Bank ID codes 1S1H and 1S1I). Color codes for the conserved rRNA core in the small subunit: domain I, red; domain II, green; domain III, yellow; domain IV, blue. Color codes for the conserved rRNA core in the large subunit: domain I, blue; domain II, cyan; domain III, yellow; domain IV, red; domain V, magenta; domain VI, green; 5S RNA, orange. h, head; bk, beak; tu, turret; h44, helix 44; h33, helix 33; h18, helix 18; h23, helix 23; DC, decoding center; CP, central protuberance; L1, L1 stalk; pr, prong; SB, stalk base; PTC, peptidyl-transferase center; SRL, sarcin–ricin loop; H34, helix 34; H38, helix 38; H68, helix 68; H69, helix 69.

structures investigated to date (24). Apart from the turret, the extra density in the 40S subunit also includes several small helical structures as part of ES6 and ES7. These helical structures observed in our density map are in accordance with the comparative analysis result based on ES6 sequences from >3,000 eukaryotes, in which several helices were identified only from kinetoplatida (25).

Besides the expansion segments ES6 and ES7, which are related to the turret structure and its subsidiaries, *T. cruzi* 18S rRNA has four additional small expansion segments, designated as ES3, ES9, ES10, and ES12, located in domains I, III, and IV of the 18S rRNA, respectively. Similarly, all of these small expansion segments are found to be associated with the unaccounted density in the *T. cruzi* 40S ribosomal subunit (Fig. 4). ES12 is located in the long penultimate helix (helix 44), and results in the helix 44 in *T. cruzi* (113 nt) being longer than in *E. coli* (103 nt), but shorter than in yeast (129 nt). Consistent with these variations in length, the span of the density attributable to helix 44 in the *T. cruzi* ribosome also has an intermediate position between *E. coli* and yeast. ES3, ES9, and ES10 are located near helices 9, 39, and 41, respectively, and are associated with three small masses in the density map of the 40S ribosomal subunit, one at the bottom of the 40S ribosomal subunit, the other two in the head region.

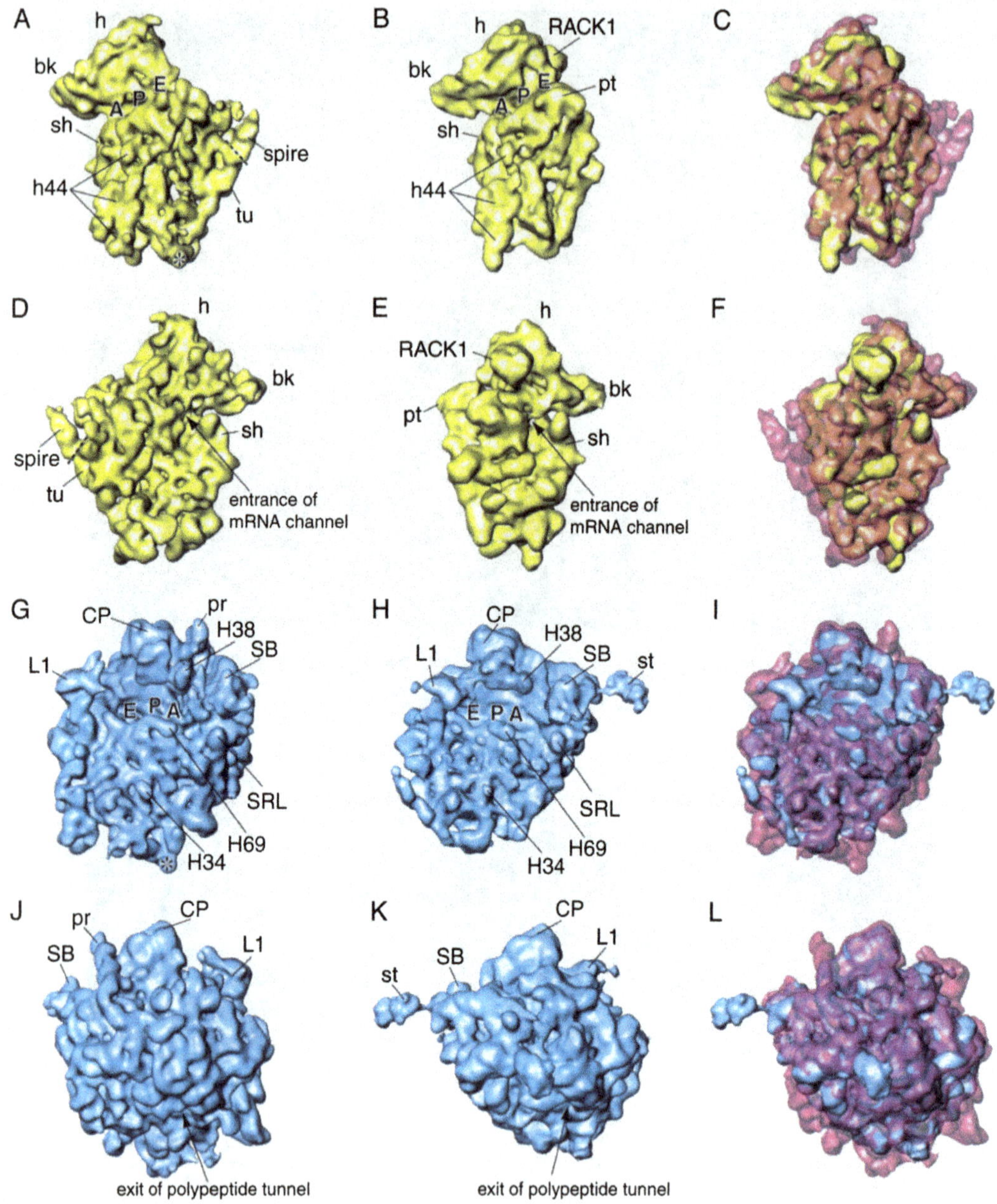

Fig. 3. Structural comparison of the ribosomal subunits from *T. cruzi* and yeast (*S. cerevisiae*). (*A* and *B*) Intersubunit view of the 40S subunits. (*D* and *E*) Solvent side view of the 40S subunits. (*C* and *F*) Superposition of the two 40S subunits (*T. cruzi* 40S is in semitransparent red, and yeast 40S in solid yellow). (*G* and *H*) Intersubunit view of the 60S subunits. (*J* and *K*) Solvent side view of the 60S subunits. (*I* and *L*) Superposition of the two 60S subunits (*T. cruzi* 60S is in semitransparent red, and yeast 60S in solid blue). RACK1, protein RACK1; st, P-protein stalk; A, aminoacyl site; P, peptidyl site; E, exit site. Other landmarks are the same as introduced in Fig. 2.

As to the large subunit, besides the common occurrence of the central protuberance (CP), L1 stalk, and stalk base of P proteins, as well as the conserved rRNA core structure, the *T. cruzi* 60S subunit contains several large extra densities located at its periphery, compared with the large ribosomal subunit in yeast (Fig. 3). Although the secondary structure of the *T. cruzi* rRNAs in the 60S subunit is not available, the locations of most of the observed extra densities are consistent with the general locations of the rRNA expansion segments, which are, as a rule, at the surface of the eukaryotic ribosomes (26). Among all of the extra densities, there is a large helical structure ("prong") located between the CP and helix 38, in the back of the 60S subunit (Figs. 1 and 3). Interestingly, a similar feature was only reported in the structure of the human ribosome, but

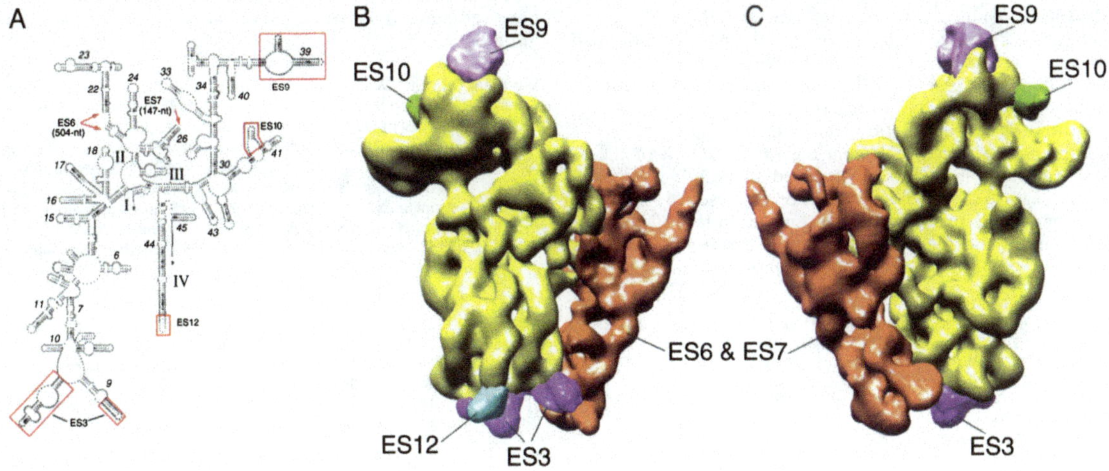

Fig. 4. RNA in the *T. cruzi* 40S ribosomal subunit. (*A*) Secondary structure diagram of the *T. cruzi* 18S rRNA (www.rna.icmb.utexas.edu). The expansion segments are marked by using the nomenclature of Gerbi (26). The regions of ES6 and ES7 are indicted by arrows and the number of encompassed nucleotides. (*B* and *C*) RNA partition of the *T. cruzi* 40S ribosomal subunit as viewed from intersubunit side and solvent side. The conserved rRNA core is rendered in yellow. The expansion segments are highlighted: ES6 and ES7, orange; ES3, dark purple; ES9, light purple; ES10, green; ES12, cyan.

not in yeast and bacterial ribosomes (27). Morphological comparison of the 60S ribosomal subunit from *T. cruzi* with those from yeast and other higher eukaryotes reveals that the *T. cruzi* 60S ribosomal subunit does not possess the universal eukaryotic feature as a planar surface near the exit site of polypeptide (24). Instead, the 60S ribosomal subunit from *T. cruzi* presents a shape that is similar to those from bacteria (Fig. 5, which is published as supporting information on the PNAS web site).

In contrast to the conserved eukaryotic rRNA core structure, the location of the L1 stalk in *T. cruzi*, which is on one side of the CP, does not match either of the two reported positions in yeast, known as the "in position" and "out position," in relation to the ratchet-like subunit rearrangement. Instead, the L1 stalk in *T. cruzi* takes an in-between position, possibly due to its high mobility. On the other side of the CP, the P protein stalk is not visible in the density map of the *T. cruzi* 60S ribosomal subunit, whereas Western blots of the ribosome preparation using monoclonal antibodies against P proteins (P0/P1/P2) showed that these proteins were present (see Fig. 6, which is published as supporting information on the PNAS web site). As homologs of the bacterial moiety $L10/(L7/L12)_4$, P proteins are known to be very flexible, and the absence of a stalk in the current EM density map is likely due to the lack of stabilization through additional ligand binding (28).

Discussion

Structural analysis of the 80S ribosome from *T. cruzi* may be of importance in detecting specific structures necessary for the process of translation of this pathogenic microorganism. Our cryo-EM result shows that one of the most notable structural features in the *T. cruzi* 80S ribosome is associated with the small subunit, as a turret structure forming a large helix with its upper end located in the vicinity of the mRNA exit channel. A comparison of the available secondary structures of the 18S rRNA among eukaryotes (29) indicates that the large expansion segments ES6 and ES7 identified to be associated with the turret structure and its subsidiaries only exist in trypanosomes, where they occur with highly conserved sequence and size (current data available for *T. cruzi*, *T. brucei*, and *L. major*, www.rna.icmb.utexas.edu). On the other hand, the 39-nt SL sequence and its associated cap structure are also highly conserved in these trypanosomes (Fig. 7, which is published as supporting information on the PNAS web site). Thirty of the 39 nucleotides in their SL sequence are identical, including the modified first four nucleotides that compose the cap-4 structure. The unique features of the *T. cruzi* ribosomal components found in our study, along with the unusually structured mRNA conserved among known trypanosomes, suggest the existence of a unique interaction between the trypanosomal ribosome and the mRNA.

In eukaryotic translation initiation, the overwhelming majority of the mRNAs recruit the ribosomal 40S subunit through their 5′-end cap, and a number of initiation factors, including eIF4E, eIF4G, and eIF4A, a process known as the cap-dependent initiation (30). A recent study of the cap-binding protein eIF4E from trypanosome showed that eIF4E in trypanosome has much lower binding affinity with the cap-4 structure, compared to the binding affinity of eIF4E in mouse with monomethylated cap (31). Furthermore, the detected cellular level of eIF4E also appears lower than that required for conducting translation sufficiently (32). In line with these findings of the eccentric behavior of eIF4E in the trypanosome, and the unusually structured mRNA 5′-end conserved among trypanosomes, we propose that the turret structure found in the *T. cruzi* 40S ribosomal subunit may be involved in the translation initiation process in *T. cruzi*.

In the absence of a tertiary structure model related to the turret structure, a legitimate guess would be that the spire region of the turret may be a target for direct interaction with the SL sequence and cap-4 structure at the 5′ end of mRNA. At the free end of the turret structure, the conformation of the spire would allow a certain freedom of movement during the interaction with mRNA. Although a direct sequence scan in ES6 and ES7 does not reveal the nucleotides that form the Watson–Crick base pairs with the cap-4, it is possible that some nucleotides in the 39-nt SL sequence are involved in the base pairing. In addition, the SL sequence itself might be recognized by the turret, perhaps by folding into a specific structure similar to a viral internal ribosome entry site sequence (33), which forms direct RNA–ribosome contacts to initiate translation. A case in point is that the 39-nt spliced leader would fit the 130-Å distance between the spire and the mRNA exit without being fully extended. In the absence of additional evidence, we cannot rule out other possibilities, e.g., that the turret may represent a special platform for binding distinct initiation factors unique to the trypanosomes or that it may substitute for some initiation factors in trypanosome.

The absence of RACK1 from the small subunit of the *T. cruzi* ribosome may also be involved in defining the very specific nature of translation initiation in these parasite cells. It is conceivable that all protozoan species of the Euglenozoa group, including free living species that possess mRNA processed by transsplicing, share the special ribosomal component observed here, as well as a unique mechanism of translation initiation.

We thank Michael Watters for assisting us with the illustrations. This work was supported by the Howard Hughes Medical Institute (HHMI), National Institutes of Health Grants GM29169 and RR01219 and grants from the National Science Foundation (to J.F.), WHO/Special Program for Research and Training in Tropical Diseases (South–South Initiative project A20363), Fondo Nacional de Investigacion Científica y Tecnològica–Argentinean National Agency for Science and Technology, and the University of Buenos Aires, Buenos Aires, Argentina (to M.J.L.). In addition, M.J.L. was supported by an International Research Scholar grant from HHMI.

1. Walder, J. A., Eder, P. S., Engman, D. M., Brentano, S. T., Walder, R. Y., Knutzon, D. S., Dorfman, D. M. & Donelson, J. E. (1986) *Science* **233,** 569–571.
2. Murphy, W. J., Watkins, K. P. & Agabian, N. (1986) *Cell* **47,** 517–525.
3. Sutton, R. E. & Boothroyd, J. C. (1986) *Cell* **47,** 527–535.
4. Perry, K. L., Watkins, K. P. & Agabian, N. (1987) *Proc. Natl. Acad. Sci. USA* **84,** 8190–8194.
5. Freistadt, M. S., Cross, G. A. & Robertson, H. D. (1988) *J. Biol. Chem.* **263,** 15071–15075.
6. Bangs, J. D., Crain, P. F., Hashizume, T., McCloskey, J. A. & Boothroyd, J. C. (1992) *J. Biol. Chem.* **267,** 9805–9815.
7. Lewdorowicz, M., Yoffe, Y., Zuberek, J., Jemielity, J., Stepinski, J., Kierzek, R., Stolarski, R., Shapira, M. & Darzynkiewicz, E. (2004) *RNA* **10,** 1469–1478.
8. Preiss, T. & Hentze, M. W. (2003) *BioEssays* **25,** 1201–1211.
9. Maroney, P. A., Denker, J. A., Darzynkiewicz, E., Laneve, R. & Nilsen, T. W. (1995) *RNA* **1,** 714–723.
10. Lall, S., Friedmann, C. C., Jankowska-Anyszka, M., Stepinksi, J., Darzynkiewicz, E. & Davis, R. E. (2004) *J. Biol. Chem.* **279,** 45573–45585.
11. Aline, R. F., Jr., Scholler, J. K. & Stuart, K. (1989) *Mol. Biochem. Parasitol.* **32,** 169–178.
12. Liou, R. F. & Blumenthal, T. (1990) *Mol. Cell. Biol.* **10,** 1764–1768.
13. Zeiner, G. M., Sturm, N. R. & Campbell, D. A. (2003) *J. Biol. Chem.* **278,** 38269–38275.
14. Gomez, E. B., Medina, G., Ballesta, J. P. G., Levin, M. J. & Tellez-Inon, M. T. (2001) *Int. J. Parasitol.* **31,** 1032–1039.
15. Gabashvili, I. S., Agrawal, R. K., Spahn, C. M. T., Grassucci, R. A., Frank, J. & Penczek, P. (2000) *Cell* **100,** 537–549.
16. Spahn, C. M. T., Penczek, P., Leith, A. & Frank, J. (2000) *Structure (London)* **8,** 937–948.
17. Frank, J., Radermacher, M., Penczek, P., Zhu, J., Li, Y., Ladjadj, M. & Leith, A. (1996) *J. Struct. Biol.* **116,** 190–199.
18. Jones, T. A., Zhou, J. Y., Cowan, S. W. & Kjeldgaard, M. (1991) *Acta Crystallogr. A* **47,** 110–119.
19. Pettersen, E. F., Goddard, T. D., Huang, C. C., Couch, G. S., Greenblatt, D. M., Meng, E. C. & Ferrin, T. E. (2004) *J. Comput. Chem.* **25,** 1605–1612.
20. Frank, J. (1996) *Three-Dimensional Electron Microscopy of Macromolecular Assemblies* (Academic, New York).
21. Spahn, C. M. T., Gomez-Lorenzo, M. G., Grassucci, G. A., Jorgensen, R., Andersen, G. R., Beckmann, R., Penczek, P. A., Ballesta, J. P. G. & Frank, J. (2004) *EMBO J.* **23,** 1008–1019.
22. Link, A. J., Eng, J., Schieltz, D. M., Carmack, E., Mize, G. J., Morris, D. R., Garvik, B. M. & Yates, J. R., III (1999) *Nat. Biotechnol.* **17,** 676–682.
23. Sengupta, J., Nilsson, J., Gursky, R., Spahn, C. M. T., Nissen, P. & Frank, J. (2004) *Nat. Struct. Mol. Biol.* **11,** 957–962.
24. Frank, J. (2003) *Genome Biol.* **4,** 237.
25. Wuyts, J., De Rijk, P., Van de Peer, Y., Pison, G., Rousseeuw, P. & De Wachter, R. (2000) *Nucleic Acids Res.* **28,** 4698–4708.
26. Gerbi, S. A. (1996) in *Ribosomal RNA. Structure, Evolution, Processing, and Function in Protein Synthesis*, eds. Zimmermann, R. A. & Dahlberg, A. E. (CRC Press, New York), pp. 71–87.
27. Spahn, C. M. T., Jan, E., Mulder, A., Grassucci, R. A., Sarnow, P. & Frank, J. (2004) *Cell* **118,** 465–475.
28. Gomez-Lorenzo, M. G., Spahn, C. M. T., Agrawal, R. K., Grassucci, R. A., Penczek, P., Chakraburtty, K., Ballesta, J. P. G., Lavandera, J. L., Garcia-Bustos, J. F. & Frank, J. (2000) *EMBO J.* **19,** 2710–2718.
29. Cannone, J. J., Subramanian, S., Schnare, M. N., Collett, J. R., D'Souza, L. M., Du, Y., Feng, B., Lin, N., Madabusi, L. V., Muller, K. M., *et al.* (2002) *BMC Bioinformatics* **3,** 2.
30. Kapp, L. D. & Lorsch, J. R. (2004) *Annu. Rev. Biochem.* **73,** 657–704.
31. Yoffe, Y., Zuberek, J., Lewdorowicz, M., Zeira, Z., Keasar, C., Orr-Dahan, I., Jankowska-Anyszka, M., Stepinski, J., Darzynkiewicz, E. & Shapira, M. (2004) *RNA* **10,** 1764–1775.
32. Dhalia, R., Reis, C. R. S., Freire, E. R., Rocha, P. O., Katz, R., Muniz, J. R. C., Standart, N. & de Melo Neto, O. P. (2005) *Mol. Biochem. Parasitol.* **140,** 23–41.
33. Hellen, C. U. T. & Sarnow, P. (2001) *Genes Dev.* **15,** 1593–1612.

Available online at www.sciencedirect.com
SCIENCE DIRECT
Journal of Structural Biology 154 (2006) 168–183
ELSEVIER
Journal of Structural Biology
www.elsevier.com/locate/yjsbi

Estimation of variance in single-particle reconstruction using the bootstrap technique

Pawel A. Penczek [a,*], Chao Yang [b], Joachim Frank [c], Christian M.T. Spahn [d]

[a] *The University of Texas–Houston Medical School, Department of Biochemistry and Molecular Biology, 6431 Fannin, MSB 6.218, Houston, TX 77030, USA*
[b] *Computational Research Division, Lawrence Berkeley National Laboratory, Berkeley, CA 94720, USA*
[c] *HHMI, Health Research, Inc., Wadsworth Center, Albany, NY 12201, USA*
[d] *Institut für Medizinische Physik und Biophysik, Universitätsklinikum Charité, Berlin, Germany*

Received 24 October 2005; received in revised form 12 January 2006; accepted 17 January 2006
Available online 13 February 2006

Abstract

Density maps of a molecule obtained by single-particle reconstruction from thousands of molecule projections exhibit strong changes in local definition and reproducibility, as a consequence of conformational variability of the molecule and non-stoichiometry of ligand binding. These changes complicate the interpretation of density maps in terms of molecular structure. A three-dimensional (3-D) variance map provides an effective tool to assess the structural definition in each volume element. In this work, the different contributions to the 3-D variance in a single-particle reconstruction are discussed, and an effective method for the estimation of the 3-D variance map is proposed, using a bootstrap technique of sampling. Computations with test data confirm the viability, computational efficiency, and accuracy of the method under conditions encountered in practical circumstances.
© 2006 Elsevier Inc. All rights reserved.

Keywords: Single-particle reconstruction; Bootstrap; Electron microscopy

1. Introduction

Cryo-electron microscopy (EM), together with associated image processing techniques, has become an established method of structural analysis of large macromolecular complexes existing in single-particle form (i.e., as isolated particles) (Frank, 2006). It has become possible to reconstruct and visualize three-dimensional (3-D) density distributions of objects ranging from individual proteins (size 10–20 nm) (Ludtke et al., 2001), to protein–RNA complexes, such as ribosomes (size range 20–30 nm) (Gabashvili et al., 2000; Spahn et al., 2001), to highly symmetric structures of viruses (size range 30–150 nm) (Böttcher et al., 1997; Conway et al., 1997; Zhou et al., 2000). The technique is based on the assumption that the macromolecules are isolated, randomly oriented, and have identical structure. A suspension of molecules is placed on a grid, rapidly frozen, and transferred to the microscope. A single-exposure picture of a section of a grid is taken, yielding a micrograph filled with hundreds of projections of macromolecules frozen in various orientations. The distribution of orientations of particle views on the support grid depends on various factors and is in most cases non-uniform. Multiple micrographs are collected using various defocus settings of the microscope. After individual projections are selected from micrographs, their relative orientations in 3-D space are determined using alignment procedures (Penczek et al., 1992; Penczek et al., 1994; Penczek et al., 1996; van Heel et al., 2000) and the 3-D density distribution is calculated using a 3-D reconstruction algorithm (Harauz and van Heel, 1986; Marabini et al., 1998; Penczek et al., 1992; Radermacher, 1992). Although in terms of the best resolution achieved (0.7 nm for asymmetric molecules), single-particle cryo-EM cannot yet match

* Corresponding author. Fax: +1 713 500 0652.
E-mail address: Pawel.A.Penczek@uth.tmc.edu (P.A. Penczek).

1047-8477/$ - see front matter © 2006 Elsevier Inc. All rights reserved.
doi:10.1016/j.jsb.2006.01.003

X-ray crystallography, there are many advantages of this technique. Thanks to the rapid freezing of the specimen, molecules are captured in their native, aqueous environment, and their native structure is preserved. The electron microscope yields true projections of the Coulomb potential[1]; thus, in principle, an accurate representation of molecular densities can be obtained (Langmore and Smith, 1992). In addition, since the native form of the molecules is preserved, it is possible to study conformational changes and to examine dynamical effects of different functional states (Frank and Agrawal, 2000; Gao et al., 2003; Lata et al., 2000; Valle et al., 2003; Zhou et al., 2001).

The extent of the reproducibility of EM maps is reported in terms of "resolution", which is evaluated using one of the available methods, such as Fourier Ring/Shell Correlation (FRC/FSC) (Saxton and Baumeister, 1982), Differential Phase Residual (DPR) (Frank et al., 1981), or Spectral Signal-to-Noise-Ratio (SSNR) (Penczek, 2002a; Unser et al., 1987). These measures, constructed in Fourier space, are used to evaluate the self-consistency of Fourier information as a function of spatial frequency. The term "resolution", although often expressed in units of length, refers to the spatial frequency limit beyond which the information in Fourier space is dominated by noise (Penczek, 1998). In real space, this limit corresponds to a finite distance, (resolution distance), i.e., the minimum distance at which two point sources can be considered distinguishable. While the underlying analysis is performed in Fourier space and yields a non-uniform distribution of SSNR (and also Fourier space variance, see (Penczek, 2002a) for details), it only allows us to deduce the mean uniform distribution of the real-space error. To obtain a per-voxel distribution of errors, one would have to calculate the 3-D variance associated with the reconstructed object.

Although Fourier-space error measures in EM have been introduced over two decades ago, there has been relatively little effort devoted to evaluation of real-space errors. One could argue that for the analysis and interpretation of 3-D maps, the latter kind of evaluation is more important and certainly more useful. A straightforward way to assess the errors is to equate them with the standard deviation of the measured quantity, which can be derived through calculation of the variance. If available, knowledge of real-space errors in 3-D maps would be helpful in (i) detection of different functional states (for example, those characterized by binding of a ligand), (ii) analysis of conformational heterogeneity of the assemblies due to fluctuations of the structure around the ground state, (iii) analysis of the significance of small details in 3-D reconstructions, (iv) analysis of the significance of details in difference maps, (v) docking of known structural domains into EM density maps. For example, for both difference maps and for analysis of conformational variability, one should be able to verify that the difference between two maps being compared exceeds the standard deviation of the mass distribution by a reasonable margin. Similarly, the uniqueness and accuracy of docking can only be asserted if the error in the EM map is known (particularly if its distribution is non-uniform, as we can reasonably expect). Also, it is worth remembering that the underlying assumption in single-particle reconstruction is that the preparation is homogenous, i.e., the sample contains multiple replicas of the same macromolecule captured in the same conformational state. It is not clear to which extent this assumption is fulfilled in practice; there is mounting evidence that macromolecules occur naturally in a mixture of conformational states (Frank and Agrawal, 2000; Gao et al., 2004; Lata et al., 2000; Zhou et al., 2001). Moreover, macromolecular complexes are to a certain extent flexible and detection of such flexible regions can itself provide important biological information. An analysis of the phenomena described would be possible if the variance of the density distribution in real space could be evaluated.

Although the need for evaluation of 3-D variance in single-particle analysis has been recognized early, the problem proved to be quite challenging (Liu and Frank, 1995). The main difficulty is that the data are available in form of single projections, i.e., the information is partial. Thus, even if the variance of 2-D projection data is available, its relation to the variance in the reconstructed 3-D density map is not obvious. In addition, in single-particle reconstruction the particle images (projections) originate from different 3-D structures. If these structures fall into different classes, due to conformational changes or non-stoichiometry of ligand binding, the obvious course of action would be to reconstruct 3-D structures from subsets of projections corresponding to the respective classes and calculate the 3-D variances directly using the resulting volumes. Unfortunately, this is not possible as we do not know the class membership of single projections. In other words, we have to recognize that *there is only one data set* of projection, while for the calculation of the variance we would have to know *which particle view corresponds to which class of macromolecules*. Finally, we have to keep in mind that the exact inversion of the projection process is impossible. Thus, the process of 3-D reconstruction itself is a source of errors.

The problem of estimating the variance in images reconstructed from sets of their projections attracted significant attention in the field of Positron Emission Tomography (PET) (Leahy and Qi, 2000). For linear reconstruction methods, it is easy to derive an analytical expression for the variance (Barrett, 1990). Due to popularity of the linear Filtered Backprojection (FBP) reconstruction algorithm in PET applications, its variance properties have been studied extensively (Carson et al., 1993; Maitra, 1997; Maitra and O'Sullivan, 1998; Palmer et al., 1985). It was demonstrated that for FBP, given estimates of the variance in projections and assuming the independence of the underlying sources of noise, the variance in the reconstructed image, including

[1] The limitations due to the curvature of Ewald's sphere can be ignored for the combination of object size, microscope voltage, and achievable resolution in single-particle reconstruction.

contributions from the FBP algorithm itself, can be calculated relatively easily and efficiently using the FFT technique (Maitra, 1997). Subsequently, attention shifted to explorations of variance properties of maximum likelihood (ML) and maximum a posteriori (MAP) image reconstruction methods. These methods can combine physical models of image formation in PET and the statistical nature of the data (Poisson distribution) with an inclusion of priors enforcing smoothness of the reconstructed image; therefore, MAP and ML reconstruction algorithms are widely used in PET applications. Unfortunately, since ML and MAP are non-linear and iterative, explicit formulae of the reconstruction variance for these algorithms are not available. Nevertheless, a number of ML and MAP analytical methods for resolution and variance estimation have been proposed based on simplifying assumptions (Barrett et al., 1994; Fessler, 1996; Fessler and Rogers, 1996; Qi and Leahy, 1999, 2000; Wilson and Tsui, 1993). To free the analysis of statistical properties of PET images from the dependence on the reconstruction algorithm used and from the assumption of the Poisson noise in projections, the use of resampling techniques was suggested (Haynor and Woods, 1989), particularly the bootstrap (Buvat, 2002; Dahlbom, 2002; Maitra, 1998). Resampling techniques are computationally intensive, but they yield excellent results irrespective of the complexities of the reconstruction process.

Despite the development of comprehensive error analysis methods for PET applications, very few of these methods can be directly used in the context of cryo-EM. Although FBP is used in single-particle reconstruction, it is implemented in a generalized form that accounts for uneven distribution of projections in EM (Radermacher, 1992). In recent years, other reconstruction algorithms were introduced and increasingly used in cryo-EM (Lanzavecchia et al., 1999, 2002; Marabini et al., 1998; Penczek et al., 1992), particularly those employing direct Fourier inversion of the 3-D ray transform (Grigorieff, 1998; Penczek et al., 2004). It is not clear whether analytical formulae for variance could be derived for all these algorithms. Bootstrap methods developed for PET take advantage of its particular mode of data collection and cannot be directly transferred to cryo-EM image collection paradigm. Finally, as we will argue later, single-particle reconstruction has its own set of requirements and constraints that have to be addressed if the analysis of variance is to yield practical results.

It is curious that estimation of real-space variance in single-particle reconstruction has attracted little attention. Despite the fact that the interpretation of 3-D maps is often performed at the limit of their resolution, little is known about the distribution of error in real space. A method for estimation of variance of a 3-D map reconstructed from projections by a general weighted backprojection (GWBP) algorithm was proposed in (Liu and Frank, 1995). The authors start by introducing two "types" of the variance: (i) type I, associated with the estimate of the variance in projection data, and (ii) type II, which could be calculated only if we could collect entire set of projections for each macromolecule so it could be reconstructed unambiguously. In this formulation, the type-II variance would be calculated according to the definition of variance, using multiple available reconstructions, and could be employed to study conformational variability of macromolecules. The authors of that study recognized that unless one can assign projections to classes that correspond to homogenous sets of macromolecules, the calculation of the type-II variance of reconstructed objects might be impossible. Similar conclusions were drawn in (Ushakov and Ushakova, 1998). To calculate the type-I variance, the authors introduced two possible methods of variance estimation in the projection data: (i) one that takes advantage of the availability of multiple projection images assigned to a given projection direction, (ii) a second one, in which the variance is estimated by comparison of projection data with reprojections of the reconstruction. The type-I variance is then calculated using GWBP algorithm by backprojecting the weighted and filtered projection variances. In general, the derivation of the method is based on rather restrictive assumptions, i.e., that the noise in projections is independent. This limits its usefulness for cryo-EM applications, particularly if conformation changes of macromolecule and the influence of the contrast transfer function (CTF) of the microscope are to be considered. In (Haley et al., 1998) it was suggested that a 3-D variance map can be calculated by backprojecting the modulus of the differences between projection data and re-projections of the structure, and the method was attributed to M. van Heel. Effectively, this is the same approach as espoused by Liu and Frank (Liu and Frank, 1995).

In this study, we will demonstrate that a meaningful estimate of the real-space variance (corresponding to type-II variance of Liu and Frank), including the variance caused by the reconstruction algorithm itself, can be calculated taking advantage of the availability of a large number of 2-D projections and by using a bootstrap technique. More importantly, the covariance of the 3-D reconstructed volume can also be calculated, making the analysis of conformational variability of macromolecules possible.

The article is organized as follows. In Section 2, we introduce the basic notation and provide an expression for the variance following a linear reconstruction method. In Section 3, we discuss various sources of errors in 3-D reconstructions from projections in single-particle reconstruction. We proceed with a discussion of the concept of the variance of a 3-D reconstruction and postulate that there are three different categories of such a variance. In Section 4, we introduce the bootstrap method and describe its efficient implementation using a variant of a 3-D reconstruction method employing direct Fourier inversion. In Section 5, we provide test results for the introduced variance estimation algorithm introduced. For the bootstrap method, we analyze the consistency and the error of the variance estimate using Monte-Carlo simulations as a function of the number of bootstrap volumes calculated. This provides a practical way of deciding how many bootstrap volumes have to be calculated to obtain the desired accuracy of the variance estimate. We proceed with tests of the bootstrap method using simulated data for various models of the variance in single-particle reconstruction and demonstrate that it yields acceptable results. We demonstrate that the expectation maximization algorithm (described in the Supplementary Material) yields erroneous results in the case of correlated noise in the structure. We also demonstrate that using the bootstrap method we can estimate the covariance (i.e., intervoxel correlations) within the reconstructed 3-D density map. Additional test results of the bootstrap method are provided in the Supplementary Material. In the closing Section 6 we point out the conceptual difficulties caused by the presence of alignment errors in the projection data and we discuss practical ways to account for the effects of the transfer function of the microscope.

The results of the application of the bootstrap method to the analysis of the 70S *Escherichia coli* ribosome complexed with EF-G and tRNAs are described in a companion paper by Penczek et al.

2. Notation

In the weak-phase approximation (Wade, 1992), a 2-D image g collected in the electron microscope is a parallel beam projection P (ray transform) of the 3-D Coulomb potential distribution of a macromolecule f. The projection in the direction of the z axis is

$$g_z(x,y) = \int f(x,y,z)\mathrm{d}z. \tag{1}$$

This integral transformation is continuous, linear, and shift-invariant. We will write the full ray transform as

$$g = Pf. \tag{2}$$

We assume that an inverse transformation $P^\dagger$ exists and that it is regularized by a 'smoothing' operator S

$$\hat{f} = SP^\dagger g, \tag{3}$$

where $\hat{f}$ is an approximate reconstruction of f from projection data g.

In the discrete case, $\mathbf{f}$ is a vector containing n arbitrarily arranged voxels of the 3-D structure, $\mathbf{g}$ is a vector containing correspondingly arranged m pixels of k 2-D projections, and $\mathbf{P}$ is an $m \times n$ matrix with entries p_{ij} whose values depend on the interpolation scheme used and on the projection directions. For the often used trilinear interpolation, the relationship $0 \leqslant p_{ij} \leqslant 1$ holds. In addition, in single-particle reconstruction, $m \gg n$ and matrix $\mathbf{P}$ is sparse with $O(\sqrt[3]{n})$ non-zero entries per row. Thus, in the discrete representation, pixels in projections are linear combinations of corresponding voxels of the 3-D structure

$$g_i = \sum_{j=1}^{n} p_{ij} f_j, \quad i = 1,\ldots,m. \tag{4}$$

By $\mathbf{P}^\dagger$ we denote any linear and shift-invariant reconstruction algorithm, for example a least-square pseudo-inverse solution

$$\hat{\mathbf{f}} = \mathbf{S}(\mathbf{P}^{\mathrm{T}}\mathbf{P})^{-1}\mathbf{P}^{\mathrm{T}}\mathbf{g} = \mathbf{S}\mathbf{P}^\dagger\mathbf{g} = \mathbf{R}\mathbf{g}, \tag{5}$$

where $\mathbf{S}$ is a matrix representation of the appropriate low-pass filter and $\mathbf{P}^\dagger$ is an $n \times m$ "reconstruction matrix". $\mathbf{R}$ denotes any linear reconstruction algorithm, for example GWBP, direct Fourier inversion, or algebraic iterative reconstruction (SIRT).

The variance of a random vector $\mathbf{f}$ is given by

$$\sigma_{\mathbf{f}}^2 = \langle \mathbf{f}^2 \rangle - \langle \mathbf{f} \rangle^2, \tag{6}$$

where $\langle\rangle$ denotes the expectation value and $\sigma_{\mathbf{f}}^2$ is a vector containing variances of components of the vector $\mathbf{f}$ as its elements, and squaring of the vector is understood componentwise. The standard deviation $\sigma_{\mathbf{f}}$ will be associated with the error of vector $\mathbf{f}$. Finally, the covariance of $\mathbf{f}$ is

$$\mathbf{C}_{\mathbf{f}} = \left\langle (\mathbf{f} - \langle\mathbf{f}\rangle)(\mathbf{f} - \langle\mathbf{f}\rangle)^{\mathrm{T}} \right\rangle = \langle \mathbf{f}\mathbf{f}^{\mathrm{T}} \rangle - \langle\mathbf{f}\rangle\langle\mathbf{f}\rangle^{\mathrm{T}}. \tag{7}$$

By combining Eq. (5) with Eq. (7), we obtain an expression for the covariance matrix of the reconstruction $\hat{\mathbf{f}}$

$$\mathbf{C}_{\hat{\mathbf{f}}} = \mathbf{R}^{\mathrm{T}}\mathbf{C}_{\mathbf{g}}\mathbf{R}, \tag{8}$$

where $\mathbf{C}_{\mathbf{g}}$ is the covariance matrix of the projection data $\mathbf{g}$.

3. Sources of variance in EM single-particle structure determination

3.1. Sources of variability

The process of image formation in electron microscopy is well understood (Wade, 1992). In the weak-phase approximation, image formation can be approximated by a linear model according to which the observed projections of ice-embedded macromolecules are linearly related to the molecules' Coulomb potential with additional modification by a (linear) transfer function. The image contrast is formed as a result of the interaction of the electrons with the specimen and the data are collected either on film, subsequently digitized, or directly in digital form in the microscope using a CCD camera. Usually, the microscope is operated with an accelerating voltage in the range of 100–300 kV. To minimize radiation damage, the pictures are taken under low-dose conditions and the dose is kept at $\sim$10 $e^-/\text{Å}^2$. With a typical pixel size ranging from 2×2 to 5×5 Å^2, the average number of electron counts per pixel is sufficient to justify a Gaussian approximation of their distribution.

We can identify six major sources of noise in electron microscopy:

(1) *The specimen*: (i) impurities in the sample, such as the presence of minor contaminants, for example extra protein; (ii) induced disorder caused by the preparation and imaging procedures, for example, from

radiation damage, dehydration effects, or stain-induced effects (i.e., variability and granularity of stains); (iii) inherent disorder or irregularities at the atomic, molecular, or macromolecular level, in particular conformational variability of the macromolecules; and (iv) non-stoichiometry of ligand binding.

(2) *Medium surrounding the specimen*: in cryo-EM it is amorphous ice.

(3) *Specimen support film*: preparation and imaging of many proteins is facilitated by addition of a thin layer of supporting carbon film, which introduces "structural noise."

(4) *The microscope*: thermal drift, mechanical vibrations, electrostatic charging, the electron statistics, and electron scattering events unrelated to the specimen.

(5) *Data collection*: when the data is collected on film, its granularity will be a source of noise; subsequent analog-to-digital conversion will introduce digitization noise that depends mainly on the dynamical range of the data relative to the number of bits in the converter; CCD cameras, in addition to digitization noise, contribute scintillator noise and noise introduced by the readout amplifier.

(6) *Image processing*: single-particle images are represented in discrete form as 2-D arrays of floating point numbers. In the course of establishing the geometrical relations between the images (alignment) and calculation of the 3-D map (3-D reconstruction), they have to be shifted by non-integer pixel values and rotated both in 2-D and in 3-D. These steps involve interpolations required to approximate image values between grid points. Interpolation will result in loss of high-frequency information and injection of spurious information. Finally, misalignment of particle images should be considered a major source of noise.

Despite the large number of sources of noise, our measurements (2-D images) basically contain two regions with distinct statistical properties: the background and the particles. The former can be considered a sample of the noise of the data collection process, while the latter contains a superposition of the signal with two kinds of noise: the background noise and the noise due to the variability of the specimen (Penczek, 2002b). In addition, some of the sources listed contribute noise that is uncorrelated between pixels in projections (for example, the noise due to granularity of the film), while others contribute correlated noise (any source affected by the CTF of the microscope will generate noise that is necessarily correlated). Note that subsequent image processing steps, prior to or during the step of 3-D reconstruction, will introduce additional correlations between pixels.

In the present work we will ignore any effects due to the CTF. This facilitates the analysis and significantly simplifies the design of the variance estimation methods, at the acceptable cost of limiting the resolution of the variance maps thus obtained.

3.2. Categories of variance

For the purpose of the current analysis, we will group the sources of noise listed according to their contributions to the noise in the 3-D map and according to our ability to recover the distribution of this noise upon the process of 3-D reconstruction from projections. We will introduce three main categories of variance and, for the purpose of easier identification, we will term them based on the broad understanding of their respective sources (see also Table 3):

1. *Solvent variance* σ^2_{Solv}: the solvent variance relates exclusively to the imaged specimen and the surrounding solvent, not to the reconstructed object. The main sources of this variance are: (i) irregularities of the medium (ice) surrounding the particle and (ii) the variation of the macromolecules that is due to radiation or thermal instabilities. These two sources are affected by the CTF. Additional components of the solvent variance are those that are conceptually different, but which cannot be distinguished from the two main components; i.e., the variance due to support film, the microscope, the data collection process, and some of the steps of image processing. Some of these sources are not affected by the CTF. If the CTF is ignored, all sources of the noise listed can be considered uncorrelated, although their spatial distribution can be non-uniform. For example, if the distribution of mass density of the macromolecule is $f(x, y, x)$, the associated variance can have the form of $\sigma^2_{\text{Solv}}(f(x, y, x))$, i.e., it is a function of the spatially varied mass density. The information about the solvent variance can be used to estimate the distribution of noise in the imaged specimen and also, to some extent, the degree of noise uniformity.
2. *Structure variance* σ^2_{SVar}: the variance in the reconstructed 3-D map is due to five possible sources: (i) reconstruction algorithm and non-uniform distribution of projections, (ii) background noise, (iii) misalignment errors, (iv) conformational variability of the macromolecule, and (v) presence or absence of a ligand. Sources (iii–v) have long-range correlations; moreover, we expect that variations in different regions of the reconstructed structure will tend to be correlated (for example, binding of ligand can cause conformational changes in distant regions of the protein). For the purpose of easier identification, we will use the following terms for two components of the structure variance:

   2a. *Reconstruction variance* σ^2_{Rec}: the variance in the 3-D map that is due to the reconstruction algorithm and the uneven distribution of projection directions. The inversion of the ray transform is an ill-posed problem and even in the absence of noise, the reconstruction process will introduce artifacts in the structure, particularly if we consider that the distribution of projection directions varies between experiments. The level and distribution of the reconstruction variance informs us about the stability of the reconstruction algorithm under practical image collection conditions.

2b. *Background variance* σ^2_{Back}: the variance in the reconstructed 3-D map that is not due to the structure. Sources of this variance include any noise in projection data, particularly those contributions listed under solvent variance. The background variance is the variance of the noise modified by the reconstruction process, i.e., it is modified by the associated filtration and averaging in Fourier space. We will assume that the background variance is well represented by the distribution of noise in the micrograph background and that the noise within the particle area has essentially the same distribution. By extension, we will assume that the resulting background variance in 3-D reconstruction is uniform (or nearly so). The information on the background variance can be used for the purpose of evaluation of docking accuracy. It also provides a reference level of error, i.e., information about the minimum possibly achievable level of noise in the reconstructed object if there were no alignment errors and the particles were truly homogenous.

3. *Variance of structures* σ^2_{Struct}: the variance due to conformational variability or non-stoichiometry of ligand binding, which could be calculated if, for each macromolecule, we had an entire set of projections, sufficient to reconstruct it unambiguously. This variance is closely related to, and derived from, the structure variance, but it differs from the latter in the exclusion of the background noise and in numerical aspects. In distinction from the structure variance, which corresponds to the variance associated with averages of mixtures of different structures, the *variance of structures* corresponds to the variance of unique structures; thus, its numerical value can be expected to be much higher (Fig. 1). Knowledge of the variance of structure can be used to guide docking experiments; it can also indicate presence of alignment errors, heterogeneity of the sample, or the location of ligands. Note that the variance of structures is not directly accessible in single-particle reconstruction, as projections originate from different macromolecules, and each projection carries only partial information about the 3-D structure.

In the nomenclature introduced by Liu and Frank (Liu and Frank, 1995), the solvent variance is of type I, while the variance of structures is of type II.

The three categories of variance listed above share the following two major distinctions: (i) whether or not the sources of errors are correlated and (ii) whether we consider the variance of the *imaged object* (or background) or the variance of the *reconstructed object* (or background). We will demonstrate that each of these categories calls for a different computational approach and each will result in quite different estimates of noise. Nevertheless, each carries important information about the statistics of the experiment, and each can therefore be used for different purposes in the variance estimation.

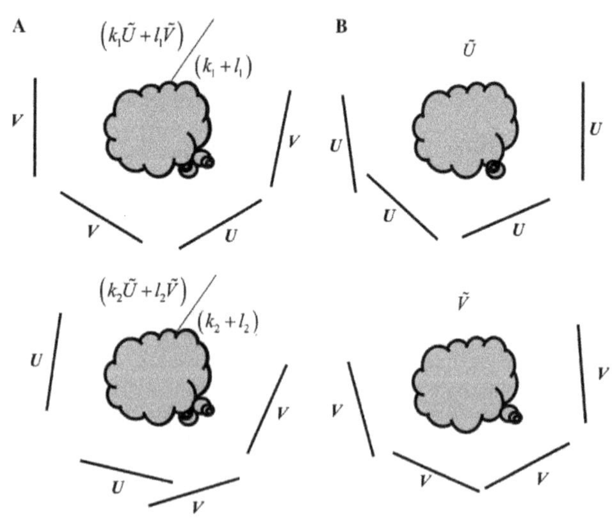

Fig. 1. Structure variance versus variance of structures in case of two conformers $\tilde{U}$ and $\tilde{V}$ of a macromolecule. (A) In single-particle reconstruction, a data set contains projections U and V of two conformers mixed in unknown quantities k and l, respectively. Repeated experiment would yield projection data in different quantities and at different angular directions. The variance calculated in this case will yield the structure variance, i.e., the variance corresponding to variability of different "average" reconstructions. (B) If the assignments of projections were available, each conformer could be reconstructed unambiguously. This would yield the variance of structures, i.e., the variance of reconstructed actual conformers. Since the assignment of projections is not known, this variance is not directly accessible in single-particle reconstruction.

4. Application of the bootstrap technique to the variance estimation

In Section 2, we showed that to calculate variance/covariance $\mathbf{C}_{\hat{\mathbf{f}}}$ of the 3-D reconstruction $\hat{\mathbf{f}}$ we would have to estimate variance/covariance $\mathbf{C}_{\mathbf{g}}$ of the projection data $\mathbf{g}$ (Eq. (8)). This is impractical because of the very large size of the matrix $\mathbf{C}_{\mathbf{g}}$ and because we do not have an efficient algorithm for calculating $\mathbf{C}_{\hat{\mathbf{f}}}$ from $\mathbf{C}_{\mathbf{g}}$.

In addition to the calculation of the variance of the reconstruction we would like to calculate the variance of the reconstructed structure. The two are not the same, as can be deduced by the following reasoning. In single-particle reconstruction, each projection image originates from a different macromolecule. If macromolecules were identical and there was no alignment error, the only source of errors in the reconstruction would be the additive noise in the projections and the reconstruction algorithm itself. In reality, macromolecules do vary in structure. Thus, if we could collect a number of projections sufficient for calculating the 3-D reconstruction of every macromolecule, Eq. (5) could be written as:

$$\hat{\mathbf{f}} = \mathbf{R}(\mathbf{g}_1 + \mathbf{g}_2 + \mathbf{g}_3 + \ldots), \tag{9}$$

where $\mathbf{g}_k$ denotes a set of projections associated with the *k-th* structure. If we could sort projections into these k homogenous sets, we could calculate 3-D reconstructions $\hat{\mathbf{f}}_k = \mathbf{R}\mathbf{g}_k$ and—from the definition – their variance $\sigma^2_{\hat{\mathbf{f}}}$. This variance we will call the *variance of structures* (recall it is the Type-II variance of Liu and Frank (Liu and Frank, 1995)). In practice we do not know which projection originated from which structure, so we can only divide the set of projection randomly into subsets and calculate the corresponding 3-D reconstructions. Since each subset contains projections from different structures, the *structure variance* σ^2_{SVar} of reconstructions thus obtained will be necessarily smaller than the "true" variance of the original, albeit unknown structures, and also smaller than the variance of structures, i.e., $\sigma^2_{\text{SVar}} < \sigma^2_{\text{Struct}}$. The two could be equal only if all macromolecules were identical and if there were no alignment errors. It is important to note that in each case the portion of variance in reconstructions that is due to uncorrelated noise in projections is the same; clearly, since the noise is uncorrelated, its contribution cannot be reduced by exchanging projections between different groups. Also, since we only consider molecules reconstructed from projections, both variances contain contribution from the variance due to the reconstruction algorithm.

In this section we will demonstrate that the variance/covariance of a 3-D reconstruction can be calculated relatively efficiently using the bootstrap technique. Moreover, we will also show that the variance of structures is related to the structure variance in a simple manner, and that its magnitude can be estimated if the level of background variance is known.

4.1. The bootstrap technique for the estimation of the variance in a 3-D structure calculated from a set of its projections

To estimate the variance/covariance in a structure calculated from the set of its projections in our earlier work (Penczek, 2002b) we proposed to employ a statistical resampling technique, the bootstrap (Davison and Hinkley, 1997). This method is particularly useful for the estimation of parameters of statistical distributions when analytical formulae are not known. The idea is to resample the original data set to create replicate sets, from which quantities of interest can be calculated. Bootstrap is based on resampling with replacements, and in our application we propose to resample the available set of K projections, as was put forward in (Penczek, 2002b).

The voxels in the reconstructed volume can be considered to be (weighted) sums of pixels in projections. If we assume that the number of 2-D projections is large and the weighting function of the reconstruction algorithm does not change much by removing or adding a subset of projections, then we propose to calculate the variance of structures σ^2_{Struct} by using a relationship between the variance of arithmetic means for sampling with replacements and the sample variance (Hansen et al., 1953)

$$\sigma^2_{\text{Struct}} = K\sigma^2_B. \tag{10}$$

In what follows, we will demonstrate that the above relationship holds only for components of the variance that are related to the structure, not for the components due to uncorrelated noise, and we will show that σ^2_B is the structure variance σ^2_{SVar}.

In the bootstrap procedure, a new set of projections is generated by randomly selecting, with replacements, projections from the original set. Thus, in the new set some of the projections will be omitted and some will be represented more than once. Using this set, a 3-D sample structure is calculated. This process is repeated B times. The resulting B bootstrap volumes are low-passed filtered to a resolution not exceeding the resolution of the analyzed structure, and the voxel-by-voxel bootstrap variances σ^2_B as well as covariances between voxels (or regions within the structure) are calculated. The covariance or, taking advantage of the fact that both variance and covariance of the mean are related to those of the sample by the same factor K, the corresponding cross-correlation coefficient can be estimated by

$$r_{ij} = \frac{\frac{1}{B}\sum_{l=1}^{B}\left(v_i^l - \bar{v}_i\right)\left(v_j^l - \bar{v}_j\right)}{\sigma_{B_i}\sigma_{B_j}}, \tag{11}$$

where i and j are indices of two different voxels/regions in a 3-D bootstrap structure, and v_i^l and $\bar{v}_i$ are the values of the i'-th voxels/regions for the l'-th structure and its average, respectively.

Application of the bootstrap technique to the variance estimation in 3-D reconstruction has the major advantage that it does not depend on the choice of a particular 3-D reconstruction algorithm; rather, the approach is general, i.e., applicable to any reconstruction algorithm as long as it is linear. This freedom of choice is important not only because the quality of reconstruction algorithms varies but, more importantly, because of considerations of computational efficiency. Since bootstrap reconstructions can be generated from an overall reconstruction simply by removing certain projections and increasing the multiplicity of others, we should employ a 3-D reconstruction in which such modifications can be performed in an efficient way. A possible approach is to use a straightforward direct Fourier inversion scheme based on the projection theorem for the 3-D ray transform, which states that the 2-D Fourier transform of a projection of a 3-D object equals the 3-D Fourier transform of this object on a central plane perpendicular to the projection direction. Thus, one can compute 2-D Fourier transforms of the available projections and add Fourier pixels to voxels of the target 3-D Fourier transform using a simple nearest-neighbor (NN) interpolation scheme (Penczek, 2002a). Before the 3-D Fourier inversion, the accumulated Fourier coefficients have to be divided by the respective numbers of added Fourier pixels (the weighting functions). Incidentally, the accuracy of this reconstruction scheme can be greatly improved if the projections are padded with zeroes in real space to increase their size fourfold (N. Grigorieff, personal communication). The padding results in a finer Fourier grid and thus reduces the nearest-neighbor interpolation errors.

The Bootstrap method for 3-D variance estimation
using direct Fourier inversion with NN-interpolation:

(i) pad K projection images with zeroes fourfold;
(ii) calculate and store FFTs of padded projections;
(iii) compute and store 3-D Fourier volume using NN-interpolation.

for $\ell = 1, 2, \ldots, B$

(a) using selection with replacements, generate K bootstrap projection indices $(i_1, i_2, \ldots, i_K)_\ell$;
(b) based on the set of projection indices $(i_1, i_2, \ldots, i_K)_\ell$, either remove or add projections to the pre-computed 3-D Fourier volume;
(c) compute inverse 3-D inverse FFT of the Fourier volume and extract the part containing the structure. This yields bootstrap volume V_ℓ.

end for
Compute 3-D bootstrap variance map:

$$\sigma_B^2 = \frac{1}{B-1}\sum_{\ell=1}^{B}\left(V_\ell - \frac{1}{B}\sum_{j=1}^{B} V_j\right)^2.$$

The approach adopted results in a particularly straightforward and efficient implementation of the bootstrap technique. In addition, due to the nature of the method, the bootstrap reconstructions can be generated entirely independently in parallel, using either the shared or distributed memory paradigm. For example, the data set described in the companion paper on the application of the bootstrap method contained 10,477 2-D particle images boxed in windows of 75×75 pixels. Using a shared memory (OpenMP) implementation we were able to generate 90 bootstrap volumes using 1 wall clock hour of SGI 3200 ORIGIN equipped with 8×400 MHz processors and 8 GB main memory.

4.2. Application of the bootstrap method to the variance estimation in a general case

In a general case, the variance in the 3-D reconstruction, which we call *the structure variance*, will contain the variance due to the reconstruction algorithm and distribution of projections σ^2_{Rec}, due to background noise in projections σ^2_{Back}, due to alignment errors σ^2_{Ali}, and due to conformational variability of the 3-D structure σ^2_{Conf} (for example due to structural variability or non-stoichiometric binding of ligands). This compound variance can be equated to the bootstrap variance and calculated using the bootstrap technique

$$\sigma^2_{\text{SVar}} = \sigma^2_B = \sigma^2_{\text{Conf}} + \sigma^2_{\text{Ali}} + \sigma^2_{\text{Rec}} + \sigma^2_{\text{Back}}. \tag{12}$$

In a series of tests described in the Supplementary Material we demonstrated that the bootstrap technique yields very good estimates of three of the components of the right hand side of Eq. (12), namely σ^2_{Ali}, σ^2_{Rec}, and σ^2_{Back}. However, if we wish to calculate the variance of structures σ^2_{Struct} using Eq. (10), we have to eliminate from σ^2_B all those components that would remain the same if we were able to calculate the variance of structures directly, i.e., by collecting an entire set of projections and calculating an independent reconstruction for each macromolecule (Eq. (9)). Clearly, the components included in Eq. (12) that fall into this category are σ^2_{Ali}, σ^2_{Rec}, and σ^2_{Back} and they have to be calculated independently. Thus, the variance of structures should be calculated using an appropriately modified version of Eq. (10), that is:

$$\sigma^2_{\text{Struct}} = K\left(\sigma^2_B - \sigma^2_{\text{Ali}} - \sigma^2_{\text{Rec}} - \sigma^2_{\text{Back}}\right). \tag{13}$$

A major difficulty is caused by the misalignment variance σ^2_{Ali}. Although, as shown in the Supplementary Material, in the absence of other components, σ^2_{Ali} can be accurately estimated using the bootstrap technique, it cannot be distinguished from the variance due to conformational variability of 3-D structure σ^2_{Conf}. In other words, σ^2_{Ali} has the same statistical properties as σ^2_{Conf}, but while the latter has to be multiplied by the number of projections K to yield the correct level of the variance of structures σ^2_{Struct} using Eq. (10), the former should not. Unfortunately it is not immediately apparent how one could design a method that would allow us to distinguish between the two components (Baldwin and Penczek, 2005).

4.3. Calculation of the background variance

If we assume that the statistical distribution of noise in micrographs is known, it should be possible, in principle, to derive from it the statistical distribution of noise in the reconstruction. For example, if the projection directions were distributed uniformly, the distribution of noise in the reconstructed volume would be modified by a filter $h(\omega) = \sqrt{\omega_x^2 + \omega_y^2 + \omega_z^2}$, where ω is the spatial frequency. This filter reflects the degree of oversampling in the reconstruction. For some 3-D distributions of projections (uniform, single-axis tilt), the appropriate filter can be calculated analytically and the resulting modification of the power spectrum of the noise can also be calculated. Unfortunately, this is not the case in single-particle reconstruction, where the distribution of projections is random and non-uniform. Therefore, we propose to estimate the background variance of the reconstruction using samples of the micrograph background noise and the distribution of projection directions that occurred in the actual reconstruction (Penczek, 2002b). One possibility would be to calculate a 3-D reconstruction using the windowed noise and in this way obtain a "sample" of a 3-D distribution of noise. However, we estimate other components of the variance in the reconstruction using the bootstrap technique, so we can also apply the same method to the estimation of the background noise, i.e., we can obtain another estimate of σ^2_{Back} by resampling, with replacements, the available projection samples of the background noise. The limitation of using the samples of background noise from micrographs to infer the level of noise in 3-D reconstruction is that the level of noise within the area occupied by the projection of the molecule is reduced by the ratio of the molecule "height" to the ice thickness. The ice thickness is very difficult to measure; moreover, rigorous treatment of the problem would introduce non-linearities into the EM image formation model, as the distribution of noise would depend in the shape of particle. Therefore, we decided to ignore this dependence while keeping in mind that the level of background noise is to some degree overestimated.

The bootstrap procedure comprises the following steps:

1. Select samples of the background noise from the micrograph. Their number has to be the same as the number of available projection images (in the companion paper it is described how to obtain these samples from noise regions of images containing windowed particles).
2. Apply the bootstrap technique described in Section 4.1 to calculate the 3-D variance map of the background noise σ^2_{Back}, which will also contain the reconstruction variance σ^2_{Rec}.
3. Calculate the average level of the background variance $\bar{\sigma}^2_{\mathrm{Back}}$ within the 3-D region corresponding to the support of the structure.

In the last step we assumed that the variance in the reconstruction that is due to the background noise in projection data is uniform, i.e., it does not depend on the location in the 3-D reconstructed volume. This follows from the fact that the noise is uniform in 2-D and the reconstruction algorithm employed is shift-invariant. An additional advantage of using the bootstrap technique for estimation of the background variance is that the result contains also the reconstruction variance. Overall, the result of the procedure is a single number, which specifies the uniform level of error (due to background noise and the reconstruction algorithm) in a 3-D reconstruction of a macromolecule. The test of the bootstrap approach and a comparison with direct estimation based on ergodic principle are included in the Supplementary Material.

Taking advantage of the bootstrap estimate of the background error, the variance of structures should be calculated using the modified version of Eq. (13)

$$\sigma^2_{\mathrm{Struct}} = K\left(\sigma^2_{\bar{B}} - \bar{\sigma}^2_{\mathrm{Back}}\right). \tag{14}$$

Note that here we disregarded the variance arising from alignment errors, as there is no method to estimate it independently.

5. Results

In this Section we provide results of tests of the bootstrap technique that help to determine the number of bootstrap volumes required for the estimation of the variance and that demonstrate that the technique works well for both noise-free and noise corrupted date. Additional tests are given in the Supplementary Material.

5.1. Monte-Carlo tests of the bootstrap estimation of the variance

To be able to use the bootstrap technique, we have to know the minimum required size of the bootstrap sample B for the method to yield reliable results. For a set of K projections, the number of possible bootstrap samples is $B = \binom{2K-1}{K-1} \cong 2\exp[2K]$. Obviously, it is impossible to generate all of them; rather, to make calculations manageable, we would like to generate as few as possible. To determine the minimum acceptable number we have to know the convergence properties of the bootstrap estimate of the variance of the sample, or the variance of the variance estimator as a function of both the number of bootstrap samples B and the size of the sample K. Unfortunately, analytical expressions do not exist (Shao and Tu, 1995); instead, we used the Monte-Carlo method to assess the relationship.

We generated K random numbers from a Gaussian distribution $N\left(1, \sigma_\gamma = \sqrt{10}\right)$. These numbers were resampled with replacements B times and for each sample set its average was calculated (this step corresponds to the calculation of a 3-D structure from a resampled set of projections). The variance σ_B^2 was calculated from this set of B averages (this step corresponds to the calculation of 3-D structure variance). We determined that the variance estimate was underbiased and that, in the case tested, it could be approximated by

$$\frac{K\sigma_B^2}{\hat{\sigma}_\gamma^2} = 1 - \frac{1}{K}, \tag{15}$$

where $\hat{\sigma}_\gamma^2$ is the actual variance of the sample. Thus, the bias is of the order $1/K$ and can be neglected in practice. Next, we similarly confirmed that the variance $\sigma_{\sigma_B^2}^2$ of the resampled variance σ_B^2 does not depend on the sample size K, and is the same as the variance of the estimated variance for a simple normally distributed random sample

$$\sigma_{\sigma_B^2}^2 = \frac{2}{B}\hat{\sigma}_\gamma^4. \tag{16}$$

In conclusion, the bootstrap variance estimator is consistent, as its variance vanishes with increased size of the bootstrap sample B, and it is negligibly underbiased. Following Eq. (16), thus the size of the bootstrap sample B depends only on the desired accuracy of the variance estimator to have relative error 0.01 it is sufficient to calculate $B = 200$ bootstrap volumes, which is perfectly manageable.

5.2. Test of the bootstrap estimation of the structure variance in a noise-free case

The tests described in this Section were designed to verify that the bootstrap method yields proper estimates of the structure variance and, on the basis of Eq. (10), the variance of structures. The basic test structure comprised five Gaussian functions: the first one (S) centered with a broad support mimicking the shape of a structure, and the remaining four (A–D) scattered and relatively compact corresponding to "features" of the structure (Fig. 2A)

$$\begin{aligned} f(x,y,z) &= 10\exp\left[0.5\left((x/8)^2 + (y/12)^2 + (z/10)^2\right)\right] \\ &+ 3\exp\left[0.5\left(((x-13)/6)^2 + ((y-7)/5)^2 + ((z-11)/4)^2\right)\right] \\ &+ 2\exp\left[0.5\left(((x-5)/6)^2 + ((y+5)/8)^2 + ((z+13)/7)^2\right)\right] \\ &- 2\exp\left[0.5\left(((x+11)/9)^2 + ((y+7)/8)^2 + ((z-11)/6)^2\right)\right] \\ &+ 1.2\exp\left[0.5\left(((x+13)/6)^2 + ((y-3)/4)^2 \right.\right. \\ &\left.\left. + ((z+5)/5)^2\right)\right] = 10S + 3A + 2B - 2C + 1.2D. \end{aligned} \tag{17}$$

(All parameters are given in voxels.) The size of the volume was 64^3 voxels with $-32 \leqslant x,y,z < 32$. Next, we modified the test structure (Eq. (17)) to randomize features A–D. Both their amplitudes and their shapes were set to vary randomly and we also introduced correlations between some of them (only feature B varied independently) (Figs. 3A and B)

$$\begin{aligned} f(x,y,z) = 10S &+ 3A\left(1 + r_1(A/3)^2\right) \\ &+ 2B\left(1 + 1.2r_2(B/2)^2\right) \\ &- 2C\left(1 + 2.5(r_1 + r_3)(C/4)^2\right) \\ &+ 1.2D\left(1 + (r_1 + 2r_4)(D/1.1)^2\right), \end{aligned} \tag{18}$$

where $r_{1\text{–}4}$ are independent Gaussian numbers $N(0,1)$. We generated 1253 quasi-evenly distributed projections of this randomized structure by selecting an independent set of numbers $r_{1\text{–}4}$ for each projection direction. In effect, each 2-D image contained a projection of the same basic 3-D shape S, but with different amplitudes of features A–D.

We calculated the structure variance using the bootstrap method with $B = 500$ as the size of the sample (Figs. 3C and D). After generating 500 bootstrap volumes we calculated the 3-D structure variance map σ_{SVar}^2, which we then

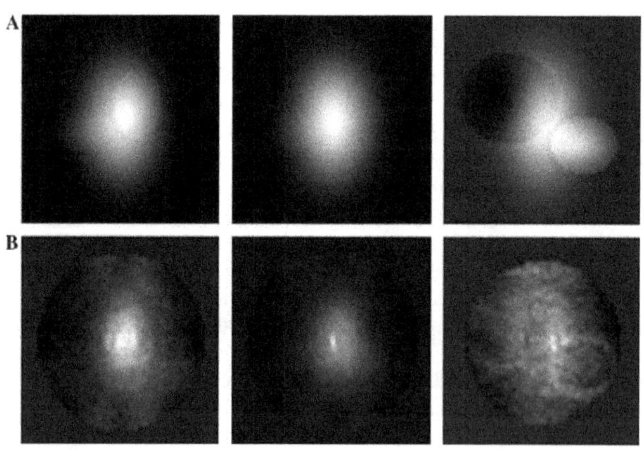

Fig. 2. Test of the variance of the reconstruction estimated using the bootstrap method (see the Supplementary Material). We show selected z-slices ($z = -5, 0, +11$) of the 3-D map masked with a centered sphere with radius $R = 27$ voxels and with the contrast within each slice adjusted independently, so the intensities do not reflect absolute values in respective slices. (A) Model structure (Eq. (17)). (B) Estimated variance.

A

B

C

D

E

F

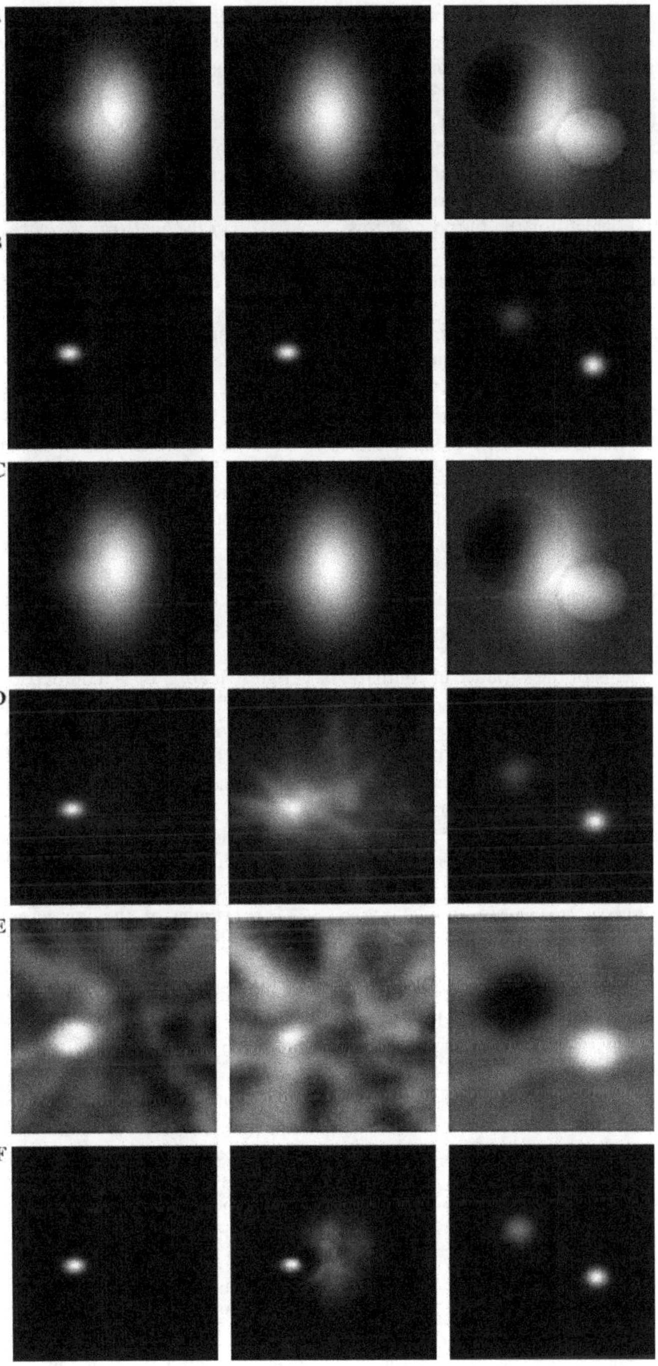

Fig. 3. Test of the estimation of the structure variance in the presence of feature variability using the bootstrap method. We show selected z-slices, left to right: $z = -5$ contains feature D, $z = 0$ with no features, $z = +11$ contains feature A. Contrast within each slice was adjusted independently, so the intensities do not reflect absolute values in respective slices. (A) Average of model structures (Eq. (18)). (B) The variance calculated using 1253 simulated model structures. (C) Average bootstrap structure. (D) Structure variance calculated using the bootstrap method. (E) Correlation map between the center of the feature A and the remaining voxels calculated using sample volumes. (F) Variance calculated using the solvent variance estimation method, i.e., the expectation maximization algorithm described in the Supplementary Material.

multiplied by the number of projections K (=1253) to obtain the 3-D map of variance of structures σ^2_{Struct} (Eq. (10)). The bootstrap variance map (Fig. 3D) corresponds closely to the variance of the test volumes (Fig. 3B). Also, numerical values in bootstrap maps match well values of variance calculated directly using test volumes (Table 1). Since the "features" in the test structure are spread, we decided to compare the average levels of variance within Regions Of Interest (ROIs), which for the purpose of this work we defined as regions around maxima of functions A–D that have densities equal to at least 60% of the maximum density of the respective functions. The values of the variances of structures within four ROIs agree very well with the values obtained for the ROIs within the original test volumes. In addition, we calculated the correlation coefficient between the center of the feature A and the remaining voxels in the volume (Fig. 3E). The average correlation coefficients (shown in Table 1) within four ROI's agree well with those in the test volume. (The correlation coefficient with the ROI A is less than one because it was averaged over the ROI support. The central value is 0.998.) Note that the full correlation function is a four-dimensional entity (64^4) which cannot be easily calculated or displayed.

The variance of structures (Fig. 3D, slice $z = 0$) clearly contains additional components of the variance in comparison with the variance of the test volumes (Fig. 3B). The magnitude of this additional variance can be evaluated based on its values within the central region of the reconstruction (*center* in Table 1), in which we do not expect any variations. We attribute this additional component to the reconstruction variance. It is important to note that the 3-D variance map does not change in a noticeable way even if the size of the bootstrap sample is increased to $B = 5000$ (results not shown).

To put the bootstrap results in a perspective, we calculated the solvent variance using the same test data set and the expectation maximization algorithm described in Supplementary Material. Following this method, correlations in the sources of the variance are ignored; in effect, the relative values of variance within ROIs are distorted.

Table 1
Tests of the bootstrap estimation of the variance of structures

	A	*B*	*C*	*D*	*Center*
σ^2_f	2.25	1.49	0.68	2.37	10^{-8}
σ^2_{Struct}	2.51	1.79	0.93	2.76	0.55
$\sigma^2_{\text{Solvent}}$	15.5	12.2	7.78	17.7	0.77
r_f	1.00	0.02	−0.71	0.48	10^{-7}
r_{Struct}	0.86	0.02	−0.55	0.37	-7×10^{-3}

A–D: ROIs within respective variance/correlation maps. *Center*: ROI defined as a centrally located ball with radius 2 pixels. σ^2_f: variance of test structures. σ^2_{Struct}: variance of structures estimated using the bootstrap technique. $\sigma^2_{\text{Solvent}}$: variance calculated using the method for the solvent variance estimation. r_f, r_{Struct}: correlation coefficients between the central center of the ROIA and the centers of all ROIs for the test and bootstrap results, respectively. Correlation coefficients were averaged within respective ROIs.

For example, the upper left peak in the rightmost slice in Fig. 3F has values that are clearly too high. Also, the variance values within the ROIs are entirely incorrect (Table 1). Finally, the method does not yield correlations within the 3-D reconstruction. Despite these inferior results, the method can still be considered useful for a fast estimation of the distribution of the variance within a reconstruction.

5.3. Test of the bootstrap variance estimation method in case of structural variability of the data corrupted by additive uncorrelated noise

In the final test we evaluated the performance of the bootstrap variance estimation method in the case of structural variability of the data corrupted by additive, uncorrelated noise. As the test data set we used the structure designed for testing the structure variance (Eq. (18)) and the set of projections generated for these tests. We corrupted the projections by independent additive Gaussian noise $N(0,30)$, which resulted in $SNR \cong 2$ in the projection data. After generating $B = 500$ bootstrap volumes, we estimated the resolution of an individual reconstruction using the Fourier shell correlation (FSC) (Saxton and Baumeister, 1982) to be 0.1 [1/voxel], applied the corresponding low-pass filtration to all sample volumes, and then calculated the 3-D bootstrap variance map. For a comparison, we applied the same low-pass filtration to the test volumes and calculated their average, variance, and correlation map with respect to the center of feature A (Figs. 4A–C). The loss of features in the average map is apparent in Fig. 4A, particularly when compared with the original structure shown in Fig. 2A. The correlation map (Fig. 4C) contains an unusual pattern, which is due to correlations introduced into the volumes by the process of low-pass filtration.

The average of the low-passed bootstrap volumes is shown in Fig. 4D, and the variance in Fig. 4E. The average bootstrap structure is indistinguishable from the average test structure (Fig. 4A) and it is similarly featureless. Nevertheless, despite the high level of noise in the data and severe low-pass filtration of the bootstrap volumes, the structure variance map contains easily distinguishable bright spots in locations corresponding to high level of variance (Fig. 4E, features D and A). The central slice $z = 0$ of the variance map (Fig. 4E) contains a typical pattern of the featureless background variance. Interestingly, neither the variance nor the correlation maps are inferior to the maps obtained in noise-free case (Figs. 3D and E).

To test the approach given by Eq. (13), we generated projections containing independent additive Gaussian noise $N(0,30)$ and calculated $B = 35$ bootstrap volumes. We applied the same low-pass filtration as used for the volumes containing the structure, calculated the noise bootstrap variance volume, and estimated from it the average background variance $K\bar{\sigma}^2_{\text{Back}} = 0.31$, which we subtracted from $K\sigma^2_{\text{SVar}}$. The values of σ^2_{Struct} obtained are very close to the values calculated for filtered test volumes (Table 2). This demonstrates that the proposed method

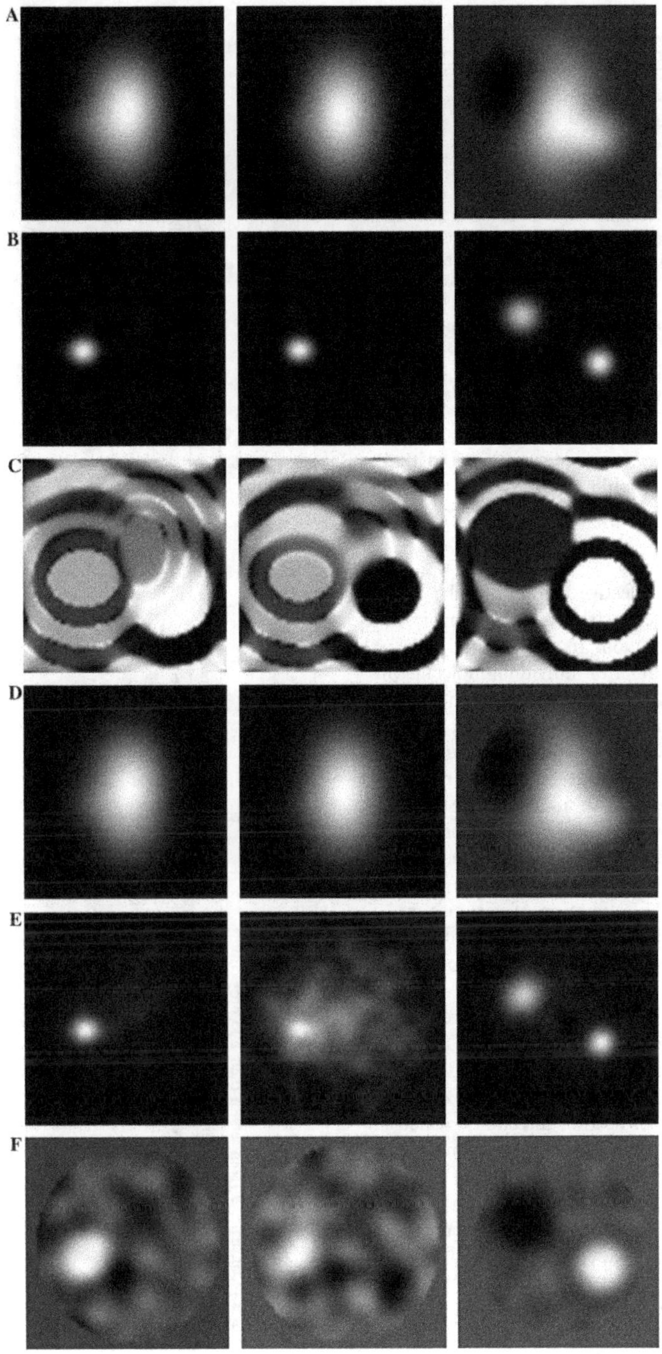

Fig. 4. Test of the estimation of the structure variance using the bootstrap method in the presence of additive independent Gaussian in projections. We show selected z-slices ($z = -5$, 0, +11) with the contrast within each slice adjusted independently, so the intensities do not reflect absolute values in respective slices. (A) Average of low-passed model structures (Eq. (18)). (B) The variance calculated using 1253 simulated low-passed model structures. (C) Correlation map between the center of the feature A and the remaining voxels calculated for simulated low-passed volumes. (D) The average of low-passed bootstrap structures. (E) Structure variance calculated using the bootstrap method and estimated from low-passed sample volumes. (F) Correlation map between the center of the feature A and the remaining voxels calculated using low-passed bootstrap volumes.

Table 2
Test of the estimation of the structure variance using the bootstrap method in the presence of additive independent Gaussian in projections

	A	*B*	*C*	*D*	*Center*
σ_f^2	1.17	1.24	0.65	1.35	4×10^{-2}
$K\sigma_{\mathrm{SVar}}^2$	1.51	1.59	1.14	1.83	0.77
$\sigma_{\mathrm{Struct}}^2$	1.19	1.28	0.82	1.52	0.46
r_f	1.00	0.00	−0.70	0.44	0.19
r_{SVar}	0.87	−0.02	−0.44	0.34	-7×10^{-3}

A–D: ROIs within respective variance/correlation maps. *Center*: ROI defined as a centrally located ball with radius 2 pixels. σ_f^2: variance of test structures. σ_{SVar}^2: structure variance estimated using the bootstrap technique. $\sigma_{\mathrm{Struct}}^2 = K(\sigma_{\mathrm{SVar}}^2 - \bar{\sigma}_{\mathrm{Back}}^2)$: the variance of structures. $\bar{\sigma}_{\mathrm{Back}}^2$: average background variance estimated using the bootstrap method. r_f, r_{SVar}: correlation coefficients between the central center of the ROI *A* and the centers of all ROIs for the test and bootstrap results, respectively. Correlation coefficients were averaged within respective ROIs.

of calculating the structure variance is sound and that it yields qualitatively correct results.

6. Discussion

We have developed an effective method for the estimation of variance in 3-D reconstructions from projections in single-particle reconstruction. In this method, we take advantage of the large number of individual molecule projections available in single-particle reconstruction and we apply the bootstrap technique to generate multiple sample sets. For each sample set we calculate the sample 3-D reconstruction. The bootstrap variance is finally calculated using a set of sample volumes. In addition, the technique makes it possible to calculate correlations either between voxels in the structure or between regions of interests, usually defined as regions that have high levels of variance. The methods related to the estimation of variance in 3-D reconstructions from projections were implemented in the SPIDER system (Frank et al., 1996) and their comprehensive list is given in Table 3.

The calculation and, in particular, the interpretation of the variance of a structure calculated from a set of projections is intricate due to conceptual and practical difficulties, particularly those caused by the presence of alignment errors. It is expected that projection directions, which are established computationally using an iterative refinement procedure, will have errors. At this point it is not clear whether the variance due to alignment errors can be distinguished from the variance due to structural heterogeneity. As we argued in our earlier paper (Baldwin and Penczek, 2005), there is ambiguity in the possible interpretation of the variance in the data, which can be thought of as caused by alignment errors or the presence of multiple classes of particles in the dataset. To reduce this variance, in the former case one would chose to correct the alignment, while in the latter one would try to classify the data into homogeneous classes. To compound the difficulty, if the structure variance was due to alignment errors it would be problematic to use Eq. (13) and interpret the varience

Table 3
Variance in 3-D structures reconstructed from sets of their 2-D projections

Variance	Notation	Variance model	Estimation method	Command in SPIDER system
Solvent	σ_{Solv}^2	Irregularities of the ice surrounding the particles; radiation/thermal instabilities of macromolecules; it is the variance of the source, ***not*** of the reconstructed map	Expectation maximization 3-D reconstruction algorithm applied to estimates of projection variances	BP EM
Structure variance	σ_{SVar}^2	The variance in the reconstructed 3-D map due to the structure, the reconstruction algorithm, the background, and the alignment errors	Bootstrap applied to 2-D projection data	VA 3Q, VA 3R
Reconstruction	σ_{Rec}^2	The variance in the 3-D map that is due to the reconstruction algorithm and the distribution of projections	Bootstrap applied to 2-D noise-free projection images	VA 3Q, VA 3R
Background	σ_{Back}^2	The variance in the reconstructed map that is due to the background noise in the data	Bootstrap applied to 2-D projection images containing samples of the background noise	VA 3Q, VA 3R
Variance of structures	$\sigma_{\mathrm{Struct}}^2$	The variance of the reconstructed 3-D structure due to conformational variability or ligand binding	Derived from the structure variance	Calculate $K(\sigma_{\mathrm{SVar}}^2 - \bar{\sigma}_{\mathrm{Back}}^2)$
Covariance (correlation)	r_{ij}	Voxel correlations in the reconstructed map	Calculated from bootstrap volumes of structure variance	VA 3C

K is the number of 2-D projections. In bold we marked three main categories of the variance, as discussed in Section 3.2. $\bar{\sigma}_{\mathrm{Back}}^2$ is the average background variance.

obtained as the variance of structures. Such an interpretation would imply that the reconstructed macromolecules have intrinsically different orientations and that it would therefore be advantageous to cluster them into classes based on their orientations. A more reasonable interpretation is to consider the alignment error as part of the error of the reconstruction, in which case it should not be interpreted in structural terms. Unfortunately, as we already pointed out, we do not have tools as of yet that would allow us to make a distinction between structural and alignment variances.

In the current work, we made a number of simplifying assumptions that allowed us to focus on the analysis on the problem of estimation of variance in reconstructions calculated from sets of their projections. In particular, we ignored the CTF effects. The main reason was that the process of CTF correction itself would constitute another source of variance and at this moment the consensus on how to perform this step computationally has not been reached. In one approach, the particle images are grouped according to associated defocus settings, separate 3-D reconstructions are calculated for each defocus group, and finally merged using a Wiener filtration approach into a single CTF-corrected map (Frank et al., 2000; Penczek et al., 1997). Because of the grouping of the data in this case one would have to modify the resampling strategy accordingly. In another approach, the CTF correction is fused with the 3-D reconstruction process (Grigorieff, 1998; Ludtke et al., 1999), in which case the variance estimation using the bootstrap technique described here could be applied with only minor modifications.

Acknowledgments

We thank Edward H. Egelman, Piotr Franaszczuk, and Anchi Cheng for helpful discussions. Supported by Grants NIH R37 GM 29169 (to J.F.) and R01 GM 60635 (to P.A.P.).

Appendix A. Supplementary material

Supplementary data associated with this article can be found, in the online version, at doi:10.1016/j.jsb.2006.01.003. The SPIDER software in which the variance estimation is implemented is available under an open source agreement at http://www.wadsworth.org/spider_doc/spider/docs/spider.html.

References

Baldwin, P.R., Penczek, P.A., 2005. Estimating alignment errors in sets of 2-D images. J. Struct. Biol. 150, 211–225.

Barrett, H.H., 1990. Objective assessment of image quality: effects of quantum noise and object variability. J. Opt. Soc. Am. A 7, 1266–1278.

Barrett, H.H., Wilson, D.W., Tsui, B.M.W., 1994. Noise properties of the EM algorithm: I. theory. Phys. Med. Biol. 39, 833–846.

Böttcher, B., Wynne, S.A., Crowther, R.A., 1997. Determination of the fold of the core protein of hepatitis B virus by electron cryomicroscopy. Nature 386, 88–91.

Buvat, I., 2002. A non-parametric bootstrap approach for analysing the statistical properties of SPECT and PET images. Phys. Med. Biol. 47, 1761–1775.

Carson, R.E., Yan, Y., Daube-Witherspoon, M.E., Freedman, N., Bacharach, S.L., Herscovitch, P., 1993. An approximation formula for the variance of PET region-of-interest values. IEEE Trans. Med. Imaging 12, 240–250.

Conway, J.F., Cheng, N., Zlotnick, A., Wingfield, P.T., Stahl, S.J., Steven, A.C., 1997. Visualization of a 4-helix bundle in the hepatitis B virus capsid by cryo-electron microscopy. Nature 386, 91–94.

Dahlbom, M., 2002. Estimation of image noise in PET using the bootstrap method. IEEE Trans. Nucl. Sci. 49, 2062–2066.

Davison, A.C., Hinkley, D.V., 1997. Bootstrap Methods and Their Application. Cambridge University Press, New York.

Fessler, J.A., 1996. Mean and variance of implicitly defined biased estimators (such as penalized maximum likelihood): applications to tomography. IEEE Trans. Image Process. 5, 493–506.

Fessler, J.A., Rogers, W.L., 1996. Spatial resolution properties of penalized-likelihood image reconstruction: space-invariant tomographs. IEEE Trans. Image Process. 5, 1346–1358.

Frank, J., 2006. Three-Dimensional Electron Microscopy of Macromolecular Assemblies. Oxford University Press, New York.

Frank, J., Agrawal, R.K., 2000. A ratchet-like inter-subunit reorganization of the ribosome during translocation. Nature 406, 318–322.

Frank, J., Penczek, P., Agrawal, R.K., Grassucci, R.A., Heagle, A.B., 2000. Three-dimensional cryoelectron microscopy of ribosomes. Methods Enzymol. 317, 276–291.

Frank, J., Radermacher, M., Penczek, P., Zhu, J., Li, Y., Ladjadj, M., Leith, A., 1996. SPIDER and WEB: processing and visualization of images in 3D electron microscopy and related fields. J. Struct. Biol. 116, 190–199.

Frank, J., Verschoor, A., Boublik, M., 1981. Computer averaging of electron micrographs of 40S ribosomal subunits. Science 214, 1353–1355.

Gabashvili, I.S., Agrawal, R.K., Spahn, C.M., Grassucci, R.A., Svergun, D.I., Frank, J., Penczek, P., 2000. Solution structure of the *E. coli* 70S ribosome at 11.5 Å resolution. Cell 100, 537–549.

Gao, H., Sengupta, J., Valle, M., Korostelev, A., Eswar, N., Stagg, S.M., Van Roey, P., Agrawal, R.K., Harvey, S.C., Sali, A., Chapman, M.S., Frank, J., 2003. Study of the structural dynamics of the *E. coli* 70S ribosome using real-space refinement. Cell 113, 789–801.

Gao, H.X., Valle, M., Ehrenberg, M., Frank, J., 2004. Dynamics of EF-G interaction with the ribosome explored by classification of a heterogeneous cryo-EM dataset. J. Struct. Biol. 147, 283–290.

Grigorieff, N., 1998. Three-dimensional structure of bovine NADH: ubiquinone oxidoreductase (complex I) at 22 Å in ice. J. Mol. Biol. 277, 1033–1046.

Haley, D.A., Horwitz, J., Stewart, P.L., 1998. The small heat-shock protein, αB-crystallin, has a variable quaternary structure. J. Mol. Biol. 277, 27–35.

Hansen, M.H., Hurwitz, W.N., Madow, W.G., 1953. Sample Survey Methods and Theory. Wiley, New York.

Harauz, G., van Heel, M., 1986. Exact filters for general geometry three dimensional reconstruction. Optik 73, 146–156.

Haynor, D.R., Woods, S.D., 1989. Resampling estimates of precision in emission tomography. IEEE Trans. Med. Imaging 8, 337–343.

Langmore, J.P., Smith, M.F., 1992. Quantitative energy-filtered electron microscopy of biological molecules in ice. Ultramicroscopy 46, 349–373.

Lanzavecchia, S., Bellon, P.L., Radermacher, M., 1999. Fast and accurate three-dimensional reconstruction from projections with random orientations via Radon transforms. J. Struct. Biol. 128, 152–164.

Lanzavecchia, S., Cantele, F., Radermacher, M., Bellon, P.L., 2002. Symmetry embedding in the reconstruction of macromolecular assemblies via the discrete Radon transform. J. Struct. Biol. 137, 259–272.

Lata, R., Conway, J.F., Cheng, N., Duda, R.L., Hendrix, R.W., Wikoff, W.R., Johnson, J.E., Tsuruta, H., Steven, A.C., 2000. Maturation dynamics of a viral capsid: visualization of transitional intermediate states. Cell 100, 253–263.

Leahy, R.M., Qi, J., 2000. Statistical approaches in quantitative positron emission tomography. Stat. Comput. 10, 147–165.

Liu, W., Frank, J., 1995. Estimation of variance distribution in three-dimensional reconstruction. I. Theory. J. Opt. Soc. Am. A–Opt. Image Sci. 12, 2615–2627.

Ludtke, S.J., Baldwin, P.R., Chiu, W., 1999. EMAN: semiautomated software for high-resolution single-particle reconstructions. J. Struct. Biol. 128, 82–97.

Ludtke, S.J., Jakana, J., Song, J.L., Chuang, D.T., Chiu, W., 2001. A 11.5 Å single particle reconstruction of GroEL using EMAN. J. Mol. Biol. 314, 253–262.

Maitra, R., 1997. Estimating precision in functional images. J. Comput. Graph. Stat. 6, 132–142.

Maitra, R., 1998. An approximate bootstrap technique for variance estimation in parametric images. Med. Image Anal. 2, 379–382.

Maitra, R., O'Sullivan, F., 1998. Variability assessment in positron emission tomography and related generalized deconvolution models. J. Am. Stat. Assoc. 93, 1340–1355.

Marabini, R., Herman, G.T., Carazo, J.M., 1998. 3D reconstruction in electron microscopy using ART with smooth spherically symmetric volume elements (blobs). Ultramicroscopy 72, 53–65.

Palmer, M.R., Bergstrom, M., Beddoes, N.P., Pate, B.D., 1985. Effects of detector wobble motion on image noise in positron emission tomography. IEEE Trans. Med. Imaging MI-4, 58–62.

Penczek, P., 1998. Measures of resolution using Fourier Shell Correlation. J. Mol. Biol. 280, 115–116.

Penczek, P., Radermacher, M., Frank, J., 1992. Three-dimensional reconstruction of single particles embedded in ice. Ultramicroscopy 40, 33–53.

Penczek, P.A., 2002a. Three-dimensional Spectral Signal-to-Noise Ratio for a class of reconstruction algorithms. J. Struct. Biol. 138, 34–46.

Penczek, P.A., 2002b. Variance in three-dimensional reconstructions from projections. Proceedings of the IEEE International Symposium on Biomedical Imaging, Washington, DC.

Penczek, P.A., Grassucci, R.A., Frank, J., 1994. The ribosome at improved resolution: new techniques for merging and orientation refinement in 3D cryo-electron microscopy of biological particles. Ultramicroscopy 53, 251–270.

Penczek, P.A., Renka, R., Schomberg, H., 2004. Gridding-based direct Fourier inversion of the three-dimensional ray transform. J. Opt. Soc. Am. A 21, 499–509.

Penczek, P.A., Zhu, J., Frank, J., 1996. A common-lines based method for determining orientations for N>3 particle projections simultaneously. Ultramicroscopy 63, 205–218.

Penczek, P.A., Zhu, J., Schröder, R., Frank, J., 1997. Three-dimensional reconstruction with contrast transfer function compensation from defocus series. Scan. Microsc. Suppl. 11, 1–10.

Qi, J., Leahy, R.M., 1999. A theoretical study of the contrast recovery and variance of MAP reconstructions from PET data. IEEE Trans. Med. Imaging 18, 293–305.

Qi, J., Leahy, R.M., 2000. Resolution and noise properties of MAP reconstructions in fully 3-D PET. IEEE Trans. Med. Imaging 19, 493–506.

Radermacher, M., 1992. Weighted back-projection methods. In: Frank, J. (Ed.), Electron Tomography. Plenum, New York, pp. 91–115.

Saxton, W.O., Baumeister, W., 1982. The correlation averaging of a regularly arranged bacterial envelope protein. J. Microsc. 127, 127–138.

Shao, J., Tu, D., 1995. The Jackknife and Bootstrap. Springer Verlag, New York.

Spahn, C.M.T., Beckmann, R., Eswar, N., Penczek, P.A., Sali, A., Blobel, G., Frank, J., 2001. Structure of the 80S ribosome from *Saccharomyces cerevisiae*-tRNA–ribosome and subunit–subunit interactions. Cell 107, 373–386.

Unser, M., Trus, B.L., Steven, A.C., 1987. A new resolution criterion based on spectral signal-to-noise ratios. Ultramicroscopy 23, 39–51.

Ushakov, N.G., Ushakova, A.P., 1998. 3D reconstruction from projections for stochastic objects (stochastic tomography). Electronics Letters 34, 512–514.

Valle, M., Zavialov, A., Sengupta, J., Rawat, U., Ehrenberg, M., Frank, J., 2003. Locking and unlocking of ribosomal motions. Cell 114, 123–134.

van Heel, M., Gowen, B., Matadeen, R., Orlova, E.V., Finn, R., Pape, T., Cohen, D., Stark, H., Schmidt, R., Schatz, M., Patwardhan, A., 2000. Single-particle electron cryo-microscopy: towards atomic resolution. Q. Rev. Biophys. 33, 307–369.

Wade, R.H., 1992. A brief look at imaging and contrast transfer. Ultramicroscopy 46, 145–156.

Wilson, D.W., Tsui, B.M.W., 1993. Noise properties of filtered-backprojection and ML-EM reconstructed emission tomographic images. IEEE Trans. Nucl. Sci. 40, 1198–1203.

Zhou, Z.H., Dougherty, M., Jakana, J., He, J., Rixon, F.J., Chiu, W., 2000. Seeing the herpesvirus capsid at 8.5 Å. Science 288, 877–880.

Zhou, Z.H., Liao, W.C., Cheng, R.H., Lawson, J.E., McCarthy, D.B., Reed, L.J., Stoops, J.K., 2001. Direct evidence for the size and conformational variability of the pyruvate dehydrogenase complex revealed by three-dimensional electron microscopy—The "breathing" core and its functional relationship to protein dynamics. J. Biol. Chem. 276, 21704–21713.

The process of mRNA–tRNA translocation

Joachim Frank*†‡§, Haixiao Gao†, Jayati Sengupta†, Ning Gao*, and Derek J. Taylor*

*Howard Hughes Medical Institute and †Health Research Incorporated, Wadsworth Center, Empire State Plaza, Albany, NY 12201-0509; and ‡Department of Biomedical Sciences, State University of New York, Albany, NY 12222.

This contribution is part of the special series of Inaugural Articles by members of the National Academy of Sciences elected on April 25, 2006.

Contributed by Joachim Frank, September 13, 2007 (sent for review August 6, 2007)

In the elongation cycle of translation, translocation is the process that advances the mRNA–tRNA moiety on the ribosome, to allow the next codon to move into the decoding center. New results obtained by cryoelectron microscopy, interpreted in the light of x-ray structures and kinetic data, allow us to develop a model of the molecular events during translocation.

protein synthesis | ribosome | translation | EF-G | cryo-EM

Synthesis of proteins from their building blocks, the amino acids, is a fundamental process in the cells of all living organisms, be it animal, plant, or bacteria. The discovery that the macromolecular assembly that facilitates this process, the ribosome, is highly conserved in all essential parts has lent additional credence to the idea of the unity of all life at the molecular level.

The ribosome is a very large (2.4 MDa in eubacteria) ribonucleic-protein complex composed of two distinct subunits, the small subunit (30S) charged with the task of decoding the genetic message carried by the messenger RNA (mRNA), the large subunit (50S) to the catalysis of peptide bond formation. Instrumental for these fundamental processes is the interaction of the ribosome with transfer RNA (tRNA), a small L-shaped molecule that embodies in its various forms the association of each amino acid with a three-base "word" of the genetic code, the codon. Translation is based on the mutual recognition, by partial Watson–Crick pairing, between the codon on the mRNA and the anticodon of the tRNA carrying the corresponding amino acid. In facilitating tRNA selection, decoding, and the stepwise formation of the polypeptide, ribosomal RNA (rRNA) acts as both a structural framework and a catalyst.

Despite the success in the elucidation of ribosomal structure by x-ray crystallography, the detailed mechanism by which translation of mRNA code into peptide proceeds is still only scantly understood. One of the obstacles we face is that although the process is complex and dynamic, x-ray crystallography represents the molecule in a static form—packed in a crystal, moreover, whose very stability depends on intermolecular contacts that are largely nonphysiological. Of crucial importance for the understanding of the multistep translation process is the knowledge of how the ribosome interacts with its ligands, notably (apart from the most crucial ligands mRNA and tRNAs) the various factors catalyzing initiation, elongation, termination, and recycling. To date, with the exception of ribosomal complexes containing eubacterial release (RF1 or RF2) (1) or recycling (RRF) (2, 3) factors, there exists no x-ray structure of a factor–ribosome complex. The crystal structures of individual subunits complexed with initiation (4–6) and recycling (7) factors have also been solved; however, crystallographic data of elongation factors bound to the ribosome are currently still not available. Moreover, to date, despite many efforts, no atomic structure is available for a eukaryotic ribosome.

Increasingly, within the past decade, cryo-electron microscopy (cryo-EM) has filled this gap; in fact, the very first three-dimensional images of the ribosome (8, 9) were obtained by this technique well before the first x-ray structure was solved. In the mean time, cryo-EM has furnished quite detailed information on the interaction of the *Escherichia coli* ribosome with initiation factors (10), elongation factors (11–14), release factors (15–18), and ribosome recycling factor (19, 20). Some of the corresponding complexes have been visualized for the 80S eukaryotic ribosome (21–23).

As the initially low-resolution maps gave way to density maps in the subnanometer range, evidence of conformational changes both in the ribosome and its ligands was uncovered, and qualitative descriptions were increasingly replaced by quantitative measurements based on fitting and docking of x-ray structures. As a result, we can begin to piece together mechanistic models that explain kinetic, genetic, and other data collected over the five decades of ribosome research in structural terms.

The Elongation Cycle, and Translocation

In the course of protein synthesis, tRNA occupies successively the universally conserved A (aminoacyl), P (peptidyl), and E (exit) sites of the mRNA-programmed ribosome. The elongation cycle of translation is a repeating, three-step process catalyzed by the ribosome and two GTPases, EF-Tu and EF-G (eEF1A and eEF2 in eukaryotes). First, in decoding, EF-Tu delivers an aminoacylated-tRNA (aa-tRNA) molecule as part of the ternary complex of aa-tRNA·EF-Tu·GTP to the A/T site. The orientation with which tRNA enters the ribosome is not favorable for codon–anticodon interaction, and the anticodon stem-loop of the tRNA needs to be strongly kinked at the A/T site to allow the anticodon to interact with mRNA (24). If the aa-tRNA is cognate—i.e., when its three-base anticodon forms complementary base pairs with the codon in the messenger RNA—GTP is hydrolyzed, and the aa-tRNA, upon dissociation of EF-Tu·GDP, is accommodated in the ribosomal A site. The second step, the peptidyl-transferase reaction, is catalyzed by the rRNA of the large subunit and occurs immediately following the accommodation of the aa-tRNA. The transfer of the nascent peptide chain, from peptidyl-tRNA in the ribosomal P site to the accommodated aa-tRNA in the A site, results in the "pretranslocational" (PRE) state of the translating ribosome, with the deacylated tRNA in the P site and a peptidyl-tRNA in the A site.

The final step of the elongation cycle is translocation, which brings the ribosome into the "posttranslocational" (POST) state, with an empty A site, peptidyl-tRNA in the P site, and deacylated tRNA in the E site. Translocation is catalyzed by the binding of EF-G and subsequent GTP hydrolysis. The movement of the tRNA molecules is concomitant with the movement of the bound mRNA chain by the length of three bases, allowing the next codon of mRNA in the ribosomal A site to be presented for decoding.

Each step of the elongation cycle of translation, namely decoding, peptidyl-transfer, and translocation, presents an enigma on its own. Whereas the mechanism of peptidyl-transfer has received particular

---

Author contributions: D.J.T. designed the animation depicting the proposed mechanism of translocation; H.G. performed real-space refinement fitting for the two conformations of the ribosome; and J.F., H.G., J.S., N.G., and D.J.T. wrote the paper.

The authors declare no conflict of interest.

See accompanying Profile on page 19668.

§To whom correspondence should be addressed. E-mail: joachim@wadsworth.org.

This article contains supporting information online at www.pnas.org/cgi/content/full/0708517104/DC1.

© 2007 by The National Academy of Sciences of the USA

attention (see ref. 25), and at least part of the tRNA selection step of decoding has been unraveled in molecular detail (6), the molecular processes of EF-G-mediated translocation have been least understood until recently. Key findings include the discovery of the tRNA hybrid states (26) and the results of kinetic experiments that GTP hydrolysis precedes translocation (27).

Important cryo-EM findings relevant to translocation are the visualization of EF-G binding to the ribosome (11, 12, 14); the visualization of tRNA in the P/E hybrid state (13, 18, 23, 28); the discovery of the ratchet motion, a rotation of the small vs. large subunit upon binding of EF-G (29) or eEF2 (22) to the ribosome; and, most recently, the elucidation of the role of GTP hydrolysis in eEF2-induced translocation of the eukaryotic ribosome (23).

Hybrid-State tRNAs and Intersubunit "Ratchet" Motion

It was originally proposed in the 1960s that translocation of tRNAs might occur independently on the two ribosomal subunits during elongation (30, 31). Later, biochemical data such as the results from direct chemical footprinting led to the hybrid-state tRNA model, which describes an intermediate step in translocation (26). Here, the acceptor ends of the A- and P-site tRNAs move into the P and E sites of the large subunit, while the anticodon stem-loops of the tRNAs remain in the A and P sites of the small subunit. This configuration results in occupation by the tRNAs of the A/P and P/E hybrid states. Cryo-EM reconstructions indeed revealed structural evidence for a position of tRNA in the P/E hybrid state (11, 13, 18, 23); however, the A/P hybrid state has thus far eluded structural analysis using cryo-EM or x-ray crystallography.

The initial report of hybrid-state tRNAs by Moazed and Noller (26) suggested that tRNAs occupy the hybrid state immediately after peptide bond formation, but structures representing the PRE ribosome determined by cryo-EM (13) and x-ray crystallography (32–35) showed that in the absence of elongation factors, the tRNAs occupy the classic A/A, P/P, and E/E sites exclusively. Potential explanations for this discrepancy are that the tRNAs are sampling conformations between classic and hybrid states in the PRE ribosome (36), or that the buffer used in the various studies stabilizes one state of binding over the other (i.e., classic over hybrid) (28).

The P/E hybrid state tRNA in the cryo-EM reconstructions was visualized only when EF-G was bound to the ribosome. In addition to bringing the tRNA into the P/E hybrid state, the binding of the factor also induces a new conformation of the ribosome, where the small subunit is in a different orientation with respect to the large subunit, related to the normal conformation by a counterclockwise rotation ("ratchet motion") (29, 37) [supporting information (SI) Fig. 6 and SI Movie 1]. Evidence for the existence of an intersubunit motion first came from small-angle neutron scatting experiments, but the precise form of this motion could not be deduced (38). It is likely that the ratchet motion of the ribosome even occurs in the absence of ribosomal factors and that it is coincident with the movement of tRNAs from the classic to the hybrid states, which facilitate translocation (39, 40). Hence, the view is emerging that, as suggested by the cryo-EM data, the binding of EF-G/eEF2 in the GTP form stabilizes the ratcheted conformation, and with it, the hybrid-state tRNAs (40, 41). The ratcheted ribosome with hybrid-state tRNAs, therefore, represents a distinct structural intermediate between the PRE and POST ribosomes. In solution, it is likely that the PRE ribosome oscillates between the ratcheted and nonratcheted states, with EF-G catalyzing translocation to the POST ribosome. The POST ribosome, with peptidyl-tRNA in the P-site, is locked in the nonratcheted state (13).

A convincing proof that the ratchet motion is instrumental for translocation is contained in a series of three recent studies. In one, the motion was inhibited by the placement of a crosslink between proteins of the two subunits facing each other in the normal conformation of the ribosome, and complete suppression of translocation was observed (42). Furthermore, two FRET studies, one bulk (40) and the other single-molecule,[¶] have provided conclusive evidence that the motion inferred from the comparison of cryo-EM maps indeed occurs in a translating ribosome.

It is noteworthy, in this context, that the kinetic parameters of tRNA movement have been investigated in detail with the use of fluorescent labels (43, 44). In addition to supporting the existence of the A/P and P/E hybrid states, these data identify distinct hybrid-state intermediates. The movements of the two tRNAs bound to the A and P sites of the small subunit occur separately, such that A/A tRNA and P/E-hybrid tRNA may exist simultaneously in an intermediate ribosomal state. In the absence of elongation factors, single-molecule FRET measurements suggest that the tRNAs transition between the classic, A/P-P/E hybrid, and A/A-P/E intermediate states but spend the majority of time ($\approx$60%) in the classic states (43).

Factor-Free Translocation

Some of the observations discussed above clearly show that the transition between two dramatically different conformations of the ribosome and, along with it, the transition between canonical and hybrid states of the tRNA occur spontaneously, in the absence of EF-G/eEF2. It is compelling to link the structural and dynamic evidence to the long-known observations of factor-free translocation.

Reports in the very early years of ribosome studies (45–47), when very little was known about its biochemistry and structure, indicated that the ribosome is capable of performing the synthesis of oligophenylalanine from a poly(U) template even in the absence of elongation factors and GTP. Subsequent work by Spirin's group (48–50) further indicated that the treatment of the ribosome with pCMB, or the removal of ribosome protein S12 (51), stimulates the rate of factor-free translocation significantly. Recent work from Rachel Green's laboratory (52) identified three target cysteine residues, in protein S12 (C104 and C27) and S13 (C87), whose modification by pCMB results in an elevated factor-free translocation rate, albeit with decreased accuracy (see Fig. 1).

Interestingly, ribosomal protein S12 is the very protein identified in cryo-EM maps as the pivot of the ratchet-like motion (13) and was shown to interact with EF-G both by cross-linking studies (53, 54) and cryo-EM (13). A number of subsequent cryo-EM studies found that other translation factors, including IF2 (10), RF3 (18, 55), and RRF (20), all shown to interact with S12, induce the same type of intersubunit movement as EF-G (Fig. 2 and SI Fig. 7). In fact, the above-mentioned C104 residue of S12 is located exactly at the interface between the 30S subunit and these factors. These cryo-EM observations thus provide a link between factor-free and factor-promoted translocation, where S12 plays an apparent regulatory role in controlling the intersubunit motion of the ribosome (see below for a similar role of S13).

One apparent benefit of introducing protein factors into the elongation cycle, in the course of evolution, is that it helps lower the energetic barriers of the various transition states, because factor-free translation occurs at a much lower rate (45, 48). Translocation itself has been shown to be an exergonic reaction (56, 57), but with a high free energy of activation [$\Delta G^{\neq} \approx 96$ kJ/mol for factor-free translocation (58, 59)]. Kinetic studies indicated that the free energy of activation is lowered by $\approx$18 kJ/mol in the presence of EF-G, and by another 10 kJ/mol with the hydrolysis of GTP, which correspond to an $\approx$1,000-fold increase in translocation rate and another 50-fold increase, respectively (27, 60).

[¶]Cornish, P. V., Ermolenko, D., Noller, H. F., Ha, T., Ribosomes Meeting, June 3–8, 2007, Cape Cod, MA, p. 69 (abstr.).

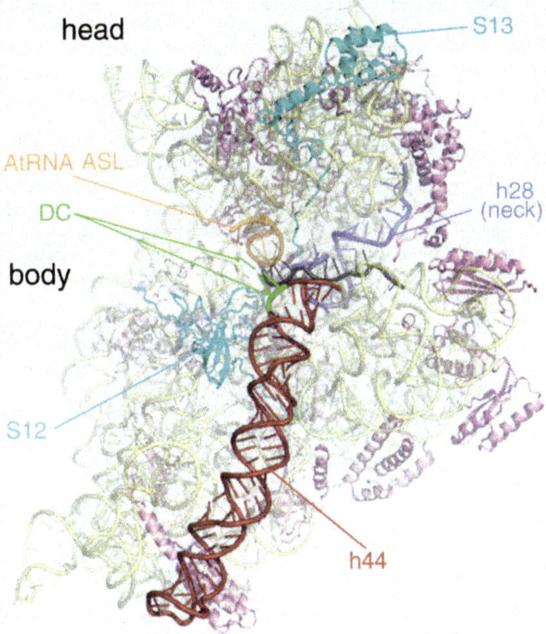

Fig. 1. Atomic model of the 30S subunit of the pretranslocational ribosome from *Thermus thermophilus* (ref. 34; Protein Data Bank ID code 2J00). This intersubunit view of the 30S highlights several important features, which are color-coded and labeled. Ribosomal proteins S12 and S13 interact directly with the 50S subunit and affect ratcheting efficiency of the ribosome. The mRNA channel passes between the head and the body of the SSU (mRNA is colored gray). Upon EF-G binding, the top of helix 44 (h44) bends from the A site toward the P site. Because helix 44 is directly connected to helix 28 (h28 or the "neck"), the movement in h44 could provide torsional force on h28. GTP hydrolysis by EF-G decouples the tether between the A-site tRNA bound to the head and the decoding center (DC) in the body of the SSU. This event would release the head to rotate about its neck and relieve the torsional force in h28. The head rotation is likely to be important for movement of the mRNA and tRNA anticodon stem-loops on the SSU.

Architecture of the Ribosome: A Molecule Made to Rock

X-ray studies show that the ribosome possesses a unique architecture with built-in instabilities, among them architectural features conducive to the ratchet motion. As we have seen, these properties are apparently required for a whole range of translational steps, not just for translocation. Among the architectural features giving rise to instability are the following.

1. The ribosome is composed of two relatively loosely coupled subunits, a fact that prompted Spirin (31) quite early on to postulate the existence of intersubunit motion as a necessary part of the translational process.
2. The coupling is such that rotational movement around a central core of connections accompanied by torsional stress of these connections is prevalent among the modes shown by normal mode analysis (61, 62). Specifically, the central bridge B2a is formed by RNA–RNA minor groove interactions (32, 63, 64). According to Valle and coworkers (13), the rotational pivot axis passes through a point between bridges B3 and B5a on h44, just above h27 of 16S rRNA (see also ref. 61).
3. The peripheral bridges (i.e., bridges located far away from the rotational pivot axis) are quite flexible and involve at least one protein [e.g., B4, involving S15 and the 715 stem loop of 23S rRNA (65); for definition of bridge nomenclature, see refs. 9 and 19]. The bridge having to absorb the largest movement, B1b, is formed by two proteins, S13 and L5, which disengage and reengage in different constellations (13). It is interesting to look at this bridge in the two constellations, using the quasi-atomic models obtained by real-space refinement (Fig. 3): in the normal configuration of the ribosome (macrostate I), the electrostatic charges of regions of S13 and L5 facing each other are of opposite polarity, lending stability to the ribosome. In the ratcheted configuration (macrostate II), in contrast, the electrostatic charges of regions facing each other are of equal polarity, conveying instability to the proteins and a propensity for gliding along their interface. By extending the motion further, we would reach a constellation where opposite polarities would again face each other, creating another stabilizing interaction. The distance traveled between the "normal" position (macrostate I) and the postulated second antipolar juxtaposition would be 30 Å, which might act as a yardstick defining the maximum amplitude of "allowed" ratchet motion.
4. There are a large number of tertiary contacts, identified by Noller (66), that are slightly outside of the range of canonical A-minor interactions (67) and hence would allow the gliding of adjacent strands of rRNA without locking them into a defined position.
5. The rRNA possesses a number of kink-turns, a new element discovered in the x-ray structure of the 50S subunit of *Haloarcula marismortui* (68). Kink-turns are elements with built-in instability, allowing extensive hinge motions of some ribosomal components, among these the L1 stalk, the A-site finger (69), and the GTPase-associated center (GAC). These elements and their biological significance are detailed in the following.

The L1 stalk, which is implicated in the transport of the tRNA, pivots around a hinge region in h76 of 23S rRNA and has been observed in different positions both by cryo-EM (13, 21, 28) and x-ray crystallography (64, 70). The position of this stalk in an "inward" position, found in macrostate II, likely blocks the exit path of tRNA. Conversely, an "outward" position, found in macrostate I, leaves the exit path open. Movement from the "inward" to the "outward" positions of the L1 stalk likely coordinates translocation of tRNAs by allowing ejection of E-site tRNA from the ribosome (see ref. 13).

The A-site finger is a long rRNA helix (h38) reaching from the 23S rRNA into the intersubunit region where it forms a contact (termed B1a) with S13 of the small subunit. In *Thermus thermophilus*, h38 possesses a kink-turn (Kt-38). In *E. coli*, it has a similar region of instability. Because the tip of the A-site finger must follow the ratchet motion of the small subunit at a peripheral location, its built-in instability may serve to reduce the activation energy for the ratcheting. Shortening of the helix, presumably leading to a dis-

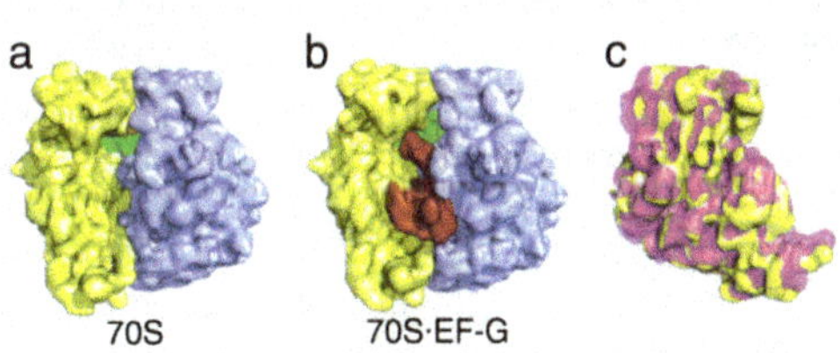

Fig. 2. Binding of EF-G stabilizes the ratcheted conformation of the ribosome. (*a*) Cryo-EM reconstruction of the pretranslocational ribosome. The large subunit is colored blue, the small subunit is colored yellow, and the P-site tRNA is colored green. (*b*) Cryo-EM reconstruction of the EF-G·70S complex, showing EF-G in red. The large subunit of the two reconstructions (*a* and *b*) was aligned and the small subunit of each was superimposed (*c*). This view, from the solvent side of the small subunit, shows that binding of EF-G induces a counterclockwise rotation of the small subunit (pink) when compared with the small subunit of the pretranslocational ribosome (yellow).

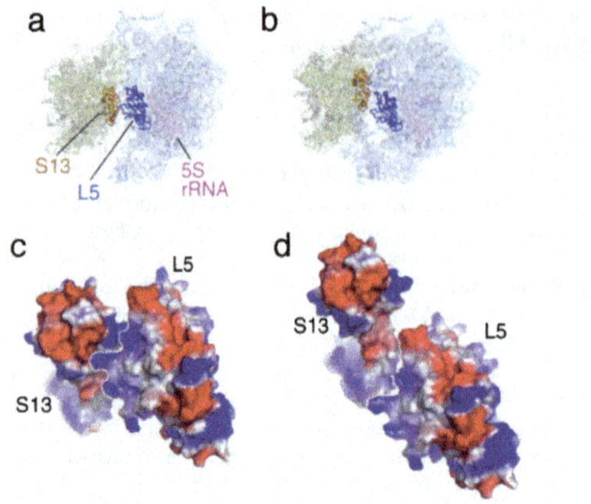

Fig. 3. The bridge B1b, formed by S13 and L5, in the two ratchet-related conformations of the ribosome. (*a* and *b*) Positions of proteins S13 and L5 in the two states. (*c* and *d*) Surface charge representation of S13-L5 in the two states. In *c* (normal conformation), the charges at the interface are of opposite polarity, leading to stabilization. In *d* (ratcheted conformation), the charges are of equal polarity, leading to instability.

ruption of B1a, results in an acceleration of translocation activity, pointing to a role of the A-site finger as a control element maintaining the reading frame (71).

Finally, the GAC, formed by 23S rRNA h43 and h44, is flexibly linked to the main body of the large subunit by h42, which contains kink-turn 42. According to cryo-EM studies, the mobility of the GAC is essential for factor binding, forming part of an "induced fit" mechanism during the binding of EF-Tu (24, 72), EF-G (11), RF3 (18), and IF2 (10).

Hence, in conclusion, the ribosome's architecture is intrinsically unstable, conducive to making the ribosome alternate between two conformations. In the context of translocation, the first of these conformations (macrostate I) is required for initial factor binding, to initiate the first step of the translocation process, whereas the second (macrostate II) is required for staging the hybrid states, triggering GTP hydrolysis, and disengaging the firm grip that holds the mRNA–tRNA moiety in the decoding center, thereby initiating the second step of the translocation process.

Observed Conformations of EF-G

EF-G-catalyzed translocation requires EF-G to assume at least four conformations: (*i*) the GTP-bound, ribosome-free form; (*ii*) the GTP-bound conformation of EF-G on the ribosome immediately before GTP hydrolysis (as binding to the ribosome induces the GTPase of EF-G); (*iii*) the GDP-bound conformation, in which EF-G remains bound to the ribosome; and (*iv*) the GDP-bound conformation after dissociation of the factor from the ribosome. All four conformations have been observed structurally—by cryo-EM for the ribosome-bound conformations and by x-ray crystallography for the unbound conformations (SI Fig. 8). Such conformational alterations, specifically a hinge-like motion between the N-terminal domains (I, II, and G′) and C-terminal domains (III–V) of EF-G, undoubtedly play a role in the physiological function of the factor.

In the 1990s, the crystal structures of GDP-bound (73) and nucleotide-free (74) EF-G were solved. Both were shown to have similar conformations, with disordered regions in domain 3 and the switch 1 loop. The ribosome-free structure of EF-G·GTP has been solved in two conformations. First, the structure of a mutant EF-G·GDPNP revealed only slight differences when compared with the GDP-bound form (75). Noteworthy differences between these structures are seen in the P-loop, the visible regions of the switch 1 loop, the switch 2 loop, and a slight shift at the tip of domain IV, where a conserved histidine resides. Whereas the γ-phosphate of GDPNP is ordered in the structure, the switch 1 loop is not. This could be due to an artifact from crystal packing or because the crystals were seeded from mutant EF-G·GDP crystals. Recently, the x-ray structure of an EF-G homolog, EF-G-2, complexed with GTP was reported (76). This structure possesses a fully ordered switch·1 loop and a conformation that is distinct from the GDP-bound EF-G structure; specifically, a reorganization of domains III–V with respect to domains I, II, and G′. This rearrangement results in a 25-Å shift at the tip of domain IV when compared with the GDP-bound EF-G, a conformation of EF-G undoubtedly favorable for binding to the ribosome as evidenced by the increased affinity for EF-G·GTP over its GDP-bound counterpart (see ref. 77 and references therein).

Attempts to find a system in the bewildering variety of observed structures for EF-G (see, for instance, ref. 78) need to account for the fact that the factor is in different environments, facing different conformational constraints when placed in a crystal context versus bound to the ribosome. Conformational differences relevant for the functioning of the factor while performing its work on the ribosome are best inferred from structures of EF-G bound to the ribosome in an authentic state. Because of the lack of x-ray structures for such a complex, the existing quasi-atomic models of EF-G computed from cryo-EM maps are currently the only source of information. A comparison of the cryo-EM reconstructions of EF-G (11–14) bound to the ribosome with the crystal structures of the factors (assuming that these are close to the solution forms) revealed that EF-G undergoes gross conformational changes when it binds to the ribosome. These changes are again described as a hinge-like joint motion of domains III, IV, and V with respect to domains I, II, and G′ (11, 12) and are coincident with the ratchet motion (29). The hinge-like conformational change in EF-G induced by ribosome binding is in the same direction as that of the GDP- to GTP-bound conformations but with even greater magnitude (≈27 Å) (SI Fig. 8).

The conformations of ribosome-bound EF-G before and after GTP hydrolysis are quite likely similar to those recently elucidated for eEF2 by cryo-EM reconstructions of 80S–eEF2 complexes (23). The cryo-EM maps of modified eEF2 bound to the 80S ribosome in the GTP-state (using GDPNP) and GDP-state (GDP plus sordarin) revealed additional, smaller conformational changes in eEF2 upon GTP hydrolysis, which include a 6-Å shift of domain IV of eEF2 toward the decoding center (23). This conformational change is likely responsible for decoupling the decoding center from the mRNA–tRNA duplex in the A site, a prerequisite for translocation.

The GTP Hydrolysis Mechanism in EF-G-Catalyzed Translocation

GTPases, including EF-Tu and EF-G and their eukaryotic counterparts, are a family of proteins characterized by a GTP-binding domain, which is responsible for the binding and hydrolysis of GTP. GTPases function as molecular switches in several processes of cellular regulation by cycling between an active, GTP-bound state and an inactive, GDP-bound state (see ref. 79). The core, or G domain, is composed of three conserved features in the nucleotide-binding pocket: the phosphate-binding loop (P-loop) and the switch 1 and switch 2 motifs. The P-loop binds the nucleotide via the α- and β-phosphates, while the switch 1 and 2 loops coordinate the γ-phosphate of GTP.

Based on the similarities in structure, particularly of the switch 1 and 2 loops and their conserved ribosomal binding site near the GAC, it is likely that the mechanism of GTP hydrolysis is conserved among the elongation factors. In fact, this mechanism may be extended to all ribosomal binding GTPases involved in initiation, elongation, and termination because of the high conservation in factor structures and the universal GAC binding site on the ribosome. However, despite the availability of several crystal structures of these factors bound with various nucleotides and antibiotics, the mechanism of GTP hydrolysis of the factors is not well understood. This is due mainly to a lack of atomic models of the elongation factors bound to the ribosome. Such a model is obviously important for the understanding of the mechanism because GTP hydrolysis by elongation factors is accelerated on the ribosome by more than seven orders of magnitude (80).

GTPase activation of the smaller G proteins is catalyzed by the binding of a GTPase-activating protein (GAP), which induces conformational changes in, and thus alters the activity of, the G proteins (79). Because there is no GAP *per se* for the elongation factors, the GAC of the ribosome apparently assumes this role. Recent cryo-EM reconstructions of EF-G·70S (76) and eEF2·80S (23) complexes revealed that the switch 1 loop of EF-G/eEF2 is in fact ordered in the GTP-bound state. After GTP hydrolysis, the switch 1 loop of the ribosome-bound eEF2 becomes disordered, similar to that of the EF-Tu·GDP·aurodox crystal structure (81) and the structure of 70S·EF-Tu·Phe-$tRNA^{Phe}$·kirromycin complex studied by cryo-EM, which also represents a post-GTPase state (24). This observation indicates a critical role of the switch 1 loop, as well as the existence of a similar mechanism of GTP hydrolysis in EF-Tu and EF-G/eEF2.

The sarcin–ricin loop (SRL) of the ribosome is a conserved tetraloop in the rRNA of the large subunit, also implicated in the activation of the GTPase activity. Cleavage of the SRL by the ribotoxin α-sarcin or ricin inhibits binding of elongation factors (82, 83). When free in solution, the SRL binds EF-G (84). Cryo-EM showed that the SRL is in close proximity of the switch 1 loop of GTP-bound EF-G and eEF2 (23, 76). Chemical protection assays revealed that the SRL is protected by EF-G in either the GTP-bound (GDPNP) or the GDP-bound (stabilized by fusidic acid) conformation (85). A conserved feature in the cryo-EM reconstructions of the ribosome-bound GTPases, IF2 (86) and eEF2 (23), is a shift in the GTP-binding domain of the factors toward the GAC of the ribosome in the transition from the GTP-bound states to GDP-bound states. This transition may alter the interactions between the SRL and the switch regions of the factors, leading to GTP hydrolysis and the disordering of the switch 1 loop.

Head Rotation of the Small Subunit: A Pivotal Step in Translocation

While the movement of the tRNAs on the large subunit (i.e., the formation of the hybrid-state tRNA) is facilitated by the binding of EF-G·GTP and the concomitant ratcheting of the ribosome, the second step of translocation, the movement of the tRNAs along with the mRNA on the small subunit by one codon, is more difficult to explain. Evidence has accumulated that supports the notion that GTP hydrolysis precedes translocation on the small subunit (44, 87). Toe-printing experiments of EF-G·70S complexes in the presence of either GTP or GDPNP demonstrated that mRNA movement relative to the small subunit occurs after GTP hydrolysis (88). Kinetic data suggest that GTP hydrolysis by EF-G leads to some kind of "unlocking" event that induces a conformational change of the ribosome, on such a scale that it limits the rate of translocation (89).

An explanation of this "unlocking" event was recently proposed on the basis of cryo-EM reconstructions of modified eEF2 bound to the 80S ribosome (23). Using the ADP moiety as a density marker, these authors showed that ADP-ribosylated eEF2 (ADPR-eEF2) undergoes conformational changes that are a direct result of GTP hydrolysis. These changes result in a 6-Å shift in the tip of domain IV of ADPR-eEF2, where a diphthamide residue resides in eEF2 (or a conserved histidine, in EF-G). Judged from the location of the tip, this movement likely severs the connection between the ribosomal decoding center in the body of the small subunit and the mRNA–tRNA duplex in the A site, bound to the head of the small subunit. Once this link is severed, the head is free to rotate around the neck of the small subunit, simultaneously translocating the mRNA–tRNA complex.

Such a head rotation of the subunit, which is independent of ratcheting, has been inferred from the difference in head positions observed both in cryo-EM reconstructions of 80S ribosomes (22, 23) and in x-ray structures of the 70S ribosome (64). Even earlier small-angle neutron scattering data predicted a head rotation of the small subunit to provide the mechanical act of translocation (90). Genetic mutations also suggest a necessary interaction between the head of the small subunit and the tRNAs during translocation, because mutations to a conserved adenine in the head-region of 16S rRNA (A1339 in *E. coli*) confer a >18-fold decrease in translation activity (91). Not surprisingly, A1339 interacts intimately with P-site tRNA in the PRE ribosome (32).

Analysis of the small subunit from the x-ray structures of vacant 70S ribosomes from *E. coli* revealed that residues G1338 to U1341 (*E. coli* numbering) in the head and residue A790 in the shoulder of the small subunit form a block between the P and E sites that opens or closes as a function of head movement (64). In the x-ray structure of the 70S ribosome in the PRE state (32) and the cryo-EM structure of the 80S ribosome in that same state (92), the distance between A1340 and A790 α-carbons is ≈18 Å and thus would impede movement of the tRNA stem from the P to E site. Binding of eEF2 opens this distance to >26 Å in the 80S ribosome (22, 23), which would accommodate the passing of the anticodon stem-loop of tRNA on the small subunit from the P to the E site (64). In the cryo-EM maps, however, eEF2 is bound and this block is "open," but the P/E-site tRNA remains in the hybrid state, interacting strongly with the G1338–U1341 region of the head (Fig. 4). The strong interaction seen between the P/E-site tRNA and the head of the small subunit contrasts with tRNAs in the classic states, where the main interactions are with the body of the small subunit (32–34). This result suggests that in a translating ribosome, the head movement physically drags the anticodon stem-loop of P/E site tRNA toward the E site through a connection with the G1338–U1341 region in the head of the small subunit after ratcheting and GTP hydrolysis. At this point, the P/E hybrid tRNA is likely to be committed to the E site through a strong interaction formed with the L1 stalk of the large subunit after ratcheting of the ribosome (13, 21, 22), an interaction that may assist the transition phase. Disruptions of the interaction between the head region of the small subunit and the tRNAs must be induced by a combination of back-ratcheting of the small subunit and back-rotation of the head, both of which occur upon dissociation of eEF2/EF-G·GDP. We speculate that it is this event that allows P/E tRNA to finally pass through the gate between the head (G1338–U1341) and body (A790) of the small subunit to fully occupy the classic E/E state.

Finally, it is likely that the binding of EF-G indirectly provides the driving force for head rotation in the small subunit: The binding of EF-G·GTP and ribosomal ratcheting are coupled events that are accompanied by a bending of h44 at a point 2/3 along its length, producing a shift of the decoding center by ≈8 Å toward the P site (93). This shift is required to fully accommodate domain IV of EF-G in the ribosomal A site. Directly connected with h44 is h28 in 16S rRNA, which makes up the neck of the small subunit (Fig. 1). It is plausible that EF-G-induced movement at the top of h44 will cause torsional stress in h28. Thus, the head of the small subunit would be primed to rotate as soon as the rigid link between the mRNA–tRNA duplex, still bound to the small subunit's head in the A site, and the decoding center in the body of the small subunit is removed upon GTP hydrolysis (see Fig. 1).

Synthesis of a Model for mRNA–tRNA Translocation

Recent evidence from structural, kinetic, and biochemical data, some of which has been reviewed above, point toward a two-step mechanism of translocation. The binding of EF-G·GTP (or EF-G·GDPNP) induces a first phase of translocation by stabilizing the

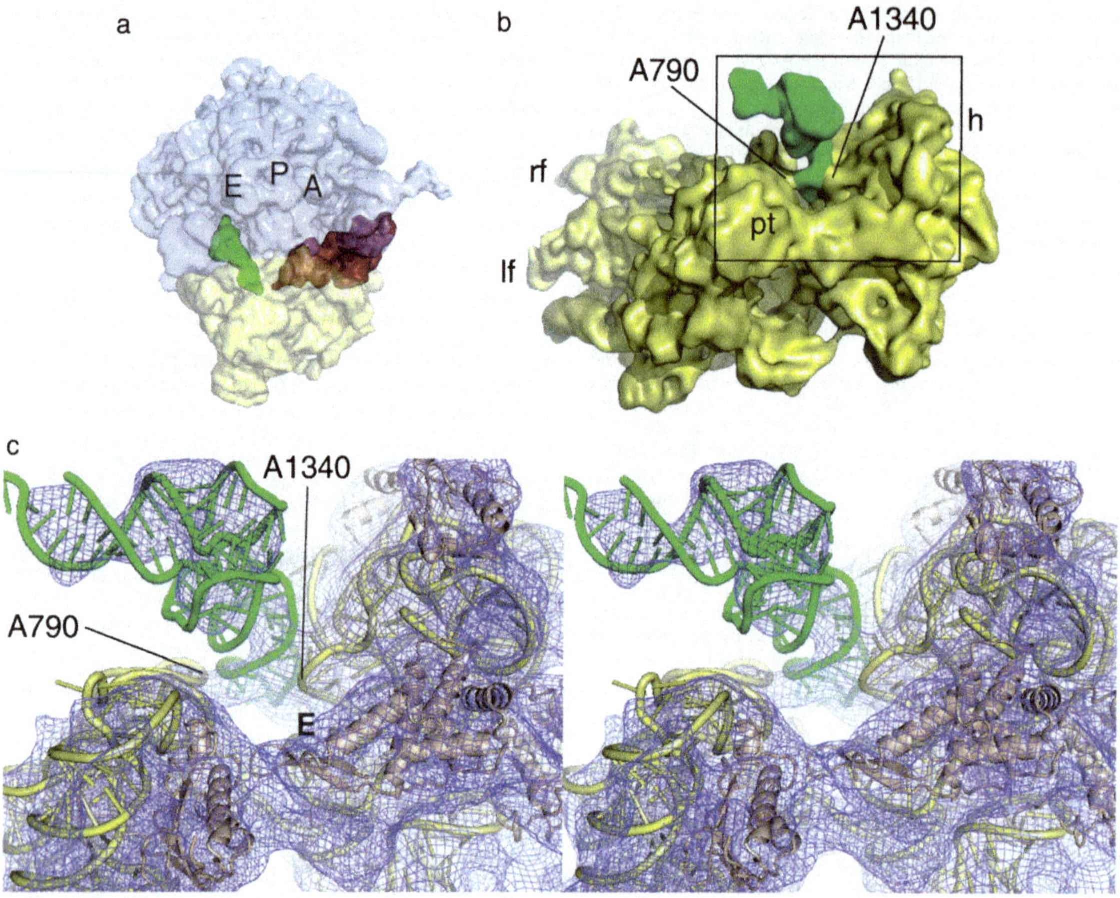

Fig. 4. The P/E-site tRNA visualized in cryo-EM reconstructions. (*a*) A side view of the 80S·ADPR-eEF2·GDPNP complex (23), showing the CCA end of the tRNA occupying the E site of the large subunit, while the anticodon stem-loop occupies the P site of the small subunit. The A, P, and E sites of the ribosome are individually labeled. (*b*) The anticodon stem-loop of the P/E-site tRNA forms a strong interaction with the small subunit head, near the G1338-U1341 ridge, and a weaker interaction with the small subunit body, via the A790 loop of the 18S rRNA. The gap between A790 and A1340 separates the ribosomal P and E sites of the small subunit. (*c*) Stereo view of the interactions between the small subunit and P/E-site tRNA described in *b*. The head of the small subunit in this complex is rotated, increasing the distance between A790 in the body and A1339 in the head of the small subunit from ≈18 Å to ≈26 Å. This distance would allow passage of the anticodon stem-loop from the P to the E site; however, the tRNA remains bound in the hybrid state. Passage of the tRNA anticodon stem-loop from the P to the E site of the small subunit therefore must occur during back-ratcheting and back-rotation of the head of the small subunit, once EF-G/eEF2 dissociates from the ribosome. E denotes the E site of the small subunit.

tRNA molecules in the hybrid state. This occurs because the deacylation of the tRNA at the P site liberates the tRNA CCA, such that it can proceed to the E site on the 50S subunit, and allows the small subunit to ratchet forward while the mRNA–tRNA remains locked with the small subunit at its P site. At the same time, the vacating of the P site on the large subunit by the P/E-tRNA creates the precondition for the formation of the A/P hybrid state (43). Once this state is attained and stabilized by EF-G·GTP, the second step of translocation will start. This step is catalyzed by GTP hydrolysis on EF-G and entails the coupled movement of the mRNA and the anticodon loops of the tRNAs with respect to the small subunit.

The recent crystal structure of EF-G-2·GTP presents a conformation of EF-G that is ideally tailored to interacting with the PRE ribosome (76). Modeling of EF-G in this conformation with the PRE ribosome suggests that the binding of domains I, II, and G′ to the large subunit of the ribosome would position the tip of domain IV precisely at the codon–anticodon binding region of A-site tRNA (SI Fig. 8). Domain IV of EF-G, specifically the conserved histidine residue at its tip, comes into intimate contact with the codon–anticodon moiety formed by the mRNA and A-site tRNA. This interaction could stabilize the base pairing of the codon–anticodon of A-site tRNA during ratcheting of the ribosome, as suggested earlier (22). This reasoning would in fact explain the adverse effects on translocation when the conserved histidine in domain IV of EF-G is mutated (94). (For the eukaryotic case, the same scenario will hold, with EF-G replaced by eEF2 and the critical histidine residue by diphthamide.)

The movement of the decoding center in the direction of mRNA–tRNA translocation also stabilizes the A-site tRNA in the A/P hybrid state. Importantly, placement of tRNAs in the hybrid state is more favorable for translocation to occur as compared with tRNAs in the classic states (39, 43). Therefore, by stabilizing tRNAs in the hybrid states, EF-G binding directs the translocation process to go forward, which explains the catalytic potential of EF-G even in the absence of GTP hydrolysis (27).

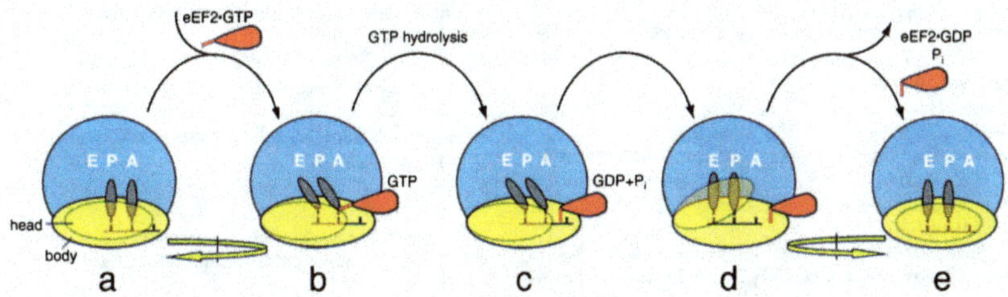

Fig. 5. Proposed sequence of events during the translocation process. Schematic depicts the top view of the translating ribosome. EF-G interacts with the pretranslocational ribosome (*a*), stabilizing the A- and P-site tRNAs into A/P and P/E hybrid sites, respectively. Binding of EF-G is coincident with the ratcheting motion of the entire small subunit (i.e., head and body ratchet together) with respect to the small subunit (*b*). GTP hydrolysis causes the tip of domain IV of EF-G to sever the connection between the small subunit head–tRNA–mRNA complex and the small subunit body (*c*). Once the link to the body is severed, the head rotates, translocating the mRNA by one codon and the anticodon stem-loop ends of the hybrid site tRNAs to the P and E sites (*d*). Translocation is completed by back-rotation of the head and reverse-ratcheting of the entire small subunit as EF-G leaves the ribosome with the tRNAs in full P/P and E/E sites (*e*).

Binding of EF-G and ratcheting of the ribosome are followed by GTP hydrolysis, causing a shift of domain IV, which in turn detaches the mRNA–tRNA complex from the decoding center (23). The mRNA–tRNA complex is temporarily freed from the body of the small subunit but likely maintains the strong interactions with the subunit's head. A head rotation in the small subunit is likely responsible for translocation of the mRNA chain by one codon but, because the tRNAs maintain their contacts with the head throughout the rotation, they technically remain in their hybrid states. Disruptions of these interactions must be induced by a combination of reverse-ratcheting of the small subunit and back-rotation of the head that both occur upon release of eEF2/EF-G. The final result is the POST-state ribosome with tRNAs in the canonical P and E sites (Fig. 5).

Conclusion

The ribosome, as we have seen, has intrinsic properties facilitating translocation. The most important one is an architecture that lends a specific kind of instability to the molecule and allows the ribosome to alternate between two distinct conformations separated by a small activation barrier. The change from one conformation to the other is along a pathway predicted as one of the highest-ranking modes by normal mode analysis of the x-ray structure (61, 62). This means, in other words, that the architecture has developed such that a minimum of energy is needed to go from the "normal" conformation to the "ratcheted" conformation, which is productive for the ultimate events of translocation.

This kind of parsimonious design is ubiquitous in all life forms. Human locomotion is an example on the macroscopic scale: during the act of walking, the legs are used as pendula, swinging with close to their innate resonance frequency and thus requiring a minimum of energy for sustenance of the movement. In the same way, energy supplied by thermal bombardment is channeled, by virtue of the ribosome's unique architecture, into a motion that is most productive in terms of translation.

Cryo-EM and the methods of flexible fitting by real-space refinement have allowed us to produce quasi-atomic models of the ribosome in both states (SI Fig. 6). An analysis of these models—particularly by watching an animation in which the two models alternate while the ribosome rotates (SI Movie 1)—makes it clear that the motion involves two interrelated events: (*i*) a conformational change in the intersubunit bridges and (*ii*) a change in the entire rRNA framework of each of the subunits, which resembles an elastic deformation, and accommodates the changes in the bridges.

The principal role of EF-G/eEF2, then, is in decreasing the energy barrier between the two ribosomal conformations. Specifically, the role of these factors is twofold: in bringing the ribosome into a conformation that is consistent with the binding of the tRNAs at the hybrid sites, and in unlocking the mRNA–tRNA moiety from the decoding center, thereby allowing the small subunit head movement to occur that is required for completion of translocation.

An "ur-ribosome" with the same basic architecture, and the same properties of instability, in a world in which elongation factors had not yet made their appearance, would already have functioned, albeit with very much decreased efficiency and accuracy. The very small rate of protein synthesis in these initial stages means that the evolution of the complex G-proteins (EF-G/eEF2, EF-Tu/eEF1A, IF2/eIF5B, and RF3/eRF3) tailored precisely to the ribosome's architecture could only proceed at a snail's pace. A similar line of reasoning holds for the impact of the addition of ribosomal proteins such as S12 and S13, which greatly improved the accuracy of factor-free translocation. Once these helper proteins had developed, however, the resultant much higher efficiency and accuracy of protein synthesis would lead to today's overflowing richness of life forms.

We thank M. Watters for assistance in generating figures and Y. Chen for help in creating SI Movie 1. This work was supported by the Howard Hughes Medical Institute and National Institutes of Health Grants R37 GM29169, R01 GM55440, and P41 RR01219.

1. Petry S, Brodersen DE, Murphy FV, IV, Dunham CM, Selmer M, Tarry MJ, Kelley AC, Ramakrishnan V (2005) *Cell* 123:1255–1266.
2. Borovinskaya MA, Pai RD, Zhang W, Schuwirth BS, Holton JM, Hirokawa G, Kaji H, Kaji A, Cate JH (2007) *Nat Struct Mol Biol* 14:727–732.
3. Weixlbaumer A, Petry S, Dunham CM, Selmer M, Kelley AC, Ramakrishnan V (2007) *Nat Struct Mol Biol* 14:733–737.
4. Carter AP, Clemons WM, Jr, Brodersen DE, Morgan-Warren RJ, Hartsch T, Wimberly BT, Ramakrishnan V (2001) *Science* 291:498–501.
5. Pioletti M, Schlunzen F, Harms J, Zarivach R, Gluhmann M, Avila H, Bashan A, Bartels H, Auerbach T, Jacobi C, *et al.* (2001) *EMBO J* 20:1829–1839.
6. Ogle JM, Brodersen DE, Clemons WM, Jr, Tarry MJ, Carter AP, Ramakrishnan V (2001) *Science* 292:897–902.
7. Wilson DN, Schluenzen F, Harms JM, Yoshida T, Ohkubo T, Albrecht R, Buerger J, Kobayashi Y, Fucini P (2005) *EMBO J* 24:251–260.
8. Frank J, Penczek P, Grassucci R, Srivastava S (1991) *J Cell Biol* 115:597–605.
9. Frank J, Zhu J, Penczek P, Li Y, Srivastava S, Verschoor A, Radermacher M, Grassucci R, Lata RK, Agrawal RK (1995) *Nature* 376:441–444.
10. Allen GS, Zavialov A, Gursky R, Ehrenberg M, Frank J (2005) *Cell* 121:703–712.
11. Agrawal RK, Penczek P, Grassucci RA, Frank J (1998) *Proc Natl Acad Sci USA* 95:6134–6138.
12. Agrawal RK, Heagle AB, Penczek P, Grassucci RA, Frank J (1999) *Nat Struct Biol* 6:643–647.
13. Valle M, Zavialov A, Sengupta J, Rawat U, Ehrenberg M, Frank J (2003) *Cell* 114:123–134.
14. Stark H, Rodnina MV, Wieden HJ, van Heel M, Wintermeyer W (2000) *Cell* 100:301–309.
15. Rawat U, Gao H, Zavialov A, Gursky R, Ehrenberg M, Frank J (2006) *J Mol Biol* 357:1144–1153.

16. Rawat UB, Zavialov AV, Sengupta J, Valle M, Grassucci RA, Linde J, Vestergaard B, Ehrenberg M, Frank J (2003) *Nature* 421:87–90.
17. Klaholz BP, Pape T, Zavialov AV, Myasnikov AG, Orlova EV, Vestergaard B, Ehrenberg M, van Heel M (2003) *Nature* 421:90–94.
18. Gao H, Zhou Z, Rawat U, Huang C, Bouakaz L, Wang C, Cheng Z, Liu Y, Zavialov A, Gursky R, *et al.* (2007) *Cell* 129:929–941.
19. Agrawal RK, Sharma MR, Kiel MC, Hirokawa G, Booth TM, Spahn CM, Grassucci RA, Kaji A, Frank J (2004) *Proc Natl Acad Sci USA* 101:8900–8905.
20. Gao N, Zavialov AV, Li W, Sengupta J, Valle M, Gursky RP, Ehrenberg M, Frank J (2005) *Mol Cell* 18:663–674.
21. Gomez-Lorenzo MG, Spahn CM, Agrawal RK, Grassucci RA, Penczek P, Chakraburtty K, Ballesta JP, Lavandera JL, Garcia-Bustos JF, Frank J (2000) *EMBO J* 19:2710–2718.
22. Spahn CM, Gomez-Lorenzo MG, Grassucci RA, Jorgensen R, Andersen GR, Beckmann R, Penczek PA, Ballesta JP, Frank J (2004) *EMBO J* 23:1008–1019.
23. Taylor DJ, Nilsson J, Merrill AR, Andersen GR, Nissen P, Frank J (2007) *EMBO J* 26:2421–2431.
24. Valle M, Zavialov A, Li W, Stagg SM, Sengupta J, Nielsen RC, Nissen P, Harvey SC, Ehrenberg M, Frank J (2003) *Nat Struct Biol* 10:899–906.
25. Barta A, Dorner S, Polacek N (2001) *Science* 291:203.
26. Moazed D, Noller HF (1989) *Nature* 342:142–148.
27. Rodnina MV, Savelsbergh A, Katunin VI, Wintermeyer W (1997) *Nature* 385:37–41.
28. Agrawal RK, Penczek P, Grassucci RA, Burkhardt N, Nierhaus KH, Frank J (1999) *J Biol Chem* 274:8723–8729.
29. Frank J, Agrawal RK (2000) *Nature* 406:318–322.
30. Bretscher MS (1968) *Nature* 218:675–677.
31. Spirin AS (1968) *Curr Mod Biol* 2:115–127.
32. Yusupov MM, Yusupova GZ, Baucom A, Lieberman K, Earnest TN, Cate JH, Noller HF (2001) *Science* 292:883–896.
33. Korostelev A, Trakhanov S, Laurberg M, Noller HF (2006) *Cell* 126:1065–1077.
34. Selmer M, Dunham CM, Murphy FV, IV, Weixlbaumer A, Petry S, Kelley AC, Weir JR, Ramakrishnan V (2006) *Science* 313:1935–1942.
35. Yusupova G, Jenner L, Rees B, Moras D, Yusupov M (2006) *Nature* 444:391–394.
36. Sharma D, Southworth DR, Green R (2004) *RNA* 10:102–113.
37. Frank J, Agrawal RK (2001) *Cold Spring Harbor Symp Quant Biol* 66:67–75.
38. Serdyuk IN, Spirin AS (1986) in *Structure, Function, and Genetics of Ribosomes*, eds Hardesty B, Kramer G (Springer, New York), pp 425–437.
39. Dorner S, Brunelle JL, Sharma D, Green R (2006) *Nat Struct Mol Biol* 13:234–241.
40. Ermolenko DN, Majumdar ZK, Hickerson RP, Spiegel PC, Clegg RM, Noller HF (2007) *J Mol Biol* 370:530–540.
41. Spiegel PC, Ermolenko DN, Noller HF (2007) *RNA* 13:1473–1482.
42. Horan LH, Noller HF (2007) *Proc Natl Acad Sci USA* 104:4881–4885.
43. Munro JB, Altman RB, O'Connor N, Blanchard SC (2007) *Mol Cell* 25:505–517.
44. Pan D, Kirillov SV, Cooperman BS (2007) *Mol Cell* 25:519–529.
45. Pestka S (1968) *J Biol Chem* 243:2810–2820.
46. Pestka S (1969) *J Biol Chem* 244:1533–1539.
47. Gavrilova LP, Smolyaninov VV (1971) *Mol Biol* 5:710–717.
48. Gavrilova LP, Spirin AS (1971) *FEBS Lett* 17:324–326.
49. Gavrilova LP, Spirin AS (1974) *FEBS Lett* 39:13–16.
50. Gavrilova LP, Kostiashkina OE, Koteliansky VE, Rutkevitch NM, Spirin AS (1976) *J Mol Biol* 101:537–552.
51. Gavrilova LP, Koteliansky VE, Spirin AS (1974) *FEBS Lett* 45:324–328.
52. Cukras AR, Southworth DR, Brunelle JL, Culver GM, Green R (2003) *Mol Cell* 12:321–328.
53. Girshovich AS, Bochkareva ES, Ovchinnikov YA (1981) *J Mol Biol* 151:229–243.
54. Nechifor R, Wilson KS (2007) *J Mol Biol* 368:1412–1425.
55. Klaholz BP, Myasnikov AG, Van Heel M (2004) *Nature* 427:862–865.
56. Makarov EM, Katunin VI, Odintsov VB, Semenkov Iu P, Kirillov SV (1984) *Mol Biol (Moscow)* 18:1342–1347.
57. Schilling-Bartetzko S, Bartetzko A, Nierhaus KH (1992) *J Biol Chem* 267:4703–4712.
58. Wintermeyer W, Savelsbergh A, Semenkov YP, Katunin VI, Rodnina MV (2001) *Cold Spring Harbor Symp Quant Biol* 66:449–458.
59. Semenkov YP, Shapkina TG, Kirillov SV (1992) *Biochimie* 74:411–417.
60. Katunin VI, Savelsbergh A, Rodnina MV, Wintermeyer W (2002) *Biochemistry* 41:12806–12812.
61. Tama'F, Valle M, Frank J, Brooks CL, III (2003) *Proc Natl Acad Sci USA* 100:9319–9323.
62. Wang Y, Rader AJ, Bahar I, Jernigan RL (2004) *J Struct Biol* 147:302–314.
63. Cate JH, Yusupov MM, Yusupova GZ, Earnest TN, Noller HF (1999) *Science* 285:2095–2104.
64. Schuwirth BS, Borovinskaya MA, Hau CW, Zhang W, Vila-Sanjurjo A, Holton JM, Cate JH (2005) *Science* 310:827–834.
65. Culver GM, Cate JH, Yusupova GZ, Yusupov MM, Noller HF (1999) *Science* 285:2133–2136.
66. Noller HF (2005) *Science* 309:1508–1514.
67. Nissen P, Ippolito JA, Ban N, Moore PB, Steitz TA (2001) *Proc Natl Acad Sci USA* 98:4899–4903.
68. Klein DJ, Schmeing TM, Moore PB, Steitz TA (2001) *EMBO J* 20:4214–4221.
69. Osswald M, Doring T, Brimacombe R (1995) *Nucleic Acids Res* 23:4635–4641.
70. Harms J, Schluenzen F, Zarivach R, Bashan A, Gat S, Agmon I, Bartels H, Franceschi F, Yonath A (2001) *Cell* 107:679–688.
71. Komoda T, Sato NS, Phelps SS, Namba N, Joseph S, Suzuki T (2006) *J Biol Chem* 281:32303–32309.
72. Frank J, Sengupta J, Gao H, Li W, Valle M, Zavialov A, Ehrenberg M (2005) *FEBS Lett* 579:959–962.
73. Czworkowski J, Wang J, Steitz TA, Moore PB (1994) *EMBO J* 13:3661–3668.
74. Aevarsson A, Brazhnikov E, Garber M, Zheltonosova J, Chirgadze Y, al-Karadaghi S, Svensson LA, Liljas A (1994) *EMBO J* 13:3669–3677.
75. Hansson S, Singh R, Gudkov AT, Liljas A, Logan DT (2005) *FEBS Lett* 579:4492–4497.
76. Connell SR, Takemoto C, Wilson DN, Wang H, Murayama K, Terada T, Shirouzu M, Rost M, Schuler M, Giesebrecht J, *et al.* (2007) *Mol Cell* 25:751–764.
77. Czworkowski J, Moore PB (1997) *Biochemistry* 36:10327–10334.
78. Liljas A (2004) *Structural Aspects of Protein Synthesis* (World Scientific, Singapore).
79. Vetter IR, Wittinghofer A (2001) *Science* 294:1299–1304.
80. Mohr D, Wintermeyer W, Rodnina MV (2002) *Biochemistry* 41:12520–12528.
81. Vogeley L, Palm GJ, Mesters JR, Hilgenfeld R (2001) *J Biol Chem* 276:17149–17155.
82. Hausner TP, Atmadja J, Nierhaus KH (1987) *Biochimie* 69:911–923.
83. Wool IG, Gluck A, Endo Y (1992) *Trends Biochem Sci* 17:266–269.
84. Munishkin A, Wool IG (1997) *Proc Natl Acad Sci USA* 94:12280–12284.
85. Moazed D, Robertson JM, Noller HF (1988) *Nature* 334:362–364.
86. Myasnikov AG, Marzi S, Simonetti A, Giuliodori AM, Gualerzi CO, Yusupova G, Yusupov M, Klaholz BP (2005) *Nat Struct Mol Biol* 12:1145–1149.
87. Wilden B, Savelsbergh A, Rodnina MV, Wintermeyer W (2006) *Proc Natl Acad Sci USA* 103:13670–13675.
88. Wilson KS, Nechifor R (2004) *J Mol Biol* 337:15–30.
89. Savelsbergh A, Katunin VI, Mohr D, Peske F, Rodnina MV, Wintermeyer W (2003) *Mol Cell* 11:1517–1523.
90. Serdyuk I, Baranov V, Tsalkova T, Gulyamova D, Pavlov M, Spirin A, May R (1992) *Biochimie* 74:299–306.
91. Abdi NM, Fredrick K (2005) *RNA* 11:1624–1632.
92. Spahn CM, Beckmann R, Eswar N, Penczek PA, Sali A, Blobel G, Frank J (2001) *Cell* 107:373–386.
93. VanLoock MS, Agrawal RK, Gabashvili IS, Qi L, Frank J, Harvey SC (2000) *J Mol Biol* 304:507–515.
94. Savelsbergh A, Matassova NB, Rodnina MV, Wintermeyer W (2000) *J Mol Biol* 300:951–961.

Disentangling conformational states of macromolecules in 3D-EM through likelihood optimization

Sjors H W Scheres[1], Haixiao Gao[2], Mikel Valle[1,5], Gabor T Herman[3], Paul P B Eggermont[4], Joachim Frank[2] & Jose-Maria Carazo[1]

Although three-dimensional electron microscopy (3D-EM) permits structural characterization of macromolecular assemblies in distinct functional states, the inability to classify projections from structurally heterogeneous samples has severely limited its application. We present a maximum likelihood–based classification method that does not depend on prior knowledge about the structural variability, and demonstrate its effectiveness for two macromolecular assemblies with different types of conformational variability: the *Escherichia coli* ribosome and Simian virus 40 (SV40) large T-antigen.

Many tasks in the living cell are performed by macromolecular assemblies encompassing various binding interactions and accompanied by conformational changes. The 3D-EM approach holds the promise of being able to visualize these 'molecular machines' in their various functional states. In the single-particle reconstruction approach (see ref. 1), individual molecules are visualized with an electron microscope, and the resulting projections—often numbering tens of thousands—are combined in a three-dimensional density map. As the technique imposes no restrictions on the range of existing conformations in the sample, information about various functional states is often available in a single experiment. However, the combination of images in a three-dimensional reconstruction requires that they represent projections of identical three-dimensional objects and that their relative orientations be known. As the problems of conformational and orientational classification are strongly intertwined, the coexistence of different conformations or ligand binding states in the sample has seriously limited the applicability of the 3D-EM approach[2].

Several strategies have been proposed for classification of heterogeneous projection data. The most potent one may be supervised classification (for example, as applied in refs. 3,4), which requires prior structural knowledge about the sample heterogeneity to separate the data according to their similarity to two or more known reference structures. Here the dependency on *a priori* available information seriously limits the general applicability of this approach. The same can be said about approaches in which the conformational variability of a molecule is predicted by normal-mode analysis from an existing density map[5]. Most other methods use recursive schemes of two-dimensional classification and three-dimensional reconstruction, in which orientational and class membership assignment are alternated (see ref. 1 and references therein). However, as these approaches typically rely on many strategic judgments, their effectiveness is difficult to predict generally.

Here we propose a classification approach based on maximum-likelihood principles for handling structurally heterogeneous data. This methodology has potentially wide applicability for mining hidden information from biological data sets that are (i) very large (in one of the experiments that we report, the number of raw data items was approximately 1,500 million), (ii) very noisy (the signal-to-noise ratios of the data sets presented are in the order of 0.1), and (iii) lack information on how the data items were generated (in our case, for any projection, we have no prior information on the associated conformation and orientation). From a theoretical point of view, maximum-likelihood estimators are particularly well suited for the problem at hand because, as the amount of experimental data increases, these methods yield less biased results with smaller variances than alternative estimators[6]. Furthermore, previous applications of the maximum-likelihood approach to related problems in the 3D-EM field have demonstrated its particular robustness to high levels of noise[7–10].

The classification technique described here aims to find the most likely set of parameters (Θ) to construct a statistical model describing structurally heterogeneous data, through optimization of the following log-likelihood function:

$$L(\Theta) = \sum_{i=1}^{I} \ln \sum_{k=1}^{K} \int_{\varphi} P(X_i|k,\varphi,\Theta)P(k,\varphi|\Theta)d\varphi,$$

where for each experimental projection X_i ($i = 1, \ldots, I$), we calculate all probabilities $P(X_i|k,\varphi,\Theta)$ of observing X_i given model Θ, position φ (that is, rotation and translation) and class k ($k = 1, \ldots, K$), also taking the prior probability of k and φ, $P(k,\varphi|\Theta)$, into account. These probability calculations are based on the assumption that all X_i are projections of one

[1]Centro Nacional de Biotecnología, CSIC, Cantoblanco, 28049, Madrid, Spain. [2]Howard Hughes Medical Institute, Health Research Inc., Empire State Plaza, Albany, New York 12201, USA. [3]The Graduate Center, City University of New York, New York, New York 10016, USA. [4]Food and Resource Economics, University of Delaware, Newark, Delaware 19716, USA. [5]Present address: CICBiogune, Parque Tecnológico de Bizkaia, 48160, Derio-Bizkaia, Spain. Correspondence should be addressed to J.-M.C. (carazo@cnb.uam.es).

RECEIVED 24 MAY; ACCEPTED 6 NOVEMBER; PUBLISHED ONLINE 10 DECEMBER 2006; DOI:10.1038/NMETH992

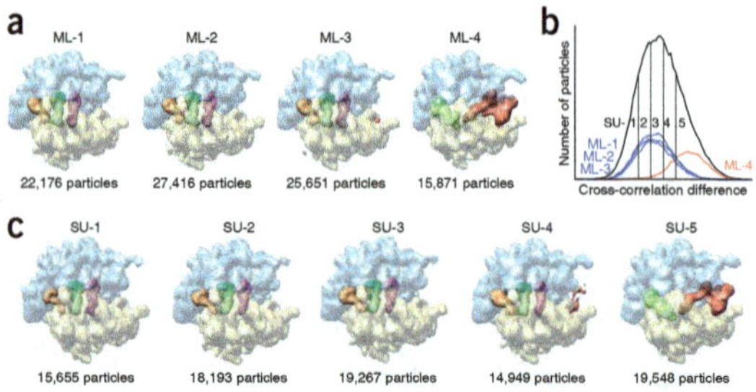

Figure 1 | Comparison of supervised and maximum-likelihood classifications for the ribosome data set. (**a**) Refined structures of the classes as determined by maximum-likelihood optimization (ML). (**b**) Supervised-classification histogram (SU, black) that served to divide the data set into five classes, and the corresponding histograms for the classes as determined by maximum-likelihood optimization (blue and red). (**c**) Refined structures of the classes as determined by supervised classification. The two classification approaches yielded similar structures, both in terms of overall quality (with resolutions in the range of 12–14 Å) and in terms of their interpretation: unratcheted ribosomes in complex with three tRNAs (orange, green, magenta) or ratcheted ribosomes in complex with one tRNA (light-green) and EF-G (red). Classes SU-5 and ML-4 overlap to a high extent, having 11,415 particles in common.

of K underlying three-dimensional objects, to which white, Gaussian noise has been added. We use an expectation-maximization algorithm[11] to optimize the log-likelihood function, and show that it is equivalent to a three-dimensional projection-matching algorithm, where all experimental images are iteratively presented to a discretely sampled library of projections of K different reference structures (for example, see ref. 12). The main difference from conventional projection-matching protocols is that the discrete assignments of projection orientation and class membership are replaced by probability-weighted integrations over all possible assignments. We describe the algorithm and its implementation in the open-source package Xmipp[13] in detail in the **Supplementary Note** online.

We show that this approach permits the use of initial reference maps that are devoid of any prior information (or bias) about the structural variability in the sample. Such bias-free seeds are obtained by performing a single iteration of likelihood optimization for K randomly drawn subsets of the data, using a low-resolution representation of the average structure as single reference. Therefore, classification based on likelihood optimization is unsupervised because it depends only on a preliminary reconstruction of the data and a rough estimate for the number of existing classes, which determines the choice of the value of K. Taken together with the intrinsically combined treatment of orientational and conformational assignment, this suggests that, in contrast to existing approaches, classification by likelihood optimization may be applicable to a broad range of structurally heterogeneous data sets.

We tested the likelihood optimization methodology on two highly challenging cryo-EM data sets of macromolecular assemblies exhibiting different types of structural variability. We obtained the first data set from a sample of 70S *E. coli* ribosome particles in a translocational state before GTP hydrolysis (H. Gao *et al.*, unpublished data; see **Supplementary Methods** online) and it represents a case of nonstoichiometric ligand binding. A preliminary reconstruction at 9.9 Å from 91,114 projection images exhibited blurred density for the 30S subunit and fragmented density owing to the binding of elongation factor G (EF-G; ~4% of the total mass of the complex), indicating problems of structural heterogeneity (**Supplementary Fig. 1** online). The second data set, comprising 5,738 projection images of SV40 large T-antigen dodecamers (see **Supplementary Methods**), represents a case of flexible domain movement, as this complex was previously observed to exhibit curvature of a varying degree along its central axis[14].

For the ribosome data, we chose $K = 4$ and performed 25 iterations of likelihood optimization using all particles (**Supplementary Fig. 2** online). The four resulting structures could be interpreted in terms of two different conformations. Class 4, comprising only 16% of the particles, clearly represents the ribosome in a ratcheted conformation (that is, the 30S subunit has rotated counter-clockwise relative to the 50S subunit[15]) with bound EF-G and a single tRNA positioned at the hybrid P/E site (ML-4; **Fig. 1a**). In contrast, the structures obtained from classes 1–3 represent the ribosome without ratcheting and without EF-G, but with three bound tRNAs at the A, P, and E sites (ML-1, ML-2, ML-3; **Fig. 1a**). (The occupancy of the E site is attributed to free deacylated tRNA that has high affinity for the ribosomal E site[15].) Slight overall rotations of the entire map are responsible for the main differences among classes 1–3. These results are in excellent agreement with those obtained by supervised classification into five subsets (**Fig. 1b,c**). Based on the assumption that the heterogeneity in the data entailed a ratchet motion, the supervised approach (**Supplementary Fig. 3** online) also yielded classes representing either ribosomes without ratcheting, lacking EF-G and with three tRNAs bound, or ratcheted ribosomes in complex with EF-G and a single tRNA. However, the unimodal character of the supervised classification histogram, which served to divide the data set, and the observed partial density in the corresponding reconstructions, suggest that the intermediate classes of this approach are likely to represent mixtures of assemblies with different ratios of both conformations. The lack of clear distinctions among the different classes without the binding of EF-G, as well as the high similarity of corresponding structures obtained using both approaches, suggest the existence of only two genuinely different classes in this data set. The high degree of structural overlap between the EF-G–containing classes using both methods (75% overlap in membership) validates the maximum-likelihood approach to classification. In contrast to the supervised approach, however, classification by likelihood optimization does not depend on prior knowledge about the nature of the structural variability in the data and may thus be applicable in situations where such knowledge is unavailable.

The large T-antigen data set represents a continuous form of structural variability, where a discrete number of classes cannot fully describe the continuum of conformations, but may serve to capture distinct representative states along the continuum. Consequently, the choice of K results from a trade-off between the degree of homogeneity desired within each class and the number of experimental images needed on statistical grounds to achieve sufficiently high resolution for its corresponding reconstruction. Because of the limited number of experimental particles in this data set, we chose $K = 3$ and performed 20 iterations of likelihood optimization (**Supplementary Fig. 4** online). The resulting maps represent dodecamers with different degrees of bending along their central

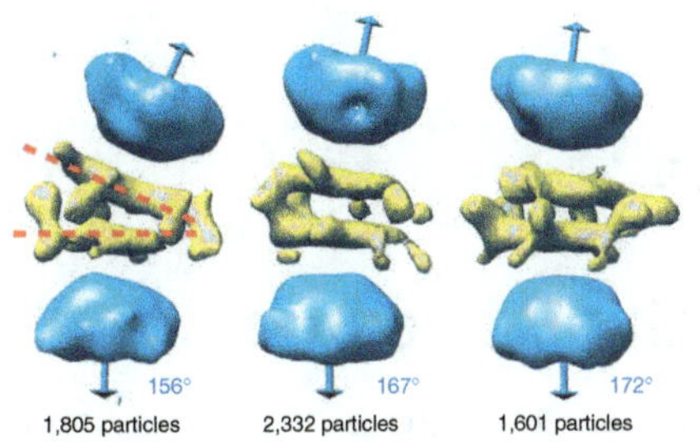

Figure 2 | Classification of the SV40 large T-antigen data set according to continuously varying bend of the molecule's long axis. The maps as obtained by maximum-likelihood optimization (with resolutions in the range of 35–38 Å) show various degrees of bending along the central axes of the dodecamers. The orientations of the sixfold symmetry axes through the hexameric C-terminal domains (blue) are indicated with arrows, and the angles between them are reported for each dodecamer. The N-terminal domains (yellow) of the most-bent dodecamer appear in an opened conformation (indicated by red lines).

axes, varying from 156 to 172° (**Fig. 2**). These results are in good agreement with random conical tilt experiments, where SV40 large T-antigen dodecamers with curvatures between 150 to 173° were observed[14]. The likelihood optimization allowed characterization of the structural variability without the need for additional tilting experiments, permitting a much simpler image acquisition scheme and opening the possibility to achieve higher resolution through the use of untilted micrographs. Moreover, the dodecamer with the highest curvature, as obtained by likelihood optimization reveals an opening of its central N-terminal domains, which was not observed for the random conical tilt reconstructions and may be implicated in the specific recognition of the SV40 origin of replication[14].

The results presented here indicate that likelihood optimization can be used to separate projection data without the need for prior knowledge about the distinct structures from which they originate, except for a rough estimate of the number of these structures. Several considerations may be relevant in the general case. First, the extensive computational cost of the likelihood optimization algorithm limits the angular sampling and the size of the images to be used. (Using downscaled images of 64 × 64 pixels and an angular sampling of 10°, classification of the ribosome data still took 6 CPU months; see **Supplementary Note**.) Consequently, the attainable resolution of the distinct reconstructions will be limited, which will in turn limit the minimum size of the structural variability that can be detected. Nevertheless, high-resolution reconstructions may still be obtained through subsequent discrete classification based on the maxima of the probability distributions for all particles. Typically, these distributions are relatively broad during the initial iterations of likelihood optimization, but tend to converge to near-delta functions (see **Supplementary Figs. 2** and **4**). Therefore, upon convergence, the maximum-likelihood estimate will be close to a discrete classification, which allows the application of standard refinement procedures with finer sampling rates and larger images.

Second, the number of different classes in the data is typically not known *a priori*, and the user may need to perform multiple calculations with different values of K. Additionally, as observed for the ribosome data, the algorithm may converge to slightly rotated but otherwise similar reconstructions, apparently as a result of accommodating particles with projection directions falling in between the angular grid used. This would imply that K should always be chosen higher than the expected number of distinct classes (and would explain why a calculation with $K = 2$ yielded only moderate separation of the structural variability in the ribosome data; data not shown). In this light, it is interesting to note that for the ribosome data, an additional calculation with $K = 6$ also resulted in a single class corresponding to EF-G–containing particles, which overlapped to 89% with the equivalent class from the run presented. We made a similar observation for the SV40 large T-antigen data, where an additional calculation with $K = 4$ resulted in classes with curvatures ranging from 159 to 172°. In this case, the classes with the highest curvature from both runs overlapped to 87% (data not shown).

Finally, as the experimental noise in cryo-EM images is typically non-white, we envisage further improvements of this technique could be made through the definition of a statistical model for colored noise, which is a topic for future research.

The successful classification of two cryo-EM data sets, representing cases of nonstoichiometric ligand binding and flexible domain movements, confirms that the proposed method may be applicable to macromolecular complexes with a broad range of structural variability. Therefore, we expect that this technique will contribute considerably to meeting the challenges posed by 3D-EM visualization of macromolecular assemblies in distinct functional states.

Note: Supplementary information is available on the Nature Methods website.

ACKNOWLEDGMENTS

We thank the Department of Computer Architecture and Electronics of the University of Almeria, and the Barcelona Supercomputing Center (Centro Nacional de Supercomputación) for providing computing resources, and T. Elfving for his help with weighted least-squares minimization. This work was supported by grants from the European Union (FP6-502828 and EGEE2-031688 to JMC), US National Institutes of Health (HL70472 to G.T.H. and J.M.C., and P41 RR01219 to J.F.), Howard Hughes Medical Institute (to J.F.) and the Spanish Comisión Interministerial de Ciencia y Tecnología (BFU2004-00217 to J.M.C.).

COMPETING INTERESTS STATEMENT

The authors declare that they have no competing financial interests.

Published online at http://www.nature.com/naturemethods/
Reprints and permissions information is available online at http://npg.nature.com/reprintsandpermissions/

1. Frank, J. *Three-dimensional Electron Microscopy of Macromolecular Assemblies* (Oxford University Press, New York, 2006).
2. Orlova, E.V. & Saibil, H.R. *Curr. Opin. Struct. Biol.* **14**, 584–590 (2004).
3. Heymann, J.B., Conway, J.F. & Steven, A.C. *J. Struct. Biol.* **147**, 291–301 (2004).
4. Gao, H., Valle, M., Ehrenberg, M. & Frank, J. *J. Struct. Biol.* **147**, 283–290 (2004).
5. Brink, J. *et al. Structure* **12**, 185–191 (2004).
6. Rice, J.A. *Mathematical Statistics and Data Analysis* (Duxbury Press, Belmost, 1995).
7. Sigworth, F.J. *J. Struct. Biol.* **122**, 328–339 (1998).
8. Yin, Z., Zheng, Y., Doerschuk, P.C., Natarajan, P. & Johnson, J.E. *J. Struct. Biol.* **144**, 24–50 (2003).
9. Scheres, S.H.W. *et al. J. Mol. Biol.* **348**, 139–149 (2005).
10. Scheres, S.H.W., Valle, M. & Carazo, J.M. *Bioinformatics* **21** (Suppl. 2), ii243–ii244 (2005).
11. Dempster, A.P., Laird, N.M. & Rubin, D.B. *J. Royal Statist. Soc. Ser. B* **39**, 1–38 (1977).
12. Valle, M. *et al. EMBO J.* **21**, 3557–3567 (2002).
13. Sorzano, C.O.S. *et al. J. Struct. Biol.* **148**, 194–204 (2004).
14. Gomez-Lorenzo, M.G. *et al. EMBO J.* **22**, 6205–6213 (2003).
15. Valle, M. *et al. Cell* **114**, 123–134 (2003).

Molecular Cell
Article

Visualization of the Hybrid State of tRNA Binding Promoted by Spontaneous Ratcheting of the Ribosome

Xabier Agirrezabala,[1,2] Jianlin Lei,[1,2] Julie L. Brunelle,[3,4] Rodrigo F. Ortiz-Meoz,[3,4] Rachel Green,[3,4] and Joachim Frank[1,5,*]
[1]Howard Hughes Medical Institute
[2]Department of Biochemistry and Molecular Biophysics
Columbia University, 630 West 168th Street, College of Physicians and Surgeons, Black Building 2-221, New York, NY 10032, USA
[3]Howard Hughes Medical Institute
[4]Department of Molecular Biology and Genetics
Johns Hopkins University School of Medicine, Baltimore, MD 21205, USA
[5]Department of Biological Sciences, Columbia University, 600 Fairchild Center, Mail Code 2402, 1212 Amsterdam Avenue, New York, NY 10027, USA
*Correspondence: jf2192@columbia.edu
DOI 10.1016/j.molcel.2008.10.001

SUMMARY

A crucial step in translation is the translocation of tRNAs through the ribosome. In the transition from one canonical site to the other, the tRNAs acquire intermediate configurations, so-called hybrid states. At this stage, the small subunit is rotated with respect to the large subunit, and the anticodon stem loops reside in the A and P sites of the small subunit, while the acceptor ends interact with the P and E sites of the large subunit. In this work, by means of cryo-EM and particle classification procedures, we visualize the hybrid state of both A/P and P/E tRNAs in an authentic factor-free ribosome complex during translocation. In addition, we show how the repositioning of the tRNAs goes hand in hand with the change in the interplay between S13, L1 stalk, L5, H68, H69, and H38 that is caused by the ratcheting of the small subunit.

INTRODUCTION

The protein elongation cycle performed by the ribosome is a multistep sequential process encompassing aminoacyl-tRNA (aa-tRNA) selection, peptide bond formation, and translocation. The aa-tRNA selection by the ribosome is based on the correct Watson-Crick base complementarity between the codon of the messenger RNA (mRNA) in the A site of the 30S subunit of the ribosome and the anticodon of the incoming tRNA. As the aa-tRNA enters the ribosome, in complex with elongation factor Tu (EF-Tu) and GTP, it assumes a deformed, high-energy conformation (A/T state of tRNA binding [Valle et al., 2003a]). During the decoding process, a signal is transmitted from the codon-anticodon recognition site and received at the GTP-binding site on EF-Tu (Ogle et al., 2002; Valle et al., 2003a; Cochella and Green, 2005). Following the conformational change of EF-Tu induced by the hydrolysis of GTP (Abel et al., 1996; Polekhina et al., 1996), EF-Tu dissociates from the aminoacyl end of the aa-tRNA and the ribosome. As the tRNA relaxes its distorted conformation, its 3′ end (acceptor arm) is accommodated into the A site of the ribosome, leading to a rapid peptide bond transfer in the peptidyl-transferase center of the 50S subunit. At this stage, the pretranslocational ribosome contains a deacylated tRNA in the P site and a peptidyl-tRNA in the A site. In the last step of the protein elongation cycle, GTPase elongation factor G (EF-G) catalyzes the translocation of the mRNA-tRNA complex through the subunit interface from the A and P sites to the P and E sites, respectively, a process coupled to the ratchet-like rotation of the 30S subunit relative to the 50S subunit (Frank and Agrawal, 2000; Frank et al., 2007). This ratcheting movement was shown to be essential for translocation (Horan and Noller, 2007).

Chemical modification experiments provided evidence that tRNAs can sample a hybrid state of binding from a postpeptidyl transfer, pretranslocation state (Moazed and Noller, 1989). These data provided support for earlier models for a hybrid state of tRNA binding proposed by Bretscher (Bretscher, 1968). Subsequent single-molecule FRET studies have detected fluctuations in the positions of tRNAs (Blanchard et al., 2004), leading to the conclusion that the classical (A/A-P/P) and hybrid states (A/P-P/E) of the tRNAs are actually in dynamic equilibrium. Kinetic analysis has also supported the relevance of such intermediates, showing that the translocation driven by EF-G is kinetically more efficient when the tRNAs are bound in the hybrid state (Dorner et al., 2006). However, the exact interaction pattern of the hybrid A site peptidyl-tRNA and P site deacylated tRNA has remained structurally uncharacterized, since the hybrid state of the pretranslocational ribosome has not been observed in previous cryo-EM or X-ray studies.

It is well known that the ribosome is very sensitive to buffer conditions, especially to Mg^{2+} concentration (Agrawal et al., 1999; Muth et al., 2001), as its high negative charge tends to make its conformation especially dependent on ionic conditions (Draper et al., 2005). Along these lines, a recent FRET analysis of the effect of Mg^{2+} on tRNA dynamics has shown that the Mg^{2+} concentration dramatically alters the dynamic equilibrium between the classical and hybrid states after peptidyl transfer in pretranslocational complexes (Kim et al., 2007). In light of this evidence, we revisit the problem analyzing the structure and dynamics of a pretranslocational ribosome in the presence of low Mg^{2+} concentration (3.5 mM) and thus more physiological-like conditions (Cromie et al., 2006). As the sample corresponds to a homogeneous ribosome preparation showing conformational variability (i.e., equilibrium distribution of conformational states), we have applied supervised classification methodology to separate the initial group of ribosomes into two different subsets. This approach has allowed us to capture both classic and hybrid states of the tRNAs, and furthermore it has enabled us to correlate these two states with the conformational changes underlying the ratchet-like rotation of the 30S subunit. In particular, we show how the reconfiguration of the tRNAs is based on the concerted action of S13, L1, L5, H68, H69, and H38, structural elements that change their relative position during the movement of the subunit interface. The structural characterization of the classical and hybrid tRNA present in our current work provides a molecular explanation for several previous findings made by other studies using biochemistry, kinetics, and structural and single-molecule techniques and also sheds light on the specific recognition of the pretranslocational ribosomal complex by EF-G. Finally, the results suggest a mechanistic explanation for the sensitivity of observed ribosome conformations to the Mg^{2+} buffer concentration as an indirect effect of E site tRNA trapped on the ribosome.

RESULTS AND DISCUSSION

Structural Characteristics of Classic and Hybrid-State Ribosomes

We collected cryo-EM data for pretranslocational ribosomes, originally with fMet-tRNA^{fMet} in the P site ("postinitiation 70S complexes"), that were incubated with cognate Trp-tRNA^{Trp}•EF-Tu•GTP ternary complexes. After Trp-tRNA^{Trp} accommodation and subsequent peptidyl transfer, the ribosomes thus obtained bear fMet-Trp-tRNA^{Trp} and deacylated tRNA^{fMet} (Experimental Procedures). Initially, a cryo-EM map at 8.75 Å resolution was calculated using the whole data set (comprising ~216,000 particles). In this initial reconstruction (see Figure S1 available online), the densities corresponding to the individual tRNAs were fused and inseparable. In addition, certain regions of the 30S subunit showed an inferior definition. The coexistence of different conformations in the initial pool of ribosomes was thus evident. Therefore, the data set was sorted based on the resemblance (as measured by the crosscorrelation coefficient) of the data to two reference maps showing the ribosome in two states related by the ratchet motion (Figure 1). In this classification strategy, two 3D references are used that represent the conformations in two different states. Then the experimental data are compared with projections from the two different references, and the class assignment is made based on the reference that gives the highest correlation value. In preparing these two reference maps, to avoid bias, the densities corresponding to tRNA and EF-G were computationally removed by the application of a soft mask so that the references differed only by the ratchet rotation of 30S subunit relative to the 50S subunit (see Gao et al., 2004). This strategy allowed us to obtain two different 3D reconstructions corresponding to the different conformations that coexist in the sample. The first map, at 10.5 Å resolution, is

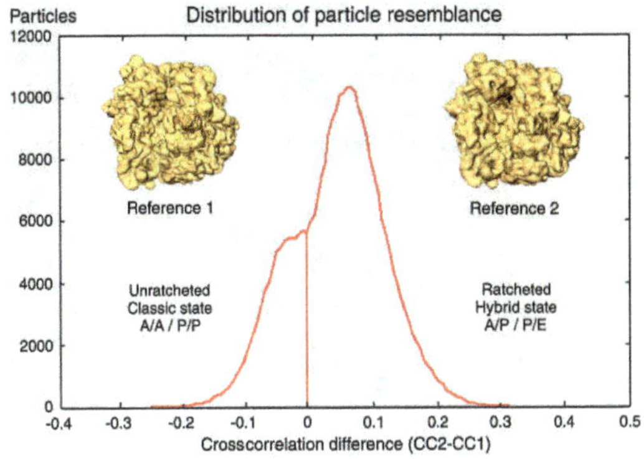

Figure 1. Statistical Distribution of Particle Resemblance
Histogram of values obtained by forming the difference between correlation coefficients of each experimental projection with respect to two references. A reference line is drawn at the selected cutoff point ($\Delta CC = 0$). The data were sorted based on their resemblance to 3D template ribosome maps that differ in the ratcheting rotation of the 30S subunit (insets). Both reference volumes were from a previous study (Valle et al., 2003b). Reference 1 was the cryo-EM density map of a pretranslocational complex (70S-tRNA^{fMet}-fMet-Phe-tRNA^{Phe}), while reference 2 was the cryo-EM map of a ribosomal complex bound with EF-G•GDPNP (70S-MFTI-tRNA^{Ile}-EF-G•GDPNP + puromycin), which displayed ratchet-like rotation. The densities corresponding to tRNAs and EF-G were computationally removed to avoid reference bias. To this end, soft Gaussian masks were used. Using this strategy, a clear bimodal distribution was obtained, and two major ribosomal populations were separated. One, composed by 57,603 particles, corresponded to unratcheted ribosomes bearing the classic A/A-P/P tRNA configuration, while the second population, consisting of 158,969 particles, showed the ratchet-like rotation of the 30S subunit hand-in-hand with the A/P-P/E hybrid tRNA configuration as shown in Figure 2.

seen to correspond to a less populated (~30%) state showing an unratcheted ribosome with tRNAs in the classic A/A-P/P configuration (Figure 2A), while the second map, visualized at 8.9 Å resolution, represents a more populated conformation (~70%) that shows a ratcheted-like rotation of the 30S subunit and tRNAs in a hybrid A/P-P/E configuration (Figure 2B).

It is worth noting that in our reconstruction of the classic state, the A/A tRNA does not show a well-defined acceptor stem end. This is clearly observed when fitting the X-ray structure of the A site tRNA into its cryo-EM density (Figure S3). As some degree of flexibility of the 3′ end of the A site tRNA is required for peptide bond formation (the reaction occurs while the tRNAs reside in their canonical A/A-P/P configuration [Schmeing et al., 2002]), the lack of visible density for the CCA end's last nucleotides in our reconstruction probably arises from conformational variability due to the mobility of this region.

Besides the different relative configuration of the tRNAs, a set of large-scale rearrangements in the ribosome can be easily noticed when comparing the two maps. A superposition of the large subunits as shown in Figure 3A reveals a clear counterclockwise rotation of the small subunit when seen from the solvent side of the 30S subunit, as the positions of the spur, beak, and head are strongly altered. This rotation causes the displacement of several components placed in the head of the small subunit and the consequent reconfiguration of several of the original

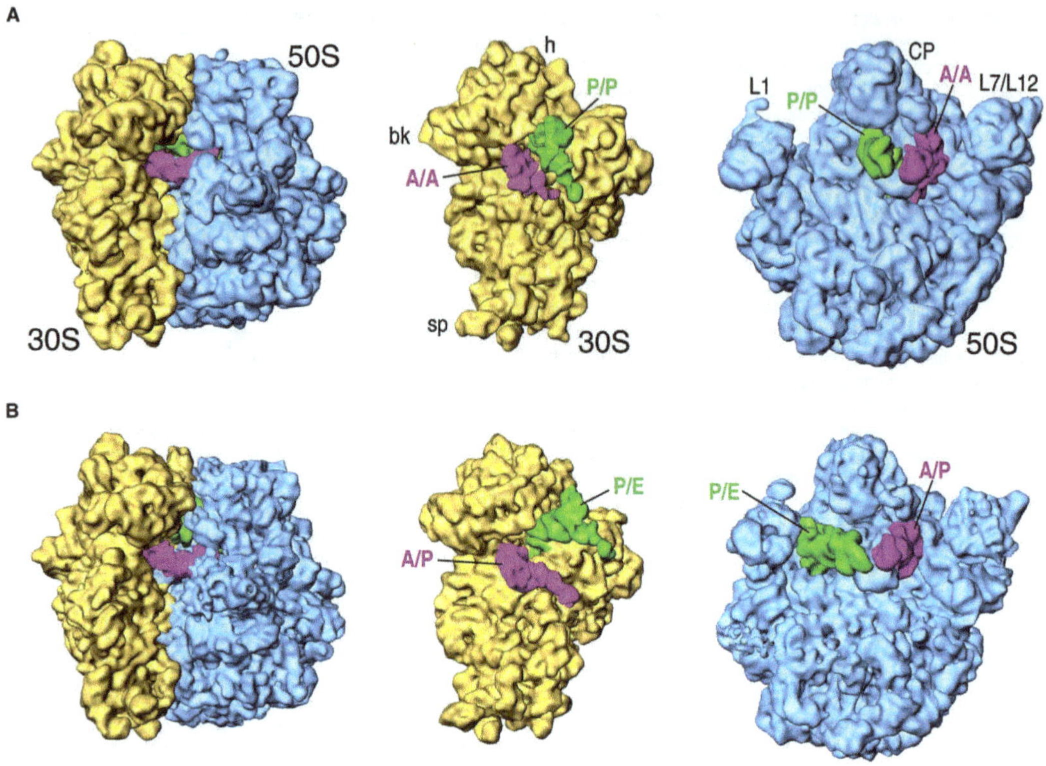

Figure 2. General Features of Classic and Hybrid Ribosomes
Cryo-EM densities are shown for the small (yellow) and large (blue) subunits and for A and P site tRNAs (magenta and green, respectively). The classic-state ribosome is shown in (A) and the hybrid-state ribosome in (B). Labels and landmarks are the following: sp (spur), bk (beak), h (head), CP (central protuberance), L1 (L1 stalk), and L7/L12 (L7/L12 stalk).

intersubunit bridges (Gabashvili et al., 2000). Bridge B1b, formed by the interaction of L5 and S13, is remodeled, and so is bridge B1a, in which H38's binding partner S13 is replaced by S19 in the ratcheted state. In this process, H38, as well as the central protuberance region where L5 is located, appears to adopt its new conformation.

Bridge B2a is formed by the interaction of the universally conserved 1915 loop of H69 and the tip of h44, just at the base of the decoding site of the small subunit. Although the bridge remains intact, slight opposite lateral movements of the tip of h44 and H69 are detected after the ratcheting movement (Figure S5). H69 is also more inclined toward the small subunit at this stage. H68, which contacts h23 of the 16S rRNA to form bridge B7a, shows a displacement toward the large subunit. In contrast to the 30S subunit, only small-scale, local changes are observed in the 50S subunit, with the exception of the L1 stalk (formed by helices 76–78 and protein L1), which moves toward the intersubunit space by ~20 Å. This rearrangement is facilitated by a large movement of H76 at the base of the stalk.

Similar conformational changes in the ribosome related to the ratcheting movement of the 30S subunit were previously observed in response to the binding of EF-G (Valle et al., 2003b), initiation factor IF2 (Allen et al., 2005), class II release factor RF3 (Klaholz et al., 2004; Gao et al., 2007), and recycling factor RRF (Gao et al., 2005). However, what distinguishes our current observation from the previous ones is that, in this case, no factor is present to facilitate ribosomal motions. We will therefore speak of spontaneous ratcheting of the ribosome, in line with data from very recent FRET studies indicating a factor-free rotation of the small subunit (Cornish et al., 2008). Spontaneous ratcheting of the ribosome has previously not been seen in cryo-EM maps of pretranslocational ribosomes, as these complexes invariably showed high occupation of canonical A, P, and E sites by tRNAs (e.g., Valle et al., 2003b).

In contrast, our current data differ in two important aspects: first, they show a mixture of two states (macrostates—see Frank et al. [2007]) in apparent equilibrium, with a preference for the ratcheted state; and second, the reconstruction of the ribosome in the canonical state from the minority population does not show a deacylated tRNA in the E site. Since FRET data have demonstrated that the P site tRNA adopting the hybrid configuration is the rate-limiting step in the transition between classical

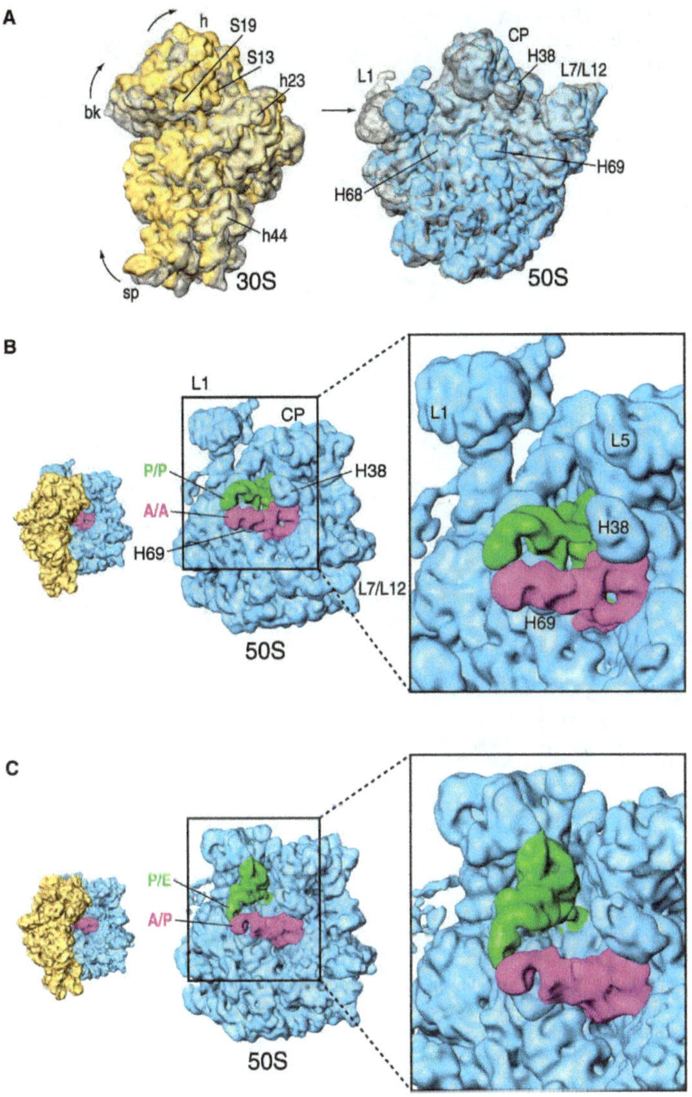

Figure 3. Conformational Changes upon Ratcheting Movement and tRNA Reconfiguration

(A) Superimposition of aligned 30S (left) and 50S (right) subunits shown from the intersubunit side. Classic-state subunits are shown in transparent gray, whereas the subunits of the hybrid-state ribosome are shown in solid yellow (30S) and blue (50S).

(B) Close-up view of the 50S subunit showing the classic tRNA configuration.

(C) Close-up view of the 50S subunit showing the hybrid tRNA configuration. The orientations of the subunits are shown as successive thumbnails on the left. Labels and landmarks are the following: S19 (small subunit/SSU protein S19), S13 (SSU protein S13), h23 (SSU helix 23), h44 (SSU helix 44), H38 (large subunit/LSU helix 38), H68 (LSU helix 68), H69 (LSU helix 69), H76 (LSU helix 76), and L5 (LSU protein L5). All other landmarks have been introduced in Figure 2.

and hybrid states (Munro et al., 2007), we argue, much in line with previous suggestions (Zavialov et al., 2005), that the striking difference in the results may result from E site binding by free deacylated tRNA, which could be in turn controlled by the Mg^{2+} concentration. The low Mg^{2+} concentration used in the present experiment could potentially prevent occupation of the E site and thus favor a more efficient transition of the classic P site tRNA to the hybrid P/E configuration.

Finally we note that the distribution of particle resemblance as measured by crosscorrelation (Figure 1) shows the presence of only two major populations, corresponding to the hybrid/ratcheted and classical/nonratcheted states of the pretranslocation complex. Along these lines, recent data (Fei et al., 2008; Cornish et al., 2008) support the view that there are no detectable intermediates beyond the two configurations visualized in this work. However, this does not necessarily exclude the possibility of a P/E-A/A intermediate, as independently described by other groups (Munro et al., 2007; Pan et al., 2007), but it does seem that this intermediate, if it exists, it is not very populated and/or stable in wild-type ribosomes.

Dynamic Structural Pattern of Interactions between the tRNAs and the Ribosome during the Ratcheting Movement, and Partial Translocation

In the transition from the classic to the hybrid state, the inferred trajectories of the tRNAs are quite different. While the elbow (D and T loops) of the P site tRNA moves by ~40 Å in the P/P-to-P/E transition, the elbow of the A site tRNA barely moves. Thus, with the exception of the CCA end, the interaction of the tRNA with the ribosome remains basically unaltered between classic A/A and hybrid A/P tRNA configurations. On the 50S subunit side, H69 (1915 loop) maintains its contact with the D stem. The elbow of the tRNA also interacts with H38 in both classic and hybrid states (Figures 3B and 3C). Nevertheless, the lateral displacement of H69, in conjunction with the conformational change of H38 accompanying the ratcheting of the small subunit, causes the pivoting of the A site tRNA around its anticodon and the reorientation of the elbow region, allowing the CCA end to reach the 50S subunit's active site in the A/P hybrid configuration. Since it is known that the peptidyl-tRNA in the A/P configuration is quite unreactive as a peptidyl donor in contrast to its high reactivity in the P/P site (Semenkov et al., 1992; Sharma et al., 2004), we can only speculate about the lower reactivity of the hybrid configuration being related to its dynamic and transient nature. On the other hand, in this A/P configuration, the tRNA is brought close to H93, where residue A2602 is situated (Figure S4). It has been shown that sparsomycin-dependent rearrangement of this residue decreases the activation energy for EF-G-independent translocation (Fredrick and Noller, 2003). These data can be rationalized if the interaction of A2602 with the CCA end stabilizes the observed A/P hybrid configuration of the tRNA that eventually leads to the repositioning of the peptidyl-tRNA from the A/P to the P/P state. Indeed, the original study by Moazed and Noller noted specific changes in the modification pattern for A2602 in the pre- and posttranslocation states, likely reporting on conformational rearrangements in the peptidyl transfer center (Moazed and Noller, 1989).

On the other hand, the extensive contacts of the P site tRNA are drastically remodeled after the P/P-to-P/E site transition. A remarkable exception is the contact with protein S13. This protein, located in the head region of 30S, undergoes a large displacement coupled to the rotation of the small subunit (Valle et al., 2003b; Frank et al., 2007). Despite the size of this movement, the C terminus of S13 does not lose its original contact with the P site tRNA after ribosomal ratcheting. Along these lines, protein S13 has been shown to be a control element in the EF-G-independent movement, as its omission promotes complete tRNA translocation (Cukras et al., 2003).

On the 50S subunit side, there is an extensive connection of the D stem of the classic-state P/P tRNA with H69, as well as between the T loop and protein L5. However, the map corresponding to the hybrid P/E tRNA indicates that it is no longer bound to H69 (see Li and Frank, 2007). The CCA stem of the P/E tRNA contacts H68 in its new configuration, and the T loop is severed from its original interaction with L5, having established a new contact with H78 situated in the L1 stalk (Figures 3B and 3C). This new interaction is only formed after the large displacement of the L1 stalk coupled to the ratcheting of the small subunit. In this hybrid P/E configuration, several residues located in H68, H74, and H11 that are known to be essential for P site tRNA translocation (Feinberg and Joseph, 2001) are placed in the vicinity of the 3′ acceptor end (Figure S4).

In summary, these observations show how the tRNA reconfiguration is promoted by the changes the ratcheting of the small subunit generates in several components involved not only in tRNA binding interactions but also in subunit association and reconfiguration of intersubunit bridges. Some of them, namely S13, L5, and H38, which form the bridges that connect the head of the 30S subunit with the central protuberance of the large subunit, may be instrumental for the control of the rotation, as they are located at the periphery where the movement is largest (Frank et al., 2007). The fact that these components are highly dynamic and change their relative position during the movement of the subunit interface reflects the direct relation between the ratcheting movement of the small subunit and the reconfiguration of the tRNAs. Thus, our data suggest that the unlocking of the ribosome (understood as the removal of the physical connection between P site tRNA and the polypeptide chain by the

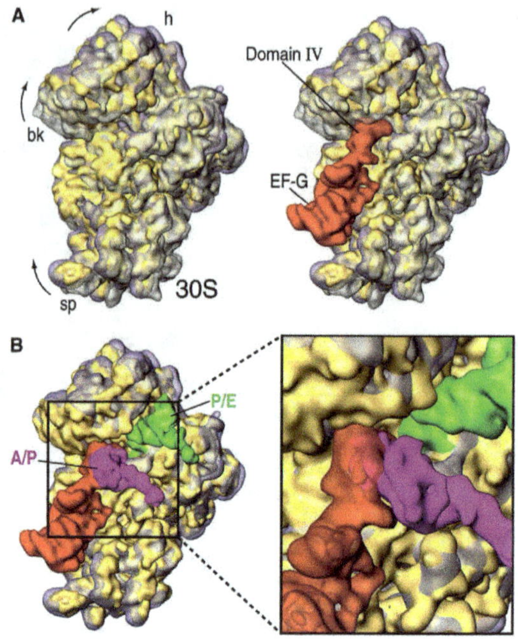

Figure 4. Hybrid Configuration of the Ribosome and Implications for EF-G-Based Translocation

(A) Superimposition of aligned 30S subunits from hybrid (solid yellow), classic (transparent gray), and EF-G-bound (transparent purple) forms (from Valle et al., 2003b). In the right panel, the binding position of EF-G (red) is shown.

(B) Close-up view of the superimposition of hybrid (solid yellow) and EF-G-bound (transparent purple) forms in the presence of EF-G and hybrid tRNAs. The subunits are shown from the intersubunit side.

deacylation of the P site tRNA [Zavialov and Ehrenberg, 2003; Valle et al., 2003b]) is what releases the inherent ratchet-related reorganization capability of the ribosome, which permits the different tRNA environments to be established.

Implications for EF-G-Based Translocation

Kinetic and thermodynamic studies have demonstrated that the hybrid state of tRNA binding makes important energetic contributions to EF-G-dependent translocation (Semenkov et al., 2000; Dorner et al., 2006). These results invite a comparison with results from previous cryo-EM reconstructions of EF-G•GDPNP-bound ribosomes (Agrawal et al., 1998; Valle et al., 2003b).

When superimposing the structures of the 30S subunit, we see that the movement of the small subunit from the hybrid ribosome goes in the same direction as the one observed in the presence of EF-G, although it is smaller in magnitude (Figure 4A, left panel). This fitting result indicates that the inherent ratcheting of the ribosome is less extensive than the one finally stabilized by the action of EF-G but that it is nevertheless sufficient to position the tRNAs in the hybrid configuration. In Figure 4A (right panel), we show EF-G and its relative position with respect to the 30S subunit in the three complexes. From this juxtaposition, it is seen that the tip of domain IV of EF-G (the contact point of the factor with h44) overlaps with

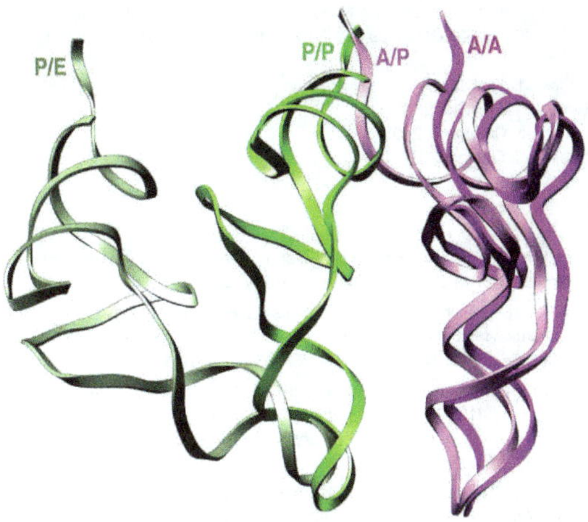

Figure 5. Comparison between the Positions and Orientations of Classic and Hybrid tRNAs Deduced from the Cryo-EM Maps

The fitting of the experimental densities with the X-ray structures of the tRNAs by real space refinement using rigid-body fitting. Although the CCA end of the classic A/A tRNA was not well-defined in the experimental density map, the complete quasiatomic structure is shown for clearer visualization and perception of the relative positions of the different configuration tRNAs. The structures are displayed in ribbons.

h34 in the classic 30S subunit configuration (shown in transparent gray). Therefore, the kinetic efficiency attributed to the hybrid-state tRNA configuration can be explained in structural terms by the fact that the small subunit is already partially ratcheted in the hybrid-state ribosome, thus no longer presenting steric constraints in the vicinity of the 30S head, which would slow down the positioning of EF-G's domain IV in the 30S subunit's A site region.

After the initial binding to the ribosome, domain IV of EF-G contacts h44 and the head of the small subunit in the vicinity of the decoding center. It is known that, in the posttranslocational ribosome, the tip of h44 is shifted toward the P site by ~8 Å (VanLoock et al., 2000). When we compare the position of the tip of h44 from classic, hybrid, and EF-G bound ribosomes (Figure S5), we see that this shift is indeed mainly caused by the direct action of EF-G. However, a displacement in h44 (~2 Å) caused by the dynamic behavior of the ribosome is also noted in the case of the hybrid-state ribosome. In Figure 5, we show the relative position of classic and hybrid-state tRNAs. From this fitting, it is clear that the anticodon stem loops of both classic and hybrid-state tRNAs do not superpose perfectly. The shift observed in the A/A-to-A/P tRNA transition would be based on this lateral displacement of the decoding center in h44, whereas the shift observed in the P/P to P/E reconfiguration (~4–5 Å) may be generated by the temporary destabilization of the codon-anticodon helix shown to occur during the movement of the tRNA into the P/E hybrid state (McGarry et al., 2005).

On the other hand, in Figure 4B we show the superposition of 30S subunits from hybrid and EF-G bound forms in the presence of hybrid A/P and P/E tRNAs. According to this fitting, it is clear that before the complete ratcheting has occurred, and h44 is further shifted by the action of EF-G, the ASL from the A/P tRNA and domain IV of EF-G occupy similar positions. As it is known that EF-G, even in the absence of GTP hydrolysis, accelerates the translocation rate by 1000-fold (Rodnina et al., 1997), the sole binding of EF-G•GTP likely facilitates the complete translocation of the hybrid A/P tRNA to the P site (Pan et al., 2007), displacing both the decoding center (the tip of h44) and the mRNA-tRNA complex toward the P site. Thus, the potential steric clash between the ASL of the A/P tRNA and EF-G explains why all previous cryo-EM reconstructions of ribosomes bearing EF-G failed to show the simultaneous presence of hybrid A/P and P/E tRNAs. Only the hybrid P/E tRNA was visualized in a ribosome bound with EF-G, as it was in the posttermination state lacking A site-bound tRNA (Valle et al., 2003b).

Concluding Remarks

Chemical footprinting experiments followed the movement of the tRNAs through the intersubunit space and showed that, after the peptidyl transfer, the tRNAs reside in intermediate hybrid positions (Moazed and Noller, 1989). The proposed driving force for this partial translocation is based on the well-established different affinities the CCA end shows for 50S subunit A, P, and E sites (Lill et al., 1989; Feinberg and Joseph, 2001; Virumae et al., 2002; reviewed in Noller et al., 2002). However, the ratcheting movement of the ribosome has never been integrated in this scheme. Based on our data, we demonstrate how the repositioning of the tRNAs in the 50S subunit is based not only on the affinity the CCA ends show for the different sites but also on the structural reconfiguration of S13, L1 stalk, L5, H68, H69, and H38 that is caused by the rotation of the small subunit and the new environment thus created, which favors transitions from A/A-P/P to A/P-P/E.

EXPERIMENTAL PROCEDURES

Sample Preparation

The ribosomal complex was prepared using an in vitro translation system in HiFi buffer (50 mM Tris-HCl [pH 7.5], 70 mM NH_4Cl, 30 mM KCl, 3.5 mM $MgCl_2$, 0.5 mM spermidine, 8 mM putrescine, 2 mM DTT) (Gromadski and Rodnina, 2004). Ribosomes from strain MRE600 were obtained as previously described (Moazed and Noller, 1986). EF-Tu and $tRNA^{Trp}$ expression and purification were performed as previously described (Cochella and Green, 2005). A 70S initiation mix (500 μl) containing 1 μM ribosomes, 1.5 μM fMet-$tRNA^{fMet}$, 1 mM GTP, 1 μM each IF (1, 2, and 3), and 8 μM MWstop-mRNA was incubated at 37°C for 70 min. After a sucrose cushion in buffer D (with $MgCl_2$ added to 10 mM final concentration), ICs were pelleted, washed once, and resuspended in HiFi buffer (100 μl). 35S-labeled fMet-$tRNA^{fMet}$-containing complexes were made in parallel with the unlabeled complexes to check the concentrations by scintillation counting. The yield of IC formation was 65%, and thus the final concentration was 3.2 μM. This complex was then diluted with the same buffer to 1 μM concentration. The ternary complex was prepared separately by incubating (in 80 μl) 2 μM Trp-$tRNA^{Trp}$, 40 μM EF-Tu, 1 mM GTP, 67 mM L-Trp, 8 μM TrpRS, and energy regeneration system (0.025 mg/ml PK, 1.5 mM ATP, 1.5 mM PEP) at 37°C for 45 min. Elongation was started by mixing equal volumes (40 μl) of the initiation and ternary complex. After 5 min incubation at 37°C, the mixtures (80 μl) were frozen in liquid nitrogen immediately. 35S-labeled fMet-$tRNA^{fMet}$-containing complexes were checked for dipetide formation and release activity (A site occupancy). The release activity was determined as the amount of peptide that could be released by RF1 in a stop codon-dependent reaction after fMet-Trp-$tRNA^{Trp}$ translocation with addition of EF-G. Elongation complexes were incubated with EF-G and RF1 for 1 min at 37°C and quenched with 10% formic acid. For peptide bond formation measurement, complexes were hydrolyzed with 0.75 N KOH. Products from both assays were separated by TLC electrophoresis in pyridine-acetic acid (pH 2.8). The complexes formed 82% fMet-Trp dipeptide and released 65% fMet-Trp upon addition of EF-G and RF1.

Electron Microscopy and Image Processing

To prepare the sample for cryo-EM, an aliquot of the ribosomal complex at 1 μM concentration was thawed on ice and diluted with HiFi buffer to a final concentration of ribosome of 30 nM. A carbon-coated Quantifoil 2/4 grid (Quantifoil Micro Tools GmbH, Jena, Germany) was prepared following standard cryo procedures (Grassucci et al., 2007). Images were recorded on a single-port 4K × 4K CCD camera (TVIPS TemCam-F415) mounted on an FEI Tecnai Polara operating at 300 kV and a nominal magnification of 59,000× in low-dose conditions (~20 $e^-/Å^2$) by using the automated data collection system AutoEMation (Lei and Frank, 2005). Because of the postmagnification ratio, the effective magnification is 100,000×, corresponding to a pixel size of 1.5 Å on the object scale. The AutoEMation session lasted 44 hr and was left completely unattended, including an overnight period, except for refilling of liquid nitrogen one to two times per day. High-magnification CCD images (2073) were acquired from 691 holes, which were automatically selected from 16 grid squares, as the system can automatically collect multiple high-magnification low-dose images in each hole. After visual inspection and evaluation of micrographs and their power spectra, 1996 CCD images were selected for further analysis. Particles were chosen via automated particle picking (Rath and Frank, 2004) followed by visual selection. The total number of particles used was 216,517. The 3D reconstructions then followed the standard SPIDER protocols for reference-based reconstruction (Frank et al., 1996), including a 3D projection alignment procedure with correction of the contrast transfer function (CTF) using defocus groups, varying the defocus of the selected images from 1.2 to 4 μm, followed by multiple rounds of angular refinement with angular increments. The map corresponding to the accommodated state of Phe-$tRNA^{Phe}$ was used as initial 3D reference (EMD-1056). Supervised classification methodology was applied as described in Figure 1. The final resolution for the maps was 8.75, 8.9, and 10.5 Å, using a cutoff of 0.5 in the Fourier shell correlation (Figure S2), and the maps were filtered accordingly.

The fitting of the cryo-EM maps with the atomic models was done using the real space refinement program RSREF (Chapman, 1995). The initial atomic model was based on the X-ray structure of an empty *E. coli* 70S ribosome (Protein Data Bank [PDB] codes 2AVY and 2AW4; Schuwirth et al. [2005]). The P site fMet-tRNA used was the one described in the 70S structure of *Thermus Thermophilus* (2j00; Selmer et al. [2006]), whereas the A site tRNA used corresponds to Phe-tRNA described in Yusupov et al. (2001) (1GIX). The 7 nt fragment from the tip of h44 (nt 1989–1996) corresponds to the X-ray structure of the 30S subunit in the presence of cognate Phe-tRNA and paramomycin (1IBK; Ogle et al. [2001]). The cryo-EM densities and the atomic models were visualized using the programs Amira (http://amira.zib.de) and Chimera (Pettersen et al., 2004), respectively.

ACCESSION NUMBERS

The EM maps have been deposited in the 3D-EM database, EMBL-European Bioinformatics Institute, Cambridge, under the corresponding accession numbers: EMD-1540 (classic-70S•fMet-$tRNA^{fMet}$•Trp-$tRNA^{Trp}$) and EMD-1541 (hybrid-70S•fMet-$tRNA^{fMet}$•Trp-$tRNA^{Trp}$).

SUPPLEMENTAL DATA

The Supplemental Data include five figures and can be found with this article online at http://www.molecule.org/supplemental/S1097-2765(08)00680-1.

ACKNOWLEDGMENTS

We thank Derek Taylor and Wen Li for discussion of the manuscript, Jayati Sengupta and Ruben L. Gonzalez for valuable suggestions, Robert A. Grassucci for assistance with the image collection, and Michael Watters for assistance with the preparation of the illustrations. This work was supported by Howard Hughes Medical Institute and National Insitutes of Health (NIH) grants R01 GM55440 and R37 GM29169 (to J.F.) and NIH grant R01 GM059425 (to R.G.).

Received: June 18, 2008
Revised: August 12, 2008
Accepted: October 1, 2008
Published: October 23, 2008

REFERENCES

Abel, K.M., Yoder, M.D., Hilgenfeld, R., and Jurnak, F. (1996). An α to β conformational switch in EF-Tu. Structure *4*, 1153–1159.

Agrawal, R.K., Penczek, P., Grassucci, R.A., and Frank, J. (1998). Visualization of elongation factor G on the Escherichia coli 70S ribosome: the mechanism of translocation. Proc. Natl. Acad. Sci. USA *95*, 6134–6138.

Agrawal, R.K., Penczek, P., Grassucci, R.A., Burkhardt, N., Nierhaus, K.H., and Frank, J. (1999). Effect of buffer conditions on the position of tRNA on the 70S ribosome as visualized by cryoelectron microscopy. J. Biol. Chem. *274*, 8723–8729.

Allen, G.S., Zavialov, A., Gursky, R., Ehrenberg, M., and Frank, J. (2005). The cryo-EM structure of a translation initiation complex from *Escherichia coli*. Cell *121*, 703–712.

Blanchard, S.C., Gonzalez, R.L., Kim, H.D., Chu, S., and Puglisi, J.D. (2004). tRNA selection and kinetic proofreading in translation. Nat. Struct. Mol. Biol. *11*, 1008–1014.

Bretscher, M.S. (1968). Translocation in protein synthesis: a hybrid structure model. Nature *218*, 675–677.

Chapman, M.S. (1995). Restrained real-space macromolecular atomic refinement using a new resolution-dependent electron density function. Acta Crystallogr. A *51*, 69–80.

Cochella, L., and Green, R. (2005). An active role for tRNA in decoding beyond codon:anticodon pairing. Science *308*, 1178–1180.

Cornish, P.V., Ermolenko, D.N., Noller, H.F., and Ha, T. (2008). Spontaneous intersubunit rotation in single ribosomes. Mol. Cell *30*, 578–588.

Cromie, M.J., Shi, Y., Latifi, T., and Groisman, E.A. (2006). An RNA sensor for intracellular Mg^{2+}. Cell *125*, 71–84.

Cukras, A.R., Southworth, D.R., Brunelle, J.L., Culver, G.M., and Green, R. (2003). Ribosomal proteins S12 and S13 function as control elements for translocation of the mRNA:tRNA complex. Mol. Cell *12*, 321–328.

Dorner, S., Brunelle, J.L., Sharma, D., and Green, R. (2006). The hybrid state of tRNA binding is an authentic translation elongation intermediate. Nat. Struct. Mol. Biol. *13*, 234–241.

Draper, D.E., Grilley, D., and Soto, A.M. (2005). Ions and RNA folding. Annu. Rev. Biophys. Biomol. Struct. *34*, 221–243.

Fei, J., Kosuri, P., MacDougall, D.D., and Gonzalez, R.L. (2008). Coupling of ribosomal L1 stalk and tRNA dynamics during translation elongation. Mol. Cell *30*, 348–359.

Feinberg, J.S., and Joseph, S. (2001). Identification of molecular interactions between P-site tRNA and the ribosome essential for translocation. Proc. Natl. Acad. Sci. USA *98*, 11120–11125.

Frank, J., and Agrawal, R.K. (2000). A ratchet-like inter-subunit reorganization of the ribosome during translocation. Nature *406*, 318–322.

Frank, J., Radermacher, M., Penczek, P., Zhu, J., Li, Y., Ladjadj, M., and Leith, A. (1996). SPIDER and WEB: processing and visualization of images in 3D electron microscopy and related fields. J. Struct. Biol. *116*, 190–199.

Frank, J., Gao, H., Sengupta, J., Gao, N., and Taylor, D.J. (2007). The process of mRNA-tRNA translocation. Proc. Natl. Acad. Sci. USA *104*, 19671–19678.

Fredrick, K., and Noller, H.F. (2003). Catalysis of ribosomal translocation by sparsomycin. Science *300*, 1159–1162.

Gabashvili, I.S., Agrawal, R.K., Spahn, C.M., Grassucci, R.A., Svergun, D.I., Frank, J., and Penczek, P. (2000). Solution structure of the *E. coli* 70S ribosome at 11.5 Å resolution. Cell *100*, 537–549.

Gao, H., Valle, M., Ehrenberg, M., and Frank, J. (2004). Dynamics of EF-G interaction with the ribosome explored by classification of a heterogeneous cryo-EM dataset. J. Struct. Biol. *147*, 283–290.

Gao, N., Zavialov, A.V., Li, W., Sengupta, J., Valle, M., Gursky, R.P., Ehrenberg, M., and Frank, J. (2005). Mechanism for the disassembly of the posttermination complex inferred from cryo-EM studies. Mol. Cell *18*, 663–674.

Gao, H., Zhou, Z., Rawat, U., Huang, C., Bouakaz, L., Wang, C., Cheng, Z., Liu, Y., Zavialov, A., Gursky, R., et al. (2007). RF3 induces ribosomal conformational changes responsible for dissociation of class I release factors. Cell *129*, 929–941.

Grassucci, R.A., Taylor, D.J., and Frank, J. (2007). Preparation of macromolecular complexes for cryo-electron microscopy. Nat. Protocols *2*, 3239–3246.

Gromadski, K.B., and Rodnina, M.V. (2004). Kinetic determinants of high-fidelity tRNA discrimination on the ribosome. Mol. Cell *13*, 191–200.

Horan, L.H., and Noller, H.F. (2007). Intersubunit movement is required for ribosomal translocation. Proc. Natl. Acad. Sci. USA *104*, 4881–4885.

Kim, H.D., Puglisi, J.D., and Chu, S. (2007). Fluctuations of transfer RNAs between classical and hybrid states. Biophys. J. *93*, 3575–3582.

Klaholz, B.P., Myasnikov, A.G., and Van Heel, M. (2004). Visualization of release factor 3 on the ribosome during termination of protein synthesis. Nature *427*, 862–865.

Lei, J., and Frank, J. (2005). Automated acquisition of cryo-electron micrographs for single particle reconstruction on an FEI Tecnai electron microscope. J. Struct. Biol. *150*, 69–80.

Li, W., and Frank, J. (2007). Transfer RNA in the hybrid P/E state: correlating molecular dynamics simulations with cryo-EM data. Proc. Natl. Acad. Sci. USA *104*, 16540–16545.

Lill, R., Robertson, J.M., and Wintermeyer, W. (1989). Binding of the 3′ terminus of tRNA to 23S rRNA in the ribosomal exit site actively promotes translocation. EMBO J. *8*, 3933–3938.

McGarry, K.G., Walker, S.E., Wang, H., and Fredrick, K. (2005). Destabilization of the P site codon-anticodon helix results from movement of tRNA into the P/E hybrid state within the ribosome. Mol. Cell *20*, 613–622.

Moazed, D., and Noller, H.F. (1986). Transfer RNA shields specific nucleotides in 16S ribosomal RNA from attack by chemical probes. Cell *47*, 985–994.

Moazed, D., and Noller, H.F. (1989). Intermediate states in the movement of transfer RNA in the ribosome. Nature *342*, 142–148.

Munro, J.B., Altman, R.B., O'Connor, N., and Blanchard, S.C. (2007). Identification of two distinct hybrid state intermediates on the ribosome. Mol. Cell *25*, 505–517.

Muth, G.W., Chen, L., Kosek, A.B., and Strobel, S.A. (2001). pH-dependent conformational flexibility within the ribosomal peptidyl transferase center. RNA *10*, 1403–1415.

Noller, H.F., Yusupov, M.M., Yusupova, G.Z., Baucom, A., and Cate, J.H. (2002). Translocation of tRNA during protein synthesis. FEBS Lett. *514*, 11–16.

Ogle, J.M., Brodersen, D.E., Clemons, W.M., Jr., Tarry, M.J., Carter, A.P., and Ramakrishnan, V. (2001). Recognition of cognate transfer RNA by the 30S ribosomal subunit. Science *292*, 897–902.

Ogle, J.M., Murphy, F.V., Tarry, M.J., and Ramakrishnan, V. (2002). Selection of tRNA by the ribosome requires a transition from an open to a closed form. Cell *111*, 721–732.

Pan, D., Kirillov, S.V., and Cooperman, B.S. (2007). Kinetically competent intermediates in the translocation step of protein synthesis. Mol. Cell *25*, 519–529.

Pettersen, E.F., Goddard, T.D., Huang, C.C., Couch, G.S., Greenblatt, D.M., Meng, E.C., and Ferrin, T.E. (2004). UCSF Chimera—a visualization system for exploratory research and analysis. J. Comput. Chem. *25*, 1605–1612.

Polekhina, G., Thirup, S., Kjeldgaard, M., Nissen, P., Lippmann, C., and Nyborg, J. (1996). Helix unwinding in the effector region of elongation factor EF-Tu-GDP. Structure *4*, 1141–1151.

Rath, B.K., and Frank, J. (2004). Fast automatic particle picking from cryo-electron micrographs using a locally normalized cross-correlation function: a case study. J. Struct. Biol. *145*, 84–90.

Rodnina, M.V., Savelsbergh, A., Katunin, V.I., and Wintermeyer, W. (1997). Hydrolysis of GTP by elongation factor G drives tRNA movement on the ribosome. Nature *385*, 37–41.

Schmeing, T.M., Seila, A.C., Hansen, J.L., Freeborn, B., Soukup, J.K., Scaringe, S.A., Strobel, S.A., Moore, P.B., and Steitz, T.A. (2002). A pre-translocational intermediate in protein synthesis observed in crystals of enzymatically active 50S subunits. Nat. Struct. Biol. *9*, 225–230.

Schuwirth, B.S., Borovinskaya, M.A., Hau, C.W., Zhang, W., Vila-Sanjurjo, A., Holton, J.M., and Cate, J.H.D. (2005). Structures of the bacterial ribosome at 3.5 Å resolution. Science *310*, 827–834.

Selmer, M., Dunham, C.M., Murphy, F.V., 4th, Weixlbaumer, A., Petry, S., Kelley, A.C., Weir, J.R., and Ramakrishnan, V. (2006). Structure of the 70S ribosome complexed with mRNA and tRNA. Science *313*, 1935–1942.

Semenkov, Y.P., Shapkina, T., Makhno, V., and Kirillov, S. (1992). Puromycin reaction for the A site-bound peptidyl-tRNA. FEBS Lett. *296*, 207–210.

Semenkov, Y.P., Rodnina, M.V., and Wintermeyer, W. (2000). Energetic contribution of tRNA hybrid state formation to translocation catalysis on the ribosome. Nat. Struct. Biol. *7*, 1027–1031.

Sharma, D., Southworth, D.R., and Green, R. (2004). EF-G-independent reactivity of a pre-translocation-state ribosome complex with the aminoacyl tRNA substrate puromycin supports an intermediate (hybrid) state of tRNA binding. RNA *10*, 102–113.

Valle, M., Zavialov, A., Li, W., Stagg, S.M., Sengupta, J., Nielsen, R.C., Nissen, P., Harvey, S.C., Ehrenberg, M., and Frank, J. (2003a). Incorporation of aminoacyl-tRNA into the ribosome as seen by cryo-electron microscopy. Nat. Struct. Biol. *10*, 899–906.

Valle, M., Zavialov, A.V., Sengupta, J., Rawat, U., Ehrenberg, M., and Frank, J. (2003b). Locking and unlocking of ribosomal motions. Cell *114*, 123–134.

VanLoock, M.S., Agrawal, R.K., Gabashvili, I.S., Qi, L., Frank, J., and Harvey, S.C. (2000). Movement of the decoding region of the 16S ribosomal RNA accompanies tRNA translocation. J. Mol. Biol. *304*, 507–515.

Virumae, K., Saarma, U., Horowitz, J., and Remme, J. (2002). Functional importance of the 3′-terminal adenosine of tRNA in ribosomal translation. J. Biol. Chem. *277*, 24128–24134.

Yusupov, M.M., Yusupova, G.Z., Baucom, A., Lieberman, K., Earnest, T.N., Cate, J.H.D., and Noller, H.F. (2001). Crystal structure of the ribosome at 5.5 Å resolution. Science *292*, 883–896.

Zavialov, A.V., and Ehrenberg, M. (2003). Peptidyl-tRNA regulates the GTPase activity of translation factors. Cell *114*, 113–122.

Zavialov, A.V., Hauryliuk, V.V., and Ehrenberg, M. (2005). Guanine-nucleotide exchange on ribosome-bound elongation factor G initiates the translocation of tRNAs. J. Biol. *4*, 9. Published online June 27, 2005. 10.1186/jbiol24.

Journal of Structural Biology 164 (2008) 24–32

Contents lists available at ScienceDirect

Journal of Structural Biology

journal homepage: www.elsevier.com/locate/yjsbi

Exploration of parameters in cryo-EM leading to an improved density map of the *E. coli* ribosome

Jamie LeBarron [a], Robert A. Grassucci [b], Tanvir R. Shaikh [a], William T. Baxter [a], Jayati Sengupta [a,1], Joachim Frank [b,c,*]

[a] *Wadsworth Center, Empire State Plaza, Albany, NY 12201-0509, USA*
[b] *Howard Hughes Medical Institute, Department of Biochemistry and Molecular Biophysics, Columbia University, 650 West 168th Street, New York, NY 10032, USA*
[c] *Department of Biological Sciences, Columbia University, USA*

ARTICLE INFO

Article history:
Received 3 December 2007
Received in revised form 16 May 2008
Accepted 17 May 2008
Available online 18 June 2008

Keywords:
Electron microscopy
Single-particle reconstruction
Ribosome
Resolution

ABSTRACT

A number of image processing parameters in the 3D reconstruction of a ribosome complex from a cryo-EM data set were varied to test their effects on the final resolution. The parameters examined were pixel size, window size, and mode of Fourier amplitude enhancement at high spatial frequencies. In addition, the strategy of switching from large to small pixel size during angular refinement was explored. The relationship between resolution (in Fourier space) and the number of particles was observed to follow a lin-log dependence, a relationship that appears to hold for other data, as well. By optimizing the above parameters, and using a lin-log extrapolation to the full data set in the estimation of resolution from half-sets, we obtained a 3D map from 131,599 ribosome particles at 6.7 Å resolution (FSC = 0.5).

© 2008 Elsevier Inc. All rights reserved.

1. Introduction

Slowly but steadily, the resolution of single-particle reconstructions (see Frank, 2006) has improved, with particles of high symmetry (icosahedral viruses, GroEL) taking the lead. For instance, GroEL, having 14-fold symmetry, is now around 4 Å (Ludtke et al., 2008), and two reoviruses have been solved to close-to-atomic resolution (Yu et al., 2008; Zhang et al., 2008). While the electron-optical resolution of modern electron microscopes is well sufficient to achieve atomic resolution, the struggle to reach this goal faces extraordinary difficulties, especially in the absence of symmetries. If we assume a well-tuned instrument with high coherence, then the limiting factors are specimen stage instability, electrostatic charging effects, conformational heterogeneity, and strong background noise due to the low dose and the large amount of inelastic scattering in ice. There is undoubtedly room for improvement in the preparation of the specimen on the EM grid. For instance, thickness of ice has been identified as one of the factors determining the resolution of the reconstruction (Stagg et al., 2006). However, once a suitable protocol has been found, and a data set has been collected, there are a large variety of ways to proceed toward a reconstruction. It is this latter problem that we have addressed with a large data set of ribosome images.

The problem can be phrased in terms of the optimization problem posed by the single-particle reconstruction from multiple projections with unknown angles (Yang et al., 2005). This optimization problem is usually addressed by an iterative processing method, called angular refinement (Penczek et al., 1994), in which alternately two different problems are solved, one being the orientation of the projections with reference to an existing density map, the other being the reconstruction of a density map from the projections set whose angles have been determined in the previous round (see the general formulation of this approach given in Yang et al., 2005). It has been pointed out (Penczek et al., 1994) that this method will not necessarily result in the global solution. For this reason, there is room and promise for a heuristic exploration.

It is difficult to explore the various factors experimentally/computationally since angular refinement, the most time-consuming step, can take days or even weeks even on a computer cluster. The approach we have taken is to explore the various strategies with a smaller subset of the data first, then apply the strategy yielding the best solution we found to the full data set available. Obviously, despite efforts to perform the exploration systematically, it is in the nature of the problem that our study must fall far short of an exhaustive treatment.

Nevertheless, the study has resulted in a much superior density map of a ribosomal complex, 70S·Phe–tRNA^{Phe}·EF–Tu·GDP·kir, than that reported before (Valle et al., 2002, 2003). The quality of the map can be characterized by comparison of various components (ribosomal proteins or RNA) with X-ray structures, and is reflected by the experimental resolution in the range of 6.7 Å.

* Corresponding author. Fax: +1 212 305 9500.
E-mail address: jf2192@columbia.edu (J. Frank).
[1] Present address: Indian Institute of Chemical Biology, Kolkata, India

1047-8477/$ - see front matter © 2008 Elsevier Inc. All rights reserved.
doi:10.1016/j.jsb.2008.05.007

J. LeBarron et al. / Journal of Structural Biology 164 (2008) 24–32 25

2. Investigation of variables limiting resolution

2.1. Principle and practice of resolution estimation

In order to investigate what processing method might lead to better results, we need to employ an appropriate measure of reconstruction quality. Conventionally, the Fourier shell correlation (FSC) curve (Saxton and Baumeister, 1982; Harauz and van Heel, 1986) is used, which is obtained by a comparison of reconstructions from randomly picked half-sets of the data set. However, we were frustrated in our initial analysis since the FSC curve as a whole (not just the figure derived from it by setting a threshold) responded only marginally to changes in processing that produced a strongly different visual appearance in the reconstruction. In other words, we lacked a quantitative confirmation for improvements in quality quite obvious to the eye.

We then discovered that the improvements were not reflected by the FSC since they affected only that part of the density map that was occupied by the reconstructed molecule. In contrast, the FSC as applied to the whole volume measured the reproducibility of the density map as a whole, including the surroundings of the molecule.

Briefly, in a density map obtained by 3D reconstruction, the reconstructed molecule is usually surrounded by speckles of density termed "clutter." The clutter becomes strongly visible at lowered density threshold. It is generated by superposition of projection density exterior to the particle itself and carries a unique signature tied to the unique set of angles of the projections entering the reconstruction. Since the set of angles is different for the two half-sets, the clutter signature is irreproducible, leading to a steep falloff of the FSC for the periphery. Since the clutter occupies a relatively large portion of the total volume, it has a large effect on the resolution estimation. When resolution is measured by comparison of two half-set reconstructions, inconsistencies in the clutter will therefore lead to under-reported resolution. Using a similar rationale—to remove peripheral, disordered portions of the molecule structure—Stewart and coworkers (2000) introduced soft masking to obtain a more realistic resolution estimate for the resolution of the icosahedrally ordered viral capsid of Ad2 virus (for another example, see Andersen et al., 2006). A smooth mask must be used, evidently, because application of a binary-valued mask to both reconstructions would lead to a false indication of reproducibility, resulting in a grossly over-reported resolution. The mask function which minimizes such artifacts for a global particle is a rotationally symmetric Gaussian mask, placed at the center of the particle, with a standard deviation corresponding to the particle radius.

The effect is demonstrated (Fig. 1) with one of the reconstructions (Fig. 2G) to be presented below. When using a reduced density threshold, the clutter outside the molecule becomes apparent. Without use of the Gaussian mask, the resolution is 12.4 Å, while with this mask in place, the resolution improves to 10.1 Å. Investigation of the part of the volume that was rejected by the mask (i.e., accepted by the mask complement) with the FSC gives a resolution of 19.7 Å, confirming that the reproducibility is much worse at the peripheral part of the volume. (The resolution of ~20 Å for the peripheral part of the volume may seem surprisingly high, but it is due to the fact that the soft mask complement inevitably catches some of the peripheral, well-reproducible ribosome mass).

A few tests with varying mask widths (±20% from the approximate radius of the ribosome, 130 Å) show that the reported resolution is to some extent, but not critically, dependent on the choice. By using the rule of Gaussian half-width = particle radius, we make sure that all nominal resolution figures for the ribosome are comparable. In the following, we therefore have used a half-width

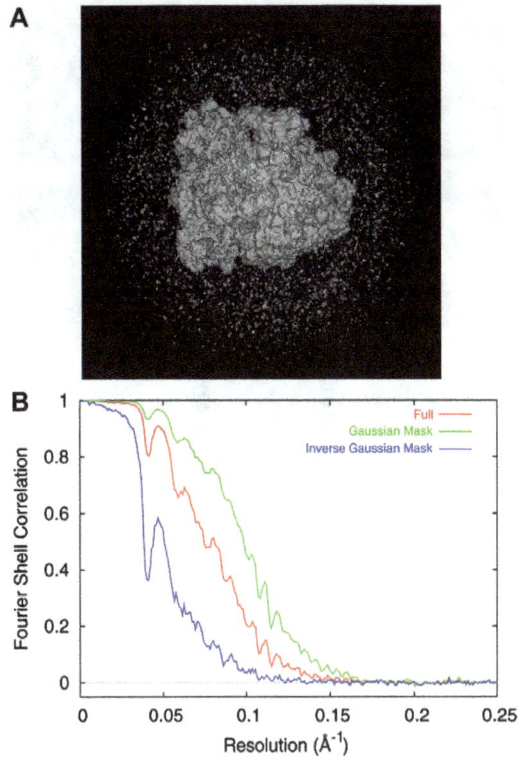

Fig. 1. (A) Clutter surrounding the ribosome in reconstruction shown in Fig. 2G. (B) FSC curves computed for the full volume (red), after soft masking (green), and for the clutter excluded by soft masking (blue).

r = 130 Å throughout. Thus, for the purpose of this study, we give two resolution figures for each density map, the first *with*, and the second, in parentheses, *without* application of the Gaussian mask to both half-set reconstructions being compared by the Fourier shell correlation. In the course of the presentation, it will become clear that only the former figure is sensitive to the various choices of parameters and strategies, and that it may best reflect the reproducible overall resolution of the particle region of the EM density map (Table 1).

2.2. Creation of the test data set

To determine which processing parameters affect resolution, a small test data set of 51,622 particles was created and processed using standard methods (Frank et al., 2000; Shaikh et al., 2008a) to produce a baseline reconstruction. A sample of the complex described in Valle et al. (2003) and known to be quite stable was prepared (Grassucci et al., 2007a) and imaged on film in the standard manner (Grassucci et al., 2007b).

To briefly summarize, the sample consisted of *E. coli* ribosome stalled in the pre-accommodation state after GTP hydrolysis with the antibiotic kirromycin. Another preparation of this sample was previously (Valle et al., 2003) found to have tRNA in both the E and P sites, with aa-tRNA·EF–Tu·GDP·kir also bound in greater than 70% of the ribosome population. Micrographs were recorded at low-dose conditions on an FEI Tecnai F30 Polara electron microscope operated with an accelerating voltage of 300 kV, a magnifica-

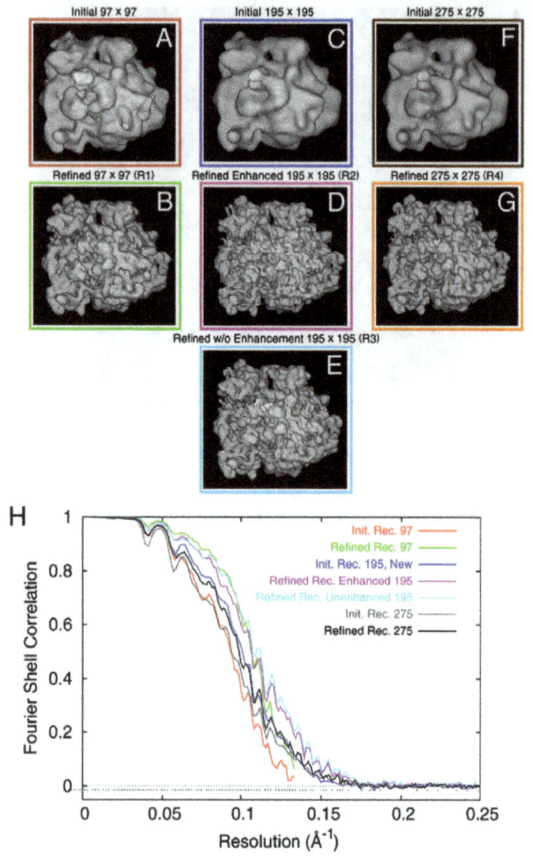

Fig. 2. Reconstructions from a 52,000-particle test data set. (A, B) Initial and refined reconstruction, with arrays 2× decimated; (C, D) initial and refined reconstruction from the non-decimated data; (E) same as (D), except that amplitude enhancement was used in each step of the refinement; (F, G) same as (C, D) but with window size enlarged from 195 × 195 to 275 × 275. (H) FSC curves for (A–G), color-coded to match color surrounding each reconstruction.

tion of ~38,000×, and with the specimen cooled to liquid-nitrogen temperature (80 K). The micrographs were scanned with a Z/I Photoscan 2000 (Z/I Imaging Corp., Huntsville, Alabama) with a step size of 7 µm, resulting in a pixel size of 1.86 Å on the object scale.

Image processing was performed as described (Shaikh et al., 2008a), with modifications described below. Prior to particle selection, all digitized micrographs were decimated by a factor of 2 using the SPIDER "DC S" command, resulting in a pixel size of 3.76 Å on the object scale. Digitized micrographs were grouped according to their defocus values into 12 defocus groups, with average defocus values ranging from 2.41 to 3.89 µm. The average range within a defocus group was 57 nm, and the maximum range 97 nm. This rather coarse grouping was chosen initially for expediency as it reduces the total number of refinement cycles in the test calculations. Defocus groups with narrower ranges were taken later on with the larger, undecimated data set (see below). Particle selection started in the standard fashion (Shaikh et al., 2008a) including automated particle windowing from digitized micrographs (Rath and Frank, 2004; Roseman, 2004). After a 3D projection alignment of windowed particles to 83 projection views of a known reference (Penczek et al., 1994; Gabashvili et al., 2000) created using an angular spacing of 15°, aligned particles were chosen by using classification (Roseman, 2004; Shaikh et al., 2008b). Briefly, particles that best match a given reference projection are low-pass filtered and are subjected to correspondence analysis. Factorial coordinates from correspondence analysis are then used to classify the filtered particles via the K-means algorithm for each reference view. Particles are then selected either individually or as an entire class. The total number of particles used was 51,622, with a window size of 97 × 97 pixels, and this data set will be referred to as *Data Set 1*. Image processing continued in the standard manner (Shaikh et al., 2008a), resulting in an initial 3D reconstruction with a resolution of 10.7 Å (11.9 Å) using a cutoff of 0.5 in the Fourier shell correlation (FSC) curve (Fig. 2A).

A cycle of angular refinement is defined in the following way: defocus group reconstructions from the previous cycle are merged and CTF-corrected, to yield an intermediate reconstruction. This reconstruction is used as a 3D reference to refine data from each defocus group. For this purpose, to maximize the correlation signal, the 3D reference is first modified (via Fourier relationship) with the CTF of the current defocus group. In this manner, a new set of alignment parameters and Eulerian angles is obtained for each defocus group, and a new reconstruction is computed for each.

Several cycles of angular refinement were performed, gradually restricting the angular search range from completely unrestricted to 5°. Midway through the refinement the step size of the angular search was reduced from 2° to 1.5° (Shaikh et al., 2008a). One of the parameters tested in the refinement was the mode of amplitude enhancement. Such enhancement is used to bring the average experimental amplitudes in each shell of the 3D Fourier transform to the level of low-angle solution scattering data (Gabashvili et al., 2000). This enhancement is normally used at the very end of the refinement (Gabashvili et al., 2000) but it has also lately been used after every iteration of the refinement (e.g., Connell et al., 2007). In this mode, the intermediate reconstruction from the previous step is first amplitude-enhanced before the individual CTFs are applied to it.

The idea of applying the enhancement after every step is that the accuracy of alignment may be enhanced due to the higher weight of high spatial frequencies in the correlation expressions.

Table 1
For resolution figures, the second number in parentheses refers to the resolution obtained before application of the soft Gaussian mask to both half-set reconstructions compared by FSC

Reconstruction	Pixel Size (Å)	Window Size (Pixels; Å)	Initial Resolution (Å)	Iterative Amplitude Enhancement	No. of Refinement Iterations	Reconstruction Resolution (Å)
1	3.76	97; 365	10.7 (11.9)	Yes	11	9.4 (10.2)
2	1.86	195; 363	9.9 (10.9)	Yes	9	9.4 (10.3)
3	1.86	195; 363	9.9 (10.9)	No	9	8.9 (10.1)
4	1.86	275; 512	10.6 (14.8)	No	9	10.1 (12.4)
High Resolution						
1st Refinement	3.6	103; 371	10.7 (13.8)	No	11	8.7 (9.3)
2nd Refinement	1.2	309; 371	8.3 (9.4)	No	10	7.7 (8.7)
Small Angle	–	–	–	No	7	7.5 (8.4)

However, a concern and possible disadvantage is that the concurrent amplification of noise may result in erratic alignment errors. For Data Set 1, we used the repeated amplitude enhancement option, each time filtering the resulting density map to the nominal resolution calculated with the FSC = 0.5 cutoff. After 11 iterations of refinement, a final reconstruction from Data Set 1 was obtained with a resolution of 9.4 Å (10.2 Å) at 0.5 FSC (Fig. 2B). In the following, this reconstruction will be referred to as *Reconstruction 1*.

2.3. Examination of the effect of pixel size

According to sampling theory, a continuous signal can be represented faithfully by a set of discrete samples if the signal is sampled at a rate of at least twice the signal's highest spatial frequency. In digital image processing, numerical errors make it necessary to use at least a factor of 3 instead (see Frank, 2006). With a sampling rate of 3.76 Å per pixel, the highest resolution possible from the decimated micrographs used in Data Set 1 is therefore somewhere in the range between 7.5 and 11.3 Å. Since the resolution of Reconstruction 1 was toward the end of this range (9.4 Å (10.2 Å)), the micrograph decimation was thought to be likely a limiting factor. The initial reconstruction of the test data set was therefore redone using the non-decimated, raw micrograph data.

The micrographs had originally been scanned at 7 μm, resulting in arrays with 1.86 Å per pixel. The positions of the final ~52,000 particles within their respective source micrographs were already known from the creation of Data Set 1 using the decimated micrographs. Using this information, and multiplying each coordinate by 2 to compensate for the initial decimation, it was possible to window each particle from the non-decimated micrographs with high precision. The new non-decimated particles were contained in arrays of 195 × 195 pixels. This data set will be referred to as *Data Set 2*. A new initial 3D reconstruction was created using Data Set 2 and the final refined alignment angles, rotations, and translations from the processing that led to Reconstruction 1. The resulting initial reconstruction (i.e., using the initial coarse angular grid of 15°) had a resolution of 9.9 Å (10.9 Å) (Fig. 2C).

Because this initial reconstruction was created using refined alignment information, the use of a large angular search range during the subsequent refinement was considered superfluous. The first two angular search ranges, unrestricted and 15°, were therefore not employed in this instance. After nine iterations of refinement, using amplitude enhancement after each iteration as described previously, the resolution of the refined density map went to 9.4 Å (10.3 Å) at FSC = 0.5 (Fig. 2D). This density map will be referred to as *Reconstruction 2*. Thus, after comparable efforts in image processing and refinement, doubling the micrograph's sampling resolution by switching to the non-decimated data set produced a nominal resolution improvement of merely 0.05 Å (0.01 Å), which is insignificant. This leads to the conclusion that with the limited ($N \sim 52{,}000$) data set used in the test, the two-fold coarser sampling, at 3.72 Å per pixel, is not a resolution-limiting factor.

2.4. Examination of the effect of the mode of amplitude enhancement

To investigate the effect of the iterative use of amplitude enhancement, the processing of Data Set 2 was repeated without employing it. The density map used as the initial reference was the 9.9-Å reconstruction from Data Set 2. Refinement processing for Data Set 2 was repeated as before, without the procedure of amplitude enhancement and low-pass filtering between each iteration that was used to obtain Reconstruction 2. The density map resulting from this processing method will be referred to as *Reconstruction 3*. According to the FSC = 0.5 cutoff, the refined map of Reconstruction 3 had a nominal resolution of 8.9 Å (10.1 Å) (Fig. 2E). Thus, with otherwise identical steps of image processing, the addition of iterative amplitude enhancement decreased the resolution achieved with this data set by ~0.5 Å (0.2 Å). We conclude from this result that the application of amplitude enhancement after each iteration of angular refinement produced no benefit; on the contrary, it reduced the final resolution significantly.

All final reconstructions were routinely amplitude-enhanced, following the protocol spelled out by Gabashvili et al. (2000): by boosting their Fourier amplitudes so that they conformed to the amplitude profile measured by low-angle X-ray solution scattering of a 70S ribosome sample. As application of a final mere multiplicative factor, this correction has no effect on resolution.

2.5. Examination of the effect of the window size

The choice of window size affects the CTF correction in a complicated way, which requires a brief explanation. There exist two different strategies for CTF correction. Following one strategy, the digitized micrographs are CTF-corrected before any additional processing is done, the other where uncorrected data are processed and corrected at the very end (Penczek et al., 1997; Zhu et al., 1997; Gabashvili et al., 2000). We routinely use the second strategy, for various reasons which mostly have to do with the numerical behavior of filters for low SNR (see Frank, 2006). This strategy requires that the window containing the particle include sufficient surrounding background, equivalent in size to the range of the point-spread function placed on the boundary of the particle. For high defocus values, the radius of the point-spread function can be 10 Å and even more. Calculations on phantom data (unpublished results) were performed to determine the minimum window size needed to capture all information required in the CTF correction.

Based on these findings another data set, Data Set 3, was created. The particles for this data set were windowed with a size of 275 × 275 pixels (corresponding to 511 × 511 Å, in contrast to 363 × 363 Å for the smaller non-decimated window size) from the non-decimated micrographs. By applying the final alignment data from the refinement that created Reconstruction 1, the initial reconstruction of Data Set 3 was created. The initial reconstruction had a nominal resolution of 10.6 Å (14.8 Å) (Fig. 2F). Using identical steps of image processing, the use of a larger particle window size led to a deterioration of the resolution by ~0.7 Å (~3.9 Å), when compared to the initial reconstruction of Reconstruction 3 (see Discussion). Refinement of this data set proceeded in a manner similar to the refinement of Reconstruction 3; one refinement of nine iterations was used without repeated amplitude enhancement. The nominal resolution (FSC = 0.5) of this refined density map, *Reconstruction 4*, was 10.1 Å (12.4 Å) (Fig. 2G).

3. Application of reconstruction strategy to the large data set

To reiterate, image processing on the non-decimated Data Set 2 without iterative amplitude enhancement produced a refined reconstruction, Reconstruction 3, with the best nominal resolution (8.9 Å (10.1 Å). To capitalize on the findings, another, larger data set of ~132,000 particles was then recorded for the same specimen with higher magnification, and image processing was used similarly as in the creation of Reconstruction 1. The larger data set was processed using two rounds of refinement without amplitude enhancement and successively smaller pixel sizes, similar to the way Reconstruction 3 had been obtained. The use of different pixel sizes within the same refinement protocol, proceeding from coarse (or decimated) to fine (or "raw"), was adopted in order to accelerate the processing. For this reason, we quote CPU times separately for both steps of the reconstruction procedure.

3.1. *Initial reconstruction and first refinement*

The biological sample used was identical to the one used in the creation of the Data Set 1 (Valle et al., 2003). Micrographs were recorded in a similar manner as for Data Set 1, except with the increased magnification of ~58,000×. In this experiment, a provision was made to reduce electron backscatter and thereby reduce the fog level, by backing the film used to record the images by an extra film. This provision requires a brief explanation: film cassettes by the manufacturer of the electron microscope were originally developed for 100 kV. These cassettes have steel backing. For this voltage, and even up to 200 kV, backscattering is minimal, and no problems due to the increase in fog level arise. Use of such cassettes becomes problematic, however, when the acceleration voltage reaches 300 kV, the voltage used in the experiment. To minimize this effect, before cassettes without metal backing were available, we placed an extra film behind the recording film.

The micrographs were scanned using a sampling of 7 microns on a Z/I Photoscan 2000 (Z/I Imaging Corp, Huntsville, Alabama) resulting in a pixel size of 1.20 Å on the object scale. Digitized micrographs were organized into 92 defocus groups with average defocus values ranging from 1.20 to 4.52 μm. The average defocus range within a defocus group was 164 nm, with a maximum range of 348 nm. The envelope function corresponding to this maximum defocus range falls off to ~50% at the spatial frequency corresponding to 6.7 Å; thus, although the choice of defocus grouping does contribute to the final resolution limitation, it is not the decisive factor.

The first steps of image processing were chosen similar to the steps used in creating Data Set 1, except for differences outlined below. A decimation factor of 3× resulted in a pixel size of 3.6 Å, which was used for the initial reconstruction and the first round of refinement. Particle picking and alignment proceeded as before for creation of the Data Set 1, resulting in 131,599 particles with a window size of 103 × 103 pixels, corresponding to 371 × 371 Å on the object scale. The initial reconstruction was obtained in the standard manner, with a resolution of 10.7 Å (13.8 Å) at FSC = 0.5 (Fig. 3A). Refinement of this initial reconstruction was continued for 11 iterations without the use of repeated amplitude enhancement. The final density map after refinement had a nominal resolution of 8.7 Å (9.3 Å) (Fig. 3B). This refinement was performed on a Beowulf cluster using 35 nodes (1.6 GHz dual-core, dual Opteron; 4 cores per node) in less than 125 h.

3.2. *Non-decimated data set and second refinement*

In a manner similar to that used to create Reconstruction 3, a new non-decimated data set was created. The ~132,000 particles were windowed from non-decimated micrographs while compensating for the previous three-fold decimation. The window size used was 309 × 309 pixels.

The initial reconstruction for this refinement was based on the non-decimated data set, and the alignment information from the final iteration of the three-fold decimated data set refinement. The initial reconstruction immediately had a resolution of 8.3 Å (9.4 Å) at FSC = 0.5 (Fig. 3C). Refinement continued through 10 iterations in total, producing a density map with a resolution of 7.7 Å (8.7 Å) at FSC = 0.5 (Fig. 3D). The refinement required less than 724 h to conclude, on a Beowulf cluster of 27 nodes.

Small-angle refinement was initiated using this 8.7-Å density map as a reference. Small-angle refinement operates functionally identical to 'standard' refinement, except that the angular search

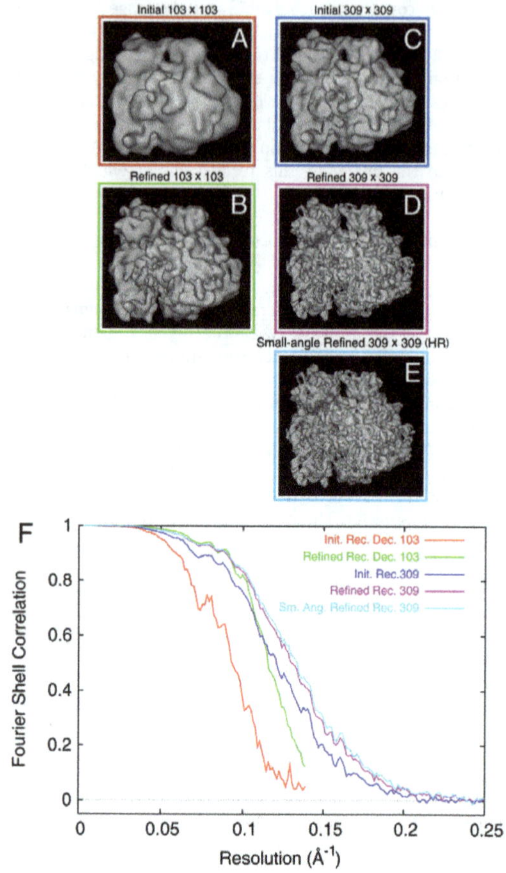

Fig. 3. Reconstructions from the full ~132,000-particle data set. (A, B) initial and refined reconstruction, with arrays 3× decimated; (C, D, E) initial, refined, and fine-angle refined reconstruction from the non-decimated data. (F) FSC curves for (A–G), color-coded to match color surrounding each reconstruction.

range is restricted to 2° and the angular step size used is 0.5°. The small-angle refinement was allowed to proceed for seven iterations, requiring approximately 920 h on a 48-node cluster, and resulting in a final map with a resolution of 7.5 Å (8.4 Å) (Fig. 3E). This final map will be referred to as the "High-resolution Reconstruction."

3.3. *Estimation of resolution using extrapolation*

The estimation of resolution via Fourier shell correlation is based on a comparison of two reconstructions from randomly picked half-sets of the total particle set. Because of this fact, the resolution is usually underestimated since the total set has better statistics than either half-set (see Frank, 2006). It is however possible to estimate the "true" resolution for the full data set by extrapolation from multiple resolution tests with increasing numbers of particles, up to the total number in the data set. We have indications, to be detailed in the following section, that this estimation by extrapolation is justified.

To estimate the "true" resolution through extrapolation, five resolution tests were performed, each with a different number of particles. Half-set reconstructions were either left unmasked or were masked using a Gaussian mask with $r = 130$ Å.

The resulting FSC = 0.5 values follow Rosenthal and Henderson's (2003) observed linear behavior when plotted as ln(*Nd*) versus $1/d^2$, where *d* denotes the resolution distance (i.e., the distance corresponding to the spatial frequency where FSC = 0.5) and *N* the number of particles. This behavior was previously confirmed in other studies (e.g., Liu et al., 2007). Tentatively, the data appear to follow two regimes, with $B = 560$ Å^2 and $B = 300$ Å^2, respectively. Extrapolation to the total number present in the data set (131,599 particles) along the latter line would yield 6.7 Å (7.6 Å) (Fig. 4). However, due to measurement errors in the reported FSC resolution, fitting of the data by a single straight line may be more realistic, and in that case the extrapolation would yield a slightly different value, 7 Å.

Based on the results of this estimation, we filtered the high-resolution reconstruction to 6.7 Å. We then applied amplitude correction using an empirical curve obtained by low-angle X-ray scattering of 70S *E. coli* ribosomes in solution which extends to 8 Å (Gabashvili et al., 2000). To deal with the extra span between 8 and 6.7 Å, we used a simple polynomial extrapolation. This empirically-guided correction should be more accurate than a correction via negative B-factor (Frank, 2006).

3.4. The ribosome at 6.7 Å resolution: a preliminary assessment

Compared to the previously published cryo-EM map of the *E. coli* ribosome from an identical specimen (Valle et al., 2003), at

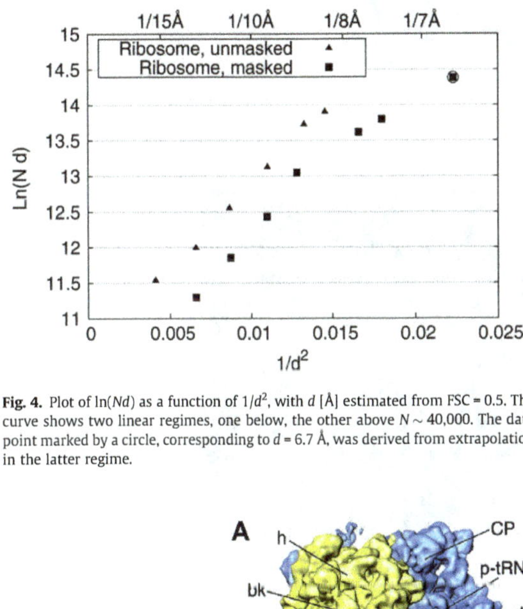

Fig. 4. Plot of ln(*Nd*) as a function of $1/d^2$, with *d* [Å] estimated from FSC = 0.5. The curve shows two linear regimes, one below, the other above $N \sim 40{,}000$. The data point marked by a circle, corresponding to $d = 6.7$ Å, was derived from extrapolation in the latter regime.

a resolution of 10.3 Å resolution (FSC = 0.5), the new map (Fig. 5A) represents a significant advance. A comparison with X-ray structures of the *E. coli* ribosome (Schuwirth et al., 2005)—albeit in a different state—enables us to assess how individual components such as peripheral ribosomal proteins, A-form RNA helices, and tRNA are rendered (Fig. 5B). Of high interest is finally the additional detail observed in the interactions among EF-Tu, GDP, aminoacyl-tRNA, and the ribosome, which advances our understanding of the tRNA selection step during decoding (see Valle et al., 2002, 2003; Frank et al., 2005; Sengupta et al., in preparation). While a more detailed evaluation of the new map using molecular-dynamics fitting (Trabuco et al., 2008) is pending (Sengupta et al., in preparation), some preliminary observations are provided in the following.

The best indicator of the quality and effective resolution of the map is provided by observation of known features. In addition to the more prominent appearance of major and minor grooves in double-helical rRNA, other secondary structural features are identifiable in the 6.7-Å map. To demonstrate some of these features, close-up views of 16S rRNA helices 16 and 17 situated at the shoulder region of the small subunit, and 5S rRNA of the large subunit are provided (Fig. 6). In addition to regions of the map filtered to the nominal resolution (Butterworth filter with start and stop at ±0.05 around the spatial frequency (in Nyquist units) corresponding to 6.7 Å), we show the same regions filtered with a Butterworth filter starting at 6.7 Å and extending to ~4 Å, with the half-width of the filter lying in the vicinity of 5 Å.

These density regions of the cryo-EM map can be compared with density maps generated by converting the coordinates of these regions taken from the crystal structure of the *E. coli* 70S ribosome (PDB code 2AVY or 2AW7 for 30S; 2AW4 or 2AWB for 50S; Schuwirth et al., 2005), and filtering them to the same resolution. All molecular surfaces are solid, without interruption by spikes or other noisy features. The fact that the molecular features are discernable with comparable definition as in the X-ray map filtered to the same resolution gives us confidence that the method of resolution estimation, by extrapolation from the half set to the full data set, is well justified.

In helix 16 of 16S rRNA, an 'interior loop' region (marked with an arrow in Fig. 7A) is clearly visible. Another example for molecular features visible with high definition is a very unique fold in the 5S rRNA structure, marked with an asterisk in Fig. 7C. Similarly, enlarged views of protein S6 at the platform of small subunit, and protein L6 near the factor-binding region of the large subunit manifest distinct features of alpha-helical and beta-sheeted regions previously not visible (Fig. 6 B and D). In addition, a comparison with the crystal structures reveals local conformational changes

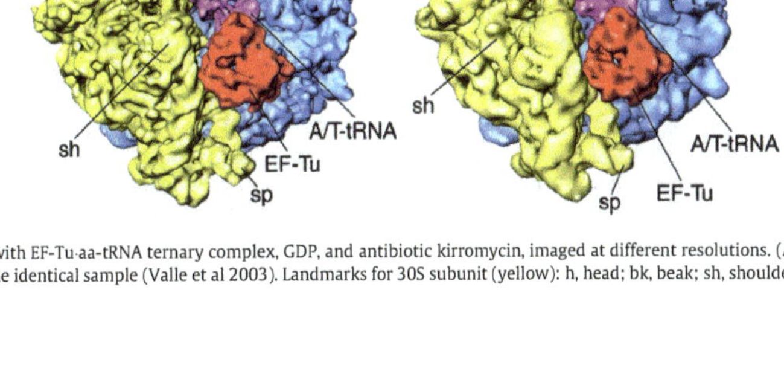

Fig. 5. 70S ribosome in complex with EF-Tu·aa-tRNA ternary complex, GDP, and antibiotic kirromycin, imaged at different resolutions. (A) 6.7 Å map obtained in the current study; (B) previous 10-Å map of the identical sample (Valle et al 2003). Landmarks for 30S subunit (yellow): h, head; bk, beak; sh, shoulder; for 50S subunit (blue): CP, central protuberance; sb, stalk base.

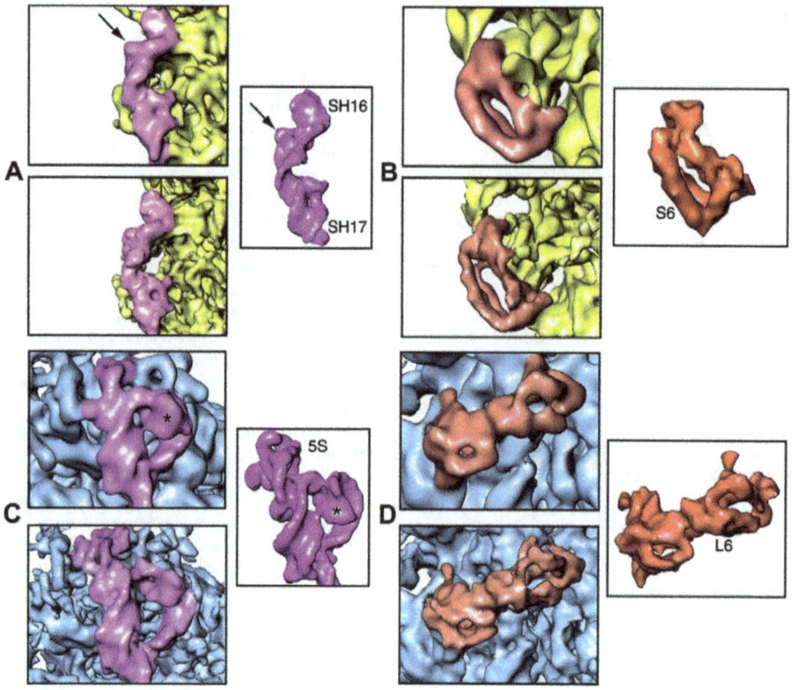

Fig. 6. Close-up views of secondary structural features of selected ribosomal proteins and rRNA regions in the 6.7-Å map. (A) Helices 16 and 17 (purple) of 16S rRNA; (B) protein S6 (red) of the 30S subunit (yellow); (C) 5S rRNA; (D) protein L6 of the 50S subunit (blue). For each of these components, an inset shows same structure from the *E. coli* X-ray coordinates (Schuwirth et al., 2005) for comparison. Atomic coordinates were converted to density maps at 6.7 Å resolution (PDB codes 30S: 2AVY, 50S: 2AW4).

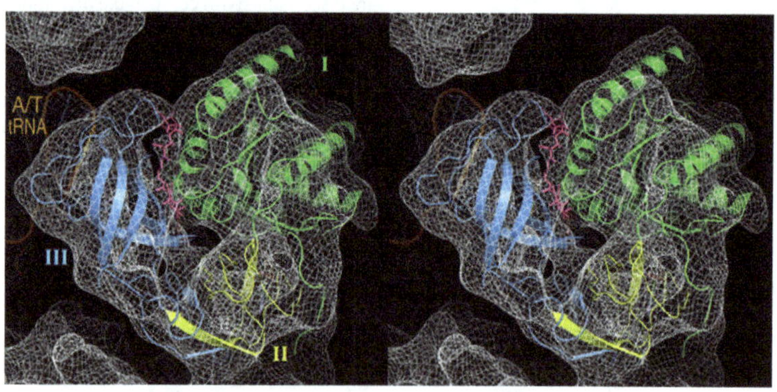

Fig. 7. Stereo view of elongation factor Tu (EF-Tu) fitted into the corresponding density in the 6.7-Å map of the ribosome complex (represented as mesh). The features of alpha-helical (see domain I, green) and beta-sheeted (see domain III, golden) regions of the protein are clearly seen. In addition, partial density corresponding to the antibiotic kirromycin (pink; molecular formula $C_{43}H_{60}N_2O_{12}$, formula weight ~800 (794.96) Da,) is distinctly visible.

in the rRNA helices and intra-domain rearrangements in individual proteins—changes that must be attributed to differences in the functional state between the ribosome in the crystal (as shown by X-ray crystallography) and in solution (by cryo-EM).

Of particular interest in this complex is the binding interaction between the ribosome and the aa–tRNA·EF–Tu·GDP·kir ternary complex. An inspection of the ligand density in the map (viewed with the crystal structure [PDB code 1OB2] previously fitted into the lower-resolution map of the identical complex (Valle et al., 2003)) (Fig. 7) indicates a much better definition of the alpha helices (see domain I) and beta-sheeted (see domain III) regions of EF-Tu. Furthermore, this map demonstrates densities with distinct features corresponding to the nucleotide (bound to domain I, close to the sarcin–ricin loop; Fig. 7A), and the antibiotic kirromycin (at the interface of domain I and III; Fig. 7B), in contrast to the lower resolution map where densities corresponding to these molecules were inseparable from the rest of the ligand mass. It is a further reflection of the high quality of the map that phosphorus atoms show up as "bumps" at the expected places along the phosphate backbones of RNA when the low-pass filter is further opened to 5 Å (Fig. 6).

4. Discussion

4.1. Resolution determination

One caveat that was reaffirmed in this study is that the FSC computed for the whole volume of the reconstruction consistently underestimates the resolution of the molecule reconstructed, and that it should be applied, instead, to a soft-masked version of the density map. Without this measure, it is virtually impossible to gauge the effects of image processing parameters on the reconstruction quality. While soft masking was introduced by Stewart et al. (2000) as a means to suppress irreproducible parts of the molecule in the resolution estimation, we would like to point out that in addition, reconstruction artifacts *unrelated* to the molecule structure are intrinsically irreproducible in a single-particle reconstruction, due to the fact that the sets of angles do not exactly match in two randomly drawn half-sets.

4.2. Standardized data set

The creation of a standardized data set is an important step in the current and future development and optimization of image processing algorithms. The biological sample was selected because it is known to be quite homogeneous, from the fact that it could be previously reconstructed to high resolution without resorting to classification (Valle et al., 2003).

4.3. Effect of pixel size

The pixel size was changed during the course of refinement, from "decimated" to "raw," in order to accelerate the computation. As expected, the non-decimated data lead to improved resolution. However, we found that the transfer of alignment parameters from a coarsely- to a finely-sampled data set can be done without a setback in resolution. Thus, the strategy of doing the refinement in 2 parts, first with a decimated version of the data set, then with the finely sampled data set and jump-starting with the alignment parameters of the first part was successful.

4.4. Effect of amplitude enhancement

Our results indicate that that refinement with amplitude enhancement in each step brings no benefit, but rather, the resolution deteriorates slightly. We used a filter that boosts the Fourier amplitudes to the values reported by low-angle X-ray scattering (Gabashvili et al., 2000) in each step. The resulting final map had resolution worse than the map in which no amplitude enhancement was used. The stated purpose of increasing Fourier amplitudes after each iteration is to increase the contribution of high spatial frequencies in the 3D projection alignment, which should translate into higher resolution of the density map. This result suggests that the concomitant amplification of noise in the high spatial frequency range counteracts the putative improvement.

4.5. Effect of window size

The final variable examined was particle window size. This was to determine if larger windows, and therefore information within the total reach of the point-spread function, would improve resolution. The standard particle window size used for Reconstructions 2 and 3 was 195 × 195 pixels, corresponding to 363 × 363 Å on the object scale. Our larger window size was 275 × 275 pixels, corresponding to 511 × 511 Å on the object scale. This window size was chosen because it would be adequate to capture high spatial frequency information contained in the side ripples of the point-spread function, according to our simulations (unpublished). In comparable refinements where the only variable modified was particle window size, Reconstructions 3 and 4, the nominal resolution deteriorated by ~2.3 Å when the larger window size was used. Even with the Gaussian mask applied to the larger window size, the resolution was significantly worse than for the smaller window. Our results show that, judged by the resolution criterion, the effect of the side ripples of the point spread function beyond the 363 × 363 Å box is negligible. We can explain the deterioration of nominal resolution both with and without soft masking by the extra amount of clutter admitted by the larger window.

4.6. Resolution by extrapolation to the full data set

The curve of $\ln(Nd)$ versus $1/d^2$ is seen to follow two regimes, with an improved B-factor toward higher particle numbers. Since the subsets were drawn randomly, with equal representations of the defocus groups, the result cannot be explained as an effect of statistical bias. However, as demonstrated by comparison with the X-ray data, the resolution estimated on the basis of the extrapolation from half to full data set is borne out by the molecular features visualized. In some portions of the ribosome, these features appear even crisper and more informative than in the X-ray map. Surprisingly, phosphate atoms are visible in a 5-Å version of the map, not just as undulations along the backbone (as expected when solely the information on the 6- to 6.5-Å distance along the backbone were transmitted) but by distinct bumps, showing that there exists valid structural information up to 5 Å.

5. Conclusions

Our exploration of parameters and processing methods that might lead to improved results has produced a density map of the ribosome with greatly improved resolution and definition, compared with previously published maps. As has been pointed out, this study falls far short of a systematic exploration of a wide range of options, which might produce even better results but would require much larger computational expenditures. The new biologically relevant information contained in the ribosome map will be expounded elsewhere. The appearance of phosphorus atoms along the phosphate backbone on the map when limited to 5 Å resolution indicates the presence of information well beyond the FSC = 0.5 cutoff, although we can say for sure that the signal-to-noise ratio at this spatial frequency falls short of the SNR = 1 cutoff (see Malhotra et al., 1998; Rosenthal and Henderson, 2003). In the course of our inquiry, we confirmed the validity of a relationship initially postulated by Rosenthal and Henderson (2003), which may allow extrapolation to the full data set as well as the prediction of the minimum data collection requirements toward a more ambitious target resolution.

6. Materials and methods

SPIDER (Frank et al., 1996) was used for all image processing. The visualization of some figures was done using IRIS Explorer (Numerical Algorithm Group, Downers, IL). For molecular modeling, Pymol (DeLano Scientific, San Carlos, CA) was used.

Acknowledgments

We would like to thank Måns Ehrenberg, Uppsala University, Sweden for the specimen of the pre-accommodated *E. coli* 70S·Phe–tRNA^{Phe}·EF–Tu·GDP·kir complex. We would like to thank Hstau Liao and Jie Fu for discussions of the extrapolation of resolution. We acknowledge the assistance of Richard Hall, UC Berkeley, for his contribution to the selection of high-quality particles, based on his analysis of micrograph quality (unpublished). We thank Michael Watters for assistance with the preparation of the illustrations. This work was supported by HHMI and NIH grants P41 RR 01219, R01 GM29169 and R01 GM55440 (to J.F.).

References

Andersen, C.B.F., Becker, T., Blau, M., Anand, M., Halic, M., Balar, B., Mielke, T., Boesen, T., Pedersen, J.S., Spahn, C.M.T., Kinzy, T.G., Andersen, G.R., Beckmann, R., 2006. Structure of eEF3 and the mechanism of transfer RNA release from the E-site. Nature 443, 663–668.

Connell, S.R., Takemoto, C., Wilson, D.N., Wang, H., Murayama, K., Terada, T., Shirouzu, M., Rost, M., Schüler, M., Giesebrecht, J., Dabrowski, M., Mielke, T., Fucini, P., Yokoyama, S., Spahn, C.M.T., 2007. Structural basis for interaction of the ribosome with the switch regions of GTP-bound elongation factors. Mol. Cell 25, 751–764.

Frank, J., 2006. Three-dimensional Electron Microscopy of Macromolecular Assemblies. Oxford University Press.

Frank, J., Radermacher, R., Penczek, P., Zhu, J., Li, Y., Ladjadj, M., Leith, A., 1996. SPIDER and WEB: processing and visualization of images in 3D electron microscopy and related fields. J. Struct. Biol. 116, 190–199.

Frank, J., Penczek, P., Agrawal, R.K., Grassucci, R.A., Heagle, A.B., 2000. Three-dimensional cryoelectron microscopy of ribosomes. Methods Enzymol. 317, 276–291.

Frank, J., Sengupta, J., Gao, H., Li, W., Valle, M., Zavialov, A., Ehrenberg, M., 2005. The role of tRNA as a molecular spring in decoding, accommodation, and peptidyl transfer. FEBS Lett. 579, 939–963.

Gabashvili, I.S., Agrawal, R.K., Spahn, C.M.T., Grassucci, R.A., Svergun, D.I., Frank, J., Penczek, P., 2000. Solution structure of the *E.coli* 70S ribosome at 11.5 Å resolution. Cell 100, 537–549.

Grassucci, R.A., Taylor, T.D., Frank, J., 2007a. Preparation of macromolecular complexes for cryo-electron microscopy. Nature Prot. 2, 3238–3246.

Grassucci, R.A., Taylor, T.D., Frank, J., 2007b. Visualization of macromolecular complexes using cryo-electron microscopy with FEI Tecnai TEMs. Nature Prot. 3, 330–339.

Harauz, G., van Heel, M., 1986. Exact filters for general geometry three-dimensional reconstruction. Optik 73, 146–156.

Liu, X., Jiang, W., Jakana, J., Chiu, W., 2007. Averaging tens to hundreds of icosahedral particle images to resolve protein secondary structure elements using multi-path simulated annealing optimization algorithm. J. Struct. Biol. 160, 11–27.

Ludtke, S.J., Baker, M., Chen, D., Song, J.-L., Chuang, D., Chiu, W., 2008. De Novo backbone trace of GroEL from single particle electron cryomicroscopy. Structure 16, 441–448.

Malhotra, A., Penczek, P., Agrawal, R.K., Gabashvili, I.S., Grassucci, R.A., Jünemann, R., Burkhardt, N., Nierhaus, K.H., Frank, J., 1998. *Escherichia coli* 70S ribosome at 15 Å resolution by cryo-electron microscopy: localization of fmet-tRNAfMet and fitting of L1 protein. J. Mol. Biol. 280, 103–116.

Penczek, P., Grassucci, R.A., Frank, J., 1994. The ribosome at improved resolution: New techniques for merging and orientation refinement in 3D cryoelectron microscopy of biological particles. Ultramicroscopy 53, 251–270.

Penczek, P.A., Zhu, J., Schroeder, R., Frank, J., 1997. Three-dimensional reconstruction with contrast transfer compensation from defocus series. Scanning Microsc. 11, 147–154.

Rath, B.K., Frank, J., 2004. Fast automatic particle picking from cryo-electron micrographs using a locally normalized cross-correlation function: a case study. J. Struct. Biol. 145, 84–90.

Roseman, A.M., 2004. FindEM—a fast, efficient program for automatic selection of particles from electron micrographs. J. Struct. Biol. 145, 91–99.

Rosenthal, P.B., Henderson, R., 2003. Optimal determination of particle orientation, absolute hand, and contrast loss in single-particle electron cryomicroscopy. J. Mol. Biol. 333, 721–745.

Saxton, W.O., Baumeister, W., 1982. The correlation averaging of a regularly arranged bacterial cell envelope protein. J. Microsc. 127, 127–138.

Schuwirth, B.S., Borovinskaya, M.A., Hau, C.W., Zhang, W., Vila-Sanjurjo, A., Holton, J.M., Cate, J.D., 2005. Structures of the bacterial ribosome at 3.5 Å resolution. Science 310, 827–834.

Shaikh, T.R., Gao, H., Baxter, W., Asturias, F., Boisset, N., Leith, A., Frank, J., 2008a. SPIDER image processing for single-particle reconstruction of biological macromolecules from electron micrographs. Nature Prot., in press.

Shaikh, T.R., Trujillo, R., LeBarron, J.S., Baxter, W.T., Frank, J., 2008b. Particle verification for single-particle, reference-based reconstruction using multivariate data analysis and classification. J. Struct. Biol., 164, 41–48.

Stagg, S.M., Lander, G.C., Pulokas, J., Fellmann, D., Cheng, A., Quispe, J.D., Mallick, S.P., Avila, R.M., Carragher, B., Potter, C.S., 2006. Automated cryo-EM data acquisition and analysis of 284 742 particles of GroEL. J. Struct. Biol. 155, 470–481.

Stewart, P.L., Cary, R.B., Peterson, S.R., Chiu, C.Y., 2000. Digitally collected cryo-electron micrographs for single particle reconstruction. Microsc. Res. Tech. 49, 224–232.

Trabuco, L.G., Villa, E., Mitra, K., Frank, J., Schulten, K., 2008. Flexible fitting of atomic structures into electron microscopy maps using molecular dynamics. Structure 16, 673–683.

Valle, M., Sengupta, J., Swami, N.K., Grassucci, R.A., Burkhardt, N., Nierhaus, K.H., Agrawal, R.K., Frank, J., 2002. Cryo-EM reveals an active role for amino-acyl-tRNA in the accommodated state. EMBO J. 21, 3557–3567.

Valle, M., Zavialov, A., Li, W., Stagg, S.M., Sengupta, J., Nielsen, R.C., Nissen, P., Harvey, S.C., Ehrenberg, M., Frank, J., 2003. Incorporation of aminoacyl-tRNA into the ribosome as seen by cryo-electron microscopy. Nat. Struct. Biol. 10, 899–906.

Yang, C., Ng, E.G., Penczek, P.A., 2005. Unified 3-D structure and projection orientation refinement using quasi-Newton algorithm. J. Struct. Biol. 149, 53–64.

Yu, X., Jin, L., Zhou, Z.H., 2008. 3.88 Å structure of cytoplasmic polyhedrosis virus by cryo-electron microscopy. Nature advance online publication doi:10.1038/nature06893.

Zhang, X., Settembre, E., Xu, C., Dormitzer, P.R., Bellamy, R., Harrison, S.C., Grigorieff, N., 2008. Near-atomic resolution using electron cryomicroscopy and single-particle reconstruction. PNAS 105, 1867–1872.

Zhu, J., Penczek, P., Schroeder, R., Frank, J., 1997. Three-dimensional reconstruction with contrast transfer-function correction from energy-filtered cryo-electron micrographs—procedures and applications to the 70S *Escherichia coli* ribosome. J. Struct. Biol. 118, 197–219.

Structure
Technical Advance

Flexible Fitting of Atomic Structures into Electron Microscopy Maps Using Molecular Dynamics

Leonardo G. Trabuco,[1,2,6] Elizabeth Villa,[1,2,6] Kakoli Mitra,[3,7] Joachim Frank,[3,4,8] and Klaus Schulten[1,5,*]
[1]Beckman Institute
[2]Center for Biophysics and Computational Biology
University of Illinois at Urbana-Champaign, Urbana, IL 61801, USA
[3]Howard Hughes Medical Institute, Health Research Incorporated, Wadsworth Center, Empire State Plaza, Albany, NY 12201, USA
[4]Department of Biomedical Sciences, State University of New York, Albany, NY 12222, USA
[5]Department of Physics, University of Illinois at Urbana-Champaign, Urbana, IL 61801, USA
[6]These authors contributed equally to this work.
[7]Present address: Skirball Institute, New York University, New York, NY 10016, USA.
[8]Present address: Howard Hughes Medical Institute, Department of Biochemistry and Molecular Biophysics and Department of Biology, Columbia University, New York, NY 10032, USA.
*Correspondence: kschulte@ks.uiuc.edu
DOI 10.1016/j.str.2008.03.005

SUMMARY

A novel method to flexibly fit atomic structures into electron microscopy (EM) maps using molecular dynamics simulations is presented. The simulations incorporate the EM data as an external potential added to the molecular dynamics force field, allowing all internal features present in the EM map to be used in the fitting process, while the model remains fully flexible and stereochemically correct. The molecular dynamics flexible fitting (MDFF) method is validated for available crystal structures of protein and RNA in different conformations; measures to assess and monitor the fitting process are introduced. The MDFF method is then used to obtain high-resolution structures of the *E. coli* ribosome in different functional states imaged by cryo-EM.

INTRODUCTION

A key to understanding how biological systems work is to look at their structures captured in their various functional states. Experimental techniques reveal different levels of macromolecular structure: high-resolution techniques such as X-ray crystallography furnish atomic detail, but structures obtained are often in functionally undefined states; techniques such as cryo-electron microscopy (cryo-EM) image systems captured in different functional states, albeit at lower resolution (Saibil, 2000). Computational techniques can help bridge the resolution gap by adapting high-resolution crystallographic structures to EM maps, thus providing atomic detail of the system in different functional states. These techniques can also be used to analyze the physical and dynamical properties of the resulting structures, revealing astonishing views of cellular processes.

Until a few years ago, typical resolutions for EM maps of biomolecules were 15–30 Å, and high-resolution crystal structures were often available only for domains of a biomolecular complex (Frank, 2002; Rossmann, 2000). This led to the development of so-called rigid-body docking techniques that fit atomic structures into density maps keeping the high-resolution structure rigid, usually by performing an exhaustive search over all rotational and translational degrees of freedom in real or reciprocal space, guided by some choice of similarity measure. Rigid-body docking has reached maturity, permitting today the independent docking of parts of the assembly into a map, thus producing a jigsaw structure of the macromolecular complex as a whole. The main limitation of this class of methods is that neither the interaction between the docked subunits nor the difference in conformation between the structures of the domains in crystal form and in the complex can be determined. A comprehensive review of the available rigid-body docking methods can be found in Wriggers and Chacón (2001).

Recent improvements in the resolution of cryo-EM structures motivated the development of methods to flexibly fit atomic structures into density maps. In these methods, many degrees of freedom are considered in the fit, allowing the atomic structures to undergo conformational changes that improve their correspondence to the EM map. Various approaches to flexible fitting have recently been employed and have provided insight into the structural mechanisms of a number of biomolecular processes. One of the first attempts to introduce flexibility in the fitting process, still in use today, consists in dividing a macromolecule into "rigid bodies" and fitting them independently (e.g., Volkmann et al., 2000). Refinement techniques can then be applied to the resulting structure, such as the real-space refinement originally developed for X-ray crystallography (Chapman, 1995; Chen et al., 2001). Other approaches include the use of so-called vector quantization, where a reduced representation is derived from both the atomic structure and the EM map and used as constraints in a molecular mechanics refinement (Wriggers et al., 2000; Wriggers and Birmanns, 2001); the use of a linear combination of low-frequency normal modes to deform the atomic structure, applying minimization techniques to maximize the correlation coefficient between atomic structure and EM map (Tama et al., 2004; Suhre et al., 2006); the combination of comparative modeling, based on alternative alignments and

 ©2008 Elsevier Ltd All rights reserved

loop conformations, and structure refinement (Topf et al., 2006); the generation of deformed structures based on the variability exhibited by the protein domains of a superfamily and subsequent selection of the best fit based on the cross-correlation coefficient (Velazquez-Muriel et al., 2006); the use of a combination of restraints imposed by the EM map and a deformable elastic network (Schröder et al., 2007); the use of Monte Carlo simulations with constraints derived from a rigidity analysis (Jolley et al., 2008); and a hierarchical approach consisting of docking the structure as one or more rigid bodies with a Monte Carlo search, followed by two refinement steps based on minimization of a scoring function and simulated annealing (Topf et al., 2008). Li and Frank (2007) recently correlated an ensemble of conformations from equilibrium molecular dynamics (MD) simulations with cryo-EM data, suggesting that a cryo-EM map can be interpreted as a conformational average and that using MD to flexibly fit structures into EM maps should yield structures representative of the conformational ensemble represented by the EM map, particularly when combined with an enhanced sampling technique.

We suggest here a novel method to perform MD simulations to flexibly fit atomic structures into EM maps, with the MD simulation incorporating EM data through an external potential. Forces proportional to the density gradient of the EM map are added in the MD simulation of the atomic structure, effectively biasing the system toward the region of conformational space of interest, i.e., one that is consistent with the density distribution of the EM map. Since large forces are applied and simulations presently are performed in vacuo, harmonic restraints must be applied to keep secondary structure elements intact, thus preventing structural distortions and "overfitting." The molecular dynamics flexible fitting (MDFF) method presented brings together several of the most desirable features of existing methods. First, MDFF takes into account all information contained in the map by avoiding the use of reduced representations or global measures of similarity to drive the fitting. Since the external potential is applied locally, it is possible to fit some components of a macromolecular assembly even when the structure of the remaining components is not available. Moreover, the degree of success of the MDFF method is expected to be independent of system size, which is an advantage over the use of optimization or Monte Carlo-based approaches that employ global-optimization criteria, where an increase in system size decreases the likelihood of successful transitions.

The two methods that MDFF is based on, namely, conventional MD simulation and 3D cryo-EM single-particle reconstruction, are introduced in the online Supplemental Data (see Molecular Dynamics Situations and 3D EM Reconstructions) that accompanies this article for the nonexpert reader. In the following, we describe how the MDFF method incorporates EM data into MD simulations and how it applies restraints to preserve the integrity of structural elements. We demonstrate MDFF by fittings into noise-free, simulated maps created from atomic structures. Finally, as an example application of MDFF, we fit atomic structures into cryo-EM maps of the *E. coli* ribosome at different resolutions.

MD Simulation with an EM-Derived External Potential

In the MDFF method, an external potential derived from the EM map is introduced into an MD simulation to steer the atoms into high-density regions. The stereochemical quality of the structure is preserved by the MD force field and also through harmonic restraints applied to enforce the integrity of secondary structure. The method, therefore, adds two extra terms to the potential energy function of an MD simulation

$$U_{total} = U_{MD} + U_{EM} + U_{SS}, \tag{1}$$

where U_{MD} is the conventional MD potential energy function (Supplemental Data, see Molecular Dynamics Situations), U_{EM} corresponds to a potential derived from the EM data, and U_{SS} is a potential that aims to preserve the secondary structure of protein and nucleic acids. We now describe the latter two terms.

The data provided by cryo-EM reconstructions represent the Coulomb potential of the macromolecule (Supplemental Data, see 3D EM Reconstructions); the dependence of this potential on the atomic number of the composing atoms makes it roughly proportional to the mass density of the macromolecule. It is then sensible to define a potential so that when the atomic structure is placed in it, the atoms are driven through application of forces into high-density areas and away from low-density areas. This potential can be defined on a grid, thus preserving all the information contained in the EM density map. The potential energy function corresponding to this map is

$$U_{EM}(\mathbf{R}) = \sum_j w_j V_{EM}(\mathbf{r}_j), \tag{2}$$

where **R** collects all atom coordinates and

$$V_{EM}(\mathbf{r}) = \begin{cases} \xi\left[1 - \frac{\Phi(\mathbf{r}) - \Phi_{thr}}{\Phi_{max} - \Phi_{thr}}\right] & \text{if } \Phi(\mathbf{r}) \geq \Phi_{thr}, \\ \xi & \text{if } \Phi(\mathbf{r}) < \Phi_{thr}. \end{cases} \tag{3}$$

Here $\Phi(\mathbf{r})$ is the Coulomb potential revealed by cryo-EM, Φ_{max} is the maximum value of all voxels in the EM map, ξ is an arbitrary scaling factor ($\xi > 0$) discussed below, Φ_{thr} is a threshold value used to disregard data not corresponding to the biomolecule, i.e., solvent density, w_i is a weight that can be varied according to the type of atom placed in this potential, typically set to the atomic mass, and $\mathbf{r}_j$ is the position of atom *j*. The global minimum of U_{EM} alone corresponds to all atoms collapsed on the density maximum; however, the other two terms in Equation 1 counterbalance this effect, preserving physically sound structures.

The threshold value Φ_{thr} is selected in accordance with the density histogram of the EM map, which reveals two peaks of different density (Frank, 2006); an example is shown in Figure 1A. The first and higher peak corresponds to solvent density, whereas the second, broader peak corresponds to protein and nucleic acid density. The density histogram thus suggests a natural threshold value at the minimum between the two density peaks; clamping all values below this threshold removes the solvent contribution and yields a flat potential in the solvent regions. For cases where the solvent and biomolecule peaks are not well resolved, the average, which generally lies at the solvent peak, is chosen as the threshold, thus removing much of the undesired density while conservatively avoiding loss of macromolecular density information, as portrayed in Figure 1C.

The force applied to an atom inside the potential defined by Equation 3 is

 ©2008 Elsevier Ltd All rights reserved

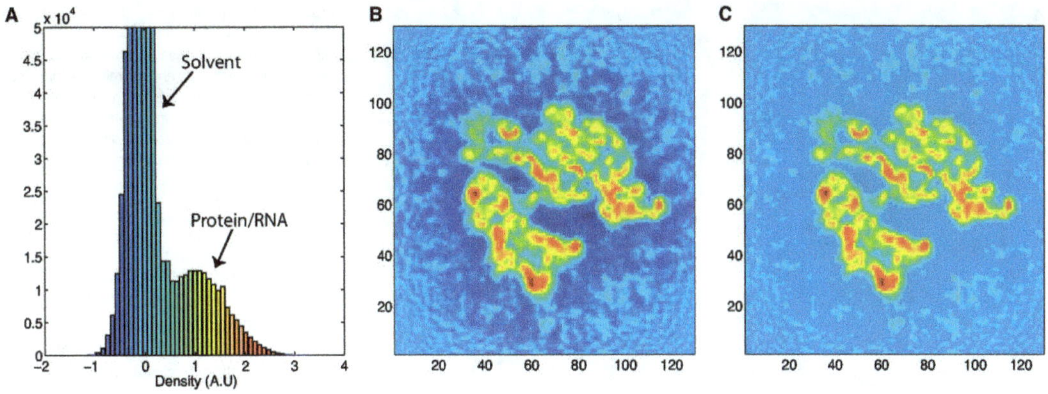

Figure 1. Reconstruction of the *E. coli* Ribosome from Cryo-EM Data at ~12.8 Å Resolution
(A) A density histogram shows two distinct peaks pertaining to the solvent and macromolecule; (B) 2D slice of the density; (C) 2D slice of the density after clamping values below the average, thus homogenizing the density corresponding to the solvent surrounding the macromolecule and the bulk solvent. Unpublished data from K.M., L.G.T., E.V., A. Zavialov, M. Ehrenberg, K.S., and J.F.

$$\mathbf{f}_i^{EM} = -\frac{\partial}{\partial \mathbf{r}_i} U_{EM}(\mathbf{R}) = -w_i \frac{\partial}{\partial \mathbf{r}_i} V_{EM}(\mathbf{r}_i) \tag{4}$$

where U_{EM} is the potential energy function introduced in Equations 2 and 3. $\mathbf{f}_i^{EM}$ can thus be tuned via the scaling factor ξ, which is the same for all atoms, and the weight w_i, which can be defined on a per-atom basis. The external forces $\mathbf{f}_i^{EM}$ are applied in MDFF using grid-steered molecular dynamics, an extension of the SMD method (Isralewitz et al., 2001; Sotomayor and Schulten, 2007) that allows an arbitrary steering potential to be defined on a grid (Wells et al., 2007). The force applied to each atom i depends only on the gradient of the potential derived from the EM density map at position $\mathbf{r}_i$; thus, the fitting in MDFF is performed locally.

In order to preserve the stereochemical quality of the structure and prevent overfitting in MDFF, harmonic restraints are applied to a set of internal coordinates relevant to the secondary structure of the macromolecule in its initial conformation—in many cases, the crystal structure. For protein structures, harmonic restraints are applied to ϕ and ψ dihedral angles of amino-acid residues in helices and β strands. For nucleic acids, the software package RNAView (Yang et al., 2003) is used to identify and classify base pairs. The following base pair types, which were observed to preserve the secondary structure of RNA in simulation when restrained, are selected: W/W, W/H, W/S, H/H, H/S, and stacked (W , Watson-Crick edge; H, Hoogsteen edge; S, sugar edge; see Figure 2A). For these base pairs, harmonic restraints are applied to the seven dihedral angles (α, β, γ, δ, ε, ζ, and χ) and two interatomic distances (d_1 and d_2) depicted in Figure 2B, the latter to preserve the planarity of the base pair. Thus, an extra term introducing secondary structure harmonic restraints is added to the potential energy function, namely

$$U_{SS} = \sum_{\mu} k_{\mu}(X_{\mu} - X_{\mu}^0)^2, \tag{5}$$

where the restraints X_{μ} stand for protein dihedral angles ϕ and ψ, RNA dihedral angles α, β, γ, δ, ε, ζ, and χ, and RNA distances d_1 and d_2. The equilibrium values X_{μ}^0 are generally taken from the initial atomic structure but can also be set to ideal values. Additional restraints can be added, such as codon-anticodon interactions in the ribosome, as discussed below.

The MDFF simulations are performed using NAMD (Phillips et al., 2005), with the CHARMM27 force field (MacKerell et al., 1998; Foloppe and MacKerrell, 2000) in vacuo, using a dielectric constant of 80. A multiple time-stepping integration scheme is used, calculating bonded interactions every 1 fs and nonbonded interactions every 2 fs; a cutoff distance of 10 Å is used for the nonbonded interactions. Temperature is maintained at 300 K using a Langevin thermostat (Brünger et al., 1984) coupled to all heavy atoms with a damping coefficient of 5 ps^{-1}. Rigid-body docking performed as the initial step of a fitting protocol in all examples presented was executed using Situs (Wriggers et al., 1999).

Flexible-Fitting Protocol

Obtaining an optimal fit in MDFF relies on a balance among the three terms in Equation 1. The first term, U_{MD}, is given by the choice of standard force field and is here not subject to alteration; the other two terms, U_{EM} and U_{SS}, contain parameters that can be tuned and represent a trade-off: higher forces derived from the EM map (U_{EM}) and lower restraints to secondary structure (U_{SS}) can yield a better fit but can also lead to distortions of the macromolecular structure, a natural concern in flexible fitting methods generally referred to as "overfitting" (Tama et al., 2004). For U_{EM}, parameters that can be varied are the scaling factor ξ and atomic weights w_i (Equations 2 and 3). The latter are typically set to the atomic masses while ξ determines the range in magnitude of forces applied to the atoms and is usually set to values around 0.3 kcal/mol, resulting in forces typically on the order of 10–15 pN (piconewton) per atom (for a carbon atom). The gradient of the potential derived from the EM density map can be calculated before the fitting to pick a value of ξ: high values can result in high accelerations of atoms, rendering the simulation unstable. The term U_{SS} requires the selection of

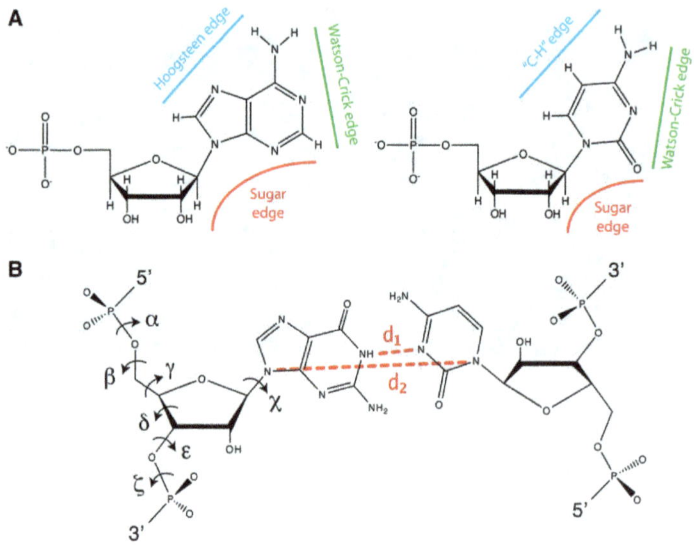

Figure 2. Harmonic Restraints Applied to Base-Paired RNA Residues

(A) RNA interaction edges for both purines (adenine is shown on the left) and pyrimidines (cytosine is shown on the right), according to Leontis and Westhof (1998); (B) dihedral angles, and the two interatomic distances (dashed lines) to which harmonic restraints are applied.

spring constants k_μ in Equation 5. Typical values used in MDFF are $k_\mu = 200$ kcal mol^{-1} rad^{-2}, or 200 kcal mol^{-1} Å^{-2} in the case of RNA distances (300 kcal mol^{-1} Å^{-2} is a typical spring constant value for a carbon-carbon bond in the CHARMM27 force field). The choice of these values renders the restrained secondary structure elements relatively stiff, thus preserving their structure throughout the simulation.

MDFF can also be performed in multiple steps, varying parameters at every step, such that the atomic structure gradually converges into the EM map. Generally, a step is considered "finished" when the MD simulation has converged as evaluated by a goodness measure, e.g., by stabilization of the root-mean-square deviation (rmsd, given by $\langle(\mathbf{r}_i - \langle\mathbf{r}_i\rangle)^2\rangle^{1/2}$, where $\mathbf{r}_i$ is the position of each atom i); one can also track the correlation between the trajectory and the EM map, as described later. Naturally, in a last step of MDFF one can switch U_{EM} and U_{SS} off and equilibrate the system subject only to the intrinsic potential U_{MD}.

In a multistep protocol of MDFF, several parameters can be varied besides those mentioned above: (1) *Temperature* can be adjusted to allow the system to overcome energy barriers, exploring a larger portion of the conformational space in less time, often done in simulated annealing simulations (Frenkel and Smit, 2002); (2) *low-pass filtering* the map in a first step can be used to avoid getting trapped in local minima, followed by use of the original map—this approach first induces overall domain motions and refines the structure subsequently on the local scale using the high-resolution information in the original map; (3) along the same line, one can initially apply strong harmonic restraints to the secondary structure and in a subsequent step *weaken the restraints* to refine the structure; (4) at any given time, any portion of the structure may be *fixed* or *restrained* to its current position (positional restraints)—this is useful, e.g., when factors or low-occupancy ligands are introduced for fitting, and the rest of the previously fitted structure can be fixed or restrained while the ligands are fitted; (5) one can also *delete the density* corresponding to parts of the structure, which is useful, for example, for fitting protein-RNA complexes, as discussed below—the deletion is done by assuming the current fit is perfect, creating a simulated map from the fitted structure as described in the next section, and using it as a mask to selectively delete a portion of the original EM map. The new map Φ' defining U_{EM} as in Equation 3 is given by

$$\Phi' = \left(1 - \frac{\Phi_{sim}}{\max(\Phi_{sim})}\right)\Phi \qquad (6)$$

where Φ is the original EM map and Φ_{sim} is the simulated map, with a maximum voxel value of $\max(\Phi_{sim})$. Equation 6 results in potentials with smoother gradients than those obtained from difference maps.

An important advantage of multistep protocols in MDFF is evident when fitting protein-RNA complexes into EM maps. The atomic number of the atoms composing RNA is on average higher than those composing protein, which may lead to incorrect fitting of proteins due to lower-energy minima attracting them into RNA density. To avoid this problem, the RNA may be fitted first, followed by deleting the RNA contribution from the EM map once it is well fitted and by then proceeding with the fitting of the proteins. Alternatively, one may positionally restrain the RNA structure to prevent the protein from being drawn into the density the RNA occupies.

Fitting Atomic Structures into Simulated Maps

In the following sections, we demonstrate application of MDFF to noise-free, simulated EM maps as a means of validation of the method. We briefly describe how simulated maps are generated from atomic structures and how we calculate cross-correlation coefficients between a fitted structure and the EM map. Using examples where X-ray crystal structures are available in two different conformations of the same molecule, we apply the method to fit one conformation into simulated maps generated from the other.

Simulated EM Maps from Atomic Structures

Noise-free simulated EM maps can be created from atomic structures using the approach described in Stewart et al. (1993), which assumes that the EM map represents the electrostatic potential of the nuclei. The atomic number of each atom is interpolated onto a grid, and the resulting 3D density map is subsequently low-pass filtered to the desired resolution. Simulated maps can be generated from a single structure or from a set of structures obtained from an MD simulation by averaging maps created from each frame of the trajectory (see Li and Frank,

Structure

Molecular Dynamics Flexible Fitting

Table 1. Effect of Resolution and Grid Spacing on Flexible Fitting into Noise-Free Simulated Maps using MDFF

System (PDB Codes)	Resolution (Å)	Initial Rmsd (Å)	Initial (Local) Correlation	Grid Spacing (Å)	Final Mean Rmsd (Å)	Final Mean (Local) Correlation
ACS/CODH (1OAO)	5.0	14.82	0.677 (0.282)	1.0	0.75	0.988 (0.934)
				2.0	1.04	0.986 (0.917)
	10.0	14.33	0.778 (0.478)	1.0	1.25	0.996 (0.977)
				2.0	1.25	0.996 (0.976)
	15.0	13.92	0.817 (0.582)	1.0	2.01	0.995 (0.980)
				2.0	2.07	0.995 (0.980)
16S rRNA (2AW7, 2AVY)	5.0	3.60	0.873 (0.611)	1.0	0.65	0.992 (0.961)
				2.0	0.84	0.988 (0.939)
	10.0	3.57	0.947 (0.812)	1.0	1.04	0.996 (0.984)
				2.0	1.21	0.995 (0.976)
	15.0	3.56	0.973 (0.901)	1.0	2.06	0.992 (0.972)
				2.0	2.21	0.991 (0.965)

For each system, maps were created computationally from a given crystal structure of conformation I and a crystal structure available in an alternative conformation, II, was fitted into the computed map. The initial backbone rmsd and cross-correlation coefficients correspond to rigid-body docked structures into computed maps with grid spacing of 2.0 Å using Situs with default options. The final mean backbone rmsd values were calculated from the last 200 ps of 500 ps trajectories. The final mean cross-correlation coefficients were calculated using computed maps generated from the average of the trajectories. Corresponding local cross-correlation coefficients that consider only the molecular envelope are also shown in parenthesis (a threshold of 0.2 σ was used in these examples).

2007). In the present work, simulated maps are used for three different purposes: (1) calculate cross-correlation coefficients between EM maps and atomic structures; (2) delete selected regions from the experimental EM map based on a fitted atomic structure; and (3) provide noise-free maps for validation purposes.

Cross-Correlation Coefficients

To quantify the goodness of the fit, a simulated map can be generated from the fitted atomic structure with the same target resolution as the EM map. The Pearson's correlation coefficient (usually referred to in the EM literature as the cross-correlation coefficient) between these two data sets, i.e., the simulated (*S*) and the experimental (*E*) 3D maps, can be used as a measure of similarity between them, and is given by

$$\rho_{SE} = \frac{\langle (S - \langle S \rangle)(E - \langle E \rangle) \rangle}{\sigma_S \sigma_E}, \quad (7)$$

where $\langle S \rangle$ and $\langle E \rangle$ correspond to the average voxel values of the simulated and experimental maps, respectively, and σ_S and σ_E correspond to their standard deviation (Frank, 2006). Note that the cross-correlation coefficient is normalized, i.e., $\rho_{SE} \in [-1, 1]$. All cross-correlation coefficients reported were computed considering either all voxel values in the density maps ("global correlation") or only voxels inside the molecular envelope of the simulated map ("local correlation") (Roseman, 2000), using a threshold that we report as the number of standard deviations (σ) above the mean of the simulated map. The global correlation, commonly quoted in the literature, depends sensitively on the box size arbitrarily selected by the electron microscopist. Larger boxes result in artificially higher correlation values, leading to overestimation of the quality of the fit; thus, local correlations should be preferred.

Validation Using Atomic Structures in Two Conformations

In order to validate the method and estimate the accuracy of the fitted structures, we use X-ray structures of molecules available in two conformations as test cases. One of the structures is fitted into noise-free simulated maps of the other at 5, 10, and 15 Å resolutions, after an initial phase of rigid-body docking. Here we present two examples: acetyl-coenzyme A synthase/carbon monoxide dehydrogenase (ACS/CODH) (Darnault et al., 2003) and a 16S rRNA (Schuwirth et al., 2005). Several other examples are presented in Supplemental Data (see Fitting Atomic Structures Into Simulated Maps).

All simulations used ξ = 0.3 kcal/mol and k_μ = 200 kcal mol^{-1} rad^{-2} (or 200 kcal mol^{-1} Å^{-2}) for the harmonic restraints (see Equation 5 and Figure 2B), except for 16S rRNA dihedral restraints that were enforced using k_μ = 50 kcal mol^{-1} rad^{-2}. The simulations converged in less than 200 ps of simulation but they were run for 500 ps to improve statistics. Table 1 lists the correlation coefficients and mean backbone rmsd between the fitted and target structures, calculated from the last 200 ps of trajectories (a plot of the rmsd through the entire trajectories for ACS/CODH can be found in the Supplemental Data [see Fitting Atomic Structures into Simulated Maps]). One can see that the rmsd decreases with higher resolution. It is important to note that the simulations result in a representative set of structures that fit the map. Since fitting atomic structures into low-resolution data is an indeterminate problem, a representative set of conformers should be considered when interpreting the data. The structures presented in this paper serve as an illustration of one such representative structure for each of the problems considered. Figure 3 shows the initial and target conformation for each of the systems, as well as representative structures from MDFF into 10 Å resolution simulated maps.

The final structures closely match the target structures used to generate the simulated maps; as expected, the match is less precise for lower resolutions or larger grid spacings. As an example, at 10 Å resolution, which represents a typical resolution for EM maps today, the final mean backbone rmsd is 1.25 Å for ACS/CODH and 1.04 Å for the 16S rRNA. The additional test cases presented in the Supplemental Data (see Fitting Atomic Structures

 ©2008 Elsevier Ltd All rights reserved

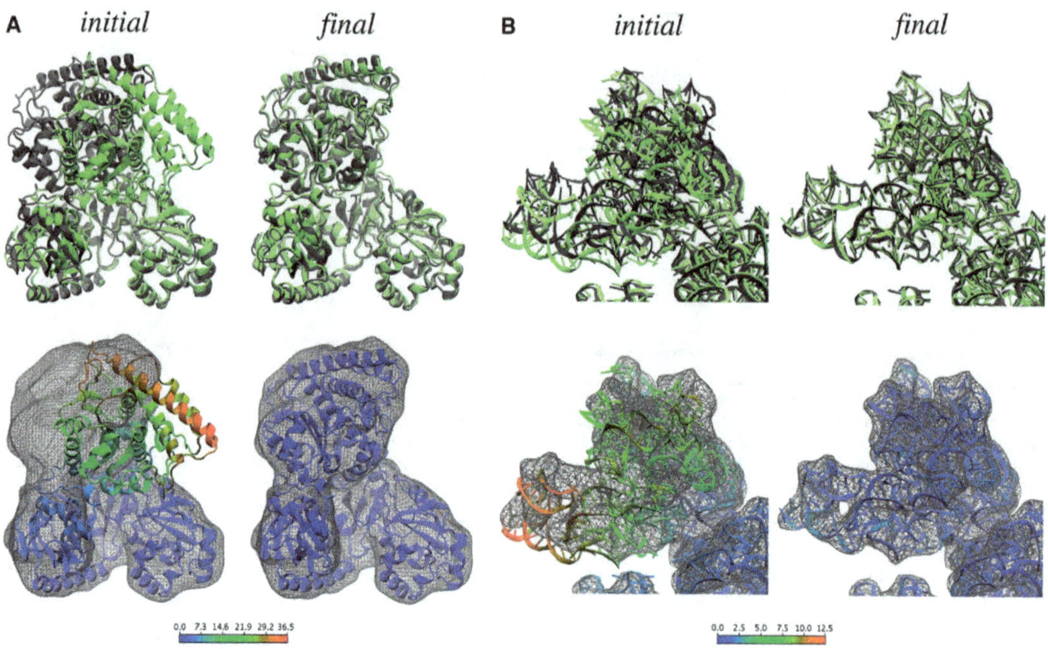

Figure 3. Validation of the MDFF Method Using X-Ray Structures in Two Conformations
(A) Acethyl-CoA synthase/carbon monoxide dehydrogenase; (B) 16S rRNA (only the head is shown for clarity, since it is the only region where the two conformations differ significantly). The target structures and simulated maps are shown in gray, whereas the initial and final fitted structures are shown in green (top) and colored by backbone rmsd per residue with respect to the target structures (bottom; color scales in Å). The final structures correspond to fittings into 10 Å simulated maps generated from the target structures. Movies of the fittings are included in the Supplemental Data (see Movies S1 and S2).

into Simulated Maps) also show that the accuracy obtained by applying MDFF is comparable to other methods (Jolley et al., 2008; Topf et al., 2008) and that MDFF can describe a large range of conformational changes, such as movements around a hinge and domain shearing. The fact that smaller grid spacings yield better results suggests that higher-order interpolation schemes in the grid-based force calculation can improve the fitting; indeed, the use of cubic interpolation in the current implementation of grid-steered molecular dynamics in NAMD (used in the calculations presented) yielded slightly better results when compared to linear interpolation (data not shown). The simulated maps presented in this section differ significantly from EM maps in that they do not contain noise emerging from the imaging process and numerical errors due to image processing and reconstruction, and in that they represent a single structure instead of an ensemble average as captured by cryo-EM. An attempt to address the latter is presented in the Supplemental Data (see Simulated Maps Generated from an Ensemble of Structures), where a protein structure is fitted into maps created from an ensemble of structures obtained from equilibrium MD simulations. For these tests, the fluctuation of atomic positions observed in MDFF reproduce reasonably well the fluctuations on the target map, especially for resolutions in the range of 10–15 Å.

Example Application: The E. coli *Ribosome*

The ribosome, a complex macromolecular machine responsible for protein synthesis in all cells, is one of the biological systems for which cryo-EM has provided much insight to date (Frank, 2003). The ribosome undergoes several conformational changes and binds a number of cofactors throughout the process of protein synthesis. Different functional states have been extensively imaged by cryo-EM at ever-increasing resolution (Frank and Spahn, 2006). The ribosome has driven the development of methods to obtain quasi-atomic structures by combining X-ray crystallographic structures with cryo-EM maps (e.g., Wriggers et al., 2000; Gao et al., 2003; Tama et al., 2004; Mitra et al., 2005, 2006; for a review, see Mitra and Frank, 2006).

We present the ribosome as an application of MDFF, using EM maps corresponding to two functional states, namely, the *E. coli* ribosome in complex with the ternary complex EF-Tu·aminoacyl-tRNA·GDP stalled by the antibiotic kirromycin (70S-fMet-tRNA^{fMet}-Phe-tRNA^{Phe}·EF-Tu·GDP·kir) at resolutions of 9 Å (Valle et al., 2003) and 6.7 Å (J. LeBaron, R.A. Grassucci, T. Shaikh, W. Baxter, J. Sengupta, and J.F., unpublished data), and a ribosome with an accommodated A-site tRNA at 9 Å resolution (Valle et al., 2003). The first functional state corresponds to the initial selection of the ternary complex (TC) in the elongation cycle, after which EF-Tu leaves the ribosome and, subsequently, the peptide bond is formed, resulting in the second functional state, with a deacylated tRNA occupying the P site and an accommodated fMet-Phe-tRNA^{Phe} in the A site (70S-tRNA^{Phe}-MF-tRNA^{Phe}). The atomic model of the ribosome

 ©2008 Elsevier Ltd All rights reserved

used in this section is described in the Supplemental Data (see Atomic Model of the *E. coli* Ribosome).

The flexible fitting was performed following a multistep protocol: (1) After rigid-body docking, the 30S and 50S subunits were flexibly fitted, using ξ = 0.3 kcal/mol, $k_{\mu,RNA}$ = 200 kcal mol^{-1} rad^{-2} or 200 kcal mol^{-1} $Å^{-2}$ for angles and distances, respectively, and $k_{\mu,\mathrm{protein}}=400$ kcal mol^{-1} rad^{-2}. In addition to the secondary structure restraints, the ϕ and ψ angles of the rest of the amino acid residues were restrained with $k_{\mu,\mathrm{protein}}=200$ kcal mol^{-1} rad^{-2}. (2) The potential corresponding to the 5S, 16S, and 23S rRNAs was deleted, and the atoms in these chains were restrained to their current positions (positional restraints) with a relatively stiff force constant of 10 kcal mol^{-1} $Å^{-2}$, effectively fixing them. This allowed the fitting of ribosomal proteins to improve, since the potential energy minima corresponding to the ribosomal RNA were removed from the simulation. (3) The harmonic restraints applied to the ribosomal proteins were relaxed to 200 kcal mol^{-1} rad^{-2}, restraining only residues in helices and β sheets. (4) The remaining ligands, namely EF-Tu, tRNAs, and mRNA, were introduced using rigid-body docking. For this step, the original EM potential (before RNA deletion) was used, all ribosomal proteins were positionally restrained, and the positional restraints on the rRNAs were lifted. The same secondary structure restraints to proteins and RNAs used in the previous step were preserved and also applied to the recently introduced ligands. In addition, equivalent RNA restraints were imposed to enforce codon-anticodon interactions between the A-site mRNA codon and the tRNA. During this last step, force scaling was increased to ξ = 1.0 kcal/mol. Convergence times and cross-correlation coefficients for each step are presented for all maps in the Supplemental Data (see Comparison between Maps at Different Resolutions). A movie of the fitting into the 6.7 Å map illustrating the multistep protocol can be found in the Supplemental Data (see Movie S3).

The MDFF method applied to different ribosome cryo-EM maps yielded quasi-atomic models that closely fit the EM densities; indeed, the cross-correlation coefficients between the maps and the fitted structures are significantly higher than those obtained from rigid-body docking (local correlations with a threshold of 0.2 σ are given in parenthesis): 0.913 (0.764) versus 0.858 (0.632) (TC-bound ribosome at 6.7 Å), 0.919 (0.739) versus 0.835 (0.503) (TC-bound ribosome at 9.0 Å), and 0.878 (0.735) versus 0.756 (0.513) (ribosome with accommodated A-site tRNA at 9.0 Å). Selected regions for which conformational changes have been previously characterized are presented in Figure 4. It can be seen that distinct structural elements such as RNA and protein helices fit well into their corresponding densities, despite internal restraints imposed to avoid structural distortions. Local cross-correlation coefficients of different elements are presented in the Supplemental Data (see Local Cross-Correlation Coefficient Map).

An overview of the fitting into the 6.7 Å map of the TC-bound ribosome is depicted in Figure 4A, along with a detailed view of the decoding center. Figure 4B shows the X-ray structure of tRNA (crystallized in complex with EF-Tu), a fitted structure, and the same crystal structure fitted manually in a previous study (Valle et al., 2003). It can be seen that the conformation of the anticodon loop (ACL) of the A/T-site tRNA is significantly different from the one adopted in the free TC. Interestingly, our fitted structure is very similar to the previously proposed one; however, the previous work assumed that the ACL of the A/T-site tRNA adopts the same conformation as that of the free ternary complex, and thus a model of the structure was built by interpolating coordinates between two manual fittings of the same crystal structure (Valle et al., 2003). With the MDFF method, we obtained the conformational change that the ACL undergoes when the TC binds to the ribosome without any assumptions, the fitting being driven only by the EM data. The structure of the ACL of the A/T-site tRNA obtained from our fitting has the same conformation as the ACL of the A-site tRNA observed by Selmer et al. (2006), as shown in Figure 4C. A comparison between the fittings into cryo-EM maps of the TC-bound ribosome at different resolutions (6.7 Å and 9.0 Å) is presented in the Supplemental Data (see Comparison between Maps at Different Resolutions).

Binding of the TC to the ribosome induces a conformational change in the GTPase-associated center (GAC) (Valle et al., 2003). Figure 4D presents a comparison of the GAC conformation in the TC-bound ribosome and the ribosome in complex with an accommodated A-site tRNA. In the first conformation (closed), the GAC approaches the 50S to interact with the TC; when EF-Tu leaves the ribosome, the second conformation (open) arises in which the GAC lobe moves back to its original position. One can see that rigid-body docking of the crystal structure to the ribosomal cryo-EM map with an accommodated A-site tRNA shows a good fit for the GAC, revealing that the crystal structure captures this conformation. Flexible fitting obtains a closer match to the EM map for this state and reveals the closed conformation of the GAC and TC in the TC-bound ribosome. The details of the atomic structures obtained from the 6.7 Å map will be discussed elsewhere (J. Sengupta, E.V., L.G.T., J. LeBaron, W.T. Baxter, T. Shaikh, R.A. Grassucci, P. Nissen, M. Ehrenberg, K.S., and J.F., unpublished data).

Even at subnanometer resolution that permits the identification of secondary structure elements, some regions are not well defined in an EM map. For example, the switch regions of EF-Tu in the 6.7 Å TC-bound ribosomal EM map are not resolved, presumably due to their high flexibility. Though information about their structure is not directly available in the EM map, MD simulations can be performed in the presence of the EM potential to assess the feasibility of different conformations of the switches: an unfeasible conformation will either change during simulation, or alter the quality of the fit of neighboring domains. The conformational dynamics of the interaction of EF-Tu with the ribosome can be deduced from the computational studies, even though a crystal structure of a ribosome-bound EF-Tu is not yet available, and density for some key elements governing the interaction is missing in the EM map (J. Sengupta, E.V., L.G.T., J. LeBaron, W.T. Baxter, T. Shaikh, R.A. Grassucci, P. Nissen, M. Ehrenberg, K.S., and J.F., unpublished data).

DISCUSSION

We have developed a novel method, MDFF, for combining atomic structures and EM maps to reveal atomic details of macromolecular complexes in functional states. High-resolution structures of complexes imaged by cryo-EM permit a better interpretation of the data, e.g., by characterizing the flexibility of

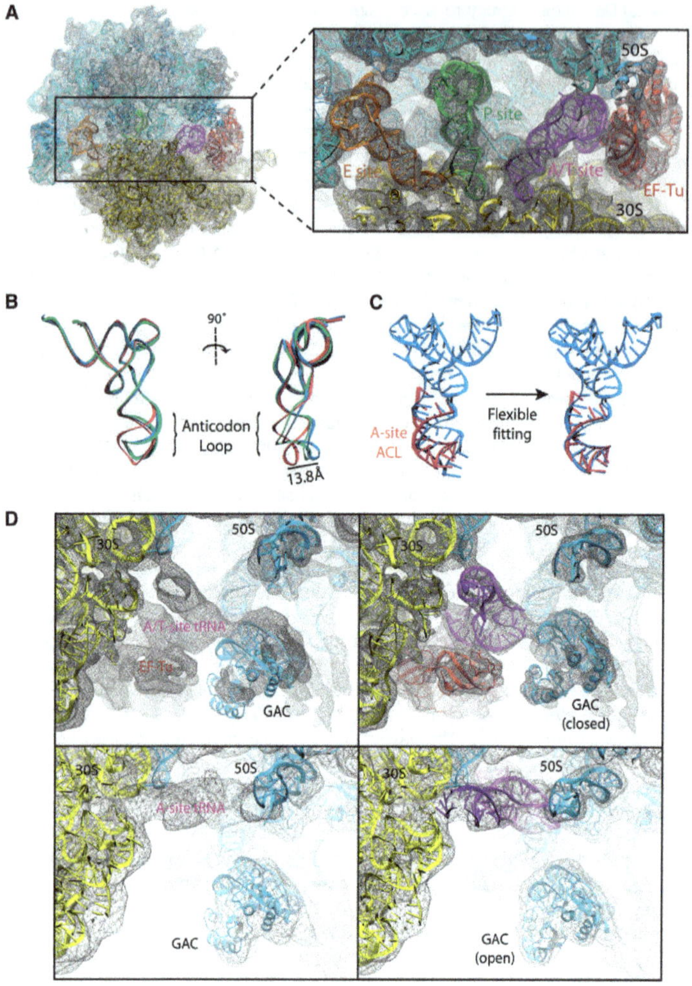

Figure 4. Fitting into the TC-Bound Ribosome Cryo-EM Map at 6.7 Å Resolution by Means of MDFF

(A) Overview of the all-atom ribosome structure fitted into the 6.7 Å map, with a close view into the decoding center (inset).

(B) Conformation of tRNA in the A/T site. The crystal structure from the free TC used as a starting point for the fitting (PDB code: 1OB2; unpublished data) is shown in red; the A/T tRNA model obtained by applying the MDFF method to the 6.7 Å map is shown in blue; the A/T tRNA model previously obtained using a 9.0 Å map constructed by interpolating two manual fittings of tRNA (PDB code: 1OB2) is shown in green (Valle et al., 2003).

(C) Conformation of tRNA in the A/T site (blue) compared to a partial crystal structure of the A-site tRNA (Selmer et al., 2006) (red). The crystal structure from the free TC used as a starting point for the fitting (PDB code: 1OB2; unpublished data) is shown on the left; the A/T tRNA model obtained by applying the MDFF method to the 6.7 Å map is shown on the right.

(D) Conformational dynamics of the GTPase-associated center. Shown are differences in the conformation of the GTPase-associated center between the TC-bound ribosome (EM map at 6.7-Å resolution, top), and the accommodated ribosome (EM map at 9 Å resolution, bottom). Rigid-body docked structures into the corresponding maps, used as initial coordinates for flexible fitting, are shown on the left; flexibly fitted structures are shown on the right.

the complex, identifying key interactions between its composing elements, or locating bound cofactors. The MDFF method takes advantage of the impressive advances both in X-ray crystallography and cryo-EM, by using MD simulations that incorporate EM data through a potential driving an atomic structure into a conformation corresponding to an EM map. The method has several characteristics that bring together the best features of previously developed methods: (1) MDFF avoids the use of *reduced representations* which necessarily discard some of the information contained in the crystal structure or the EM map. (2) MDFF considers information contained in the map through a potential, such that the fitting is performed *locally*, i.e., independently of other regions of a molecule. (3) MDFF uses *global measures* of the fit only to assess convergence and not to drive the fit. Methods that rely on global measures to obtain a good fit become less efficient as the *size of the system* increases. (4) MDFF can be used to fit both *protein and nucleic acids*, as well as systems composed of both. (5) MDFF can fit parts of the complex independently when complete atomic structures are not available. (6) MDFF does not require *user input* to divide a molecule into pieces to flexibly fit them into the map; rather, the flexibility is indigenous to the molecular structure, with additional restraints dictated by the secondary structure. (7) MDFF follows *multistep protocols* that permit adjusting the method to face various challenges; e.g., in the case of systems composed of protein and nucleic acids, fitting can be performed on nucleic acids first and the protein component second. (8) MDFF ensures *stereochemical correctness* during the fitting process, obviating the need for post-fitting refinement, which often results in deviations from the EM map. In fact, the structures obtained by MDFF can be used as initial coordinates for further simulation studies, in particular, for an equilibration testing the stability of the model arrived at. (9) MDFF uses restraints to preserve *secondary structure* elements and other structural features in order to prevent overfitting; in this respect it bears some similarity to real-space refinement (Chapman, 1995), but MDFF is more automated and more adaptive. (10) MDFF can represent the structural variability present in the experimental map by providing a representative set of fitted structures instead of a single one. (11) MDFF can be extended to take advantage of other enhanced-sampling techniques

 ©2008 Elsevier Ltd All rights reserved

Structure

Molecular Dynamics Flexible Fitting

such as the use of elastic network models (Tama et al., 2004; Schröder et al., 2007).

A natural concern about flexible fitting methods is overfitting. We have shown that the harmonic restraints proposed are enough to prevent overfitting into noise-free maps. However, the degree of overfitting into experimental EM maps prevented by applying the restraints presented here cannot be evaluated at this time, and thus users are discouraged from softening the tight restraints suggested in this paper. The choice of restraints in MDFF might be improved further, e.g., through incorporation of restraints based on rigidity analysis (Jacobs et al., 2001; Jolley et al., 2008). A limitation intrinsic to our method is that rotation of structural elements are difficult to capture; this may be addressed, e.g., by enhancing the lowest frequency normal modes in the simulation (Zhang et al., 2003) or by refining with MDFF models generated with a flexible fitting method based on normal modes (Tama et al., 2004; Suhre et al., 2006). Naturally, with the current restraints it is not possible to capture conformational changes that involve unfolding of secondary structure elements. Even though the fitting in MDFF is performed locally, proteins that interact with nucleic acids should not be fitted in the absence of the latter, since they might be attracted to the higher nucleic acid density. An interesting alternative involves the use of local, but not global, cross-correlation maps to define U_{EM}.

Naturally, MDFF results cannot be better than the EM data used permit. Any artifact in the molecular density reported in the EM map can be propagated to the fitted structure. The quantitative use of the density map requires high-quality data collection and efficient 3D reconstruction algorithms, as extensively discussed in Frank (2006). The fitting and the interpretation of the resulting structures must be done taking into account factors that prevent the 3D reconstruction from accurately representing the molecular density. An additional concern that the MDFF technique might raise is computational cost. Indeed, the necessary calculations are extensive, akin to other available flexible-fitting methods (Schröder et al., 2007; Wriggers and Birmanns, 2001). However, with the prospect of automated cryo-EM data eventually being able to match the speed of throughput achieved by other structural methods (Zhu et al., 2001), the advantages of not requiring ad hoc user input and of incorporating as much of information from the original data as possible, together with expected increases in computer power, will make MDFF a feasible and attractive tool for obtaining atomic structures from the wealth of EM data. Moreover, by using NAMD, the computational cost of the MDFF method scales linearly with system size, permitting the application of MDFF to large macromolecular complexes.

The success of the MDFF method is evident from the quality of the high-resolution structures obtained for cryo-EM maps of ribosomal complexes. The ribosome represents one of the greatest challenges to fitting methods due to its sheer size, its lack of symmetry, and its mixed composition of protein and RNA. We have recently fitted atomic structures to several ribosomal cryo-EM maps to study the conformational changes that take place during the decoding process. (K.M., L.G.T., E.V., A. Zavialov, M. Ehrenberg, K.S., and J.F., unpublished data; J. Sengupta, E.V., L.G.T., J. LeBaron, W.T. Baxter, T. Shaikh, R.A. Grassucci, P. Nissen, M. Ehrenberg, K.S., and J.F., unpublished data). Typical global cross-correlation coefficients obtained improve from ~0.8 after rigid-body docking, to ~0.9 after completion of the flexible fitting.

In recent years, it has become evident that cellular functions are carried out by assemblies of interacting macromolecules (Alberts, 1998), many of them existing only transiently in the cell. In order to provide a comprehensive description of such complexes, spanning the atomic and the system level, data obtained from various structural techniques must be combined into high-resolution structures with the aid of theoretical approaches (Sali et al., 2003; Alber et al., 2007). MD is a method of choice, being increasingly used to refine macromolecular structures, and is an established tool for studying structural dynamics of large biomolecules. The conceptual grounds of the methodology presented here can easily be extended to other sources of structural data by including them in MD simulations of atomic or coarse-grained structures (Shih et al., 2007; Arkhipov et al., 2006) as external potentials that merge several levels of description into high-resolution structures.

As structural methods become more prolific, automated software capable of interpreting intermediate-resolution structures at the atomic level is becoming crucial; indeed, it seems likely that in years to come this type of approach will yield the main, and perhaps only, source of atomic structures for large macromolecular complexes in functional conformations. The software should be widely available and should not require very detailed technical expertise or ad hoc input, but rather should be easy to use and general. These requirements are met by the MDFF method. The method is currently implemented in the molecular dynamics software package NAMD (Phillips et al., 2005), and the automated setup and analysis of the simulations is done through the molecular visualization software package VMD (Humphrey et al., 1996).

SUPPLEMENTAL DATA

Supplemental Data include an overview of MD simulations and 3D EM reconstructions; several additional validation cases for MDFF, including density maps generated from an MD simulation; a description of the atomic model of the *E. coli* ribosome used in the current work; an analysis of the effect of resolution on the fitting into experimental cryo-EM maps; a local cross-correlation coefficient map for a fitted ribosome structure; and three movies illustrating the MDFF method. Supplemental Data can be found with this article online at http://www.structure.org/cgi/content/full/16/5/673/DC1/.

ACKNOWLEDGMENTS

The authors would like to thank David Wells, Aleksei Aksimientiev, and Jim Phillips for help with the implementation of the MDFF method in NAMD; Jordi Cohen for stimulating discussions; and Emma Falck for contributing to initial stages of this project. This work was supported by HHMI (J.F.), the Burroughs Wellcome Fund (K.M.), and grants NIH P41-RR05969 (K.S.), NIH P41-RR01219, NIH R37-GM29169, and NIH R01-GM55440 (J.F.). Computer time was provided through LRAC grant MCA93S028 (K.S.).

Received: January 16, 2008
Revised: March 3, 2008
Accepted: March 22, 2008
Published: May 6, 2008

REFERENCES

Alber, F., Dokudovskaya, S., Veenhoff, L.M., Zhang, W., Kipper, J., Devos, D., Suprapto, A., Karni-Schmidt, O., Williams, R., Chait, B.T., et al. (2007). Determining the architectures of macromolecular assemblies. Nature *450*, 683–694.

 ©2008 Elsevier Ltd All rights reserved

Alberts, B. (1998). The cell as a collection of protein machines – preparing the next generation of biologists. Cell *92*, 291–294.

Arkhipov, A., Freddolino, P.L., and Schulten, K. (2006). Stability and dynamics of virus capsids described by coarse-grained modeling. Structure *14*, 1767–1777.

Brünger, A.T., Brooks, C.L., III, and Karplus, M. (1984). Stochastic boundary conditions for molecular dynamics simulations of ST2 water. Chem. Phys. Lett. *105*, 495–498.

Chapman, M.S. (1995). Restrained real-space macromolecular atomic refinement using a new resolution-dependent electron-density function. Acta Crystallogr. A *51*, 69–80.

Chen, L.F., Blanc, E., Chapman, M.S., and Taylor, K.A. (2001). Real space refinement of acto-myosin structures from sectioned muscle. J. Struct. Biol. *133*, 221–232.

Darnault, C., Volbeda, A., Kim, E.J., Legrand, P., Vernède, X., Lindahl, P.A., and Fontecilla-Camps, J.C. (2003). Ni-Zn-[Fe-S] and Ni-Ni-[Fe-S] clusters in closed and open α subunits of acetyl-CoA synthase/carbon monoxide dehydrogenase. Nat. Struct. Biol. *10*, 271–279.

Foloppe, N., and MacKerell, A.D., Jr. (2000). All-atom empirical force field for nucleic acids: I. Parameter optimization based on small molecule and condensed phase macromolecular target data. J. Comput. Chem. *21*, 86–104.

Frank, J. (2002). Single-particle imaging of macromolecules by cryo-electron microscopy. Annu. Rev. Biophys. Biomol. Struct. *31*, 309–319.

Frank, J. (2003). Electron microscopy of functional ribosome complexes. Biopolymers *68*, 223–233.

Frank, J. (2006). Three-Dimensional Electron Microscopy of Macromolecular Assemblies (New York: Oxford University Press).

Frank, J., and Spahn, C.M.T. (2006). The ribosome and the mechanism of protein synthesis. Rep. Prog. Phys. *69*, 1383–1417.

Frenkel, D., and Smit, B. (2002). Understanding Molecular Simulation From Algorithms to Applications (California: Academic Press).

Gao, H., Sengupta, J., Valle, M., Korostelev, A., Eswar, N., Stagg, S.M., Van Roey, P., Agrawal, R.K., Harvey, S.C., Sali, A., et al. (2003). Study of the structural dynamics of the *E. coli* 70S ribosome using real space refinement. Cell *113*, 789–801.

Humphrey, W., Dalke, A., and Schulten, K. (1996). VMD – Visual Molecular Dynamics. J. Mol. Graph. *14*, 33–38.

Isralewitz, B., Gao, M., and Schulten, K. (2001). Steered molecular dynamics and mechanical functions of proteins. Curr. Opin. Struct. Biol. *11*, 224–230.

Jacobs, D.J., Rader, A.J., Kuhn, L.A., and Thorpe, M.F. (2001). Protein flexibility predictions using graph theory. Proteins *44*, 150–165.

Jolley, C.C., Wells, S.A., Fromme, P., and Thorpe, M.F. (2008). Fitting low-resolution cryo-EM maps of proteins using constrained geometric simulations. Biophys. J. *94*, 1613–1621.

Leontis, N.B., and Westhof, E. (1998). Conserved geometrical base-pairing patters in RNA. Q. Rev. Biophys. *31*, 399–455.

Li, W., and Frank, J. (2007). Transfer RNA in the hybrid P/E state: Correlating molecular dynamics simulations with cryo-EM data. Proc. Natl. Acad. Sci. USA *104*, 16540–16545.

MacKerell, A., Jr., Bashford, D., Bellott, M., Dunbrack, R.L., Jr., Evanseck, J., Field, M.J., Fischer, S., Gao, J., Guo, H., Ha, S., et al. (1998). All-atom empirical potential for molecular modeling and dynamics studies of proteins. J. Phys. Chem. B *102*, 3586–3616.

Mitra, K., and Frank, J. (2006). Ribosome dynamics: insights from atomic structure modeling into cryo-electron microscopy maps. Annu. Rev. Biophys. Biomol. Struct. *35*, 299–317.

Mitra, K., Schaffitzel, C., Shaikh, T., Tama, F., Jenni, S., Brooks, C.L., Ban, N., and Frank, J. (2005). Structure of the *E. coli* protein-conducting channel bound to a translating ribosome. Nature *438*, 318–324.

Mitra, K., Schaffitzel, C., Fabiola, F., Chapman, M., Ban, N., and Frank, J. (2006). Elongation arrest by SecM via a cascade of ribosomal RNA rearrangements. Mol. Cell *22*, 533–543.

Phillips, J.C., Braun, R., Wang, W., Gumbart, J., Tajkhorshid, E., Villa, E., Chipot, C., Skeel, R.D., Kale, L., and Schulten, K. (2005). Scalable molecular dynamics with NAMD. J. Comput. Chem. *26*, 1781–1802.

Roseman, A.M. (2000). Docking structures of domains into maps from cryo-electron microscopy using local correlation. Acta Crystallogr. D Biol. Crystallogr. *56*, 1332–1340.

Rossmann, M.G. (2000). Fitting atomic models into electron-microscopy maps. Acta Crystallogr. D Biol. Crystallogr. *56*, 1341–1349.

Saibil, H.R. (2000). Macromolecular structure determination by cryo-electron microscopy. Acta Crystallogr. D Biol. Crystallogr. *56*, 1215–1222.

Sali, A., Glaeser, R., Earnest, T., and Baumeister, W. (2003). From words to literature in structural proteomics. Nature *422*, 216–225.

Schröder, G.F., Brunger, A.T., and Levitt, M. (2007). Combining efficient conformational sampling with a deformable elastic network model facilitates structure refinement at low resolution. Structure *15*, 1630–1641.

Schuwirth, B.S., Borovinskaya, M.A., Hau, C.W., Zhang, W., Vila-Sanjurjo, A., Holton, J.M., and Cate, J.H.D. (2005). Structures of the bacterial ribosome at 3.5 Å resolution. Science *310*, 827–834.

Selmer, M., Dunham, C.M., Murphy, F.V., IV, Weixlbaumer, A., Petry, S., Kelley, A.C., Weir, J.R., and Ramakrishnan, V. (2006). Structure of the 70S ribosome complexed with mRNA and tRNA. Science *313*, 1935–1942.

Shih, A.Y., Freddolino, P.L., Arkhipov, A., and Schulten, K. (2007). Assembly of lipoprotein particles revealed by coarse-grained molecular dynamics simulations. J. Struct. Biol. *157*, 579–592.

Sotomayor, M., and Schulten, K. (2007). Single-molecule experiments in vitro and in silico. Science *316*, 1144–1148.

Stewart, P.L., Fuller, S.D., and Burnett, R.M. (1993). Difference imaging of adenovirus: Bridging the resolution gap between X-ray crystallography and electron microscopy. EMBO J. *12*, 2589–2599.

Suhre, K., Navaza, J., and Sanejouand, Y.H. (2006). NORMA: a tool for flexible fitting of high-resolution protein structures into low-resolution electron-microscopy-derived density maps. Acta Crystallogr. D Biol. Crystallogr. *62*, 1098–1100.

Tama, F., Miyashita, O., and Brooks, C.L., III. (2004). Normal mode based flexible fitting of high-resolution structure into low-resolution experimental data from cryo-EM. J. Struct. Biol. *147*, 315–326.

Topf, M., Baker, M.L., Marc-Renom, M.A., Chiu, W., and Sali, A. (2006). Refinement of protein structures by iterative comparative modeling and cryoEM density fitting. J. Mol. Biol. *357*, 1655–1668.

Topf, M., Lasker, K., Webb, B., Wolfson, H., Chiu, W., and Sali, A. (2008). Protein structure fitting and refinement guided by cryo-EM density. Structure *16*, 295–307.

Valle, M., Zavialov, A., Li, W., Stagg, S.M., Sengupta, J., Nielsen, R.C., Nissen, P., Harvey, S.C., Ehrenberg, M., and Frank, J. (2003). Incorporation of aminoacyl-tRNA into the ribosome as seen by cryo-electron microscopy. Nat. Struct. Biol. *10*, 899–906.

Velazquez-Muriel, J.A., Valle, M., Santamaría-Pang, A., Kakadiaris, I.A., and Carazo, J.M. (2006). Flexible fitting in 3D-EM guided by the structural variability of protein superfamilies. Structure *14*, 1115–1126.

Volkmann, N., Hanein, D., Ouyang, G., Trybus, K.M., DeRosier, D.J., and Lowey, S. (2000). Evidence for cleft closure in actomyosin upon ADP release. Nat. Struct. Biol. *7*, 1147–1155.

Wells, D.B., Abramkina, V., and Aksimentiev, A. (2007). Exploring transmembrane transport through α-hemolysin with grid-steered molecular dynamics. J. Chem. Phys. *127*, 125101–125110.

Wriggers, W., and Birmanns, S. (2001). Using Situs for flexible and rigid-body fitting of multiresolution single-molecule data. J. Struct. Biol. *133*, 193–202.

Wriggers, W., and Chacón, P. (2001). Modeling tricks and fitting techniques for multiresolution structures. Structure *9*, 779–788.

Wriggers, W., Milligan, R.A., and McCammon, J.A. (1999). Situs: A package for docking crystal structures into low-resolution maps from electron microscopy. J. Struct. Biol. *125*, 185–195.

 ©2008 Elsevier Ltd All rights reserved

Wriggers, W., Agrawal, R.K., Drew, D.L., McCammon, A., and Frank, J. (2000). Domain motions of EF-G bound to the 70S ribosome: Insights from a hand-shaking between multi-resolution structures. Biophys. J. *79*, 1670–1678.

Yang, H., Jossinet, F., Leontis, N., Chen, L., Westbrook, J., Berman, H., and Westhof, E. (2003). Tools for the automatic identification and classification of RNA base pairs. Nucleic Acids Res. *31*, 3450–3460.

Zhang, Z., Shi, Y., and Liu, H. (2003). Molecular dynamics simulations of peptides and proteins with amplified collective motions. Biophys. J. *84*, 3583–3593.

Zhu, Y., Carragher, B., Kriegman, D.J., Milligan, R.A., and Potter, C.S. (2001). Automated identification of filaments in cryoelectron microscopy images. J. Struct. Biol. *135*, 302–312.

 ©2008 Elsevier Ltd All rights reserved

Ribosome-induced changes in elongation factor Tu conformation control GTP hydrolysis

Elizabeth Villa[a,b,1], Jayati Sengupta[c,1,2], Leonardo G. Trabuco[a,b], Jamie LeBarron[c], William T. Baxter[c], Tanvir R. Shaikh[c], Robert A. Grassucci[d], Poul Nissen[e], Måns Ehrenberg[f], Klaus Schulten[a,g], and Joachim Frank[d,h,3]

[a]Beckman Institute, University of Illinois at Urbana-Champaign, Urbana, IL 61801; [b]Center for Biophysics and Computational Biology, University of Illinois at Urbana-Champaign, Urbana, IL 61801; [c]Wadsworth Center, Empire State Plaza, Albany, NY 12201; [d]Howard Hughes Medical Institute, Department of Biochemistry and Molecular Biophysics, Columbia University, 650 West 168th Street, New York, NY 10032; [e]Department of Molecular Biology, University of Aarhus, Gustav Wieds vej 10C, DK-8000 Aarhus, Denmark; [f]Department of Cell and Molecular Biology, Biomedical Center, Box 596, S-75124 Uppsala, Sweden; [g]Department of Physics, University of Illinois at Urbana-Champaign, Urbana, IL 61801; and [h]Department of Biological Sciences, Columbia University, 600 Fairchild Center, Mail Code 2402, 1212 Amsterdam Avenue, New York, NY 10027

Contributed by Joachim Frank, November 12, 2008 (sent for review October 10, 2008)

In translation, elongation factor Tu (EF-Tu) molecules deliver aminoacyl-tRNAs to the mRNA-programmed ribosome. The GTPase activity of EF-Tu is triggered by ribosome-induced conformational changes of the factor that play a pivotal role in the selection of the cognate aminoacyl-tRNAs. We present a 6.7-Å cryo-electron microscopy map of the aminoacyl-tRNA·EF-Tu·GDP·kirromycin-bound *Escherichia coli* ribosome, together with an atomic model of the complex obtained through molecular dynamics flexible fitting. The model reveals the conformational changes in the conserved GTPase switch regions of EF-Tu that trigger hydrolysis of GTP, along with key interactions, including those between the sarcin-ricin loop and the P loop of EF-Tu, and between the effector loop of EF-Tu and a conserved region of the 16S rRNA. Our data suggest that GTP hydrolysis on EF-Tu is controlled through a hydrophobic gate mechanism.

decoding | GTPase | flexible fitting | cryo-EM | ternary complex

In the elongation cycle of protein synthesis, each aminoacyl-tRNA (aa-tRNA) is delivered to the mRNA-programmed ribosome as a ternary complex with elongation factor Tu (EF-Tu) and GTP. On binding to the ribosome, aa-tRNA enters the A/T state, where it occupies the A site on the 30S ribosomal subunit while still interacting with EF-Tu. When the codon–anticodon interaction is cognate, GTPase activity of EF-Tu is greatly stimulated and GTP hydrolysis occurs rapidly, inducing a large conformational change of EF-Tu followed by its dissociation from the ribosome and subsequent accommodation of aa-tRNA into the ribosomal A site. The irreversible GTP hydrolysis separates 2 kinetic steps of aa-tRNA incorporation, namely initial selection and proofreading, that jointly establish the overall accuracy of tRNA selection by the codon-programmed ribosome (1). Much effort has been devoted to elucidate the mechanism by which codon recognition in the 30S subunit leads to GTP hydrolysis by EF-Tu >75 Å away. Biochemical and structural studies identified some key interactions, but the atomic details of the mechanism have remained elusive. In particular, the transduction of the codon recognition signal and subsequent ribosome-induced EF-Tu conformational changes leading to GTPase activation are not well understood.

GTP hydrolysis in EF-Tu takes place through an in-line, direct attack on the γ-phosphate by a water molecule (2). His-84 of EF-Tu has been proposed to activate a water molecule for nucleophilic attack on the γ-phosphate by extracting one of its protons (3, 4). This proposal agrees with studies identifying His-84 as the catalytic residue that stabilizes the transition state of GTP hydrolysis by hydrogen bonding to the attacking water molecule or the γ-phosphate group of GTP (5). Crystal structures of EF-Tu in the GTP form revealed a "hydrophobic gate" that appears to control access of His-84 to the GTP binding pocket (3, 6). For His-84 to carry out its catalytic role and activate the water molecule, the gate would have to open on interaction with the ribosome (3, 6). So far, gate opening has only been observed in the crystallographic structure of *Thermus thermophilus* EF-Tu in complex with GDP and the GTPase-stimulating antibiotic methyl-kirromycin (aurodox), outside the ribosome.

In the present work, we focus on the ribosome-induced EF-Tu conformational changes that trigger GTP hydrolysis on the factor as revealed by cryo-electron microscopy (cryo-EM) studies. We provide structural evidence suggesting that the ribosome induces the opening of the hydrophobic gate and the repositioning of His-84 of EF-Tu. Our results reconcile a large body of experimental data and explain how the ribosome controls the GTPase activity of EF-Tu in the ternary complex.

Results

We obtained a 6.7-Å cryo-EM map of the preaccommodated *Escherichia coli* 70S ribosome bound to the Phe-tRNA^{Phe}·EF-Tu·GDP ternary complex stalled by the antibiotic kirromycin (kir) (7). We applied the recently developed molecular dynamics flexible fitting (MDFF) method (8) to obtain an atomic model of the complex that enables the interpretation of the cryo-EM data in unprecedented detail. The map, along with the fitted atomic model of the 70S ribosome, the ternary complex, P- and E-site tRNAs, and mRNA are shown in Fig. 1*A*. This map presents a substantial advancement in resolution over previously reported cryo-EM data (9–11), allowing us to corroborate previous observations in greater detail, and to observe new contacts and changes in conformation of the complex. In particular, with the improvement in resolution, the structural features of EF-Tu and its interactions with the ribosome are clearly discernible in the density map.

The relative orientation between the domains of the ribosome-bound EF-Tu with kirromycin (Fig. 1*B*) is most similar to the "closed," GTP-bound form of EF-Tu (3, 6), as previously

---

Author contributions: M.E., K.S., and J.F. designed research; E.V., J.S., L.G.T., J.L., W.T.B., T.R.S., and R.A.G. performed research; M.E. contributed new reagents/analytic tools; E.V., J.S., L.G.T., J.L., W.T.B., T.R.S., P.N., and K.S. analyzed data; and E.V., J.S., L.G.T, K.S., P.N., and J.F. wrote the paper.

The authors declare no conflict of interest.

Freely available online through the PNAS open access option.

Data deposition. The atomic coordinates have been deposited in the Protein Data Bank, www.pdb.org [PDB ID codes 3FIH (30S and factors) and 3FIK (50S)]. The cryo-EM density map has been deposited in the 3D-EM database, EMBL-European Bioinformation Institute (code EMD-5036).

[1]E.V. and J.S. contributed equally to this work.

[2]Present address: Structural Biology and Bioinformatics Division, Indian Institute of Chemical Biology, Calcutta, 700032, India.

[3]To whom correspondence should be addressed. E-mail: jf2192@columbia.edu.

This article contains supporting information online at www.pnas.org/cgi/content/full/0811370106/DCSupplemental.

© 2009 by The National Academy of Sciences of the USA

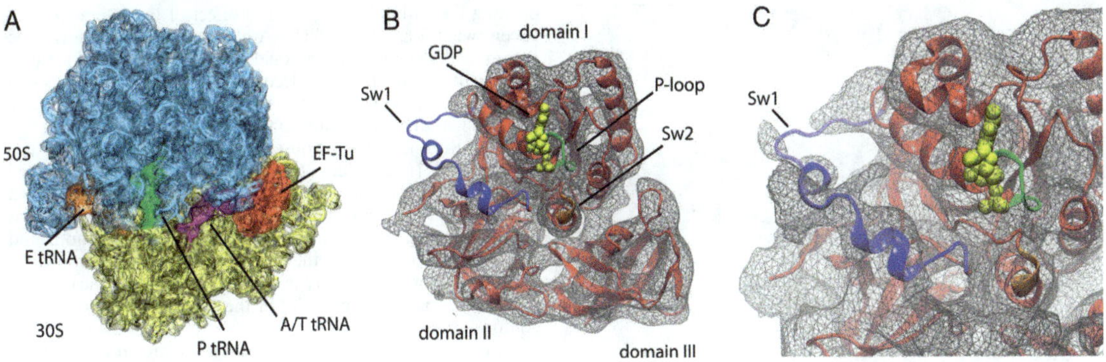

Fig. 1. Overview of EF-Tu structure bound to the ribosome. (*A*) Cryo-EM map of the 70S·fMet-tRNA^{fMet}·Phe-tRNA^{Phe}·EF-Tu·GDP·kir complex at a resolution of 6.7 Å shown in transparent surface. The atomic model obtained by molecular dynamics flexible fitting (MDFF) is shown in cartoon representation. (*B*) Atomic model of the ribosome-bound EF-Tu showing a domain orientation most similar to the closed, GTP-bound form. The switch regions are highlighted: switch I (Sw1) in blue, switch II (Sw2) in orange, and P-loop in green. The same coloring scheme is used in all figures. (*C*) Closeup of the GTPase domain of EF-Tu. The cryo-EM density map is displayed at a lower threshold to make the switch I density visible. Density thresholds are 2.6 σ (*A*), 2.2 σ (*B*), and 1.4 σ (*C*).

observed outside of the ribosome. The structure of EF-Tu·GDP·kir bound to the ribosome has been proposed to correspond to a state immediately after GTP hydrolysis, yet preceding the conformational change into the free, GDP-bound structure and, hence, preceding EF-Tu release (11). Our data show that this EF-Tu structure indeed corresponds to an activated state for the hydrolysis of the nucleotide, as detailed below. Several conserved regions playing prominent roles in the working cycle of GTPases, namely switch I (EF-Tu residues 40–62; *E. coli* numbering is used throughout this article), switch II (80–100) and P loop (18–23), are found to undergo characteristic conformational changes in EF-Tu. We describe these changes in the context of the ribosome-bound state of EF-Tu captured by the map.

GTPase Activation Is Controlled by a Hydrophobic Gate. The hydrophobic gate of EF-Tu, represented schematically in Fig. 2*A*, is formed by residues Val-20 (P loop) and Ile-60 (switch I). The gate in its closed form is present in many crystal structures of GDP- and GTP-bound EF-Tu both bound and unbound to tRNA, indicating that gate opening occurs only in the presence of the ribosome. Fig. 2*B* shows the hydrophobic gate present in the crystal structure of the free *E. coli* ternary complex (PDB ID code 1OB2; unpublished data). The aurodox-bound structure, shown in Fig. 2*D*, displays an open gate, where Ile-60 and the remainder of switch I are disordered and His-84 is repositioned toward the nucleotide (12), hinting that the increased GTPase activity induced by aurodox is achieved by gate opening. One wing of the hydrophobic gate is formed by residue Val-20, which forms part of the P loop, a conserved element in GTPases important in guanine nucleotide release (13, 14). Our structure shows that the P-loop interacts with the sarcin-ricin loop (SRL; nucleotides 2646–2674 of the 23S rRNA). It was established that SRL interacts with EF-Tu (see ref. 11 and references therein), but available cryo-EM data until now lacked the resolution needed

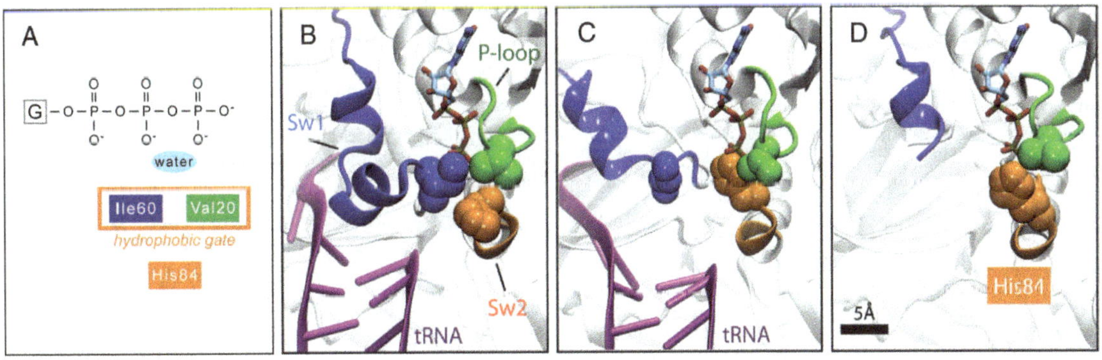

Fig. 2. Hydrophobic gate. (*A*) Schematic representation of the hydrophobic gate formed by EF-Tu residues Ile-60 and Val-20, which prevents His-84 from activating the water molecule and thus catalyzing GTP hydrolysis (G, guanisine). (*B*) The gate shown in the crystal structure of the ternary complex bound to kirromycin (PDB ID code 1OB2; unpublished data). (*C*) The ribosome-bound ternary complex model obtained through MDFF displaying conformational changes in the nucleotide binding area. Switch I has moved away, opening the gate on the Ile-60 side; the acceptor stem of the tRNA has moved significantly closer to switch II; and the side chain of His-84 has been repositioned toward the nucleotide, resulting in catalysis of GTP hydrolysis. (*D*) Open hydrophobic gate seen in the crystal structure of the aurodox-bound EF-Tu [PDB ID code 1HA3 (12)]. The structure suggests that an opening of the hydrophobic gate is necessary for hydrolysis, and exhibits a conformation of the gate very similar to that obtained by cryo-EM and MDFF. A GTP molecule is shown in *B–D* for comparison purposes, to illustrate the spacial rearrangements in the hydrophobic gating, although the structures depicted were obtained in presence of GDPNP (*B*) and GDP (*C* and *D*). The coordinates of the GTP molecule correspond to those of GDPNP in the crystal structure depicted in *B*; the coordinates of the nucleotides present in the structures depicted in *C* and *D* are very similar.

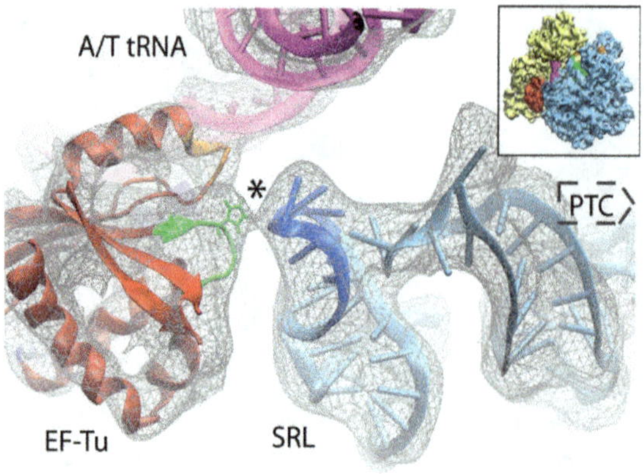

Fig. 3. Interaction (marked with an asterisk) between His-19 in the P loop of EF-Tu (green) and A2260 in the tetraloop of the SRL (cyan; tetraloop shown in blue). This interaction presumably establishes an anchor point holding one of the wings of the hydrophobic gate (compare Fig. 2) in place. (*Inset*) The ribosome is shown in the same orientation. The map is contoured at 2.5 σ. This figure is rotated 90° with respect to the view in Fig. 2.

to identify the residues of EF-Tu involved in this interaction. Fig. 3 shows the residues as resolved in our MDFF analysis, namely His-19 of EF-Tu and A2660 in the tetraloop of SRL. Because of the importance of the P loop for nucleotide exchange in EF-Tu (14), one may expect that interaction with SRL pulls the P loop away from the nucleotide binding pocket, thereby opening the gate through a mechanism similar to nucleotide exchange. However, our analysis shows that the contact with SRL stabilizes the P loop in its position, acting as an anchor point for this wing of the gate, whereas the other wing opens as described below. The P loop-SRL interaction might also be involved in signal transduction after GTP hydrolysis from the G domain of EF-Tu to the peptidyltransferase center (PTC), inducing conformational changes in the PTC required for complete accommodation and efficient peptide bond formation (see refs. 15 and 16 and references therein).

The other wing of the hydrophobic gate (Ile-60) is part of switch I (also known as effector loop), which assumes an α-helix conformation in the closed form of EF-Tu and changes to a β-hairpin conformation in the open form (17, 18). The large flexibility of this region has prevented switch I from being resolved in EM density maps and some crystallographic structures. At low-density threshold, the present map displays clear density corresponding to switch I; i.e., all molecular surfaces are solid, without disruption by spikes or other noisy features. The map allowed us to build an atomic model of switch I interacting with the ribosome (Fig. 4*A*; see *Materials and Methods*). In the state captured by the map, switch I interacts with the junction formed by helices h8 and h14 of the 16S rRNA (Fig. 4 *A* and *B*). It is interesting to note that a similar interaction was recently observed between the eukaryotic 40S subunit and the elongation factor eEF2 in a cryo-EM map of a transition-state complex (19), suggesting that this interaction is already present before GTP hydrolysis. The conformation of switch I in the ribosome-bound EF-Tu is different from that in the closed (GTP) and open (GDP) forms of free EF-Tu. The conformational change from the closed form involves repositioning of the wing containing Ile-60 leading to the opening of the hydrophobic gate, as depicted in Fig. 2*C*. This scenario receives support from the previous observation that cleavage of the peptide bond between Arg-58 and Gly-59 in switch I of EF-Tu abolishes ribosome-induced GTPase activation (20), because interruption of this peptide bond should prevent the movement of switch I from opening the hydrophobic gate. The effect of the interaction between switch I and the 16S rRNA h8/h14 junction remains to be determined, but this contact could serve as a pathway for signal transduction to and from the decoding center through helix h44 of the 16S rRNA; this helix interacts with the decoding center at one end and with the h8/h14 junction at the other (Fig. 4*C*).

At low threshold, we also observe a bulged density in the S turn motif of SRL (nucleotides 2653–2656 of the 23S rRNA, marked with an asterisk in Fig. 4*A*). The S turn motif has been shown to be critical for EF-G binding (21), and thus it is likely that this density represents an interaction between this motif and switch I that occurs before the movement of the latter toward the h8/h14 junction; the density of the bulged region indicates that only a fraction of the particles in the sample are found in this conformation.

As depicted in Fig. 2, the third effector region of EF-Tu is switch II, which contains the catalytic residue His-84 (5). On gate opening, the side chain of His-84 is free to reach the GTP binding pocket and to activate hydrolysis. Our atomic model resolves the activating conformation of His-84 (Fig. 2*C*), which is indeed similar to the one seen in the aurodox-bound structure of EF-Tu outside of the ribosome (12), shown in Fig. 2*D*. In our model, the His-84 side chain is positioned ≈4Å from the nucleotide, i.e., much closer than the ≈8Å seen in the crystal structure of the free ternary complex that exhibits the closed hydrophobic gate (PDB ID code 1OB2, unpublished data; see also ref. 22). A comparison of all of the above structures is presented in supporting information (SI) Fig. S1. Reorientation of His-84 to form the activated state is supported by mutational data showing that replacement of Gly-83 by Ala abolishes ribosome-induced GTPase activation (23), likely because any side chain bulkier than a hydrogen in position 83 prevents the reorientation of His-84 toward the GTP binding pocket (3, 6). A similar mechanism is observed in the G protein $G\alpha_{i1}$, where rotation of the catalytic residue Gln-204 to stabilize the transition state for GTP hydrolysis requires the presence of Gly at position 203 (24).

Interaction Between EF-Tu and Ribosomal Protein L12. Another element known to interact with EF-Tu is ribosomal protein L12, which binds to the G domain of EF-Tu at helix D (25) (Table S1). L12 greatly contributes to the GTPase activity of EF-Tu on the ribosome (26), which led to the speculation that L12 induces conformational rearrangements of EF-Tu, in particular, the reorientation of His-84 (26). However, this claim could not be corroborated from structural data. In our map, helix D of EF-Tu has clear density, from which we see no evidence for conformational changes of EF-Tu in the region binding to L12. It is therefore likely that L12 stabilizes the transition state complex on the ribosome, without actively inducing the required structural rearrangements of EF-Tu.

Role of tRNA in Signal Transduction. tRNA plays a central role in GTPase activation induced by codon recognition, as suggested by several mutations (1) and by the fact that an intact tRNA is required to trigger GTP hydrolysis by EF-Tu on the ribosome (27). Previous cryo-EM data showed that the elbow of tRNA interacts with the L11-binding region of the 23S rRNA, which, together with ribosomal protein L11, constitutes the GTPase-associated center (GAC) (9, 11). Cryo-EM data at lower resolution suggested that the interaction occurs at A1067 (11); this claim was supported by mutational studies of A1067 that impair EF-Tu function (28). The present map corroborates these findings, showing that the interaction is established between a flipped A1067 of the GAC and C56 in the elbow region of the tRNA (Fig. S2). Previous molecular dynamics (MD) studies showed that A1067 can exist in 2 different conformations in solution, suggesting that the flipped-out conformation might

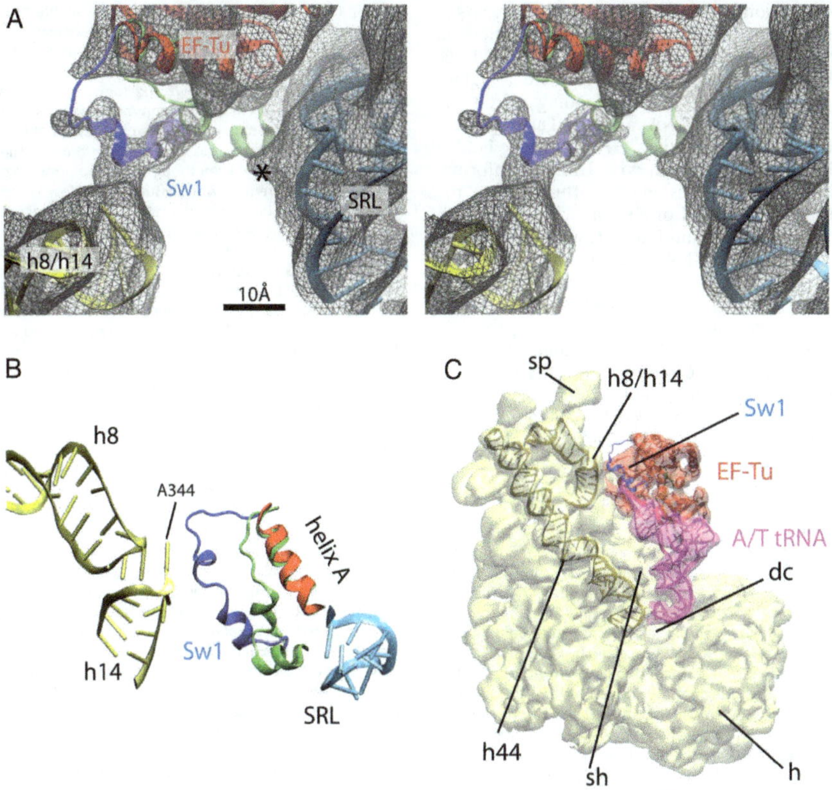

Fig. 4. Structure of switch I in the ribosome-bound EF-Tu. (*A*) Stereoview illustrating the movement of switch I (blue) toward the junction of helices h8 and h14 of 16S rRNA (yellow). Rigid-body fitting of the crystal structure (green; PDB ID code 1OB2) places switch I close to the SRL (cyan). At low-density threshold, shown as gray mesh, the density corresponding to SRL exhibits a protuberance (marked with an asterisk) around the S turn motif, possibly corresponding to an interaction with switch I (blue) during its movement toward the h8/h14 junction (yellow). The map is contoured at 1.25 σ. (*B*) Detail of the interaction between switch I and the 16S rRNA. Elements of EF-Tu (red and blue) and the 30S (yellow) and 50S (cyan) subunits are shown along with the crystal structure of switch I (green; PDB ID code 1OB2). (*C*) Overview of the ternary complex bound to the 30S subunit (yellow). switch I (blue) interacts with the h8/h14 junction; h44 interacts with the latter on one side, and with the decoding center on the other, possibly serving as part of a signal transduction pathway. Landmarks: sp, spur; sh, shoulder; dc, decoding center; h, head.

occur during aa-tRNA binding and selection (29). Ribosomal protein S12 contacts the tRNA both at the acceptor stem and the anticodon loop, connecting the movement of the acceptor stem with the reorganization of the decoding center on codon recognition (9, 10). Fig. S2 shows the residues involved in this direct link between the codon-anticodon helix and the tRNA acceptor stem through ribosomal protein S12.

Cryo-EM data had already demonstrated that the relative orientation of tRNA and EF-Tu changes on binding to the ribosome; our model reveals that even though the overall movement of tRNA is directed away from EF-Tu (Fig. S3), this rearrangement results in a closer interaction between tRNA's acceptor stem and EF-Tu's switch II, specifically with Asp-86 on switch II (Figs. 2*C*, S1, and S3). The universally conserved residue Asp-86 was proposed to play a role in stabilization of the protonated His-84 (3), and was observed to interact strongly with the tRNA in MD simulations of the free ternary complex (30). These findings, together with our structural evidence, strongly suggest that repositioning of the tRNA leading to interaction with switch II assists in the required conformational change of the latter, which results in the repositioning of His-84.

Discussion

Our data represent a significant improvement in resolution over previously reported cryo-EM structures of the ternary complex-bound ribosome. The advance toward higher resolution has allowed us to use a systematic method to obtain atomic models that explain the cryo-EM density. By using force fields that represent physical-chemical properties of biomolecules, similar to those used in X-ray crystallography structure refinement, the interpretation of the data can be pushed beyond the nominal resolution. An extensive comparison of the model obtained through the use of MDFF and models obtained through other methods is offered in ref. 8, along with a discussion of the role of resolution on the model accuracy.

Here, we presented the interactions between EF-Tu and both the ribosome and tRNA in atomic detail. The interactions induce the conformational changes necessary to trigger GTP hydrolysis on cognate codon–anticodon interaction. Our data suggest that interaction of the ternary complex with the ribosome opens a hydrophobic gate formed by residues Ile-60 and Val-20, allowing His-84 to act as a general base activating an active-site nucleophilic water molecule, and thus triggering hydrolysis.

Based on the available structural and biochemical data, we propose a refined model of aa-tRNA incorporation into the ribosome. The interaction between tRNA and GAC occurs at an early stage (11). The geometry of the codon–anticodon helix is monitored by 23S rRNA nucleotides A1492, A1493, and G530, leading to a 30S domain closure and a tighter binding of the anticodon loop (1). This binding affects the relative orientation between tRNA and EF-Tu, a process influenced by S12. While tRNA moves, EF-Tu must be held in place, likely aided by its interaction with L12. A conformational change in switch II takes place, influenced by the acceptor stem of tRNA. Switch I of EF-Tu moves away from the nucleotide binding pocket and interacts with the junction of helices h8/h14 of the 16S rRNA, opening one wing of the hydrophobic gate, while the other wing is held fixed by the interaction between SRL and P loop. His-84 reorients toward the nucleotide, which leads to activation of a water molecule and subsequent GTP hydrolysis. After inorganic phosphate release, EF-Tu undergoes a large conformational change to an "open," GDP-bound form, leading to its dissociation from the ribosome, after which tRNA can be accommodated into the A site. The conformational change of EF-Tu can be sensed by SRL and communicated to the PTC, inducing changes in PTC required for peptide bond formation.

Materials and Methods

Cryo-EM. The sample consisted of preaccommodated *E. coli* 70S ribosome bound to the Phe-tRNA^{Phe}·EF-Tu·GDP ternary complex stalled by the antibiotic kirromycin (kir) (11) by using a standard preparation (31). The map was obtained from ≈130,000 particles, recorded with ≈58,000× magnification on a FEI Tecnai F30 Polara electron microscope operated at 300 kV at liquid-nitrogen temperature (80 K) and processed by using SPIDER. The resulting resolution of the map was estimated to be 6.7 Å with the 0.5 FCS criterion, extrapolated to the full dataset with a pixel size of 1.2 Å. A version of the map filtered to 5 Å proved to show the positions of phosphorus atoms along exposed RNA helices (7). The map represents a significant improvement in resolution over the previously published 10.3-Å map obtained from an identical sample preparation (11). The improvement in resolution was realized through optimization of image processing parameters, described in detail in ref. 7.

Flexible Fitting. The atomic model of the aminoacyl-tRNA·EF-Tu·GDP·kirromycin ternary complex-bound *E. coli* ribosome was obtained by applying the recently developed MDFF method (8). MDFF incorporates the EM data into molecular dynamics simulation as a potential proportional to the density gradient of the EM map. The potential applies forces on the atoms that actively drive the model into the density. The initial coordinates of the model were taken from available structural data. For the ribosome, an atomic model of the *E. coli* ribosome based on a 3.22-Å crystal structure (PDB ID code 2I2U/2I2V) (32) was built, as described in ref. 8. The initial structure of the ternary complex was taken from the crystal structure of *E. coli* EF-Tu in complex with kirromycin, a GTP analog, and yeast Phe-tRNA^{Phe} (PDB ID code 1OB2; unpublished data). For details on MDFF of these models to the map presented in this report, the reader is referred to ref. 8. An additional step was required to fit the structure of the switch I of EF-Tu into the density. Even though the map presents clear density corresponding to this region when visualized at low threshold, a model for switch I could not be obtained by using MDFF alone, because the gradient of the EM map in this area is not steep enough to drive the structure into the density unequivocally. Therefore, we manually fitted the helices of switch I in the GTP conformation into the density, and used the positions of the C_α atoms in this conformation as a target location for the respective atoms during MDFF simulations. The C_α atoms were driven into the target positions by introducing an additional potential term equivalent to a targeted molecular dynamics simulation (33), in which the root-mean-squared deviation between the simulated and the target positions of the C_α atoms is minimized along the simulation. In this way, the C_α atoms of switch I became located close to the target positions, whereas the rest of the atoms were still driven into the measured density, the stereochemical correctness of the model being enforced by the molecular dynamics force field.

The model produced by MDFF is useful to interpret the EM data in atomic detail; however, the atomic structure should always be analyzed together with the original data to ensure consistency and proper interpretation. Uncertainty in the coordinates arises both from features of the EM data, e.g., resolution and flexibility of certain regions, and from the conservative secondary structure restraints used in MDFF to avoid overfitting (see ref. 8 for a detailed discussion). A cryo-EM map represents an ensemble average of structures in the same functional state; certain regions can only be seen at low threshold, implying that, in the represented functional state, these regions are flexible and exist in 2 or more conformations. Combining MDFF with cryo-EM maps at resolutions like the one presented here permits the location of certain elements beyond the nominal resolution. An example is the acceptor stem of the tRNA where, even though the deviation from the crystal structure is only a few angstroms, it is clear that an overall movement of the tRNA away from the density places this regions closer to switch II of EF-Tu (Fig. S3).

All figures were made with the molecular graphics program VMD (34).

ACKNOWLEDGMENTS. We thank Andrew Carter, Zan Luthey-Schulten, and members of the Frank and Schulten labs for stimulating discussions, and Lila Iino-Rubenstein for assistance with the illustrations. This work was supported by Howard Hughes Medical Institute (J.F.), the Swedish Research Council (M.E.), National Science Foundation Grant PHY0822613 (to K.S.), and National Institutes of Health Grants P41-RR05969 (to K.S.), P41-RR01219, R37-GM29169, and R01-GM55440 (to J.F.), and R01-GM070768 (to M.E). Computer time for MDFF was supported by Large Resources Allocation Committee Grant MCA93S028 (to K.S.).

1. Ogle JM, Ramakrishnan V (2005) Structural insights into translational fidelity. *Annu Rev Biochem* 74:129–177.
2. Eccleston JF, Webb MR (1982) Characterization of the GTPase reaction of elongation factor Tu. *J Biol Chem* 257:5046–5049.
3. Kjeldgaard M, Nissen P, Thirup S, Nyborg J (1993) The crystal structure of elongation factor EF-Tu from *Thermus aquaticus* in the GTP conformation. *Structure* 1:35–50.
4. Limmer S, Reiser COA, Schirmer NK, Grillenbeck NW, Sprinzl M (1992) Nucleotide binding and GTP hydrolysis by elongation factor Tu from *Thermus thermophilus* as monitored by proton NMR. *Biochemistry* 31:2970–2977.
5. Daviter T, Wieden HJ, Rodnina MV (2003) Essential role of histidine 84 in elongation factor Tu for the chemical step of GTP hydrolysis on the ribosome. *J Mol Biol* 332:689–699.
6. Berchtold H, *et al.* (1993) Crystal structure of active elongation factor Tu reveals major domain rearrangements. *Nature* 365:126–132.
7. LeBarron J, *et al.* (2008) Exploration of parameters in cryo-EM leading to an improved density map of the *E. coli* ribosome. *J Struct Biol* 164:24–32.
8. Trabuco LG, Villa E, Mitra K, Frank J, Schulten K (2008) Flexible fitting of atomic structures into electron microscopy maps using molecular dynamics. *Structure* 16:673–683.
9. Valle M, *et al.* (2002) Cryo-EM reveals an active role for aminoacyl-tRNA in the accommodation process. *EMBO J* 21:3557–3567.
10. Stark H, *et al.* (2002) Ribosome interactions of aminoacyl-tRNA and elongation factor Tu in the codon-recognition complex. *Nat Struct Biol* 9:849–854.
11. Valle M, *et al.* (2003) Incorporation of aminoacyl-tRNA into the ribosome as seen by cryo-electron microscopy. *Nat Struct Biol* 10:899–906.
12. Vogeley L, Palm GJ, Mesters JR, Hilgenfeld R (2001) Conformational change of elongation factor Tu (EF-Tu) induced by antibiotic binding. *J Biol Chem* 276:17149–17155.
13. Vetter IR, Wittinghofer A (2001) The guanine nucleotide-binding switch in three dimensions. *Science* 294:1299–1304.
14. Dahl LD, Wieden HJ, Rodnina MV, Knudsen CR (2006) The importance of the P-loop and domain movements in EF-Tu for guanine nucleotide exchange. *J Biol Chem* 281:21139–21146.
15. Chan YL, Dresios J, Wool IG (2006) A pathway for the transmission of allosteric signals in the ribosome through a network of RNA tertiary interactions. *J Mol Biol* 355:1014–1025.
16. Blanchard SC, Gonzalez RL, Kim HD, Chu S, Puglisi JD (2004) tRNA selection and kinetic proofreading in translation. *Nat Struct Mol Biol* 11:1008–1014.
17. Abel K, Yoder MD, Hilgenfeld R, Jurnak F (1996) An α to β conformational switch in EF-Tu. *Structure* 4:1153–1159.
18. Polekhina G, *et al.* (1996) Helix unwinding in the effector region of elongation factor EF-Tu–GDP. *Structure* 4:1141–1151.
19. Sengupta J, *et al.* (2008) Visualization of the eEF2–80S ribosome transition-state complex by cryo-electron microscopy. *J Mol Biol* 382:179–187.
20. Zeidler W, *et al.* (1996) Limited proteolysis and amino acid replacements in the effector region of *Thermus thermophilus* elongation factor Tu. *Eur J Biochem* 239:265–271.
21. Munishkin A, Wool IG (1997) The ribosome-in-pieces: Binding of elongation factor EF-G to oligoribonucleotides that mimic the sarcin/ricin and thiostrepton domains of 23S ribosomal RNA. *Proc Natl Acad Sci USA* 94:12280–12284.
22. Nissen P, *et al.* (1995) Crystal structure of the ternary complex of Phe-tRNA^{Phe}, EF-Tu, and a GTP analog. *Science* 270:1464–1472.
23. Knudsen C, Wieden HJ, Rodnina MV (2001) The importance of structural transitions of the switch II region for the functions of elongation factor Tu on the ribosome. *J Biol Chem* 276:22183–22190.
24. Berghuis AM, Lee E, Raw AS, Gilman AG, Sprang SR (1996) Structure of the GDP-P_i complex of Gly203→Ala $G_{i\alpha1}$: A mimic of the ternary product complex of G_α-catalyzed GTP hydrolysis. *Structure* 4:1277–1290.

25. Kothe U, Wieden HJ, Mohr D, Rodnina MV (2004) Interaction of helix D of elongation factor Tu with helices 4 and 5 of protein L7/12 on the ribosome. *J Mol Biol* 336:1011–1021.
26. Mohr D, Wintermeyer W, Rodnina MV (2002) GTPase activation of elongation factors Tu and G on the ribosome. *Biochemistry* 41:12520–12528.
27. Piepenburg O, *et al.* (2000) Intact aminoacyl-tRNA is required to trigger GTP hydrolysis by elongation factor Tu on the ribosome. *Biochemistry* 39:1734–1738.
28. Saarma U, Remme J, Ehrenberg M, Bilgin N (1997) An A to U transversion at position 1067 of 23S rRNA from *Escherichia coli* impairs EF-Tu and EF-G function. *J Mol Biol* 272:327–335.
29. Li W, Sengupta J, Rath BK, Frank J (2006) Functional conformations of the L11-ribosomal RNA complex revealed by correlative analysis of cryo-EM and molecular dynamics simulations. *RNA* 12:1240–1253.
30. Eargle J, Black AA, Sethi A, Trabuco LG, Luthey-Schulten Z (2008) Dynamics of recognition between tRNA and elongation factor Tu. *J Mol Biol* 377:1382–1405.
31. Grassucci RA, Taylor DJ, Frank J (2007) Preparation of macromolecular complexes for cryo-electron microscopy. *Nat Protoc* 2:3239–3246.
32. Berk V, Zhang W, Pai RD, Cate JHD (2006) Structural basis for mRNA and tRNA positioning on the ribosome. *Proc Natl Acad Sci USA* 103:15830–15834.
33. Schlitter J, Engels M, Krüger P, Jacoby E, Wollmer A (1993) Targeted molecular dynamics simulation of conformational change—Application to the T → R transition in insulin. *Mol Sim* 10:291–308.
34. Humphrey W, Dalke A, Schulten K (1996) VMD—Visual Molecular Dynamics. *J Mol Graphics* 14:33–38.

Structural characterization of mRNA-tRNA translocation intermediates

Xabier Agirrezabala[a,1], Hstau Y. Liao[b,1], Eduard Schreiner[c,d,2], Jie Fu[b], Rodrigo F. Ortiz-Meoz[e], Klaus Schulten[c,d,f], Rachel Green[e], and Joachim Frank[b,g,3]

[a]Structural Biology Unit, CIC-bioGUNE, Bizkaia Technology Park, Derio 48160, Basque Country, Spain; [b]Howard Hughes Medical Institute, Department of Biochemistry and Molecular Biophysics, Columbia University, 630 168th Street, P&S BB 2-221, New York, NY; [c]Beckman Institute for Advanced Science and Technology, [d]Center for Biophysics and Computational Biology, University of Illinois at Urbana-Champaign, Urbana, IL; [e]Howard Hughes Medical Institute, Department of Molecular Biology and Genetics, Johns Hopkins University School of Medicine, Baltimore, MD 21205; [f]Department of Physics, University of Illinois at Urbana-Champaign, Urbana, IL; and [g]Department of Biological Sciences, Columbia University, New York, NY

Contributed by Joachim Frank, January 31, 2012 (sent for review November 21, 2011)

Cryo-EM analysis of a wild-type *Escherichia coli* pretranslocational sample has revealed the presence of previously unseen intermediate substates of the bacterial ribosome during the first phase of translocation, characterized by intermediate intersubunit rotations, L1 stalk positions, and tRNA configurations. Furthermore, we describe the domain rearrangements in quantitative terms, which has allowed us to characterize the processivity and coordination of the conformational reorganization of the ribosome, along with the associated changes in tRNA ribosome-binding configuration. The results are consistent with the view of the ribosome as a molecular machine employing Brownian motion to reach a functionally productive state via a series of substates with incremental changes in conformation.

cryo-EM | E. coli | ribosome | translation

Changes in ribosome conformation during protein synthesis are substantial, the most pronounced ones occurring during mRNA-tRNA translocation along the A (aminoacyl), P (peptidyl), and E (exit) tRNA binding sites of the ribosome, as postulated early on by Spirin (1) and Bretscher (2) and shown in recent studies by cryo-EM, X-ray crystallography, and smFRET [see (3)]. These changes go along with changes in binding configurations the ribosome forms with the tRNAs and elongation factor G (EF-G) in the process of translocation.

Translocation can be broadly divided into two phases [see (4)]: during the first phase, the tRNAs move with respect to the large (50S) subunit, and in the second, the mRNA and the tRNAs affixed to it move with respect to the small (30S) subunit. After accommodation of the incoming aminoacyl-tRNA into the A/A site, and peptide transfer from the peptidyl-tRNA residing in the P/P site, the tRNAs proceed from the classical (A/A, P/P) to the hybrid (A/P, P/E) binding configuration (5): while the anticodon stem loops (ASLs) of tRNAs stay in the small subunit's A and P sites, the acceptor ends move to the large subunit's P and E sites, respectively. As EF-G binds to the complex, the ribosome undergoes a ratchet-like motion ("intersubunit rotation")—the 30S subunit rotates with respect to the 50S subunit (6). This rotation is accompanied by a movement of the L1 stalk of the 50S subunit toward the main body of the ribosome and a rotation of the head domain of the 30S subunit around its long axis (7–10). These movements separate two distinct states during the first phase of translocation, termed macrostate I (MS I) and II (MS II) (4). The fact that the conformational changes of the ribosome and classical-hybrid transitions of tRNAs occur spontaneously in a pretrans-locational (PRE) ribosome (11–16) has confirmed the view of the ribosome as a Brownian machine (17). In this view, the role of ribosomal factors is to modulate the free-energy landscape, promoting or controlling structural and kinetic routes underlying functional dynamics of translation (18, 19). In the case of translocation, smFRET has provided rich detail on the way EF-G promotes and controls the reaction [e.g., (13, 14, 20–27)].

Cryo-EM (11, 15) of factor-free PRE WT *Escherichia coli* samples revealed a large structural reorganization of the ribosome and associated classical → hybrid changes of tRNA ribosome-binding configuration in going from MS I to MS II. The size of these changes made it very likely that structural intermediates exist. Evidence for such intermediates was found by smFRET (28), by cryo-EM of the back-translocation process (29), of the second phase of translocation (8), of ribosomes with a P-loop mutation (30), and by X-ray crystallography of 70S ribosomes with ASL mimics (31), prompting us to reexamine the WT *E. coli* PRE ribosome complex. We employed ML3D (32), a technique of maximum-likelihood based classification requiring no prior knowledge of the nature of the structural variability.

Results

Summary of Reconstructions. We applied ML3D to an existing dataset of the factor-free WT *E. coli* PRE complex (15), and obtained six classes. The class-1 reconstruction was discarded as it showed evidence for bias in particle orientations (see Fig. S1). Class 4 showed indications of residual heterogeneity, especially in the features of the tRNAs. Subsequent ML3D analysis subdivided this class into two more homogenous classes, bringing the total back to six. The outcome of the classification was evaluated by visual inspection and 3D variance analysis of the final six reconstructions (see Fig. S2).

Classes 2 (approximately 16%; see Fig. S3*A*) and 4A (10%; Fig. S3*C*) both correspond to ribosomes bearing classical A/A and P/P tRNAs. Class 3 (16%) corresponds to ribosomes bearing a single tRNA in the P site (Fig. S3*B*). Classes 4B (10%), 5 (17%), and 6 (18%) (Fig. S3 *D–F*, respectively) correspond to ribosomes that have undergone intersubunit rotation and bear hybrid-like tRNAs in A and P sites. The conformations of class-5 and -6 ribosomes are quite similar at the current resolution, and in turn similar to the one previously described as MS II (11, 15). (For a comparison of performance of the ML3D algorithm with that of the supervised approach, see Table S1).

Author contributions: X.A. and J. Frank designed research; X.A., H.Y.L, and E.S. performed research; R.F.O.-M. and R.G. contributed new reagents/analytic tools; X.A., H.Y.L., E.S., J. Fu, K.S., and J. Frank analyzed data; and X.A. and J. Frank wrote the paper.

The authors declare no conflict of interest.

Data deposition: The EM reconstructions were deposited in the 3D-EM database under the following accession numbers: EMD-5361, -5360, -5359, -5364, -5363, and -5362 for class 2, 3, 4A, 4B, 5, and 6, respectively. The atomic coordinates generated by MDFF were deposited with PDB ID codes 3J0U-3J0T, 3J13-3J11, 3J0V-3J0W, 3J0X-3J0Y, 3J0Z-3J12 and 3J10-3J14, for the small and large subunits, respectively.

[1]X.A. and H.L. contributed equally to the work.

[2]Present address: BASF SE, Materials Modeling, 67056 Ludwigshafen, Germany.

[3]To whom correspondence should be addressed. E-mail: jf2192@columbia.edu.

This article contains supporting information online at www.pnas.org/lookup/suppl/doi:10.1073/pnas.1201288109/-/DCSupplemental.

www.pnas.org/cgi/doi/10.1073/pnas.1201288109

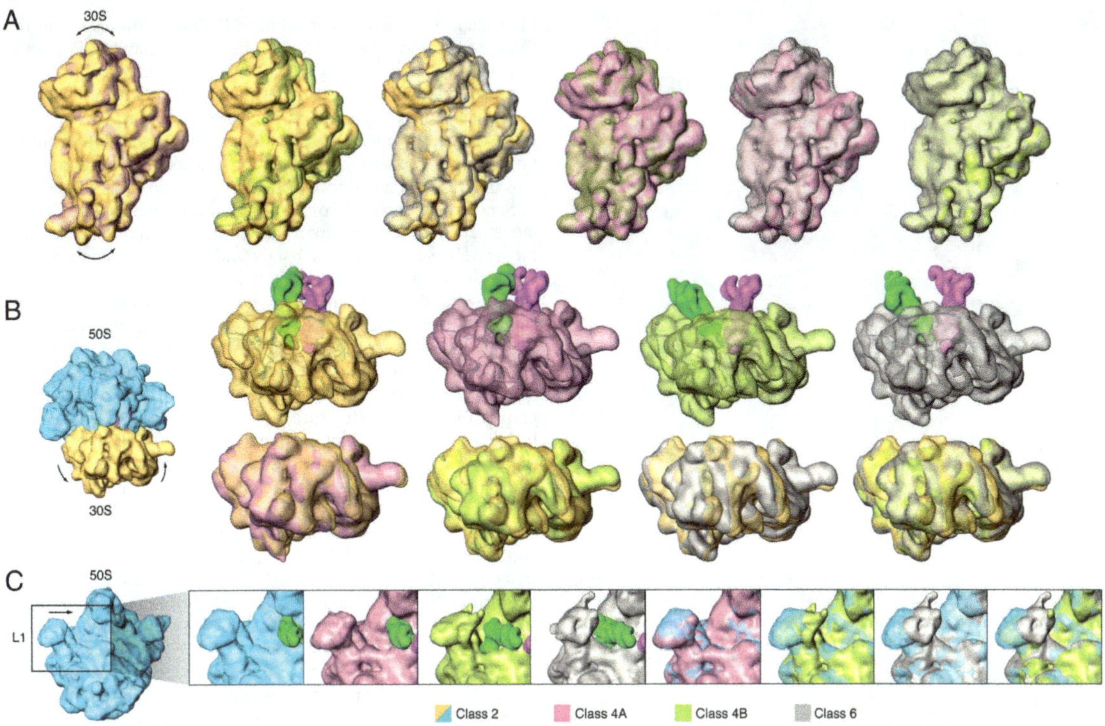

Fig. 1. Movement of domains and tRNAs. (*A*) Left to right, different superimposed cryo-EM densities of the 30S subunits from class 2 (yellow), class 4A (violet), class 4B (olive), and class 6 (gray). (*B*) Left to right, different reconstructions (upper row) and superimposed densities (lower row) of the 30S subunits; color-coded as in (*A*). The new orientation of the subunit is shown as a thumbnail on the left. The A- and P-site tRNAs are in magenta and green. (*C*) Left to right, close-ups of different reconstructions and superimposed cryo-EM densities of the 50S subunits; color-coded as in (*A*) and (*B*). The orientation of the subunits is shown as a thumbnail on the left. The arrows in (*A*), (*B*), and (*C*) indicate the directions of intersubunit rotation, head swivel, and L1 stalk movement, respectively. Colored labels identify the different class reconstructions.

Progression of Ribosomal Dynamics and tRNA Movements. When analyzing the structural movements implicated in tRNA translocation, including intersubunit rotation (Fig. 1*A*), swivelling of head domain (Fig. 1*B*), and inward movement of L1 stalk (Fig. 1*C*), a progression is seen as we go from class 2 (unrotated, classic state) over classes 4A and 4B to classes 5 and 6 (fully rotated, hybrid). Thus the new classes 4A and 4B represent two distinct structural intermediates during the first phase of translocation. While class 4A still shows classical-like tRNA positions, it displays distinct positions in the intersubunit rotation and L1 stalk movement. Class 4B is further along in both movements, and in addition shows a unique hybrid-like tRNA configuration: the deacylated P-site tRNA contacts the large subunit's E-site region via the acceptor stem, but it shows a distinct position of the elbow, approximately halfway through its transition from class 2 to classes 5 and 6.

To gain further insights, atomic models were built using MD-based flexible fitting (MDFF) as previously described (34). Movies S1 and S2 depict the conformational transitions between classes 2, 4B, and 6 by interpolation of the atomic coordinates generated for each of these classes, and represent (in this simplified model) a feasible pathway for forward translocation. Note that recent X-ray data show the reconfiguration of the intersubunit bridges that link the ribosomal subunits during the transition from the unrotated, classical to the rotated, hybrid state (35, 36). It may be possible to analyze the rearrangements of bridges, notably B1b (4), during the intermediate transitions by comparing the fitted structures in atomic detail, but the current resolution does not support a detailed fitting in the intersubunit region, as the gradients in the interface region distort the structure.

Atomic models generated for the tRNAs (Fig. 2; also see Fig. S4) show that, coupled to the intersubunit rotation, the P-site tRNA ASL-mRNA complex has moved laterally for class 4B and classes 5 and 6 compared to its position in class 2. In addition to the intersubunit rotation, a concomitant lateral movement of the tip of h44 with respect to its classical position is responsible for the shift in the A-site tRNA. In both fully classical and fully hybrid states (i.e., MS I and II), the A-site tRNA interacts with H69 and H38 via its D stem and its T arm, respectively. A slight reorientation of the ASL and elbow region, coupled with the straightening of the AS, is responsible for placing the CCA end of the A-site tRNA in the vicinity of the P site of the peptidyl-transferase center (PTC) in classes 5 and 6. In contrast, ribosomes in class 4B display the A/A configuration, although the location of the 3′ CCA end cannot be assigned unequivocally. As to the hybrid-like P/E-configuration tRNA, it is characterized in both intermediate (class 4B) and fully hybrid (classes 5 and 6) states by (*i*) a distortion in the vicinity of the ASL, enabling it to reach the E site with its 3′ end, and (*ii*) by loss of contact with H69. However, while going in the same direction, the magnitude of the rotation of the anticodon region is smaller for the new class-4B hybrid-like intermediate; this results in a different elbow position when compared to classes 5 and 6. Note that the acceptor stem of the hybrid-like P/E tRNA model in class 4B appears contracted, but it fits the density well.

Regarding the L1 stalk, we observe different conformations that are apparently coupled with the intersubunit rotation and—except for class 4A—changes in tRNA positions (see Fig. 1*C* and Fig. S4). In class 2, the L1 stalk is fully open. Classes 4A and 4B show (slightly distinct) intermediate positions for the L1 stalk

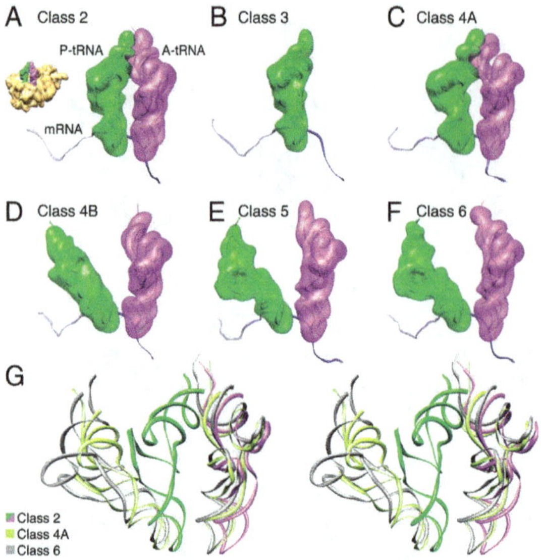

Fig. 2. Atomic models of tRNA obtained by flexible fitting. A gallery of images is shown for tRNA-mRNA complexes for classes as indicated in (*A*)–(*F*). Thumbnail in (*A*) shows the orientation of the tRNAs with respect to the 30S subunit (yellow). Cryo-EM densities for A- and P-site tRNAs are in transparent magenta and green. Fitted structures are in ribbons. Modeled mRNA sequence is shown in blue. (*G*) Stereo-views of fitted A and P-site tRNAs from class 2 (magenta and green), class 4B (olive), and class 6 (gray), aligned with respect to the 70S ribosomes (see colored labels). See also Fig. S4.

(class 4A being more proximal to the open position). Finally, classes 5 and 6 present a fully closed stalk along with the fully hybrid tRNA configurations and a fully rotated 30S subunit. A similar half-closed intermediate configuration of the L1 stalk as in class 4B was recently proposed on the grounds of smFRET analysis [see (22)] in the posttranslocational (POST) ribosome, which apparently presented E/E tRNAs and a nonrotated small subunit. We also note that in the analysis of the back-translocation process by Fischer and coworkers (29), only three L1 positions were found (open and closed in PRE complexes, half-closed in POST complexes), which may indicate that our class-4A and -4B intermediates are not visited during back-translocation, and that the two processes follow different preferred pathways. On the other hand, examination of the reconstructions shows that the L1 stalk does not participate in the stabilization of the class-4B intermediate hybrid position of the P/E tRNA, as it does for classes 5 and 6. However, due to the possibility that class 4B might contain some residual contamination as a consequence of merging particles from other classes—potentially some from classes 5 and 6 (see Fig. S2 for details)—it is still uncertain whether the new intermediate tRNA position is achieved independently of the action of the L1 stalk.

Quantification of Conformational Changes. To quantify the relative movements among the structures derived from the conformational classes, the displacement angles and distances for each structural feature relative to a reference structure were determined by tensor analysis (33). Tables S2 and S3 show the angles formed between selected vectors of the inertia tensors from two different components [for definition of angles between domains, see Suppl. Fig. 4 in Ref. (33)]. In addition, the different values in Table S4 track the progress of the tRNAs through the intersubunit corridor and the gradual closing of the L1 stalk. As the tRNAs travel through the ribosome, from the entrance to the exit of intersubunit space, we follow their movements by measuring the distances of their centers of mass (COMs) with respect to the COM of the very first tRNA; i.e., the A/A as manifest in class 2. The approach of measuring distances between centers of mass of a large selection of atoms was taken because distances between individual atoms, especially in the peripheral part, can be quite noisy, given the moderate resolution and the resulting fitting uncertainty. (The same approximation was taken in the tensor analysis that resulted in the rotation angles shown in Table S2 and S3).

Specifically, Table S2 contains quantitative descriptions of the movements of different ribosomal domains within the 30S subunit, as well as those of the L1 stalk and the tRNAs. In extending the analysis, Table S3 contains data describing relevant, published structures and include the X-ray crystal structures of 70S ribosomes in the presence of ASL tRNA mimics (31), which revealed intermediate states of intersubunit rotation (see 3I1Q-3I1R and 3I1Z-3I2O). PDB entries 2WRI-2WRJ and 2WRK-2WRL correspond to ribosome-EF-G complexes reported by X-ray crystallography (37). These structures, arrested in the POST state by fusidic acid, display no intersubunit rotation. On the other hand, the cryo-EM structures of ribosome-EF-G complexes obtained by Spahn and coworkers using image sorting (8), also in the presence of fusidic acid, display intersubunit rotation uncoupled to swiveling movements, and are considered PRE (2XSY-2XTG) and POST (2XUY-2XUX) states, respectively. The recently described structures of fully rotated ribosomes bound to RRF (35) are also included in the analysis (3R8N-3R8S and 3R8O-3R8T for rotated/hybrid and unrotated/classical ribosomes, respectively). Finally, the structures denoted as NR, IRS1, IRS2, and RS correspond to the nonrotated, the intermediate rotated structures 1 and 2, and the rotated structures, respectively, as obtained by cryo-EM of PRE complexes of mutant ribosomes (30).

From the analysis, it can be concluded that the cryo-EM structures 2XSY-2XTG and X-ray structures 3R8N-3R8S, obtained in the presence of RRF, most closely resemble our classes 5 and 6, while the structure 2XUY-2XUX is a state which we do not observe, characterized by a large swiveling angle of the head. In the latter structure, the tRNA is in fact more similar to the classical E/E tRNA than to a hybrid P/E tRNA. The X-ray crystal structures of 70S ribosomes in the presence of ASL tRNA mimics (31) are very similar to our class-4B reconstruction with regards to the intersubunit rotation. Note that the structure described by PDB entry 3I1M (31) also shows a considerable degree of swiveling of the head, although still falling short when compared to structure 2XUY-2XUX. On the other hand, when considering the body/platform opening values in hybrid forms, the structures can be divided in two groups: (*i*) the present cryo-EM structures and the X-ray structures obtained in the presence of ASL mimics and (*ii*) the hybrid structures obtained in the presence of EF-G (2XUY-2XUX, 2XSY- 2XTG) or RRF (3R8N-3R8S). We also note that structure 2XUY-2XUX shows a distinctive position of the head relative to the shoulder, as well as a strong tilt of the head domain toward the subunit interface. Structure 3I1M, and, to a lesser extent, the hybrid cryo-EM classes 5 and 6 show a similar trend, although the magnitude of the movement is less pronounced.

Of special interest are measurements for the structures obtained by Fu et al. (30). This work, a follow-up analysis of a smFRET study of P-loop mutant ribosomes (28), also described two intermediate states. We applied our tensor analysis to unpublished RSREF-fitted coordinates for these maps and observed that, as a function of intersubunit rotation, all domain movements show the same trends as in the current models, with the exception of the tilt of the small subunit head domain toward the subunit interface. A large difference is only seen in the position of the acceptor stem of the P-site tRNA for both IRS1 and IRS2, presumably the position most affected by the mutation. In terms of ribosome conformation alone, including intersubunit angle, IRS1 evidently corresponds to class 4A, and IRS2 to class 4B. (RSREF-fittings for the A-site tRNA were not considered in this comparison as they show large inconsistencies).

Finally, we note that several X-ray structures of unrotated ribosomes in our table show intersubunit rotation angles that are smaller than those for our EM reconstructions of unrotated complexes. This could be explained by the uncertainty of classification in the EM data analysis, which may allow "adjacent" populations with larger rotation to contribute to our reconstructions. However, some X-ray structures even show negative values; i.e., less rotated than 2I2U/2I2V (38), the structure that was used as reference. Thus, it is possible that the subunit position is influenced by crystal packing effects or uncontrolled experimental effects (e.g., buffer conditions and temperature).

Coordination of the Dynamics Between the Ribosome and the tRNAs. We followed the way angles or distances of ribosomal components change as we move through the states, ordered according to "time" parameterized by intersubunit rotation. The rationale in singling out intersubunit rotation as an "independent" variable is that the movements of the massive subunits are likely to occur on a timescale that is considerably longer compared to the movements of all subunit domains and especially the tRNA. We envision that, being on a longer timescale, subunit positions will define the environment for the tRNAs, and thus dictate the range of domain and tRNA motions within energetic reach at any point in time. We therefore plotted the different measurements (ordinates) from Tables S2, S3, and S4 against the angle of intersubunit rotation (abscissa). Connecting lines between measurements have a meaning since they indicate, via interpolation, an estimate of where a given measurement would be if we had been able to sample the states more finely. We interpret the states with intermediate intersubunit rotations as intermediates in the transition of the ribosome from the classical to the final, fully rotated configuration.

To investigate the degree to which the movements of structural components are coupled, we also computed the correlation coefficients between all pairs of variables shown in Tables S2, S3, and S4 (see Fig. S6). These coefficients measure how closely any two variables are tied to each other by a linear relationship. Although causal relations cannot be proven based on correlation coefficients, a high correlation between quantities that measure the varying positions of ribosome components (notably L1 stalk versus intersubunit rotation, see below) indicates that we observe the gradual change of an entire tightly connected structure through different states.

Fig. 3*A* displays the trends of several angles measured on the small subunit in graphic form, while the progression of the L1 stalk closing and movement of tRNAs is displayed in Fig. 3 *B–E*. As seen in Fig. 3*A*, the intersubunit rotation occurs in concert with structural rearrangements within the 30S subunit. The closing of the body and platform runs through a maximum for the intermediate classes 4A and 4B. This is in qualitative agreement with the classical as well as partially (31) and fully rotated (35)

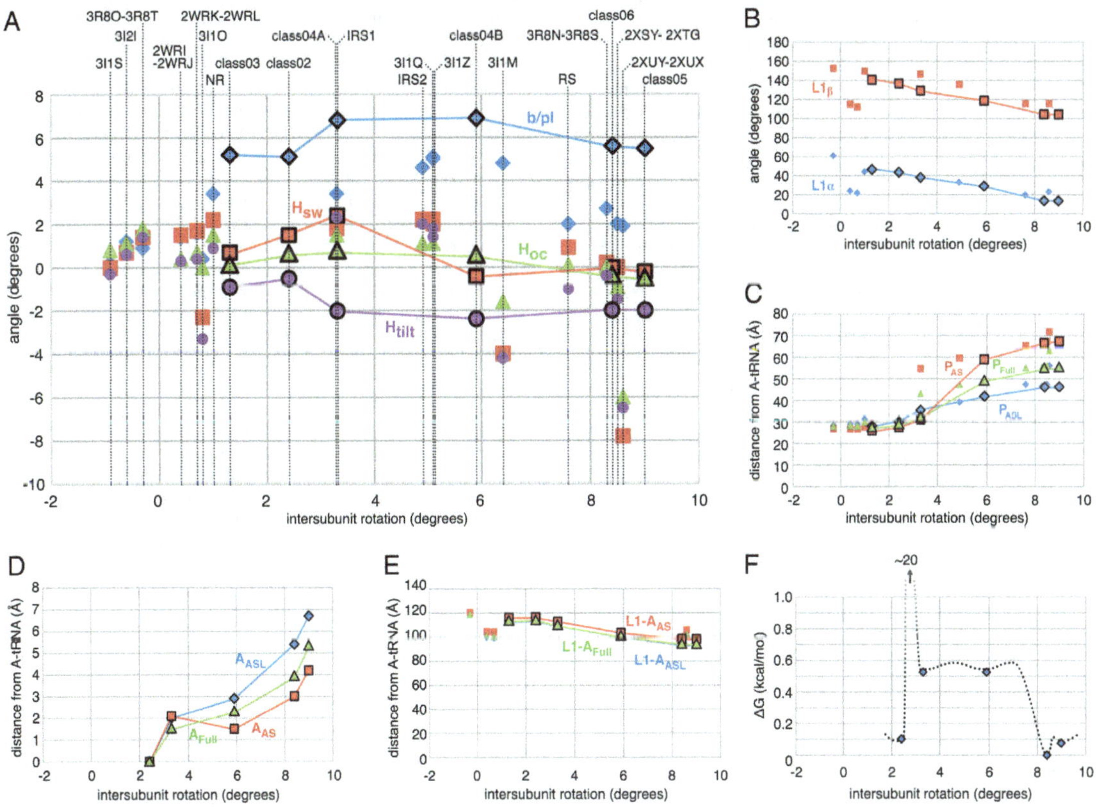

Fig. 3. Tensor analysis of individual domains. (*A*) Trends of measured angles on the small subunit for this study (connected points), those obtained by Fu et al. (32) (NR, IRS1, IRS2, RS) and those by X-ray crystallography (PDB codes indicated). The values plotted, b/pl, H_{tilt}, H_{oc}, and H_{sw}, are as defined in Tables S2 and S3. (*B*) Progression of the L1 stalk closing. ($L1_\alpha$ and $L1_\beta$ correspond to the same angle in an oblique plane). (*C, D, E*) Progression of tRNA and L1 movements. (*F*) Free-energy landscape along the intersubunit rotation coordinate, computed from occupancies of states in equilibrium. The numbers describe the relative stabilities of the various ground states. Dotted line represents the largely unknown topology of the energy landscape between the sample points obtained by cryo-EM. It includes a peak at 20 kcal/mol, going off scale in our diagram, indicating the barrier separating MS I and MS II as calculated from tRNA-tRNA smFRET measurements.

structures obtained by X-ray crystallography; note that the apo forms (3I1O and 3I1M) behave differently due to the lack of tRNAs. A very similar behavior can be observed for the closing of the head domain relative to the shoulder of the 30S subunit and the tilt of the head with respect to the subunit interface. However, while the motion relative to the shoulder closes the small subunit during the rotation, a larger rotation goes along with a tilt of the head away from the interface. The latter observation may be rationalized by the potential strain caused by the P-site tRNA moving between the head and the apical part of h44. However, the crystal structures of partially rotated ribosomes show a larger tilt *toward* the interface for tRNA-containing complexes, while a large tilt *away* from the interface is observed for the apo form. The origin of this discrepancy is not clear. A somewhat different trend is observed for the swiveling of the head domain: the head rotates away from the subunit interface reaching the maximum amplitude at class 4A and rotates back subsequently. This behavior can be observed in X-ray crystal structures (31, 35). Interestingly, however, the partially rotated structures of the ribosome in complex with tRNAs shows larger swiveling, while the apo complex shows significantly smaller swiveling than presently observed. Obviously, the presence of tRNAs does not allow the head domain to rotate too far into the intersubunit space. Note that the head already swivels back at the point when the body and platform are still closed and the head domain is closed relative to the shoulder (class 4B).

One of the most striking observations in the graphs is that the motion of the L1 stalk as a function of intersubunit rotation is reflected by a strict (negative) proportionality between stalk and intersubunit angles (Fig. 3*B*). The rationale for the high correlation may be found in the normal-mode analysis by Tama and coworkers (39), who found these movements to be coupled in one of the principal modes of the mechanical system. The only outliers on this graph belong to data obtained by X-ray crystallography, presumably since the component is peripheral, and thus strongly affected by crystal contacts.

The progression of tRNA and L1 movements is shown in Fig. 3 *C–E*. For both tRNAs, a continuous progression is seen as the intersubunit rotation angle increases. The classic-to-hybrid transition is reflected in the motions of AS and ASL for A- and P-site tRNAs, as well as in the change of angle formed by the two arms of the tRNA. Looking at the angles characterizing the tRNA structures and constellations, it is now seen that classes 5 and 6, which are similar with regards to ribosomal conformation, are in fact slightly different as far as the tRNAs are concerned. In contrast to the P-site tRNA (panel 4*C*), the graphs B and E show anomalies in the X-ray measurements—they do not fit into the smooth trajectory of the EM data. In contrast, the reconstructions from Spahn and coworkers (8) fit into our data quite well. The correlation analysis (Table in Fig. S6) shows similar trends as the plots in Fig. 3. For example, the negative correlation between L1 stalk motion and intersubunit rotation is so high that it almost suggests the existence of a gear.

The analysis also reveals the participation of the head swiveling in the movement of the P-site tRNA [via A-minor interactions between the head and the P-site-tRNA ASL, see (31)]. Something similar may also go on in the case of the correlation between the A-site ASL and the opening/closing of the small subunit, as it is conceivable that the closing of the body "pushes" the tRNA. Interestingly, there is a strong correlation of the intersubunit rotation with the distance of the ASL of the A-site tRNA but not with the corresponding angle. Conversely, the opening/closing of body/platform is correlated with the angle, but not the distance, of the ASL of the A-site tRNA.

Translating Relative Occupancies of the PRE States into Free-Energy Differences. The classification has given us numbers of molecules in each state as a by-product of the analysis. In a freely equilibrating system, these numbers reflect relative stabilities of the corresponding states, and can be converted into free-energy differences ΔG, reflecting the topology of the energy landscape (18, 19, 29). In this conversion we followed the analysis by Fischer et al. (29) (see *Supporting Information*). We take the energetics of the intersubunit rotation as the most interesting quantity to evaluate (see Fig. 3*F*), even though similar plots can be obtained for any of the quantities measured and tabulated in Tables 1–3.

Evidently (Fig. 3*F*), the intermediate states represented by classes 4A and 4B are energetically unfavorable relative to classes 2, 5, and 6, representing local minima within a relatively flat, plateau-like barrier that molecules traversing from the MS I ground state (class 2) to the MS II ground state (classes 5 and 6) must cover. According to the calculated landscape, the lowest free-energy minimum appears at the position of the hybrid, fully rotated forms, in line with FRET experiments showing that excursions to the classical state are less likely under the given experimental conditions (40). From the analysis, it can be concluded that classes 4A+4B and 5+6 represent two "superstates," the latter corresponding to MS II. In both cases, the energy minima of the constituent states are closely matched, suggesting local equilibration in the corresponding regions of the energy landscape.

Discussion

As shown by our work, the ribosome PRE complex exists as an ensemble of substates which span a large conformational space. Understanding the correlated dynamics of this system is essential for mapping structure to function. To this end, we have studied the multidimensional range of conformations sampled by the ribosome, by combining cryo-EM and MD techniques with the quantification of domains rearrangements. Our premise has been that extending the analysis to include correlations among movements of subunit domains and tRNAs contributes to defining the framework of functional interactions within the translating ribosome. Among other unresolved questions, the results of our study may contribute to the discussion concerning temporal uncoupling between movements of tRNA and the L1 stalk on the one hand (41) and between movements of the L1 stalk and intersubunit rotation on the other, as suggested by smFRET (22).

There is also currently a debate over the extent to which intermediates are sampled in wild-type PRE ribosomes during translocation. For example, a recent tRNA-tRNA smFRET study (26) found no evidence for intermediates in WT ribosomes, even when employing hidden Markov modeling methods similar to those used in Munro et al. (28). All existing crystallographic and cryo-EM studies that identified intermediate structures of ribosome-tRNA complexes used mutant ribosomes, ASL mimics, an empty A site or other modifications that impair or prohibit translocation. To date, our study is the only one to identify intermediate states of a PRE ribosome-tRNA complex, prepared using WT bacterial ribosomes and intact tRNAs, and competent for unimpaired forward translocation.

The observation of a progression in movements of structural components and tRNAs in our study suggests that the PRE ribosome oscillates between two extreme conformations going through at least two intermediates with distinct conformations. It is in the nature of such pathway intermediates that they represent stepping stones without commitment of the molecule to proceed further in the same direction; i.e., ribosomes captured in classes 4A and 4B may either be on pathway from one extreme to the other, or represent unsuccessful attempts of a ribosome in the MS-I or MS-II state to acquire the alternative end conformation. On the other hand, since in cryo-EM we only "see" molecules that occur in sufficient numbers, settled in local minima, the actual profile of the free-energy landscape is expected to be more complex (e.g., as suggested by a dotted line in Fig. 3*F*), but its trajectory outside the minima is only accessible by alternative experimental methods, such as smFRET (18, 19). In this context we note that for simplicity, Movies S1 and S2 display the conformational changes as a single, continuous, unidirectional process; in reality the changes are discontinuous, with long "pauses" in the wells, and—as the true free-energy landscape is likely very complex—multiple pathways connecting the observed conformational states may exist.

The calculated free-energy landscape suggests that all observed states are interchangeable, with little energetic difference (<1 kcal/mol). In contrast, smFRET studies report that the activation energy required to initiate the movement of the 30S subunit rotation from one macrostate to the other is considerably larger, in the order of 20 kcal/mol (12, 19). It is tempting to suggest that the required reconfiguration of intersubunit bridges, notably those in the periphery of the intersubunit space (4, 35, 36) presents the activation barrier, with one or more poorly occupied, short-lived high-energy transition states, unlikely to be captured by cryo-EM. As to the location of the barrier on our free-energy plot, a plausible placement is at the very beginning of the intersubunit rotation (Fig. 3*F*), where proteins L5 and S13, the two components of intersubunit bridge B1b, are locked by juxtaposition of opposing charges (4). Additional free-energy barriers may exist in the whole range.

All in all, our data support the idea that the intersubunit motion necessary for translocation occurs in several steps, and that each step involves a conformational reorganization of the ribosome and each of its subunits. These changes determine the corresponding configurations of the tRNAs. The progression we see implies that the different motions are coupled, if not kinetically then certainly thermodynamically. Evidently, the intermediate steps of incremental conformational changes and mRNA/tRNA binding positions are necessary to ensure processivity during translocation, as the binding partners are not allowed to disengage from the ribosome, and we can assume that all other phases of translation follow the same principle.

Note: As this manuscript was finalized, we became aware of a study of the PRE mammalian ribosome with the same methods of cryo-EM and classification (42). The results of this study, including the existence of intermediate positions of the tRNAs, are in broad agreement with our study of the *E. coli* system.

Materials and Methods

Image Processing. The ribosomal complex was prepared as described (15). To separate the heterogeneous dataset into structurally homogeneous subsets, a maximum-likelihood (ML)-based classification approach (32), was applied (see *SI Methods*).

Fitting of Crystallographic Structures into Electron Microscopy Densities. Structural models for the different conformational states of the ribosome were obtained using MDFF (34) (see *SI Methods*).

ACKNOWLEDGMENTS. We thank Ruben L. Gonzalez for a critical reading of the manuscript, Melissa Thomas for assistance with the preparation of the illustrations, and Wen Li for assistance with the deposition of the atomic models. X.A. is a recipient of a "Ramon y Cajal" fellowship from the Spanish Government (RYC-2009-04885). Supported by Grant PI2011-23 from the Department of Education, Universities and Research of the Basque Country (to X.A.), Howard Hughes Medical Institute and National Institutes of Health Grants R01 GM55440 and R37 GM29169 (to J.F.), National Institutes of Health Grant R01 GM059425 (to R.G.), and National Institutes of Health P41-RR005969 and National Science Foundation PHY0822613 (to K.S). Computer time for MDFF simulations was provided by National Science Foundation through the National Resources Allocation Committee (MCA93S028).

1. Spirin AS (1968) How does the ribosome work? A hypothesis based on the two subunit construction of the ribosome. *Curr Mod Biol* 2:115–127.
2. Bretscher MS (1968) Translocation in protein synthesis: A hybrid structure model. *Nature* 218:675–677.
3. Agirrezabala X, Frank J (2009) Elongation in translation as a dynamic interaction among the ribosome, tRNA, and elongation factors EF-G and EF-Tu. *Q Rev Biophys* 42:159–200.
4. Frank J, Gao H, Sengupta J, Gao N, Taylor DJ (2007) The process of mRNA-tRNA translocation. *Proc Natl Acad Sci USA* 104:19671–19678.
5. Moazed D, Noller HF (1989) Intermediate states in the movement of transfer RNA in the ribosome. *Nature* 342:142–148.
6. Frank J, Agrawal RK (2000) A ratchet-like inter-subunit reorganization of the ribosome during translocation. *Nature* 406:318–322.
7. Schuwirth BS, et al. (2005) Structures of the bacterial ribosome at 3.5 A resolution. *Science* 310:827–834.
8. Ratje AH, et al. (2010) Head swivel on the ribosome facilitates translocation by means of intra-subunit tRNA hybrid sites. *Nature* 468:713–716.
9. Taylor DJ, et al. (2007) Structures of modified eEF2 80S ribosome complexes reveal the role of GTP hydrolysis in translocation. *EMBO J* 26:2421–2431.
10. Spahn CM, et al. (2004) Domain movements of elongation factor eEF2 and the eukaryotic 80S ribosome facilitate tRNA translocation. *EMBO J* 23:1008–1019.
11. Julián P, et al. (2008) Structure of ratcheted ribosomes with tRNAs in hybrid states. *Proc Natl Acad Sci USA* 105:16924–16927.
12. Blanchard SC, Kim HD, Gonzalez RL, Jr., Puglisi JD, Chu S (2004) tRNA dynamics on the ribosome during translation. *Proc Natl Acad Sci USA* 101:12893–12898.
13. Fei J, Kosuri P, MacDougall DD, Gonzalez RL, Jr. (2008) Coupling of ribosomal L1 stalk and tRNA dynamics during translation elongation. *Mol Cell* 30:348–359.
14. Cornish PV, Ermolenko DN, Noller HF, Ha T (2008) Spontaneous intersubunit rotation in single ribosomes. *Mol Cell* 30:578–588.
15. Agirrezabala X, et al. (2008) Visualization of the hybrid state of tRNA binding promoted by spontaneous ratcheting of the ribosome. *Mol Cell* 32:190–197.
16. Ermolenko DN, et al. (2007) Observation of intersubunit movement of the ribosome in solution using FRET. *J Mol Biol* 370:530–540.
17. Spirin AS (2009) The ribosome as a conveying thermal ratchet machine. *J Biol Chem* 284:21103–21119.
18. Munro JB, Sanbonmatsu KY, Spahn CM, Blanchard SC (2009) Navigating the ribosome's metastable energy landscape. *Trends Biochem Sci* 34:390–400.
19. Frank J, Gonzalez RL, Jr. (2010) Structure and dynamics of a processive Brownian motor: The translating ribosome. *Annu Rev Biochem* 79:381–412.
20. Pan D, Kirillov SV, Cooperman BS (2007) Kinetically competent intermediates in the translocation step of protein synthesis. *Mol Cell* 25:519–529.
21. Fei J, et al. (2009) Allosteric collaboration between elongation factor G and the ribosomal L1 stalk directs tRNA movements during translation. *Proc Natl Acad Sci USA* 106:15702–15707.
22. Cornish PV, et al. (2009) Following movement of the L1 stalk between three functional states in single ribosomes. *Proc Natl Acad Sci USA* 106:2571–2576.
23. Munro JB, Altman RB, Tung CS, Sanbonmatsu KY, Blanchard SC (2010) A fast dynamic mode of the EF-G-bound ribosome. *EMBO J* 29:770–781.
24. Munro JB, Wasserman MR, Altman RB, Wang L, Blanchard SC (2010) Correlated conformational events in EF-G and the ribosome regulate translocation. *Nat Struct Mol Biol* 17:1470–1477.
25. Ermolenko DN, Noller HF (2011) mRNA translocation occurs during the second step of ribosomal intersubunit rotation. *Nat Struct Mol Biol* 18:457–462.
26. Chen C, et al. (2011) Single-molecule fluorescence measurements of ribosomal translocation dynamics. *Mol Cell* 42:367–377.
27. Ermolenko DN, et al. (2007) The antibiotic viomycin traps the ribosome in an intermediate state of translocation. *Nat Struct Mol Biol* 14:493–497.
28. Munro JB, Altman RB, O'Connor N, Blanchard SC (2007) Identification of two distinct hybrid state intermediates on the ribosome. *Mol Cell* 25:505–517.
29. Fischer N, Konevega AL, Wintermeyer W, Rodnina MV, Stark H (2010) Ribosome dynamics and tRNA movement by time-resolved electron cryomicroscopy. *Nature* 466:329–333.
30. Fu J, Munro JB, Blanchard SC, Frank J (2011) Cryoelectron microscopy structures of the ribosome complex in intermediate states during tRNA translocation. *Proc Natl Acad Sci USA* 108:4817–4821.
31. Zhang W, Dunkle JA, Cate JH (2009) Structures of the ribosome in intermediate states of ratcheting. *Science* 325:1014–1017.
32. Scheres SH, et al. (2007) Disentangling conformational states of macromolecules in 3D-EM through likelihood optimization. *Nat Methods* 4:27–29.
33. Agirrezabala X, et al. (2011) Structural insights into cognate versus near-cognate discrimination during decoding. *EMBO J* 30:1497–1507.
34. Trabuco LG, Villa E, Mitra K, Frank J, Schulten K (2008) Flexible fitting of atomic structures into electron microscopy maps using molecular dynamics. *Structure* 16:673–683.
35. Dunkle JA, et al. (2011) Structures of the bacterial ribosome in classical and hybrid states of tRNA binding. *Science* 332:981–984.
36. Jin H, Kelley AC, Ramakrishnan V (2011) Crystal structure of the hybrid state of ribosome in complex with the guanosine triphosphatase release factor 3. *Proc Natl Acad Sci USA* 108:15798–15803.
37. Gao YG, et al. (2009) The structure of the ribosome with elongation factor G trapped in the posttranslocational state. *Science* 326:694–699.
38. Berk V, Zhang W, Pai RD, Cate JH (2006) Structural basis for mRNA and tRNA positioning on the ribosome. *Proc Natl Acad Sci USA* 103:15830–15834.
39. Tama F, Valle M, Frank J, Brooks CL, 3rd (2003) Dynamic reorganization of the functionally active ribosome explored by normal mode analysis and cryo-electron microscopy. *Proc Natl Acad Sci USA* 100:9319–9323.
40. Kim HD, Puglisi JD, Chu S (2007) Fluctuations of transfer RNAs between classical and hybrid states. *Biophys J* 93:3575–3582.
41. Munro JB, et al. (2010) Spontaneous formation of the unlocked state of the ribosome is a multistep process. *Proc Natl Acad Sci USA* 107:709–714.
42. Budkevich T, et al. (2011) Structure and dynamics of the Mammalian ribosomal pretranslocation complex. *Mol Cell* 44:214–224.

LETTER

doi:10.1038/nature11872

High-resolution cryo-electron microscopy structure of the *Trypanosoma brucei* ribosome

Yaser Hashem[1]*, Amedee des Georges[1]*, Jie Fu[2], Sarah N. Buss[3], Fabrice Jossinet[4], Amy Jobe[5], Qin Zhang[6], Hstau Y. Liao[2], Robert A. Grassucci[1], Chandrajit Bajaj[6], Eric Westhof[4], Susan Madison-Antenucci[3] & Joachim Frank[1,2,5]

Ribosomes, the protein factories of living cells, translate genetic information carried by messenger RNAs into proteins, and are thus involved in virtually all aspects of cellular development and maintenance. The few available structures of the eukaryotic ribosome[1–6] reveal that it is more complex than its prokaryotic counterpart[7,8], owing mainly to the presence of eukaryote-specific ribosomal proteins and additional ribosomal RNA insertions, called expansion segments[9]. The structures also differ among species, partly in the size and arrangement of these expansion segments. Such differences are extreme in kinetoplastids, unicellular eukaryotic parasites often infectious to humans. Here we present a high-resolution cryo-electron microscopy structure of the ribosome of *Trypanosoma brucei*, the parasite that is transmitted by the tsetse fly and that causes African sleeping sickness. The atomic model reveals the unique features of this ribosome, characterized mainly by the presence of unusually large expansion segments and ribosomal-protein extensions leading to the formation of four additional inter-subunit bridges. We also find additional rRNA insertions, including one large rRNA domain that is not found in other eukaryotes. Furthermore, the structure reveals the five cleavage sites of the kinetoplastid large ribosomal subunit (LSU) rRNA chain, which is known to be cleaved uniquely into six pieces[10–12], and suggests that the cleavage is important for the maintenance of the *T. brucei* ribosome in the observed structure. We discuss several possible implications of the large rRNA expansion segments for the translation-regulation process. The structure could serve as a basis for future experiments aimed at understanding the functional importance of these kinetoplastid-specific ribosomal features in protein-translation regulation, an essential step towards finding effective and safe kinetoplastid-specific drugs.

The uniqueness of the *T. brucei* ribosomes, and those of other kinetoplastids, was highlighted by the discovery that the LSU rRNA chain is processed into six pieces[10–12]. The rRNA secondary structure of *T. brucei* (Supplementary Fig. 1a, b) pointed to the existence of several interesting features, such as the size of several expansion segments. A comparison of *T. brucei* ribosomal protein sequences with those of eukaryotic homologues[13] revealed the presence of *T. brucei*-specific large protein extensions. However, the arrangement and the structure of these features have remained unknown, although a low-resolution cryo-electron microscopy (cryo-EM) study of the *Trypanosoma cruzi* ribosome has been reported previously[14]. It is worth noting that gene expression in trypanosomes is primarily regulated post-transcriptionally (see ref. 15), at the protein-translation level. Several observations suggest that there are differences in both translation and translational regulation between *T. brucei* (and kinetoplastids in general) and other eukaryotes. For example, kinetoplastid messenger RNAs have a 35-nucleotide conserved spliced leader that is trans-spliced onto their 5′ ends (see ref. 16) and that has a markedly modified cap. In addition, unique protein factors are involved in the maturation of the *T. brucei* ribosomes[17], whereas several eukaryotic initiation factors (eIF-4B, eIF-3 subunit P135 and eIF-3 subunit 3) and the eukaryotic ribosomal recycling factor (eRRF) are absent[18].

Here we present a high-resolution cryo-EM structure of the *T. brucei* ribosome, and with this information we model the atomic structure of the ribosome using a combination of map segmentation[19,20], homology modelling, *ab initio* rRNA modelling based on covariation analysis with S2S and Assemble[21,22], and the molecular dynamics flexible fitting (MDFF) method[23] (see Methods).

The cryo-EM map was reconstructed from approximately 164,000 particles (see Methods). The *T. brucei* ribosome is in the inter-subunit-rotated state, with a transfer RNA at the exit (E) site (Fig. 1a, b). A measurement of local resolution (see Methods) shows that this is approximately 5 Å in most parts of the map (Supplementary Fig. 2a) but that there are small regions of high variance where it reaches 9 Å. Consistent with this, the three-dimensional variance[24] of the *T. brucei* ribosome map (Supplementary Fig. 2b) shows little structural variability, with the exception of the inter-subunit space where tRNA and translation-factor binding takes place, and areas around the L1 and P stalks. At the tRNA binding sites, the three-dimensional variance points mostly to the intermittent presence of tRNAs at the A and P sites. As our ribosomes are purified directly from *T. brucei* cell extracts, and therefore include a large proportion of translating ribosomes, such heterogeneity is expected. Data classification (see Methods) was attempted on a subset of the particles, but presumably because of the local nature of the variability it was not possible to identify distinct classes. Most of the *T. brucei* ribosome particles seem to exist in the same state. Periodically arranged densities for the phosphate groups along the rRNA strands and indications of protein side chains are visible (Supplementary Fig. 2c) in most regions of the map, and a comparison of the experimental cryo-EM map with simulated density maps generated from the atomic model, at a resolution of 5.0 Å (Supplementary Fig. 2d), validate our resolution estimation.

The first notable observation is the unusual size of several expansion segments (Fig. 1). In the small ribosomal subunit (SSU), expansion segments ES3S, ES6S, ES7S and ES9S seem to be several times larger than in other 80S ribosomes of known structure (Fig. 1c). Expansion segments ES4L, ES7L, ES19L, ES27L and ES31L of the LSU (Fig. 1d) are also considerably larger than in other known eukaryotic ribosomes. An additional domain on the LSU, which we refer to as the kinetoplastid-specific domain (KSD), is situated in the vicinity of ES7L (Fig. 1, red). These observations are consistent with the rRNA sequence and secondary structure in both ribosomal subunits[25] (Supplementary Fig. 1). We were able to model all proteins of the *T. brucei* ribosome (Fig. 2 and Supplementary Fig. 2d), except for some unstructured tails, and we were able to model the rRNAs in their entirety, with the exception of parts of ES27L and ES9S, which have poorly resolved densities due to their dynamic behaviour.

[1]Howard Hughes Medical Institute (HHMI), Department of Biochemistry and Molecular Biophysics, Columbia University, New York, New York 10032, USA. [2]Department of Biochemistry and Molecular Biophysics, Columbia University, New York, New York 10032, USA. [3]Division of Infectious Diseases, Wadsworth Center, New York State Department of Health, Albany, New York 12201, USA. [4]Architecture et Réactivité de l'ARN, Université de Strasbourg, Institut de Biologie Moléculaire et Cellulaire (CNRS), Strasbourg 67084, France. [5]Department of Biological Sciences, Columbia University, New York, New York 10027, USA. [6]Department of Computer Science, Institute for Computational Engineering and Sciences, University of Texas, Austin, Texas 78712, USA.
*These authors contributed equally to this work.

©2013 Macmillan Publishers Limited. All rights reserved

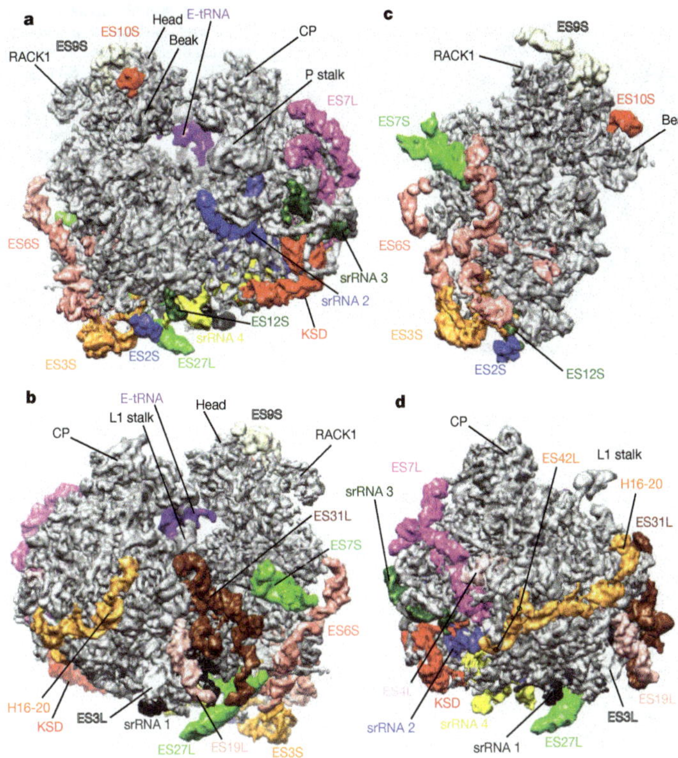

Figure 1 | High-resolution cryo-EM structure of the *T. brucei* ribosome. The density map was filtered with a locally varying bandpass, according to the local resolution measurements (see Methods and Supplementary Fig. 2A). In all panels, the largest rRNA expansion segments (ESs) are shown in different colours. **a**, The *T. brucei* ribosome viewed from the front (P stalk and beak side). **b**, The *T. brucei* ribosome viewed from the back (L1 stalk and platform side). **c**, The solvent side of the SSU. **d**, The solvent side of the LSU. CP, central protuberance; E-tRNA, exit-site RNA.

Our atomic model of the *T. brucei* ribosome reveals the structures of its expansion segments and its additional rRNA insertions (in both subunits) (Fig. 2a, b). These expansion segments and insertions represent a substantial additional mass compared to other known eukaryotic ribosomes (Fig. 2c, d). In spite of this addition and the change in the rRNA arrangement at several sites (Supplementary Fig. 3a, c), the structures and binding sites of ribosomal proteins are conserved (Supplementary Fig. 3b, d), as exemplified by proteins S7e on the SSU and L19e on the LSU (Fig. 3a). The high homology of these proteins with their counterparts in yeast contrasts with the differences in the arrangement of the surrounding rRNA context, formed by expansion segments ES6S, ES27L and ES31L (Fig. 3a).

The largest expansion segments are concentrated mostly in the regions of the LSU L1 stalk and the platform of the SSU. These large expansion segments form four additional inter-subunit bridges that are not seen in either of the two available 80S structures[1–4] (Fig. 3a). We refer to these bridges as B_{Tb}-1 B_{Tb}-2 B_{Tb}-3 and B_{Tb}-4 (Tb, *T. brucei*). B_{Tb}-1 is formed by three contacts (Fig. 3a and Supplementary Fig. 4a); B_{Tb}-1a, SSU ES7S helix A with LSU ES31L helix C. B_{Tb}-1b, SSU ES7S helix A with L34e carboxy-terminal extension accompanying ES31 helix C. B_{Tb}-1c, SSU ES6S helix F with LSU ES31L helix C. This first bridge seems to be stronger in the *T. cruzi* density map published previously[14] (Supplementary Fig. 4a). B_{Tb}-2 is formed by SSU ES6S helix G and LSU ES27L helix A (Fig. 3a and Supplementary Fig. 4b). This bridge is not established in *T. cruzi* despite the presence of its components. B_{Tb}-3 is formed by SSU ES12S and a trypanosome-specific expansion segment grafted onto one of four short ribosomal RNAs in *T. brucei* (we refer to this here as srRNA-4-ES; expansion segment on short rRNA 4) (Fig. 3a and Supplementary Fig. 4c). This bridge is the strongest among the four, as seen from the *T. brucei* and *T. cruzi* density maps. B_{Tb}-4 is formed by SSU ES3S helix C and LSU ES27L helix B (Fig. 3a and Supplementary Fig. 4d), but the cryo-EM density of this bridge is poorly resolved and can be observed only at a low-density threshold in *T. brucei*, probably owing to the inherent flexibility of ES3S and ES27L. In contrast, in *T. cruzi* this bridge is visible at a relatively high-density threshold.

In terms of the enigmatic processing of LSU rRNA into six pieces, two large ribosomal RNAs, LSU-α and LSU-β, and four short ribosomal RNAs (which we refer to here as srRNA 1, srRNA 2, srRNA 3 and srRNA 4), our structure shows the arrangement of these pieces. Although the processing of the LSU rRNA into two large pieces has been observed in other organisms (see ref. 10), only trypanosomes and other kinetoplastids[10] show such extensive processing. srRNA 1, which is composed of approximately 220 nucleotides[11,12], is generated from the main immature rRNA of LSU by the removal of two flanking sequences (Supplementary Fig. 5a), by splitting the main immature rRNA chain of the LSU into two components, LSU-α and LSU-β (Fig. 3a). These flanking sequences are localized at the positions of LSU ES20L and ES26L, respectively, which are also found in other eukaryotic ribosomes[9] (next to H54 on LSU domain III)[12]. It is possible that LSU expansion segments ES20L and ES26L in kinetoplastids have evolved to become cleavage sites.

srRNA 1 is structurally very similar to its homologous region on domain III in yeast (Supplementary Fig. 5b), except that it bears an expansion segment, which we name srRNA-1-ES. It is interesting to note that LSU ES19L is two to three times larger in *T. brucei* than it is in other 80S ribosomes (Fig. 3). srRNA 1 cleavage in *T. brucei* seems to remove a piece of rRNA that would otherwise collide with ES19L. It is possible that this cleavage was an expedient evolutionary solution to allow room for the unusually large ES19L. However, the functional importance of ES19L is unknown.

©2013 Macmillan Publishers Limited. All rights reserved

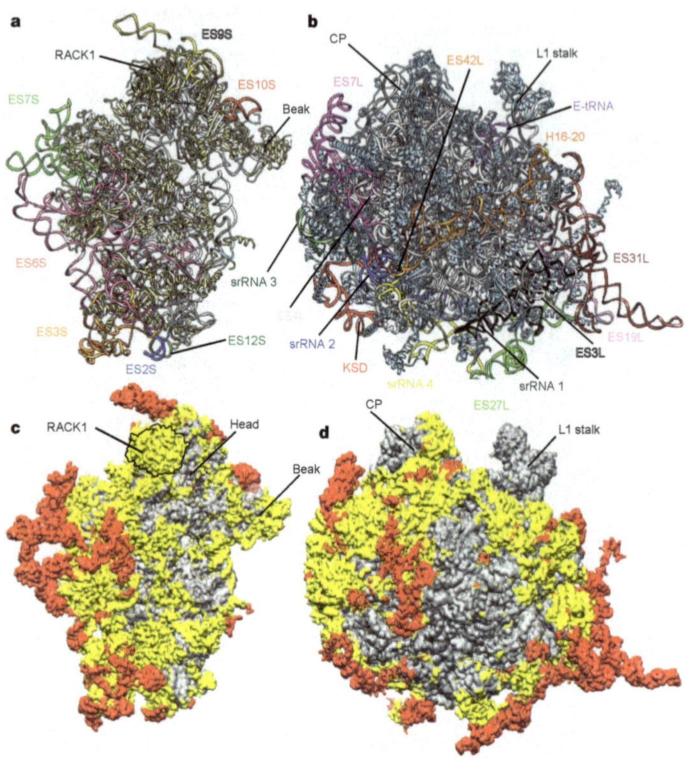

Figure 2 | Atomic model of the *T. brucei* ribosome. a, b, Atomic models of SSU (**a**) and LSU (**b**) with the expansion segments coloured and annotated as in Fig. 1. **c, d,** Atomic models of SSU (**c**) and LSU (**d**) in surface presentations. Grey regions indicate the location of conserved common core elements of all (prokaryotic and eukaryotic) ribosomes. Yellow regions highlight eukaryote-specific conserved elements, including those for trypanosomes. Red regions indicate trypanosome-specific elements that do not exist in other 80S ribosomes.

The remaining srRNAs (srRNA 2, srRNA 3 and srRNA 4 (180, 70 and 140 nucleotides, respectively)) were suggested previously, based on sequence conservation and two-dimensional modelling, to interact with one another and mimic the missing last domain of the LSU-β rRNA in the *T. brucei* ribosome, corresponding to domain VI in yeast and other 80S ribosomes[12]. Our structure confirms this hypothesis and shows, in addition, that srRNA 2, srRNA 3 and srRNA 4 interact directly with the KSD (Fig. 4a, left and middle) to form one single stable domain similar to domain VI in other 80S ribosomes (Fig. 4a, right). Similar to the processing of srRNA 1, the processing of srRNA 2, srRNA 3 and srRNA 4 may be, at least in part, a structural necessity. Indeed, the existence of KSD in the structure and conformation observed is possible only if several pieces of rRNA are cleaved, as they would otherwise collide with KSD. The cleavage sites correspond to LSU expansion segments ES36L and part of ES39L in other eukaryotic ribosomes, and the cleavage yields the three srRNAs. The bundle formed by srRNA 2, srRNA 3, srRNA 4 and KSD is anchored in place through two specific protein extensions on L14e and L31e that hug the rRNA bundle (Fig. 4a, middle). There is also a unique interaction between srRNA 2 and domain I of LSU-α rRNA (Fig. 4a, arrow). This interaction occurs between a proximal stem-loop on srRNA 2 and another stem-loop, part of a *T. brucei* expansion segment between helices 19 and 20 on the LSU-α rRNA domain I, which we call ES42L. Although the effect of this connection is unknown, we speculate that it may contribute to maintaining the structure of the srRNAs bundle. *T. brucei* ribosome 5S and 5.8S rRNAs are very similar to their counterparts in yeast, except that ES3L on the 5.8S rRNA is larger.

No trypanosome-specific proteins were found, based on the analysis of the density map. However, several *T. brucei* ribosomal proteins have evolved specific helical extensions that interact with the additional rRNA mass and contribute towards stabilizing the extended expansion segments in their observed conformations (Fig. 4b). In other cases, ribosomal protein tails have adopted different structures. On the LSU, L4 presents a long amino-terminal helical tail, different in the other available structures, which is probably due to the presence of a large, trypanosome-specific rRNA helical insertion in LSU ES7L, helix A (Fig. 4b). On the SSU, the trypanosomal RACK1 protein (also known as guanine nucleotide-binding protein subunit β-like protein) is found in its conserved binding site (Fig. 2a), which contrasts with *T. cruzi* ribosome[14], in which trypanosomal RACK1 was missing from the density map, for a reason that is unresolved[26].

An interesting protein–rRNA interaction is formed by ribosomal protein S6e with expansion segments ES3S and ES6S (Fig. 3a, black circle). Ribosomal protein S6e possess an unstructured C-terminal tail, unresolved in both yeast 80S[1,2] and *Tetrahymena thermophila* 40S subunit structures[5]. This tail is involved in translation regulation by phosphorylation of several serine residues recognized by S6e kinases (RSK and S6K) in higher eukaryotes and many lower eukaryotes (see ref. 27). The *T. brucei* S6e C-terminal tail contains a conserved RXXSS motif, recognizable by S6 kinases, and seems to have a helical structure (Fig. 3a, black circle, and Supplementary Fig. 6). The S6eC-terminal tail is surrounded by additional parts of SSU expansion segments ES3S and ES6S shielding the phosphorylation site, Ser 233 and Ser 234 (Supplementary Fig. 6), and could render it inaccessible to kinases on the ribosome. This is suggestive of a difference in translation regulation between trypanosomes and higher eukaryotes.

The role of the kinetoplastid-specific expansion segments in the regulation of the translation process remains unknown. In fact, very little is known about the translation regulation in kinetoplastids. However, the localization and the arrangement of the specific expansion segments, revealed by our structure, suggest the importance of several elements. In *T. brucei*, SSU expansion segments ES6S and ES7S are localized near the mRNA channel exit and are poised for a potential role in the regulation of translation initiation. The localization of

©2013 Macmillan Publishers Limited. All rights reserved

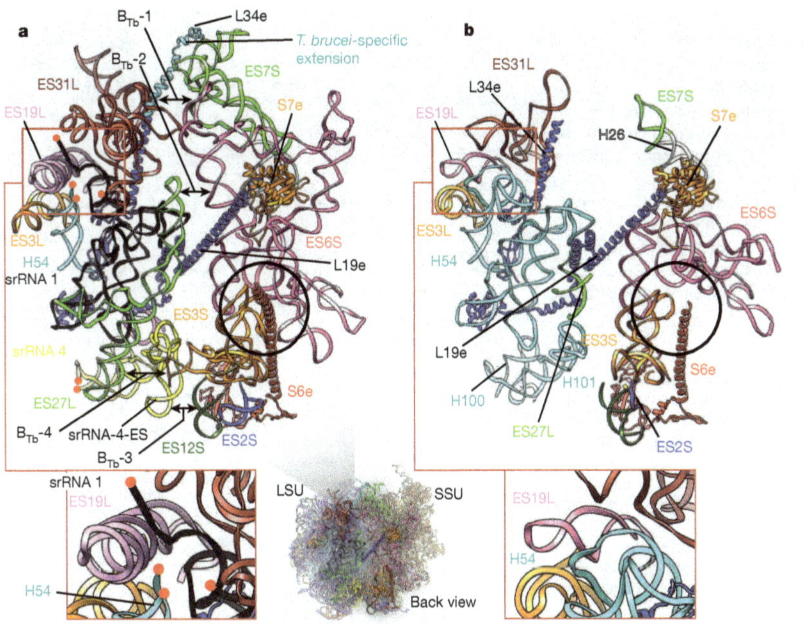

Figure 3 | Comparison between *T. brucei* and yeast ribosomes. **a, b,** Back view of rRNA expansion segments at the back of the *T. brucei* (**a**) and yeast (**b**) ribosomes. *T. brucei*-specific inter-subunit bridges are indicated by double arrows in **a.** Insets show enlargements of the srRNA 1 cleavage site in *T. brucei* and its counterpart region in yeast. Black circles highlight the interaction of S6e with ES3S and ES6S in both ribosomes. Red spheres in **a** denote srRNA 1 and srRNA 4 cleaved ends.

expansion segments ES6S and ES7S on the SSU, near the mRNA channel exit, suggests their possible interaction with the mRNA spliced leader and its unusually modified cap, but also with eIF3 (refs 28, 29). It is interesting to note that SSU expansion segments ES6S and ES7S collide with the proposed binding site of eIF3 (ref. 29) (Fig. 4c). The initiation factor eIF3 does exist in trypanosomes[18] (Supplementary Fig. 7); in *T. brucei* it contains at least seven or eight subunits. The structure here suggests that the proposed binding site of eIF3 (ref. 29) may require reconsideration. The *T. brucei* ribosome structure may alternatively suggest that the SSU interaction with eIF3 is different from that in other eukaryotes, and that expansion segments ES6S and ES7S may have an important role in this interaction.

It was found that the *T. brucei* ribosome interacts with specific factors involved in translation regulation (such as L34 and L37 (not to be confused with the LSU proteins of the same name))[17] that are required in the maturation and cytoplasmic exportation of the LSU. It is possible that kinetoplastid-specific expansion segments are involved in the binding of such specific factors that regulate the translation process. This hypothesis

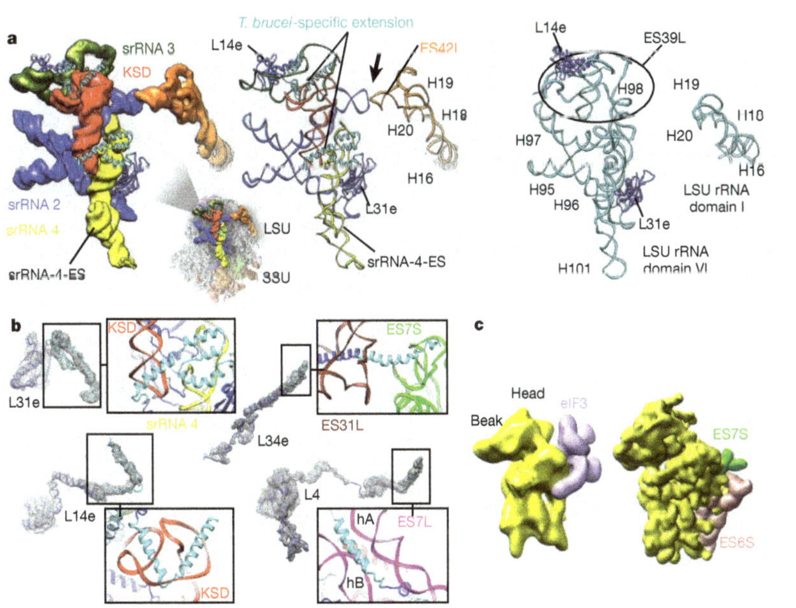

Figure 4 | *T. brucei* srRNAs and protein extensions. a, Left and middle, srRNA 2, srRNA 3, srRNA 4 and KSD cryo-EM density segmentations with their atomic models. Right, yeast LSU rRNA domain I and LSU rRNA domain VI. Arrow shows the kissing-loop interaction. Black circle surrounds ES39L in the yeast ribosome. **b,** Ribosomal proteins presenting specific extensions in *T. brucei* (cyan ribbon). Insets show enlargements of the interactions of their extensions with the surrounding rRNA. hA, helix A, hB, helix B. **c,** Left, cryo-EM-derived model of the eIF3-binding site (recreated according to ref. 21). Right, segmented map of the SSU from the *T. brucei* ribosome, filtered to 20 Å.

©2013 Macmillan Publishers Limited. All rights reserved

is supported by the observation that *T. brucei* expansion segments contain several pyrimidine-rich internal loops, that have been suggested to be important for protein recognition or binding to other RNA elements[30].

In conclusion, our cryo-EM study reveals that the *T. brucei* ribosome has an unusual structure and shows the unique arrangement of its expansion segments. These features may have consequences for the regulation of translation in trypanosomes. Our results may explain, in structural terms, the rationale for the processing of LSU rRNA into six pieces. More generally, our structure highlights the extraordinary diversity of ribosome structures among eukaryotes. We believe that these findings provide a basis for future investigations of translation regulation in kinetoplastids and of the function of the specific expansion segments, opening the door to the development of new drugs.

METHODS SUMMARY

Trypanosome culture and ribosome isolation. Pro-cyclic forms of *T. brucei* strain TREU 667 were grown at 27 °C. Cell pellets were washed and resuspended in SKS buffer containing 5 mM EGTA, 0.5% Triton X-100 and protease inhibitor cocktail (0.5 mM TLCK (*N*-tosyl-L-leucine chloromethyl ketone), 0.1 mM PMSF (phenylmethylsulphonyl fluoride), 25 μg ml^{-1} aprotinin, 10 μg ml^{-1} leupeptin). Cells were freeze-thawed. Several centrifugation steps were used to obtain a ribosome-enriched pellet that was resuspended in a buffer containing 20 mM HEPES, pH 7.2, 10mM $MgCl_2$, 500 mM KCl and 5 mM β-mercaptoethanol. The suspension was clarified and ribosomes were purified.

Electron microscopy and image processing. Three microlitres of each sample were applied to holey carbon grids (carbon-coated Quantifoil 2/4 grid, Quantifoil Micro Tools, GmbH) containing an additional continuous thin layer of carbon, and glow-discharged. Grids were blotted for 6 s at 4 °C in 100% humidity and vitrified by plunging rapidly into liquid ethane at −180 °C with a Vitrobot (FEI). Cryo-EM data were acquired under low-dose conditions (25 electrons (e^-) per Å^2) on a Tecnai F30 Polara (FEI) operating at 300 kV. The micrographs were taken at a nominal magnification of ×59,000 on Kodak SO-163 film. The data were processed using SPIDER. The data windows were inspected manually. The contrast transfer function (CTF) of the reconstructions was corrected by phase-flipping the particles using the defocus value estimated for each micrograph and a single reconstruction was obtained from the entire data set using conjugate gradients with regularization (BP CG command in SPIDER).

Ribosomal RNA and protein modelling and fitting. The *T. brucei* SSU and LSU rRNA were modelled by homology, and rRNA expansion segments were modelled mostly *ab initio* by performing a covariation analysis against a number of orthologous rRNA sequences found in other related kinetoplastids. The ribosomal proteins were modelled mostly by homology. The resulting models were fitted using the MDFF method[23].

Full Methods and any associated references are available in the online version of the paper.

Received 3 October; accepted 21 December 2012.
Published online 10 February 2013.

1. Ben-Shem, A. *et al.* Crystal structure of the eukaryotic ribosome. *Science* **330,** 1203–1209 (2010).
2. Ben-Shem, A. *et al.* The structure of the eukaryotic ribosome at 3.0 Å resolution. *Science* **334,** 1524–1529 (2011).
3. Armache, J. P. *et al.* Localization of eukaryote-specific ribosomal proteins in a 5.5 Å cryo-EM map of the 80S eukaryotic ribosome. *Proc. Natl Acad. Sci. USA* **107,** 19754–19759 (2010).
4. Armache, J. P. *et al.* Cryo-EM structure and rRNA model of a translating eukaryotic 80S ribosome at 5.5 Å resolution. *Proc. Natl Acad. Sci. USA* **107,** 19748–19753 (2010).
5. Rabl, J., Leibundgut, M., Ataide, S. F., Haag, A. & Ban, N. Crystal structure of the eukaryotic 40S ribosomal subunit in complex with initiation factor 1. *Science* **331,** 730–736 (2011).
6. Klinge, S., Voigts-Hoffmann, F., Leibundgut, M., Arpagaus, S. & Ban, N. Crystal structure of the eukaryotic 60S ribosomal subunit in complex with initiation factor 6. *Science* **334,** 941–948 (2011).
7. Klinge, S., Voigts-Hoffmann, F., Leibundgut, M. & Ban, N. Atomic structures of the eukaryotic ribosome. *Trends Biochem. Sci.* **37,** 189–198 (2012).
8. Wilson, D. N. & Doudna Cate, J. H. The structure and function of the eukaryotic ribosome. *Cold Spring Harb. Perspect. Biol.* **4,** a011536 (2012).
9. Yokoyama, T. & Suzuki, T. Ribosomal RNAs are tolerant toward genetic insertions: evolutionary origin of the expansion segments. *Nucleic Acids Res.* **36,** 3539–3551 (2008).
10. White, T. C., Rudenko, G. & Borst, P. Three small RNAs within the 10 kb trypanosome rRNA transcription unit are analogous to domain VII of other eukaryotic 28S rRNA. *Nucleic Acids Res.* **14,** 9471–9489 (1986).
11. Cordingley, J. S. & Turner, M. J. 6.5 S RNA; preliminary characterization of unusual small RNAs in *Trypanosoma brucei. Mol. Biochem. Parasitol.* **1,** 91–96 (1980).
12. Campbell, D. A., Kubo, K., Graham Clark, C. & Boothroyd, J. C. Precise identification of cleavage sites involved in the unusual processing of trypanosome ribosomal RNA. *J. Mol. Biol.* **196,** 113–124 (1987).
13. Berriman, M. *et al.* The genome of the African trypanosome *Trypanosoma brucei. Science* **309,** 416–422 (2005).
14. Gao, H., Juri Ayub, M., Levin, M. L. & Frank, J. The structure of the 80S ribosome from *Trypanosoma cruzi* reveals unique rRNA components. *Proc. Natl Acad. Sci. USA* **102,** 10206–10211 (2005).
15. Clayton, C. & Shapira, M. Post-transcriptional regulation of gene expression in trypanosomes and leishmanias. *Mol. Biochem. Parasitol.* **156,** 93–101 (2007).
16. Michaeli, S. Trans-splicing in trypanosomes: machinery and its impact on the parasite transcriptome. *Future Microbiol.* **6,** 459–474 (2011).
17. Ciganda, M. & Williams, N. Characterization of a novel association between two trypanosome-specific proteins and 5S rRNA. *PLoS ONE* **7,** e30029 (2012).
18. Ivens, A. C. *et al.* The genome of the kinetoplastid parasite, *Leishmania major. Science* **309,** 436–442 (2005).
19. Zhang, Q., Bettadapura, R. & Bajaj, C. Macromolecular structure modeling from 3D EM using VolRover 2.0. *Biopolymers* **97,** 709–731 (2012).
20. Pintilie, G., Zhang, J., Goddard, T., Chiu, W. & Gossard, D. Quantitative analysis of cryo-EM density map segmentation by watershed and scale-space filtering, and fitting of structures by alignment to regions. *J. Struct. Biol.* **170,** 429–438 (2010).
21. Jossinet, F. & Westhof, E. Sequence to Structure (S2S): display, manipulate and interconnect RNA data from sequence to structure. *Bioinformatics* **21,** 3320–3321 (2005).
22. Jossinet, F., Ludwig, T. E. & Westhof, E. Assemble: an interactive graphical tool to analyze and build RNA architectures at the 2D and 3D levels. *Bioinformatics* **26,** 2057–2059 (2010).
23. Trabuco, L. G., Villa, E., Mitra, K., Frank, J. & Schulten, K. Flexible fitting of atomic structures into electron microscopy maps using molecular dynamics. *Structure* **16,** 673–683 (2008).
24. Liao, H. Y. & Frank, J. Classification by bootstrapping in single particle methods. *IEEE Int. Symp. Biom. Imaging* 169–172 (2010).
25. Cannone, J. J. *et al.* The comparative RNA web (CRW) site: an online database of comparative sequence and structure information for ribosomal, intron, and other RNAs. *Bioinformatics* **3,** 2 (2002).
26. Ayub, M. J., Atwood, J., Nuccio, A., Tarleton, R. & Levin, M. J. Proteomic analysis of the *Trypanosoma cruzi* ribosomal proteins. *Biochem. Biophys. Res. Commun.* **382,** 30–34 (2009).
27. Meyuhas, O. Physiological roles of ribosomal protein S6: One of its kind. *Int. Rev. Cell Mol. Biol.* **268,** 1–37 (2008).
28. Srivastava, S., Verschoor, A. & Frank, J. Eukaryotic initiation factor 3 does not prevent association through physical blockage of the ribosomal subunit-subunit interface. *J. Mol. Biol.* **226,** 301–304 (1992).
29. Siridechadilok, B., Fraser, C. S., Hall, R. J., Doudna, J. A. & Nogales, E. structural roles for human translation factor eIF3 in initiation of protein synthesis. *Science* **310,** 1513–1515 (2005).
30. Lescrinier, E. M. H. P. *et al.* Structure of the pyrimidine-rich internal loop in the poliovirus 30′-UTR: The importance of maintaining pseudo-2-fold symmetry in RNA helices containing two adjacent non-canonical base-pairs. *J. Mol. Biol.* **331,** 759–769 (2003).

Supplementary Information is available in the online version of the paper.

Acknowledgements This work is dedicated to the memory of Mariano Levin, who collaborated with J.F. and S.M.A. on the ribosomes from *T. cruzi* and *T. brucei*. We thank G. Cardone for assistance in the local resolution computation, and M. Thomas for her assistance with the preparation of figures. We wish to thank N. Williams for an useful discussion about the *T. brucei* LSU rRNA processing. This work was supported by the Howard Hughes Medical Institute (HHMI) and the National Institutes of Health (NIH) R01 GM29169 (to J.F.), L'Agence Nationale de la recherche (ANR) project AMIS ARN ANR-09-BLAN-0160 (E.W. and F.J.), as well as NIH R01-EB004873 and R01-GM074258 (to Q.Z. and C.B.). S.N.B. was supported by a Centers for Disease Control (CDC) Emerging Infectious Diseases (EID) fellowship program.

Author Contributions Y.H., A.d.G., S.N.B., F.J., Q.Z., C.B., S.M.-A., E.W. and J.F. interpreted the data and wrote the manuscript. S.N.B. purified the *T. brucei* ribosomes. Y.H., J. Fu and R.A.G. carried out the cryo-EM experiments. H.Y.L. performed the three-dimensional variance estimation. Y.H., A.J. and Q.Z. performed the density-map segmentations. Y.H., A.d.G., J. Fu, A.J. and H.Y.L. carried out the cryo-EM data processing. Y.H. and F.J. modelled the rRNA. Y.H. and Q.Z. modelled the ribosomal proteins. J.F. directed research.

Author Information The electron microscopy map has been deposited in the European Molecular Biology Laboratory (EMBL) European Bioinformatics Institute Electron Microscopy Data Bank (EMDB) under accession code EMD-2239. Coordinates of electron-microscopy-based model have been deposited in the RCSB Protein Data Bank under accession numbers 3ZEQ, 3ZEX, 3ZEY and 3ZF7. Reprints and permissions information is available at www.nature.com/reprints. The authors declare no competing financial interests. Readers are welcome to comment on the online version of the paper. Correspondence and requests for materials should be addressed to J.F. (jf2192@columbia.edu).

©2013 Macmillan Publishers Limited. All rights reserved

METHODS

Trypanosome culture and ribosome isolation. Pro-cyclic forms of *T. brucei* strain TREU 667 were grown at 27 °C in a semi-defined medium supplemented with 10% heat-inactivated fetal bovine serum and 25 mM HEPES, as described previously[31]. For ribosomal isolation, cells were grown to a density of $0.5–1.2 \times 10^7$ cells per ml in a 4 l volume, and collected by 10 min centrifugation at 6,000*g*.

Ribosomes were prepared using a method described previously[32], but with slight modifications. In brief, cell pellets were washed three times with SKS buffer (0.25 M sucrose, 5 mM KCl) and resuspended in SKS buffer containing 5 mM EGTA, 0.5% Triton X-100 and protease inhibitor cocktail (0.5 mM TLCK, 0.1 mM PMSF, 25 μg ml^{-1} aprotinin, 10 μg ml^{-1} leupeptin). The cells were freeze-thawed five times and ruptured further using 10 strokes of a Dounce homogenizer. Centrifugation steps described previously[1] were used to obtain a ribosome-enriched pellet, and the pellet was resuspended in a buffer containing 20 mM HEPES, pH 7.2, 10 mM $MgCl_2$, 500 mM KCl and 5 mM β-mercaptoethanol. The suspension was clarified and ribosomes were purified using methods described previously[1], however, 20 mM HEPES, pH 7.2, was substituted for Tris in all buffers to reduce ribosomal aggregation.

Electron microscopy. Three microlitres of each sample was applied to holey carbon grids (carbon-coated Quantifoil 2/4 grid, Quantifoil Micro Tools GmbH) containing an additional continuous thin layer of carbon[33], and glow-discharged. Grids were blotted for 6 s at 4 °C in 100% humidity and vitrified by plunging rapidly into liquid ethane at −180 °C with a Vitrobot (FEI)[34,35]. Data were acquired under low-dose conditions (25 e^- per $Å^2$) on a Tecnai F30 Polara (FEI) operating at 300 kV. The micrographs were taken at a nominal magnification of ×59,000 on Kodak SO-163 film using the automated data collection system AutoEMation[36], and scanned on a Nikon Super Coolscan 9000 ED with a step size of 6.35 μm, resulting in a pixel size of 1.09 Å on the object scale.

Image processing. The data were processed using SPIDER[37]. Approximately 700,000 particle windows were automatically extracted from 1,100 film-recorded micrographs using the lfcpick.spi spider batch file[8] (a batch file of SPIDER commands that extracts potential particles by template matching) with a window size of 500 pixels and a pixel size of 1.09 Å. The windows were inspected manually; only particles with high contrast were selected, and those in contact with other particles were eliminated. This conservative manual verification step yielded approximately 164,000 particles in total. Reconstruction was carried out following standard SPIDER protocols for reference-based reconstruction[38], except that contrast transfer function of the reconstructions was corrected by phase-flipping the particles using the defocus value estimated for each micrograph and a single reconstruction was obtained from the entire data set using conjugate gradients with regularization (BP CG command in SPIDER). The resolution at FSC = 0.5 for the final reconstruction was estimated at 5.57 Å (Supplementary Fig. 2C). However, this resolution is an average of the local resolutions over all of the ribosome and it underestimates most of the ribosome volume that is resolved at approximately 5.0 Å (see below).

Maximum-likelihood classification[39], on a subset of 40,000 particles taken from the full data set, was attempted using the XMIPP package[40]. The full *T. brucei* reconstruction was used as a seed for four classes. The resulting classes were all very similar, as expected based on the three-dimensional variance calculation (see below) and the resolution of the map, probably as a consequence of most of the ribosomes being in the same state. No further classification was attempted.

We attribute the high resolution obtained in our study to several factors; first, the CTF correction procedure by phase flipping, which has the advantage of correcting the CTF more precisely at the micrograph level; second, carrying out the three-dimensional reconstructions (including all the phase-flipped particles) at once, instead of summing different reconstructions issued from different defocus groups; third, the global homogeneity of the sample in our study; and finally, the very carful particle selection.

Local resolution estimation and filtering. The local resolution was estimated on the final map using the 'blocres' module, part of Bernard's Software Package (Bsoft; http://lsbr.niams.nih.gov/bsoft/)[41,42]. The three-dimensional map depicting local resolutions showed that the ribosome volume has a resolution of approximately 5.0 Å (at FSC = 0.5) that concerns mostly the core (Supplementary Fig. 2A), and a resolution between 5 and 6 Å at the outer shell, whereas regions of higher variability have resolutions ranging between 7 and 9 Å. The *T. brucei* cryo-EM map was filtered according to the estimated local resolution using the following procedure: the map was segmented in accordance with the local resolution using the Segger module[43] implemented in the Chimera UCSF program[44]. The segments obtained were filtered according to their local resolution and then merged together into one map.

Map segmentation. A detailed description of the segmentation protocol can be found in the Supplementary Information section. A preliminary segmentation of the *T. brucei* map was carried out using Segger[43], now implemented in Chimera[44], and VolRover[45] (see also protocols described in refs 46–48). These preliminary segments were used during the atomic modelling process. After the atomic model was obtained, we used it to re-segment the *T. brucei* map more accurately (see Supplementary Information).

Three-dimensional variance estimation. The particle windows were binned six times for the three-dimensional variance estimation step, to reduce the large amount of time and memory required for this operation, resulting in a pixel size of 6.54 Å. The three-dimensional variance map was computed using the bootstrapping method, developed by several groups[49–52], in the following way: 40,000 bootstrap reconstructions were generated, each of which was reconstructed from *n* particle projections that were randomly sampled with replacement from the total set of approximately 164,000 particles. To enhance the signal-to-noise ratio of the three-dimensional variance, the bootstrap volumes were filtered to a resolution corresponding to approximately half the frequency of the first zero-crossing of the CTFs. The structural variance was estimated as the sample variance of the bootstrap volumes minus the variance of the noise, and the difference was then multiplied by *n* (ref. 50). In this estimation, the noise variance is assumed to be uniform across the map[50]. Ideally, the three-dimensional variance map should reflect only regions of high conformational changes, including presence versus absence of a ligand such as a protein or a tRNA. Here, the low-pass filtration was necessary to reduce the noise variance. However, the filtering also introduces spatial correlation of the noise, and the resulting variance map is distorted to some extent by the noise variance (as verified using simulations).

Ribosomal RNA modelling. The *T. brucei* rRNAs were modelled using a workflow combining *ab initio* and homology approaches and can be found in the online Supplementary Information section, schematized in Supplementary Fig. 8. The 18S rRNA (SSU rRNA) was modelled based on the *T. thermophila* 40S crystal structures[5]. The LSU rRNA was modelled based on the yeast ribosome crystal structure[2]. The LSU and SSU rRNAs were modelled by homology using the S2S tool[21]. *T. brucei* and other kinetoplastid rRNA sequences were fetched from the SILVA webserver (http://www.arb-silva.de/)[53]
and the comparative RNA website and project (CRW site) (http://www.rna.icmb.utexas.edu/)[54]. *T. brucei* expansion segments were modelled mostly *ab initio* by performing a covariation analysis on each one of these expansion segments sequences against a number of orthologous rRNA sequences found in other related kinetoplastids, using the mlocarna algorithm[55]. This allowed deriving a consensus secondary structure that was used in the Assemble tool[22] and three-dimensional models were generated for each expansion segment and fitted into the EM map.

Modelling of ribosomal proteins. The modelling workflow can be found in the Supplementary Information, and it is schematized in Supplementary Fig. 10. The ribosomal proteins were modelled by homology to mainly three crystal structures, the yeast ribosome[2], the *T. thermophila* 40S subunit in complex with eIF1 (ref. 5), and the 60S subunit with eIF6 (ref. 6). The *T. brucei* ribosomal protein sequences, used as inputs for the homology modelling, were taken from the National Center for Biotechnology Information (NCBI) protein databases (http://www.ncbi.nlm.nih.gov/). Swiss-Model web server was used for protein structure homology modelling (http://swissmodel.expasy.org/)[56,57]. We used the Phyre and the Phyre2 web servers[58] for the *ab initio* modelling of the missing fragments and extensions into the density map. We assessed the structure of each model using the MolProbity web server (http://molprobity.biochem.duke.edu/)[59,60].

Molecular dynamics flexible fitting method. A detailed description of the MDFF method can be found in the Supplementary Information. The initial system was prepared for MDFF[88] using Visual Molecular Dynamics (VMD)[61] and run in the Not Another Molecular Dynamics (NAMD) program[62] for 1.5 ns of simulation time.

31. Kaminsky, R., Beaudoin, E. & Cunningham, I. Cultivation of the life cycle stages of *Trypanosoma brucei* sspp. *Acta Trop.* **45,** 33–43 (1988).
32. Gómez, E. B., Medina, G., Ballesta, J. P., Levin, M. J. & Téllez-Iñón, M. T. Acidic ribosomal P proteins are phosphorylated in *Trypanosoma cruzi*. *Int. J. Parasitol.* **31,** 1032–1039 (2001).
33. Grassucci, R. A., Taylor, D. J. & Frank, J. Preparation of macromolecular complexes for cryo-electron microscopy. *Nature Protocols* **2,** 3239–3246 (2007).
34. Dubochet, J. *et al.* Cryo-electron microscopy of vitrified specimens. *Q. Rev. Biophys.* **21,** 129–228 (1988).
35. Wagenknecht, T., Frank, J., Boublik, M., Nurse, K. & Ofengand, J. Direct localization of the tRNA–anticodon interaction site on the *Escherichia coli* 30 S ribosomal subunit by electron microscopy and computerized image averaging. *J. Mol. Biol.* **203,** 753–760 (1988).
36. Lei, J. & Frank, J. Automated acquisition of cryo-electron micrographs for single particle reconstruction on an FEI Tecnai electron microscope. *J. Struct. Biol.* **150,** 69–80 (2005).
37. Frank, J. *et al.* SPIDER and WEB: processing and visualization of images in 3D electron microscopy and related fields. *J. Struct. Biol.* **116,** 190–199 (1996).
38. Shaikh, T. R. *et al.* SPIDER image processing for single-particle reconstruction of biological macromolecules from electron micrographs. *Nature Protocols* **3,** 1941–1974 (2008).

©2013 Macmillan Publishers Limited. All rights reserved

39. Scheres, S. H. W. *et al.* Disentangling conformational states of macromolecules in 3D-EM through likelihood optimization. *Nature Methods* **4,** 27–29 (2007).
40. Scheres, S. H. W., Nuñez-Ramirez, R., Sorzano, C. O. S., Carazo, J. M. & Marabini, R. Image processing for electron microscopy single-particle analysis using Xmipp. *Nature Protocols* **3,** 977–990 (2008).
41. Heymann, J. B. Bsoft: image and molecular processing in electron microscopy. *J. Struct. Biol.* **133,** 156–169 (2001).
42. Heymann, J. B., Cardone, G., Winkler, D. C. & Steven, A. C. Computational resources for cryo-electron tomography in Bsoft. *J. Struct. Biol.* **161,** 232–242 (2008).
43. Pintilie, G., Zhang, J., Goddard, T., Chiu, W. & Gossard, D. Quantitative analysis of cryo-EM density map segmentation by watershed and scale-space filtering, and fitting of structures by alignment to regions. *J. Struct. Biol.* **170,** 427–438 (2010).
44. Pettersen, E. F. *et al.* UCSF Chimera—a visualization system for exploratory research and analysis. *J. Comput. Chem.* **13,** 1605–1612 (2004).
45. Baker, M. L., Yu, Z., Chiu, W. & Bajaj, C. Automated segmentation of molecular subunits in electron cryomicroscopy density maps. *J. Struct. Biol.* **156,** 432–441 (2006).
46. Yu, Z. & Bajaj, C. Automatic ultrastructure segmentation of reconstructed cryoEM maps of icosahedral viruses. *IEEE Trans. Image Process.* **14,** 1324–1337 (2005).
47. Zhang, Q., Bettadapura, R. & Bajaj, C. Macromolecular structure modeling from 3D EM using VolRover 2.0. *Biopolymers* **97,** 709–731 (2012).
48. Zeyen, Y. & Bajaj, C. Computational approaches for automatic structural analysis of large biomolecular complexes. *IEEE/ACM Trans. Comput. Biol. Bioinform.* **5,** 568–582 (2008).
49. Penczek, P. A., Yang, C., Frank, J. & Spahn, C. M. Estimation of variance in single-particle reconstruction using the bootstrap technique. *J. Struct. Biol.* **154,** 168–183 (2006).
50. Zhang, W., Kimmel, M., Spahn, C. M. & Penczek, P. A. Heterogeneity of large macromolecular complexes revealed by 3D cryo-EM variance analysis. *Structure* **16,** 1770–1776 (2008).
51. Liao, H. Y. & Frank, J. Classification by bootstrapping in single particle methods. *IEEE Int. Symp. Biom. Imaging* 169–172 (2010).
52. Simonetti, A. *et al.* Structure of the 30S translation initiation complex. *Nature* **455,** 416–420 (2008).
53. Pruesse, E. *et al.* SILVA: a comprehensive online resource for quality checked and aligned ribosomal RNA sequence data compatible with ARB. *Nucleic Acids Res.* **35,** 7188–7196 (2007).
54. Cannone, J. J. *et al.* The Comparative RNA Web (CRW) Site: an online database of comparative sequence and structure information for ribosomal, intron, and other RNAs. *Bioinformatics* **3,** 2 (2002).
55. Will, S., Joshi, T., Hofacker, I. L., Stadler, P. F. & Backofen, R. LocARNA-P: accurate boundary prediction and improved detection of structural RNAs. *RNA* **18,** 900–914 (2012).
56. Arnold, K., Bordoli, L., Kopp, J. & Schwede, T. The SWISS-MODEL workspace: a web-based environment for protein structure homology modelling. *Bioinformatics* **22,** 195–201 (2006).
57. Kiefer, F., Arnold, K., Künzli, M., Bordoli, L. & Schwede, T. The SWISS-MODEL Repository and associated resources. *Nucleic Acids Res.* **37,** D387–D392 (2009).
58. Kelley, L. A. & Sternberg, M. J. E. Protein structure prediction on the Web: a case study using the Phyre server. *Nature Protocols* **4,** 363–371 (2009).
59. Chen, V. B. *et al.* MolProbity: all-atom structure validation for macromolecular crystallography. *Acta Crystallogr. D* **66,** 12–21 (2010).
60. Davis, I. W. *et al.* MolProbity: all-atom contacts and structure validation for proteins and nucleic acids. *Nucleic Acids Res.* **35,** W375–W383 (2007).
61. Humphrey, W., Dalke, A. & Schulten, K. VMD: visual molecular dynamics. *J. Mol. Graph. Model.* **1,** 33–8–27-8 (1996).
62. Phillips, J. C. *et al.* Scalable molecular dynamics with NAMD. *J. Comput. Chem.* **26,** 1781 (2005).

©2013 **Macmillan Publishers Limited. All rights reserved**

Trajectories of the ribosome as a Brownian nanomachine

Ali Dashti[a,1], Peter Schwander[a,1], Robert Langlois[b], Russell Fung[a], Wen Li[b], Ahmad Hosseinizadeh[a], Hstau Y. Liao[b], Jesper Pallesen[c,2], Gyanesh Sharma[b,3], Vera A. Stupina[d], Anne E. Simon[d], Jonathan D. Dinman[d], Joachim Frank[b,c,4], and Abbas Ourmazd[a,1,4]

[a]Department of Physics, University of Wisconsin, Milwaukee, WI 53211; [b]Department of Biochemistry and Molecular Biophysics, and [c]Howard Hughes Medical Institute, Columbia University, New York, NY 10032; and [d]Department of Cell Biology and Molecular Genetics, University of Maryland, College Park, MD 20742

Contributed by Joachim Frank, October 8, 2014 (sent for review September 10, 2014)

A Brownian machine, a tiny device buffeted by the random motions of molecules in the environment, is capable of exploiting these thermal motions for many of the conformational changes in its work cycle. Such machines are now thought to be ubiquitous, with the ribosome, a molecular machine responsible for protein synthesis, increasingly regarded as prototypical. Here we present a new analytical approach capable of determining the free-energy landscape and the continuous trajectories of molecular machines from a large number of snapshots obtained by cryogenic electron microscopy. We demonstrate this approach in the context of experimental cryogenic electron microscope images of a large ensemble of nontranslating ribosomes purified from yeast cells. The free-energy landscape is seen to contain a closed path of low energy, along which the ribosome exhibits conformational changes known to be associated with the elongation cycle. Our approach allows model-free quantitative analysis of the degrees of freedom and the energy landscape underlying continuous conformational changes in nanomachines, including those important for biological function.

cryo-electron microscopy | elongation cycle | manifold embedding | nanomachines | translation

Ideally, one would like to "see" the conformational changes of a Brownian machine as it traverses its work cycle trajectory over the energy landscape. This information is particularly relevant for a biologically important molecular machine such as the ribosome, which is responsible for protein synthesis in all living cells. During the so-called elongation process, the ribosome repeatedly links an amino acid carried in by transfer RNA (tRNA) to the nascent polypeptide chain, with the choice of amino acid in each cycle dictated by the genetic message on the mRNA. In the eukaryotic ribosome, this process is facilitated by elongation factors eEF1A and eEF2, both GTPases.

It is believed that many intermediate conformational states must be involved in the elongation cycle of the ribosome (1), but the evidence is inferred, albeit from an impressive array of experimental techniques. Both cryogenic electron microscopy (cryo-EM) (2) and X-ray crystallographic approaches (3) have been used to determine the structures of several biochemically "trapped" states along the conformational trajectory. However, it has been pointed out that these likely represent only a fraction of the relevant conformational states, that each biochemically trapped state may correspond to more than one conformational state, and that the observed intermediate structures may be affected by the trapping process itself (1). Powerful algorithms (4, 5) have been used to sort cryo-EM snapshots into a small number of discrete classes, each presumed to represent an intermediate state (6). In some cases, however, snapshots of major ribosomal regions with large conformational flexibility have defied classification into discrete states altogether, even by the most advanced analytical methods (7). Single-molecule FRET experiments have yielded evidence for discrete conformational changes in single, freely equilibrating pretranslocational ribosomes, and provided ensemble averages for such changes, but have been unable to provide data for short-lived intermediates (8, 9).

In a groundbreaking study, Fischer and coworkers (10) used cryo-EM to determine the structure and occupancy of different ribosomal conformational states as a function of time, obtaining the free-energy landscape through the Boltzmann factor (11). The specific process studied was that of back-translocation, a slow process (on the order of 30 min), in which the elongation cycle is partially reversed through interaction with another GTPase: LepA. Cryo-EM snapshots were classified in a hierarchical series of supervised (reference-based) steps to yield multiple conformations, differing mainly in specific preselected features: degree of intersubunit rotation, tRNA positions, and degree of head swivel of the small subunit—all changes known to be associated with the elongation work cycle of the ribosome.

Significance

Many functions in the cell are performed by Brownian machines, macromolecular assemblies that use energy from the thermal environment for many of the conformational changes involved in their work cycles. Here we present a new approach capable of mapping the continuous motions of such nanomachines along their trajectories in the free-energy landscape and demonstrate this capability in the context of experimental cryogenic electron microscope snapshots of the ribosome, the nanomachine responsible for protein synthesis in all living organisms. We believe our approach constitutes a universal platform for the analysis of free-energy landscapes and conformational motions of molecular nanomachines and their dependencies on temperature, buffer conditions, and regulatory factors.

Author contributions: A.D., P.S., and A.O. designed research; R.F., A.H., V.A.S., A.E.S., and J.D.D. contributed new reagents/analytic tools; A.D., P.S., R.L., W.L., H.Y.L., J.P., J.F., and A.O. analyzed data; J.F. and A.O. wrote the paper; A.D. and A.O. implemented algorithms; P.S. implemented geometry transformations and algorithms; R.L. helped prepare movies and geometry transformations; R.F. and A.H. provided NLSA and parameter-fitting codes and advice; W.L. interpreted movies; H.Y.L. and J.P. preprocessed ribosome data; G.S. performed EM experiments and particle verification; V.A.S. purified yeast ribosomes; A.E.S. and J.D.D. directed ribosome purification; and J.F. directed EM experiments and interpreted movies.

The authors declare no conflict of interest.

Freely available online through the PNAS open access option.

Data deposition: The 50 3D maps generated along the free-energy trajectory (clockwise in Fig. 3B) have been deposited in the Electron Microscopy Data Bank (accession no. EMD-6044).

[1]A.D., P.S., and A.O. contributed equally to this work.

[2]Present address: 35 W. 127th St., Apt. 1, New York, NY 10027.

[3]Present address: Sir Mortimer B Davis Jewish General Hospital, Montreal, QC, Canada H3T 1E2.

[4]To whom correspondence may be addressed. Email: jf2192@columbia.edu or ourmazd@uwm.edu.

This article contains supporting information online at www.pnas.org/lookup/suppl/doi:10.1073/pnas.1419276111/-/DCSupplemental.

There is general agreement, however, that the value of any approach capable of revealing conformational changes along specific trajectories is determined by the extent to which it provides unbiased quantitative insight into the conformational degrees of freedom, the ensemble kinetics, and the free-energy landscape traversed by the system under examination (1, 12, 13).

Here, we present a new analytical approach capable of mapping continuous conformational changes of nanomachines along any trajectory in the energy landscape, without timing information, supervision, or templates. Our unbiased approach determines the number of degrees of freedom exercised during the observations, the energy landscape explored by the nanomachine, the energy trajectory traversed during the work cycle, and the continuous conformational changes associated with the work cycle. These novel capabilities constitute a powerful platform for quantitative study of the conformational and energy trajectories of nanomachines, including those engaged in a wide range of important biological processes. As reported below, the first application of this approach to an experimental cryo-EM dataset of ribosomes purified from yeast not engaged in translation reveals a closed trajectory of minimum energy along which the ribosomes appear to idle with conformational changes reminiscent of those observed during the protein elongation cycle.

Conceptual Outline

A flow diagram is shown in Fig. 1, with a more technical description following the conceptual outline. Details are presented in the *SI Text*. As used, for example, in early approaches to image classification in electron microscopy (14), a snapshot can be represented as a vector in high-dimensional space by regarding the intensity value at each pixel of the snapshot as a component of the vector. Distance is a measure of similarity between the snapshots in this space. A collection of snapshots produces one or more data clouds, with discrete conformations producing separate clouds. We have previously shown that this approach is able to determine 3D structure from a collection of ultra-low signal 2D snapshots of unknown orientation (15–18), and, in the presence of orientational heterogeneity and defocus variations, distinguish between discrete conformations with best-in-class performance (17).

Here, we are concerned with continuous conformational changes, which produce correlations among the points representing the snapshots. These correlations define a hypersurface—a manifold—a concept found useful also in a previous approach to heterogeneity in cryo-EM (19). Using advanced machine-learning methods, we have demonstrated the ability to identify such manifolds in the presence of overwhelming noise and defocus variations (17).

Our approach begins with determining the orientations of the snapshots by any of a number of approaches (17), in this instance by a standard method used in electron microscopy (20). As shown previously (17, 19), this can be achieved without taking conformational heterogeneity into account, because the effects of orientational change dominate. Snapshots lying within a tight orientational aperture are then selected, and the manifold spanned by them is determined. This step in the analysis yields the conformational manifold (spectrum) in the selected viewing (projection) direction. Of course, the conformational changes may be governed by more than one parameter. In general, therefore, the conformational manifold is best described by a set of orthogonal coordinates. Such a description can be achieved by one of many well-established machine-learning techniques, which also reveal the intrinsic dimensionality of the manifold, and hence the number of degrees of freedom exercised by the system under observation.

Unfortunately, it is not possible to determine the conformational changes from such a description, because the local rates of change in multidimensional manifolds obtained from machine-learning techniques are, in general, unknown (21, 22) and cannot be easily related to the underlying changes in the system under observation (15, 23). To overcome this well-known difficulty, we introduce an additional step, in which the cloud of points is mapped to another coordinate system, where the local rates of change can be determined exactly, and related to the underlying conformational changes. This step leads to a representation of conformational change in terms of a universal parameter

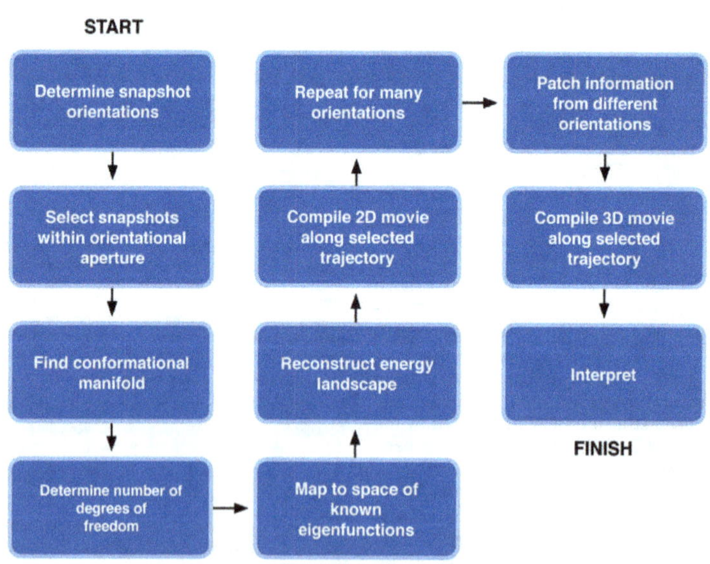

Fig. 1. Flowchart representation of the approach used to determine the free-energy landscape, work-cycle trajectory, and associated continuous conformational changes from experimental snapshots of nanomachines in unknown orientational and conformational states.

(metric). The density of points in this space can now be related to the energy landscape sampled by the system through the Boltzmann factor $e^{-\Delta G/k_B T}$ (10), with ΔG denoting the change in the Gibbs free energy, k_B the Boltzmann constant, and T the temperature. The locus of minimum energy in this landscape represents the trajectory traversed by the machine during its work cycle. In each viewing direction, 2D movies can be compiled to reveal the conformational changes along the path of minimum energy or indeed any chosen trajectory. 3D movies can be compiled by stepping along such a trajectory and, in each step, performing a 3D reconstruction by integrating the information from many different viewing directions.

Analytical Procedure

The analysis is illustrated below in more technical terms with reference to a set of 849,914 experimental cryo-EM snapshots of 80S ribosomes from yeast, obtained in the course of a study of translational initiation by a plant virus (*SI Text*). The procedure of purification rendered the ribosome free of mRNA and most of the tRNAs.

Manifolds defined by data clouds are, in general, nonlinear: the data lie on intrinsically curved rather than flat hypersurfaces. Such manifolds can be identified and described (embedded) by graph-theoretic machine-learning techniques often used for dimensionality reduction. The so-called diffusion map embedding algorithm used in our work yields a description of a curved manifold in terms of the orthogonal eigenfunctions (more precisely eigenvectors) of known operators, specifically the Laplace–Beltrami operator (21, 22, 24) (*SI Text*). We use a specially developed kernel to deal with the substantial defocus variations encountered in cryo-EM data (17) (*SI Text*). In Fig. 2, we show representative cryo-EM snapshots (Fig. 2*A*) and a typical data manifold (Fig. 2*B*) described in terms of the first two eigenvectors of the Laplace–Beltrami operator. [In general, there is no simple correspondence between the eigenvectors obtained from this nonlinear analysis, and those obtained with linear approaches, such as principal component analysis (PCA) and singular value decomposition (SVD).] The manifold produced by the experimental snapshots is, in fact, nonlinear, with an intrinsic dimensionality of five (*SI Text*).

Fig. 2 describes the data in terms of the eigenfunctions of the Laplace–Beltrami operator with respect to an unknown metric (22). The absence of information on the metric precludes a consistent description of the conformational changes in terms of a known universal parameter. We solve this problem by mapping the manifold to another space, in which the eigenfunctions are known exactly. For this, we use an approach used in nonlinear Laplacian spectral analysis (NLSA) (25), a technique capable of performing SVD on nonlinear manifolds (nonlinear SVD for short) (*SI Text*). Briefly, one considers a collection of supervectors formed by concatenating snapshots falling within a window moving over the data vectors. The snapshots within each supervector are ordered according to the projections of the points representing them on a line through the origin, making an angle θ with, say, the horizontal axis (Fig. 2 and *SI Text*). By virtue of this projection, the arrangement picks out the conformational evolution along the selected line characterized by θ, with the random ordering of conformational changes along other directions assuming the character of noise. Nonlinear SVD (25) is then used to extract characteristic images (topos) and their evolutions (chronos) from these supervectors (*SI Text*). Each topo/chrono pair constitutes an element of a biorthogonal decomposition of the conformational changes along the selected line. Noise-reduced snapshots can be reconstructed from the topo/chrono pairs with significant (above-noise) singular values and embedded to obtain the manifold characteristic of the conformational changes along the selected line (*SI Text*). By construction, this manifold is one-dimensional, described in terms of known eigenfunctions, viz. $\cos(k\pi\tau)$, $k = 1, 2, 3, \ldots$, and governed by the single parameter τ (*SI Text*).

Given a sufficiently dense collection of radial lines, each making an angle θ with the horizontal axis, the conformational changes can be described in any direction in the multidimensional space of conformations by the parameter $\tau(\theta)$. The mapping from a space of unknown metric and hence eigenfunctions to one characterized by known eigenfunctions governed by a single parameter $\tau(\theta)$ allows a consistent description of the conformational changes (*SI Text* and Fig. S1). This description of conformational changes in a given projection direction can be related to descriptions in other projection directions by assuming that the same conformational spectrum is viewed in all projection directions. In other words, the histograms of occupancy vs.

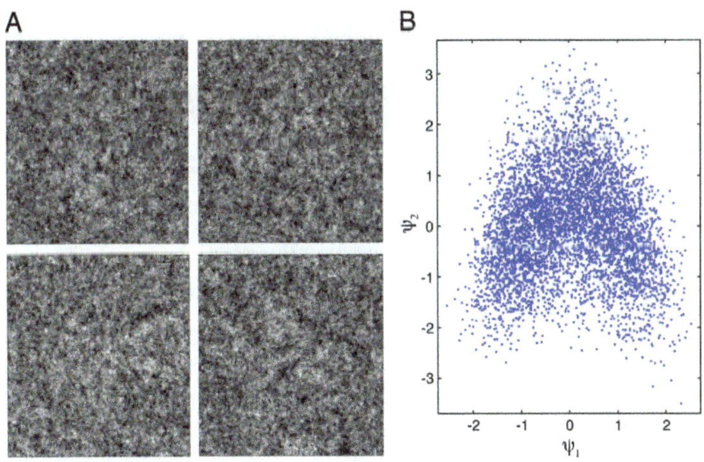

Fig. 2. (*A*) Representative cryo-EM snapshots, and (*B*) 2D view of a typical conformational manifold. The manifold is derived from a measure of similarity among ~1,500 cryo-EM snapshots of ribosome particles viewed within a tight orientational aperture. The axes ψ_1,ψ_2 represent the first two eigenvectors obtained by the diffusion map algorithm with a kernel able to deal with defocus variations.

conformational parameter $\tau(\theta)$ for different projection directions all represent the same spectrum of conformations and can thus be equalized. This equalization allows a universal description of the conformational spectrum across all projection directions. For any selected projected direction, the energy landscape along any line can be determined from the density of points along that line via the Boltzmann factor. Energy landscapes of higher dimension (with the dimensionality given by the number of eigenvectors) can be reconstructed by a tomographic extension of this approach (Fig. 3 and *SI Text*). Here, we concentrate on the first two eigenvectors, which contain some of the most interesting biological information (see below).

With this information in hand, the structural evolution of the system, including 2D movies of the conformational changes and the system's thermodynamic properties, can be quantitatively investigated in any viewing direction (*SI Text* and Movie S1). For any point on the energy landscape, a 3D structure map can be compiled by integrating the 2D information from many viewing directions into a 3D representation (Fig. 4) (*SI Text* and Movies S2–S4). Thus, 3D movies can be compiled by stepping along any selected trajectory in the energy landscape and reconstructing a 3D structure map in each step. The conformational changes along the closed minimum-energy trajectory of Fig. 3*B* are summarized in Fig. 3 and exemplified in Fig. 4. The associated movies (Movies S2–S4) present the ribosome as it evolves, with each movie corresponding to a different viewing direction selected for optimal visibility of the biologically relevant domain motions (Fig. 3*A*). We note that, by construction, these movies are based on similarity rather than time. Accordingly, identical trajectories in opposite directions cannot be distinguished.

Results

The trough of minimum free energy is seen to form a closed, roughly triangular path (Fig. 3*B*), with variations of <0.9 kcal/mol ($1.5k_BT$ at room temperature) in the energy positions of the deepest points. The energy difference between the lowest and highest points of the entire landscape, corresponding to states with the highest and lowest nonzero occupancy in the experiment, amount to 3.8 ± 0.65 kcal/mol ($6.5 \pm 1.0k_BT$ at room temperature), with the uncertainty stemming chiefly from the low occupancy of high-energy states. The energy range covered nonetheless indicates that so-called transition states with energies several times the thermal energy have been probed in our experiment.

We examine the 3D movies showing the conformational changes along the closed triangular trajectory, analyzing in detail seven structure maps along the way (*SI Text* and Movie S5). Using rigid-body fitting of domains while making use of a published cryo-EM map of the yeast ribosome (26) for reference, we observe combinations of four motions previously described in the literature and associated with the elongation cycle, as follows: (*i*) ratchet-like intersubunit rotation, a counterclockwise rotation of the small subunit about an axis normal to the subunit interface (Fig. 3*A*, solvent view) (27, 28); (*ii*) rotation (closing movement) of

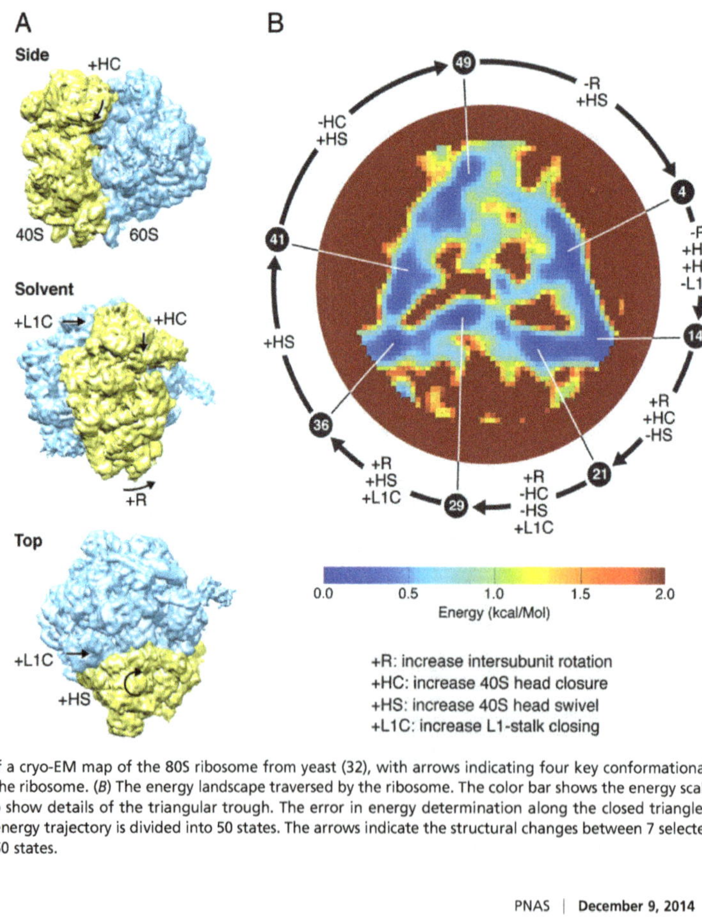

Fig. 3. (*A*) Three views of a cryo-EM map of the 80S ribosome from yeast (32), with arrows indicating four key conformational changes associated with the elongation work cycle of the ribosome. (*B*) The energy landscape traversed by the ribosome. The color bar shows the energy scale. The energy range has been truncated at 2 kcal/mol to show details of the triangular trough. The error in energy determination along the closed triangle is 0.05 kcal/mol. The roughly triangular minimum free-energy trajectory is divided into 50 states. The arrows indicate the structural changes between 7 selected states, each identified by its place in the sequence of 50 states.

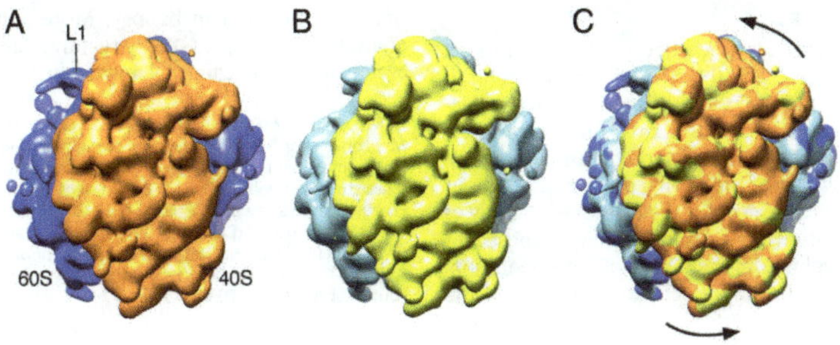

Fig. 4. Example of conformational changes along the trajectory: ratchet-like motion. (*A*) Unrotated ribosome, map 14 in Fig. 3*B*. (*B*) Maximally rotated ribosome, map 36. (*C*) Superposition of the two maps. The full set of frames showing continuous conformational changes is shown in three common viewing directions in Movies S2–S4.

the L1 stalk toward the intersubunit space (Fig. 3*A*, solvent and top views) (29); (*iii*) small subunit head swivel, a rotation of the subunit head about its long axis (Fig. 3*A*, top view) (30); and (*iv*) small subunit head closure, a "nodding" motion of the head (Fig. 3*A*, side and solvent views) (31). The first three motions are correlated and known to be associated with the process of mRNA–tRNA translocation (30, 32), whereas the fourth is known to accompany the selection of cognate aminoacyl tRNA during the process of decoding (31).

Proceeding clockwise along the minimum free-energy pathway, these motions occur in the sequence encountered during the elongation cycle: starting at the apex of the triangle and moving to the right side of the base (Fig. 3*B*), we see the small subunit head closure as a prominent motion, corresponding to the tRNA selection step. Next, proceeding along the base of the triangle, three coordinated movements of small subunit body, head swivel, and L1 stalk occur—all expected during translocation—while the small subunit head remains in the closed position. Finally, moving up from the left side of the base to the apex, we observe the reversal of head closing (the conformational change associated with the disengagement of the mRNA from the decoding center), as well as the reversal of the intersubunit motion.

Discussion

The energy landscape constructed by the analysis of the ribosome ensemble shows multiple interconnected paths of relatively low energy. Of these, a path of roughly triangular shape clearly stands out, encompassing roughly one third of the molecules. Comparison of 3D reconstructions from data at selected points along this path reveals conformational changes previously observed in the elongation cycle of fully programmed, translating ribosomes. Because the ribosomes captured in our snapshots were not engaged in protein synthesis, lacking mRNA and aminoacylated tRNAs, we must assume that their idle motions in the thermal environment sample the conformational space permitted by the degrees of freedom, thus exhibiting the conformational changes that would be productive in the presence of the ligands of the translational machinery (mRNA, aa-tRNA, eEF2, and eEF1A). Indeed, idling of the pretranslocational ribosome in the absence of elongation factors along the direction of the most prominent conformational changes (intersubunit motion and opening/closing of L1 stalk) has been previously observed by single-molecule FRET (8, 9) and cryo-EM (33). Idling of the empty ribosome along the same path has also been inferred from a series of X-ray structures (34). It remains to be established by the approach outlined in this paper whether fully programmed ribosomes follow the same path identified here. As to the other paths traversing the free-energy landscape (Fig. 3*B*), it is tempting to speculate that they may represent alternative routes of the molecular machine under different buffer and temperature conditions.

We now discuss the salient features of our approach. First, the usefulness of movies of Brownian machines has been rightly questioned, because each trajectory of a single machine is strongly influenced by stochastic factors and thus is unique (1). The movies presented here, however, integrate information from a large ensemble of snapshots, each stemming from an object viewed only once. By using manifolds to capture the properties of the entire dataset, and nonlinear singular value analysis to suppress noise, our approach offers an efficient means for extracting the ensemble kinetics, i.e., the information common to a collection of objects, each viewed in an initially unknown orientational and conformational state.

Most successful experimental studies of the conformational spectra of biological machines have been hitherto restricted to sorting snapshots into a small number of classes (10, 11), using templates in some form, relying on timing information, or a combination of these tools (10). In contrast, our approach naturally yields detailed conformational trajectories and energy landscapes without a priori information or assumptions and at moderate computational expense (*SI Text*).

Of course, the maximum number of detectable conformational states is limited by various factors. These factors include the frequency (via the Shannon–Nyquist theorem) with which the conformational spectrum has been sampled, the information content of the individual snapshots, and, ultimately, the atomic nature of the object itself. In our analysis, the conformational signal stems primarily from the most heavily sampled projection directions, in each of which the conformational spectrum is randomly sampled by up to 5,000 snapshots. The number of meaningfully distinct conformational states is governed by the signal-to-noise ratio, which determines the statistical confidence with which neighboring states can be distinguished. In this study, the requirement for neighboring conformational states to be separated by 3 SDs (3 σ) means that ~50 conformational states can be distinguished (*SI Text*). This number is about an order of magnitude larger than previously achieved without timing information or templates. In combination with the recent availability of large cryo-EM datasets with near-atomic resolution (7, 35), our approach promises the possibility to extract conformational information with unprecedented detail.

In recent years, increasing computing power has fueled efforts to simulate the conformational trajectories of molecular machines, particularly the ribosome (36), by molecular dynamics. Experimental results obtained with the new approach presented here offer the promise to guide these efforts and provide the means for verifying important modeling assumptions.

Conclusions

Given our increasingly detailed knowledge of the structure of biological machines in general, and the ribosome in particular, it has been suggested that the "heroic age" of biostructure determination is drawing to a close and that further progress requires the study of ensemble kinetics and conformational energy landscapes (1). As shown by the results presented here, our approach offers a powerful platform for the quantitative study of continuous conformational motions over the energy landscapes traversed by biological machines in the course of their operation. This development may significantly accelerate the dawn of the predicted new era in biology. More generally, the way to the quantitative study of the structure, kinetics, and operational cycles of nanomachines is now open.

ACKNOWLEDGMENTS. We thank D. Giannakis, K. Sanbonmatsu, and M. Schmidt for helpful comments on an early version of the manuscript and M. Thomas-Baum for assistance in the preparation of the illustrations. This research was supported by US Department of Energy, Office of Science, Basic Energy Sciences Award DE-FG02-09ER16114 (to A.O.); the Howard Hughes Medical Institute and National Institutes of Health Grants R01 GM29169 and GM55440 (to J.F.); and National Science Foundation MCB Grant 1157906 (to A.E.S.). The publication of this work was supported by US National Science Foundation Award STC 1231306.

1. Moore PB (2012) How should we think about the ribosome? *Annu Rev Biophys* 41(1): 1–19.
2. Villa E, et al. (2009) Ribosome-induced changes in elongation factor Tu conformation control GTP hydrolysis. *Proc Natl Acad Sci USA* 106(4):1063–1068.
3. Korostelev A, Ermolenko DN, Noller HF (2008) Structural dynamics of the ribosome. *Curr Opin Chem Biol* 12(6):674–683.
4. Scheres SHW (2010) Maximum-likelihood methods in cryo-EM. Part II: Application to experimental data. *Methods Enzymol* 482:295–320.
5. Scheres SH (2012) A Bayesian view on cryo-EM structure determination. *J Mol Biol* 415(2):406–418.
6. Frank J (2012) Intermediate states during mRNA-tRNA translocation. *Curr Opin Struct Biol* 22(6):778–785.
7. Amunts A, et al. (2014) Structure of the yeast mitochondrial large ribosomal subunit. *Science* 343(6178):1485–1489.
8. Fei J, Kosuri P, MacDougall DD, Gonzalez RL, Jr (2008) Coupling of ribosomal L1 stalk and tRNA dynamics during translation elongation. *Mol Cell* 30(3):348–359.
9. Cornish PV, Ermolenko DN, Noller HF, Ha T (2008) Spontaneous intersubunit rotation in single ribosomes. *Mol Cell* 30(5):578–588.
10. Fischer N, Konevega AL, Wintermeyer W, Rodnina MV, Stark H (2010) Ribosome dynamics and tRNA movement by time-resolved electron cryomicroscopy. *Nature* 466(7304):329–333.
11. Agirrezabala X, et al. (2012) Structural characterization of mRNA-tRNA translocation intermediates. *Proc Natl Acad Sci USA* 109(16):6094–6099.
12. Frank J, Gonzalez RL, Jr (2010) Structure and dynamics of a processive Brownian motor: The translating ribosome. *Annu Rev Biochem* 79(9):381–412.
13. Munro JB, Sanbonmatsu KY, Spahn CMT, Blanchard SC (2009) Navigating the ribosome's metastable energy landscape. *Trends Biochem Sci* 34(8):390–400.
14. van Heel M, Frank J (1981) Use of multivariate statistics in analysing the images of biological macromolecules. *Ultramicroscopy* 6(2):187–194.
15. Giannakis D, Schwander P, Ourmazd A (2012) The symmetries of image formation by scattering. I. Theoretical framework. *Opt Express* 20(12):12799–12826.
16. Schwander P, Giannakis D, Yoon CH, Ourmazd A (2012) The symmetries of image formation by scattering. II. Applications. *Opt Express* 20(12):12827–12849.
17. Schwander P, Fung R, Ourmazd A (2014) Conformations of macromolecules and their complexes from heterogeneous datasets. *Philos Trans R Soc Lond B Biol Sci* 369(1647): 20130567.
18. Hosseinizadeh A, et al. (2014) High-resolution structure of viruses from random diffraction snapshots. *Philos Trans R Soc Lond B Biol Sci* 369(1647):20130326.
19. Fu J, Gao H, Frank J (2007) Unsupervised classification of single particles by cluster tracking in multi-dimensional space. *J Struct Biol* 157(1):226–239.
20. Penczek PA, Grassucci RA, Frank J (1994) The ribosome at improved resolution: New techniques for merging and orientation refinement in 3D cryo-electron microscopy of biological particles. *Ultramicroscopy* 53(3):251–270.
21. Belkin M, Niyogi P (2003) Laplacian eigenmaps for dimensionality reduction and data representation. *Neural Comput* 15(6):1373–1396.
22. Coifman RR, et al. (2005) Geometric diffusions as a tool for harmonic analysis and structure definition of data: Diffusion maps. *Proc Natl Acad Sci USA* 102(21):7426–7431.
23. Coifman RR, Shkolnisky Y, Sigworth FJ, Singer A (2010) Reference free structure determination through eigenvectors of center of mass operators. *Appl Comput Harmon Anal* 28(3):296–312.
24. Coifman RR, Kevrekidis IG, Lafon S, Maggioni M, Nadler B (2008) Diffusion maps, reduction coordinates, and low dimensional representation of stochastic systems. *Multiscale Model Simulation* 7(2):842–864.
25. Giannakis D, Majda AJ (2012) Nonlinear Laplacian spectral analysis for time series with intermittency and low-frequency variability. *Proc Natl Acad Sci USA* 109(7):2222–2227.
26. Taylor DJ, et al. (2009) Comprehensive molecular structure of the eukaryotic ribosome. *Structure* 17(12):1591–1604.
27. Frank J, Agrawal RK (2000) A ratchet-like inter-subunit reorganization of the ribosome during translocation. *Nature* 406(6793):318–322.
28. Budkevich T, et al. (2011) Structure and dynamics of the mammalian ribosomal pretranslocation complex. *Mol Cell* 44(2):214–224.
29. Valle M, et al. (2003) Locking and unlocking of ribosomal motions. *Cell* 114(1):123–134.
30. Ratje AH, et al. (2010) Head swivel on the ribosome facilitates translocation by means of intra-subunit tRNA hybrid sites. *Nature* 468(7324):713–716.
31. Ogle JM, Carter AP, Ramakrishnan V (2003) Insights into the decoding mechanism from recent ribosome structures. *Trends Biochem Sci* 28(5):259–266.
32. Frank J, Gao H, Sengupta J, Gao N, Taylor DJ (2007) The process of mRNA-tRNA translocation. *Proc Natl Acad Sci USA* 104(50):19671–19678.
33. Agirrezabala X, et al. (2008) Visualization of the hybrid state of tRNA binding promoted by spontaneous ratcheting of the ribosome. *Mol Cell* 32(2):190–197.
34. Zhang W, Dunkle JA, Cate JHD (2009) Structures of the ribosome in intermediate states of ratcheting. *Science* 325(5943):1014–1017.
35. Liao M, Cao E, Julius D, Cheng Y (2013) Structure of the TRPV1 ion channel determined by electron cryo-microscopy. *Nature* 504(7478):107–112.
36. Whitford PC, Blanchard SC, Cate JHD, Sanbonmatsu KY (2013) Connecting the kinetics and energy landscape of tRNA translocation on the ribosome. *PLOS Comput Biol* 9(3):e1003003.

Supporting Information

Dashti et al. 10.1073/pnas.1419276111

SI Text

Cryo-EM Dataset.

Description. The dataset consisted of 849,914 projection images ("particles") of the yeast 80S ribosome embedded in a thin layer of ice, obtained with a transmission electron microscope, as detailed below.

Ribosome preparation. Purification of active yeast 80S ribosomes using cysteine-charged sulfolink columns was performed as described in Leshin et al. (1).

Electron microscopy. A carbon-coated Quantifoil 2/4 grid (Quantifoil Micro Tools) was prepared following standard cryo-EM procedures (2). Grids were glow-discharged for 25 s in an H_2, O_2 atmosphere, with 15 W and 25 s using a plasma cleaner (Gatan Solarus Model 950 Advanced Plasma System; Gatan) to render them hydrophilic. An aliquot of 4 µL containing the sample at a concentration of 50 nM was pipetted onto the carbon-coated grid. The grid was blotted in 100% humidity at 4 °C for 5 s and plunge-frozen into liquid ethane cooled by liquid nitrogen using the Vitrobot (FEI). Images were recorded with a 4k × 4k Tietz CCD detector mounted on an FEI Tecnai Polara electron microscope operating at 300 kV and a nominal magnification of 80,000× under low-dose conditions (~20 $e^-/Å^2$). The resulting pixel size was 1.5 Å on the object scale. For automated data collection, the program AutoEMation (3) was used.

Image preprocessing and orientation determination. After visual inspection and evaluation of the micrographs and their power spectra, ~4,700 micrographs were selected for further analysis. The total number of computer-selected particle candidates was ~1,100,000, brought down to 849,914 by manual verification. Orientations of particles were determined by iterative projection matching using the pySPIDER (4) software package with an angular step size decreasing from 15° to 0.5°. Electron Microscopy Data Bank (EMDB) 1067 (5), low pass-filtered to 70Å, was used as the initial reference. Contrast transfer functions (CTFs) were determined for each micrograph in the standard way (6). Before further analysis, the snapshots within a small aperture over the two-sphere of projection directions were brought into in-plane orientational registry.

Manifold Embedding by Diffusion Map.

Outline. Common to all manifold embedding algorithms is the representation of manifolds in a low-dimensional Euclidean space (7–12). The family of diffusion maps, introduced by Coifman and coworkers (11, 12), establishes a rigorous link between the eigenfunctions (more precisely eigenvectors) ψ_i of the Laplace–Beltrami operator with respect to the Riemannian metric g

$$\Delta_g = -\frac{1}{\sqrt{\det(g)}}\partial_i\left[\sqrt{\det(g)}g^{ij}\partial_j\right] \quad i,j=1\ldots n, \qquad \text{[S1]}$$

and the similarity between snapshots, as measured by the diffusion distance between them. The intrinsic dimensionality of the manifold, determined by well-established procedures (13), corresponds to the number of degrees of freedom exercised by the system under observation.

Defocus-tolerant kernel. The information content of a cryo-EM snapshot depends on the defocus at which it is obtained. The effect of defocus on the similarity measure between two otherwise identical snapshots must be eliminated. We achieve this by a double-filtering kernel, which corrects the Euclidean distance D_{ij}, as follows:

$$\begin{aligned} D_{ij}^2 &= \sum_{pixels}|\mathrm{PSF}_j\otimes I_i - \mathrm{PSF}_i\otimes I_j|^2 \\ &= \sum_{pixels}|\mathrm{PSF}_j\otimes(\mathrm{PSF}_i\otimes P_i) - \mathrm{PSF}_i\otimes(\mathrm{PSF}_j\otimes P_j)|^2. \\ &= \sum_{pixels}|\mathrm{PSF}_i\otimes\mathrm{PSF}_j\otimes\{P_i-P_j\}|^2 \end{aligned} \qquad \text{[S2]}$$

For each snapshot i, I_i represents the image intensity distribution, PSF_i the microscope point-spread function, P_i the projected potential, and ⊗ the convolution operator. This scheme ensures a zero Euclidean distance between two snapshots differing in defocus only. For computational efficiency, the distances are calculated in Fourier space so that convolution becomes multiplication. Let $\tilde{I}_i$ be the Fourier transform of the image and CTF_i the Fourier transform of the point-spread function. The application of Parseval's theorem (14) yields

$$D_{ij}^2 = \sum_q|\mathrm{CTF}_j\cdot\tilde{I}_i - \mathrm{CTF}_i\cdot\tilde{I}_j|^2. \qquad \text{[S3]}$$

One-dimensional manifolds. Snapshots of a system exercising only one degree of freedom give rise to a 1D Riemannian manifold described in terms of the eigenfunctions of the operator

$$\Delta_g = -\frac{1}{\sqrt{g}}\partial\left(\frac{1}{\sqrt{g}}\partial\right). \qquad \text{[S4]}$$

For such manifolds, it is always possible to select a coordinate system in which the metric $g(\tau)=1$. The eigenvalue equation then assumes the simple form

$$\frac{\partial^2\psi(\tau)}{\partial\tau^2} = -\lambda\psi(\tau), \qquad \text{[S5]}$$

whose normalized eigenfunctions satisfying the Neumann boundary conditions are of the form $\sqrt{2}\cos(k\pi\tau)$. It can be easily seen that a representation of a bounded 1D manifold in terms of the first two eigenfunctions ($k=1,2$) is a parabola. Unbounded 1D manifolds are characterized by eigenfunctions of the form $\sqrt{2}\cos(2k\pi\tau)$ and $\sqrt{2}\sin(2k\pi\tau)$. The representation of an unbounded 1D manifold in terms of the first two eigenfunctions is thus a circle. For both types of 1D manifold, each snapshot is associated with a value of τ (modulo 2π for the unbounded type), regardless of the dimensionality of the representation (number of eigenfunctions used).

SVD on Nonlinear Manifolds.

Outline. SVD is a data-analytical approach capable of extracting characteristic images (topos) and their time evolutions (chronos) from large noisy datasets (15, 16). Like PCA, SVD assumes a linear structure in the data. Geometrically, this is tantamount to the data lying on a flat hyperplane (16). Often, however, the data have a nonlinear intrinsic structure and thus lie on a curved hyperplane (nonlinear manifold) (7), precluding a straightforward application of SVD. This limitation has been overcome by a machine-learning technique (NLSA), which is able to perform SVD on nonlinear manifolds (17, 18).

Concatenation. NLSA analysis requires a series of snapshots ordered according to a single parameter. In our approach, the snapshot order is determined by the projected position of the embedded point representing it on a trajectory. This trajectory can constitute a straight line, for example, the ψ_1 or ψ_2 axis above, or any other trajectory of interest. The analysis proceeds as follows. First, ordered (time-lagged) embedding (19–22) is used to obtain c-fold concatenated supervectors X from a dataset consisting of vectors x, with a typical supervector taking the form

$$X_t = \left(x_t, x_{t-\delta t}, \ldots x_{t-(c-1)\delta t}\right). \quad \text{[S6]}$$

The timestamp assigned to each supervector corresponds to the mean of the timestamps of its constituent vectors. Second, graph-based machine learning (9, 11, 23, 24), here the diffusion map algorithm (11, 12), is used to identify the nonlinear data manifold formed by the collection of supervectors. This analysis provides a Euclidean description of the manifold in terms of the eigenfunctions of the Laplace–Beltrami operator, collectively referred to as Ψ below.

Nonlinear SVD. The supervectors X_t are projected onto the manifold to obtain the matrix A

$$A = \mathbf{X}\mu\Psi, \quad \text{[S7]}$$

with $\mathbf{X}$ representing the matrix of supervectors X_t, μ the Riemannian measure of the manifold, and the (empirical) eigenfunction Ψ is a truncated set of the eigenfunctions of the Laplace–Beltrami operator on the manifold (17). This Euclidean description of the nonlinear manifold allows one to analyze the matrix A by standard SVD.

Reconstruction and parameter extraction. The modes (specifically the chronos) obtained from SVD are projected from the space spanned by Ψ back to the time domain, and the topos corresponding to the supervectors are unwrapped to obtain individual snapshots (17). These snapshots reflect the conformational changes along the selected trajectory (e.g., a line in embedding space making an angle θ with the horizontal axis). The embedding of these snapshots by diffusion map produces an excellent 1D manifold, accurately described by eigenfunctions of the form $\sqrt{2}\cos(k\pi\tau)$. The conformational parameter $\tau(\theta)$ is extracted from the 1D manifold using the first three eigenvectors of the 1D manifold (Fig. S1).

The mapping from a description of the data in terms of unknown eigenfunctions of the Laplace–Beltrami operator ψ_i to one in terms of known eigenfunctions of the type $\cos(k\pi\tau)$ is a key feature of our analysis. The resulting conformational parameterization in terms of τ allows a consistent approach to reconstructing the energy landscape and compiling movies along any trajectory, including the minimum-energy path.

Free-Energy Landscape. We use a tomographic approach to retrieve the 2D free-energy landscape from snapshots obtained by NLSA. The dimensionality here refers to the number of eigenfunctions ψ_i considered in the analysis. The density of points $n(\tau,\theta)$ on the 1D manifold stemming from concatenation along a straight line at an angle θ to the horizontal axis is the result of projecting the data cloud onto that line. Thus, $n(\tau,\theta)$ is a measure of the density of points perpendicular to the line at a point characterized by τ. Thus, $n(\tau,\theta)$ is, in effect, the Radon transform of the embedded data cloud in the direction $(\theta+\pi/2)$. The set of $n(\tau,\theta)$ compiled over the range $(0\leq\theta\leq\pi)$ thus represents a collection of Radon transforms of the cloud of points in the space of ψ_i. The inverse Radon transform reproduces the cloud of points, described in terms of the known eigenfunctions above.

This approach was implemented using 180 concatenation directions at 1° intervals. The projected density $n(\tau,\theta)$ was deduced from the $\tau(\theta)$ values as follows. For each direction, the $\tau(\theta)$ values were centered to zero-mean to align the different Radon transforms. The density was obtained by counting the number of points in each of 50 classes. Inversion was performed by filtered back-projection with a Shepp–Logan filter. The energy landscape was derived from the density of points via the Boltzmann factor, $n(\tau,\theta)=e^{-\Delta G/k_BT}$, with ΔG the change in Gibb's free energy and the usual meaning for the other parameters. It can be readily shown that the error in this approach is determined by counting statistics. In our case, this is about 0.05 kcal/mol along the minimum-energy trajectory.

To retrieve n-dimensional energy landscapes (with n being the number of ψ_i included), concatenation is performed along directions defined by an $(n-1)$ sphere.

Compiling Movies and 3D Volumes.

2D movies. We now address the compilation of movies showing the continuous conformational changes as seen in a given projection direction (2D movies). These movies consist of a sequence of NLSA reconstructed snapshots, ordered along the minimum-energy path in the energy landscape (Movies S1–S5). To identify the raw snapshots in the neighborhood of this path, the trajectory is mapped back from space of known eigenfunctions to the space of ψ_i using Eq. **S7** and the procedure described below. NLSA is then applied to the snapshots along the path to generate NLSA reconstructed snapshots as frames of the 2D movie.

The movies were compiled using the MATLAB video writer object, with the color axis for all frames normalized to the minimum and maximum intensities of the ensemble; 12.5 frames are shown per second to produce a 50-frame, 4-s movie.

Patching together information from different orientations. Because the minimum energy trajectory is a closed path (Fig. 3), embedding reconstructed snapshots along this path results in a circular manifold, characterized by a single angular parameter $0\leq\tau<2\pi$, which serves as a conformational parameter. In principle, the parameter τ can be used to establish equivalent conformational classes across projection (viewing) directions and thus reconstruct a 3D volume for each conformational class. However, the zero of the parameter τ in each direction is arbitrary. The resulting ambiguity can be resolved as follows:

i) At each point along a great circle, all snapshots spanning the conformational spectrum are concatenated into a supervector ordered according to increasing τ in that projection direction.

ii) The correlation coefficient between two consecutive supervectors along a great circle is calculated.

iii) The same correlation coefficient is calculated with the snapshot sequence for all possible starting (zero-τ) positions, with and without reversing the sense of the sequence.

iv) The largest correlation identifies the mutually consistent sense and starting point for adjacent conformational spectra. The results are propagated along the great circle by repeating the process above.

v) The conformational sense established along one great circle is propagated to other great circles via their points of intersection with the first.

The above approach is robust and computationally fast.

The rate at which the conformational parameter τ changes along the 1D manifold can vary slightly from projection direction to projection direction due to orientational inhomogeneity of the metric g (Eq. **S4**). We compensate for this effect by equalizing the histogram of the τ parameter in each projection direction to that corresponding to the union of 10 projection directions with the most active conformational spectra (measured by the magnitude of the singular values in the NLSA reconstruction along the selected trajectory). The efficacy of this approach has been validated by simulation.

3D volumes and movies. 3D movies were compiled using University of California, San Francisco (UCSF) Chimera (25), taking orthographic surface representations of output volumes for the 50 reconstructed classes used as input frames. Each volume was first segmented into large and small subunits using Chimera. A Python script was used to capture the sequence of frames from three standard points of view (side view, solvent view of the 40S subunit, and top view; Movies S2–S4, respectively). Movie S1 shows the conformational changes between the seven maps of Fig. 3 as seen from the standard viewing directions.

Computational Details.

Hardware. Computations were performed on the following cluster: 16 CPU nodes, each consisting of two Intel Xeon E5620 CPUs @2.4 GHz with four cores; 48 GB of memory; and 200 GB of local hard drive.

Software. Snapshots were picked and preprocessed using SPIDER (4). The NLSA pipeline, energy landscape reconstruction, and compilation of 2D movies were implemented in MATLAB. 3D reconstruction was performed with the Arachnid software package (4). UCSF Chimera was used for visualization and compilation of 3D movies.

Computational expense and scaling. The NLSA pipeline, an embarrassingly parallelizable problem for different projection directions, was implemented using the MATLAB parallel processing toolbox. NLSA data analysis with 32 MATLAB workers requires ~10 h for 320 projection directions with a total of 184,500 snapshots. The number of snapshots in the most densely populated projection direction is ~6,500. Reconstruction of the energy landscape for all 320 projection directions requires ~7 d on the same cluster. 3D reconstruction of volumes for 50 different conformations can be completed in ~1 h using one cluster node. Due to the nonuniform density of snapshots over the two-sphere, the computational expense of the NLSA varies across projection directions. The data analysis process is thus asynchronous between nodes, with the projection direction with the largest number of snapshots determining the total computation time. Generation of a (k-nearest neighbor) k-NN graph of all snapshots in each projection direction is the most time-consuming part of NLSA and thus a good measure of algorithmic scaling. This task has $O(n_s^2 \times d)$ complexity, with n_s as the number of snapshots in each projection direction and d as the number of pixels in each snapshot.

Number of Conformational Classes. We consider the effect of SVD on noise and the number of distinguishable conformational classes with reference to the process used to extract continuous conformations:

i) Embed data by diffusion map. This step produces a cloud of points, viewed, e.g., in projection on the (ψ_1, ψ_2) plane.
ii) Reconstruct 2D energy landscape using NLSA.
iii) Reconstruct NLSA snapshots along a selected trajectory and embed to obtain a 1D manifold.

This process is designed to extract the conformational variations along the selected trajectory, e.g., the minimum-energy path.

The extent to which two conformations can be distinguished is determined by two factors: (*i*) the relevant conformational signal, as manifested in the extent of the data cloud along the trajectory, hereon designated Δ; and (*ii*) The noise as manifested by the distribution of the data cloud perpendicular to the trajectory, characterized, for a normal distribution, by an SD σ. The noise characterized by σ determines the confidence with which two conformations can be regarded as distinct. The number of distinct conformational states is the extent of the data cloud along the trajectory, measured in units of $p\sigma$, $(1 \le p \le 6)$, depending on the desired level of confidence.

Consider now the noise-reduction capability of the SVD step. Henry and Hofrichter (16) discussed the effect of noise on SVD analysis of a matrix consisting of n vectors, each with m components. They show that with input data characterized by noise of variance σ^2_{input}, SVD produces an output, whose variance due to noise is $(\frac{1}{m}+\frac{1}{n})\sigma^2_{input}$. When, e.g., $m \gg n$, this yields $\sigma_{output} = \sigma_{input}/\sqrt{n}$, a conceptually appealing scaling behavior.

Consider next a single-mode reconstruction, ignoring the singular value. As before, let the range spanned by the resulting conformations be Δ. Different conformational classes must be separated by $p\sigma_{output}$. The number of permissible conformational classes is therefore

$$N_{class} = \frac{\Delta}{p\sigma_{output}} = \frac{\Delta}{p\sigma_{input}\sqrt{\left(\frac{1}{m}+\frac{1}{n}\right)}} \approx \frac{\Delta\sqrt{n}}{p\sigma_{input}}, \quad m \gg n, \qquad \textbf{[S8]}$$

with p depending on the required confidence level. The parameter m is replaced by cm for c-fold concatenated snapshots. In the present work, $\Delta \approx 5$, $\sigma_{input} \approx 1$, $p = 3$, $n \approx 10^3$, $cm \gg n$, yielding $N_{class} \sim 50$.

The above analysis ignores noise reduction due to projection onto the eigenvectors of the Laplace–Beltrami operator. Taking this into account suggests that up to ~300 conformational classes can be distinguished, subject, of course, to the other limiting factors mentioned in the main text. In this paper, we use the more conservative estimate of 50 conformational classes based on the noise-reduction capability of SVD only.

Pseudocode.

End-to-end analysis.

Input.

Cryo-EM micrographs.

Outputs.

Spectrum of conformational changes along a selected trajectory.

2D and 3D movies of conformation spectrum.

Steps.

i) Particle picking and orientation recovery (using SPIDER).
ii) Locate n_{GC} great circles (one Shannon angle width) in the space of orientations that are highly populated with experimental snapshots.
iii) Divide each great circle into n_P projections each spanning about one Shannon angle (the ratio of resolution to object diameter).
iv) For great circle i_{GC} = 1 to n_{GC} do.
v) For projection i_P = 1 to n_P do.
vi) Identify snapshots that fall into great circle i_{GC} and projection i_P.
vii) Bring snapshots into in-plane registry.
viii) Embed with diffusion map with defocus-tolerant kernel to obtain eigenvectors ψ of Laplace–Beltrami operator.
ix) Order snapshots along a line making an angle θ with the horizontal axis in ψ_1-ψ_2 space.
x) Perform NLSA (see pseudocode below).
xi) Extract the parameter $\tau(\theta)$ in 1° intervals for $0 \le \theta < \pi$.
xii) Reconstruct 2D energy landscape.
xiii) Perform NLSA on a selected trajectory on the landscape.
xiv) Embed NLSA reconstructed snapshots with diffusion map.
xv) Extract conformation parameter τ (conformation spectrum).
xvi) Establish consistent sense and starting points for all conformation spectra.
xvii) Compile 2D movie along selected trajectory on energy landscape.
xviii) endfor.
xix) endfor.
xx) Bin snapshots into N_{class} classes based on conformation parameter τ.
xxi) Perform 3D reconstruction of each class using Arachnid.
xxii) Compile 3D movie.

NLSA.

Input.

Set of n_s snapshots (each with d pixels) along the selected direction (ψ_1, ψ_2 or a line in between).

Output.

Set of NLSA noise-reduced reconstructed snapshots.

Steps.

i) Generate supervectors X of c-fold concatenated snapshots. $\mathbf{X}$ is a matrix with dimensions $cd \times (n_s - c + 1)$, $X[(k-1)d + 1{:}\ kd, j]$ is the $j + k - 1$th snapshot.
ii) Embed X with diffusion map to obtain truncated set of eigenvectors of the Laplace–Beltrami operator $\Psi = \{\psi_i | i = 1, \cdots, l\}$ and Riemannian measure μ.
iii) Project X onto $\Psi : A = \mathbf{X}\mu\Psi$.
iv) Perform SVD: $A = USV^T$.
v) Project V from manifold back to time domain: $V'^T = V^T\Psi^T$.
vi) Reconstruct with p modes: $\tilde{X} = \sum_1^p X_k$, where $X_k = U_k S_{kk} V_k'^T$ (U_k is the kth column of the matrix U; similarly, for V_k', S_{kk} is the kth element of the diagonal matrix S).
vii) c reconstructions are available for each time point t: $\tilde{X}[(k-1)d+1 \ldots kd,\ t-k+1]$ $1 \le k \le c$. These reconstructions are averaged to obtain noise-reduced reconstructed snapshots.

Reconstruct NLSA snapshots along the selected trajectory.

Input.

Selected trajectory on the map of density of points in τ space.

Output.

Set of NLSA reconstructed snapshots along the trajectory.

Steps.

i) On the map of density of points, identify two lines passing through each point on the trajectory, with the first line connecting the point to the center of the density map and the second normal to the first.
ii) Find class of raw snapshots at the intersection of the two lines using Eq. **S7**.
iii) Repeat steps *i* and *ii* to identify all raw snapshots contributing to selected trajectory.
iv) Perform NLSA on the raw snapshots to obtain a 1D compact manifold (circle).
v) Extract universal conformational parameter τ for each NLSA reconstructed snapshot.

1. Leshin JA, Rakauskaitė R, Dinman JD, Meskauskas A (2010) Enhanced purity, activity and structural integrity of yeast ribosomes purified using a general chromatographic method. *RNA Biol* 7(3):354–360.
2. Grassucci RA, Taylor DJ, Frank J (2007) Preparation of macromolecular complexes for cryo-electron microscopy. *Nat Protoc* 2(12):3239–3246.
3. Lei J, Frank J (2005) Automated acquisition of cryo-electron micrographs for single particle reconstruction on an FEI Tecnai electron microscope. *J Struct Biol* 150(1):69–80.
4. Langlois R, et al. (2014) Automated particle picking for low-contrast macromolecules in cryo-electron microscopy. *J Struct Biol* 186(1):1–7.
5. Spahn CM, et al. (2004) Domain movements of elongation factor eEF2 and the eukaryotic 80S ribosome facilitate tRNA translocation. *EMBO J* 23(5):1008–1019.
6. Shaikh TR, et al. (2008) SPIDER image processing for single-particle reconstruction of biological macromolecules from electron micrographs. *Nat Protoc* 3(12):1941–1974.
7. Tenenbaum JB, de Silva V, Langford JC (2000) A global geometric framework for nonlinear dimensionality reduction. *Science* 290(5500):2319–2323.
8. Roweis ST, Saul LK (2000) Nonlinear dimensionality reduction by locally linear embedding. *Science* 290(5500):2323–2326.
9. Belkin M, Niyogi P (2003) Laplacian eigenmaps for dimensionality reduction and data representation. *Neural Comput* 15(6):1373–1396.
10. Donoho DL, Grimes C (2003) Hessian eigenmaps: Locally linear embedding techniques for high-dimensional data. *Proc Natl Acad Sci USA* 100(10):5591–5596.
11. Coifman RR, et al. (2005) Geometric diffusions as a tool for harmonic analysis and structure definition of data: Diffusion maps. *Proc Natl Acad Sci USA* 102(21):7426–7431.
12. Coifman RR, Lafon S (2006) Geometric harmonics: A novel tool for multiscale out-of-sample extension of empirical functions. *Appl Comput Harmon Anal* 21(1):31–52.
13. Coifman RR, Shkolnisky Y, Sigworth FJ, Singer A (2008) Graph Laplacian tomography from unknown random projections. *IEEE Trans Image Process* 17(10):1891–1899.
14. Champeney DC (1973) *Fourier Transforms and Their Physical Applications* (Academic Press, London), p 256.
15. Aubry N, Guyonnet R, Lima R (1991) Spatiotemporal analysis of complex signals: Theory and applications. *J Stat Phys* 64(3-4):683–739.
16. Henry E, Hofrichter J (1992) Singular value decomposition: Application to analysis of experimental data. *Methods Enzymol* 210:129–192.
17. Giannakis D, Majda AJ (2012) Nonlinear Laplacian spectral analysis for time series with intermittency and low-frequency variability. *Proc Natl Acad Sci USA* 109(7): 2222–2227.
18. Giannakis D, Majda AJ (2012) Nonlinear Laplacian spectral analysis: Capturing intermittent and low-frequency spatiotemporal patterns in high-dimensional data. *Stat Anal Data Mining* 6(3):1–15.
19. Packard N, Crutchfield J, Farmer J, Shaw R (1980) Geometry from a time series. *Phys Rev Lett* 45(9):712–716.
20. Takens F (1981) *Dynamical Systems & Turbulence, Warwick 1980* (Springer, Berlin).
21. Sauer T, Yorke JA, Casdagli M (1991) Embedology. *J Stat Phys* 65(3):579–616.
22. Deyle ER, Sugihara G (2011) Generalized theorems for nonlinear state space reconstruction. *PLoS ONE* 6(3):e18295.
23. Coifman R, Lafon S (2006) Diffusion maps. *Appl Comput Harmon Anal* 21(1):5–30.
24. Jones PW, Maggioni M, Schul R (2008) Manifold parametrizations by eigenfunctions of the Laplacian and heat kernels. *Proc Natl Acad Sci USA* 105(6):1803–1808.
25. Pettersen EF, et al. (2004) UCSF Chimera—a visualization system for exploratory research and analysis. *J Comput Chem* 25(13):1605–1612.

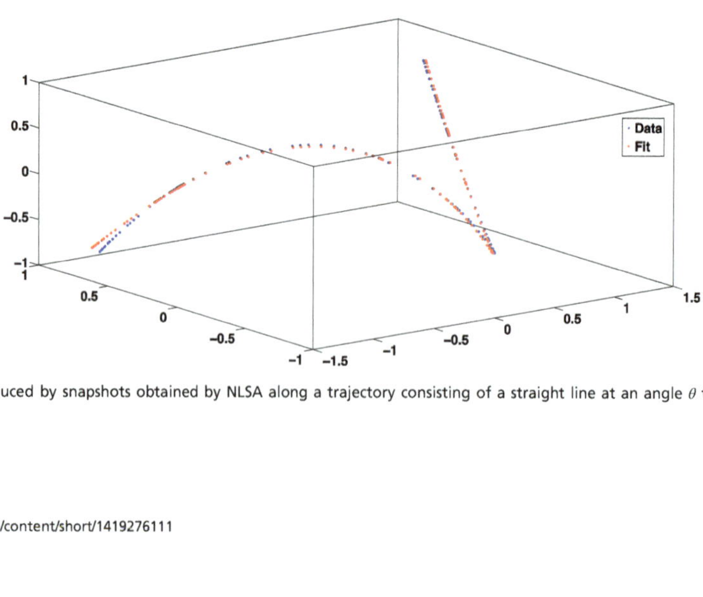

Fig. S1. The manifold produced by snapshots obtained by NLSA along a trajectory consisting of a straight line at an angle θ to the horizontal axis in the (ψ_1,ψ_2) plane.

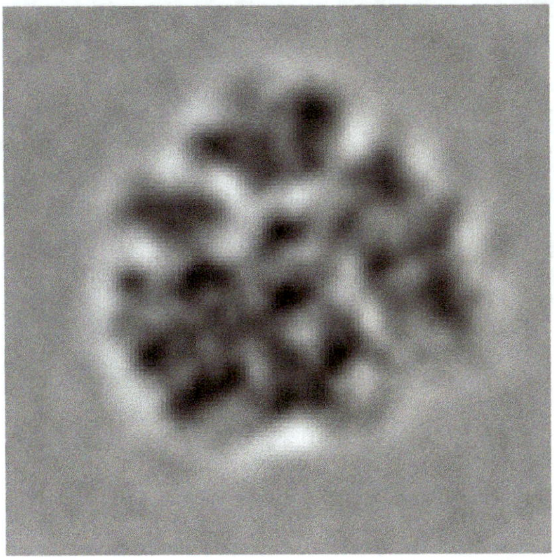

Movie S1. Changes in reconstructed cryo-EM snapshots from a selected viewing direction, as the path of lowest energy is traversed. The position along the path is indicated by the purple dot in the right panel.

Movie S1

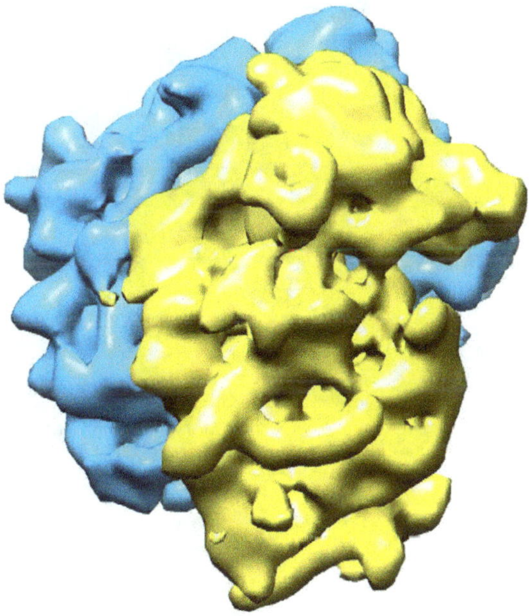

Movie S2. Variations in the ribosome structure along the path of lowest energy, as indicated by the position of the purple dot in the right panel. The sequence of 3D structures is identical in Movies S2 through S4, but is shown in three different standard viewing directions. Viewing direction here: from the factor-binding side.

Movie S2

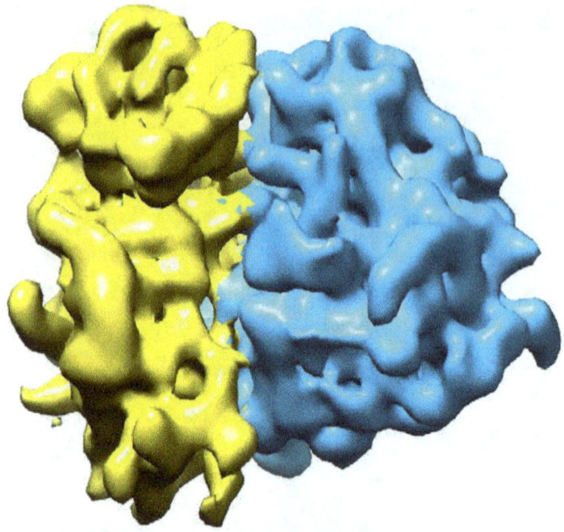

Movie S3. Variations in the ribosome structure along the path of lowest energy, as indicated by the position of the purple dot in the right panel. The sequence of 3D structures is identical in Movies S2 through S4, but is shown in three different standard viewing directions. Viewing direction here: from the solvent side.

Movie S3

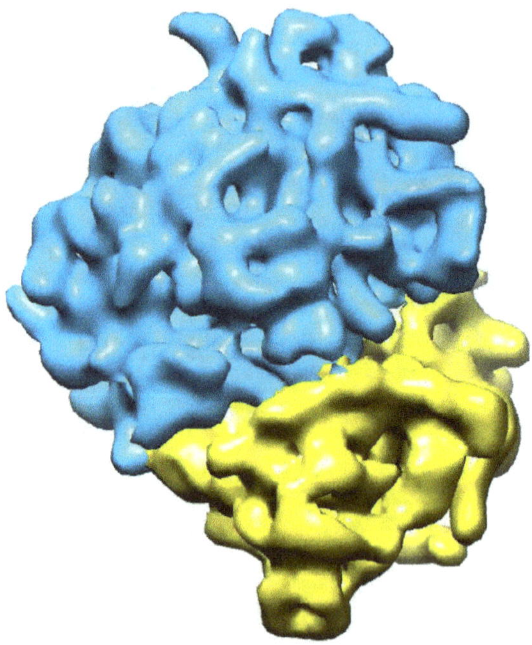

Movie S4. Variations in the ribosome structure along the path of lowest energy, as indicated by the position of the purple dot in the right panel. The sequence of 3D structures is identical in Movies S2 through S4, but is shown in three different standard viewing directions. Viewing direction here: from the top.

Movie S4

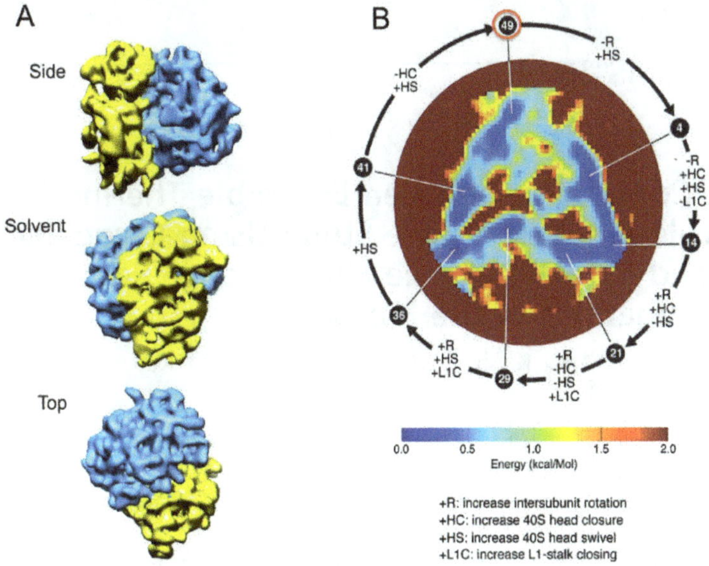

Movie S5. The ribosome structure, in three standard views, as it varies between the seven successive points along the path of lowest energy indicated by the position of the red circle in the right panel.

Movie S5

THE JOURNAL OF PHYSICAL CHEMISTRY B

Article

pubs.acs.org/JPCB

Quantitative Connection between Ensemble Thermodynamics and Single-Molecule Kinetics: A Case Study Using Cryogenic Electron Microscopy and Single-Molecule Fluorescence Resonance Energy Transfer Investigations of the Ribosome

Colin D. Kinz-Thompson,[†,∥] Ajeet K. Sharma,[‡,∥,¶] Joachim Frank,[⊥,§] Ruben L. Gonzalez, Jr.,[†] and Debashish Chowdhury[*,‡]

[†]Department of Chemistry, Columbia University, New York, New York 10032, United States

[‡]Department of Physics, Indian Institute of Technology, Kanpur 208016, India

[⊥]Howard Hughes Medical Institute, Department of Biochemistry and Molecular Biophysics, Columbia University, New York, New York 10027, United States

[§]Department of Biology, Columbia University, New York, New York 10027, United States

ABSTRACT: At equilibrium, thermodynamic and kinetic information can be extracted from biomolecular energy landscapes by many techniques. However, while static, ensemble techniques yield thermodynamic data, often only dynamic, single-molecule techniques can yield the kinetic data that describe transition-state energy barriers. Here we present a generalized framework based upon dwell-time distributions that can be used to connect such static, ensemble techniques with dynamic, single-molecule techniques, and thus characterize energy landscapes to greater resolutions. We demonstrate the utility of this framework by applying it to cryogenic electron microscopy (cryo-EM) and single-molecule fluorescence resonance energy transfer (smFRET) studies of the bacterial ribosomal pre-translocation complex. Among other benefits, application of this framework to these data explains why two transient, intermediate conformations of the pre-translocation complex, which are observed in a cryo-EM study, may not be observed in several smFRET studies.

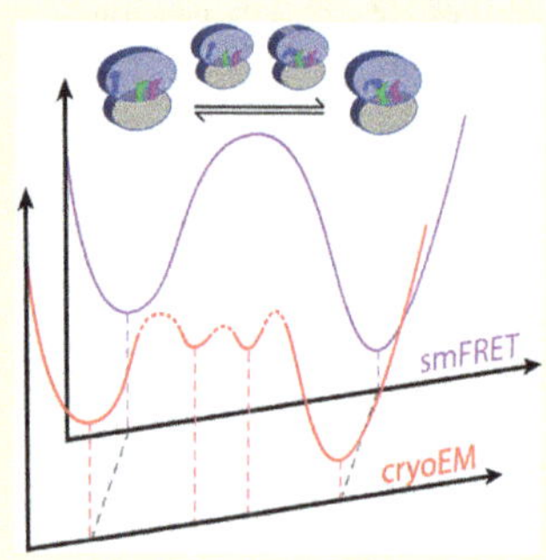

1. INTRODUCTION

Biomolecular machines operate on energy landscapes with transition-state energy barriers which range from $\sim k_B T$ to the energy of covalent bonds.[1−3] Characterizing the wells and barriers which comprise these energy landscapes is important for understanding the thermodynamics and kinetics of biomolecular machines, and how these thermodynamics and kinetics can be modulated in order to regulate the activities of these machines.[4−9] However, due to the stochasticity inherent to these processes, as well as the transient and/or rare nature of states separated by low transition-state energy barriers, extremely sensitive techniques are often required to obtain the level of detail necessary to adequately describe such systems.[10,11] Sufficiently sensitive ensemble techniques, such as cryogenic electron microscopy (cryo-EM), can measure static, equilibrium-state populations, and this provides information on the relative energy differences between distinct states. However, because of the vanishingly small probability of observing a transition state, techniques such as cryo-EM are not able to characterize the transition-state energy barriers responsible for much of the regulation of biomolecular processes.[12] Fortunately, dynamic, time-dependent, single-molecule techniques, such as single-molecule fluorescence resonance energy transfer (smFRET), can directly monitor the kinetics of these processes, and allow the characterization of the transition-state energy barriers with theories such as transition-state theory or Kramers's theory.[13] smFRET is a particularly powerful technique for connecting single-molecule kinetics to ensemble thermodynamics obtained from cryo-EM in that the FRET efficiency (E_{FRET}) obtained from the smFRET experiments can be correlated to structures obtained from the cryo-EM experiments. Despite this significant advantage over many other single-molecule techniques, like all techniques, smFRET approaches often suffer from limitations to spatial and temporal resolution, and also often require structural information to develop biologically informative signals.[14] Therefore, for any particular system, static, equilibrium-state, ensemble techniques and dynamic, time-dependent, single-molecule techniques provide complementary approaches for studying the underlying biological processes. Nonetheless, given the current limitations in their application, the pictures they provide may not always be congruous.

Special Issue: Biman Bagchi Festschrift

Received: December 25, 2014
Revised: March 15, 2015
Published: March 18, 2015

 © 2015 American Chemical Society

DOI: 10.1021/jp5128805
J. Phys. Chem. B 2015, 119, 10888−10901

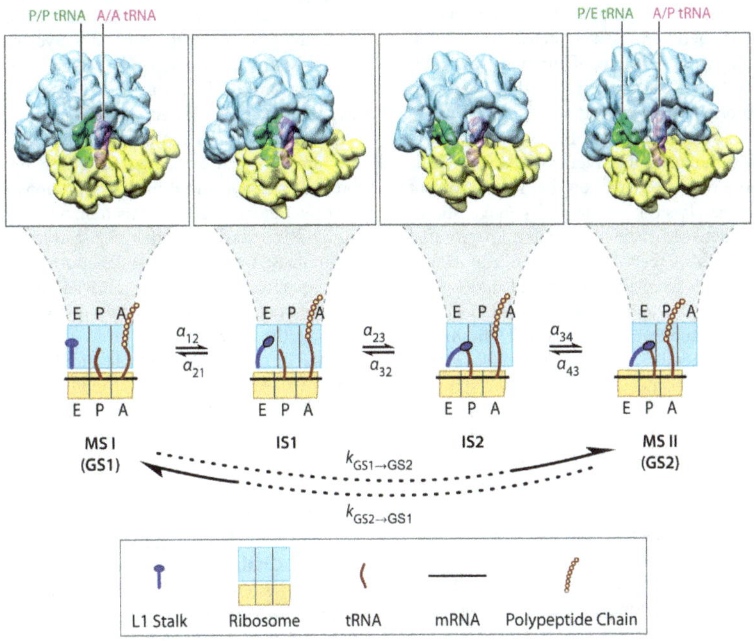

Figure 1. Cartoon schematic mechanism of PRE complex fluctuations. After peptide-bond formation, the PRE complex fluctuates between MS I/GS1 and MS II/GS2, passing through IS1 and IS2, until EF-G-catalyzed translocation occurs.

The bacterial ribosome is one example of a biological system that has been well studied by both ensemble and single-molecule techniques, although the associated energy landscape remains only coarsely defined.[15,16] Responsible for translating messenger RNAs (mRNAs) into their encoded proteins, the ribosome is composed of a large and a small subunit (50S and 30S in bacteria, respectively). During the elongation stage of translation, the ribosome undergoes consecutive rounds of an elongation cycle, in which it successively adds amino acids to the nascent polypeptide chain in the order dictated by the sequence of the mRNA. In the first step of the elongation cycle, the mRNA-encoded aminoacyl-transfer RNA (aa-tRNA) is delivered to the aa-tRNA binding (A) site of the ribosome in the form of a ternary complex (TC) that is composed of the ribosomal guanosine triphosphatase (GTPase) elongation factor (EF) Tu, guanosine triphosphate (GTP), and aa-tRNA.[17−20] Upon delivery of the mRNA-encoded aa-tRNA into the A site, peptide bond formation results in transfer of the nascent polypeptide chain from the peptidyl-tRNA bound at the ribosomal peptidyl-tRNA binding (P) site to the aa-tRNA at the A site, generating a ribosomal pre-translocation (PRE) complex carrying a newly deacylated tRNA at the P site and a newly formed peptidyl-tRNA, extended by one amino acid, at the A site.[21−23] Subsequently, the ribosome must translocate along the mRNA, moving the newly deacylated tRNA from the P site to the ribosomal tRNA exit (E) site and the newly formed peptidyl-tRNA from the A site to the P site.[15,18,24−30] While translocation can occur spontaneously, albeit slowly, *in vitro*,[31] it is accelerated by orders of magnitude *in vivo* through the action of EF-G, another ribosomal GTPase.[18,30]

Prior to translocation and in the absence of EF-G, at least three individual structural elements of the PRE complex undergo thermally driven conformational fluctuations: (i) the P- and A-site tRNAs fluctuate between their classical P/P and A/A configurations and their hybrid P/E and A/P configurations (where, relative to the classical P/P and A/A configurations, the hybrid P/E and A/P configurations are characterized by the movement of the acyl acceptor ends of the P- and A-site tRNAs from the P and A sites of the 50S subunit into the E and P sites of the 50S subunit, respectively); (ii) the ribosome fluctuates between its non-rotated and rotated subunit orientations (where, relative to the non-rotated subunit orientation, the rotated subunit orientation is characterized by a counterclockwise rotation of the 30S subunit relative to the 50S subunit when viewed from the solvent-accessible side of the 30S subunit);[32] and (iii) the L1 stalk of the 50S subunit fluctuates between its open and closed conformations (where, relative to the open L1 stalk conformation, the closed L1 stalk conformation is characterized by movement of the L1 stalk into the inter-subunit space such that it can make a direct contact with the hybrid P/E-configured tRNA) (Figure 1).[33] Because of the stochastic nature of thermally driven processes, the tRNAs, ribosomal subunits, and L1 stalk within an ensemble of PRE complexes will asynchronously fluctuate between these transiently populated states in the absence of EF-G. While this structural heterogeneity impedes ensemble studies of these dynamics, they have been successfully characterized by single-molecule methods.[29,34−44]

Remarkably, smFRET studies performed by Fei and co-workers have observed PRE complexes fluctuating between two discrete states: (i) global state 1 (GS1), characterized by classically configured tRNAs, non-rotated subunits, and an open L1 stalk, and (ii) global state 2 (GS2), characterized by hybrid-configured tRNAs, rotated subunits, and a closed L1 stalk.[34,35] The observation by Fei and co-workers that the PRE complex fluctuates between just two states in the smFRET studies is consistent with numerous subsequent smFRET studies from several other groups in which the tRNAs,

DOI: 10.1021/jp5128805
J. Phys. Chem. B 2015, 119, 10888−10901

ribosomal subunits, or L1 stalk elements of PRE complexes are also observed to fluctuate between just two states, corresponding to the classical and hybrid tRNA configurations, the non-rotated and rotated subunit orientations, or the open and closed L1 stalk conformations, respectively.[34,35,37,38,40,44,45] Furthermore, the initial observation by Fei and co-workers that the PRE complex fluctuates between GS1 and GS2 is consistent with the more recent observation that fluctuations of the tRNAs between their classical and hybrid configurations, the ribosomal subunits between their non-rotated and rotated orientations, and the L1 stalk between its open and closed conformations are physically coupled, and coordinated by the ribosome in order to maximize and regulate the efficiency of translocation.[44] smFRET studies reveal that the thermodynamics and kinetics of the equilibrium between GS1 and GS2 are sensitive to (i) the presence, identity, and acylation status of the P-site tRNA;[29,34−42] (ii) the presence and acylation status of the A-site tRNA;[29,34−42] (iii) the binding of EF-G;[34,35,37−40] (iv) Mg^{2+} concentration;[41] (v) temperature;[43] (vi) the binding of ribosome-targeting antibiotic inhibitors of translocation;[41,44,46] and (vii) the perturbation of inter-subunit rotation via disruption of specific ribosomal inter-subunit interactions.[42,44] Collectively, these studies have provided deep insights into the roles that the P- and A-site tRNAs, EF-G, antibiotics, cooperative conformational changes, and allostery play in regulating translocation.

Cryo-EM studies of PRE complexes performed by Agirrezabala and co-workers have directly observed two states that are presumably the structural equivalents of GS1 and GS2, termed macrostate 1 (MS I) and macrostate 2 (MS II), respectively.[32,47−49] More recent molecular dynamics simulations of cryo-EM-derived structural models of MS I and MS II support this presumption, and found that, in the studies by Fei and co-workers, the E_{FRET} values observed in GS1 and GS2 are consistent with the simulations of MS I and MS II, respectively.[50] However, in addition to MS I and MS II, a more recent analysis of the cryo-EM data set reported by Agirrezabala et al. has revealed the presence of two additional states that are presumably intermediate between MS I/GS1 and MS II/GS2 along the reaction coordinate.[51] Neither of these two intermediate states, referred to here as intermediate state 1 (IS1) and intermediate state 2 (IS2), nor any others, were detected in the smFRET studies by Fei and co-workers[34,35,44,45] or several other groups.[37,38,40] In contrast, we note that smFRET studies of PRE complexes by Munro and co-workers identified and characterized two additional states that presumably lie along the reaction coordinate between MS I/GS1 and MS II/GS2.[29,42] However, these studies relied heavily on the use of smFRET data collected using ribosomes in which a substitution mutation disrupts a critical ribosome−tRNA interaction, and consequently causes the P-site tRNA in the resulting intermediate states to adopt conformations that are very different from those observed in either IS1 or IS2 in PRE complexes formed using wild-type ribosomes.[51]

Since the smFRET experiments by Fei and co-workers and the cryo-EM experiments by Agirrezabala and co-workers interrogate PRE complexes composed of wild-type ribosomes (i.e., without mutations that disrupt ribosome−tRNA interactions), it is highly likely that IS1 and IS2 are also present in the smFRET data, but that, given the spatial resolution, time resolution, and/or signal-to-background ratio (SBR) of the smFRET experiments, IS1 and IS2 do not produce large enough changes in E_{FRET} to be distinguished from GS1 or GS2. By connecting the results from static, equilibrium-state, ensemble experiments, such as cryo-EM, with the results from dynamic, time-dependent, single-molecule experiments, such as smFRET, through a theoretical framework, these hypotheses can be tested, and the energy landscape where the PRE complex exists can be characterized more precisely. Here, we present such a framework based upon equilibrium-state probabilities and dwell-time distributions (see refs 52 and 53, and references therein). This framework is general and can be applied to various other ensemble and single-molecule techniques; we use a linear kinetic model but emphasize that the equations can also be derived for other models. As an illustrative case study, we apply this generalized framework to analyze the data obtained from the cryo-EM and smFRET studies by Agirrezabala et al. and Fei et al., respectively. In doing so, we connect the distribution of the MS I, MS II, IS1, and IS2 states of the PRE complex observed by cryo-EM to the transition rates between the GS1 and GS2 states observed by smFRET.

2. METHODS

2.1. Dwell-Time Distribution Framework for *N*-State Markov Chain.

Consider two distinct chemical states, 1 and N, of the system connected linearly by a number of on-pathway intermediate states, 2 through $N - 1$, with transitions from state i to state j occurring at rate α_{ij},

$$1 \underset{\alpha_{21}}{\overset{\alpha_{12}}{\rightleftharpoons}} 2 \underset{\alpha_{32}}{\overset{\alpha_{23}}{\rightleftharpoons}} \cdots \underset{\alpha_{N-1,N-2}}{\overset{\alpha_{N-2,N-1}}{\rightleftharpoons}} N-1 \underset{\alpha_{N,N-1}}{\overset{\alpha_{N-1,N}}{\rightleftharpoons}} N \tag{1}$$

If $P_\mu(t)$ is the probability of finding the system in chemical state μ at time t, then the time evolution of these probabilities is governed by a set of coupled master equations. The steady state of this set of equations, corresponding to the constraints $\partial P_\mu/\partial t = 0$, yields the equilibrium-state occupation probabilities P_μ^{eq} of populating each state. The distribution of times taken by the system to reach one terminus from the other can be calculated by modifying the original kinetic scheme, imposing an absorbing boundary at the destination state. If the destination state is N, then the corresponding modified kinetic scheme would be

$$1 \underset{\alpha_{21}}{\overset{\alpha_{12}}{\rightleftharpoons}} 2 \underset{\alpha_{32}}{\overset{\alpha_{23}}{\rightleftharpoons}} \cdots \underset{\alpha_{N-1,N-2}}{\overset{\alpha_{N-2,N-1}}{\rightleftharpoons}} N-1 \xrightarrow{\alpha_{N-1,N}} N \tag{2}$$

By writing master equations for this new scheme, and adopting the method of Laplace transform, we analytically calculated the probability, $f^{p}(t)\, dt$, that the system, initially at state 1, reaches state N in the time interval between t and $t + dt$. By evaluating the first moment of this distribution, the mean time for this transition, $\langle t_p \rangle$, can be calculated. An analogous process can be performed to calculate the probability, $f^{r}(t)\, dt$, that the system, initially in state N, reaches state 1 in the time interval between t and $t + dt$, and hence obtain the mean time for this transition, $\langle t_r \rangle$.

The two expressions for the mean transition times between the termini, $\langle t_p \rangle$ and $\langle t_r \rangle$, form a system of equations with all $2N - 2$ α_{ij} as variables. Ratios of the P_μ^{eq} define relationships between the rate constants α_{ij}; so, if the equilibrium-state probabilities are known, substitution of these ratios into the expressions for $\langle t_p \rangle$ and $\langle t_r \rangle$ reduces the number of degrees of freedom in the system of equations. With an experimental measure of the mean transition time between the terminal states, the system of equations can be solved for the α_{ij} rate

DOI: 10.1021/jp5128805
J. Phys. Chem. B 2015, 119, 10888−10901

constants. The four-state model (two intermediate states) is solved in Appendix A, as is the derivation of the expression for the variances of t_p and t_r. Equivalent expressions for the three-state model (one intermediate state) are given in Appendix B.

2.2. smFRET Simulations. We simulated 100 E_{FRET} versus time trajectories with a linear, three-state kinetic scheme. Dwell-times prior to transitions to other states were exponentially distributed according to the appropriate rate constants. In each E_{FRET} versus time trajectory, the value of E_{FRET} corresponding to each state was randomized by choosing r from a normal distribution, and calculating $E_{FRET} = (1 + (r/R_0)^6)^{-1}$, where R_0 is the Förster radius. For each E_{FRET} versus time trajectory, R_0 was also randomly chosen from a normal distribution. Noise reflecting a reasonable SBR for the total-internal reflection fluorescence (TIRF) microscope used in the smFRET experiments (*i.e.*, $\sigma = 0.05$) was also added to each E_{FRET} versus time trajectory. More details can be found in Appendix C.

3. RESULTS AND DISCUSSION

3.1. General Framework. Because of their complexity, biomolecular systems are often investigated with multiple techniques—each with its individual strengths and weaknesses. However, different techniques occasionally yield disparate mechanistic pictures that must ultimately be resolved. One situation in which this problem manifests itself is when a static, equilibrium-state technique such as cryo-EM detects on-pathway intermediates, but a dynamic, time-dependent technique such as smFRET does not. This situation could arise if the dynamic technique is not sensitive enough to distinguish the intermediate state from other states of the biomolecular system. In order to reconcile such contrasting measurements, we need to estimate the lifetime of the transient intermediates if these, indeed, exist. In an effort to get these estimates we consider a linear kinetic pathway with on-pathway intermediates, such as in eq 1, though the framework presented here can easily be extended to include off-pathway intermediates. Note that for an N-state linear kinetic scheme there are $2N - 2$ rate constants α_{ij}. Therefore, in principle, the numerical values of all the individual α_{ij} could be obtained if $2N - 2$ independent algebraic equations satisfied by these rate constants were available. As we argue now, except for some small values of N, the rate constants α_{ij} are usually underdetermined by the available experimental information.

Time-dependent smFRET experiments are typically analyzed with a hidden Markov model (HMM).[14,54−57] Among other things, such an analysis yields HMM-idealized state versus time trajectories from which a distribution of lifetimes in a particular state can be calculated. If sufficiently transient, on-pathway intermediates between the initial and final state exist, the distribution of idealized lifetimes will not appear significantly different from what it would be in the absence of the intermediate states (e.g., an exponential distribution for a random transition with a time-independent probability of occurrence). In such a case, the simplest model that the smFRET data supports is that of a transition with no intermediate states; so, assuming Markovian transitions, the mean lifetimes obtained from the HMM-idealized trajectories would be taken to be the inverses of the effective rate constants for the transitions between the initial and final states. In contrast, the analytical expressions for the mean lifetimes spent traveling between the terminal states, via intermediate states, of an N-state kinetic scheme contain the rate constants α_{ij} that describe the direct transitions between the intermediate states (see Appendices A and B). Therefore, equating the mean lifetimes for the forward and reverse transitions between the terminal states that were inferred from dynamic, single-molecule experiments with the corresponding theoretically calculated mean lifetimes yields two algebraic equations that involve $2N - 2$ rate constants α_{ij}.

Thus, in practice, except for the trivial case of $N = 2$, the information available in the form of the effective rates of forward and reverse transitions between the two terminal states would be inadequate to determine all the $2N - 2$ rate constants α_{ij} that describe the full kinetic mechanism. Obviously, for larger values of N, the number of degrees of freedom must be reduced further by acquiring additional experimental information. This extra information comes from the equilibrium-state experiments. Including information about the equilibrium-state probabilities for the N states provides $N - 1$ additional independent equations (the constraint of normalization of the probabilities, i.e., their sum must be equal to unity, reduces the number from N to $N - 1$). So, for $N = 3$, one would have just enough information to write down four independent equations satisfied by the four rate constants in the three-state model. However, for $N = 4$, we have fewer equations than the number of unkowns and, therefore, in the absence of any other information, one of the rate constants would remain a free parameter. Any one of the six rate constants can be selected as the free parameter. Then, as we will show later in this section, varying the selected free parameter allows one to enumerate all the solutions which are consistent with the data, and thereby impose lower and/or upper bounds on the magnitudes of the rates. In case of higher values of N, the analytical expressions for the variance of t_p and t_r, reported in the Appendices, can be utilized for further reduction of the number of degrees of freedom if the corresponding experimental data becomes available in the future.

3.2. Model System: The Bacterial Pre-translocational Complex. Agirrezabala and co-workers collected cryo-EM data on PRE complexes containing $tRNA^{fMet}$ in the P site, and fMet-Trp-$tRNA^{Trp}$ in the A site.[49] Using ML3D, a maximum-likelihood-based classification method, particles from this data set were more recently grouped into six classes.[51] The conclusions of this study strongly suggest that three of the classes represent MS I and MS II–MS II, being comprised of two structurally similar classes. Additionally, the authors propose that two of the other classes represent on-pathway intermediate states (IS1 and IS2, respectively) between MS I and MS II. As the remaining class represents PRE complexes that are missing a tRNA in the A site, it is therefore ignored. Thus, the model of PRE dynamics proposed by this study is

$$1 \underset{\alpha_{21}}{\overset{\alpha_{12}}{\rightleftharpoons}} 2 \underset{\alpha_{32}}{\overset{\alpha_{23}}{\rightleftharpoons}} 3 \underset{\alpha_{43}}{\overset{\alpha_{34}}{\rightleftharpoons}} 4 \quad (3)$$

where states 1, 2, 3, and 4 represent MS I, IS1, IS2, and MS II, respectively, and are distributed as shown in Table 1.

Similarly, smFRET experiments performed by Fei and co-workers monitored PRE complexes as they transitioned, driven by thermal energy, between two global conformational states, GS1 and GS2, which correspond structurally to MS I and MS II, respectively.[34,35] By monitoring the relative change in the distance between the P-site tRNA and the ribosomal protein L1 within the L1 stalk of the 50S subunit, the smFRET signal developed by Fei and co-workers probably reports upon the tRNA motions along the pathway proposed by Agirrezabala

DOI: 10.1021/jp5128805
J. Phys. Chem. B 2015, 119, 10888−10901

Table 1. Summary of the Ribosomal PRE Complexes Observed by Agirrezabala and Co-workers[51]

state index μ	state	class	P_μ^{eq}
1	MS I	2	0.231
2	IS1	4A	0.131
3	IS2	4B	0.140
4	MS II	5/6	0.498

and co-workers. However, these smFRET measurements were performed for several PRE complexes of variable composition. Of these complexes, perhaps the one most relevant to the work performed by Agirrezabala and co-workers is the $PRE_{fM/F}$ complex carrying a $tRNA^{fMet}$ in the P site and a fMet-Phe-$tRNA^{Phe}$, rather than a fMet-Trp-$tRNA^{Trp}$, in the A site. Since the identities of the A-site dipeptidyl-tRNA in the two experiments differ, this could potentially lead to tRNA-dependent differences in the populations and lifetimes of the various states and, consequently, the rates of transitions between these states. Indeed, smFRET studies have shown that the lifetimes of GS1 and GS2 do depend on the presence[34,35] and acylation status (i.e., deacylated $tRNA^{Phe}$ versus Phe-$tRNA^{Phe}$ versus fMet-Phe-$tRNA^{Phe}$) of the A-site dipeptdyl-tRNA.[36,41] It should be noted, however, that the effect of the identity of the A-site dipeptidyl-tRNA itself (i.e., tRNAs other than $tRNA^{Phe}$) has not yet been tested by smFRET. In addition to the difference in the identities of the A-site dipeptidyl-tRNA in the cryo-EM and smFRET studies, the Mg^{2+} concentrations employed in the two studies differ. The cryo-EM studies were performed at $[Mg^{2+}]$ = 3.5 mM, and the smFRET studies were performed at $[Mg^{2+}]$ = 15 mM. Previously, smFRET studies have demonstrated that changes to the Mg^{2+} concentration over this range affect the populations and lifetimes of the GS1 and GS2 states.[41] With this in mind, it is likely that the equilibrium-state populations observed in the cryo-EM experiments and the corresponding state occupancies in the smFRET experiments are disparate. Nonetheless, despite their experimental differences, these cryo-EM and smFRET studies are the most experimentally similar cryo-EM and smFRET studies of wild-type bacterial PRE complexes that have been reported in the literature. Therefore, as a case study, we have chosen to quantitatively compare these two particular studies in order to demonstrate the application of the general framework developed in section 3.1.

The transition rates between GS1 and GS2 reported using the $PRE_{fM/F}$ complex for the L1-tRNA donor–acceptor labeling scheme were $k_{GS1\to GS2} = 2.8 \pm 0.2\ s^{-1}$ and $k_{GS2\to GS1} = 3.0 \pm 0.4\ s^{-1}$.[35] Given that no evidence of intermediate states was observed, this suggests that any intermediates states, if they exist, might be very transient relative to the time resolution with which the smFRET data were acquired. Indeed, there is a limitation to the time resolution with which smFRET data can be acquired with the electron-multiplying charge-coupled device (EMCCD) cameras that are typically used as detectors in TIRF microscopy-based smFRET experiments. Transitions that are faster than the EMCCD camera's acquisition rate (20 s^{-1} in the work of Fei et al.)[35] result in time averaging of the E_{FRET} and the recording of a single, artifactual data point that appears at the time-averaged value of the E_{FRET} between the states involved in the rapid fluctuations. This is a well-documented feature of smFRET data analysis, which we term "blurring".[54] This effect is further compounded by the fact that current state-of-the-art computational methods used to analyze the smFRET data cannot distinguish between artificial, short-lived (i.e., one data point) "states" resulting from blurring and actual, short-lived (i.e., one data point) states resulting from the sampling of true intermediate states.[54] With such an analysis, the true molecular states become hidden among the "blurred" states.

Reanalysis of the original $PRE_{fM/F}$ data using the software package ebFRET—a state-of-the-art, HMM-based analysis method for smFRET data[56,57]—yields a better estimate of the transition rates. This is because ebFRET uniquely enables analysis of the entire ensemble of individual E_{FRET} versus time trajectories, instead of analyzing them in the traditional, isolated, one-by-one manner. The two-state rates inferred by means of ebFRET are similar to, though perhaps more accurate than, those reported originally by Fei and co-workers: $k_{GS1\to GS2} = 2.0 \pm 0.2\ s^{-1}$ and $k_{GS2\to GS1} = 2.8 \pm 0.1\ s^{-1}$ (see Table 2).

Table 2. Summary of the Rates of Transition between GS1 and GS2 Observed by Fei and Co-workers[35] and in This Work

$k_{GS1\to GS2}$ (s^{-1})	$k_{GS2\to GS1}$ (s^{-1})	ref
2.8 ± 0.2	3.0 ± 0.4	35
2.0 ± 0.2	2.8 ± 0.1	this work

Interestingly, application of ebFRET reveals that the smFRET $PRE_{fM/F}$ data are best described by a five-state model. However, further analysis indicates that the three additional states are probably artifacts of blurring, because they are negligibly populated, have extremely transient lifetimes, and occur at an E_{FRET} that is in between the E_{FRET} values of the two well-defined states.

3.3. Four-State Model of PRE Complex Dynamics. The dynamics of PRE complexes were analyzed using the general framework presented in section 3.1, where the experimental data summarized in Tables 1 and 2 were used as constraints for the linear, four-state kinetic scheme shown in eq 3. For this kinetic scheme, the mean time needed for the forward transition from the terminal state 1 to the terminal state 4, $\langle t_p \rangle$, is given by

$$\langle t_p \rangle = \frac{1}{\alpha_{12}}\left[1 + \frac{\alpha_{21}}{\alpha_{23}} + \frac{\alpha_{21}\alpha_{32}}{\alpha_{23}\alpha_{34}}\right] + \frac{1}{\alpha_{23}}\left[1 + \frac{\alpha_{32}}{\alpha_{34}}\right] + \frac{1}{\alpha_{34}} \tag{4}$$

(see Appendix A for the full derivation). The corresponding mean time for the reverse transition from state 4 to state 1, $\langle t_r \rangle$, is given by

$$\langle t_r \rangle = \frac{1}{\alpha_{43}}\left[1 + \frac{\alpha_{34}}{\alpha_{32}} + \frac{\alpha_{34}\alpha_{23}}{\alpha_{32}\alpha_{21}}\right] + \frac{1}{\alpha_{32}}\left[1 + \frac{\alpha_{23}}{\alpha_{21}}\right] + \frac{1}{\alpha_{21}} \tag{5}$$

(see Appendix A for the full derivation). The probabilities for the occupation of the four states at equilibrium are given by

$$P_1^{eq} = \frac{\alpha_{43}\alpha_{32}\alpha_{21}}{\alpha_{43}\alpha_{32}\alpha_{21} + \alpha_{12}\alpha_{43}\alpha_{32} + \alpha_{12}\alpha_{23}\alpha_{43} + \alpha_{12}\alpha_{23}\alpha_{34}} \tag{6}$$

$$P_2^{eq} = \frac{\alpha_{12}\alpha_{43}\alpha_{32}}{\alpha_{43}\alpha_{32}\alpha_{21} + \alpha_{12}\alpha_{43}\alpha_{32} + \alpha_{12}\alpha_{23}\alpha_{43} + \alpha_{12}\alpha_{23}\alpha_{34}} \tag{7}$$

DOI: 10.1021/jp5128805
J. Phys. Chem. B 2015, 119, 10888–10901

$$P_3^{eq} = \frac{\alpha_{12}\alpha_{23}\alpha_{43}}{\alpha_{43}\alpha_{32}\alpha_{21} + \alpha_{12}\alpha_{43}\alpha_{32} + \alpha_{12}\alpha_{23}\alpha_{43} + \alpha_{12}\alpha_{23}\alpha_{34}} \quad (8)$$

$$P_4^{eq} = \frac{\alpha_{12}\alpha_{23}\alpha_{34}}{\alpha_{43}\alpha_{32}\alpha_{21} + \alpha_{12}\alpha_{43}\alpha_{32} + \alpha_{12}\alpha_{23}\alpha_{43} + \alpha_{12}\alpha_{23}\alpha_{34}} \quad (9)$$

with the normalization condition $\sum_{\mu=1}^{4} P_\mu^{eq} = 1$ (see Appendix A for the full derivation). Using the equations for $\langle t_p \rangle$ and $\langle t_r \rangle$ (eqs 4 and 5, respectively), and the equations for P_μ^{eq} (eqs 6–9), a plot was generated of all the rate constants α_{ij} as functions of an independent α_{43} (Figure 2).

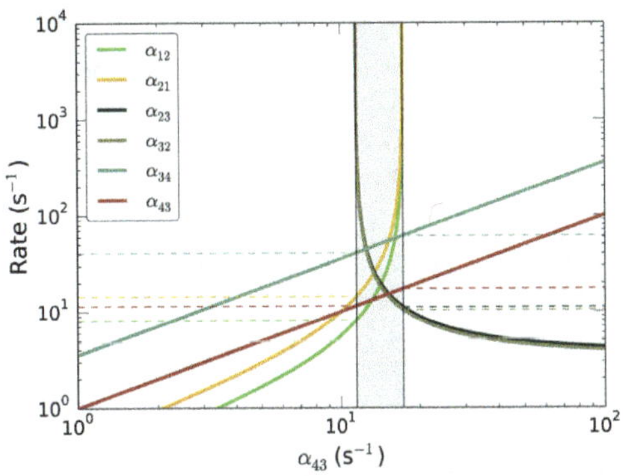

Figure 2. Rate constants for the four-state model as a function of α_{43}. The gray region contains the solutions where all rate constants are positive. Both axes are log-scaled, so any incompatible rates (negative valued) are not shown. The points where rate constants switch from being negative to positive valued are denoted with a black vertical line. Horizontal dashed lines denote upper and lower bounds for particular rate constants.

Notably, for some values of the independent rate constant, solutions for the dependent rate constants are negative. While this is a consistent solution of the model, only the values where all rate constants are positive are physically relevant solutions. The boundaries to this region where all rate constants are positive therefore represent the upper and lower bounds on the rate constants for the four-state model that are consistent with both the cryo-EM and the smFRET studies.

Interestingly, Figure 2 depicts the upper and lower bounds for α_{34} and α_{43}, but only lower-bound cutoffs for the other α's. These other rate constants all asymptotically converge to positive infinity; α_{12} and α_{21} increase with increasing α_{43}, while α_{23} and α_{32} compensate by decreasing to their lower bound. Plotting these results as a function of an independent α_{12} or α_{23} yields the same bounds—there is only a narrow window where all α's are consistent with the cryo-EM and smFRET data. With such boundaries on the individual rate constants, one can estimate the EMCCD camera acquisition rate that would be needed to distinctly observe the transient states of interest.

3.4. Three-State Pre-translocation Model. Since the number of equations available in our four-state model is five, whereas the number of unknown rate constants is six, we could only express five rate constants in terms of the sixth one. In contrast, because we can reduce the number of states from four to three (*vide infra*), in this subsection we use the four corresponding independent equations to extract the absolute values of the four rate constants associated with the three-state model,

$$1 \underset{\alpha_{21}}{\overset{\alpha_{12}}{\rightleftharpoons}} 2 \underset{\alpha_{32}}{\overset{\alpha_{23}}{\rightleftharpoons}} 3 \quad (10)$$

As explained in Appendix C, structural analysis strongly suggests that the L1–tRNA distance in IS1 is insufficiently different from that of MS I so as to result in an E_{FRET} that is significantly different than that of MS I. Thus, MS I and IS1 can be combined into a single state, state 1, thereby reducing the four-state model into a three-state kinetic scheme as shown in eq 10, where states 2 and 3 correspond to IS2 and MS II, respectively.

The expressions for $\langle t_p \rangle$, $\langle t_r \rangle$, and P_μ (μ = 1, 2, 3) are given by

$$\langle t_p \rangle = \frac{1}{\alpha_{12}}\left[1 + \frac{\alpha_{21}}{\alpha_{23}}\right] + \frac{1}{\alpha_{23}} \quad (11)$$

$$\langle t_r \rangle = \frac{1}{\alpha_{32}}\left[1 + \frac{\alpha_{23}}{\alpha_{21}}\right] + \frac{1}{\alpha_{21}} \quad (12)$$

$$P_1^{eq} = \frac{\alpha_{21}\alpha_{32}}{\alpha_{21}\alpha_{32} + \alpha_{12}\alpha_{32} + \alpha_{23}\alpha_{12}} \quad (13)$$

$$P_2^{eq} = \frac{\alpha_{12}\alpha_{32}}{\alpha_{21}\alpha_{32} + \alpha_{12}\alpha_{32} + \alpha_{23}\alpha_{12}} \quad (14)$$

$$P_3^{eq} = \frac{\alpha_{23}\alpha_{12}}{\alpha_{21}\alpha_{32} + \alpha_{12}\alpha_{32} + \alpha_{23}\alpha_{12}} \quad (15)$$

where the normalization condition is $\sum_{\mu=1}^{eq} P_\mu = 1$ (see Appendix B for detailed derivations). These expressions, together with the corresponding experimental data, are utilized to write down four independent equations. The rate constants computed by solving those four equations are shown in Table 3.

Table 3. Rate Constants for $PRE_{fM/F}$ Using a Linear, Three-State Kinetic Scheme Where MS I and IS1 Have Been Combined into the First State[a]

α	max $k_{L1-tRNA}$ (s^{-1})	mean $k_{L1-tRNA} \pm 1\sigma$ (s^{-1})
α_{12}	18.1	23.3 ± 22.7
α_{21}	46.8	52.5 ± 33.6
α_{23}	5.90	5.89 ± 0.42
α_{32}	1.66	1.66 ± 0.12

[a]Error from the ebFRET-estimated and smFRET-determined rate constants, and the counting error from the cryo-EM study, were propagated into distributions of the α. These distributions are strictly not normal distributions, although α_{23} and α_{32} are approximately normal.

Interestingly, this calculation suggests that the rate-limiting steps for the foward and reverse reactions are IS2→MS II and MS II→IS2, respectively, while interconversion between MS I/IS1 and IS2 occurs relatively rapidly. These rates for MS I/IS1 to IS2 interconversion are approximately the same as or faster than the 20 s^{-1} EMMCD camera acquisition rate of the original smFRET data from Fei et al.; additionally, the change in the L1–tRNA distance between the MS I/IS1 and IS2 states is relatively small (~80 Å to 64 Å), resulting in a correspondingly small difference in E_{FRET} (~0.15 to 0.40), so any separation of MS I/IS1 and IS2 that might have been observed in the

DOI: 10.1021/jp5128805
J. Phys. Chem. B 2015, 119, 10888–10901

The Journal of Physical Chemistry B Article

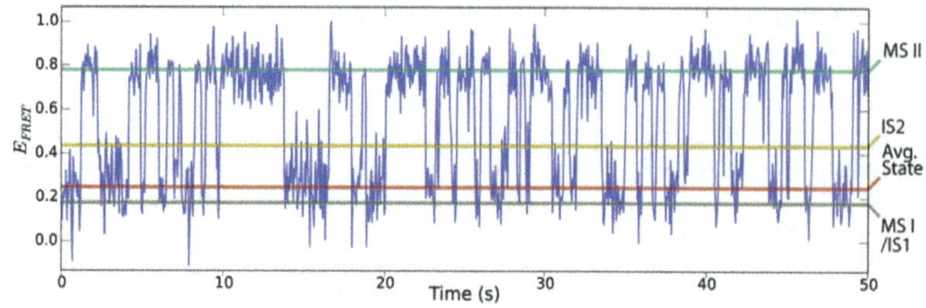

Figure 3. Example synthetic E_{FRET} versus time trajectory. The gray, yellow, and blue horizontal lines denote the E_{FRET} means used to simulate MS I/IS1, IS2, and MS II for this time series, respectively. The red horizontal line is the equilibrium-state-weighted average of the means of the MS I/IS1 and IS2 states.

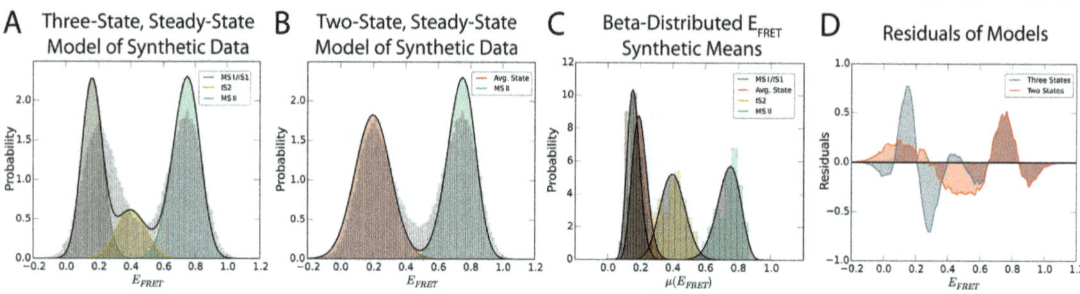

Figure 4. (A) Histogram of synthetic PRE complex data (gray) modeled with three non-blurred states. (B) Histogram of synthetic PRE complex data (gray) modeled with two non-blurred states. The "Avg. State" is a weighted combination of MS I/IS1 and IS2. (C) Histograms and probability distributions of the static E_{FRET} value means generated for the synthetic PRE complex data set. (D) Plots of the model probability densities minus the normalized histograms from panels A and B. The sum-of-squares values of these residuals are 10.03 and 4.02 for the three-state and two-state models, respectively, suggesting that the blurred, synthetic, PRE complex data are better represented by a two-state model.

smFRET data would likely have been obscured in the HMM analysis process by camera blurring arising from interconversion rates that are similar to the acquisition rate. The fast interconversion and small expected changes in E_{FRET} suggest that MS I, IS1, and IS2 might have originally been interpreted as a "single", averaged state in the analysis of the smFRET data.

3.5. Synthetic smFRET Time Series. To investigate how ebFRET would treat this "single", averaged smFRET state, synthetic time series simulating a three-state $\mathrm{PRE}_{\mathrm{fM/F}}$ complex were constructed guided by the analysis above. Estimates for E_{FRET} were based upon the cryo-EM structures reported by Agirrezabala and co-workers (see Appendix C), and the kinetic scheme and associated rate constants employed are those in section 3.4. Since the rate constants for the transition between states MS I/IS1 and IS2 are of the same order of magnitude as the frame rate of this simulation, traditional smFRET data analysis of these synthetic data provides insight into whether blurring could have obscured any transient, intermediate states in the data reported by Fei and co-workers. Typically, such obfuscation begins to manifest when dwell-times in a state of interest approach the same order of magnitude as the EMCCD camera acquisition time, because of errors in estimating the lengths of the dwell-times.[54]

The synthetic smFRET data set was constructed by carrying out simulations of E_{FRET} versus time trajectories where each "single ribosome" had randomized simulation parameters as described in Appendix C. This probabilistic approach accounts for experimental variation (e.g., uneven illumination in the field-of-view), as well as ensemble variations (e.g., static disorder from a small sub-population of ribosomes lacking an A-site dipeptidyl-tRNA). An example of a synthetic E_{FRET} vs time trajectory is shown in Figure 3, where the ensemble mean values of E_{FRET} were ~0.16, 0.40, and 0.74 for states MS I/IS1, IS2, and MS II, respectively.

With regard to the distribution of E_{FRET} values observed from any ensemble of E_{FRET} versus time trajectories, blurring would result in a shift of some of the density of the equilibrium-state occupancy probability distribution to an intermediate, averaged value between the blurred states. Deviation of a histogram of the observed, simulated E_{FRET} in the synthetic data set from the distribution predicted by the equilibrium-state occupancies of the linear, three-state model therefore can be used to characterize the amount of blurring present in the synthetic data. We modeled the normalized histogram of the synthetic ensemble with normal distributions weighted by their respective equilibrium-state probability, P_{μ}^{eq} (Figure 4A). The mean of each state was distributed according to the distribution of static E_{FRET} for that state in each of the synthetic time series (Figure 4C). This approach accurately reflects a non-blurred histogram of E_{FRET} (Figure 5). Deviations that occur are therefore due to blurring or, if they had been simulated in this synthetic data set, could have been due to the presence of unaccounted-for states. Notably, a large portion of the MS I/IS1 density in Figure 4A is relocated into the region between MS I/IS1 and IS2. This is a direct manifestation of blurring. By collapsing MS I/IS1 and IS2 into one averaged state (Figure 4B,C), we find that the data are much better described by only two states (Figure 4D). In this case, the artifactual, blurred, averaged state overwhelms any distinction between the MS I/IS1 and IS2 states, whereas when the simulation is performed

DOI: 10.1021/jp5128805
J. Phys. Chem. B 2015, 119, 10888–10901

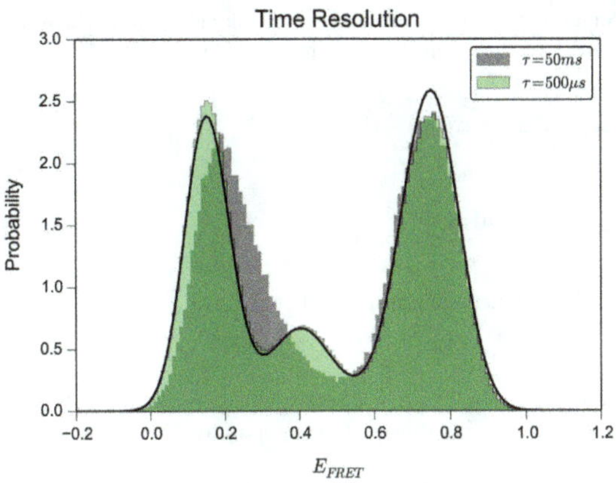

Figure 5. Histograms of the same ensemble of synthetic E_{FRET} vs time trajectories rendered with different time resolutions. The left-most peak in the gray histogram, which was used in Figure 4, is well-resolved into two separate peaks when the synthetic data are rendered with a 100× faster acquisition rate (green histogram). The three-state model distribution (black curve) is that from Figure 4, which was modeled using only the initial simulation parameters. Analysis of the faster time resolution time series (green histogram) using ebFRET accurately estimated all four transition rates ($k_{12} = 17.8 \pm 0.6$ s^{-1}, $k_{21} = 47.2 \pm 1.6$ s^{-1}, $k_{23} = 6.4 \pm 0.6$ s^{-1}, $k_{32} = 1.82 \pm 0.17$ s^{-1}), as well as the distribution of E_{FRET} means ($\mu_1 = 0.16$, $\sigma_1 = 0.035$, $\mu_2 = 0.41$, $\sigma_2 = 0.067$, $\mu_3 = 0.74$, $\sigma_3 = 0.058$) and the noise parameter ($\sigma_{noise} = 0.050$).

with an acquisition rate that is significantly faster than the transitions of interest (2000 s^{-1}), these states are well-resolved (Figure 5).

This blurred, synthetic, three-state ensemble of E_{FRET} versus time trajectories was then analyzed with ebFRET. Interestingly, ebFRET overestimated the true number of kinetic states. Most likely this is due to the fact that the dwell-times in states MS I/IS1 and IS2 are too transient relative to the simulation's "acquisition rate", because ebFRET is able to accurately infer the model parameters from the same data with a more appropriate acquisition rate of 2000 s^{-1} (Figure 5). As a result, the blurred, synthetic data points are modeled by ebFRET as distinct "states"—even though these states do not actually exist on the "energy landscape" of the simulated $PRE_{fM/F}$ complex.

4. CONCLUSION

We have established a general, theoretical framework that integrates equilibrium-state cryo-EM observations with dynamic, time-dependent smFRET observations. This framework is not limited exclusively to cryo-EM and smFRET; any technique that provides static, equilibrium-state populations can be integrated with any other technique that provides dynamic, time-dependent transition rates. The analysis reported here suggests that states IS1 and IS2, previously identified by cryo-EM, are not detected by smFRET because their short lifetimes result in transitions that are too fast to be characterized, given the acquisition rate of the EMCCD camera and the limitations of the manner in which hidden Markov models are implemented. Moreover, structural analyses indicate that the L1–tRNA smFRET signal is unable to yield distinguishable signals for MS I and IS1, given the typical SBR of TIRF microscope-based smFRET measurements. Quantitative analysis of the experimental data, based on the analytical theory presented here, provides a possible explanation for why the PRE complex intermediates observed by cryo-EM (i.e., IS1 and IS2) escaped detection in the smFRET studies by Fei et al.[34,35] and, possibly, other groups.[37,38,40] Conversely, application of this framework to smFRET data in which intermediate states have been detected, but have been difficult to assign to specific PRE complex structures,[29,42] should allow researchers to determine whether and how such intermediates correspond to IS1 and/or IS2. Perhaps more importantly, we can now predict the lifetimes and corresponding rates of transitions into and out of IS1 and IS2, which are crucial for designing future kinetic experiments. This work therefore resolves a discrepancy in the field and opens a path for performing and analyzing future experiments. Furthermore, we hope to use this theoretical framework to make predictions regarding future experiments in which the cryo-EM and smFRET data would be collected under more comparable conditions. This theoretical framework will also be useful in guiding future experimental explorations which include, for example, stabilization of IS1 or IS2 (or any other intermediates that may be ultimately identified) using different tRNAs or mutant ribosomes.

■ APPENDICES

A. Four-State Model

The theoretical results reported here are based on approaches similar to those followed in refs 52 and 53 for analytical calculations of the distribution of the dwell-times of a ribosome.

We use the kinetic scheme

$$1 \underset{\alpha_{21}}{\overset{\alpha_{12}}{\rightleftharpoons}} 2 \underset{\alpha_{32}}{\overset{\alpha_{23}}{\rightleftharpoons}} 3 \underset{\alpha_{43}}{\overset{\alpha_{34}}{\rightleftharpoons}} 4 \tag{16}$$

where integer indices 1, 2, 3, and 4 represent a discrete chemical state and α_{ij} denotes the transition probability per unit time (i.e., rate constant) for the $i \to j$ transition. If $P_\mu(t)$ is the probability of finding the system in chemical state μ at time t, then the time evolution of these probabilities is governed by the following master equations:

$$\frac{dP_1(t)}{dt} = -\alpha_{12}P_1(t) + \alpha_{21}P_2(t) \tag{17}$$

$$\frac{dP_2(t)}{dt} = \alpha_{12}P_1(t) - (\alpha_{23} + \alpha_{21})P_2(t) + \alpha_{32}P_3(t) \tag{18}$$

$$\frac{dP_3(t)}{dt} = \alpha_{23}P_2(t) - (\alpha_{32} + \alpha_{34})P_3(t) + \alpha_{43}P_4(t) \tag{19}$$

$$\frac{dP_4(t)}{dt} = \alpha_{34}P_3(t) - \alpha_{43}P_4(t) \tag{20}$$

Now we calculate the time-independent occupation probability, P_μ^{eq}, of each of these states by finding the equilibrium-state solutions of eqs 17–20,

$$P_1^{eq} = \frac{\alpha_{43}\alpha_{32}\alpha_{21}}{\alpha_{43}\alpha_{32}\alpha_{21} + \alpha_{12}\alpha_{43}\alpha_{32} + \alpha_{12}\alpha_{23}\alpha_{43} + \alpha_{12}\alpha_{23}\alpha_{34}} \tag{21}$$

$$P_2^{eq} = \frac{\alpha_{12}\alpha_{43}\alpha_{32}}{\alpha_{43}\alpha_{32}\alpha_{21} + \alpha_{12}\alpha_{43}\alpha_{32} + \alpha_{12}\alpha_{23}\alpha_{43} + \alpha_{12}\alpha_{23}\alpha_{34}} \tag{22}$$

$$P_3^{eq} = \frac{\alpha_{12}\alpha_{23}\alpha_{43}}{\alpha_{43}\alpha_{32}\alpha_{21} + \alpha_{12}\alpha_{43}\alpha_{32} + \alpha_{12}\alpha_{23}\alpha_{43} + \alpha_{12}\alpha_{23}\alpha_{34}} \tag{23}$$

DOI: 10.1021/jp5128805
J. Phys. Chem. B 2015, 119, 10888–10901

$$P_4^{eq} = \frac{\alpha_{12}\alpha_{23}\alpha_{34}}{\alpha_{43}\alpha_{32}\alpha_{21} + \alpha_{12}\alpha_{43}\alpha_{32} + \alpha_{12}\alpha_{23}\alpha_{43} + \alpha_{12}\alpha_{23}\alpha_{34}} \quad (24)$$

We also calculate the distribution of the time spent transitioning from chemical states 1–4 for the first time by modifying the original kinetic scheme into

$$1 \underset{\alpha_{21}}{\overset{\alpha_{12}}{\rightleftharpoons}} 2 \underset{\alpha_{32}}{\overset{\alpha_{23}}{\rightleftharpoons}} 3 \overset{\alpha_{34}}{\longrightarrow} 4$$

and writing the master equations according to this new scheme:

$$\frac{dP_1(t)}{dt} = -\alpha_{12}P_1(t) + \alpha_{21}P_2(t) \quad (25)$$

$$\frac{dP_2(t)}{dt} = \alpha_{12}P_1(t) - (\alpha_{21} + \alpha_{23})P_2(t) + \alpha_{32}P_3(t) \quad (26)$$

$$\frac{dP_3(t)}{dt} = \alpha_{23}P_2(t) - (\alpha_{32} + \alpha_{34})P_3(t) \quad (27)$$

$$\frac{dP_4(t)}{dt} = \alpha_{34}P_3(t) \quad (28)$$

These equations can be re-written in terms of the following matrix notations:

$$\frac{d\mathbf{P}(t)}{dt} = \mathbf{MP}(t) \quad (29)$$

where $\mathbf{P}(t)$ is a column matrix whose elements are $P_1(t)$, $P_2(t)$, and $P_3(t)$, and

$$\mathbf{M} = \begin{bmatrix} -\alpha_{12} & \alpha_{21} & 0 \\ \alpha_{12} & -(\alpha_{21} + \alpha_{23}) & \alpha_{32} \\ 0 & \alpha_{23} & -(\alpha_{34} + \alpha_{32}) \end{bmatrix} \quad (30)$$

Now, by introducing the Laplace transform of the probability of kinetic states,

$$\tilde{P}_\mu(s) = \int_0^\infty P_\mu(t)\, e^{-st}\, dt \quad (31)$$

the solution of eq 29 in Laplace space is

$$\tilde{\mathbf{P}}(s) = (s\mathbf{I} - \mathbf{M})^{-1}\mathbf{P}(0) \quad (32)$$

The determinant of the matrix $s\mathbf{I}-\mathbf{M}$ is a third-order polynomial,

$$(s\mathbf{I} - \mathbf{M})^{-1} = a_3s^3 + a_2s^2 + a_1s + a_0 \quad (33)$$

where

$$a_3 = 1 \quad (34)$$

$$a_2 = \alpha_{12} + \alpha_{21} + \alpha_{23} + \alpha_{32} + \alpha_{34} \quad (35)$$

$$a_1 = \alpha_{21}\alpha_{34} + \alpha_{21}\alpha_{32} + \alpha_{23}\alpha_{34} + \alpha_{12}\alpha_{23} + \alpha_{12}\alpha_{34} + \alpha_{12}\alpha_{32} \quad (36)$$

$$a_0 = \alpha_{12}\alpha_{23}\alpha_{34} \quad (37)$$

We can solve eq 32 by using the initial conditions

$$P_1(0) = 1 \quad \text{and} \quad P_2(0) = P_3(0) = P_4(0) = 0 \quad (38)$$

Suppose that the probability of transitioning from chemical state 1 to state 4 in the time interval between t and $t + \Delta t$ is $f^P(t)\Delta t$. Then,

$$f^P(t)\Delta t = \Delta P_4(t) \quad (39)$$

Therefore, we find

$$f^P(t) = \frac{dP_4(t)}{dt} = \omega_{34}P_3(t) \quad (40)$$

Taking the Laplace transform of eq 40 gives

$$\tilde{f}^P(s) = \omega_{34}\tilde{P}_3(s) \quad (41)$$

Now we can calculate an analytical expression for $\tilde{P}_3(s)$ by solving eq 32. Inserting this expression into eq 41 yields

$$\tilde{f}^P(s) = \frac{\alpha_{12}\alpha_{23}\alpha_{34}}{(s + \omega_1)(s + \omega_2)(s + \omega_3)} \quad (42)$$

where ω_1, ω_2, and ω_3 are the solutions of the equation

$$\omega^3 - a_2\omega^2 + a_1\omega - a_0 = 0 \quad (43)$$

Taking the inverse Laplace transform of eq 42 gives

$$f^P(t) = \frac{\alpha_{12}\alpha_{23}\alpha_{34}}{(\omega_1 - \omega_2)(\omega_1 - \omega_3)}e^{-\omega_1 t} + \frac{\alpha_{12}\alpha_{23}\alpha_{34}}{(\omega_2 - \omega_1)(\omega_2 - \omega_3)}e^{-\omega_2 t} + \frac{\alpha_{12}\alpha_{23}\alpha_{34}}{(\omega_3 - \omega_1)(\omega_3 - \omega_2)}e^{-\omega_3 t} \quad (44)$$

Now, solving for the first moment of this distribution,

$$\langle t_p \rangle = \int_0^\infty t f^P(t)\, dt \quad (45)$$

gives

$$\langle t_p \rangle = \frac{1}{\alpha_{12}}\left[1 + \frac{\alpha_{21}}{\alpha_{23}} + \frac{\alpha_{21}\alpha_{32}}{\alpha_{23}\alpha_{34}}\right] + \frac{1}{\alpha_{23}}\left[1 + \frac{\alpha_{32}}{\alpha_{34}}\right] + \frac{1}{\alpha_{34}} \quad (46)$$

Similarly, the second moment is

$$\langle t_p^{\,2} \rangle = \int_0^\infty t^2 f^P(t)\, dt = \frac{2(a_1^{\,2} - a_0a_2)}{a_0^{\,2}} \quad (47)$$

Analogously, one can also obtain the exact formula for the distribution of the time spent transitioning from chemical state 4 to 1:

$$f^r(t) = \frac{\alpha_{43}\alpha_{32}\alpha_{21}}{(\Omega_1 - \Omega_2)(\Omega_1 - \Omega_3)}e^{-\Omega_1 t} + \frac{\alpha_{43}\alpha_{32}\alpha_{21}}{(\Omega_2 - \Omega_1)(\Omega_2 - \Omega_3)}e^{-\Omega_2 t} + \frac{\alpha_{43}\alpha_{32}\alpha_{21}}{(\Omega_3 - \Omega_1)(\Omega_3 - \Omega_2)}e^{-\Omega_3 t} \quad (48)$$

Here, Ω_1, Ω_2, and Ω_3 are the solutions of the equation

$$\Omega^3 - \Omega^2 b_2 + \Omega b_1 - b_0 = 0 \quad (49)$$

where

$$b_0 = \alpha_{43}\alpha_{32}\alpha_{21} \quad (50)$$

DOI: 10.1021/jp5128805
J. Phys. Chem. B 2015, 119, 10888–10901

$$b_1 = \alpha_{34}\alpha_{21} + \alpha_{34}\alpha_{23} + \alpha_{32}\alpha_{21} + \alpha_{43}\alpha_{32} + \alpha_{43}\alpha_{21} + \alpha_{43}\alpha_{23} \quad (51)$$

$$b_2 = \alpha_{43} + \alpha_{34} + \alpha_{32} + \alpha_{21} + \alpha_{23} \quad (52)$$

Now, solving for the first moment of this distribution gives

$$\langle t_r \rangle = \int_0^\infty t f^r(t)\, dt = \frac{b_1}{b_0} = \frac{1}{\alpha_{43}}\left[1 + \frac{\alpha_{34}}{\alpha_{32}} + \frac{\alpha_{34}\alpha_{23}}{\alpha_{32}\alpha_{21}}\right] + \frac{1}{\alpha_{32}}\left[1 + \frac{\alpha_{23}}{\alpha_{21}}\right] + \frac{1}{\alpha_{21}} \quad (53)$$

and solving for the second moment gives

$$\langle t_r^2 \rangle = \int_0^\infty t^2 f^r(t)\, dt = \frac{2(b_1^2 - b_0 b_2)}{b_0^2} \quad (54)$$

Assuming that the experimentally observed, fractional population of chemical state μ, χ_μ, represents the equilibrium-state solutions, P_μ^{eq}, ratios of χ can be used to write relationships between several α's:

$$\alpha_{21} = \frac{\chi_1}{\chi_2}\alpha_{12} \qquad \alpha_{32} = \frac{\chi_2}{\chi_3}\alpha_{23} \qquad \alpha_{34} = \frac{\chi_4}{\chi_3}\alpha_{43} \quad (55)$$

The expectation values for the time spent transitioning from chemical states 1 to 4 ($\langle t_p \rangle$), and transitioning from chemical states 4 to 1 ($\langle t_r \rangle$), are assumed to be equivalent to the inverses of the experimentally observed transition rates between the two states of a two-state model (k_{12} and k_{21}). Using these expressions and the experimentally observed two-state rates, making substitutions with eq 55, and rearranging yields the following system of equations:

$$\frac{1}{k_{12}} = 1\cdot\frac{1}{\alpha_{12}} + C_1\cdot\frac{1}{\alpha_{23}} + C_2\cdot\frac{1}{\alpha_{43}}; \quad C_1 = \left(\frac{\chi_1}{\chi_2} + 1\right),\; C_2 = \left(\frac{\chi_1 + \chi_2 + \chi_3}{\chi_4}\right) \quad (56)$$

$$\frac{1}{k_{21}} = C_3\cdot\frac{1}{\alpha_{12}} + C_4\cdot\frac{1}{\alpha_{23}} + 1\cdot\frac{1}{\alpha_{43}}; \quad C_3 = \left(\frac{\chi_4 + \chi_3 + \chi_2}{\chi_1}\right),\; C_4 = \left(\frac{\chi_4 + \chi_3}{\chi_2}\right) \quad (57)$$

which has fewer constraints than degrees of freedom. To proceed, we solve the system of equations keeping one degree of freedom independent, moving that term to the left-hand side of the equation, and treating it as part of the constraints.

Independent α_{12}. We solve the system of equations for an independent α_{12} by matrix inversion:

$$\mathbf{B} = \mathbf{AX} \rightarrow \begin{bmatrix} \left(\frac{1}{k_{12}} - 1\cdot\frac{1}{\alpha_{12}}\right) \\ \left(\frac{1}{k_{21}} - C_3\cdot\frac{1}{\alpha_{12}}\right)\end{bmatrix} = \begin{bmatrix} C_1 & C_2 \\ C_4 & 1 \end{bmatrix}\begin{bmatrix} \frac{1}{\alpha_{23}} \\ \frac{1}{\alpha_{43}}\end{bmatrix} \Rightarrow \mathbf{X} = \mathbf{A}^{-1}\mathbf{B} \quad (58)$$

where

$$\mathbf{A}^{-1} = \frac{1}{(C_1\cdot 1 - C_2C_4)}\cdot\begin{bmatrix} 1 & -C_2 \\ -C_4 & C_1 \end{bmatrix}$$

which yields the following rate constants:

$$\alpha_{12} = \text{independent}$$

$$\alpha_{21} = \left(\frac{\chi_1}{\chi_2}\right)\alpha_{12}$$

$$\alpha_{23} = \frac{(C_1 - C_2C_4)(k_{12}k_{21}\alpha_{12})}{[1\cdot(k_{21}\alpha_{12} - k_{12}k_{21}) - C_2\cdot(k_{12}\alpha_{12} - C_3k_{12}k_{21})]}$$

$$\alpha_{32} = \left(\frac{\chi_2}{\chi_3}\right)\alpha_{23}$$

$$\alpha_{34} = \left(\frac{\chi_4}{\chi_3}\right)\alpha_{43}$$

$$\alpha_{43} = \frac{(C_1 - C_2C_4)(k_{12}k_{21}\alpha_{12})}{[-C_4\cdot(k_{21}\alpha_{12} - k_{12}k_{21}) + C_1\cdot(k_{12}\alpha_{12} - C_3k_{12}k_{21})]}$$

Independent α_{23}. Similarly, for an independent α_{23},

$$\alpha_{12} = \frac{(1 - C_2C_3)(k_{12}k_{21}\alpha_{23})}{[1\cdot(k_{21}\alpha_{23} - C_1k_{12}k_{21}) - C_2\cdot(k_{12}\alpha_{23} - C_4k_{12}k_{21})]}$$

$$\alpha_{21} = \left(\frac{\chi_1}{\chi_2}\right)\alpha_{12}$$

$$\alpha_{23} = \text{independent}$$

$$\alpha_{32} = \left(\frac{\chi_2}{\chi_3}\right)\alpha_{23}$$

$$\alpha_{34} = \left(\frac{\chi_4}{\chi_3}\right)\alpha_{43}$$

$$\alpha_{43} = \frac{(1 - C_2C_3)(k_{12}k_{21}\alpha_{23})}{[-C_3\cdot(k_{21}\alpha_{23} - C_1k_{12}k_{21}) + 1\cdot(k_{12}\alpha_{23} - C_4k_{12}k_{21})]}$$

Independent α_{43}. Finally, for an independent α_{43},

$$\alpha_{12} = \frac{(C_4 - C_1C_3)(k_{12}k_{21}\alpha_{43})}{[C_4\cdot(k_{21}\alpha_{43} - C_2k_{12}k_{21}) - C_1\cdot(k_{12}\alpha_{43} - k_{12}k_{21})]}$$

$$\alpha_{21} = \left(\frac{\chi_1}{\chi_2}\right)\alpha_{12}$$

$$\alpha_{23} = \frac{(C_4 - C_1C_3)(k_{12}k_{21}\alpha_{43})}{[-C_3\cdot(k_{21}\alpha_{43} - C_2k_{12}k_{21}) + 1\cdot(k_{12}\alpha_{43} - k_{12}k_{21})]}$$

$$\alpha_{32} = \left(\frac{\chi_2}{\chi_3}\right)\alpha_{23}$$

$$\alpha_{34} = \left(\frac{\chi_4}{\chi_3}\right)\alpha_{43}$$

DOI: 10.1021/jp5128805
J. Phys. Chem. B 2015, 119, 10888–10901

The Journal of Physical Chemistry B Article

α_{43} = independent

B. Three-State Model

Given the following linear, three-state model,

$$1 \underset{\alpha_{21}}{\overset{\alpha_{12}}{\rightleftharpoons}} 2 \underset{\alpha_{32}}{\overset{\alpha_{23}}{\rightleftharpoons}} 3$$

the time evolution of the probability, $P_\mu(t)$, will be governed by

$$\frac{dP_1(t)}{dt} = -\alpha_{12}P_1(t) + \alpha_{21}P_2(t) \tag{59}$$

$$\frac{dP_2(t)}{dt} = \alpha_{12}P_1(t) - (\alpha_{21} + \alpha_{23})P_2(t) + \alpha_{32}P_3(t) \tag{60}$$

$$\frac{dP_3(t)}{dt} = \alpha_{23}P_2(t) - \alpha_{32}P_3(t) \tag{61}$$

We calculate the equilibrium-state probabilities,

$$P_1^{eq} = \frac{\alpha_{21}\alpha_{32}}{\alpha_{21}\alpha_{32} + \alpha_{12}\alpha_{32} + \alpha_{23}\alpha_{12}} \tag{62}$$

$$P_2^{eq} = \frac{\alpha_{12}\alpha_{32}}{\alpha_{21}\alpha_{32} + \alpha_{12}\alpha_{32} + \alpha_{23}\alpha_{12}} \tag{63}$$

$$P_3^{eq} = \frac{\alpha_{23}\alpha_{12}}{\alpha_{21}\alpha_{32} + \alpha_{12}\alpha_{32} + \alpha_{23}\alpha_{12}} \tag{64}$$

Then, we also calculate the distribution of the time spent by a molecule transitioning from chemical states 1 to 3. For this purpose, we modify the scheme into

$$1 \underset{\alpha_{21}}{\overset{\alpha_{12}}{\rightleftharpoons}} 2 \overset{\alpha_{23}}{\longrightarrow} 3$$

and write the master equations according to this new scheme:

$$\frac{dP_1(t)}{dt} = -\alpha_{12}P_1(t) + \alpha_{21}P_2(t) \tag{65}$$

$$\frac{dP_2(t)}{dt} = \alpha_{12}P_1(t) - (\alpha_{21} + \alpha_{23})P_2(t) \tag{66}$$

$$\frac{dP_3(t)}{dt} = \alpha_{23}P_2(t) \tag{67}$$

As in the four-state result, we can solve these equations with the Laplace transform method to yield

$$f^{p}(t) = \frac{\alpha_{12}\alpha_{23}}{\omega_2 - \omega_1} e^{-\omega_1 t} + \frac{\alpha_{12}\alpha_{23}}{\omega_1 - \omega_2} e^{-\omega_2 t} \tag{68}$$

where ω_1 and ω_2 are the solutions of the equation

$$\omega^2 - \omega(\alpha_{12} + \alpha_{21} + \alpha_{23}) + \alpha_{12}\alpha_{23} = 0 \tag{69}$$

Now, solving for the first moment,

$$\langle t_p \rangle = \frac{1}{\alpha_{12}}\left[1 + \frac{\alpha_{21}}{\alpha_{23}}\right] + \frac{1}{\alpha_{23}} \tag{70}$$

and the second moment,

$$\langle t_p^{\,2} \rangle = \frac{2(c_1^{\,2} - c_0)}{c_0^{\,2}} \tag{71}$$

where

$$c_0 = \alpha_{12}\alpha_{23} \tag{72}$$

$$c_1 = \alpha_{12} + \alpha_{21} + \alpha_{23} \tag{73}$$

Similarly, we can also calculate the distribution of the time spent transitioning from chemical states 3 to 1,

$$f^{r}(t) = \frac{\alpha_{21}\alpha_{32}}{\Omega_2 - \Omega_1} e^{-\Omega_1 t} + \frac{\alpha_{21}\alpha_{32}}{\Omega_1 - \Omega_2} e^{-\Omega_2 t} \tag{74}$$

where Ω_1 and Ω_2 are the solutions of the equation

$$\Omega^2 - \Omega(\alpha_{32} + \alpha_{23} + \alpha_{21}) + \alpha_{32}\alpha_{21} = 0 \tag{75}$$

In this case, we calculate the first moment,

$$\langle t_r \rangle = \frac{1}{\alpha_{32}}\left[1 + \frac{\alpha_{23}}{\alpha_{21}}\right] + \frac{1}{\alpha_{21}} \tag{76}$$

and the second moment,

$$\langle t_r^{\,2} \rangle = \frac{2(d_1^{\,2} - d_0)}{d_0^{\,2}} \tag{77}$$

where

$$d_0 = \alpha_{32}\alpha_{21} \tag{78}$$

$$d_1 = \alpha_{32} + \alpha_{23} + \alpha_{21} \tag{79}$$

Assuming that the experimentally observed, fractional population of state μ, χ_μ, represents the equilibrium-state solutions, P_μ^{eq}, ratios of χ can be used to write relationships between several α:

$$\alpha_{21} = \frac{\chi_1}{\chi_2}\alpha_{12} \quad \text{and} \quad \alpha_{23} = \frac{\chi_3}{\chi_2}\alpha_{32} \tag{80}$$

The expectation values for the time spent transitioning from chemical states 1 to 3 ($\langle t_p \rangle$), and from chemical states 3 to 1 ($\langle t_r \rangle$), are assumed to be equivalent to the inverses of the experimentally observed transition rates between the two final states of a two-state model, k_{12} and k_{21}. Using these expressions and experimentally observed rates, substituting eq 80, and then rearranging yields the following system of equations:

$$\langle t_p \rangle = \frac{1}{k_{12}} = 1\cdot\frac{1}{\alpha_{12}} + C_1\cdot\frac{1}{\alpha_{32}}; \quad C_1 \equiv \left(\frac{\chi_1}{\chi_3} + 1\right) \tag{81}$$

$$\langle t_r \rangle = \frac{1}{k_{21}} = C_2\cdot\frac{1}{\alpha_{12}} + 1\cdot\frac{1}{\alpha_{32}}; \quad C_2 \equiv \left(\frac{\chi_3 + \chi_2}{\chi_1}\right) \tag{82}$$

which can be solved as for the four-state model,

$$\mathbf{B} = \mathbf{AX} \rightarrow \begin{bmatrix} \frac{1}{k_{12}} \\ \frac{1}{k_{21}} \end{bmatrix} = \begin{bmatrix} 1 & C_1 \\ C_2 & 1 \end{bmatrix}\begin{bmatrix} \frac{1}{\alpha_{12}} \\ \frac{1}{\alpha_{32}} \end{bmatrix} \Rightarrow \mathbf{X}$$

$$= \begin{bmatrix} \frac{1}{\alpha_{12}} \\ \frac{1}{\alpha_{32}} \end{bmatrix} = \frac{1}{(1 - C_1C_2)}\cdot\begin{bmatrix} 1\cdot\frac{1}{k_{12}} - C_2\cdot\frac{1}{k_{21}} \\ -C_2\cdot\frac{1}{k_{12}} + 1\cdot\frac{1}{k_{21}} \end{bmatrix} \tag{83}$$

to yield the three-state model rate constants,

$$\alpha_{12} = \frac{(1 - C_1C_2)(k_{12}k_{21})}{(1\cdot k_{21} - C_1\cdot k_{12})}$$

DOI: 10.1021/jp5128805
J. Phys. Chem. B 2015, 119, 10888–10901

$$\alpha_{21} = \left(\frac{\chi_1}{\chi_2}\right)\alpha_{12}$$

$$\alpha_{23} = \left(\frac{\chi_3}{\chi_2}\right)\alpha_{32}$$

$$\alpha_{32} = \frac{(1 - C_1C_2)(k_{12}k_{21})}{(-C_2 \cdot k_{21} - 1 \cdot k_{12})}$$

C. smFRET Simulations

In order to simulate PRE complex dynamics with transient intermediates, we estimated the values of E_{FRET} for each PRE complex conformational state. We estimated E_{FRET} for the L1–tRNA labeling scheme by measuring the distances between the β-carbon of the threonine at position 202 of the L1 protein of the 50S ribosomal subunit and the sulfur of the thiouridine at position 8 of tRNA^{fMet} from the atomic-resolution, molecular dynamics flexible fitting of the classes of PRE complexes deposited in the Protein Data Bank by Agirrezabala and co-workers (Table 4).[51] While the absolute accuracy of these estimates is likely imprecise, it is reasonable to interpret the relative distances as an informative measure of the relative E_{FRET} values. As the distances measured ignore any foreshortening due to the space occupied by the fluorophore and its hydrocarbon linker, subsequent analysis compensated by overestimating R_0 = 60 Å. Notably, all the classes measured yielded distinct values of E_{FRET} except for classes 2 and 4A (MS I and IS1, respectively). Since the distances between the labeling sites on the L1 protein and the P-site tRNA are 78 Å ($E_{FRET} \approx 0.17$) and 81 Å ($E_{FRET} \approx 0.14$) for classes 2 and 4A, respectively, MS I and IS1 are most likely indistinguishable, given the SBR of the TIRF-based smFRET measurements used by Fei and co-workers.[34,35] As such, we chose to group MS I and IS1 together into state 1, while IS2 corresponded to state 2, and MS II corresponded to state 3.

Table 4. Distances and Approximate FRET Efficiencies of PRE$_{fM/F}$ Ribosomes from Cryo-EM Structures

class	r(L1–tRNA) (Å)	E_{FRET}, R_0 = 60 Å
2	78	0.17
4A	81	0.14
4B	64	0.41
5	43	0.89
6	55	0.62

From these E_{FRET} estimates, Markovian transitions along a linear, three-state kinetic scheme were then simulated for 100 state versus time trajectories. Each state versus time trajectory was 50 s in length, and they were eventually transformed into discrete E_{FRET} versus time trajectories, where each data point is the mean E_{FRET} value during a 50 ms time period. For each state versus time trajectory, the distances between the donor and acceptor fluorophores in the ith state, r_i, were randomized for each time series with a normal distribution, $\mathcal{N}(\mu = r_i, \sigma = 2$ Å). The E_{FRET} of each state was then calculated as $E_{FRET} = (1 + (r/R_0)^6)^{-1}$, where R_0 is the Förster radius (a parameter dependent upon the identity of donor and acceptor fluorophores, as well their local environments). R_0 was randomized for each time series within a reasonable range for the Cy3-Cy5 FRET donor–acceptor pair with a normal distribution, $\mathcal{N}(\mu = 60$ Å, $\sigma = 2$ Å). The E_{FRET} versus time trajectory was then discretized by calculating the average value of E_{FRET} during sequential 50 ms long time periods. Noise was added to the E_{FRET} versus time trajectories that was normally distributed at each data point with a standard deviation of 0.05—a reasonable SBR for data collected on the TIRF microscope used in the smFRET experiments.

In order to model the histogram of the simulated E_{FRET} versus time trajectories, we used a Gaussian mixture model where each state is modeled to contribute as a normal distribution centered at the respective mean E_{FRET} value for that state, and is weighted by the equilibrium-state probability for that state (see Appendix B). To account for the simulated heterogeneity in the ensemble of synthetic E_{FRET} versus time trajectories, the mean E_{FRET} value of each state was marginalized out by integrating over the joint-probability distribution of the normal distribution of E_{FRET} and a beta distribution of the mean E_{FRET} observed in that state (Figure 4C) with parameters determined by a maximum likelihood estimate from the exact simulated E_{FRET} means. For the "Avg. State" distribution (cf. Figure 4B), the distribution of E_{FRET} means that was employed was beta distributed with a linear combination of the parameters of the mean E_{FRET} distributions from states 1 and 2 (Figure 4C). The standard deviation used for the normal distribution of E_{FRET} values for each state was taken exactly as the standard deviation used to add noise to the synthetic E_{FRET} versus time trajectories.

As described in section 3.5, this model of the observed E_{FRET} value histograms is for a temporally-resolved histogram without any blurring present. Performing the same simulation described above, but with a 0.5 ms acquisition time period yields an accurately modeled set of E_{FRET} versus time trajectories (Figure 5). Furthermore, analyzing this data with ebFRET yields an accurately estimated number of states, rate constants, distribution of E_{FRET} means, and noise parameter.

Analysis of the 50 ms time resolution data with ebFRET found the most evidence for a five-state kinetic model—all of which were significantly populated. Since the synthetic data was simulated with a three-state kinetic model, the ebFRET analysis is not consistent with the original simulation. Additionally, the rate constants inferred by ebFRET for a three-state model from the simulated data do not match the original simulation parameters. The rate constants inferred by ebFRET for the two-state model were $k_{GS1\rightarrow GS2} = 1.41 \pm 0.05$ s^{-1} and $k_{GS2\rightarrow GS1} = 1.41 \pm 0.05$ s^{-1}, but these differ from those learned from the data of Fei and co-workers (cf. section 3.2). This suggests that the original simulation parameters are not consistent with the experimental data from the smFRET study of Fei and co-workers.[34,35] Most likely, the discrepancy is due to experimental differences between the smFRET and cryo-EM studies that were compared using the general framework presented here.

■ AUTHOR INFORMATION

Corresponding Author

*E-mail: debch@iitk.ac.in.

Present Address

¶A.K.S.: Chemistry Department, University Park, PA 16802, United States.

Author Contributions

‖C.D.K.-T. and A.K.S. contributed equally.

Notes

The authors declare no competing financial interest.

DOI: 10.1021/jp5128805
J. Phys. Chem. B 2015, 119, 10888–10901

■ ACKNOWLEDGMENTS

The authors thank Dr. Jingyi Fei for comments on the manuscript, and Ms. Melissa Thomas and Dr. Wen Li for assistance in the preparation of the figures. This work is supported in part by the U.S. Department of Energy Office of Science Graduate Fellowship Program (DE-AC05-06OR23100) (C.D.K.-T.); the Council of Scientific and Industrial Research (A.K.S.); the Howard Hughes Medical Institute and NIH-NIGMS grants (R01 GM29169 and R01 GM55440) (J.F.); a Burroughs Wellcome Fund CABS Award (CABS 1004856), an NSF CAREER Award (MCB 0644262), an NIH-NIGMS grant (R01 GM084288), an American Cancer Society Research Scholar Grant (RSG GMC-117152), and a Camille Dreyfus Teacher-Scholar Award (R.L.G.); and the Prof. S. Sampath Chair Professorship and a J.C. Bose National Fellowship (D.C).

■ REFERENCES

(1) Astumian, R. D. Thermodynamics and Kinetics of a Brownian Motor. *Science* **1997**, *276*, 917–922.

(2) Alberts, B. The Cell as a Collection of Protein Machines: Preparing the Next Generation of Molecular Biologists. *Cell* **1998**, *92*, 291–294.

(3) Peskin, C. S.; Odell, G. M.; Oster, G. F. Cellular Motions and Thermal Fluctuations: The Brownian Ratchet. *Biophys. J.* **1993**, *65*, 316–324.

(4) Frank, J. Introduction. In *Molecular Machines in Biology: Workshop of the Cell*; Frank, J., Ed.; Cambridge University Press: New York, 2011; pp 1–3.

(5) Bustamante, C. In singulo biochemistry: when less is more. *Annu. Rev. Biochem.* **2008**, *77*, 45–50.

(6) Chowdhury, D. Statistical Mechanical Treatment of Molecular Machines. In *Molecular Machines in Biology: Workshop of the Cell*; Frank, J., Ed.; Cambridge University Press: New York, 2011; pp 38–58.

(7) Chowdhury, D. Modeling stochastic kinetics of molecular machines at multiple levels: from molecules to modules. *Biophys. J.* **2013**, *104*, 2331–2341.

(8) Kolomeisky, A. B. Motor proteins and molecular motors: how to operate machines at the nanoscale. *J. Phys.: Condens. Matter* **2013**, *25*, No. 463101.

(9) Tinoco, I.; Wen, J.-D. Simulation and analysis of single-ribosome translation. *Phys. Biol.* **2009**, *6*, No. 025006.

(10) Tinoco, I.; Gonzalez, R. L. Biological mechanisms, one molecule at a time. *Genes Dev.* **2011**, *25*, 1205–1231.

(11) Xie, S. Single-Molecule Approach to Enzymology. *Single Mol.* **2001**, *2*, 229–236.

(12) Frank, J. Story in a sample—the potential (and limitations) of cryo-electron microscopy applied to molecular machines. *Biopolymers* **2013**, *99*, 832–836.

(13) Zwanzig, R. *Nonequilibrium Statistical Mechanics*;Oxford University Press: Oxford, 2001; pp 67–82.

(14) Kinz-Thompson, C. D.; Gonzalez, R. L. smFRET studies of the 'encounter' complexes and subsequent intermediate states that regulate the selectivity of ligand binding. *FEBS Lett.* **2014**, *588*, 3526–3538.

(15) Frank, J.; Gonzalez, R. L. Structure and dynamics of a processive Brownian motor: the translating ribosome. *Annu. Rev. Biochem.* **2010**, *79*, 381–412.

(16) Munro, J. B.; Sanbonmatsu, K. Y.; Spahn, C. M. T.; Blanchard, S. C. Navigating the ribosome's metastable energy landscape. *Trends Biochem. Sci.* **2009**, *34*, 390–400.

(17) Schmeing, T. M.; Ramakrishnan, V. What recent ribosome structures have revealed about the mechanism of translation. *Nature* **2009**, *461*, 1234–1242.

(18) Voorhees, R. M.; Ramakrishnan, V. Structural basis of the translational elongation cycle. *Annu. Rev. Biochem.* **2013**, *82*, 203–236.

(19) Johansson, M.; Lovmar, M.; Ehrenberg, M. N. Rate and accuracy of bacterial protein synthesis revisited. *Curr. Opin. Microbiol.* **2008**, *11*, 141–147.

(20) Rodnina, M. V. Quality control of mRNA decoding on the bacterial ribosome. In *Advances in Protein Chemistry and Structural Biology*, 1st ed.; Elsevier Inc.: Amsterdam, 2012; Vol.*86*, pp 95–128.

(21) Rodnina, M. V. Mechanisms of decoding and peptide bond formation. In *Ribosomes Structure, Function, and Dynamics*; Rodnina, M. V., Wintermeyer, W., Green, R., Eds.; Springer-Verlag: Berlin, 2011; pp 199–212.

(22) Simonović, M.; Steitz, T. A. A structural view on the mechanism of the ribosome-catalyzed peptide bond formation. *Biochim. Biophys. Acta* **2009**, *1789*, 612–623.

(23) Beringer, M.; Rodnina, M. V. The ribosomal peptidyl transferase. *Mol. Cell* **2007**, *26*, 311–321.

(24) Moazed, D.; Noller, H. F. Intermediate states in the movement of transfer RNA in the ribosome. *Nature* **1989**, *342*, 142–148.

(25) Chen, J.; Tsai, A.; O'Leary, S. E.; Petrov, A.; Puglisi, J. D. Unraveling the dynamics of ribosome translocation. *Curr. Opin. Struct. Biol.* **2012**, *22*, 804–814.

(26) Rodnina, M. V.; Wintermeyer, W. The ribosome as a molecular machine: the mechanism of tRNA-mRNA movement in translocation. *Biochem. Soc. Trans.* **2011**, *39*, 658–662.

(27) Cooperman, B. S.; Goldman, Y. E.; Chen, C.; Farrell, I.; Kaur, J.; Liu, H.; Liu, W.; Rosenblum, G.; Smilansky, Z.; Stevens, B.; et al.Mechanism and dynamics of the elongation cycle. In *Ribosomes Structure, Function, and Dynamics*; Rodnina, M. V., Wintermeyer, W., Green, R., Eds.; Springer-Verlag: Berlin, 2011; pp 339–348.

(28) Noller, H. F.; Ermolenko, D. N.; Korostelev, A.; Laurberg, M.; Zhu, J.; Asahara, H.; Lancaster, L.; Horan, L.; Hirschi, A.; Donohue, J. P.; et al. Studies on the mechanisms of translocation and termination. In *Ribosomes Structure, Function, and Dynamics*; Rodnina, M. V., Wintermeyer, W., Green, R., Eds.; Springer-Verlag: Berlin, 2011; pp 349–360.

(29) Munro, J. B.; Altman, R. B.; Tung, C.-S.; Cate, J. H. D.; Sanbonmatsu, K. Y.; Blanchard, S. C. Spontaneous formation of the unlocked state of the ribosome is a multistep process. *Proc. Natl. Acad. Sci. U.S.A.* **2010**, *107*, 709–714.

(30) Shoji, S.; Walker, S. E.; Fredrick, K. Ribosomal translocation: one step closer to the molecular mechanism. *ACS Chem. Biol.* **2009**, *4*, 93–107.

(31) Gavrilova, L. P.; Kostiashkina, O. E.; Koteliansky, V. E.; Rutkevitch, N. M.; Spirin, A. S. Factor-free ("non-enzymic") and factor-dependent systems of translation of polyuridylic acid by Escherichia coli ribosomes. *J. Mol. Biol.* **1976**, *101*, 537–552.

(32) Frank, J.; Agrawal, R. K. A ratchet-like inter-subunit reorganization of the ribosome during translocation. *Nature* **2000**, *406*, 318–322.

(33) Frank, J. Intermediate states during mRNA-tRNA translocation. *Curr. Opin. Struct. Biol.* **2012**, *22*, 778–785.

(34) Fei, J.; Kosuri, P.; MacDougall, D. D.; Gonzalez, R. L. Coupling of ribosomal L1 stalk and tRNA dynamics during translation elongation. *Mol. Cell* **2008**, *30*, 348–359.

(35) Fei, J.; Bronson, J. E.; Hofman, J. M.; Srinivas, R. L.; Wiggins, C. H.; Gonzalez, R. L. Allosteric collaboration between elongation factor G and the ribosomal L1 stalk directs tRNA movements during translation. *Proc. Natl. Acad. Sci. U.S.A.* **2009**, *106*, 15702–15707.

(36) Blanchard, S. C.; Kim, H. D.; Gonzalez, R. L.; Puglisi, J. D.; Chu, S. tRNA dynamics on the ribosome during translation. *Proc. Natl. Acad. Sci. U.S.A.* **2004**, *101*, 12893–12898.

(37) Cornish, P. V.; Ermolenko, D. N.; Noller, H. F.; Ha, T. Spontaneous intersubunit rotation in single ribosomes. *Mol. Cell* **2008**, *30*, 578–588.

(38) Cornish, P. V.; Ermolenko, D. N.; Staple, D. W.; Hoang, L.; Hickerson, R. P.; Noller, H. F.; Ha, T. Following movement of the L1 stalk between three functional states in single ribosomes. *Proc. Natl. Acad. Sci. U.S.A.* **2009**, *106*, 2571–2576.

(39) Wang, Y.; Qin, H.; Kudaravalli, R. D.; Kirillov, S. V.; Dempsey, G. T.; Pan, D.; Cooperman, B. S.; Goldman, Y. E. Single-molecule

DOI: 10.1021/jp5128805
J. Phys. Chem. B 2015, 119, 10888–10901

structural dynamics of EF-G-ribosome interaction during translocation. *Biochemistry* **2007**, *46*, 10767–10775.
(40) Chen, C.; Stevens, B.; Kaur, J.; Cabral, D.; Liu, H.; Wang, Y.; Zhang, H.; Rosenblum, G.; Smilansky, Z.; Goldman, Y. E.; et al. Single-molecule fluorescence measurements of ribosomal translocation dynamics. *Mol. Cell* **2011**, *42*, 367–377.
(41) Kim, H. D.; Puglisi, J. D.; Chu, S. Fluctuations of transfer RNAs between classical and hybrid states. *Biophys. J.* **2007**, *93*, 3575–3582.
(42) Munro, J. B.; Altman, R. B.; O'Connor, N.; Blanchard, S. C. Identification of two distinct hybrid state intermediates on the ribosome. *Mol. Cell* **2007**, *25*, 505–517.
(43) Wang, B.; Ho, J.; Fei, J.; Gonzalez, R. L.; Lin, Q. A microfluidic approach for investigating the temperature dependence of biomolecular activity with single-molecule resolution. *Lab Chip* **2011**, *11*, 274–281.
(44) Ning, W.; Fei, J.; Gonzalez, R. L. The ribosome uses cooperative conformational changes to maximize and regulate the efficiency of translation. *Proc. Natl. Acad. Sci. U.S.A.* **2014**, *111*, 12073–12078.
(45) Fei, J.; Richard, A. C.; Bronson, J. E.; Gonzalez, R. L. Transfer RNA-mediated regulation of ribosome dynamics during protein synthesis. *Nat. Struct. Mol. Biol.* **2011**, *18*, 1043–1051.
(46) Wang, L.; Pulk, A.; Wasserman, M. R.; Feldman, M. B.; Altman, R. B.; Cate, J. H. D.; Blanchard, S. C. Allosteric control of the ribosome by small-molecule antibiotics. *Nat. Struct. Mol. Biol.* **2012**, *19*, 957–963.
(47) Valle, M.; Zavialov, A.; Sengupta, J.; Rawat, U.; Ehrenberg, M. N.; Frank, J. Locking and unlocking of ribosomal motions. *Cell* **2003**, *114*, 123–134.
(48) Frank, J.; Gao, H.; Sengupta, J.; Gao, N.; Taylor, D. J. The process of mRNA-tRNA translocation. *Proc. Natl. Acad. Sci. U.S.A.* **2007**, *104*, 19671–19678.
(49) Agirrezabala, X.; Lei, J.; Brunelle, J. L.; Ortiz-Meoz, R. F.; Green, R.; Frank, J. Visualization of the hybrid state of tRNA binding promoted by spontaneous ratcheting of the ribosome. *Mol. Cell* **2008**, *32*, 190–197.
(50) Trabuco, L. G.; Schreiner, E.; Eargle, J.; Cornish, P.; Ha, T.; Luthey-Schulten, Z.; Schulten, K. The role of L1 stalk-tRNA interaction in the ribosome elongation cycle. *J. Mol. Biol.* **2010**, *402*, 741–760.
(51) Agirrezabala, X.; Liao, H. Y.; Schreiner, E.; Fu, J.; Ortiz-Meoz, R. F.; Schulten, K.; Green, R.; Frank, J. Structural characterization of mRNA-tRNA translocation intermediates. *Proc. Natl. Acad. Sci. U.S.A.* **2012**, *109*, 6094–6099.
(52) Garai, A.; Chowdhury, D.; Chowdhury, D.; Ramakrishnan, T. Stochastic kinetics of ribosomes: Single motor properties and collective behavior. *Phys. Rev. E* **2009**, *80*, No. 011908.
(53) Sharma, A. K.; Chowdhury, D. Distribution of dwell times of a ribosome: effects of infidelity, kinetic proofreading and ribosome crowding. *Phys. Biol.* **2011**, *8*, No. 026005.
(54) Bronson, J. E.; Fei, J.; Hofman, J. M.; Gonzalez, R. L.; Wiggins, C. H. Learning rates and states from biophysical time series: a Bayesian approach to model selection and single-molecule FRET data. *Biophys. J.* **2009**, *97*, 3196–3205.
(55) Bronson, J. E.; Hofman, J. M.; Fei, J.; Gonzalez, R. L.; Wiggins, C. H. Graphical models for inferring single molecule dynamics. *BMC Bioinformatics* **2010**, *11*, S2.
(56) van de Meent, J.-W.; Bronson, J. E.; Wood, F.; Gonzalez, R. L.; Wiggins, C. H. Hierarchically-coupled hidden Markov models for learning kinetic rates from singlemolecule data. In *Proceedings of the 30th International Conference on Machine Learning*, Atlanta, GA; Dasgupta, S., McAllester, D., Eds.; Microtome Publishing: Brookline, MA, 2013; Vol. *28*, Cycle 2 Papers, pp 361–369.
(57) van de Meent, J.-W.; Bronson, J. E.; Wiggins, C. H.; Gonzalez, R. L., Jr. Empirical Bayes Methods Enable Advanced Population-Level Analyses of Single-Molecule FRET Experiments. *Biophys. J.* **2014**, *106*, 1327–1337.

DOI: 10.1021/jp5128805
J. Phys. Chem. B 2015, 119, 10888–10901

RESEARCH ARTICLE

BIOCHEMISTRY

2015 © The Authors, some rights reserved; exclusive licensee American Association for the Advancement of Science. Distributed under a Creative Commons Attribution NonCommercial License 4.0 (CC BY-NC).
10.1126/sciadv.1500169

Activation of GTP hydrolysis in mRNA-tRNA translocation by elongation factor G

Wen Li,[1]* Zheng Liu,[1]* Ravi Kiran Koripella,[2] Robert Langlois,[1] Suparna Sanyal,[2] Joachim Frank[1,3,4]†

During protein synthesis, elongation of the polypeptide chain by each amino acid is followed by a translocation step in which mRNA and transfer RNA (tRNA) are advanced by one codon. This crucial step is catalyzed by elongation factor G (EF-G), a guanosine triphosphatase (GTPase), and accompanied by a rotation between the two ribosomal subunits. A mutant of EF-G, H91A, renders the factor impaired in guanosine triphosphate (GTP) hydrolysis and thereby stabilizes it on the ribosome. We use cryogenic electron microscopy (cryo-EM) at near-atomic resolution to investigate two complexes formed by EF-G H91A in its GTP state with the ribosome, distinguished by the presence or absence of the intersubunit rotation. Comparison of these two structures argues in favor of a direct role of the conserved histidine in the switch II loop of EF-G in GTPase activation, and explains why GTP hydrolysis cannot proceed with EF-G bound to the unrotated form of the ribosome.

INTRODUCTION

The ribosome translates the genetically encoded sequence of codons along mRNA into a polypeptide chain, which is subsequently folded into a protein. The ribosome has three binding sites for transfer RNA (tRNA): the aminoacyl (A) site, peptidyl (P) site, and exit (E) site. In each cycle of the peptide chain elongation, an amino acid is delivered to the ribosome at the A site by a cognate aminoacylated tRNA (aa-tRNA) bound with elongation factor Tu (EF-Tu). After EF-Tu separates from the ribosome and the peptide is transferred from the P-site tRNA to the amino acid on the A-site tRNA, the ribosome equilibrates between two conformations, termed "rotated" and "nonrotated," distinguished by a ~9° rotation between the two subunits (*1–5*). Efficient translocation of mRNA-tRNAs, a process required to move a new codon into the A site, requires the binding of elongation factor G (EF-G) to the ribosome and subsequent guanosine triphosphate (GTP) hydrolysis (*6, 7*). A recent x-ray study reported a compact conformation of EF-G bound with the pretranslocational ribosome, indicating that the translocation depends on the structural transition of EF-G, starting from a compact conformation (*8*). In spite of significant progress by x-ray crystallography (*9–12*) and cryogenic electron microscopy (cryo-EM) (*13, 14*), and studies by single-molecule fluorescence resonance energy transfer (smFRET) (*15–18*) to characterize the binding interactions of EF-G with the ribosome, the role of EF-G in translocation is still not well understood. An smFRET study showed that EF-G•GTP is able to bind to the ribosome in both rotated and nonrotated states, but that only the former is productive in leading to GTP hydrolysis and translocation (*19*). We reasoned that structural investigation of these two different binding states might give us clues on the mechanism of guanosine triphosphatase (GTPase) activation. Imaging of EF-G•GTP–bound complexes for structural visualization is challenged by their very short life span because GTP hydrolysis is a very fast process and results in rapid conformational changes downstream. A solution to circumvent this problem in the past was to use nonhydrolyzable GTP analogs, such as GDPNP (*9, 10*) or GMPPCP (*12*), as a substitute for GTP, thereby blocking the complex in the state before GTP hydrolysis. The structures obtained in this way showed that the His[87] residue in *Thermus thermophilus* EF-G (His[91] in *Escherichia coli* EF-G) is situated one water molecule away from the GTP molecule. The role of this histidine residue in the activation of GTP hydrolysis was proposed to either be direct, through catalytic activation of the water molecule for nucleophilic attack (*20–22*), or indirect, through stabilization of the structural elements surrounding GTP; see review in (*23*). In an attempt to resolve the question of the role of H91 in GTPase activation and to shed further light on the interactions between EF-G and the ribosome, we used cryo-EM to image a sample of translating ribosomes in the presence of EF-G in which His[91] is mutated to Ala (see Materials and Methods).

RESULTS AND DISCUSSION

Biochemical analysis shows that a mutant of EF-G, H91A, hydrolyzes GTP at a substantially slower rate ($k_{GTP\ H91A} = 28 \pm 4.5\ s^{-1}$) than the wild type ($k_{GTP\ WT} = 202 \pm 29\ s^{-1}$) does on the ribosome (Fig. 1). The same EF-G mutant has been reported earlier as being completely inactive in GTP hydrolysis under different experimental conditions (*24, 25*). Also, in the case of EF-Tu, the corresponding H84A mutation was shown to be defective in GTP hydrolysis by five orders of magnitude (*26, 27*). These results, along with the smFRET results cited above (*19*), suggest that with H91A, stable EF-G–ribosome complexes suitable for imaging can even be obtained in the presence of authentic GTP; this is confirmed by co-sedimentation analysis by ultracentrifugation followed by SDS–polyacrylamide gel electrophoresis (SDS-PAGE) (Fig. 2).

Refined density maps were obtained for four major classes (fig. S1 and Fig. 3), with resolutions in the range of 3.6 to 5.7 Å (Table 1, movies S1 to S3, and fig. S2), whose major differences are seen in a combination of presence or absence of EF-G and presence or absence of intersubunit rotation. The atomic details were extracted in terms of existing x-ray crystallographic structures of the 70*S E. coli* ribosome

[1]Department of Biochemistry and Molecular Biophysics, Columbia University, 2-221 Blackwell, 165 West 168th Street, New York, NY 10032, USA. [2]Department of Cell and Molecular Biology, Uppsala University, Box-596, BMC, 75124 Uppsala, Sweden. [3]Howard Hughes Medical Institute, Columbia University, New York, NY 10032, USA. [4]Department of Biological Sciences, Columbia University, New York, NY 10027, USA.
*These authors contributed equally to this work.
†Corresponding author. E-mail: jf2192@cumc.columbia.edu

RESEARCH ARTICLE

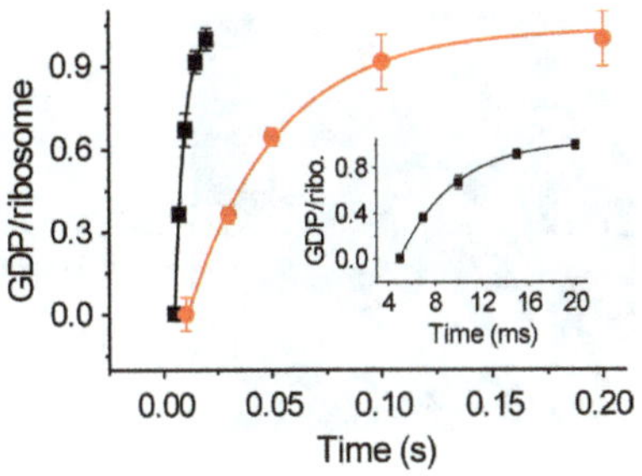

Fig. 1. Single-turnover GTP hydrolysis. Time course of single-turnover GTP hydrolysis by wild-type (WT; black) and H91A (red) EF-G. The amount of [^{3}H]GDP produced is plotted against time and fitted with a single exponential function. The rate constants are 202 ± 29 s^{-1} for WT and 28 ± 4.5 s^{-1} for H91A EF-G. Error bars represent SDs obtained from three independent experiments. Ribo., ribosome.

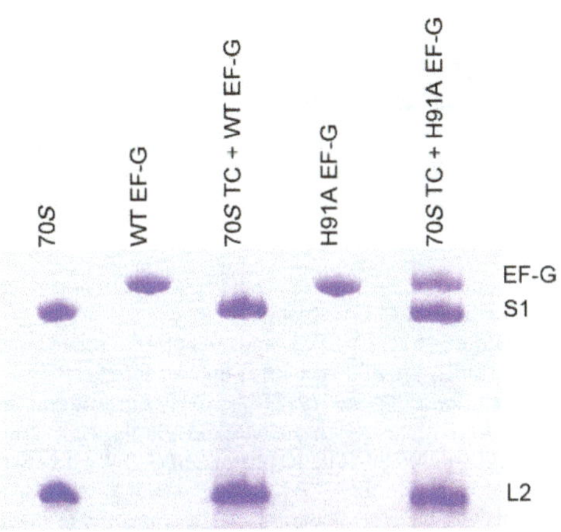

Fig. 2. Occupancy of H91A EF-G in the ribosomal translocation complex. Translocation complexes (70S TC) containing WT or H91A EF-G are ultracentrifuged at 100,000*g* at 4°C with 30% sucrose cushion. The ribosomal pellet was carefully isolated, washed, and subjected to SDS-PAGE analysis. Whereas no EF-G band was seen in the 70S TC + WT EF-G complex, a clear band of H91A EF-G, in almost 1:1 stoichiometry with the ribosomal protein S1, could be seen in the 70S TC + H91A EF-G complex.

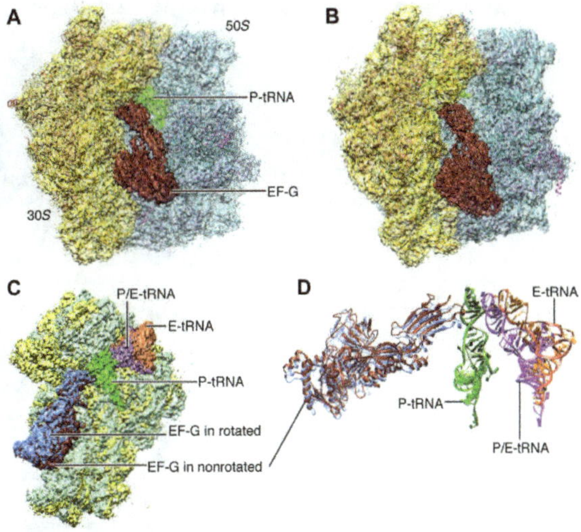

Fig. 3. Cryo-EM maps of the EF-G-bound complexes. (**A**) Nonrotated ribosome (transparent) fitted with atomic models for both ribosome and EF-G. (**B**) Rotated ribosome (transparent) fitted with atomic models. (**C**) Superimposition of the maps of the 30S subunit (green for rotated) when two maps are aligned on the 50S subunits. (**D**) Map-fitted structures of all tRNAs and EF-Gs at their respective positions shown in (C).

Table 1. Summary of cryo-EM maps. r70S, rotated ribosome; nr70S, nonrotated ribosome.

Maps	Annotation of states	Resolution (Å)	No. of particles
1	nr70S•P-tRNA•E-tRNA	4.0	~50,000
2	nr70S•P-tRNA•E-tRNA•EF-G	3.6	~90,000
3	r70S•P/E-tRNA•EF-G	4.2	~35,000
4	r70S•P/E-tRNA	5.7	~15,000

using the flexible fitting program MDFF (see Materials and Methods), and yielding representative structural models for four states, as summarized in Table 1. Here, we report our results on two structures showing EF-G bound to the rotated (Fig. 3B) and nonrotated (Fig. 3A) state of the ribosome, the former with one tRNA in the hybrid P/E position and the latter with tRNAs in the P and E positions (Fig. 3, C and D). The map resolution in the region surrounding the mutation H91A in EF-G is close to 3.2 and 3.4 Å in the unrotated and rotated state, respectively (Fig. 4). These two structures appear to correspond to the two GTP states observed in Chen *et al.*'s smFRET study (*19*). In both density maps, the size of the nucleotide bound to EF-G, singled out within the density of EF-G when displayed at a high threshold, is too large for a guanosine diphosphate (GDP) but just large enough to accommodate a GTP or GDP-Pi (inorganic phosphate) (Fig. 5 and fig. S3). Thus, the complexes could have been trapped in the GTP or GDP-Pi state. Furthermore, EF-G in both structures exhibits the domain constellation characteristic for its GTP form, distinguished from the GDP form (*28, 29*) by a rotation of the superdomain formed by domains II, IV, and V.

The rotated, EF-G–bound structure (Fig. 3B) has the characteristics of a pre-translocational ribosome in the GTP state, captured at the precise point where GTP hydrolysis is initiated in the wild-type situation. As in several x-ray structures (*5, 10–12*), no A/P-tRNA is present in this complex. This common observation is explained by the small size of the peptide; in an smFRET study (*30*) on the dynamics of EF-G in the pre-translocational complex showed an increase in the occupancy of tRNA in the hybrid A/P configuration as the peptide length was increased. As in these previous crystallography studies, the authenticity of the pre-translocational complex missing A/P is confirmed

RESEARCH ARTICLE

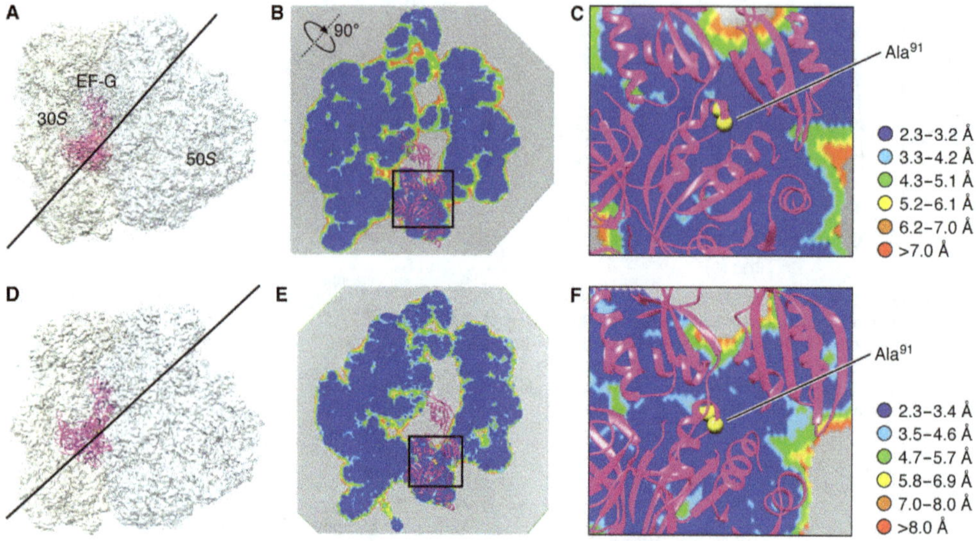

Fig. 4. Map resolution in a plane adjacent to residue Ala^{91}. (**A**) Map of the nonrotated ribosome complex. The line indicates the position of the section of the map in which the local resolution map is viewed. (**B**) Local resolution map in the section at the position shown in (A). (**C**) Zoomed-in view of the boxed region in (B). Resolution ranges are color-coded. (**D** to **F**) Illustrations of the rotated complex, corresponding to (A) to (C), respectively.

by the fact that the conformation of the ribosome is closely comparable with the authentic, A/P-tRNA–containing PRE complexes obtained by previous cryo-EM studies (*4, 14, 15*). The structure shows that the nonrotated EF-G–bound structure (Fig. 3A) was the result of the binding of EF-G in its GTP state to the post-translocational ribosome, which is weak and transient (*19*), but may be frequently sampled on account of the high concentration of EF-G•GTP in our sample. This structure is distinct from the structure of the post-translocational ribosome bound with EF-G in the GDP state, which can be trapped by fusidic acid (*28*) and is only reached after a round of translocation.

We first come to a close examination of the GTPase activation region. In both density maps, the region encompassing switch I, switch II, and the P-loop is virtually identical, showing the characteristics of the GTP state: switch I is ordered (Fig. 5) as in all x-ray structures of eubacterial ribosomes bound with the EF-G and GTP analogs GDPNP (*10*), GDPCP (*9, 11*), or GMPPMP (*12*). An overall structural comparison of the current EF-G and the x-ray structures is summarized in Table 2. However, our structures differ from previous wild-type EF-G structures in the distances between the Cα atom of H91A and the γ phosphate of GTP, between the Cα atom of H91A and the Cα atom of Thr^{63}, between the Cα atom of H91A and the Cα atom of Asp^{22}, and between the Cα atom of Thr^{63} and the Cα atom of Thr^{63} by about 1 to 2 Å. The slightly greater distances, compared with the corresponding distances in the x-ray structures of *T. thermophilus* [Table 3 and Fig. 6; (*9–12, 20*)], must be attributed to the mutation from the charged histidine to the nonpolar alanine residue. Hence, despite the very similar positions and conformations of the local structures in the immediate environment of GTP, the change in the distance between residue 91 and the γ phosphate, due to absence of the positive charge and the bulky side chain of His, renders the GTPase center of EF-G defective in triggering GTP hydrolysis. This result shows that the positioning of the side chain of His^{91} is critical for the activation of GTP hydrolysis.

That the mutant H91A EF-G binds to both the rotated and unrotated ribosomes, yet wild-type EF-G only leads to GTP hydrolysis in the rotated ribosome [see (*19, 25*)], strongly suggests that the position of His^{91} (His^{87} in *T. thermophilus*) seen in the x-ray structures of ribosomes bound to EF-G in the presence of GTP is a necessary but not sufficient condition for GTP hydrolysis. Clues for the additional requirements will therefore come from an examination of the differences in the binding interactions between EF-G•GTP in the two conformations of the ribosome. As will be shown in detail below, this comparison reveals that the additional action effected by the rotation is the stabilization of domain III, which is poised to interact with the switch I loop. Because domain III is part of a superdomain that it forms with domains IV and V, stabilization of these latter two domains is required as well. Examination of the contacts between EF-G and the ribosome shows that in the rotated complex, the superdomain is stabilized by flanking interactions with domain III on one side and with domain V on the other side (Fig. 7), as follows.

In the present rotated pre-translocational complex, switch I and domain III are in contact through residues Glu^{57}-Arg^{474}, Arg^{58}-Asp^{466} in *E. coli* EF-G (*12*). This result agrees with the identification, in the x-ray structures, of the interacting pairs Glu^{56}-Arg^{468}, Gln^{57}-Asp^{464}, and Arg^{61}-Glu^{460} for EF-G of *T. thermophilus* (*11*). We also find in the rotated complex that the 75-loop of protein S12 directly contacts domain III of EF-G, whereas residue 76 inserts itself into the middle of the helix (residues 424 to 440) and the β-turn (around residue 452) in domain III (Figs. 7, A and B, and 8), similarly as seen in the x-ray structure by Pulk and Cate (*12*). In sharp contrast, the 75-loop is positioned outside of domain III in the present nonrotated EF-G–bound complex. It is worth noting that in spite of the different positions of the 75-loop

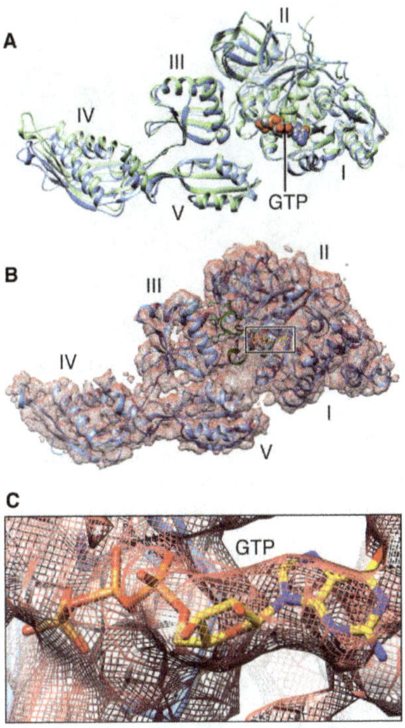

Fig. 5. Structures of 70S-bound EF-G in its GTP-bound state. (**A**) EF-G structure obtained from our cryo-EM maps of the nonrotated (green) and rotated complexes (blue), aligned on domains I and II. (**B**) Map-fitted structure in the rotated complex (see corresponding fig. S3 for the nonrotated complex). (**C**) Map in (B) displayed at a very high threshold, singling out the density for GTP.

relative to domain III in our present two complexes, the conformation of protein S12 remains virtually the same, and the difference is only introduced by the intersubunit rotation. This inserted position is maintained during translocation because it also appears in the x-ray structure of the POST complex in which the nonrotated ribosome is bound with EF-G•GDP [fig. S4; (*28*)]. In conclusion, the insertion of the 75-loop of protein S12 into domain III of EF-G in the rotated ribosome is one of the critical elements for stabilization of domain III ensuring efficient activation of GTP hydrolysis.

We now proceed to the stabilization of domain V from the other side of the superdomain (Fig. 7C). The interactions between the elements of domain V (side chains of Leu^{641} and Gln^{644}) and nucleotides A1067 and A1096 in helices 43 and 44 are only found in the rotated complex, in which the N-terminal domain of L11 is also rotated toward domain V, compared with both the EF-G–unbound and EF-G–bound nonrotated ribosomes. The closed position of the L11 lobe appears to stabilize domain V, whereas domain III is restricted in its orientation by protein S12's insertion on the opposite side of the structure (Fig. 7C). The interactions with protein S12 and the L11 lobe produce a pair of flanking restrictions on EF-G, which, in contrast, are not formed in the nonrotated state.

Table 2. Comparison of EF-G conformations in terms of root mean square deviation.

	EF-G in 4KIY.pdb (from *E. coli* PRE; resolution: 2.9 Å)	EF-G in 4BTC.pdb (from *T. thermophilus* PRE; resolution: 2.95 Å)
EF-G in nonrotated complex	1.3 Å	1.3 Å
EF-G in rotated complex	1.2 Å	1.3 Å

Table 3. Geometries for region of switch I and II and P-loop. Distances d1 (between Cα of His^{87} and the Pγ atom of GTP), d2 (between Cα atoms of His^{87} and Thr^{64}), d3 (between Cα atoms of His^{87} and Asp^{22}), and d4 (between Cα atoms of Thr^{64} and Asp^{22}) were obtained by averaging these distances from four x-ray structures (4JUW, 4KIY, 4KBT, 4KCY). For EF-Tu, the x-ray structure 4QDS was used. The x-ray POST used PDB 2WRI. "rt" and "nrt" denote rotated and nonrotated structures, respectively (see also Fig. 3).

Distances (Å)	X-ray EF-G average	X-ray EF-Tu	rt	nrt	X-ray POST
d1	7.0	6.6	7.3	7.3	9.0 (to P_β)
d2	9.1	9.2	8.8	9.1	19.0
d3	8.8	9.2	10.3	11.3	6.7
d4	10.2	9.5	10.4	11.7	18.5

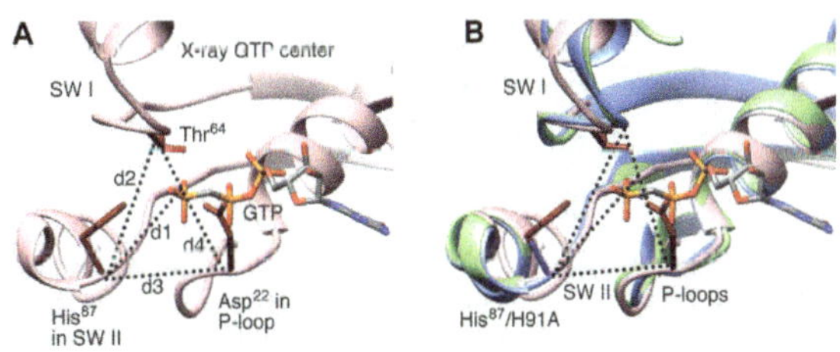

Fig. 6. Structures for region of switch I and II and P-loop. (**A**) Structure (x-ray PDB; 4JUW) around GTP showing the residues directly involved in GTP hydrolysis. Distances d1 (between Cα of His^{87} and the Pγ atom of GTP), d2 (between Cα atoms of His^{87} and Thr^{64}), d3 (between Cα atoms of His^{87} and Asp^{22}), and d4 (between Cα atoms of Thr^{64} and Asp^{22}), listed in Table 2 for our cryo-EM structures and x-ray structures. SW I, switch I; SW II, switch II. (**B**) Superimposition of our cryo-EM structures in the rotated (blue) and nonrotated (green) complexes on a representative x-ray structure of GTP state (PDB: 4JUW, pink).

We also see in our maps that the linker between domains III and IV is stabilized in the rotated conformation, and domains II and V are connected (at the displayed threshold), which is probably attributable to the aforementioned flanking restrictions (fig. S5). Additionally, in the rotated state, domain IV of EF-G, which is directly involved in the translocation of mRNA and the anticodon

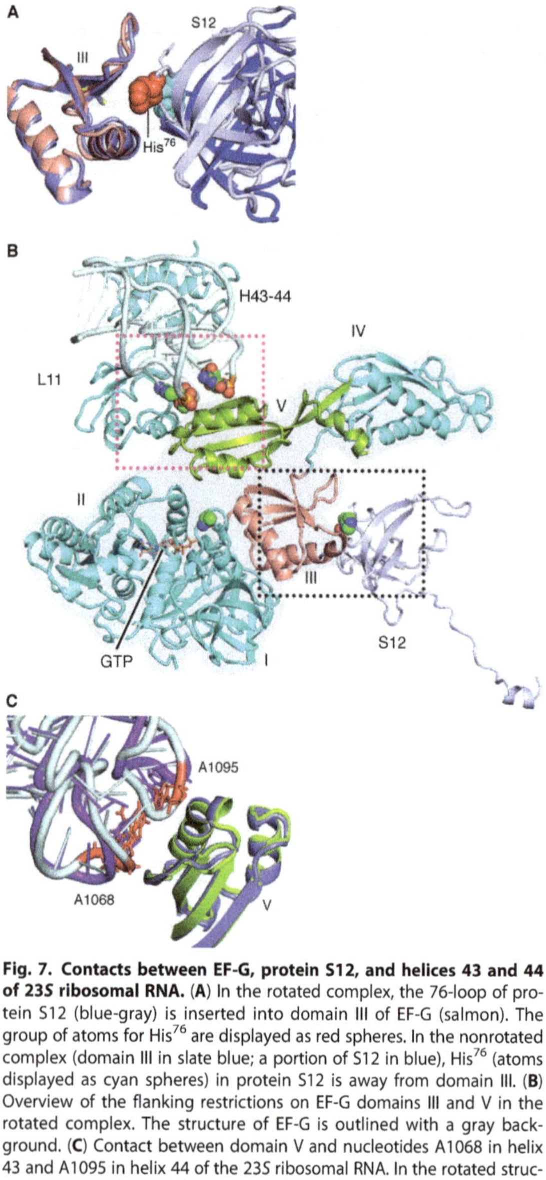

Fig. 7. Contacts between EF-G, protein S12, and helices 43 and 44 of 23*S* ribosomal RNA. (**A**) In the rotated complex, the 76-loop of protein S12 (blue-gray) is inserted into domain III of EF-G (salmon). The group of atoms for His[76] are displayed as red spheres. In the nonrotated complex (domain III in slate blue; a portion of S12 in blue), His[76] (atoms displayed as cyan spheres) in protein S12 is away from domain III. (**B**) Overview of the flanking restrictions on EF-G domains III and V in the rotated complex. The structure of EF-G is outlined with a gray background. (**C**) Contact between domain V and nucleotides A1068 in helix 43 and A1095 in helix 44 of the 23*S* ribosomal RNA. In the rotated structure, domain V is green and helices (H) 43 and 44 are blue-gray, in which A1068 and A1095 are red. In the nonrotated complex, domain V is slate blue and helices are purple.

loop of tRNAs (*31*), interacts more strongly with bridge B2a, which is formed between helix 69 of 23*S* RNA and helix 44 of 16*S* RNA.

It is finally worth asking, given the evidence that EF-G•GTP competes with the aa-tRNA•EF-Tu•GTP complex for binding to the POST ribosome, what are the structural features that favor the ternary complex? In the smFRET experiments of Chen *et al.* (*19*), EF-G•GTP was seen to only visit the nonrotated POST ribosome intermittently, indicating a weak interaction. In contrast, these authors found much longer lifetimes of the complex formed between EF-G•GTP and the ribosome in the rotated state. Comparison of our two EF-G–bound cryo-EM density maps shows that in the nonrotated state, EF-G's interaction with the ribosome is unstable, whereas in the rotated state, the factor establishes strong contacts with protein S12 and the L11-lobe. Thus, the overall weak binding of EF-G to the nonrotated state stands in stark contrast to multiple binding interactions with aa-tRNA•EF-Tu•GTP, explaining the results of Chen *et al.* (*19*) in structural terms.

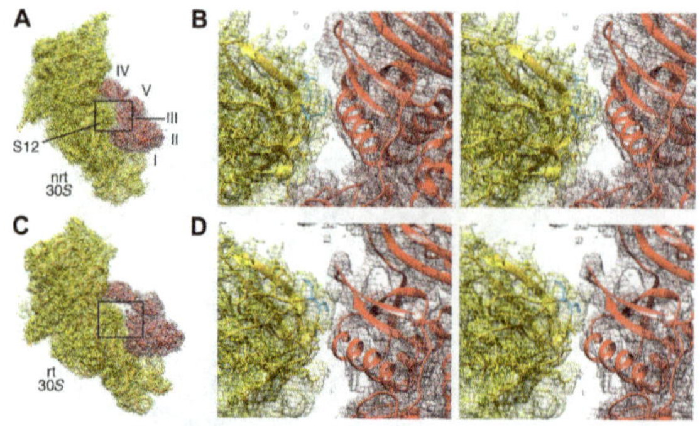

Fig. 8. Positions of residues 75 to 76 in protein S12 relative to domain III of EF-G. (**A**) Density maps (mesh) of the 30*S* subunit and EF-G in the nonrotated (nrt) state, with structures shown as ribbon. The domains of EF-G are labeled. (**B**) A stereo view of the boxed region in (A). Residues 75 to 76 are in cyan. (**C** and **D**) Illustrations corresponding to (A) and (B) respectively, but for the rotated state.

MATERIALS AND METHODS

Buffer and translation components

All experiments were performed using Hepes polymix buffer (pH 7.5) at room temperature (*32*). *E. coli* MRE600 70*S* ribosomes, His-tagged initiation factors (IF1, IF2, and IF3) and elongation factors (EF-Tu, EF-Ts, and EF-G), XR7-Met-Leu-Phe-stop mRNA, $tRNA^{Leu}$, and $[^3H]fMet\text{-}tRNA^{fMet}$ were purified following standard laboratory protocols (*32, 33*). H91A EF-G was created by mutating histidine (H) at position 91 to alanine (A) in wild-type EF-G construct (C-terminally His-tagged) (a gift from K. S. Wilson, Oklahoma State University) by site-directed mutagenesis (*34*). Nucleotide triphosphates [adenosine triphosphate (ATP) and GTP] were purchased from GE Healthcare.

Single-turnover GTP hydrolysis

GTP hydrolysis by EF-G (wild type and H91A) was measured under single-turnover conditions by mixing 70*S* ribosomes (3 μM) and $[^3H]GTP$ (15 μM) with EF-G (15 μM) in a quench-flow device (RQF-4, KinTek Corp.) at 37°C. The reaction was quenched at different time points with 25% formic acid. The amount of $[^3H]GDP$ formed was estimated by separating the $[^3H]GTP$ and $[^3H]GDP$ fractions on a Mono Q column attached to a high-performance liquid chromatography system. The rate of GTP hydrolysis (k_{GTP}) was obtained by fitting the data with a single exponential function using Origin 8.0 (OriginLab).

Preparation of the translocation complex with H91A EF-G

An initiation mixture (IM) containing 70*S* ribosomes (0.5 μM), all the three initiation factors [IF1, IF2, and IF3 (0.5 μM each)], XR7-Met-Leu-Phe mRNA (2 μM), [^{3}H]fMet-tRNA^{fMet} (1 μM), ATP (1 mM), GTP (1 mM), and phosphoenolpyruvate (PEP; 10 mM), and an elongation mixture (EM) containing elongation factors EF-Tu (0.5 μM), EF-Ts (0.2 μM), and EF-G H91A (5 μM), leucine (100 μM), tRNA^{Leu} 0.5 μM), Leu-tRNA synthatase (0.2 μM), ATP (1 mM), GTP (1 mM), and PEP (10 mM) were incubated separately at 37°C for 15 min. Both reaction mixtures were cooled down by leaving them at room temperature for 10 min. To initiate peptide elongation, 2 μl of IM was mixed with 8 μl of EM to achieve a final concentration of 70*S* ribosomes of 100 nM. The reaction was done for 15 s at room temperature (between 20° and 23°C). Aliquots of 4 μl of the final mixture were dropped on carbon-coated and glow-discharged holey carbon grids (Quantifoil 2/4 grids), which were incubated for 30 s in the Vitrobot chamber at a temperature of 4°C and a humidity of 100%. The grids were blotted for 1 s using an FEI Vitrobot, and the grids were immediately freeze-plunged in liquid ethane at the temperature of liquid nitrogen.

Co-sedimentation analysis to determine the occupancy of H91A EF-G in the translocation complex

The H91A EF-G–containing translocation complex was analyzed by co-sedimentation assay to determine the occupancy of EF-G in the complex. First, the translocation complex (prepared as above) was ultracentrifuged at 100,000*g* at 4°C with 30% sucrose cushion in an analytical ultracentrifuge. Then, the ribosomal pellet was carefully separated from the supernatant, washed twice with 1× Hepes polymix buffer (*32*), and subjected to SDS-PAGE analysis. A control experiment with wild-type EF-G was done in parallel.

Cryogenic electron microscopy

The specimens were transferred to a Titan Krios electron microscope (FEI) under cryo-conditions. The microscope was operated at 300 kV. Images were recorded using an FEI Falcon I direct electron detection camera with the Leginon program (*35*) at a calibrated magnification of ×59,000, yielding a pixel size of 1.09 Å. Defocus values for the contrast transfer function were calculated using CTFFIND3 (*36*) and were in the range of 1.5 to 3.5 μm in the final data set. A total of 8800 micrographs were collected, 6800 of which were used for particle selection after excluding those micrographs that had either poor power spectra or thick ice, using the method described by Jiang *et al.* (*37*). Program EMAN was used to select ribosome-sized single-particle images and then manually examine the quality of these images, yielding a data set of 380,000 single-particle images. The images were subjected to three-dimensional (3D) classifications and then map refinements, all performed using program RELION 1.2 (*38*). The process of hierarchical 3D classification is detailed in fig. S6. The refined maps were corrected for the modulation transfer function of the detector, and the high spatial frequencies were boosted using *B*-factor sharpening (*39*). Local resolutions of the maps were calculated using ResMap [fig. S7; (*40*)].

Model building and map fitting

With the program Modeller, the sequence of EF-G from *E. coli* was aligned to the sequence of EF-G from *T. thermophilus*, and then a homology model was built using the x-ray structure of EF-G from *T. thermophilus* [Protein Data Bank (PDB) ID: 2WRI] as the template. This template structure was bound with GDP and fusidic acid in the original form. The atomic structures of the 70*S* ribosome from *E. coli*, which included an fMet-tRNA in the P site (PDB IDs: 3J0U and 3J9T) of the nonrotated ribosome, were used for the starting model of the map-fitting of the nonrotated states. The atomic model (PDB IDs: 3J0Z and 3 J12) of the rotated ribosome including P/E-tRNA was used for the starting model of the fitting for the maps of the rotated states. The starting structures for the 70*S* ribosome and EF-G were fitted separately into the segmented maps, and then the fitted structures were combined together. The flexible fitting was done using the program MDFF (*41*) with implicit solvent, the generalized-Born implicit solvent model as implemented in NAMD. The final product fitting for each map was run for a 0.5-ns simulation, which was followed by 5000 steps of energy minimization. The cryo-EM maps and the atomic structures have been deposited (EMD-6315, EMD-6316, 3J9Z.pdb, and 3JAl.pdb.

SUPPLEMENTARY MATERIALS

Supplementary material for this article is available at http://advances.sciencemag.org/cgi/content/full/1/4/e1500169/DC1

Fig. S1. Gallery of four cryo-EM maps obtained by classification.
Fig. S2. Examples for map fittings.
Fig. S3. Density for GTP in the nonrotated complex.
Fig. S4. Interaction of protein S12 with domain III.
Fig. S5. Densities linking domain III to IV and domain II to V in our maps.
Fig. S6. Three-stage (I, II, III) classification of the entire data set using RELION.
Fig. S7. Color code for the resolution values in the movie showing the local resolution map.
Movie S1. Overview and local resolution estimation for the 3.6-Å map.
Movie S2. Examples of high-resolution protein features in the 3.6-Å map.
Movie S3. Examples of high-resolution RNA features in the 3.6-Å map.

REFERENCES AND NOTES

1. D. Moazed, H. F. Noller, Intermediate states in the movement of transfer RNA in the ribosome. *Nature* **342**, 142–148 (1989).
2. J. Frank, R. K. Agrawal, A ratchet-like inter-subunit reorganization of the ribosome during translocation. *Nature* **406**, 318–322 (2000).
3. J. Frank, Intermediate states during mRNA–tRNA translocation. *Curr. Opin. Struct. Biol.* **22**, 778–785 (2012).
4. X. Agirrezabala, J. Lei, J. L. Brunelle, R. F. Ortiz-Meoz, R. Green, J. Frank, Visualization of the hybrid state of tRNA binding promoted by spontaneous ratcheting of the ribosome. *Mol. Cell* **32**, 190–197 (2008).
5. W. Zhang, J. A. Dunkle, J. H. Cate, Structures of the ribosome in intermediate states of ratcheting. *Science* **325**, 1014–1017 (2009).
6. P. C. Spiegel, D. N. Ermolendo, H. F. Noller, Elongation factor G stabilizes the hybrid-state conformation of the 70S ribosome. *RNA* **13**, 1473–1482 (2007).
7. M. V. Rodnina, A. Savelsbergh, V. I. Katunin, W. Wintermeyer, Hydrolysis of GTP by elongation factor G drives tRNA movement on the ribosome. *Nature* **385**, 37–41 (1997).
8. J. Lin, M. G. Gagnon, D. Bulkley, T. A. Steitz, Conformational changes of elongation factor G on the ribosome during tRNA translocation. *Cell* **160**, 219–227 (2015).
9. D. S. Tourigny, I. S. Fernández, A. C. Kelley, V. Ramakrishnan, Elongation factor G bound to the ribosome in an intermediate state of translocation. *Science* **340**, 1235490 (2013).
10. J. Zhou, L. Lancaster, J. P. Donohue, H. F. Noller, Crystal structures of EF-G–ribosome complexes trapped in intermediate states of translocation. *Science* **340**, 1236086 (2013).
11. Y. Chen, S. Feng, V. Kumar, R. Ero, Y. G. Gao, Structure of EF-G–ribosome complex in a pretranslocation state. *Nat. Struct. Mol. Biol.* **20**, 1077–1084 (2013).
12. A. Pulk, J. H. Cate, Control of ribosomal subunit rotation by elongation factor G. *Science* **340**, 1235970 (2013).
13. D. J. Ramrath, L. Lancaster, T. Sprink, T. Mielke, J. Loerke, H. F. Noller, C. M. Spahn, Visualization of two transfer RNAs trapped in transit during elongation factor G-mediated translocation. *Proc. Natl. Acad. Sci. U.S.A.* **110**, 20964–20969 (2013).
14. A. F. Brilot, A. A. Korostelev, D. N. Ermolenko, N. Grigorieff, Structure of the ribosome with elongation factor G trapped in the pretranslocation state. *Proc. Natl. Acad. Sci. U.S.A.* **110**, 20994–20999 (2013).
15. D. Pan, S. V. Kirillov, B. S. Cooperman, Kinetically competent intermediates in the translocation step of protein synthesis. *Mol. Cell* **25**, 519–529 (2007).

RESEARCH ARTICLE

16. J. Fei, P. Kosuri, D. D. MacDougall, R. L. Gonzalez Jr., Coupling of ribosomal L1 stalk and tRNA dynamics during translation elongation. *Mol. Cell* **30**, 348–359 (2008).
17. J. B. Munro, R. B. Altman, N. O'Connor, S. C. Blanchard, Identification of two distinct hybrid state intermediates on the ribosome. *Mol. Cell* **25**, 505–517 (2007).
18. D. N. Ermolenko, Z. K. Majumdar, R. P. Hickerson, P. C. Spiegel, R. M. Clegg, H. F. Noller, Observation of intersubunit movement of the ribosome in solution using FRET. *J. Mol. Biol.* **370**, 530–540 (2007).
19. J. Chen, A. Petrov, A. Tsai, S. E. O'Leary and J. D. Puglisi, Coordinated conformational and compositional dynamics drive ribosome translocation. *Nat. Struct. Mol. Biol.* **20**, 718–727 (2013).
20. R. M. Voorhees, T. M. Schmeing, A. C. Kelley and V. Ramakrishnan, The mechanism for activation of GTP hydrolysis on the ribosome. *Science* **330**, 835–838 (2010).
21. A. Liljas, M. Ehrenberg, J. Aqvist, Comment on "The mechanism for activation of GTP hydrolysis on the ribosome". *Science* **333**, 37 (2011).
22. G. Wallin, S. C. Kamerlin, J. Aqvist, Energetics of activation of GTP hydrolysis on the ribosome. *Nat. Commun.* **4**, 1733 (2013).
23. R. M. Voorhees, V. Ramakrishnan, Structural basis of the translational elongation cycle. *Annu. Rev. Biochem.* **82**, 203–236 (2013).
24. B. R. Cunha, F. Peske, W. Holtkamp, W. Wintermeyer, M. V. Rodnina, Dual use of GTP hydrolysis by elongation factor G on the ribosome. *Translation* **1**, e24315 (2013).
25. W. Holtkamp, C. E. Cunha, F. Peske, A. L. Konevega, W. Wintermeyer, M. V. Rodnina, GTP hydrolysis by EF-G synchronizes tRNA movement on small and large ribosomal subunits. *EMBO J.* **33**, 1073–1085 (2014).
26. R. H. Cool, A. Parmeggiani, Substitution of histidine-84 and the GTPase mechanism of elongation factor Tu. *Biochemistry* **30**, 362–366 (1991).
27. G. Scarano, I. M. Krab, V. Bocchini, A. Parmeggiani, Relevance of histidine-84 in the elongation factor Tu GTPase activity and in poly(Phe) synthesis: Its substitution by glutamine and alanine. *FEBS Lett.* **365**, 214–218 (1995).
28. Y. G. Gao, M. Selmer, C. M. Dunham, A. Weixlbaumer, A. C. Kelley, V. Ramakrishnan, The structure of the ribosome with elongation factor G trapped in the posttranslocational state. *Science* **326**, 694–699 (2009).
29. J. Frank, H. Gao, J. Sengupta, N. Gao, D. J. Taylor, The process of mRNA–tRNA translocation. *Proc. Natl. Acad. Sci. U.S.A.* **104**, 19671–19678 (2007).
30. J. B. Munro, M. R. Wasserman, R. B. Altman, L. Wang, S. C. Blanchard, Correlated conformational events in EF-G and the ribosome regulate translocation. *Nat. Struct. Mol. Biol.* **17**, 1470–1477 (2010).
31. J. Zhou, L. Lancaster, J. P. Donohue, H. F. Noller, How the ribosome hands the A-site tRNA to the P site during EF-G–catalyzed translocation. *Science* **345**, 1188–1191 (2014).
32. C. S. Mandava, K. Peisker, J. Ederth, R. Kumar, X. Ge, W. Szaflarski, S. Sanyal, Bacterial ribosome requires multiple L12 dimers for efficient initiation and elongation of protein synthesis involving IF2 and EF-G. *Nucleic Acids Res.* **40**, 2054–2064 (2012).
33. C. Huang, C. S. Mandava, S. Sanyal, The ribosomal stalk plays a key role in IF2-mediated association of the ribosomal subunits. *J. Mol. Biol.* **399**, 145–153 (2010).
34. R. K. Koripella, Y. Chen, K. Peisker, C. S. Koh, M. Selmer, S. Sanyal, Mechanism of elongation factor-G-mediated fusidic acid resistance and fitness compensation in *Staphylococcus aureus*. *J. Biol. Chem.* **287**, 30257–30267 (2012).
35. B. Carragher B, N. Kisseberth, D. Kriegman, R. A. Milligan, C. S. Potter, J. Pulokas, A. Reilein, Leginon: An automated system for acquisition of images from vitreous ice specimens. *J. Struct. Biol.* **132**, 33–45 (2000).
36. J. A. Mindell, N. Grigorieff, Accurate determination of local defocus and specimen tilt in electron microscopy. *J. Struct. Biol.* **142**, 334–347 (2003).
37. W. Jiang, F. Guo, Z. Liu, A graph theory method for determination of cryo-EM image focuses. *J. Struct. Biol.* **180**, 343–351 (2012).
38. S. H. Scheres, RELION: Implementation of a Bayesian approach to cryo-EM structure determination. *J. Struct. Biol.* **180**, 519–530 (2012).
39. P. B. Rosenthal, R. Henderson, Optimal determination of particle orientation, absolute hand, and contrast loss in single-particle electron cryomicroscopy. *J. Mol. Biol.* **333**, 721–745 (2003).
40. A. Kucukelbir, F. J. Sigworth, H. D. Tagare, Quantifying the local resolution of cryo-EM density maps. *Nat. Methods* **11**, 63–65 (2014).
41. L. G. Trabuco, E. Villa, K. Mitra, J. Frank, K. Schulten, Flexible fitting of atomic structures into electron microscopy maps using molecular dynamics. *Structure* **16**, 673–683 (2008).

Acknowledgments: We thank Z. Yu and J. de la Cruz for their assistance in the data collection at Janelia Farm Research Campus of Howard Hughes Medical Institute (HHMI). We also thank A. des Georges for his assistance with the use of the Relion program, and M. Thomas-Baum for her assistance in preparing the illustrations. **Funding:** This research was supported by HHMI and NIH grants R01 GM29169 and GM55440 (to J.F.); the Swedish Research Council [2011-6088 (NT) and 2008-6593 (Linnaeus grant to Uppsala RNA Research Center)] and the Knut and Alice Wallenberg Foundation (KAW 2011.0081, RiboCORE) grants (to S.S.). **Competing interests:** The authors declare that they have no competing interests.

Submitted 12 February 2015
Accepted 3 April 2015
Published 22 May 2015
10.1126/sciadv.1500169

Citation: W. Li, Z. Liu, R. K. Koripella, R. Langlois, S. Sanyal, J. Frank, Activation of GTP hydrolysis in mRNA-tRNA translocation by elongation factor G. *Sci. Adv.* **1**, e1500169 (2015).

Structural Basis for Gating and Activation of RyR1

Amédée des Georges,[1,2,7,8] Oliver B. Clarke,[1,7] Ran Zalk,[3,7,9] Qi Yuan,[3] Kendall J. Condon,[1,2] Robert A. Grassucci,[1,2] Wayne A. Hendrickson,[1,3,*] Andrew R. Marks,[3,4,5,*] and Joachim Frank[1,2,6,10,*]
[1]Department of Biochemistry and Molecular Biophysics
[2]Howard Hughes Medical Institute
[3]Department of Physiology and Cellular Biophysics
[4]Department of Medicine
[5]Wu Center for Molecular Cardiology, College of Physicians and Surgeons
[6]Department of Biological Sciences
Columbia University, New York, NY 10032, USA
[7]Co-first author
[8]Present address: CUNY Advanced Science Research Center, 85 Saint Nicholas Terrace, New York, NY 10031, USA
[9]Present address: The National Institute for Biotechnology in the Negev, Ben-Gurion University of the Negev, Beer Sheva 84105, Israel
[10]Lead Contact
*Correspondence: wayne@xtl.cumc.columbia.edu (W.A.H.), arm42@cumc.columbia.edu (A.R.M.), jf2192@cumc.columbia.edu (J.F.)
http://dx.doi.org/10.1016/j.cell.2016.08.075

SUMMARY

The type-1 ryanodine receptor (RyR1) is an intracellular calcium (Ca^{2+}) release channel required for skeletal muscle contraction. Here, we present cryo-EM reconstructions of RyR1 in multiple functional states revealing the structural basis of channel gating and ligand-dependent activation. Binding sites for the channel activators Ca^{2+}, ATP, and caffeine were identified at interdomain interfaces of the C-terminal domain. Either ATP or Ca^{2+} alone induces conformational changes in the cytoplasmic assembly ("priming"), without pore dilation. In contrast, in the presence of all three activating ligands, high-resolution reconstructions of open and closed states of RyR1 were obtained from the same sample, enabling analyses of conformational changes associated with gating. Gating involves global conformational changes in the cytosolic assembly accompanied by local changes in the transmembrane domain, which include bending of the S6 transmembrane segment and consequent pore dilation, displacement, and deformation of the S4-S5 linker and conformational changes in the pseudo-voltage-sensor domain.

INTRODUCTION

Ryanodine receptors (RyRs) are homotetrameric intracellular calcium (Ca^{2+}) release channels of extraordinary size and complexity located in the sarco/endoplasmic reticula (SR/ER) of most metazoan cell types. RyR1 is the predominant isoform in mammalian skeletal muscle, where SR Ca^{2+} release via RyR1 is required for excitation-contraction (EC) coupling. Paracrystalline RyR1 arrays on the terminal cisternae interact directly with, and are regulated by, $Ca_V1.1$ voltage-gated Ca^{2+} channels on the transverse (T)-tubule membrane (Franzini-Armstrong et al., 1999). Every other RyR1 channel associates with a cluster of four $Ca_V1.1$ channels (also known as dihydropyridine receptors, or DHPRs), whereas the intervening RyR1 channels are not associated with DHPRs. RyR1 channels in these arrays exhibit coupled gating (Marx et al., 1998).

In planar lipid bilayers, the open probability (P_o) of RyR1 channels has a biphasic dependence on cytosolic Ca^{2+} concentration, with a maximal P_o in the 10- to 100-μM range and decreased activity at both higher and lower levels of Ca^{2+} (Bezprozvanny et al., 1991). RyR2, the predominant isoform in cardiac muscle, is activated primarily by Ca^{2+} entry through $Ca_V1.2$ via Ca^{2+}-induced Ca^{2+}-release (CICR) (Fabiato, 1983). The plant alkaloid ryanodine is a specific RyR channel modifier. It locks the channel in a reduced conductance state at low concentration and blocks it at higher concentrations (Nagasaki and Fleischer, 1988). Ryanodine receptors are also potently activated by ATP, which binds at a relatively low-affinity, non-hydrolyzing site. Given the affinity of this site and the high cytosolic concentration of ATP in muscle (Kushmerick et al., 1992), it is likely that RyR1 in skeletal muscle is constitutively bound to ATP or one of its analogs.

A number of proteins bind and regulate RyR activity, including calmodulin (Hamilton and Reid, 2000) and the immunophilins FKBP12/FKBP12.6, also called calstabin1/calstabin2, which are subunits of the RyR1/RyR2 channel complexes, respectively. FKBP12 and FKBP12.6 regulate the coupled gating of adjacent receptors (Marx et al., 1998) and also modulate the open probability of both RyR1 (Brillantes et al., 1994) and RyR2 (Kaftan et al., 1996) in planar lipid bilayer experiments but do so without altering gating properties with respect to Ca^{2+}, ATP, or ryanodine. The high affinity and specificity of the RyR-FKBP interaction has been exploited here for purification of the receptor (Zalk et al., 2015).

Recently, three single-particle cryoelectron microscopy (cryo-EM) studies have described the architecture of the closed state of RyR1 (Efremov et al., 2015; Yan et al., 2015; Zalk et al., 2015). In this study, we have sought to understand Ca^{2+}-dependent gating and activation of RyR1 and its modulation by key ligands by analyzing structures in multiple functional states. Earlier structural studies of RyR1 showed that gating involves dilation

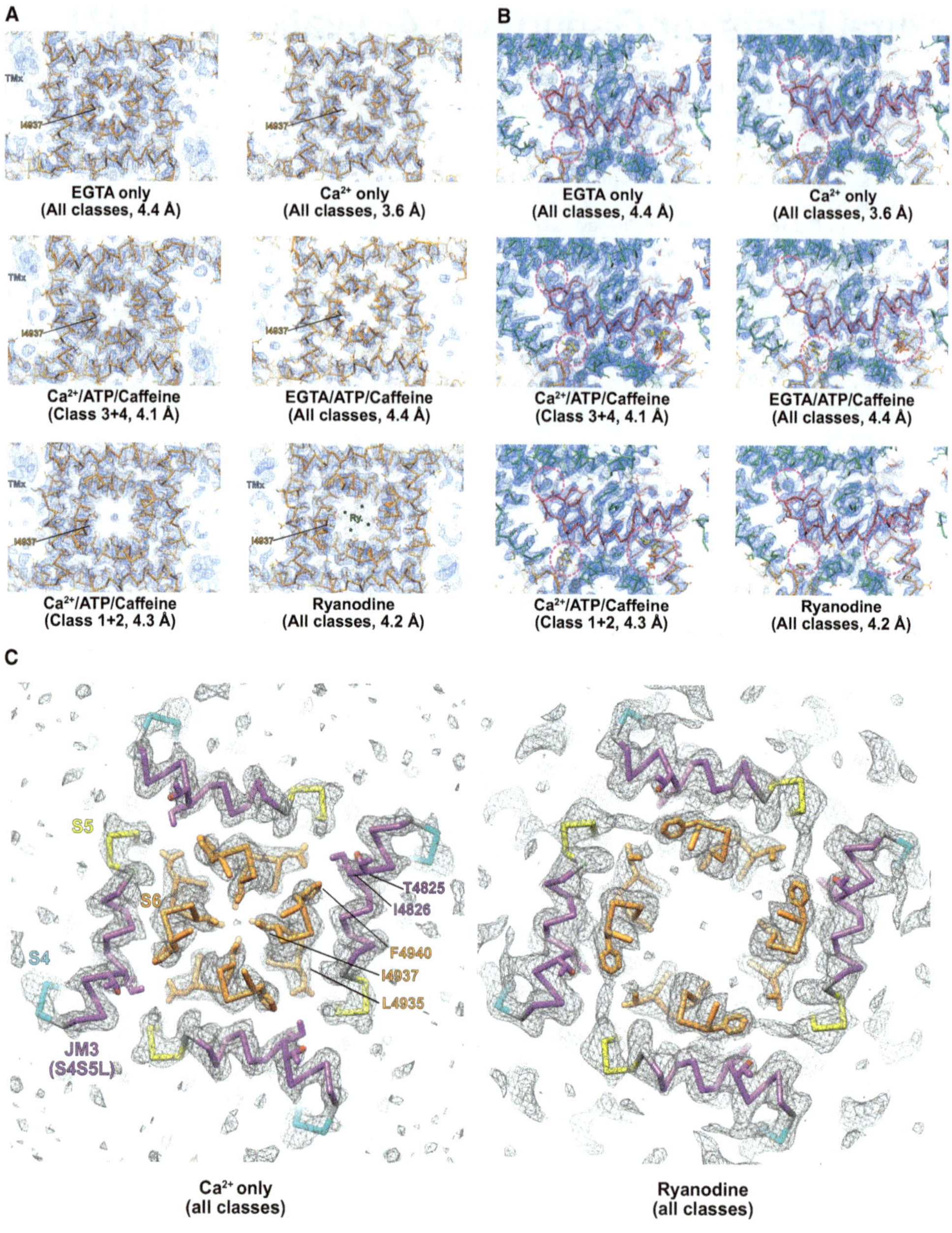

Figure 1. Structures of RyR1 in Multiple Functional States Reveal Conformational Changes and Ligand-Binding Sites

(A and B) For six representative refined reconstructions (identity as labeled), the atomic model, shown as a Cα trace with side chains in stick representation, is depicted along with the density map, which is contoured at 4 σ and represented as a dark blue mesh. In (A), a 7-Å-thick slab is shown which is orthogonal to, and centered upon, the channel axis at the narrowest part of the pore, at the height of residue I4937 (labeled). In (B), an orthogonal view is represented, showing the cytoplasmic extension of S6 and the C-terminal domain, as well as the surrounding regions of the core solenoid and TaF domain. In all panels, the transmembrane region is colored in orange, the C-terminal domain in dark red, and the core solenoid and TaF domains in sea green. The three ligand-binding sites around the C-terminal domain are highlighted with dashed magenta circles. The position of a transmembrane helix of uncertain provenance

(legend continued on next page)

of the transmembrane pore accompanied by rearrangement of the cytosolic assembly with displacement of the peripheral regions of the latter toward the SR membrane (Samsó et al., 2009; Serysheva et al., 1999). In this report, we define "activation" as the increase in open probability induced by the addition of a channel agonist, while "gating" refers to the structural transition between the closed (non-conducting) and open (conducting) states of the channel, which may occur under equilibrium conditions. We present structures of RyR1 in the presence and absence of channel modulators (such that open probability is expected to range from near-zero to near-unitary), providing detailed insights into the molecular mechanisms of RyR1 activation and gating, and we identify the binding sites for Ca^{2+}, ATP, caffeine, and ryanodine. In the presence of Ca^{2+}, ATP, and caffeine, we obtained structures of both open and closed states of the channel from the same sample using three-dimensional classification, allowing detailed characterization of the local and global structural changes associated with RyR1 gating.

RESULTS

We have determined the structure of the RyR1/Calstabin2 (RyR1/Cs2) complex by cryo-EM under the following five conditions: (1) nominally Ca^{2+}-free state (denoted "EGTA only"); (2) activating [Ca^{2+}] (denoted "Ca^{2+} only"); (3) activating Ca^{2+} plus ATP and caffeine (denoted "Ca^{2+}/ATP/Caffeine"); (4) nominally Ca^{2+}-free plus ATP and caffeine (denoted "EGTA/ATP/Caffeine"); and (5) activating Ca^{2+} plus ryanodine (denoted "ryanodine"). Conditions for each dataset are summarized in Table S1. Conditions 3 and 5 were intended to maximize open probability during data collection. Condition 5 was intended to lock the channel in a stable state characteristic of ryanodine-modified purified channels (Imagawa et al., 1987). Although we know that in single-channel experiments a mixture of micromolar Ca^{2+} and millimolar ATP is sufficient to achieve near-unitary open probability, it was unclear whether the same would hold true for rapidly vitrified detergent-solubilized RyR1 imaged under cryo-conditions. To maximize a highly populated open state, we included the RyR channel activator caffeine in condition 3.

Single-channel recordings of purified RyR1/Cs2 reconstituted in planar lipid bilayers reveal that the purified protein is functional (Figure S1), with properties similar to those recorded from SR microsomes. No open channels could be detected in the EGTA-only condition. Under the Ca^{2+}-only and EGTA/ATP/Caffeine conditions, the open probabilities of representative channels were 0.22 and 0.13, respectively, while under the Ca^{2+}/ATP/Caffeine condition, the open probability was 0.91 (for the purified RyR1) or 0.92 (for RyR1 in SR microsomes). The addition of 10 μM ryanodine locked RyR1 in the characteristic sub-conductance state.

After initial 2D and 3D classification to eliminate poor-quality particles, all cryo-EM datasets were subjected to 3D classification into four classes with identical classification parameters to assess conformational heterogeneity (see Star Methods). Tables S1 and S3 summarize the numbers of particles and domain-specific resolutions for each dataset; Table S4 gives particulars for each class.

Improved resolution compared to published structures has allowed us to revise our previous model (Zalk et al., 2015) and extend sequence registration significantly. The overall resolution (using "gold-standard" protocol and Fourier shell correlation = 0.143) ranges from 3.8 Å for the Ca^{2+}-only state to 4.5 Å for the EGTA-only state. In all cases, the highest resolutions were obtained in the core of the channel (C-terminal 1,400 residues). Thus, for the Ca^{2+}/ATP/caffeine state (classes 3 and 4), the overall resolution is 4.5 Å with a resolution of 4.1 Å in the core, while for the best reconstruction (Ca^{2+} only, all classes combined) the respective values are 3.8 and 3.6 Å. We show portions of the density maps at the narrowest part of the pore (I4937) and from the CTD in Figures 1A and in 1B, respectively, for each condition after refinement of all particles together for each dataset, giving the gold-standard core resolution for each reconstruction. Two maps are represented from the Ca^{2+}/ATP/Caffeine dataset since both open and closed pore conformations are highly populated.

We fitted a largely complete RyR1 atomic model (4,170 of 5,037 residues) to the best map and then adapted it to the other states. We identified previously undescribed features, including a polyproline segment in S2S3, localization of calmodulin-binding peptides, and two additional putative transmembrane helices. Moreover, conformational changes between the different states led us to divide RyR1 into a cytosolic shell that comprises N-terminal domains through the end of the bridging solenoid from each protomer, and a channel and activation core from the C-terminal remainders. This core includes the core solenoid and its intimate associations with cytosolic extensions from the tetrameric pore. Because of hinge flexing now evident between BrA (residues 1657–2144) and BrB of the bridging solenoid, but relative rigidity at the BrB-BrC junction, we renamed BrA as the junctional solenoid (JSol) and BrB-BrC as the bridging solenoid (BSol). JSol is named for its joining of the N-terminal solenoid (NSol) with BSol and Csol (Figure S2). RyR1 domain definitions are in Table 1 and Figure S2.

Architectural Organization

The transmembrane pore shares the overall architecture of the six-transmembrane (6TM) cation channel family. A pseudo-voltage-sensor domain (pVSD) spanning helices S1–S4 is linked via a short amphipathic helix to the pore domain, composed of helices S5 and S6. RyR1 pore domain possesses unique structural features including the S2S3 helical bundle (S2S3 domain) inserted in the cytosolic loop between S2 and S3, a 30-Å cytosolic extension of S6 (S6c) terminating in a zinc-finger containing C-terminal domain (CTD), and an intrinsically disordered (glycine-rich) and highly acidic luminal loop between S1 and S2.

(TMx) is labeled in light blue when present. In the ryanodine map, putative ryanodine densities in the transmembrane pore are asterisked and labeled in dark green.

(C) Identical views parallel to the channel axis are shown for the Ca^{2+}-bound closed state (Ca^{2+} only, all classes) and the ryanodine-bound open state (all classes), with specific residues at the S6/JM3 (S4S5L) interface labeled. Notably, F4940 reorients during opening, coincident with straightening and displacement of S4S5L and expansion of the pore.

Table 1. RyR1 Domain Definitions

Domain Symbol	Identification	Residue Span
Cytosolic Shell		(1–3613)
NTD-A	N-terminal domain A	(1–208)
NTD-B	N-terminal domain B	(209–392)
Nsol	N-terminal solenoid	(393–627)
SPRY1	SP1a/ryanodine receptor domain 1	(628–849)
RY1&2	RYR repeats 1 and 2	(850–1054)
SPRY2	SP1a/ryanodine receptor domain 2	(1055–1241)
SPRY3	SP1a/ryanodine receptor domain 3	(1242–1656)
JSol	junctional solenoid	(1657–2144)
BSol	bridging solenoid	(2145–3613)
(RY3&4)[A]	RYR repeats 3 and 4	(2735–2938)
Channel and Activation Core		(3614–5037)
SCLP	shell-core linker peptide, CaM, and JSol binding sites	(3614–3666)
CSol	core solenoid	(3667–4174)
(EF1&2)[A]	EF-hand pair	(4060–4134)
TaF	thumb and Forefingers domain	(4175–4253)
TMx	auxiliary transmembrane helices	(4322–4370)
pVSD	pseudo voltage sensor domain	(4541–4819)
(S2S3)[A]	helical-bundle domain between S2 and S3	(4666–4786)
Pore	channel pore domain	(4820–4956)
(S6c)[A]	cytoplasmic extension of S6	(4938–4956)
CTD	C-terminal domain	(4957–5037)
Cs2	calstabin-2/FKBP12.6	

[A]Parentheses indicate subdomains of the preceding entries.

The S2S3 domain is unique to the RyR family, as it is absent even from the closely related inositol 1,4,5-trisphosphate receptor (IP3R) channels (Fan et al., 2015). The majority of the cytosolic assembly forms an extensively interwoven scaffold of α-solenoid repeats, interspersed with compact domains of other folds, including two N-terminal β-trefoil domains (NTD-A and NTD-B) (Tung et al., 2010), three SPRY domains (SPRY1–SPRY3) (Lau and Van Petegem, 2014), two pairs of RYR repeats (RY1and 2 and RY3 and 4) (Sharma et al., 2012; Yuchi et al., 2015) and a pair of EF-hands (EF1and 2) (Xiong et al., 2006) inserted in CSol (Zalk et al., 2015).

Interprotomer Interfaces in the Cytosolic Shell

The inter-protomer contacts in the homotetrameric cytoplasmic shell can be divided into three main interaction sites. We denote interactions as X:X′ or X:X″, where X′ and X″ refer to domain X in the clockwise and anti-clockwise adjacent protomer as viewed from the cytosol, respectively. The three major sets of contacts coupling adjacent protomers are thus NTD-A:NTD-B′, NTD-A:BSol′, and SPRY2:BSol′. All have previously been implicated in RyR channel gating.

Contact between the NTDs and a region in the bridging solenoid including the DP4 peptide (2242–2477) is suggested as important in gating, and disruption by addition of synthetic DP4 sensitizes channels to activation by Ca^{2+} and caffeine (Yamamoto et al., 2000). In our model, this region forms an extensive interface with NTD-A, making contacts with the 121–132 and 174–178 loops, and the 75–84 α helix. Interestingly, the latter also contacts the CSol.

Auxiliary Transmembrane Helices

We previously noted an unassigned helical density in the TM region of RyR1 (Zalk et al., 2015). Now, with an improved map for the nominally Ca^{2+}-free state, we can identify this density as auxiliary transmembrane helices (TMx) for a single transmembrane helix, a short luminal extension lacking definable secondary structure, and an associated density that could be from a partially unfolded second helix. This segment, located between the pVSD and pore domains of adjacent protomers, is not contiguous with other densities in the TM region. Such density was not present in other RyR1 reconstructions (Efremov et al., 2015; Yan et al., 2015), suggesting that TMx presence or stability depends on differences in protein purification procedures. Indeed, although it is clearly present in our Ca^{2+}/ATP/caffeine, Ca^{2+}-free, and ryanodine-bound datasets, it is absent in the 30 μM Ca^{2+} and EGTA/ATP/caffeine datasets (Figure 1A). TMx could be an unidentified RyR1-binding protein; however, we could not detect any additional proteins in our RyR1 preparations. TMx is more likely part of RyR1, presumably corresponding to the predicted TM hairpin in the region 4322–4370 (Du et al., 2002). Assuming that TMx is intrinsic to RyR1, we propose a revised TM-domain topology that has two TMx helices separated from the pVSD and pore domains by a long disordered cytoplasmic segment.

An RyR1-Specific Polyproline Motif in the S2S3 Domain

Improved resolution in the RyR1 core allows complete sequence assignment of the S2S3 domain in pVSD, a domain unique to RyRs consisting of a bundle of five α helices, the last of which is an amphipathic juxtamembrane helix (JM2). JM2 connects to S2S3 through a short, proline-rich linear motif, 4756-RKPDPPPGLL-4765 that is divergent among RyR isoforms, but conserved in RyR1 from different species. Having a conserved, proline-rich linear motif proximal to the core-gating machinery suggests a regulatory role, possibly as an SH3-domain binding site.

Binding Sites for Activating Ligands: Ca^{2+}, ATP, and Caffeine

It was apparent from inspection of density maps in the core of RyR1 that additional features, correlated with the presence and absence of Ca^{2+}, ATP, and caffeine, were located at CTD interdomain interfaces (Figure 1B). By comparing RyR1 structures in different states, we find that the binding sites for activating ligands are located at interfaces between the CTD and, depending on the ligand, specific interacting domains (CSol, TaF, S2S3, and/or S6c) from the same protomer. Locally aligned difference density maps calculated between the reconstructions obtained for the Ca^{2+}-free closed state and that in the presence of Ca^{2+}, ATP, and caffeine reveal three extra densities, clustered around

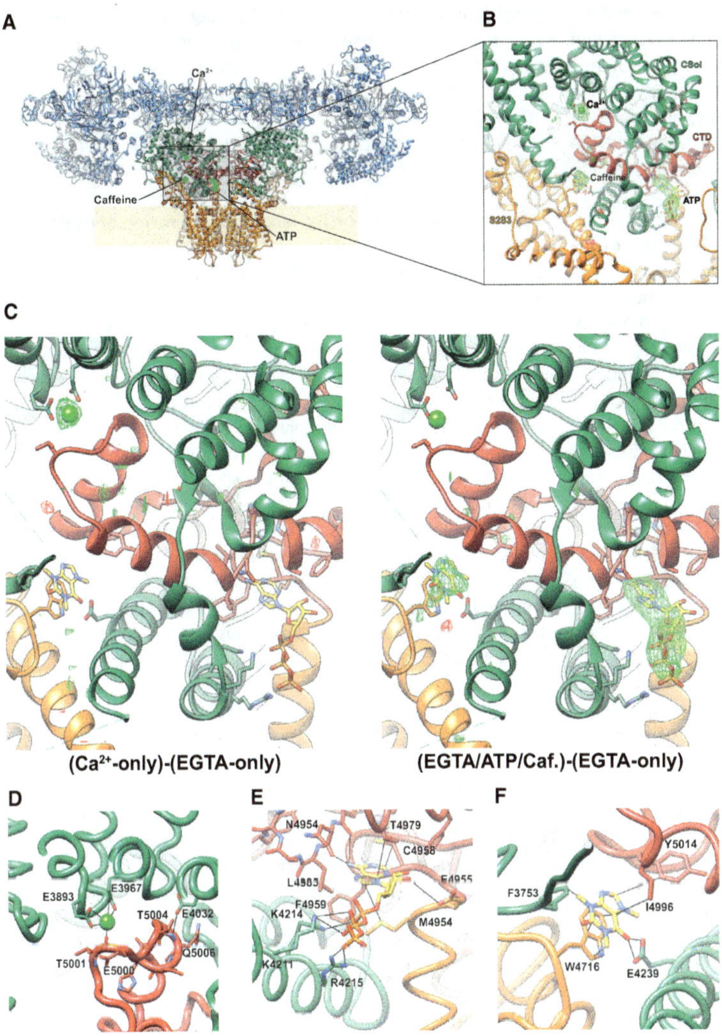

Figure 2. Ca^{2+}, ATP, and Caffeine Bind to the C-Terminal Domain of RyR1

(A) The architecture of RyR1/Cs2 is depicted in ribbon representation, with the shell (residues 1–3666) colored light blue, the core solenoid (residues 3667–4253) dark green, the transmembrane region and S6c (residues 4540–4956) orange, and the CTD (residues 4957–5037) dark red. In the boxed area, a locally aligned difference density map calculated between the EGTA-only reconstruction (all classes combined) and the reconstruction obtained in the presence of Ca^{2+}, ATP, and caffeine (classes 3 and 4) is depicted in green/red mesh, contoured at 5σ. Putative binding sites for Ca^{2+}, ATP, and caffeine are labeled.

(B) Enlarged view of the inset in (A), with ligands and interacting residues on RyR1 depicted in stick representation.

(C) Locally aligned difference maps, all contoured at 5σ, between the following respective pairs of structures: (Ca^{2+} only) - (EGTA only) (left) and (ATP/Caffeine/EGTA) - (EGTA only) (right).

(D–F) The binding sites of Ca^{2+}, ATP, and caffeine, respectively, with likely interacting residues labeled and depicted in stick representation, and putative RyR-ligand interactions indicated as thin black sticks. The RyR1 backbone is shown in licorice/worm representation. One plausible orientation of caffeine is depicted, as it cannot be determined with certainty from the density alone.

the CTD (Figures 2A and 2B). Comparison of difference maps calculated from reconstructions that have Ca^{2+}, but lack ATP/caffeine, and those that have ATP/caffeine, but lack Ca^{2+}, allow us to uniquely assign a peak corresponding to Ca^{2+}; density fit considerations allow us to infer which of the two remaining peaks is ATP and which is caffeine (Figure 2C; Star Methods). Ca^{2+} binds at the interface of the CTD and CSol (Data S1, 1), ATP binds at the mutual junction of S6c, the CTD and TaF domains (Data S1, 2), and caffeine binds between the CTD and the S2S3 domain (Data S1, 3; Figures 2D–2F).

The putative Ca^{2+}-binding site is between the carboxylate side chains of E3893 and E3967 of the core solenoid, and the backbone carbonyl of T5001 from the CTD (Figures 2D and S3A; Data S1, 1). In addition, Q3970 and H3895 contribute to the second coordination sphere of Ca^{2+} (Figure S3A; Data S1, 1). These five seeming Ca^{2+}–interacting residues are conserved in both the RyR and IP3R families (Figure S3B), consistent with a possible role for this site in Ca^{2+}-dependent activation of RyR and IP3R channels, both of which display a similar biphasic response to Ca^{2+} (Bezprozvanny et al., 1991). Mutation of residue E4032 severely impairs Ca^{2+} activation in both RyR1 and RyR2 (Fessenden et al., 2001; Li and Chen, 2001). E4032 appears to stabilize the CTD-CSol interface by directly H-bonding to the amide nitrogens at the end of one of the CTD helices, but it does not directly contribute to the Ca^{2+}-binding site (Figures 2D and S3; Data S1, 1), and E4032 is not conserved in IP3R channels (Figure S3B).

The ATP-binding site of RyR1 is located at the interface formed by the S6c-CTD junction and the C-terminal half of the TaF motif, with the entrance of the pocket facing away from the channel axis, not directly accessible from the permeation pathway. The adenine base is buried in a hydrophobic cleft lined by M4954, F4959, T4979, and L4985, while the ribose ring and triphosphate tail interact with S6c and the "forefingers" of the TaF domain (Figure 2E; Data S1, 2). Three positively charged residues (K4211, K4214, and R4215) from the first TaF finger are positioned to interact with the triphosphate tail of the nucleotide, while E4955 appears likely to interact with the ribose ring. The adenine-binding site is located directly adjacent to the recently identified Zn-binding site in the CTD (Yan et al., 2015).

The caffeine-binding site is situated between the S2S3 domain and the CTD. The orientation is ambiguous, given the resolution of the data, but it appears to be stacking parallel to the aromatic side chain of W4716 from S2S3, with I4996 forming hydrophobic interactions with the other face of the molecule, and possible hydrogen bonds between one of the two carbonyls and the carboxylate moiety of E4239 of the TaF domain (Figure 2F; Data S1, 3).

Concerted Cytosolic Shell Movements in All States of Activation

RyR1 exhibits substantial conformational variability in the EGTA-only dataset (Figure 3A; Data S2, 1), characterized primarily by the presence of multiple states of the cytosolic shell up (away from) or down (toward) with respect to the plane of the SR membrane, in addition to a small degree of variability in the displacement of the Csol and TaF domains away from the channel vertical axis. The population of particles can be classified into an arbitrary number of classes with distinct conformations, suggesting a continuum of conformational states due to oscillation of the shell in solution. These movements can be modeled by rotations of multidomain units (comprising NTDABC, SPRY123-RY1&2, JSol, Cs2, and BSol′-RY3&4′), each being rigid to a first approximation, about axes parallel to the membrane (Figure S4). In the absence of activating ligands the observed motion is confined exclusively to the cytosolic region (Figures 1A, 1C, 3A, and 3C). Similar cytosolic shell conformations persist upon additions of activating ligands (Figure 3B), but the variation in shell conformation are then over different and more limited ranges (Figure S4; Data S2, 2–5).

Activation Module Changes that Prime the Channel for Opening

In contrast to the Ca^{2+}-free state, where the core is largely fixed despite shell movements, the addition of either Ca^{2+} or ATP/caffeine results in characteristic changes in the core (Figures 3C–3F;Data S1, 4 and 5, and Data S2, 6–10). Associated with binding of these activating ligands, cytoplasmic portions of the core adopt a new conformation while leaving the transmembrane portion of the channel closed, not significantly changed from the Ca^{2+}-free state (Figures 1A and 1C). We call this cytoplasmic portion of the core the activation module, which comprises cytoplasmic core domains (CSol-EF1 and 2, TaF, and CTD) and sub-domains that extend into the cytoplasm from the 6TM channel pore itself (S2S3 and S6c). Changes in the activation module include a plastic deformation of the core solenoid, presumably due to contacts with the CTD mediated by Ca^{2+} or ATP (Figures 2B–2E). The CSol deformations lead to re-orientation of the TaF/CTD unit relative to CSol (Figures 3C and 3D; Data S2, 7) and also a change in the disposition of the EF1 and 2 pair (Figures 3C and 3E; Data S2, 10), which is displaced in the opposite direction from the TaF/CTD unit. Movement of the EF1 and 2 pair is particularly noticeable in the tetramer where it then directly contacts S2S3 of a neighboring protomer, whereas these features are separated in the absence of Ca^{2+} or ATP/caffeine (Figure 3F; Data S2, 10–15).

Our observation that the channels remain closed under the conditions of Ca^{2+} alone or ATP/caffeine alone is qualitatively consistent with the single-channel properties (Figure S1A). The pore remains constricted at I4937 to an aperture insufficient to allow passage of a hydrated monovalent or divalent cation. We consider the common conformational state obtained with 30 μM Ca^{2+} or ATP/caffeine alone to be "primed" for activation but held from opening by a barrier that can be overcome by the partner ligand.

Channel Opening upon Synergistic Activation of RyR1 by Ca^{2+} and ATP/Caffeine

Binding of Ca^{2+}, ATP, and caffeine results in four distinct conformations (Data S2, 3); two of these have a constricted pore with an aperture similar to that observed in the closed states (classes 3 and 4; Figures 1A, 4C [left], and 4D), and two have a dilated pore consistent with passage of hydrated ions (open state) (classes 1 and 2; Figures 1A, 4C [right], and 4D; Data S1, 6 and 7), each comprising approximately 50% of the particles compatible with the high (but not unitary) open probability of RyR1/Cs2 in the presence of Ca^{2+}, ATP, and caffeine (Figure S1A).

The dilated-pore and constricted-pore classes each independently segregate into classes in which the cytosolic assembly is either up or down with respect to the plane of the SR membrane, as also occurs in other states (Figures S4A–S4C; Data S2, 3), without accompanying changes in the channel core. Moreover, the two constricted-pore classes are similar to the up and down states from the Ca^{2+}-alone and ATP/caffeine conditions, respectively.

The conformational change between open and closed classes involves concerted movement of the channel core and the cytosolic shell in which a rigid unit comprising shell residues 1657–3613 (JSol and BSol-RY3&4) plus the primed activation module (core residues 3614–4253, 4666–4786, and 4938–5037) tilts outward from the vertical axis of the channel, resulting in dilation of the pore (Figure 4A). The rigid units responsible for pore dilation can be superposed very closely, with a root-mean-square deviation (RMSD) of 0.93 Å over 2,300 aligned Cα positions (Figure 4E). Concomitantly, the remainder of the cytosolic shell (residues 1–1656) rotates about an axis parallel to the vertical axis of the channel to accommodate the tilting of the rest of the protomer. This movement disrupts interprotomer contacts between NTD-A and NTD-B (Data S1, 8) at a previously identified "hotspot" for malignant hyperthermia (MH) mutations (Kimlicka et al., 2013). Dilation of the pore also disrupts the inter-protomer contact that forms between EF1 and 2 and S2S3′ upon binding of Ca^{2+} or addition of ATP/caffeine (Data S2, 25).

Dilation of the pore does not occur via bending of S6 at a single "hinge" alone, but rather by bowing and distortion of the S6 helix from a point midway through the membrane (Figure 4B), near the position of G4934 hinge, previously identified to influence gating of both RyR1 and RyR2 (Euden et al., 2013; Mei et al., 2015). Three other notable changes occur in the transmembrane region coincident with pore dilation. First, the "cinch" formed by the four S4S5L helices of the tetramer loosens, with S4S5L straightening and its contacts with S6 altering. Most notably, F4940 from S6c switches from interacting with L4935 on S6 of the adjacent protomer to interacting with T4825, I4826, and S4829 from S4S5L on the same protomer (Figure 1C). Second, the displacement of S4S5L away from the channel vertical axis in turn pushes

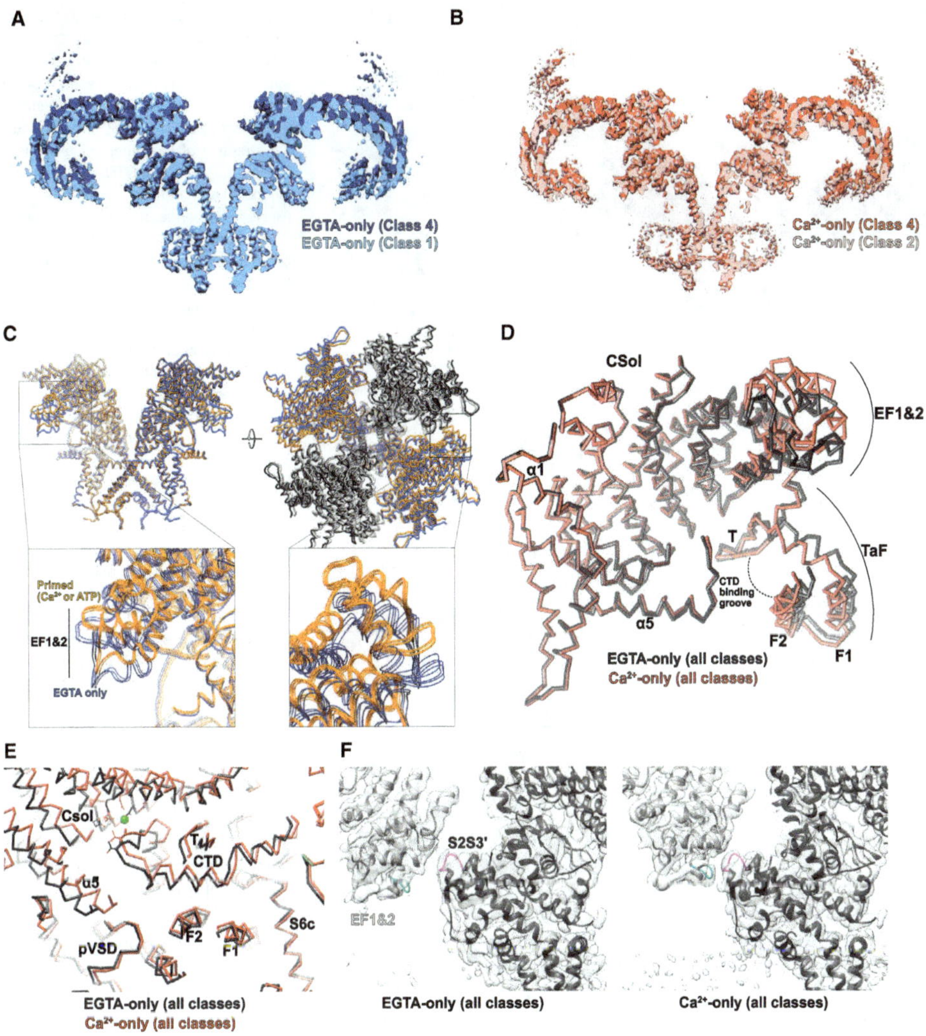

Figure 3. Ligand-Induced Conformational Changes in the RyR1 Closed State

(A and B) Conformational variability of maps derived from three-dimensional classes in, respectively, the EGTA-only dataset (blue/light blue density, contoured at 3σ), and the Ca^{2+}-only dataset (orange/red density, contoured at 3σ). In both cases, only the two extremes are represented, as labeled; a comparison of all classes is provided for both datasets by, respectively, Data S2, 1 and Data S2, 2.

(C) Conformational changes of the RyR1 core in the presence of Ca^{2+} alone or ATP/caffeine alone. The EGTA-only (all classes combined) structure (residues 3667–5037 of two opposing protomers; in the top view, the two alternating protomers are depicted in gray/charcoal to show the change in interprotomer interactions) is depicted in thin blue ribbons, while the Ca^{2+}-only (all classes combined) structure is depicted in orange ribbons. In the inset, an overlay of all four EGTA-only classes (thin blue ribbons; EGTA-only classes 1–4) and all eight "primed" structures (thin orange ribbons; EGTA/ATP/caffeine classes 1–4 and Ca^{2+}-only classes 1–4) in the EF1 and 2 region are shown to illustrate the consistent conformational change observed between the apo- and primed states of the activation core.

(D) The activation core undergoes flexible changes during priming. Here, residues 3639–4253 of the EGTA-only (charcoal) and Ca^{2+}-only (dark red) structures are shown aligned on the core solenoid. The TaF and EF1 and 2 domains are displaced in opposite directions, changing their disposition with respect to the body of the core solenoid.

(E) Change in orientation of the CTD with ligand binding. Here, a Cα trace of the atomic model for the EGTA-only reconstruction (all classes combined) and the Ca^{2+}-only reconstruction (all classes combined) are shown in charcoal and dark red, respectively. The bound Ca^{2+} ion is represented as a green sphere, and the side chains of coordinating residues are shown in stick representation. The view is the same as represented in Figure 1B.

(F) The EF1 and 2/S2S3′ interface, viewed orthogonal to the channel axis, for the EGTA-only (left) and Ca^{2+}-only (right) reconstructions, with density maps shown as transparent surfaces, contoured at 3σ. A loop in the EF1 and 2 domain is colored cyan, and a loop in the S2S3 domain is colored magenta, to improve visibility of the conformational change.

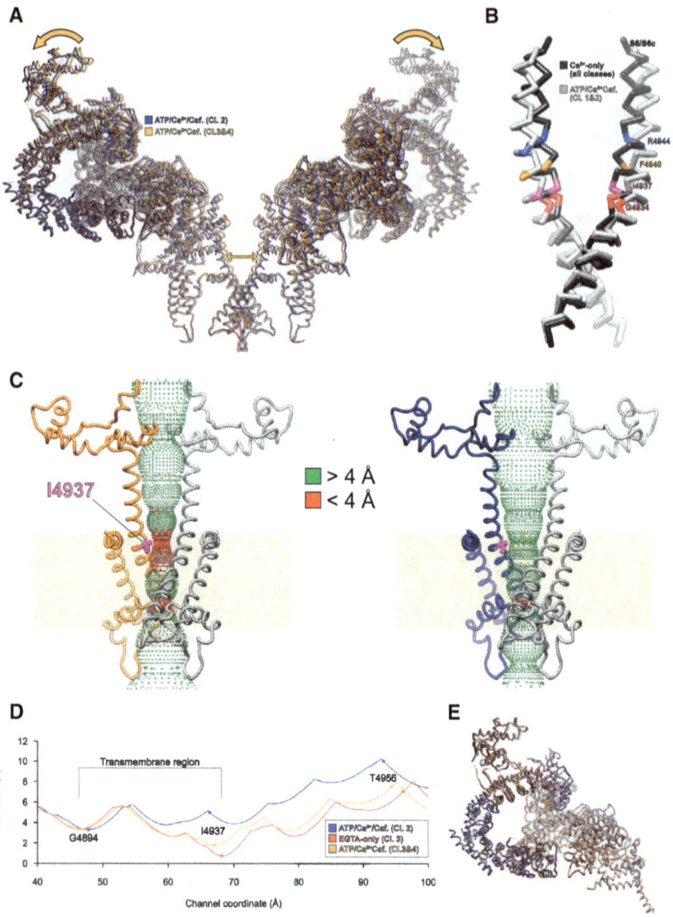

Figure 4. Ryanodine Receptor Gating in the Presence of Ca^{2+}, ATP, and Caffeine

(A) Global conformational changes between the closed (orange, Ca^{2+}/ATP/Caf. classes 3&4) and the open (blue, Ca^{2+}/ATP/Caf. classes 1&2) states identified in the presence of Ca^{2+}, ATP, and caffeine. Arrows show the direction of movement of a semi-rigid entity encompassing S6c, the CTD, the core solenoid and the junctional and bridging solenoids.

(B) Helix S6 bends during gating. Here, two opposite S6/S6c helices (residues 4910–4956) are represented as Cα traces viewed orthogonal to the channel axis for both the closed state (dark gray) and the open state (light gray), with several marker residues colored on both models.

(C) A side view of the transmembrane pore (residues 4820–5037) is depicted in ribbon representation with the hydrophobic gate residue, I4937, highlighted in magenta for the closed state (left, Ca^{2+}/ATP/Caf. classes 3&4) and the open state (right, Ca^{2+}/ATP/Caf. classes 1&2). A dotted representation of the accessible inner surface of the channel as calculated by HOLE (Smart et al., 1996) is colored green (where accessible to a sphere of greater than 4 Å radius) and red (where inaccessible to such a sphere)

(D) 1D graph of the pore radius in the Ca^{2+}-free closed state (dark red), activated closed (orange), and activated open (dark blue) structures.

(E) Superposition of the rigid units (residues 1657–4253 and residues 4934–5037) between the closed (orange) and open (dark blue) activated structures.

on the cytosolic end of S4, which is displaced internally within the pVSD, resulting in the apparent formation of a salt bridge between R4563 on S1 and D4815 on S4 (Figure 5D), and the reorientation of S4. A network of hydrogen bonds involving two histidines in pVSD, H4650 from S2 and H4812 from S4, also rearranges during this process. Both R4563 and D4815 are conserved in the IP3R family, highlighting the potential importance of this interaction. Last, the interactions between S6c and S6c′ are altered. R4944 shifts from forming an apparent salt bridge with E4942 on S6c′ in the closed state (Figure 5A) to interacting with D4938 on S6c′ when opened (Figure 5B), one turn closer to the membrane. All three residues involved in this interaction are highly conserved in both RyR and IP3R families (Figure 5C). Concomitantly, R4944 rotates outward to leave an electronegative ion conduction pathway lined by E4942 and other carboxylate residues.

Ryanodine Locks the Pore in the Dilated State

In contrast to the closed state that predominates in the Ca^{2+}-only dataset, the addition of 10 μM ryanodine together with 300 μM Ca^{2+} results in a map in which both major conformational classes have the pore maximally opened (Figures 1A, 1C, and 6A). The ryanodine-modified pore structure is very similar to that of the open-pore reconstruction (classes 1 and 2) of the Ca^{2+}/ATP/caffeine-activated state (Figure 1A), except for extra density features in the ryanodine-modified pore (Figure 6B). We attribute these features to one ryanodine molecule averaged over four symmetrically equivalent binding sites. Besides locking the channel in a partially open state, even a single ryanodine molecule is likely to occlude part of the pore, consistent with reduced conductance in the presence of ryanodine (Nagasaki and Fleischer, 1988). This putative ryanodine position is directly adjacent to Q4933 (Figure 6B), mutation of which specifically reduces ryanodine binding without affecting channel function (Fessenden et al., 2001). The equivalent residue in IP3R channels, which have a similar transmembrane pore but do not respond to ryanodine, is a phenylalanine (Figure 5C).

DISCUSSION

Mechanism of Ryanodine Receptor Activation

We have developed a structural description of RyR1 activation and gating (Figure S5) based on high-resolution cryoEM reconstructions of multiple RyR functional states (Figure S6–S8). In the absence of Ca^{2+}, ATP, or caffeine, the RyR1 core as a whole is extremely rigid and non-permissive of pore dilation. The minor

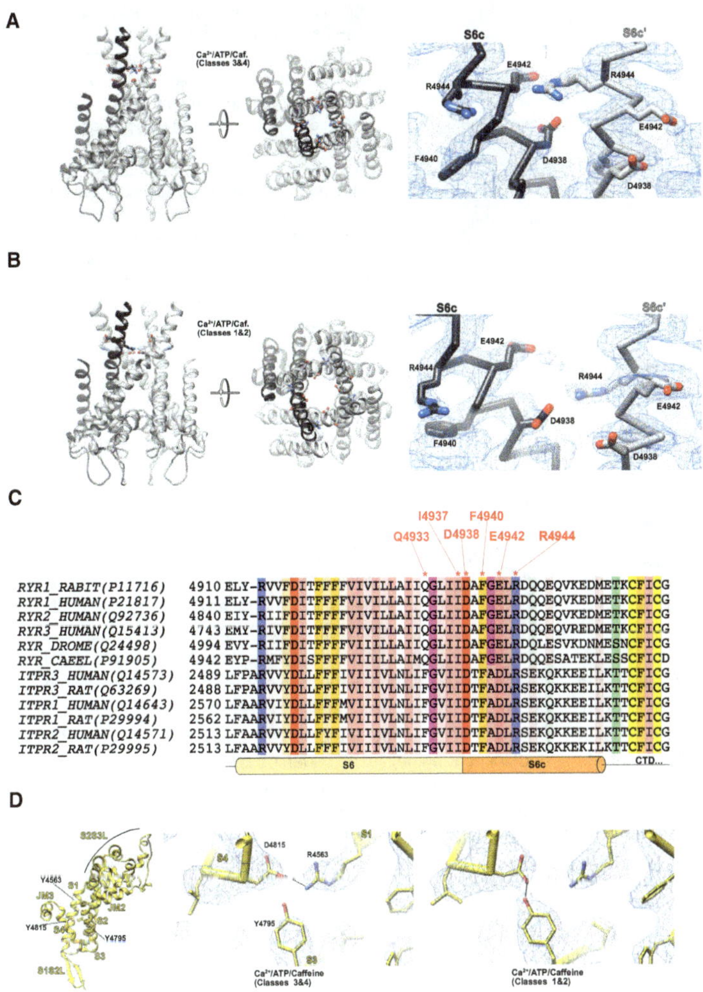

Figure 5. A Conserved Network of Interactions between S6c and S6c′ Rearranges during Gating

(A) In the closed state (here Ca^{2+}/ATP/caffeine classes 3 and 4), R4944 is in close contact with E4942 of the adjacent protomer, and the guanadino moiety of the arginine side-chain lines the conduction pathway of the pore. Density map is shown at right (blue mesh, contoured at 2.5σ) with fitted model (Cα trace with selected side chains shown and labeled).

(B) In the open state (here Ca^{2+}/ATP/caffeine classes 1 and 2), S6 splays out away from the channel axis, and R4944 rotates out of the conduction pathway, approaching the side chain of D4938 from the adjacent protomer. Density map is shown at right (blue mesh, contoured at 2.5σ) with fitted model (Cα trace with selected side chains shown and labeled).

(C) Portion of a sequence alignment of representative RyR and IP3R genes, prepared using MUSCLE (Edgar, 2004) with default parameters, visualized using Jalview (Waterhouse et al., 2009). The alignment is colored by conservation using a modified version of the Zappo coloring scheme, with a conservation cutoff of 60%. Residues of particular interest are asterisked and labeled, including R4944, D4938, and E4942. The Uniprot accession numbers of the relevant sequences are given in parentheses.

(D) An apparent salt bridge between R4563 and D4815, in the core of the pVSD (left), rearranges during channel opening, due to the conformational change of S4S5L and the rotation of S5. Here, the fitted atomic models of the Ca^{2+}/ATP/caffeine closed (center, classes 3&4) and open (right, classes 1&2) structures are shown with the density maps (blue mesh, contoured at 3σ).

displacements of the CTD and CSol that do occur in classes 1 and 3 (Data S2, 1) are not accompanied by pore dilation. In contrast, the interfaces between the core and the shell are flexible allowing for a large range of cytosolic shell conformations without affecting dilation of the pore (Data S2, 1). Binding of activators (Ca^{2+}, ATP, and caffeine) at the CTD interfaces initiates a rearrangement of the RyR1 core ("priming"), involving a reorientation of the CTD and flexible changes in the activation module (CSol-EF1 and 2, TaF, and CTD from the cytoplasmic core and S2S3 and S6c from the 6TM channel pore) (Data S2, 6). The activation module now adopts a conformation and orientation poised for pore dilation. In the presence of a single activator (either Ca^{2+} or ATP), the population of open channels is too low to isolate by our 3D classification; addition of ryanodine prevents these channels from closing, by intercalating between the pore-lining S6 helices and locking the channel in the open state. Multiple activators (Ca^{2+}, ATP, and caffeine), binding at different sites on the CTD, increase the channel open probability such that open channels are observed (Data S2, 3). For our Ca^{2+}/ATP/caffeine sample, the populations with open and closed pores are nearly equal.

The comparison of open and closed states under activating conditions allows us to describe the structural changes that occur in the channel upon stochastic channel gating at equilibrium. These closed-state structures are essentially the same as those from the single-activator conditions. Cytoplasmic portions of the channel protomers, including the activation modules, move as rigid bodies and interface with the 6TM pore domain from this closed state to open the pore. An unanswered question concerns the activation barrier that restricts channels from opening freely whenever activators are present.

The dilation of the opened transmembrane pore is particularly evident at the S6 helix-bundle crossing, where I4937 forms a hydrophobic seal in the closed state (Data S1, 9). Pore dilation occurs due to outward bowing and rotation of the transmembrane S6 helices such that I4937 no longer points directly toward the channel axis. S6 bowing is also coincident with a conformational change in which S4S5L ,which otherwise curves inward toward the outer edge of S6, straightens (Figure 1C; Data S1, 9), accommodating both the increased aperture of the open channel and

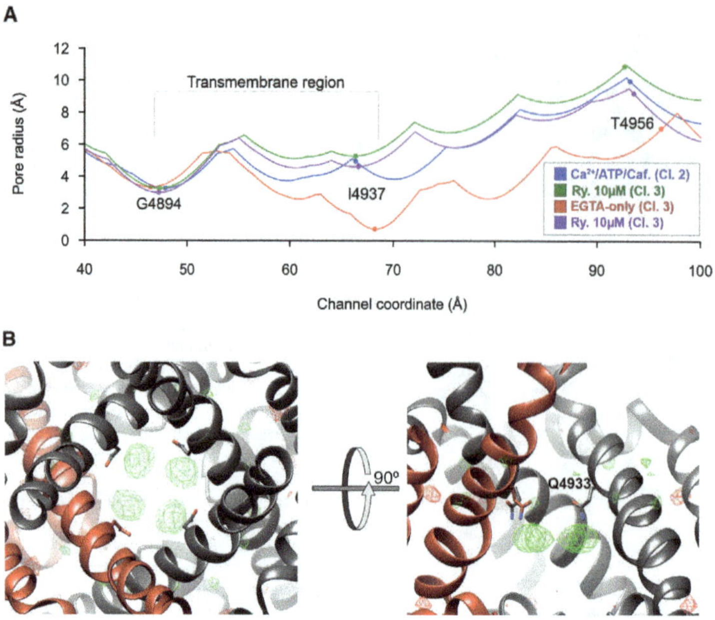

Figure 6. Ryanodine Binding Site and Mechanism of Channel Modulation

(A) 1D graph of the pore radius in the EGTA-only (red), Ca^{2+}/ATP/caffeine- activated open (dark blue), and ryanodine-locked (purple and green) structures, prepared using the HOLE software package. Classes are as labeled.

(B) Difference map between Ca^{2+}/ATP/caffeine-activated open-pore reconstruction and 10 μM ryanodine-locked open-pore reconstruction, superimposed onto the 10 μM ryanodine-locked open model (gray ribbons; one protomer colored dark red) and shown in two orientations (left, from the cytosol; right, parallel to the membrane). The difference map is represented as green/red mesh and contoured at 5σ. Q4933 is depicted in stick representation.

associated changes in pVSD and in the interfaces between the S6c pore helices. The structurally equivalent S4-S5 linker of Kv channels plays a role in coupling conformational changes at the activation gate to those in the voltage sensor (Labro et al., 2008), and mutations in this region of RyR1 have been functionally characterized (Murayama et al., 2011). The bowing in S6 coincides with displacement of a region that encompasses CTD and the BSol, JSol, and CSol solenoids away from the pore axis.

Ca^{2+}-Binding Site

Previous work on Ca^{2+}-dependent activation of RyR1 has suggested that the channel contains a moderate-affinity Ca^{2+}-dependent activation binding site (μM K_D), and a lower-affinity Ca^{2+}-dependent inhibition site at high (millimolar) Ca^{2+} (Bezprozvanny et al., 1991). Here, we have identified a Ca^{2+}-binding site that is highly occupied at a concentration of 30 μM Ca^{2+}, consistent with a Ca^{2+}-activation site. All residues involved in the Ca^{2+} coordination are conserved in both RyRs and IP3Rs (Figure S3B). Moreover, the putative Ca^{2+} activation site is located in proximity to two other channel activators (ATP and caffeine). Finally, this site is directly adjacent to an interface involving E4032, mutation of which inhibits Ca^{2+}-dependent activation of the channel. Although E4032 has previously been proposed to form part of the activating Ca^{2+}-binding site (Fessenden et al., 2001), this not so in the Ca^{2+}-bound structures. Rather E4032 forms an interface with CTD in the Ca^{2+}-bound state, most likely stabilizing a conformation that permits Ca^{2+} binding.

ATP-Binding Site

Adenine-derived nucleosides and nucleotides are competitive partial agonists of both skeletal (Laver et al., 2001) and cardiac (Kermode et al., 1998) RyRs. ATP increases the maximal P_O of the channel, without altering conductance (Kermode et al., 1998). Given that the cytosolic concentration of ATP in a myocyte is ~5 mM (Kushmerick et al., 1992), ATP or one of its analogs will likely be constitutively bound to the channel. The adenine ring appears to be critically important for affinity, as neither GTP (Chan et al., 2000) nor inosine monophosphate (Laver et al., 2001) activates RyRs. Channel activation as assessed by the maximal P_O in the presence of varying [Ca^{2+}] correlates with the number of phosphates attached to the ribose ring (ATP > ADP > AMP > adenosine ≈ adenine), with adenine and adenosine exhibiting minimal effects on the open probability of RyR2, whereas ATP increases the maximal open probability of RyR2 to >0.9 (Chan et al., 2000).

Studies using spin-labeled ATP analogs have shown that each protomer of RyR1 possesses one to two ATP-binding sites (Dias and Vogel, 2009). It has been suggested that the last 100 residues of the C-terminal "tail" contribute to the ATP-binding pocket (Popova et al., 2012). We now confirm that this C-terminal segment contains the ATP-binding pocket of RyR1.

Caffeine-Binding Site

The possibility that natural nucleotide metabolites may regulate the caffeine site in vivo remains to be explored. The putative caffeine-binding site identified here is at the interface of the CTD, S2S3, and TaF domains. The co-localization of binding sites for ATP, caffeine, and Ca^{2+} at interfaces of the CTD with other channel domains highlights the importance of this region in channel activation.

Ryanodine-Binding Site

At low (nM-μM) concentrations ryanodine binds to the open channel and locks it in a partial conductance state of exceptionally long duration, the fractional conductance of which varies with the structural features of the ryanoid (Welch et al., 1997). Increased ryanoid concentrations (~100 μM) lock the channel in a non-conducting state (Zimányi et al., 1992). RyRs possess one high-affinity ryanodine-binding site per tetramer, and up to three lower-affinity sites (Carroll et al., 1991). Several lines of evidence suggest that the high-affinity ryanodine-binding site is located within the transmembrane pore of the channel. First, in photoaffinity labeling experiments, an azido derivative of ryanodine specifically labels the C-terminal 76-kDa fragment of the channel (Witcher et al., 1994). Concordant results were obtained by proteolysis of the channel labeled with tritiated ryanodine (Callaway et al., 1994). Second, ryanodine and the impermeant cation tetraethylammonium (TEA) compete for the same binding site, and ryanodine-binding results in relocation of the TEA-binding site toward the cytoplasm (Tanna et al., 2006). Third, mutation of a pore-lining glutamine residue (Q4863 in RyR2; Q4933 in RyR1) located just luminal to the cytoplasmic gate (I4937) dramatically affects ryanoid-binding kinetics (Ranatunga et al., 2007). Comparison of RyR1 with structures of potassium and sodium channels complexed with quaternary ammonium ions demonstrates that the expected TEA-binding site is directly adjacent to Q4933, consistent with the aforementioned report that ryanoids and quaternary alkyl ammonium ions compete for the same binding site (Tinker and Williams, 1993). Last, ryanoid-binding kinetics are positively correlated with prior open probability, such that RyR activators, including Ca^{2+} and ATP, increase the rate of ryanoid association, while ryanoids do not bind when the pore is closed (Chu et al., 1990).

We show here that a ryanodine-binding site is located within the pore, proximal to Q4933, and that binding of ryanodine to this site locks the transmembrane pore in a dilated state. The structure of ryanodine-modified RyR1 has several implications for understanding the mechanism of ryanodine binding and its modulation of RyR channel activity. The four equivalent ryanodine-binding sites within the pore are sufficiently close to one another that the association of a ryanodine molecule to any one of them likely adversely impacts ryanodine binding at the three remaining sites, consistent with the observation that ryanodine binds to a single site with high affinity, followed by slow association at lower-affinity sites. The ryanodine-binding site is located within the permeation pathway of the channel, but displaced from the vertical axis of the channel, suggesting that association of a single ryanoid may lock the channel in a fractional conductance state by partially occluding the permeation pathway, while association of multiple ryanoids would likely result in complete occlusion of Ca^{2+} permeation. One implication of these observations is that the high- and low-affinity ryanodine-binding sites may be one and the same, as previously proposed (Callaway et al., 1994).

Conservation of Shared Functional Motifs in RyRs and IP3Rs

IP3Rs are the other members of the intracellular Ca^{2+} release channel family and are also tetrameric, Ca^{2+}-modulated channels. IP3Rs share significant sequence and structural similarity with RyR1 in the TM, CSol, and NTD-A/B/C regions (Fan et al., 2015). IP3Rs also exhibit a biphasic response to Ca^{2+} and sensitivity to ATP but lack sensitivity to ryanodine (Bezprozvanny et al., 1991). Mapping sequence conservation across the RyR/IP3R family onto our structure, we see that the residues comprising the putative Ca^{2+}-binding site at the CTD-CSol interface are conserved, as are many residues in the ATP-binding site. Notably, both the EF1 and 2 and S2S3 domains are absent in IP3R, suggesting that any function these two domains (and the interactions between them) have is likely to be of a regulatory, RyR-specific nature, rather than due to involvement in Ca^{2+} activation. This hypothesis is supported by the observations that the EF1 and 2 pair can be deleted without substantially affecting RyR2 activation by Ca^{2+} (Guo et al., 2016), and that the EF hands of the equivalent region in RyR1 can be sequence-scrambled without abolishing Ca^{2+} activation of the channel (Fessenden et al., 2004). Last, the key residues in the ryanodine-binding site, most notably Q4933, are not conserved in IP3Rs, supporting the hypothesis that this is the functionally relevant ryanodine-binding-site.

STAR★METHODS

Detailed methods are provided in the online version of this paper and include the following:

- KEY RESOURCES TABLE
- CONTACT FOR REAGENT AND RESOURCE SHARING
- METHOD DETAILS
  - Rabbit Skeletal Muscle RyR1 Purification
  - Single-Channel Recordings
  - Electron Microscopy
  - Image Processing
  - 3D Classification
  - Resolution Estimation
  - Model Building
  - Ligand Identification and Assignment
  - Pore Radius Calculations
  - Movie Preparation

SUPPLEMENTAL INFORMATION

Supplemental Information includes eight figures, four tables, and two data files and can be found with this article online at http://dx.doi.org/10.1016/j.cell.2016.08.075.

AUTHOR CONTRIBUTIONS

Conceptualization, A.d.G., O.B.C., R.Z., W.A.H., A.R.M., and J.F.; Investigation, A.G., O.B.C., R.Z., Q.Y., K.J.C., and R.G.; Writing, A.G., O.B.C., R.Z., K.J.C., R.G., W.A.H., A.R.M., and J.F.

ACKNOWLEDGMENTS

This work was supported by grants from the National Institutes of Health (R01AR060037 and R01HL061503 to A.R.M., P41GM116799 to W.A.H., R01GM29169 to J.F.) and by HHMI (to J.F.). R.Z. was a fellow of the American Heart Association (0625919T), K.J.C. was supported by the HHMI EXROP-Capstone Program, and O.B.C. was supported by a Charles H. Revson Senior Fellowship. A.R.M. is a consultant to, and owns shares in, ARMGO Pharma, Inc., a biotech company targeting RyR channels for therapeutic purposes.

Received: February 29, 2016
Revised: June 8, 2016
Accepted: August 30, 2016
Published: September 22, 2016

REFERENCES

Afonine, P.V., Headd, J.J., Terwilliger, T.C., and Adams, P.D. (2013). Computational Crystallography Newsletter. http://phenix-online.org/newsletter/CCN_2013_07.pdf.

Barad, B.A., Echols, N., Wang, R.Y.-R., Cheng, Y., Dimaio, F., Adams, P.D., and Fraser, J.S. (2015). EMRinger: side chain-directed model and map validation for 3D cryo-electron microscopy. Nat. Meth *12*, 943–946.

Bers, D.M., Patton, C.W., and Nuccitelli, R. (2010). A practical guide to the preparation of Ca(2+) buffers. Methods Cell Biol *99*, 1–26.

Bezprozvanny, I., Watras, J., and Ehrlich, B.E. (1991). Bell-shaped calcium-response curves of Ins(1,4,5)P3- and calcium-gated channels from endoplasmic reticulum of cerebellum. Nature *351*, 751–754.

Brillantes, A.B., Ondrias, K., Scott, A., Kobrinsky, E., Ondriasová, E., Moschella, M.C., Jayaraman, T., Landers, M., Ehrlich, B.E., and Marks, A.R. (1994). Stabilization of calcium release channel (ryanodine receptor) function by FK506-binding protein. Cell *77*, 513–523.

Callaway, C., Seryshev, A., Wang, J.-P., Slavik, K.J., Needleman, D.H., Cantu, C., 3rd, Wu, Y., Jayaraman, T., Marks, A.R., and Hamilton, S.L. (1994). Localization of the high and low affinity [3H]ryanodine binding sites on the skeletal muscle Ca^{2+} release channel. J. Biol. Chem. *269*, 15876–15884.

Carroll, S., Skarmeta, J.G., Yu, X., Collins, K.D., and Inesi, G. (1991). Interdependence of ryanodine binding, oligomeric receptor interactions, and Ca^{2+} release regulation in junctional sarcoplasmic reticulum. Arch. Biochem. Biophys. *290*, 239–247.

Chan, W.M., Welch, W., and Sitsapesan, R. (2000). Structural factors that determine the ability of adenosine and related compounds to activate the cardiac ryanodine receptor. Br. J. Pharmacol. *130*, 1618–1626.

Chen, S., McMullan, G., Faruqi, A.R., Murshudov, G.N., Short, J.M., Scheres, S.H.W., and Henderson, R. (2013). High-resolution noise substitution to measure overfitting and validate resolution in 3D structure determination by single particle electron cryomicroscopy. Ultramicroscopy *135*, 24–35.

Chu, A., Díaz-Muñoz, M., Hawkes, M.J., Brush, K., and Hamilton, S.L. (1990). Ryanodine as a probe for the functional state of the skeletal muscle sarcoplasmic reticulum calcium release channel. Mol. Pharmacol. *37*, 735–741.

Dias, J.M., and Vogel, P.D. (2009). Effects of small molecule modulators on ATP binding to skeletal ryanodine receptor. Protein J. *28*, 240–246.

Du, G.G., Sandhu, B., Khanna, V.K., Guo, X.H., and MacLennan, D.H. (2002). Topology of the Ca^{2+} release channel of skeletal muscle sarcoplasmic reticulum (RyR1). Proc. Natl. Acad. Sci. USA *99*, 16725–16730.

Dubochet, J., Adrian, M., Chang, J.J., Homo, J.C., Lepault, J., McDowall, A.W., and Schultz, P. (1988). Cryo-electron microscopy of vitrified specimens. Q. Rev. Biophys. *21*, 129–228.

Edgar, R.C. (2004). MUSCLE: multiple sequence alignment with high accuracy and high throughput. Nucleic Acids Res. *32*, 1792–1797.

Efremov, R.G., Leitner, A., Aebersold, R., and Raunser, S. (2015). Architecture and conformational switch mechanism of the ryanodine receptor. Nature *517*, 39–43.

Emsley, P., Lohkamp, B., Scott, W.G., and Cowtan, K. (2010). Features and development of Coot. Acta Crystallogr. D Biol. Crystallogr. *66*, 486–501.

Euden, J., Mason, S.A., and Williams, A.J. (2013). Functional characterization of the cardiac ryanodine receptor pore-forming region. PLoS ONE *8*, e66542.

Fabiato, A. (1983). Calcium-induced release of calcium from the cardiac sarcoplasmic reticulum. Am J Physiol *245*, C1–C14.

Fan, G., Baker, M.L., Wang, Z., Baker, M.R., Sinyagovskiy, P.A., Chiu, W., Ludtke, S.J., and Serysheva, I.I. (2015). Gating machinery of InsP3R channels revealed by electron cryomicroscopy. Nature *527*, 336–341.

Fessenden, J.D., Chen, L., Wang, Y., Paolini, C., Franzini-Armstrong, C., Allen, P.D., and Pessah, I.N. (2001). Ryanodine receptor point mutant E4032A reveals an allosteric interaction with ryanodine. Proc. Natl. Acad. Sci. USA *98*, 2865–2870.

Fessenden, J.D., Feng, W., Pessah, I.N., and Allen, P.D. (2004). Mutational analysis of putative calcium binding motifs within the skeletal ryanodine receptor isoform, RyR1. J. Biol. Chem. *279*, 53028–53035.

Franzini-Armstrong, C., Protasi, F., and Ramesh, V. (1999). Shape, size, and distribution of Ca(2+) release units and couplons in skeletal and cardiac muscles. Biophys. J. *77*, 1528–1539.

Grant, T., and Grigorieff, N. (2015). Measuring the optimal exposure for single particle cryo-EM using a 2.6 Å reconstruction of rotavirus VP6. eLife *4*, e06980.

Guo, W., Sun, B., Xiao, Z., Liu, Y., Wang, Y., Zhang, L., Wang, R., and Chen, S.R.W. (2016). The EF-hand Ca^{2+} Binding Domain Is Not Required for Cytosolic Ca^{2+} Activation of the Cardiac Ryanodine Receptor. J. Biol. Chem. *291*, 2150–2160.

Hamilton, S.L., and Reid, M.B. (2000). RyR1 modulation by oxidation and calmodulin. Antioxid. Redox Signal. *2*, 41–45.

Imagawa, T., Smith, J.S., Coronado, R., and Campbell, K.P. (1987). Purified ryanodine receptor from skeletal muscle sarcoplasmic reticulum is the Ca^{2+}-permeable pore of the calcium release channel. J. Biol. Chem. *262*, 16636–16643.

Kaftan, E., Marks, A.R., and Ehrlich, B.E. (1996). Effects of rapamycin on ryanodine receptor/Ca(2+)-release channels from cardiac muscle. Circ. Res. *78*, 990–997.

Kermode, H., Williams, A.J., and Sitsapesan, R. (1998). The interactions of ATP, ADP, and inorganic phosphate with the sheep cardiac ryanodine receptor. Biophys. J. *74*, 1296–1304.

Kimlicka, L., Lau, K., Tung, C.-C., and Van Petegem, F. (2013). Disease mutations in the ryanodine receptor N-terminal region couple to a mobile intersubunit interface. Nat. Commun. *4*, 1506.

Kucukelbir, A., Sigworth, F.J., and Tagare, H.D. (2014). Quantifying the local resolution of cryo-EM density maps. Nat. Methods *11*, 63–65.

Kushmerick, M.J., Moerland, T.S., and Wiseman, R.W. (1992). Mammalian skeletal muscle fibers distinguished by contents of phosphocreatine, ATP, and Pi. Proc. Natl. Acad. Sci. USA *89*, 7521–7525.

Labro, A.J., Raes, A.L., Grottesi, A., Van Hoorick, D., Sansom, M.S.P., and Snyders, D.J. (2008). Kv channel gating requires a compatible S4-S5 linker and bottom part of S6, constrained by non-interacting residues. J. Gen. Physiol. *132*, 667–680.

Lang, P.T., Brozell, S.R., Mukherjee, S., Pettersen, E.F., Meng, E.C., Thomas, V., et al. (2009). DOCK 6: combining techniques to model RNA-small molecule complexes. RNA *15*, 1219–1230.

Lau, K., and Van Petegem, F. (2014). Crystal structures of wild type and disease mutant forms of the ryanodine receptor SPRY2 domain. Nat. Commun. *5*, 5397.

Laver, D.R., Lenz, G.K., and Lamb, G.D. (2001). Regulation of the calcium release channel from rabbit skeletal muscle by the nucleotides ATP, AMP, IMP and adenosine. J. Physiol. *537*, 763–778.

Lee, H.B., Xu, L., and Meissner, G. (1994). Reconstitution of the skeletal muscle ryanodine receptor-Ca^{2+} release channel protein complex into proteoliposomes. J. Biol. Chem. *269*, 13305–13312.

Li, P., and Chen, S.R. (2001). Molecular basis of Ca(2)+ activation of the mouse cardiac Ca(2)+ release channel (ryanodine receptor). J. Gen. Physiol. *118*, 33–44.

Li, X., Mooney, P., Zheng, S., Booth, C.R., Braunfeld, M.B., Gubbens, S., Agard, D.A., and Cheng, Y. (2013). Electron counting and beam-induced motion correction enable near-atomic-resolution single-particle cryo-EM. Nat. Methods *10*, 584–590.

Marx, S.O., Ondrias, K., and Marks, A.R. (1998). Coupled gating between individual skeletal muscle Ca^{2+} release channels (ryanodine receptors). Science *281*, 818–821.

Mei, Y., Xu, L., Mowrey, D.D., Mendez Giraldez, R., Wang, Y., Pasek, D.A., Dokholyan, N.V., and Meissner, G. (2015). Channel gating dependence on pore lining helix glycine residues in skeletal muscle ryanodine receptor. J. Biol. Chem. *290*, 17535–17545.

Murayama, T., Kurebayashi, N., Oba, T., Oyamada, H., Oguchi, K., Sakurai, T., and Ogawa, Y. (2011). Role of amino-terminal half of the S4-S5 linker in type 1 ryanodine receptor (RyR1) channel gating. J. Biol. Chem. *286*, 35571–35577.

Nagasaki, K., and Fleischer, S. (1988). Ryanodine sensitivity of the calcium release channel of sarcoplasmic reticulum. Cell Calcium *9*, 1–7.

Pettersen, E.F., Goddard, T.D., Huang, C.C., Couch, G.S., Greenblatt, D.M., Meng, E.C., and Ferrin, T.E. (2004). UCSF Chimera–a visualization system for exploratory research and analysis. J. Comput. Chem. *25*, 1605–1612.

Popova, O.B., Baker, M.R., Tran, T.P., Le, T., and Serysheva, I.I. (2012). Identification of ATP-binding regions in the RyR1 Ca^{2+} release channel. PLoS ONE *7*, e48725.

Ranatunga, K.M., Chen, S.R.W., Ruest, L., Welch, W., and Williams, A.J. (2007). Quantification of the effects of a ryanodine receptor channel mutation on interaction with a ryanoid. Mol. Membr. Biol. *24*, 185–193.

Rohou, A., and Grigorieff, N. (2015). CTFFIND4: Fast and accurate defocus estimation from electron micrographs. J. Struct. Biol. *192*, 216–221.

Rosenthal, P.B., and Henderson, R. (2003). Optimal determination of particle orientation, absolute hand, and contrast loss in single-particle electron cryomicroscopy. J. Mol. Biol. *333*, 721–745.

Rubinstein, J.L., and Brubaker, M.A. (2015). Alignment of cryo-EM movies of individual particles by optimization of image translations. J. Struct. Biol. *192*, 188–195.

Russo, C.J., and Passmore, L.A. (2014). Electron microscopy: Ultrastable gold substrates for electron cryomicroscopy. Science *346*, 1377–1380.

Samsó, M., Feng, W., Pessah, I.N., and Allen, P.D. (2009). Coordinated movement of cytoplasmic and transmembrane domains of RyR1 upon gating. PLoS Biol. *7*, e85.

Scheres, S.H.W. (2012). RELION: implementation of a Bayesian approach to cryo-EM structure determination. J. Struct. Biol. *180*, 519–530.

Scheres, S.H.W. (2015). Semi-automated selection of cryo-EM particles in RELION-1.3. J. Struct. Biol. *189*, 114–122.

Schoenmakers, T.J., Visser, G.J., Flik, G., and Theuvenet, A.P. (1992). CHELATOR: an improved method for computing metal ion concentrations in physiological solutions. BioTechniques *12*, 870–874 and 876–879.

Serysheva, I.I., Schatz, M., van Heel, M., Chiu, W., and Hamilton, S.L. (1999). Structure of the skeletal muscle calcium release channel activated with Ca^{2+} and AMP-PCP. Biophys. J. *77*, 1936–1944.

Sharma, P., Ishiyama, N., Nair, U., Li, W., Dong, A., Miyake, T., Wilson, A., Ryan, T., MacLennan, D.H., Kislinger, T., et al. (2012). Structural determination of the phosphorylation domain of the ryanodine receptor. FEBS J. *279*, 3952–3964.

Smart, O.S., Neduvelil, J.G., Wang, X., Wallace, B.A., and Sansom, M.S. (1996). HOLE: A program for the analysis of the pore dimensions of ion channel structural models. J. Mol. Graph *14*, 354–60–376.

Suloway, C., Pulokas, J., Fellmann, D., Cheng, A., Guerra, F., Quispe, J., Stagg, S., Potter, C.S., and Carragher, B. (2005). Automated molecular microscopy: The new Leginon system. J. Struct. Biol. *151*, 41–60.

Tanna, B., Welch, W., Ruest, L., Sutko, J.L., and Williams, A.J. (2006). The interaction of an impermeant cation with the sheep cardiac RyR channel alters ryanoid association. Mol. Pharmacol. *69*, 1990–1997.

Tinker, A., and Williams, A.J. (1993). Using large organic cations to probe the nature of ryanodine modification in the sheep cardiac sarcoplasmic reticulum calcium release channel. Biophys. J *65*, 1678–1683.

Tung, C.-C., Lobo, P.A., Kimlicka, L., and Van Petegem, F. (2010). The amino-terminal disease hotspot of ryanodine receptors forms a cytoplasmic vestibule. Nature *468*, 585–588.

Wagenknecht, T., Grassucci, R., and Frank, J. (1988). Electron microscopy and computer image averaging of ice-embedded large ribosomal subunits from Escherichia coli. J. Mol. Biol. *199*, 137–147.

Waterhouse, A.M., Procter, J.B., Martin, D.M.A., Clamp, M., and Barton, G.J. (2009). Jalview Version 2–a multiple sequence alignment editor and analysis workbench. Bioinformatics *25*, 1189–1191.

Welch, W., Williams, A.J., Tinker, A., Mitchell, K.E., Deslongchamps, P., Lamothe, J., Gerzon, K., Bidasee, K.R., Besch, H.R., Jr., Airey, J.A., et al. (1997). Structural components of ryanodine responsible for modulation of sarcoplasmic reticulum calcium channel function. Biochemistry *36*, 2939–2950.

Witcher, D.R., McPherson, P.S., Kahl, S.D., Lewis, T., Bentley, P., Mullinnix, M.J., Windass, J.D., and Campbell, K.P. (1994). Photoaffinity labeling of the ryanodine receptor/Ca^{2+} release channel with an azido derivative of ryanodine. J. Biol. Chem. *269*, 13076–13079.

Xiong, L., Zhang, J.-Z., He, R., and Hamilton, S.L. (2006). A Ca^{2+}-binding domain in RyR1 that interacts with the calmodulin binding site and modulates channel activity. Biophys. J. *90*, 173–182.

Yamamoto, T., El-Hayek, R., and Ikemoto, N. (2000). Postulated role of interdomain interaction within the ryanodine receptor in Ca(2+) channel regulation. J. Biol. Chem. *275*, 11618–11625.

Yan, Z., Bai, X.-C., Yan, C., Wu, J., Li, Z., Xie, T., Peng, W., Yin, C.-C., Li, X., Scheres, S.H.W., et al. (2015). Structure of the rabbit ryanodine receptor RyR1 at near-atomic resolution. Nature *517*, 50–55.

Yuchi, Z., Yuen, S.M.W.K., Lau, K., Underhill, A.Q., Cornea, R.L., Fessenden, J.D., and Van Petegem, F. (2015). Crystal structures of ryanodine receptor SPRY1 and tandem-repeat domains reveal a critical FKBP12 binding determinant. Nat. Commun. *6*, 7947.

Zalk, R., Clarke, O.B., des Georges, A., Grassucci, R.A., Reiken, S., Mancia, F., Hendrickson, W.A., Frank, J., and Marks, A.R. (2015). Structure of a mammalian ryanodine receptor. Nature *517*, 44–49.

Zimányi, I., Buck, E., Abramson, J.J., Mack, M.M., and Pessah, I.N. (1992). Ryanodine induces persistent inactivation of the Ca^{2+} release channel from skeletal muscle sarcoplasmic reticulum. Mol. Pharmacol. *42*, 1049–1057.

STAR★METHODS

KEY RESOURCES TABLE

REAGENT or RESOURCE	SOURCE	IDENTIFIER
Chemicals, Peptides, and Recombinant Proteins		
GST-Calstabin1	Zalk et al., 2015	
GST-Calstabin2	Zalk et al., 2015	
Calf intestinal alkaline phosphatase	NEB	Cat# M0290L
CHAPS	Anatrace	Cat# C316S; CAS: 75621-03-3
DOPC	Avanti Polar Lipids	Cat# 850375C; CAS: 4235-95-4
DOPE	Avanti Polar Lipids	Cat# 850725C; CAS: 4004-05-1
Na_2ATP	Sigma-Aldrich	Cat# A2383; CAS: 34369-07-8
Ryanodine	Abcam	Cat# ab120083; CAS:15662-33-6
Caffeine	Sigma-Aldrich	Cat# C0750; CAS: 58-08-2
Critical Commercial Assays		
GSTrap	GE Healthcare	Cat# 17-5281-01
Ni Sepharose 6 fast flow	GE Healthcare	Cat# 17-5318-01
Thrombin	Biopharm	Cat# 91-035
TSK-GEL G4SWxl	Tosoh	Cat# 20025
Deposited Data		
N-terminal 3 domain of ryr1	Tung et al., 2010	PDB: 2XOA
RyR1-Cs2 EGTA closed-state	Zalk et al., 2015	PDB: 3J8E
RyR1-Cs2 EGTA closed-state	Yan et al., 2015	PDB: 3J8H
SPRY1 domain of RyR2	Yuchi et al., 2015	PDB: 5C33
SPRY2 domain of RyR1	Lau and Van Petegem, 2014	PDB: 4P9J
RY1&2 domain of RyR1	Yuchi et al., 2015	PDB: 5C30
RyR1-Cs2, Ca^{2+} only (all particles)	This study	PDB:5T15; EMDB: EMD-8342
RyR1-Cs2, Ca^{2+} only (Class 1)	This study	PDB: 5T9M; EMDB: EMD-8372
RyR1-Cs2, Ca^{2+} only (Class 2)	This study	PDB: 5T9N; EMDB: EMD-8373
RyR1-Cs2, Ca^{2+} only (Class 3)	This study	PDB: 5T9R; EMDB: EMD-8374
RyR1-Cs2, Ca^{2+} only (Class 4)	This study	PDB: 5T9S; EMDB: EMD-8375
RyR1-Cs2, EGTA (all particles)	This study	PDB: 5TB0; EMDB: EMD-8391
RyR1-Cs2, EGTA (Class 1)	This study	PDB: 5TB1; EMDB: EMD-8392
RyR1-Cs2, EGTA (Class 2)	This study	PDB: 5TB2; EMDB: EMD-8393
RyR1-Cs2, EGTA (Class 3)	This study	PDB: 5TB3; EMDB: EMD-8394
RyR1-Cs2, EGTA (Class 4)	This study	PDB: 5TB4; EMDB: EMD-8395
RyR1-Cs2, ATP/Caf. (all particles)	This study	PDB: 5TAP; EMDB: EMD-8381
RyR1-Cs2, ATP/Caf. (Class 1)	This study	PDB: 5TAS; EMDB: EMD-8383
RyR1-Cs2, ATP/Caf. (Class 2)	This study	PDB: 5TAT; EMDB: EMD-8384
RyR1-Cs2, ATP/Caf. (Class 3)	This study	PDB: 5TAU; EMDB: EMD-8385
RyR1-Cs2, ATP/Caf. (Class 4)	This study	PDB: 5TAV; EMDB: EMD-8386
RyR1-Cs2, Ca^{2+}/ATP/Caf. (Class 1)	This study	PDB: 5T9V; EMDB: EMD-8376
RyR1-Cs2, Ca^{2+}/ATP/Caf. (Class 2)	This study	PDB: 5TA3; EMDB: EMD-8377
RyR1-Cs2, Ca^{2+}/ATP/Caf. (Class 1&2)	This study	PDB: 5TAL; EMDB: EMD-8378
RyR1-Cs2, Ca^{2+}/ATP/Caf. (Class 3)	This study	PDB: 5TAN; EMDB: EMD-8380
RyR1-Cs2, Ca^{2+}/ATP/Caf. (Class 4)	This study	PDB: 5TAM; EMDB: EMD-8379

(*Continued on next page*)

Continued		
REAGENT or RESOURCE	SOURCE	IDENTIFIER
RyR1-Cs2, Ca^{2+}/ATP/Caf. (Class 3&4)	This study	PDB: 5TAQ; EMDB: EMD-8382
RyR1-Cs2, ryanodine (all particles)	This study	PDB: 5TAW; EMDB: EMD-8387
RyR1-Cs2, ryanodine (Class 1)	This study	PDB: 5TAX; EMDB: EMD-8388
RyR1-Cs2, ryanodine (Class 2)	This study	PDB: 5TAY; EMDB: EMD-8389
RyR1-Cs2, ryanodine (Class 3)	This study	PDB: 5TAZ; EMDB: EMD-8390
Experimental Models: Organisms/Strains		
Male New Zealand rabbits skeletal muscle	Sierra for Medical Science	By request
Software and Algorithms		
MaxChelator	Bers et al., 2010	version 2.50; http://maxchelator.stanford.edu
Axoscope 10.1	Axon Instruments	http://mdc.custhelp.com/app/answers/detail/a_id/18779
Clampfit 10.1	Molecular Devices	http://mdc.custhelp.com/app/answers/detail/a_id/18779
MotionCorr	Li et al., 2013	http://cryoem.ucsf.edu/software/driftcorr.html
UNBLUR	Grant and Grigorieff, 2015	http://grigoriefflab.janelia.org/unblur
RELION 1.4	Scheres, 2015	http://www2.mrc-lmb.cam.ac.uk/relion
CTFFIND4	Rohou and Grigorieff, 2015	http://grigoriefflab.janelia.org/ctf
alignparts_lmbfgs	Rubinstein and Brubaker, 2015	https://sites.google.com/site/rubinsteingroup/direct-detector-align_lmbfgs
UCSF Chimera	Pettersen et al., 2004	https://www.cgl.ucsf.edu/chimera/
ResMap	Kucukelbir et al., 2014	http://resmap.sourceforge.net
EMringer	Barad et al., 2015	http://fraserlab.com/2015/02/18/EMringer
Coot	Emsley et al., 2010	http://www2.mrc-lmb.cam.ac.uk/personal/pemsley/coot/
phenix.real_space_refine	Afonine et al., 2013	https://www.phenix-online.org/documentation/reference/real_space_refine.html
HOLE	Smart et al., 1996	http://www.smartsci.uk/hole/
DOCK-PREP	Lang et al., 2009	https://www.cgl.ucsf.edu/chimera/docs/ContributedSoftware/dockprep/dockprep.html
VideoBlend	Videoblend	http://www.videoblend.org
Affinity Designer	Serif	https://affinity.serif.com
Gnu Image Manipulation Program	GIMP	https://www.gimp.org
Other		
C-flat CF-1.2/1.3-2C-T grids	Protochips Inc, NC	*CF-1.2/1.3-2C-T*
Quantifoil R1.2/1.3 grids	Quantifoil Micro Tools GmbH, Germany	*R1.2/1.3*

CONTACT FOR REAGENT AND RESOURCE SHARING

Further information and requests for reagents may be directed to, and will be fulfilled by the Lead Contact Joachim Frank (jf2192@cumc.columbia.edu).

METHOD DETAILS

Rabbit Skeletal Muscle RyR1 Purification

To obtain highly purified RyR1, 100 g of rabbit skeletal muscle flash frozen in liquid nitrogen were blended for 90 s in 500 ml of cold buffer containing 10 mM Tris-Maleate (pH 6.8), 1 mM DTT, 1 mM EDTA, 150 μM PMSF and 1 mM benzamidine. The mixture was then centrifuged for 10 min at 8000 × g. The supernatant was centrifuged for 20 min at 40,000 × g. Pellets were solubilized in 50 ml buffer containing 10 mM HEPES (pH 7.5), 1% CHAPS, 1 M NaCl, 2 mM EGTA, 2 mM TCEP, and a protease inhibitors cocktail (Roche). The solubilized membranes were diluted 1:1 in the same buffer without NaCl and centrifuged for 30 min at 100,000 × g. The supernatant was filtered using a 0.2 μm filter and bound overnight at 4 °C to a pre-equilibrated 5 ml GSTrap (GE healthcare) column with bound GST-Calstabin1. The column was washed with 10 column volumes of cold modified solubilization buffer (0.5%CHAPS and 0.5M NaCl) and RyR1 was eluted with two column volumes of 10 μM Calstabin2 in the same buffer (expressed in 1L IPTG induced transformed BL-21 cells and thrombin cleaved for 1 hr at room temperature; Calstabin2 was used for elution because of its higher affinity for RyR1 compared to Calstabin1). The eluent was pre-cleared with 0.5 ml glutathione beads to remove traces of GST-calstabin and treated with calf intestinal alkaline phosphatase (CIP, NEB, 100 U/mL final concentration) for 3 hr at room temperature to dephosphorylate RyR1 (Zalk et al., 2015). The CIP-treated RyR1 was concentrated on a 100 kDa cut-off centrifugation filter (Millipore) to less than 500 μl, loaded on a size-exclusion column (Tosoh TSKgel G4SWxl) and washed in modified solubilization buffer with 1 mM EGTA, 0.25% CHAPS and 0.001% (w/v) DOPC (Avanti). The peak fractions were combined and modulating regents were added and allowed to bind (Ca^{2+}, ATP and caffeine for 10 min at room temperature, ryanodine for 2 hr at room temperature followed by overnight at 4 °C, See Table S1 for a summary of the final solution conditions of each sample). Final free [Ca^{2+}] was determined using MaxChelator (Schoenmakers et al., 1992). The samples were concentrated on a 100 kDa cut-off centrifugation filter (Millipore) to 5-10 mg/ml determined using spectroscopy (Abs at 280 nm, 1% (w/v) = 1.018).

Single-Channel Recordings

To prepare SR vesicles, muscles were homogenized using a tissue vitamizer (Fisher Scientific) at the highest speed for 1 min with 2 volumes of 20 mM Tris-maleate (pH 7.4), 1 mM EDTA and protease inhibitors (Roche). Homogenate was centrifuged at 4,000 × g for 15 min at 4°C and the supernatant was centrifuged at 40,000 × g for 30 min at 4°C. The final pellet, containing the SR fractions, was resuspended and aliquoted in: 250 mM sucrose, 10 mM MOPS (pH 7.4), 1 mM EDTA and protease inhibitors. Samples were frozen in liquid nitrogen and stored at −80°C.

Purified RyR1/Cs2 was reconstituted into microsomes as described by (Lee et al., 1994). Briefly, the purified receptor complex was brought to 5mg/mL of lipid mixture (5:3 DOPC:DOPE) and placed in a dialysis cassette (10k MWCO). The sample was dialyzed overnight in 1L of buffer containing 0.5M NaCl, 0.1mM EGTA, 0.2mM Ca, 150uM PMSF, 1mM DTT, and 10mM HEPES (pH 7.4) to remove free detergent. Following dialysis, samples were snap frozen in liquid nitrogen and stored at −80°C until use.

SR vesicles containing RyR1 or microsomes reconstituted with purified RyR1/Cs2 complex were fused to planar lipid bilayers formed by painting a lipid mixture of phosphatidylethanolamine and phosphatidylcholine (Avanti Polar Lipids) in a 5:3 ratio in decane across a 200-μm hole in polysulfonate cups (Warner Instruments) separating 2 chambers. The trans chamber (1.0 ml), representing the intra-SR (luminal) compartment, was connected to the head stage input of a bilayer voltage clamp amplifier. The *cis* chamber (1.0 ml), representing the cytoplasmic compartment, was held at virtual ground. Symmetrical solutions used were as follows (in mM): 1 mM EGTA, 250/125 mM HEPES/Tris, 50 mM KCl (pH 7.35) as *cis* solution and 53 mM Ca(OH)2, 50 mM KCl, 250 mM HEPES (pH 7.35) as trans solution. The concentration of free Ca^{2+} in the *cis* chamber was calculated with WinMaxC program (v.2.50; http://web.stanford.edu/~cpatton/maxc.html). SR vesicles or microsomes containing purified RyR1/Cs2 complex were added to the *cis* side and fusion with the lipid bilayer was induced by making the *cis* side hyperosmotic by the addition of 400-500 mM KCl. After the appearance of potassium and chloride currents, the *cis* side was perfused with the *cis* solution. For the nominally Ca^{2+}-free condition, EGTA was added to the *cis* side to get a final concentration of 5mM EGTA. For the EGTA/ATP/Caffeine condition, ATP (2mM final concentration) and caffeine (5mM final concentration) were added to the *cis* side. Ca^{2+} (30 μM final concentration) was added to *cis* side to get the Ca^{2+}-only condition. ATP (2mM final concentration), caffeine (5mM final concentration) and Ca^{2+} (30 μM final concentration) was added to *cis* side to get the ATP/Caffeine/Ca^{2+} condition. All ligands and additives were prepared as 100X stock solutions solubilized in *cis* buffer. At the end of each experiment, Ryanodine (10 μM) was added to block the RyR channel. Single-channel currents were recorded at 0 mV using a Bilayer Clamp BC-525C (Warner Instruments), filtered at 1 kHz using a Low-Pass Bessel Filter 8 Pole (Warner Instruments), and digitized at 4 kHz. All experiments were performed at room temperature (23°C). Data acquisition was performed using Digidata 1322A and Axoscope 10.1 software (Axon Instruments). The recordings were analyzed by using Clampfit 10.1 (Molecular Devices) and Graphpad Prism software.

Electron Microscopy

Grid preparation and imaging were carried out in two different ways (see Table S1). For grid preparation, early samples (RyR-EGTA and RyR-30 μM Ca^{2+}-ATP-caffeine) were prepared as previously described (Zalk et al., 2015). Briefly, 3 μl of each sample was applied to holey carbon grids (C-flat CF-1.2/1.3-2C-T, Protochips Inc, NC) without prior plasma cleaning. Grids were blotted for ~1 s and vitrified by rapidly plunging into liquid ethane at −180°C (Dubochet et al., 1988; Wagenknecht et al., 1988) with a Vitrobot (FEI).

For subsequent experiments (RyR-30 μM Ca^{2+}, RyR-300 μM Ca^{2+}-10 μM Ryanodine), holey-gold grids were prepared as described by Russo and Passmore (Russo and Passmore, 2014) from either C-flat CF-1.2/1.3-2C-T (Protochips Inc, NC) or from Quantifoil R1.2/1.3 (Quantifoil Micro Tools GmbH, Germany). 3 μl of each sample was applied to the holey-gold grids after plasma cleaning with a mixture of H_2 and O_2. Grids were blotted for 3.5 to 4 s and vitrified by rapidly plunging into liquid ethane at −180°C (Dubochet et al., 1988; Wagenknecht et al., 1988) with a Vitrobot (FEI). Optimal blotting time was reproducible across samples.

Data acquisition was done on a FEI Tecnai F30 Polara (FEI, Eindhoven) operating at 300 kV. Datasets were collected with the automated data collection system Leginon (Suloway et al., 2005) at a nominal magnification of 31,000x in electron counting mode on a K2 Summit direct electron detector camera (Gatan, Pleasanton, CA) with a camera pixel size of 5 μm, corresponding to a calibrated pixel size of 1.255 Å on the specimen scale.

After obtaining the initial reconstruction, the pixel size was calibrated by real-space correlation with a crystal structure of the N-terminal 3 domains of the receptor (PDB ID: 2XOA). A clear maximum was observed at 1.255Å, and this value was found to be consistent across multiple different reconstructions, data collections and samples, and the same value was obtained using crystal structures of other fragments of the receptor. Applying the same approach to the reconstructions reported in (Yan et al., 2015) gave a calibrated pixel size for that map of 1.38Å, compared to the nominal pixel size of 1.34Å. This discrepancy highlights the importance of always performing pixel size calibration with a known crystal structure – a 3% difference in magnification could easily stymie interpretation of global conformational changes between structures determined of the same molecule by different groups. This does not have to be carried out on the molecule under investigation – the value determined here proved accurate for other datasets collected at this magnification on the same microscope for structurally unrelated assemblies.

For the early experiments on holey carbon grids, images were recorded in dose-fractionated mode, each image being fractionated into 20 frames, with an accumulation time of 0.2 s per frame. The dose rate on the camera was set to 8 counts per physical pixel per second. The total exposure time was 4 s, leading to a total accumulated dose of 25 electrons per $Å^2$ on the specimen. The beam diameter was set at approximately 500nm in order to capture 2 images per hole using image shift.

For subsequent experiments using holey gold grids, data acquisition was done under the same conditions except that the dose was increased from 25 to 50 $e^-/Å^2$ (100 $e^-/Å^2$ for the 10 μM ryanodine dataset)(Figure S6), with images fractionated into 34 frames with 1.5 $e^-/Å^2$/frame. The total exposure time was 8 s (16 s for the 10 μm ryanodine dataset). The additional contrast provided by the doubling of the electron dose improved the accuracy of orientation assignment estimated by RELION (Scheres, 2012) from 1.3 degrees to 1 degree. The beam diameter was also increased from ~0.5 μm to approximately 2 μm in order to obtain more parallel illumination and to expose gold on all sides of the hole to further limit beam-induced movement of the ice layer(Russo and Passmore, 2014). Therefore, only one image per hole was collected. Images were collected every second hole to reduce the possibility of double exposure of the imaged areas, which we found to be frequent with automated acquisition procedures.

Image Processing

The image processing followed the same protocol for every dataset. Datasets initially processed with a different protocol were subsequently reprocessed to match the established protocol. Dose-fractionated image stacks were aligned using either MotionCorr (Li et al., 2013) or UNBLUR (Grant and Grigorieff, 2015), depending on the dataset, and the sum of aligned frames was used for further preprocessing. The first two frames were omitted in the early datasets collected on C-flat holey-carbon grids, as they contained motions greater than 4Å per frame on average. For the datasets collected on holey-gold grids all frames were kept, as the motion was measured as less than 1Å per frame in the first frames. The particles were picked with the reference-based automated particle picking procedure implemented in RELION 1.3 (Scheres, 2015) and their defocus values estimated by CTFFIND4 (Rohou and Grigorieff, 2015). Those particles were subjected to 3D classification using RELION with the number of classes, K, varying between 8 and 10. The particles from the highest-resolution classes were then pooled and their frame stacks realigned using alignparts_lmbfgs (Rubinstein and Brubaker, 2015) with optimized alignment parameters (See Figure S7). Radiation-damage weighting following the procedure described by Grant and Grigorieff (Grant and Grigorieff, 2015) was used as incorporated in alignparts_lmbfgs, as follows. For each dataset, individual frames were filtered according to the optimal exposure curve calculated for Rotavirus VPC (Grant and Grigorieff, 2015), modulated by a constant (the effective dose per frame), chosen based on the overall resolution of the resulting reconstruction. The resulting per-frame resolution-dependent weighting factors are presented in Table S2. In the case of the ryanodine dataset, for which the largest dose of $100e^-/Å^2$ was collected, the dose-filtered reconstruction using the full $100e^-/Å^2$ dose reached a resolution of 4.7Å for the core, while a resolution of 4.2Å in the core was obtained for a reconstruction using only the frames up to $60e^-/Å^2$ total dose.

All pooled particles were then refined using RELION 1.4 with C_4 symmetry imposed. The combination of per-particle frame alignment, radiation-damage weighting and use of CTFFIND4 brought the resolution of our RyR-EGTA closed-state dataset (Zalk et al., 2015) from 4.8 to 4.4Å in the core (see Tables S1 and S4). This procedure was applied to all subsequent datasets. On the later datasets, the combination of holey-gold grids, higher total electron dose and parallel illumination greatly reduced the number of particles needed to reach resolutions of 4.5Å and better (see Table S4 and Figure S8), with the highest resolution dataset reaching 3.6Å resolution in the core from only 65,000 particles. Plots of the orientation distribution for each dataset reveals that preferred orientations of the particles vary between datasets, independently of the grid support material (Figure S6B).

3D Classification

The first one or two rounds of 2D/3D classification were used to remove low-quality particles or non-particle contaminants from the dataset and to obtain a first assessment of the conformational variability in the dataset. The particles retained were then classified following the same protocol for each dataset in order to minimize bias in the classification results. The results were also compared with those of earlier non-standardized rounds of classification and showed consistent results. The remaining factors varying between experiments were the number of particles and the imaging conditions.

For each dataset, the orientation parameters of the retained particles were first refined together as a single class (Table S4). The orientations were then kept for a classification with C4 symmetry imposed, a local orientation search and a number of classes set at K = 4. Maps from the previous combined refinement filtered at 25Å served as initial references. Orientational sampling was set to 0.9 degrees until convergence (~5 to 10% of particles changing class assignment) and then to 0.5 degrees for an additional 10 iterations

(1% or less particles changing class assignment). Each class was then refined using RELION autorefine procedure with C4 symmetry imposed (Table S4). Deviations from C4 symmetry were observed in non-symmetrized reconstructions. These deviations were difficult to interpret due to the constraints introduced by the local nature of the particles alignment search at fine increment, which prevents particles from changing between the 4 possible orientations about the symmetry axis. This prompted us to impose C4 symmetry during classification to simplify interpretation of the results and to maintain consistency with subsequent refinements, which reached higher resolution when carried out with C4 symmetry imposed.

Those classifications gave very similar ranges of motions of the shell and similar radius for the maximally open and the maximally closed pores as in the primary classifications. To measure the reproducibility of those classifications, we ran the Ca^{2+}/ATP/Caffeine dataset classification 3 times with the number of classes constant at K = 4, but with randomly generated seeds and with slightly different initial model resolutions (25, 30 and 35Å resolution) to introduce more variability in the starting parameters. The resulting classifications gave similar results with differences in min and max shell height between each run of 0.9 and 1.2Å SD, respectively. We have not refined the classes of all the runs to high resolution; we therefore cannot estimate the SD in pore aperture as precisely as for the shell height. But based on the ~7Å resolution classification volumes obtained, we do not see significant differences in pore aperture among the maximally open classes nor among the maximally closed classes of the Ca^{2+}/ATP-Caffeine dataset.

Resolution Estimation

Reported resolutions are based on the 'gold-standard' protocol with the FSC = 0.143 criterion using soft masks with a 8 pixel soft edge, and were corrected for the effects of the mask on the FSC curve using high-resolution noise substitution (Chen et al., 2013).

Because of the intrinsic flexibility of the channel, in particular the cytosolic assembly, resolution varies substantially between domains (Figure S6C). To characterize the local resolution on a domain level, FSCs were calculated domain by domain. The fitted atomic model of the ryanodine receptor was segmented following domain boundaries described in Figure S2 to create simulated density maps for each domain in UCSF Chimera (Pettersen et al., 2004). These simulated maps were then used to create masks for each domain of the receptor using relion_mask_create, with mask edges having an 8 pixel falloff. The FSC of each domain was then calculated by masking the two half-sets with the corresponding masks using relion_postprocess. Domain-by-domain resolutions are reported in Table S3 for each dataset. For the highest-resolution dataset (30 μM Ca^{2+}) resolutions vary among domains from 8.5Å (RY34) to 3.5Å (CTD). Resolutions reported by this method agreed well, in most cases, with features visible in the corresponding domains and with the local resolution estimated by ResMap (Kucukelbir et al., 2014) (Figure S6B). However, in lower-resolution domains, local resolutions determined by domain masking tend to be artificially high due to the width of the mask edge overlapping with neighboring domains of much higher local resolution.

For each class of every dataset, average resolution for the whole protomer, the core, and the shell regions, as well as class distributions and shell height are reported in Table S4. Shell height is measured from the membrane plane (approximated by a plane intersecting with the S4-S5 linkers) to the centers of mass of the RY1&2 domains at the corner of the shell.

Prior to visualization, all density maps were sharpened by applying a negative B-factor estimated according to Rosenthal and Henderson (Rosenthal and Henderson, 2003) on solvent-flattened maps. An additional negative B-factor, empirically determined based on visual inspection of the map in Coot, was applied for visualization and model fitting in areas of particularly good local resolution within the core of the molecule. In preparation for model building and map comparison, all maps were aligned (using the 'fitmap' module of UCSF Chimera) to a single reference map, comprised of the transmembrane region of the highest-resolution map, such that all maps are effectively aligned on the transmembrane pore.

Model Building

Model building proceeded in two stages. First, an optimal model was built and refined into the highest-resolution map (Ca^{2+}-only classes 1-4 combined), and then, in the second stage, this model was used for fitting the other states – Ca^{2+}/ATP/Caffeine (classes 1&2 and classes 3&4), EGTA/ATP/Caffeine (all classes combined), EGTA-only (all classes combined), and Ca^{2+}/ryanodine (all classes combined). The Ca^{2+}-only model had an EMringer score of 2.3, consistent with a well-refined model.

In the first stage, fitting began by placing the highest-resolution model of the Ca^{2+}-free closed state (PDB ID: 3J8H; (Yan et al., 2015)), into the Ca^{2+}-30μM map and rigid-body fitting each domain using a Python script launched from within Coot (Emsley et al., 2010). Recently determined crystal structures of SPRY1 (PDB ID: 5C33), SPRY2 (PDB ID: 4P9J) and the RY1&2 tandem repeat (PDB ID: 5C30) were incorporated into the model (Lau and Van Petegem, 2014; Yuchi et al., 2015), and the model was manually rebuilt and corrected in Coot based on the improved map, resulting in an increase in the overall number of placed and sequenced residues and a reduced number of chain breaks. Global real-space refinement of the model was carried out using phenix.real_space_refine, and this model became the initial model for subsequent fits to maps of other samples and conformational classes. Two regions of unassigned sequence remain. These are a large portion of SPRY3 (1275-1573) and the second half of BSol (2487-2735 and 2939-3613). Either a cryoEM reconstruction of substantially improved resolution, or in the case of SPRY3 a crystal structure of this region, will be required to address these remaining ambiguities in the structure. Residues with uncertain sequence assignment have been given the residue name "UNK"; The sequence numbering still reflects our best guess at the most likely assignment of the residue in question. All maps used in model fitting were first aligned to a reference map consisting of density corresponding to the RyR1 transmembrane region alone, and resampled upon the grid of this map, allowing facile comparison of the different structures. The map used in model fitting has had B-factor sharpening applied to a level appropriate for optimal visualization of high-resolution features such as sidechains in the best resolved parts of the structure. In order to have a better view of connectivity, model fitting, and interpretation used two maps in conjunction – the aforementioned map, and a version of it that had been filtered at 4.5 Å, for which connectivity and low resolution features are more clearly defined due to the absence of high-resolution noise.

In the second stage, the refined atomic model was rigid-body fit to each map, domain by domain, using the same Python script as used in the initial stages of building the reference model. The entire chain was manually inspected, and sections of the main chain that were flexibly deformed (for example, S6/S6c and interdomain interfaces) were corrected using the real-space refine zone tool within Coot, with Ramachandran restraints applied. In the case of the EGTA-only, Ca^{2+}/ATP/Cafeine and ryanodine sets, the putative transmembrane helix hairpin was modeled in, although sequence was not assigned. After identification and placement of ligands (see below), a final round of global real space refinement was carried out using phenix.real_space_refine.

Ligand Identification and Assignment

Candidate ligand binding sites (for ATP, caffeine, ryanodine and Ca^{2+}) were identified using locally aligned difference maps, calculated as follows for each domain (implemented as a command line alias in UCSF Chimera). First, the density within a 5Å border of the atoms is extracted (using the 'vop zone' command) for both map/model pairs. Then, the two map fragments are aligned, using the 'fitmap' command, and a difference map is calculated in real space using the 'vop subtract' command. This procedure allows for the identification of ligand densities from a difference map in the presence of domain reorientation, which would otherwise introduce too much noise into the difference map to visualize the bound ligands. Most of the locally aligned difference maps generated by this procedure were essentially featureless, but two had strong peaks of distinctive shape – those calculated around the CTD, and those calculated around the transmembrane region. The presence and morphology of these density features corresponded to the addition of specific ligands – a spherical density feature at the interface of the CTD and CSol was exclusively present when Ca^{2+} was included, while two larger, more elongated features at the S6c/CTD boundary and the CTD-S2S3 interface only appeared in the presence of ATP and caffeine. The larger of the pair of features correlating with the addition of ATP/caffeine could readily be assigned as ATP based on density fit considerations and the local chemical environment at the S6c/CTD/TaF interface. The smaller feature was more ambiguous, but was assigned as caffeine based on size, shape, and the fact that it was not present in a map obtained in the presence of Ca^{2+} and ATP alone (unpublished data). A peak was also located within the transmembrane pore, above Q4933 specifically in the presence of ryanodine, and was assigned as such.

Pore Radius Calculations

Pore radius calculations were carried out on the final atomic models of the core region of the channel (residues 3639-5037) using HOLE (Smart et al., 1996), with the default set of atomic radii; graphs of the pore radius versus the channel coordinate were generated using gnuplot. Prior to HOLE, incomplete sidechains were replaced using the DOCK-PREP tool as implemented in UCSF Chimera, which picks the most probable rotamer from the Dunbrack rotamer library that introduces the least severe steric clashes. HOLE calculations carried out on models with sidechains omitted yielded qualitatively consistent results (in terms of the differences between conformational classes) to those performed using the full atomic models with sidechains intact.

Movie Preparation

The movies included in Data S1, which are intended to convey the nature of the conformational changes observed between different states of the channel based solely on linear interpolation of the aligned density maps, were prepared in UCSF Chimera as follows: The "Morph Map" tool in UCSF Chimera was used to create a trajectory consisting of a linear interpolation at each voxel of the density values of the pair of maps being compared. Each resulting trajectory was displayed as a maximum intensity projection using the following command line alias,

alias ^volume_project background solid black; volume $1 step 1 sdlevel 0,0 color white sdlevel 20,1 color white style solid projectionMode 3d maximumIntensityProjection true btCorrection true linearInterpolation true; unset depthCue

such that the intensity of each voxel is related to the density values (higher density is white, while zero density is black). In each case, five key density slabs of 5-10Å thickness were chosen to best represent the observed conformational changes. These consisted of two slabs parallel to and intersecting with the channel axis, spaced 45 degrees apart about the channel axis; and three slabs orthogonal to the channel axis, one of which is at the putative activation gate (I4937), the second of which is at the height of the EF1&2 domains, and the third of which is in the plane of NTD-A&B. The location and thickness of each slab is depicted on an orthogonal view of the entire volume as a thumbnail in the lower right-hand corner of the movie frame. Movies were saved in UCSF Chimera as MP4, converted to Graphics Interchange Format (GIF) using VideoBlend, labeled and annotated frame-by-frame using Affinity Designer and Gnu Image Manipulation Program (GIMP), and finally converted back to MP4 format.

Structure
Article

Key Intermediates in Ribosome Recycling Visualized by Time-Resolved Cryoelectron Microscopy

Ziao Fu,[1,6] Sandip Kaledhonkar,[2,6] Anneli Borg,[3,6] Ming Sun,[4] Bo Chen,[4] Robert A. Grassucci,[2,5] Måns Ehrenberg,[3] and Joachim Frank[2,4,5,7,*]
[1]Integrated Program in Cellular, Molecular, and Biomedical Studies, Columbia University College of Physicians and Surgeons, New York, NY 10032, USA
[2]Department of Biochemistry and Molecular Biophysics, Columbia University, New York, NY 10027, USA
[3]Department of Cell and Molecular Biology, Uppsala University, Uppsala 751 24, Sweden
[4]Department of Biological Sciences, Columbia University, New York, NY 10027, USA
[5]Howard Hughes Medical Institute, Columbia University, New York, NY 10032, USA
[6]Co-first author
[7]Lead Contact
*Correspondence: jf2192@cumc.columbia.edu
http://dx.doi.org/10.1016/j.str.2016.09.014

SUMMARY

Upon encountering a stop codon on mRNA, polypeptide synthesis on the ribosome is terminated by release factors, and the ribosome complex, still bound with mRNA and P-site-bound tRNA (post-termination complex, PostTC), is split into ribosomal subunits, ready for a new round of translational initiation. Separation of post-termination ribosomes into subunits, or "ribosome recycling," is promoted by the joint action of ribosome-recycling factor (RRF) and elongation factor G (EF-G) in a guanosine triphosphate (GTP) hydrolysis-dependent manner. Here we used a mixing-spraying-based method of time-resolved cryo-electron microscopy (cryo-EM) to visualize the short-lived intermediates of the recycling process. The two complexes that contain (1) both RRF and EF-G bound to the PostTC or (2) deacylated tRNA bound to the 30S subunit are of particular interest. Our observations of the native form of these complexes demonstrate the strong potential of time-resolved cryo-EM for visualizing previously unobservable transient structures.

INTRODUCTION

Ribosome recycling is one of four major stages of protein synthesis, following initiation, elongation, and termination (Ramakrishnan, 2002). After termination, an mRNA-containing ribosome complex with a deacylated tRNA in the P site (post-termination complex, PostTC) is targeted for recycling by the concerted action of ribosome-recycling factor (RRF) and elongation factor G (EF-G). During recycling the PostTC is split into its large (50S) and small (30S) subunits in a guanosine triphosphate (GTP) hydrolysis-dependent manner for a new round of initiation (Hirashima and Kaji, 1973; Zavialov et al., 2005). Regarding the molecular mechanism of ribosome recycling, it was found that the shape of the crystal structure of RRF from *Thermotoga maritima* superimposes almost perfectly with that of a tRNA (Selmer et al., 1999). This finding prompted speculations that RRF binds to the A site of the ribosome as a peptidyl-tRNA mimic and is then translocated by EF-G by a mechanism akin to canonical mRNA-tRNA translocation. However, site-directed hydroxyl radical cleavage assays (Lancaster et al., 2002), cryo-electron microscopy (cryo-EM) (Agrawal et al., 2004; Barat et al., 2007; Gao et al., 2005; Yokoyama et al., 2012), and X-ray crystallography (Borovinskaya et al., 2007; Dunkle et al., 2011; Wilson et al., 2004) of ribosome-RRF complexes indicated that the position of RRF bound to the ribosome is distinct from that of an A-site tRNA. Binding of RRF stabilizes a fully rotated state of the ribosome, with the P-site tRNA moved to the hybrid P/E binding state (Dunkle et al., 2011).

Domain I of RRF is important for its binding to the ribosome. Domain II of RRF is connected to domain I by two flexible linkers, which permit relative domain rotation (Weixlbaumer et al., 2007). Comparison of the various crystal structures of RRF (Kim et al., 2000; Nakano et al., 2003; Toyoda et al., 2000) with its solution structure (Yoshida et al., 2001), NMR relaxation experiments (Yoshida et al., 2003), and a study of mutations around the hinge region (Toyoda et al., 2000) have demonstrated that such a rotation does take place and suggested that the flexibility of the hinge region is functionally important. In a cryo-EM study it has been shown that when *Thermus thermophilus* RRF (ttRRF) binds together with *Escherichia coli* EF-G to the *E. coli* ribosome (Yokoyama et al., 2012), domain II of ttRRF is rotated by a large angle. The physiological relevance of these findings is uncertain because this heterologous combination does not have authentic recycling activity either in vitro (Raj et al., 2005) or in vivo (Fujiwara et al., 1999; Ito et al., 2002; Toyoda et al., 2000). Another player in ribosome recycling is initiation factor 3 (IF3). After separation of the ribosomal subunits this factor binds to the 30S subunit to prevent its rebinding to the 50S subunit. According to one model of ribosome recycling (Hirokawa et al., 2006), in which the tRNA already leaves the PostTC upon action by RRF and EF-G, this is the only role of IF3. In another model, the deacylated tRNA stays bound to the 30S subunit until the ribosome has been split, and IF3 has the additional role to accelerate dissociation of deacylated tRNA, which in turn speeds up dissociation of mRNA from the 30S subunit (Karimi et al., 1999).

 © 2016 Elsevier Ltd.

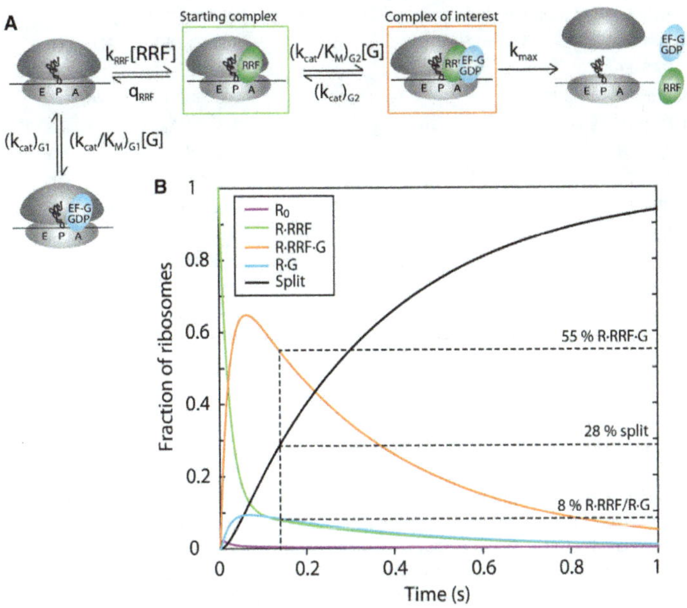

Figure 1. The Recycling Process

(A) Simplified scheme for EF-G- and RRF-dependent recycling of the post-termination complex. The 26°C rate constant estimates were obtained and interpolated from experimental estimates of the corresponding rate constants at 20°C and 37°C using the Arrhenius equation (Table S2). [RRF], concentration of RRF; [G], concentration of EF-G; $k_{RRF} = 6.6\ \mu M^{-1}\ s^{-1}$, $q_{RRF} = 14\ s^{-1}$, $(k_{cat}/K_M)_{G1} = 23\ \mu M^{-1}\ s^{-1}$, $(k_{cat})_{G1} = 15\ s^{-1}$, $(k_{cat}/K_M)_{G2} = 3.8\ \mu M^{-1}\ s^{-1}$, $(k_{cat})_{G2} = 4.5\ s^{-1}$ and $k_{max} = 3.7\ s^{-1}$. R_0, vacant ribosome; R·RRF, RRF-bound ribosome; R·RRF·G, RRF and EF-G-bound ribosome; R·G, EF-G-bound ribosome. Split, the fraction of split ribosomes.

(B) Simulation of the recycling reaction at 26°C, starting from the RRF-bound post-termination complex. The distinct ribosome fractions were predicted from the reaction mechanism and its rate constants (A), and plotted as functions of time. The dashed lines indicate the sample composition at 140 ms incubation time.

The mechanism of action of RRF and EF-G in the recycling of the bacterial ribosome poses a major puzzle that has defied exploration by existing techniques of structure research. Ribosome recycling entails transitions between structural intermediates of the ribosome, driven by domain movements of EF-G and RRF, which, in turn, are driven by the free energy liberated upon EF-G-catalyzed GTP hydrolysis (Savelsbergh et al., 2009). This sequence of events implies the existence of short-lived ribosome structures that unlock the interactions of the ribosomal subunits and allow their rapid separation by the concerted action of EF-G and RRF. It would, in principle, be possible to visualize one of these structures by locking the recycling process in an intermediate state, e.g., by replacing GTP with one of its non-hydrolyzable analogs. The caveat here is that the relevance of the visualized structure would be difficult to assess in this process. Furthermore, all such efforts to artificially lock the ribosome in an intermediate recycling state with active EF-G and RRF of the same species have failed. The only 70S ribosome visualized in complex with both of these translation factors is the above mentioned complex with EF-G from *E. coli* and RRF from *T. thermophilus* (Yokoyama et al., 2012).

This short outline may serve to underline the urgent need for an experimental method that allows both the determination of structures that appear in sequential transients and the examination of the kinetic order in which they appear. If successful, this method would be applicable not only to the study of ribosome recycling but also of a variety of other fast processes presently out of reach for structural and dynamics studies such as initiation of mRNA translation, peptide elongation, termination of translation, and recycling of class-1 release factor by a class-2 release factor.

Recent kinetic experiments (Borg et al., 2016) have shown that ribosome recycling occurs in the subsecond time range, offering hope that reaction intermediates can be captured by a recently established technique of time-resolved cryo-EM, which operates in the same time frame (Chen et al., 2015; Lu et al., 2009; Shaikh et al., 2014). In this mixing-spraying setup of time-resolved cryo-EM, a monolithic silicon chip is used to mix and react two solutions for pre-set times, followed by spraying the reaction product onto an EM grid that is being plunged into liquid cryogen. In the present study we have used this method to study the ribosome recycling driven by native RRF, EF-G, and GTP. Our results, obtained by single-particle reconstruction aided by 3D maximum-likelihood classification, unequivocally show RRF and EF-G bound together to an authentic PostTC within the time window predicted from the kinetic analysis.

RESULTS

Biochemical Characterization and Computer Simulations

In a recent study we performed real-time ribosome splitting experiments by the stopped-flow technique with Rayleigh light scattering detection at 37°C (Borg et al., 2016). By complementing these experiments with measurements of EF-G-dependent GTP hydrolysis, the complete kinetic mechanism of ribosome recycling was identified and its rate constants at 37°C were estimated (Table S1 and Figure S1). In these experiments it was found that EF-G competes with RRF for binding to the PostTC, leading to inhibition of recycling and futile GTP hydrolysis cycles at comparatively high EF-G concentration. Ribosome splitting is efficient and takes place in the milliseconds range. According to the ribosome splitting model proposed by Borg et al. (2016), RRF binding to the PostTC is followed by EF-G binding to the PostTC·RRF complex, which in turn triggers the splitting of the subunits (see the simplified reaction scheme in Figure 1A). As the time-resolved cryo-EM apparatus is operated at room temperature, we repeated the previous kinetic experiments (Borg et al., 2016) also at 20°C (Table S2 and Figure S2). Most

Table 1. List of the Cryo-EM Structures Obtained in this Study

Experiment	Ribosome and Other Components (Rot or Non-rot)	Abbreviation (Resolution, Å)
Control	70S ribosome, RRF, P/E tRNA, mRNA (Rot)	PostTC·$RRF_{control}$ (10)
140 ms	70S ribosome, RRF, P/E tRNA, mRNA (Rot)	PostTC·RRF_{140} (15)
	70S ribosome, RRF, mRNA (Non-rot)	NR-PostTC·RRF_{140} (16)
	70S ribosome, RRF, P/E tRNA, EF-G, mRNA (Rot)	PostTC·RRF·EF-G_{140} (7.4)
	50S subunit, RRF, EF-G	50S·RRF·EF-G_{140} (12)
	50S subunit, RRF, EF-G, E tRNA	50S·RRF·EF-G·$tRNA_{140}$ (16)
	30S subunit, P/I tRNA, mRNA	30S·$tRNA_{140}$ (10)
	30S subunit, IF3, mRNA	30S·$IF3_{140}$ (22)
Long incubation	50S subunit, RRF, EF-G	50S·RRF·EF-G_{long} (14)
	50S subunit, RRF, EF-G, E tRNA	50S·RRF·EF-G·$tRNA_{long}$ (12)
	50S subunit, RRF, E tRNA	50S·RRF·$tRNA_{long}$ (16)
	50S subunit, E tRNA	50S·$tRNA_{long}$ (16)
	30S subunit, IF3, mRNA	30S·$IF3_{long}$ (10)

In the control experiment, EF-G and IF3 were left out. Structures obtained at 140 ms are subscripted with "140." RRF, ribosome-recycling factor; EF-G, elongation factor G; IF3, initiation factor 3. In the long-incubation experiment (subscripted "long"), samples are incubated first, then forced through the time-resolved apparatus for spraying. Note that all the experiments were done with the same time-resolved apparatus.

association and dissociation rate constants were between three- and ten-fold smaller at 20°C than at 37°C, and the maximal splitting rate (k_{max}) was estimated as 1.3 s^{-1} at 20°C and 25 s^{-1} at 37°C. As it turned out, the actual cryo-EM experiments were performed at ~26°C, and we used the Arrhenius equation (see Equations 1 and 2 in the Supplemental Experimental Procedures) to estimate the rate constants at this temperature from the 20°C and 37°C data (Tables S1 and S2). These rate constants show that the highest value of the maximal PostTC·RRF·EF-G fraction can be obtained by pre-incubating the PostTC with high concentration of RRF. In this way, idling of EF-G on the factor-free PostTC can be curbed by reducing the frequency with which ribosomes enter the vertical branch of the reaction scheme (Figure 1A). At 45 μM [RRF] and 10 μM [EF-G] the fraction of the PostTC·RRF·EF-G complex peaked at around 100 ms, constituting almost 70% of the ribosomes (see Figure 1). We therefore used the closest available reaction chip, with 140 ms total reaction time, for the time-resolved cryo-EM experiment. To confirm the completeness of the recycling reaction, we performed a long-incubation experiment (see Experimental Procedures).

Altogether in this study we obtained 13 cryo-EM structures of the 30S subunit, 50S subunit, and 70S ribosome complexes from 3D classification of data collected under three experimental conditions: control experiment (details are in the next section), at the 140 ms time point, and after long (30 min) incubation. They are listed with their abbreviations in Table 1. Henceforth, we refer to the structures with these abbreviations.

Control Experiment Confirming the Stability of the RRF-Bound Post-termination Complex

We performed a control experiment in the absence of EF-G and IF3 to validate the purity and stability of the RRF-bound PostTC during its passage through the mixing and reaction channel of the microfluidic device. For this experiment we employed a device with a reaction time of 560 ms, having the longest reaction channel among the presently available devices. The preparation of a time-resolved grid followed the procedure of Chen et al. (2013). We injected a mixture containing PostTC (2 μM) incubated with RRF (90 μM) in polymix buffer, supplemented with components for energy regeneration, into inlet 1 and only polymix buffer along with the components for energy regeneration into inlet 2 of the microfluidic device. Equal volumes of the two reaction solutions were rapidly mixed inside the mixing channel and sprayed onto the EM grid, which was immediately plunged into liquid ethane. Details of data collection and processing can be found in the Supplemental Experimental Procedures.

In the dataset obtained from the control experiment, no 50S or 30S subunits were observed, and only a single 70S ribosome structure class was identified with ~9,000 particles (Figure 2), which yielded a 10 Å-density map of the 70S ribosome in rotated conformation carrying a P/E tRNA and an additional mass of density in the inter-subunit space that could be readily attributed to RRF (Borovinskaya et al., 2007) (Figure 2B). Henceforth we refer to this map obtained in the control experiment as the PostTC·$RRF_{control}$ complex (see Table 1).

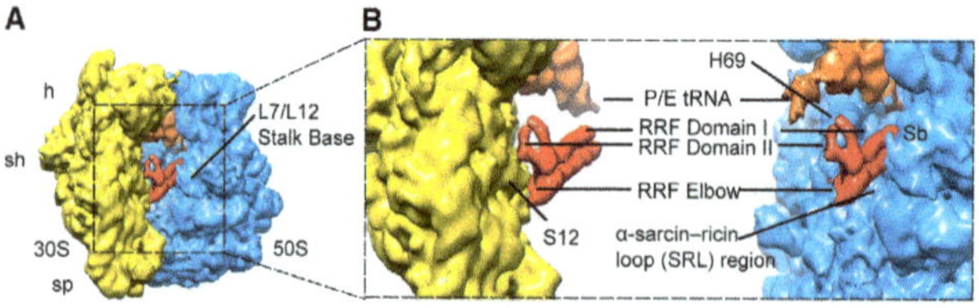

Figure 2. Post-termination 70S Ribosome in Complex with RRF and P/E tRNA

(A) The PostTC·RRF complex. Gold, 30S subunit; blue, 50S subunit; red, RRF; and orange-red, P/E-site tRNA. h, head; sh, shoulder; sp, spur.

(B) Orientation and interaction of RRF and P/E tRNA with the 30S and 50S subunits, respectively.

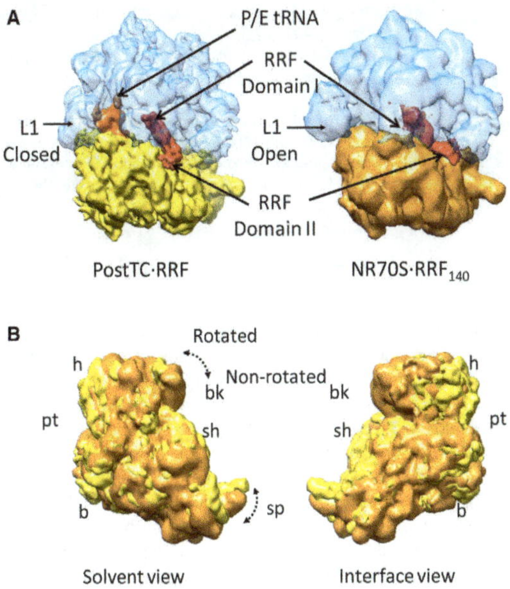

Figure 3. Interaction of RRF and tRNA in the PostTC·$RRF_{control}$ and NR-PostTC70S·RRF_{140} Complexes

(A) PostTC·RRF and NR70S·RRF_{140} complexes in top view.

(B) 30S subunit of PostTC·RRF (gold) and NR70S·RRF_{140} (orange) in solvent and interface views. L1, L1-stalk; h, head; pt, platform; b, body; sp, spur; sh, shoulder; bk, beak. The rotational movement is indicated by dashed lines.

In the PostTC·$RRF_{control}$ complex (10 Å) obtained in the control experiment, RRF adopts the same conformation (Figure 2B) as in the structures published previously (Agrawal et al., 2004; Gao et al., 2005) (EMD: 1128 and 1077). However, a closer analysis of one of the maps (Agrawal et al., 2004) (EMD: 1077), where no 3D classification was employed, shows that domain II actually appears in two positions; a majority density, equivalent to the presently observed one, and a minority density, the latter only seen when the density threshold was lowered in their map. In our studies domain I of RRF in the PostTC·$RRF_{control}$ complex is seen to reach into the interface canyon of the 50S subunit and to make contact with several parts of the 23S rRNA (Figure 2B). The elbow between the two RRF domains faces the L7/L12 stalk base and the α-sarcin-ricin loop region of the 50S subunit. Domain II of RRF is found in a single conformation, pointing toward the shoulder of the 30S subunit, and contacting protein S12 (Figure 2B). No density is observed to extend toward the stalk base of the 50S subunit, even when the density threshold is reduced. We note that this conformation is quite different from that observed by Yokoyama et al. (2012). In their ttRRF-bound PostTC, although domain I of RRF binds similarly to the 50S subunit as in our structure, domain II is rotated about the long axis of domain I toward the 50S subunit (see Figure S3).

Observed Structures of the 70S Recycling Complex at 140 ms

In the time-resolved cryo-EM experiment capturing the recycling reaction at the 140 ms time point, after mixing the RRF-bound PostTC with EF-G and IF3, we identified three major classes of 70S ribosome particles by 3D classification of 45,000 particles. The first class yields a map quite similar to the one in the first control experiment, except that the resolution is 15.5 Å. Henceforth we refer to this complex as PostTC·RRF_{140} (Table 1).

The second class yields an RRF-bound non-rotated 70S complex (NR-PostTC·RRF_{140}). Compared with PostTC·RRF, this complex displays three striking differences. First, while the 70S ribosome adopts the non-rotated state in NR-PostTC·RRF_{140}, it adopts the rotated state in PostTC·$RRF_{control}$ (Figure 3). Related to this change, the L1 stalk of the 50S subunit is half-closed in NR-PostTC·RRF_{140}, but fully closed in PostTC·$RRF_{control}$. Second, NR-PostTC·RRF_{140} contains no tRNA density, while PostTC·$RRF_{control}$ contains P/E tRNA (Figure 3). The third difference is related to the position of RRF: in NR-PostTC·RRF_{140}, compared with PostTC·$RRF_{control}$, the whole domain I of RRF is shifted toward the peptidyl-transfer center by 5 Å, and domain II makes contact with the stalk base of the 50S subunit. Clearly, in NR-PostTC·RRF_{140} there is no interaction between domain II of RRF and the 30S subunit, whereas in PostTC·$RRF_{control}$, domain II of RRF is in contact with protein S12.

The third class of 70S ribosome particles yields a complex that has both RRF and EF-G bound to the ribosome in the rotated state, and contains density for P/E tRNA (PostTC·RRF·EF-G_{140}; Figure 4). Here domain II of EF-G make contact with the 30S subunit while domain III of EF-G appears disordered in this map, indicating residual unresolved heterogeneity of this class. Domain IV of EF-G contacts domain II of RRF, while domain V interacts with the 50S subunit. Regarding the mechanism of ribosome splitting, it is believed that it involves a movement of RRF domain II with respect to domain I upon EF-G binding (Gao et al., 2005, 2007). Domain II is initially (in PostTC·$RRF_{control}$) in contact with S12 on the small subunit. Due to the low local resolution of the factor binding site on PostTC·RRF·EF-G_{140}, the density mass of domain II of RRF is fused with that of EF-G. Still, it is apparent that domain II is no longer in contact with S12, and the angle between domain I and II of RRF can be determined. Comparison with PostTC·$RRF_{control}$ shows that upon binding of EF-G the angle between domains I and II of RRF decreases by approximately 20° so that domain II points toward h44 (Figure 5). By comparison, for the heterologous *E. coli* PostTC·ttRRF complex Yokoyama et al. (2012) obtained results that are quite different. In that study, domain II is initially in contact with the stalk base of the 50S subunit and, upon EF-G binding, a large decrease in the angle between domains I and II of RRF (75°) takes place as domain I moves from a position where it contacts the 50S subunit side to one where it points toward the side of the 30S subunit. In addition, they observe a shift of RRF by 8 Å toward the E-site tRNA.

50S Subunit Complexes Observed at the 140 ms Time Point

We found two classes of RRF- and EF-G-containing 50S subunits at 140 ms, distinguished by the presence or absence of an E-site tRNA ("50S·RRF·EF-G·$tRNA_{140}$" and "50S·RRF·EF-G_{140}"). RRF on the 50S subunit is in the same position as observed previously by our group (Gao et al., 2005). We compared the conformation of RRF in our PostTC·RRF_{140} with 50S·RRF·EF-G_{140},

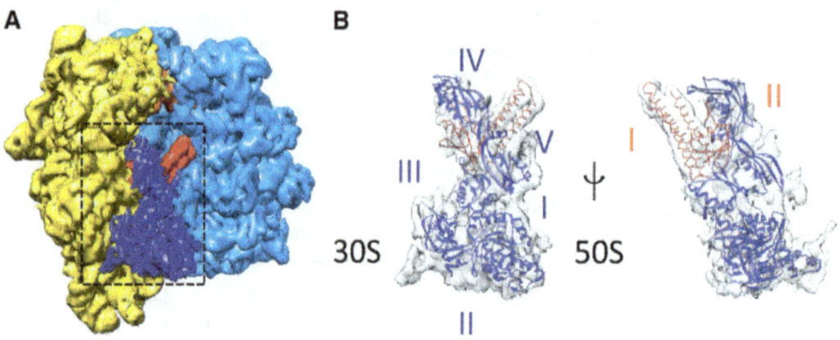

Figure 4. Segmented Map of the PostTC·RRF·EF-G$_{140}$ Complex

(A) PostTC·RRF·EF-G$_{140}$ complex. Gold, 30S subunit; blue, 50S subunit; red, RRF; orange-red, P/E tRNA; dark blue, EF-G.

(B) Zoom-in views of (A) showing EF-G and RRF interaction. Density is rendered 50% transparent to show the rigid-body fitting of the EF-G and RRF structures. Five EF-G domains and two RRF domains are labeled.

NR-PostTC·RRF$_{140}$, and PostTC·RRF·EF-G$_{140}$ (Figure 6). Domain I of RRF binds in the same position in all four complexes; in contrast, domain II in the 50S·RRF·EF-G$_{140}$ complex is extensively rotated (~60°) compared with its orientation in the PostTC·RRF$_{140}$ complex. Domain II of RRF in the PostTC·RRF·EF-G$_{140}$ complex lies between its positions in 50S·RRF·EF-G$_{140}$ and PostTC·RRF$_{140}$. Domain II of RRF in NR-PostTC·RRF$_{140}$ extensively interacts with the stalk-base region of the 50S subunit. Detailed comparison of the conformations of RRF and EF-G in our complexes with those in published structures can be found in the Supplemental Experimental Procedures.

In the other 50S complex, 50S·RRF·EF-G·tRNA$_{140}$, we see E-site tRNA on the 50S subunit, matching the position of E-site tRNA density from *T. thermophilus* or *E. coli* 70S ribosomes (Schmeing et al., 2003). The L1 stalk is in the half-closed conformation, as it is in the non-rotated 70S ribosome in the presence of the E-site tRNA (Réblová et al., 2012). The elbow region of the tRNA is interacting with the L1 stalk, and its CCA end is in contact with H88 of 23S rRNA. The anticodon loop of the tRNA becomes visible when the density threshold level is reduced.

30S Subunit Observed at the 140 ms Time Point

In our tRNA-bound 30S complex map (30S·tRNA, 11 Å) obtained at the 140 ms time point, a mass of density for tRNA is observed (Figure 7A). Superimposition of our map with the 30S subunit portion of the initiation complex of Allen et al. (2005) (EMD: 3523) locates the tRNA to the P/I position, with the tRNA elbow shifted toward the E site (Figure 7A). Thus, even in the absence of initiation factor 2 (IF2), tRNA alone on the 30S subunit occupies the P/I site. The same observation was made by Simonetti et al. (2008) for fMet-tRNA^{fMet} in the initiation complex from *T. thermophilus*. Density for mRNA is also observed (not shown in Figure 7A for simplicity). In the other 30S complex obtained by classification (30S·IF3) we find density for IF3 but not tRNA. However, the resolution of this class (22 Å) is limited because of the small number of particles (data not shown).

Observed Structures of the 30S and 50S Subunits after a Long Reaction Time

To quantify the splitting reaction and obtain a better resolution for the 30S·IF3 complex, we collected additional data with a reaction time of 30 min, by mixing the RRF-bound PostTC with EF-G and IF3 in a test tube. To make the results strictly comparable with those in the time-resolved experiment, the reaction product was forced through the 140 ms chip for spraying and plunging of the EM grid. As we expected, by that time all 70S ribosomes have split, and all 30S subunits are bound with IF3. With about 9,000 particles in the 30S·IF3 class, we solved the structure of this complex to 9 Å (Figure 7B). The binding location of IF3 to the 30S subunit has been controversial in the field even though studies on IF3 have been carried out with cryo-EM (Julián et al., 2011; McCutcheon et al., 1999), chemical and site-directed probing (Dallas and Noller, 2001; Fabbretti et al., 2007), and crystallography on 30S subunits soaked in IF3 solution (Pioletti et al., 2001). In our 30S structure bound with IF3, we see a clear density for the C-terminal domain of IF3 at the h44 location and its N-terminal domain in the platform region (Figure 7B). Unexpectedly, we still observe mRNA density in this complex as well (Figure 7B).

For the 50S subunit, we observe four classes. Two of these (50S·RRF·EF-G$_{long}$ and 50S·RRF·EF-G·tRNA$_{long}$) match the 50S subunit classes observed at 140 ms, while the others lack EF-G and are new (50S·RRF·tRNA$_{long}$ and 50S·tRNA$_{long}$). In 50S·RRF·tRNA$_{long}$, domain I of RRF is bound at the same position as in the 50S·RRF·EF-G·tRNA and 50S·RRF·EF-G complexes. Domain II of RRF is in contact with the stalk-base region of the 50S subunit.

DISCUSSION

In this study we present a snapshot of the transient structure of an *E. coli* post-termination ribosome on the pathway to its splitting into subunits by *E. coli* RRF and EF-G in the presence of GTP. The RRF- and EF-G-bound post-termination ribosome has previously been visualized only when locked by the presence of fusidic acid and with heterologous translation factors, RRF from *T. thermophilus* and EF-G from *E. coli*, not active in ribosome splitting. In our case the PostTC·RRF·EF-G$_{140}$ complex, which is a key intermediate in the ribosome-recycling process, was formed during the active splitting reaction by homologous translation factors, in the presence of GTP, with no antibiotics and without the help of any chemical intervention or mutation. The mixing-spraying method of time-resolved cryo-EM, which allows choosing reaction times in the range of tens of milliseconds to one second, was used to trap the short-lived complex on the EM grid. The complete mechanism of the ribosome-recycling reaction was biochemically characterized at 37°C in previous work and now at 20°C in an *E. coli*-based in vitro translation system with in vivo-like function. These data were used to identify the complete kinetic mechanism for ribosome recycling along with its kinetic parameters and their temperature dependencies. The experiments enabled us to estimate kinetic rates

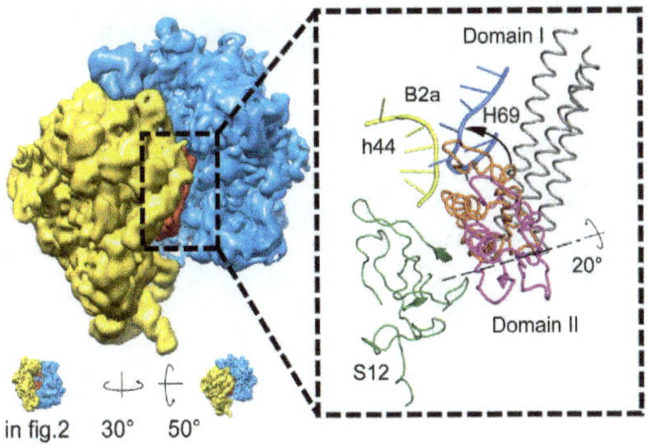

Figure 5. Movement of Domain II of RRF in PostTC·RRF·EF-G_{140} Compared with its Position in PostTC·RRF_{140}

The central bridge between 30S and 50S, B2a, involves h44 (yellow) and H69 (blue). S12 protein is shown in green. RRF domain II in orange shown by the fitting of PDB structure (PDB: 1EH1) into its density in the PostTC·RRF·EF-G_{140} complex. For comparison with the RRF domain II in PostTC·RRF_{140} (red), the rotational movement of domain II toward bridge B2a is shown with an arrow. The position of domain I (gray) is the same in these two complexes. h, head; sh, shoulder; sp, spur; Sb, L7/L12 stalk base. Thumbnails in the lower left depict the relationship of the views presented here to the view shown in Figure 2.

of ribosome splitting at 26°C, the temperature actually prevailing in our time-resolved cryo-EM experiments, and to optimize the experimental setup for high yield of the transient ribosome complex containing both EF-G and RRF.

At a reaction temperature of 26°C and a reaction time of 140 ms, the model predicted fractions of ribosomes in complex with EF-G and RRF or split into subunits of 55% or 28%, while the measured fractions were 20% or 40%, respectively. Part of the explanation for this kinetic discrepancy might be that some ribosomes form the tRNA-lacking NR-PostTC·RRF complex (Table 1). This ribosome class is, we suggest, not an authentic intermediate on the ribosome splitting pathway. It was not identified by our biochemical experiments nor is it present in our kinetic model (Supplemental Experimental Procedures), which proposes that the P-site tRNA remains ribosome-bound until subunit separation (Borg et al., 2016). As can be seen from the long-time incubation experiment, this class of ribosomes will also eventually be split. The latter result is in line with a previous observation that tRNA-lacking ribosomes can be split by RRF and EF-G, albeit more slowly than their tRNA-containing counterparts (Pavlov et al., 2008).

Among the complexes identified by classification, the most interesting ones, in terms of their potential for providing new information on the recycling process, are PostTC·RRF_{140} and PostTC·RRF·EF-G_{140}, interpreted as successive complexes in which the latter is a short-lived (~100 ms) intermediate. A comparison between these complexes yields information about a rotation of RRF domain II toward domain I, into a position close to inter-subunit bridge B2a, that is potentially triggered by one of three events: binding of EF-G, GTP hydrolysis on EF-G, or Pi release, or a combination thereof. While EF-G binding and GTP hydrolysis closely follow each other, Pi release may follow with some delay. It is therefore likely that the map PostTC·RRF·EF-G_{140} depicts a majority population of ribosomes that are in a state prior to Pi release. As to the study by Yokoyama et al. (2012), there are major discrepancies compared with our results in both PostTC·RRF_{140} and PostTC·RRF·EF-G_{140} complexes, precluding a detailed comparison of their findings with ours. These discrepancies may originate, in part, with the construction of their complex from heterologous components. The conformational change between PostTC·RRF and PostTC·EF-G·RRF complexes indicated by the two maps suggests that EF-G assists the movement of domain II of RRF toward the bridge B2a and jointly acts with RRF to split the PostTC into the two subunits.

A related issue of biological importance is the disposition of deacylated tRNA and mRNA at the end of the recycling process, and the role of IF3. Our data from the 140 ms time point, showing two classes of the 30S subunit, bound either with tRNA or IF3, suggest that the post-termination deacylated tRNA remains initially with the 30S subunit and is subsequently removed by competitive binding of IF3. This conclusion is supported by the results of the long-time incubation study (Figure 6) in which a sole class of 30S subunit, bound with IF3 is found, interpreted as the final state after all initially bound tRNAs have been displaced. Our results thus seemed to argue in favor of the model advocated by Zavialov et al. (2005) and against the view favored by Hirokawa et al. (2006), according to which the tRNA leaves the PostTC already during the splitting reaction. On the other hand, our finding that mRNA remains with the IF3-bound 30S complex in the long-time incubation study indicates that for our choice of mRNA the recycling process is not finished. It was very recently suggested to us that the scenario advocated by Zavialov et al. may be one of two possible pathways, the selection of which is conditional on the strength of the interaction of the Shine-Dalgarno (SD) sequence of the mRNA and the anti-SD sequence of the 16S rRNA (B. Cooperman and A. Kaji, personal communication). In fact, the mRNA used in our experiment has a strong SD sequence, and could be responsible for the selection of the observed pathway. Resolving this issue with the present method will require novel experiments with variegated strength of the SD sequence to allow detection of the presence or absence of mRNA.

Concerning the fate of the 50S subunit, its RRF- and EF-G-bound fraction is 100% at 140 ms and then decreases to 79% at the long reaction time. At 140 ms, 44% of the RRF- and EF-G-bound fraction contains and 56% lacks an E-site-bound tRNA. At long reaction time the fraction of RRF- and EF-G-bound ribosomes that contains an E-site tRNA increases to 82%. We tentatively interpret this to mean that splitting of the post-termination ribosome results in an RRF- and EF-G-containing 50S subunit lacking an E-site tRNA, and that this is the dominating 50S complex also at the 140 ms reaction time. With increasing reaction time RRF, EF-G, and deacylated tRNA will eventually reach the steady state for their 50S-bound and free forms that is observed at the long reaction time. By hypothesis, this would mean that the 50S subunit fraction with E-site-bound tRNA increases over time, in line with the recycling model by Zavialov et al. (2005).

The present cryo-EM study of a very short-lived ribosome complex on the pathway to its splitting into subunits provides a proof of principle for cryo-EM visualization of short-lived, native

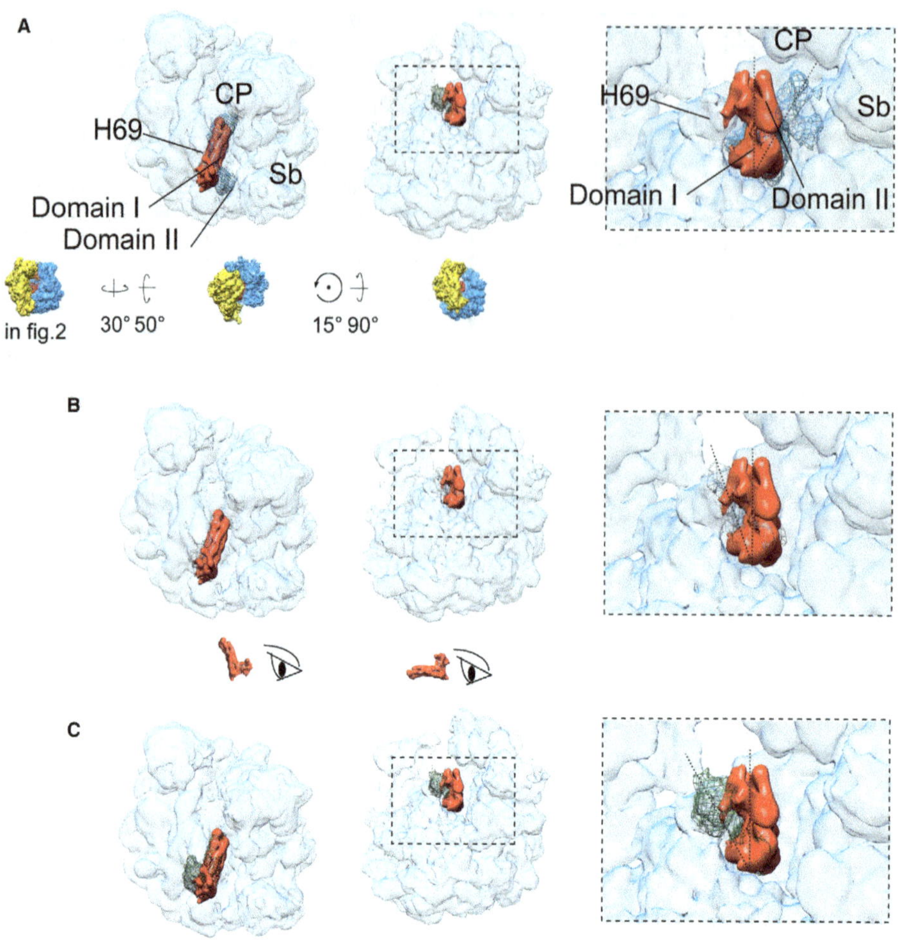

Figure 6. Domain II of RRF Is in Different Positions in PostTC·$RRF_{control}$, NR-PostTC·RRF_{140}, PostTC·RRF·EF-G_{140}, and 50S·RRF·EF-G_{140}

(A) Comparison of domain II of RRF in PostTC·$RRF_{control}$ and NR-PostTC·RRF_{140}. CP, central protuberance; Sb, stalk base. Helix H69, and domains I and II of RRF are indicated. The position of RRF in PostTC·RRF_{140} is shown in red. RRF in NR70S·RRF_{140} is shown in blue mesh. Rotational movement is shown by the dashed lines.

(B) Comparison of domain II of RRF in PostTC·$RRF_{control}$ and PostTC·RRF·EF-G_{140}. RRF in PostTC·RRF·EF-G_{140} is shown as a gray mesh.

(C) Comparison of domain II of RRF in PostTC·$RRF_{control}$ and 50S·RRF·EF-G_{140}. RRF in 50S·RRF·EF-G_{140} is shown as a green mesh.

reaction intermediates in biological processes. This, we suggest, will open the door to high-resolution structural analysis of a wide range of previously inaccessible large biological complexes in their well-defined functional contexts. We also note that time-resolved cryo-EM can be used as a very powerful experimental technique for kinetic modeling of complex biochemical processes along with precise estimation of the kinetic parameters of these models. With this technique, and by virtue of the classification methods, essentially all intermediate large molecular complexes of a reaction pathway can be identified from their high-resolution cryo-EM structures, and their time evolutions can be precisely monitored by single-particle counting that is the byproduct of classification.

Currently time-resolved cryo-EM is still a somewhat cumbersome technique because of limitations to data collection yield, but we are optimistic that technological advances will soon allow the full potential of this method, with its deep integration of structural and functional approaches, to manifest in a multitude of experimental applications. It should be noted that a previous study using time-resolved cryo-EM of the back-translocation process (Fischer et al., 2010) had a time resolution in the minute range. It is the achievement of a time resolution in the more important 10 ms to 1 s range that has the potential of revolutionizing structural biology by facilitating studies of transient large molecular complexes operating in their proper biological contexts and time scales.

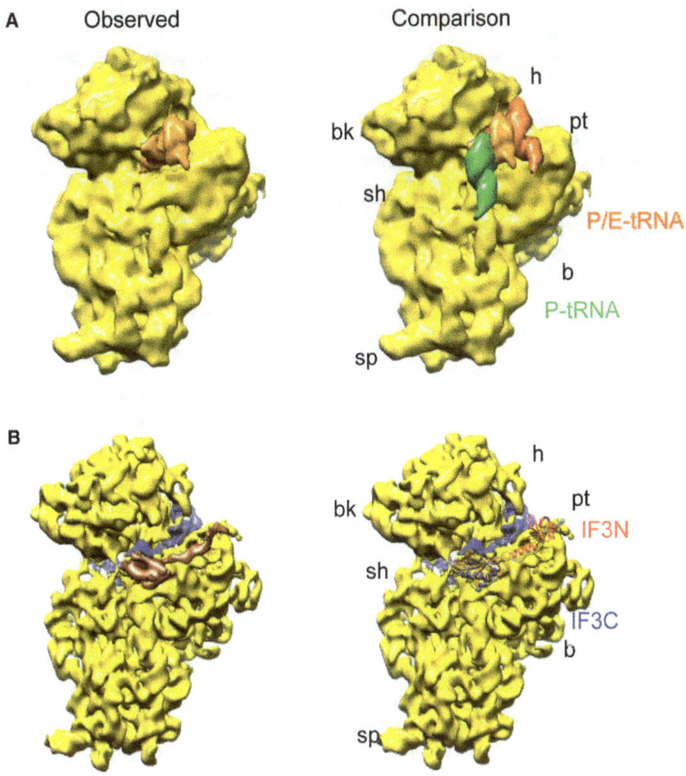

Figure 7. 30S Subunit Classes

(A) Interaction of P/I tRNA with the 30S subunit at 140 ms. Yellow, 30S subunit; green, P-tRNA; orange, P/I tRNA; red, P/E tRNA. The mRNA density is not shown for simplicity.

(B) Interaction of IF3 with the 30S subunit at 30 min. h, head; pt, platform; b, body; bk, beak; sh, shoulder; sp, spur. The mRNA density is shown in dark blue. The density attributed to IF3 is segmented out in brown and fitted with the NTD (1TIF, red) and CTD (1TIG, purple) of the IF3 structure. Panels on the left show observed densities, those on the right show the comparison with positions of tRNA (Allen et al., 2005) and IF3 (McCutcheon et al., 1999).

In most of the studies carried out either with cryo-EM (Barat et al., 2007; Gao et al., 2005; Hirashima and Kaji, 1973; Yokoyama et al., 2012) or X-ray crystallography (Selmer et al., 1999; Weixlbaumer et al., 2007; Wilson et al., 2004) to understand the ribosome splitting mechanism, each complex was prepared separately using a chemical intervention. By employing time-resolved cryo-EM, in contrast, we have captured most of the transient structures that exist during ribosome splitting with a single biochemical preparation, which proves to be of advantage in terms of experimental control.

Concerning future directions of the present work on ribosome recycling, we will attempt to closely monitor the conformational dynamics, in the ribosome and in the translation factors, that lead to ribosome splitting. This type of study will have a good chance to answer the long-standing question of how EF-G and RRF interact both with each other and with the ribosome to achieve separation of the ribosomal subunits, and the role of the thermodynamic driving force provided by GTP hydrolysis in the process.

EXPERIMENTAL PROCEDURES

Preparation of Time-Resolved Cryo-EM Grids

Quantifoil R1.2/1.3 300-mesh Cu EM grids were carbon coated and glow-discharged following standard procedures (Grassucci et al., 2007). The ribosome-recycling reaction was performed using a mixing-spraying device (Lu et al., 2009) with an environmental chamber as described previously (Chen et al., 2015), with a few minor alterations. During the experiment, the ambient conditions were maintained at 24°C–26°C and 80%–90% relative humidity. In the mixing-spraying chips equal volumes of two mixtures were injected, each at a flow rate of 3 μL/s.

Materials

Ribosomes (*E. coli* MRE600) and mRNA, encoding the peptide fMet-Phe-Thr(GGGAAUUCGGGCCC UUGUUAACAAUUAAGGAGGUAUUAA**AUGUUU ACGUAA**UUGCAGAAAAAAAAAAAAAAAAAAAAA [ORF in bold, SD underlined]), were prepared as described previously (Borg et al., 2016). His-tagged EF-G and RRF were overexpressed in *E. coli* and purified by nickel-affinity chromatography. $tRNA^{Phe}$ was overexpressed in, and purified from, *E. coli*. All experiments were performed in polymix buffer, containing 95 mM KCl, 5 mM NH_4Cl, 0.5 mM $CaCl_2$, 8 mM putrescine, 1 mM spermidine, 5 mM potassium phosphate (pH 7.5), 1 mM dithioerythritol, and 5 mM $Mg(OAc)_2$. All reaction mixtures also contained components for energy supply: GTP (1 mM), ATP (1 mM), phosphoenolpyruvate (10 mM), pyruvate kinase (50 μg/mL), and myokinase (2 μg/mL).

Control Experiment

A PostTC mixture was prepared containing 70S ribosomes (2 μM), RRF (90 μM), $tRNA^{Phe}$ (5 μM), and MFT mRNA (10 μM). A second mixture was prepared containing only polymix buffer and components for energy regeneration. The two mixtures were incubated for 15 min at 37°C and then kept on ice. They were then centrifuged for 3 min at 20,800 × *g* before being loaded into the mixing-spraying device. Equal volumes of the two mixtures were rapidly mixed in a reaction chip with 560 ms total incubation time, i.e., from mixing to freezing on the EM grid.

Time-Resolved Cryo-EM Experiment

A PostTC mixture was prepared containing 70S ribosomes (2 μM), RRF (90 μM), $tRNA^{Phe}$ (5 μM), and MFT mRNA (10 μM). A recycling mixture was prepared containing IF3 (16 μM) and EF-G (20 μM) and components for energy regeneration. The two mixtures were incubated for 15 min at 37°C and then kept on ice. They were centrifuged for 3 min at 20,800 × *g* before being loaded into the mixing-spraying device. Equal volumes of the two mixtures were rapidly mixed in a reaction chip with 140 ms total incubation time, from mixing to freezing on the EM grid. In the long-incubation experiment, equal volumes of the two mixtures were mixed and incubated for 25 min at 26°C. Then the reaction mixture was split into two halves and loaded into the mixing-spraying device. The EM grid was made in the same way as in 140 ms time-resolved cryo-EM experiment. The total reaction time was roughly estimated to be 30 min (25 min incubation and 5 min preparation of EM grid with the mixing-spraying device).

Data Acquisition

The grids were imaged using an FEI Tecnai Polara transmission electron microscope operated at 300 kV and at a nominal magnification of ×31,000. Datasets were collected with Leginon (Potter et al., 1999) on a K2 Summit direct electron detector (Gatan) with a physical pixel size of 5 μm, corresponding to 1.255 Å per pixel at the specimen in electron-counting mode. The dose rate was set to eight counts per physical pixel per second. The total exposure time was 10 s. Fifty frames were recorded. All images were taken with a defocus in the range of 1.5–3 μm.

Image Processing

For the control experiment dataset, after assessment of the micrographs, 963 micrographs were selected for subsequent processing. For the 140 ms experiment and long-incubation dataset, 947 and 881 micrographs were selected, respectively, for subsequent processing. To correct for stage movement and beam-induced movements, all 50 frames of each micrograph were aligned (Li et al., 2013). Particle picking was done with RELION-1.3 (Scheres, 2012) and manually checked.

For the control experiment, all picked particles were classified into four classes. One class that did not conform to the known shapes of ribosomes was rejected. Good classes were pooled and reconstructed with auto-refinement in RELION.

For the 140 ms experiment, classification was done in four steps. In a first step, it is performed with low-pass-filtered reference maps and four-times binned particle images to separate the data into 30S subunits, 50S subunits, and 70S ribosomes. The next level of classification resolved conformational differences, and two classes were obtained, "non-rotated 70S" (NR70S) and "rotated 70S" (RT70S), based on the presence or absence of inter-subunit rotation. The RT70S class was further divided into two subclasses, one (PostTC·RRF_{140}) identical (save for the difference in resolution) to the control complex (PostTC·$RRF_{control}$), and the other one (PostTC·RRF·EF-G_{140}). Using focused classification, by applying a soft-edged mask to the EF-G and RRF binding region, we found no evidence of subclasses in PostTC·RRF·EF-G_{140}. The 50S subunit class was resolved into two subclasses of RRF- and EF-G-containing 50S subunits, distinct by the presence or absence of an E-site tRNA ("50S·RRF·EF-G·$tRNA_{140}$" and "50S·RRF·EF-G_{140}"). The 30S subunit class was resolved into two subclasses, one bound with tRNA in the P/I position ("30S·$tRNA_{140}$") and the other with IF3 ("30S·$IF3_{140}$").

For the long-incubation experiment, in the first step of classification, we found two classes, 50S and 30S subunits. At the next level, particles of pooled particles of 30S subunits and 50S subunits were classified into subclasses. All of the 30S subunit was found to be bound with IF3. For 50S subunits, four classes were found, 50S·RRF·EF-G_{long}, 50S·RRF·EF-G·$tRNA_{long}$, 50S·RRF·$tRNA_{long}$, and 50S·$tRNA_{long}$.

ACCESSION NUMBERS

The electron microscopy maps have been deposited in the EMBL-European Bioinformatics Institute EM Databank under accession codes EMD: 8411, 8412, 8413, 8415, 8416, 8417, and 8418 (PostTC·$RRF_{control}$, NR-PostTC·RRF_{140}, PostTC·RRF·EF-G_{140}, 50S·RRF·EF-G_{140}, 50S·RRF·EF-G·$tRNA_{140}$, 30S·$tRNA_{140}$, and 30S·$IF3_{long}$ maps, respectively).

SUPPLEMENTAL INFORMATION

Supplemental Information comprises Supplemental Experimental Procedures, four figure, and two tables and can be found with this article online at http://dx.doi.org/10.1016/j.str.2016.09.014.

AUTHOR CONTRIBUTIONS

Conceptualization, Z.F., S.K., A.B., M.S., B.C., M.E., and J.F.; Formal Analysis, Z.F., S.K., A.B., and M.S.; Investigation, Z.F., S.K., and A.B.; Resources, M.E. and J.F.; Writing – Original Draft, Z.F., S.K., A.B., M.E., and J.F.; Writing – Review & Editing, Z.F., S.K., A.B., M.S., B.C., M.E., and J.F.; Visualization, Z.F., S.K., and A.B.; Supervision, R.A.G., M.E., and J.F.; Funding Acquisition, Z.F., S.K., A.B., M.E., and J.F.

ACKNOWLEDGMENTS

This work was supported by HHMI and NIH R01 GM55440 and GM29169 (to J.F.) and by the Swedish Research Council (2015-04682), the Knut and Alice Wallenberg Foundation, RiboCORE (KAW 2009.0251) (to M.E.), and a Liljewalchs travel stipend (Uppsala University) (to A.B.).

Received: June 7, 2016
Revised: September 21, 2016
Accepted: October 7, 2016
Published: November 3, 2016

REFERENCES

Agrawal, R.K., Sharma, M.R., Kiel, M.C., Hirokawa, G., Booth, T.M., Spahn, C.M.T., Grassucci, R.A., Kaji, A., and Frank, J. (2004). Visualization of ribosome-recycling factor on the *Escherichia coli* 70S ribosome: functional implications. Proc. Natl. Acad. Sci. USA *101*, 8900–8905.

Allen, G.S., Zavialov, A., Gursky, R., Ehrenberg, M., and Frank, J. (2005). The cryo-EM structure of a translation initiation complex from *Escherichia coli*. Cell *121*, 703–712.

Barat, C., Datta, P.P., Raj, V.S., Sharma, M.R., Kaji, H., Kaji, A., and Agrawal, R.K. (2007). Progression of the ribosome recycling factor through the ribosome dissociates the two ribosomal subunits. Mol. Cell *27*, 250–261.

Borg, A., Pavlov, M., and Ehrenberg, M. (2016). Complete kinetic mechanism for recycling of the bacterial ribosome. RNA *22*, 10–21.

Borovinskaya, M.A., Pai, R.D., Zhang, W., Schuwirth, B.S., Holton, J.M., Hirokawa, G., Kaji, H., Kaji, A., and Cate, J.H.D. (2007). Structural basis for aminoglycoside inhibition of bacterial ribosome recycling. Nat. Struct. Mol. Biol. *14*, 727–732.

Chen, S., McMullan, G., Faruqi, A.R., Murshudov, G.N., Short, J.M., Scheres, S.H.W., and Henderson, R. (2013). High-resolution noise substitution to measure overfitting and validate resolution in 3D structure determination by single particle electron cryomicroscopy. Ultramicroscopy *135*, 24–35.

Chen, B., Kaledhonkar, S., Sun, M., Shen, B., Lu, Z., Barnard, D., Lu, T.-M., Gonzalez, Ruben L., Jr., and Frank, J. (2015). Structural dynamics of ribosome subunit association studied by mixing-spraying time-resolved cryogenic electron microscopy. Structure *23*, 1097–1105.

Dallas, A., and Noller, H.F. (2001). Interaction of translation initiation factor 3 with the 30S ribosomal subunit. Mol. Cell *8*, 855–864.

Dunkle, J.A., Wang, L., Feldman, M.B., Pulk, A., Chen, V.B., Kapral, G.J., Noeske, J., Richardson, J.S., Blanchard, S.C., and Cate, J.H.D. (2011). Structures of the bacterial ribosome in classical and hybrid states of tRNA binding. Science *332*, 981–984.

Fabbretti, A., Pon, C.L., Hennelly, S.P., Hill, W.E., Lodmell, J.S., and Gualerzi, C.O. (2007). The real-time path of translation factor IF3 onto and off the ribosome. Mol. Cell *25*, 285–296.

Fischer, N., Konevega, A.L., Wintermeyer, W., Rodnina, M.V., and Stark, H. (2010). Ribosome dynamics and tRNA movement by time-resolved electron cryomicroscopy. Nature *466*, 329–333.

Fujiwara, T., Ito, K., Nakayashiki, T., and Nakamura, Y. (1999). Amber mutations in ribosome recycling factors of *Escherichia coli* and *Thermus thermophilus*: evidence for C-terminal modulator element. FEBS Lett. *447*, 297–302.

Gao, N., Zavialov, A.V., Li, W., Sengupta, J., Valle, M., Gursky, R.P., Ehrenberg, M., and Frank, J. (2005). Mechanism for the disassembly of the posttermination complex inferred from cryo-EM studies. Mol. Cell *18*, 663–674.

Gao, N., Zavialov, A.V., Ehrenberg, M., and Frank, J. (2007). Specific interaction between EF-G and RRF and its implication for GTP-dependent ribosome splitting into subunits. J. Mol. Biol. *374*, 1345–1358.

Grassucci, R.A., Taylor, D.J., and Frank, J. (2007). Preparation of macromolecular complexes for cryo-electron microscopy. Nat. Protoc. *2*, 3239–3246.

Hirashima, A., and Kaji, A. (1973). Role of elongation factor g and a protein factor on the release of ribosomes from messenger ribonucleic acid. J. Biol. Chem. *248*, 7580–7587.

Hirokawa, G., Demeshkina, N., Iwakura, N., Kaji, H., and Kaji, A. (2006). The ribosome-recycling step: consensus or controversy? Trends Biochem. Sci. *31*, 143–149.

Ito, K., Fujiwara, T., Toyoda, T., and Nakamura, Y. (2002). Elongation factor G participates in ribosome disassembly by interacting with ribosome recycling factor at their tRNA-mimicry domains. Mol. Cell *9*, 1263–1272.

Julián, P., Milon, P., Agirrezabala, X., Lasso, G., Gil, D., Rodnina, M.V., and Valle, M. (2011). The cryo-EM structure of a complete 30S translation initiation complex from *Escherichia coli*. PLoS Biol. *9*, e1001095.

Karimi, R., Pavlov, M.Y., Buckingham, R.H., and Ehrenberg, M. (1999). Novel roles for classical factors at the interface between translation termination and initiation. Mol. Cell *3*, 601–609.

Kim, K.K., Min, K., and Suh, S.W. (2000). Crystal structure of the ribosome recycling factor from *Escherichia coli*. EMBO J. *19*, 2362–2370.

Lancaster, L., Kiel, M.C., Kaji, A., and Noller, H.F. (2002). Orientation of ribosome recycling factor in the ribosome from directed hydroxyl radical probing. Cell *111*, 129–140.

Li, X., Mooney, P., Zheng, S., Booth, C., Braunfeld, M.B., Gubbens, S., Agard, D.A., and Cheng, Y. (2013). Electron counting and beam-induced motion correction enables near atomic resolution single particle cryoEM. Nat. Methods *10*, 584–590.

Lu, Z., Shaikh, T.R., Barnard, D., Meng, X., Mohamed, H., Yassin, A., Mannella, C.A., Agrawal, R.K., Lu, T.-M., and Wagenknecht, T. (2009). Monolithic microfluidic mixing-spraying devices for time-resolved cryo-electron microscopy. J. Struct. Biol. *168*, 388–395.

McCutcheon, J.P., Agrawal, R.K., Philips, S.M., Grassucci, R.A., Gerchman, S.E., Clemons, W.M., Ramakrishnan, V., and Frank, J. (1999). Location of translational initiation factor IF3 on the small ribosomal subunit. Proc. Natl. Acad. Sci. USA *96*, 4301–4306.

Nakano, H., Yoshida, T., Uchiyama, S., Kawachi, M., Matsuo, H., Kato, T., Ohshima, A., Yamaichi, Y., Honda, T., Kato, H., et al. (2003). Structure and binding mode of a ribosome recycling factor (RRF) from mesophilic bacterium. J. Biol. Chem. *278*, 3427–3436.

Pavlov, M.Y., Antoun, A., Lovmar, M., and Ehrenberg, M. (2008). Complementary roles of initiation factor 1 and ribosome recycling factor in 70S ribosome splitting. EMBO J. *27*, 1706–1717.

Pioletti, M., Schlünzen, F., Harms, J., Zarivach, R., Glühmann, M., Avila, H., Bashan, A., Bartels, H., Auerbach, T., Jacobi, C., et al. (2001). Crystal structures of complexes of the small ribosomal subunit with tetracycline, edeine and IF3. EMBO J. *20*, 1829–1839.

Potter, C.S., Chu, H., Frey, B., Green, C., Kisseberth, N., Madden, T.J., Miller, K.L., Nahrstedt, K., Pulokas, J., Reilein, A., et al. (1999). Leginon: a system for fully automated acquisition of 1000 electron micrographs a day. Ultramicroscopy *77*, 153–161.

Réblová, K., Šponer, J., and Lankaš, F. (2012). Structure and mechanical properties of the ribosomal L1 stalk three-way junction. Nucleic Acids Res. *40*, 6290–6303.

Raj, V.S., Kaji, H., and Kaji, A. (2005). Interaction of RRF and EF-G from *E. coli* and *T. thermophilus* with ribosomes from both origins—insight into the mechanism of the ribosome recycling step. RNA *11*, 275–284.

Ramakrishnan, V. (2002). Ribosome structure and the mechanism of translation. Cell *108*, 557–572.

Savelsbergh, A., Rodnina, M.V., and Wintermeyer, W. (2009). Distinct functions of elongation factor G in ribosome recycling and translocation. RNA *15*, 772–780.

Scheres, S.H.W. (2012). RELION: implementation of a Bayesian approach to cryo-EM structure determination. J. Struct. Biol. *180*, 519–530.

Schmeing, T.M., Moore, P.B., and Steitz, T.A. (2003). Structures of deacylated tRNA mimics bound to the E site of the large ribosomal subunit. RNA *9*, 1345–1352.

Selmer, M., Al-Karadaghi, S., Hirokawa, G., Kaji, A., and Liljas, A. (1999). Crystal structure of *Thermotoga maritima* ribosome recycling factor: a tRNA mimic. Science *286*, 2349–2352.

Shaikh, T.R., Yassin, A.S., Lu, Z., Barnard, D., Meng, X., Lu, T.-M., Wagenknecht, T., and Agrawal, R.K. (2014). Initial bridges between two ribosomal subunits are formed within 9.4 milliseconds, as studied by time-resolved cryo-EM. Proc. Natl. Acad. Sci. USA *111*, 9822–9827.

Simonetti, A., Marzi, S., Myasnikov, A.G., Fabbretti, A., Yusupov, M., Gualerzi, C.O., and Klaholz, B.P. (2008). Structure of the 30S translation initiation complex. Nature *455*, 416–420.

Toyoda, T., Tin, O.F., Ito, K., Fujiwara, T., Kumasaka, T., Yamamoto, M., Garber, M.B., and Nakamura, Y. (2000). Crystal structure combined with genetic analysis of the *Thermus thermophilus* ribosome recycling factor shows that a flexible hinge may act as a functional switch. RNA *6*, 1432–1444.

Weixlbaumer, A., Petry, S., Dunham, C.M., Selmer, M., Kelley, A.C., and Ramakrishnan, V. (2007). Crystal structure of the ribosome recycling factor bound to the ribosome. Nat. Struct. Mol. Biol. *14*, 733–737.

Wilson, D.N., Schluenzen, F., Harms, J.M., Yoshida, T., Ohkubo, T., Albrecht, R., Buerger, J., Kobayashi, Y., and Fucini, P. (2004). X-ray crystallography study on ribosome recycling: the mechanism of binding and action of RRF on the 50S ribosomal subunit. EMBO J. *24*, 251–260.

Yokoyama, T., Shaikh, T.R., Iwakura, N., Kaji, H., Kaji, A., and Agrawal, R.K. (2012). Structural insights into initial and intermediate steps of the ribosome-recycling process. EMBO J. *31*, 1836–1846.

Yoshida, T., Oka, S., Uchiyama, S., Nakano, H., Kawasaki, T., Ohkubo, T., and Kobayashi, Y. (2003). Characteristic domain motion in the ribosome recycling factor revealed by 15N NMR relaxation experiments and molecular dynamics simulations. Biochemistry *42*, 4101–4107.

Yoshida, T., Uchiyama, S., Nakano, H., Kashimori, H., Kijima, H., Ohshima, T., Saihara, Y., Ishino, T., Shimahara, H., Yoshida, T., et al. (2001). Solution structure of the ribosome recycling factor from *Aquifex aeolicus*. Biochemistry *40*, 2387–2396.

Zavialov, A.V., Hauryliuk, V.V., and Ehrenberg, M. (2005). Splitting of the post-termination ribosome into subunits by the concerted action of RRF and EF-G. Mol. Cell *18*, 675–686.

Structure and assembly model for the *Trypanosoma cruzi* 60S ribosomal subunit

Zheng Liu[a,1], Cristina Gutierrez-Vargas[b,1], Jia Wei[b,1], Robert A. Grassucci[a,c], Madhumitha Ramesh[d], Noel Espina[e], Ming Sun[b], Beril Tutuncuoglu[d], Susan Madison-Antenucci[e], John L. Woolford Jr.[d], Liang Tong[b], and Joachim Frank[a,b,c,2]

[a]Department of Biochemistry and Molecular Biophysics, Columbia University, New York, NY 10032; [b]Department of Biological Sciences, Columbia University, New York, NY 10027; [c]Howard Hughes Medical Institute, Columbia University, New York, NY 10027; [d]Department of Biological Sciences, Carnegie Mellon University, Pittsburgh, PA 15213; and [e]Parasitology Laboratory, Wadsworth Center, New York State Department of Health, Albany, NY 12208

Contributed by Joachim Frank, September 1, 2016 (sent for review July 12, 2016; reviewed by Shulamit Michaeli and James R. Williamson)

Ribosomes of trypanosomatids, a family of protozoan parasites causing debilitating human diseases, possess multiply fragmented rRNAs that together are analogous to 28S rRNA, unusually large rRNA expansion segments, and r-protein variations compared with other eukaryotic ribosomes. To investigate the architecture of the trypanosomatid ribosomes, we determined the 2.5-Å structure of the *Trypanosoma cruzi* ribosome large subunit by single-particle cryo-EM. Examination of this structure and comparative analysis of the yeast ribosomal assembly pathway allowed us to develop a stepwise assembly model for the eight pieces of the large subunit rRNAs and a number of ancillary "glue" proteins. This model can be applied to the characterization of *Trypanosoma brucei* and *Leishmania* spp. ribosomes as well. Together with other details, our atomic-level structure may provide a foundation for structure-based design of antitrypanosome drugs.

ribosome structure | *Trypanosoma cruzi* | biogenesis | multiply fragmented rRNA | antitrypanosome drug design

Ribosomes share a universally conserved core that carries out the fundamental processes of protein synthesis (1). Outside of this core, ribosome composition varies considerably. The main differences among eukaryotic ribosomes are due to rRNA expansion segments (ESs) and variations of ribosomal proteins (r-proteins) (2). An extreme case is presented by the ribosome of trypanosomatids, a family of kinetoplastid protozoans the members of which cause human diseases such as Chagas disease (*Trypanosoma cruzi*), sleeping sickness (*Trypanosoma brucei*), and Leishmaniasis (*Leishmania* spp.) (3–5). Trypanosomatid ribosomes have unusually large ESs and r-protein variations (6). Moreover, their large subunit contains eight pieces of rRNA, six of which result from unique cleavages of a precursor rRNA. Upon assembly, these rRNAs fulfill the functions of the 28S rRNA found in other eukaryotes (7, 8). This unique multiple fragmentation strongly suggests that trypanosomatid ribosomes have pronounced differences that could be exploited for drug design.

Results and Discussion

Here, we report the structure of the 60S large ribosomal subunit of *T. cruzi* at a 2.5-Å resolution (Fig. 1; *SI Appendix*, Fig. S2). Our large subunit structure represents a significant improvement over earlier trypanosomatid ribosome models (6, 9) (*SI Appendix*, Fig. S14) and provides extensive details of the constitutive molecular interactions that stabilize the eight pieces of rRNA (Fig. 1 *C* and *D*; *SI Appendix*, Fig. S4). The quality of the EM map is indicated by the clearly resolved density of the nucleotides and the base-pairing and base-stacking in the helical segments of rRNA (Fig. 1*B*; *SI Appendix*, Fig. S3 *A–D*). Purine bases can be readily distinguished from pyrimidine bases, and it is also routinely possible to distinguish between G and A (Fig. 1*B*; *SI Appendix*, Fig. S3). Well-defined EM density is also observed for most protein side chains (*SI Appendix*, Fig. S7 and Table S7). In addition, the high resolution allowed us to identify magnesium ions as well as water molecules (Fig. 1*B*; *SI Appendix*, Fig. S3). Importantly, we located many rRNA covalent modifications (Fig. 1*B*). The number of 2′-O methylations (66 sites) remarkably exceeds that in other eukaryotes, which is consistent with bioinformatics analysis (9) and the modifications on the *Leishmania* ribosome structure recently revealed (10).

In general outline, the 60S ribosomal subunit of *T. cruzi* resembles that of yeast. However, the total size of the six rRNA pieces is about one-fourth larger than the 25S rRNA in yeast (4,225 vs. 3,392 nts) (11, 12). This increase is accounted for by the larger size of ESs and the addition of trypanosome-specific ESs. We refer to the six pieces analogous to 28S rRNA as LSU-α, LSU-β, and srRNAs1–4 (13). The corresponding six pieces across the three pathogenic trypanosomatids are shown in *SI Appendix*, Fig. S15. LSU-α roughly corresponds to domains I and II of the yeast 25S RNA, srRNA1 to domain III, LSU-β to domains IV and V, and srRNA2–4 to domain VI (*SI Appendix*, Fig. S6). LSU-α and LSU-β are located on the solvent and interface side, respectively, whereas srRNA1 is located at the bottom of the large subunit (solvent-side view) (Fig. 1*D*; *SI Appendix*, Fig. S4). The srRNAs2–4 are in close mutual contact and situated on the left side of the large subunit under the P stalk (solvent-side view) (Fig. 1; *SI Appendix*, Fig. S4). Most r-proteins are conserved between yeast and *T. cruzi* with exceptions typically occurring near the srRNAs. Comparative analysis of trypanosome-specific rRNA–rRNA and r-protein–rRNA interactions with those of other eukaryotes, especially yeast, has allowed us to identify the

Significance

The pathogenic trypanosomatids—*Trypanosoma cruzi, Trypanosoma brucei*, and *Leishmania* spp.—are the causative agents of Chagas disease, African trypanosomiasis, and leishmaniasis, respectively. These diseases, with high morbidity and mortality rates, affect millions of people worldwide. Current treatments typically use drugs with high toxicity and marginal efficacy. Here we present, a 2.5-Å structure of the *T. cruzi* ribosome large subunit by single-particle cryo-EM. Our structure highlights distinctive trypanosome interactions and has allowed us to propose a tentative model for assembly of the 60S large ribosomal subunit. These atomic details highlighting trypanosome-specific interactions and the differences between *T. cruzi* and the human ribosome can be used directly for structure-based drug design of antitrypanosome drugs.

Author contributions: Z.L. and J.F. designed research; Z.L., C.G.-V., J.W., R.A.G., and L.T. performed research; N.E. and S.M.-A. contributed new reagents/analytic tools; Z.L., C.G.-V., J.W., M.R., M.S., B.T., J.L.W., L.T., and J.F. analyzed data; and Z.L., C.G.-V., J.W., M.R., B.T., S.M.-A., J.L.W., L.T., and J.F. wrote the paper.

Reviewers: S.M., Faculty of Life Sciences, University Ramat-Gan Israel; and J.R.W., The Scripps Research Institute.

The authors declare no conflict of interest.

Data deposition: The data reported in this article have been deposited in The Electron Microscopy Data Bank (accession no. EMD-8361) and the Protein Data Bank (PDB ID code: 5T5H).

[1]Z.L., C.G.-V., and J.W. contributed equally to this work.

[2]To whom correspondence should be addressed. Email: jf2192@cumc.columbia.edu.

This article contains supporting information online at www.pnas.org/lookup/suppl/doi:10.1073/pnas.1614594113/-/DCSupplemental.

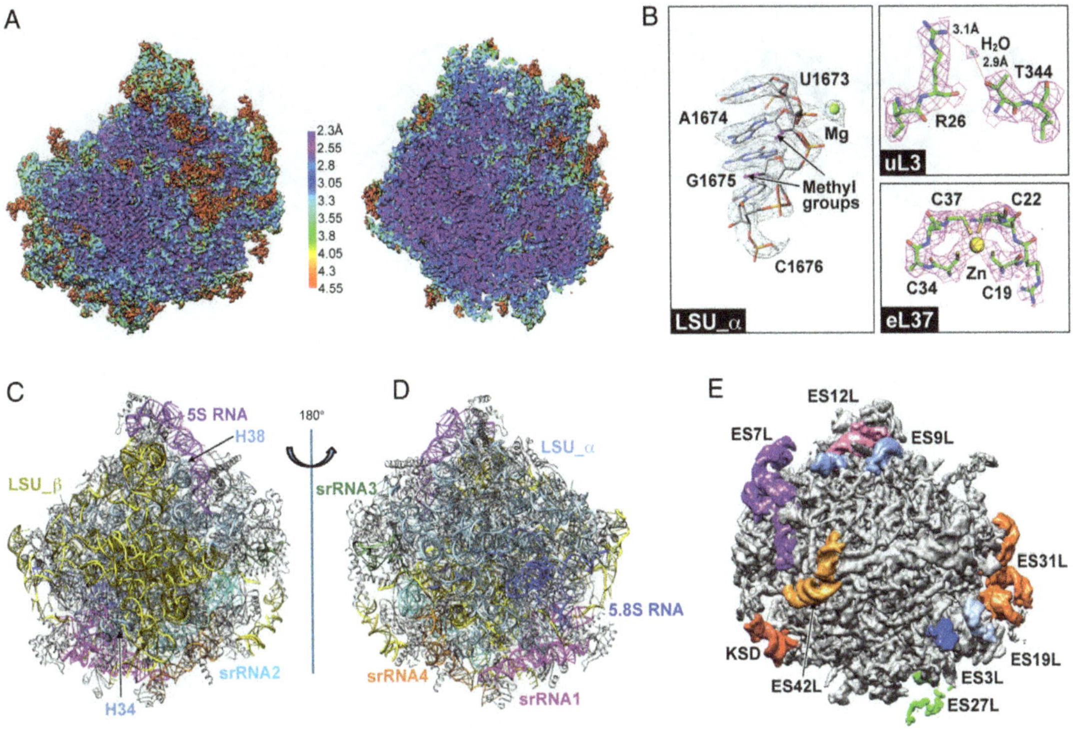

Fig. 1. Structure of the *T. cruzi* large subunit ribosome. (*A*) Cryo-EM map of the 60S subunit after sharpening, colored by local resolution and viewed from the subunit interface. (*Left*) Surface view. (*Right*) Central cut-away view. (*B*) Selected views of density for rRNA and proteins with associated ions and water molecule. (*C*) Some expansion segments in the unsharpened map of a large subunit, viewed from the solvent side. (*C* and *D*) rRNA architecture of the large subunit: interface (*C*) and solvent (*D*) view. (*E*) Some expansion segments in the unsharpened map of a large subunit, viewed from the solvent side.

interactions stabilizing the eight pieces of rRNA to form a mature large subunit. Using our structure as the endpoint of assembly in combination with this comparative analysis has allowed us to arrive at a plausible model for assembly of the eight trypanosomal rRNA pieces.

The rRNA Scaffold. Examination of rRNA–rRNA interactions in our structure reveals that the 5.8S rRNA and the conserved LSU-α and LSU-β domains form the backbone of the rRNA scaffold with two rigid blocks acting as stabilizing core. The first block is formed by the whole 5.8S rRNA and nucleotides in the 5′-end of LSU-α. The second rigid block is constructed by nucleotides in the 3′-end of LSU-α and the 5′-end of LSU-β. Both of the rigid blocks are conserved in yeast; however, in trypanosomes they interact with each other via ES3L of 5.8S rRNA (Fig. 2*A*). Clearly, the 5.8S serves as a critical organizing center for rRNA folding and assembly. The stability of the LSU-α–LSU-β junction is further reinforced by the insertion of two helices, H38 and H34 from LSU-α, into the interface from the solvent side anchoring LSU-α to LSU-β. Mg^{2+} ions and modified nucleotides also mediate contacts in the scaffold (*SI Appendix*, Fig. S11). Along with rRNA–rRNA interactions, r-proteins play important roles in stabilizing this scaffold (*SI Appendix*, Fig. S7).

Structure and Interactions Stabilizing the Small rRNAs. The srRNA1 consists of five helices corresponding to yeast H55-59 and a trypanosome-specific ES, named srRNA1-ES, which adopts a similar position in *T. brucei* (13) and shows highly dynamic behavior, preventing us from modeling it (*SI Appendix*, Fig. S9*A*). The srRNA1 is anchored to the scaffold with the help of srRNA2 via base-pairing interactions. In addition to the contacts with srRNA2, two anchoring proteins, eL19 and eL34, pin H58 and H55 of srRNA1 to the scaffold (Fig. 2*F*). Protein eL34 has adopted an insertion and protein eL19 has a C-terminal extension of about 170 aa, crossing through to the SSU, as already revealed in the structure of *T. brucei*, which suggests that the small subunit bears a role in further stabilizing srRNA1 (13).

The srRNA2 has a scissor-shape architecture and is composed of four helices corresponding to H94-97 in yeast (*SI Appendix*, Fig. S9*B*). H94-95 and H96-97 line up to form the two prongs of the scissor. We found in our structure that srRNA2 has five contact sites with LSU-β. In addition, srRNA2 forms various contacts with LSU-α (Fig. 2*B*). Apart from direct interactions with the scaffold, we found srRNA2 to be further anchored by uL3 onto the scaffold. Specifically, protein uL3 forms a C-shaped cavity, which facilitates the stable assembly of srRNA2 onto the scaffold by accommodating H94 and H96 (Fig. 2*B*; *SI Appendix*, Figs. S8*B* and S9*B*).

The srRNA3 of *T. cruzi* corresponds to yeast ES39L, which is made up of three helices, ES39L1–3, threaded together by single rRNA stretches (*SI Appendix*, Fig. S9*C*). In contrast, srRNA3 is composed of a stretch of single-stranded rRNA plus a helix corresponding to helix ES39L-h2 (*SI Appendix*, Fig. S12*A*). The vacancy resulting from the missing ES39L-h1 adjacent to the sRNA2–4 cleavage sites is filled partially by ES42L, a trypanosome-specific ES. The absence of ES39L-h3 makes room for the kinetoplastid-specific domain (KSD) stretch, the extension of the C-terminal portion of eL14, and the N-terminal domain of eL33. Protein eL33 anchors srRNA3 to the scaffold and contains an

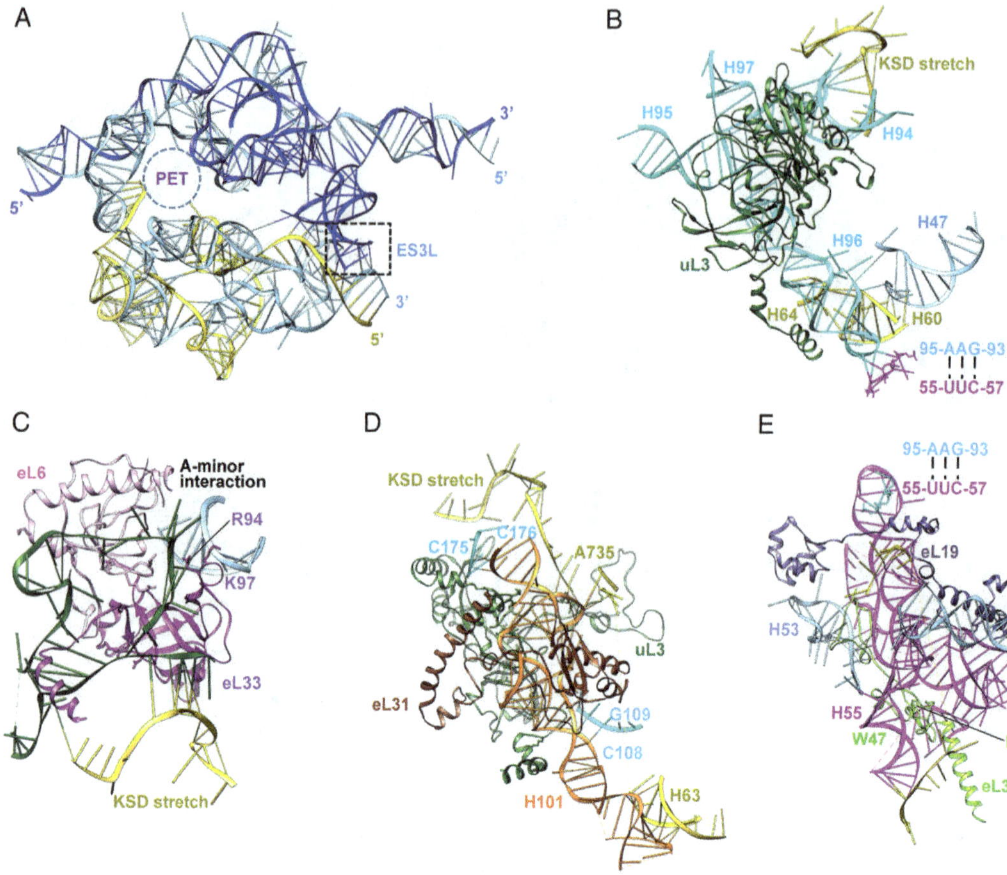

Fig. 2. Interactions stabilizing the srRNAs. (*A*) Scaffold formed by 5.8S rRNA, LSU-α, and LSU-β. PET, peptide exit tunnel. (*B*) srRNA2 (cyan). The components interacting with srRNA2 include the KSD stretch, H60, H64, H47, and uL3. (*C*) srRNA3 (forest green). The rRNA-contacting residues of the anchoring protein, eL6 and eL33, are shown. R94 and K97 are in the trypanosome-specific insertion of eL33. (*D*) srRNA4 (orange). The interacting residues are from the KSD stretch, H63, srRNA2, eL31, and uL3. (*E*) srRNA1 (magenta). W47 and H51 are from the trypanosome-specific insertion of eL34.

insertion in its globular domain that provides extra binding sites for both srRNA3 and LSU-α (Fig. 2*C*; *SI Appendix*, Figs. S8*D*, S9*C*, and S11*A*). Moreover, srRNA3 forms an A-minor interaction with ES7L (Fig. 2*C*) (14), indicating that ES7L also facilitates the assembly of srRNA3 onto the scaffold.

Finally, the srRNA4 is composed of a long helix, corresponding to H101 in yeast, and two short helices (H99 and H100 in yeast) located on one end of H101, as well as an ES, named srRN4-ES, emerging from H101 (*SI Appendix*, Fig. S9*D*). Nucleotides 106–108 of srRNA2 and the exposed positively charged residues of uL3 establish a binding surface for srRNA4 to stack onto (Fig. 2*D*). Furthermore, several residues of the N-terminal extension of eL31 cover H101 of srRNA4. These additions on the N-terminal portion of eL31 exhibit more flexibility compared with the C-terminal region of eL31, which results in only C-A modeling for this region. Also, *T. cruzi* srRNA4 adopts a shorter loop at the base of H100 than its counterpart in yeast, thus avoiding conflict with the N terminus of eL31 (*SI Appendix*, Fig. S9*D*). In addition, residues 1561–1564 of the KSD stretch provide two binding sites with H99 from srRNA4 (Fig. 2*D*).

From the arrangement of the srRNAs revealed in our structure, we see that both srRNA1 and srRNA4 require srRNA2 to be in its final position to provide binding sites for their stable assembly onto the scaffold (Fig. 2 *B*, *D*, and *E*). Interestingly, the known relative time points of assembly of the r-proteins anchoring these small pieces correlates with this arrangement. Studies in yeast have shown that r-proteins, localized in specific positions within the mature 60S subunit, can be classified as early, middle-, or late-acting proteins, thus providing a temporal hierarchy of 60S assembly (15–17). As described earlier, we found in our structure that srRNA2 is anchored by uL3 and srRNA3 is anchored by eL6 and eL33 (Fig. 2 *B* and *C*), all of which are early acting proteins whereas proteins eL19, eL34, and eL31 (Fig. 2 *D* and *E*; *SI Appendix*, Fig. S8 *E–G*), which are identified as middle-acting proteins in yeast, interact with srRNA1 and srRNA4 (Fig. 2 *D* and *E*) (17, 18). Thus, the arrangement of the srRNAs as well as their anchoring proteins help define the likely order of assembly of the srRNAs, such that stable association of srRNA2 and srRNA3 occurs before that of srRNA1 and srRNA4. Indeed, deletion of the counterpart of srRNA3 in yeast (ES39L) leads to an early processing defect (19). Strikingly, although srRNA1 is the first small rRNA transcribed, it is stably assembled after srRNA2 and srRNA3, suggesting that its RNA processing occurs during or after assembly of the ribosome instead of the 5′ to 3′ direction following transcription. Furthermore, this

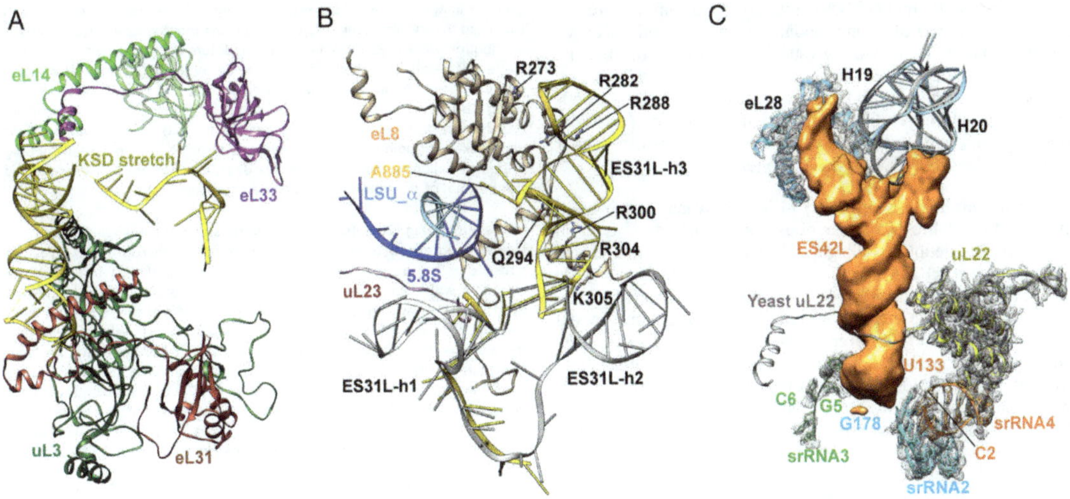

Fig. 3. *T. cruzi*-specific expansion segments. (*A*) KSD contacts uL3, eL31, eL33, and eL14. (*B*) ES31L (yellow) in *T. cruzi* has the specific helix ES31L-h3 contacting eL8 and the 5′ end of 5.8S rRNA. ES31L-h1 and -h2 are only partially modeled. For comparison, ES31L from yeast is also shown here (in gray). (*C*) ES42L, emerging from H19, covers the cleavage sites of srRNA2–4 and blocks the extension of uL22.

assembly order is also consistent with the proposed order of assembly of the three focal points on the *Leishmania* ribosome structure (10) (*SI Appendix*, Fig. S6).

Prominent Expansion Segments and the 5S rRNA. The trypanosome-specific ESs, ES42L and the KSD, both located on the surface of the ribosome, bundle srRNAs2–4. The KSD is a binding partner of the sRNA2- and srRNA4-anchoring proteins, uL3 and eL31, respectively (Fig. 3*A*). This places KSD in a pivotal position interacting with ribosomal functional centers (*SI Appendix*, Table S6) (20, 21). ES42L covers the hub region encompassing the cleavage sites of srRNAs2–4 (Fig. 3*C*; *SI Appendix*, Fig. S12), where the extension of uL22, a component of the peptide exit tunnel (PET), would reach toward ES39L in yeast. The location of ES42L suggests that it may also play a role in the remote regulation of PET by uL22 (22) (Fig. 3*C*; *SI Appendix*, Fig. S12). Both ES42L and KSD envelope small rRNAs2–4 and make additional contacts to stabilize these small rRNA pieces (Figs. 1*E* and 3 *A* and *C*), which suggests that their assembly occurs after that of srRNAs2–4.

Another prominent ES, ES31L, emerges from the base of the L1 stalk and contains an additional helix compared with yeast (Fig. 3*B*). This helix contacts extensions of eL8 and interacts with the 3′-end of the 5.8S rRNA. In yeast, deleting ES31L or the N-terminal extension of eL8 blocks ribosome assembly at later stages, suggesting that ES31L may be involved in subsequent stabilization steps of the previously established scaffold (23). Also associated with the later stages is the final positioning of the 5S rRNA. As in other eukaryotes, the *T. cruzi* 5S rRNA is independently transcribed and associates with uL5 and uL18 forming the 5S ribonucleoprotein particle (RNP) (24). In yeast, this 5S RNP associates with the body of the large subunit during early steps of biogenesis, but is essential only for much later pre-rRNA processing. During this late nuclear stage, the 5S RNP undergoes a 180° rotation before nuclear export (25, 26). The conservation of 5S RNA and its associated proteins uL5 and uL18 (*SI Appendix*, Figs. S10 and S13), as well as identification of trypanosome homologs for 5S RNP assembly factors (27), lead us to hypothesize that similar pathways may exist for maturation of the 5S RNP in trypanosomes at late stages.

Implications for Trypanosomatid Ribosome Assembly. Ribosome biogenesis is a complex, multistep process, requiring the actions of hundreds of assembly factors and small nucleolar RNAs (26, 28). The multiply fragmented rRNAs increase the complexity of *T. cruzi* ribosome biogenesis (27). Based on the above analysis of interactions of individual rRNA components with r-proteins, and on a comparison with the hierarchy of yeast ribosome biogenesis, we propose here a step-wise assembly model for rRNAs of the large subunit (Fig. 4). First, the earlier transcribed components, 5.8S rRNA, LSU-α (domains I, II and part of domain III in yeast), and LSU-β (domains IV, V, and part of domain III), form two large intermingled blocks, and with the addition of the associated r-proteins, the scaffold is constructed. Next, srRNA2 and srRNA3 (both part of domain VI) are positioned in part by their anchoring r-proteins, which are already bound to the scaffold. ES7L is stably assembled concurrently with srRNA3. Subsequently, srRNA1 (the

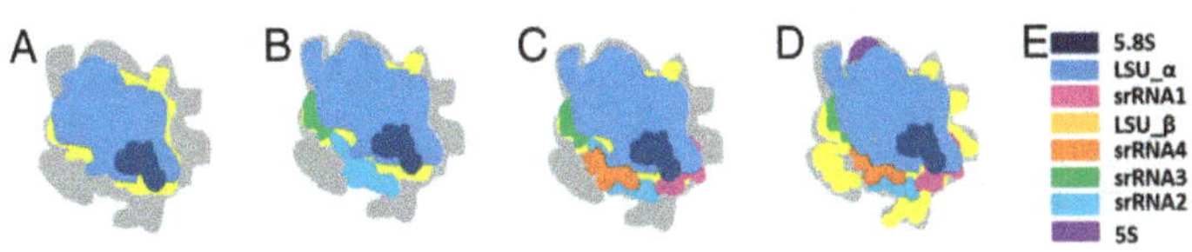

Fig. 4. A model of the proposed *T. cruzi* large subunit rRNAs assembly pathway. (*A*) 5.8S, LSU-α, and LSU-β form a scaffold. The gray color indicates the outline of the completely assembled large subunit. (*B*) srRNA2, -3, and ES7L are assembled. (*C*) srRNA4 and srRNA1 are assembled next. (*D*) 5S rRNA, ES42L, ES31L, and the KSD properly assemble to complete the large subunit. (*E*) Color key for the eight pieces of rRNA.

rest of domain III) and srRNA4 (the rest of domain VI) are stabilized with the aid of strong binding interactions with srRNA2. ES31L stabilizes the previously established scaffold, and the trypanosome-specific ESs (ES42L and the KSD) stabilize srRNAs2–4. The 5S rRNP containing 5S RNA, having assembled with the body of the large subunit earlier, is then rotated to its final position, enabling subsequent maturation steps, concurrent with proper folding of the large ESs on the periphery.

Conclusion. Our high-resolution cryo-EM density map has revealed the structure of the 60S large ribosomal subunit in *T. cruzi* in unprecedented detail, including the rRNA nucleotides, amino acid side chains, and rRNA modifications as well as small molecules (metal ions, and water molecules). We present an analysis of distinctive trypanosome interactions and propose a tentative model for assembly of the 60S large ribosomal subunit. These results from trypanosomatid ribosomes can be used to further understand how different rRNA domains are compacted as assembly proceeds in yeast and other eukaryotes. The multiple fragmentation of the *T. cruzi* ribosome is conserved within the whole trypanosomatid family (*SI Appendix*, Fig. S15 and Table S1), and thus our assembly model can be applied to the characterization of *T. brucei* and *Leishmania* spp. ribosomes as well. Moreover, trypanosomatids, as a group of protozoa early diverged from the eukaryotic lineage, are a compelling model to study the variability of highly conserved processes such as protein translation. Our structure thus serves as a basis for future studies of trypanosome-specific features in protein synthesis. Most importantly, the atomic details highlighting the differences between the *T. cruzi* and the human ribosome can be directly used for structure-based design of antitrypanosome drugs (*SI Appendix*, Tables S9 and S10).

Materials and Methods

***T. cruzi* Culture and Ribosome Purification.** *T. cruzi* Y-strain epimastigote cultures were grown at 27 °C in American Type Culture Collection 1029 liver infusion trypsose medium (9 mg/ml liver infusion broth, 5 mg/ml tryptose, 17 mM NaCl, 56 mM Na_2HPO_4, 5 mM KCl, 5 mM glucose, 10% heat inactivated fetal bovine serum, 0.01 mg/ml hemin) to a concentration of 3 × 107/mL in 1-L glass bottles. A total of 12–18 L of culture was chilled on ice, and cells were harvested by centrifugation at 1,000 × *g* for 20 min. Cell pellets from 5 L of culture were pooled and washed three times in ice-cold buffer containing 0.25 M sucrose and 5 mM KCl. Washed pellets were resuspended in 30 mL of buffer containing 0.25 M sucrose, 5 mM KCl, 5 mM EGTA, 1× protease inhibitor mixture (Sigma P2714), and 0.5% Triton X-100. Lysis was achieved by five freeze–thaw cycles using a dry ice/ethanol bath and warm water. Following lysis the sample was frozen at −80 °C.

Ribosomes were prepared based on previous purification protocols (29) with slight modifications. Cell lysate was loaded onto a 30% (wt/vol) sucrose cushion prepared in buffer I (containing 20 mM Hepes·KOH, pH 7.5), 100 mM KOAc, 4 mM Mg(OAc)2, 20% glycerol, 2 mM DTT, and Roch EDTA-free protease inhibitor mixture) and was centrifuged at 33,800 × *g* for 18 h at 4 °C. The resulting ribosome-enriched pellets were further purified using the 20-K PEG precipitation methods described previously (29). The final pellets were suspended in buffer II (10 mM Hepes·KOH, pH 7.4, 50 mM KOAc, 10 mM NH4Cl, 5 mM Mg(OAc)2, 2 mM DTT, and protease inhibitor mixture) and kept at −80 °C for further use.

Electron Microscopy. Three microliters of ~80 nM (0.3 mg/mL) purified ribosomes were applied onto a holey carbon grid (carbon-coated Quantifoil 2/2 grid, Quantifoil Micro Tools). The grids were coated by an additional prefloated continuous thin layer of carbon and glow-discharged before use. The sample-loaded grids were blotted for 3 s at 4 °C in 100% humidity and vitrified by plunging rapidly into liquid ethane cooled by liquid nitrogen at −180 °C with a Vitrobot (FEI). Prepared grids were quickly transferred to a storage box and stored in liquid nitrogen for imaging.

Micrographs were acquired with a back-thinned FEI Falcon II detector on a TITAN KRIOS (FEI) operating at 300 kV with the automatic imaging software FEI EPU. The defocus ranged from −1.0 to −3.0 μm, and the nominal magnification was 143,000× (133,970 after calibration), resulting in a final pixel size of 1.045 Å. Sixteen frames per second were collected with a total dose of about 32e-/Å2 in integrating mode—the only mode available for this camera. To select the grid for imaging, four grids were loaded into the microscope column and subjected to careful screening. Only grids presenting thin ice and good particle distribution were used for data collection. Before data collection, the microscope was carefully aligned including the Cs corrector to minimize electron optical aberrations, linear geometrical distortions, and coma. During the imaging session, micrographs were monitored for evidence of bad pixels, unusually large drift, beam shift, or astigmatism, in which case corrective actions were taken including regaining reference, skipping the squares exhibiting drift, and realigning the microscope. About 11,000 micrographs were collected from a total of two grids during 5 consecutive days.

Image Processing. Dose-fractionated frame stacks were aligned using the program dosefgpu_driftcorr (30). The averaged micrographs were first screened with Arachnid (31), a python version of SPIDER (32), to exclude micrographs with obvious drift, ice contamination, or sparse coverage with particles. The contrast transfer function (CTF) parameters of the remaining good micrographs were calculated using CTFFIND3 (33). Particle selection was performed using the program e2boxer.py (34). About 700,000 selected ribosome-like single particles were subjected to 3D classifications with RELION (35) using a 70S ribosome map (EMD-6315) (36) filtered to 60 Å as the starting reference (*SI Appendix*, Fig. S1). About 400,000 particles were kept after excluding bad particles. One further classification with 10 classes was performed to separate ribosomes in different binding states in the dataset, yielding 60S subunits, E-tRNA–bound ribosomes, and empty ribosomes as the only relevant distinctions. No other tRNAs or factors such as eEF2 or eEF1A were found bound to the ribosomes. One resulting dominant class of 80S-E-tRNA including 79,000 particles was auto-refined and yielded a 3.1-Å resolution map showing a rigid 60S subunit and a less ordered 40S subunit, which indicated residual heterogeneity in the dataset.

To improve the resolution of the reconstruction, we used several image-processing strategies. First, all of the 60S subunit-containing particles from the first classification were recombined into a pool of about 345,000 particles. These were subjected to focused classification using a large-subunit mask. Before the refinement, the particle dataset was subjected to 2D reference-free classification to eliminate heterogeneous particles. Focused classification and reconstruction were used to overcome heterogeneity in the dataset and even in the classified subset (37). To further improve the homogeneity, three sets of particle positional parameters were obtained by refining each particle with three independent references, which were generated from 3D classification, and only the particles having consistent views were kept (38). In this way, about 4% of the particles were excluded. Second, the micrographs and their power spectra were visually reassessed to exclude those with contamination by ice and ethane, uncorrected drift, and astigmatism, ensuring that only particles extracted from high-quality micrographs were kept. Third, CTF values were recalculated using another program, *S2focus* (39), and the resultant values were cross-validated with those previously determined by CTFFIND3. Micrographs with discrepancies in defocus larger than 500 Å were excluded—in our case, a total of 480 micrographs (4.4%). Fourth, particles for final reconstruction were extracted from good micrographs averaged with weighting from frames 3–10, taking into account the fact that the earlier frames have more drift and the later frames experience more radiation damage.

In this manner, about 235,000 particles were used to reconstruct the large subunit to 2.5 Å. Resolution was estimated using the 0.143 Fourier shell correlation criterion computed from two independently reconstructed maps, following the "gold standard" protocol (40) (*SI Appendix*, Fig. S2). In addition, the two independently reconstructed maps were also used to estimate the distribution of local resolution using Resmap (41) (Fig. 1*A* and *SI Appendix*, Fig. S2 *B–D*). The resolution estimates were in agreement with visual assessment of map quality for secondary structure elements (e.g., appearance of α-helices and β-sheets), individual amino acid residues and nucleotides, small molecules such as water, and methyl groups (*SI Appendix*, Fig. S2 *B–D*). The accuracies of rotation and translation from final reconstruction reported by RELION were 0.52° and 0.49 Å, respectively. The resolution of the final reconstruction without masking was 3.0 Å. RELION postprocessing with auto_mask, auto_bfac, and the inimask_threshold 0.035 (which yields a masked map of the 60S subunit approximately enclosing the atomic model) determined the B-factor as 58.5 $Å^2$ and the resolution of the masked region as 2.5 Å. The resolution of the whole map of the 80S ribosome including the disordered small subunit was determined as 2.8 Å (*SI Appendix*, Fig. S2).

Modeling. A multimap modeling strategy was used in model building (42). The crystal structure of the yeast large ribosomal subunit (PDB: 4V88) (12) and the high-resolution cryo-EM model of the *T. brucei* large ribosomal subunit (PDB: 4V8M) (13) were used as references for building atomic models of rRNA and ribosomal proteins, respectively. The references were initially fitted as rigid bodies into the negative B-factor–sharpened density map using Chimera (43). The rRNA sequences were obtained from TriTrypDB Kinetoplastid Genomics Resource (tritrypdb.org/tritrypdb/) (11). Regions of rRNA conserved between *T. cruzi* and yeast were used as starting points, followed by extensive manual building using Coot (44). The high-resolution density map and the base-pairing principle allowed us to identify most of the bases. The polypeptide chains in *T. brucei* were mutated to match their sequences in *T. cruzi* taken from the National Center for Biotechnology Information protein databases (www.ncbi.nlm.nih.gov) and manually adjusted to fit the density map. The atomic model of the large ribosomal subunit was refined in real space using PHENIX (45). In addition to Mg^{2+} and Zn^{2+} ions, a number of water molecules were identified as density globules in the distance range of 2.8–3.4 Å from neighboring amino acids, a typical range for a hydrogen bond, occurring reproducibly in two independent reconstructions. The statistics for reconstruction and model refinement are summarized in *SI Appendix*, Table S2. The modeled parts and unmodeled residues are summarized in *SI Appendix*, Tables S3–S5.

ACKNOWLEDGMENTS. We thank Amédée des Georges for helpful suggestions on image processing; Harry Kao for assistance with computer hardware; Nadia Severina for help in ribosome purification; Melissa Thomas-Baum (Buckyball Design) for assistance with the preparation of figures; and Z. Yu, C. Hong, R. Huang, and H. He for their assistance in the data collection at the Janelia Farm Research Campus of Howard Hughes Medical Institute (HHMI). This study was supported by HHMI and NIH Grants R01 GM29169 (to J.F.), R35GM118093 (to L.T.), and R01 GM028301 (to J.L.W.) and by National Science Foundaton Grant GRFP DGE-11-44155 (to C.G.-V).

1. Schmeing TM, Ramakrishnan V (2009) What recent ribosome structures have revealed about the mechanism of translation. *Nature* 461(7268):1234–1242.
2. Yusupova G, Yusupov M (2014) High-resolution structure of the eukaryotic 80S ribosome. *Annu Rev Biochem* 83:467–486.
3. El-Sayed NM, et al. (2005) The genome sequence of Trypanosoma cruzi, etiologic agent of Chagas disease. *Science* 309(5733):409–415.
4. El-Sayed NM, et al. (2005) Comparative genomics of trypanosomatid parasitic protozoa. *Science* 309(5733):404–409.
5. Ivens AC, et al. (2005) The genome of the kinetoplastid parasite, Leishmania major. *Science* 309(5733):436–442.
6. Gao H, Ayub MJ, Levin MJ, Frank J (2005) The structure of the 80S ribosome from Trypanosoma cruzi reveals unique rRNA components. *Proc Natl Acad Sci USA* 102(29):10206–10211.
7. Gray MW (1981) Unusual pattern of ribonucleic acid components in the ribosome of Crithidia fasciculata, a trypanosomatid protozoan. *Mol Cell Biol* 1(4):347–357.
8. White TC, Rudenko G, Borst P (1986) Three small RNAs within the 10 kb trypanosome rRNA transcription unit are analogous to domain VII of other eukaryotic 28S rRNAs. *Nucleic Acids Res* 14(23):9471–9489.
9. Liang XH, et al. (2005) A genome-wide analysis of C/D and H/ACA-like small nucleolar RNAs in Trypanosoma brucei reveals a trypanosome-specific pattern of rRNA modification. *RNA* 11(5):619–645.
10. Shalev-Benami M, et al. (2016) 2.8-Å Cryo-EM ctructure of the large ribosomal subunit from the eukaryotic parasite Leishmania. *Cell Reports* 16(2):288–294.
11. Aslett M, et al. (2010) TriTrypDB: A functional genomic resource for the Trypanosomatidae. *Nucleic Acids Res* 38(Database issue):D457–D462.
12. Ben-Shem A, et al. (2011) The structure of the eukaryotic ribosome at 3.0 Å resolution. *Science* 334(6062):1524–1529.
13. Hashem Y, et al. (2013) High-resolution cryo-electron microscopy structure of the Trypanosoma brucei ribosome. *Nature* 494(7437):385–389.
14. Nissen P, Ippolito JA, Ban N, Moore PB, Steitz TA (2001) RNA tertiary interactions in the large ribosomal subunit: The A-minor motif. *Proc Natl Acad Sci USA* 98(9):4899–4903.
15. Woolford JL, Jr, Baserga SJ (2013) Ribosome biogenesis in the yeast Saccharomyces cerevisiae. *Genetics* 195(3):643–681.
16. de la Cruz J, Karbstein K, Woolford JL, Jr (2015) Functions of ribosomal proteins in assembly of eukaryotic ribosomes in vivo. *Annu Rev Biochem* 84:93–129.
17. Gamalinda M, et al. (2014) A hierarchical model for assembly of eukaryotic 60S ribosomal subunit domains. *Genes Dev* 28(2):198–210.
18. Pöll G, et al. (2009) rRNA maturation in yeast cells depleted of large ribosomal subunit proteins. *PLoS One* 4(12):e8249.
19. Ramesh M, Woolford JL, Jr (2016) Eukaryote-specific rRNA expansion segments function in ribosome biogenesis. *RNA* 22(8):1153–1162.
20. García-Ortega L, Alvarez-García E, Gavilanes JG, Martínez-del-Pozo A, Joseph S (2010) Cleavage of the sarcin-ricin loop of 23S rRNA differentially affects EF-G and EF-Tu binding. *Nucleic Acids Res* 38(12):4108–4119.
21. Mailliot J, et al. (2016) Crystal structures of the uL3 mutant ribosome: Illustration of the importance of ribosomal proteins for translation efficiency. *J Mol Biol* 428(10 Pt B): 2195–2202.
22. Nilsson OB, et al. (2015) Cotranslational protein folding inside the ribosome exit tunnel. *Cell Reports* 12(10):1533–1540.
23. Ramesh M, Woolford JL, Jr (2016) Eukaryote-specific rRNA expansion segments function in ribosome biogenesis. RNA 22(8):1153–1162.
24. Michaeli S, Agabian N (1990) A Trypanosoma brucei small RNP particle containing the 5S rRNA. *Mol Biochem Parasitol* 41(1):7–15.
25. Leidig C, et al. (2014) 60S ribosome biogenesis requires rotation of the 5S ribonucleoprotein particle. *Nat Commun* 5:3491.
26. Wu S, et al. (2016) Diverse roles of assembly factors revealed by structures of late nuclear pre-60S ribosomes. *Nature* 534(7605):133–137.
27. Michaeli S (2011) rRNA biogenesis in Trypanosomes. *RNA Metabolism in Trypanosomes (Nucleic Acids and Molecular Biology)*, ed Bindereif A (Springer, Berlin), Vol 28, pp 123–148.
28. Gamalinda M, Woolford JL, Jr (2015) Paradigms of ribosome synthesis: Lessons learned from ribosomal proteins. *Translation (Austin)* 3(1):e975018.
29. Khatter H, et al. (2014) Purification, characterization and crystallization of the human 80S ribosome. *Nucleic Acids Res* 42(6):e49.
30. Li X, et al. (2013) Electron counting and beam-induced motion correction enable near-atomic-resolution single-particle cryo-EM. *Nat Methods* 10(6):584–590.
31. Langlois R, et al. (2014) Automated particle picking for low-contrast macromolecules in cryo-electron microscopy. *J Struct Biol* 186(1):1–7.
32. Frank J, et al. (1996) SPIDER and WEB: Processing and visualization of images in 3D electron microscopy and related fields. *J Struct Biol* 116(1):190–199.
33. Mindell JA, Grigorieff N (2003) Accurate determination of local defocus and specimen tilt in electron microscopy. *J Struct Biol* 142(3):334–347.
34. Tang G, et al. (2007) EMAN2: An extensible image processing suite for electron microscopy. *J Struct Biol* 157(1):38–46.
35. Scheres SH (2015) Semi-automated selection of cryo-EM particles in RELION-1.3. *J Struct Biol* 189(2):114–122.
36. Li W, et al. (2015) Activation of GTP hydrolysis in mRNA-tRNA translocation by elongation factor G. *Sci Adv* 1(4):e1500169.
37. Brown A, et al. (2014) Structure of the large ribosomal subunit from human mitochondria. *Science* 346(6210):718–722.
38. Liu Z, Guo F, Wang F, Li TC, Jiang W (2016) 2.9 Å resolution Cryo-EM 3D reconstruction of close-packed virus particles. *Structure* 24(2):319–328.
39. Jiang W, Guo F, Liu Z (2012) A graph theory method for determination of cryo-EM image focuses. *J Struct Biol* 180(2):343–351.
40. Henderson R, et al. (2012) Outcome of the first electron microscopy validation task force meeting. *Structure* 20(2):205–214.
41. Kucukelbir A, Sigworth FJ, Tagare HD (2014) Quantifying the local resolution of cryo-EM density maps. *Nat Methods* 11(1):63–65.
42. Brown A, et al. (2015) Tools for macromolecular model building and refinement into electron cryo-microscopy reconstructions. *Acta Crystallogr D Biol Crystallogr* 71(Pt 1): 136–153.
43. Pettersen EF, et al. (2004) UCSF chimera: A visualization system for exploratory research and analysis. *J Comput Chem* 25(13):1605–1612.
44. Emsley P, Cowtan K (2004) Coot: Model-building tools for molecular graphics. *Acta Crystallogr D Biol Crystallogr* 60(Pt 12 Pt 1):2126–2132.
45. Adams PD, et al. (2002) PHENIX: Building new software for automated crystallographic structure determination. *Acta Crystallogr D Biol Crystallogr* 58(Pt 11):1948–1954.

THE JOURNAL OF PHYSICAL CHEMISTRY B

Article

pubs.acs.org/JPCB

Quantitative Characterization of Domain Motions in Molecular Machines

Suvrajit Maji,[†] Rezvan Shahoei,[‡,§,∥] Klaus Schulten,*[,‡,§,∥,∇] and Joachim Frank*[,†,⊥,#]

[†]Department of Biochemistry and Molecular Biophysics, Columbia University, New York, New York 10027, United States

[‡]Department of Physics, University of Illinois at Urbana−Champaign, Champaign, Illinois 61801, United States

[§]Center for the Physics of Living Cells, Department of Physics, University of Illinois at Urbana−Champaign, Champaign, Illinois 61801, United States

[∥]Beckman Institute for Advanced Science and Technology, University of Illinois at Urbana−Champaign, Champaign, Illinois 61801, United States

[⊥]Howard Hughes Medical Institute, Columbia University, New York, New York 10027, United States

[#]Department of Biological Sciences, Columbia University, New York, New York 10027, United States

Supporting Information

ABSTRACT: The work of molecular machines such as the ribosome is accompanied by conformational changes, often characterized by relative motions of their domains. The method we have developed seeks to quantify these motions in a general way, facilitating comparisons of results obtained by different researchers. Typically there are multiple snapshots of a structure in the form of pdb coordinates resulting from flexible fitting of low-resolution density maps, from X-ray crystallography, or from molecular dynamics simulation trajectories. Our objective is to characterize the motion of each domain as a coordinate transformation using moments of inertia tensor, a method we developed earlier. What has been missing until now are ancillary tools that make this task practical, general, and biologically informative. We have provided a comprehensive solution to this task with a set of tools implemented on the VMD platform. These tools address the need for reproducible segmentation of domains, and provide a generalized description of their motions using principal axes of inertia. Although this methodology has been specifically developed for studying ribosome motion, it is applicable to any molecular machine.

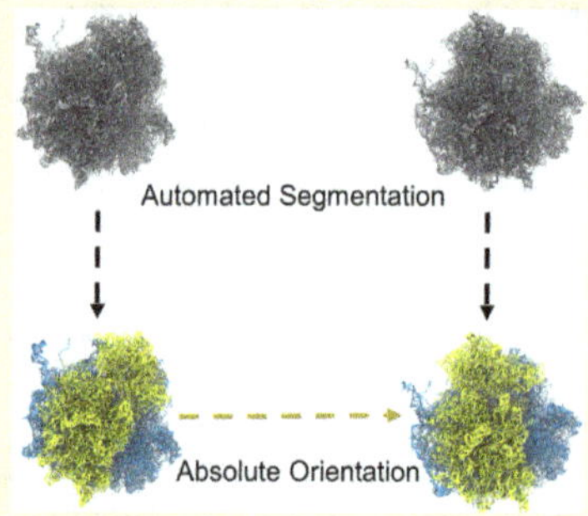

■ INTRODUCTION

To analyze the mechanism underlying a biophysical process it is often necessary to characterize the dynamics of the macromolecular machine responsible for this process, using simple and biologically meaningful motion parameters. Our study was motivated by the need for a common framework for describing domain motions in the ribosome during the process of mRNA-tRNA translocation, but the problem we are addressing is quite general and of increasing importance as structural knowledge on molecular machines in various states rapidly accumulates. For this reason we will state the problem and the approach we have taken in a general way, and then proceed to exemplify it with the ribosome.

The structures of macromolecules to be characterized may arise from a variety of sources: X-ray crystallography, cryo-EM, or molecular dynamics (MD) simulations. They are represented in the form of atomic coordinates and are usually stored as pdb files for individual models derived from experiments, or dcd files for MD trajectories. In the case of cryo-EM the coordinates may be obtained by flexible fitting[1,2] of existing structures, or, resolution permitting, by *de novo* modeling.[3] We opted to use VMD,[4] a popular software package for visualization and analysis of biomolecular systems, as a platform to develop a toolset to analyze the domain motions of molecular machines.

There are numerous studies in the literature describing motions in molecular machines, for instance ribosomes,[5−19] RNA polymerase,[20−22] ATP-synthase,[23,24] and GroEL,[25,26] but only few of these make use of a general method or tool.[14−19] In this work we sought to address some shortcomings of existing methods, such as the need for a domain segmentation tool that makes allowance for inconsistencies in the residues, and the lack of an integrated platform for segmentation, computation, and display of coordinate transformations.

First, in order to quantify domain motions, it is necessary to properly define the domains in the molecule that are structurally and functionally distinct, i.e., domains that at least in approximation behave like rigid bodies. Given the lack of tools in VMD to accomplish this task in a straightforward way, we have

Special Issue: Klaus Schulten Memorial Issue

Received: October 25, 2016
Revised: February 14, 2017
Published: February 15, 2017

 © 2017 American Chemical Society

DOI: 10.1021/acs.jpcb.6b10732
J. Phys. Chem. B 2017, 121, 3747−3756

developed an automated segmentation method on the VMD platform for extracting the major domains from any macromolecular structure whose atomic coordinates are given. The challenge posed by the comparison of two realizations of a domain in two different structures has been noted by Whitford et al.:[15] The coordinate transformation for a given residue is usually not representative for other residues in the domain. Rather, positions of residues are subject to random local fluctuations but also additional changes that are functionally relevant. For expediency, these local fluctuations are often ignored, and the motion is approximated by a collective rigid-body motion with a set of global parameters. These authors developed an approach where global motion of a domain from one structure to the other is calculated from the average of the local displacements between corresponding atoms in the two structures. Compared to the problem Whitford et al.[15] addressed—the comparison of structures along MD trajectories—experimentally derived structures often present an additional problem because of gaps in coverage, as residues in one structure may not be all matched to those in the other. Here, we developed a segmentation approach that disregards such inconsistencies, and expanded a general method to describe domain position based on inertia tensor with principal axes, and further, to describe a domain motion by a transformation between one set of axes with another in a global way. This method was introduced earlier by us to characterize the domain motion of the ribosome[27] and later on applied in two additional studies of its dynamics.[28,29]

We aim to describe the macromolecular domain motion in a compact and precise manner. Typically, a rigid body motion is measured with the aid of a set of three Euler angles obtained from the transformation or rotation matrices. The problem with such representation is that different orders in a sequence of rotations will generate different sets of Euler angles and so the reported results may be inconsistent and not comparable, although one can circumvent this problem by using the angle-axis representation derived from the rotation matrix.[12,14] We propose an alternative approach for a unique representation and characterization of the macromolecular rotations, by using unit quaternions. Some of the interesting features of unit quaternions for 3D rotations are smooth interpolation between two quaternions and smaller round-off errors with unit normalization.

■ THE APPROACH

A molecule may exist in a series of conformations, such that each of its domains (modeled in approximation as rigid bodies) undergoes a series of coordinate transformations. We sought to develop a toolset aimed at defining the domains and describing these transformations. For the purpose of this exposition, it is sufficient to consider the case of a molecule in two states, A and B, whose domains undergo coordinate transformations in the process of going from A to B.

The following description is divided into three sections. One deals with the practical problem of defining the domains in a structure given by its atomic coordinates. The second sketches out the method of determining the principal axes of the inertia tensor for a domain thus defined. The third describes the way to use these systems of principal axes to track the motion of the domain in 3D.

1. Domain Segmentation—Overall Strategy. Our aim has been to develop a general method for segmentation of structural models represented by pdb coordinates that will not depend on the type of structures and residues (protein or RNA), nor, to some extent, on the presence of gaps or other inconsistencies as we compare the structures encountered in conformation A or B. Our approach to this problem is to perform the segmentation not on the atomic structure, but on a 3D density map generated from that structure. This step renders the segmentation virtually insensitive to minor inconsistencies in residue counts and relative gaps. The boundaries of the resulting volume segments are then used as masks for extracting the enclosed residues for the corresponding domain.

1.1. Automated 3D Volume Segmentation. For segmentation, we use a method known as the Watershed Algorithm which is widely used in computer vision for segmenting 2D images and 3D volumetric images (or density maps). We will first briefly describe the basic concept of the Watershed Algorithm[30−32] and then summarize the segmentation method making use of it.[33]

Watershed Algorithm. The algorithm we use in the implementation of our segmentation method is part of the Matlab toolbox (see Text S8), so we will only briefly mention the general working principle.[33] Its schematic for the 1D case is illustrated in the Supporting Information (Figure S1), but the underlying idea remains the same for a 2D image or a 3D density map. A density map is treated as a topographical surface with peaks and valleys, or "catchment basins" where the elevation level is given by the density values in each voxel. There are two ways to perform Watershed segmentation. One is the top-down approach, where the water flow is simulated from regions of higher elevation, or peaks, to the lower level, or valleys, following the path of steepest descent. The other way, which we have used, is the bottom-up approach where we first find the local minima or catchment basins. Each of the local minima is treated as an independent water source which floods the associated catchment basin up to the point that it spills into the neighboring basin. This level defines the boundary between the two basins, and is considered as the boundary between two segments. Instead of the original density maps, we have applied the Watershed Algorithm to the Distance Transform[30,32,34] of the binary map (i.e., the thresholded density map in Section 1.2). The idea to apply the Watershed method to the Distance Transform of the binary map is that the peaks (regional maxima) of the Distance Transform map provide the initial markers for the foreground objects of interest (domains in our case) we wish to identify.

Distance Transform. The Distance Transform (see general description provided in Text S1) of a density map is defined as the distance of every voxel X from its nearest nonzero valued voxel, and is given by

$$D_T(X) = \min\{d(X, Y) | Y \in O, X \in O \cup O^c\} \quad (1.1)$$

where O is the set of object voxels, and O^c is the set of background voxels. The distance function is given by

$$d(X, Y) = \sum_{k=1}^{3} |x_k - y_k| \quad (1.2)$$

where $X = (x_1, x_2, x_3)$, $Y = (y_1, y_2, y_3)$ are voxel coordinates.

To obtain a proper initial segmentation one can try different distance functions, combined with a proper region merging method, as will be sketched out below.

Region Merging. Application of the Watershed Algorithm to real density maps often produces severe oversegmentation due to the existence of numerous regional minima. Therefore, it is necessary to take additional steps either in preprocessing (by smoothening the map) or postprocessing, or both, to produce

DOI: 10.1021/acs.jpcb.6b10732
J. Phys. Chem. B 2017, 121, 3747−3756

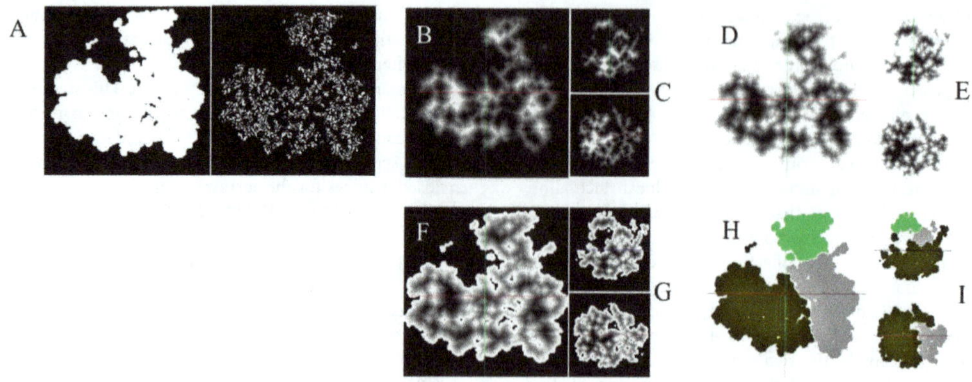

Figure 1. Density map segmentation. (A) Binary maps, generated by application of a threshold to the density map (central *Z*-axis slice shown); threshold = 0 (left) is appropriate while threshold = 0.5 (right) is unsuited for segmentation. (B) Distance Transform (D_T) of the complementary binary map ~ M_{bin} (central *Z*-axis slice shown). (C) Same map as in (B) but showing central *X*-axis slice (top) and *Y*-axis slice (bottom). (D) Inverted distance transform map $D_T := -D_T$. (E) Same map as in (D) but showing central *X*-axis slice (top) and *Y*-axis slice (bottom). (F) Distance Transform map with voxels corresponding to the background voxels of M_{bin}, labeled as−infinity, so that the background voxels becomes inaccessible. (G) Same map as in (F) but showing central *X*-axis slice (top) and *Y*-axis slice (bottom). (H) Segmentation of the volume maps after applying the Watershed Algorithm with region merging. (I) Same map as in (F) but showing central *X*-axis slice (top) and *Y*-axis slice (bottom).

fewer and more meaningful segments using region merging methods.[35−41] The basic approach is to use some type of region similarity or dissimilarity metric in a statistical sense based on the intensity or contour variation and then merge the neighboring regions, for example using a region adjacency graph, in a hierarchical manner. As we go higher in the hierarchical order (merging level), more regions are merged with one another and we obtain fewer oversegmented regions.

1.2. Process of Segmentation. We first convert the input pdb structure into a 3D density map and then apply a filter to obtain a smooth map. Next we apply a threshold on the smooth map to obtain a binary map and verify the binary map for proper thresholding (Figure 1A). Next we convert the binary map into a distance transform map (Figure 1B) and, subsequently, an inverted distance transform map (Figure 1D,F) is obtained by inverting the sign of the density values. Finally, we employ the Watershed Algorithm on the distance transform map obtained from the last step and also apply a region merging step as discussed above, to reduce the oversegmentation. Once we have the proper density map volume segments, we can segment out the corresponding pdb structure enclosed within the individual volume segments (Figure 2).

All steps of the segmentation method are outlined in Supporting Text S1.

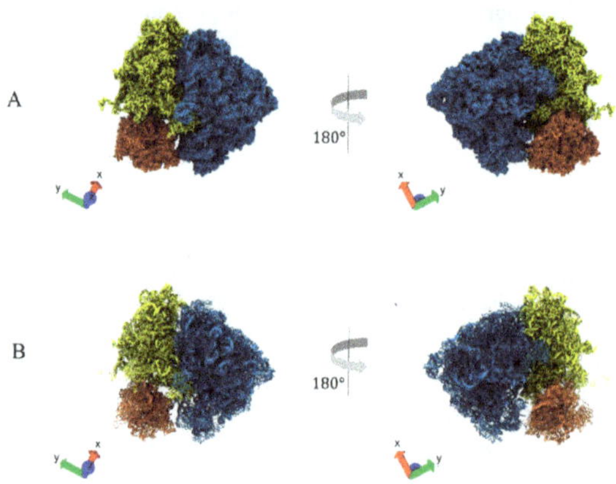

Figure 2. Segmentation of ribosome structure. (A) Segmentation of the 3D density map of a ribosome structure into its three relevant domains: large subunit (LSU) (blue), small subunit (SSU) body (yellow), and SSU head (orange). (B) Segmented atomic model of the ribosome. The segments were obtained by extracting the residues contained within the individual volume segments obtained from (A). VMD axes X(red), Y(green), Z(blue) are shown in the bottom left inset of each panel.

2. Inertia Tensor and Principal Axes. The original idea of using moment of inertia tensor and principal axes to characterize the motions of ribosomal domains was first introduced in a cryo-EM study, focused on the role of L1 stalk and tRNA interaction, whose aim was to analyze the dynamics of the translation elongation cycle.[27] We have used this method as part of our toolset as it is a general method to compute the principal axes for any macromolecular structure represented by a set of atomic coordinates. We have provided a brief recapitulation of the concepts of moment of inertia and principal axes in the Supporting Information (Text S2), since they form the basis of our approach.

2.1. Domain Motion Using Principal Axes. Once we have the individual domains segmented and selected, we can compute the inertia tensors and hence the principal axes for each individual domain, as outlined in the previous section. The task of characterizing the domain motions is now reduced to one of quantifying the transformation (i.e., rotation, translation) of the principal axes coordinate system of a particular domain from one state to the other. We can measure the domain rotation in different ways as implemented in the toolset (Text S6). In addition to the conventional Euler angles, another method we have employed here to describe the domain motions is to compute the absolute orientation using unit quaternion which characterizes the rotation unambiguously between two coordinate systems with one unique rotation axis and the corresponding rotation angle. The details are illustrated in Section 3.

2.2. Example. In the following we go through the basic steps for setting up the descriptor using principal axes for domain motions, using the ribosome as an example.

DOI: 10.1021/acs.jpcb.6b10732
J. Phys. Chem. B 2017, 121, 3747−3756

The Journal of Physical Chemistry B Article

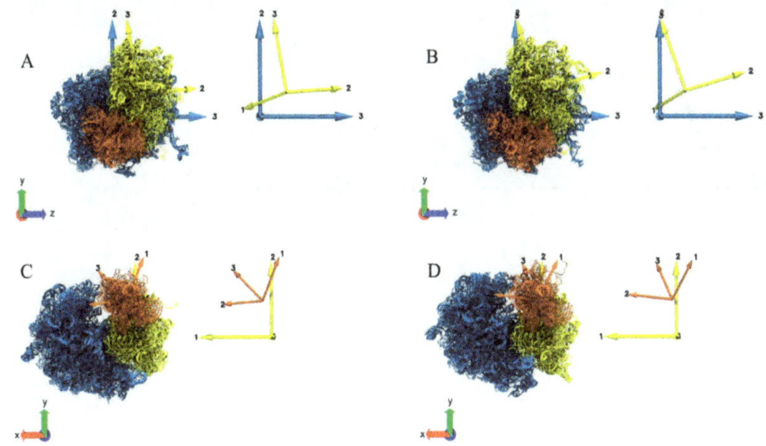

Figure 3. Principal axes computation for the segmented domains (Figure 2) of the ribosome structures in two states, "A" and "B". The principal axes for individual domains are shown with the same color as the corresponding domain. The inset on the right of each panel shows the principal axes only. (A) Principal axes of LSU (blue) and SSU body (yellow) for the structure in state "A". (B) Principal axes of LSU (blue) and SSU body (yellow) for the structure in state "B". (C) Principal axes of SSU body (yellow) and SSU head (orange) for the structure in state "A". (D) Principal axes of SSU body (yellow) and SSU head (orange) for the structure in state "B". For panels (A) and (B), the LSU is fixed and aligned to the Cartesian VMD coordinate system axes XYZ. For panels (C) and (D), the SSU body is fixed and aligned to the Cartesian coordinate system axes XYZ.

a. For calculating the domain motion of the small subunit (SSU) with respect to the large subunit (LSU) going from State A to State B, we align the principal axes of the LSU of structure #1 with the Cartesian coordinate system XYZ (Figure 3). First the principal axis 3 is aligned with the Z axis, then the principal axis 2 is aligned with the Y axis and, since both XYZ and the principal axes form orthogonal coordinate systems, the principal axis 1 would be already aligned to the X axis at this stage. The principal axes of the LSU for the subsequent structures are also aligned to XYZ and consequently the LSU of both structures are aligned to one another. Then we compute the transformation of the principal axes of the SSU body from State A to B.
b. For computing the domain motion of the SSU head, we align the principal axes of the SSU body to XYZ in the same way as mentioned above (a) and calculate the rotation for the head domain.

There is an option to either use all atoms for computing the inertia tensors for all the domains or just a subset of particular atoms such as phosphorus. Using all atoms takes typically several minutes on a single processor per ribosome model compared to using just phosphorus atoms, which just takes several seconds, at the expense of accuracy.

3. Characterization of Domain Motion with Absolute Orientation. Once we have the principal axes for each domain, the transformation between the principal axes coordinate system for a particular domain from one state to the next (Figure 4A,B) will determine the unique rotation (Figure 4C) for that domain with a single rotation axis and rotation angle (Figure 4D). A similar idea of characterizing the molecular motion was presented in a recent method[12,14] which uses the Euler–Rodrigues (E–R) formula to calculate the interdomain rotation relative to a reference model. This method provides a single E–R rotation axis and rotation angle for one pair of molecules at a time. However, these authors measure the rotation by computing the transformation matrix for the whole domain, whereas we are dealing with the transformation of the domain principal axes only.

The problem of describing the transformation of coordinate axes and, in general, the least-squares fitting of two sets of points has several closed-form solutions.[42−44] We have chosen to use the more classical approach for absolute orientation algorithm proposed by Horn.[42] In this approach, a closed-form solution to the least-squares problem of absolute orientation is determined using unit quaternions. Quaternions have been previously used in biomolecular research and structural biophysics for molecular modeling.[45−48] The basics of quaternions and the least-squares problem of absolute orientation using unit quaternion are briefly outlined in the Supporting Information (Text S3 and S4, respectively). Here we merely present the implementation steps, in Section 3.1 below.

3.1. Absolute Orientation Algorithm. Suppose we have the transformation from coordinate system 1 to coordinate system 2 and the set of points in the two systems be $\{\boldsymbol{r}_{1,i}\}_{i=1\cdots n}$ and $\{\boldsymbol{r}_{2,i}\}_{i=1\cdots n}$, respectively. Then we can summarize the absolute orientation algorithm for coordinate axes transformation in simple implementable steps as follows.

a. Compute the centroids of the input coordinate systems 1 and 2, as shown in Supporting equation S4.4

$$\bar{r}_1 = \frac{1}{n}\sum_{i=1}^{n} \boldsymbol{r}_{1,i},\ \bar{r}_2 = \frac{1}{n}\sum_{i=1}^{n} \boldsymbol{r}_{2,i}$$

b. Next we subtract the centroids from the respective coordinate systems so that we only deal with measurements relative to the centroids (eq S4.5).

$$r'_{1,i} = r_{1,i} - \bar{r}_1,\ r'_{2,i} = r_{2,i} - \bar{r}_2$$

c. Then we compute the matrix $\boldsymbol{M}$ (S4.25) .

$$\boldsymbol{M} = \sum_{i=1}^{n} r'_{1,i} {r'}^{T}_{2,i}$$

d. Having the matrix $\boldsymbol{M}$, we can use its element values to assign the elements S_{xx}, S_{xy}, S_{xz} and so on (eq S4.26).

DOI: 10.1021/acs.jpcb.6b10732
J. Phys. Chem. B 2017, 121, 3747−3756

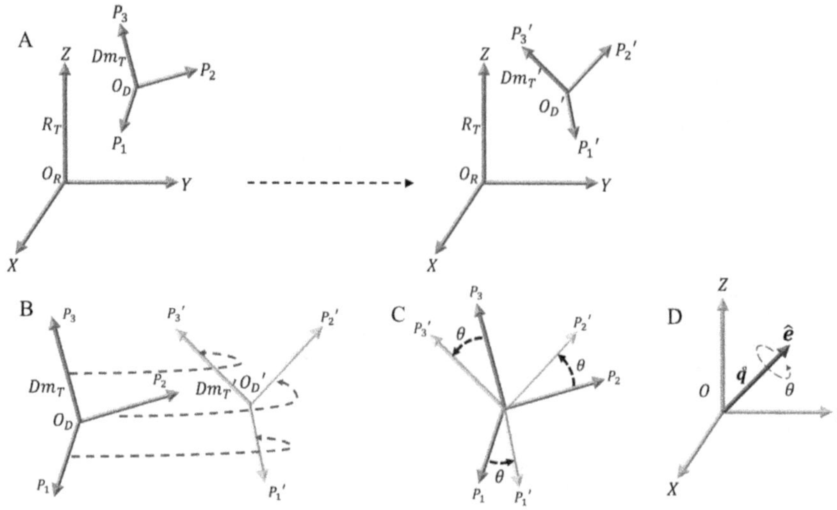

Figure 4. Coordinate axes transformation using Absolute Orientation. (A) Fixed reference frame denoted by the principal axes of inertia tensor R_T. The domain motion relative to the fixed reference frame is denoted by the principal axes of inertia tensors Dm_T and Dm'_T on the left-hand side (lhs) and right-hand side (rhs), respectively. The principal axes P_1, P_2, and P_3 (lhs) move to P'_1, P'_2, and P'_3(rhs), respectively. (B) Schematic illustrating the Absolute Orientation problem for coordinate axes transformation between Dm_T and Dm'_T. There is a unique rotation and also a translation (given by the difference between the corresponding centers $O_D - O'_D$) to obtain the complete coordinate axes transformation. (C) Special case where the origins for Dm_T and Dm'_T coincide and the coordinate axes transformation is achieved by just rotating Dm_T by an angle θ about an axis perpendicular to the plane of rotation, to be transformed into D'_T. (D) Angle-axis representation for the unit quaternion $\mathring{q}$. The solution to the rotation for the coordinate axes transformation in B or C is given by $\mathring{q}$. The axis of rotation is given by $\hat{\mathbf{e}}$ and the angle of rotation is given by θ.

e. Using the matrix elements from step (d), we compute the matrix $\mathbf{N}$ (eq S4.27).

f. Finally, we perform eigen-decomposition of $\mathbf{N}$ and determine the eigenvector $\boldsymbol{e}_{max}$ with the largest positive eigenvalue. The unit quaternion $\mathring{q}$ (representing the rotation) in the direction of $\boldsymbol{e}_{max}$ is the required solution to the absolute orientation problem.

3.2. Component of Rotation Around a Specified Axis. The unit quaternion (Section 3.1) along with a unique rotation axis and a rotation angle (Text S3), provides us with precise information about the domain's rigid-body rotation. However, in certain situations, it may be desirable to obtain the component of the rotation around a specified axis $\vec{v}$ that is different from the true rotation axis. This can be achieved through the decomposition[49] (see Text S5 for detailed description) of the unit quaternion $\mathring{q}$ into two components, $\mathring{q}_s$ and $\mathring{q}_v$, such that $\mathring{q} = \mathring{q}_s\mathring{q}_t$.

Let the unit quaternion $\mathring{q} = (q_w, q_x, q_y, q_z)$ represent the complete rotation and the specified axis be the vector $\vec{v} = (v_x, v_y, v_z)^T$. The rotation axis component of $\mathring{q}$ is $\vec{r}_q = (q_x, q_y, q_z)^T$. The projection of the rotation axis $\vec{r}_q$ along the vector $\vec{v}$ is given by

$$\vec{P}_{r_q v} = \frac{(\vec{r}_q \cdot \vec{v})\vec{v}}{\|\vec{v}\|^2} = (\vec{r}_q \cdot \hat{v})\hat{v} \tag{3.1}$$

where $\hat{v} = \frac{\vec{v}}{\|v\|}$ is a unit vector along $\vec{v}$.

Using the projection vector $\vec{P}_{r_q v}$ and the scalar part of $\mathring{q}$, we construct (Text S5, equation S5.3) the unit quaternion $\mathring{q}_t$ as

$$\mathring{q}_t = \frac{1}{\sqrt{{q_w}^2 + \|\vec{r}_q \cdot \hat{v}\|^2}}(q_w, (\vec{r}_q \cdot \hat{v})\hat{v}) \tag{3.2}$$

Therefore, $\mathring{q}_v$ as defined above, is a unit quaternion and represents the component of the rotation around the specified vector $\vec{v}$. We can determine the other component as $\mathring{q}_s = \mathring{q}\mathring{q}_t^{-1}$.

■ RESULTS AND DISCUSSION

We have applied our method to a cryo-EM data set from a sample of translating eukaryotic ribosomes. The details of the results and detailed biological implications will be provided in a separate research article (Shahoei, Maji et al., in preparation). Here we aim to illustrate the application of our toolset to compute the domain rotation between two structures representing the ribosome in two different states (labeled as A(class2.pdb) and B(class5.pdb); see movies MS1, MS2, MS3 in Supporting Information).

Segmentation. We demonstrate the segmentation method (Section 1.2) with an example of the ribosome structure. To begin with, we generated a simulated density map (M_{vol}) from the ribosome pdb structure and applied a Gaussian filter to the density map in order to obtain a smooth map (M_f) (see description of the filter parameter used in Text S8). Next, we converted the smooth map into a binary map using thresholding (Section 1.2 step (c)). Since the initial density map does not contain any background noise, we chose a threshold of zero to obtain the binary map (M_{bin}). Through visual inspection of the 3D volume, we can confirm that the binary map is contiguous and the thresholding was performed properly. Here we just show the 3D binary map with representative central slice along Z-axis (Figure 1A left). In contrast, a choice of higher threshold produced a severely fragmented binary map (central Z-axis slice shown in Figure 1A right).

In the next step, we converted the complement of the binary map into a Distance Transform map (D_T) as shown with a

DOI: 10.1021/acs.jpcb.6b10732
J. Phys. Chem. B 2017, 121, 3747–3756

representative central volume slice along the Z-axis (Figure 1B). The Distance Transform map makes it simpler, compared to the original density map, to find the regional minima or catchment basins. We also show central slices along X-axis (Figure 1C top) and Y-axis (Figure 1C bottom). Next, we computed the inverted Distance Transform map ($-D_T$) (representative central Z-axis slice, Figure 1D) and the corresponding two other orthogonal slices (Figure 1E). Then we labeled the voxels belonging to the set of background voxels in the original binary map (M_{bin}) as inaccessible (Section 1.2 step (f)) in the inverted Distance Transform map obtained in the previous step. The resulting Distance Transform map is shown with representative orthogonal central slices (Figure 1F,G).

Next, we applied the Watershed Algorithm (Section 1.2 step (g) and (h)) to the final Distance Transform map obtained in the last step. Using our segmentation approach on this map, together with the specific set of parameter values (Text S8), the Watershed method initially produced 396 segments before the application of region merging (i.e., merging level is set to 0). The maximum merging size was set to 50^3. Higher merging levels produced less oversegmentation and we settled at a merging level that generated three volume segments corresponding to the three major domains in the ribosome. We can inspect representative orthogonal slices of the segmented density map (Figure 1H,I) to determine if the segmentation was performed properly. The map (Figure 2A) shows the three relevant domains, large subunit (blue), small subunit body (yellow), and small subunit head (orange). Finally, using the individual segments as mask for extracting the residues from the whole ribosome structure, we obtained the segmented pdb structure (Figure 2B).

Inertia Tensor and Principal Axes Calculation. After we had the relevant domains segmented out, we proceeded to set up the platform for characterizing the motion of the relevant domains of the ribosome. For each of the domains of interest (approximated as a rigid body) we computed the inertia tensor and principal axes (Section 2 and Text S2) using all atoms. As described in the Supporting Information (Text S2), we can observe from the example of ribosome structures (Figures 3, S2C) how the distribution of mass of a rigid body determines the order of the three principal axes. From the eigen-decomposition of the inertia tensor matrix for any domain of the ribosome structure, we obtained the eigenvalues (principal moments), I_{11}, I_{22}, I_{33} which are ordered as $I_{11} > I_{22} > I_{33}$ (numerical data not shown). This implies that the distribution of mass should be most elongated along principal axis 3, the eigenvector corresponding to the smallest moment I_{33} and the least elongated along principal axis 1. We see that for each of the domains (LSU, SSU, SSU body, etc.) (Figure 3, Figure S2C), the distribution of mass is indeed most elongated along principal axes 3, followed by principal axis 2, and the least elongated along principal axis 1. For any ribosome domain approximated as a rigid body, the order of the three principal axes and their orientation should be consistent at all points during the motion. By consistent orientation, we mean that if the rigid body has only translation and no rotation moving from state A to B, then the pair of principal axes at A and B should align with each other only by translation.

For the purpose of demonstrating the characterization of domain motion with principal axes, we measured two types of motion for the pair of structures A and B—(a) intersubunit motion, which is motion of the entire SSU or SSU body (target) relative to the LSU (reference), and (b) head rotation, which is motion of the SSU head (target) relative to the SSU body (reference).

For measuring the intersubunit rotation we first computed the principal axes of the reference LSU of structure A (R_T in Figure 4A left) and aligned it to the Cartesian coordinate axes XYZ (see Section 2.2 for the detailed alignment procedure). Then we computed the principal axes of the SSU body of A (Dm_T in Figure 4A, Figure 3A). After that, we computed and aligned the principal axes of the LSU of structure B (R_T in Figure 4A right) with XYZ and, as a result, the principal axes of the LSU of both structures A and B were aligned to each other. Next, we computed the principal axes of the SSU body of B (Dm'_T in Figure 4A, Figure 3B). The coordinates of the principal axes of the SSU body for both A and B are now expressed relative to the fixed coordinate system (principal axes) of the LSU of A and B.

For the head rotation we followed the same alignment procedure for the principal axes of the SSU body as in case above with LSU as the reference. In this case, the coordinates of the principal axes of the SSU head for A and B are now expressed relative to the fixed coordinate system of the SSU body of A and B (Figure 3C,D).

Measuring the rotation of a domain (SSU, SSU body, SSU head) at this stage is equivalent to tracking the associated principal axes from one state to the other (Figure 4B). The rotation of principal axes is estimated by solving the least-squares problem of absolute orientation, i.e., by calculating the transformation of the coordinate axes between state A and B (Figure 4C) using unit quaternions. The angle-axis representation (Figure 4D) of the unique quaternion gives us the rotation axis and the corresponding rotation angle.

Coordinate Axes Transformation/Absolute Orientation. We demonstrate the working principle of the Absolute Orientation algorithm (Section 3.1) using a simple example in the Supporting Information (Text S9). Here we show its application by calculating the transformation of the principal axes computed for SSU body and SSU head from state A (class2) to B (class5) and the rotation measurements are tabulated below (Table 1). The rotation of SSU body from state A (Figure 3A) to B (Figure 3B) (movie MS1) is characterized by the unit quaternion $\mathring{q}_{body}$ and the corresponding rotation angle $\boldsymbol{\theta}_{body}$ around the native rotation axis given by the unit vector $\hat{\mathbf{e}}_{body}$.

Table 1. Unit Quaternion $\mathring{q}$ and the Corresponding Axis–Angle Estimates Obtained by Solving the Absolute Orientation Problem for the Principal Axes Computed for the Domains (SSU Head, SSU Body, Full SSU, and the Reference SSU) in the Two States A and B

domain		$q^{\circ} = (q_w, q_x, q_y, q_z)$			
SSU head	$\mathring{q}_{head}$	0.989429	−0.007936	−0.100939	0.103817
SSU body	$\mathring{q}_{body}$	0.992263	−0.091447	−0.073468	0.040654
SSU	$\mathring{q}_{SSU}$	0.996006	−0.085232	−0.007273	0.025572
SSU ref	$\mathring{q}_{SSU-ref}$	0.995708	−0.089049	−0.005387	0.024636
	$\hat{\mathbf{e}} = (e_x, e_y, e_z)$				$\boldsymbol{\theta}$ (degree)
SSU head	$\hat{\mathbf{e}}_{head}$	−0.054727	−0.696057	0.715898	16.68
SSU body	$\hat{\mathbf{e}}_{body}$	−0.736592	−0.591778	0.327462	14.26
SSU	$\hat{\mathbf{e}}_{SSU}$	−0.9554637	−0.081456	0.286415	10.24
SSU ref	$\hat{\mathbf{e}}_{SSU-ref}$	−0.962162	−0.058203	0.266189	10.62

We calculated the rotation of SSU head, full SSU, and reference SSU from state A (Figure 3C) to B (Figure 3D)

DOI: 10.1021/acs.jpcb.6b10732
J. Phys. Chem. B 2017, 121, 3747–3756

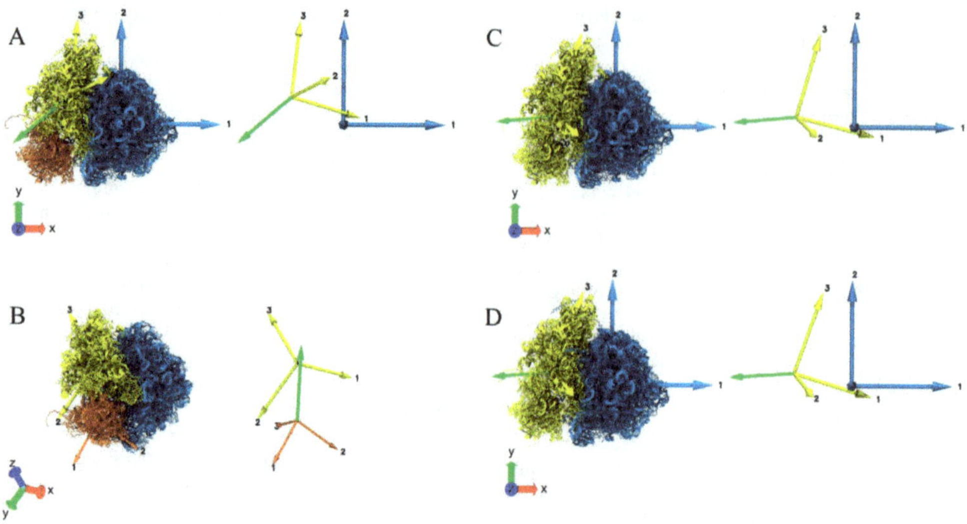

Figure 5. Characterization of domain motion between two structures "A" and "B" (Figure 3) using the Absolute Orientation algorithm for the principal axes transformation from "A" to "B". The resulting unit quaternion describing the rotation is expressed in its axis-angle representation. The principal axes for individual domains are shown with the same color as the corresponding domain, and the rotation axis is shown in green. The inset on the right of each panel shows the principal axes and the rotation axis. (A) Rotation of SSU body (yellow) from state "A" to "B". The LSU remains fixed here. The green arrow represents the axis of rotation of the SSU body between states "A" and "B" and is drawn overlaid on model "B". (B) Rotation of SSU head (orange) from state "A" to "B". The SSU body remains fixed. The green arrow represents the axis of rotation of the SSU head between the states "A" and "B" and is drawn overlaid on model "B". (C) Rotation for the full SSU (yellow) from state "A" to "B". The LSU remains fixed. The green arrow represents the axis of rotation of the SSU body between states "A" and "B" and is drawn overlaid on model "B". (D) Rotation of the full SSU (yellow) from state "A" to "B". The LSU remains fixed here. The LSU and SSU, in this case, are obtained from a reference segmentation. The green arrow represents the axis of rotation of the SSU body between states "A" and "B" and is drawn overlaid on model "B".

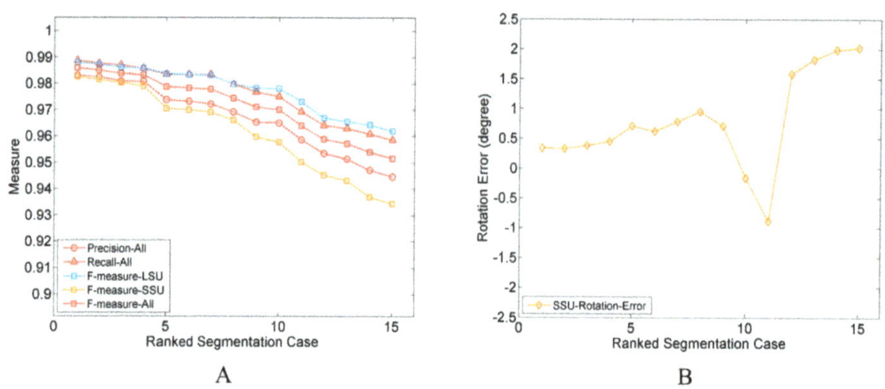

Figure 6. Characterizing the error in domain segmentation and domain rotation. (A) Segmentation quality measures, with Precision, Recall calculated for the full structure, and *F*−measure calculated for both the individual domains and the full structure, in all the segmentation cases listed in Table S1. The plotted values are ranked based on the of the SSU (Table S4). (B) Errors in rotation angle of the SSU for the segmentation cases and ranked in the same manner as in A.

(movies MS2, MS3) in the same way. The graphical visualization of the rotation axis for the SSU body, SSU head, full SSU (head and body combined), and the reference SSU is shown as a green arrow (Figure 5A,B,C,D). One can validate the above rotation axis and angle by visual inspection of the change in the two models (movies MS1, MS2, MS3).

The rotation axis not only tells us about the directionality (clockwise or anticlockwise sense) of the rotation, but also provides us with a way to categorize the rotation into different types, as described in the literature for macromolecular motion, depending on the alignment of the rotation axis and the principal axes of the reference molecule. For example, in our case (Figure 5C), the acute angles between the rotation axis $\hat{e}_{SSU}$ and the three principal axes, $(1, 0, 0)^T$, $(0, 1, 0)^T$, $(0, 0, 1)^T$ of the fixed LSU are 17.3, 85.3, and 73.4°, respectively. We can see that the rotation axis $\hat{e}_{SSU}$ is more closely aligned with the principal axis 1 of LSU compared to the other two axes. Since principal axis 1 is approximately perpendicular to the intersubunit interface (Figure 5C), we can refer to this SSU rotation as "intersubunit rotation". If the rotation axis $\hat{e}_{SSU}$ was closely aligned with

DOI: 10.1021/acs.jpcb.6b10732
J. Phys. Chem. B 2017, 121, 3747−3756

principal axis 2 of LSU, which is approximately parallel to the principal axis 3 of SSU, then the SSU rotation would be described as "rolling", and so on. Similar motions (head rotation, swiveling, etc.) can be described for the rotation of the SSU head relative to SSU body. In general, the true rotation axis may not be closely aligned to a prespecified axis of rotation and, in certain cases, it may be necessary to know the rotation around such chosen axis.

Given a unit quaternion representing the rotation of a domain with a unique rotation axis and rotation angle, we can obtain the component of the rotation around a specified axis using the decomposition of the unit quaternion (Section 3.2). As an illustration, we chose to find the component of SSU body rotation ($\mathring{q}_{\text{body}}$), around the principal axis 1 of LSU, i.e., $\vec{v} = (1, 0, 0)^T$. From the decomposition (eq 3.2) of $\mathring{q}_{\text{body}}$, we obtained the component $\mathring{q}_t = (0.995780, -0.091771, 0.0, 0.0)^T$. The corresponding rotation angle $\theta_t = 10.53°$ and the rotation axis is $\hat{e}_t = (-1, 0, 0)^T = -\vec{v}$; i.e, $\hat{e}_t$ is aligned with $\vec{v}$. These measurements will help us to validate our rotation estimates with those reported in the literature where the rotation is calculated around a manually chosen axis.

Error Analysis. Since the absolute orientation provides a closed-form solution to the problem of determining coordinate axes transformations, the major source of error in the measurement of the rotation is the uncertainty in the segmentation of the domains as it affects the subsequent computation of their principal axes. In order to characterize the magnitude of errors, we have investigated several instances of segmentations with small variations in the parameters, and in each instance evaluated the segmentation quality with regard to a reference structure using Precision, Recall, and *F*—measure, calculated based on the atoms included or excluded in the segmentation, (see Supporting Text S10). As there is no differentiation between SSU body versus head in the coordinates of the reference model, we only used the SSU and LSU domains in this error assessment. Since the segmented density map has no background noise, the error in the segmentation stems primarily from errors in assignment at the interface of the two subunits and the neighboring regions. The incorrectly segmented atoms, in the above-mentioned regions, constitute a very small fraction (Text S10, Table S3) of the total number of atoms in the two subunits. Consequently, the Precision, Recall, and *F*—measure values are quite high ~0.90–0.99 (Figures 6A, S6) for the two individual subunits and overall structure. The resulting error in the rotation angle of the SSU is lower than ±0.5°(Tables S1, S3; Figure 6B) for the highest-quality segmentation cases 1–4 (Table S1, S4) with small segmentation error (in terms of all three quality measures including Precision ≥ 0.97, for individual subunits and combined). Indeed, visual assessment (Figure S3) confirms that segmented domains for the cases 1–4 appear to be closest to the corresponding reference domains. However, one additional case (case 6) with low rotation angle error presents an anomaly as the segmentation error is much larger (Precision = 0.92) (Tables S3, S4; Figures S5, S6). A possible explanation for this discrepancy is that the error in the segmentation from different regions of the intersubunit interface (Figure S5) is being averaged out to a considerable extent due to the nature of principal axes calculation over a large number of atoms.

The rotation angles are close for cases 1–4, when we measured the SSU body and SSU head separately (Figure S6B, Table S2). As references for the SSU body and head domains are not available, the error in their respective rotation angles could not be calculated. Therefore, it was not possible to compare the error in the SSU body and head rotation angles with the error in rotation angles for the full SSU, which could have provided us with an estimate about the sensitivity (see Text S10, S10.1) of the SSU rotation measurements to segmentation error.

From our earlier discussion, we know that the error in segmentation is mainly at the periphery of the subunits. We should note that this is the worst-case scenario where the wrongly assigned atoms are farthest away from the axis and thus have larger contribution in the calculation of the principal axes. Consequently, the rotation measurements are more strongly affected, compared to contributions by the inner atoms.

As our calculations show, the computation of rotation angles is robust to small segmentation errors (Precision > 0.97) (Table S4), but the domain segmentation should be carefully verified prior to the principal axes calculations in order to accurately measure small rotations.

■ CONCLUSIONS

We have developed an integrated platform for segmentation and computation of coordinate system transformations to characterize the motion of individual domains of biomolecules. We have demonstrated the application of our toolset on a specific example of ribosome data. The characterization of the rotational motion of a domain using a single rotation axis and the associated rotation angle through the use of quaternions is straightforward and unambiguous. This choice of expressing motion circumvents the confusing proliferation of definitions and gimbal ambiguities that come with the choice of Eulerian angles.

As we have pointed out, there have been a variety of approaches to delineate domains and measure domain motions, but they do not utilize an integrated framework as we have aimed to do in our toolset. Unlike the previous methods for determining the domains directly from atomic models, our domain-segmentation method uses density maps derived from these models to obtain domain boundaries. Hence, it is insensitive to the nature of the residues, minor inconsistencies in residue counts, and gaps in the structure, rendering the method quite general. However, this advantage comes at a price. A shortcoming of our toolset is that the initial segmentation step—since the information on the underlying atomic structure is lost—entails a judgment on the validity of the segments, and may require an adjustment. Nevertheless, once the domain of interest is specified by either using our segmentation method or from another source, the subsequent determination of transformations is automated and straightforward.

It should also be noted that we have opted to use a structure-based domain decomposition approach here to characterize motion. An alternative approach is the use of motion-based domain decomposition methods which define the dynamics of domains by computing the local motion of atoms from an ensemble of (at least two) structures with different conformations.[50−52] A disadvantage of this approach is that it requires structures that are complete, and that it cannot deal with sets of structures that have gaps or that are inconsistent in some parts—in other words, inconsistencies we are trying to circumvent by fleshing out atomic models into densities.

Even though the methodology we have presented here was developed and demonstrated with data from ribosomes, it should be pointed out that similar properties of density maps, domain definitions and domain motions are found in a wide range of molecular machines. The adaptation of the toolset and its conventions for measuring and reporting domain motions by the structural biology community would make the characterization of macromolecular domain motions biologically informative, compact, and quantitatively comparable.

DOI: 10.1021/acs.jpcb.6b10732
J. Phys. Chem. B 2017, 121, 3747−3756

■ ASSOCIATED CONTENT

Supporting Information

The Supporting Information is available free of charge on the ACS Publications website at DOI: 10.1021/acs.jpcb.6b10732.

Details of the various methods used in the toolset (PDF)
Movie of fixed LSU with SSU body rotation from state A to B (AVI)
Movie of fixed SSU body with SSU head rotation from state A to B (AVI)
Movie of fixed LSU with full SSU rotation from state A to B (AVI)

■ AUTHOR INFORMATION

Corresponding Authors

*Tel: 217-244-1604; E-mail: kschulte@ks.uiuc.edu.
*Tel: (212) 305 9510; E-mail: jf2192@cumc.columbia.edu.

ORCID

Joachim Frank: 0000-0001-5449-6943

Notes

Software will be made available upon request. After sufficient testing, arrangements will be made for distribution along with the VMD package.

The authors declare no competing financial interest.

∇Deceased.

■ ACKNOWLEDGMENTS

We would like to thank Wen Li, Amedee Des Georges and Hstau Liao for helpful discussions. This work was supported by HHMI and NIH R01 GM29169 to J.F.; NSF PHY1430124 and NIH 9P41GM104601 to K.S. The authors also acknowledge the computer time provided by the National Science Foundation (NSF)-funded Extreme Science and Engineering Discovery Environment (XSEDE) MCA93S028, and a Blue Waters Illinois allocation, which is part of the Blue Waters sustained-petascale computing project supported by the NSF (awards OCI-0725070 and ACI-1238993) and the state of Illinois.

■ REFERENCES

(1) Trabuco, L. G.; Villa, E.; Schreiner, E.; Harrison, C. B.; Schulten, K. Molecular dynamics flexible fitting: A practical guide to combine cryo-electron microscopy and x-ray crystallography. *Methods* **2009**, *49* (2), 174–180.

(2) Trabuco, L. G.; Villa, E.; Mitra, K.; Frank, J.; Schulten, K. Flexible fitting of atomic structures into electron microscopy maps using molecular dynamics. *Structure* **2008**, *16* (5), 673–683.

(3) Noel, J. K.; Levi, M.; Raghunathan, M.; Lammert, H.; Hayes, R. L.; Onuchic, J. N.; Whitford, P. C. Smog 2: A versatile software package for generating structure-based models. *PLoS Comput. Biol.* **2016**, *12* (3), e1004794.

(4) Humphrey, W.; Dalke, A.; Schulten, K. Vmd: Visual molecular dynamics. *J. Mol. Graphics* **1996**, *14* (1), 33–38 27–28..

(5) Behrmann, E.; Loerke, J.; Budkevich, T. V.; Yamamoto, K.; Schmidt, A.; Penczek, P. A.; Vos, M. R.; Burger, J.; Mielke, T.; Scheerer, P.; et al. Structural snapshots of actively translating human ribosomes. *Cell* **2015**, *161* (4), 845–857.

(6) Spahn, C. M.; Gomez-Lorenzo, M. G.; Grassucci, R. A.; Jorgensen, R.; Andersen, G. R.; Beckmann, R.; Penczek, P. A.; Ballesta, J. P.; Frank, J. Domain movements of elongation factor eef2 and the eukaryotic 80s ribosome facilitate trna translocation. *EMBO J.* **2004**, *23* (5), 1008–1019.

(7) Wriggers, W.; Agrawal, R. K.; Drew, D. L.; McCammon, A.; Frank, J. Domain motions of ef-g bound to the 70s ribosome: Insights from a hand-shaking between multi-resolution structures. *Biophys. J.* **2000**, *79* (3), 1670–1678.

(8) Agirrezabala, X.; Lei, J.; Brunelle, J. L.; Ortiz-Meoz, R. F.; Green, R.; Frank, J. Visualization of the hybrid state of trna binding promoted by spontaneous ratcheting of the ribosome. *Mol. Cell* **2008**, *32* (2), 190–197.

(9) Ratje, A. H.; Loerke, J.; Mikolajka, A.; Brunner, M.; Hildebrand, P. W.; Starosta, A. L.; Donhofer, A.; Connell, S. R.; Fucini, P.; Mielke, T.; et al. Head swivel on the ribosome facilitates translocation by means of intra-subunit trna hybrid sites. *Nature* **2010**, *468* (7324), 713–716.

(10) Agirrezabala, X.; Frank, J. Elongation in translation as a dynamic interaction among the ribosome, trna, and elongation factors ef-g and ef-tu. *Q. Rev. Biophys.* **2009**, *42* (3), 159–200.

(11) Taylor, D. J.; Nilsson, J.; Merrill, A. R.; Andersen, G. R.; Nissen, P.; Frank, J. Structures of modified eef2 80s ribosome complexes reveal the role of gtp hydrolysis in translocation. *EMBO J.* **2007**, *26* (9), 2421–2431.

(12) Guo, Z.; Noller, H. F. Rotation of the head of the 30s ribosomal subunit during mrna translocation. *Proc. Natl. Acad. Sci. U. S. A.* **2012**, *109* (50), 20391–20394.

(13) Ermolenko, D. N.; Majumdar, Z. K.; Hickerson, R. P.; Spiegel, P. C.; Clegg, R. M.; Noller, H. F. Observation of intersubunit movement of the ribosome in solution using fret. *J. Mol. Biol.* **2007**, *370* (3), 530–540.

(14) Mohan, S.; Donohue, J. P.; Noller, H. F. Molecular mechanics of 30s subunit head rotation. *Proc. Natl. Acad. Sci. U. S. A.* **2014**, *111* (37), 13325–13330.

(15) Whitford, P. C.; Blanchard, S. C.; Cate, J. H.; Sanbonmatsu, K. Y. Connecting the kinetics and energy landscape of trna translocation on the ribosome. *PLoS Comput. Biol.* **2013**, *9* (3), e1003003.

(16) Poornam, G. P.; Matsumoto, A.; Ishida, H.; Hayward, S. A method for the analysis of domain movements in large biomolecular complexes. *Proteins: Struct., Funct., Genet.* **2009**, *76* (1), 201–212.

(17) Matsumoto, A.; Ishida, H. Global conformational changes of ribosome observed by normal mode fitting for 3d cryo-em structures. *Structure* **2009**, *17* (12), 1605–1613.

(18) Tama, F.; Valle, M.; Frank, J.; Brooks, C. L., 3rd Dynamic reorganization of the functionally active ribosome explored by normal mode analysis and cryo-electron microscopy. *Proc. Natl. Acad. Sci. U. S. A.* **2003**, *100* (16), 9319–9323.

(19) Paci, M.; Fox, G. E. Centers of motion associated with ef-tu binding to the ribosome. *RNA Biol.* **2016**, *13* (5), 524–530.

(20) Nudler, E. Rna polymerase active center: The molecular engine of transcription. *Annu. Rev. Biochem.* **2009**, *78*, 335–361.

(21) Steitz, T. A. Visualizing polynucleotide polymerase machines at work. *EMBO J.* **2006**, *25* (15), 3458–3468.

(22) Hahn, S. Structure and mechanism of the rna polymerase ii transcription machinery. *Nat. Struct. Mol. Biol.* **2004**, *11* (5), 394–403.

(23) Yoshida, M.; Muneyuki, E.; Hisabori, T. Atp synthase–a marvellous rotary engine of the cell. *Nat. Rev. Mol. Cell Biol.* **2001**, *2* (9), 669–677.

(24) Boyer, P. D. The atp synthase–a splendid molecular machine. *Annu. Rev. Biochem.* **1997**, *66*, 717–749.

(25) Horwich, A. L.; Farr, G. W.; Fenton, W. A. Groel-groes-mediated protein folding. *Chem. Rev.* **2006**, *106* (5), 1917–1930.

(26) Horwich, A. L.; Fenton, W. A.; Chapman, E.; Farr, G. W. Two families of chaperonin: Physiology and mechanism. *Annu. Rev. Cell Dev. Biol.* **2007**, *23*, 115–145.

(27) Trabuco, L. G.; Schreiner, E.; Eargle, J.; Cornish, P.; Ha, T.; Luthey-Schulten, Z.; Schulten, K. The role of l1 stalk-trna interaction in the ribosome elongation cycle. *J. Mol. Biol.* **2010**, *402* (4), 741–760.

(28) Agirrezabala, X.; Schreiner, E.; Trabuco, L. G.; Lei, J. L.; Ortiz-Meoz, R. F.; Schulten, K.; Green, R.; Frank, J. Structural insights into cognate versus near-cognate discrimination during decoding. *EMBO J.* **2011**, *30* (8), 1497–1507.

(29) Agirrezabala, X.; Liao, H. Y.; Schreiner, E.; Fu, J.; Ortiz-Meoz, R. F.; Schulten, K.; Green, R.; Frank, J. Structural characterization of mrna-trna translocation intermediates. *Proc. Natl. Acad. Sci. U. S. A.* **2012**, *109* (16), 6094–6099.

(30) Meyer, F. Topographic distance and watershed lines. *Signal Process.* **1994**, *38* (1), 113–125.

DOI: 10.1021/acs.jpcb.6b10732
J. Phys. Chem. B 2017, 121, 3747–3756

(31) Vincent, L.; Soille, P. Watersheds in digital spaces - an efficient algorithm based on immersion simulations. *IEEE Trans. Pattern Anal. Mach. Intell.* **1991,** *13* (6), 583–598.
(32) Meyer, F.; Beucher, S. Morphological segmentation. *J. Vis. Comm. Img. Rep.* **1990,** *1* (1), 21–46.
(33) Roerdink, J. B. T. M.; Meijster, A. The watershed transform: Definitions, algorithms and parallelization strategies. *Fundam. Inf.* **2000,** *41* (1), 187–228.
(34) Paglieroni, D. W. Distance transforms: Properties and machine vision applications. *CVGIP: Graph. Models Image Process* **1992,** *54* (1), 56–74.
(35) Bleau, A.; Leon, L. J. Watershed-based segmentation and region merging. *Comput. Vis. Image. Und.* **2000,** *77* (3), 317–370.
(36) Haris, K.; Efstratiadis, S. N.; Maglaveras, N. Watershed-based image segmentation with fast region merging. *International Conference on Image Processing - Proceedings* **1998,** *3,* 338–342.
(37) Pires, R. L. V. P. M.; De Smet, P.; Philips, W. Watershed segmentation and region merging. *Proc. SPIE* **2004,** *5308,* 1127–1135.
(38) Wang, J.; Lu, H. Q.; Eude, G.; Liu, Q. S. A fast region merging algorithm for watershed segmentation. *7th International Conference on Signal Processing - Proceedings* **2004,** *1–3,* 781–784.
(39) Beucher, S. Watershed, hierarchical segmentation and waterfall algorithm. In *Mathematical morphology and its applications to image processing;* Serra, J.; Soille, P., Eds.; Springer: Netherlands, 1994; Vol. 2, pp 69–76.
(40) Crespo, J.; Schafer, R. W.; Serra, J.; Gratin, C.; Meyer, F. The flat zone approach: A general low-level region merging segmentation method. *Signal Process.* **1997,** *62* (1), 37–60.
(41) Sezgin, M.; Sankur, B. Survey over image thresholding techniques and quantitative performance evaluation. *J. Electron. Imaging.* **2004,** *13* (1), 146–168.
(42) Horn, B. K. P. Closed-form solution of absolute orientation using unit quaternions. *J. Opt. Soc. Am. A* **1987,** *4* (4), 629–642.
(43) Umeyama, S. Least-squares estimation of transformation parameters between two point patterns. *IEEE Trans. Pattern Anal. Mach. Intell.* **1991,** *13* (4), 376–380.
(44) Arun, K. S.; Huang, T. S.; Blostein, S. D. Least-squares fitting of two 3-d point sets. *IEEE Trans. Pattern Anal. Mach. Intell.* **1987,** *PAMI-9* (5), 698–700.
(45) Coutsias, E. A.; Seok, C.; Dill, K. A. Using quaternions to calculate rmsd. *J. Comput. Chem.* **2004,** *25* (15), 1849–1857.
(46) Kneller, G. R. Superposition of molecular structures using quaternions. *Mol. Simul.* **1991,** *7* (1–2), 113–119.
(47) Hanson, R. M.; Kohler, D.; Braun, S. G. Quaternion-based definition of protein secondary structure straightness and its relationship to ramachandran angles. *Proteins: Struct., Funct., Genet.* **2011,** *79* (7), 2172–2180.
(48) Skone, G.; Cameron, S.; Voiculescu, I. Doing a good turn: The use of quaternions for rotation in molecular docking. *J. Chem. Inf. Model.* **2013,** *53* (12), 3367–3372.
(49) Dobrowolski, P. *Swing-twist decomposition in clifford algebra* [Online], 2015. https://arxiv.org/abs/1506.05481.
(50) Dziubinski, M.; Daniluk, P.; Lesyng, B. Resicon: A method for the identification of dynamic domains, hinges and interfacial regions in proteins. *Bioinformatics* **2015,** *32* (1), 25–34.
(51) Romanowska, J.; Nowinski, K. S.; Trylska, J. Determining geometrically stable domains in molecular conformation sets. *J. Chem. Theory Comput.* **2012,** *8* (8), 2588–99.
(52) Bernhard, S.; Noe, F. Optimal identification of semi-rigid domains in macromolecules from molecular dynamics simulation. *PLoS One* **2010,** 5 (5), e10491.

DOI: 10.1021/acs.jpcb.6b10732
J. Phys. Chem. B 2017, 121, 3747–3756

ARTICLE

doi:10.1038/nature23479

Channel opening and gating mechanism in AMPA-subtype glutamate receptors

Edward C. Twomey[1,2], Maria V. Yelshanskaya[1], Robert A. Grassucci[1,3], Joachim Frank[1,3,4] & Alexander I. Sobolevsky[1]

AMPA (α-amino-3-hydroxy-5-methyl-4-isoxazole propionic acid)-subtype ionotropic glutamate receptors mediate fast excitatory neurotransmission throughout the central nervous system. Gated by the neurotransmitter glutamate, AMPA receptors are critical for synaptic strength, and dysregulation of AMPA receptor-mediated signalling is linked to numerous neurological diseases. Here we use cryo-electron microscopy to solve the structures of AMPA receptor-auxiliary subunit complexes in the apo, antagonist- and agonist-bound states and determine the iris-like mechanism of ion channel opening. The ion channel selectivity filter is formed by the extended portions of the re-entrant M2 loops, while the helical portions of M2 contribute to extensive hydrophobic interfaces between AMPA receptor subunits in the ion channel. We show how the permeation pathway changes upon channel opening and identify conformational changes throughout the entire AMPA receptor that accompany activation and desensitization. Our findings provide a framework for understanding gating across the family of ionotropic glutamate receptors and the role of AMPA receptors in excitatory neurotransmission.

Excitatory neurotransmission is tightly governed by the AMPA subtype of the ionotropic glutamate receptors (iGluRs)[1,2]. Gated by the neurotransmitter glutamate, AMPA receptors (AMPARs) activate on a sub-millisecond timescale and rapidly depolarize the post-synaptic membrane. AMPAR-mediated neuronal signalling has profound effects on synaptic strength, and many neurological diseases have been directly linked to changes in AMPAR synaptic signalling[3–5].

AMPARs are tetrameric, two-fold symmetric membrane proteins with a three-layer, 'Y'-shaped architecture[6,7]. Extracellularly, the amino-terminal domains (ATDs) form the top of the Y. Below the ATDs are the ligand-binding domains (LBDs) that bind the neurotransmitter glutamate. Two polypeptide stretches (S1 and S2) make up each of the four AMPAR subunit LBDs, which are tethered to the transmembrane domain (TMD) at the base of the Y by flexible linkers. The nearly four-fold symmetric TMD is made up of three membrane-spanning helices (M1, M3 and M4) and a re-entrant pore loop (M2) between helices M1 and M3. The cation-selective ion channel is lined by the M3 helices with their bundle crossing forming a tight gate to regulate permeation.

Since the first, closed-state intact AMPAR structure[8], many studies have pursued different receptor conformations using both X-ray crystallography[9–12] and cryo-electron microscopy (cryo-EM)[13–17]. However, the structure of the ion channel is yet to be resolved to high resolution, and the open-state of the receptor has yet to be seen. This is a fundamental barrier to understanding AMPAR biology and structurally informed design of therapeutic agents that target AMPARs in neuropathological conditions. Here, we use cryo-EM to solve the structure of an AMPAR ion channel at high resolution and resolve the opening of the AMPAR ion channel in response to binding of glutamate.

Ion channel structure

We used cryo-EM to analyse the structure of the homomeric GluA2 flip splice variant AMPAR in complex with the auxiliary subunit germline-specific gene 1-like (GSG1L), which we surmised would stabilize the closed-state structure of the ion channel as it lowers the ion channel open probability compared to the transmembrane AMPAR regulatory protein (TARP)-γ2 or stargazin (STZ)[15,16]. Indeed, using a covalent fusion construct (Methods) between truncated forms of GluA2 and GSG1L, we solved structures of the digitonin-solubilized receptor complex bound to the antagonist ZK200775 in two states, termed GluA2–GSG1L$_{ZK\text{-}1}$ (4.6 Å) and GluA2–GSG1L$_{ZK\text{-}2}$ (4.4 Å) (Extended Data Table 1; Extended Data Fig. 1), with local qualities of the maps showing higher resolution details in the ion channel pore (Extended Data Fig. 2).

The GluA2–GSG1L$_{ZK\text{-}1}$ complex has a typical Y-shaped GluA2 tetramer in the centre (Fig. 1a), with two GSG1L subunits assembling between protomers A/B and C/D on the periphery of the GluA2 TMD (Fig. 1b). The GluA2 TMD is completely resolved and allows building of the entire ion channel pore-forming region (Fig. 1c, d). The M3 bundle crossing, formed by T617, A621, T625 and M629, occludes cation permeation, as previously observed in the original GluA2 crystal structure[8]. Below T617, the M3 bundle crossing opens up into a hydrophobic cavity in the middle of the channel pore. Below this cavity is a second constriction formed by the extended regions of the M2 re-entrant loops. Similar to other tetrameric ion channels, this constriction might serve as a lower gate of the channel as well as a selectivity filter[18–20]. At the re-entrant loop tip, the glutamine/arginine (Q/R)-site glutamine residues (Gln586) protrude towards the centre of the ion channel pore, occluding the permeation pathway and forming a lower gate. Their location is consistent with the observation that channels composed of edited GluA2 subunits (Gln586Arg) have reduced Ca^{2+} permeation and polyamine block, probably due to electrostatic repulsion[21,22]. The pore loop, under the Q/R-site, appears to be more flexible in our closed-state structures, compared to the rest of the channel, but we expect it to become more ordered upon channel opening to form a selectivity filter along the permeation pathway.

The α-helical portion of M2 is largely hydrophobic (Fig. 1c) and contributes to cross-subunit interfaces within the TMD (Extended Data

[1]Department of Biochemistry and Molecular Biophysics, Columbia University, 650 West 168th Street, New York, New York 10032, USA. [2]Integrated Program in Cellular, Molecular, and Biomedical Studies, Columbia University, 650 West 168th Street, New York, New York 10032, USA. [3]Howard Hughes Medical Institute, 650 West 168th Street, New York, New York 10032, USA. [4]Department of Biological Sciences, Columbia University, 650 West 168th Street, New York, New York 10032, USA.

© 2017 Macmillan Publishers Limited, part of Springer Nature. All rights reserved.

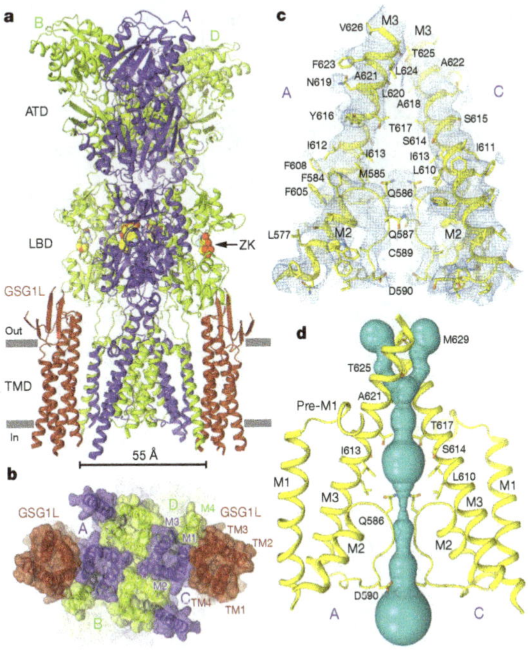

Figure 1 | GluA2–GSG1L and ion channel structure. a, b, Model of GluA2–2×GSG1L$_{ZK-1}$ viewed parallel to (**a**) or from the intracellular side of (**b**) the membrane with GluA2 subunits A and C coloured purple, B and D in green and GSG1L in red. The competitive antagonists ZK200775 (ZK) are shown as space-filling models. **c,** Close-up view of the pore-lining domains M2 and M3 in subunits A and C with cryo-EM density shown as blue mesh. **d,** Ion conduction pathway (violet) with pore-lining residues in the M2 and M3 segments of subunits A and C shown as sticks.

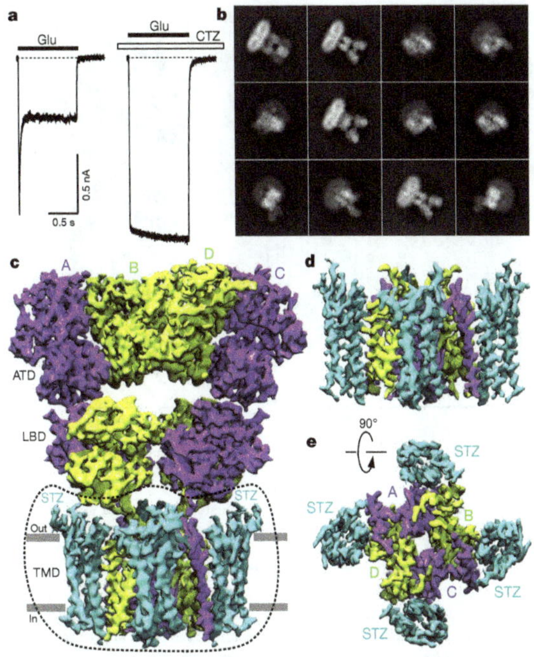

Figure 2 | Cryo-EM of an activated GluA2–STZ complex. **a,** Representative whole-cell currents recorded at −60 mV membrane potential from a HEK293 cell expressing GluA2–STZ in response to 1 s applications of Glu alone or in the continuous presence of 30 μM CTZ. **b,** Two-dimensional class averages of GluA2–STZ$_{Glu+CTZ}$ particles, showing diverse orientations. **c,** 4.2 Å cryo-EM reconstruction of the entire GluA2–STZ$_{Glu+STZ}$ complex, with GluA2 subunits coloured green and purple, and STZ in blue; viewed perpendicular to the membrane. The dashed outline highlights the area focused on in directed refinement to improve the TMD density. **d, e,** Density of the GluA2–STZ$_{Glu+CTZ}$ TMD at 4.0 Å resolution from directed refinement, viewed parallel (**d**) and perpendicular (**e**) to the membrane.

Fig. 3b). These hydrophobic interactions involve Leu540 and Tyr533 from M1, as well as Trp605 and Phe608 from the neighbouring M3, each from the same subunit as the M2, and Trp606 and Ile600 from M3 of the adjacent subunit. The M2 helix contributes to these hydrophobic interfaces through Leu577, Trp578, Leu581 and Phe584. For a helix that has often evaded structural studies, the M2 helix has quite an extensive network of hydrophobic interactions.

The second closed state structure, GluA2–GSG1L$_{ZK-2}$, has the same ion channel architecture as in GluA2–GSG1L$_{ZK-1}$, although with higher resolution details throughout the TMD (Extended Data Fig. 2). While the TMD structure is the same, the ECD in GluA2–GSG1L$_{ZK-2}$ is markedly different, being shorter by 6 Å (Extended Data Fig. 4). The ATD rotates 19° as a rigid body around the overall two-fold axis of symmetry (Extended Data Fig. 4c, d), and the LBD dimers rotate by 4°, creating a separation between the proximal subunits A and C (Extended Data Fig. 4e, f). An elongated density at the ATD–LBD interfaces inside subunits A and C perfectly matches a digitonin molecule (Extended Data Fig. 4b, inset). Thus, digitonin binding might help to trap GluA2 in a distinct closed state conformation. We hypothesize that the ATD–LBD interface represents a new binding site for hydrophobic molecules with a possible therapeutic importance and may bind to fatty acids that have recently been identified as non-competitive AMPAR inhibitors[23,24].

Cryo-EM of activated GluA2–STZ complex

To understand how the native neurotransmitter glutamate (Glu) activates and opens the AMPAR ion channel, we used a fusion construct between GluA2 and STZ (Methods) that promotes GluA2–STZ complex formation and expression while maintaining its functional properties. In contrast to GSG1L, which stabilizes the AMPAR inactive states, STZ favours the open state[25–27], as signified by increased steady-state current in response to Glu application in whole-cell patch-clamp recordings (Fig. 2a). Additional stabilization of the open state is provided by the positive allosteric modulator cyclothiazide (CTZ)[28], which makes AMPAR responses non-desensitizing (Fig. 2a).

We solved a cryo-EM structure of the GluA2–STZ complex in the presence of Glu and CTZ to 4.2 Å resolution (Extended Data Fig. 1), with two-dimensional class averages showing the secondary structure features and diverse orientations (Fig. 2b). The reconstruction (Fig. 2c) shows the entire GluA2–STZ complex in the agonist-bound state, with high-resolution features in the GluA2 TMD core (Extended Data Fig. 2n, o) but lower local resolution for STZ. To achieve unambiguous side chain definition in the TMD, we performed directed refinement (Fig. 2c) that resulted in a 4.0 Å cryo-EM reconstruction (Fig. 2d) and distinct side-chain densities (Extended Data Fig. 5e). The TMD cryo-EM map shows clear differences in the central pore (Fig. 2e) compared to the closed-state GluA2–GSG1L complex (Fig. 1b) and four STZ molecules bound. Also, glutamate and CTZ are easily discernible in the cryo-EM density (Extended Data Fig. 5a, d), suggesting that we trapped the complex in an activated state.

The original, closed-state cryo-EM structures of the GluA2–STZ complex suggested that acidic residues in the STZ β4-TM2 loop mediate electrostatic interactions with positively charged residues in the lower

© 2017 Macmillan Publishers Limited, part of Springer Nature. All rights reserved.

RESEARCH ARTICLE

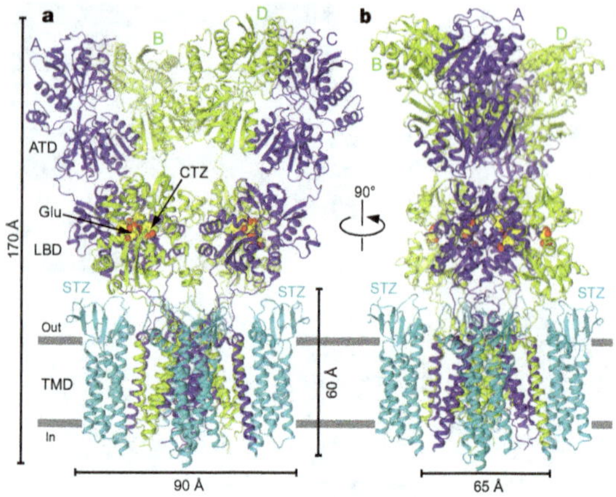

Figure 3 | Structure of the GluA2–STZ complex. a, b, Model of GluA2–4xSTZ$_{\text{Glu+CTZ}}$ viewed parallel to the membrane with GluA2 subunits A and C coloured purple, B and D in green and STZ in cyan. The molecules of agonist Glu and positive allosteric modulator CTZ are shown as space-filling models.

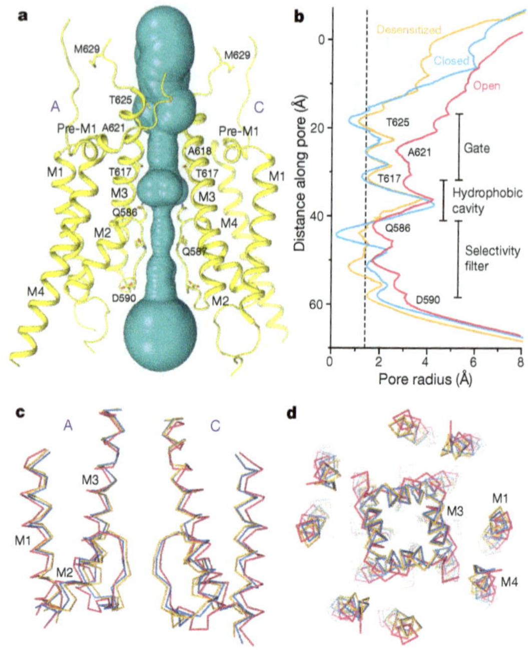

Figure 4 | Ion channel pore in open, closed and desensitized states. a, The GluA2–STZ$_{\text{Glu+CTZ}}$ ion conduction pathway (cyan) with pore-lining residues in M2 and M3 segments of subunits A and C shown as sticks. **b,** Pore radius calculated using HOLE for GluA2–STZ$_{\text{Glu+CTZ}}$ in the open state (pink), GluA2–GSG1L$_{\text{ZK-1}}$ in the closed state (blue) and GluA2–2×GSG1L$_{\text{Quis}}$ in the desensitized state (orange). **c, d,** TMD ribbon diagrams for the structures in **b** viewed parallel to the membrane (**c**) or extracellularly (**d**). In **c**, subunits B and D as well as M4 segments are removed for clarity.

GluA2 LBD and S1-pre-M1 linker to enhance receptor activation[17,29]. In our active state cryo-EM reconstruction, this region of STZ seems to be disordered. However, based on well-defined adjacent regions of STZ (Extended Data Fig. 5e), the same acidic residues are juxtaposed to the GluA2 basic patch. Perhaps rather than directly interacting in the activated state, the β4-TM2 loop has an important role in the structural transitions necessary for receptor activation.

Our density, and the corresponding GluA2–STZ$_{\text{Glu+CTZ}}$ model (Fig. 3, Methods) show four STZ molecules bound around the AMPAR core (Fig. 2e, Extended Data Fig. 6c). We used the same fusion design to solve the first structure of the GluA2–STZ complex in the closed state[15] but observed only one or two STZ molecules bound to the AMPAR core. Here we solubilized the complex in digitonin instead of dodecyl maltoside (DDM) detergent and observed maximum occupancy of STZ binding sites around the AMPAR core, similar to GluA2–STZ co-expression[17]. Accordingly, instead of the four-layer particle architecture in DDM[15], we now see a clearly discernible three-layer architecture in the two-dimensional (2D) class averages (Extended Data Fig. 6a). Interestingly, the GluA2–GSG1L complex in digitonin, which showed a four-layer architecture when solubilized in DDM[16], also showed the three-layer particle architecture (Extended Data Fig. 6b) but with only two, not four, molecules of GSG1L bound to GluA2 core (Fig. 1a, Extended Data Fig. 6d). As the nature of the fusion construct design does not preclude full occupancy of the auxiliary subunit binding sites, we hypothesize that STZ and GSG1L use different stoichiometries to assemble with AMPARs.

Open ion channel pore

As indicated by the markedly different architecture of the AMPAR core in GluA2–STZ$_{\text{Glu+CTZ}}$ compared to the closed state, the width of the TMD increases by 10 Å from 55 Å in the closed state structures to 65 Å in GluA2–STZ$_{\text{Glu+CTZ}}$, and the ion channel pore undergoes widening (compare Figs 1b and 3b). We probed the ion conduction pathway in GluA2–STZ$_{\text{Glu+CTZ}}$ (Fig. 4a) and found it much wider than in the closed state GluA2–GSG1L structures (Fig. 1d). Measurements of the pore radius (Fig. 4b) confirmed that in contrast to the pore-occluded conformations in the closed GluA2–GSG1L$_{\text{ZK-1}}$, GluA2–GSG1L$_{\text{ZK-2}}$ and desensitized GluA2–GSG1L$_{\text{Quis}}$[16] states, the GluA2–STZ$_{\text{Glu+CTZ}}$ pore is wide open for ion conductance. The distances between the C_α atoms of diagonal Gln586, Thr617 and Ala621 pairs increased almost uniformly by about 3 Å in GluA2–STZ$_{\text{Glu+CTZ}}$ (Fig. 4b, c), indicating nearly four-fold symmetric, iris-type opening of the main portion of the pore (Fig. 4d), reminiscent of other tetrameric ion channels[19,30,31]. In stark contrast, the distance between diagonal Thr625 pairs, which is similar for the A/C and B/D subunit pairs in the closed state (9.5 and 11 Å), increases to 16 Å for the A/C pair and to 32 Å for the B/D pair in GluA2–STZ$_{\text{Glu+CTZ}}$. This 6.5 Å versus 21 Å separation upon channel opening indicates that at the level of Thr625, the AMPAR pore loses its pseudo four-fold symmetry and becomes two-fold symmetric, consistent with previous functional experiments[32]. We also compared our recent molecular dynamics model of the open state[33] and the GluA2–STZ$_{\text{Glu+CTZ}}$ structure. The narrowest portions of the pore in the molecular dynamics model, estimated as C_α distances between the diagonal Gln586 (9.6 Å) and Thr617 (12.6 Å) pairs, are narrower than the corresponding portions of the GluA2–STZ$_{\text{Glu+CTZ}}$ pore (10.2 Å and 14.1 Å, respectively). Yet, we clearly observed ion conducting events through our molecular dynamics model[33], an observation strongly supporting the conducting conformation of the GluA2–STZ$_{\text{Glu+CTZ}}$ pore. In addition, a clear density in the centre of the pore (Extended Data Fig. 5b, c), probably representing a sodium ion, also suggests that the selectivity filter of the channel is indeed in a conducting state.

The selectivity filter located at and below the Q/R site (Fig. 4a, b), which is not visible in the closed state, is clearly resolved in GluA2–STZ$_{\text{Glu+CTZ}}$ and is approximately four-fold symmetric, similar to K^+ channels[30,31]. Just below the Q/R site, the backbone carbonyls of Gln587, Gly588 and Cys589 line the selectivity filter, while the side chains of Gln587 and Cys589 point away from the central pore axis. At the bottom of the selectivity filter is Asp590 that creates a negative entry point into the cytoplasm from the channel pore, perhaps having a role in cation versus anion selectivity of AMPAR channels.

© 2017 Macmillan Publishers Limited, part of Springer Nature. All rights reserved.

Gating mechanism

To understand the conformational changes that are associated with AMPAR gating, we compared the open-state GluA2–STZ$_{Glu+CTZ}$ with the closed-state GluA2–GSG1L$_{ZK-1}$ and the desensitized-state GluA2–2×GSG1L$_{Quis}$ structures (Fig. 5, Supplementary Video 1). Since gating starts with binding of glutamate to an unliganded AMPAR, we first wanted to see whether GluA2–GSG1L$_{ZK-1}$ represents the resting, apo state of the receptor. We solved cryo-EM structures of the digitonin-solubilized GluA2–GSG1L complex in the absence of ligands, which, similar to the ZK200775-bound structures, are represented by two different states. The apo state structures, GluA2–GSG1L$_{apo-1}$ and GluA2–GSG1L$_{apo-2}$, turned out to be very similar to GluA2–GSG1L$_{ZK-1}$ and GluA2–GSG1L$_{ZK-2}$ (Extended Data Figs 3, 4), superposing on each other with root mean square deviation (r.m.s.d.) values of 0.526 Å and 0.701 Å, respectively. We therefore used the higher resolution GluA2–GSG1L$_{ZK-1}$ structure that is nearly identical to GluA2–GSG1L$_{apo-1}$ as an accurate representation of the closed resting state.

In the first approximation, activation, which includes ion channel opening, is represented by the GluA2–GSG1L$_{ZK-1}$ to GluA2–STZ$_{Glu+CTZ}$ transition. At the level of a single LBD, glutamate binding

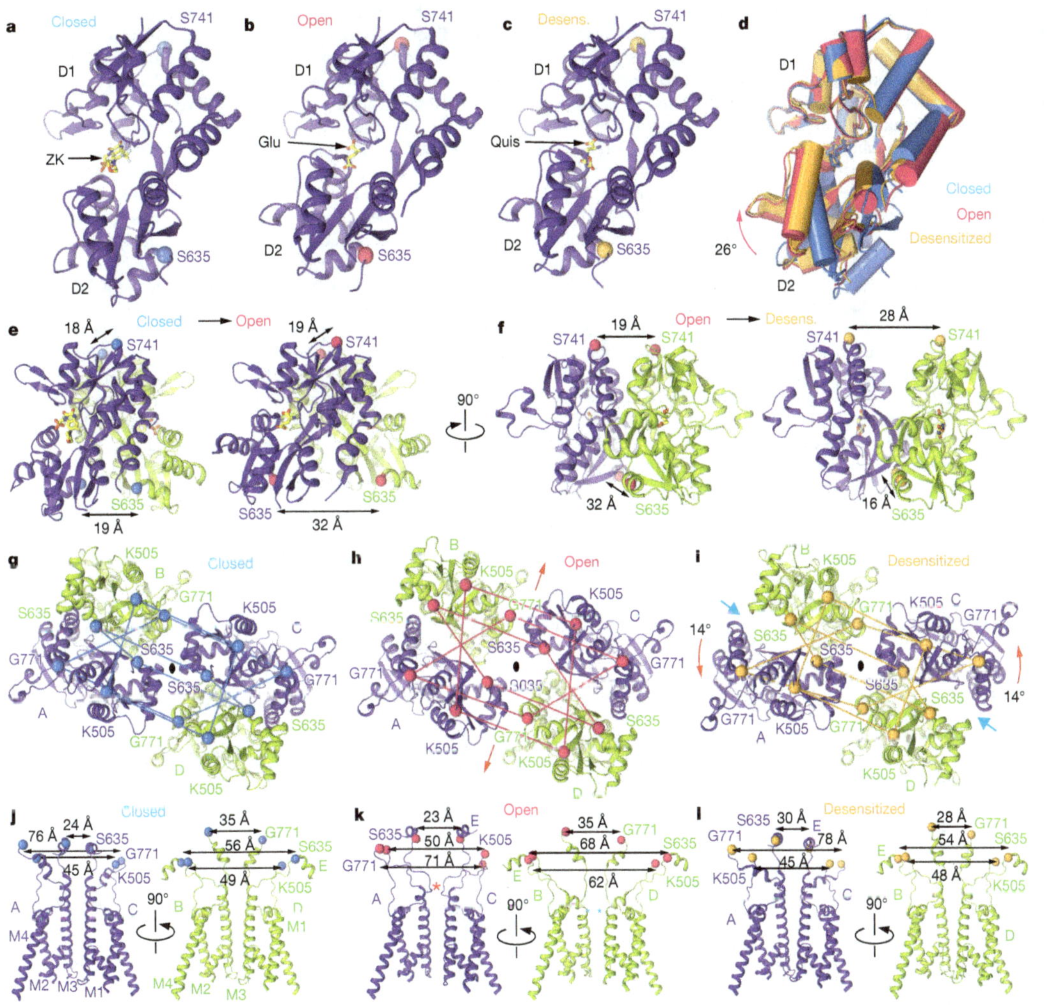

Figure 5 | Structural rearrangements during gating. GluA2 domains are shown for structures of GluA2–GSG1L$_{ZK-1}$ in the closed state (blue), GluA2–STZ$_{Glu+CTZ}$ in the open state (pink) and GluA2–2×GSG1L$_{Quis}$ in the desensitized state (orange). **a–d**, LBD monomers shown individually (**a–c**) or in superposition (**d**). ZK200775, Glu and quisqualate (Quis) molecules are shown in sticks. **e, f**, Changes in LBD dimer conformation upon transition from closed to open (**e**) and open to desensitized (**f**) states. Separation of upper D1 and lower D2 lobes is indicated by arrows. **g–i**, LBD tetramers viewed from the ion channel. Broadening of the LBD layer in the open state and rotation of the A and C monomers in the desensitized state are indicated by red arrows. Blue arrows in **i** point to the cleft between the desensitized state LBD protomers signifying the loss of local LBD dimer two-fold symmetry and threefold reduction of intradimer interface[16]. **j–l**, S1–M1, M3–S2 and S2–M4 linkers viewed parallel to membrane. The A/C and B/D subunit pairs are shown separately and viewed orthogonally. Distances between C_α atoms of Lys505, Ser635 and Gly771 are indicated. Red and blue asterisks in **k** indicate the S2–M4 linkers extended towards the pore and one helical turn-unwound M3 helices in A/C subunits and kinked M3 helices in the B/D subunits of the open state, respectively.

© 2017 Macmillan Publishers Limited, part of Springer Nature. All rights reserved.

causes 26° LBD clamshell closure (Fig. 5a–d). Consistent with studies on isolated LBDs[34–36], this closure leaves the D1–D1 interface intact and leads to a 12 Å separation of the D2 lobes (Fig. 5e). The D2 separation results in widening of the LBD tetramer and expansion of the LBD gating ring in the middle of the LBD layer (Fig. 5g, h and Extended Data Fig. 7g, h), which translates to pore opening by the LBD–TMD linkers (Fig. 5j, k).

We monitored conformational rearrangements in the S1–M1, M3–S2 and S2–M4 linkers by measuring distances between C_α atoms of the diagonal Lys505, Ser635 and Gly771 pairs. The M3–S2 linker in subunits B and D has the central role in transmitting conformational changes in LBDs to opening of the ion channel, as signified by an 11 Å increase in Ser635 separation and a distinct kink in the M3 helix at Ala618 (Fig. 5j, k and Extended Data Fig. 5e), which turns the upper gate at the bundle crossing away from the central pore axis. The M3 helices in subunits A and C unwind and become one helical turn shorter, but the S635 separation in subunits A and C decreases by 1 Å, suggesting a weaker role in gating of the corresponding linkers. The S1–M1 linkers in subunits B and D make a much more notable contribution to channel opening than the S1–M1 linkers in subunits A and C, with separation between Lys505 increasing by 13 and 5 Å, respectively. Unexpected changes were observed in the S2–M4 linkers in subunits A and C, which included complete unwinding of the pre-M4 helices (Fig. 5j, k) and stretching these linkers towards the centre of the ion channel pore, while reducing the distance between G771 by 5 Å. These changes are in stark contrast to the S2–M4 linkers in subunits B and D, which maintain their relative positioning. The unprecedented conformational changes in the subunits A/C S2–M4 linkers that occur during channel opening support the importance of the M4 segments for iGluR function[37–39] and suggest a possible contribution of these linkers to the open pore extracellular vestibule.

We next mapped the changes between GluA2–$STZ_{Glu+CTZ}$ and GluA2–2×$GSG1L_{Quis}$[16] to assess conformational rearrangements during desensitization. Upon desensitization, individual LBDs maintain their closed clamshell conformation (Fig. 5a–d), while the LBD dimers undergo marked rearrangements (Fig. 5f), signified by rupture of the D1–D1 interface (increase in the S741 distance by 9 Å) allowing D2 lobes to come closer together (by 15 Å, measured as a change in the S635 distance) and permit channel closure. Rupture of the D1–D1 interface parallels loss of two-fold symmetry by individual LBD dimers, with 14° rotation of the A/C LBDs away from their B/D dimer partners and a more four-fold symmetric LBD tetramer arrangement[16] signified by appearance of a cleft between local LBD dimers (Fig. 5i). Correspondingly, the D2 attachments to the LBD–TMD linkers in this new LBD tetramer arrangement acquire positions similar to the closed state, compressing the expanded open-state LBD gating ring (Fig. 5g–i). This results in the LBD–TMD linkers (Fig. 5l) returning to positions (Fig. 5j) that confer a closed, non-conducting ion channel. To return back to the resting state, the receptor must undergo conformational rearrangement from the desensitized to the closed, resting state, in a process termed recovery from desensitization[16].

The ATD local dimers maintain their relative positioning between different gating states, resulting in rigid-body movements of the ATD tetramer (Extended Data Fig. 7d–f). During opening, the ATD tetramer rotates by 11° clockwise as viewed extracellularly (Extended Data Fig. 7e). Upon desensitization, the ATD further rotates clockwise by an additional 7° (Extended Data Fig. 7f). To recover from desensitization back to the resting state, the ATD must then rotate anticlockwise by 18°. Maintenance of the ATD tetrameric arrangement in different gating states[29,40] emphasizes importance of the ATDs for iGluR assembly[41–45], while enormous rigid body rotations provide a way to fine tune synaptic transmission via altering cross-synaptic interactions[46].

Discussion

The structures of AMPAR complexes allow better understanding of the iGluR gating mechanism (Extended Data Fig. 8). Channel opening in GluA2–$STZ_{Glu+CTZ}$ is accompanied by the outward flipping of Gln586, which occludes the channel in the closed state, away from the central pore axis. Therefore, the Q/R site at the tip of M2 seems to function as a lower gate. We propose that individual AMPAR subunit contributions to the lower gate result in the sub-conductance states observed in single-channel AMPAR studies[47,48]. AMPARs occupy their highest conductance state in the presence of glutamate[49,50] and STZ[26]. We expect that the highest conductance state results from all subunits being engaged by glutamate[51,52] when channel occlusion is relieved at both gates. STZ also promotes the higher conductance states by stabilizing the open conformation of the lower gate through interaction with the M2 helices. We therefore believe that our GluA2–$STZ_{Glu+CTZ}$ structure represents the maximum-conductance state of the channel.

In addition, the prominent placement of Gln586 of the Q/R-site in the ion channel pore explains why AMPARs containing the R-edited GluA2 subunits show markedly reduced ion channel conductance[53], probably through adding a bulky, charged residue into the permeation pathway. At the level of glutamate receptors, this demonstrates how compartment-specific mRNA editing alters ion channel conductance and AMPAR-mediated excitatory neurotransmission[2]. Importantly, identification of the earlier-predicted[32] M3 kink at Ala618 in subunits B/D upon channel opening suggests a wider role for Ala618 across gating in the iGluR family, as it is part of the highly conserved SYTANLAAF motif[20]. Also in this motif is the Lurcher mutation Ala622Thr, which notably changes function across iGluR subtypes, resulting in constitutively activated receptors and markedly different gating properties[54]. The proximity of this site to Ala618 suggests that the Lurcher mutation may change iGluR gating through altering the properties of M3 kinking upon channel opening. Better understanding of AMPAR gating and availability of the high-resolution open-state structure will help to overcome barriers in the design of therapeutic agents and the treatment of neurodegenerative diseases.

Online Content Methods, along with any additional Extended Data display items and Source Data, are available in the online version of the paper; references unique to these sections appear only in the online paper.

Received 15 June; accepted 14 July 2017.
Published online 24 July 2017.

1. Kumar, J. & Mayer, M. L. Functional insights from glutamate receptor ion channel structures. *Annu. Rev. Physiol.* **75,** 313–337 (2013).
2. Traynelis, S. F. *et al.* Glutamate receptor ion channels: structure, regulation, and function. *Pharmacol. Rev.* **62,** 405–496 (2010).
3. Bowie, D. Ionotropic glutamate receptors & CNS disorders. *CNS Neurol. Disord. Drug Targets* **7,** 129–143 (2008).
4. Chater, T. E. & Goda, Y. The role of AMPA receptors in postsynaptic mechanisms of synaptic plasticity. *Front. Cell. Neurosci.* **8,** 401 (2014).
5. Henley, J. M. & Wilkinson, K. A. Synaptic AMPA receptor composition in development, plasticity and disease. *Nat. Rev. Neurosci.* **17,** 337–350 (2016).
6. Sobolevsky, A. I. Structure and gating of tetrameric glutamate receptors. *J. Physiol. (Lond.)* **593,** 29–38 (2015).
7. Mayer, M. L. Structural biology of glutamate receptor ion channel complexes. *Curr. Opin. Struct. Biol.* **41,** 119–127 (2016).
8. Sobolevsky, A. I., Rosconi, M. P. & Gouaux, E. X-ray structure, symmetry and mechanism of an AMPA-subtype glutamate receptor. *Nature* **462,** 745–756 (2009).
9. Chen, L., Dürr, K. L. & Gouaux, E. X-ray structures of AMPA receptor-cone snail toxin complexes illuminate activation mechanism. *Science* **345,** 1021–1026 (2014).
10. Dürr, K. L. *et al.* Structure and dynamics of AMPA receptor GluA2 in resting, pre-open, and desensitized states. *Cell* **158,** 778–792 (2014).
11. Yelshanskaya, M. V., Li, M. & Sobolevsky, A. I. Structure of an agonist-bound ionotropic glutamate receptor. *Science* **345,** 1070–1074 (2014).
12. Yelshanskaya, M. V. *et al.* Structural bases of noncompetitive inhibition of AMPA subtype ionotropic glutamate receptors by antiepileptic drugs. *Neuron* **91,** 1305–1315 (2016).
13. Herguedas, B. *et al.* Structure and organization of heteromeric AMPA-type glutamate receptors. *Science* **352,** aad3873 (2016).
14. Meyerson, J. R. *et al.* Structural mechanism of glutamate receptor activation and desensitization. *Nature* **514,** 328–334 (2014).
15. Twomey, E. C., Yelshanskaya, M. V., Grassucci, R. A., Frank, J. & Sobolevsky, A. I. Elucidation of AMPA receptor-stargazin complexes by cryo-electron microscopy. *Science* **353,** 83–86 (2016).

© 2017 Macmillan Publishers Limited, part of Springer Nature. All rights reserved.

16. Twomey, E. C., Yelshanskaya, M. V., Grassucci, R. A., Frank, J. & Sobolevsky, A. I. Structural bases of desensitization in AMPA receptor–auxiliary subunit complexes. *Neuron* **94,** 569–580.e5 (2017).
17. Zhao, Y., Chen, S., Yoshioka, C., Baconguis, I. & Gouaux, E. Architecture of fully occupied GluA2 AMPA receptor–TARP complex elucidated by cryo-EM. *Nature* **536,** 108–111 (2016).
18. Doyle, D. A. *et al.* The structure of the potassium channel: molecular basis of K^+ conduction and selectivity. *Science* **280,** 69–77 (1998).
19. Cao, E., Liao, M., Cheng, Y. & Julius, D. TRPV1 structures in distinct conformations reveal activation mechanisms. *Nature* **504,** 113–118 (2013).
20. Huettner, J. E. Glutamate receptor pores. *J. Physiol. (Lond.)* **593,** 49–59 (2015).
21. Wilding, T. J., Lopez, M. N. & Huettner, J. E. Radial symmetry in a chimeric glutamate receptor pore. *Nat. Commun.* **5,** 3349 (2014).
22. Lopez, M. N., Wilding, T. J. & Huettner, J. E. Q/R site interactions with the M3 helix in GluK2 kainate receptor channels revealed by thermodynamic mutant cycles. *J. Gen. Physiol.* **142,** 225–239 (2013).
23. Chang, P. *et al.* Seizure control by decanoic acid through direct AMPA receptor inhibition. *Brain* **139,** 431–443 (2016).
24. Chang, P. *et al.* Seizure control by derivatives of medium chain fatty acids associated with the ketogenic diet show novel branching-point structure for enhanced potency. *J. Pharmacol. Exp. Ther.* **352,** 43–52 (2015).
25. Priel, A. *et al.* Stargazin reduces desensitization and slows deactivation of the AMPA-type glutamate receptors. *J. Neurosci.* **25,** 2682–2686 (2005).
26. Tomita, S. *et al.* Stargazin modulates AMPA receptor gating and trafficking by distinct domains. *Nature* **435,** 1052–1058 (2005).
27. MacLean, D. M., Ramaswamy, S. S., Du, M., Howe, J. R. & Jayaraman, V. Stargazin promotes closure of the AMPA receptor ligand-binding domain. *J. Gen. Physiol.* **144,** 503–512 (2014).
28. Patneau, D. K., Vyklicky, L. Jr & Mayer, M. L. Hippocampal neurons exhibit cyclothiazide-sensitive rapidly desensitizing responses to kainate. *J. Neurosci.* **13,** 3496–3509 (1993).
29. Yelshanskaya, M. V., Saotome, K., Singh, A. K. & Sobolevsky, A. I. Probing intersubunit interfaces in AMPA-subtype ionotropic glutamate receptors. *Sci. Rep.* **6,** 19082 (2016).
30. Jiang, Y. *et al.* The open pore conformation of potassium channels. *Nature* **417,** 523–526 (2002).
31. Long, S. B., Campbell, E. B. & Mackinnon, R. Crystal structure of a mammalian voltage-dependent Shaker family K^+ channel. *Science* **309,** 897–903 (2005).
32. Sobolevsky, A. I., Yelshansky, M. V. & Wollmuth, L. P. The outer pore of the glutamate receptor channel has 2-fold rotational symmetry. *Neuron* **41,** 367–378 (2004).
33. Yelshanskaya, M. V., Mesbahi-Vasey, S., Kurnikova, M. G. & Sobolevsky, A. I. Role of the ion channel extracellular collar in AMPA receptor gating. *Sci. Rep.* **7,** 1050 (2017).
34. Armstrong, N. & Gouaux, E. Mechanisms for activation and antagonism of an AMPA-sensitive glutamate receptor: crystal structures of the GluR2 ligand binding core. *Neuron* **28,** 165–181 (2000).
35. Sun, Y. *et al.* Mechanism of glutamate receptor desensitization. *Nature* **417,** 245–253 (2002).
36. Armstrong, N., Jasti, J., Beich-Frandsen, M. & Gouaux, E. Measurement of conformational changes accompanying desensitization in an ionotropic glutamate receptor. *Cell* **127,** 85–97 (2006).
37. Salussolia, C. L. *et al.* Interaction of the M4 segment with other transmembrane segments is required for surface expression of mammalian α-amino-3-hydroxy-5-methyl-4-isoxazolepropionic acid (AMPA) receptors. *J. Biol. Chem.* **286,** 40205–40218 (2011).
38. Salussolia, C. L. *et al.* A eukaryotic specific transmembrane segment is required for tetramerization in AMPA receptors. *J. Neurosci.* **33,** 9840–9845 (2013).
39. Gan, Q., Dai, J., Zhou, H. X. & Wollmuth, L. P. The transmembrane domain mediates tetramerization of α-Amino-3-hydroxy-5-methyl-4-isoxazolepropionic acid (AMPA) receptors. *J. Biol. Chem.* **291,** 6595–6606 (2016).
40. Shaikh, S. A. *et al.* Stargazin modulation of AMPA receptors. *Cell Reports* **17,** 328–335 (2016).
41. Leuschner, W. D. & Hoch, W. Subtype-specific assembly of α-amino-3-hydroxy-5-methyl-4-isoxazole propionic acid receptor subunits is mediated by their N-terminal domains. *J. Biol. Chem.* **274,** 16907–16916 (1999).
42. Papadakis, M., Hawkins, L. M. & Stephenson, F. A. Appropriate NR1-NR1 disulfide-linked homodimer formation is requisite for efficient expression of functional, cell surface *N*-methyl-D-aspartate NR1/NR2 receptors. *J. Biol. Chem.* **279,** 14703–14712 (2004).
43. Ayalon, G., Segev, E., Elgavish, S. & Stern-Bach, Y. Two regions in the N-terminal domain of ionotropic glutamate receptor 3 form the subunit oligomerization interfaces that control subtype-specific receptor assembly. *J. Biol. Chem.* **280,** 15053–15060 (2005).
44. Ayalon, G. & Stern-Bach, Y. Functional assembly of AMPA and kainate receptors is mediated by several discrete protein-protein interactions. *Neuron* **31,** 103–113 (2001).
45. Jin, R. *et al.* Crystal structure and association behaviour of the GluR2 amino-terminal domain. *EMBO J.* **28,** 1812–1823 (2009).
46. Watson, J. F., Ho, H. & Greger, I. H. Synaptic transmission and plasticity require AMPA receptor anchoring via its N-terminal domain. *eLife* **6,** e23024 (2017).
47. Cull-Candy, S. G. & Usowicz, M. M. Multiple-conductance channels activated by excitatory amino acids in cerebellar neurons. *Nature* **325,** 525–528 (1987).
48. Jahr, C. E. & Stevens, C. F. Glutamate activates multiple single channel conductances in hippocampal neurons. *Nature* **325,** 522–525 (1987).
49. Rosenmund, C., Stern-Bach, Y. & Stevens, C. F. The tetrameric structure of a glutamate receptor channel. *Science* **280,** 1596–1599 (1998).
50. Smith, T. C. & Howe, J. R. Concentration-dependent substate behavior of native AMPA receptors. *Nat. Neurosci.* **3,** 992–997 (2000).
51. Poon, K., Ahmed, A. H., Nowak, L. M. & Oswald, R. E. Mechanisms of modal activation of GluA3 receptors. *Mol. Pharmacol.* **80,** 49–59 (2011).
52. Prieto, M. L. & Wollmuth, L. P. Gating modes in AMPA receptors. *J. Neurosci.* **30,** 4449–4459 (2010).
53. Swanson, G. T., Kamboj, S. K. & Cull-Candy, S. G. Single-channel properties of recombinant AMPA receptors depend on RNA editing, splice variation, and subunit composition. *J. Neurosci.* **17,** 58–69 (1997).
54. Kohda, K., Wang, Y. & Yuzaki, M. Mutation of a glutamate receptor motif reveals its role in gating and δ2 receptor channel properties. *Nat. Neurosci.* **3,** 315–322 (2000).

Supplementary Information is available in the online version of the paper.

Acknowledgements We thank H. Kao for computational support, A. des Georges, I. S. Fernandez, M. Fislage and A. K. Singh for processing advice. E.C.T. is supported by National Institutes of Health (NIH) F31 NS093838. A.I.S. is supported by the NIH (R01 NS083660, R01 CA206573), the Pew Scholar Award in Biomedical Sciences, and the Irma T. Hirschl Career Scientist Award. J.F. is supported by the Howard Hughes Medical Institute and the NIH (R01 GM029169). Cryo-EM data were collected at the Columbia University Medical Center cryo-EM facility and at the Simons Electron Microscopy Center and National Resource for Automated Molecular Microscopy located at the New York Structural Biology Center, supported by grants from the Simons Foundation (349247), NYSTAR, and the NIH National Institute of General Medical Sciences (GM103310).

Author Contributions E.C.T. designed the constructs, prepared the protein samples, carried out cryo-EM data collection and processing, built models, analysed data and wrote the manuscript. M.V.Y. carried out electrophysiology experiments, assisted in protein production and edited the manuscript. R.A.G. assisted in cryo-EM data collection. J.F. advised on the cryo-EM workflow and provided funding. A.I.S. supervised the project, built models, analysed data, wrote the manuscript and provided funding. E.C.T. and A.I.S. designed the project.

Author Information Reprints and permissions information is available at www.nature.com/reprints. The authors declare no competing financial interests. Readers are welcome to comment on the online version of the paper. Publisher's note: Springer Nature remains neutral with regard to jurisdictional claims in published maps and institutional affiliations. Correspondence and requests for materials should be addressed to A.I.S. (as4005@cumc.columbia.edu).

Reviewer Information *Nature* thanks M. Mayer, S. Traynelis and the other anonymous reviewer(s) for their contribution to the peer review of this work.

© 2017 Macmillan Publishers Limited, part of Springer Nature. All rights reserved.

RESEARCH ARTICLE

METHODS

Data reporting. No statistical methods were used to predetermine sample size. The experiments were not randomized, and investigators were not blinded to allocation during experiments and outcome assessment.

Construct. The fusion constructs used for cryo-EM were prepared by introducing a single GT linker between rat GluA2$_{flip}$ construct described in literature[11] and mouse GSG1L truncated after Phe237[16] or STZ truncated after Leu207[15], in the pEG BacMam vector[55], with a C-terminal enhanced green fluorescent protein (eGFP) for fluorescence-detection size-exclusion chromatography (FSEC)[56] profiling and monitoring during expression, in addition to a streptavidin affinity tag for purification purposes. The amino acids Thr-Gly-Gly were included between the C terminus of GSG1L or STZ and a thrombin cleavage site (Leu-Val-Pro-Arg-Gly-Ser) before the C-terminal eGFP and affinity tag.

Electrophysiology. DNA encoding wild-type GluA2 or GluA2–STZ was introduced into a plasmid for expression in eukaryotic cells that was engineered to produce green fluorescent protein via a downstream internal ribosome entry site[11]. Human embryonic kidney 293 (HEK293) cells (ATCC CRL-1573) grown on glass coverslips in 35 mm dishes were transiently transfected with 1–5 μg of plasmid DNA using Lipofectamine 2000 Reagent (Invitrogen). Recordings were made 24–96 h after transfection at room temperature. Currents from whole cells, typically held at a −60 mV potential, were recorded using Axopatch 200B amplifier (Molecular Devices, LLC), filtered at 5 kHz and digitized at 10 kHz using low-noise data acquisition system Digidata 1440A and pCLAMP software (Molecular Devices, LLC). The external solution contained (in mM): 140 NaCl, 2.4 KCl, 4 $CaCl_2$, 4 $MgCl_2$, 10 HEPES pH 7.3 and 10 glucose; 7 mM NaCl was added to the extracellular activating solution containing 3 mM L-glutamate (Glu). The internal solution contained (in mM): 150 CsF, 10 NaCl, 10 EGTA, 20 HEPES pH 7.3. Rapid solution exchange was achieved with a two-barrel theta glass pipette controlled by a piezoelectric translator. Typical 10–90% rise times were 200–300 μs, as measured from junction potentials at the open tip of the patch pipette after recordings. Data analysis was performed using the computer program Origin 9.1.0 (OriginLab Corp.).

Expression and purification. The GluA2–GSG1L and GluA2–STZ bacmid and baculovirus were made using standard methods[55], and the following methods were applied to the expression and purification of both constructs. P2 virus, produced from Sf9 cells (GIBCO 12659017), was added to HEK293S GnTI$^-$ cells (ATCC CRL-3022) at 37 °C and 5% CO_2. At 12 h after infection, 10 mM sodium butyrate was added the temperature was changed to 30 °C. At 72 h after infection, cells were collected by low-speed centrifugation (5,500g, 15 min). Cells were then washed with PBS pH 8.0 and pelleted with low speed centrifugation (5,500g, 15 min). The washed cells were lysed with a Misonix sonicator (18 × 10 s, power level 7) in 150 mM NaCl, 20 mM Tris-HCl (pH 8.0) containing protease inhibitors (0.8 μM aprotinin, 2 μg ml^{-1} leupeptin, 2 μM pepstatin A and 1 mM phenylmethysulfonyl fluoride); 50 ml was used per 1 l HEK293 cells. Next, the lysate was clarified after centrifugation (9,900g, 15 min), and the membranes were collected by ultracentrifugation (186,000g, 40 min). The membranes were collected and mechanically homogenized, then solubilized for 2 h in 150 mM NaCl, 20 mM Tris-HCl pH 8.0, 1% digitonin (Cayman Chemical Company, 14952). Insoluble material was removed by ultracentrifugation (186,000g, 40 min). Streptavidin-linked resin was added to the soluble material (1 ml resin per 1 l cells), and left to bind for 10–14 h. The resin was washed with 10 column volumes of 150 mM NaCl, 20 mM Tris-HCl pH 8.0, 0.1% digitonin (0.05% digitonin was used in the GluA2–STZ purification). Elution was done with the same buffer but supplemented with 2.5 mM D-desthiobiotin. The sample was then concentrated and digested with thrombin (1:200 mass ratio of thrombin to eluted protein) for 2 h at 22 °C. The sample was then loaded onto a size-exclusion chromatography column (Superose-6) equilibrated with 150 mM NaCl, 20 mM Tris-HCl pH 8.0, 0.1% digitonin (0.05% digitonin was used in the GluA2–STZ purification). The peak fractions were pooled for cryo-EM specimen preparation, and concentrated to 5 mg ml^{-1} (4 mg ml^{-1} for GluA2–STZ). All steps, unless otherwise noted, were performed at 4 °C.

Cultured cells. HEK293 cells were used for patch-clamp electrophysiology. Cells were obtained from ATCC and were not further authenticated. Cells were grown and maintained using standard protocols at 37 °C with 5% CO_2, in DMEM media (Corning) supplemented with 10% heat inactivated fetal bovine serum (Crystalgen). Before recordings, cells were plated on glass coverslips in 35 mm dishes and transiently transfected with 1–5 mg of plasmid DNA using Lipofectamine 2000 Reagent (Invitrogen). Cells were passed twice a week and were used until the 30th passage.

HEK293S GnTI$^-$ cells were used to express protein for cryo-EM. Cells were obtained from ATCC and were not further authenticated. Cells were grown and maintained at 37 °C and 5% CO_2 using standard protocols, in Freestyle 293 media (GIBCO) supplemented with 2% heat inactivated fetal bovine serum (Crystalgen). Cells were passed twice a week and were used until the 30th passage.

Sf9 insect cells were used to make baculovirus for infecting HEK293S GnTI$^-$ cells. Cells were obtained from GIBCO and were not further authenticated. Cells were grown and maintained at 27 °C using standard protocols, in Sf-900 III SFM (GIBCO). Cells were passed twice a week and were used until the 30th passage. All cell lines were tested for mycoplasma contamination by the providers (mycoplasma test negative) and have not been retested again.

Cryo-EM sample preparation and data collection. C-flat (Protochips, Inc., Morrisville, NC) CF-1.2/1.3-2Au 200 mesh holey carbon grids were coated with gold using an Edwards Auto 306 evaporator. Carbon was removed using Ar/O_2 (6 min, 50 W, 35.0 sccm Ar, 11.5 sccm O_2) plasma treatment with a Gatan Solarus 950 advanced plasma system, making Au 1.2/1.3 holey on Au mesh grids[57]. To make the surface hydrophilic before sample application, the grid surface was plasma-treated with H_2O_2 (20 s, 10 W, 6.4 sccm H_2, 27.5 sccm O_2). Frozen-hydrated grids were prepared using a Vitrobot Mark IV (FEI). 3 μl of sample (5 mg ml^{-1} GluA2–GSG1L with or without 0.3 mM antagonist ZK200775) was applied to the plasma-treated grids using 3.0 s blotting time, 3 blot force, 30.0 s wait time and 100% humidity at 22 °C. For the open state, 4 mg ml^{-1} GluA2–STZ was incubated for 15 min with 100 μM CTZ, then 3 μl of this was quickly added and pipetted up-and-down in a mixture of 0.5 μl 700 mM glutamate (final concentration 100 mM). 3 μl of this mixture was added to the grid, blotted as described above.

The GluA2–GSG1L data (apo/ZK) were collected on a Tecnai F30 Polara (FEI) microscope operating at 300 kV, using Leginon[58] with a Gatan K2 Summit electron detection (DED) camera (Gatan) in counting mode with a pixel size of 0.98 Å. Data were collected across 40 frames (0.2 s per frame), with a dose rate of 8 e$^-$ pixel^{-1} s^{-1} (total dose of approximately 67 e$^-$ Å^{-2}), within defocus range −1.5 μm to −3.5 μm, C_s 2.26 mm. The GluA2–STZ data were collected on a Titan Krios microscope (FEI) operating at 300 kV, using Leginon with a post-GIF quantum energy filter (20 eV slit) and Gatan K2 Summit direct camera DED camera in counting mode operating at a pixel size of 1.08 Å pixel^{-1}. A dose rate of 8 e$^-$ pixel^{-1} s^{-1} (total dose of approximately 55 e$^-$ Å^2) was used across 40 frames (0.2 ms per frame), within the defocus range −1.0 μm to −3.0 μm.

Image processing. All frame alignment and dose-weighting was carried out in MotionCor2[59]. CTF correction, with gCTF[60], was performed on non-dose-weighted micrographs, with all further processing on dose-weighted micrographs using RELION 2.0[61]. From 2,938 micrographs for GluA2–GSG1L bound to ZK200775, 541,261 particles were picked. The data were binned to 3.92 Å pixel^{-1} and subject to 2D classification, which resulted in cleaning the particle pool to 513,406 particles. These particles were then subject to 3D classification without symmetry into ten classes and the GluA2–2×GSG1L$_{ZK}$ model[16] low-pass filtered to 40 Å as a reference. Three classes, with two showing different ECD conformations (that is, state 1 and state 2) and two auxiliary subunits bound were chosen for additional 3D sub-classification. This resulted in a group of 234,426 particles, which were unbinned to a pixel size of 1.96, and subject to classification with the GluA2–2×GSG1L$_{ZK}$ model (40 Å filter) as the reference. This resulted in two well-structured groups of particles: GluA2–GSG1L$_{ZK\text{-}1}$ (26,971 particles) and GluA2–GSG1L$_{ZK\text{-}2}$ (41,926 particles). The particles were unbinned to the original 0.98 Å pixel^{-1} size, and refined with C2-symmetry. Of note is that we did not observe any stoichiometric heterogeneity, as was observed in cryo-EM performed on the GluA2–GSG1L complex solubilized in DDM[16].

For the GluA2–GSG1L collection in the absence of ligand, 242,078 particles were automatically picked from 2,593 micrographs using six 2D classes as references (generated from manual picking as described above). The particle images were binned to 3.92 Å pixel^{-1}, and 235,543 particles remained following 2D classification. Next, 3D classification was carried out without symmetry into ten classes with the 40 Å low-pass filtered GluA2–GSG1L$_{ZK\text{-}1}$ map. Six classes, numerating 115,120 particles showed structural details that warranted further image processing. The particles were unbinned to 0.98 Å pixel^{-1} and classified in 3D into 10 classes with the GluA2–GSG1L$_{ZK\text{-}1}$. Resulting were two prominent classes, GluA2–GSG1L$_{apo\text{-}1}$ (20,392 particles) and GluA2–GSG1L$_{apo\text{-}2}$ (18,926 particles), which were refined with C2-symmetry, and resemble the two states observed for GluA2–GSG1L in the presence of ZK200775 (Extended Data Figs 3, 4).

A total of 4,116 micrographs were collected for GluA2–STZ, and 595,889 particles were automatically picked, with manually picked particles used to generate reference classes as described above. The particles were binned twice to 2.16 Å pixel^{-1} and subject to 2D classification, which resulted in 581,495 particles being subject to further image processing. The original GluA2–2xSTZ antagonist-bound map[15], low-pass filtered to 40 Å, was used as a reference model for 3D classification into ten classes without symmetry imposed. 3D classification resulted in 278,454 particles, from four classes, being chosen for further image processing, without observation of stoichiometric heterogeneity as in the DDM-solubilized complex[15], and the particles were unbinned to 1.08 Å pixel^{-1} and classified in 3D to ten classes. Particles from three classes, numerating 69,207

© 2017 Macmillan Publishers Limited, part of Springer Nature. All rights reserved.

particles and showing structurally similar features, were chosen for refinement. Initial refinement, with C2-symmetry, resulted in a 4.2 Å map, which showed side-chain features in the GluA2 core but was lacking details for STZ. To align particles better according to the GluA2 TMD and STZ, we generated a soft spherical mask around the detergent micelle (which encompassed all of the GluA2 and STZ TMDs and STZ extracellular head). This resulted in an overall 4.0 Å map, with distinct side-chain features throughout the TMD for model building of the GluA2 and STZ TMDs. We did not observe a digitonin-bound state of GluA2–$STZ_{Glu+CTZ}$, as we did for GluA2–GSG1L in the resting states (GluA2NGSG1L_{ZK-2}, GluA2–GSG1L_{apo-2}).

The resolution for all reconstructions was estimated using the Fourier shell correlation (FSC) = 0.143 criterion[62] between independent half maps on corrected FSC curves in which the influences of the mask were removed[63] (Extended Data Fig. 1). All maps were post-processed using a softmask in RELION, and B-factors for map sharpening were automatically estimated (Extended Data Table 1). All visualization of electron microscopy densities was done in UCSF Chimera[64]. Local resolution for each map (Extended Data Fig. 2) was calculated with unfiltered half maps using ResMap[65].

Model building. To build the state 1 ZK200775-bound and GluA2–GSG1L models, we individually isolated the ATD and LBD dimers, as well as the TMD, from the GluA2–2×GSG1L_{ZK} complex structure[16] using rigid-body fitting in COOT[66]. Side chains in the TMD were defined based on local resolution, as was the structure of the M2 helix and pore loop. The resulting model was refined against an unfiltered half map (work) in real space with constraints using PHENIX[67]. The refined model was tested for overfitting (Extended Data Fig. 2) by shifting the coordinates with shake in PHENIX and building a density in EMAN2[68] from the shaken model. FSC values were calculated between the densities from the shaken model, the map used in PHENIX refinement (work), second half map (free) and unfiltered sum map, using EMAN2. The resulting GluA2–GSG1L_{ZK-1} model was used as a basis to build GluA2–GSG1L_{apo-1}. Correspondingly, the ATD and LBD local dimers were moved as rigid bodies to define the state 2 models (GluA2–GSG1L_{ZK-2} and GluA2–GSG1L_{apo-2}), and features were adjusted manually according to high resolution features in GluA2–GSG1L_{ZK-2}. Secondary structure features of the M3 pore were further defined in the GluA2–GSG1L_{ZK-2} M3 density. The resulting models were also real space refined in PHENIX and tested for overfitting (Extended Data Fig. 2). State 2 models were refined in the absence of digitonin.

For GluA2–$STZ_{Glu+CTZ}$, we rigid-body fit the high-resolution structures of the ATD[45] and glutamate/CTZ-bound LBD[35]. We individually rigid body fit helices from the TMD of our GluA2–GSG1L_{ZK-1} refined structure, and rigid-body fit STZ subunits from the GluA2–2xSTZ_{ZK} structure[15] in COOT for an initial model. After increasing the resolution in the TMD map, we built STZ and the GluA2 TMD *de novo*. The resulting models were also real space refined in PHENIX and tested for overfitting (Extended Data Fig. 2), of which there are no signs of over-fit models. Structures were visualized and figures were prepared in Pymol[67].

Data availability. Cryo-EM density maps have been deposited in the Electron Microscopy Data Bank (EMDB) under accession numbers EMD-8819 (GluA2–GSG1L_{ZK-1}), EMD-8820 (GluA2–GSG1LL_{ZK-2}), EMD-8821 (GluA2–GSG1L_{apo-1}), EMD-8822 (GluA2–GSG1L_{apo-2}), and EMD-8823 (GluA2–$STZ_{Glu+CTZ}$, including directed TMD map). Model coordinates have been deposited in the Protein Data Bank (PDB) under accession numbers 5WEK (GluA2–GSG1L_{ZK-1}), 5WEL (GluA2–GSG1LL_{ZK-2}), 5WEM (GluA2–GSG1L_{apo-1}), 5WEN (GluA2–GSG1L_{apo-2}), and 5WEO (GluA2–$STZ_{Glu+CTZ}$). All other data are available from the corresponding author upon reasonable request.

55. Goehring, A. *et al.* Screening and large-scale expression of membrane proteins in mammalian cells for structural studies. *Nat. Protocols* **9,** 2574–2585 (2014).
56. Kawate, T. & Gouaux, E. Fluorescence-detection size-exclusion chromatography for precrystallization screening of integral membrane proteins. *Structure* **14,** 673–681 (2006).
57. Russo, C. J. & Passmore, L. A. Electron microscopy: ultrastable gold substrates for electron cryomicroscopy. *Science* **346,** 1377–1380 (2014).
58. Suloway, C. *et al.* Automated molecular microscopy: the new Leginon system. *J. Struct. Biol.* **151,** 41–60 (2005).
59. Zheng, S. Q. *et al.* MotionCor2: anisotropic correction of beam-induced motion for improved cryo-electron microscopy. *Nat. Methods* **14,** 331–332 (2017).
60. Zhang, K. Gctf: Real-time CTF determination and correction. *J. Struct. Biol.* **193,** 1–12 (2016).
61. Kimanius, D., Forsberg, B. O., Scheres, S. H. & Lindahl, E. Accelerated cryo-EM structure determination with parallelisation using GPUs in RELION-2. *eLife* **5,** e18722 (2016).
62. Scheres, S. H. & Chen, S. Prevention of overfitting in cryo-EM structure determination. *Nat. Methods* **9,** 853–854 (2012).
63. Chen, S. *et al.* High-resolution noise substitution to measure overfitting and validate resolution in 3D structure determination by single particle electron cryomicroscopy. *Ultramicroscopy* **135,** 24–35 (2013).
64. Pettersen, E. F. *et al.* UCSF Chimera—a visualization system for exploratory research and analysis. *J. Comput. Chem.* **25,** 1605–1612 (2004).
65. Kucukelbir, A., Sigworth, F. J. & Tagare, H. D. Quantifying the local resolution of cryo-EM density maps. *Nat. Methods* **11,** 63–65 (2014).
66. Emsley, P., Lohkamp, B., Scott, W. G. & Cowtan, K. Features and development of Coot. *Acta Crystallogr. D* **66,** 486–501 (2010).
67. Afonine, P. V. *et al.* Towards automated crystallographic structure refinement with phenix.refine. *Acta Crystallogr. D* **68,** 352–367 (2012).
68. Tang, G. *et al.* EMAN2: an extensible image processing suite for electron microscopy. *J. Struct. Biol.* **157,** 38–46 (2007).

© 2017 Macmillan Publishers Limited, part of Springer Nature. All rights reserved.

RESEARCH ARTICLE

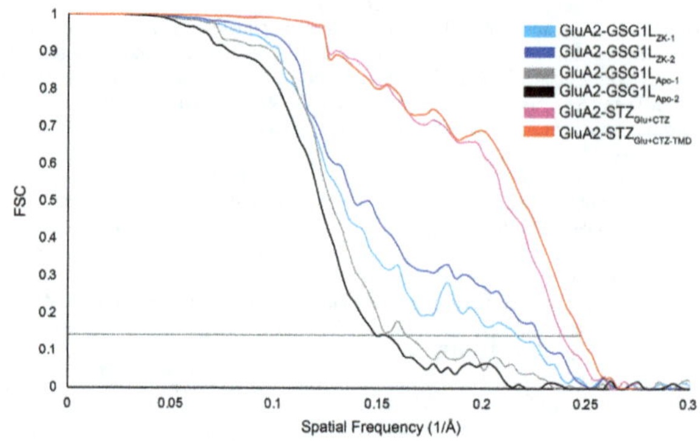

Extended Data Figure 1 | FSC curves for cryo-EM reconstructions. FSC curves calculated between half-maps for GluA2–GSG1L$_{ZK\text{-}1}$, GluA2–GSG1L$_{ZK\text{-}2}$, GluA2–GSG1L$_{apo\text{-}1}$, GluA2–GSG1L$_{apo\text{-}2}$ and GluA2–STZ$_{Glu+CTZ}$ cryo-EM reconstructions, as well as for the GluA2–STZ$_{Glu+CTZ}$ TMD reconstruction from directed refinement. The dashed line indicates FSC = 0.143.

© 2017 Macmillan Publishers Limited, part of Springer Nature. All rights reserved.

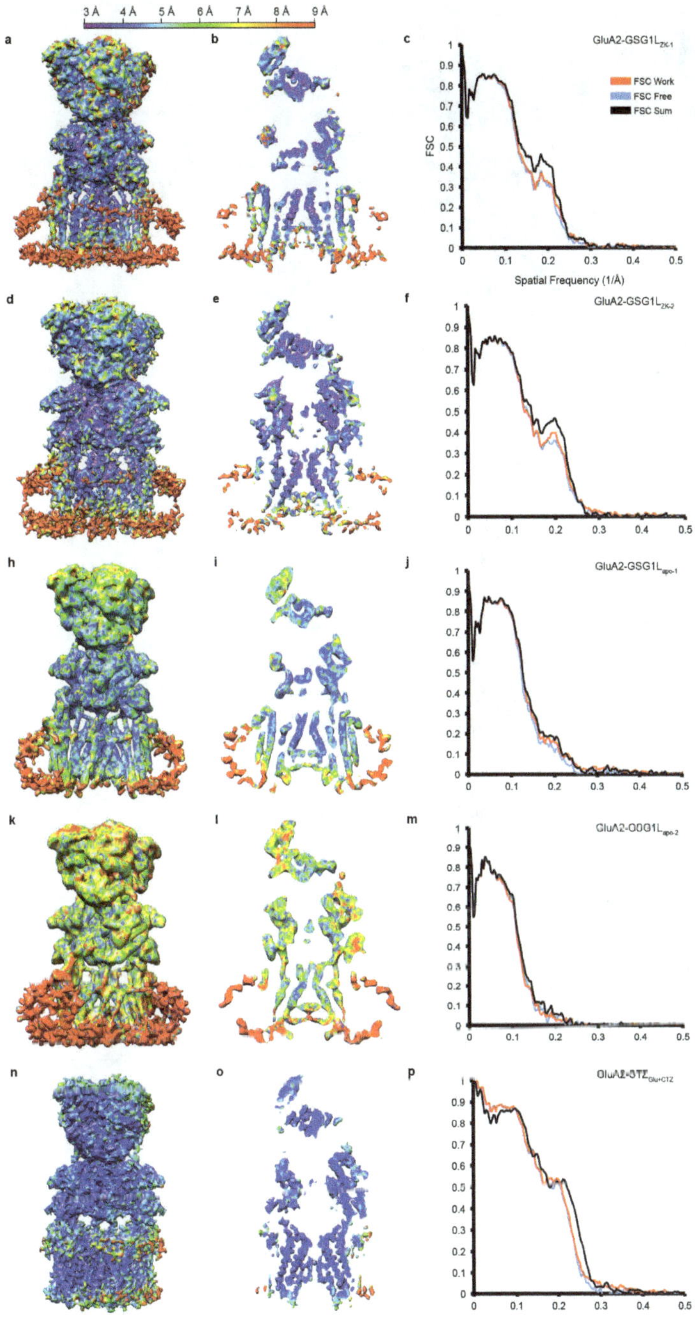

Extended Data Figure 2 | Local resolution and fitting of cryo-EM maps. a–p, Local resolution calculated using Resmap and two unfiltered halves of the reconstruction for GluA2–GSG1L_{ZK-1}, GluA2–GSG1L_{ZK-2}, GluA2–GSG1L_{apo-1}, GluA2–GSG1L_{apo-2} and GluA2–$STZ_{Glu+CTZ}$ structures viewed parallel to the membrane as a surface (**a**, **d**, **h**, **k**, **n**) and slice through the centre of the receptor (**b**, **e**, **i**, **l**, **o**), with the cross-validation FSC curves for the refined model versus unfiltered half maps (one used in the refinement, work, and another one not, free) and the unfiltered summed maps shown on the right (**c**, **f**, **j**, **m**, **p**).

© 2017 Macmillan Publishers Limited, part of Springer Nature. All rights reserved.

RESEARCH ARTICLE

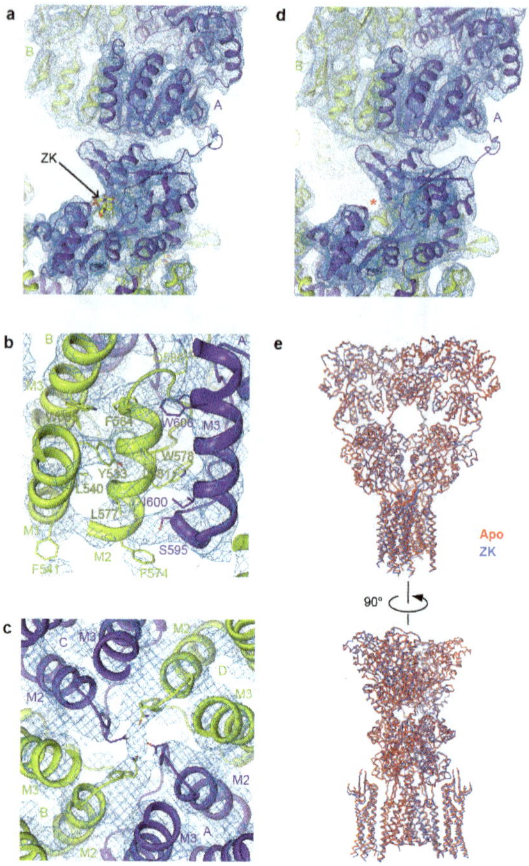

Extended Data Figure 3 | Closed state 1 cryo-EM density and comparison of ZK200775-bound and apo states. a–d, Fragments of GluA2–GSG1$L_{ZK\text{-}1}$ and GluA2–GSG1$L_{apo\text{-}1}$ with the corresponding cryo-EM density. **a, d,** ATD and LBD of subunit A in GluA2–GSG1$L_{ZK\text{-}1}$ (**a**) and GluA2–GSG1$L_{apo\text{-}1}$ (**d**) with density for ZK200775 indicated in the GluA2–2×GSG1$L_{ZK\text{-}1}$ structure. **b, c,** M2 helix (**b**) and selectivity filter (**c**) with the Q/R-site Gln586 side chains pointing towards the centre of the pore in GluA2–GSG1$L_{ZK\text{-}1}$. **e,** Superposition of GluA2–GSG1$L_{ZK\text{-}1}$ (blue) and GluA2–GSG1$L_{apo\text{-}1}$ (red) viewed parallel to the membrane. Note that the structures are almost indistinguishable (r.m.s.d. = 0.526 Å). Densities are shown at 6σ.

© 2017 Macmillan Publishers Limited, part of Springer Nature. All rights reserved.

ARTICLE RESEARCH

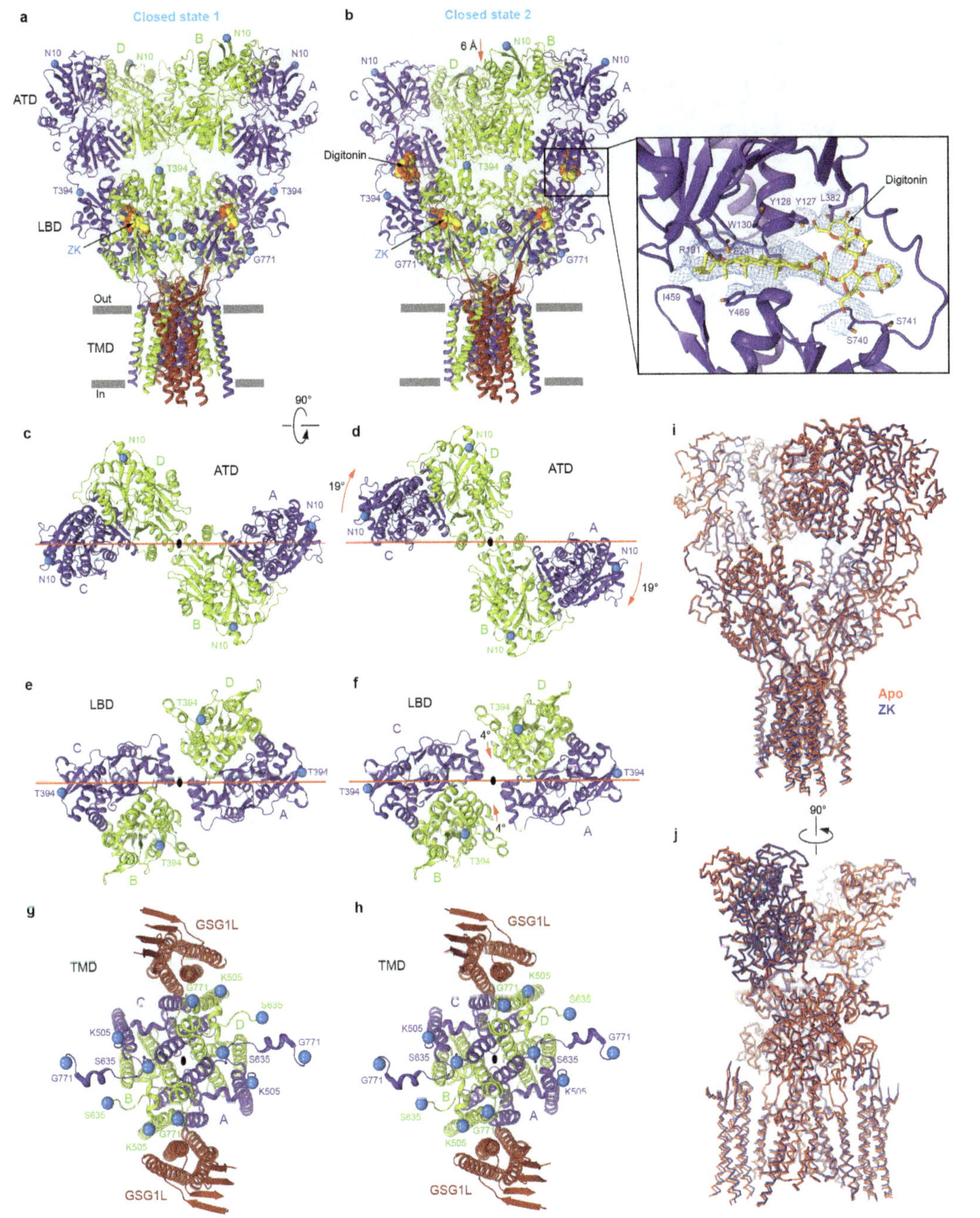

Extended Data Figure 4 | Closed state 2 structure and digitonin-binding pocket. a, **b**, Structures of GluA2–GSG1L$_{ZK-1}$ (**a**) and GluA2–GSG1L$_{ZK-2}$ (**b**) viewed parallel to the membrane. The GluA2 subunits A and C are coloured purple, B and D in green and GSG1L in red. The competitive antagonist ZK200775 and digitonin are shown as space-filling models. In **b**, inset shows expanded view of the boxed region, demonstrating cryo-EM density for digitonin (blue mesh, 4σ). Digitonin and the surrounding residues in the inset are shown in stick representation. **c**–**h**, Top down views along the axis of the overall two-fold rotational symmetry on the ATD (**c**, **d**), LBD (**e**, **f**) and TMD (**g**, **h**) layers. Rigid-body rotation of the ATD tetramer in **d** and rotation of LBD dimers in **f** are indicated by red arrows. **i**, **j**, Superposition of GluA2–GSG1L$_{ZK-2}$ (blue) and GluA2–GSG1L$_{apo-2}$ (red) viewed parallel to the membrane. Note that the structures are almost indistinguishable (r.m.s.d. = 0.701 Å).

© 2017 Macmillan Publishers Limited, part of Springer Nature. All rights reserved.

RESEARCH ARTICLE

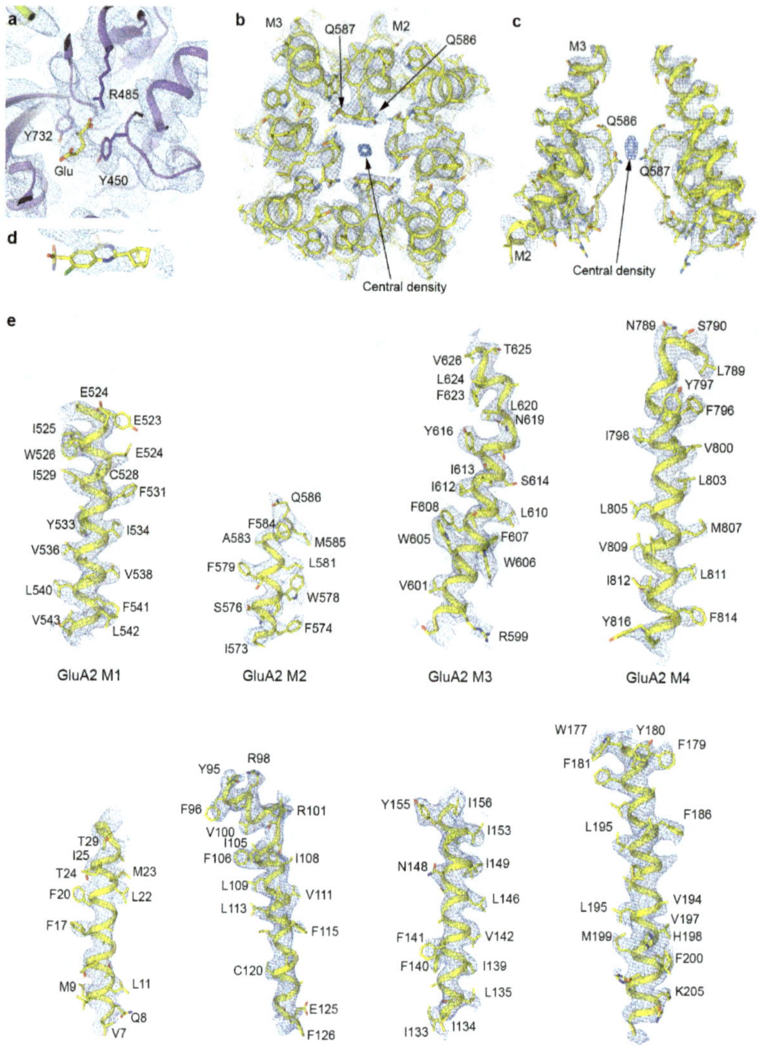

Extended Data Figure 5 | Cryo-EM density for the open state. **a–e**, Fragments of GluA2–STZ$_{Glu+CTZ}$ with the corresponding cryo-EM density. **a**, Zoomed view of the glutamate-binding pocket. **b**, **c**, The ion channel pore with a central density at the selectivity filter, probably for a sodium ion that is hydrated based on the pore diameter, viewed from the top of the selectivity filter looking down into the cytoplasm (**b**) or parallel to the membrane with two (front and back) GluA2 subunits removed (**c**). **d**, Density for CTZ. **e**, Transmembrane domain segments for GluA2 (top row) and STZ (bottom row).

© 2017 Macmillan Publishers Limited, part of Springer Nature. All rights reserved.

ARTICLE RESEARCH

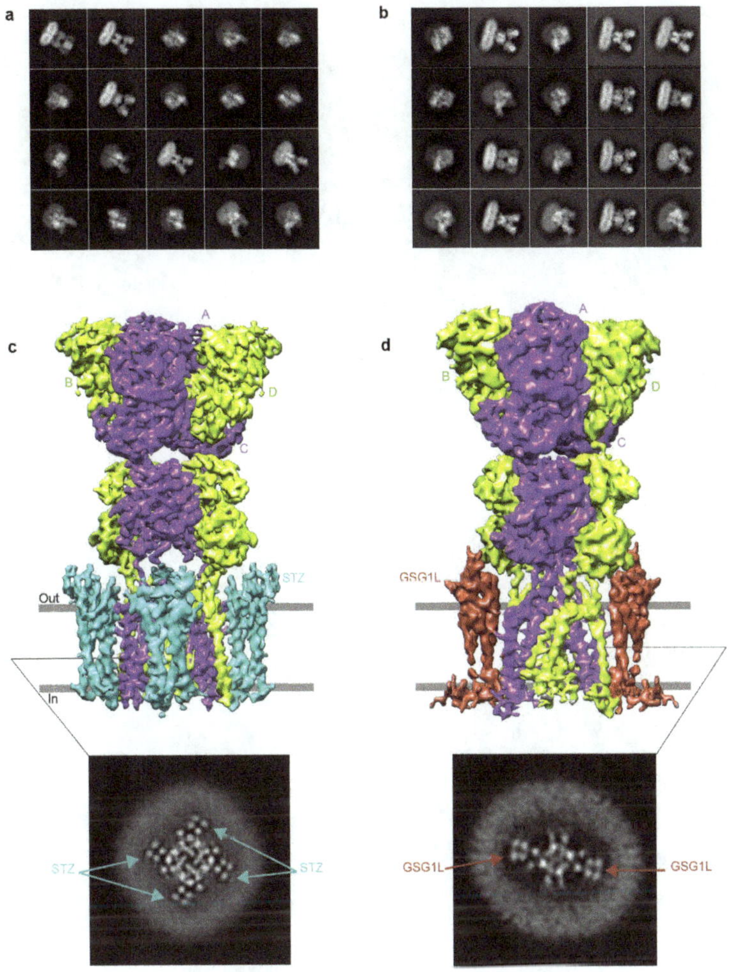

Extended Data Figure 6 | Overview of single-particle cryo-EM and stoichiometry for GluA2–STZ and GluA2–GSG1L solubilized in digitonin. a, b, Two-dimensional class averages for GluA2–$STZ_{Glu+CTZ}$ (**a**) and GluA2nGSG1L_{ZK} (**b**) indicating three-layer architecture of the particles. **c, d,** Final densities for GluA2–$STZ_{Glu+CTZ}$ (**c**) and GluA2–GSG1L_{ZK-1} (**d**) with the GluA2 subunits A and C coloured purple, B and D in green, STZ in cyan and GSG1L in red. Insets show 2D slices made parallel to the membrane through the refined, nonfiltered map. Note, although four STZ molecules bind one receptor, only two copies of GSG1L can bind per GluA2 tetramer.

© 2017 Macmillan Publishers Limited, part of Springer Nature. All rights reserved.

RESEARCH ARTICLE

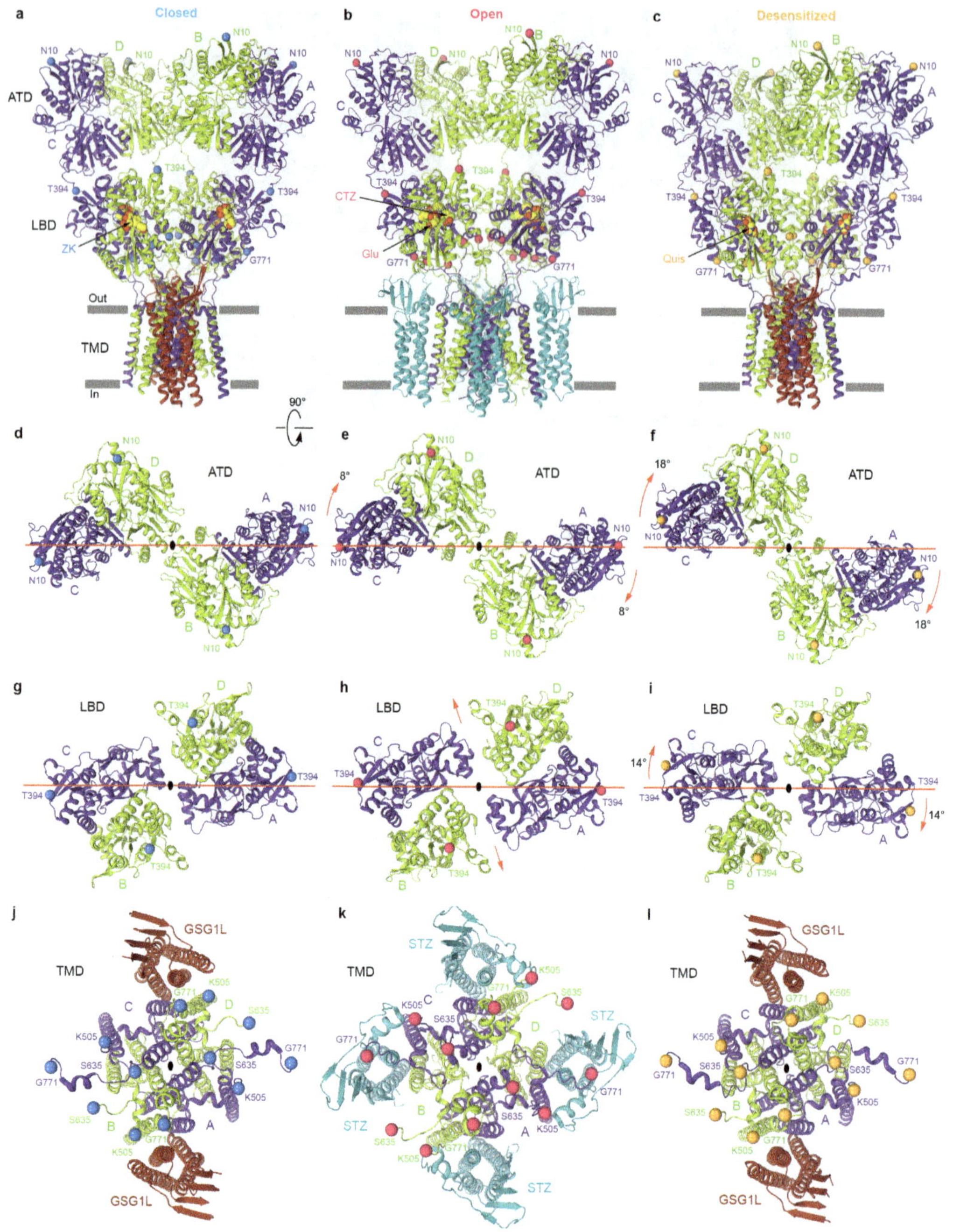

Extended Data Figure 7 | Conformational differences between the closed, open and desensitized states. a–c, Structures of GluA2–GSG1L$_{ZK-1}$ in the closed state (**a**), GluA2–STZ$_{Glu+CTZ}$ in the open state (**b**) and GluA2–2×GSG1L$_{Quis}$ in the desensitized state (**c**), viewed parallel to the membrane. The GluA2 subunits A and C are coloured purple, B and D in green, GSG1L in red and STZ in cyan. The competitive antagonist ZK200775, agonists Glu and Quis and positive allosteric modulator CTZ are shown as space-filling models. **d–l,** Top down views along the axis of the overall two-fold rotational symmetry on the layers of ATD (**d–f**), LBD (**g–i**) and TMD (**j–l**). Rigid-body rotation of the ATD tetramers in **e** and **f**, broadening of LBD layer in **h** and rotation of subunit A/C LBDs in **i** are indicated by red arrows. Note the dramatic opening in the middle of the LBD layer (**h**) and pore dilation (**k**) in the open state.

© 2017 Macmillan Publishers Limited, part of Springer Nature. All rights reserved.

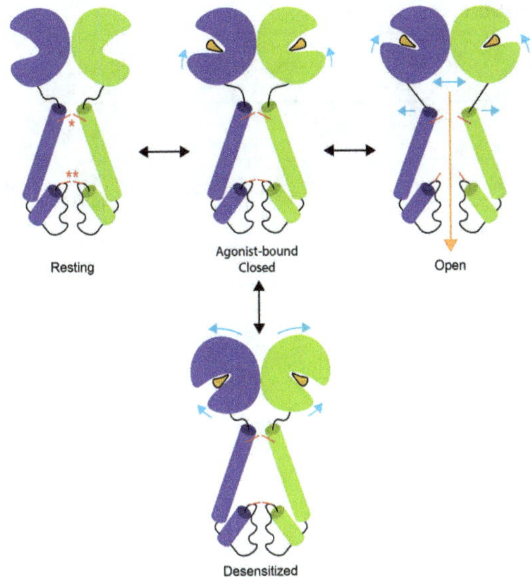

Extended Data Figure 8 | iGluR gating mechanism. Two out of four iGluR subunits are shown with the ATDs omitted. Four basic states of iGluR gating are illustrated: resting, represented by apo (GluA2–GSG1L$_{apo-1}$) or antagonist-bound closed state (GluA2–GSG1L$_{ZK-1}$) structures; closed, agonist-bound (pre-active state crystal structures[10,11]); open (GluA2–STZ$_{Glu+CTZ}$); and desensitized (GluA2–2×GSG1L$_{Quis}$ complex[16]). Transitions between the states are indicated by black arrows, conformational rearrangements by blue arrows, and ionic current through the open channel by an orange arrow. Upper and lower gates are indicated by one and two red asterisks, respectively, with red sticks at the upper gate representing channel occluding residues at the bundle crossing and the Q/R site at the lower gate. Glutamate molecules are illustrated by orange wedges. The receptor sits in a resting, closed state, with its LBD clamshells in the maximally open conformations, unoccupied by the neurotransmitter glutamate. Upon glutamate binding, the LBD clamshells close, as described in the pre-activated crystal structures, to an intermediate state that does not put enough strain on the LBD–TMD linkers to open the channel. The LBDs then transition to their maximally closed state, which strains the LBD–TMD linkers, causing the channel pore to open and conduct ions. Most AMPARs, however, quickly desensitize, transitioning to the desensitized state from the open state via the agonist-bound, closed state. Desensitization is accompanied by the rupture of the upper LBD interfaces, with the LBDs adapting their maximally closed clamshell conformations, as described in the desensitized-state GluA2–GSG1L complex.

© 2017 Macmillan Publishers Limited, part of Springer Nature. All rights reserved.

RESEARCH ARTICLE

Extended Data Table 1 | Cryo-EM data collection, refinement and validation statistics

	GluA2-GSG1L$_{ZK-1}$ (EMDB-8819) (PDB 5WEK)	GluA2-GSG1L$_{ZK-2}$ (EMDB-8820) (PDB 5WEL)	GluA2-GSG1L$_{apo-1}$ (EMDB-8821) (PDB 5WEM)	GluA2-GSG1L$_{apo-2}$ (EMDB-8822) (PDB 5WEN)	GluA2-STZ$_{Glu+CTZ}$ (EMDB-8823) (PDB 5WEO)
Data collection and processing					
Magnification	39,000x		39,000x		130,000x
Voltage (kV)	300		300		300
Electron exposure (e⁻/Å²)	67		67		55
Defocus range (μm)	-1.5 to -3.5		-1.5 to -3.5		-1.0 to -3.0
Pixel size (Å)	0.98		0.98		1.08
Symmetry imposed	C2		C2		C2
Initial particle images (no.)	541,261		242,078		595,889
Final particle images (no.)	26,971	41,926	20,392	18,926	69,207
Map resolution (Å) FSC=0.143	4.6	4.4	6.1	6.8	4.2
Map resolution range (Å)	3 to 9	3 to 9	4 to 9	4 to 9	3 to 9
Refinement					
Initial model used (PDB code)	5VHY	5VHY	5VHY	5VHY	5KBU
Model resolution (Å) FSC=0.143	4.6	4.4	6.1	6.8	4.2
Model resolution range (Å)	3 to 9	3 to 9	4 to 9	4 to 9	3 to 9
Map sharpening *B* factor (Å²)	-172	-196	-225	-653	-160
Model composition					
Non-hydrogen atoms	27,468	27,468	27,360	27,360	30,940
Protein residues	27,360	27,360	27,360	27,360	30,804
Ligands	108	108	N/A	N/A	136
B factors (Å²)					
Protein	263.37	248.89	337.97	409.74	240.10
Ligand	217.60	88.03	N/A	N/A	113.71
R.m.s. deviations					
Bond lengths (Å)	0.01	0.01	0.01	0.01	0.01
Bond angles (°)	1.18	1.25	1.37	1.25	1.16
Validation					
MolProbity score	1.69	1.81	1.85	1.86	2.00
Clashscore	6.12	6.79	7.83	7.84	9.25
Poor rotamers (%)	0.23	0.34	0.14	0.37	0.59
Ramachandran plot					
Favored (%)	93.11	93.22	93.48	93.34	91.10
Allowed (%)	5.48	6.63	6.34	6.55	8.20
Outliers (%)	1.42	0.14	0.17	0.12	0.70

© 2017 Macmillan Publishers Limited, part of Springer Nature. All rights reserved.

Corpus Acknowledgments

We wish to thank the authors and the following publishers for their permission to reproduce the articles found in this volume.

	Article	*Publisher*
1.	**Frank**, J. (1969). (in German) Nachweis von Objektbewegungen im lichtoptischen Diffraktogramm von elektronenmikroskopischen Aufnahmen [Detection of object movement in the optical diffractograms of electron micrographs]. *Optik* **30**: 171–180. [Translated version of paper included.]	Elsevier
2.	**Frank**, J. (1973). The envelope of electron microscopic transfer functions for partially coherent illumination. *Optik* **38**: 519–539.	Elsevier
3.	**Frank**, J. (1975). Averaging of low exposure electron micrographs of non-periodic objects. *Ultramicroscopy* **1**: 159–162.	Elsevier
4.	**Frank**, J., and Al-Ali, L. (1975). Signal-to-noise ratio of electron micrographs obtained by cross correlation. *Nature* **256**: 376–379. https://doi.org/10.1038/256376a0	Nature Publishing Group
5.	**Saxton**, W.O., and Frank, J. (1977). Motif detection in quantum noise-limited electron micrographs by cross-correlation. *Ultramicroscopy* **2**: 219–227.	Elsevier
6.	**Frank**, J., Goldfarb, W., Eisenberg, D., and Baker, T.S. (1978). Reconstruction of glutamine synthetase using computer averaging. *Ultramicroscopy* **3**: 283–290.	Elsevier
7.	**Frank**, J., Shimkin, B., and Dowse, H. (1981). SPIDER — A modular software system for electron image processing. *Ultramicroscopy* **6**: 343–358.	Elsevier
8.	**Frank**, J., Verschoor, A., and Boublik, M. (1981). Computer averaging of electron micrographs of 40S ribosomal subunits. *Science* **214**: 1353–1355.	The American Association for the Advancement of Science

9. **van Heel**, M., and Frank, J. (1981). Use of multivariate statistics in analysing the images of biological macromolecules. *Ultramicroscopy* **6**: 187–194.	Elsevier
10. **Frank**, J., and van Heel, M. (1982). Correspondence analysis of aligned images of biological particles. *J. Mol. Biol.* **161**: 134–137.	Elsevier
11. **Bretaudiere**, J.-P., and Frank, J. (1986). Reconstitution of molecule images analysed by correspondence analysis: a tool for structural interpretation. *J. Microsc.* **144**: 1–14.	John Wiley and Sons
12. **Radermacher**, M., Wagenknecht, T., Verschoor, A., and Frank, J. (1987). Three-dimensional reconstruction from a single-exposure, random conical tilt series applied to the 50S ribosomal subunit of *Escherichia coli*. *J. Microsc.* **146**: 113–136.	John Wiley and Sons
13. **Verschoor**, A., and Frank, J. (1990). Three-dimensional structure of the mammalian cytoplasmic ribosome. *J. Mol. Biol.* **214**: 737–749.	Elsevier
14. **Penczek**, P., Radermacher, M., and Frank, J. (1992). Three-dimensional reconstruction of single particles embedded in ice. *Ultramicroscopy* **40**: 33–53.	Elsevier
15. **Srivastava**, S., Verschoor, A., and Frank, J. (1992). Eukaryotic initiation factor 3 does not prevent association through physical blockage of the ribosomal subunit–subunit interface. *J. Mol. Biol.* **226**: 301–304.	Elsevier
16. **Penczek**, P.A., Grassucci, R.A., and Frank, J. (1994). The ribosome at improved resolution: new techniques for merging and orientation refinement in 3D cryo-electron microscopy of biological particles. *Ultramicroscopy* **53**: 251–270.	Elsevier
17. **Frank**, J., Zhu, J., Penczek, P., Li, Y., Srivastava, S., Verschoor, A., Radermacher, M., Grassucci, R., Lata, R.K., and Agrawal, R.K. (1995). A model of protein synthesis based on cryo-electron microscopy of the *E. coli* ribosome. *Nature* **376**: 441–444. https://doi.org/10.1038/376441a0	Nature Publishing Group
18. **Agrawal**, R.K., Penczek, P., Grassucci, R.A., Li, Y., Leith, A., Nierhaus, K.H., and Frank, J. (1996). Direct visualization of A-, P-, and E-site transfer RNAs in the *Escherichia coli* ribosome. *Science* **271**: 1000–1002.	The American Association for the Advancement of Science

19. **Beckmann**, R., Bubeck, D., Grassucci, R., Penczek, P., Verschoor, A., Blobel, G., and Frank, J. (1997). Alignment of conduits for the nascent polypeptide chain in the ribosome-Sec61 complex. *Science* **278**: 2123–2126.	The American Association for the Advancement of Science
20. **Penczek**, P.A., Zhu, J., Schröder, R., and Frank, J. (1997). Three dimensional reconstruction with contrast transfer compensation from defocus series. *Scanning Microsc.* **11**: 147–154.	Scanning Microscopy International
21. **Agrawal**, R.K., Penczek, P., Grassucci, R.A., and Frank, J. (1998). Visualization of elongation factor G on the *Escherichia coli* 70S ribosome: the mechanism of translocation. *Proc. Natl. Acad. Sci. USA* **95**: 6134–6138.	National Academy of Sciences, USA
22. **Ban**, N., Freeborn, B., Nissen, P., Penczek, P., Grassucci, R.A., Sweet, R., Frank, J., Moore, P.B., and Steitz, T.A. (1998). A 9Å resolution X-ray crystallographic map of the large ribosomal subunit. *Cell* **93**: 1105–1115.	Elsevier
23. **Gabashvili**, I.S., Agrawal, R.K., Spahn, C.M.T., Grassucci, R.A., Svergun, D.I., Frank, J., and Penczek, P. (2000). Solution structure of the *E. coli* 70S ribosome at 11.5 Å resolution. *Cell* **100**: 537–549.	Elsevier
24. **Frank**, J., and Agrawal, R.K. (2000). A ratchet-like inter-subunit reorganization of the ribosome during translocation. *Nature* **406**: 318–322. https://doi.org/10.1038/35018597	Nature Publishing Group
25. **Beckmann**, R., Spahn, C.M.T., Eswar, N., Helmers, J., Penczek, P.A., Sali, A., Frank, J., and Blobel, G. (2001). Architecture of the protein-conducting channel associated with the translating 80S ribosome. *Cell* **107**: 361–372.	Elsevier
26. **Spahn**, C.M.T., Beckmann, R., Eswar, N., Penczek, P.A., Sali, A., Blobel, G., and Frank, J. (2001a). Structure of the 80S ribosome from *Saccharomyces cerevisiae* — tRNA-ribosome and subunit-subunit interactions. *Cell* **107**: 373–386.	Elsevier
27. **Spahn**, C.M.T., Kieft, J.S., Grassucci, R.A., Penczek, P.A., Zhou, K., Doudna, J.A., and Frank, J. (2001b). Hepatitis C virus IRES RNA-induced changes in the conformation of the 40S ribosomal subunit. *Science* **291**: 1959–1962.	The American Association for the Advancement of Science

28. **Valle**, M., Sengupta, J., Swami, N.K., Grassucci, R.A., Burkhardt, N., Nierhaus, K.H., Agrawal, R.K., and Frank, J. (2002). Cryo-EM reveals an active role for aminoacyl-tRNA in the accommodation process. *EMBO J.* **21**: 3557–3567. — John Wiley and Sons

29. **Gao**, H., Sengupta, J., Valle, M., Korostelev, A., Eswar, N., Stagg, S.M., Van Roey, P., Agrawal, R.K., Harvey, S.C., Sali, A., Chapman, M.S., and Frank, J. (2003). Study of the structural dynamics of the *E. coli* 70S ribosome using real-space refinement. *Cell* **113**: 789–801. — Elsevier

30. **Rawat**, U.B.S., Zavialov, A.V., Sengupta, J., Valle, M., Grassucci, R.A., Linde, J., Vestergaard, B., Ehrenberg, M., and Frank, J. (2003). A cryo-electron microscopic study of ribosome-bound termination factor RF2. *Nature* **421**: 87–90. https://doi.org/10.1038/nature01224 — Nature Publishing Group

31. **Valle**, M., Gillet, R., Kaur, S., Henne, A., Ramakrishnan, V., and Frank, J. (2003). Visualizing tmRNA entry into a stalled ribosome. *Science* **300**: 127–130. — The American Association for the Advancement of Science

32. **Valle**, M., Zavialov, A., Li, W., Stagg, S.M., Sengupta, J., Nielsen, R.C., Nissen, P., Harvey, S.C., Ehrenberg, M., and Frank, J. (2003). Incorporation of aminoacyl-tRNA into the ribosome as seen by cryo-electron microscopy. *Nat. Struct. Biol.* **10**: 899–906. https://www.nature.com/articles/nsb1003 — Nature Publishing Group

33. **Valle**, M., Zavialov, A., Sengupta, J., Rawat, U., Ehrenberg, M., and Frank, J. (2003). Locking and unlocking of ribosomal motions. *Cell* **114**: 123–134. — Elsevier

34. **Yarus**, M., Valle, M., and Frank, J. (2003). A twisted tRNA intermediate sets the threshold for decoding. *RNA* **9**: 384–385. — RNA Society

35. **Spahn**, C.M.T., Gomez-Lorenzo, M.G., Grassucci, R.A., Jørgensen, R., Andersen, G.R., Beckmann, R., Penczek, P.A., Ballesta, J.P.G., and Frank, J. (2004). Domain movements of elongation factor eEF2 and the eukaryotic 80S ribosome facilitate tRNA translocation. *EMBO J.* **23**: 1008–1019. — John Wiley and Sons

36. **Allen**, G.S., Zavialov, A., Gursky, R., Ehrenberg, M., and Frank, J. (2005). The cryo-EM structure of a translation initiation complex from *Escherichia coli*. *Cell* **121**: 703–712. — Elsevier

37. **Gao**, H., Ayub, M.J., Levin, M.J., and Frank, J. (2005). The structure of the 80S ribosome from *Trypanosoma cruzi* reveals unique rRNA components. *Proc. Natl. Acad. Sci. USA* **102**: 10206–10211. — National Academy of Sciences, USA

38. **Penczek**, P.A., Yang, C., Frank, J., and Spahn, C.M.T. (2006). Estimation of variance in single-particle reconstruction using the bootstrap technique. *J. Struct. Biol.* **154**: 168–183. — Elsevier

39. **Frank**, J., Gao, H., Sengupta, J., Gao, N., and Taylor, D.J. (2007). The process of mRNA–tRNA translocation. *Proc. Natl. Acad. Sci. USA* **104**: 19671–19678. — National Academy of Sciences, USA

40. **Scheres**, S.H.W., Gao, H., Valle, M., Herman, G.T., Eggermont, P.P.B., Frank, J., and Carazo, J.-M. (2007). Disentangling conformational states of macromolecules in 3D-EM through likelihood optimization. *Nat. Methods* **4**: 27–29. https://doi.org/10.1038/nmeth992 — Nature Publishing Group

41. **Agirrezabala**, X., Lei, J., Brunelle, J.L., Ortiz-Meoz, R.F., Green, R., and Frank, J. (2008). Visualization of the hybrid state of tRNA binding promoted by spontaneous ratcheting of the ribosome. *Mol. Cell* **32**: 190–197. — Elsevier

42. **LeBarron**, J., Grassucci, R.A., Shaikh, T.R., Baxter, W.T., Sengupta, J., and Frank, J. (2008). Exploration of parameters in cryo-EM leading to an improved density map of the *E. coli* ribosome. *J. Struct. Biol.* **164**: 24–32. — Elsevier

43. **Trabuco**, L.G., Villa, E., Mitra, K., Frank, J., and Schulten, K. (2008). Flexible fitting of atomic structures into electron microscopy maps using molecular dynamics. *Structure* **16**: 673–683. — Elsevier

44. **Villa**, E., Sengupta, J., Trabuco, L.G., LeBarron, J., Baxter, W.T., Shaikh, T.R., Grassucci, R.A., Nissen, P., Ehrenberg, M., Schulten, K., and Frank, J. (2009). Ribosome-induced changes in elongation factor Tu conformation control GTP hydrolysis. *Proc. Natl. Acad. Sci. USA* **106**: 1063–1068. — National Academy of Sciences, USA

45. **Agirrezabala**, X., Liao, H.Y., Schreiner, E., Fu, J., Ortiz-Meoz, R.F., Schulten, K., Green, R., and Frank, J. (2012). Structural characterization of mRNA–tRNA translocation intermediates. *Proc. Natl. Acad. Sci. USA* **109**: 6094–6099. — National Academy of Sciences, USA

46. **Hashem**, Y., des Georges, A., Fu, J., Buss, S.N., Jossinet, F., Jobe, A., Zhang, Q., Liao, H.Y., Grassucci, R.A., Bajaj, C., Westhof, E., Madison-Antenucci, S., and Frank, J. (2013). High-resolution cryo-electron microscopy structure of the *Trypanosoma brucei* ribosome. *Nature* **494**: 385–389. https://doi.org/10.1038/nature11872
Nature Publishing Group

47. **Dashti**, A., Schwander, P., Langlois, R., Fung, R., Li, W., Hosseinizadeh, A., Liao, H.Y., Pallesen, J., Sharma, G., Stupina, V.A., Simon, A.E., Dinman, J.D., Frank, J., and Ourmazd, A. (2014). Trajectories of the ribosome as a Brownian nanomachine. *Proc. Natl. Acad. Sci. USA* **111**: 17492–17497.
National Academy of Sciences, USA

48. **Kinz-Thompson**, C.D., Sharma, A.K., Frank, J., Gonzalez, R.L., and Chowdhury, D. (2015). Quantitative connection between ensemble thermodynamics and single-molecule kinetics: a case study using cryogenic electron microscopy and single-molecule fluorescence resonance energy transfer investigations of the ribosome. *J. Phys. Chem. B* **119**: 10888–10901.
American Chemical Society

49. **Li**, W., Liu, Z., Koripella, R.K., Langlois, R., Sanyal, S., and Frank, J. (2015). Activation of GTP hydrolysis in mRNA-tRNA translocation by elongation factor G. *Science Adv.* **1**: e1500169.
The American Association for the Advancement of Science

50. **des Georges**, A., Clarke, O.B., Zalk, R., Yuan, Q., Condon, K.J., Grassucci, R.A., Hendrickson, W.A., Marks, A.R., and Frank, J. (2016). Structural basis for gating and activation of RyR1. *Cell* **167**: 145–157.
Elsevier

51. **Fu**, Z., Kaledhonkar, S., Borg, A., Sun, M., Chen, B., Grassucci, R.A., Ehrenberg, M., and Frank, J. (2016). Key intermediates in ribosome recycling visualized by time-resolved cryoelectron microscopy. *Structure* **24**: 2092–2101.
Elsevier

52. **Liu**, Z., Gutierrez-Vargas, C., Wei, J., Grassucci, R.A., Ramesh, M., Espina, N., Sun, M., Tutuncuoglu, B., Madison-Antenucci, S., Woolford, J.L., Tong, L., and Frank, J. (2016). Structure and assembly model for the *Trypanosoma cruzi* 60S ribosomal subunit. *Proc. Natl. Acad. Sci. USA* **113**: 12174–12179.
National Academy of Sciences, USA

53.	**Maji**, S., Shahoei, R., Schulten, K., and Frank, J. (2017). Quantitative characterization of domain motions in molecular machines. *J. Phys. Chem. B* **121**: 3747–3756.	American Chemical Society
54.	**Twomey**, E.C., Yelshanskaya, M.V., Grassucci, R.A., Frank, J., and Sobolevsky, A.I. (2017). Channel opening and gating mechanism in AMPA-subtype glutamate receptors. *Nature* **549**: 60–65. https://doi.org/10.1038/nature23479	Nature Publishing Group

www.ingramcontent.com/pod-product-compliance
Lightning Source LLC
LaVergne TN
LVHW080751300926
849162LV00008BA/1790